TAILINGS AND MINE WASTE '99

PROCEEDINGS OF THE SIXTH INTERNATIONAL CONFERENCE ON TAILINGS AND MINE WASTE '99/FORT COLLINS/COLORADO/USA/24-27 JANUARY 1999

Tailings and Mine Waste '99

A.A. BALKEMA/ROTTERDAM/BROOKFIELD/1999

Organized by Geotechnical Engineering Program, Department of Civil Engineering, Colorado State University

ORGANIZING COMMITTEE

John D. Nelson	Conference Chair, Department of Civil Engineering, Geotechnical Engineering Program, Colorado State University
Debora Cave	HSI Geotrans, Boulder, Colorado
William A. Cincilla	Golder Associates, Lakewood, Colorado
Cary L. Foulk	McCulley, Frick and Gilman, Boulder, Colorado
R. Craig Ford	Cyprus Climax Metals Company, Tempe, Arizona
Linda L. Hinshaw	Department of Civil Engineering, Geotechnical Engineering Program, Colorado State University
David C. Keith	Shepherd Miller, Inc., Fort Collins, Colorado
Robert L. Medlock	Engineering Consultant, Denver, Colorado
Debora J. Miller	ESA Consultants, Fort Collins, Colorado
Sean C. Muller	Brown and Caldwell, Denver, Colorado
Charles D. Shackelford	Department of Civil Engineering, Geotechnical Engineering Program, Colorado State University
William P. Van Liew	Engineering Consultant, Fort Collins, Colorado
Dirk J.A. VanZyl	TRC Environmental Corp., Englewood, Colorado

The texts of the various papers in this volume were set individually by typists under the supervision of each of the authors concerned.

Published by
A.A. Balkema, P.O. Box 1675, 3000 BR Rotterdam, Netherlands
Fax: +31.10.413.5947; E-mail: balkema@balkema.nl; Internet site: www.balkema.nl
A.A. Balkema Publishers, Old Post Road, Brookfield, VT 05036-9704, USA
Fax: 802.276.3837; E-mail: info@ashgate.com

ISBN 90 5809 025 6

Printed in the Netherlands

Tailings and Mine Waste '99 © 1999 Balkema, Rotterdam, ISBN 90 5809 025 6

Table of contents

Liners, covers and barriers

Groundwater and geochemistry

Remediation and reclamation

Reprocessing, utilization, and treatment

Case histories

Tailings and Mine Waste '99 © 1999 Balkema, Rotterdam, ISBN 90 5809 025 6

Preface

This is the sixth annual Tailings and Mine Waste Conference held at Colorado State University in Fort Collins, Colorado. The purpose of these conferences is to provide a forum for discussion and establishment of dialogue among all people in the mining industry and environmental community regarding tailings and mine waste. Previous conferences have been successful in providing opportunities for formal and informal discussion, exhibits by equipment and instrumentation companies, technical exhibits, and general social interaction.

This year's conference proceedings include 88 papers. These papers address the important issues faced by the mining industry today. This book will provide a record of the discussions at the conference that will remain of value for many years.

Site characterization, investigation and monitoring

Tailings and Mine Waste '99 © 1999 Balkema, Rotterdam, ISBN 90 5809 025 6

The benefits of sonic core drilling to the mining industry

T.Oothoudt
Environmental Drilling Division, Boart Longyear Company, Little Falls, Minn., USA

ABSTRACT: In this age of ever increasing costs of operation and environmental regulation it has never been more critical to obtain highly representative, good quality subsurface samples for geohydrologic, geochemical and geotechnical studies, and for assessment of mineral content for initial or secondary mining purposes. Conventional drilling and sampling methods commonly used to drill test borings for these studies and assessments may not and often do not provide the high quality sample data which is necessary to accurately characterize the subsurface environment, make accurate assessments, make good engineering decisions, and to make correct containment or cleanup plans, etc.

Sonic drilling is now available offering many unique capabilities and advantages to the mining industry, the mineral exploration market and the environmental industry. By using a special drill head, or oscillator, that imparts high frequency vibrations into a dual string of steel drill pipe, continuous, very representative, relatively undisturbed core samples can be recovered from virtually any unconsolidated formation and from most bedrock formations. The sonic method requires no air, water, mud, or additives of any kind to collect core samples from unconsolidated formations. Sonic penetration is fast, drills at angles, and yields very little or no drill cuttings or mess. Case histories of sonic use for numerous mining and exploration projects in the U.S. and Canada over the past 14 years have demonstrated excellent results.

1 INTRODUCTION

Over the past 20 years the environmental contamination assessment and cleanup industry has become a highly regulated multi-billion dollar business. Because of the immense costs both to private industry and government entities involved in complying with all of the regulatory demands there has been strong interest in development of new technologies for doing this work in a more cost effective and efficient manner.

A task that is nearly always needed for this work, and therefore represents one of the major cost areas, is the drilling of borings to obtain subsurface data, install wells, and other borehole devises and instruments. Most, if not all, of the conventional drilling methods that have been and are being used for environmental work are carry over methods designed for other purposes that have been altered or adapted to attempt to comply with environmental drilling requirements which are much more stringent and different than most other types of drilling.

For these reasons, sonic drilling has been introduced to the environmental industry as a new and better way to satisfy many of the complicated drilling requirements. The demand for sonic drilling has increased dramatically over the past 5-7 years as more and more people become aware of these rigs, what they are, and what they will do.

Sonic or Rotasonic is a relatively new drilling method that got it's commercial contracting start in the U.S. 14 years ago. As of today sonic rigs are being widely utilized in the environmental industry and are now being used more and more frequently in the mining industry throughout the U.S and Canada.. This increased use is because of the larger quantity and better quality of the subsurface samples (soil cores, bedrock cores, soil vapor and groundwater) as well as for quick, accurate, and controlled installations of piezometers, monitoring wells, and geotechnical borehole instrumentation.

2 HISTORICAL OVERVIEW

The concept behind sonic drilling is not new. The original development work was started in 1946 in the U.S. by a firm called Drilling Research, Inc. (DRI) which was formed and supported by the petroleum industry. Some of their goals were to examine ways to speed up oil drilling operations by adding vibrations to the rotary motion of the drill bits and to investigate the use of high frequency vibration as a way to develop and redevelop oil wells. This company went out of operation in 1958. It is also known that the Russians were working on vibratory drilling in the late 1940's and early 1950's. The extent of their research or the results of their work are not known. In the early 1960's Shell Oil Company supported a group led by Dr. Alfred Bodine that continued to research and develop vibratory drilling technology for several years. Their major focus was on sonic pile driving equipment. A few years later when Shell cut off funding, Dr. Bodine started his own company to carry on the development. His company applied for and received many patents related to sonic drilling during this time. In the early 1970's Dr. Bodine entered into a licensing agreement with Hawker Siddeley , a British manufacturing company that has numerous Canadian divisions, initially to build sonic pile driving equipment and later to build sonic drill rigs. Between 1975 and 1981 approximately 10 drill rigs and 15 sonic drill heads were built and tested with limited success because of numerous mechanical problems with the vibratory drill heads, the rigs, and the tooling. The rigs were too expensive and far too unreliable for successful operation. Their intended markets for the rigs were seismic drilling and placer gold exploration which both took downturns in the early 1980's, hence Hawker Siddley decided to discontinue their sonic operations.

Midwest Drilling, a Canadian core drilling firm from Winnipeg was the major player in keeping the technology alive in the early 1980's. They recognized the benefits and potential for the sonics' unique ability to collect a continuous core sample of any unconsolidated and most bedrock formations. They built their own drill rigs, tools and drilling systems, and were the first company able to successfully operate the sonic rigs on a commercial basis. Ray Roussy, a former Hawker Siddley engineer also worked on his own to build sonic parts and a rig of his own in the early 1980's. North Star Drilling, a Minnesota drilling firm started leasing rigs from Midwest in 1985 and operated them for mineral exploration and environmental drilling in the U.S. Since that time, major changes and improvements have been made in the vibratory drill heads, the drilling tools, operating systems, etc. Today, the rigs are a very reliable piece of equipment.

3 SAMPLING AND COMPARISONS

Sonic drilling combines the primary force of high frequency vibration with hydraulic pulldown and slow rotation [when or if needed] to advance a dual string of drill rod and drill casing into the ground both vertically and at angles. The inner drill pipe has a core barrel attached to it and is advanced ahead of the outer casing into the undisturbed formations to collect a continuous large diameter core sample. The core sample is pulled to the surface and extracted into a plastic sleeve or a sample tray, or if coring with a split core barrel containing a poly or lexan liner, the liner is simply taken out of the barrel and the ends are capped to preserve the core sample for examination and testing . The outer casing is then advanced down the previously core sampled borehole in order to maintain the borehole wall, prevent caving or sloughing from above, and to help in preventing cross contamination.

Not having to use air, water, mud, or additives of any type when coring unconsolidated formations eliminates the mixing and disturbance of the sample that is caused by the addition of these circulation mediums. Sonic core samples provide excellent lithological data from which an accurate characterization of the subsurface can be made. A sonic rig will core softer bedrock formations very well, and will core hard bedrock. However, penetration rates in hard rock are much slower and bit and tool costs are higher than with a standard diamond wireline hard rock coring rig, so if any large amounts of hard rock need to be cored a standard wireline core rig is much more cost effective. Sonic rigs are generally used to core into competent hard rock from 10 to 50 feet, or far enough to accurately identify it. The sonic rig utilizes water or air when coring hard bedrock to remove the rock cuttings from the bit face, the same as with any other rock coring or sawing process.

The sonic subsurface samples provide accurate data for location of perched water, location of the water table(s) and water bearing units, and collection of in-situ borehole vapor and/or water samples. All of this allows the consultant to find and identify contamination if present, and to positively evaluate contaminate distribution and transport zones.

When compared to hollow stem auger drilling with continuous split spoon sampling, some of the advantages of a sonic rig are: its ability to drill much faster, deeper, and through virtually any type of formation or obstruction without meeting refusal or having to convert to mud rotary drilling in order to advance the boring to depth, collection of excellent core samples of silts, sands, gravel, cobbles, or boulders, the ability to collect samples below the water table with no problem. Specially designed "trap" bits are utilized to hold core samples of loose sand, silts, mud, and saturated formations.

When compared to reverse air drilling the sonic is usually more expensive on a per foot drilled basis because it is slower. However the sonic core sample is not drilled up and mixed with air, therefore is much less disturbed and more representative. The sonic core sample also allows for field screening for volatile organic compounds and for borehole soil vapor sampling. Both methods allow for temporary screening and pumping for groundwater sampling, draw down tests, etc. The sonic allows for more accurate location of perched water, and generally crates less borehole wall disturbance which translates into less well development and purging time. Today's sonic rigs are limited maximum to depths of 600 to 800 ft. with 4 inch x 6 inch diameter borings, and maximum borehole diameters of 10 - 12 inches to lesser depths.

4 VIBRATION AND RESONANCE

The key to the operation of the sonic rig is the hydraulically powered sonic drill head or oscillator which produces the high frequency vibrations. This oscillator contains two balanced, eccentric, counter rotating weights or rollers that are timed to oppose each other and cancel out each others radial centrifugal force until they reach the 0 and 180 degree positions, at which point they are perfectly synchronized delivering 100% of their combined energy to create vertical vibration. The rotation speed of the rollers [or vibratory frequency] can be infinitely controlled with the hydraulic drive system from 0 - 10,800 rpm. The faster the rollers are turned , the more centrifugal [vibratory]energy they produce.

The drill head contains either an air or mechanical spring that insulates the drill rig from the vibration. It also contains a hydraulic rotation system, and an air/water swivel for drilling conditions where air or water are needed and for easy conversion to air or mud rotary drilling when or if advantageous.

The drill head is threaded to the drill rod or outer casing and the vibrations are imparted directly down the drill steel to the circular cutting face of the drill bit which contains carbide buttons or cutters. Vibrational frequency is controlled to suit operating conditions to achieve optimum penetration rates and maximum core recovery. The vibration at the face of the bit creates three different drilling actions which are:

a. Displacement. In silts, sands, gravel and small cobbles the soil particles do not vibrate in harmony with the bit and pipe or each other, as a result they vibrate in all different directions which has a fluidizing effect on the particles, moving them away from the bit and reducing friction on the bit and pipe.
b. Shearing. In tight clays or cohesive silt type formations, the bit shears the material.
c. Fracturing. In boulders and consolidated formations the bit tends to break up or fracture the material.

When the combination of the length of the drill pipe and the frequency of the vibrational stress waves being induced into the top of the drill pipe coincide with the steel pipes' natural ability to accept and reflect each wave, a natural phenomenon called resonance occurs. A detailed description or discussion of resonance is beyond the scope of this paper, however a brief or simplified explanation of the effect it has on sonic drilling is as follows: When the steel drill pipe is in resonance it accepts maximum amounts of hydraulically generated vibrational energy. Energy is stored in the elastic properties of the steel, it expands and contracts in time with the vibrational stress waves being induced into it, causing it to act like a flywheel or a spring delivering tremendous amounts of energy directly to the cutting face of the bit with very minimal energy loss in the process. While in resonance the system reaches maximum efficiency.

5 SONIC USES IN THE MINING INDUSTRY

Sonic core drilling has been utilized in the mining industry for several different purposes. Some of them are as follows:

1. Environmental characterization investigations in and around existing mines and around former mine sites.
2. Geotechnical studies for design and planning of coffer dams and tailing ponds.
3. Core sampling heap leach pads for geochemical purposes and to make assessments of the mineral content of the samples.
4. Core sampling of tailings for geochemical and other testing.
5. Core sampling old rock dumps for geochemical testing and mineral content assessments for possible secondary recovery operations.
6. Installation of piezometers, monitoring wells, inclinometers, and installation of various other types of borehole instrumentation.
7. Pre-collaring boreholes through overburden or fill areas for wireline core drilling, with the added advantage of obtaining the sonic core sample to be used for environmental, geotechnical, or other studies.

6 EXAMPLES OF COMPLETED MINING PROJECTS

Iron mines in Northern Minnesota have used sonic rigs for numerous environmental investigations in and around rock dumps because they were able to core through the large boulders and broken rock and collect good core samples. They were also used for another dual purpose; for drilling through the difficult overburden formations to pre-collar boreholes into the competent bedrock for a wireline core drilling rig, while at the same time providing a continuos core sample which could be examined for contamination, used for lithologic information, and could be assessed for mineral content.

Consultants leading the environmental investigation at the Summitville Mine near Del Norte, Colorado employed sonic rigs to provide continuous core samples of former rock dumps, tailing ponds, and the overburden formation around the mine, and to install monitoring wells.

Over the past year sonic rigs have been used at 6 or 8 different mines in the western U.S. for core sampling overburden around mines, heap leach pads, tailing areas, and rock dumps for both environmental investigations and to assess the mineral content.

A sonic rig was used to conduct test borings at a property which was known from previous core drilling programs to contain a sizable manganese deposit. The manganese was contained within an iron deposit between 184 and 420 feet below surface. The previous wireline core holes indicated the presence of the manganese but because it is a soft clay like deposit, most of the core washed away and very sketchy information was provided. The sonic was used to continuously core sample the 236 foot ore zone without the use of any air or water. Except for four feet of a soft, running, mud layer, core sample recovery was 100%. Core had been recovered from this formation where nothing but washed cuttings were obtained before and was described by U.S. Bureau of Mines geologist, Linda Dahl, as "amazing".

Sonic rigs have been utilized for numerous mineral exploration projects to core sample the overburden and into the bedrock looking for gold, diamonds, manganese, phosphate, kaolin, and other minerals. They are also widely used for accurate definitions of gravel deposits for the aggregate industry.

Sonic rigs have also been used in Northern Canada to core sample for phosphate deposits, and to sample lake bottom silts, on down through boulders and till, and into the bedrock for geotechnical studies associated with the construction of a coffer dam. And in a neighboring mine under similar drilling conditions to provide core samples for geotechnical studies, lithologic information and mineral assessment.

7 SAFETY

Operational safety is another positive feature of sonic rigs. The rigs have elevated hydraulically leveled drilling platforms which provide a safe, controlled work area. They are made of expanded and corrugated metal to prevent slipping. They keep the drill crew up off the ground, out of drill cuttings, water, or ice and snow in cold conditions. The drill rods and outer casing are flush pipe with no auger flights, breakout flats or bars to catch clothing on. While drilling, they are either not rotated at all, or very slowly, which makes it very unlikely that anyone could get wrapped up in the pipe. In addition the rigs have rotation guards in place while the rigs are drilling. The drill pipe and casing are made up and taken apart with a fully automated hydraulic breakout wrench and hydraulic rod spinner which eliminates the need to use hand tools. Ten foot drill rods are stacked on a rack at hip level for ease of handling. The drill head pivots out 90 degrees toward the back of the rig to facilitate easy connection of, and unloading of, the drill rods.

There are hydraulic jib booms on the rig masts and hydraulic cranes on the tender trucks for lifting heavy pipe and other supplies. There are four shutoffs located on each machine in case of an emergency. One is in the truck cab or on the drive motor, one at the operators control station, an emergency crew shut down on the back of the work deck, and an emergency shut down on the side of the rig which can be accessed from the ground.

Elevated work areas have safety rails and the ladder to the work deck has corrugated steps and a safety rail. When drilling in contaminated materials the crew is up and away from borehole vapors, air and fluids are not circulating or carrying the contamination to the surface, the samples are extruded into plastic sleeves without being touched by the crew, or the core can be collected in specially fitted acrylic, lexan, or poly core barrel liners which encase the core in the liner below surface as the core barrel is advanced into the formation.

CONCLUSION

The conclusion I've reached is that sonic drilling provides consultants and mining companies with access to a very beneficial new tool to use whenever high quality subsurface samples are useful or necessary for; environmental investigations, lithologic definition, location of perched water, location of water tables, location of water bearing formations, field screening of samples for volatile organic compounds, borehole vapor or water sampling, draw down tests, assessment of the mineral content, geotechnical engineering studies, and for installation of monitoring wells, injection or extraction wells, piezometers, observation wells, inclinometers and various other borehole instrumentation.

The quality and quantity of the sonic core sample yields accurate subsurface data which provide the basis for faster and more accurate characterization of sites, better design and planning, and better assessment of the mineral content. All of these items can lead to savings in several ways, by needing to drill less boreholes in order to obtain adequate information, by drilling continuously core sampled borings much faster, by creating much less mess and drill cuttings, and from potential huge savings in the future by providing data which allows for correct design and planing up front, thereby eliminating costly mistakes later on.

Tailings and Mine Waste '99 © 1999 Balkema, Rotterdam, ISBN 90 5809 025 6

Image is everything

R.J. Miller
10:1 Solution Group of Golden, Colo., USA
M.J. Culig
COLOG Incorporated of Golden, Colo, USA
J. Viellenave
teg-Rocky Mountain of Golden, Colo., USA

ABSTRACT: The "image" referred to is the ability to "see" inside the tailings structure itself. A suite of subsurface exploration technologies is available that allows for the imaging of physical and chemical characteristics of earth materials. Types of information typically obtained include: subsurface structure, material types, chemistry and moisture content. Other properties that can be measured include density, velocity, electromagnetics, temperature, soil gases, and even the in-situ strength of the material. Flow paths within, under or around tailings can be identified, and the effects of remediation/stabilization efforts quantified.

1 INTRODUCTION

Important to the evaluation of any tailings structure is to know what the conditions are inside the structure. An open excavation is one way of inspecting the subsurface, but it is obviously very intrusive, expensive, and often not even possible in key areas of the structure. Surface geophysics can provide some information, but is critically dependent on site conditions, and often lacks the resolution required at depth. What is needed is to get the measurement tools close to the zones of interest. An effective, minimally-intrusive way to do this is to log holes placed in critical areas of the structures. In this way the whole section of interest to total depth can be characterized. Also, if a direct-push method is employed to create the holes to be logged, disturbance is further minimized and the handling of potentially hazardous materials greatly reduced. A structure can be quickly sampled, and a 3-D image created of the subsurface. Before-and-after logging can evaluate the effectiveness of stabilization or remediation efforts by identifying changes in subsurface conditions.

Logging has already been successfully used in the characterization of tailings, but recent new approaches and variations of techniques promise to expand the value of logging to many geotechnical applications. Following is a review of available technologies, their applicability to tailings characterization, and some examples of their successful use.

2 TECHNOLOGIES

Direct-Push Sampling - This method represents an alternative to conventional drilling, and offers the advantages of not only being less intrusive (smaller boreholes), but produces no cuttings or other hazardous waste needing disposal. Direct push equipment can be mounted on a variety of vehicles ranging from small ATV's to 15,000 pound trucks, and can penetrate 100 feet or greater into tailings. Other than obstruction by large cobbles and boulders, the rigs can easily push, using hydraulic and vibratory methods, through mine/mill tailings and spoil piles. Samples of earth materials (soil), soil gases, and liquids are also easily acquired for analysis and testing.

Logging-Geophysical logging techniques provide continuous, depth correlated measurements along the axis of a borehole. Many different measurements can be made to investigate formation and borehole properties such as density, moisture content, naturally occurring radioactivity, acoustic velocity, electric and electromagnetic properties, fluid flow, water quality, and others.

Both the well completion and drilling method limit the suite of geophysical measurements obtainable. The geophysical methods must be able to detect physical variations in the surrounding media given the constraints provided by the borehole or well environment.

Geo-Chemistry - A wide variety of geochemical tests can be performed on soil, gas, and liquid samples acquired from tailings and spoils. Mine types fall into two broad classifications: Hard rock (metal) mines and mills; and fuel (coal, oil shale) mines. Regardless of type, the analytical approaches described can readily be conducted in either mobile or fixed laboratories, depending on how quickly the analysis is needed.

Metal mines and mills produce wastes that can contain base metals that are regarded as threats to surface and ground water. In addition, it is common to analyze for temperature, pH, Eh, conductivity, a variety of cations and anions, sulfur compounds, and occasional radionuclides and organics. Fuel mines produce a variety of organic materials in addition to many of the waste products mentioned above. Further, there are often volatile gases (methane, H_2S) associated with the mines themselves which can be hazardous to surface occupants.

3 APPLICATIONS

Several direct applications of these technologies to the characterization of tailings structures have been identified, and others surely exist. Some that have been demonstrated to be effective include:

Subsurface Structure - borehole imaging logs: acoustic televiewer (ATV), Borehole Image Processing System (BIPS), borehole video, sonic logs, and borehole radar can be used to identify structural features such as layer contacts, fill-to-native contacts, voids, and other anomalous underground conditions.

Material Types - deep/shallow normal resistivities, induced polarization (IP), spontaneous potential (SP), gamma, single point resistance (SPR), and total magnetic field provide information used to determine differing soil/fill types and thus the potential for contaminant transport.

Geo-Chemistry - can be used for measuring chemical composition of the soil and soil fluids.

Moisture Content - neutron logs measure the moisture content in the vicinity of the borehole by detecting the interactive response from a neutron emitting radioactive source.

Material Density - gamma-gamma density logs use radioactive sources to determine bulk density of materials in specific zones around the borehole or cased wells.

Seismic Velocities - can be measured within and between holes and can be helpful in determining relative compaction over larger areas.

Electromagnetic Properties - are determined with EM induction and magnetic susceptibility logs which can help in material characterization and the location of features such as metal or conductive fluids.

Flow Paths - using flow meters and hydrophysical logging techniques, in-situ flow paths and associated quantities can be measured within and around the structure.

4 CASE HISTORIES

Soils Logging - The authors have recently become involved in the application of existing technologies used in new ways. One of the new markets being served is logging ahead of excavation, such as microtunneling, to identify ground conditions and potential hazards to excavation. The technology has successfully determined: changes in soil types; the presence of existing pipes or utilities; the location of boulders, trees and general trash; and the presence of contaminated ground. Direct-push and density logging have been used to determine relative compaction in back-filled soils, which should translate directly to tailings and waste piles. Measurements of changes in density from compaction-pile grouting suggests that relative stabilization or strength increases can be determined for tailings and waste piles as well.

Moisture Monitoring with Nuclear Borehole Logs - Applications of nuclear borehole logs to moisture studies in the vadose zone have included measurement of infiltration rates due to seasonal rainfall and artificial ponding, and the monitoring of leachate migration near landfills and beneath chemical storage lagoons. In general, these studies were designed to evaluate such parameters as specific retention and unsaturated permeability, as well as provide field information on contaminant migration in the unsaturated zone.

At a recent project conducted in the Mojave Desert in California, a study was done to monitor a ponded wetting front, and track the infiltration of seasonal rainfall. The purpose of this study was to characterize unsaturated flow in terms of infiltration rate, flow pattern, and ultimate moisture content. Gamma-gamma density, neutron, and induction logs were run over time in order to identify the infiltration front. This technology can be directly applied to determine moisture content and acid infiltration in heap leach piles.

Imaging - Borehole imaging techniques provide a means of looking at the in-situ conditions of the borehole wall. Developments in optical, acoustic, and electrical imaging devices have improved the usefulness of these techniques for mining and ground water studies. By characterizing the borehole wall face, information regarding cobble size, flow paths and orientation, hole rugosity or diameter can be obtained. This method can also be used to orient core.

Oriented core drilling has traditionally been used by gold mines for fracture analysis in slope stability characterizations. This method necessitates an angle drill hole 50 to 70° from horizontal and is limited by the difficulty of drilling through unconsolidated surface overburden. At a Nevada mine site, traditional oriented core drilling was inappropriate for their low angle (near horizontal) bedding. By using borehole imaging logs in combination with reverse circulation (RC) drilling methods, oriented fracture data were obtained quickly and at a much lower cost than the oriented core drilling process.

HydroPhysical Logging™(HpL™) - This new technique provides precise measurements of flow and water quality over the vertical extent of a well. HpL™ identifies producing zones and quantifies the flow from these zones more quickly and economically than conventional packer tests. This method of testing has been used successfully to identify flow paths and hydraulic conductivities in the pre-site evaluations for in-situ copper mines operations.

A project at an in-situ mining operation in Arizona was performed to accurately describe the characteristics of groundwater flow and the underlying aquifer under several stressed (pumping) conditions and evaluate the subsurface hydraulic connections between extraction and injections wells. Cross hole testing was utilized and hydraulic connections between wells were identified and quantified. This data was instrumental in the design and construction of the pilot study.

5 SUMMARY

True, image is not everything, but certainly in tailings and waste pile engineering, knowing conditions inside the structure greatly facilitates evaluation and control. It has been demonstrated that technologies exist and function for application to tailings and waste pile evaluation. Existing and new logging technologies provide engineering-level data critical to evaluation, stabilization, and remediation.

Tailings and Mine Waste '99 © 1999 Balkema, Rotterdam, ISBN 90 5809 025 6

Lysimeter design in theory and practice

B.E. Bews & S.L. Barbour
Department of Civil Engineering, University of Saskatchewan, Sask., Canada
B.Wickland
Clifton Associates Limited, Saskatoon, Sask., Canada

ABSTRACT: Lysimeters are often used in waste containment facilities to provide accurate measurements of the net flux of water through soil covers or liners. In many cases, these lysimeters are constructed without a full appreciation of the effect that unsaturated conditions within the cover and underlying waste material will have on performance. When lysimeters are used in conjunction with a site monitoring program to determine whether a liner or cover is performing as required, an inaccurate reading could lead to over or under estimates of net infiltration and consequently misrepresent the actual performance. It is therefore very important that lysimeters be designed properly with unsaturated flow theory in mind. In this paper, the performance of a lysimeter under unsaturated flow conditions is illustrated through the use of a laboratory prototype and numerical modelling of lysimeter performance.

1 INTRODUCTION

Soil covers and liners are commonly used to control the influx of water through mine waste such as tailings or waste rock. Lysimeters are often used as part of a monitoring system to measure moisture fluxes through these barriers in order to determine if they are performing correctly. Monitoring the water fluxes through mine waste is often complicated, however, by the fact that the soil or mine waste beneath the cover or liner is frequently unsaturated. When designing such lysimeters, a number of issues must be addressed in order for the instrumentation to function properly. Included in this list of design issues are the geometry of the lysimeter, the anticipated infiltration rate, and the hydraulic properties of the backfill.

In this paper, the theory behind lysimeter design will be illustrated with reference to the lysimeters designed for the storage and release test covers at the semi-arid Kidston Gold Mine in north-central Queensland, Australia. The results of a laboratory column study are also used to emphasize the key role that the properties of the backfill have on lysimeter performance. The results of the numerical modelling of unsaturated flow compare quite favorably with the measured results from the laboratory study.

2 UNSATURATED FLOW THEORY FOR LYSIMETER DESIGN

Lysimeters are currently one of the only ways to collect and measure flow through unsaturated soil (Wickland 1998). Many lysimeters do not, however, provide accurate measurements of infiltration because of poor design. A fundamental understanding of flow through unsaturated soils is required in the design of properly functioning lysimeters. This section provides a brief overview of unsaturated flow theory as it relates to lysimeters designed under steady state infiltration.

The flow of water through a soil is governed by Darcy's law:

$$q = -k * i \tag{1}$$

where q = flux (L/T); k = hydraulic conductivity (L/T); and i = hydraulic gradient. The hydraulic gradient has two components - the pressure head gradient and the elevation gradient. In vertical flow, the elevation gradient (dz/dz) is always equal to unity. For an unsaturated soil, the hydraulic conductivity is not constant but is expressed as a non-linear function of water content and/or suction (i.e. negative water pressure).

Consider the case of an infinitely high column of soil with a water table at its base. If a steady state flux which is less than the saturated hydraulic conductivity of the soil is applied to this column, then the vertical hydraulic gradient in the soil will become unity at some distance above the water table (Kisch 1959, Barbour 1990). When this occurs, the pressure gradient is equal to zero and the hydraulic gradient is equal to unity. Consequently, the unsaturated hydraulic conductivity of the soil is equal to the applied flux:

$$q = -k * 1 \tag{2}$$

Three pressure head distributions are shown in Figure 1 in order to illustrate this behavior. The pore water pressures at the water table are zero. In a soil with zero infiltration, the pressure head will decrease linearly (e.g. hydrostatically) with elevation above the water table (a). Where the infiltration rate is equal to the saturated hydraulic conductivity of the soil, the pressure head remains at zero for all elevations (b). For infiltration into an unsaturated profile, as shown in (c), the pressure curve initially follows a hydrostatic line. Just above the water table, the pressures are only slightly negative and the hydraulic conductivity is much higher than the applied flux. As a result, the gradient is very small, that is, the pressure gradient is just greater than –1. As the pressure becomes more negative, the hydraulic conductivity begins to decrease and the negative pressure gradient begins to decrease, allowing the hydraulic gradients to increase. At some point, the hydraulic gradient reaches a value of unity and the curve breaks upward to a constant value of pressure head dictated by the infiltration rate and the soil's hydraulic conductivity function. Therefore, the pore water pressure distribution above the water table may be determined using Darcy's law, the soil's hydraulic conductivity function, and the applied flux rate (Barbour 1990).

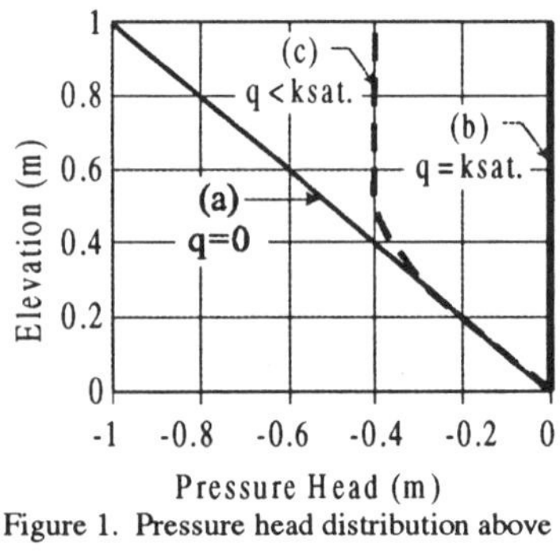

Figure 1. Pressure head distribution above the water table for a soil under different rates of infiltration

Different negative pore water pressures (i.e. suctions) will be reached for different soils under the same infiltration rate because hydraulic conductivity functions vary between soils. The hydraulic conductivity functions of the two soils used in the laboratory experiment in this paper are shown in Figure 2. For these two soils subjected to a flux rate of 1e-07 m/s, the coarse soil would reach a constant pressure of approximately 2 kPa, whereas the fine soil would reach a constant pressure of approximately 5.5 kPa. If you assume the pressure profile to be hydrostatic prior to the break to a constant pressure, then the height at which this break occurs would be approximately 20 cm for the coarse soil and 55 cm for the fine soil. It is interesting to note that there is only one possible flux rate at which both soils would produce the same pressure, that being where the two hydraulic conductivity functions cross.

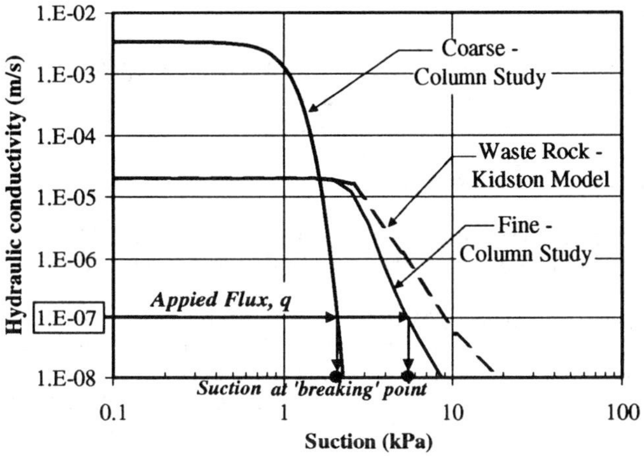

Figure 2. Hydraulic conductivity functions for the two materials used in the laboratory experiment and the waste rock modelled at Kidston.

A fundamental feature for the design of a lysimeter in unsaturated soil is in this relationship between the expected infiltration and the hydraulic conductivity function of the lysimeter backfill and the material in which the lysimeter is placed. The maximum suction developed in the waste rock and lysimeter backfill can be estimated from the 'breaking' point as shown in Figure 2. A lysimeter causes an artificial water table to be established within the lysimeter which may be different from that in the soil outside the lysimeter (O'Kane et al. 1996). In order for the lysimeter to give an accurate measurement of the flux from the base of the cover, the lysimeter itself must not influence that flux. To achieve this, the suction, A_1, inside the top of the lysimeter wall must equal the suction, A_2, outside the lysimeter; that is, there must be no horizontal gradient between the two points which would result in divergence of flow towards or away from the lysimeter (Figure 3). The pressure head profile in the soil follows an approximately hydrostatic line to Ψ_1, and then goes vertically upwards. A zero gradient between points A_1 and A_2 can be achieved by forcing the breaking point, A, on the pressure profile to occur within the lysimeter.

The use of a backfill material that is different than what the lysimeter is buried in will create measurement errors because different soils produce different suctions for the same infiltration rate. In addition, the depth of the lysimeter walls must be greater than the suction (m) at the breaking point in order to be on the vertical portion of the curve (for example, if the suction is 20 kPa then the lysimeter walls must be greater than 2 m deep). Because the suction at this breaking point is where the applied flux equals the hydraulic conductivity, by simply knowing the applied flux and the hydraulic conductivity function, the suction is known and the required lysimeter depth can be estimated.

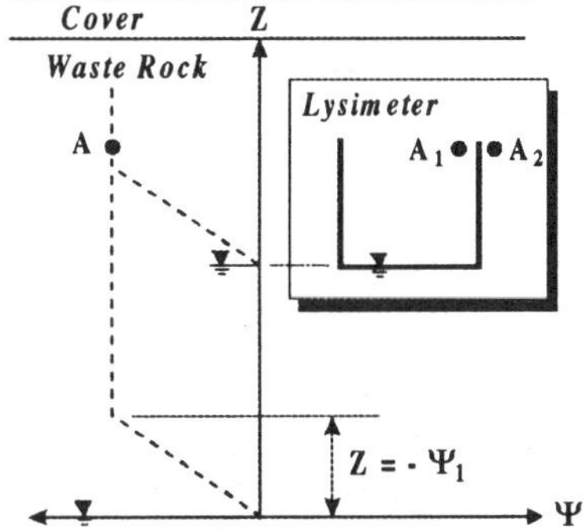

Figure 3. Cross-section of covered waste rock showing the breaking point in the suction profile.

3 KEY LYSIMETER DESIGN ISSUES

The geometry (e.g. wall height), applied infiltration rate, and the hydraulic properties of the material used as backfill are all significant issues for the design of a lysimeter. In the following sections, backfill and geometry are the key issues that will be discussed.

3.1 *Illustration of the importance of lysimeter wall height*

A lysimeter design to accurately measure the net flux through the test cover plots at Kidston was required to ensure that the previously designed storage and release test covers were performing as required (Bews et al. 1997).

The two-dimensional finite element model SEEP/W (GEO-SLOPE International 1995) was used as a tool to aid in the design of the lysimeters at Kidston. The hydraulic conductivity function for the waste rock is shown in Figure 2. The wall height was varied from a minimum of 15 cm high to a maximum where it was extended up to the base of the cover system in 50 cm increments. The lower boundary condition for all of the models was set as a water table.

The simulations illustrated that a pan type lysimeter placed at depth within the waste rock under the soil cover system will not function properly (Figure 4a). The 2-D model predicted that water entering the pan from above will be drawn over the sides of the pan if the height of the lysimeter wall is not sufficient. This syphoning effect is due to the hydraulic gradient between the material inside the pan and the waste rock material outside the pan. As the water enters the material in the pan, the suction decreases. However, higher suctions are maintained outside the pan. Therefore, at a given elevation, a hydraulic gradient develops which draws water from the area of lower suction within the pan to the area of higher suction outside the pan.

The modelling demonstrated that if the base of the lysimeter is 2 m below the base of the cover system, the wall height of the lysimeter must be extended to at least 1.5 m above the base of the lysimeter (Figure 4b). This reduces the effect of the water being drawn out of the lysimeter. The residual value of suction for the backfill material is approximately 10-15 kPa. At suctions greater than the residual, the hydraulic conductivity of the soil is essentially zero, consequently the movement of water in the liquid phase is greatly reduced. This corresponds to the 1.5 m minimum wall height required.

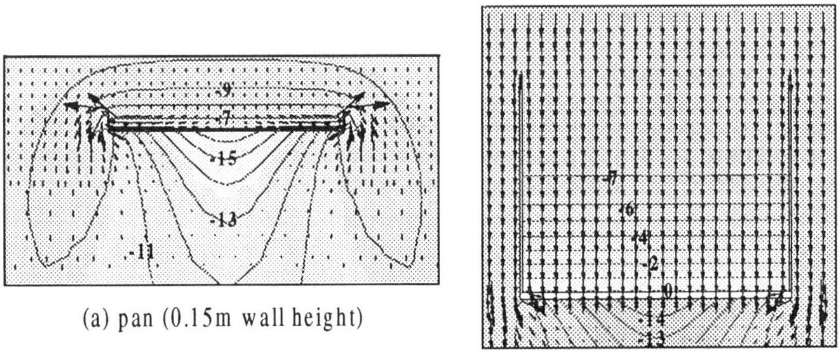

(a) pan (0.15m wall height)

(b) 1.5 m wall height

Figure 4. Pressure contours (kPa) for (a) a poorly designed lysimeter with diverging flow (pan), and (b) a lysimeter with limited deflection of flow (1.5 m walls).

The simulations carried out show that in order to essentially eliminate the effect of the elevated water table at the base of the lysimeter, the lysimeter wall must extend to heights greater than 1.5m. This was confirmed by calculating the gradient at the top of the lysimeter (between points A_1 and A_2 in Figure 3). Figure 5 shows that the gradients are dramatically reduced at 1 m but that extending the lysimeter walls to a depth of 2 m increased the lysimeter efficiency and minimized the lateral gradient at the top of the lysimeter as required to prevent water loss.

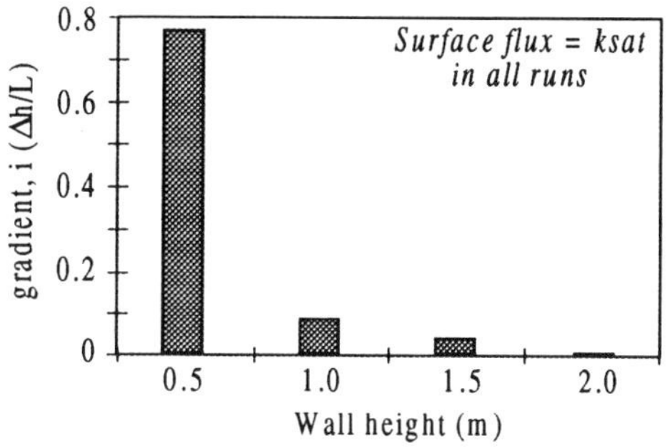

Figure 5. Gradients calculated at the top of the modelled lysimeters.

3.2 *Illustration of the importance of lysimeter backfill material*

Another key design feature for lysimeters is the selection of backfill material. A common design error is to use the wrong material to backfill the lysimeter. Sands and gravels are often used as lysimeter backfills because they are considered to be 'drainage materials'. A column study was undertaken in order to illustrate the impact of this key design feature on lysimeter performance. Various combinations of coarse and fine material were used to represent the unsaturated mine waste or the lysimeter backfill. The lysimeter collection efficiency was measured under a variety of flux conditions for these various combinations. Lysimeter performance was also predicted through the use of numerical seepage analyses (Wickland 1998).

In each case, a lysimeter was buried within a column of soil. It was assumed that the sands used in the experiment were isotropic and homogeneous. Boundary conditions and geometry were similar for all three columns and included water tables at the base of the lysimeter and the column. Measurements of flux were taken after steady state flow was achieved. Numerical modelling using SEEP/W was undertaken to verify the experimental results.

3.2.1 *Laboratory Experiment*

Three lysimeter designs were tested in three soil filled columns:

- Column 1 contained fine sand, while the buried lysimeter contained coarse sand,
- Column 2 contained coarse sand, while the buried lysimeter contained fine sand, and
- Column 3 contained coarse sand, and the buried lysimeter also contained coarse sand.

The coarser sand was a manufactured 12/20 silica sand. The fine sand was collected from a natural aeolin deposit near Beaver Creek, Saskatchewan. A 35.56 cm (14 inch) diameter, strippable concrete form was used to construct the columns. Each column was sealed at one end with a plywood disc and wax and contained a flexible tubing drain at the base. Vents were installed at three levels in the lysimeters to ensure that if the overlying soils became saturated, the soil in the lysimeter would still be open to atmospheric pressure. Porous ceramic tensiometer tips were installed through the walls of the columns to measure soil suctions just inside and outside of the mouth of the lysimeter. A series of uniformly distributed fluxes were delivered to the surface of the column using a constant head reservoir and a peristaltic pump to supply water to a constructed rainmaker system using drip heads on a 2.54 x 2.54 cm spacing. Figure 6 shows the column design.

3.2.2 *Numerical Modelling*

Each of the three physical lysimeter designs was simulated using a steady state, axisymmetric finite element model, SEEP/W. Inputs to the model included the geometry of the columns and lysimeters, the flux rates applied, and the hydraulic conductivity functions of the two materials.

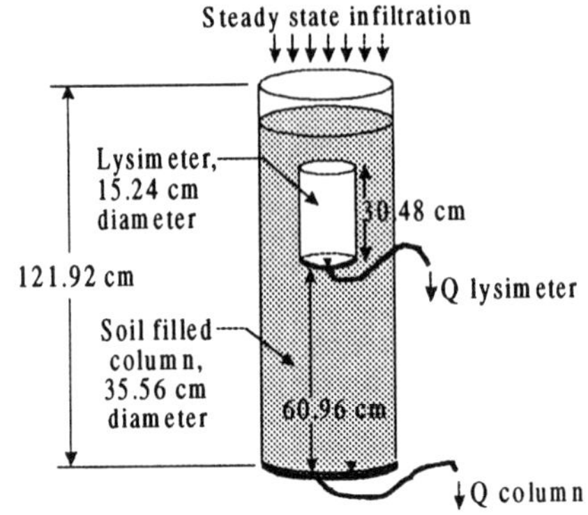

Figure 6. Laboratory experimental apparatus.

The fine sand had not been previously characterized, so its hydraulic conductivity function was estimated. The soil water characteristic curve was measured by a Tempe cell test (Fredlund and Rahardjo 1993). The Fredlund and Xing method was then used to estimate the hydraulic conductivity function of the fine sand (Fredlund and Xing 1997). The hydraulic conductivity function of the coarse material was previously determined (Newman et al. 1997) and is shown along with that of the fine sand in Figure 2.

3.2.3 *Comparison of laboratory and numerical modelling results*

The experimental and numerical modelling results are summarized in graphical form to compare the measured and applied fluxes in Figures 7 to 9. If a lysimeter is working as designed, the measured flux should equal the applied flux, and all data points should fall on the diagonal line of equality.

Column 1 represents the case where the lysimeter backfill is coarser than the surrounding material, such as a lysimeter containing gravel being used to monitor fine tailings (Figure 7). It is evident that when the applied flux is high, the lysimeter collects more water than is being applied. In this condition, the suctions are low within the waste material and the gravel in the lysimeter is at a higher value of hydraulic conductivity than is present in the waste material. Consequently the lysimeter tends to concentrate flow. When the fluxes are low, however, the suctions increase and the hydraulic conductivity in the gravel backfill is lower than that outside of the lysimeter. Consequently flow is diverted away from the lysimeter.

The lysimeter in Column 2 contains a backfill that is finer than the material surrounding the lysimeter, such as a lysimeter containing sand buried within a coarse waste rock (Figure 8). Under high applied fluxes (i.e. low suctions) the hydraulic conductivity of the lysimeter is lower than the coarser waste material and flow is diverted away from the lysimeter. At low applied fluxes the reverse should be true, and in fact the measured fluxes are slightly above the line of equality.

The lysimeter in Column 3 contains coarse sand, and is buried in coarse sand (Figure 9). This design demonstrates the effectiveness of using the same material inside the lysimeter as outside, assuming the lysimeter walls are sufficiently high and the 'pressure break' is contained within the lysimeter. The lowest applied flux in this case was just below 1e-05 m/s, therefore, the equilibrium suction at this infiltration would be less than 2 kPa (Figure 2). This would mean that the break point would be at an elevation of no more than 20 cm above the artificial water table in the lysimeter, well within the height of the lysimeter used in the laboratory column.

All the results obtained in the laboratory experiment and in the numerical modelling can be explained with reference to the plot of the hydraulic conductivities for the two materials used (Figure 2). Columns 1 and 2 work properly only when fluxes are applied which are equal to the hydraulic conductivity at the point where the two material's functions cross (e.g. 1.8e-5 m/s). It may be seen that the soils have the same pressure for only this one infiltration rate. At this rate,

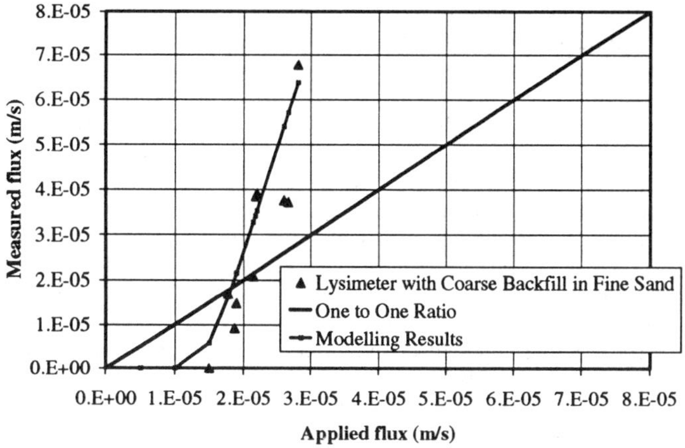

Figure 7. Measured versus applied fluxes for Column 1.

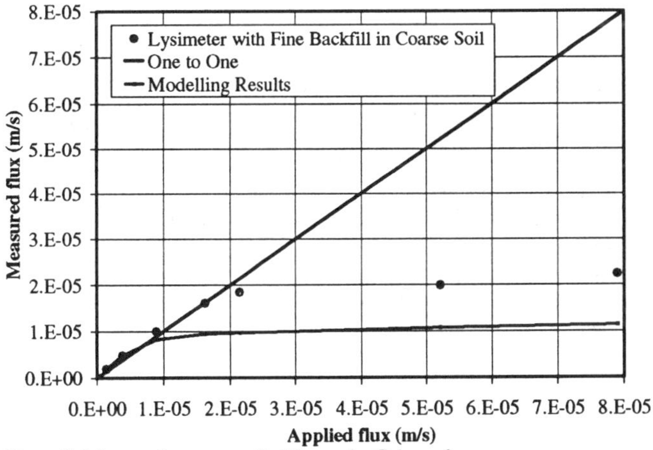

Figure 8. Measured versus applied fluxes for Column 2.

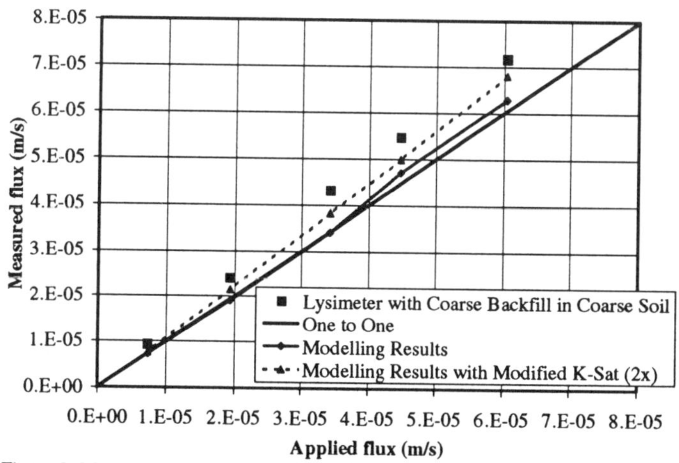

Figure 9. Measured versus applied fluxes for Column 3.

the pressure heads in the materials inside and outside of the lysimeter are equal, and the lysimeter collects only flux from directly above the lysimeter. For Column 3, the relation of measured to applied flux is linear. This is reasonable considering that for all fluxes applied, the pressures and hydraulic conductivity of the material both inside and outside of the lysimeter are the same.

The modelling results follow the same pattern as the experimental ones. Differences between the experimental and modelling results are likely due to inaccurate estimates of the saturated hydraulic conductivity used in the model. It is important to note that all properties were obtained from independent tests and no 'calibration' was done to improve the agreement between measured and predicted performance. The hydraulic conductivity values of the two materials can vary by as much as one order of magnitude because of density changes that occur as a result of different methods of placement. The sand outside and below the lysimeter was placed dry, whereas the material inside and above the lysimeter was placed and compacted wet. It was suspected that material placed wet would have a lower density and higher saturated hydraulic conductivity than the material placed dry. This hypothesis was tested in Column 3 by increasing the saturated hydraulic conductivity of the material above and within the lysimeter by a factor of two in a second modelling run. The results of the second run show much better agreement with the measured results confirming that the variation was likely due to changes in density caused by differing placement methods (Figure 9).

4 DISCUSSION

The following recommendations are given for the design and installation of lysimeters (Wickland 1998). The results from Columns 1 and 2 demonstrate that using a backfill that is different from the material in which the lysimeter is buried will cause gross measurement errors. Different materials develop different pressure head profiles under different infiltration rates causing horizontal gradients across the mouth of the lysimeter. Infiltrating water moves according to the gradient, either into or away from the lysimeter. To accurately measure infiltration, a lysimeter must be backfilled with the same material in which it is buried.

The depth of the lysimeter will also affect its performance as illustrated in the design of the lysimeter for the Kidston mine. A lysimeter must be deep enough to develop a constant pressure head for the anticipated rates of infiltration.

Density and water content significantly affect the hydraulic conductivity of a soil. The density of a lysimeter backfill should be as close as possible to that of the surrounding material. The same is true for water content. At the very least, measurements for water content and density of both the lysimeter backfill and surrounding material should be taken during construction.

Other suggestions include the use of vents at several levels to allow the backfill access to atmospheric pressure. Saturation of the soil above the lysimeter may cause the lysimeter to trap air. Without access to atmospheric pressure, the performance of the lysimeter may become unpredictable, causing measurement errors. In addition, tensiometers and water content probes should be installed in and around a lysimeter. A lysimeter make take years to collect a measurable amount of water because the rates of infiltration to be measured may be quite small. Water content probes allow the tracking of wetting fronts moving through the soil within the lysimeter. Tensiometers measure pore water pressure, and provide information on hydraulic gradients. These probes should be installed within the lysimeter, and at some distance away from the lysimeter at the same elevations.

5 CONCLUSIONS

A number of design issues must be addressed in order to obtain a properly functioning lysimeter. The geometry of the lysimeter, hydraulic properties of the backfill, and the flux from the base of the cover must be integrated into the design in order to ensure that the suction at the entrance to the lysimeter is the same both inside and outside, thus preventing the development of lateral gradients under unsaturated infiltration and ultimately measurement errors. The results of the laboratory column and numerical modelling study illustrated the key role that backfill selection and infiltration rate has on lysimeter performance, while a numerical modelling program for the design of lysimeters at Kidston showed that lysimeter wall height is also of significant importance. It was found that only the lysimeter which was filled with material the same as that in which the lysimeter was placed and with sufficient wall height would perform properly.

6 REFERENCES

Barbour, S.L. 1990. Reduction of acid generation in mine tailings through the use of moisture-retaining cover layers as oxygen barriers: Discussion. *Canadian Geotechnical Journal.* 27: 398-401.

Bews, B.E., Barbour, S.L., Wilson, G.W. and O'Kane, M.A. 1997. The Design of Lysimeters for a Low Flux Cover System over Acid Generating Waste. *Canadian Geotechnical Golden Jubilee Conference*, Pre-print Vol. 1: 26-33.

Fredlund, D.G. and Rahardjo, H. 1993. *Soil mechanics for unsaturated soils.* Toronto: John Wiley & Sons Inc.

Fredlund, M.D. 1997. Design of a knowledge-based system for unsaturated soil properties, Master's Thesis, University of Saskatchewan, Saskatoon, Saskatchewan, Canada.

GEO-SLOPE International Ltd. 1995. SEEP/W User's Manual. Geo-Slope International Ltd., Calgary, AB.

Kisch, M. 1959. The theory of seepage from clay-blanketed reservoirs. *Géotechnique*, 9:9-21.

Newman, L.L., Herasymuik, G.M., Barbour, S.L., Fredlund, D.G. and Smith, T. 1997. The hydrogeology of waste rock dumps and a mechanism for unsaturated preferential flow. *Proceedings International Conference on Acid Rock Drainage*, Vol. II: 551-565, Vancouver.

O'Kane, M., Bews, B.E., and Wilson, G.W. 1996. Laboratory Characterization of SoilCover and Waste Materials, Preliminary Numerical Modelling of Soil Cover and Lysimeter Design, and Field Work for the South Dump Research Program at Kidston Gold Mines. Progress Report 1, Prepared for Mr. Nick Currey, Senior Environmental Officer and Mr. Paul Ritchie, Environmental Officer of Kidston Gold Mine, Placer Pacific Ltd., Queensland, Australia.

Wickland, B. 1998. Lysimeter Backfill Design. CE 499.3 Project Report, Department of Civil Engineering, University of Saskatchewan, Saskatoon, Saskatchewan, Canada.

Tailings and Mine Waste '99 © 1999 Balkema, Rotterdam, ISBN 90 5809 025 6

Field performance monitoring of the Kimberley Operations siliceous tailings test plots

M.O'Kane & L.Ryland
O'Kane Consultants Incorporated, Saskatoon, Sask., Canada
R.T.Gardiner
Cominco Limited, Kimberley, B.C., Canada

ABSTRACT: The Cominco Ltd., Kimberley Operations constructed a total of seven field test plots on the siliceous tailings in 1993 and 1994 to assess alternate soil cover systems for acid rock drainage abatement and determine if a suitable growth medium could be established. The construction of the field test plots was a continuation of research on soil covers for tailings pond closure which began in 1978 and was initially focused on the establishment of a self-sustaining vegetation cover. Operation of a collection/treatment system and application of a soil cover system is part of the reclamation plan for the tailings ponds. The purpose of the soil cover system is to reduce the volume of tailings seepage requiring treatment by limiting infiltration due to precipitation and to provide a growth medium capable of sustaining vegetation that achieves land use and productivity objectives. Climate conditions at the Kimberley Operations are a key factor controlling performance of the test plots. The time of year that precipitation occurs appears to significantly control the percentage of the incident precipitation percolates to the tailings. In general, precipitation that contributes to snowpack or rainfall during the winter, will increase net percolation to the tailings while summer rainfall is buffered by the presence of cover material and does not percolate to the underlying tailings.

1 INTRODUCTION

The Cominco Ltd. Sullivan Mine is scheduled for closure in the year 2001 after more than 90 years of continuous operation. The iron-lead-zinc sulphide deposit is located in Kimberley, British Columbia, Canada. Cominco Ltd. developed a comprehensive decommissioning and closure plan that was submitted to the Ministry of Energy, Mines, and Petroleum Resources for British Columbia (Cominco Ltd., 1991). The plan, which is the subject of an intensive review involving public as well as federal and provincial regulatory agencies, describes measures for controlling acid rock drainage (ARD) generated in underground mine workings, waste rock dumps, and tailings ponds and for establishing vegetation that achieves land use and productivity objectives (Gardiner *et al.*, 1997).

This paper focuses on the field performance of siliceous tailings test plots constructed in 1993 and 1994. Background information on the Sullivan Mine is presented. The design of the field test plots is also presented through a discussion on the cover system design philosophy and a review of material properties. Field performance of the test plots including meteorological conditions during the monitoring period, is presented and a discussion on performance is provided.

2 BACKGROUND

Cominco Ltd. acquired the Sullivan ore body in 1909. The iron-lead-zinc sulphide deposit was discovered in 1892. Tailings disposal began in 1923 following development of a differential flotation process capable of separating complex ore into lead, zinc, and iron concentrates. A total of 90 million tonne of tailings containing pyrrhotite and pyrite have been discharged since 1923 to ponds occupying 373 hectares of land. Approximately 140 million tonne of ore and 15 million tonne of waste rock have been removed from underground during the 87 years the mine has operated.

Table 1 shows the mineral composition and metal concentrations of typical tailings in the siliceous tailings pond. The siliceous tailings are acid generating and oxidize rapidly following exposure to ambient conditions. The depth of oxidation is generally less than 0.5m with pH in the 2.0 to 2.5 range.

Table 1. Mineral composition and metal content of siliceous tailings (after Gardiner *et al.*, 1997).

Mineral	Composition (%)	Metal	Content (%)
Pyrrhotite	20	Iron	21
Pyrite	2	Sulphur	21
Sphalerite	< 1	Lead	0.6
Galena	< 1	Zinc	0.5
Quartz	5 – 10	Tin	0.03
Calcite	10 – 15		
Mica	0 – 5		
Chlorite	0 – 5		
Feldspar	2		

The siliceous tailings ponds are located in a relatively flat area with bedrock and till deposits forming topographic highs to the north and south. A thin veneer of till rests on the majority of the bedrock highs. However, relatively continuous till deposits exist between the bedrock highs and range in thickness up to 30m. Discontinuous glaciofluvial deposits of sand and gravel cover the till surface. Prior to tailings discharge, the pond area was a marshy flat that probably represents a groundwater discharge area (Harpley and Day, 1993).

2.1 *Climate*

Average annual precipitation is 409mm with 42% as snow, based on measurements recorded approximately 2.5km from the siliceous tailings test plot area over a forty year period up to 1987. The recording station was reactivated in 1992. Mean monthly annual temperature recorded over the forty-year period was 5°C with a maximum mean monthly temperature of 37°C and a minimum of –40°C. The site can be characterized on an annual basis as arid to semi-arid since an annual moisture deficit occurs. However, the site typically experiences hot dry summer conditions and can experience humid fall and winter conditions. It is not uncommon for rainfall to occur during the winter months. Spring freshet contributes significantly to flow in surface drainage courses.

2.2 *Acid rock drainage abatement*

A number of ARD abatement technologies are incorporated into the reclamation plan for the tailings pond and include: collection and treatment of seepage and surface water; interception and diversion of an up gradient surface water course; and construction of a soil cover system.

A system of ditches, pumps, and pipelines are in place to collect surface seepage from the tailings pond dykes. The seepage is combined with mine water and tailings effluent before being pumped to a treatment plant. The main source of ARD following closure, in terms of both volume and metal loadings, will be drainage from underground workings (Gardiner *et al.*, 1997). Hence, the need to treat mine drainage following closure was a key factor in determining the optimum soil cover system design for the tailings ponds. The drainage water treatment plant uses lime in a high-density sludge that neutralizes the acid and precipitates metal into a free draining, non-leaching sludge (Murdock *et al.*, 1994). The plant is designed to treat a normal maximum of 18,000L/min., and can be forced to 27,000L/min. Sludge is discharged to an exfiltration pond with a life expectancy of 150 years at expected post closure production rates. Mine drainage and seepage will be treated during a six to seven month period each year during winter and spring. Improvement to the effectiveness of seepage and groundwater collection in the tailings pond area is a continuous effort. Construction and monitoring of the siliceous tailings field test plots is part of an effort to evaluate alternate soil cover systems to control the ingress of oxygen and infiltration of water to the underlying tailings while also providing a medium for sustainable vegetation.

3 FIELD TEST PLOT DESIGN AND CONSTRUCTION

The test plots were constructed using a coarse low-density reject rock from the mill (float rock) and a cobbly non-plastic till. A total of seven test plots were constructed, as shown schematically in Figure 1, four in 1993 and three in 1994. The three additional test plots were constructed in 1994 since the field lysimeters constructed to monitor infiltration from the 1993 test plots were damaged during placement of the cover materials. The design of these three plots were identical to Plots 2, 3, and 4 and were designated Plots 2s, 3s, and 4s.

3.1 *Soil cover system design philosophy*

The philosophy used for the design of engineered soil cover systems based on current technology attempts to (from Barbour *et al.*, 1996):

1. Minimize the flux of oxygen and water to potentially acid generating mine waste,
2. Ensure that the soil cover system will maintain long term physical and ecological stability, where "long term" refers to the environmental liability associated with acid rock drainage, and
3. Use a design that integrates all aspects of the ecosystem, climate, weathering effects, soil cover layers, and waste material.

The reduction of oxygen diffusion coefficients, and as a result the control of oxygen flux, is the primary factor controlling acid generation. This reduction, together with a reduction of the water flux and in turn ARD migration from the waste facility, is the objective sought when engineered soil cover systems are placed over sulphidic waste material. It is difficult and usually not economically feasible in arid and semi-arid climates to construct a soil cover system which contains a layer that remains highly saturated thereby reducing oxygen transport. The soil cover system will be subjected to extended dry periods and therefore the effect of evapotranspiration will be significant. However, subjecting the cover system to evaporative demands can be beneficial in arid climates and result in a reduction of infiltration to the underlying sulphidic waste material. An upper cover surface layer possessing high storage capacity can be used to retain water during a precipitation event or freshet. Subsequent to the increase in storage in the upper layer, it would release a significant portion of pore water to the atmosphere by evapotranspiration during extended dry periods, thereby reducing the net infiltration across the soil cover system.

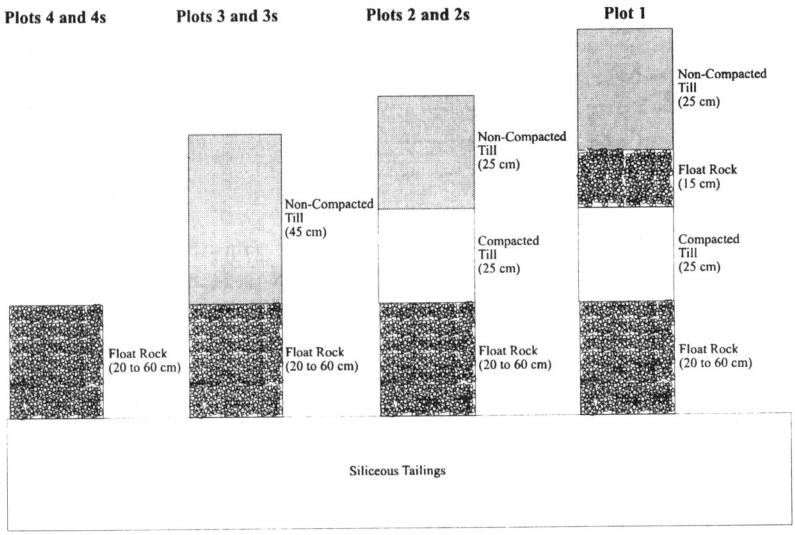

Figure 1. Schematic diagram of siliceous tailings test plot profiles.

3.2 *Cover and tailings material properties*

The till cover material is a non-plastic well graded material with a relatively low air entry value in the non-compacted condition, as shown in Figure 2. Compacting the till cover material increased the air entry value from 10 kPa to 30 kPa and decreased the saturated hydraulic conductivity by at least 2 orders of magnitude, as shown in Figure 3 and Table 1.

The consolidated and oxidized tailings and the un-oxidized tailings have air entry values equal to approximately 5 kPa. However, the un-oxidized tailings have more moisture storage at suctions less than the air entry value and less moisture storage at suctions greater than the residual suction, as compared to the consolidated and oxidized tailings. The upper 45cm of siliceous tailings consist of layers of material possessing varying degrees of oxidation and

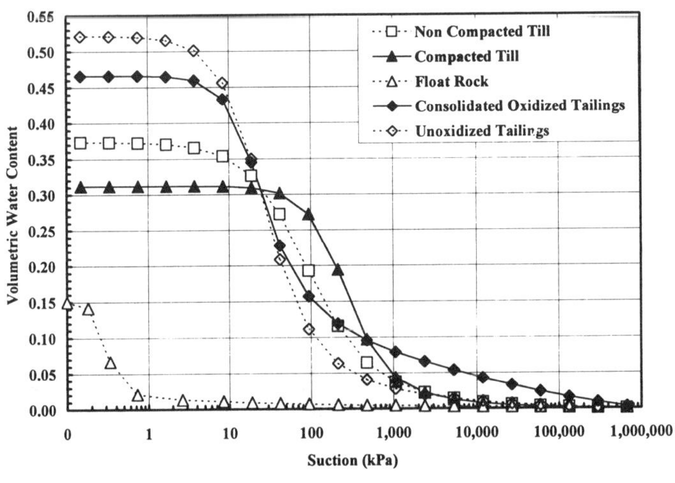

Figure 2. Soil water characteristic curves for the cover and tailings materials (OKC, 1996).

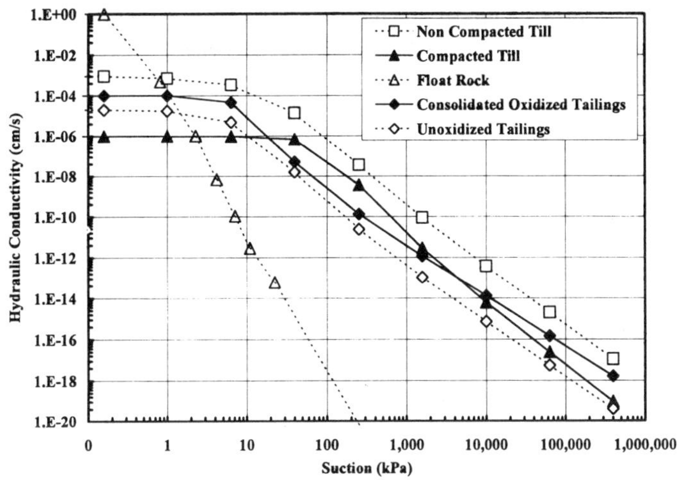

Figure 3. Hydraulic conductivity functions for the cover and tailings materials (OKC, 1996).

Table 2 Summary of cover and tailings material laboratory measured properties (from OKC, 1996)

Material	Porosity	Saturated hydraulic conductivity (cm/s)	Air Entry Value (kPa)
Non-compacted till	0.38	1×10^{-3}	10 kPa
Compacted till	0.32	1×10^{-6}	30 kPa
Float rock	0.15	1×10^{0}	0.1 kPa
Oxidized tailings	0.47	1×10^{-4}	5 kPa
Non-oxidized tailings	0.52	2×10^{-5}	4 kPa

consolidation. The material below the initial 45cm is un-oxidized siliceous tailings. The float rock is characterized as having little or no liquid moisture storage at even low levels of suction, as shown in Figure 2.

The siliceous tailings test plots were constructed on pond 3 of the siliceous tailings area. The tip of the piezometer nearest to the test plot area is approximately 20m below the tailings surface and is dry, as are the majority of the piezometers installed in the pond 3 siliceous tailings area.

4 FIELD TEST PLOT PERFORMANCE MONITORING

Meteorological conditions monitored in the test plot area included air temperature, precipitation, and relative humidity. Information on global radiation and sunshine hours are recorded at nearby regional meteorological stations and was used to supplement site measurements. Snow coarse measurements were taken subsequent to the onset of winter conditions and up to spring freshet. The suction and temperature at the mid-depth of the non-compacted and compacted till layer of each test plot, as well as the underlying tailings, was monitored since construction. Sensors were installed to measure the pH and redox potential of the siliceous tailings underlying plots 1, 2, 3, and 4. The field lysimeters monitored infiltration from the base of the cover systems and the gaseous oxygen concentration at the base of the cover system. The temperature, pH, and redox potential of water collected from the lysimeters was measured. The monitoring data presented and discussed in this report is focused on evaluating the reduction of infiltration and ingress of oxygen to the underlying tailings.

Gardiner *et al.* (1997) reported that biomass production on similar cover systems established in 1989 have consistently exceeded productivity of an adjacent undisturbed native range site.

4.1 *Climate*

The 1995, 1996, and 1997 rainfall and total precipitation, as compared to average monthly rainfall and total precipitation, are shown in Figures 4, 5, and 6 respectively. The average rainfall and total precipitation were based on mill site historical data adjusted to an October to September monitoring period. 1995, 1996, and 1997 can be characterized as being wet years in comparison to the historic average. Approximately 230mm more, or 57%, precipitation was recorded during the 1997 monitoring period as compared to the historic average. Approximately 95mm (24%) and 175mm (44%) more precipitation was recorded than the historic average for 1995 and 1996, respectively. However, rainfall recorded during the 1997 monitoring period was nearly one-third less than recorded during the 1995 and 1996 monitoring periods. Rainfall during the 1997 monitoring period was approximately equal to the historic average. Hence, water equivalent snowfall was more than 135% higher during the 1997 monitoring period, as compared to the historic average. Rainfall during the 1995 and 1996 monitoring periods was more than 100mm above average. The increase in annual rainfall recorded during 1996 was due to unseasonable significant rainfall in February and April 1996. In addition, rainfall during May, June, and July 1996 was above average. Water equivalent snowfall during the 1995 monitoring period was approximately equal to the historical average and 35% more than the historical average during 1996.

Siliceous tailings test plot snow coarse measurements during the 1997 monitoring period occurred on January 8, 1997 (144mm), January 29, 1997 (180mm), March 1, 1997 (169mm), April 1, 1997 (62mm), and April 15, 1997 (0mm). The snow course measurements at the test plot area illustrated the data presented in Figures 4, 5, and 6 since the highest spring 1997 water equivalent snow coarse measurement was more than 25% greater than that measured for 1996.

4.2 *In situ performance*

Identical pairs of instruments (designated A and B) were installed to obtain duplicate measurements from each test plot (NDM, 1996). Thermocouples and gypsum moisture blocks

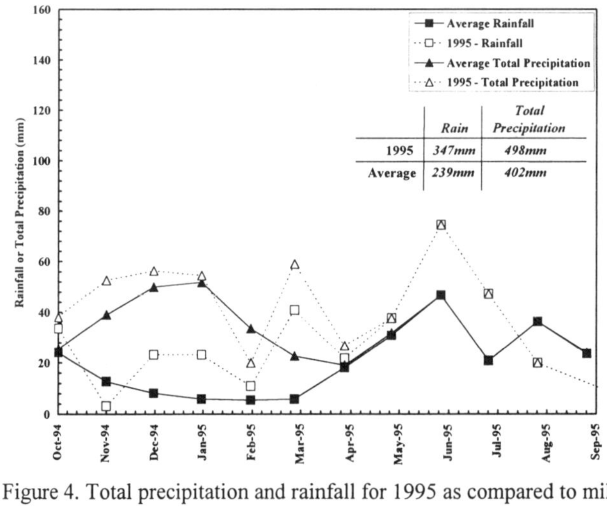

Figure 4. Total precipitation and rainfall for 1995 as compared to mill site averages (OKC, 1998).

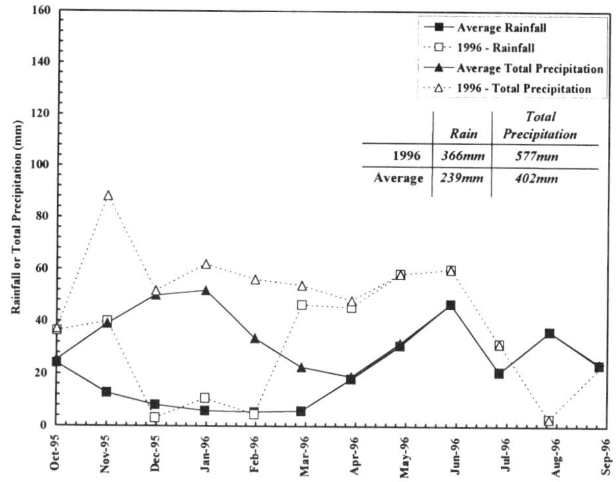

Figure 5. Total precipitation and rainfall for 1996 as compared to mill site averages (OKC, 1998).

(measurement of matric suction) were installed in each till layer and the underlying tailings.

Permanent pH and redox potential meters were installed in the tailings underlying the 1993 test plots. Lysimeters were installed to monitor infiltration.

Figure 7 presents *in situ* suction of the tailings underlying the test plot 3 and 3s cover systems and at the mid-depth of the non-compacted layer. Suction measured at the mid-depth of the non-compacted layer of plots 3 and 3s increased rapidly past the sensor measurement range in May, 1997 and remained at the high level throughout the summer of 1997. The suction was approximately 20 kPa during November, 1996 to April, 1997 inclusive. The maintenance of a low suction, and hence near saturated conditions, within the non-compacted till during the winter and spring was consistent to observations for the 1995 and 1996 monitoring periods. In addition, the sharp increase in suction at the beginning and end of summer was also consistent to that observed in 1995 and 1996.

Hydraulic performance of the test plot 3 and 3s non-compacted layer material was typical of the test plot 1, 2, and 2s non-compacted material. In addition, the comparatively consistent suction measured in the tailings underlying test plots 3 and 3s is typical of the tailings underlying the

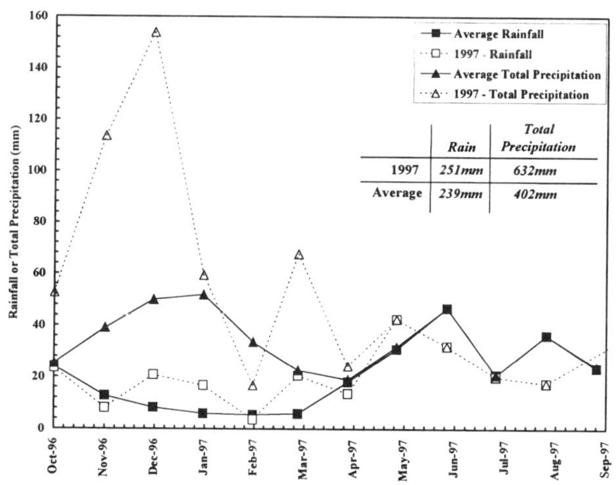

Figure 6. Total precipitation and rainfall for 1997 as compared to mill site averages (OKC, 1998).

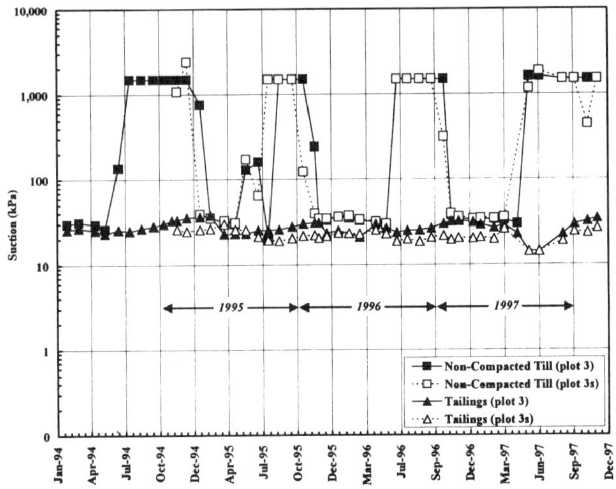

Figure 7. Measured suction in the non-compacted till and tailings of plots 3 and 3s for 1994, 1995, 1996, and 1997 (OKC, 1998).

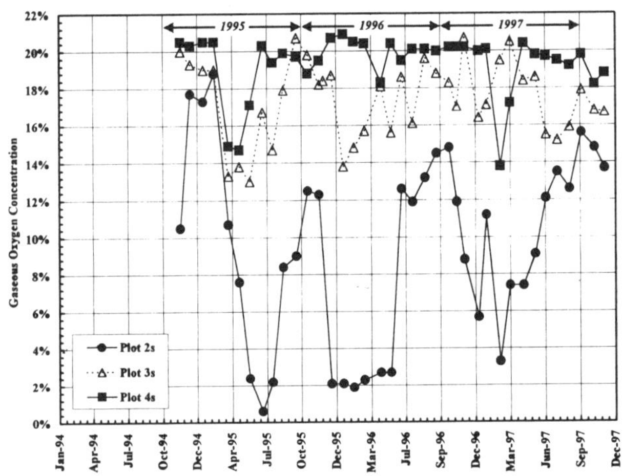

Figure 8. Measured gaseous oxygen concentration in the lysimeters underlying plots 2s, 3s, and 4s for 1995, 1996, and 1997 (OKC, 1998).

plot 1 and 2 cover designs. Suction in the tailings underlying the control test plot was approximately 20kPa during the monitoring period.

The gaseous oxygen concentration measured throughout the monitoring period for plots 2s, 3s, and 4s are presented in Figure 8. Atmospheric conditions were measured at the base of the float rock material of plot 4s during the 1997 monitoring period, similar to previous monitoring years. The gaseous oxygen concentration at the base of the plot 3s test plot ranged from 15% to near atmospheric conditions during the 1997 monitoring period. Plot 2s did limit oxygen ingress during the months of January to June 1997 inclusive as was observed during the same period in 1996, as shown in Figure 8. The measured gaseous oxygen concentration then increased from 3% to more than 16% during the summer of the 1997 monitoring period. It is reasonable to assume that the plot 1 cover profile would function in a similar manner since similar hydraulic performance was measured for the two test plots.

5 DISCUSSION

The float rock layer was intended to limit migration of salts to the root zone and function as a capillary break below a layer within the soil cover system that remained at a high degree of saturation. In general the diffusion of oxygen across a cover system is minimized if the degree of saturation of a layer of the soil cover system is at or near saturation. In short, oxygen will diffuse through air at a rate approximately four orders of magnitude greater than through water, given a similar gradient.

Evapotranspiration during the summer of 1997 led to a significant decrease in moisture content in the till layers of all test plots, including the compacted layer underlying the upper float rock layer of plot 1. Moisture within the cover layers during the spring of 1997 was a result of fall and winter 1996 infiltration as well as the 1997 spring freshet. The seasonal variation in moisture content during the 1997 monitoring period that characterized the hydraulic performance of the soil cover systems was consistent with that observed in 1995 and 1996. The measured gaseous oxygen concentration was reduced to 3% at the base of plot 2s for the month of February, 1997 and subsequently increased to almost 16% by the end of the 1997 monitoring period. Reliable data was not available for the plot 1 cover design, although it is reasonable to assume similar performance. The oxygen concentration at the base of the plot 3 and plot 4 cover designs varied between 14% and atmospheric conditions (≈ 20%) during the 1997 monitoring period which matched trends observed in 1995 and 1996.

Figure 9 and Table 3 compares the cumulative flux that reported to the field lysimeters during 1995, 1996, and 1997. The plot 2 cover profile reduced the net infiltration to the tailings by 5 times in 1995 and 4 times in 1996, as compared to the plot 3 cover profile. However, essentially the same volume of percolation reported to the lysimeters underlying plots 2s and 3s during the 1997 monitoring period. This did not agree with the two previous monitoring periods. The net infiltration to the plot 4s field lysimeter was approximately 73% of the 1997 incident precipitation, more than three times that measured for plots 3s and 2s.

6 SUMMARY

The test plots did not control the ingress of oxygen during the 1997 monitoring period, which continued a trend observed since their construction. In general, the cover material decreased in

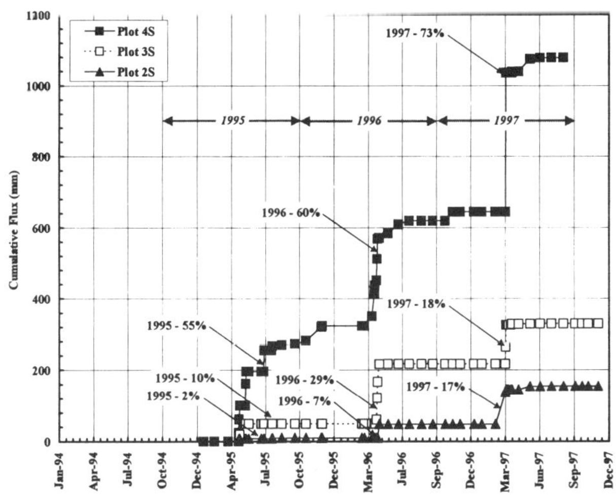

Figure 9. Cumulative measured net infiltration to the tailings underlying plots 4s, 3s, and 2s for 1995, 1996, and 1997 and net infiltration as a percentage of annual precipitation (OKC, 1998).

moisture content during the summer following moist winter and spring conditions. The plot 2, 2s, and 1 compacted layers did not possess sufficient moisture retention to prevent migration of moisture to the soil-atmosphere interface for subsequent evapotranspiration during the summer. Hence, measured oxygen concentration at the base of each test plot during the 1997 monitoring period was a function of climate conditions and *in situ* tailings conditions rather than being controlled by the presence of a tension saturated component of the cover system.

The volume of water entering the siliceous tailings underlying the test plots is a function of three site specific factors: the test plot profile (geometry, material type and preparation), the amount of rainfall during the fall (i.e. year round climate), and the amount of spring snowmelt (hence snowpack). This was illustrated when comparing the performance of the test plots during the 1996 and 1997 monitoring periods when snowfall was 80% more in 1997 as compared to 1996. The increased 1997 snowpack resulted in an increase in percolation to the underlying tailings during the 1997 spring freshet that was more than the over all increase in total precipitation for 1997. The difference between the rainfall and snowfall characteristics observed during 1996 and 1997 illustrates the impact of annual changes in climate and in particular when the changes occur during the year. It is also important to note that this variable cannot be controlled nor accurately predicted, while seemingly obvious is a certainty that is often overlooked when evaluating long term cover performance. Clearly, annual monitoring is a "snapshot" in time and performance of the test plots was a direct function of climate conditions. Cominco Ltd. supplemented the annual monitoring with field response and predictive soil-atmosphere numerical modelling.

The site can be characterized as experiencing an arid to semi-arid climate on an annual basis since an annual moisture deficit is prevalent. However, while arid conditions are prevalent during the summer the site typically experiences semi-humid to humid conditions during late fall, winter, and spring. The consequences of climatic variation at the site on soil cover system design and performance are significant because a system that maintains a tension saturated layer during dry climate conditions while also controlling water infiltration is not technically feasible using the available cover materials.

The increase in precipitation during the 1997 monitoring period led to an increase in percolation to the tailings underlying each test plot. Significant winter precipitation resulted in the increase in precipitation observed in 1997, which exacerbated the percolation increase to the tailings. It is anticipated significantly less percolation would have reported to the underlying tailings if the increase in 1997 precipitation was due to summer rainfall, even though the total precipitation for the year would have been the same. Summer rainfall is stored and later released as evapotranspiration during the typically hot dry summer months.

In summary, climate conditions continue to be a key factor controlling performance of the test plots. Higher annual incident precipitation will generally lead to an increase in percolation to the tailings underlying a given test plot. In general, precipitation that contributes to snowpack, or occurs during the winter, will increase net percolation to the tailings while summer rainfall is buffered by the presence of cover material and does not percolate to the underlying tailings.

REFERENCES

Barbour, S.L., Vanapalli, S., and O'Kane, M., 1996. A critical review of the Waite Amulet and Heath Steele Cover Systems. A MEND project report.

Cominco Ltd., 1991. Decommissioning and closure plan, Kimberley Operations. April, 1991.

Day, S.J. and Harpley, D.P., 1993. Survey of wells in the Kimberley-Cranbrook Region interpretation report. Norecol, Dames & Moore Inc., Vancouver, B.C.

Gardiner, R.T., Dawson, D.B., and Gray, G.G., 1997. Application of ARD abatement technology in reclamation of tailings ponds at Cominco Ltd., Sullivan Mine. *In: Proceedings of the Fourth International Conference on Acid Rock Drainage, Volume 1, pp. 47-63. May 31 – June 6, 1997, Vancouver, B.C.*

Murdock, D.J., Fox, J.R.W., and Bensley, J.G., 1994. Treatment of acid mine drainage by the high density sludge process. *In: Proceedings of the Third International Conference on the Abatement of Acidic Drainage, Volume 1, pp. 241-249. April 24 - 29, 1994, Pittsburgh, PA.*

Norecol, Dames & More, 1996. Evaluation of Cover Design for Tailings Ponds, Phase 3 - Site Test Plots, 1995 Monitoring Report, Cominco Ltd. Job No. 25141-031-310. March 26, 1996.

O'Kane Consultants Inc., 1996. Soil Cover Characterization and Numerical Modelling of the Siliceous Tailings Test Plots at Cominco Ltd., Kimberley Operations - Final Report. Prepared for Cominco Ltd., Kimberley Operations, December, 1996.

O'Kane Consultants Inc., 1998. 1997 Annual monitoring report on performance of the siliceous tailings test plots at Cominco Ltd., Kimberley Operations. Prepared for Cominco Ltd., Kimberley Operations, March, 1998.

Tailings and Mine Waste '99 © 1999 Balkema, Rotterdam, ISBN 90 5809 025 6

Some geophysical methods for tailings/mine waste work

David L.Campbell, Robert J.Horton, Robert J.Bisdorf, David L.Fey, Michael H.Powers & David V.Fitterman
US Geological Survey, MS 964, Denver Federal Center, Colo., USA

ABSTRACT: The geophysical methods of most help for addressing typical mine tailings and mine waste problems are generally geoelectrical in nature. These methods mainly measure the electrical conductivity of the ground, a property that reflects lithology, porewater saturation, and porewater chemistry. For dumps that are fairly uniform and somewhat moist, and as long as the pH of the porewater is 5 or less, conductivity can be a good indicator of acid generating potential. Induced polarization (IP) measurements have been especially useful in distinguishing mineralogy in mine dumps; in particular, they can help outline concentrations of sulfide minerals therein. Besides helping find concentrations of acid-generating materials inside mine dumps, geoelectrical methods also can help locate plumes of acidic/high TDS groundwater outside the dumps. For tracing plumes, electromagnetic (EM) techniques are often adequate and more cost effective than other methods. Ground-penetrating radar (GPR) methods give the most resolution where the ground is resistive enough to allow wave propagation, but they often fail at sites in the West where evaporation processes have concentrated salts near the surface of the mine dump or tailings deposit.

1. INTRODUCTION

The geophysical methods of most help for addressing typical mine tailings and mine waste problems are generally geoelectrical in nature. These geoelectrical methods (Telford et al. 1976) include direct current (DC), electromagnetic (EM), induced polarization (IP), and ground-penetrating radar (GPR) methods. Of these, DC, EM, and IP methods mainly measure the electrical conductivity of the ground, whereas GPR responds to a combination of conductivity and other physical properties. The results typically reflect a combination of lithology, porosity, and characteristics of the porewater, including porewater saturation, amount of total dissolved solids (TDS) in the porewater, and pH of the porewater. Saturated, high-porosity ground is more conductive than unsaturated, low-porosity ground. Water with high concentrations of TDS is more conductive than low TDS water. The presence of conductive minerals, such as clays (e.g. montmorillonite) or sulfides (e.g. pyrite) can significantly increase the electrical conductivity of the ground.

Within a mine waste pile, acidic water is produced by sulfide oxidation. This acidic water can have very high TDS concentrations. Conductivity anomalies are produced by acidic ground water plumes high in TDS. Zones of sulfide-rich waste contained within a waste rock pile may produce a conductivity anomaly due to the conductive nature of the sulfide mineral(s). Furthermore, acidic conditions may break down feldspar minerals in mine waste to produce clay and dissolved ions, also leading to increased conductivity. Unfortunately, most mine waste sites also contain buried metallic junk such as steel cable, pipes, rails, and corrugated steel roofing, which produce additional "false" conductive anomalies.

2. DC EXAMPLES

Both DC and EM methods are usually set up to obtain data either as "soundings", in which electrical conductivity is determined as a function of depth below a particular site, or as "profiles", in which lateral variations of conductivity are measured along surveyed lines or grids of lines. Some kinds of DC and EM methods, such as dipole-dipole DC or multifrequency EM, are able to do both sounding and profiling simultaneously. Most DC and EM transmit currents into the ground and use receivers to measure the voltages that result. Typically, the DC methods transmit and receive signals via electrodes that are in contact with the ground, whereas the EM methods use induction coils that need not touch the ground. Results from DC work are usually reported as electrical resistivity, in units of ohm-m, whereas those from EM work are reported as electrical conductivity, in units of milliSiemens/m (mS/m).

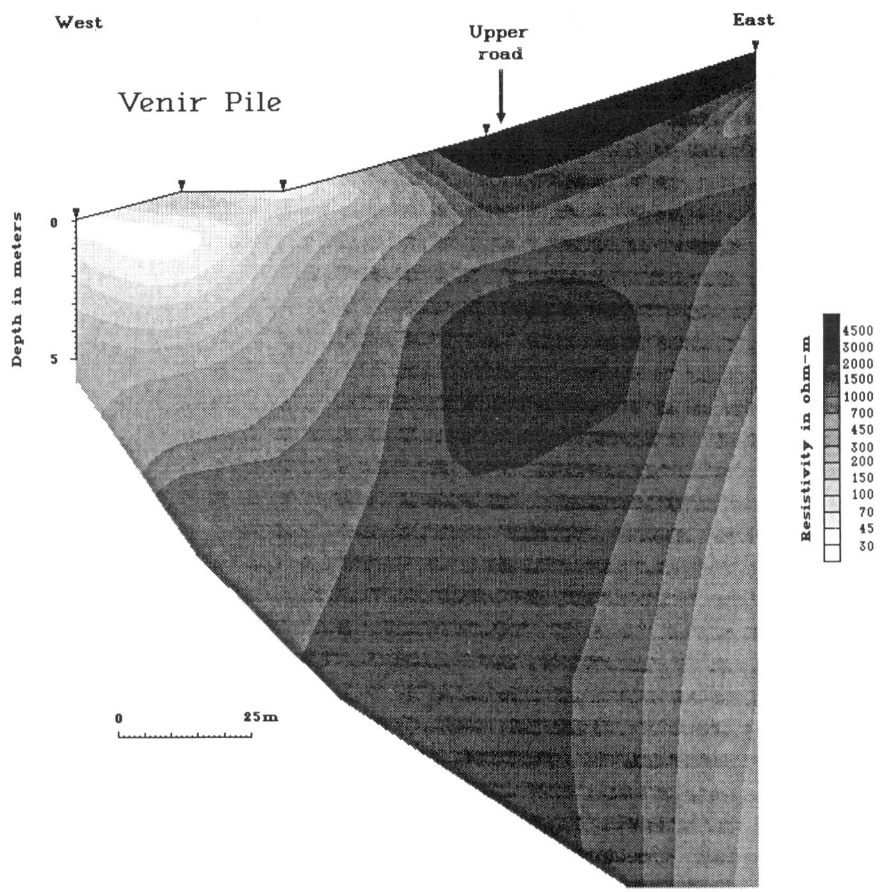

Figure 1. DC resistivity cross section across Venir mine pile.

2.1 The Venir mine pile

Figure 1 shows a resistivity cross section across the Venir mine pile, near Leadville CO. This cross section was made by contouring resistivity-depth results from five individual DC soundings, three of which were on the Venir pile itself, and two more of which were located on the hillslope above the Venir pile. The section shows that the waste pile material is considerably less resistive (more conductive) than the surrounding host rocks -- material on or near the waste pile had resistivities on the order of 10s of ohm-m, typically under 70 ohm-m. These low resistivity values probably reflect electrically conducting porewater in the relatively-high-porosity dump material; indeed, a slightly higher resistivity is seen at the crest of the pile where the material may be somewhat drier. A low-resistivity channel extends under the pile that may reflect groundwater reacting with the dump material as it moves along its base. Resistivities of the surrounding undisturbed areas are on the order of 1000 ohm-m and more. Resistivities on the section in the intermediate range, say 100-1000 ohm-m, may result in this case from our conservative interpretation approach, which avoids assigning abrupt resistivity contrasts to precise depths when a more smoothed-out transition (as shown here) fits the data just as well. The general shape of the pile, or at any rate its associated electrically-conducting porewater, is nevertheless apparent.

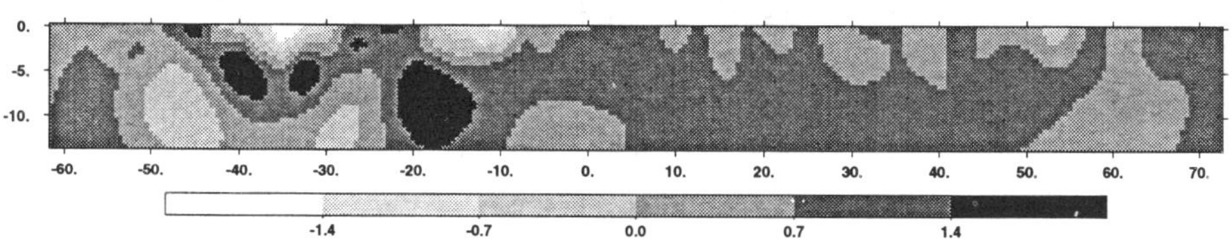

Figure 2. DC conductivity cross section along a portion of the face of the Mayday mine dump.

2.2 The Mayday mine dump

Figure 2 shows a conductivity cross section along a portion of the dump face of the Mayday mine dump, near Silverton CO. The data shown was collected using a pole-dipole array, so as to do simultaneous sounding and profiling. Individual pockets of both high conductivity (>~10 mS/m) and low-conductivity (<~1 mS/m) material are apparent. For reasons explained below, we interpreted the Mayday dump to be quite uniform in composition, with no high concentrations of sulfides present. If that is so, it implies in turn that the conductive pockets on Figure 2 reflect not lithology so much as porewater conductivity. Such porewater should have a pH appropriate to, and contain solids produced by, interactions with nearby grains of dump material.

Figure 3 shows an empirical correlation between electrical conductivity of leachate from samples of mine waste and the pH of these waters. The data was obtained using the EPA1312 protocol (Environmental Protection Agency 1995). We see that for acid leachate (pH < ~ 5) conductivity correlates inversely with pH. Note that the conductivity values

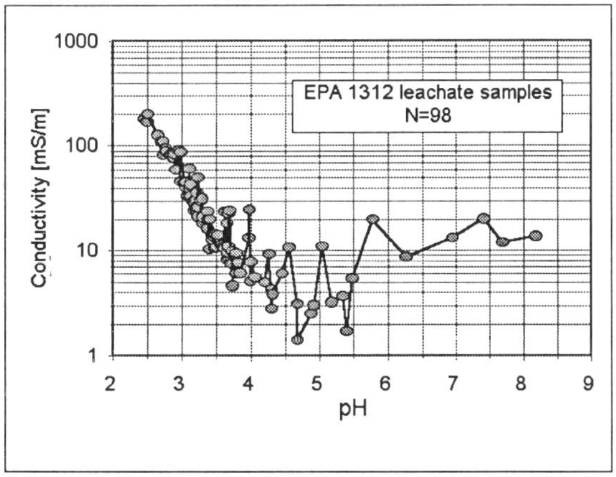

Figure 3. Conductivity vs. pH for leachates from 98 samples of tailings and mine waste from the western United States.

in Figure 3 are for leachate alone, not porewater in a porous matrix of grains. The bulk conductivity of such a porewater-matrix mixture is typically 1/5 - 1/20[1] of that of the porewater alone.

Water that drains from the Mayday dump falls in the pH<5 range. We therefore interpret that the conductivity section in Figure 2 reflects locations within the dump where the porewater in equilibrium with nearby grains of dump material has especially low pH. Figure 2, in other words, is a sort of "acid-generating potential" map for the Mayday dump.

3. EM EXAMPLES

In terms of the final product, EM methods accomplish about the same ends as do DC methods -- both give soundings, profiles, or both that map conductivity of the ground. Depending on the job in hand, either one or the other method may have the advantage. Because EM surveys can be done by only one or two crew members, and because most EM gear does not require the crew to emplace electrodes in the ground, EM surveys can often be done more quickly and cheaply than DC ones.

3.1 Fossil plume at Sunday #2 pile

Figure 4 is a conductivity profile across the slope downhill from the Sunday #2 pile, near Leadville CO. The profile crosses a denuded zone where acid runoff from the waste pile has killed vegetation. When the survey was made, the ground was moist but no runoff was observed. The data was taken by one person using standard EM gear that directly displays conductivity values. The profile shows higher conductivities over the path of a fossil plume of acid leachate from the mine pile. We think the plume has now dried up, and that the EM unit is responding to residual evaporate salts that the plume left behind. EM methods are known to work well to map plumes of contaminated groundwater near landfills (Lucius & Bisdorf 1995), and we think they should be similarly effective in mapping plumes of mine waste leachate.

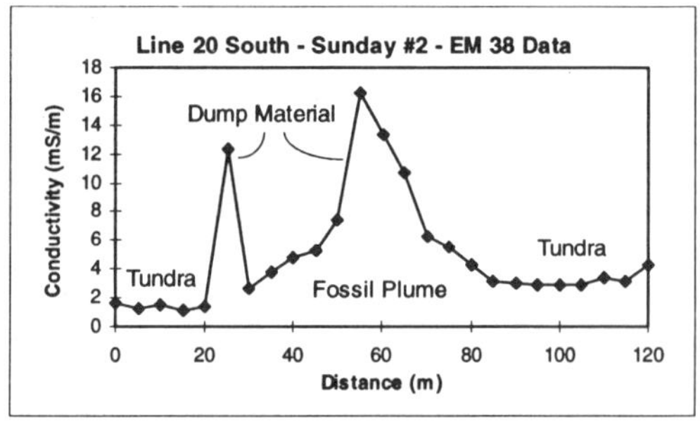

Figure 4. Conductivity profile across a fossil plume.

4. IP EXAMPLES

The IP method measures charge transport (conduction) and charge storage (polarization) processes. Polarization effects are produced by current-induced, electron transfer between ions in the groundwater and minerals composing the rock or sediment. When the IP transmitter sets up electric fields in the ground, the ions present are redistributed, producing a temporary polarization effect. When the electric field is turned off, the ions return to their normal distribution.

One kind of IP work involves "spectral" (SIP) measurements, in which resistivity and phase are measured as a function of frequency. Frequency dependent electrical properties can be characteristic of a particular substance. For two materials, such as sulfide-bearing and nonsulfide-bearing rock, the frequency dependent electrical properties may be different enough to distinguish them by their IP signatures. Figure 5 contrasts two samples of sulfide-bearing material with two samples of material from the Mayday dump in which the original sulfides have

---

[1]This "Archie's law" coefficient depends on porosity, grain shape, and other factors.

mostly been altered to sulfates (Campbell et al 1998). It appears that these two kinds of material can be distinguished by their different phase values at low frequencies.

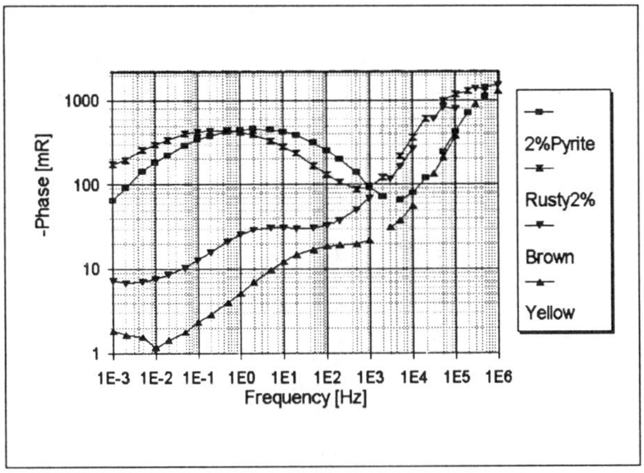

Figure 5. Laboratory measurements of SIP phases for 4 samples. 2% Pyrite sample is fresh; Rusty 2% is same sample with limonite alteration; lower 2 samples are material taken directly from Mayday dump.

4.1 Mayday and Yukon mine dumps

Figure 6 is a cross section showing interpreted phase values at the frequency of 0.125 Hz at the Mayday dump (above) and the nearby Yukon mine dump (below). All phase values in the Mayday are low, indicating a lack of polarization processes. This implies that any sulfides that originally may have been there have now been altered to sulfates. The Yukon mine dump, by contrast, still contains pockets of material having high phases like those on Figure 5 that characterize sulfides. We accordingly interpret that the Yukon dump may contain yet-unaltered sulfide concentrations.

One cautionary note: using a magnetometer we checked the locations of the high-phase anomalies at the Yukon, and found one of them to be very magnetic. Just under the surface at this location was a crumbling cement foundation containing rusty iron rebar. So not all high-phase anomalies are due to sulfides!

As stated above, the uniform low-phase values we found in the Mayday dump led us to interpret that there is little sulfide material now left therein. Bulk samples at the Mayday contained less than 1% pyrite, while those at the Yukon had about 6% pyrite (Sutley, written commun. 1998) The Mayday dump nevertheless generates acidic drainwater. USGS geochemists are currently investigating mechanisms by which acidic waters may be produced from low-sulfide dumps like the Mayday (Leinz et al. 1998, Smith et al. 1998).

5. GPR EXAMPLE

A good introduction to GPR work is given by Daniels (1989). GPR is useful, among other things, for finding buried artifacts and outlining stratigraphic details in fluvial and aeolian deposits. Its drawback is that it cannot penetrate materials with high electrical conductivity, such as mineralogical clays. (GPR waves often *will* propagate well in "clays" made merely of fine particles, if those particles do not consist of true mineralogical clay.) Experience has taught us to expect a poor GPR response anywhere that near-surface conductivity is greater than about 30 mS/m. Our attempts to use standard GPR methods to investigate mine dumps and fluvial tailings deposits near Leadville CO were unsuccessful, apparently because the GPR signals would not propagate into the ground. We speculate that evaporative processes that concentrate electrically conducting salts near the surfaces of such bodies, and acidic or high-TDS porewater inside them, result in high near-surface conductivity values. We think it unlikely that standard GPR methods will work well on such targets anywhere in the West.

5.1 Coeur d'Alene River

GPR did help outline possible stratigraphic locations where fluvial mine tailings were redeposited on the floor of the Coeur d'Alene River (Campbell et al. 1997). Figure 7 shows interpreted slump blocks in the river floor that we think may have been formed during the great

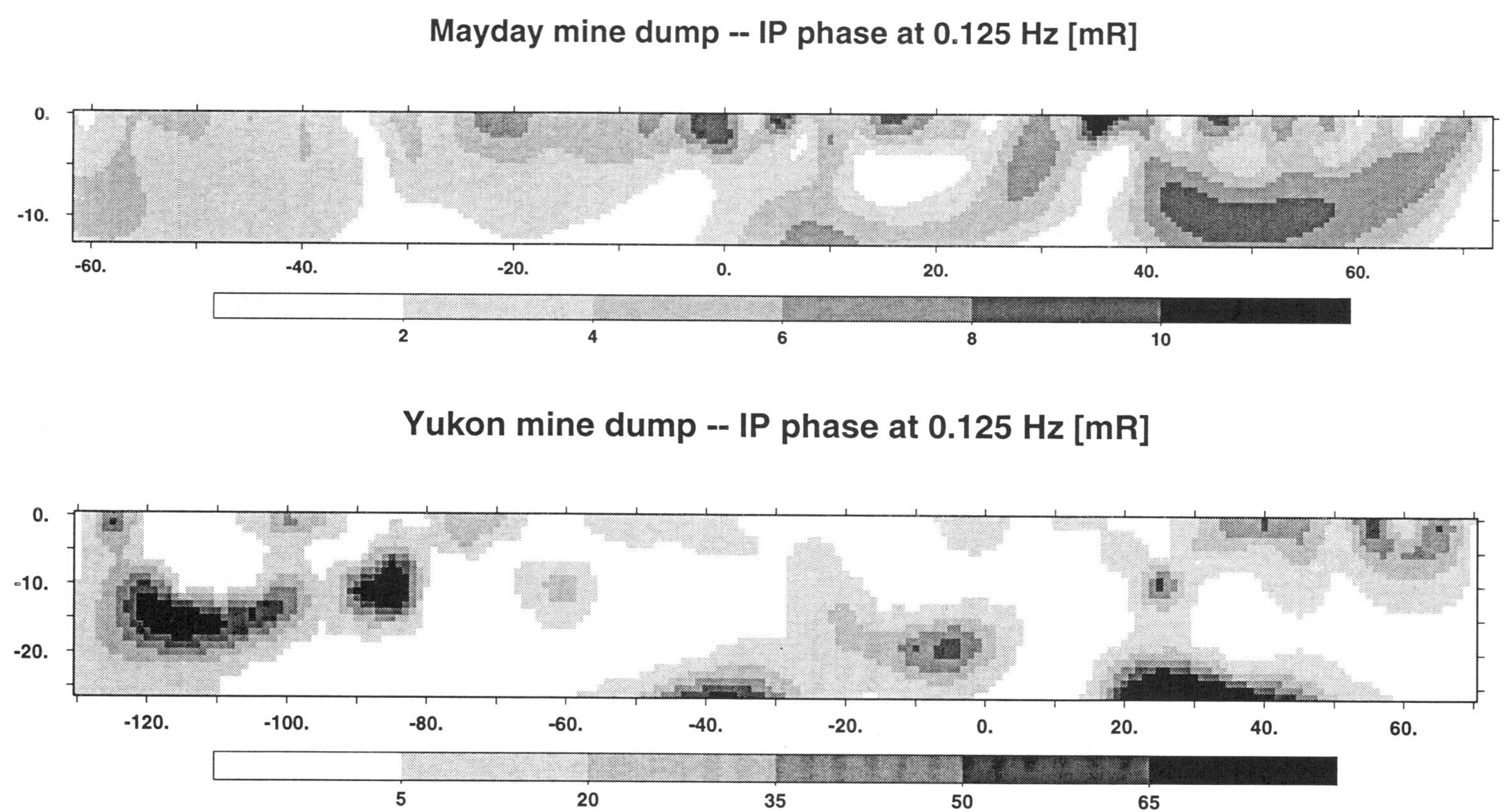

Figure 6. Interpreted SIP phase values in milliradians (mR) at the Mayday (above) and Yukon (below) dumps. Distance coordinates are in feet. Note the different phase scales.

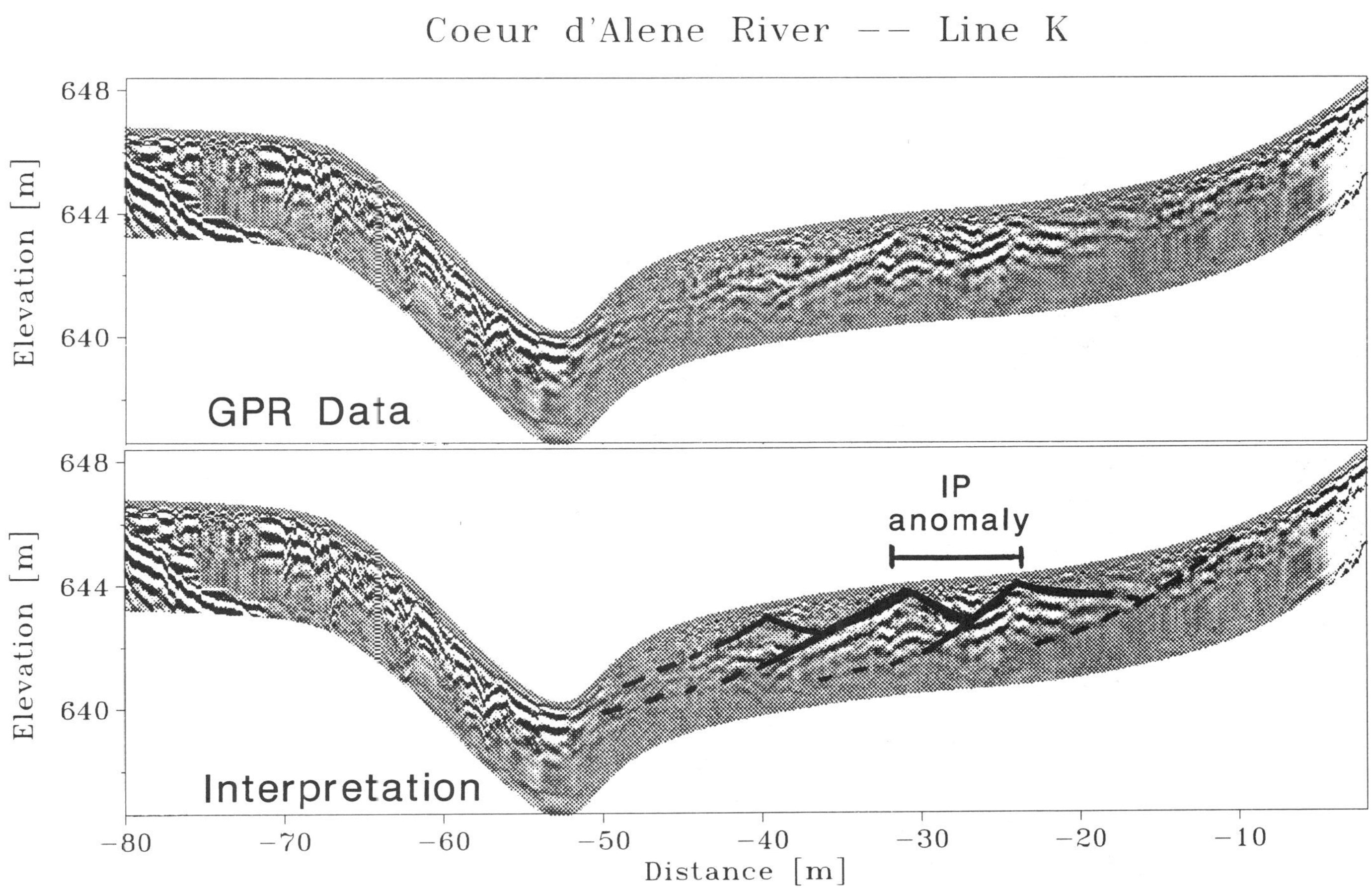

Figure 7. GPR data and interpretation of a profile across the Coeur d'Alene River.

flood of 1974. We think high-metal-concentration tailings exhumed from locations further upriver were redeposited in one of the Vs formed at the tops of these slump blocks. Such concentrations were indicated by high IP responses over the V.

6. OTHER GEOPHYSICAL METHODS

6.1 Self-potential

The self-potential (SP) method measures electrical potentials generated from naturally occurring electrochemical, electrokinetic, and thermoelectric processes. Electrochemical potentials can be generated by several processes including, but not limited to, ion diffusion due to chemical gradients produced by dissolution of waste rock and electron transfer associated with oxidation-reduction reactions involving sulfide minerals. Electrokinetic potentials are generated by fluid flow. Runoff and ground water movement through fractured or porous media, such as a waste rock pile, could produce electrokinetic potentials. Thermoelectric potentials are generated by diffusion of ions and electrons due to thermal gradients. Exothermic chemical reactions, such as sulfide oxidation, could produce thermal gradients and generate thermoelectric potentials. We are now doing SP surveys at mine piles and in the laboratory to try to detect places where such processes, particularly sulfide oxidation, may be taking place (Vaughn et al. 1998a,b).

6.2 Magnetics

Magnetic surveys respond to the presence of ferromagnetic minerals, such as magnetite, and can detect buried metallic objects composed of iron and steel. Metallic objects buried at shallow depths generate high-amplitude magnetic anomalies that are easily identified. For mine waste studies, the primary use of the magnetic data is to identify metallic objects. Buried iron and steel objects generate conductivity anomalies. The corrosion of iron may produce "false" SP and IP anomalies. The magnetic data helps to discriminate conductivity, IP, and SP anomalies produced by buried metal from those produced by "natural" causes. (Cf. Section 4.2, above.)

There are, however, natural processes that may vary the magnetic character of the mine waste. Magnetite is stable at relatively high pH. Acidic conditions may reduce magnetite to a non-magnetic iron oxide. If the mine waste contains magnetite, or other magnetic minerals susceptible to acid reduction, the magnetic signature of the waste may change over time due to acidic conditions generated by the oxidation of sulfide minerals. In order to observe such changes in the waste's magnetic signature, measurements would have to be repeated periodically.

7. SUMMARY AND CONCLUSIONS

Geophysical surveys can be used to help characterize mine waste sites. The geophysical methods that measure the conductivity of the ground, DC, EM, and IP, all indicate the waste rock can be more conductive than the surrounding undisturbed areas. The high conductivity of the waste rock is attributed to its high porosity, relative to the host rock; acidic and high TDS porewater; and high concentrations of conductive minerals including sulfide minerals and clay. EM methods probably can be used to trace acidic/high-TDS groundwater plumes from mine waste and tailings. High polarizability observed in IP data can indicate the presence of sulfide minerals, and/or clay minerals in the waste rock.

Magnetic surveys provide an effective means of identifying buried metallic objects that can complicate the interpretation of conductivity and SP data by producing "false" anomalies. Active oxidation-reduction reactions within mine waste piles should produce observable SP anomalies. Standard GPR methods generally have not worked well for direct investigations of mine waste and tailings deposits in the arid West, but may have a limited role in outlining fluvial and aeolian stratigraphy relevant to particular mine waste problems.

REFERENCES

Campbell, D.L., D.V. Fitterman, A.S. Hein, & D.P. Jones 1998. Spectral induced polarization studies of mine dumps near Silverton, Colorado, *Symposium for the Application of Geophysics to Engineering and Environmental Problems 1998*, Wheat Ridge CO: Environmental and Engineering Geophysics Society, p. 761-769.

Campbell, D.L., J.C. Wynn, S.E. Box, A.A. Bookstrom & R.J. Horton 1997. Tests of ground penetrating radar and induced polarization for mapping fluvial mine tailings on the floor of Coeur d'Alene River, Idaho. *Symposium for the Application of Geophysics to Engineering and Environmental Problems 1997*, Wheat Ridge CO: Environmental and Engineering Geophysics Society, p. 81-87.

Daniels, J.J. 1989. Fundamentals of ground penetrating radar. *Symposium for the Application of Geophysics to Engineering and Environmental Problems 1989*, Wheat Ridge CO: Environmental and Engineering Geophysics Society, p. 61-142.

Environmental Protection Agency 1995. Revision 28, dated April 4, 1995, to *Test methods for evaluating solid waste, Vols I and II (SW-846)*, 3rd Ed., November, 1986.

Leinz, R.W., S.J. Sadly & P.H. Briggs 1998. The use of sequential extractions for the chemical speciation of mine wastes. *This volume.*

Lucius, J.E. & R.J. Bisdorf 1995. Results of geophysical investigations near the Norman, Oklahoma, municipal landfill, *U.S. Geological Survey Open-File Report OF95-0825.*

Smith, K.S., J.F. Ranville, P.J. Lamothe, A.L. Meier & K. Walton-Day 1998. Metal leaching through a fluvial tailings deposit along the upper Arkansas River, Colorado. *This volume.*

Telford, W.M., L.P. Geldart, R.E. Sherriff & D.A. Keys 1976. *Applied Geophysics*, New York: Cambridge University Press.

Vaughn, R.B., R.J. Horton, M.R. Stanton, D.L. Fey, S, Sutley & P. Theodorakos 1998a. Antfarms: mine dump simulation studies in the laboratory. *This volume.*

Vaughn, R.B., M.R. Stanton & R.J. Horton 1998b. A year in the life of a mine dump: A diachronic case study. *This volume.*

Tailings and Mine Waste '99 © 1999 Balkema, Rotterdam, ISBN 90 5809 025 6

Undrained shear strength determination using dynamic pore pressure measurements from CPT data

D.R.Jenevein & J.M.Johnson
Golder Associates Incorporated, Denver, Colo., USA

S.Vahdani
Golder Associates Incorporated, Oakland, Calif., USA

ABSTRACT: A method is presented using cone penetration testing (CPT) data to determine undrained shear strength in milled mine tailings. The method uses the dynamic pore pressure measured during penetration and calculates the effective stress at failure. This method produces a relationship of undrained shear strength for the tailings versus depth which is believed to be more accurate and that can the be used to better predict behavior of the tailings.

1 INTRODUCTION

The traditional method of assigning shear strength parameters to milled tailings is to use correlations with tip resistance and sleeve friction without reference to measured pore pressures to estimate total stress strength parameters. This method is conservative and may be reasonable if the conditions or planned loading sequence is not critical. However, where more critical circumstances exist, a more detailed or accurate model describing the tailings shear strength may be useful.

The method of deposition of the tailings can have significant bearing on the shear strength of the tailings. For a tailings impoundment that has been filled using sub-aqueous deposition, the material would be expected to posses behavior consistent with a normally consolidated, soft soil deposit. However, for impoundments employing sub-aerial deposition, the tailings are allowed to desiccate prior to re-deposition. This causes the material to achieve an overconsolidated state prior to the subsequent lift(s) of tailings being deposited. Therefore, the material would be expected to behave like overconsolidated soil with higher shear strength, lower void ratio and lower induced pore pressures during shearing.

For this paper, a strength model is presented using field and laboratory data obtained for milled gold tailings that uses a measure of this overconsolidated behavior to better describe the shear strength of the tailings.

2 INVESTIGATION

A field and laboratory investigation was carried out to determine the physical properties of the tailings materials. As part of the field investigation, 37 CPT borings and 5 geotechnical borings were drilled. From the geotechnical borings, laboratory testing was performed to determine index, consolidation, and shear strength properties of the tailings material. The laboratory testing was used to calibrate the CPT data for material classification and shear strength model verification.

3 MATERIAL PROPERTIES

The materials contained in the tailings were roughly grouped into two categories, sands and slimes. The sands are basically medium-to-fine grained uniform sands with varying amounts of silt-to-clay sized fines. The slimes are a low plasticity silty-clay to clayey-silt, and contain trace to little amounts of fine sands. The index testing from the investigation program are summarized in Table 1.

Table 1. Material Properties of Tailings Material

	Misture Content	Atterberg Limits		Grain Size Analysis		Unit Weight		Specific Gravity
				%Passing	%Passing	Wet	Dry	
	%	LL	P.I.	#4	#200	pcf	pcf	
Total Number of Tests	100	17	17	64	90	37	37	10
Max. Value	56.2	40	22	100	100	129.8	104.3	2.85
Min. Value	5.2	27	7	100	13	95.2	79.9	2.7
Ave. Value	28	32	12	100.0	62.0	121.9	93.7	2.8

4 INTERPRETATION

The CPT data were initially separated using the Classification Index, Ic developed by Jefferies and Davies, (1993). From the laboratory data taken from the companion geotechnical borings, the materials were classified into two groups: 1) very fine materials with classification index (IC) greater than 2.54 and 2) material containing various amounts of sand with IC less than 2.54. This cutoff value was determined to be the boundary at which the material changed from sandy silts or clays to pure silts or clays. The later materials would be expected to behave essentially in an undrained manner when sheared.

After discretizing the CPT data by material type, the undrained shear strength of silt tailings layers (IC > 2.54) was then calculated using the following equation:

$$S_u = \frac{Qt - \sigma_v}{Nk} \tag{1}$$

Where Qt is the corrected cone tip resistance and σv is the total overburden pressure. Nk was taken to be 8 as determined from the back analysis of the triaxial shear strength data. The Su values computed for silts using this procedure indicated that the shear strength of the material varied considerably at different locations and depths. A plot of shear strength versus depth for 7 of the 37 CPT soundings for materials with Ic values greater than or equal to 2.54 are shown in Figure 1.

To identify the cause of this behavior, the variation of pore water pressure generated by the CPT probe with undrained shear strength was then carefully examined. It was observed that the high shear strength values corresponded with low induced pore water pressure values and low shear strength values corresponded with high induced pore water pressure values. This correlation was found to be very strong.

To further investigate the relationship between Su and pore water pressure, the Su values were plotted as a function of induced effective stresses which were computed by subtracting the instantaneous (hydrostatic plus induced) pore water pressures (as opposed to hydrostatic) from the total overburden pressure as shown in Figure 2.

After observing this relationship, it appeared that the Su values are independent of effective confining pressure where effective stresses are close to zero or negative, whereas they increase with increasing positive effective stresses. This behavior indicates that the fine grained tailings (silts) have both cohesion (portion of strength which is independent of effective stresses) and a

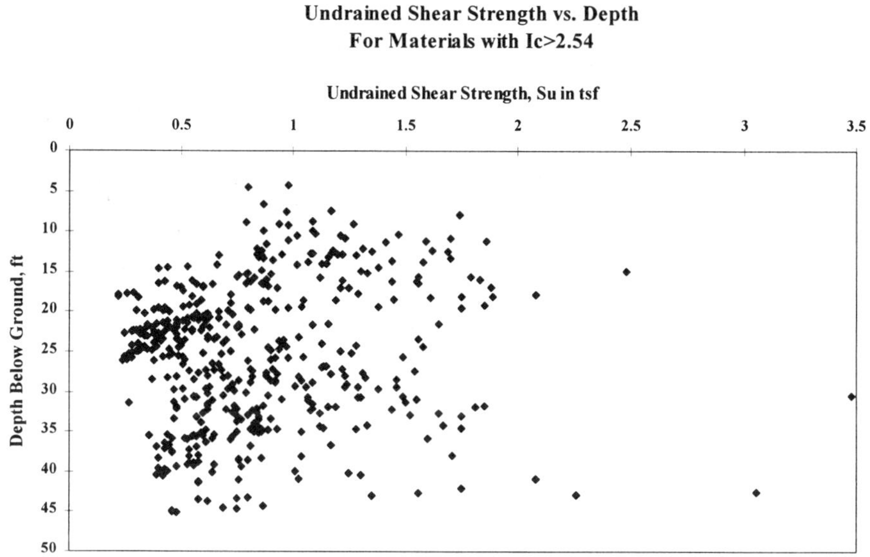

Figure 1. Undrained shear strength versus depth below ground for fine grained material

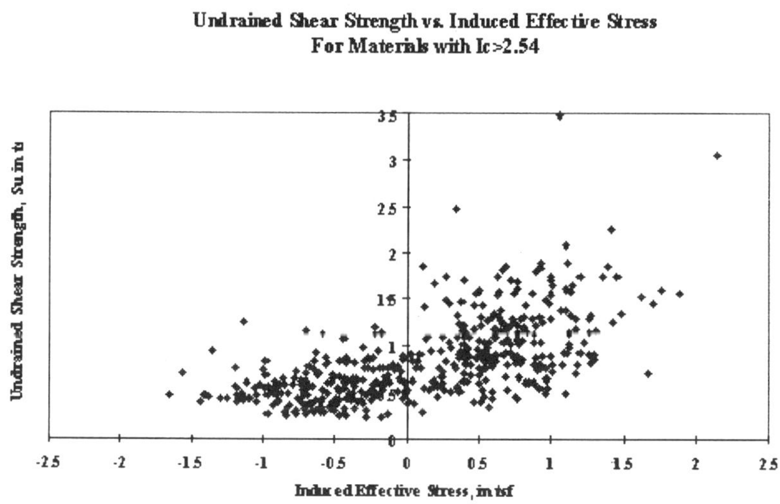

Figure 2. Undrained shear strength versus induced effective stress for fine grained material.

non-zero friction angle (portion of strength which depends on the effective stresses). Therefore, the most appropriate model would be one that recognizes both components.

5 SHEAR STRENGTH MODEL

After recognizing that the shear strength of the fine grained tailings material possess a behavior that has both a cohesive and frictional component, the soil deposits were divided into 5 foot thick layers to evaluate the relationship at various depths. It was then observed that the shear strength data indicated that the cohesion component (Su values corresponding to the negative effective stresses) did vary with depth of deposit or effective consolidation pressure. From the examination of data presented in Figure 2, we have developed the following undrained strength parameters for the fine grained tailings (silts):

$$Su = 0.31\ \sigma cv + \sigma v' * \text{Tan}\ (20°) \quad (2)$$

Where σcv is the vertical effective consolidation pressure and σv' is the instantaneous effective stresses. Figures 3 and 4 show the undrained shear strength measured from CPT data versus induced effective stress for two zones of material along with the shear strength envelope as given by equation 2.

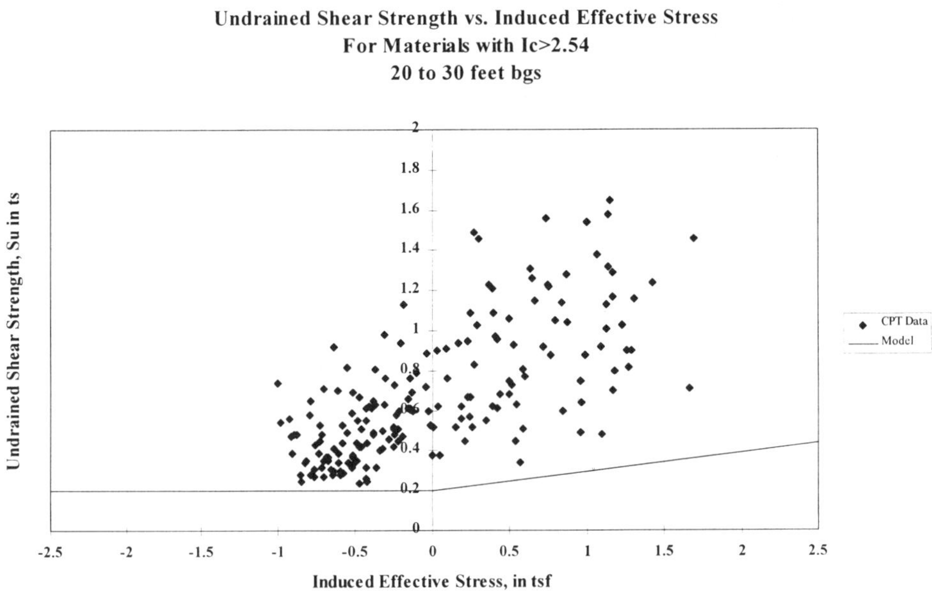

Figure 3. Undrained shear strength versus induced effective stress for fine grained materials from 20 to 30 feet below ground surface and model prediction.

Equation 2 represents the undrained shear strength of fine grained tailings on a horizontal plane. However, since the effective consolidation pressure varies with the orientation of plane of failure, the material possesses anisotropy. For example, the undrained shear strength Su on a plane of failure which is 45 degrees to the horizontal can be calculated as follows:

$$Su = 0.23\ \sigma cv + \sigma v' * \text{Tan}\ (20°) \quad (3)$$

Similarly, for a vertical plane of failure:

$$Su = 0.16\ \sigma cv + \sigma v' * \text{Tan}\ (20°) \quad (4)$$

For a normally consolidated tailings where pore water pressure is hydrostatic the equation reduces to:

$$Su = 0.31\ \sigma cv + \sigma v' * 0.35, \text{ or} \quad (5)$$

$$Su = 0.66\ \sigma v', \text{ or} \quad (6)$$

$$Su = \sigma v' * \text{Tan}\ (34°) \quad (7)$$

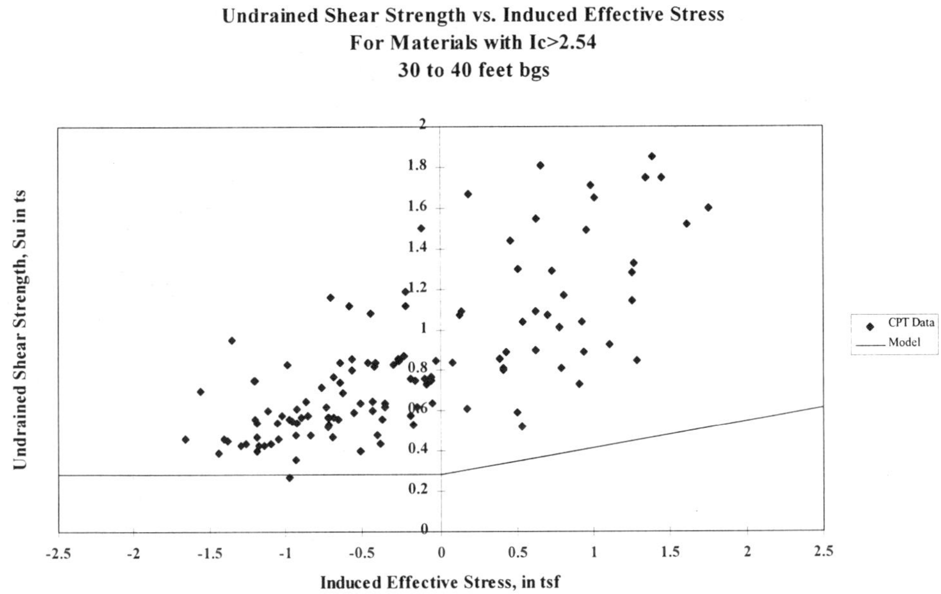

Figure 4. Undrained shear strength versus induced effective stress for fine grained materials from 30 to 40 feet below ground surface and model prediction.

In other words, the strength could also be represented by an effective stress friction angle of 34 degrees and a cohesion of zero.

To further examine the validity of this model, the shear strength values computed using this equation were compared with shear strength data obtained from TXCU tests. There is a very good correlation between the strength modeling presented by the equation and test results and is shown in Figure 5.

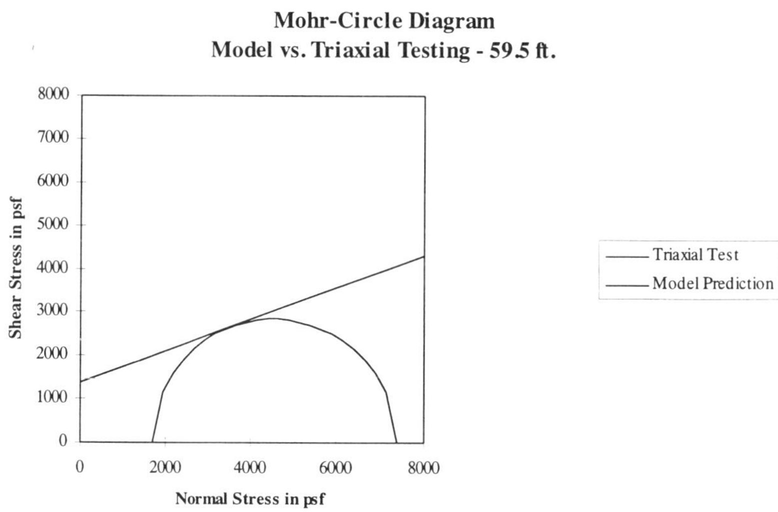

Figure 5. Mohr-Circle stress diagram for TXCU testing data compared with model prediction.

It should be noted that the effective stress parameters computed by TXCU data also indicate an effective friction angle of 34 degrees.

The shear strength model based on this equation also indicates that the shear strength of fine grained tailings depends partially on the induced or instantaneous pore water pressure values which in turn depend on the rate of loading. This indicates that had the CPY's been performed with slower or faster rates of penetration, the interpreted shear strengths would have been different if interpreted in the standard manner without the use of this model.

Because of the high rate of loading and the resultant large soil strains associated with dynamic loading conditions (earthquake), it is appropriate to neglect the frictional component of the shear strength, thereby assuming generation of large pore water pressures during dynamic loading. The tailings shear strength will then consist solely of the cohesion component.

6 CONCLUSIONS

It has been found that for the tailings investigated, the undrained shear strength of the fine grained materials has both a cohesive and a frictional component. It is also apparent that the cohesive component is a function of the effective consolidation stress. This cohesive component is believed to be due to an overconsolidation effect attributed to sub-aerial deposition scheme which allowed the tailing to desiccate prior to the deposition of the subsequent layer. This behavior has been investigated for other tailings deposits, and while the components of the model differ, the same phenomenon has been observed.

7 REFERENCES

Jefferies, M.G. and Davies, M.P., "Use of CPTu to Estimate Equivalent SPT N60", Geotechnical Testing Journal, Vol. 16, No. 4, December 1993, pp. 458-466. American Society for Testing and Materials.

Tailings and Mine Waste '99 © 1999 Balkema, Rotterdam, ISBN 90 5809 025 6

An alternative approach to characterizing existing waste rock dumps

M.Wickham & E.Swanson
Golder Associates Inc., Lakewood, Colo., USA
M.Logsdon
Geochimica, Inc., Ojai, Calif., USA
J.Clark
Claypool, Ariz., USA

ABSTRACT: The use of existing mine records in conjunction with an appropriate classification scheme can be used successfully to identify and characterize the composition of existing waste rock dumps where traditional approaches based on sampling may be impractical or costly. This paper describes an alternative approach used in an ARD investigation at a porphyry copper mine in Arizona with numerous waste rock dumps of varying ages and composition. The approach integrated detailed lithologic and mineralogic information about the ore body and overburden materials with records of pit development and dispatch. A waste rock characterization matrix was developed for the site to meaningfully classify lithology and mineralogic composition of the waste rock. The approach resulted in a focused geochemical testing program and facilitated interpretation and application of the test results.

1 INTRODUCTION

Knowledge of lithologic and mineralogic composition of existing waste rock dumps is critical to characterizing their acid rock drainage (ARD) or leachate-generating potential. Even the most rigorous geochemical testing program has limited value if knowledge of the composition of the dumps cannot be demonstrated. A costly discrete sampling program, consisting of drilling and testing, may be used to determine existing dump composition if the dumps are of limited mass and complexity. However, such programs can be inconclusive or cost prohibitive if the dumps are numerous, large, and highly variable in composition. Under these conditions, investigators can effectively use existing mine records to estimate the types and quantities of waste rock in existing dumps. When used in conjunction with an appropriate waste rock classification scheme, historical records can also reduce the uncertainty and cost of the characterization program over strategies that rely on discrete sampling.

This paper describes an ARD investigation that relied on historical mine records to identify waste rock composition. The investigators were confronted with the challenge of developing a sampling and analysis program for 13 distinct waste rock dumps containing over 350 Mt of waste rock at a porphyry copper mine in central Arizona. The lithologic and mineralogic composition of the waste rock at the site is highly variable. An approach was necessary that could accurately and precisely characterize and classify the composition of existing waste rock dumps without an extensive and costly sampling and testing program. To this end, an approach was developed that integrated detailed lithologic and mineralogic information about the ore body and overburden materials with records of pit development and dispatch. A compositional model was developed to integrate the records and allowed the lithologic and mineralogic composition of each dump to be determined with reasonable certainty. A waste rock characterization matrix was developed for the site to meaningfully classify lithology and geochemical facies of waste rock within the compositional model.

Prior knowledge of dump composition narrowed the focus of the sampling program to obtaining and testing representative samples from each category of the waste rock characterization matrix. Samples from the various lithologies and geochemical facies were collected in proportions to the known composition of waste rock at the site. Samples were collected from test pits and pre-existing core. Known dump compositions were also used to estimate the net acid-generating potential of each dump. The result was a defensible waste rock characterization program at a considerable cost savings to the client.

2 WASTE ROCK CLASSIFICATION

Waste rock was classified using a characterization matrix developed for the site. The matrix was based on the major lithologic types and geochemical facies (Table 1). The purpose of the characterization matrix was to provide a method of grouping waste rock into categories expected to exhibit similar geochemical behavior, thereby ensuring that sampling and testing of site waste rock was representative. In addition, the matrix provided the primary means by which the estimated waste dump compositions were reported. Classification of waste rock using the matrix enabled integration of the compositional model to the results of the laboratory geochemical testing.

The characterization matrix is composed of rows and columns corresponding to the principal lithologies and geochemical facies, respectively, present in waste rock at the site. The principal lithologies are granite, schist, dacite, and conglomerate. An additional matrix category, unknown/other, was necessary to account for data deficiencies, anomalies, or uncertainties regarding lithology. The ore body occurs in both granite porphyry and quartz sericite-amphibole schist, which are simply termed granite and schist in the characterization matrix. Virtually all mineralization is contained within these two lithologies.

Table 1. Waste Rock Characterization Matrix. Shaded areas indicate rarely occurring or not applicable categories.

	GEOCHEMICAL FACIES				
LITHOLOGY	Supergene Enriched (Sulfide Ore)	Hypogene Enriched (Protore)	Supergene Depleted (Leach Cap)	Unmineralized	Unknown/ Other
Schist					
Granite					
Dacite	[shaded]	[shaded]	[shaded]		[shaded]
Conglomerate	[shaded]	[shaded]	[shaded]		[shaded]
Unknown/Other					

Geochemical facies were defined as rock with similar mineral assemblages resulting from similar geochemical processes. Most of the productive deposits in the mining district are of hydrothermal origin subsequently enriched by supergene-processes. Therefore, the geochemical facies used in the matrix include supergene-enriched, supergene-depleted, hypogene-enriched, unmineralized, and unknown/other. The mineral assemblages and approximate juxtaposition of the various facies are shown in Figure 1. The unknown/other category was adopted for situations where the mineralogy of the waste rock was not known, the geochemical facies did not fall within the established facies, or the geochemical facies could not be identified with the information available.

Implausible categories in the matrix are the hypogene-enriched conglomerate and dacite, the supergene-enriched conglomerate and dacite, and the supergene-depleted conglomerate and dacite. These lithologies are younger than the mineralizing processes and are unmineralized.

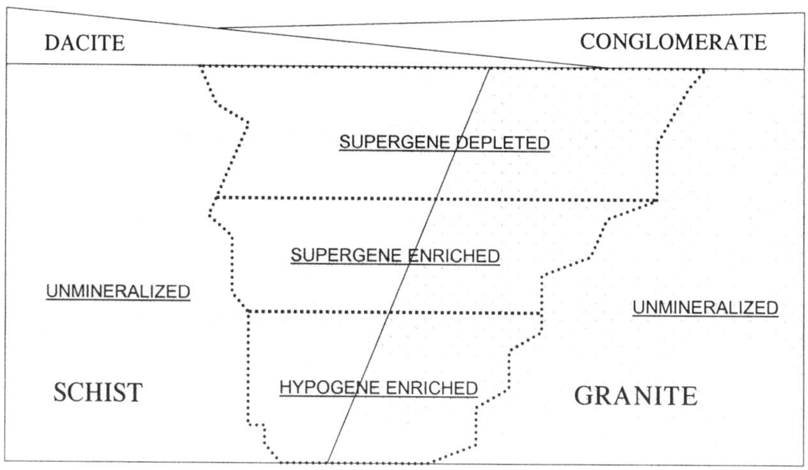

Figure 1. Conceptualization of site lithology and mineralogy in relation to the mineralized zone.

3 COMPOSITIONAL MODEL

Defining the composition of each waste dump required assimilation of large databases, classification of waste rock according to the characterization matrix, and summation and reporting of results in a meaningful format. To this end, a waste rock compositional model was developed. The model was developed in FORTRAN and integrated pit development, ore handling, and dispatch records with the mine exploratory corehole database. For the period of available records, the model performed accounting for tonnage by lithologic type, geochemical facies, and waste rock containing sulfide minerals. This accounting was performed for the entire site and for the individual waste rock dumps, each as a function of time. The model logic is shown schematically in Figure 2 and described below.

Daily ore records (DORs) were provided by the mine for the period January 1984 through May 1996. Daily ore records contained, in part, coordinates and bench elevation of the origin of ore and waste rock, destination of the material as dump name or number, and total (Cu) and acid-soluble (AsCu) copper content. The DOR database was first processed to remove records for ore that reported to a leach dump, leaving a total of 22,000 DORs for waste rock.

The compositional model relied on the mine's exploratory corehole database to characterize the lithology and mineralogy of each DOR. The corehole database contained an estimated 3,600 coreholes, each with lithology and mineralogy descriptions for discrete intervals ranging from 1 foot to 10 feet. Each DOR was assigned the properties of the core interval corresponding to the bench and within a radius of 400 feet from the DOR coordinates. Core was located using a 3-dimensional nearest-neighbor approach. Waste rock was identified in the corehole logs as material with a total copper content of less than 0.1 percent by weight. Material with a copper content 0.1 percent or greater was assumed to be ore. Waste rock for each waste rock core interval within the bench was assigned to the DOR. This often involved numerous waste rock categories for a single DOR.

The waste rock was then classified using the characterization matrix. The assignment of lithology was relatively straightforward. An algorithm, using the following criteria evaluated in this order, was used to assign the geochemical facies, based on mineral composition:

- Supergene-enriched facies: chalcocite, bornite, or covellite present as accessory minerals;
- Hypogene-enriched facies: chalcopyrite, and pyrite or molybdenite present as accessory minerals;
- Supergene-depleted facies: chrysocolla, limonite, hematite, malachite or azurite present as accessory minerals;

TABLE2a. MINE TOTAL, TONNAGE OF WASTE ROCK

LITHOLOGY	SUPERGENE ENRICHED (SULFIDE ORE) TONS	%	HYPOGENE ENRICHED (PROTORE) TONS	%	SUPERGENE DEPLETED (LEACH CAP) TONS	%	UNMINERALIZED TONS	%	UNKNOWN/ OTHER TONS	%	GRAND TOTAL TONS	%
SCHIST	5,606,000	1.58%	7,327,000	2.07%	71,580,000	20.23%	22,874,000	6.47%	4,499,000	1.27%	111,886,000	31.62%
GRANITE	5,845,000	1.65%	7,436,000	2.10%	57,445,000	16.24%	28,870,000	8.16%	2,844,000	0.80%	102,440,000	28.95%
DACITE	⊠	⊠	⊠	⊠	⊠	⊠	12,079,000	3.41%	⊠	⊠	12,079,000	3.41%
CONGLOMERATE	⊠	⊠	⊠	⊠	⊠	⊠	92,444,000	26.13%	⊠	⊠	92,444,000	26.13%
UNKNOWN/OTHER									34,956,000	9.88%	34,956,000	9.88%
TOTALS	11,451,000	3.24%	14,763,000	4.17%	129,025,000	36.47%	156,267,000	44.17%	42,299,000	11.96%	353,805,000	100.00%

TABLE 2b. SUBTOTAL, TONNAGE CONTAINING SULFIDE MINERALS

LITHOLOGY	SUPERGENE ENRICHED (SULFIDE ORE) TONS	%	HYPOGENE ENRICHED (PROTORE) TONS	%	SUPERGENE DEPLETED (LEACH CAP) TONS	%	UNMINERALIZED TONS	%	UNKNOWN/ OTHER TONS	%	GRAND TOTAL TONS	%
SCHIST	5,606,000	1.58%	7,327,000	2.07%	2,458,000	0.69%	3,550,000	1.00%	4,281,000	1.21%	23,222,000	6.56%
GRANITE	5,845,000	1.65%	7,436,000	2.10%	4,063,000	1.15%	1,374,000	0.39%	2,268,000	0.64%	20,986,000	5.93%
DACITE	⊠	⊠	⊠	⊠	⊠	⊠	0	0.00%	⊠	⊠	0	0.00%
CONGLOMERATE	⊠	⊠	⊠	⊠	⊠	⊠	768,000	0.22%	⊠	⊠	768,000	0.22%
UNKNOWN/OTHER									549,000	0.16%	549,000	0.16%
TOTALS	11,451,000	3.24%	14,763,000	4.17%	6,521,000	1.84%	5,692,000	1.61%	7,098,000	2.01%	45,525,000	12.87%

TABLE 2c. SUBTOTAL - TONNAGE CONTAINING PYRITE

LITHOLOGY	SUPERGENE ENRICHED (SULFIDE ORE) TONS	%	HYPOGENE ENRICHED (PROTORE) TONS	%	SUPERGENE DEPLETED (LEACH CAP) TONS	%	UNMINERALIZED TONS	%	UNKNOWN/ OTHER TONS	%	GRAND TOTAL TONS	%
SCHIST	5,023,000	1.42%	7,228,000	2.04%	2,458,000	0.69%	3,550,000	1.00%	4,281,000	1.21%	22,540,000	6.37%
GRANITE	3,468,000	0.98%	6,248,000	0.98%	2,974,000	0.84%	1,346,000	0.38%	2,253,000	0.64%	16,289,000	0.64%
DACITE	⊠	⊠	⊠	⊠	⊠	⊠	0	0.00%	⊠	⊠	0	0.00%
CONGLOMERATE	⊠	⊠	⊠	⊠	⊠	⊠	4,000	0.00%	⊠	⊠	4,000	0.00%
UNKNOWN/OTHER									126,000	0.04%	126,000	0.04%
TOTALS	8,491,000	2.40%	13,476,000	3.02%	5,432,000	1.54%	4,900,000	1.38%	6,660,000	1.88%	38,959,000	7.04%

Tonnages are cumulative through May 1996

Tonnages have been rounded.

⊠ = Rarely Occuring or Not Applicable

• Unmineralized facies: no accessory minerals present or, if not previously classified, total copper content less than 0.05 percent; or
• Unknown/other facies: could not be classified based on available information.

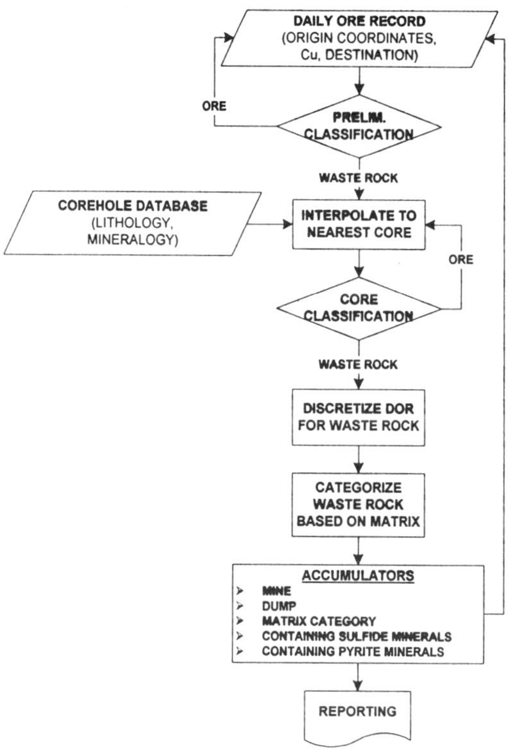

Figure 2. Flow diagram of the waste rock compositional model.

It is worth emphasizing that the "unmineralized facies" is used in the mining sense, that is, it is based on the absence of accessory minerals or a threshold copper content. The facies should not be construed as unmineralized in a strict mineralogical and geochemical sense and may exhibit a potential for adverse geochemical impacts.

Accounting of waste rock tonnages occurred at numerous levels including mine scale, dump scale, lithologic type, geochemical facies, and individual matrix category. Each level of accounting was performed monthly, annually, and for total period of record. In addition, the model could distinguish waste rock with sulfide-accessory minerals noted in the corehole logs. The model reported total tonnage with any sulfide mineral noted as an accessory mineral and total tonnage with pyrite noted as an accessory mineral. These reported tonnages should not be interpreted as sulfidic or pyritic waste rock. They are tonnages of waste rock in which sulfide minerals were noted without specification of the amount or occurrence (i.e., weight percent of rock mass, fine-grained and disseminated).

Total waste dump tonnage for pre-existing dumps was computed using geometric volumes estimated from present day and ancestral topography. Pre-record waste rock tonnage for certain dumps was assumed to be the difference between the total geometric tonnage estimate and compositional model estimates for some dumps.

4 RESULTS

The compositional model was found to be accurate in the processing of the mine records, waste rock classification, and accounting. The model culled out waste rock from the DORs (which include both waste rock and ore records) with 99.96 percent accuracy (by mass) for the 29-month model verification period of January 1994 through May 1996. The algorithm for assigning geochemical facies was also found to be accurate; the percentage of "unknown/other" assignments was less than 0.4 percent of all DORs or less than 0.3 percent of the total waste rock mass for the model verification period. The algorithm provided correct assignments for all coreholes checked during model development and verification.

The DORs and the model were found to provide accurate accounting for total tonnage reporting to the various waste rock dumps at the mine. Verification was performed by comparing total tonnages for each dump as calculated by the model with tonnages estimated

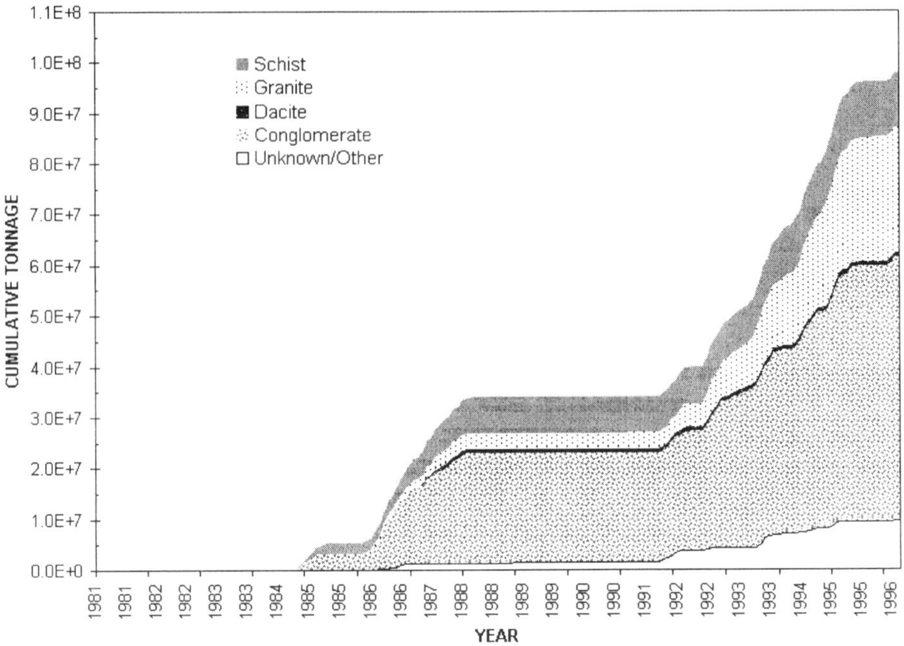

Figure 3. Waste rock tonnage by lithology, 22/23 Dump.

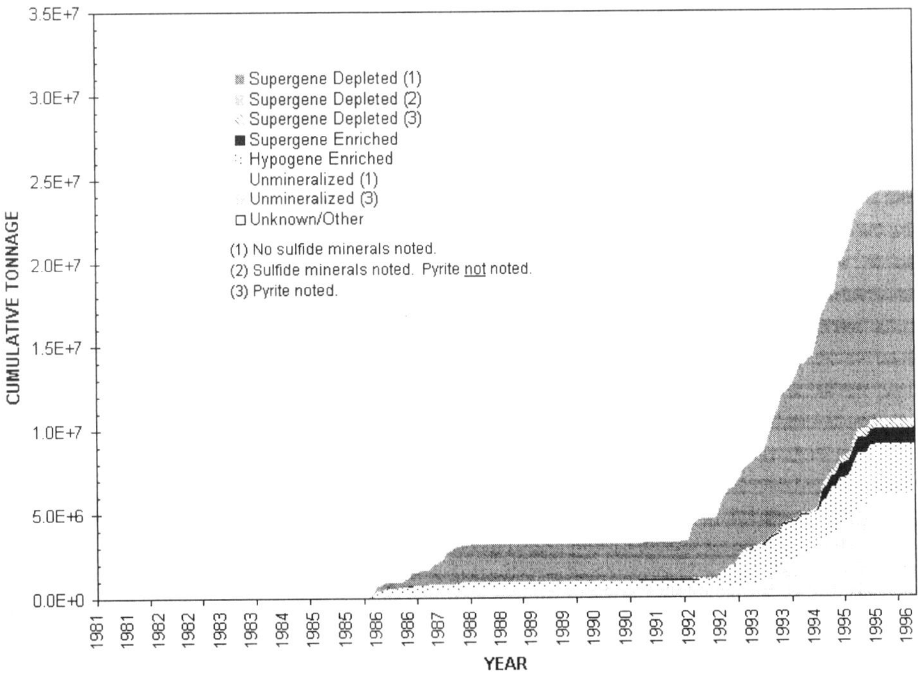

Figure 4. Waste rock tonnage by geochemical facies, schist lithology, 22/23 Dump.

from volume calculations. Volume calculations were made for each dump using ancestral and current topography maps for the site. Generally, the model outputs matched the estimated volume calculations where accurate ancestral topography was available. Discrepancies were attributed to inaccurate definition of ancestral topography within the dump footprint areas.

Approximately 88% of all existing waste rock was fully characterized using this approach. Of the remaining 12% classified as "Unknown/Other", 2% could be classified by lithology but not by geochemical facies. Most of the remaining 10% could not be classified because the waste rock was generated prior to the period of record available to the investigators. Other minor contributors included undetected database errors and rock categories not represented by the characterization matrix.

Results were presented in different formats to satisfy the informational requirements of the geochemical testing program. Results were reported as total tonnage for each matrix category for each dump, as shown for the mine in Table 2. The model differentiated between total, sulfide-, and pyrite-bearing materials for each dump. Results were also reported as a function of time. Any reporting time interval was possible. Examples of the temporal distribution of lithology and geochemical facies for a specific dump are provided in Figures 3 and 4. This information was available to compare against aerial photography and dump maps to determine the approximate placement of waste rock materials for each category in the matrix.

The approach supported a representative sampling and testing program to assess ARD potential at each waste rock dump. Based on the results of the geochemical testing program, the classification scheme accurately classified waste rock exhibiting similar geochemical behavior as determined by static and kinetic testing. Samples for testing were collected from existing core and from test pits constructed in each of the dumps specifically for material characterization. The number of samples tested was proportional to the total tonnage at the site as determined by the compositional modeling and the variability in geochemical behavior of waste rock within each category. The test results were then interpreted based on waste rock composition to estimate the net ARD potential for each dump.

5 CONCLUSIONS

An appropriate characterization scheme and mine records can be used successfully to identify and characterize the composition of existing waste rock dumps under circumstances where traditional approaches are impractical. This alternative approach can provide the following advantages:

- provides an efficient means to identify and characterize the composition of existing waste rock dumps;
- supports demonstration of representative sampling and analysis for geochemical testing;
- substantially reduces sampling and testing costs; and
- facilitates interpretation and application of geochemical testing results.

Tailings and Mine Waste '99 © 1999 Balkema, Rotterdam, ISBN 90 5809 025 6

Environmental ARD baseline at a high altitude gold mine project in the Central Andes

B.R. Keller – *ARCADIS Geraghty and Miller, Santa Bárbara, Calif., USA*
F.J. Johns, II & W.W. Huskie – *ARCADIS Geraghty and Miller, Denver, Colo., USA*
C. Tore & S. Castro – *ARCADIS Geotécnica, Santiago, Chile*
F. Ramírez & E. Kelleher – *Barrick Chile, Santiago, Chile*

ABSTRACT: Potential acid rock drainage (ARD) impacts from waste management at a proposed gold mine, located at approximately 5,000 meters elevation at the crest of the Andes (29° south latitude) has prompted a review of natural ARD hydrogeologic conditions. Data reviewed include geologic mapping, historical surface water flow and quality data, groundwater elevation and quality data, field reconnaissance of hydrogeologic features, and ARD of waste rock at nearby mines. Recharge of surface water and groundwater is dominated by snowmelt, with some contribution of melting of glacial ice. Poorly consolidated near-surface material includes irregular permafrost bodies. A conceptual groundwater model of the rounded mountain top location includes a "skin" aquifer parallel to the topographic surface, in which natural ARD generation occurs due to contact with sulfide minerals, modified by fault systems which were involved in Neogene ore body mineralization. Discharge of ARD water occurs at springs structurally related to the fault zones.

1. INTRODUCTION

Acid rock drainage (ARD) refers to the occurrence of acidic groundwater and surface water, produced by the decomposition of sulfide minerals. ARD has been documented as a significant environmental concern at many sulfide ore mining locations throughout the world, both as a naturally occurring condition and as a result of mining activities. In many locations with lengthy histories of mining activity, it is difficult to assess the natural ARD conditions that existed prior to mining. An understanding of the natural conditions would be useful in evaluating potential environmental impacts from subsequent mining activities. A case history to-date of an ongoing project in the central Andes of Chile and Argentina illustrates a situation in which an evaluation of the natural conditions is being performed. This hydrogeologic and geochemical investigation uses data that, in many instances, were collected for other purposes; a situation that may be common as increasing emphasis is put on environmental aspects of mining waste management.

2. PURPOSE AND SETTING OF INVESTIGATION

A pre-mine ARD assessment is being performed for the Pascua Mine project of Barrick Chile. The purpose of the pre-mine ARD assessment is to establish baseline water quality conditions in the vicinity of the proposed mine and to collect site-specific data to assess the potential of stockpiled ore, waste rock, and tailings to generate ARD. These site-specific ARD data will be compared to the baseline conditions to determine if, and to what extent, mine operations will result in impacts to water quality.

2.1 *Project, Site, and Mine Description*

The Pascua Mine project is located at the crest of the Andes at approximately 29° south latitude. The Pascua Mine hill and mountains within a few kilometers to the southeast are called the Cerros Nevados, and the area is also known as the Nevada mineral zone. The Pascua Mine hill, in the pre-mine configuration (Figure 1), is an east-west trending ridge with a high point of 5,171 meters (m) above mean sea level (amsl) on the west side of the proposed pit area and an elevation of 5,259 m amsl on the east side, at the Chile / Argentina border.

Three surface water courses drain the Pascua Mine hill. These are Río Del Estrecho on the northwest side, Río El Toro on the southwest side, and Arroyo Turbio on the east side. The divide between Río Del Estrecho and Río El Toro on the west, and Arroyo Turbio on the east is the continental divide of South America. Both Río Del Estrecho, which flows into Río Tránsito, and Río El Toro, which flows into Río del Carmen, are ultimately tributaries of Río Huasco, which flows west to the Pacific Ocean. The valleys of Río Huasco and its lower tributaries are important agricultural areas. Arroyo Turbio is a tributary of Río de las Taguas, which flows into Río de la Palca, then Río Jachal and Río Desaguadero, which is normally dry and does not reach the Atlantic Ocean, but infiltrates in La Pampa on the east side of the Andes several hundred kilometers (km) to the southeast of the Pascua Mine location.

2.2 *Geologic Setting*

The Pascua ore body is located in the Nevada hydrothermal alteration zone, which is part of a regional area of gold-bearing hydrothermal alteration zones known as the El Indio or Eastern belt. The country rock is characterized by north-south longitudinal bands of volcanic, volcaniclastic, sedimentary, and plutonic Paleozoic to Mesozoic or Paleogene basement rocks. Mineralization is associated with overlying Neogene volcanic rocks, whose location is related to north-south trending fault zones. Most notable of these is the Baños del Toro fault zone (Martin et al, 1995, Nasi et al, 1990), which separates a horst to the west from a graben to the east. The Neogene volcanism and mineralization is associated with the east side of the fault zone.

In the Pascua ore body, the altered host rocks are upper Paleozoic to lower Triassic granitic rocks and rhyolites, as opposed to Neogene volcanic rocks in much of the rest of the belt. The most abundant alteration minerals are silica, alunite, gypsum, limonite, jarosite, opal, and finely disseminated pyrite. Of these minerals, the sulfide mineral, pyrite, contributes most strongly to the generation of natural ARD. Waste rock which will ultimately be disposed in waste rock dumps during projected mining operations will include both hydrothermally altered material and unaltered rock. Laboratory tests are currently underway to better understand the ARD potential of the waste rock.

In the vicinity of the proposed Pascua mine, faults within the general Baños del Toro fault zone trend include the Nevada fault, approximately 2 km west of the ore body, and the Esperanza fault zone within the ore body and extending north and south of it (Figure 1). Locally, in the vicinity of the ore body, very detailed three dimensional mapping of the Esperanza fault zone has been performed as part of the ore body exploration, while elsewhere the mapping is restricted to regional surface studies. Both of these fault zones are associated with natural ARD.

The area near Pascua Mine hill was heavily glaciated during Pleistocene time, with characteristic "U" shaped valleys and lateral and terminal moraines within the valleys. The valley alluvial fill is a combination of glacio-fluvial deposits, talus from the modern hillsides, and modern stream deposits. Modern glacial ice occurs locally at elevations of about 5000 m. Erratic lenses of permafrost ice occur within poorly consolidated volcanic deposits at such altitude, and are reported to reach dimension of as much as tens of meters in drill holes.

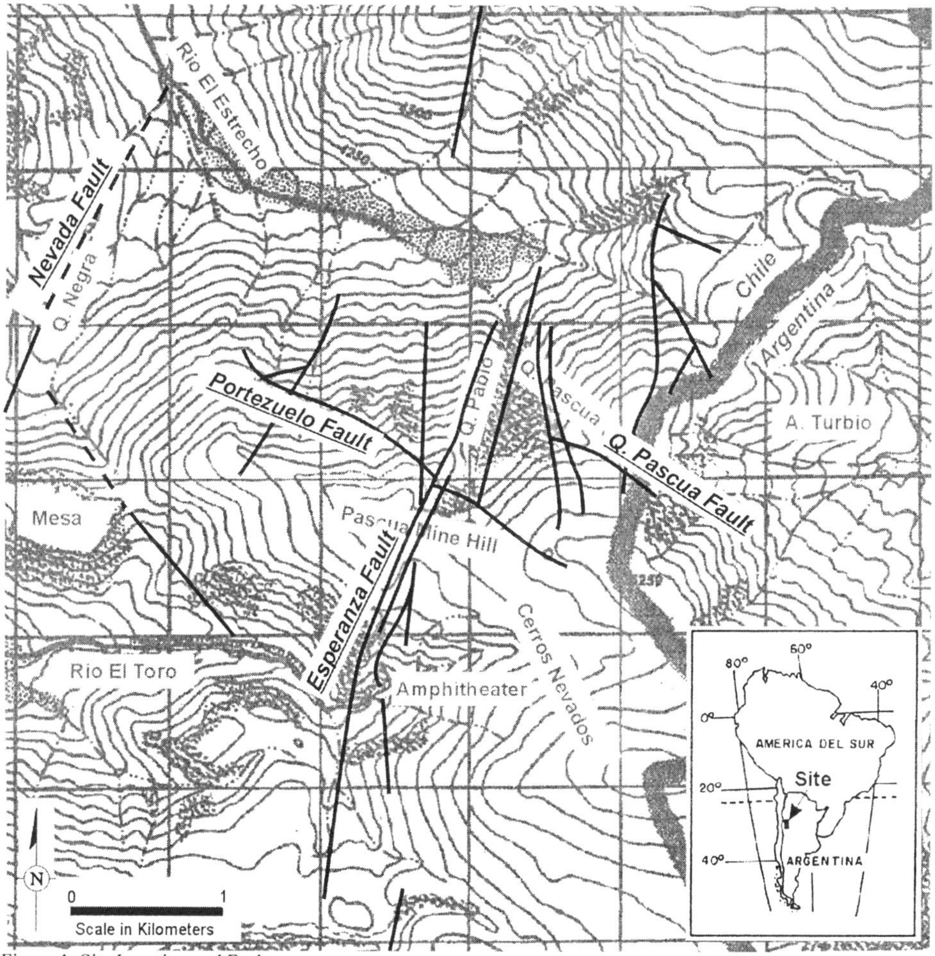

Figure 1. Site Location and Faults

3. MINE AREA HYDROGEOLOGY AND WATER QUALITY

Both surface water and groundwater occur in the three drainages at Pascua Mine hill. On the ridge itself, limited data indicate the existence of a near-surface aquifer in fractured rock, referred to herein as a "skin" aquifer. The data on all of these water occurrences are based either on field observations as part of this study or on data collected historically for water resource purposes, rather than for environmental or waste management purposes.

3.1 *Surface Water Hydrology and Quality*

Surface water data which have been obtained to date are more plentiful for the drainage on the west side of the Andes than on the east side, so this description focuses chiefly on the west side. It is possible that many of the general characteristics of the west side may also apply to the east side, although the influence of "Invierno Altiplánico" summer thunderstorm precipitation and of orographic differences could produce locally important variations. For regional scale information, long term hydrographs based on stream gauge records maintained by the Chilean federal water agency, Dirección General de Aguas (DGA) are available for the

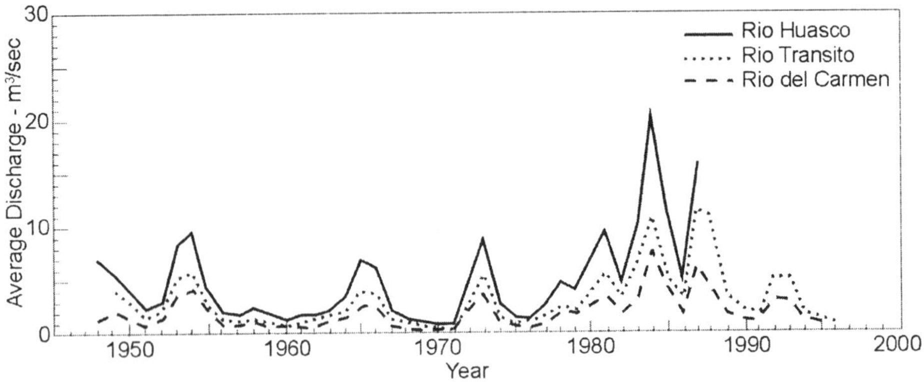

Figure 2. River flow hydrographs for three rivers at Alto del Carmen, the confluence of Rio Transito, which drains the northwest side of the project area, and Rio del Carmen, which drains the southeast side, to form Rio Huasco.

two rivers that drain the northwest and southwest sides of Pascua Mine hill, and Río Huasco, which is formed by their confluence. A hydrograph of approximately 50 years duration, based on daily measurements for these stations, is shown in Figure 2. This remains to be updated with the very high flows experienced in the 1997-98 El Niño season.

The hydrographs of the gauging stations are very similar in their temporal patterns of variations. This indicates that the precipitation and runoff regimes of the Río del Tránsito and Río del Carmen catchments are very similar, and general conclusions about the drainages on the west side of the continental divide apply to both streams, allowing conclusions to be drawn which are applicable to areas without long hydrographic records. This part of the Andes receives mainly austral winter precipitation from the Pacific Ocean, which may be in contrast to mining districts farther north in the Andes, where the "Invierno Altiplánico" summer thunderstorm precipitation, carrying Atlantic Ocean moisture, may produce notable local spatial variations.

The variations between individual years in the hydrographs is quite pronounced, typical of desert rivers. Prolonged dry periods were more common in the decades of the 1950s to 1970s than in the 1980s and 1990s. Hydrographic information in the vicinity of the mine is available only from the 1980s and 1990s, so the earlier prolonged dry periods are not represented by the data available in the proposed mine vicinity, but might recur during the operational life of a mine. Hydrographs have been intermittently recorded in the mine vicinity since about 1980, but data gaps, especially during the periods of highest flow in the hydrographs of Figure 3, make correlation difficult. However, the historical flow records in the Río del Estrecho valley indicate that there is generally some flow even during drought periods, while limited records and inspection of historical aerial photos indicate that in the upper part of Río El Toro, near the proposed mine area, significant periods may pass with no surface water flow. Since the natural ARD is carried in surface water, these observations mean that periods of much lower natural ARD generation may occur in the future.

Surface water quality data representative of natural ARD conditions include analytical values of historic water sampling and field measurement of pH values. The pH was measured in the field with restricted range pH paper (resulting in values expressed as 0.5 pH unit ranges. The pH values in the vicinity of Pascua Mine hill range from 2.0 to 6 (Figure 3). Springs with low pH values are associated with both the Esperanza fault zone and the Nevada fault. The Esperanza fault is represented by pH = 2.0 -2.5 in an exploratory tunnel and pH = 4.0 - 4.5 observed downstream of springs in the amphitheater area south of Pascua Mine hill. The Nevada fault is represented by pH = 4.0 - 4.5 measured in the field at flowing monitoring well BT-3, and pH between 3.05 and 3.43 reported from laboratory tests.

A comparison of surface water quality analytical data along the stream draining the northwest side of Pascua Mine hill shows natural ARD characteristics (Table 1). Analytical results for some typical ARD indicator parameters are presented for locations that are

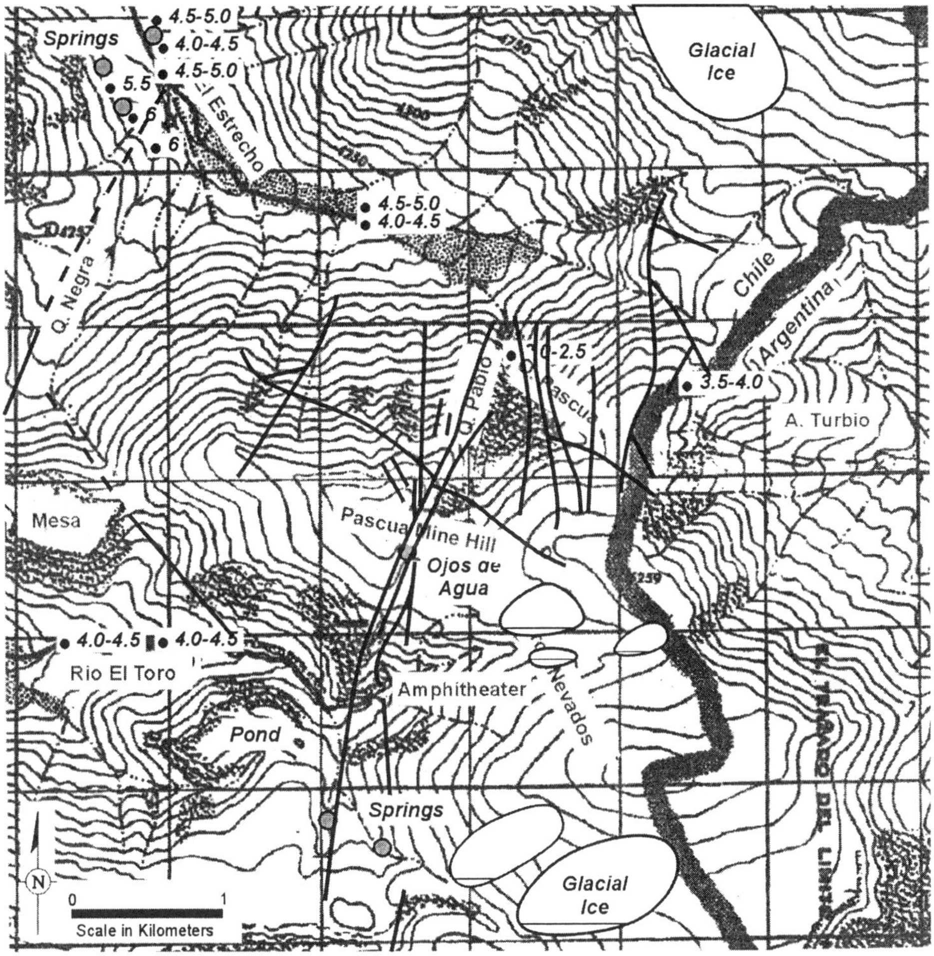

Figure 3. Locations of field pH measurements, springs, and faults.

Table 1 Laboratory Analytical Values of Surface Water ARD Indicator Parameters, Río del Estrecho Valley. All values in mg/l, except pH.

Constituent	NE-5	BT-3	NE-2	NE-1	NE-3	NE-4	WRD
Arsenic (total)	<0.0001	<.001	0.006	0.0225	0.008	<0.0001	0.42
Cadmium	0.0007	0.13	0.015	0.005	0.12	0.008	0.035
Copper	0.4	3	0.41	0.013	0.216	0.15	28
Iron	2.4	116	1.1	0.78	2.4	1.1	143
Manganese	6.3	36	6.6	0.026	4	3.3	26
Lead	0.075	2	0.09	0.185	0.06	0.01	26
Sulfate	206	1850	432	132	281	258	2100
Zinc	3.1	14.1	3	0.2	1.9	1.6	5.3
pH - lab	4.15	3.05	4.26	7.04	4.48	4.58	2.66

Notes: - Results are for dissolved concentrations unless noted progressively further downstream from Pascua Mine hill in the Río del Estrecho drainage (northwest side) for March, 1997, a date during a short-term drought period. The data for this date are characteristic of long-term averages at stations for which long term data are available.

The stations are:

- NE-5 - headwaters seep in the stream bed near proposed mine area;
- BT-3 - spring on stream bank in Nevada fault trend;
- NE-2 - stream approximately 6 km downstream from proposed mine area;
- NE-1 - side canyon stream of good quality water entering below NE-2;
- NE-3 - stream approximately 2 km below NE-2, with little other inflow than NE-1 and NE-2;
- NE-4 - approximately 5 km further down stream, below the confluence with another, larger stream that does not drain the proposed mine area; and
- WRD - outflow from an existing waste rock dump at a mine approximately 30 km south in the same mineralized belt, which has similar ore mineralogy.

The surface water closest to the mine site (NE-5) exhibits low pH, the primary characteristic of ARD, but does not exhibit notably elevated concentrations of ARD indicator parameters, relative to either the existing waste rock dump (WRD) or to the fault zone spring (BT-3). This is interpreted to indicate that although generation of acidic water by decomposition of sulfide minerals is ongoing in the vicinity of the proposed mine, the exposed rock surfaces through which this water passes do not have appreciable residual concentrations of soluble metals. The concentrations at the fault zone spring are elevated for some constituents, particularly manganese and zinc, indicating active geochemical processes in the subsurface. This water is volumetrically insignificant to the stream, since these high concentrations are not observed further downstream. The concentrations of metals appear to be slightly diluted downstream, although the diluting influence of the large stream entering above NE-4 is less than that of the smaller NE-1 stream entering above NE-3. Thus, metals concentrations downstream of NE-4 probably represent contributions from regional geologic materials, rather than the influence of the area of the proposed mine. The concentration of sulfate, which is the dominant anion, does not decrease downstream, indicating that this parameter also represents a regional value.

3.2 *Groundwater*

Three types of occurrence of groundwater have been observed in the vicinity of Pascua Mine hill. These are groundwater in alluvial aquifers in the valley floor, water issuing from springs, and the skin aquifer of Pascua Mine hill.

Groundwater in the alluvial aquifers in the valley floors near Pascua Mine hill has been characterized by monitoring wells installed for water resources exploration, for evaluation of a potential tailings dam site, and recently for environmental monitoring. With exception of the flowing wells in the Nevada fault zone, these have good quality water. Some deeper wells exhibit confined conditions, probably due to clay layers within the glacio-fluvial deposits. In both the Río del Estrecho and Río El Toro valleys, downstream of the proposed mine area, are reaches with exposed bedrock in the bottom, so that the alluvial aquifers are terminated and essentially all groundwater flow discharges to surface water flow. This will allow simple monitoring of any potential impacts to groundwater migrating away from the proposed mine area, to supplement monitoring well data.

Springs that discharge groundwater include ARD springs, which in the valley are restricted to the area near flowing well BT-3 in the Nevada fault zone, and many springs that discharge good quality water. The perennial springs of good quality water support meadow grass vegetation ("vegas" in Spanish), while the ARD springs have no vegetation or very stunted vegetation, and iron/manganese oxide staining of rocks. No vegetation is observed in aerial photographs of springs in the upper part of Río El Toro, attributable either to the ARD nature of the water, to the intermittency of flow, or to the fact that they are at higher altitude than vegas in Río del Estrecho valley.

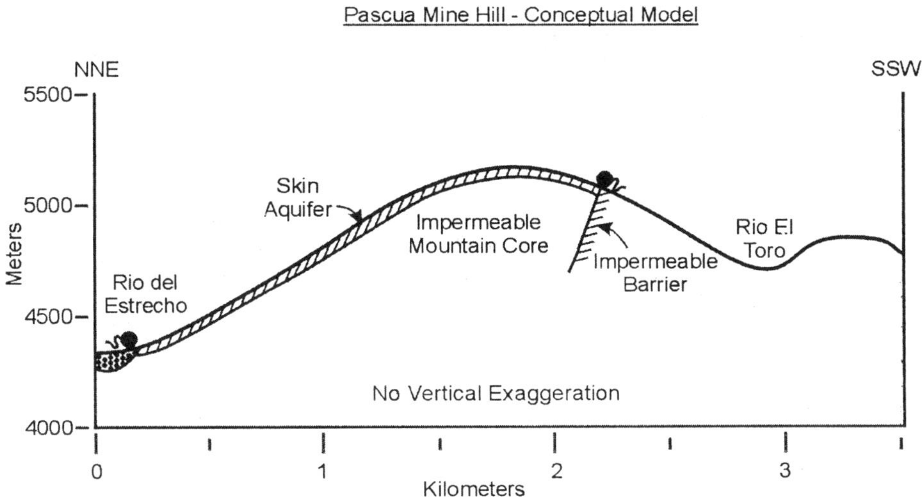

Figure 4. Schematic Diagram of Conceptual Model

A conceptual model of the preexisting condition at Pascua Mine hill is a dome-shaped hill consisting of rock that is extensively and openly fractured in an outer surface layer, permitting infiltration and migration of water which infiltrates principally from snowmelt, as well as some infiltration from melting glacial ice, and an inner core of rock which is sealed with respect to infiltration. This fractured surface layer is referred to as the skin aquifer. The conceptual model is that groundwater migrates downward in a locally saturated condition within the skin aquifer until it encounters either a topographic low point or an impermeable barrier, at which locations the flow may discharge to surface water flow or continue as alluvial fill groundwater, or both. This is shown schematically in Figure 4.

The thickness of the skin aquifer is depicted as approximately 40 meters, based on observations in an exploration tunnel. This tunnel is within the Esperanza fault zone, and, as noted, springs in the upper part of the Río El Toro drainage appear to be related to the Esperanza fault zone, so the detailed configuration of the skin aquifer may vary depending on location relative to fault zone features.

In its relation to topography, this conceptual groundwater regime would be of similar scale to a "local" groundwater system in the terminology of Toth (1963), with the difference that the impermeable base is within the vertical dimension of the topographic relief, instead of being a flat-lying surface at greater depth. Similar to the Toth (1963) local groundwater regime, groundwater flow is away from local topographic highs and toward local topographic lows, except where blocked by impermeable barriers. Thus, both topographic highs and lows constitute symmetrical, no-flow barriers, with groundwater flow originating at the highs and discharging to either surface water flow or alluvial groundwater flow at the lows.

4. SUMMARY AND CONCLUSIONS

An evaluation of natural ARD conditions in the vicinity of a proposed gold mine project at about 5,000 m elevation in the central Andes indicates that generation of acidic water is related to fault zones, which in some cases are also related to the ore mineralization. Low pH water is characteristic of surface water flow leaving the ore zone area, but with relatively low metals concentrations, as compared to water flowing from a waste rock dump at an existing mine in the region. An exception, which has both low pH and elevated metals, is water sampled from a flowing monitoring well in a zone of ARD springs located in a fault zone, indicating localized subsurface geochemical activity. This evaluation can be used to compare potential impacts from proposed mining and waste disposal activities against naturally occurring ARD conditions.

REFERENCES

Martin, M., J. Clavero, C. Mpodozis & L. Cuitiño. 1995. *Estudio Geológico Regional De La Franja El Indio, Cordillera De Coquimbo*. SERNAGEOMIN.

Nasi, C., R. Moscoso & V. Maksaev. 1990. *Hoja Guanta - Región de Coquimbo. Carta Geológica de Chile*. SERNAGEOMIN.

Toth, J. 1963. *A Theoretical Analysis of Groundwater Flow in Small Drainage Basins*. Journal of Geophysical Research. V. 68. 4795 - 4812.

Tailings and Mine Waste '99 © 1999 Balkema, Rotterdam, ISBN 90 5809 025 6

A groundwater mass loading balance for the Apache Tailings area, Leadville, Colorado

Gregory S. Davis, Cary L. Foulk & Bence V. Close
McCulley, Frick and Gilman Incorporated, Boulder, Colo., USA

ABSTRACT: A groundwater loading mass balance model was developed to assess the relative contributions of metals to groundwater from various potential sources and the mechanisms of metals loading at the Apache Tailings site. Results of the loading mass balance are consistent with site hydrochemical data, and the model can be used as a semi-quantitative tool to assess the relative reduction in the mass load of metals to groundwater downgradient of the Apache Tailings area that may be achieved by various remedial alternatives. Mass balance modeling of various groundwater constituents, in conjunction with site observations, water quality monitoring, and geochemical evaluations of the various types/sources of tailing materials, confirms that most of the metals loading to groundwater results from stormwater runoff from the Main Impoundment embankments. Metals loading to groundwater from percolation through the various tailing impoundments at the site is highly variable due to the variable nature of the tailing materials.

1 INTRODUCTION

Site characterization is being performed as part of a Remedial Investigation and Feasibility Study for the Apache Tailings - Operable Unit 7 (OU7), located within the California Gulch Superfund Site, Leadville, Colorado. Included in this assessment is an evaluation of the nature and extent of contamination within OU7 and an assessment of anticipated reductions in metals loading downgradient of tailing impoundments that may be achieved by the final remedial action selected for OU7. To assist in the evaluation of the site and remedial actions, a mass-balance model was developed to assess the relative contributions of metals to groundwater from various potential sources and the mechanisms of loading.

1.1 *Site Description*

Several types of ores were mined in the Leadville district, including placer and secondary oxidized or enriched ores which were mined early in the districts history. Later in the district's history, blanket, or manto, ores and vein ores were mined. The blanket ore bodies are closely associated with the porphyry intrusives and dolomites in the district and represent the majority of the district's base-metal production. Most of the vein deposits of the district are hosted in the igneous and sedimentary rocks of the district, including the Precambrian rocks. The veins were typically most productive where they intersected dolomite beds and certain quartzite and grit beds (Tweto, 1968). Most of the vein deposits were highly pyritic, some almost pure pyrite, and contained notable amounts of gold and silver. Tailings from the milling of both manto and vein ores appear to be present in the Apache Tailings impoundments.

OU7 consists of four distinct tailings impoundments, which include the Main Impoundment, Tailing Pond No. 2, Tailing Pond No. 3 and the North Impoundment (Figure 1). The Main

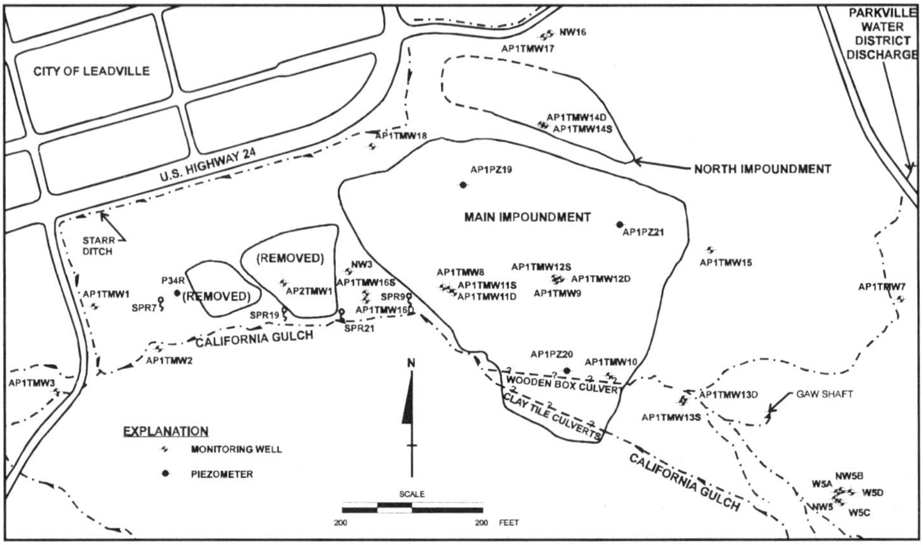

Figure 1 - Site Location Map

Impoundment is the farthest to the east, extends across the gulch, and covers approximately 11.3 acres. Tailing Ponds Nos. 2 and 3, which were removed in 1997 and placed on the Main Impoundment, were located west and downstream of the Main Impoundment, directly north of and adjacent to water flowing in the California Gulch channel, and were approximately 1.5 and 0.5 acres in size, respectively. The North Impoundment is separate from the Main Impoundment and covers an area of approximately 1.8 acres. The volumes of tailings in the Main Impoundment and North Impoundment are estimated to be approximately 630,000 cubic yards and 14,500 cubic yards, respectively. The volumes of material in Tailing Ponds Nos. 2 and 3 were approximately 36,200 cubic yards and 5,200 cubic yards, respectively.

The California Gulch channel is the main surface water feature in the Apache Tailings area. This constructed stream channel conveys discharge from the Yak Tunnel Water Treatment Plant and also, during high-flow periods, runoff from the Upper California Gulch watershed. The constructed portion of the California Gulch channel follows the south side of the valley upstream of the Main Impoundment, with most of this reach elevated above the valley floor. The channel intersects the southern edge of the Main Impoundment where flow is routed through two 24-inch-diameter clay-tile culverts which are bedded in alluvial materials beneath the Main Impoundment. Flow exiting the clay-tile culverts follows a small valley formed by the intersection of the embankment toe of the Main Impoundment and the hillside to the south. Approximately 50 feet downstream from the outlet for the clay-tile culverts, flow in the California Gulch channel combines with flow from a wooden box culvert which represents a major source of tributary inflow associated with surface water sources upgradient of the Main Impoundment and not associated with the California Gulch channel. Surface water sources upstream of the Main Impoundment include discharge from several springs, the Gaw Shaft, and the Parkville Water District storage tank. Discharges from these sources flow toward the Main Impoundment and coalesce just upstream of the impoundment. This flow is then conveyed under the Main Impoundment through the wooden box culvert located at the approximate tailings/alluvium interface. Below former Tailings Ponds Nos. 2 and 3, the California Gulch channel is joined by Starr Ditch. Starr Ditch approximately parallels the northern and western boundaries of the Apache Tailings area and conveys runoff from the Stray Horse Gulch watershed and the eastern portion of Leadville.

1.2 *Tailing Impoundments and Foundation Soils*

Tailings in the Main Impoundment range in size from medium-grained sand to clay. The maximum observed tailings thickness (59 feet) was noted in the south-western portion of the impoundment. In general, and as typical in tailing impoundments built using the upstream construction method, sand-sized material predominates along the constructed edges of the impoundment, and silt to clay-sized material predominates toward the center of the impoundment.

Three geochemically distinct types of tailings were observed in boreholes drilled in the Main Impoundment. The uppermost weathered layer consists of oxidized, weathered sulfidic tailings to a depth of approximately six inches to two feet below the tailings surface. The weathered sulfidic tailings, which also occur in the North Impoundment, are the most leachable and acid-generating tailings. Their total metal concentrations may be relatively low because of prior metals leaching. Underlying the uppermost weathered layer is a layer of dark gray tailings with generally high sulfide content. The dark gray sulfidic tailings make up slightly less than half of the total volume of the Main Impoundment and the majority of the volume of the North Impoundment and Tailing Ponds Nos. 2 and 3. The brown oxide tailings, which occur only in the lower portion of the Main Impoundment, have oxide minerals characteristic of the milled ore and, with the exception of lead, have metal concentrations significantly less than the overlying dark gray sulfidic tailings. The brown oxide tailings appear to have been derived from a different ore type than the overlying sulfidic tailings. Sulfides are much less abundant (typically less than 1% to 5%) than in the upper dark gray tailings (20% to 30%). The brown oxide tailings also have a significant average net neutralization potential (Table 1).

Table 1 - Tailings Geochemical Summary

Impoundment	Summary Statistics	Zinc Total (mg/kg)	Zinc SPLP (mg/L)	Total Sulfur (%)	Pyritic Sulfur (%)	Net Acid/Base Account (tons/1000T)	Acid Generation Potential (tons/1000T)	Acid Neutralization Potential (tons/1000T)	ANP/AGP Ratio
WEATHERED SULFIDIC TAILINGS									
Main	Mean[1]	2,294	21.1	11.8	4.8	-150	150	<1	0.007
	Minimum	677	12.10	NA	NA	NA	NA	NA	NA
	Maximum	5,850	51.3	NA	NA	NA	NA	NA	NA
	n	6	3	1	1	1	1	1	1
North	Mean	1,345	5.99	18.85	8.72	-271	272	1	0.005
	Minimum	485	1.94	6.46	<0.01	-455	<1	<1	NA
	Maximum	4,640	24	33.78	14.70	0	458	3	0.007
	n	6	6	6	6	6	6	6	6
GRAY SULFIDIC TAILINGS									
Main	Mean	7,567	4.187	15.50	7.02	-164	219	56	0.253
	Minimum	2,720	0.200	13.60	6.73	-164	210	46	0.219
	Maximum	16,000	75.2	17.40	7.30	-163	228	65	0.285
	n	8	8	2	2	2	2	2	2
North	Mean	21,852	58.62	21.96	9.41	-251	294	44	0.148
	Minimum	5,220	32	17.40	3.66	-387	114	30	0.263
	Maximum	59,700	81.9	31.28	13.60	-40	425	74	0.174
	n	8	8	8	8	8	8	8	8
BROWN OXIDE TAILINGS									
Main	Mean	4,230	0.503	0.97	0.36	92	11.3	97	8.53
	Minimum	1,150	<0.02	0.26	0.02	-15	1	2	2.0
	Maximum	17,600	12.3	1.61	0.70	266	22	244	11.1
	n	17	17	6	6	6	6	6	6

Notes: [1] Zinc concentrations are presented as geometric means, all other parameters are presented as arithmetic means. One-half of the detection limit is used to replace undetected values for statistical calculations.
Net Acid/Base Account = Acid Neutralization Potential (ANP) - Acid Generation Potential (AGP)
SPLP = Synthetic Precipitation Leaching Procedure
NA = Calculation not possible or appropriate due to nondetected value, single value or negative value.

In the North Impoundment, tailings are exposed in a topographic depression in the eastern portion of the impoundment, whereas tailings in the western portion of the impoundment are covered by as much as 8 to 12 feet of fill material. The tailings thickness is greatest (12.5 feet) in the south-central portion of the impoundment, and decreases radially. The base of the tailings appears to reflect the topography of the native ground surface which generally slopes upward to the north and northeast. The grain size of the tailings becomes finer from east to west. Tailings in the east end of the impoundment are generally sand size, whereas tailings in the west half of the impoundment consist of clay- and silt-sized particles with some fine sand (silty/sandy slimes). With the exception of a relatively thin (approximately one-foot thick) oxidized layer on top of the tailings in the exposed eastern portion, the tailings are dark gray, unoxidized sulfidic tailing, with a very high pyritic sulfur content (9.4 % average) that results in high acid generating potentials and very negative net acid-base accounts (Table 1). The dark gray sulfidic tailings in the North Impoundment have a sulfide content that is typically higher and total metals concentrations that are substantially higher (approximately 22,000 mg/Kg average zinc) than the dark gray sulfidic tailings in the Main Impoundment (approximately 7,600 mg/Kg average zinc). Sandy tailings are present and exposed in a topographic depression that receives runoff from the north and east in addition to incident precipitation and snowmelt.

Tailing Ponds Nos. 2 and 3 contained reprocessed surficial weathered tailings and dark gray sulfidic tailings originally from the northwest portion of the Main Impoundment. The maximum tailings thicknesses observed in Tailing Ponds Nos. 2 and 3 during the removal action were approximately 14 feet and 6 feet, respectively.

1.3 *Tailings Pore Water and Groundwater Conditions*

Groundwater in the Apache Tailings area occurs within tailings, unconsolidated sedimentary (glacial till and alluvium) deposits (the "Alluvial Aquifer"), and consolidated bedrock. The occurrence and extent of saturation in the various tailing impoundments and the Alluvial Aquifer are briefly summarized below.

The occurrence and extent of saturation in the different tailing impoundments are variable. Pore water within the Main Impoundment appears to be perched on top of relatively low permeability materials consisting of silty clay tailings (i.e., slimes) and, in some locations, a low-permeability, stiff, organic clay layer at the base of the impoundment. The geometric mean of horizontal hydraulic conductivity values for tailings in the Main Impoundment is 1.5×10^{-4} cm/sec (0.42 ft/day). With few exceptions, only minor (less than 2.3 feet) fluctuations in water level elevations have been observed in the Main Impoundment monitoring wells since November 1996. In former Tailing Ponds Nos. 2 and 3, the occurrence and extent of saturation were evaluated based on very limited information, but observations during the tailings removal suggested that saturated conditions may have existed within portions of these two tailing impoundments. In contrast, tailings in the North Impoundment are unsaturated, and sandy tailings exposed in a topographic depression in the eastern portion of the impoundment have a relatively high vertical permeability (1.5×10^{-3} cm/sec).

Alluvial Aquifer sediments are primarily composed of a very poorly sorted, unstratified glacial till with a stiff clay matrix interbedded with laterally discontinuous coarser-grained outwash deposits of silts, sands and angular to subrounded gravels. Gravels, pebbles and occasionally cobbles composed of granite, metamorphic and sedimentary rocks are present in the Alluvial Aquifer sediments. Alluvial Aquifer sediments are at least 260 feet thick beneath the Apache Tailings impoundments.

Depth to groundwater in the Alluvial Aquifer ranges from approximately 34 feet below ground surface upgradient of the North Impoundment to less than approximately 1 foot below ground surface downgradient, near the California Gulch channel. Recharge of the Alluvial Aquifer results primarily from direct infiltration of precipitation and surface water, including snowmelt (Golder, 1996). In some areas, groundwater in the Alluvial Aquifer is unconfined; however, confined and/or semi-confined conditions commonly occur as a result of the heterogeneity of the aquifer materials and the interbedded, laterally discontinuous layers of fine-grained and coarser-grained materials. Groundwater generally flows from east to west in a direction that roughly follows the surface topography along the axis of California Gulch. The geometric mean of hydraulic conductivity values for the Alluvial Aquifer is 9.8 x 10^{-4} cm/sec (2.8 ft/day). Based on an assumed effective porosity of 22.5% and a horizontal hydraulic gradient of 0.040 ft/ft, the calculated average linear velocity for the Alluvial Aquifer in the Apache Tailings area is 0.5 ft/year.

Water levels in the Alluvial Aquifer have shown only minor seasonal variation with less than three feet of variation observed at most locations during the November 1996 to October 1997 measurement period. The most significant exceptions were observed at monitoring wells AP1TMW14S, AP1TMW18, AP1TMW7, AP1TMW15 and W5A (Figure 1). At well AP1TMW14S, water levels exhibit a rapid response to infiltration from precipitation events and/or snowmelt. At well AP1TMW18, water levels appear to respond to periodic flow in Starr Ditch. Wells AP1TMW7, AP1TMW15 and W5A, upgradient of the Main Impoundment, show water level variations consistent with changes in water levels in the underlying bedrock. Bedrock water levels have increased by approximately 35 to 40 feet between 1991 and 1997.

Unsaturated alluvium underlies all of the North Impoundment and all or a majority of the Main Impoundment. The potentiometric surface in the Alluvial Aquifer periodically appears to be above the base of tailings in a relatively small area under the southwestern portion of the Main Impoundment; however, apparent confined or semi-confined conditions observed while drilling beneath the impoundment in the Alluvial Aquifer in this area indicate that water level elevations measured in the wells represent confined pressure levels and not contact with tailings. Therefore, the area of actual hydrologic contact between tailings and the Alluvial Aquifer may be non-existent or less extensive than the area of potential contact that would be estimated assuming unconfined aquifer conditions. During the winter months of 1996/97, the potentiometric water level in well AP1TMW11D was below the base of the tailings and in the spring and summer of 1997 was as much as 0.5 feet above the base of the tailings. Water levels in other wells screened in the Alluvial Aquifer beneath the tailings have remained below the base of tailings.

2 GROUNDWATER MASS LOADING EVALUATION

2.1 *Site Conceptual Model*

The site conceptual model (Figure 2) is the basis for evaluation of: (1) the mass loading contributions to groundwater from the various sources and inputs in the Apache Tailings area; and (2) the fate and transport of potential constituents of concern (COCs) within the Apache Tailings area hydrologic system. The site conceptual model incorporates loading from multiple tailings impoundments that may release COCs to varying degrees to groundwater and surface water and other sources upgradient/upstream of the impoundments and also considers groundwater/surface water interactions and geochemical reactions.

2.2 *Overview of the Loading Mass Balance Methodology*

The loading mass balance for groundwater was evaluated by comparing the inflow and outflow loads of the COCs for the hydrologic system. For OU7, the estimated constituent loads in groundwater upgradient of the Apache Tailings and loading to groundwater from individual sources within the Apache Tailings area (referred to as hydrologic-system inflows) were compared to the estimated loads in downgradient groundwater and discharges to surface water (referred to as hydrologic-system outflows). Of the COCs, which includes arsenic, cadmium, lead and zinc, dissolved zinc is the most useful in developing the loading mass balance and in evaluating the relative contributions from various sources because zinc is present at the highest concentrations in groundwater and tailings and it is the most geochemically conservative (i.e., non-reactive) of the site COCs. Additionally, chloride was used to calibrate the loading mass balance model. Even though the absolute magnitude of the calculated mass loads are subject to some uncertainty,

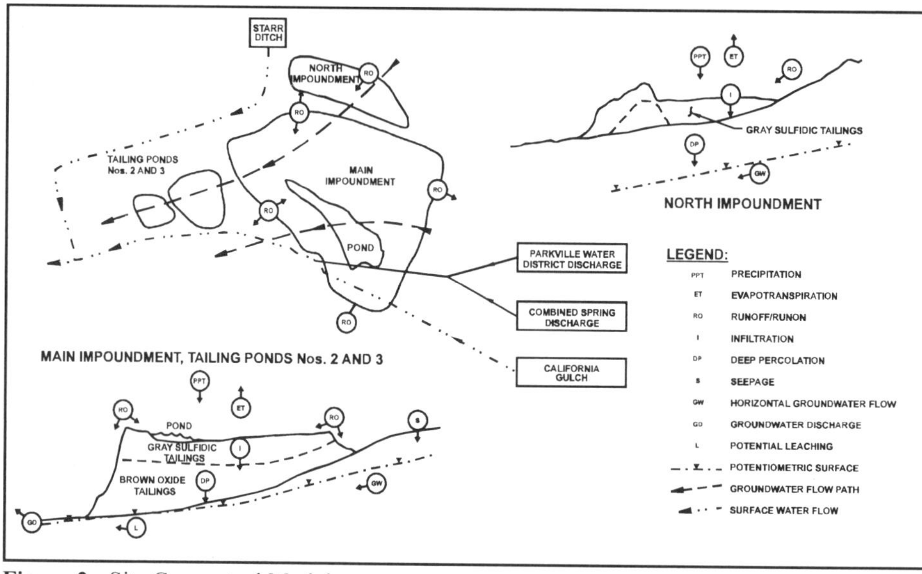

Figure 2 - Site Conceptual Model

modification of the input parameters within a range of values that are reasonable for the Apache Tailings area did not substantially change the relative contributions associated with each of the identified sources/release mechanisms. Components of the loading mass balance are described below.

Upgradient Groundwater Inflow And Loading

Average annual groundwater flow and chemical loading entering the site were based on the estimated flow and load through two upgradient segments representing the inflow ends of groundwater streamtubes. The groundwater velocity through the upgradient segments was estimated based on: average hydraulic conductivity testing results from wells in the area; hydraulic gradients calculated along perpendicular transects spaced equally across the segments; and an assumed effective porosity. Aquifer thickness in the discharge calculation was estimated based on an evaluation of the boring logs for wells in the vicinity of the segments. Discharge through each upgradient segment was multiplied by a weighted average of chemical concentrations in nearby wells to obtain mass loading estimates through each segment.

Runoff Loading To Groundwater

Outward runoff is only significant for the Main Impoundment with its extensive steeply sloped faces. Runoff from the Main Impoundment flows to either surface water or ponds in areas with limited drainage. Ponded runoff either evaporates or infiltrates relatively rapidly into the surrounding alluvial materials. Evaporation, while possibly being significant, has no apparent effect on infiltration loadings. Heavy salt crusts are not observed in areas where ponded runoff water occurs after precipitation events. This suggests that most of the metals load infiltrates. In developing the mass loading model, portions of the metals load associated with runoff were assigned to either infiltration or flow to surface water drainage.

Focused recharge and rapid infiltration have been observed along the northern and southwestern portions of the Main Impoundment as a result of topographically-controlled drainage patterns and the presence of coarser-grained, more permeable alluvial materials. These observations are consistent with soil sample tests and slug test results in wells AP1TMW16S/D and AP1TMW14D, which indicate higher hydraulic conductivities than most other areas of the site. The impoundment faces were divided into five segments based on runoff characteristics (i.e., similar slopes, contribution to surface water, or ponding areas). For each segment, a horizontal area receiving precipitation was measured and a percentage of runoff reaching groundwater was estimated based on topographic considerations and field observations. A volume of precipitation based on annual totals was then used to estimate the amount of runoff from the slopes on an annual basis. Runoff from the other impoundments was considered to be insignificant because only a very small outward exposure of tailings exists at the North Impoundment, and Tailing Ponds Nos. 2 and 3 had berms constructed primarily of alluvial material. The estimation of loadings from Main Impoundment runoff to groundwater was made by multiplying the estimated volume of runoff that infiltrates to groundwater through more permeable materials adjacent to the impoundment by the observed dissolved metals concentrations in runoff water collected during a storm event.

Impoundment Percolation To Groundwater

Percolation through the Main Impoundment may originate in two areas: (1) the exposed impoundment surface; and (2) the impoundment pond. The HELP model (EPA, 1994) was used to estimate infiltration and deep percolation through the base of that portion of the Main Impoundment not covered by ponded water, whereas Darcy's Law was used to estimate percolation from the impoundment pond. To be conservative, the HELP model simulation was performed without considering the slime layers that are present at and near the base of the Main Impoundment. The loading to groundwater was calculated by multiplying the percolation from each portion of the impoundment by average chemical concentration data from monitoring wells screened in the tailings.

The HELP model was also used to estimate infiltration and percolation through the eastern portion of the North Impoundment where acid-generating tailings are exposed in a topographic depression. Water infiltrating this portion of the North Impoundment can react with the tailings and percolate relatively rapidly through the impoundment without encountering significant neutralizing components or low permeability layers. Percolation through the finer-grained, fill-covered tailings in the western portion of the North Impoundment was assumed to be 33% of the percolation through the exposed tailings.

The loading to groundwater from former Tailing Ponds No. 2 and 3 was calculated by multiplying the percolation to groundwater based on Darcy's Law calculations by the pond water concentrations. The assumption that the chemical character of this discharge approximates that of the water ponded on the impoundments is supported by the chemistry of spring SPR7 and piezometer P34R, located immediately downgradient of the ponds, which have chemistry very similar to the pond water. Infiltration through the unponded portions of these impoundments was not considered to be significant because of the limited area.

Potential Groundwater Loading From Main Impoundment/ Groundwater Contact Zone

Although groundwater may not actually contact the base of the tailings because of confining conditions in the alluvial aquifer, for conservatism an evaluation of loading from a potential contact area was included. Loading from the potential contact area was calculated initially by multiplying the flow through the potential contact area by the leachable zinc concentrations (i.e., geometric mean SPLP results) for the brown oxide tailings and for the potentially impacted foundation soils. As part of a sensitivity analysis of input parameters, loading was also estimated assuming that leachable metals concentrations are two orders of magnitude greater than the geometric mean of the SPLP results (which is equivalent to the dissolved zinc concentration in well AP1TMW11S).

Downgradient Groundwater Outflows And Loadings

Mechanisms for transport of COCs away from the Apache Tailings area include groundwater flow, spring flow (primarily SPR7), and diffuse groundwater discharge to the California Gulch channel. Groundwater flow leaving the Apache Tailings area was estimated for two segments corresponding to the downgradient ends of the two groundwater streamtubes for the two upgradient segments. Groundwater flow through the north segment is largely derived from northern portions of the Apache Tailings area, including the North Impoundment and a portion of the Main Impoundment. This flow generally leaves the Apache Tailings area as spring discharge and groundwater flow through the center of California Gulch. The southern segment includes flow from upgradient California Gulch and sources in the center and southern portions of the Apache Tailings area. This groundwater flow leaves the Apache Tailings area along the southern edge below the Main Impoundment and appears to partially discharge to California Gulch with some possible flow beneath the hill to the south of California Gulch. The groundwater velocity through the segments was estimated based on: average hydraulic conductivity testing results from nearby wells; hydraulic gradients across each segment; and an assumed effective porosity. Weighted average concentration data from wells near each segment were used for estimating loading.

3 GROUNDWATER MASS LOADING BALANCE RESULTS

Based on the mass loading evaluation, the three most significant sources of loading to groundwater in the Apache Tailings area, in decreasing order of magnitude, are discussed below and are summarized in Table 2.

- Infiltration of runoff from the Main Impoundment embankments is estimated to account for approximately 9,300 lbs/year or 64% of the total estimated zinc load to groundwater downgradient of the Apache Tailings. The metals loading to groundwater appears to be derived primarily from stormwater contact with relatively thin weathered sulfidic tailings. Because the acid generating potential of the surficial tailings exceeds the neutralizing potential (which is substantially depleted due to weathering), the tailings readily produce low pH and metals-enriched runoff.
- Percolation through Tailing Ponds Nos. 2 and 3 (prior to removal) is estimated to account for approximately 2,300 lbs/year or 16% of the total estimated zinc load downgradient of the Apache Tailings. Reprocessed tailings in these impoundments (which were largely removed from surficial portions of the Main Impoundment) appeared to be acid generating and contributing to the metals loading in groundwater. The pond water quality from Tailing Ponds Nos. 2 and 3 is very similar to the groundwater and surface water quality downgradient of the impoundments, and to lesser extents, other springs and wells in the immediate area.
- Percolation through the North Impoundment tailings is estimated to account for approximately 2,300 lbs/year or 16% of the total estimated zinc load downgradient of the Apache Tailings. The eastern portion of the impoundment contains high-permeability, sulfide-rich tailings, which also contain some of the highest total metals concentrations and acid generating potential of the tailings impoundments at OU7 (Table 1). Groundwater beneath the impoundment is a concentrated metalliferous, sulfate-rich groundwater with very low pH (less than 3) that represents the North Impoundment leachate. Groundwater from this area flows under the Main Impoundment to the Tailing Ponds Nos. 2 and 3 area.

Approximately 3% (420 lbs/year) of the remaining zinc load appears to come from upgradient groundwater flowing into the Apache Tailings area. The zinc is associated with a metals-enriched groundwater with some bicarbonate and near neutral pH (above 6). The bicarbonate water type is representative of deep bedrock groundwater that is recharging the deeper portions of the alluvial aquifer and upwelling along faults and mine workings to the shallow alluvial aquifer.

Percolation through the Main Impoundment appears to account for 0.8% (110 lbs/year) of the zinc load, and thus, does not appear to be a major contributor to the metals loading in groundwater. This result is consistent with the hydrochemistry at the site. Pore water in the Main Impoundment has a unique magnesium/potassium chemical signature that is not evident in the alluvial aquifer

Table 2 - Summary of Groundwater Mass Loading Evaluation

SYSTEM COMPONENT	Discharge (gpm)	Dissolved Zinc (mg/L)	Zinc Loading (lbs/year)	Zinc Loading (% of total)	Chloride Loading (lbs/year)	Chloride Loading (% of total)
HYDROLOGIC SYSTEM INFLOWS						
I. Upgradient Alluvial Groundwater			524	3.6%	1,967	86.3%
Upgradient of Main Impoundment	27	4.4				
Upgradient of North Impoundment	25	0.04				
II. Source Loading						
A. Runoff Infiltration to Groundwater - Main Impoundment	1.15	1,840	9,268	63.9%	116	5.1%
B. Percolation Through North Impoundment			2,290	15.8%	9	0.4%
Exposed Tailings	0.26	1,530				
Fill-covered Tailings	0.08	1,530				
C. Percolation Through Main Impoundment			110	0.8%	50	2.2%
Brown Oxide Tailings	0.75	20.4				
Foundation Soils	0.485	20.4				
D. Potential Horizontal Flow Through Base of Tailings and Foundation Soils			2.5	0.02%	NA[(1)]	0.0%
Brown Oxide Tailings	0.003	0.503				
Foundation Soils	0.294	1.9				
E. Potential Percolation Through Lower Impoundments			2,316	15.9%	138	6.1%
Tailings Pond No. 2	1.1	269				
Tailings Pond No. 3	0.9	246				
Sum of Groundwater Inflows and Source Loading	57.0	NA	14,510	100%	2,280	100%
HYDROLOGIC SYSTEM OUTFLOWS						
I. Downgradient Groundwater						
A. North Segment Downgradient Flow	21	60.9	5,599	36.5%	1,886	81.9%
B. South Segment Downgradient Flow	13.7	84.0	5,043	32.9%	179	7.8%
C. Spring Discharge (SPR7, SPR9, SPR19 and SPR21)	2.5	NA[(2)]	1,887	12.3%	90	3.9%
D. Diffuse Groundwater Flow to California Gulch Channel	20	32.0	2,805	18.3%	149	6.5%
Sum of Downgradient Groundwater Outflow and Loading	57.2	NA	15,334	100%	2,304	100%
UNACCOUNTED GAIN (LOSS)[(3)]	0.2	NA	824	5.7%	24	1.0%

Notes: (1) NA means not applicable.
(2) Dissolved zinc loading from SPR7 represents 75% of the total load associated with downgradient spring discharge. The average dissolved zinc concentration for SPR7 is 1,417 mg/L, and the average discharge rate is 1.3 gpm.
(3) Due to rounding, column values presented in the table may not sum precisely.

directly beneath or downgradient of the impoundment.

Potential contact between shallow alluvial groundwater and the base of tailings, if any, would represent less than 0.02% (2.5 lbs/year) of the zinc load to groundwater downgradient of the Apache Tailings area. Even if one assumes that the concentration of zinc leached from the base of the tailings due to groundwater contact is two orders of magnitude greater than the geometric mean of the SPLP results for brown-oxide tailings used in the leaching analysis (which is equivalent to the dissolved zinc concentrations observed in well AP1TMW11S), the zinc mass loading associated with contact would still represent only 0.02% of the total zinc load in groundwater downgradient of OU7. Alternatively, if the potential contact area is increased by an order of magnitude, the zinc mass loading associated with potential contact with the base of tailings would still represent only 0.02% of the zinc total mass load in groundwater downgradient of OU7.

Starr Ditch is also a source of metals loading to groundwater; however, the magnitude of this source is variable and not readily quantified due to the intermittent nature of flow and limited number of monitoring points in the vicinity of the site.

4 CONCLUSIONS

The following conclusions are based on historic and recent site characterization information.

- Collectively, runoff from the Main Impoundment embankments and percolation through the North Impoundment and Tailing Ponds Nos. 2 and 3 account for approximately 96% of the mass loading of dissolved zinc to groundwater in the Apache Tailings area. The remaining sources of mass loading to groundwater include upgradient groundwater flowing into the Apache Tailings area (3%), percolation through the Main Impoundment (0.8%) and potential contact, if any, between shallow alluvial groundwater and the base of tailings (<0.1%).
- Unsaturated alluvium underlies all of the North Impoundment and all or a majority of the Main Impoundment. The potentiometric surface in the alluvial aquifer periodically appears to be above the base of tailings in a relatively small area under the southwest portion of the Main Impoundment. However, confined and semi-confined conditions occur in the Alluvial Aquifer, and therefore, water levels above the base of tailings appear to represent confined pressure levels and not contact with the tailings.
- The mass loading model provides a basis for comparing the effectiveness of various remedial alternatives for the site. The model suggests that actions to reduce contact and subsequent runoff from the Main Impoundment tailings and infiltration through the North Impoundment and Tailings Ponds Nos. 2 and 3 would help to remediate up to 96% of the COC loadings to groundwater.

ACKNOWLEDGMENTS

The authors acknowledge ASARCO Incorporated, for which these studies were performed, for permission to present and publish this paper.

REFERENCES

Golder and Associates (Golder). 1996. Final Hydrogeologic Remedial Investigation Report, California Gulch Site, Leadville, Colorado. May.

Tweto, Ogden. 1968. Leadville District, Colorado; in John D. Ridge (ed.), Ore Deposits of the United States, 1933-1967, American Institute of Mining, Metallurgical, and Petroleum Engineers, Inc., New York, pp. 681-705.

U.S. Environmental Protection Agency (EPA). 1994. The Hydrologic Evaluation of Landfill Performance (HELP) Model: Engineering Documentation for Version 3.05. EPA/600/9-94/xxx, U.S. Environmental Protection Agency Risk Reduction Engineering Laboratory, Cincinnati, Ohio.

Tailings and Mine Waste '99 © *1999 Balkema, Rotterdam, ISBN 90 5809 025 6*

Digital photogrammetry and GIS in tailings and mine waste management

Chr. Fischer
Institute of Mine Surveying, Technical University of Clausthal, Germany

D. Hermsmeyer
Institute of Soil Science, University of Hannover, Germany

ABSTRACT: Mine waste mounds frequently are little accessible, irregularly shaped bodies of appreciable size that are physically difficult to manage. Digital terrain models (DTMs) are useful management tools, especially when integrated as a data layer in Geographic Information Systems (GIS). Frequently, DTMs of waste mounds are derived from stereographic analysis of aerial photographs. In our contribution, we discuss increasingly applied softcopy or digital photogrammetry versus analytical measurement as a method to generate DTMs. Issues of DTM accuracy are considered. Results from ongoing studies in the German coal and potash mining industries are presented. As an example, it is shown how we used DTMs to calculate short wave solar radiation and potential evaporation from a potash mine residue mound.

1 INTRODUCTION

By-production of solid wastes is unavoidable in many mining operations. In German coal mines, currently as much as 50 % of all mined materials are non-combustible solid by-products. Only 25 % of these can be used in construction industry or disposed of underground. In potash mining, as much as 60 % of all residues are returned to caverns. In both cases, remaining wastes are piled in mounds. According to European and German environmental regulations, mining companies have to supervise waste heaping and reclamation of tailing mounds. Storage area management in accordance with these regulations requires reliable and up-to-date information on mound topography. To gather such data, stereoscopic analysis of aerial photographs is commonly applied. Resulting terrain models can help to answer a wide range of management questions, some of which are discussed in the sample application section of this paper.

1.1 *Softcopy or digital versus analytical generation of DTMs*

Analytical stereoplotters are widespread devices in DTM generation. Conceptually, they operate through the formation of a mathematical model adjusting the geometry of a cartographic camera to the terrain formation as depicted by a pair of partially overlapping aerial photographs (*stereopair*). Positions in the model can then be mathematically transformed into ground coordinates and elevations. Analytical stereoplotters are known to be extremely accurate in that they have little optical or mechanical limitations. However, manual (point by point) measurement of DTMs is time consuming and cumbersome.

Today, these operations are increasingly being performed by softcopy or digital photogrammetric systems. Fundamental distinctions between traditional analog systems and softcopy systems are (i) the use of digital raster images instead of photographs in the softcopy approach (often produced by computer scanning of aerial photographs) and (ii) the central role that mathematical modeling plays in softcopy systems to computerize several steps of DTM generation. Measurement of the interior orientation as described by Kersten and Haering (1997), of tie points for automated aerial triangulation, as discussed by Tsingas (1995), Dörstel (1996), and Krystek et al. (1996), and of homologuous points in stereoscopic imagery, as described by Heipke (1996) and Gülch (1994) are all automated steps that in the analytical approach need to be performed by a human operator. Not only does digital photogrammetry incorporate the functionality of analytical stereoplotters (i.e. measurement of points), but it also provides for automated DTM generation, computation of digital orthophotos, preparation of animated visualizations from perspective views to 'fly-through' sequences, and capture of data for direct entry into a GIS. Digital systems result in the integration of photogrammetry, image processing, and GIS on a single computer platform. It is therefore expected that data handling is improved and time is saved, particularly when large volumes of data need to be processed to generate high quality mass data.

On the other hand, it is difficult for the end user to evaluate and control the accuracy of digitally generated DTMs. At present, commercially available programs that are widely used for softcopy DTM production offer only limited control tools for quality assessment. When DTM generation turns from analytical to softcopy systems, the loss of accuracy that might be connected with this technological change is a matter of concern.

As softcopy sytems allow for a faster and easier way to generate DTMs, an increasing number of papers on storage area management include a terrain analysis section. However, the issues of DTM quality are often not fully discussed. For this paper, DTM generation of a coal mine waste mound was conducted both analytically and digitally. Results of the two methods are compared. As the mound under consideration is subdivided into sections of different surface topography and vegetation cover, the quality of the softcopy approach is evaluated with respect to the analytical method for various surface conditions.

1.2 *Sample application of DTMs in tailings and mine waste management: evaporation study*

When integrated as a data layer into GIS computer software, DTMs can be used as information source in many applications of tailing storage area management.

Water balance calculations of mine tailing mounds are an important aspect of reclamation and water quality enhancement programs. Residue mounds of the German coal mining industry are frequently revegetated after waste deposition has stopped. Large amounts of money are spent on the planting of trees and shrubs, on fertilizers, water, and irrigation systems.

Mining operation residues frequently contain chemicals that can contaminate the environment. In potash mine residues, such chemicals are NaCl, $MgSO_4 \cdot H_2O$, $CaSO_4$, $CaSO_4 \cdot H_2O$, and KCl. Over decades, the German potash mining industry has taken comprehensive measures to prevent contamination of ground and surface water. Where technically and geologically practicable, solid mine wastes are disposed of underground. If underground disposal is not possible, mounds are constructed over impermeable, man-made clay layers. Briny discharge collected from mounds is conveyed through pipelines into rivers of low stream order and high runoff to relieve smaller local streams and to accomplish maximum dilution ratios. Yet, under Germany's humid climate, potash tailing mounds will continue to produce contaminated runoff. Therefore, further measures are necessary to protect the environment against contamination.

Because of the high salt content, potash mine tailings usually are bare. Covering such tailings with soil or soil-like material and establishing a plant cover seems to be one possibility to reduce the amount of discharge. It is expected that a plant cover will enhance evapotranspiration. Instead of valuable soil material, it has been suggested to use industrial by-products, such as ashes, slags or drosses as cover materials. The use of such wastes has the additional advantage of relieving the available landfill capacity, which is limited in densely populated Central Europe.

In both cases (coal and potash mine wastes) evaporation from the exposed surface of tailing mounds is an important aspect of water balance calculations. A common approach to describe the evaporative demand of the atmosphere is the combined energy balance and aerodynamic concept, first proposed by Penman (1948). To estimate the daily potential evaporation or evapotranspiration, Priestley and Taylor (1972) published an empirically derived simplification of the Penman equation:

$$Ep = v \frac{s}{s+\gamma} \frac{Rn-G}{L}, \tag{1.1}$$

in which expression *Ep* (mm d^{-1}) stands for the daily evaporation rate, v is a constant that lumps the aerodynamic or wind function of the Penman equation and is proposed to be 1.26 for agricultural crops, $s/(s+\gamma)$ is a dimensionless temperature function, *Rn* is the net radiation balance of the evaporating surface, *G* is the soil heat flux, and *L* is the specific heat for evaporation of water. For further details, the reader is referred to Priestley and Taylor (1972).

In the expression for *Ep*, *Rn*, and both air and ground temperature must be known. For a few single locations, *Rn* can be measured. However, since tailing mound slopes are steep and face different compass directions, it is expected that in mid-latitudes solar irradiance, *Rn*, and thus evaporation vary significantly with slope aspect. It is therefore necessary to estimate *Rn* for varying surface locations from other meteorological data, e.g. net global (shortwave) radiation, duration of bright sunshine hours per day, atmospheric turbidity, and the percentage of cloud cover. Several models exist to calculate solar radiation on mountain slopes. Commonly, information on slope steepness and direction is required as input parameters to these models. Using GIS software, such data can conveniently be derived in high spatial resolution from DTMs of tailing mounds.

As a DTM sample application in mine waste management, results from an ongoing study on the evaporation of a potash mine waste mound that possibly will be covered with an industrial by-product are presented. In this paper, a comprehensive model for *Rn* is used (VDI, 1994).

2 MATERIALS AND METHODS

2.1 *Evaluation of softcopy DTM accuracy*

Three stereopairs of standard aerial color positive transparencies (23 × 23 cm^2) were used to study the geometry of a terraced and partly vegetated, actively operated coal waste mine in the Rhine-Ruhr area in western Germany. For flight parameters see Table 1, column two. DTMs from these photographs were generated both analytically and digitally (Figure 1).

Table 1. Parameters for test flights over a coal and a potash waste mound.

	Coal waste mound	Potash residue mound
Camera	Leica, RC 30	Leica, RC 30
Focal length, mm	153.23	153.23
Date of flight	March 23, 1995	August 06, 1996
Scale	1 : 6,280	1 : 8,000
Height, m	1030	1230
Image overlap, %	60	60
Film	color	color

Deposition of coal mine wastes on the mound under consideration progresses from south to north. Because older parts of the mound's surface are already covered with different stages of vegetation, the DTM was subdivided into four characteristic parts for further analysis.

On the day of the test flight, the northern part was awaiting deposition within an earlier deposited outer dam of waste material. Inside this dam, topography was characterized by small undulating surface features and only few edges of markedly changing slope steepness. The middle part contains on its eastern slope the steepest inclinations within the entire test site and on the day of the flight was only partially vegetated. The south-western part was covered with a lush vegetation of herbs and shrubs close to the mound's base line, and with single shrubs and grassland at higher elevations. The south-eastern part was similar, but lush shrub vegetation reached higher than in the south-west.

In digital photogrammetry and DTM generation, a number of technical terms is used. Short definitions of these shall be given here for the convenience of the reader. *Ground control points* are points flagged on the ground before flight. Coordinates of ground control points are known planimetrically in a cartesian coordinate system from terrestrial survey. Elevations are known with respect to a level datum (e.g. sea level). *Fiducial marks* are built-in devices of cartographic cameras that are projected to the middle of the sides and/or the corners of the photograph. Usually, a central perspective projection is assumed with aerial photographs. To determine geometric parameters of a photograph, its *interior orientation* has to be established. This is done by measuring the fiducial marks and calculating the *principal point*, which is the intersection of the lines connecting opposite fiducial marks. Any measured points in the photograph are then registered with respect to its principal point in a relative coordinate system. During the next step, the *relative orientation* of a stereopair is established. A number of points is measured within the overlapping area and used as *tie points*. From tie points, the plane is defined that is covered by both images of the stereopair.

Figure 1. Digitally generated DTM of the coal waste mound from a test flight on March 23, 1995.

This plane contains all points to be possibly measured three-dimensionally, including the projection centers of both photographs. At this stage, there is no relation between the relative coordinate system of the stereo model and the real-world coordinate system. To adjust this orientation of the stereopair to the ground control points, parameters of the *absolute orientation* need to be calculated during a process called *bundle block adjustment*.

Mass points and *breaklines* are surface feature types used during the process of DTM generation. During this process an approximated surface is generated by constructing a mesh of triangles with triangle nodes being tied to mass points. In this paper, we used a triangulation method that satisfies the Delaunay criterion. With Delaunay triangulation, triangle nodes are tied to the mass points in such a way that (i) triangles are mostly isosceles and (ii) a circle drawn through the three nodes of any triangle does not contain other mass points. This means that all mass points are connected with their two nearest neighbours to form triangles. Moreover, triangle edges are forced to follow surface features that have been defined as breaklines. This additional criterion yields a much improved approximation of the resulting DTM in the vicinity of surface features that form strong changes in slope steepness, e.g. the edge of a terrace or the base line of a waste mound.

Analytical (manual) photogrammetric measurement of the parameters of the interior, relative, and absolute orientation, as well as mass points and breaklines was conducted using a Zeiss P3 Planicomp plotter (Carl Zeiss, Inc.). Edges of terraces were measured and coded as breaklines.

Bundle block adjustment was conducted with help of the computer software package BLUH (Jacobsen, 1997). The Triangulated Irregular Network (TIN) module of ArcInfo GIS software (ESRI, 1996a) was used to generate a DTM from measured points and breaklines. Also by using TIN, the mound's unknown base plane was approximated from its three-dimensionally recorded base line, and the volume of the mound was calculated. Using ArcInfo GIS software, the TIN-generated DTMs were translated to grid format with a ground resolution (cell size) of 1 m^2.

According to Konecny and Lehmann (1984), errors in planimetric locality (σ_{xy}) and in elevation (σ_z) for analytically measured points can be calculated with the following equations:

$$\sigma_{xy} = \pm \sigma_{px} m_b \tag{2.1}$$

$$\sigma_z = \pm \sigma_{xy} \frac{h}{b} = \pm \sigma_{px} \frac{h^2}{cb}, \tag{2.2}$$

in which expression σ_{px} is the focussing error of the analytical plotter, h is the elevation or height of flight above sea level, c is the camera focal length, m_b is the scale number and is the inverse of the photo scale ($m_b = h / c$), and b is the base of the photograph which is an approximation of the distance between projection centers of two neighbouring photographs. Using the Zeiss P3 Planicomp and adjusting measured points by bundle block adjustment yields $\sigma_{px} = \pm 15$ µm. Thus, together with flight parameters from Table 1 (coal waste mound), from Eq.(2.2) an error of elevation in analytical measurement of approximately ± 0.18 m is obtained.

For softcopy photogrammetric analysis, the before-mentioned transparencies were digitized in a black and white and, for visualization purposes, also a three-color (red, green, blue) model. This was done with a high resolution photoscanner (RM-1, Wehrli & Assoc., Inc.), set to a resolution of 24 µm (1000 dpi or 0.151 m ground resolution, respectively). Absolute orientation was established from identical parameters as in the analytical approach. The software package OrthoMax (Vision International Inc.) was used for the automated measurement of *homologuous points*. These are points in stereopairs which correspond to the same point on the ground. Conceptually, homologuous points are selected by marked changes in shades of gray within a local edit mask that is moved over the stereopair. For a detailed description of the program OrthoMax, including mathematical algorithms, see ERDAS (1994). Neither special treatment of terrace edges and the mound base line as breaklines, nor TIN triangulation was conducted. The DTM was directly generated in raster format from homologuous points and was set to a ground cell resolution equal to that used with the analytical method (1 m^2). From the resulting grid, the analytically generated grid was subtracted cell by cell. The resulting new grid was converted to ascii format with cells of a value of zero (for cells with identical softcopy and analytical elevations) being excluded. From the ascii file, a new TIN was generated and its volume (i.e. the volumetric deviation of the softcopy from the analytical DTM) was calculated.

Two different approaches were followed to evaluate softcopy DTM accuracy with respect to reference data from analytical photogrammetry. In both cases, results from analytical measurements and triangulation were assumed to be free of error. As a first approach, simply results of volumetric calculations in DTMs from the analytical and the softcopy method were compared. If DTMs from both methods would match exactly, volumes should be identical. However, if the DTM constructed by one method yields higher elevations (i.e. volumes) under certain surface conditions (e.g. slope steepness or vegetation cover), and lower elevations under other conditions, potentially significant local errors in 'softcopy elevation' could average out in calculation of the total volume of the mound.

For a more sophisticated approach, analytically measured points were used as independent control points. Statistical analysis was conducted to compare elevations of analytically measured points with those of digitally correlated points in their proximity. The cartesian coordinates of analytically measured points were fed into the software programm PIDENT, which allowed for the selection of points from the softcopy DTM within a user defined search radius around each analytical point. Search radii of 0.05, 0.10, 0.30, and 0.49 m on the ground were used, and xy-coordinates of 'softcopy points' around analytical mass points and points on breaklines for the northern, middle, south-eastern and south-western parts were selected separately. Statistical analysis was then carried out with help of the software program BLAN. Results of BLAN include the square mean of differences, the systematic, and the non-systematic error. Both PIDENT and BLAN are subroutines of BLUH (Jacobsen, 1997). The difference in elevation between corresponding 'analytical' and 'softcopy points' was calculated by use of the following equation:

$$SX_{cp} = \sqrt{\frac{(\sum_{i=1}^{N} dh_i)^2}{N}} \tag{2.2}$$

In this expression, SX_{cp} is the total deviation or error in elevation, dh_i is the deviation in elevation of a single 'softcopy point' from the corresponding 'analytical point', and N is the number of 'softcopy points' found within the current search radius. The total error in elevation consists of a systematic error (also called systematic shift, or drift) and a random error. BLAN distinguishes between these by use of the expression:

$$SX_{cp} = \sqrt{(systematic\ error^2 + random\ error^2} \tag{2.3}$$

The systematic shift is calculated from the sum of elevation differences. If a standard normal distribution is assumed for these differences, the systematic shift is the arithmetic mean of elevation differences. When the systematic shift is subtracted from the sum of differences, it can be assumed that the distribution of the remaining random errors is also a standard normal or Gaussian distribution. Therefore, calculated differences in elevation at the control points are the Root Mean Square Errors (RMSE) of this distribution.

2.2 *DTM sample application*

For evaporation and water balance studies, a DTM of an actively operated potash mine residue mound near Hannover in northern Germany was generated from digitized aerial photographs. Mound slopes are cone shaped because waste material is deposited by conveyor belts. Because the material consists 96 % of sodium chloride, the mound is unvegetated. Flight parameters are listed in column three of Table 1. For DTM generation, the softcopy approach as described before was followed, with a grid spacing (ground resolution) set to 5 m^2.

Net radiation Rn is required as part of the energy balance for every DTM grid cell (see Eq. 1.1). For a horizontal surface with no limitation of the horizon, Rn can be given as:

$$Rn = (1 - \alpha) S_g + L_i - L_o \tag{2.4}$$

where S_g is global shortwave irradiance, L_i is longwave sky irradiance, L_o is longwave radiation emitted by the surface (all in W m^{-2}), and α is the surface albedo. It is well known that daily variation is much larger for shortwave than for longwave irradiance, and that longwave irradiance integrated over a day is almost balanced. Therefore, evaporation is mainly controlled by shortwave irradiance. For simplicity only shortwave radiation is discussed here (i.e. L_i and L_o are ignored).

According to the model used (VDI, 1994), S_g for an arbitrarily oriented surface at variable atmospheric conditions is the sum of the direct (I) and diffuse (D) shortwave radiation fluxes under these conditions:

$$S_g(\beta,\eta;n) = I(\beta,\eta;n) + D(\beta,\eta;n)\,, \tag{2.5}$$

in which expression the suffix ($\beta,\eta;n$) designates that all components are functions of slope steepness β, slope aspect η, and cloud cover n. For a detailed discussion of the calculation of $S_g(\beta,\eta;n)$, see VDI (1994), or any textbook on solar radiation (e.g. Iqbal, 1983).

Conceptually, direct solar irradiance and global radiation onto an inclined surface are derived from an average extra-terrestrial solar radiation (1367 W m^{-2}), corrected for the varying sun to earth distance, the total extinction to which the solar radiation is subjected in the atmosphere, cloud cover n in sky octas, and the angular variation between the surface normal and the direction towards the sun. This variation is obtained from geographical latitude, orientation of the surface, declination of the sun, and true local day time. Total atmospheric extinction (scattering and absorption) is estimated from the Linke turbidity factor T_L. This factor specifies the optical thickness of a turbid and moist atmosphere as a multiple of a clean, dry atmosphere. It is estimated from visibility (in m). Both visibility and cloud cover are routinely recorded on an hourly basis at many airfields and are available from the German Weather Service upon request. For this paper, data from June 17 to December 31, 1997 from Wunstorf military airfield, located six kilometers to the north-east of the mound, was used. Diffuse solar radiation for a horizontal plane under a cloudless sky is calculated as the difference between global radiation and direct solar radiation under that sky. Results are reduced for inclined surfaces and separated into an isotropic and an anisotropic component for partly cloudy or overcast conditions, because maximum values of diffuse solar radiation under these conditions occur in the vicinity of the sun.

Values of surface albedo for various materials have been listed in the literature (e.g. by Iqbal, 1983), but apparently not for NaCl surfaces. Therefore, an average albedo of 0.25 was calculated from measurements of shortwave global radiation and reflected radiation. These were taken from June 17, 1997 to June 18, 1998 with two CM3 pyranometers integrated into a CNR-1 net radiometer (Kipp & Zonen, Delft, The Netherlands). The radiometer was installed on a south-easterly facing slope of the mound (appr. 160°, 0° being north). To this aspect, the CNR-1 was tilted to an angle of 34°, which is equal to slope steepness in that location.

Using Avenue object oriented programming language (ESRI, 1996c), an extension to ArcView GIS computer software (ESRI, 1996b) was developed to solve equations for direct, diffuse, global, and net shortwave radiation components. The extension program reads values of cloud cover and visibility from a data base file and calculates radiation components for either a single point of user-defined location and orientation, or every grid cell of a DTM within a user-defined time span on an hourly basis (in W m^{-2}) or as daily integrals (in MJ m^{-2}). Calculated hourly values of solar altitude and azimuth are used to conduct standard ArcView GIS hillshade analysis, i.e. to select grid cells that do not receive direct sunlight. Values of net shortwave radiation integrated over a day are used to estimate potential evaporation according to Eq.(1.1). Measured data from the CNR-1 radiometer were used to evaluate calculated values for global shortwave radiation for a single point.

3 RESULTS AND DISCUSSION

3.1 *DTM accuracy*

Calculation of volumes resulted in 36,356,069 m^3 for the analytical and 36,946,805 m^3 for the softcopy DTM, i.e. a deviation less than 2 %. This indicates that use of softcopy photogrammetry is suitable for volumetric analysis. Results from statistical analysis of individual points in subareas of the coal waste mound to characterize deviation of the softcopy from the analytical DTM are listed in Table 2. Results indicate that the degree of accuracy commonly known from analytical measurements cannot be achieved with the softcopy system. This can be seen from calculated differences in elevation for all parts of the mound, and especially from those of points on breaklines. Deviation of the analytical from the softcopy DTM seems to be positively correlated with slope steepness. Also, shrub vegetation is a relevant source of error.

Table 2. Results of DTM accuracy evaluation.

	Points on breaklines				Masspoints			
Search radius, m	0.05	0.10	0.30	0.49	0.05	0.10	0.30	0.49
				North				
Total no. of points selected	58	250	2013	5331	279	438	742	1291
Square mean of differences, m	± 1.737	± 0.892	± 0.690	± 0.571	± 0.212	± 0.218	± 0.255	± 0.412
Systematic error, m	0.120	- 0.095	- 0.126	- 0.139	- 0.113	- 0.122	- 0.133	- 0.104
Non-systematic error, m	± 1.733	± 0.877	± 0.679	± 0.554	± 0.179	± 0.180	± 0.229	± 0.399
				Middle				
Total no. of points selected	38	123	1144	2949	132	178	469	982
Square mean of differences, m	± 3.300	± 1.978	± 2.470	± 2.062	± 1.909	± 1.807	± 1.994	± 2.017
Systematic error, m	0.701	0.078	0.071	- 0.014	0.127	0.127	0.056	0.006
Non-systematic error, m	± 3.225	± 1.976	± 2.469	± 2.062	± 1.906	± 1.802	± 1.993	± 2.017
				Southwest				
Total no. of points selected	13	47	414	1050	205	219	328	531
Square mean of differences, m	-	± 0.564	± 0.630	± 0.613	± 0.600	± 0.598	± 0.528	± 0.532
Systematic error, m	-	- 0.315	- 0.367	- 0.354	- 0.377	- 0.372	- 0.332	- 0.320
Non-systematic error, m	-	± 0.468	± 0.512	± 0.500	± 0.466	± 0.468	± 0.411	± 0.425
				Southeast				
Total no. of points selected	16	54	455	1226	279	438	742	1291
Square mean of differences, m	-	± 0.917	± 0.875	± 0.942	± 0.212	± 0.218	± 0.255	± 0.412
Systematic error, m	-	- 0.600	- 0.558	- 0.541	- 0.113	- 0.122	- 0.133	- 0.104
Non-systematic error, m	-	± 0.694	± 0.674	± 0.771	± 0.179	± 0.180	± 0.229	± 0.399

Negative values of the systematic error for the (most densely vegetated) south-eastern part of the mound indicate that automatically correlated points are predominantly higher than those from analytical measurement. For many waste mound management applications, such deviations need to be corrected by manual editing. During this process, single bushes can be detected by a plausibility check of their elevation with respect to the mean elevation of their local surroundings.

3.2 *Evaporation*

Figure 2 shows the calculated potential evaporation from the potash residue mound for September 21, 1997 (autumnal equinox). From net shortwave radiation, S_g, the salt water evaporation equivalent was estimated for this raster image by use of Eq.(1.1). A visibility of 40,000 m and a fractional cloud cover of 1 octa were recorded for the entire day. Average ground temperature during daytime hours (i.e. hours between sunrise and sunset) was 15.1° Celcius. From this temperature, the term $s / (s + \gamma)$ and the specific heat for evaporation of water, L, were calculated to be 0.62632 and 2.46154 MJ m^{-2} mm^{-1}, resp. Using these values and daily integrated values for Rn as calculated with the extension to ArcView GIS described in section 2.2, potential evaporation for every grid cell in the DTM was calculated. Energy partitioning at the mound's surface was ignored, i.e. it was assumed that all energy was consumed by evaporation. Equation (1.1) was derived for the evaporation of fresh water.

Figure 2. Evaporation from the potash residue mound on September 21, 1997.

Because evaporation of salt water is less than evaporation of fresh water under the same environmental conditions, resulting values from Eq.(1.1) were devided by the density of a saturated NaCl solution (1.21 Mg m^{-3}; Weast, 1986), as suggested by Wittenberg (1985). Thus, results shown in Figure 2 are valid for the mound's surface only and are invalid for the surrounding landscape. On September 21, 1997, calculated potential evaporation varied significantly between slopes facing different aspects. It was about 0.4 mm d^{-1} on north facing slopes (darkest shades of gray), 5.4 mm d^{-1} on south facing slopes (lightest shades of gray), and 3.4 mm d^{-1} on the mound's plateau and on slopes facing east or west (intermediate shades of gray in Figure 2). From June 17 to December 31, 1997, measured CNR-1 global radiation as well as visibility and cloud cover data from Wunstorf airfield were available. From the airfield data, global radiation was also calculated for the CNR-1 location for hours between sunrise and sunset by use of the ArcView GIS extension. Measured and calculated values were separated for the nine classes of cloud cover (0 to 8 octas), and the square of the Pearson correlation coefficient (r) was calculated for every class. Together with the number of hours per class, results are shown in Table 3.

Results suggest that the model performance for shortwave radiation is best under clear sky conditions, and is less accurate for high fration of cloud cover and overcast conditions. This may be due to the fact that a low fraction of cloud cover is easier to estimate than a high fraction of cloud cover, because the observer is able to look vertically upwards into the sky in only one point (zenith), and will see the rest of the sky with an angle increasing towards the horizon. Thus, because of an error of parallax, the observer might not be able to see holes in the clouds of a seemingly overcast sky. Also, a high fraction of cloud cover might consist of either thick or thin clouds, absorbing quite different amounts of direct solar radiation. Cloud thickness, however, is not accounted for by the shortwave radiation model used. Table 3 indicates that during most hours of the measuring period a high fraction of cloud cover prevailed, and that model performance was somewhat weak for this majority of hours. However, under mostly sunny conditions, accuracy of the model was high. Because of high energy input, it is expected that evaporation mainly occurs under such conditions. The advantage of the radiation model used here as compared to other models is the availability of required input data from airfields (i.e. in relatively high spatial resolution).

Table 3. Accuracy of the radiation model at different cloud cover fractions.

Cloud cover octas	r^2 -	Absolute frequency -	Relative frequency %
0 (clear sky)	0.91	32	1.43
1	0.93	209	9.33
2	0.93	132	5.90
3	0.90	133	5.94
4 (partly cloudy)	0.75	141	6.30
5	0.66	147	6.57
6	0.54	211	9.42
7	0.56	781	34.88
8 (overcast)	0.52	453	20.23
total		2239	100

4 CONCLUSIONS

When integrated as a data layer in GIS systems, DTMs are particularly useful in tailings and mine waste management. Today, digital systems integrate functionalities of photogrammetry and photogrammetric measurement, image processing, and GIS on a single computer platform. Computation and visualization of digital orthophotos, DTMs, and derivation of data from those can easily be conducted. In tailings and mine waste applications, DTMs can conveniently be used as input data for cut-and-fill type analysis. When used in connection with GIS-based models, environmental processes as -for example- evaporation and its spatial and temporal variability can also be analyzed and quantified in high resolution.

It is shown that DTM generation and application in waste mound management include many contiguous steps. Any of these may introduce error. Generation of DTMs is a first and fundamental step, and special care should be taken to provide for DTM accuracy. Commercially available software packages currently offer only restricted control tools for quality assessment. In practice, time initially saved through automated steps of softcopy DTM production often is spent later to manually measure additional information, e.g. along breaklines. Therefore, an essential aspect of future developments has to be the implementation of extended functions into software programs to control and edit automatically generated DTMs. It is expected that use of encoded three-dimensional line measurement (i.e. automated measurement of breaklines) by digital photogrammetric software will greatly enhance softcopy DTM accuracy.

Quantification of environmental processes with models often involves empirical parameters or parameters that can only be estimated. In case of the solar radiation model used here, such a parameter is cloud cover. Because measurement of solar radiation over an extended period of time is currently only possible for a few single locations, analysis of model sensitivity to empirical or estimated input data cannot be conducted for extended areas. Yet, management decisions in tailings and mine waste mound operation based on limited knowledge shall lead to better results than those ignoring this knowledge.

ACKNOWLEDGEMENTS

Chr. Fischer's work on digital photogrammetry in mine waste management is supported by Ruhrkohle Bergbau AG, Bottrop, Germany. The work of D. Hermsmeyer on the radiation and water balance of potash tailing mounds is supported by Kali & Salz GmbH, Wunstorf, Germany, and by the Deutsche Forschungsgemeinschaft (German Research Association, DFG). Both authors were supported by the DFG through a travel grant. We also wish to thank Dr. R.R. van der Ploeg and Dr. W. Busch for help with preparation of this manuscript, and Dr. K. Jacobsen for use of his software package BLUH and the Wehrli RM-1 photoscanner.

REFERENCES

Dörstel, C. 1996. PHODIS Innovations. *Proceedings of the OEEPE-Workshop on Applications of Digital Photogrammetric Workstations, Lausanne* 4-6 March 1996:445-453.

ERDAS (Earth Resources Data and Analysis System, Inc.) 1994. *Imagine OrthoMax user's guide.* Version 8.2. Atlanta, Ga.

ESRI (Environmental Systems Research Institute, Inc.) 1996a. *ArcInfo Geographic Information System.* Version 7.0.4. and online manuals. Redlands, CA.

ESRI (Environmental Systems Research Institute, Inc.) 1996b. *ArcView GIS.* Version 3.0a and online manuals. Redlands, CA.

ESRI (Environmental Systems Research Institute, Inc.) 1996c. *Avenue. Customization and application development for ArcView.* Redlands, CA.

Gülch, E. 1994. *Generation of Digital Terrain Models by automated image orientation.* Deutsche Geodätische Kommision, Reihe C. Munich: C.H.Beck (in German).

Heipke, C. 1996. Overwiev of Image Matching Techniques. *Proceedings of the OEEPE-Workshop on Applications of Digital Photogrammetric Workstations, Lausanne*, 4-6 March 1996:173-192.

Iqbal, M. 1983. *An introduction to solar radiation.* Toronto: Academic Press.

Jacobsen, K. 1997. *Program System BLUH (Bundle block adjustment University of Hannover).* Institute for Photogrammetry and Engineering Surveys, University of Hannover.

Kersten, Th. and S. Haering 1997. Automatic interior orientation of digital aerial images. *Photogrammetric Engineering & Remote Sensing.* 63:1007-1011.

Konecny, G. and G. Lehmann 1984. *Photogrammetrie*, 4th edition. Berlin: deGruyter (in German).

Krystek, P.; Heuchel, T.; Hirt, U.; Petran, F. 1996. An Integral Approach to Automatic Aerial Triangulation and Automatic DEM Generation. *Int. Archieves of Photogrammetry and Remote Sensing, Vol. XXXI, Part B3:*, Vienna.

Penman, H.L. 1948. Natural evaporation from open water, bare soil and grass. *Proc. Roy. Soc.* London A 193:120-145.

Priestley, C.H.B. and R.J. Taylor.1972. On the assessment of surface flux and evaporation using large-scale parameters. *Month. Weath. Rev.* 100:81-92.

Tsingas, V. 1995. Operational Use and Empirical Results of Automatic Aerial Triangulation. In D. Fritsch & D. Hobbie (eds), *Photogrammetric Week*: 207-214. Heidelberg: Wichmann.

VDI, Verein Deutscher Ingenieure (Association of German Engineers) 1994. *Environmental Meteorology. Interactions between atmosphere and surfaces. Calculation of shortwave and longwave radiation.* VDI Guideline 3789, Part 2. Berlin: Beuth Verlag (in German and English).

Weast, R.C. (ed.) 1986. *CRC Handbook of chemistry and physics.* 67th edition. Boca Raton, FL: CRC Press.

Wittenberg, H. 1985. *Evaporation from salt water lakes. Studies for hydro-solar energy projects.* Beiträge zur Hydrologie, Sonderheft 5.2, Kirchzarten (Germany).

Design, operation and disposal

Tailings and Mine Waste '99 © 1999 Balkema, Rotterdam, ISBN 90 5809 025 6

Behavior of sub-aerial deposited tailings and design of sub-aerial deposition in arid regions

Yunxin Qiu & Dave C. Sego
Department of Civil and Environmental Engineering, University of Alberta, Edmonton, Alb., Canada

ABSTRACT: The sub-aerial tailings disposal method is a promising method to effectively and economically dispose of tremendous volumes of the tailings. To improve the basic understanding of the behavior of sub-aerial deposited tailings, the appropriate engineering properties of the tailings must be ascertained. First, this paper presents the engineering properties of four different tailings which represent diverse mine tailings. Then, the theoretical formulas for modeling the physical processes of the disposal method are described. Finally, the considerations leading to the optimum deposition design to achieve the maximum availability of water for recycle are presented.

1 INTRODUCTION

The mining industry produces various types and tremendous volumes of mine tailings every day. Most of the tailings have to be treated and disposed of on the earth surface in the most environmentally and economically acceptable manner. On the other hand, mineral processing requires large amounts of water. The demand on water resources is becoming a critical issue, especially in arid regions. The sub-aerial technique is a promising tailings disposal method (Qiu and Sego 1998a). Knight and Haile (1983) outlined that the sub-aerial technique involves the systematic deposition of tailings in thin layer thus allowing each layer to settle, drain and partially air dry prior to covering it with a fresh layer of tailings. This paper will present the behavior of the sub-aerial deposited tailings and evaluate the optimum deposition design for sub-aerial tailings disposal with the intent of maximizing the amount of water available for recycle. In this research, copper mine tailings, gold mine tailings, coal wash plant tailings and oil sand composite/consolidated tailings (CT) were chosen to represent diverse types of mine tailings for use in this study.

2 BEHAVIOR OF SUB-AERIAL DEPOSITED TAILINGS

To understand the tailings behavior, a series of laboratory tests were carried out (Qiu and Sego 1998b). The appropriate engineering properties of the tailings have been investigated (Qiu and Sego 1998c) and are presented as follows.

2.1 *Basic physical and shear strength properties*

The basic physical and shear strength properties of the mine tailings are presented in Table 1, and the grain size distributions are shown in Figure 1.

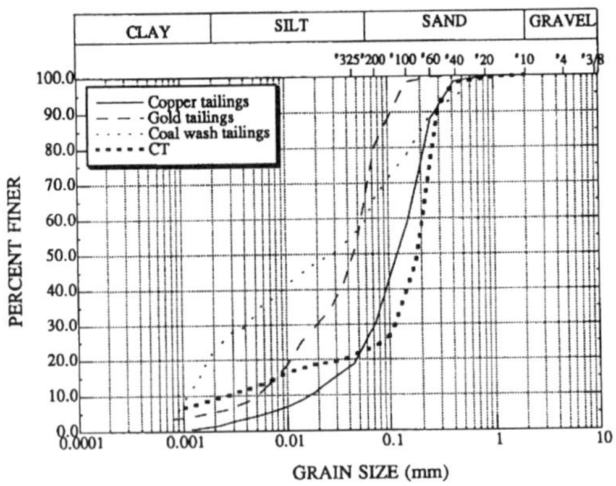

Figure 1. Tailings grain size distribution (after Qiu and Sego, 1998c)

Table 1. The basic physical and shear strength properties of mine tailings (modified from Qiu and Sego, 1998c)

Tailings	Copper	Gold	Coal	CT
Specific Gravity, Gs	2.75	3.17	1.94	2.60
Liquid limit, (%)	----	----	40	----
Plasticity Index, (%)	----	----	16	----
Clay size particles(< 2μm), (%)	1.3	5.3	22.5	8.9
Sand content (>0.06mm), (%)	74.5	33.3	40	77
Fines content (<74 μm), (%)*	31.3	81.3	66.4	21.2
Unified Soil Classification	SM	ML	CL	SM
c' (kPa)	0	0	10	3
φ'	34°	33°	32°	30°

Note: fines are referred to the particle size less than 45 μm for CT

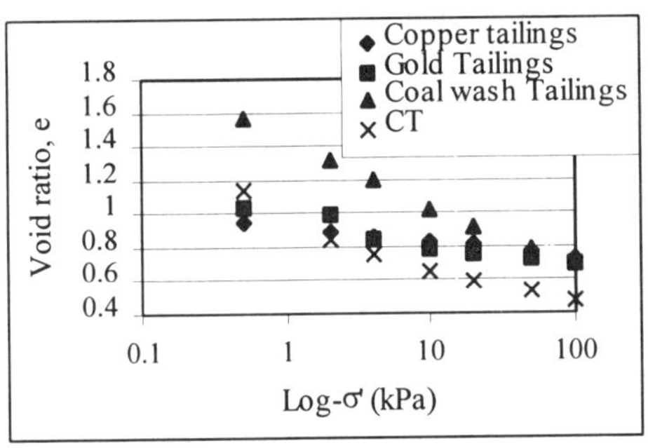

Figure 2. Void ratio versus log-σ' relationship

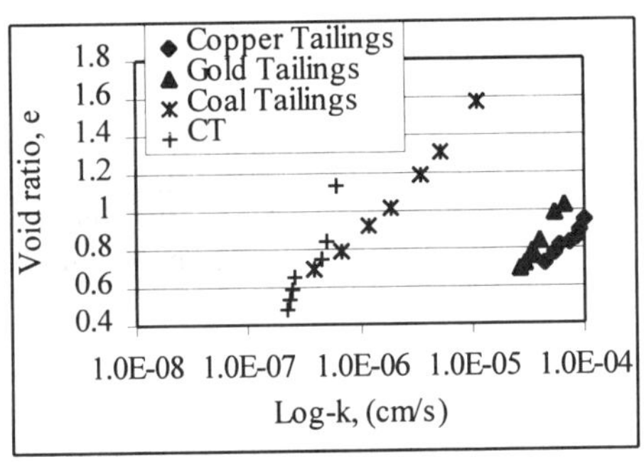

Figure 3. Void ratio versus Hydraulic conductivity plot

2.2 *Consolidation and permeability behavior*

The consolidation and constant head permeability tests were conducted over an effective stress range of 0.5 to 100kPa. The tests were carried out in a large strain consolidation apparatus (Qiu and Sego 1998b). Figure 2 and Figure 3 present the void ratio versus vertical effective stress and hydraulic conductivity relationships, respectively. The compression indexes for the copper tailings, gold tailings, coal tailings and CT are 0.09, 0.16, 0.38 and 0.27, respectively.

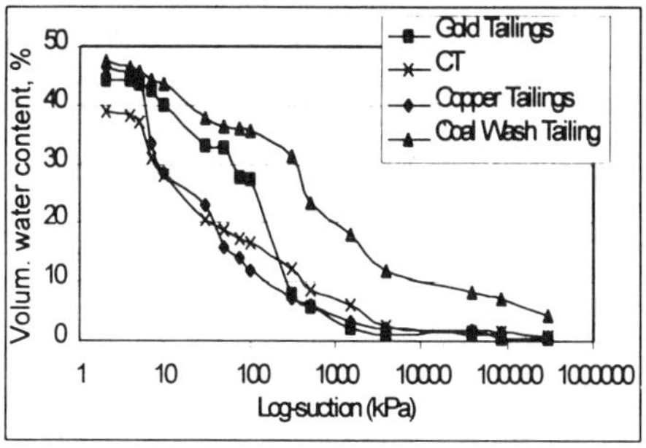

Figure 4. Tailings-water retention curve

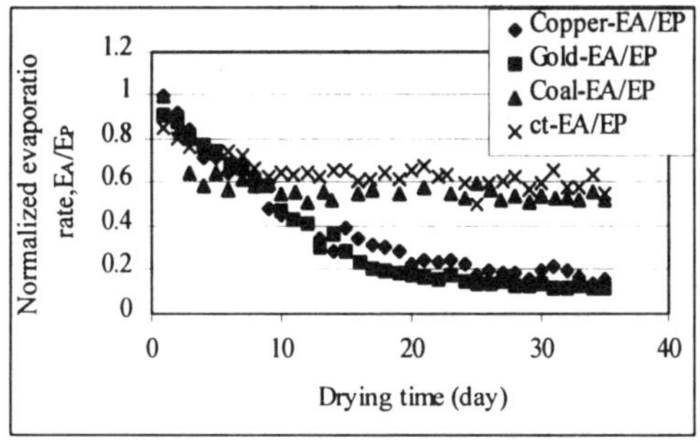

Figure 5. Normalized evaporation rate versus drying time plot

2.3 *Tailings-water retention characteristics and desiccation behavior*

To obtain the tailings-water retention curve over the entire suction range, both a pressure plate extractor test and a glass desiccator test were carried out (Qiu and Sego 1998b). The tailings-water retention curves are illustrated in Figure 4.

Column drying tests on the tailings were carried out (Qiu and Sego 1998b). The ratios of the actual evaporation rate of the tailings to the potential evaporation rate of the environment, i.e. normalized evaporation rates are shown in Figure 5.

3 MODELING SUB-AERIAL TAILINGS DEPOSITION

The physical process of sub-aerial tailings disposal after deposition are sedimentation, consolidation and desiccation. Therefore, to model the process of the sub-aerial tailings disposal, a unified theory is needed.

3.1 *Theories of sedimentation and consolidation*

To establish the theoretical formula for modeling sedimentation and consolidation, some assumptions are required: (1) the relationship between void ratio and effective stress is unique; (2) the relationship between pore pressure and saturation is unique; (3) the soil particles are incompressible; (4) the flow is one dimension; and (5) isothermal conditions exist within the profile.

In modeling large strain consolidation, the Lagrangian coordinate system and a reduced coordinate z are often used. The definitions of the coordinate systems have been described in the literatures by Gibson et al (1967), Schiffman et al (1988) and Swarbrick (1992). The relationship between z and Lagrangian coordinate is as follows.

$$\xi(z,t) = \int_0^{H_z}(1 + e(z,t))dz \tag{1}$$

$$\frac{\partial \xi}{\partial z} = 1 + e \tag{2}$$

where e is the void ratio and H_z is the total height of solids or the total volume of solids per unit area.

Considering a tailings element with a thickness of $\partial\xi$ at some point in the Lagrangian coordinate system, then the governing equations can be derived based on the equilibrium and continuity of mass.

• The equilibrium of the tailings (solids and water):

$$\frac{\partial\sigma}{\partial a}+\frac{\partial\xi}{\partial a}\left(\theta\gamma_w+\frac{\gamma_s}{1+e}\right)=0 \tag{3}$$

where σ is total stress, θ is the volumetric water content, γ_w is the unit weight of water, γ_s is the unit weight of the tailings particles and e is the void ratio of the tailings.

• Continuity of the fluid phase:

$$\frac{\partial}{\partial a}\left[\theta\gamma_w\left(v_w-v_s\right)\right]+\frac{\partial}{\partial t}\left[\frac{\partial\xi}{\partial a}\theta\gamma_w\right]=0 \tag{4}$$

where v_w and v_s are the velocities of water and solids, respectively, and t is time.

• Continuity of the solids phase:

$$\frac{\partial\xi}{\partial a}=\frac{(1+e)}{(1+e_0)} \tag{5}$$

where e_0 is initial void ratio profile.

• Darcy's Law:

$$\theta\left(v_w-v_s\right)=-k\left[\frac{\partial h}{\partial\xi}+1\right] \tag{6}$$

where h is the pressure head, and k is the hydraulic conductivity.

Using equation (2) to covert to the z coordinate leads to the following two governing equations.

$$\frac{\partial}{\partial z}\left\{\frac{k}{(1+e)}\left[\frac{\partial h}{\partial z}+(1+e)\right]\right\}=\frac{\partial[(1+e)\theta]}{\partial t} \tag{7}$$

and

$$\frac{\partial\sigma}{\partial z}+\gamma_w\left[\theta(1+e)+G_s\right]=0 \tag{8}$$

where G_s is the specific gravity of the tailings particles.

The above theoretical formulas are similar to those presented by Swarbrick (1992).

For saturated consolidation, introduction of the effective stress principle into Equations 7 and 8, and assuming that the relationships between k and e, and σ' and e are unique result in the following expression:

$$\frac{\partial e}{\partial t}=\frac{\partial}{\partial z}\left\{\frac{k}{(1+e)}\left[\left(-\frac{1}{\gamma_w}\frac{d\sigma'}{de}\right)\frac{\partial e}{\partial z}+(1-G_s)\right]\right\} \tag{9}$$

Equation 9 is the governing equation of the non-linear large strain consolidation theory proposed by Gibson et al (1967).

These equations can be solved by numerical methods, e.g. the finite difference method, and can be programmed to model the sedimentation and saturated and unsaturated consolidation when different boundary conditions are applied.

3.2 *Hydraulic conductivity of the tailings*

Hydraulic conductivity is one of the most important parameters which affect the behaviors of the sub-aerial tailings disposal. During the disposal process, the tailings first undergo sedimentation, then they undergo self-weight consolidation and finally they dry. They change from a saturated state to an unsaturated state. Therefore, the hydraulic conductivity changes dramatically.

3.2.1 *Saturated hydraulic conductivity*

Based upon the laboratory measured results, the saturated hydraulic conductivity of the tailings can be expressed using an exponential relationship with the void ratio (Qiu and Sego 1998c).

$$k = A\, e^{B} \tag{10}$$

where A and B are constants, e is the void ratio.

3.2.2 *Unsaturated hydraulic conductivity*

Due to desiccation and drying at the surface, the tailings will desaturate, and therefore the hydraulic conductivity decreases. Fredlund et al (1994) proposed a function of relationship of hydraulic conductivity which depends on the suction versus water content curve.

$$k(\psi)=k_s\frac{\int_{\ln(\psi)}^{b}\frac{\theta(e^y)-\theta(\psi)}{e^y}\theta'(e^y)dy}{\int_{\ln(\psi_{aev})}^{b}\frac{\theta(e^y)-\theta_s}{e^y}\theta'(e^y)dy} \tag{11}$$

where $k(\psi)$ is hydraulic conductivity of the unsaturated soil, k_s is saturated hydraulic conductivity of the soil, b equals $\ln(10^6)$, e is the natural number (2.71828), y is a dummy variable of integration representing the logarithm of suction, θ is volumetric water content of the soil, θ_s is the saturated water content , θ' is the derivative of θ function, and ψ_{aev} is the air-entry value of the soil under consideration.

3.3 *Compressibility of the tailings*

The consolidation characteristics of the tailings can be evaluated from the laboratory tests. It is necessary to obtain data such as compressibility for purposes of settlement analysis.

Furthermore, for modeling consolidation, the relationship between void ratio and effective stress is important. Based on the tests carried out in this study, the relationship can be expressed as (Qiu and Sego, 1998c):

$$e = a \log(\sigma_v') + b \tag{12}$$

where e is void ratio, σ_v' is the vertical effective stress, a and b are constants. The different mine tailings has different a and b values.

3.4 *Tailings-water retention curve*

The tailings-water retention curve is the relation between suction within the tailings and volumetric water content of the tailings. Aubertin et al (1998) proposed a modified Kovács Model (MK) to describe the tailings-water curve as follows.

$$\theta = \theta_s [S_c + S_a (1 - S_c)] \tag{13}$$

with

$$S_c = 1 - \left[\left(\frac{0.4}{eD_{10}\Psi}\right)^2 + 1\right] \exp\left[-m\left(\frac{0.4}{eD_{10}\Psi}\right)^2\right] \tag{13a}$$

with

$$S_a = \frac{a}{e^{1/3}\Psi^{1/6}}\left(\frac{0.4}{eD_{10}}\right)^{2/3}\left[1 - \frac{\ln\left(1 + \frac{\Psi}{\Psi_r}\right)}{\ln\left(1 + \frac{\Psi_0}{\Psi_r}\right)}\right] \tag{13b}$$

where θ_s is the saturated volumetric water content, e is the void ratio, D_{10} is the diameter for which 10% by weight passed the sieve having an opening size D (in mm), Ψ is the matric suction (in cm water), m and a are material parameters, Ψ_0 equal to 10^7 cm water and Ψ_r is the suction corresponding to the residual water content. Figure 6 shows the application of MK model to the copper tailings test data.

3.5 *Evaporation of the tailings*

Surface drying of deposited tailings is mainly caused by evaporation. To evaluate the water balance in the sub-aerial tailings disposal, equations for estimating both potential and actual evaporation rates are needed.

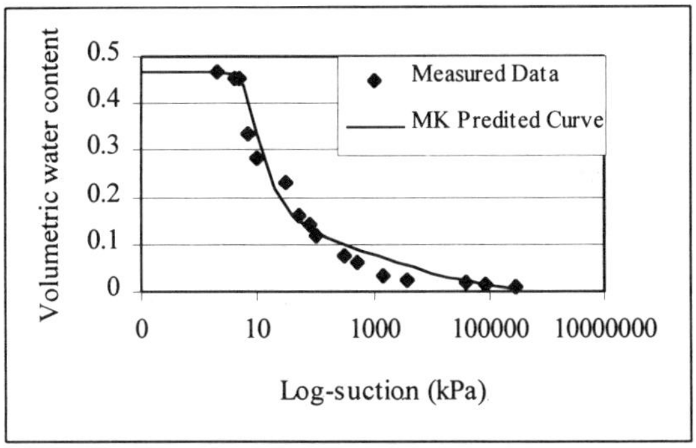

Figure 6. Application of MK model to copper tailings

3.5.1 *Potential evaporation rate (E_p)*

Chow et al (1988) proposed an aerodynamic method to estimate potential evaporation rate as:

$$E_p = B(e_s - e) \qquad \text{(mm/day)} \tag{14}$$

where, e_s and e are the saturated and the actual vapor pressure, respectively, B is a parameter related to wind velocity.

3.5.2 *Actual evaporation rate of tailings (E_A)*

The actual evaporation rate of the deposited tailings decreases with time after the layer of tailings begin to dry from the surface downward. Qiu and Sego (1998c) proposed the following equation to account for the normalized evaporation rate (E_n) of laboratory measured drying rate, or the ratio of actual evaporation rate to potential evaporation rate (E_A/E_P):

$$E_n = \frac{E_A}{E_p} = A_r\left[\frac{1}{1+(\alpha T)^p}\right]^q + B_r \tag{15}$$

where A_r and B_r are positive constants related to the residual evaporation rate, α is a constant related to the air-entry value, p and q are the positive constants related to the material properties, and T is the drying time (day).

4 OPTIMUM DEPOSITION DESIGN

To implement the sub-aerial method, the tailings should be deposited cyclically on the same area in the tailings facilities. In general, mines produce tailings continuously, and consequently generate a continuous discharge of tailings. To maintain continuous deposition in the sub-aerial method, the tailings facilities are usually divided into several sub-areas (i.e. deposition cells) to allow for tailings to be continually deposited but into different cells with time.

For the optimum deposition design, the optimum deposition parameters, such as number and dimensions of the cell, deposition layer thickness and the cycle time (or drying time), should be determined.

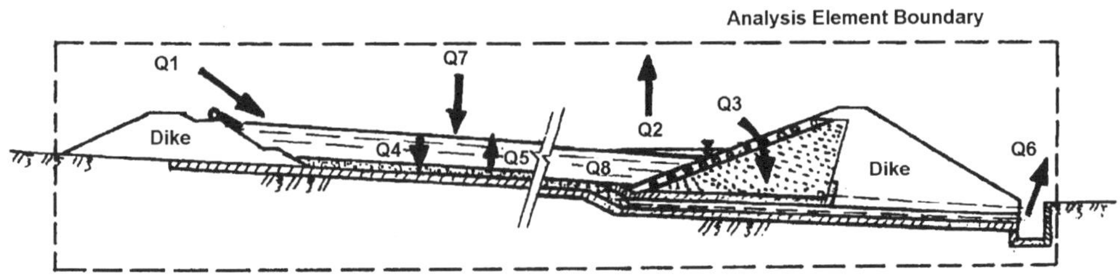

Figure 7. Water balance components during the tailings pond operation

4.1 *Water balance evaluation*

To evaluate the water balance in the sub-aerial method, an identification of all components of the water balance in the tailings facility is needed. Some assumptions are made to simplify the problem. They are as follows: (1) the flow across the bottom seal within the tailings facility is negligible; (2) there is no surface runoff from outside of the facility onto the tailings surface; (3) there is no seepage through the dikes; (4) the infiltration across the dikes is negligible. The water balance components during the operation of the facility are as follows (Figure 7).

- Tailings discharge water: this water is from the tailings discharge system (Q_1)
- Evaporation: this is lost water due to evaporation (Q_2)
- Drainage through the free drainage zone (Q_3)
- Drainage through the sand filter blanket (Q_4)
- Discharge due to consolidation (Q_5)
- Recyclable water from drainage collection ditch (Q_6)
- Precipitation (rain or snow) (Q_7)
- Storage water in the tailings layers (Q_8)

The amount of recyclable water per unit area can be determined using following equation.

$$Q_6 = Q_3 + Q_4 = Q_1 + Q_7 - Q_2 - Q_8 \tag{16}$$

The evaporation amount and the storage amount can be calculated using following equations.

$$Q_2 = \int_0^{t_d} E_A(t)dt \tag{17}$$

where $E_A(t)$ is the actual evaporation rate, it is a function of the time when the potential evaporation is constant, and t_d is the drying time.

$$Q_8 = \int_0^{H(t)} \theta(z,t)dz \tag{18}$$

where H(t) is the total thickness of the tailings layers at time t, $\theta(z, t)$ is the volumetric water content, and a function of both the depth and time, and can be obtained from the model.

4.2 *Calculation of the tailings volume*

Assuming the deposition area is constant, then, the accumulated layer thickness change can represent the volume change of the tailings. Based on the definition of the reduced coordinate z, the total height of the tailings, or, the total volume of the tailings per unit area, can be calculated using Equation (19) (Shiffman et al 1988).

$$H(t) = \int_0^{h_z(t)} [1 + e(z,t)] dz \tag{19}$$

where, H(t) is the total height of the tailings, or the total volume of the tailings per unit area, $h_z(t)$ is the total height of the solids at time t, or total volume of solids per unit area at time t, and e(z,t) is the void ratio at time t.

4.3 *Factors influencing the optimum design*

There are many factors affecting the optimum design of the sub-aerial method used to optimize the release water for recycle. The main factors are hydraulic conductivity (k), suction, drying time, evaporation and layer thickness.

4.3.1 *Hydraulic conductivity and suction*

Hydraulic conductivity is very important in the sub-aerial method. It not only affects drainage and consolidation, but also affects the actual evaporation rate. The higher the hydraulic conductivity, the quicker the consolidation occurs. On the other hand, the higher the hydraulic conductivity, the greater the evaporation rate, thus, the more water lost, therefore, the less water available for recycle. This issue is critical for design in the arid regions. As a result, from the optimum design point view, the larger the hydraulic conductivity, the better the disposal results as a rule.

In the saturated state, the hydraulic conductivity of the tailings has a exponential relationship with its void ratio and the higher the void ratio, the higher the hydraulic conductivity.

In the unsaturated state, the suction created by the surface drying reduces the hydraulic conductivity dramatically after the air-entry value. Therefore, limiting the suction to being lower than the air-entry value can prevent the dramatic decrease of the hydraulic conductivity.

4.3.2 *Drying time and evaporation*

The longer the drying time, the more water evaporates, and the higher suction which develop, and consequently, the more decrease in the hydraulic conduction. However, once the suction is high enough, cracks form on the tailings surface. This may result in horizontal flow channel when a fresh layer is deposited. In this case, the macro hydraulic conductivity increases with little further change in void ratio. Evaporation influences not only the amount of the recyclable water, but also the hydraulic conductivity of the deposit. The more water evaporates, the more salt is deposited on the tailings surface. This thin salt layer has the potential of blocking the water flow through it as the salt crust seals the surface.

4.3.3 *Layer thickness*

The thickness of each deposition layer is an important operating factor. The thinner the layer, the more rapid the drying. This process causes a reduction of hydraulic conductivity earlier. Thus it results in reduced downward flow when a new layer is deposited on its surface. On the other hand, if the layer is too thick, it takes a long time to consolidation and to reach the air-entry value. Therefore, there must exist an optimum deposition layer thickness under specific mine condition to achieve the best disposal results.

Combining all aspects discussed above, the optimum deposition parameters can be obtained.

5 CONCLUSIONS

The copper tailings and gold tailings are non-plastic cohesionless soils. They behavior like sandy soils. However, the CT is a low-plastic cohesive soil, and coal wash tailings is a plastic cohesive soils due to the presence of clay minerals.

There are eight identifiable components of the water balance in the sub-aerial tailings deposition. Meanwhile, the main factors that affect the sub-aerial deposition design for optimizing the release water for recycle are hydraulic conductivity, suction, drying time, evaporation and layer thickness.

There exists a unified theory to model the sub-aerial deposition and the optimum deposition parameters can be obtained using the model.

ACKNOWLEDGEMENT

The authors would like to thank the Natural Sciences and Engineering Research Council of Canada (NSERC) for providing the financial support for this research.

REFERENCES

Aubertin, M., Richard J. & Chapuis, R.P. 1998. A predictive model for the water retention curve: application to tailings from hard-rock mines. *Canadian Geotechnical Journal*,35:55-69.

Chow, V.T., Maidment, D.R., & Mays, L.W. 1988. *APPLIED HYDROLOGY*. McGraw-Hill Book Company. 572pp.

Fredlund, D.G., Xing, A. & Huang, S. 1994. Predicting the permeability function for unsaturated soils using the soil-water characteristic curve. *Canadian Geotechnical Journal*, 31: 533-546.

Gibson, R.E., England, G.L., & Hussey, M.J.L. 1967. The theory of one-dimensional consolidation of saturated clays, I. finite nonlinear consolidation of thin homogeneous layers. *Geotechnique*, 17(3), pp261-273.

Knight, R.B. & Haile, J.P. 1983. Sub-aerial tailings deposition. *Proceedings of 7th Pan-American Soil Mechanics Conference*, Vol.2, Vancouver, Canada. Canadian Geotechnical Society, pp 627-639.

Qiu, Y. & Sego. D.C. 1998a. Design of sub-aerial tailings deposition in arid regions. *Proceedings of 51st Canadian Geotechnical Conference*. Edmonton, Canada, October 4-7.

Qiu, Y. & Sego, D.C. 1998b. Laboratory test methods on mine tailings. *Proceedings of 51st Canadian Geotechnical Conference*. Edmonton, Canada, October 4-7.

Qiu, Y. & Sego, D.C. 1998c. Engineering properties of mine tailings. *Proceedings of 51st Canadian Geotechnical Conference*. Edmonton, Canada, October 4-7.

Schiffman, R.L., Vick, S.G. & Gibson, R.E. 1988. Behavior and properties of hydraulic fills. In *Hydraulic Fill Structures*, Ed(s): D.J. Van Zyl & S.G. Vick, American Society of Civil Engineers, Colorado State University, pp. 166-202.

Swarbrick, G.E. 1992. Transient unsaturated consolidation in desiccating mine tailings, Unpublished Ph.D. thesis, University of New South Wales, Australia.

Tailings and Mine Waste '99 © 1999 Balkema, Rotterdam, ISBN 90 5809 025 6

Site selection approach for mine tailings

A.J. Krause & D.L. Dwire
Montgomery Watson, Mining Group, Steamboat Springs, Colo., USA

ABSTRACT: Tailings management is generally a significant cost in the overall development of a new mining property. The selection of tailings sites for a new mine typically involves the independent evaluation of various components including at a minimum efficiency ratios, watershed to impoundment ratios, distance to process facilities, ability to permit, and borrow source availability. Some components can be quantitatively defined (i.e. efficiency ratio) while other components can only be defined qualitatively (i.e. ability to permit). In each case, it is important to be able to compare and weigh components to effectively select the most appropriate tailings site.

Montgomery Watson has developed an innovative decision analysis tool for tailings site selection. This tool incorporates a methodology to weight components and assign uncertainty levels for those components which are not easily quantified. Utilizing stochastic risk assessment approaches, this method has been effectively employed for tailings management issues for mines in North America and abroad. This paper presents the input parameters and methodologies used to effectively complete site selection and provides working examples of several actual mining case studies.

A. INTRODUCTION

The design, construction and operations of tailings facilities generally represent a significant portion of the overall capital and operational costs for a mining development. While costs vary depending on the disposal technique (e.g. cycloned tailings may be 1/10th the cost of conventional wet tailings), it is not uncommon to see tailings disposal costs exceed $1(US) per tonne for conventional wet tailings disposal. These costs may increase for tailings sites located in high elevations with restrictive topography and steep basal slopes.

While the costs of constructing and operating tailings are a significant consideration, accomplishing environmental compliance and permitting may be an even greater obstacle. Most countries today have promulgated rigorous environmental regulations designed to protect the environment from tailings dam failures and associated adverse environmental impacts. For example, prior to 1960, most of the tailings facilities built in Chile were constructed through upstream, hydraulic fill techniques. In 1928, the 65 meter upstream constructed Barahona tailings dam at the El Teniente Mine failed as a consequence of a major earthquake registering >7.5M, resulting in several deaths and the contamination of the Cachapoal River. A second catastrophe occurred as the result of a large magnitude earthquake when the upstream constructed El Cobre tailings dam failed, released a significant volume of tailings downstream and resulted in the loss of over 200 lives. Because of these failures, the Chilean government adopted Supreme Decree 86 in 1970 entitled "Regulation on the Construction and Operation of Tailings Dams" which has in effect, solely referenced and recommended downstream and centerline construction techniques. Similar regulations have been enacted in other areas of the world where seismic issues impact tailings dam performance.

As a result of the costs and regulatory issues affecting tailings management, it is imperative that a clearly established site selection process be used to effectively identify the best location for long-term tailings storage. With environmental effectiveness and cost efficiency as the goal, it is often difficult to fully evaluate each potential site without exhaustive field investigations, data collection programs, adequate schedule and budget. The data helpful in evaluating options may not be available or complete or may not be cost effective to obtain. As a result, uncertainty plays a role in the final site selection process. This paper presents an integrated approach to tailings site selection using proven techniques and tools to understand the real issues, systematically analyze specific selection criteria, manage uncertainty, evaluate sensitivity, simulate scenarios, and arrive at defensible selection alternatives. This process involves identifying the site selection criteria which are truly important to the ultimate selection of a tailings dam site. The results of the analysis include an objective ranking of sites based on pre-established priorities which are determined interactively with the mining client. The paper utilizes a practical software tool, Criterium Decision Plus (CDP), to accomplish this systematic evaluation.

B. PAST PRACTICES

Past practices in tailings site selection have been based on a variety of criteria. Constrained by the location of the orebody, site selection is essentially a screening process which initially involves an infinite array of possible tailings impoundment sites. Conventional site selection involves the application of various constraints to the initial array, which ultimately results in a much smaller subset of reasonable siting possibilities. Constraints include factors related to topography, mill location, hydrology, geology and groundwater factors. Each factor is briefly discussed below.

Topography

Topography may be considered the single most important site selection criteria. The general aim is to identify sites where the maximum storage capacity can be achieved with the least amount of embankment fill material while satisfying the storage volume requirements. To numerically evaluate topographic advantages the site efficiency ratio is often calculated. This value estimates the ratio of impoundment storage to embankment volume. For a given tailings storage capacity, the higher the efficiency ratio, generally the lower the cost per tonne.

Distance and Elevation Relative to the Mill

Construction and operating costs for the transport of tailings and reclaim pipelines may be significant. These costs may increase due to pipeline dimensions and remote and steep terrain. Considering that the initial cost of tailings and reclaim water pipelines may exceed $0.5 million per mile, a significant cost penalty quickly accrues for a site located a greater distance from the mill. In addition to the distance to the mill, it is generally desirable that the impoundment be situated downgradient of the mill to accommodate gravity flow of the tailings slurry.

Hydrology

Closely related to topography are surface-water hydrology considerations. To the extent that surface-water inflow is directly proportional to drainage catchment area, minimizing inflow requires that the impoundment be sited as close as possible to the head of a drainage basin. Depending on the climatic conditions at the site, the control and management of surface water runoff can provide significant design challenges.

Geology

Geologic factors impact both the embankment and the impoundment through seepage and foundation conditions. While the site geology may not play as critical of a role as the factors listed above, geologic conditions can often present fatal flaws (e.g. active faulting) which may preclude an originally selected site from further development.

Groundwater

Groundwater also plays a role in site selection. For example, in areas with high groundwater conditions, seepage from an unlined pond may adversely impact downgradient groundwater quality. Therefore, permanent impoundment lining systems intended to minimize the seepage of tailings solutions to the environment may be required in the design due to groundwater conditions.

The criteria listed above are commonly used to evaluate potential tailings sites. Often, sites perform well for some categories, but tradeoff poor performance in other categories. Therefore, the best site to select is not always obvious and may be difficult to determine. In the past, the selection of an appropriate site has been subjective and driven primarily by cost considerations. This approach has often resulted in the selection of tailings sites or tailings disposal techniques which do not best serve the overall environmental and cost objectives. Today, environmental factors must be considered. They play as large a role in site selection as costs and are often a differentiating factor used by regulatory agencies when approving permits for mining development.

C. PROPOSED METHODOLOGY FOR TAILINGS SITE SELECTION

The proposed site selection approach builds on the practices established historically, but offers a more detailed, objective and systematic approach to evaluating various tailings sites. The objective of this approach is to provide a defensible selection of the tailings site, with consideration for all the important factors, site-specific conditions, and level of uncertainty in the available data.

A quantitative evaluation is accomplished through the use of a decision management software tool, Criterium Decision Plus (CDP). A model created through CDP facilitates the tailings site selection process by implementing the following logical sequence of activities.

- Identification of the reasonable site alternatives;
- Identification of the factors or criteria important in selecting a site;
- Determination of the relative importance of the criteria to site selection ; and
- Quantitatively rating the potential sites for each of the criteria, including levels of uncertainty.

A reasonable set of siting possibilities is identified through an initial, subjective screening process. Typically, a senior engineer or geologist experienced in identifying potential storage locations in remote areas reviews site topographic conditions to identify all potential sites. These possible tailings sites are then screened according to past practices to eliminate sites which are not considered to be acceptable or reasonable options.

Identifying and weighting the criteria on which to evaluate the potential sites is an important step in the analysis. The criteria must be clearly defined and independent of one another to avoid duplicate representation in the analysis. The criteria presented previously (i.e. past practices) may be used as a starting point, but may be further defined and broken into more detailed categories. Input from the owner/operator of the mine is crucial to insure all the factors important in selecting a site are represented with appropriate emphasis. The key to successful site selection lies in how each criteria is weighted against the others. Questions should be asked, such as "what is a reasonable cost to achieve a specific improvement in environmental performance?" These types of numerical tradeoffs must be defined to allow comparison of the criteria on a relative basis.

Once the potential tailings sites are identified and criteria has been defined and weighted, each site is evaluated for each criteria and assigned a rating. Because comprehensive quantitative information is often not available or cost-effective to obtain for all criteria, uncertainty is commonly a significant factor. The uncertainty in rating the site alternatives against the criteria can be represented by overlaying an appropriate statistical distribution around the rating. Typical distribution types used to represent uncertainty are shown in Figure 1 below.

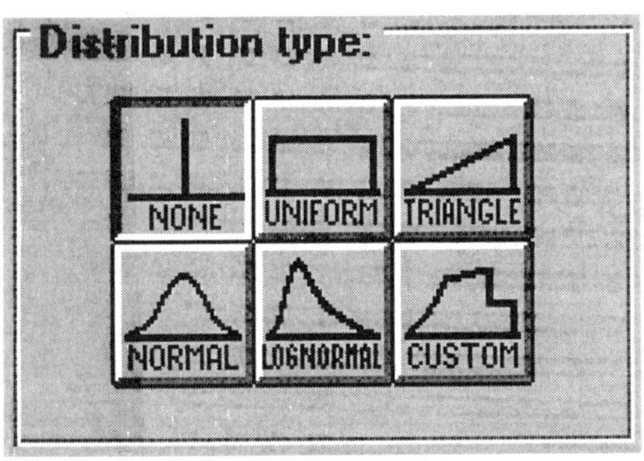

Figure 1 – Distribution types used to represent uncertainty.

Following completion of the systematic steps described above, results of the model may be analyzed. The model calculates a score for each of the potential sites based on the quantitative ratings and the tradeoffs accepted for the criteria. The scores of all potential sites may be ranked to determine the preferred location, as well as provide a comparison among all sites considered in the model. A simulation on the underlying uncertainty distributions provides an outcome illustrating the percentage of time that each site scores higher than the rest of the sites.

An additional benefit provided by CDP is a sensitivity analysis. A sensitivity analysis illustrates the robustness of the site selection result and its sensitivity to changes in the relative importance of the criteria. These analyses also identify the most cost effective areas for additional focus and resource allocation.

D. CASE STUDY

The proposed methodology for tailings site selection was recently applied to a large tailings facility located in South America. Through the review of topographical maps and available geologic and hydrologic data for the site, twelve potential tailings sites were identified. Consecutive letters of the alphabet were assigned to distinguish the sites (Sites A-L).

The specific criteria used in the site selection evaluation are listed below and a brief description of each is provided. Note that the criteria were defined specifically for a site situated in a high Andes mountain physiographic setting in South America and, therefore, are not intended for universal application.

- Efficiency Ratio - A high efficiency ratio is desirable and represents the impoundment storage capacity to the volume of embankment fill required.
- Watershed Area per Impoundment Area - Comparisons of watershed areas to impoundment areas illustrate the expected level of surface water inflows into the tailings, which should be minimized. In addition to the entire watershed area, the watershed area below 3000 masl was calculated because higher elevations are considered to be in the snowfall zone and, therefore, do not provide an adverse impact for the Probable Maximum Precipitation (PMP) event.
- Distance and Elevation Difference from Process Plant to Impoundment - Distance and elevation from the process plant represent costs associated with pipeline and potential pumping requirements.
- Risk of Avalanche - Sites which are aligned perpendicular to the prevailing wind direction and on the leeward side are considered to be susceptible to the development of avalanche cornices. The elevation of the snowshed area and historic evidence of avalanches were also considered in evaluating avalanche risk.
- Effect on Wetlands/Agricultural Land
- Permitability - The permitability of a site was determined through evaluation for environmental fatal flaws, such as water rights issues, the presence of threatened or endangered species, archaeological or cultural significance.
- Condemnation Status - Condemnation status is determined by whether a site contains mineral development potential.
- Borrow Source Location - Impoundment sites with clearly identified alluvial deposits provide a desirable borrow source, as opposed to sites which are primarily bedrock and would require importation of material to construct an earthfill tailings dam[1].
- Foundation Conditions and Seepage Potential - The potential for seepage is often dependent on whether the geologic foundation is composed of alluvial deposits or bedrock.
- Depth to Groundwater and Pre-mining Groundwater Quality - The level of potential water treatment or tailings neutralization is directly related to potential

[1] Rockfill dams would be evaluated differently. The presence of alluvium in the footprint area may require stripping and removal prior to rockfill placement.

groundwater impacts. The greater the depth to groundwater, the more likely the water quality will not be impacted.

- Seismic Activity - Seismic activity, and specifically bedrock acceleration, directly impacts the embankment section and design criteria developed to control the internal phreatic level.

For many of the criteria, quantitative data was not available for all of the potential sites considered in the evaluation. In some cases, a more general subjective rating of high, medium, or low was translated to a numerical scale. In cases where no data was available at all, an average rating was used. Appropriate ranges of uncertainty were applied in both these cases.

The criteria were weighted in relation to one another to represent their relative importance in the selection of a tailings site. The relative weight percentages are shown in Figure 2.

relative weight	Name
100%	Efficiency Ratio
25%	Watershed/Impound
25%	Watershed(<3000m)/Impound
50%	Distance to Plant
30%	Elevation Change to Plant
20%	Avalanche
10%	Wetlands/Agricultural
35%	Permitability
20%	Condemnation Status
20%	Borrow Source Location
25%	Foundation Conditions/Seepage
10%	Depth to Groundwater
10%	Pre-mining Groundwater Quality
15%	Seismic

Figure 2 – Relative weight percentages for all criteria.

Figure 2 shows that efficiency ratio was considered by far the single most important factor in selecting a tailings site, given twice the weighting of any other criteria. Distance to the plant was considered the second most important consideration, followed by the permitability of the site and the elevation change from the plant. At first glance, it appears that cost related issues are vastly outweighing the environmental concerns in this model. However, there are many more criteria relating to environmental issues and the overall weighting is spread among more individual criteria. In actuality, cost is only given approximately 15% more emphasis than environmental issues as a whole.

The decision analysis results for all the sites considered in the evaluation are presented in Figure 3. The calculated value score for each site is presented both numerically and graphically with a maximum possible score of 1.00. The value score for each alternative is based on its ratings and the relative weighting of the criteria. Uncertainty incorporated into the ratings is represented at the end of each bar graph, showing the percentage of time a given site is the best option.

The results of the evaluation show that Site K is the best site for tailings storage. Even given the uncertainties in the data, Site K is the preferred choice 85% of the time. Site A scored a close second, only one percentage point behind the leading site, and is the best option 8% of the time. The rest of the sites were further behind in score and represented the top choice less than 5% of the time.

Alternatives	Value	% of times alternative is better than all others
Site K	0.53	85%
Site A	0.52	8%
Site D	0.49	<5%
Site I	0.49	<5%
Site E	0.49	<5%
Site G	0.48	<5%
Site C	0.47	<5%
Site B	0.46	<5%
Site J	0.43	<5%
Site F	0.42	<5%
Site H	0.40	<5%
Site L	0.36	<5%

Figure 3 – Results of site selection evaluation showing ranking of site alternatives and uncertainty analysis.

Performing a sensitivity analysis tests the robustness of these decision results, showing their susceptibility to rank reversal due to a change in the weighting of the criteria. Sensitivity analysis results for the heaviest weighted criteria, efficiency ratio, is presented in Figure 4. The vertical marker represents the current weighting assigned to efficiency ratio. Moving the marker to the right (increasing the weight) or left (decreasing the weight) illustrates any change in the ranking of the alternatives shown.

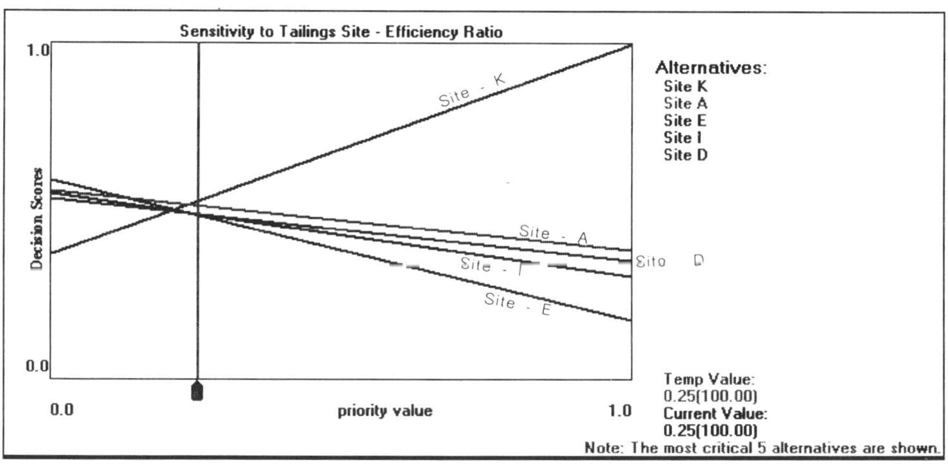

Figure 4 – Sensitivity analysis results for 'Efficiency' criteria.

This figure shows that the model is fairly sensitive to the relative importance assigned to efficiency ratio. However, only if the relative importance of efficiency ratio is reduced will a different site be selected. With all other factors remaining constant, a slight reduction in the weight of efficiency ratio (2.0%) would enable Site A to receive the highest score. A more drastic reduction in weight causes Site E to rise to the top.

The results of the site screening evaluation enabled the owner/operator of the site to prioritize the type of additional data to be collected. For example, the collection of additional data to better define groundwater conditions for the sites would likely be costly and would have minimal impact on the final selection of the preferred site. Conversely, additional information on avalanche conditions, condemnation status, effects on wetlands/agricultural land and efficiency ratios could alter the final selection of the site. The focus of additional investigations should be placed on narrowing uncertainty only for the criteria which significantly impact the results and for sites which are reasonable contenders.

E. CONCLUSIONS

The methodology for selection of a tailings site presented herein provides many advantages. The systematic, quantitative approach removes much of the subjectivity inherent in conventional practices used in the past. This process of evaluation allows the owner to participate in identifying and weighting the key criteria to ensure that owner's priorities are represented appropriately.

Criterium Decision Plus is an interactive program which is readily modified to explore what-if scenarios. In addition to meeting the overall objective of selecting the best site for tailings, additional beneficial analyses are provided. The results provide the information necessary to allocate resources most efficiently and focus the collection of additional data on that which is most valuable to selecting a tailings site.

Tailings and Mine Waste '99 © 1999 Balkema, Rotterdam, ISBN 90 5809 025 6

Integrating environmental requirements into project planning

R.W. Bleil
MACTEC-ERS, US Department of Energy, Grand Junction, Colo., USA

ABSTRACT: Proper integration of environmental compliance requirements into all phases of project management will result in an expedited, cost-effective approach. Integration should begin at the conceptual design phase of project planning. The approach presented in this paper mirrors the process effectively implemented by the U.S. Department of Energy's Grand Junction Office for the 24 sites identified in the Uranium Mill Tailings Radiation Control Act of 1978. The approach includes a system of identifying state, federal, and tribal regulatory requirements, negotiation issues and strategies, records requirements, training requirements, and budgeting techniques. Most importantly, the approach emphasizes a coordinated team effort among environmental scientists, regulatory agencies, project management, and the customer. This results in environmental regulatory requirements being accomplished on a "fast-track," in a cost-effective manner that meets project schedules and results in minimal risks to the environment and public health.

INTRODUCTION

Environmental regulation in the United States began in earnest in the late 1960s and resulted in the enactment of the National Environmental Policy Act of 1969. This act set the stage for regulating any physical, chemical, biological, or radioactive impacts that may affect our land, air, surface water, and groundwater resources. In the 1970s, Love Canal and other notable incidents convinced the public that regulation was necessary to ensure responsible management and conservation of natural resources; contaminated media meant potential risks to our environment and human health. Environmental regulation can be divided into three general categories: protection of natural and cultural resources (e.g., endangered species, wilderness, timber); pollution prevention (e.g., pesticides, PCBs, asbestos, chemical discharges) and cleanup; and remediation (e.g., mine waste, mill tailings, NPL sites). The first category is the focus of numerous federal land and water management agencies such as the U.S. Forest Service, the Bureau of Land Management, the Army Corps of Engineers, and the National Park Service. The last two categories are the focus of the U.S. Environmental Protection Agency and, to a lesser extent, other federal agencies such as the Department of Defense and the Department of Energy.

By 1985, largely because of pressure from conservationists, preservationists, and other interest groups, hundreds of federal and state regulations were implemented. When the objective is to address virtually every possible effect or concern, the regulatory path to compliance is not always clear. In some cases the regulations have gone too far, and rather than providing an extra degree of environmental protection, are actually impeding responsible management of environmental resources and pollutants. In other cases the science or technology behind the regulation is weak or nonexistent. Some regulations are media specific, some are industry specific, some are contaminant specific, and some are nonspecific. Tailings and mine waste fall into all of these.

Several approaches to compliance management have been tested since the 1970s, including Total Quality Management (Marguglio 1991) and, most recently, the International Environmental Management System (ISO 14000/14004) (Technical Committee 1996). However, the following key elements have persisted:

- Environmental policy and commitment
- Planning
- Implementation
- Monitoring (e.g., auditing, sampling, measuring)
- Re-evaluation and change

This paper discusses the planning phase of project and compliance management, which touches all phases of project management. Overlooking this relatively simple process can result in noncompliance, shutdowns, fines, civil penalties, and a poor image with stakeholders.

PLANNING ELEMENTS

When the elements outlined above are integrated into a company's overall management system (usually outlined in a brief manual), the last four elements (planning, implementation, monitoring, and re-evaluation) must also be implemented at a project level to adequately manage compliance requirements applicable to each project task. For companies and industries who budget and schedule tasks annually, this can be accomplished superficially during the budgeting process. However, detailed compliance planning should be accomplished at least a few months in advance of the project start date. The following planning elements are required for successful integration of environmental requirements into project planning.

- Establish a lead environmental consultant (LEC).
  —LEC reviews project scope with client/management/engineers.
  —LEC determines the required expertise (NEPA, RCRA, SMCRA, etc.).
  —LEC and team develop compliance strategy and approach.
  —LEC initiates the compliance plan.
- Develop Environmental Issues Checklist.
  —Team identifies key regulations and requirements.
  —Team completes the compliance plan for management review.
  —LEC and management establishes budgets and schedules.
  —Team completes regulatory compliance requirements.

ESTABLISHING AN LEC

The LEC serves as a focal point for all environmental issues related to the project. This eliminates the need for the project manager to deal with several individuals handling various aspects of environmental compliance. Although not necessarily in a management position, the LEC does have the expertise that will accomplish key environmental requirements. For companies that manage several projects simultaneously, it is desirable to have several LECs reporting to the environmental manager. The LEC should have broad expertise and negotiating skills in the environmental regulation field. The LEC should also have basic project management skills, including budgeting and scheduling. Good communication skills and rapport with regulatory agencies are essential. Most important, the LEC should be an individual who constantly seeks compliance solutions (in cooperation with regulators) that comply with the regulations in the most timely, cost-efficient manner possible. Environmental regulations cannot address every possible compliance situation. Innovative techniques to attain compliance are essential and in fact are being supported by federal, state, and tribal regulatory agencies. In some cases, particularly where violations or noncompliance become an issue, legal counsel may serve as a co-lead. However, legal counsel's role should be limited to legal aspects of the project and should not routinely include serving as the liaison with regulatory agencies. This function is best performed by the LEC and team members (in consultation with legal counsel) who are prepared to discuss the scientific and regulatory issues related to the project.

LEC scope review

Project management should provide the scope for the proposed project to the LEC early in the planning phase, often referred to as the conceptual design phase. At this stage the LEC should be able to provide initial direction to the project manager on the basis of suspected risks, liabilities, budget, and schedules. This also allows the project manager to consider less costly, more expedient methods of compliance before beginning detailed project design. The LEC should be included as part of the project design team.

Required expertise

Once the LEC and project manager have identified the general tasks and subtasks required to begin the project, the LEC can determine the regulatory expertise needed to complete compliance requirements. Team members must have an understanding of current and proposed regulations that apply to the project.

COMPLIANCE STRATEGY AND APPROACH

Developing the compliance strategy and approach and recognizing that it is an evolving process are the keys to effectively integrating compliance requirements and tasks into project planning. The strategy must consider several factors, including the schedule to complete the project, site conditions, budget, available expertise, required training, agencies involved in the process, stakeholders, and client concerns. An effective means to develop the approach is the "kick-off" meeting with the LEC, environmental team, and project manager. The project manager provides an overview of the project scope and tasks, and the LEC discusses the schedule and relevant regulatory requirements and tasks. This establishes and promotes communication early in the project. Team members are assigned compliance tasks on the basis of their areas of expertise and are expected to be innovative in their compliance approach. This includes developing cost-effective, fast-track options to comply with regulatory requirements and proposing them to regulators. Too often, consultants and legal counsel contact regulating agencies, describe their project, and then ask, "What do we have to do?" This places an unnecessary burden on regulators and often results in compliance requirements beyond those that are actually required. The agencies have no choice but to take this approach because they do not have the detailed knowledge of the project that the LEC and team do. It is the team's responsibility to research regulatory requirements and relevant guidance, establish a reasonable approach with regulators, and discuss options. My experience has been that regulators are very open to this approach.

Compliance plan initiation

Initiating the compliance plan is where the detailed work begins, and every task and activity is evaluated using the Compliance Process Overview (Figure 1).

Environmental issues checklist

The Environmental Issues Checklist (Figure 2) allows the LEC and team to develop a baseline of site conditions before the proposed project and compare it to anticipated effects (both beneficial and

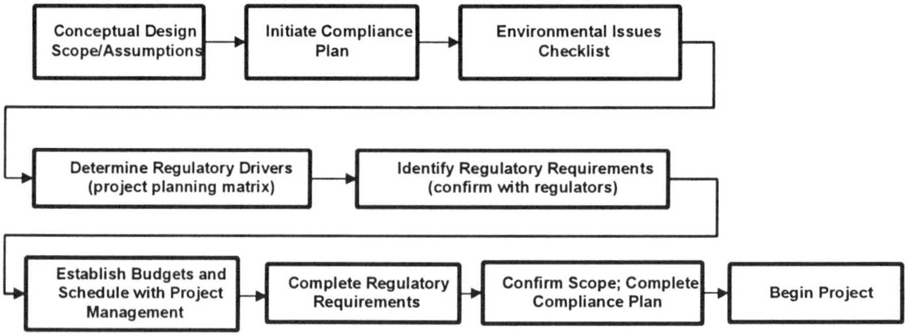

Figure 1. Compliance planning process overview

Project

Site:

Proposed actions/activities:

Location/ownership/agency info

Legal description:

Current land use/zoning:

Site use description:

Site features:

EFFECTS

	Physical	Chemical	Biological	Radiological
Land issues				
Surface disturbance	X	X		
Haz. waste disposal				
Waste storage				
T&E species	X			
Visual resources				
Wilderness				
Sensitive areas				
Pesticide use				
Air issues				
Stationary emissions				
Mobile emissions				
Fugitive dust				
Water				
Direct discharges				
Storm water				
Chemical use				
POTW discharges				
Floodplains/wetlands				
Subsurface				
Groundwater injection				

Figure 2. Sample environmental issues checklist

adverse) of the proposed project. This example checklist is not comprehensive; a more comprehensive checklist would be developed for a specific project. Remember to consider physical, chemical, biological, and radiological effects when determining activities that may be regulated.

Determining regulatory drivers and requirements

The purpose of this step in the process is to list the project actions and to develop the key regulatory drivers and compliance actions required to support each action, using the example Project Planning Matrix in Table 1. The planning matrix has been abbreviated for illustration. Note that the regulatory drivers are grouped into the three general categories outlined earlier. One of the primary assumptions in completing this section is that regulatory experts are aware of both current and proposed federal and state regulations and applicable guidance documents. Books and software are available for staying current with regulatory change. After listing all activities in the Activity/Product/Service column, evaluate the regulations for applicability. The LEC should review the matrix with management and the client before discussing compliance options with regulators. Consider stakeholder expectations when developing the matrix. Discussing options with regulators will help establish a more reasonable budget and schedule and will also reduce the number of assumptions.

Completing the compliance plan

The compliance plan incorporates all the relevant issues and regulatory drivers that have been determined in the previous steps of the process and is issued as a formal document to the project manager. Complex plans should not exceed 10 pages; most plans are 3 to 5 pages. The formal compliance plan should include at least the following elements:

- Project scope and tasks
- Key reference materials and documents
- Key contacts, stakeholders, and team members
- A summary of regulatory drivers and requirements
- Budget, schedule, and assumptions
- Status reporting frequency and report format
- Records requirements
- Training requirements

Table 1. Sample project planning matrix

Activity, product, or service	Resource Protection		Pollutant Regulation		Cleanup		
	NEPA	ESA	CWA	CAA	CERCLA	UMTRA	SMCRA
Land clearing, excavation	EA	Survey	Storm-water permit	Permit			
Chemical storage	EA						
Dredging in floodplain	EA		404 permit				
Radiological contaminants	EA		NPDES			Standards	
Coal extraction	EA	Survey?		Permit			Permit

For short-duration projects involving a limited number of tasks, a stand-alone plan may not be required, and the elements can be incorporated into another project document such as a work plan or statement of work. In cases where documentation is not deemed necessary, the same thought process outlined in Figure 1 can still be used.

Budgeting

Maintaining a core staff and subcontracting out small tasks (e.g., a threatened and endangered species survey or cultural resources survey) is critical to a cost-effective approach. Some companies attempt to maintain expertise in all areas at high cost, and other companies subcontract everything. The best approach is somewhere in between. Determine the regulatory drivers that your company deals with on a routine basis and maintain staff to support those functions. As a general rule, a company should maintain a least one full-time regulatory specialist for every 5 million dollars of business, or a cost of approximately one percent of gross revenue. However, for companies that specialize in environmental remediation, it is not uncommon to maintain one regulatory specialist for every one million dollars in gross revenue. In some cases, an experienced LEC supported by less expensive technical and administrative staff can reduce costs dramatically. Costs associated with administration and obtaining permits (e.g., well permits, fees, and surveys) cannot be overlooked.

Completing regulatory requirements

Once the compliance plan has been completed, regulatory experts need to accomplish the compliance requirements identified in the plan. An effective approach should be developed for each requirement (e.g., permits, surveys, and studies). Again, early and ongoing communication with project management, regulatory agencies, and the client is essential.

There are several methods of fast-tracking compliance tasks. Technology over the past decade has given us the Internet, fax machines, E-mail, and more, but surprisingly few people use these innovations to expedite compliance requirements. I have found E-mail the most effective means, both internally and externally, to expedite tasks. Copies of all E-mail messages should be maintained in the project file.

A common tendency of consultants is to constantly re-invent the wheel. Working smart, not hard, is a key to expedited compliance. On many occasions I have requested a recently approved agency permit or document (E-mailed if possible) that I can use as a template to complete a particular requirement. It's fast, simple, cost-effective, and gives the LEC an idea of what information is acceptable to the agency. Some type of compliance requirement tracking system or status report is essential for the LEC to keep a handle on several tasks simultaneously. Although this is basic organization to provide good customer service, it is often overlooked. Changes in tasks and work scope need to be renegotiated with all parties.

Weighing risks to human health and the environment, knowing the science and rationale behind environmental regulation, and having the ability to negotiate the methods of compliance with regulators are other integral elements of fast-track accomplishment of requirements. These go hand in hand with balancing the concerns of the clients, interest groups, and stakeholders.

CONCLUSIONS

Although there are many approaches to integrating compliance issues into project management, adopting a simple, consistent, common-sense approach that works for you is all that is required. Identifying and negotiating regulatory requirements is critical to being competitive. Therefore, when selecting a consulting firm or LEC, consider the extent to which their experience applies to your specific needs. Also consider their network of professional contacts and their relationship with regulatory agencies. Most important, the firm or LEC must be an organized, effective communicator.

The recommendations presented in this paper should result in a compliance approach that will save time and money and promote a positive public image.

REFERENCES

Marguglio, B.W. 1991. *Environmental management systems*. Milwaukee: ASQC Quality Press.

Technical Committee ISO/TC207 1996. *ISO 14004, Environmental management systems—general guidelines on principles, systems, and supporting techniques*. Geneva: International Organization for Standardization.

Tailings and Mine Waste '99 © 1999 Balkema, Rotterdam, ISBN 90 5809 025 6

Improved waste rock pile design at the Kennecott Greens Creek Mine

P.Condon
Kennecott Greens Creek Mining Company, Juneau, Alaska & University of Oregon, Eugene, Oreg., USA

ABSTRACT: Information from samples of waste rock piles and underground drill core was used to improve waste rock site design at the Greens Creek Mine near Juneau, Alaska. Waste rock pile drilling provided samples on which acid-base accounting (ABA) analyses, geochemical analyses and shake flask extraction tests were performed. ABA analyses were also performed on samples from underground drill core and development drifts. These data were used to evaluate how the rocks behave under site-specific conditions and served as the basis for improving waste site design strategies. Design improvements include encapsulation of sulfide-rich material away from final pile surfaces; placement of rock in thin, compacted lifts to improve stability and reduce water and oxygen ingress; development of a composite soil cover to further limit water and oxygen ingress; and improvement of water management around and under the pile.

1 BACKGROUND

The Kennecott Greens Creek Mine (KGCMC) is an underground mine located on the northern end of Admiralty Island approximately 18 miles southwest of Juneau, Alaska. The mine produces concentrates containing silver, gold, zinc and lead from volcanogenic massive sulfide ore which is hosted by Triassic (220 Ma) calcareous argillite and metavolcanic rock (phyllite). The deposit was discovered in 1975, and full-scale mine development commenced in 1987. The mine began full production in 1989 and operated until 1993 when production was suspended due to depressed metal prices. Following discovery of new reserves, the mine reopened in 1996 and currently processes approximately 1300 tonnes of ore per day. As of 1997, reserves were 8 million tonnes of 4.7 mg/kg Au, 600 mg/kg Ag, 4 percent Pb and 12 percent Zn with an anticipated mine life of approximately 17 years. Tailings from the mill concentrator process are filter pressed and approximately 60 percent are placed as backfill underground. The remaining tailings are disposed of on the surface in a dry tailings pile.

Initial waste rock characterization analyses were not definitive with respect to predicting the potential for acid generation. Subsequent analyses (static and kinetic tests) indicated that some lithologies had the potential to generate acid rock drainage and that high carbonate concentrations in several lithologies would provide a significant lag time to acid generation. To further characterize waste rock at the mine, existing waste dumps were drilled in 1994 and samples were sent out for acid base accounting (ABA) analyses, geochemical inductively coupled plasma mass spectrometry (ICP) analyses and shake flask extraction tests. Pre-production drill core was also sampled and sent out for similar analyses.

The purpose of this paper is to highlight the methods used to characterize both previously placed waste rock and pre-production rock and to describe the improvements in waste site design and construction that resulted from the investigation.

2 ROCK CHARACTERIZATION

Sampling methods and analytical tests used in this study are discussed below. Five surface waste rock piles were drilled and sampled in 1994, and approximately 100 samples from definition and pre-production drill core were collected in 1995. ABA analyses, ICP analyses and shake flask extraction tests (surface samples only) provided the bulk of the information used to improve waste site design at Greens Creek.

2.1 *Drilling and sampling methods*

A track-mounted Nodwell (Acker Core Max) drill rig was used for drilling and sampling surface waste rock piles. The mobility of the track-mounted rig was extremely helpful, enabling drilling in locations that were not accessible by truck. The rig was equipped with a 20 cm (o.d.) hollow stem auger and samples were recovered by driving a 6.4 cm (i.d.) split spoon below the auger bit with a 63.5 kg hammer. Maximum drilling depths approached 15 m. The drilling and sampling method worked surprisingly well considering the medium. Only on rare occasions was the auger refused within the waste rock. In some cases the auger was able to drill through large rocks, usually at the expense of a few teeth and repeated hardfacing of the bit. The decision to use the auger/split spoon combination was made to enable dry recovery of relatively intact samples. Recovery in the split spoon was good, typically 60 to 80 percent, and the sample size was large enough to meet analytical requirements. Drilling cost was approximately $100 per meter not including moves, mobilization-demobilization and consumables.

Samples from definition and pre-development drill core (NQ_{TK}, 4.78 cm) were used to predict the relative amounts and acid generation potential of the different lithologies that would be encountered while mining. Sections of core that intersected planned development headings were segregated by lithotype and sent out for ABA and geochemical ICP analyses. Chip sampling of the rib in active headings augmented the underground sampling program, providing additional samples for lithotype characterization. Chip sampling also helped with verification of predictions made from drill core and visual classification calls made at the face.

2.2 *Analytical methods*

Acid-base accounting analyses and shake flask extraction tests were performed by Cominco Engineering Services Limited (CESL) in Vancouver, British Columbia. Analytical procedures as described by CESL are presented below.

The modified Sobek method (Sobek et al. 1978) used for acid-base accounting is as follows: the sample was ground to 80 percent passing 200 mesh. Paste pH was determined by saturating a small amount of pulverized sample with distilled water and recording the pH. For neutralization potential, 2.0 grams of sample were placed into an Erlenmeyer flask to which standardized hydrochloric acid was added to obtain a target pH of 1.5-2.0. The flask was then placed on an oscillating table and agitated for 24 hours. Periodic pH measurements and adjustments with hydrochloric acid ensured that pH remained in the target range. The procedure was repeated if the pH after 24 hours was not in the target range. After 24 hours of agitation the contents of the flask were titrated to pH 8.3 with standardized sodium hydroxide. Neutralization potential (NP in kg $CaCO_3$/tonne) was calculated based on the volume of standardized base added. Acid potential (AP in kg $CaCO_3$/tonne) was derived by multiplying the amount of sulfidic sulfur (total sulfur - sulfate sulfur) by 31.25. The difference between NP and AP yields net neutralization potential (NNP) [NP - AP = NNP].

The procedure for the shake flask extraction test is as follows: 200 grams (wet weight) of sample (as received) were placed in a 1.0 liter glass beaker to which 400 ml of distilled water were added. The beaker was placed on an oscillating table and shaken for 24 hours. Conductivity and pH of the mixture were recorded after three hours of agitation. After 24 hours of agitation the beaker was removed and the contents were filtered through a 0.45 micron membrane filter. A sample of the filtrate was preserved and sent, along with representative rock samples, to Min-En Laboratories for geochemical ICP analyses. The remainder of the filtrate was tested for conductivity, pH, redox potential, alkalinity, acidity and sulfate.

Mineralogical studies on select samples were performed at the University of Oregon. Transmitted and reflected light microscopy, X-ray diffraction spectrometry and electron spectrometry were the primary techniques for mineralogical and physical characterization of fresh and weathered samples.

2.3 *Results and interpretations*

The mineralogy and physical characteristics of the rocks at Greens Creek are complex. The structural hanging wall of the deposit consists primarily of altered mafic volcanic rocks which are generally called phyllites. Phyllite lithologies predominately contain sericite, chlorite, quartz, dolomite, magnesite, calcite and pyrite. Serpentine, talc, mariposite, leucoxene, epidote and graphite are locally significant. Sphalerite, galena, tetrahedrite and chalcopyrite are only significant in mineralized zones. The phyllitic units are well-foliated and pyrite occurs as disseminations and along foliation planes. Pyrite morphology ranges from sub-millimeter-sized framboids to centimeter-sized euhedral crystals and aggregates. Millimeter-sized aggregates and subhedral to euhedral crystals are most common. Carbonate minerals also occur as disseminations and along fractures and foliation planes. Disseminated dolomite and iron-rich magnesite are common in the phyllite units, typically comprising up to ten percent of the rock. Calcite, when present, usually occurs as fracture filling with quartz. Phyllitic rocks that do not contain abundant quartz exfoliate rapidly upon weathering and expose pyrite readily. The pyrite aggregates oxidize rapidly when exposed, whereas the euhedral crystals develop a tarnish and oxidize more slowly. The high iron content of some carbonates is evident as a reddish-brown patina on weathered surfaces.

The structural footwall of the deposit consists of calcareous marine sedimentary rocks which are collectively called argillites. The argillite units are grey to black, reflecting their high graphite content. Quartz, dolomite and calcite predominate, but sericite and pyrite are locally significant. Dolomite is disseminated throughout the rock but also occurs as fracture filling with quartz and calcite. Pyrite is often present as disseminated sub-millimeter-sized framboids and subhedral crystals and is locally concentrated along foliation planes. In most samples of argillite the pyrite content is one to three percent, and the carbonate content is approximately 50 percent. The texture of the argillite units ranges from slatey to massive, and when present the foliation is less well developed than in the phyllite units. The pyrite in most argillite units reacts readily upon exposure, but the acidity is quickly neutralized by adjacent carbonate grains. A less common form of argillite has very little carbonate, a higher pyrite content and well-developed slatey cleavage. The reactive pyrite and lack of carbonate make this rock type particularly susceptible to ARD formation.

The results of 135 ABA analyses from surface and underground definition and pre-production drilling are shown in Figure 1. The data indicate that there is a considerable range in acid generation potential (AP) and neutralization potential (NP) and that lithotypes plot in distinct but overlapping zones. The majority of argillite samples are net acid neutralizers (mean NNP 441 kg $CaCO_3$/t). Most phyllite samples are potential acid generators (mean NNP -143 kg $CaCO_3$/t) but have relatively high buffering capacities, which extend the lag time to acid generation. Samples from surface drilling plot intermediate to the two main lithotypes (mean NNP -48 kg $CaCO_3$/t), demonstrating that they are generally a blend of rock types. Samples of DSA (termed dolomitic slatey argillite but actually siliceous slatey argillite) are less common but are significantly acid generating (mean NNP -399 kg $CaCO_3$/t).

The results of ABA analyses and shake flask extraction tests on samples from representative surface drilling program holes are shown in Table 1. The ABA analyses indicate the three piles contain rock that has significant buffering capacity (NP values from 22 to 529 kg $CaCO_3$/t) but in most cases is potentially acid generating (negative NNP values). The results suggest that prolonged weathering would eventually lead to acid generation but that a significant lag time can be expected. The actual lag time to acid generation, however, can not be determined without kinetic tests and field monitoring (Steffen Robertson & Kirsten Inc. 1989, Price et al. 1997). Results of kinetic testing and field monitoring at Greens Creek (unpublished company documents) indicate that the lag time to acid generation is generally on the order of 10 to 15 years. This stated

Table 1. Selected ABA and shake flask extraction test results.

Hole	Footage	Lithology	AP	NP	NNP	3hr. pH	Sulfate	Acidity	Alkalinity	Ca	Mg	Fe	Mn	Cu	Pb	Zn
			kg $CaCO_3$/t				mg/L	mg/L $CaCO_3$		mg/L	mg/L	mg/L	mg/L	mg/L	mg/L	mg/L
13-4	0-2	phyllite	103	64	-39	6.4	325	2.0	15.5	90	46	0.62	0.13	0.002	0.12	0.03
13-4	2-4	phyllite	87	49	-38	6.5	298	2.0	16.5	85	50	0.65	0.04	0.004	0.11	0.00
13-4	4-8	phyllite	93	26	-68	6.6	717	2.0	15.0	150	101	1.31	0.05	0.002	0.25	0.01
13-4	8-12	phyllite	119	36	-83	7.0	298	2.0	19.0	60	61	0.83	0.02	0.002	0.16	0.01
13-4	12-16	phyllite	172	22	-150	7.1	304	2.0	17.0	54	55	0.74	0.03	0.002	0.13	0.00
E-2	0-3	arg/phyl	75	296	221	8.0	165	1.5	39.0	62	21	0.47	0.23	0.004	0.11	0.02
E-2	3-5	phyllite	72	187	115	7.9	179	3.4	30.0	53	25	0.47	0.27	0.002	0.08	0.01
E-2	8-10	arg/phyl	112	529	417	7.5	300	4.4	27.0	74	40	0.50	0.21	0.048	0.13	0.00
E-2	13-15	phyl/arg	163	83	-81	6.4	596	12.3	7.0	154	70	0.95	3.12	0.002	1.32	7.43
E-2	18-20	phyllite	230	73	-157	6.7	560	4.9	11.0	179	54	0.76	1.95	0.002	0.38	2.54
E-2	23-24	phyllite	88	196	108	7.2	292	1.0	29.0	103	30	0.01	0.32	0.004	0.07	0.02
E-2	26-28	phyllite	218	84	-134	7.2	469	2.9	16.0	107	42	0.01	1.41	0.004	0.15	0.37
D-1	0-2	phyl/arg	123	154	30	7.0	607	2.0	20.5	194	31	0.01	0.65	0.006	0.16	1.14
D-1	4-6	phyl/arg	90	236	146	7.1	516	1.0	11.5	137	41	0.01	1.12	0.004	0.16	0.43
D-1	7-9	phyllite	149	138	-11	7.0	549	2.0	10.5	136	62	0.23	2.01	0.004	0.38	0.91

ABA by modified Sobek method.

Acidity titration to pH 8.3.

Extraction tests involved stirring 200g of material with 400 mL of D.I. water for 24h and analyzing filtrate.

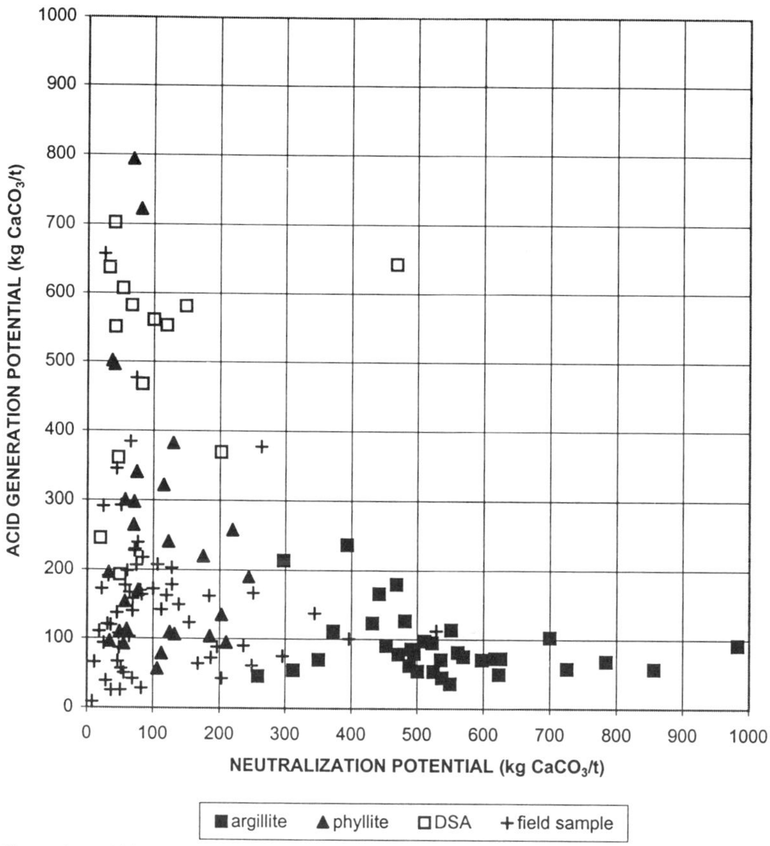

Figure 1. Acid-base accounting data.

lag time takes site-specific conditions into account but is a generalization. Actual lag times vary depending upon specific lithologic characteristics and environmental conditions such as temperature and exposure to oxygen and water. Of the 60 samples upon which ABA analyses and shake flask extraction tests were performed, only three had initial shake flask pH values of less than 6.0 (4.9, 5.8, 2.9). Paste pH values for these three samples were 6.4, 5.9 and 6.9, respectively, which demonstrates that paste pH measurement on powdered samples is not always an effective indicator of acidification of the medium. The few samples that showed signs of acid generation had been exposed the longest (approximately 10 years) and were in close proximity to materials with NNP values of negative several hundred kg $CaCO_3$/tonne. All other samples had neutral to alkaline paste pH values and neutral initial shake flask pH values, indicating that neutralization reactions are currently buffering the pH of the pore water.

Shake flask extraction test results indicate that weathering reactions have produced a significant stored soluble load, consisting primarily of calcium and magnesium sulfate salts. Dolomite and iron-rich magnesite are common in Greens Creek rocks and are the source of magnesium in the secondary salts. The data in Table 1 indicate that in some cases stored acidity exceeds alkalinity in the soluble load, but most samples produced filtrate with higher alkalinity. Iron concentrations are low, resulting from low solubility of iron oxy-hydroxides at near-neutral pH under oxidizing conditions. Manganese mobility is more variable than iron (typically less than 3.0 mg/l). Most samples produced measurable amounts of copper, lead and zinc, the latter having the highest concentrations but rarely exceeding a few milligrams per liter.

When combined with drill hole information and ABA data, the results of the shake flask extraction tests indicate that oxidation is occurring throughout the pile profile. The combined information shows that areas with high pyrite content and high water and oxygen flux, regardless

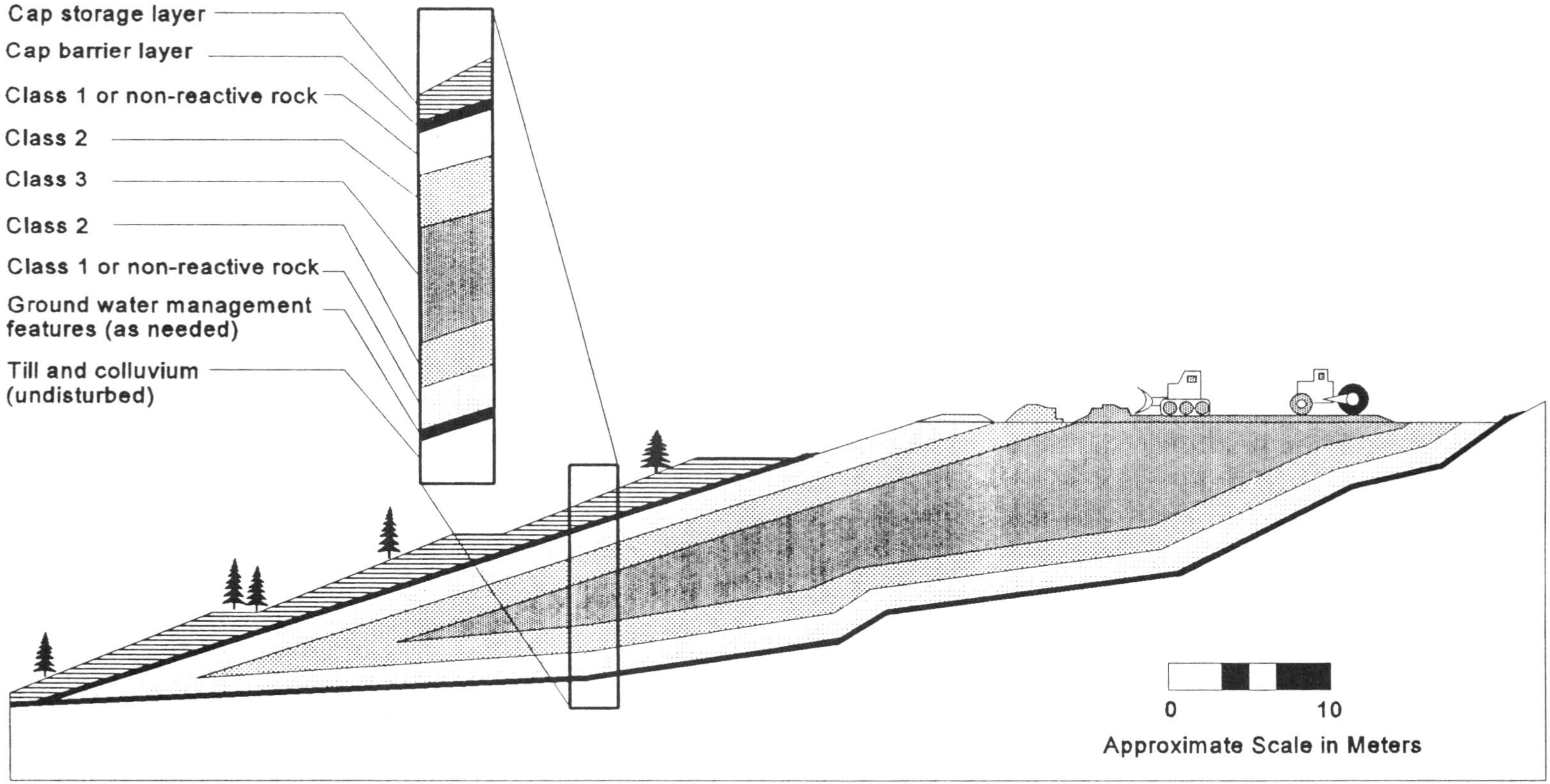

Figure 2. Development rock disposal site conceptual design.

of position in the pile, produce more abundant weathering products. With continued weathering, the high pyrite-high flux areas would likely be loci of initial acidification. This illustrates that while initial acidification generally starts at the surface of the pile where water and oxygen flux is highest, hot spots may form within the pile as well. Once acidic conditions develop, bacterially catalyzed oxidation of pyrite by ferric iron can commence, increasing the rate of acidity and metal production exponentially (Singer & Stumm 1970, Nordstrom & Southam 1997). Such an occurrence is obviously undesirable and measures to prevent acidification of the waste piles are discussed in Section 4.

3 ROCK CLASSIFICATION

KGCMC developed an improved classification system that maximizes ARD prevention strategies while allowing flexibility with waste rock handling. The system integrates ABA data, lithologic descriptions and three-dimensional mine plan imaging with visual identification methods and enables effective production-paced classification. Initially, KGCMC used the distribution of approximately 100 ABA results to define the four-part classification strategy shown in Table 2 below.

Table 2. Waste rock classification strategy.

Classification	Net Neutralizing Potential
Class 1	NNP greater than 100 kg $CaCO_3$/tonne
Class 2	NNP between 100 and -100 kg $CaCO_3$/tonne
Class 3	NNP between -100 and -300 kg $CaCO_3$/tonne
Class 4	NNP less than -300 kg $CaCO_3$/tonne

To create a production forecast of the relative amounts of each class, a three-dimensional wireframe image of the proposed mine workings was constructed using AUTOCAD™ and DATAMINE™ software. Information from definition and pre-production drilling was then used to assign proper lithologic units to respective areas of the wireframe image. The results of 111 ABA analyses were grouped by rock type and were the basis for numerical classification in the forecast. Where insufficient data density precluded using NNP values to directly classify areas on the wireframe image, the average NNP for the lithologic unit was used to determine the class of the rock from a given area.

As the workings advance, mine geologists visually inspect the active face or blasted round to determine if the rock matches the classification identified in the forecast. Once confirmed or reclassified, the rock is either used as underground backfill or hauled to the surface and placed according to the methods discussed in Section 4. Leco carbon/sulfur analyses and ABA analyses of samples from the active headings are periodically used to verify the visual classification determinations made by the geologists.

4 WASTE ROCK DISPOSAL SITE DESIGN

Design modifications utilized by KGCMC are a combination of several ARD prevention strategies. The conceptual design for a new waste site commissioned in 1995 is illustrated in Figure 2. The design includes management of water around, below and on the surface of the pile. Diversion ditches are constructed up-slope of active pile development areas, and underdrains are placed at the base of the pile to prevent ingress of runoff and groundwater. Underdrains consist of perforated 15 cm HDPE pipe bedded in screened 5 to 15 cm drain rock surrounded by filter fabric. A 0.3 m thick, approximately 45 m wide blanket drain comprised of screened 5 to 10 cm rock was also constructed at the toe of the pile. Bottom-up construction methodology is used at the site.

Waste rock is hauled directly from the face to the waste site using rubber-tired underground haul trucks. Placards on the active placement surface indicate where haul truck drivers are to dump each rock type. Encapsulation of pyritic material within the center of the pile is achieved by placing acid neutralizing rock near the outer and inner edges of the active placement surface (Figure 2). The objective is to keep potentially acid generating rock away from final pile surfaces (including the base of the pile) where water and oxygen flux is highest. Class 2 and Class 3 materials are placed at least 1.5 m and 3 m, respectively, from final pile surfaces. Class 4 rock remains underground and is used as backfill. The relative amounts and availability of the different waste rock types preclude full-time blending of potentially acid generating rock with acid consuming waste rock to create a net acid neutralizing mass. Layers of Class 1 rock were initially placed within the center of the pile, but recent studies (Day 1994, Donovan & Ziemkiewicz 1994, Mehling et al. 1997) indicate that layering of rock types may not be effective in preventing ARD. Because Class 1 rock is sometimes in limited supply, its primary function is to keep pyritic materials away from final pile surfaces. If there is a surplus of Class 1 rock beyond that which is required for encapsulation, it is blended with Class 2 and Class 3 material. Although a constant supply of all three rock types is ideal, it is not required for pile construction. Once the foundation layers are placed, portions of the active surface can, to a limited extent, be raised at different rates. If there is a temporary shortage of Class 1 rock, Class 2 and Class 3 zones can be elevated prior to bringing up the Class 1 "shell". Similar encapsulation concepts were utilized at the Montana Tunnels open pit mine in southwest Montana (Wyllie 1994).

After the rock is dumped by the haul trucks, it is spread into 0.3 m thick lifts with a bulldozer and compacted with a vibratory roller. Construction using thin, compacted lifts improves pile stability and minimizes water infiltration and air convection. Runoff from the pile is directed to the toe of the pile where it is incorporated into the facility's site water management system.

Design of a composite soil cover is an integral component of KGCMC's ARD prevention strategy. Figure 2 illustrates the cover design concepts. A 30 cm, compacted, clay-rich barrier layer is maintained at saturation by 90 cm of overlying loose storage layer material. The storage layer minimizes evaporative loss from the underlying barrier layer and is a growth medium for vegetation. The low permeability and saturated condition of the barrier layer minimizes both water and oxygen infiltration into the pile (Swanson et al. 1997, Wilson et al. 1997). An alternative design currently under investigation utilizes 60 cm of barrier layer overlain by 20 cm of gravel which is itself overlain by 60 cm of storage layer material. The gravel layer acts both as a drain and capillary break, minimizing evaporative loss from the barrier layer during dry periods and preventing the build up of pore pressure during heavy rain events. The alternative design also has no benches, which greatly reduces the volume of required capping material. Glacial sediments and colluvium derived on-site or nearby meet design specifications and will be used for cap construction. Concurrent capping with material derived from the pile construction site will minimize the need for double handling. The pile is constructed with 3:1 (H:V) slopes to provide a stable substrate for the soil cover.

The older piles, upon which analyses for this study were performed, will be recontoured, treated with alkaline amendments and capped prior to final reclamation. In some cases geotechnical, geochemical and hydrologic constraints may require selective backfilling of material later in the mine life. Such remediative efforts are costly and illustrate the importance of proactive planning and design as demonstrated by KGCMC with its new disposal site.

5 CONCLUSIONS

Mine planning and characterization of surface and underground samples at the Kennecott Greens Creek Mine led to improved waste rock pile design and construction. The mine currently uses an integrated approach to prevent acid rock drainage. Waste rock segregation, encapsulation, compaction and capping strategies are employed in conjunction with proactive water management efforts to isolate pyrite from water and oxygen. The pile design described above has several advantages over more conventional, historic pile designs. The pile is constructed with outer slopes that will accept a soil cover without further earthwork. Enough room is left around the perimeter of the pile to allow construction of the cover and water management features. Because the rock is not end-dumped, there is no significant particle size sorting, which often leads to air entry at the pile toe (Robertson & Barton-Bridges 1990, Shafer & Associates 1995). Pyritic material is encapsulated away from final pile surfaces, minimizing contact with air and water. Runoff and groundwater management is addressed prior to pile construction. Beneficial materials (topsoil and glacial sediments) are removed from the site during pile construction and incorporated with concurrent reclamation activities.

6 ACKNOWLEDGMENTS

Foremost, credit must be given to the KGCMC employees who, through determination and hard work, turned what were mere lines on paper into a state-of-the-art waste rock disposal site. B.C. Research Inc. performed initial ABA analyses and kinetic testing. Cominco Engineering Services Ltd. performed ABA analyses and shake flask extraction tests. Unsaturated Soils Engineering Ltd. designed the soil cover system. Steffen Robertson & Kirsten Inc. provided initial geotechnical support. Conners Drilling Inc. and Wink International Geo Tech Inc. performed underground and surface drilling, respectively. I am grateful to Jennifer Perez who helped with document preparation and formatting and Dr. Mark Reed and KGCMC for their critical review and interest in the project.

REFERENCES

Day, S.J. 1994. Mixing of acid generating and acid consuming rock to prevent acid generation. In *Proceedings of the Third International Conference in the Abatement of Acidic Drainage, Pittsburgh, 24-29 April, 1994*: 83-106.

Donovan, J.J. & Ziemkiewicz, P.F. 1994. Early weathering behavior of pyritic coal spoil piles interstratified with chemical amendments. In *US Dept. of the Interior Bureau of Mines Special Publication SP 06A-94: International Land Reclamation and Mine Drainage Conference and Third International Conference of the Abatement of Acidic Drainage Proceedings, Pittsburgh, 24-29 April, 1994*: 119-128.

Mehling, P.E., Day, S.J. & Sexsmith, K.S. 1997. Blending and layering waste rock to delay, mitigate or prevent acid generation: a case study review. In *Proceedings of the Fourth International Conference on Acid Rock Drainage, Vancouver, B.C., 31 May - 6 June, 1997*: 953-969.

Nordstrom, K.D. & Southam, G. 1997. Geomicrobiology of sulfide mineral oxidation. In J.F. Banfield & K.H. Nealson (eds.), *Reviews in Mineralogy, Geomicrobiology: Interactions between Microbes and Minerals*: 361-390. Washington, D.C.: Mineralogical Society of America.

Price, W., Morin, K. & Hutt, N. 1997. Guidelines for the prediction of acid rock drainage and metal leaching for mines in British Columbia: Part II. Recommended procedures for static and kinetic testing. In *Proceedings of the Fourth International Conference on Acid Rock Drainage, Vancouver, B.C., 31 May - 6 June, 1997*: 15-30.

Robertson, A. & Barton-Bridges, J. 1990. Cost effective methods of long term acid mine drainage control from waste rock piles. In *Acid Mine Drainage, Designing for Closure: Papers presented at the GAC/MAC Joint Annual Meeting, Vancouver, B.C., 16-18 May, 1990*: 153-175. Vancouver, B.C.: BiTech.

Schafer & Associates 1995. *Kennecott Corporation Acid Rock Drainage Source Book.* Bozeman, MT: Schafer and Associates.

Singer, P.C. & Stumm, W. 1970. Acidic mine drainage: the rate determining step. *Science* 167: 1121-1123.

Sobek, A.A., Schuller, W.A., Freeman, J.R. & Smith, R.M. 1978. *Field and laboratory methods applicable to overburden and mine soils, US Environmental Protection Agency EPA-600/Z-78-054.* Springfield, VA: National Technical Information Service.

Steffen Robertson & Kirsten Inc. 1989. Draft acid rock drainage technical guide, volume one. Prepared for *British Columbia Acid Mine Drainage Task Force Report.* Vancouver, B.C.: BiTech.

Swanson D.A., Barbour, S.L. & Wilson, G.W. 1997. Dry-site versus wet-site cover design. In *Proceedings of the Fourth International Conference on Acid Rock Drainage, Vancouver, B.C., 31 May - 6 June, 1997*: 1595-1609.

Wilson, G.W., Newman, L., Barbour, S.L., O'Kane, M., Eng, P. & Swanson, D.A. 1997. The cover research program at Equity Silver Mine Ltd. In *Proceedings of the Fourth International Conference on Acid Rock Drainage, Vancouver, B.C., 31 May - 6 June, 1997*: 197-210.

Wyllie, R.J.M. (Ed.) 1994. Setting a new environmental standard, a pro-active attitude helps Pegasus' Montana Tunnels reduce acid drainage. *Engineering and Mining Journ.* 195 (9): 44-45.

Tailings and Mine Waste '99 © 1999 Balkema, Rotterdam, ISBN 90 5809 025 6

Thickened tailings discharge: A review of Australian experience

M.P.A.Williams & K.D.Seddon
MPA Williams and Associates, Melbourne, Vic., Australia

ABSTRACT:

Thickened tailings discharge facilities have been operating in Australia since 1981. There are now eight operating facilities with others under construction or being planned. This paper summarizes the application of the method to Australian conditions, including aspects such as staging, perimeter bunding, and raising or extending. Multiple cone and dry stacking techniques are also covered. The design issues of seismic liquefaction stability and acid drainage at closure are covered in the context of arid to semi-arid environments.

1 INTRODUCTION

The concept of central thickened discharge (CTD) of tailings to create a low conical stack and avoid the need for major containment embankments was developed by Robinsky in the 1970's (Robinsky 1975, 1978). The method was first applied at Kidd Creek at Timmins in Ontario and has successfully operated there since. Selbaie in Quebec followed soon after but the design beach slopes were not achieved in practice and this required a major re-think of the tailings disposal system.

This setback appears to have stalled the widespread adoption of the method, at least so far as the authors are aware, in Canada and the USA. Not so in Australia. The first CTD scheme was commissioned in Australia in 1981 and there are now eight operating mines that use stacked tailings schemes, one further scheme is under construction, and three more in the feasibility or planning stage.

This paper updates an earlier review of this experience (Williams 1992) and also deals with two specific technical issues, behaviour in response to seismic shaking and the issue of oxidation and acid generation in the long term.

2 THE CTD METHOD

If non-segregating tailings are discharged onto level ground, experience shows that they will build up a beach at a low but consistent angle. Conventional thickening of tailings usually produces a non-segregating slurry that can still be pumped by centrifugal pumps and the beach slope in this instance will probably be in the range 0.5-2.0%. If discharge is from a single central point, the result will be a low conical hill (stack) of tailings at the relevant slope. In due course the storage capacity in such a cone is considerable. This configuration is called central thickened discharge (CTD).

But it is also possible to discharge tailings downslope from a hillside or from the head of a gully running out onto flat land. The resulting stack infills against the side of the hill or in the gully (Figure 1).

Table 1. Summary of thickened tailings storage facilities

Facility	Location	Type	Operator	Status	Ore Type	Start Date	Current / Planned Production Mtpa	Planned Total Capacity Mt	Discharge Slurry Density % Solids	Beach Slope %	Average In-situ Density t/m3
ELURA	NSW	CTD Seg	Pasminco	Operational	LEAD / ZINC	1981	1.0	15	60	1.7	2.37
ELURA EXTENDED	NSW	CTD	Pasminco	Planned	LEAD / ZINC	~2002	1.25	55	"	"	"
GOVE	NT	CTD mp	Nabalco	Operational	ALUMINA	?	?	?	?	?	?
ARGYLE	WA	CTD Seg/DV	Argyle	Terminated	DIAMOND	1985	3.3 - 6.5	160	72	~10.0	?
PEAK	NSW	CTD	Rio Tinto	Operational	GOLD	1992	0.55	~4	58-60	1.8	1.65
UNION REEFS #1	NT	DV	Acacia	Operational	GOLD	1994	2.5	15	52-55	0.9	1.40
UNION REEFS #2	NT	DV	Acacia	Planned	GOLD	~2002	3.0	15	"	"	"
McARTHUR RIVER	NT	CTD Seg	MIM	Operational	LEAD / ZINC	1995	2.4	120	60	1.0	1.50
ERNEST HENRY	QLD	CTD Seg	MIM/Savage	Operational	COPPER / GOLD	1997	7.0	127	75	(1.1)	?
MT. KEITH	WA	CTD mp	WMC	Operational	NICKEL	1997	10.5	250	44	(~2.0)	?
PILLARA	WA	CTD	Western Metal	Operational	ZINC / LEAD	1998	1.5	15	65	(1.5)	?
CENTURY	QLD	DV	Pasminco	Under Construction	ZINC	1999	4.2	84	40	(0.5)	n/a
SUNRISE DAM	WA	CTD	Acacia	Feasibility Study	GOLD	n/a	~2.0	~25	>48	(0.8)	n/a
BODDINGTON / WANDOO	WA	DV	BGM	Feasibility Study	GOLD	n/a	~8.0	~160	(55)	(0.4)	n/a
MURRIN MURRIN	WA	CTD	Anaconda	Pre-feasibility	NICKEL	n/a	~10.0	?	>40	?	n/a

CTD	Central thickened discharge	( )	Design estimate
CTD Seg	Segmented central thickened discharge	?	Data not known
CTD mp	Multiple cone discharge	n/a	Not applicable

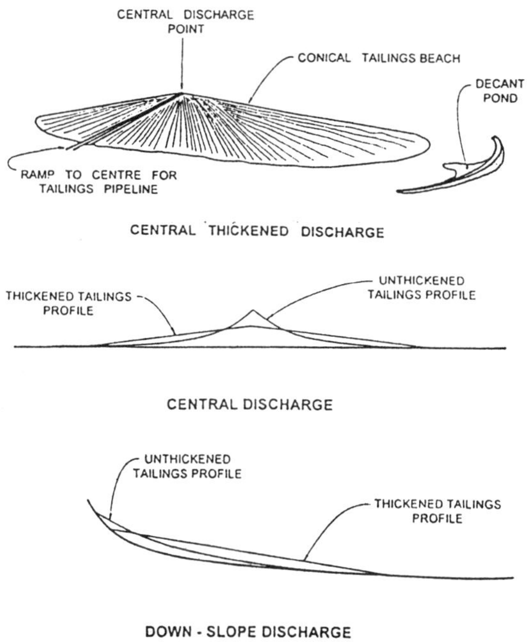

Fig.1. Thickened Tailings Disposal

The emphasis on non-segregating tailings is important. Non-segregating tailings means that the slurry consistency is such that hydraulic sorting does not occur during deposition, i.e. there is no tendency for the coarse sand sized particles to deposit close to the discharge point leaving the fine fraction (slimes) to run down to the toe of the beach. Instead, the slurry remains as a uniform blend of the full range of particle sizes. Subsequent sampling of the beach will show the same particle size distribution regardless of the distance from the discharge point. This behaviour is critical to the success of tailings stacking.

The potential cost savings from CTD are self-evident since no retaining embankments, or at least greatly reduced embankments, are required. There are no raising costs and negligible operating costs.

There are also environmental benefits. Runoff from the stack, both excess supernatant and rainfall, collects at the toe and may be directed around to a suitable recovery point for return to the plant. The absence of a decant pond on the tailings means that there is no source of constant driving head for seepage.

Furthermore, if the stack is sized correctly the tailings will end up unsaturated. This occurs if there is sufficient elapse of time between periods of deposition of fresh tailings at any given location on the stack for evaporative drying to achieve unsaturated conditions. Some infiltration will occur during rainfall but it will be slow. Most rainfall will run off, and evaporative drying can then once more restore the unsaturated state. Needless to say, the design process needs to take note of the climate conditions at each site, but the arid climate that exists over much of Australia is advantageous in this regard.

Achieving unsaturated tailings is of significance because drainage of pore water will be minimal from unsaturated tailings, i.e. there will be no seepage from the stack. Also, the increase in density associated with shrinkage during drying, combined with non-saturation, lead to high strength. There can thus be immediate trafficability onto the tailings for spreading cover materials for rehabilitation and revegetation on closure.

In the long term, the gentle slopes of the stack greatly reduce the problem of erosion protection and, aesthetically, are more likely to be in keeping with the surrounding topography than the steep slopes of an embankment.

3 AUSTRALIAN THICKENED DISCHARGE SCHEMES

Australian thickened tailings schemes, both operating and planned are summarised in Table 1.

3.1 *Thickening strategy and staging*

The Robinsky concept at Kidd Creek was to commence discharge at a relatively low slurry density, achieve full coverage of the proposed stack footprint early, and then gradually steepen the slope by reducing the underflow percent solids from the thickener to provide long term storage capacity. Australian practice by contrast has always aimed at achieving a steady thickener performance from the outset and allowing the stack area to progressively increase.

At Elura, the circular footprint of the cone has been subdivided by radial embankments. The objective was to create a No.1 segment that would be filled first, then No. 2, etc. The radial embankments have been constructed from local borrow material since it is an underground mine and the production of waste rock is limited. One of the radial embankments is the ramp for the tailings pipeline from the plant. The No. 4 (final) segment is currently being filled. The intention of the subdivision was initially to allow re-treatment of the early mine life tailings if recovery rates in the plant had improved enough to warrant this by the end of mine life. At the time of writing this is not planned. A side-benefit of the arrangement is that it permits progressive rehabilitation. Ernest Henry and McArthur River also employ segmented stacks.

3.2 *Perimeter bunding*

A low perimeter bund is employed at all of the CTD schemes for practical reasons and for efficiency. If, for example, the bund is designed to hold 1 m depth of tailings, then for a cone with 1% beach slope and a diameter of 2000 m, the reduction in base footprint area would be about 20% but the stored tonnage at, say 1.5 t/m³ dry density, would only come down from 15.71 Mt to 15.27 Mt, less than 3%. The reduced area means less site clearing and less area to rehabilitate. Such a low bund has negligible engineering significance from a design or stability standpoint, is not expensive to construct from local borrow materials, and fulfills a useful runoff control function.

3.3 *Toe embankments*

Substantial toe embankments can be employed if the topography warrants this but this will usually only apply to down valley (DV) schemes, see Table 1. The final embankment at Union Reefs will be 22 in maximum height and at Century 28 m. Notwithstanding, the schemes are still more efficient than conventional up-valley discharge from such an embankment. The head of the beach will be approximately 10 m above the embankment crest level at Union Reefs and 15 m at Century. In both cases the decant recovery and/or evaporation ponds are located immediately downstream of the embankment. The embankments must be able to cope with water standing against them but it is not necessary for them to be designed as "watertight" structures. Indeed, both dams are designed to leak, at Union Reefs by being constructed of pervious weathered rockfill (partly mine waste) and at Century by the construction of a drainage system on the upstream face. The objective is to minimise ponded water on the tailings and to maximise drainage and consolidation of the tailings adjacent to the embankment. Overflow decant outlets are also provided plus emergency spillways.

3.4 *Method of operation and raising*

Provision for raising may or may not be part of thickened tailings stacking schemes. At Elura and Peak rockfill ramps to the centre of each cone were constructed to full final design height at the outset. Subsequently, the total tonnage forecasts at both mines have increased significantly, as is often the case. The Peak scheme will be expanded by simply raising the ramp. The beach surface is dry and quite trafficable so that a centreline raise is feasible without any stability problems. The toe of the cone will spread out further in all directions and the existing perimeter contour drains will be pushed back.

At Elura the cone cannot expand in all directions since there are infrastructure constraints such as a railway line and the process plant to the north and north-west. The required increase in total capacity is also considerable, from 15 Mt to 55 Mt. The solution is to advance the ramp out and up. This will result in a translation of the location of the final cone peak but, surprisingly, by not very much. The original cone design has a height at the centre of 12 m, a diameter of roughly 1.3 km and an area of 130 ha. To accommodate the extra 40 Mt the ramp will only need to be extended about 300 m and upward by 4 m. The final area covered will be close to 300 ha. These figures demonstrate the efficiency of stacking schemes as they get larger.

It is not necessary for the discharge point to be at the final location from the beginning. A relatively short ramp can be constructed only part way towards the centre to begin with. This can then be advanced in stages. Once again, the high strength of the stack beach means that advancing over the tailings is a straight forward matter. This is the approach being adopted at Pillara.

There are also alternative methods for running the tailings pipeline to the centre of the disposal area. For the schemes described so far this has involved a surface pipeline on a ramp or causeway. However, at Mt. Keith and at Gove the pipelines have been laid on the ground to the centre and then terminate in vertical risers. The tailings overflow the top of the riser and cascade down to form the conical deposit. Access to the riser can only be obtained across the tailings beach. Such a method overcomes a shortage of waste rock for ramp or causeway construction.

3.5 *Multiple cones*

At Gove there are some twenty risers in a series of paddock style cells. Rather than being run continuously, each riser discharges for only about 3-4 hours then is rested for 2 or 3 days. This produces steeper beach slopes than for continuous operation.

Mt. Keith also uses multiple risers, nine in all, eight of them configured in a circle around a central riser. This produces multiple cones. It is claimed that this results in enhanced runoff recovery (Robinson et al. 1997). A disadvantage is the concentration of runoff in the gullies that are formed between intersecting cones. This could lead to erosion problems in the long term after closure. Such features are absent from a single cone which is a purely dissipating landform so far as runoff is concerned.

3.6 *Dry stacking*

A technique known in the bauxite industry as "dry stacking" is also worth mentioning in the context of this paper. The method involves discharging thickened red mud in even thin sheets across a long rectangular paddock and allowing it to dry out. Discharge is rotated to other paddocks during the drying period. The rate of drying can be enhanced by windrowing the tailings surface layer to and fro with a bulldozer as occurs at Alcoa's bauxite refineries in Western Australia (Cooling 1997a). Once satisfactorily dried, a fresh sheet of tailings is discharged.

4. RESPONSE TO SEISMIC SHAKING

4.1 *Seismic liquefaction at Argyle*

The concept of thickened tailings storage deposits, comprising many millions of tonnes of material stacked above the level of any confining perimeter structure has raised some concerns on the part of operators and regulatory authorities regarding liquefaction stability.

These concerns have been reinforced following the Argyle experience (McMahon et al. 1996). At this site, a thickened discharge operation was terminated at considerable cost following an assessment of seismicity and liquefaction risk.

It is the authors' opinion that the problems at Argyle were more in the detail than in the concept. However, in any event this site was non-typical for a number of reasons, including particle size, slope, rate of placement and seismicity of the site.

Table 2. Summary of tailings material properties

Facility	Typical Particle SG	Plasticity / Classification LL	PI	USC	Particle Size - 0.075mm %	Reference
ELURA	4.42	-	-	-	90	Williams & Ennis 1997
WAGERUP	2.90	-	-	-	-	Cooling, 1997b
ARGYLE	-	-	-	-	25	McMahon et al. 1996
PEAK	2.78	17	4	ML	56	Williams & Ennis 1996
UNION REEFS	2.78	25	6	CL-ML	81 - 87	" "
McARTHUR RIVER	3.00	24	7	CL-ML	90	" "
CENTURY	2.79	23	7	CL-ML	86	Seddon et al. 1999

- *Particle Size* The Argyle tailings were formed by recombining a coarse sand-gravel tailings, with a fine sandy silt tailings. The resulting all-in mix was a well graded gravelly silty sand. However, this mix proved to be difficult to pump at the minimum design slurry density of 75% Solids, and lower densities were used. This resulted in a degree of segregation on the beach, with the coarser particles dropping out near the top. The resulting particle size distribution of the critical central beach section was a silty sand, which type of materially is classically susceptible to liquefaction. In comparison, the data presented in Table 2 indicate that a more typical tailings material is a fine grained, low plasticity silt or silty clay.

- *Beach Slope* Principally because of the particle size distribution, the achieved beach slope at Argyle was exceptionally steep, at an average slope of about 10% - 12%. This is in the range originally proposed by Robinsky (1978), but much steeper than any other slope adopted in Australia before or since.

- *Rate of Placement* McMahon et al. (1996) report that the original facility was designed for a production rate of 3.3 Mtpa. The Argyle storage is essentially a down-valley scheme, and combined with the steep beach slope this limits the area of deposition, and results in a rate of rise conservatively estimated to be about 4-5 m/year for the mature scheme.

  In 1998 production was increased from 3.3 Mtpa to 6.8 Mtpa. The resulting rate of rise would then have been of the order of 10 m/year. At this rate of rise it is not surprising that the deposit was loose and saturated. The rate of rise would have been higher again if the placement method had not been modified before the planned increase to 9.0 Mtpa in 1994.

- *Seismicity of Site* Standards Australia (1993) gives a 1 in 500 year seismic acceleration coefficient for the Argyle site of around 0.11 g. This value is high, but not excessive for Australian conditions. McMahon et al. (1996) report the results of additional seismicity studies carried out for the site. These studies appear to have been at least partly deterministic (based on identified active fault structures), and resulted in the adoption of a design earthquake having a Richter magnitude of 7.5, a return period of 500 years, and an estimate horizontal acceleration at the site of 0.3 g. This acceleration is therefore three times the value given in AS1170.4, for the same return period. If correct, it suggests that seismic accelerations in the Argyle area are twice that of any other area in Australia. McMahon et al. do not provide any reconciliation between these two values. It is possible however that given the condition of the deposit it was felt that a conservative approach to the assessment of the probability of the design earthquake was justified.

4.2 *Liquefaction assessment*

With respect to seismic stability, it is the authors' opinion that there are likely to be large numbers of existing conventional tailings dams, constructed by up-stream raising in Australia, which have never been properly assessed for stability. Because of the steeper external slopes and therefore energy associated with failure, these need to be evaluated at least as closely as CTD schemes. This, however, does not obviate the need for the designer to properly address the issue of liquefaction stability of a thickened tailings scheme.

At the design stage, liquefaction assessments of stacked tailings schemes have generally followed reasonably conventional lines. For new projects there is the usual limitation on the quantity of materials available for testing; in some cases only a few kilograms of prototype tailings are available, which can place constraints on a testing program.

Tailings materials are typically silt sized (Table 2) and therefore the relative density approach as used for sands is not applicable. Density predictions may be based on finite strain consolidation theory, using Rowe-cell oedometer data, modified in the light of considerations of evaporative drying and the shrinkage limit density. In some cases direct empirical data may also be available.

Static undrained effective strength triaxial tests have been carried out to assess the overall contractive/dilative nature of the materials, and stress strain characteristics. Test results suggest that on these fine grained materials large strains, probably in excess of 15%, are necessary to generate the full reduction in strength from peak to steady-state undrained. Vane shear tests to large strains, following the practice of Poulos et al. (1985), have been used to evaluate residual/steady-state undrained strength.

A number of the larger recently commissioned thickened tailings schemes are committed to ongoing monitoring and testing to further evaluate seismic liquefaction resistance. This will include direct measurements of in-situ density and moisture content/degree of saturation. Additional sampling, followed by cyclic triaxial testing to evaluate the number of shaking cycles required to generate initial liquefaction is also proposed.

4.3 *Flow Slide Velocity*

The design requirement to provide containment for potential liquefaction flow slide debris in the early years of life of a storage facility has raised questions regarding the velocity of sliding, and whether any additional freeboard may be required to provide containment.

Some documented cases of liquefaction flow slide failures have shown relatively high velocities (Jeyapalan et al. 1983b), however these relate to dams constructed by the upstream method. These had steep downstream slopes, and a high potential energy release on failure. In the case of thickened discharge stacks, with slopes of the order of 0.5%- 2%, such high flow velocities would seem to be unlikely.

Liquefied tailings in a flow slide can be considered to act as a Bingham plastic fluid, characterized by two parameters : yield stress (τ_y), and plastic velocity (η_p) (Jeyapalan 1983a). This approach has been used to develop a predictive equation for the maximum flow velocity for a flow slide occurring on a semi-infinite planar slope (Seddon et al., 1999):-

$$V_{max} = (\sigma_v \sin \beta - \tau_y)\, h / 1.5\, \eta_p \qquad (1)$$

where V_{max} = maximum equilibrium flow velocity, h = depth of sliding, σ_v = total vertical stress at depth h, β = beach slope angle (degrees), τ_y = yield stress, and η_p = plastic viscosity

For the Century facility, testing and trial calculations gave the following critical combination of parameters for the predicted beach slope of 0.5%:-

τ_y = 80 Pa. (= 0.08 kPa)
η_p = 155 Pa. sec
h = 2.2 m
σ_v = 39 kPa, and hence
V_{max} = 1.05 m/sec

The velocity head at this flow rate is only 0.05 m, and has been taken as confirmation that there is no need to allow for additional free-board to cover a "run-up" velocity, or to design for impact forces on toe retaining embankments.

Analyses completed to date indicate that thickened discharge deposits are most susceptible to liquefaction flow slides in the initial stages of placement, when the depositional areas are relatively small, and the rate of rise high. At these times consolidation and evaporative drying effects are not significant, and low density deposits occur. One design response to this situation is to include perimeter embankments with sufficient capacity for horizontal containment of the slumped stack. Seddon et al. (1999) provide a case study for the Century project where this approach was adopted.

It has generally been found that after the critical initial stages (typically 1-2 years) the geometry of deposition is such that the rate of rise reduces markedly. Evaporative drying becomes significant at the surface, and consolidation catches up at depth. This results in significantly higher densities, and a reduced risk of liquefaction.

It is concluded that liquefaction stability is a manageable issue in the Australian context. On the other hand, the authors have recently been involved in a preliminary study for a CTD scheme located on the island of New Guinea. New Guinea has a level of seismicity and rainfall substantially greater than typical Australian conditions. On-going studies proposed for this site included detailed evaluation of evaporative drying and likely degree of saturation, and the liquefaction resistance of saturated tailings.

5 ACID DRAINAGE

5.1 Field Data

At Elura the tailings have a high pyrrhotite and pyrite content and are strong acid producers. The pH in the runoff pond from the No.1 area beach a year after deposition ceased was 2.9 (Williams 1992). There is, however, some buffering before the onset of acidic conditions. Two boreholes were drilled to depths of 5 m in the stack and samples were tested for moisture-content and pH. The results are shown on Figures 2 and 3. They are surprising for two reasons. Firstly, the saturation moisture content for the Elura stack density is 18.4%. The in-situ moisture content was typically between 11-13% over the full depth of each hole. Borehole 2 was drilled to within less than a metre of the original ground surface. The inference is that the whole stack is at a fairly uniform moisture content around 65% of saturation. This clearly occurs as a consequence of

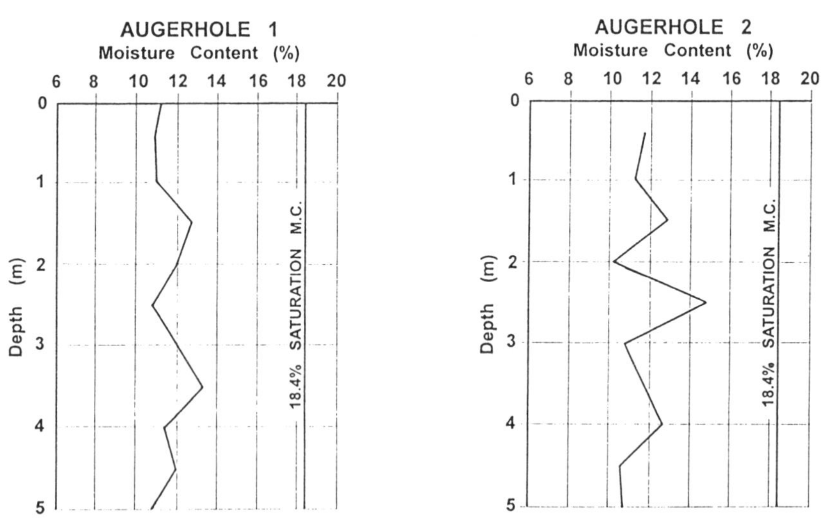

Fig.2 Elura stack moisture contents

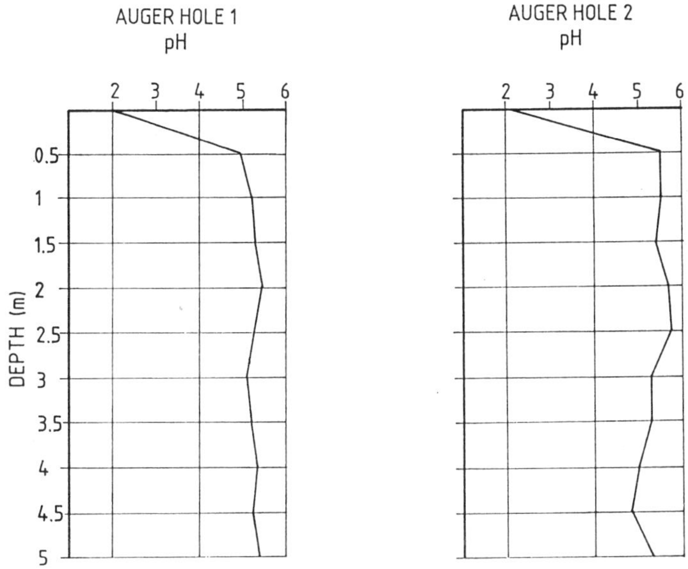

Fig.3 Elura stack pH versus depth

evaporative drying. If the period of drying is long enough before inundation with fresh tailings, they will dry back beyond the shrinkage limit density and become unsaturated. Despite some re-wetting from a fresh layer of deposition, the progress of the wetting front is it is too slow to infiltrate far before being reversed as evaporation once more takes over.

The second surprise is that the tailings pH is relatively high at between 5.0-5.5. It is presumed that the oxygen in the entrained air in the pore spaces has been consumed and the inflow of fresh air is sufficiently inhibited by the overlying tailings. Only the surface tailings have gone acid as a result of prolonged exposure.

5.2 Cover Design

A recognised strategy for the prevention of oxidation of mine wastes is to keep them permanently saturated. The most effective method is to establish and maintain a water cover. Unfortunately in much of Australia which experiences an arid or strongly seasonal rainfall pattern this is simply not possible. The emphasis must therefore shift to the design of a suitable soil cover with the objective of minimising the flux of moisture at the cover to tailings interface.

The self-shedding shape of a tailings stack and absence of ponded water on the surface become critical factors in achieving this objective. Computer programs such as SOILCOVER can be used to analyse the balance between rainfall runoff and infiltration, evaporation, and evapotranspiration by plants. The thickness and composition of a cover will depend on the availability of materials as well as the local climate. A trial area of cover has been established at Elura that consists of 0.5 m of clayey gravel/gravelly clay overlain by 0.15 m of topsoil. A sheet of black plastic film has been placed at the tailings to cover interface. Mean rainfall at Elura is 340 mm/yr and Class A pan evaporation is 2560 mm/yr. The performance of the cover will be monitored.

By avoiding the infiltration of rainfall, there will be no long term seepage or drainage from the fine grained non-segregated tailings in the stack. If the climate is such that the tailings are saturated at the end of mine life, as is anticipated at Kidd Creek, then so much the better. Non-segregated tailings, with its wider range of particle sizes, is seen as having better moisture retaining properties in this regard. But the unsaturated conditions at Elura are also satisfactory. In the absence of seepage there is no transport mechanism for oxidation products to be leached to the environment.

6 SUMMARY

Considerable experience has been gained with the discharge of thickened non-segregating tailings at Australian mining operations to form conical (CTD) stacks or down-valley in-fill. There are currently eight operating schemes. Others are either under construction or are being studied.

The Peak tailings disposal scheme at Cobar in New South Wales is featured as a case study in the Federal EPA publications entitled Best Practice Environmental Management in Mining.

The advantages of tailings stacking are:-

- the absence of, or minimal, embankments and hence low construction costs,
- low operating costs,
- the large beach area maximises the opportunity for evaporative drying and hence the density and strength of the tailings will be high,
- the decant pond is located off the tailings and can be smaller thus resulting in reduced scope for seepage to groundwater,
- the final profile is self shedding (of rainfall) on closure,
- the flat slopes are more stable and erosion resistant than those associated with the embankments needed for impoundment schemes, and
- the final landform is more likely to be in keeping with the surrounding topography.

Disadvantages of the method are:-

- a large area is required, possibly two or three times that needed for an impoundment scheme,
- possible higher rehabilitation costs as a result of the larger area, and
- the cost of installing and running a thickener.

Drilling and sampling at one site has shown that the tailings stack is unsaturated. Long term seepage will not occur from unsaturated tailings, which has obvious environmental advantages, particularly for tailings with the potential to generate acid through oxidation. Cover designs need to have as their objective the minimisation of flux of soil moisture at the tailings to cover interface.

REFERENCES

Cooling, D.J. Nov. 1997a. Dry Stacking of Bauxite Tailings. *Proceedings of the AIC Conference, Tailings Disposal Management Summit.* Sydney.

Cooling, D.J. 1997b. Tailings Management within Alcoa of Australia - Two Decades of Change. *Workshop Proceedings, Tailings Management for Decision Markers.* Australian Centre for Geomechanics. Perth.

Jeyapalan, J.K., Duncan, J.M., & Seed, H.B. 1983a. Analysis of Flow Failures of Mine Tailings Dams. *ASCE Journal of Geotechnical Engineering,* V109(2) 150-171.

Jeyapalan, J.K., Duncan, J.M., & Seed, H.B. 1983b. Investigation of Flow Failures of Tailings Dams. *ASCE Journal of Geotechnical Engineering,* V109(2) 172-189.

McMahon, B., Phillips, J.T. & Hutton, W.A. 1996. The Evaluation of Thickened Tailings in a Seismically Active Area. *Tailings and Mine Waste 96.* . Fort Collins CO:133-143. Brookfield/ Rotterdam, AA Balkema.

Poulos, S.J., Robinsky, E.I. & Keller, T.O. 1985. Liquefaction resistance of thickened tailings. *ASCE Journal of Geotechnical Engineering*, 111(12).

Robinsky, E.I. 1975. Thickened Discharge - a new approach to tailings disposal. *Bull. Can. Inst. Min. Metall.,* 68, 764, 47-53.

Robinsky, E.I. 1978. Tailings Disposal by the Thickened Discharge Method for Improved Economy and Environmental Control. *Proc. 2nd Int. Tailings Symp.* Denver.

Robinson, J.E., Joyce, D.C. & Bentel, G.M. 1997. Mt. Keith Nickel Operation, Central Thickened Discharge Tailings Storage Facility. *Proceedings of the IMR Conference, Successful Tailings Management,* Sydney.

Seddon, K.D., Murphy, S.D. & Williams, M.P.A. 1979. Assessment of Liquefaction Flow Stability of a Thickened Tailings Stack. *Proc. 8th Australia New Zealand Conference on Geomechanics*. Hobart.

Standards Australia 1993. *SAA Load Code, Part 4 : Earthquake Loads*. AS1170.4 - 1993.

Williams, M.P.A. 1992. Australian Experience with the Central Thickened Discharge Method for Tailings Disposal. *Proc. Second Int. Conf. on Environmental Issues and Waste Management in Energy and Mineral Production.* Calgary. Alberta, 1, 567-577.

Williams, M.P.A. & Ennis, P.C.E 1996. Suitability of the Central Thickened Discharge Method for Nickel tailings disposal in Western Australia. *Proceedings of the AusIMM Conference, Nickel '96,* Kalgoorlie.

Tailings and Mine Waste '99 © 1999 Balkema, Rotterdam, ISBN 90 5809 025 6

Optimization of water management for an alluvial gold mine in Ghana

R.-J. Berg, H. van Muijen & J.J. de Ruiter
Faculty of Applied Earth Sciences, Department of Raw Materials, Delft University of Technology, Netherlands,

ABSTRACT: Alluvial mining is rather small compared to the large-scale open pit and underground mining and thus of low financial capacity. This paper deals with a low cost option to cleanse and re-circulate the water for a skid mounted alluvial gold washplant, operating in Ghana.

One of the most important parameters dictating the success of an alluvial mine is the presence of clean water for the washplant throughout the life of the mine. The major (and only) pollution of the used water will be the suspension of fines. These fines do not settle by itself and have to be treated with chemicals to allow the water to be suitable for re-circulation. This option is expensive and thus unwanted.

A cheaper option is found by the mixing of present terrace gravel with the mined river gravel. Testing, both in practice as in laboratories, has given a comparison between this rather natural flocculent and a chemical flocculent, which proved to be the best alternative for cleaning the tailings water.

1 INTRODUCTION

Midras Mining Ltd. is an alluvial mining company registered in Ghana. The company is jointly owned by Blue Water International (BWI) and Ashanti Goldfields Company Limited (AGC). BWI is an American investors group based in North Carolina. AGC is a Ghanaian gold mining company and an established name in the gold mining industry. The mine concession is situated in the town of Kibi, 100 km Northeast of the main capital Accra.

1.1 Local geology

The alluvial gold mineralization in the area contains fluvial conglomerates with a fine grained sand-clay matrix. The gold bearing gravel directly overlies weathered and clay rich bedrock and is overlain itself by a layer of silt and clay overburden. This overburden is in sharp contact with the gravel layers.

Along the Birim river flats, the gold mineralized units described in the field as "gravel" can be found. Gravel is made up of deposits of the predominantly clast supported conglomerates with a matrix of clay and sand sized material. There is not much variation in the composition or size of clast. In most areas, the gravel consists of clays and quartz, shales, mafic rocks and laterite ironstones. In other areas, clast is predominantly quartz with very minor amounts of other rock types. Clast dimensions are variable with the abundance of cobble and coarse pebble sizes. Matrix material is quartz sand, silt and clay with minor amounts of heavy minerals.

The top layers of gravel are often made up of sub-angular quartz fragments and latteritic pebbles set in a mottled laterised clay matrix. The contact between this unit and the underlying gravel is transitional and is usually marked by a gradually increase of rounded white quartz pebbles.

Within the area of the concession a distinction can be made between two types of gravel. First a type of gravel, referred to as terrace gravel, has been deposited. In time a second type of gravel, referred to as river gravel, has worn out parts of the terrace gravel and has been deposited. The area consists of a line of river gravel, indicating the flow of the ancient river, with on both sides a line of terrace gravel.

Due to excessive exploration a distinction in reserve estimation can be made between relative "high-grade" river gravel and "low-grade" terrace gravel. These materials can be found in a ratio river gravel: terrace gravel = 2,1:1.

1.2 Treatment plant

The heart of the alluvial operation is a skid mounted alluvial gold treatment plant. Since the gold in the gravel occurs as free gold, a gravity separation method is employed using water only. The process is designed on a requirement of 432 m^3/hr of fresh water, which is initially taken from a local river, the Birim. A freshwater pump takes water from the river.

The treatment plant consists of a two stages feed hopper, realizing a constant feeding to the plant after which the material is screened prior to entering a drum by a grizzly with a screen diameter of 200 mm. A gravity driven drum scrubber/screen liberates the gold from the gravel and screens the material off at 9.5 mm. The oversize is hauled by a short conveyor-belt and dumped behind the plant, the undersize is collected in a bin and pumped into the jigging section.

A two-stage jig system pre-concentrates the gold. The jigs are of the trapezoidal type with a sawtooth drive: the primary jig system consisting of six cells, the secondary jig system of only one. They are arranged in a closed circuit, the tailings of the secondary jig returns into the primary jig. A 12" Knelson concentrator is concentrating the gold. The Knelson is fed with the concentrate of the second jigging step. A tailing disposal system pumps the tailings of the primary jig into the tailing pond.

Based on the design processing capacity of the plant, which is 150-m^3/hr loose gravel, a production rate of the plant of 100 m^3/hr bank volume is induced.

The gold recovery process in this plant is only gravity driven and therefore the only pollution will be in the suspension of very fines in the water.

2 QUANTITY AND QUALITY OF WATER SOURCES

The water, needed for the recovery of the gold, is drawn from one of the rivers, the river Birim. This river is used by the locals, mostly for drinking. Most of the time this is their only source of potable water and therefore disturbance should be minimized. It is of its most importance that the locals will have enough clean water at all times.

The source of the Birim river is only 10 km upstream, so the width, within the boundaries of the concession, is small and varies between 4 m and 10 m. The flow rates range from 1,800 m^3/hr in the dry season to 4,000 m^3/hr in the wet season. A quick calculation shows that the withdrawal of water for the processing has no effect on the situation for the downstream users since the withdrawal represents only 11-24 % of the river flow.

However, from practical experiences it can be concluded that the possibility of the river running dry is plausible, although the measurements suggest otherwise. In the dry season the source of the Birim river may run dry and the river contains less than 432 m^3/hr needed by the washplant. Therefore there will be no water going downstream. The fact that no or a limited amount of water is running downstream through the Birim will have different effects; social, environmental and economic:

- The fact that no fresh water will run downstream from the Birim is socially unacceptable. The local population depends on the Birim as their source of drinking water. Practical experience from nearby mines has shown that a change in the river flow rate will give tremendous problems among the local population and can even lead to the closure of the whole mining operation.
- The full use of the flow rate of the river can be compared with cutting off approximately 3 km of the river. This can have influences on the ecosystem. The entire fauna from upstream, irrespective of size or shape, has only the possibility to go through the plant. For the big species this is not possible, for the micro-fauna the action of the washplant on them is uncertain.
- Insufficient delivery of water by the source of the river will mean that the plant has to be shut down if the water is not re-circulated. Therefore the decision to re-circulate is inevitable to assure that there is enough water to run the mine. An extra criterion is to have a variable amount of surplus water and to be able to provide a total re-circulation of the water. This eliminates all kinds of environmental problems.

A baseline study was undertaken related to the quality of the rivers Birim and Ben, the two rivers in the concession area to be affected by the mining operation.

It is clear from the baseline study that there is deterioration in water quality of the rivers Birim and Ben after rain. In particular high levels of iron, total suspended solids and turbidity. The quality of both rivers, even before rain, is not suitable for potable purposes in their raw state according to the GWSc (Ghana Water and Sewerage company) guidelines, because of high levels of turbidity and the presence of suspended solids.

The wastewater in the tailing ponds is also tested and sampled. The tests indicate a huge increase in Total Suspended Solids (TSS) and turbidity compared to the values from the river Birim. TSS increases from average 18 to 774 mg/l at the efflux after considerable settling time. Therefore it can be stated that, although the quality of the water from the river Birim is not sufficient to allow for potable water, the processing in the plant will result in an unacceptable increase in pollution of the water and should therefore be cleansed.

The aim for the mine is to re-circulate the tailing water and to cleanse the water to a quality equal or better than the quality of the water in the river Birim, preferable according to the GWSc guidelines.

3 WATER MANAGEMENT

A mixture of solids and water is pumped from the washplant into the tailing ponds. During this transport the particles suspend in water. These particles need to settle from the water in order to cleanse the water from possible pollution clung to the fines of the material.

When the density of a particle differs from the density of the fluid, the particles can be removed from the water. The location where a particles settles in the settling pond depends on the settling velocity of the particle and the stream velocity of the water. As a result of changing the stream characteristics, variation in the texture and concentration of solids at the influx, the settling location of particles with the same size will vary.

Stokes has summarized all these theory in "Stokes law" *{MacDonald, 1983}*. Based on this settling theory an indication of the required settling length can be made. However, the settling performance of the very small fines (<10 μm) can never be predicted. These particles of fines are too small to be affected by gravity and do not settle fast enough (up to 63 years) *{Argall, 1978}*. On this part of the grainsize distribution our investigation is focussed.

In order to cleanse the water, the required settling time for the particles is guaranteed by the capacity of tailing ponds. In case of the very fine particles the settling time required equals infinity and therefore the settling has to be catalyzed by flocculent.

In this chapter the settling behavior of a particle will be reviewed in two ways: The settling behavior with the help of flocculent and the natural settling of a particle.

3.1 Settling tests description

The standard settling test is done as follows:

- The material is stirred for 15 min. at 10,000 rpm.
- The slurry is poured into a 1,000 ml cylinder and allowed to cool till the temperature of the slurry is lowered to 25 ^{0}c, approximately the temperature of the freshwater in the Birim.
- After each fixed time period a sample is taken from 10 cm under the surface. This depth is also used in the monitoring of the river Birim, so that the results may be compared.
- The tests are done with the material <75-µm from the grain size analyses. Larger particles settle immediately on a formed tailing dump in the tailing pond. The ratio between the water and the mixture for the tests is equal to the average feed of the plant with a constant throughput of freshwater; 76 m^3/hr of solids with 432 m^3/hr of water.

Thereafter:

***A** For the flocculent settling*:

- After adding flocculent, the column is stirred for 1 min. to get the flocculent mixed with the solids. The speed is between 800-1000 rpm.
  Turbidity measurements are taken after 5, 10, 30 and 60 min.
  The turbidity is measured with the Hanna HI 93703 Microprocessor Turbidity meter.

***B** For the natural settling*:

- The sample is filtered and after drying in the oven the difference between the single filter paper and the settled solids plus filter paper is measured. This gives an indication of the Total Suspended Solids (TSS).

The difference between the turbidity and TSS measurement is in the fact that the TSS will not be affected by the very fine particles, since these particles will pass the sieve of the filter paper. The small particles will still influence the color of the water and the quality. So the TSS provides an indication of the amount of particles, while the turbidity reflects the quality of the drink water. In practice, the TSS is checked first each test before the turbidity is measured.

3.2 Flocculent test results

A flocculent is a long chained water-soluble polymer, called polyelectrolyte, used to clean water. A number of flocculent, named Sedipur and provided by Chemproha B.v.-Dordrecht; The Netherlands, are tested and the results from these tests are reviewed.

The Sedipur polyelectrolytes can be classified in two solid grain types. These types are divided in the A-type, weak acid and strong alcalitic suspension, and the C-type, with cationic characteristics.

The test is done with seven of the Sedipur products. Also the influence of mixing both types of polyelectrolytes, the A-type and C-type, is checked. The polyelectrolytes that are tested are CF 404, CF 504, CF 604, CF 803, AF 100, AF 403 and AF 900.

Figure 1 shows the instantaneous influence of the flocculent. It is a summation of the behavior of all the tested flocculent in time. The time effect (the different lines indicate minutes after starting the test) will have influence on the value of Total Suspended Solids, but this is of negligible importance.

The most important observation is that the lines of the different sample times are showing the same pattern and thus it can be stated that the effect of a dosage will not change in time. The cleaning effect is purely dependent on the amount of flocculent added. The different lines indicate the fact that for each polyelectrolyte the performance increases with increasing dose to a maximum level in an optimal dose range, above which the performance measure decreases with increasing dose.

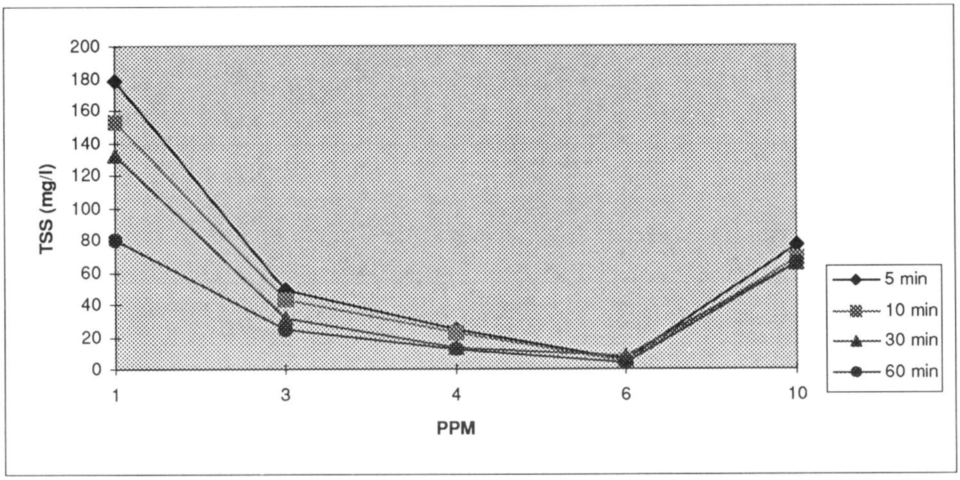

Figure 1 Behavior of the polyelectrolytes in time as a function of the dosage.

Figure 2a and 2b compare the different flocculent. The anionic flocculent do not give a satisfactory result.

All cationic polyelectrolytes reach the environmental regulations level of 5 NTU (Turbidity) but the lines show different paths. The CF 803 has a long low base, but never reaches the low turbidity level of the other three cationic polymers. CF 404 and CF 504 do reach a low level of turbidity; CF 504 reaches even the lowest level, but has a small base. The inclination on these lines at low and high amounts of dosage is steep and thus the success of these flocculent is based on the right dosage. The CF 604 is low in value and has the longest base, therefore is less dependent on the amount added. Its optimum of the line is around the same point as the CF 404 and CF 504 and will thus give no economical disadvantages. CF 604 is most effective and most economical for the cleansing of the water. A mixture of AF 100-type and CF 604-type will also cleanse the water but the overall costs are higher.

To increase the reliability more of these tests have been undertaken. They give an indication of the possibilities of the coagulation of the fines with flocculent and the global amount of polyelectrolytes that might be needed.

Further testing on the CF 604 indicated that its optimum point on the cleaning curve was reached at a dosage of 5.4 PPM.

It can be concluded that a flocculent can be found that will cleanse the water most effectively and satisfy the set requirements for the quality of the water, e.g. within the limits of the WHO-values.

The financial consequence of this solution, at a dosage of 5.4 PPM, an average water consumption of 328.1 m^3/hr and the costs for the flocculent at $5,80/kg, will be that an appreciable amount of polyelectrolytes has to be bought. The costs for this solution will be approx. $250, - per day or $90,000, - yearly. More specific, the operating costs are increased by $10/ounce of gold. Although this option is economical acceptable, the surge has to continue in finding a cheaper solution.

3.3 Results of settling tests with gravel

At the mine site two types of gravel are present, the more "high-grade" river gravel and the "low-grade" terrace gravel. During production it was noticed that both types of gravel behave differently after being pumped into the tailing ponds, resulting in a visually difference in quality of the wastewater. Therefore a suspended solids settling test is undertaken at AGC's Environmental Laboratories. The influence of mixing the two types of gravel on the settling behavior of the material is also studied.

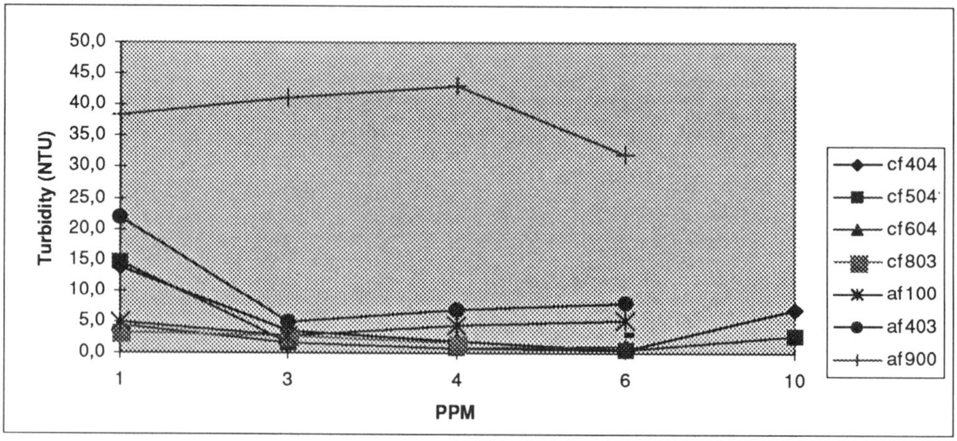

Figure 2a Behavior of the polyelectrolytes on the cleansing of the water, as a function of the dosage.

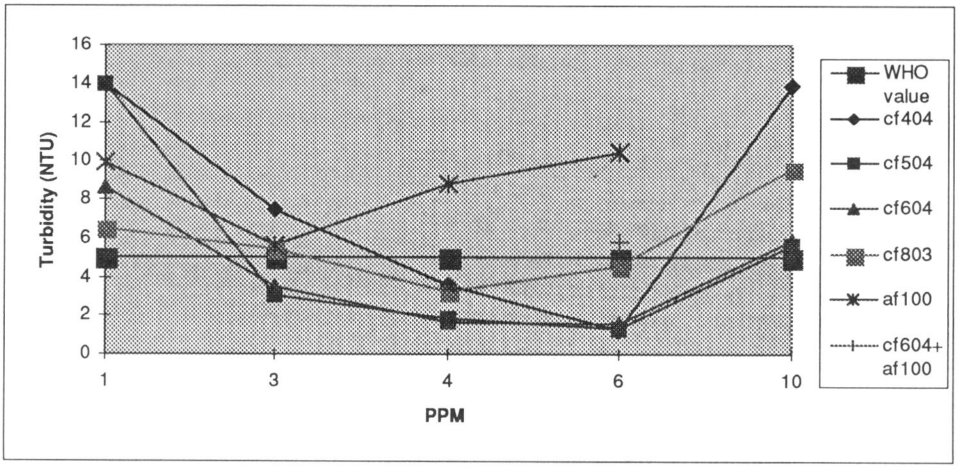

Figure 2b Behavior of the polyelectrolytes on the cleansing of the water, as a function of the dosage.

The mixtures that were investigated are: 100% river gravel, 90% river, 80% river, 70% river and 100% terrace gravel.

The tests are done conform the described standard method (Option 1) of paragraph 3.1.
To simulate the conditions in the pond the temperature of the water, turbulence in the pipes and horizontal velocity in the pond has to be considered.
Two changes on this settling test are:

- The cooling time. No cooling time is allowed. The temperature of the slurry is 40 ^{0}c, approximately the temperature of the water in the tailings ponds. (Option 2).
- The stirring time. The stirring time is changed to 16 hrs at 800 rpm. In order to simulate the conditions of turbulence in the pipe (Option 3).

The general observations from all the tests are:

- The terrace gravel will settle more quickly than the river gravel in all circumstances. The terrace gravel will settle almost immediately, while the river gravel will stay in suspension. This can be seen in figure 5 and figure 6.

- The mixing of both types of gravel has a good influence on the settling behavior. A small (10%) addition of terrace gravel to the river gravel reduces the TSS by more than half of its value (compare 100% river and 90% river). Within the mixtures there is a distinction in TSS based on the amount of terrace gravel added, but beyond 70% river mixture the results go to the same value of TSS. This can be seen in figure 3.
- From the different test options, Option 2 gives the best settling results (figure 4). The increase of the water temperature will have a small, but negligible, positive effect on the settling. This is due to the decrease in viscosity at a higher temperature, resulting in a higher settling velocity. The more the slurry has been mixed, the more particles stay in suspension, even till the end of the test.

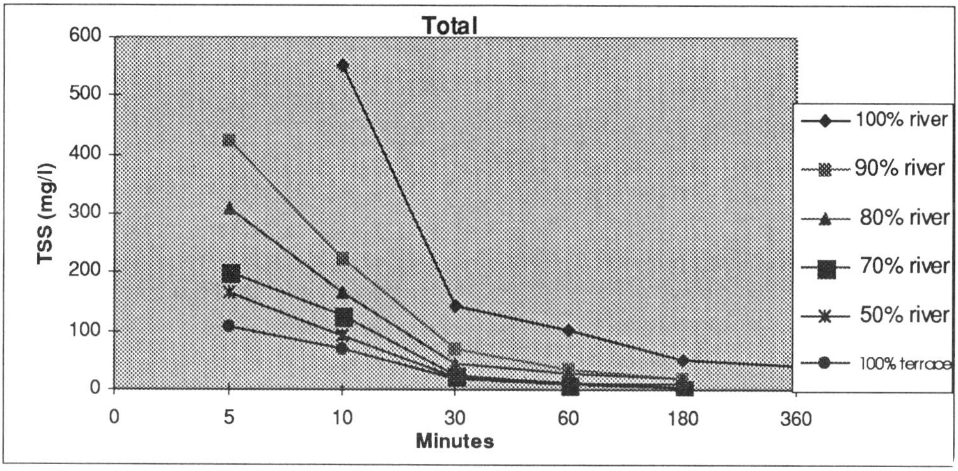

Figure 3 Settling test results for all mixture as a summation of Option 1 and Option 2.

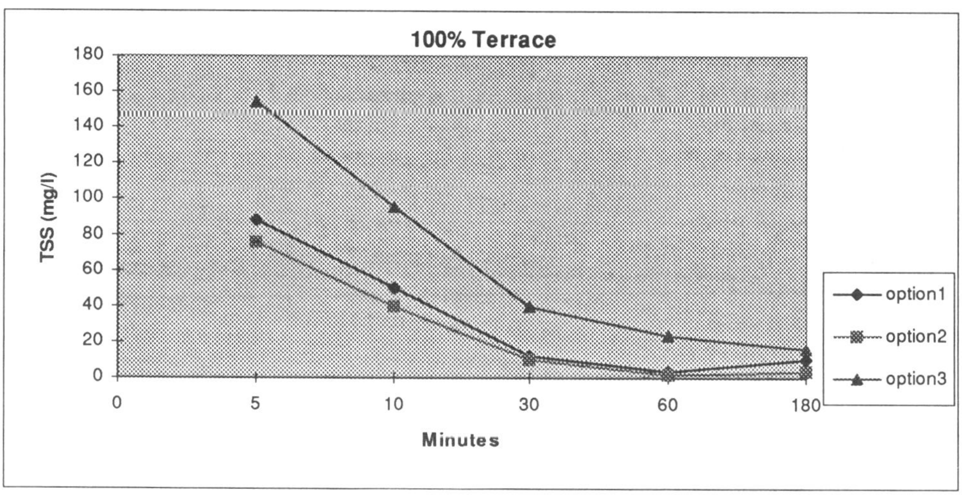

Figure 4 Settling test results for the terrace 100% for the different options.

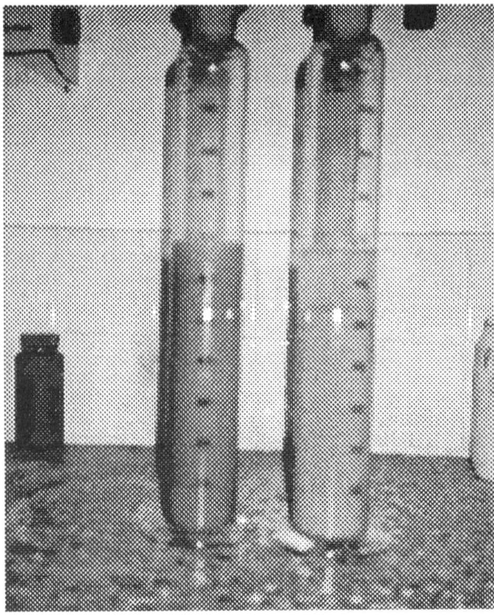

Figure 5 Settling performance of river and terrace gravel after 5 minutes

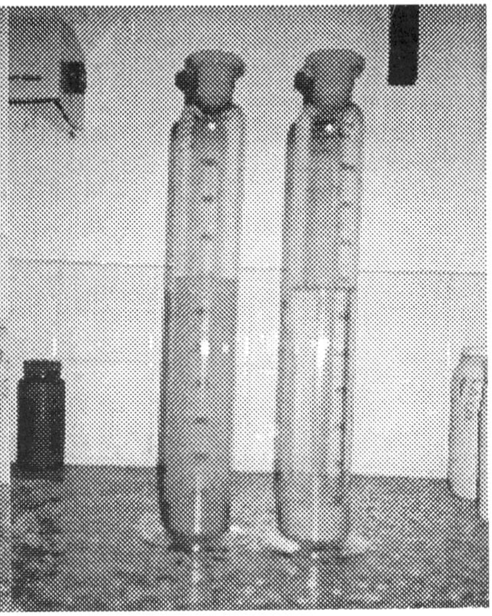

Figure 6 Settling performance of river and terrace gravel after 3 hours.

3.4 Implementation of the settling test results with gravel

As a result of the settling tests a mixing of both terrace and river gravel, to meet the criteria for clean water throughout the life of the mine, can be considered.

With a mixture of both deposits to feed the plant, a ratio of mixing of 2,1:1 (river: terrace) is possible (according to the exploration results of paragraph 1.1).

A mixture of 2:1 has been practiced for 9 days in reality in February 1998, after a first test was conducted in January 1998. These results can be compared with results from monitoring the water from the river Birim. For this comparison a quality-value of the river before and after rain is needed. The rain will cause suspension of material from the banks of the river and will therefore pollute the river naturally. The first 5 days of the test (21 Feb.-25 Feb.) both the river material and terrace material were taken from a stockpile near the treatment plant. The next 4 days (26 Feb.-2 March) the material from the mining was directly dumped into the plant and mixed with terrace from the stockpile. The results are given in table 1.

Table 1. Results of the settling tests in comparison with the water quality in the river Birim.

	Birim After rain	Birim Before rain	22-2'98	23-2'98	25-2'98	27-2'98	2-3'98
TSS (mg/l)	42	10	27	7	26	24	142
Turbidity (NTU)	52.5	5.5	4.36	4.09	1.84	6.85	23.77

From the tests it can be concluded that at the given ratio between terrace and river gravel the water in the pond will be as clean (or cleaner) than the water in the Birim after raining. The quality of the water in the river Birim before raining can only be met if material is used that has been allowed to dry. The direct dumping of the material from the mined out blocks will have a negative effect on the cleansing process of the wastewater. Therefore it may be advised to dump the material at the stockpile to allow it to dry.

The fact that the river material directly from the mine is more difficult to cleanse by the terrace gravel (compare values 1,2 and 3 with values 4 and 5 from table 1) can be attributed to the amount of very small fines (< 10μm). Allowing the material to dry at the stockpile will have a positive effect on the clay structure.

The results of the tests are hopeful. However, a hundred percent secured prediction can never be given based on these results. In order to do so, the local circumstances at the mine site should be investigated by laboratory tests and therefor more tests need to be done. Maybe with the input of water from the bottom of the cuvet the same situation as in reality can be simulated. However, at the moment, the values can be taken as a further input in the construction of the tailing pond and the prediction of the settling behavior in the pond.

3.5 Explanation of settling results

The explanation of the cleaning effect of the terrace gravel should be found in a physical or a chemical difference between the two types of gravel. There might be a physical difference in the distribution of either grainsizes or a difference in texture. Also it might be possible that a chemical reaction between the suspended fine particles and another component of the terrace gravel is resulting in natural flocculation.

Previous geological studies indicate that there is not much variation in the clast composition of the gravel. In most areas, the gravel consists of clays and quartz, shales, mafic rocks and laterite ironstones. Clast dimensions are variable with the abundance of cobble and coarse pebble sizes. There is however no significant difference in composition and texture between the terrace gravel and river gravel.

A grainsize distribution test has been undertaken on behalf of Midras Mining by PTP of AGC. The results (figure 7) show only a slight difference in the grainsize distribution of both types of gravel. The terrace gravel has a higher percentage of coarse material, which can influence the settling of the smaller particles positively, and a relative low percentage of very fine material. However, this difference is not enough to explain the huge influence of a small amount of terrace gravel on the total cleansing process. Based on this knowledge it can be said that the difference in behavior of both types of gravel cannot be explained physically.

Therefore the answer should be found in a chemical distinction between both types of gravel. In order to find the reason why the terrace gravel settle and the river gravel does not, a XRD analysis on both gravel is done at the Faculty of Applied Earth Sciences; Delft University of Technology. A sample of the settled material from tailing pond 2 and tailing pond 4 is also examined. The composition of these samples is given in the table below (table 2).

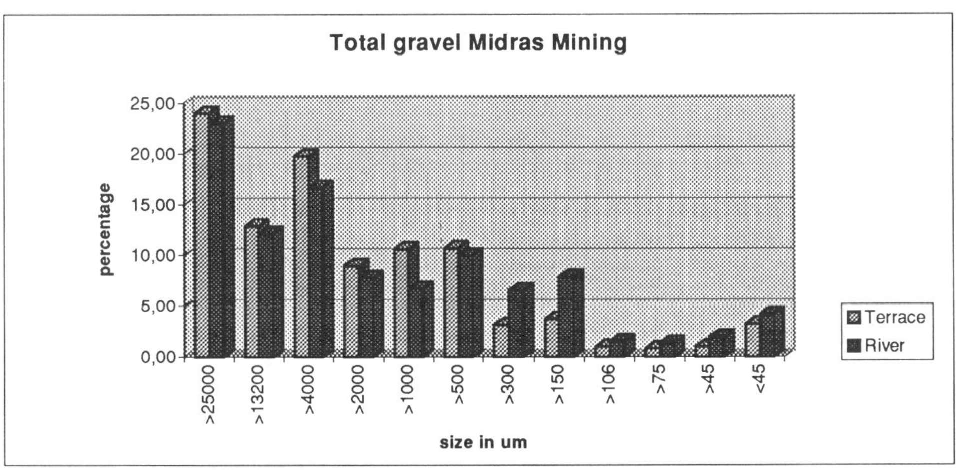

Figure 7 Grain-size distribution according to PTP.

Table 2. Results from XRD-analysis.

River + 45 µm	Terrace + 45 µm	Pond 2	Pond 4
α-Quartz (SiO_2)	α-Quartz (SiO_2)	α-Quartz (SiO_2)	α-Quartz (SiO_2)
Kaolonite $Al_2Si_2o_5(OH)_4$	Kaolonite $Al_2Si_2o_5(OH)_4$	Kaolonite $Al_2Si_2o_5(OH)_4$	Kaolonite $Al_2Si_2o_5(OH)_4$
Muscovite* $Kal_2(Si_3Al)O_{10}(OH,F)_2$	Muscovite $KAl_2(Si_3Al)O_{10}(OH,F)_2$	Muscovite* $KAl_2(Si_3Al)O_{10}(OH,F)_2$	Rutile* (TiO_2))
Arsenopyrite* (FeAsS)	Arsenopyrite* (FeAsS)	Arsenopyrite* (FeAsS)	
	Chlorite Mg-Al-silicate hydroxide		
	Feldspar* ($NaAlSi_3O_8$)		
* = Presumable Present.			

Table 3. Results from testing in practice.

	Pond 1 Terrace	Pond 2 Start terrace	Pond 2 12hr terrace	Pond 2 24hr terrace	Pond 4 No terrace
TSS (mg/l)	10.1	494.3	93.5	33.8	774
Turbidity (NTU)	1.2	115	16.8	5.3	213.5

From the results of the XRD-analysis only one major difference in the components of both types of gravel can be found. The terrace gravel consists of chlorite and the river gravel does not. Maybe the presence of chlorite in gravel will work as a catalysor resulting in the settling of the very small fines in the settling pond. This thought is accompanied by the results from both pond 2 and pond 4. The settled material in pond 4, which is totally filled with wastewater resulting from processing of only river gravel, consists only of quartz and kaolonite. However, the settled material after mixing both types of gravel (pond 2) indicates two extra components, namely muscovite and arsenopyrite. An explanation might be that the chlorite assures that the suspended components muscovite and arsenopyrite will settle. The influence of chlorite on the settling of particles can be part of a next study, but is yet unknown.

The mixed tailings material was directly pumped into pond 2 and not into pond 4, giving another explanation for the presence of the Muscovite and Arsenopyrite in pond 2 and the absent in pond 4. Apparently a direct contact of the slurry with the water in the tailing ponds is needed to assure the positive effect of the terrace gravel on the wastewater.

To check this assumption some samples are taken from the different ponds. The sampling results of pond 1 and pond 2 during the first practical test undertaken in January are given in table 3. The mixture of both types of gravel was first pumped into pond 1. At a certain time the inflow of material changes from pond 1 to pond 2 and a sample was taken from both ponds (value 1 and 2). After changing the inflow to pond 2, this pond was sampled in time and compared with the values in pond 4 (value 3,4 and 5). Comparing of the first two values and the difference in results between pond 2 and pond 4 confirm the given assumption that a direct contact with the water is needed to allow for the cleansing effect of the terrace gravel.

4. DISCUSSION

More and more the tendency in the world is to subject industrial activities to regulations. Especially the environmental regulations have been increased over the last years. The Mining industry has been heavily affected by the strictness of the new environmental criteria.

The alluvial gold mining has always been a type of mining with a high degree of interaction with the environment. Large areas of land are affected by the mining. The whole ecosystem in the region could be temporarily disturbed. In the past this was of minor interest to the Government in the mostly underdeveloped countries where mines are situated. Today the government of Ghana states strict environmental regulations concerning the quality of water and land reclamation.

These stricter regulations require the mining companies to surge for more environmental friendly, but still economical attractive solutions for cleansing waste water. One of the most effective ways to cope with this problem is to use polyelectrolytes. The negative aspect of this solution is that every different site requires excessive laboratory tests to find the most economical polyelectrolyte. This involves a lot of costs.

Another aspect is that polyelectrolytes are chemicals and the use of these chemicals can result in a lot of skepticism among the local population considering the environmental friendly aspect of the product. Therefore excessive sampling and testing is needed for quality control.

In the described case a better solution for the problem could be found in the usage of "low-grade" gravel present. Using this natural product, the skepticism against the environmental friendliness of the cleaning product will disappear.

Beside this environmental and social benefit there is also an economical benefit. It is likely that the cut-off grade of the whole concession can be increased. The extra costs involved in the usage of the alternative, the polyelectrolyte, can be used to lower the cut-off grade. A balance has to be found between the increase of ore reserve due to the mining of the "low-grade" gravel, involving the cost of the mining of this gravel, and the decrease in cut-off grade due to less expenses. This decrease in cut-off grade can be up to 5-10 %.

This new solution is easy applicable. Due to the usage of terrace gravel, the settling of very small particles is instantaneous and therefore the application of it will not affect the size and volume of the tailing ponds.

The question however is till which extend the use of "low-grade" terrace gravel is applicable to other mines. For this specific case it has been a solution, but more information is needed in general about the role of chlorite in the cleansing process of water. At the same time it has to be investigated if, and in which extend, there are sites with gravel with a composition comparable to the composition of the terrace gravel at this specific concession.

5 CONCLUSION:

For this specific concession in Ghana a solution has been found to the wastewater problem. Although the quality of water in the river before it enters the washplant is not sufficient for drinking purposes, the turbidity of the water can be reduced to below the required 5 NTU (GWSc and WHO regulations) in two ways.

The first option is using the flocculent CF 604 in the tailing pond at 5.4 PPM throughout the life of the mine. The second option is mixing terrace gravel with river gravel in a ratio of river: terrace 2:1 or more for the throughput of the plant

The first option is a reliable solution. However, it involves high costs and the process involves environmental unfriendly chemicals, which has to be tested and restricted to certain dosage. The second option is much more environmental friendly while it involves the use of natural products.

The method for cleansing water resulting from this study has been found by empirical research. Independent research is needed to determine if these results are applicable in other situations. The presence or absence of chlorite in the gravel is a major parameter for these studies. Especially the impact of the presence of chlorite in gravel in different mining situations and its effect on the cleansing of wastewater requires further investigation.

If satisfactory answers to remaining questions can be given, it is expected that a new, cheap and environmental friendly solution for the wastewater problem of most alluvial mines has been found.

6 REFERENCES

Argall, G.O. Jr et al. 1978. Tailing disposal today, volume 2: Proceeding of the Second International Tailing Symposium. Denver.

Dalmijn, Prof. Ir. W.L. Dalmijn. 1991. *Grondstoffen Technologie I*. Technische Universiteit Delft.

GHR. 1994. *Grootschalige praktijkproef zand uit baggerspecie*. Gemeentelijk Havenbedrijf Rotterdam.

Groot, de M.B. , F.T. Heezen, D.R. Matbergen, H. stefess. 1988: Slopes and densities of hydraulic placed sands. *Hydraulic Fill Structures*. Special Conference, GT Div/ASCE, Fort Collins.

Huisman. ir. L. 1965. *Theoretische grondslagen der bezinking*. Technische Universiteit Delft.

Keersemaker, M. 1996. *Environmental impact statement of the Asikam alluvial gold mining project(addendum)*.

Keersemaker, M. 1996.: *Supplementary Environmental Impact statement of the Asikam alluvial gold mining project*.

Keersemaker, M. 1997. *Supplemetary environmental impact statement of the asikam alluvial gold mining project (addendum)*.

Macdonald, Eoin. H. 1983. Alluvial Mining: The geology, technology and economics of placers. Chapman and Hall.

PHB. 1996. *Deel 1, Zandwinning met behulp van scheidingtechnieken P-DWW-94-503*. Delft: Dienst Weg en Waterbouwkunde.

PHB. 1996. *Deel 2, Plan van aanpak voor de uitvoering van proefprojecten P-DWW-94-528*. Delft; Dienst Weg en Waterbouwkunde.

PHB. 1996. *Deel 6, kennisinventarisatie depotscheiding W-DWW-95-348*. Delft: Dienst Weg en Waterbouwkunde.

Roffel, dr.ir. B. 1994. *Grondstoffenverwerking 2*. TuDelft.

Smith, J.M, E. Stammers, L.P.B.M. Janssen. 1981. *Fysische transportverschijnselen*. Delft: Vereniging voor studie- en studentenbelangen.

Van den Akker, Prof.dr.ir. H.E.A , dr.ir. R.F. Mudde. 1992. *Fysische transportverschijnselen 1* TuDelft.

Welch, D.E. , L.C. Botham, J.M. Johnson. 1995. Prediction of tailings effluent flows. *Tailings and Mine waste'95*. Fort Collins.

Zwakhals, W. , I. Deibel, et al. 1995. *Zand uit baggerspecie tweede gebruikssessie sedimentatiebekken (1994)*. Gemeentelijk Havenbedrijf Rotterdam.

Tailings and Mine Waste '99 © 1999 Balkema, Rotterdam, ISBN 90 5809 025 6

Tailings management: Adapting to more stringent environmental standards in Chile

M.D.Davis
Department of Civil and Environmental Engineering, University of California, Davis, Calif., USA
M.Andía K.
Centro de Minería, Pontificia Universidad Católica de Chile, Santiago, Chile

ABSTRACT: Mining is a key industry in Chile, accounting for $8.5 billion in 1997 and representing 49.5% of exports in the period 1988-97. In the 1990s, the mining industry has been confronted with more demanding environmental standards, including passage of the Environmental Framework Law (EFL) in 1994. The EFL delegated the National Environmental Commission (CONAMA) to oversee a system of environmental impact assessment (SEIA); consolidate existing and establish new environmental emission, quality, and protective standards; and manage renewable natural resources. New regulations will impact tailings and waste management. Other important environmental issues involve risk assessment and mitigation measures, as well as the effectiveness and best manner to comply with new environmental standards. As a result, the mining industry is in a process of transition to become more environmentally responsible in its operations.

1 INTRODUCTION

During the past three decades, Chile's economic growth has been based largely upon export markets, taking advantage of the country's vast natural resources. Mining is a significant component, accounting for $8.5 billion in export revenues in 1997, of which copper exports totaled $7.1 billion (BCC, 1997a). Mining made up 49.5% of exports in the period 1988-97 (BCC, 1997a). The sector is the primary target of foreign investors, receiving $1.7 billion in 1997 and 44.4% of total investments in Chile for the period 1988-97 (CIE, 1997).

Since the late 1980s, there has been a growing concern in regard to environmental degradation. In the 1990s, the mining industry has been confronted with more demanding environmental standards and enforcement. In the first half of the 1990s, public agencies were established to address environmental issues and regulation. The National Environmental Commission (CONAMA), an independent entity, was established in 1990; in 1992 the Department of Engineering and Environmental Management was created at the National Geologic and Mining Service, part of the Ministry of Mining, whose responsibilities include minimizing environmental impact risk and coordination with CONAMA. The Environmental Framework Law (EFL - L 19,300) of 1994 delegated CONAMA to oversee a system of environmental impact assessment (SEIA); consolidate existing and establish new environmental emission, quality, and protective standards; and manage renewable natural resources. New regulations and tighter enforcement of existing regulations will impact tailings and waste management. Other important issues include risk assessment and mitigation measures, as well as the effectiveness and best manner to comply with new environmental standards.

As a result of more stringent standards, the mining industry is in a process of transition to become more environmentally responsible in its operations. At many mining companies, adaptation will require establishment of environmental departments, targeted and regular environmental monitoring programs, and environmental education of employees. Some large copper mining operations have voluntarily incorporated improved environmental measures in their operations in the early 1990s; smaller mining operations may encounter more difficulty to comply fully with new standards due to cost constraints. In this paper, new environmental regulations are presented in context of tailings and waste management in Chile.

2 MINING SECTOR

The mining sector in Chile can be divided into three groups: (1) large scale operations run by the autonomous state institution CODELCO, (2) large scale operations usually run by a consortium of foreign companies, and (3) small to medium scale operations typically managed by private Chilean firms. CODELCO operates four copper mines that currently account for approximately 40% of copper production in Chile; copper production, currently 3.4 million metric tons (mmt) in 1997 (SERNAGEOMIN, 1998a) is anticipated to reach 4 mmt by 2000 (Claude, 1997). Foreign-run operations are expected to triple production to 2.3 mmt by 2000 (Claude, 1997). Mineral extraction, reserves, and tailings are presented in Table 1. The spatial distribution of mineral extraction in the country during 1997 can be found in Table 2.

3 ENVIRONMENTAL REGULATIONS

Until the early-1990s, environmental regulations passed since the 1900s reflected piecemeal legislation that addressed a specific issue when enacted by a particular government institution or legislative body: as of December 1992, 782 legislative texts were identified that directly or indirectly deal with environmental issues (CONAMA, 1992 and 1993). This has resulted in

Table 1. Mineral extraction, reserves, and tailings in Chile (in metric tons, mt, or kilograms, kg)

Material		1988	1989	1990	1991	1992	1993	1994	1995	1996	1997
Copper	extraction[1]	1.47	1.63	1.62	1.86	1.97	2.08	2.23	2.51	3.14	3.44
(10^6 mt)	reserves[2]	120	120	129	100	102	131	185	182*	179*	176*
(10^6 dry mt)	tailings[3]	166	185	168	195	207	219	233	244	289	314
Gold	extraction[1]	20.6	22.6	27.5	28.9	34.5	33.6	38.8	44.6	53.2	49.4
(10^3 kg)	reserves[2]	193	249	299	310	358	341	583	538*	485*	436*
Coal	extraction[1]	2.47	2.40	2.73	2.74	2.11	1.79	1.66	1.48	1.44	1.41
(10^3 mt)	reserves[2]	84	88	86	67	65	63	110	109*	107*	106*
Iron	extraction[1]	2.9	3.5	2.8	3.2	2.3	2.2	2.8	2.6	2.9	5.8
(10^6 mt)	reserves[2]	46	43	40	37	34	32	29	27*	24*	15*
Potassium Nitrate	extraction[4]	1.1	1.2	1.2	1.3	1.4	1.4	1.4	NA	NA	NA
(10^3 mt)	reserves[2]	34	33	32	31	29	27	87	NA	NA	NA
Calcium Carbonate	extraction[1]	3.65	3.75	3.78	4.00	4.89	5.65	6.31	5.91	6.01	5.62
(10^3 mt)	reserves[2]	159	154	166	178	178	175	170	164*	158*	152*
Lithium	extraction[1]	1.4	1.5	1.7	1.6	2.0	1.9	2.0	2.4	2.7	4.6
(10^3 mt)	reserves[2]	196	194	193	191	189	187	205	203*	200*	195*
Manganese	extraction[1]	0.01	0.01	0.01	0.01	0.01	0.01	0.02	0.02	0.02	0.02
(10^6 mt)	reserves[2]	0.7	0.7	0.7	1.3	1.3	1.3	1.3	1.3	1.3	1.3
Molybdenum (10^3 mt)	extraction[1]	15.5	16.6	13.8	14.4	14.8	14.9	16.0	17.9	17.4	21.6
Silver (10^6 kg)	extraction[1]	0.51	0.55	0.65	0.67	1.02	0.97	0.98	1.04	1.17	1.09
Lead (10^3 mt)	extraction[1]	1.36	1.24	1.12	1.05	0.30	0.34	1.01	0.94	1.37	1.26
Zinc (10^3 mt)	extraction[1]	19.2	18.4	25.1	31.0	29.7	29.4	31.0	25.4	36.0	33.9

NA: Not available.

*: Reserves were determined for the period 1988-94 (BCC, 1997). Reserves for the period 1995-97 reflect reserves as of 1994 minus production; they do not account for newly discovered reserves during 1995-97.

Sources: [1] SERNAGEOMIN, 1998a; [2] BCC, 1997b; [3] CCC, 1998; and [4] Claude, 1997.

Table 2. Spatial distribution of mining activity during 1997 in Chile (in metric tons, mt, or kilograms,kg)

Region	Copper 10^3 mt	Gold 10^3 kg	Iron 10^3 mt	Manganese 10^3 mt	Molybdenum 10^3 mt	Silver 10^3 kg	Lead 10^3 mt	Zinc 10^3 mt
I	144.6	0.2	---	---	---	1.5	---	---
II	2,082.0	12.2	---	---	12.6	339.1	---	---
III	378.6	20.3	3,190.5	---	1.4	456.3	---	---
IV	88.0	11.3	2,576.7	18.1	---	63.8	---	---
V	227.8	2.2	---	---	1.8	66.9	---	---
MR	155.8	1.4	---	---	2.5	61.0	---	---
VI	361.3	0.2	---	---	3.2	25.2	---	---
VII	---	0.4	---	---	---	2.2	1.0	2.3
VIII	---	---	---	---	---	---	---	---
IX	---	---	---	---	---	---	---	---
X	---	---	---	---	---	---	---	---
XI	---	1.2	---	---	---	75.3	0.2	31.6
XII	---	---	---	---	---	---	---	---
Total	3,438.1	49.4	5,767.2	18.1	21.5	1091.3	1.3	33.9

MR: Metropolitan Region.
Source: SERNAGEOMIN, 1998.

fragmented protection, confusion regarding applicability, and noncompliance. In the 1990s, several regulatory measures have been enacted that are directed toward public participation, management goals, and, in general, a more systematic approach to environmental protection. Recent environmental legislation is a result of (a) requirements by foreign lending agencies for Chile to receive infrastructure loans, (b) assurance of access of export goods to foreign markets that require higher environmental standards, (c) legislative goals for consolidation and comprehensive application of existing regulations, and (d) societal sentiment to prevent environment degradation.

3.1 *Litigation and Public Participation*

In the late 1980s, a protective appeal (petition to cease and desist) was filed to prevent further contamination produced by CODELCO's copper mine El Salvador. Tailings were dumped directly into the Salado river, seriously damaging the river ecosystem and eliminated fishing and tourism in the Chañaral Bay. In 1990 the court ordered CODELCO to cease dumping tailings into the river and to construct a pond to house the tailings (Claude, 1997). Prior to 1990, this law was not utilized successfully because proof is required that the offensive action is damaging and intentional, which is difficult to prove legally. Furthermore, less public involvement in environmental protection occurred, which has changed following transition from a military dictatorship to democracy in 1990.

This case perhaps marks a transition in Chile for the mining sector as well as other industrial activities; more environmentally responsible operations will be required. Existing regulations, many of which were routinely ignored, will be more readily enforced. In addition, the public will likely become more involved in protecting the environment.

3.2 *Existing Legislation*

Currently, the principal regulations that tailings and waste management at mining operations are subject to include the system of environmental impact assessment (SEIA) for new or significant modification to existing facilities, tailings pond construction and operation approvals, water and air quality standards, land use restrictions, and conservation measures. The principal laws affecting construction and operation of tailings and waste in Chile are contained in Table 3.

The list of existing laws in Table 3 exemplifies the multitude of regulations that must be adhered to. This list also illustrates that many items, such as water quality, are referred to several times in the existing legislation and can be regulated and enforced by distinct government agencies.

Table 3. Principal legislation affecting construction and operation of tailings and waste ponds in Chile

Item	Description	Regulatory Institution	Legislation
Environmental protection	System of environmental impact assessment (SEIA)	President of the Republic through the National Environmental Commission (CONAMA)	DL 30, 1997 LN 19,300, 1994
Construction and operation of tailings ponds	(1) Report initiation of mining activities	(1) National Geologic and Mining Service	(1) DN 72, 1986
	(2) Construction of reservoirs and hydrologic risk	(2) National Water Bureau	(2) DL 1122, 1981 - Water Code
	(3) Industrial construction outside urban limits	(3) (a) Ministry of Housing and Development & (b) Agriculture and Cattle Service	(3) DFL 458, 1975
	(4) Construction and operation	(4) National Geologic and Mining Service	(4) DN 86, 1970
Management and disposal of tailings and waste	(1) Management and disposal of liquid and solid waste	(1) Ministry of Heath	(1) DSN 5333, 1990
	(2) Regulation of landfills	(2) Ministry of Heath	(2) RN 024444, 1980
Water quality / emissions	(1) Neutralization of wastewater discharged to water bodies or sewer systems	(1) Superintendency of Sanitary Services	(1) DSN 351, 1992
	(2) Prohibition of discharging untreated wastewater from industrial activities to water bodies that can be used for potable, irrigation, or recreation water	(2) Ministry of Heath	(2) DSN 553, 1990
	(3) Prohibition of discharging industrial or mining waste into water bodies	(3) Ministry of Heath	(3) DFL 725, 1967 - Sanitary Code
	(4) Prohibition of discharging untreated wastewater from industrial activities to water bodies that can be used for potable water or irrigation	(4) Ministry of Heath	(4) LN 3133, 1916
Air quality / emissions	(1) Sulfur dioxide, arsenic, and particulate matter emissions	(1) Ministry of Mining	(1) DSN 185, 1991
	(2) Noise levels	(2) Ministry of Heath	(2) DN 268, 1984
	(3) Ambient air quality for suspended particulate matter, sulfur dioxide, carbon monoxide, ozone, and nitrogen dioxide	(3) Ministry of Heath	(3) RN 1215, 1978
	(4) Emissions from factory processes	(4) Ministry of Heath	(4) DN 144, 1961
Land use	Local zoning ordinances	Ministry of Housing and Development	DN 458, 1976
Conservation and protection of cultural heritage	(1) Protection of architecture, culture, and indigenous heritage	(1) President of the Republic	(1) LN 19253, 1993
	(2) Protection of national monuments, including archeological sites and artifacts	(2) President of the Republic	(2) LN 17288, 1970

Continued next page.

Table 3. Principal legislation affecting construction and operation of tailings and waste ponds in Chile cont.

Item	Description	Regulatory Institution	Legislation
Conservation and protection of natural resources	(1) Biological diversity conservation	(1) President of the Republic	(1) International agreement signed by Chile, 1994
	(2) Restrictions on cutting native forest	(2) Ministry of Agriculture	(2) DS 87, 1992 and DS 259, 1980
	(3) Wildlife conservation	(3) President of the Republic	(3) International agreement signed by Chile, 1987
	(4) Obligation of state to preserve nature	(4) President of the Republic	(4) Constitution, Art. 19, No. 8, 1980
	(5) Restrictions on cutting native forest	(5) Forestry Agency	(5) DFL 701, 1974
	(6) Protection of flora, fauna, and scenic beauty	(6) President of the Republic	(6) Washington Agreement - international agreement signed by Chile, 1967

Source: DICTUC, 1997.

4 ENVIRONMENTAL RISK AND MITIGATION

An important aspect of producing a revised set of coherent environmental regulations and the ability to review the environmental impact assessment of a tailings and waste pond project is to identify environmental risks and to minimize these risks. Environmental risk assessment must include impacts on human and natural environments, as well as risks generated during normal operations and extreme events (natural hazards).

4.1 *Existing regulations of tailings and waste ponds*

Current laws regulating construction and operation of tailings and waste ponds are based on structural integrity. This criterion encompasses design standards that are well established and straightforward. Typically, the design standards are dependent upon the material that is deposited, with different designs existing for materials categorized as dangerous, for example. The governing regulations typically are minimum design standards. Current standards, however, do not address environmental risk directly nor are there any incentives for mining companies to minimize this risk. Under existing regulations, mining operators locate tailings and waste ponds irrespective of important environmental risk issues such as population centers, water bodies, and downstream agricultural activity. In the following paragraphs, a general discussion is presented in regard to environmental risk and mitigation of tailings and waste ponds.

4.2 *Environmental Risk*

There are certain factors to consider when assessing the environmental risk of a mining operation. One important factor is location of the activity and its spatial relationship to human and natural environments. In Chile mining occurs predominantly in Regions I-VI (Table 2). Operations are typically located in desert areas and in the upper portions of watersheds in the Andean foothills. In addition, there are isolated mining activities that are close to urban populations (i.e. Andacollo Cobre and Andacollo Oro adjacent to the town Andacollo) and a few located at sea (i.e. Lota Schwager). Thus, most cases involve sites in relatively deserted areas where there is little competition for land with other human activities. At the same time, there are mining activities near human settlements and activities where negative impacts can be observed. Likewise, the spatial relationship with the natural environment should be examined. Thus, a general approach to environmental impact assessment must take into account diverse conditions.

Other important factors are environmental management and type of mining activity. The manner in which tailings are managed has an obvious impact on the environment, which is the main point of assessing environmental risk. The most important management aspect of tailings and waste management is capturing the tailings in a controlled area. The majority of mining operations make use of tailings ponds; some operations, nonetheless, discharge tailings into rivers or the ocean (SERNAGEOMIN, 1989), as occurred previously in the case of the El Salvador mine.

The type of material mined and *in situ* constituents at the mining operation contribute to environmental risk as well. For example, iron and copper ores are the minerals of most production and tailings by volume in Chile. However, impacts of iron mining are less harmful due to its low toxicity in the environment. Copper poses the largest threat because of its toxicity, magnitude of operations, and human activity downstream of mining operations (González, 1995).

4.3 *Environmental Risk - Normal Operations*

Environmental risks associated with normal operation of mining tailings and waste ponds include impacts on human settlements, downstream agriculture, and the natural environment. Important impacts, for example, include suspension of particulate matter, contamination of soil and water, and destruction of habitat.

In regard to tailings management, suspension of particulate matter is likely one of the easier environmental impacts to mitigate. For instance, deposition of tailings down wind of a cement mining operation at Polpaico was prevalent in the early 1990s. This impact was mitigated by placement of nets on the dried portions of the existing and older tailings ponds (DICTUC, 1997).

The environmental impact of mining activities on downstream agriculture is twofold: air and water. Impacts are a result of the toxicity of material emitted directly and indirectly from the mining operations. Emissions from mining activities degrade soils downwind, although this aspect is more associated with ore processing than material suspended from tailings ponds. The effect, however, can be very significant. Mining operations can emit elements that accumulate and are of high toxicity in the soil matrix, which can impact not only agricultural production but also the natural environment (González, 1995).

The principal conflict between mining and agriculture involves copper. Copper poses problems in Regions I-VII, where nearly all of the produce in the country is grown. As mentioned, many of the mining operations are located in the Andean foothills, whereas agricultural areas are typically located downstream in the flatter portions of the transverse valleys. Water, of degraded quality and containing mining residues, returns to rivers, which in turn is used in agriculture downstream. In addition, water extractions for mining lessen the dilution capability of receiving water bodies to assimilate return flows from mining activity (González, 1995). However, at many mining plants, water recuperation and reuse are employed, principally due to water scarcity. Recirculation systems can minimize negative surface water quality impacts associated with mining operations. Groundwater, nevertheless, could still be impacted.

Perhaps the least developed issue in terms of evaluation and mitigation is the impact on the flora and fauna of the natural environment. Knowledge of the natural environment is limited. For instance, the life cycle is known for approximately 50% of indigenous fresh water fish in Chile (Arratia, 1981). Less is known about other fresh water fauna. Thus, evaluation of the environmental risk of a project or operation posed on the natural environment is difficult to assess tangibly. Mining operations can destroy habitat. Also, operations can alter the quality of habitat beyond the limits of the operating area. An initial guideline is to minimize release of toxic elements while observing cost constraints. In addition, continual monitoring of emissions can be conducted while noting any adverse impacts on neighboring areas. Last, mitigation measures can be employed, such as when habitat is damaged at the mining site. Equivalent habitat can be protected, restored, and/or created at another site, such as exists with wetland trading in the United States. Initial mitigation measures could target special protections for threatened and endangered species - species that have been identified in two studies (Benoit, 1989 and Glade, 1993).

4.4 *Environmental Risk - Natural Hazards*

Natural hazard risks are listed in Table 4 for regions where most mining takes place in Chile. The effect of such hazards is possible failure of tailings ponds. Besides possible loss of life and economic loss, the effect is an injection of a large concentration of potentially harmful constituents that were stored in the tailings pond into the natural environment.

Failures of tailings ponds due to seismic activity are presented in Table 5 for Region V. The failure of numerous tailings ponds due to seismic activity indicates that it is possible that many existing ponds do not meet current structural standards. Furthermore it is possible that seismic standards for tailings ponds may need to be upgraded so as to account for and minimize environmental impacts of such failures, in addition to other design criteria already considered, such as economic loss or loss of life.

Additionally, there have been failures due to flooding in Region V. Flood waters washed away three tailings ponds during an intense rainstorm in 1987 (SERNAGEOMIN, 1989), which again exemplifies the important issue of the location of the tailings pond and environmental risk.

4.5 *Issues in Incorporating Environmental Risk in New Regulations*

There is interest in creating a regulatory system that addresses social cost and self-regulation. Thus, the cost of mining would not only include direct operation, maintenance, and capital costs but also the social cost of conducting business. The social cost component would reflect the internalization of costs associated with degradation caused by emissions or the environmental risk posed by locating a tailings pond near a dry stream bed, for instance, which in large part are not currently paid by the mining operators but rather externalized. Another good example is in regard to the eventual fate of tailings ponds and mining sites. In evaluating a new site for a tailings pond, future land use and environmental compatibility following its useful life must be weighed, especially upon closure of the mining operation. Such items to contemplate include

Table 4. Natural hazard risks in Chile

Region	Principal Natural Disaster Risks
I	Bolivian winter with annual floods in high plains area of the Andes. Coastal zone subject to tidal waves. Presence of active volcanoes. High seismic activity.
II	Bolivian winter with annual floods in high plains area of the Andes. Presence of active volcanoes. High seismic activity.
III	Bolivian winter with annual floods in high plains area of the Andes. Flooding on Copiapó and Huasco rivers. Presence of active volcanoes. High seismic activity.
IV	Flooding and landslides in interior gullies and small valleys. Presence of active volcanoes. High seismic activity.
V	Flooding on Aconcagua, La Ligua and Petorca rivers. Presence of active volcanoes. High seismic activity.
MR	Flooding on Maipo and Mapocho rivers and large urban drainage and irrigation canals. Presence of active volcanoes. High seismic activity.
VI	Flooding on Claro, Tinguiririca and Cachapoal rivers. High seismic activity.

MR: Metropolitan Region.
Source: Prepared by the authors.

Table 5. Earthquakes and number of tailings ponds that collapsed in Region V

Year	Magnitude (Richter)	Number of Tailings Ponds that Collapsed
1965	7.1	8
1971	7.75	4
1981	6.8	2
1985	7.8	4

Source: Aliste, 1966 and SERNAGEOMIN, 1989.

alteration of topography, impact on surface runoff or stream flow, and aesthetics. Last, effort is being made to craft a system where the industry regulates itself. These are issues that many countries and fields are considering in efforts toward sustainable growth. They are also issues that are difficult to design and implement.

5 CASE EXAMPLE - TAILINGS POND 5 AT POLPACIO CEMENT FACTORY

Experience at a small to medium mining operation is presented to indicate the dynamics regarding implementation of environmental regulations and the transition in the industry. Polpaico Cement Plant, located 38 km north of Santiago, is a small to medium producer of cement and lime. Polpaico submitted to the system of environmental impact assessment (SEIA) to evaluate the construction of tailings pond 5 in 1996 (DICTUC, 1997).

Polpaico's case is, perhaps, typical of small to medium mining operations. Advanced method and management techniques are utilized in production. However, in regard to environmental management at the time of the environmental impact assessment (EIA), there was no department dedicated specifically to environmental management at the central office, no one dedicated to environmental management at the cement plant, and no habitual environmental monitoring taking place at the plant in 1996.

Prior to the tailings pond project, Polpaico contracted an environmental audit of the operations at the cement plant. Polpaico voluntarily submitted to the SEIA since participation was not mandatory until 1997. The company followed through with the EIS, despite the preliminary ruling by CONAMA that Polpaico could receive a waiver due to low environmental impact and a case load that exceeded CONAMA's ability to review the EIA. Additionally, in 1997 a person was hired to deal with environmental management at the cement plant. Polpaico's actions indicate the transition period in which the mining sector is faced, as well as foresight by the company to anticipate more environmental regulations, which is probably not typical of all mining operations in Chile.

6 DISCUSSION AND CONCLUSION

Existing environmental regulations currently reflect a mixture of specific environmental mandates as well as systematic methods, which taken as a whole are still somewhat fragmented. Additionally, there are overlaps in institutional responsibilities. Moreover, there are several items that are not addressed directly by current legislation. Some of the more important items not dealt with in existing legislation include (1) no regulatory oversight to deal with existing and abandoned tailings and waste ponds, (2) no specific regulatory procedure nor obligation to mitigate lands upon completion of the useful operating life of a tailings pond, (3) no specific regulatory procedure to monitor air and water quality, such as particulate matter near the tailings pond or seepage from the pond and possible affects on groundwater, and (4) no specific procedure to evaluate the environmental risk that a tailings and waste pond poses. Due to gaps in existing environmental law, in some cases, the only recourse to prevent negative environmental impacts is litigation.

The Polpaico case exemplifies how both regulations and regulatory agencies are in transition, as CONAMA and other government agencies are in the process of establishing coherent legislation and also creating operating protocol within their agencies. The creation of CONAMA and the EFL are the first steps in a coordinated environmental management approach. The SEIA, for example, serves not only to evaluate the environmental impact of a new project but also coordinates regulatory and institutional approval, which can involve many government entities and allows for public participation.

Initially, there were some complaints regarding EIA evaluations that differed throughout the county. (Regional CONAMA offices (COREMAs) are in charge of evaluating projects in their regions). CONAMA has made concerted efforts to be consistent in EIA evaluations, conducting internal workshops and developing specific evaluation procedures. Furthermore, CONAMA has established a program of certified individuals/private entities that can assist CONAMA so that all projects subject to SEIA can be duly processed.

Existing environmental guidelines, however, are still in a state of flux, as one of CONAMA's responsibilities is to produce a streamlined, straightforward, and comprehensive set of environmental regulations. It is likely that there will be a continuing process of revision of regulations. The administration has developed an environmental agenda for the year 2000 which focuses on six broad topics: (1) recuperate and improve environmental quality, (2) prevent environmental deterioration, (3) advocate protection of environmental heritage and sustainable use of natural resources, (4) increase public participation in environmental management, (5) strengthen national and regional institutions, and (6) improve environmental legislation (La Época, 1998). These items will build on the transitions that have occurred in the 1990s. Special focus will be on improving the SEIA, involving more public participation in the SEIA, and improving environmental legislation to comply with the free trade agreement reached with Canada. Also, specifically targeted areas for recuperating and improving environmental quality include three mining areas where environmental degradation has occurred due to mining activity (principally due to copper foundry emissions): Chuquicamata, Paipote, and Ventanas. However, it is important to note that there is no consensus on environmental policy. One of many topics debated but left out of the environmental agenda is a specific mining policy in regard to management of nonrenewable natural resources.

Therefore, it is probable that the mining sector will confront more environmental regulations. Some of these regulatory measures may be enacted quickly so as to conform to international agreements. These changes will require adaptation in the mining industry - something that is already beginning to occur. Adaptation will likely include creation of departments dedicated to environmental management, environmental training of personnel, and a program of environmental monitoring at mining facilities. Large mining operations will be more readily able to adapt to increased environmental standards. At small to medium locally run operations, adaptation may be more difficult due to cost constraints.

Rapid changes and/or stringent regulatory measures are not anticipated to occur, however, as decisive environmental policy has not been established while Chile grapples with tradeoffs between economic development versus environmental protection and private initiative versus government regulation. The current administration and legislative bodies are reluctant to enact restrictive environmental controls that might impact economic growth, especially in light of an economic downturn due to economic instability in Asia and other export markets.

REFERENCES

Aliste, J. 1966. "Efectos del sismo de marzo de 1965," *Boletín 20 del Instituto de Investigaciones Geológicas* (IIG). Santiago, Chile: IIG.

Arratia F., G. 1981. *Géneros de peces de aguas continentales de Chile,* Publicación Ocasional 34. Santiago, Chile: Museo Nacional de Historia Natural.

Banco Central de Chile (BCC) 1997a. *Boletín del Banco Central de Chile,* July 1997 and previous months editions. Santiago, Chile: BCC.

Banco Central de Chile and Servicio Nacional de Geología y Minería de Chile (BCC) 1997b. *Cuantificación de los principales recursos minerales de Chile (1985-1994).* Santiago, Chile: BCC.

Benoit C., I.L., ed. 1989. *Libro rojo de la flora terrestre de Chile.* Santiago, Chile: Corporación Nacional Forestal de Chile (CONAF).

Claude, M. 1997. *Una vez más la miseria - ¿Es Chile en país sustentable?* Santiago, Chile: LOM Ediciones Ltda.

Comisión Chilena del Cobre (CCC) 1998. *Estadísticas del cobre 1997.* Santiago, Chile: CCC.

Comisión Nacional del Medio Ambiente (CONAMA) 1992. *Repertorio de la legislación de relevancia ambiental vigente en Chile.* Santiago, Chile: CONAMA.

Comisión Nacional del Medio Ambiente (CONAMA) 1993. *Repertorio de la legislación de relevancia ambiental vigente en Chile, suplemento 1.* Santiago, Chile: CONAMA.

Comité de Inversiones Extranjeras (CIE) 1998. *Inversiones extranjeras 1997.* Santiago, Chile: CIE.

Dirección Técnica de la Universidad Católica S.A. (DICTUC), 1997. *Estudio de impacto ambiental: proyecto tranque relaves No. 5, Cemento Polpaico S.A.* Santiago, Chile: DICTUC.

Figueroa, E. et al 1996. "Sustentabilidad ambiental del sector exportador chileno." In *Sustentabilidad ambiental del crecimiento económico chileno,* edited by O. Sunkel. Santiago, Chile: ANDROS Productora Gráfica.

Glade, A.A., ed. 1993. *Libro rojo de los vertebrados terrestres en Chile,* 2nd edition. Santiago, Chile: Corporación Nacional Forestal de Chile (CONAF).

Gonzáles M., S. 1995. "Estado de la contaminación de los suelos en Chile." In *Perfil Ambiental de Chile,* edited by G. Espinoza G., et al. Santiago, Chile: Comisión Nacional del Medio Ambiente.

Hudson, R.A. 1994. *Chile, a Country Study.* Washington, D.C.: U.S. Government Printing Office.

La Época 1998. "Gobierno se fijó siete objetivos en materia ambiental hasta el 2000," *La Época.* Page 18, Column 2, January 20.

Le-Bert M., J. 1995. "Descripción de las actividades económicas de Chile." In *Perfil Ambiental de Chile,* edited by G. Espinoza G., et al. Santiago, Chile: Comisión Nacional del Medio Ambiente.

O'Ryan, R. and A. Ulloa, 1996. "Instrumentos de regulación ambiental en Chile." In *Sustentabilidad ambiental del crecimiento económico chileno,* edited by O. Sunkel. Anatiago, Chile: ANDROS Productora Grafica.

Praus G., S. 1995. "Derecho y medio ambiente en Chile." In *Perfil Ambiental de Chile,* edited by G. Espinoza G., et al. Santiago, Chile: Comisión Nacional del Medio Ambiente.

Publicaciones Editorial Gestión (PEG), 1996. *Informe sobre Chile 1996.* Santiago, Chile: Imprenta Salesianos S.A.

Romero A.H. 1995. "Información ambiental en el sector público chileno." In *Perfil Ambiental de Chile,* edited by G. Espinoza G., et al. Santiago, Chile: Comisión Nacional del Medio Ambiente.

Servicio Nacional de Geología y Minería de Chile (SERNAGEOMIN) 1998a. *Anuario de la Minería de Chile, 1997.* Santiago, Chile: SERNAGEOMIN.

Servicio Nacional de Geología y Minería de Chile (SERNAGEOMIN) 1998b. *Web Page - Department Descriptions and Mining Statistics.* http://www.sernageomin.cl/. Downloaded September 6, 1998.

Servicio Nacional de Geología y Minería de Chile (SERNAGEOMIN) 1989. *Levantamiento Catastral de los Tranques de Relave de Chile.* Santiago, Chile: SERNAGEOMIN.

Tailings and Mine Waste '99 © 1999 Balkema, Rotterdam, ISBN 90 5809 025 6

A new development in paste backfill technology

L.J.Chen, D.G.Millette & A.Annor
Natural Resources Canada, CANMET, Sudbury, Ont., Canada

ABSTRACT: Paste backfill has been introduced and practised in underground mines since the early 1980s by which the mill tailings are disposed in the mine openings to provide a competent ground support. However, the technology is still in its development. Current paste backfill systems are based on either filtration or a blending technique which both involve expensive equipment and high energy consumption which has led the mines to search for a cost-effective technology in recent years. Alternative approaches to the problem are now being developed whereby the full plant tailings can be made into a paste by a sedimentation and fluidization process. This new development demonstrates that a tailings paste can be produced simply by flluidizing the settled solids in a silo system. The fluidization mechanism, experimental system and pilot plant test results are described and presented in this paper.

INTRODUCTION

The use of mill tailings as a mine backfill is aimed at reducing waste disposal on surface and to provide stability to underground openings. It is ideal if this waste can be totally placed underground to produce a competent support, and this is possible if dilute tailings as discharged from the mill can be prepared into a paste prior to placement. Filtration is a well known technique for dewatering the fine solid particles, however, the high cost involved in equipment and energy consumption has encouraged mining companies to search for a more cost-effective paste backfill technology. Through extensive research and experiments, significant progress has been made in recent years. A fluidization technique has been developed in recent experiments with a pilot plant and a paste tailings fill has been manufactured simply within a silo. The whole tailings mass is agitated into uniform suspension in terms of particle size and pulp density. The material produced is discharged from the silo by gravity.

The objective of this paper is to present the operating principle of a new technology for full plant tailing paste backfill and test results obtained from pilot plant experiments and full scale trials. Tailings paste behaves as a time-dependent non-Newtonian fluid during its stay in the silo and during discharge. Silo geometry, the fluidization device and the discharge system have been tested and their effectiveness proven. A paste backfill plant can be designed by scaling up the silo model from data obtained. The potential application of this new development is explained by its simplicity and cost-effectiveness in operation compared to existing technologies.

FLUIDIZATION MODEL AND MECHANISM

Tailings pumped from the mill and stored in a silo are allowed to settle. As a result, a "settled bed " is formed, and is thickened and consolidated under the load of the solids in upper layers and static pressure of the water. To get the tailings into suspension and therefore movable, the solid particles must be fluidized. The fluidization process takes place between the settled bed and the area where tailings particles are still moving freely[6] . Fluidization in the silo is completed through the use of fluid jets. The model in Figure 1 shows the mechanism used. Tailings deposited in the silo are in static balance when no external force is applied. As the fluid flows through the voids between particles, turbulent flow among the particles and friction among the surfaces of the particles produces resistance to the flow. The pressure in the lower layers of the settled bed will be greater than that in the upper layers if the fluid flows upwards from the bottom. When the tailings are in settled bed , the pressure $\Delta P1$ increases with the flow velocity of the fluid, but the tailings' porosity remains unchanged. When the velocity and pressure reach a certain value, the bed expands uniformly to take up the increased flow, and the particles space themselves, so that the tailings will be supported by friction force and not by the silo bottom and they will no longer remain in their original position. Losing mechanical support, every particle can move freely in the fluidized bed.

When fluid under pressure is introduced, the "settled bed" expends and porosity changes from E_0 to E as shown in the model. The particles up to a certain height are in a fluidized state. The lowest velocity, V_f, which changes tailings from a settled bed into fluidized particles is called the critical fluidization velocity[5]. The value of the pressure loss ΔP_2 in this case is equal to particles weight/unit area:

$$\Delta P_2 = L_o(\rho_s - \rho_f)(1 - E_o)$$

Where: L_o = the height of the tailings in settled bed,

ρs = the specific gravity of tailings,

ρ_f = the specific gravity of fluid,

E_o = the porosity of tailings in the settled bed,

In an air fluidized bed, slurry is moved upwards by action of bubbles via two mechanism: wake and drift. The solids motion is completed through wake transport as well as drift in the wake fraction[8].

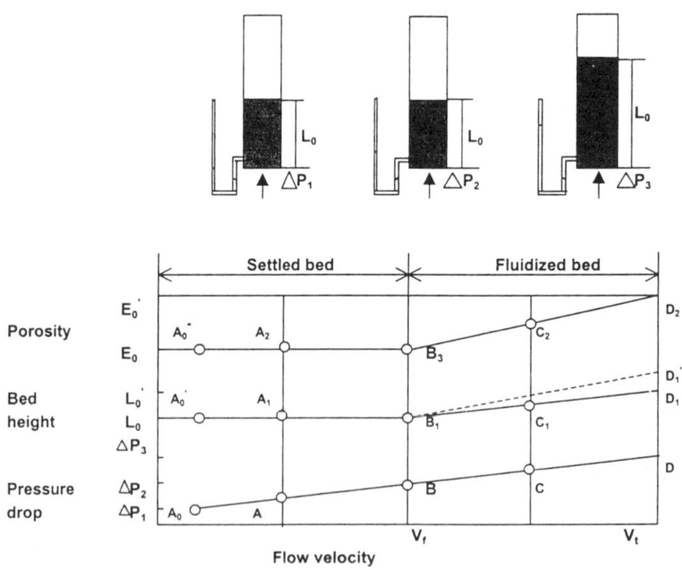

Figure 1. Fluidization model

PILOT PLANT TEST SYSTEM

The experimental system used to fluidize the tailings is shown in Figure 2 and comprises the following sections: dilute tailings feed, dewatering, fluidization, slurry discharge and monitoring system. Tailings received from the mines are dumped into a portable mixer where water is added to prepare a dilute slurry having the same pulp density as discharged from the particular mine's mill, generally 30% solids, then it is delivered to the silo by a mud pump. In the silo the dilute slurry is densified by a dewatering device during the settling process. When the tailings are sufficiently de-watered, the fluidization process commences by introducing fluid through an array of nozzles. Tailings in the silo move up and down under the action of the fluid and are in suspension after a certain period of time. Paste manufactured in the silo is then discharged by opening the valve attached to silo bottom into a receiver/mixing tank where water is added to repulp the paste into a dilute suspension which is then sent back by a sand pump to the silo to repeat the experiments. Fluidization parameters are controlled during the experiments and the pulp density and flow rate of the paste are constantly monitored. The silo is a cylindrical steel structure with a conical bottom, and the experimental system operates as a closed-loop tailings cycling system that is capable of handling 3 to 5 tonnes of paste per test cycle.

TEST MATERIAL

Different tailings from base metal and gold mines were tested, and their typical size distributions are presented in Figure 3. Classified tailing contained coarse particles sizing from 10μm to 600μm, while full plant tailing is a complex of fines, ranging from 5μm to 200μm. The specific gravity of the tailings was measured between 2.65 to 3.6 depending on their mineralogical composition. Particle size has a profound affect on solids settling and behaviour in the slurry. Coarse particles settle down rapidly, soon after the pour is completed and a dense mass is formed in the bottom of the tank, while fines have a low settling rate and tend to remain in suspension forming a less dense mass. Classified tailings from base metal mines settle to 76-78% pulp density after 12 hours and then consolidate to 82% within 24 hours, while the full plant tailings from gold mines settle to 70% and consolidate to 76% solids[(3)]. Porosity in a compacted tailing mass is a function of size gradation. Generally, well graded tailings have a low porosity under compression. Minimum porosity of 40% and 45% have been measured for classified and full plant tailings in the settled bed.

SEDIMENTATION

The water contained in a settled tailings mass is much less than that in a suspension, thus solids sedimentation has become one of the major means to remove the water and densify the slurry.

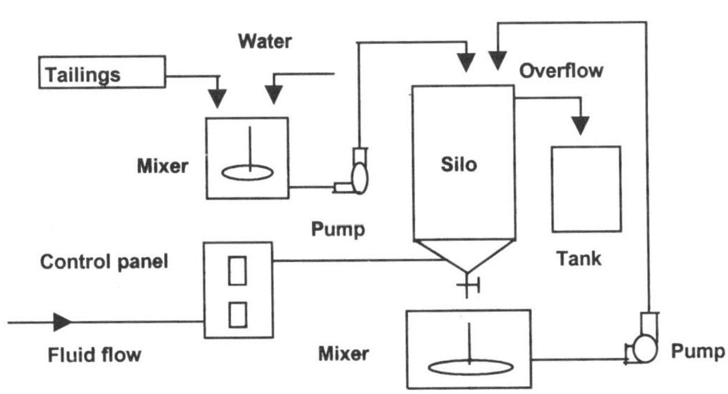

Figure 2. Pilot plant test system

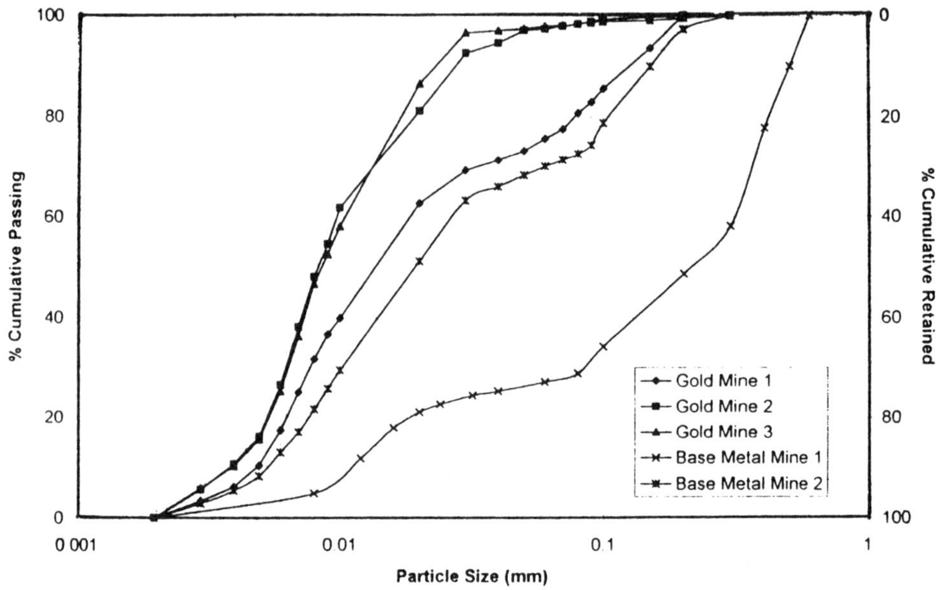

Figure 3. Particle size distribution

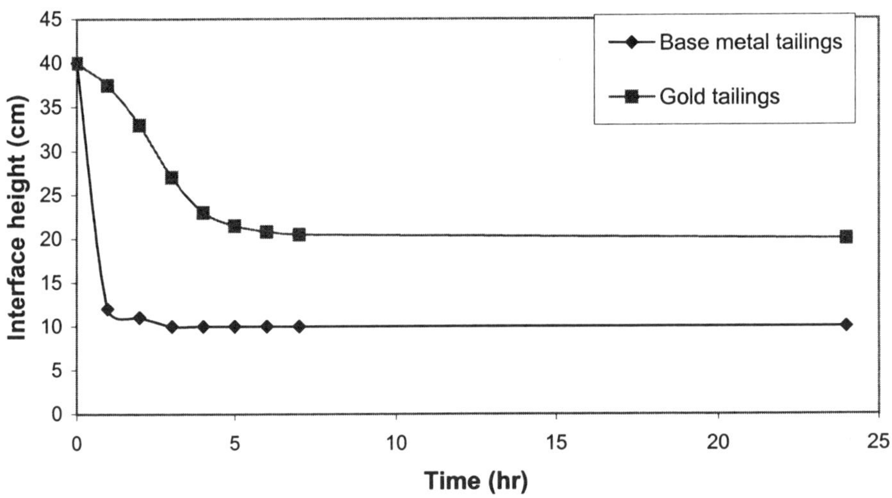

Figure 4. Sedimentation of the tailings

Bench scale tests are applied to define the settling pulp density, settling rate and paste regime of the material. As the solids settle in a dilute slurry a corresponding saturation is measured in the container. The tailings from base metal and gold mines present different settling rates (Figure 4) and settling pulp density due to their particle size distribution and specific gravity. The fine tailings from gold mines settle at a low rate and present a lower settling pulp density compared to base metal tailings. As measured, the settling pulp density for gold and base metal tailings are 68-70% and 76-78% solid respectively for a 12 hours settling period under the same condition[(3)].

Particle size has a profound effect on settling rate , as can be seen in Figure 4. A classified tailings, from a base metal mine for example, can settle within 3 hours to compose a saturation over 75% solids, while tailings from a gold mine must settle for at least 6 hours to produce a 70% solids suspension. It has been found, that the tailings settle at different rate in suspension

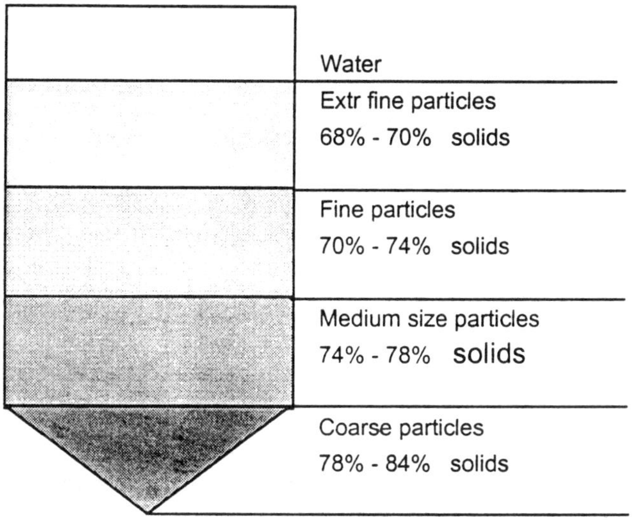

Figure 5. Profile of settled tailings inside the silo

and thus segregation always occurs in the settled bed. A thin layer of fines forms on the top of the saturation and it hinders water from further exudation thus explaining why a total tailings paste has a lower settling pulp density in comparison to a classified tailings mixture. In a silo system, the coarse particles accumulate at the bottom and become compressed, giving a highly densified material. The particle size distributions and pulp densities at different elevations in a settled bed within the silo have been investigated as shown in Figure 5.
A dewatering device has been developed to continuously remove the water from the silo during and after the feeding of the dilute slurry. This installation operates on percolation/decantation basis and allows both the interstitial and supernatant water to be removed as soon as the slurry enters the silo and solids begin to settle. Sedimentation together with extensive dewatering, causes the tailings to become a dense solid mass and there is no water left above the saturated solids in the silo before the fluidization commences.

FLUIDIZATION

Tailings settled in the silo are unable to move under gravity due to their high viscosity and shear stress. To make the tailings mass dischargeable from the silo, a fluidization device has been employed which allows fluid to flow through the voids between the particles making the solids saturation into a paste-like suspension. The fluidization process is strictly controlled in order to effectively produce a paste and so, the parameters are defined by testing a specific material. The fluidization technology and a silo system are evaluated by the following criteria[(1)].

a. High pulp density: The tailings slurry produced must be in the paste regime.
b. Continuous discharge: The tailings slurry must be dischargeable continuously from the silo at a constant flow rate.
c. High silo volume utilization: The silo fluidization system should be able to discharge as much of the slurry as possible. Generally, at least 75 per cent of the total volume of the slurry should be removed from the silo.
d. Uniformity of the slurry: The paste like solids mixture as prepared, should have uniform particle size distribution and pulp density.

TEST RESULTS

Pulp density

The unique feature of the system is that the air fluidizes the settled mass making it into a suspension without any changes in water content. The pulp densities obtained in pilot plant experiments are 72% to 75% solids for gold mine tailings, and 76% to 82% solids for base metal mine tailings which are higher than settling pulp density of these solids. It has been found that the tailings' paste regime can exist in a wide pulp density range, however the pulp density must be controlled strictly to keep the paste movable in the pipeline, since the slurry's flowability decreases dramatically with increasing pulp density. It has been observed that the discharge flow becomes impossible when the pulp density is over the range as mentioned above.

Discharge flow

The tailings paste prepared in the silo is discharged into the mixing/receiver by opening the valve attached to the silo bottom. Due to agitation in the fluidized bed, the particles space themselves increasing porosity, so the viscosity and shear stress are largely reduced which makes the paste flowable under the gravity(2). In addition, the pressure gradient inside the silo assists the flow and therefore a continuous and stable discharge has been observed. The paste is a non-Newtonian fluid and its flow obeys a general rule for a flow in a vertical gravitational system(9). It has been found that if the settled tailings are sufficiently fluidized, the mass flow will dominate the discharge, otherwise a channel flow and a rat-holing could occur(1). The discharge rate is a function of the discharge hole size and height of the fluidized bed and it relates to physical properties of the paste. The material prepared with base metal mine tailings generally has a higher discharge rate than that with gold mine tailings and this is because of the differences in physical properties of the materials.

Silo volume utilization

The silo, along with the fluidization system, should be designed and operated in such a way that not only produces a paste but also allows removal of as much of the paste from the silo as possible. It is ideal if the paste can be totally discharged, however due to the high viscosity of the material, part of the paste will inevitably remain in the silo.

It has been measured that about 75% to 95% of the paste prepared can be removed and this is because:

A. The silo is designed with conical bottom which has a smaller dead zone during discharge compared to a silo with a hemispherical bottom.
B. All tailings are fully fluidized from the bottom to the top of the silo.
C. A mass flow dominates the discharge and rate-holing is automatically avoided.

In a fluidized bed, the entire contents are in motion during discharge. A mass flow allows most of the solids to be discharged smoothly and only a small dead zone could build up near the silo wall. To maximum the silo volume usage, two aspects are considered as important: the solids should be sufficiently fluidized and the discharge system should be designed properly to avoid the dead zone inside the silo(4).

Particle size distribution in the paste

Particle size analysis of the paste is aimed at assessing the fluidization effectiveness in terms of the uniformity of size composition in the fill. If the settled bed is not fully fluidized, the coarse tailings will discharge first from the silo followed by the fines. The changes in size gradation due to segregation could cause problems in transportation and reduce the strength of the entire filled body. So it is ideal if the coarse and fine particles are mixed together and distributed uniformly in the pastefill and this has been realized in the fluidization system. The vertical and lateral movement during circulation suspend and mix the particles together, so a uniform paste in terms

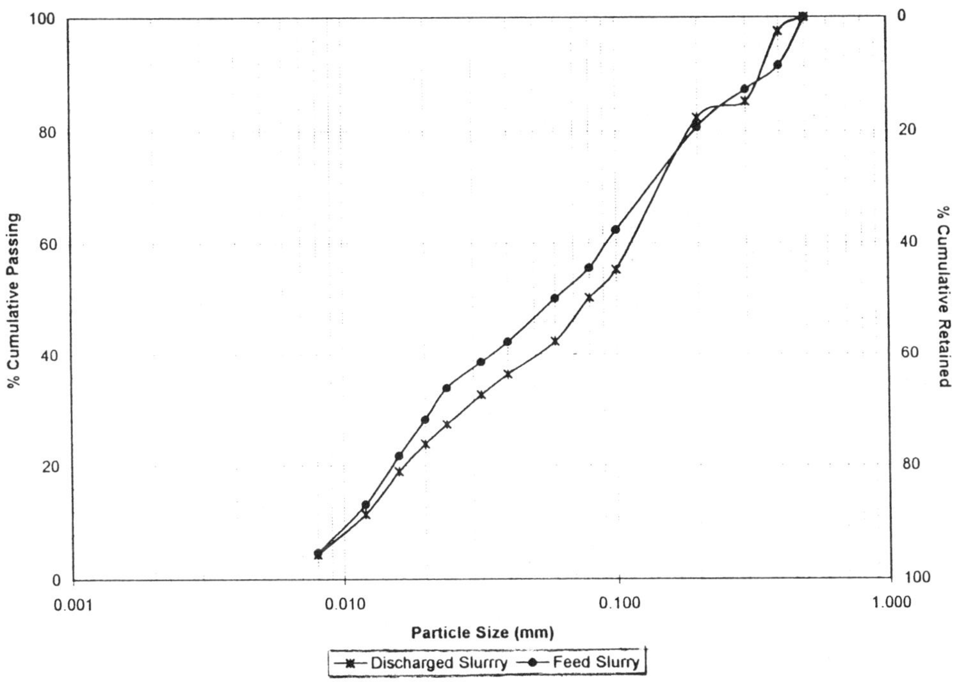

Figure 6. Particle size distribution in the feed and discharge samples

of particle size and pulp density is produced in a fluidized bed. The consistency of the particle size distribution has been confirmed by analyzing the samples taken from the feed and discharge as shown in Figure 6, which indicates that the paste has the same particle size distribution as that in feed slurry from the mill.

FULL-SCALE TRIALS

Early in 1995, CANMET began to actively look for a mining company to help fund the research and development of this technology and later in the year, made an agreement with a mine to provide the mine with paste fill technology on a step by step experimental development basis. In 1996, it was agreed that one of their two existing hydraulic fill tanks would be converted to a fluidization system, so that paste could be poured underground. A full-scale trial was then conducted in the existing backfill plant in September, 1996 at the mine site. The existing backfill plant consisting of two storage tanks equipped with Marcona-jets, a cement silo and a mixer was designed to prepare hydraulic fill with classified mill tailings. The storage tanks are tube-shaped steel structures with flat bottoms and sized 12m high and 12m in diameter. The tailings slurry produced by operating the Marcona-jet is discharged into a mixer by a sand pump. Upon sufficient mixing of the slurry with cement, the hydraulic fill is distributed underground through bore-hole, pipeline systems (Figure 7).

One of the existing tanks was converted by using CANMET's fluidization system to produce paste fill with full plant tailings. An array of nozzles was installed inside the tank to replace the Marcona-jet and a control system was designed to monitor the fluidization parameters. The tank was fed with two dilute tailings streams namely cyclone underflow and slimes and the supernatant above the settled solids was then removed by a dewatering device. A paste fill was produced by flluidizing the settled tailings and it behaved like a non-Newtonian fluid. Pulp density was measured in the range of 78% to 82% solids and the paste fill was delivered at a rate of $100m^3/hr$ to $160m^3/hr$ underground.

Figure 7. Existing backfill plant at test mine

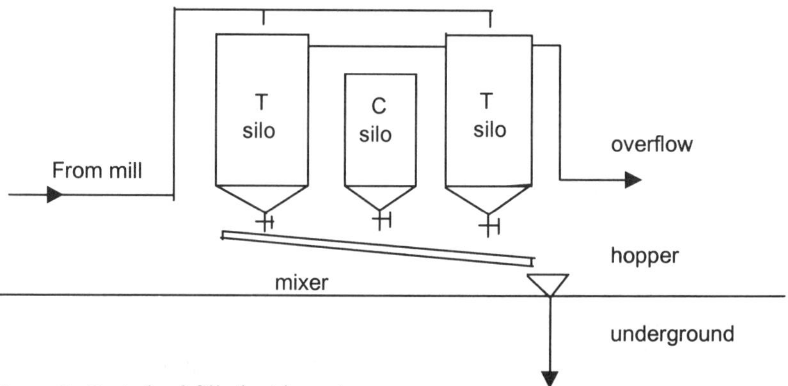

Figure 8. Paste backfill plant layout

PASTE BACKFILL PLANT

A paste backfill plant can be designed by scaling up the model silo and experimental data. Although production rate and operation are site specific, a general plant layout is applied as shown in Figure 8. The storage silo along with the fluidization and dewatering devices are the key components required for receiving and dispensing the hydraulic tailings material in paste fill production. Therefore, it has been decided to use the silo itself as a thickener. Dilute tailings, generally 30% solids is either pre-densified in a tank or sent directly to the silo by pump, where the solids settle and free water is drained off through a dewatering device. A fluidization system composed of non-plugging nozzles allows the fluid to agitate the entire solids mass from the bottom to the top of the silo. All solids are in suspension and mixed together producing a uniform paste material which is then discharged into a screw-feed mixer where the binder (cement) is added. From the mixer, the paste is delivered underground through a pipeline by gravity or positive displacement pump[7].

ADVANTAGES OF THE NEW PASTE BACKFILL SYSTEM

In comparison to the existing technology, this new system has the following advantages:
- simplicity in operation and less equipment
- less capital investment and operating costs
- very few mechanical components meaning less maintenance

- nozzles can be installed from outside silo ensuring continuous silo operation and reducing interruption due to maintenance
- no additional facilities required for the plant if a silo already exists

CONCLUSION

A paste to pulp density of 82% solids with full plant tailings has been produced in a silo equipped with a fluidization device. The dilute tailings poured into the silo is densified by a dewatering device, then the saturated solids mass is fluidized, so a uniform paste has been achieved. The solids fluidization process and silo operation are secured by non-plugging nozzles through which the fluid can be constantly introduced. A conical bottom silo has been successful in paste discharge allowing high utilization of the silo's storage capacity in pastefill preparation. The simplicity of the system is explained by nature of the technology which produces a paste from slurry in a silo system. So less equipment and energy are involved and consumed in paste backfill operation.

ACKNOWLEDGMENTS

The authors would like to thank Mr. R. Powers, Chief Engineer and the engineering staff at Westmin Resources Ltd., Myra Falls Operations for their co-operation in the research and development of the new pastefill technology.

REFERENCES

1. Chen, L .J . 1984. Studies of silo working performance and tailings movement pattern. *Mining Science and Technology*, 2, pp. 33-44.
2. Chen,L. J. 1991. Fluidization technique for mine backfill with slurry of high density. *Mining Science and Technology*, 12, pp. 325-332.
3. Chen, L. J. and Annor., A. 1995. A study on the properties of gold mine tailings fills, *97th Annual General Meeting of C.I.M.*, Halifax, Nova Scotia.
4. Chen, L. J. 1997. Criteria for silo design in backfill operation. *CIM Bulletin*, Vol. 90, pp. 95-99.
5. Davidson, J. F. and Harrison, D. 1963. Fluidized particles. *Cambridge University Press*, G. Britain.
6. Go, M.S. 1963. Fluidization-motion of homogeneous spheroid and fluid in vertical system. *Science Publication*, China.
7. Millette, D.G., Chen, L.J., Annor, A. 1995. Preparation of paste backfill using full stream tailings with the CANMET system. *Paste Backfill Seminar, Research Committee - Mining Techniques and Processing*, Val d'Or, P.Q.
8. Shen, L. H., Zhang, M. G. and Xu, Y. Q., 1995. Solids mixing in fluidized beds, *Powder Technology*, 84, 207-212.
9. Wasp, E. J. 1977. Solid-liquid flow, slurry pipeline transportation. *Trans Tech Publication*, Germany.

Tailings and Mine Waste '99 © 1999 Balkema, Rotterdam, ISBN 90 5809 025 6

Innovative tailings backfill using HSM

Yunxin Qiu, Dave C. Sego & Jozef Szymanski
Department of Civil and Environmental Engineering, University of Alberta, Edmonton, Alb., Canada

ABSTRACT: Cemented tailings backfill for underground disposal is an effective ground control method used widely in underground mining operations. Cement binder is usually mixed in the backfill to enhance its strength and reduce the excess water in the backfill. A new type of binder named High-percent-water Solidifying Material (HSM) is described. Laboratory test results on both paste and hydraulic backfill using both HSM and conventional Portland cement are described. The test results show that HSM is a promising binder for the backfill which can reduce not only the binder consumption, but also the mining cycle time.

1 INTRODUCTION

Backfill in mine operations has been practiced for centuries. The main objective of the backfill is to support mined-out openings to prevent surface subsidence, caving, rock bursting, or excessive earth pressure on the working face. Tremendous volumes and diversity of characteristic of mine tailings are produced worldwide every day. Tailings backfill for underground disposal is an effective ground control method widely used in underground mining operations. To enhance the strength of the backfill and consume excess water within the backfill, cement is commonly used as a binder in the backfill. In Canada, almost all of underground hard rock mines use the backfill, and the backfill costs account for 7% to 29 % of the total mining costs (Canada Mining Journal, 1998). Therefore, the reduction of the cement consumption is a major challenge for cemented backfill especially in the mining operations at remote sites with excessive transportation costs. This paper presents a new type of cementing binder named High-percent-water Solidifying Material (HSM), and its potential applications for both hydraulic backfill and paste backfill.

2 NEW CEMENTING MATERIAL----HSM

In the early 1990s, a new type of cementing material named High-percent-water Solidifying Material (HSM) was developed by the first author in collaboration with his former colleagues at the Beijing Graduate School, China University of Mining and Technology. The new cementing binder, HSM, is a non-combustible powder, and has the following characteristics (Qiu et al, 1992):

(1) Strong water-solidifying capability,
- when HSM is mixed with water along at water content up to 90% by volume, the mixture sets quickly.

(2) Rapid strength development (Figure 1),

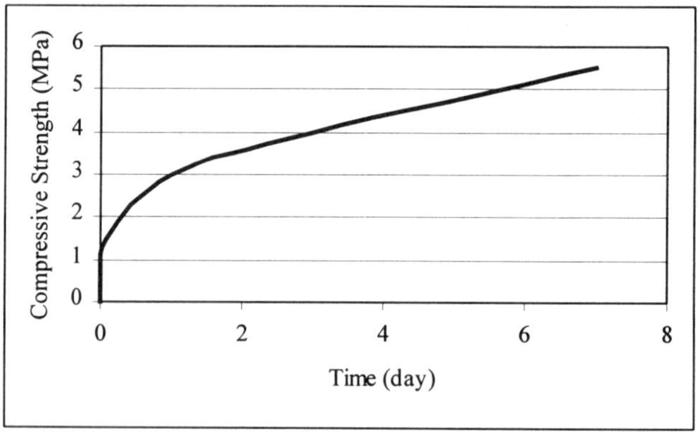

Figure 1. HSM strength development at 90% of volumetric water content

(3) Addition of aggregates or sands is unnecessary,
(4) The backfill has a high residual strength.

HSM has been successfully used in the underground backfilling in China (Qiu et al 1992). The applications of HSM are not only in mine backfill, but also in other applications such as sealing mine openings and firefighting in mines.

3 LABORATORY TEST PROGRAM FOR BACKFILL USING HSM

In order to evaluate the applications of HSM as a cemented tailings backfill, laboratory evaluation has been conducted at the University of Alberta since June 1995.

3.1 *Test program for hydraulic cemented backfill*

The hydraulic cemented backfill is widely used in the mine operations. To assess the feasibility of using HSM with tailings, a sample of INCO tailings was obtained and a series of laboratory tests was carried out.

3.1.1 *Specimen preparation*

Classified mill tailings from Thompson T-1 mine of INCO's Manitoba Division were used in the hydraulic cemented backfill tests. The tailings had a solids content of 60% by weight. For comparison, both HSM and the conventional binder were used as binders in the test program. The conventional binder was supplied by the mine and consisted of 70 per cent of type 10 Portland cement and 30 per cent of flyash. The binders were directly mixed into the tailings at the ratio of 25:1 tailings to binder by weight. Then, the mixture was cast into molds. Finally, the specimens were stored in a moisture room to cure prior to their strength being measured. The specimens had the dimensions of 50 x 50 x 50mm, as shown in Figure 2.

3.1.2 *Compression test equipment and test procedure*

Due to the low load resistance of backfill specimens, a standard soil compression test machine was used to conduct the unconfined compression tests on the specimens. The machine was a 1000 kgf Wykeham Farrance machine.

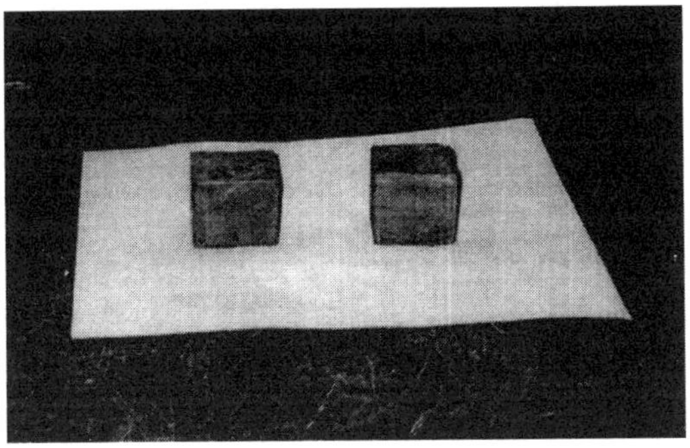

Figure 2. Hydraulic cemented backfill specimens

In order to quantify the water release, the specimens cured for the appropriate length of time were weighed, and their dimensions measured. Then, the specimens were tested at a deformation rate of 0.79mm/min. After testing to failure, the specimens were returned to the moisture room to allow for continue curing. After recuring, the previous failed specimen were tested again to measure their "residual strength" at the previously outlined deformation rate.

3.2 *Test program for paste backfill*

Paste backfill is becoming more popular due to its higher placement solids content and use of the total tailings stream in addition to the much reduced release water from the backfill. This study assessed the feasibility of changing the cementitious additive to develop a new paste backfill with superior characteristics. The Lupin Mine tailings were used in this study. A series of laboratory tests was carried out after September 1997.

The dewatered tailings with a solids content of 78 per cent were used as this matched the mix design used at the mine site. Both HSM and conventional binder, i.e., the type 10 Portland cement were used. To optimize the mix design using the new binder, the binder was added at rates of 1.5, 2.0, 3.0, 4.0 and 5.0% of total tailings by weight for this study. Finally, the compression strengths of the specimens were measured after the samples cured.

In this study, the design strength requirements specified by the mine operations were 280kPa to 350kPa after 7 days with a Portland cement addition at a rate of 2.5%, and 770kPa to 1050kPa after 7 days with a Portland cement addition at a rate of 5%.

3.2.1 *Specimen preparation*

The binders were mixed with the dewatered tailings in a mixer. Then, the slump of the mixture was measured within 20 minutes. Next, the mixture was placed in cylinder molds with dimensions of 76mm diameter x 150mm height. The specimens were stored in the moisture room for curing until they were loaded in compression.

3.2.2 *Test equipment and procedure*

The mixer used in this study was HOBART Mixer and the compression test machine was the Tritest-50 (Figure 3).

Figure 3. Paste backfill specimen compression test

Table 1. Test results for hydraulic cemented backfill

Binder Type	Uniaxial Compression Strength (kPa)										Water Release (%)
	1 day		3 days		7 days		14 days		28 days		
HSM	280	244	413	415	535	496	574	559	651	642	24
Conventional	87	88	135	137	184	190	235	246	292	322	48

Table 2. Peak strength and residual strength of HSM Backfill

Curing Age	14 days	28 days
Intact strength (σ_c) (kPa)	559	642
Residual strength (σ_r) (kPa)	461	417
σ_r / σ_c (%)	83	65

The specimens at the desired curing ages were removed from the moisture room. They were weighed, and their dimensions were measured again. Then, all cylinder specimens were capped with sulfur mortars using ASTM C617-87 capping procedure. After capping, the specimens were placed in the test machine and loaded to failure at a deformation rate of 0.75 mm per minute. The paste backfill specimen in the compression test apparatus is shown in Figure 3.

4 LABORATORY TEST RESULTS

4.1 *Hydraulic cemented backfill test results*

The uniaxial compression strengths after different curing times along with the measured water released during curing are summarized in Table 1. Table 2 presents a comparison between the intact strengths and residual strengths measured for the HSM samples which were allowed to continue curing for 3 month after having been failed. The stress-strain relationships for HSM treated backfill are shown in Figure 4.

4.2 *Paste backfill test results*

The strengths of the paste backfill specimens are presented in Table 3. The 7 day stress-strain curves for the paste backfill are shown in Figure 5.

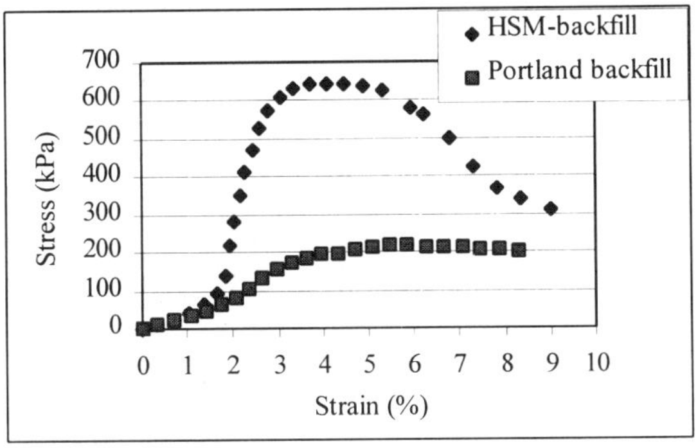

Figure 4. Stress-strain relationship at 28th day for hydraulic backfill

Table 3. Strength of the paste backfill

Cement Content		1.5%	2.0%	3.0%	4.0%	5.0%
HSM Paste Backfill	1 Day Strength	138	306	585	967	1574
		125	295	581	950	1547
		140	310	588	948	1550
		135	298	583	964	1570
	7 Day Strength	176	382	733	1021	1648
		165	330	718	948	1647
		231	386	763	1153	1703
		222	379	737	1107	1695
Portland Cement Paste Backfill	1 Day Strength	90	119	323	310	424
		71	103	272	275	383
		64	116	210	245	427
		67	71	297	237	361
	7 Day Strength	241	324	566	756	999
		176	314	557	725	916
		206	341	557	748	954
		209	298	524	729	925

5 RESULTS AND DISCUSSIONS

Table 1 and Figure 6 show that the strengths of both the HSM backfill and modified Portland cement backfill increase with curing time. However, the strengths of the HSM backfill develops more rapidly than that of the conventional binder. The strengths of the HSM backfill are two times the conventional backfill at the same age. Meanwhile, one-day strengths of the HSM backfill nearly reach the 28-day strengths of the conventional backfill. This indicates that with the same curing time and strength requirement, the ratio of the binder to tailings can be reduced by using HSM instead of the conventional binder. The cement consumption will be dramatically reduced in the backfill operation at existing mines. On the other hand, with the same ratio, the HSM backfill can reach the desired strength much earlier than the conventional backfill, thus, it can reduce the mining cycle time.

The excess water drained from the backfill after placement and as it cures must be pumped. The test results (Table 1) show that the excess water from the HSM backfill is only half of that in the conventional backfill. Therefore, the pumping cost can be reduced by using HSM instead of the conventional binder in a hydraulic backfill operation.

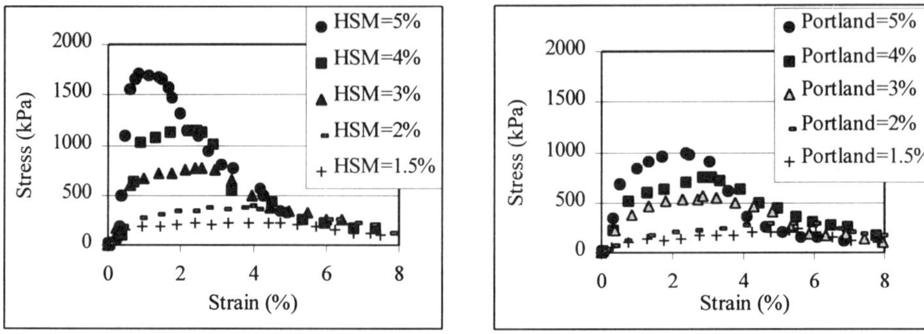

(a) HSM paste backfill

(b) Portland cement paste backfill

Figure 5. Stress-strain cures after 7 day curing of paste backfill

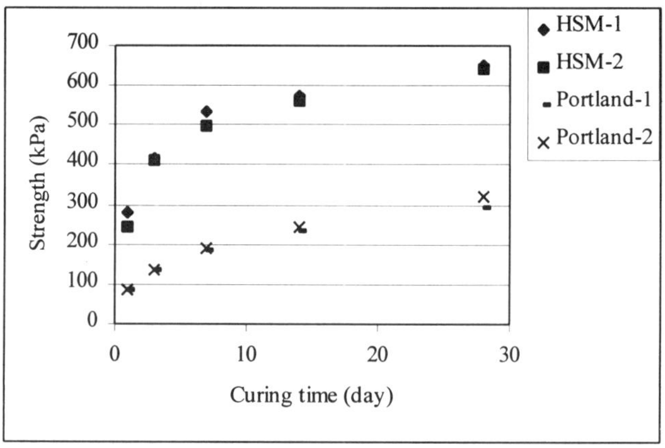

Figure 6. Strength development for the hydraulic backfill

Table 2 shows that the HSM backfill also has high residual strength. This is due to the re-crystallization ability of HSM. The test results reveal that the backfill specimens first tested to failure on the 14th day can redevelop as much as 83 per cent of this strength. This feature is very useful in the ground control operations.

The paste test results presented in Table 3 also show that the HSM paste backfill has higher early strength than Portland cement backfill. As mentioned before, the designed strength requirements for the paste backfill after 7 days are 280kPa to 350kPa with a Portland cement addition at a ratio of 2.5% and 770kPa to 1050kPa with a Portland cement addition at a ratio of 5%. Based on this design requirement, the test results (Figure 7) show that the HSM backfill prepared with addition rates of 1.75% and 3.4% of HSM can meet the low strength (280kPa to 350kPa) and high strength (770kPa to 1050kPa) requirements, respectively. Therefore, the optimized mix design for the HSM paste backfill is 1.75% of HSM in low strength zone and 3.4% of HSM in high strength zone.

Compared to the present mix design for Portland cement paste backfill, i.e. 2.5% of type 10 Portland cement for the low strength and 5% of type 10 Portland cement for the high strength, the optimized mix design for the HSM paste backfill can reduce the binder consumption by as

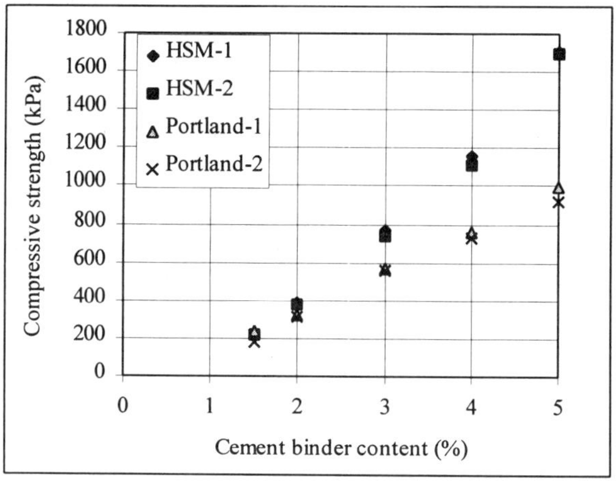

Figure 7. Strength versus binder content relationship after seven day curing

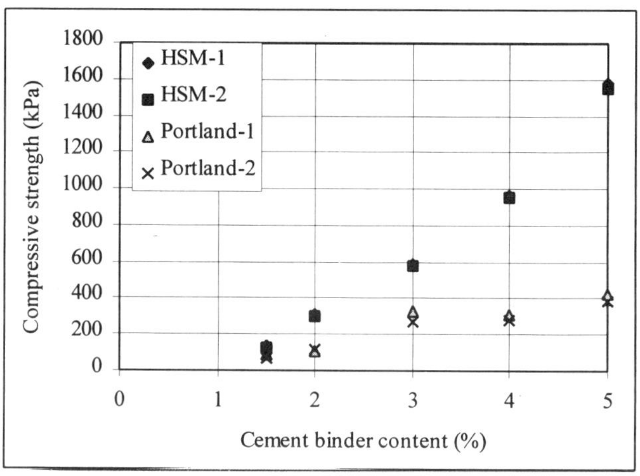

Figure 8. Strength versus binder content relationship after one day curing

much as 30%. This is significant, especially for the mining operation at remote sites where transportation costs can be significant.

Figure 8 shows the relationship between binder content and backfill strength after one day of curing. The plot illustrates that the HSM paste backfill with 2% and 3.75% of HSM after one day of curing can meet the strength requirements of 7 days. Comparing to the present mix design for the Portland cement paste fill, the HSM paste fill can reduce not only the mining cycle time by 6 days, but also the binder consumption by 20% to 25%.

Figures 4 and 5 show that both HSM and conventional binder backfill have the strain softening characteristics as these backfills are subjected to strains of greater than 1.5%.

6 CONCLUSIONS

HSM is a promising binder for use in both hydraulic cemented and paste backfill.

For the hydraulic cemented backfill, the HSM backfill has two times of the early strength of the conventional binder backfill. Using HSM as a binder in the backfill can reduce not only the mining cycle time, but also the binder consumption. In addition, HSM backfill can reduce the release water pumping cost dramatically.

For the paste backfill, compared to the Portland cement paste backfill, using HSM as the cementing binder can reduce binder consumption by 30% while developing the required 7 day design strength. Furthermore, with the HSM, the mining cycle time can be reduced by 6 days and the binder consumption can be decreased by 20% to 25%.

ACKNOWLEDGEMENT

The authors would like to thank the Natural Sciences and Engineering Research Council of Canada (NSERC) for providing the financial support for this research.

REFERENCES

Canada Mining Journal, 1998. 1998 Mining sourcebook, pp33-34.

Qiu, Y., Sun, H. and Yin, J. 1992. A study on the High-water Solidifying Material for gob-along roadside instead of coal pillars. *Proceedings of 11th International Conference on Ground Control in Mining*. The University of Wollongong, Wollongong, Australia, July 1992. PP371-375.

Geotechnical considerations

Tailings and Mine Waste'99 © 1999 Balkema, Rotterdam, ISBN 90 5809 025 6

Investigation of flow failures by means of a flume model

G. Papageorgiou, A. B. Fourie & G. E. Blight
University of the Witwatersrand, Johannesburg, South Africa

ABSTRACT: The paper presents a current investigation into flow slides with the aid of flume models. Runout distances and profiles for different surface conditions, materials, densities and height of model are presented. An evaluation of the factors affecting the run out profile of the model flow slides is carried out. Observations of the flow behaviour and development of the model flow slide was made with the aid of video recordings. The master profile principle for hydraulic beach profiles was applied to the flume model runout profiles. A comparison between the model flow slides and the Merriespruit flow slide is performed.

1 INTRODUCTION

On the night of 22 February 1994, the Merriespruit gold tailings dam in South Africa failed, resulting in a flow slide. In the wake of the disaster 17 people died, 80 houses were demolished and 200 severely damaged. The events leading too and the cause of the flow slide have been dealt with extensively by Wagner et al (1997 & 1998).

Following the inquiry into the flow slide, it became apparent that very little is understood about the effect on flow of; surface conditions, trigger mechanisms, initial dam height, breach width and soil properties and in particular on their runout distance and lateral spread. A research project was initiated to investigate flow slides with the aid of model tailings constructed in a flume.

Attempts to predict the extent of flow slides are not new. Blight et al (1985) developed a numerical model for the runout distance following the failure of the Bafokeng platinum tailings dam. Lucia (1981) and Hungr (1995) among others, have also proposed numerical models to simulate flow slides. Due to flow slides being rare occurrences, with little or no prior warning, very few flow slides have actually being witnessed. During the investigation of flow failures, the emphasis is usually on establishing blame and, pertinent data required for flow failure numerical modelling and future analyses or investigations is usually not obtained. In all cases, the pre-failure state and condition of the material involved in the flow slide is not known. The modelling of flow slide by means of flume modelling is therefore an ideal means of reproducing numerous flow slides under controlled conditions and documenting their behaviour.

2 FLUME MODELLING OF FLOW SLIDES

It is acknowledged that simulated flow slides in a flume model differ from the prototypes in various aspects. Scaling has a significant influence on modelled flow slides. In the prototype the energy (potential, kinetic and momentum) is the overriding factor whereas in the model, viscosity and capillary forces play a larger role during failure. The state and condition of the material involved in the prototype cannot be duplicated exactly by the model. Flume modelling of flow slides provides a basis upon which numerical models can be evaluated and gives an insight into the complexities involved in flow slides.

2.1 *Objectives of flume modelling of flow slides*

The present flume modelling aims to provide a qualitative evaluation of flow slides in terms of their flow behaviour, runout distance and profiles relative to various conditions ie ground slope and surface condition onto which the flow slide occurs, the height of the tailings dam, obstacles in the path of the runout material and the state and condition of the material prior to failure.

For the Merriespruit flow slide, a comparison of the runout profiles of the prototype and flume models can be made.

2.2 *Description of flume testing procedure*

Two flumes have been used in modelling the flow slides, the "small" and "big" flumes. Figure 1 and 2 show schematics of the respective features and dimensions.

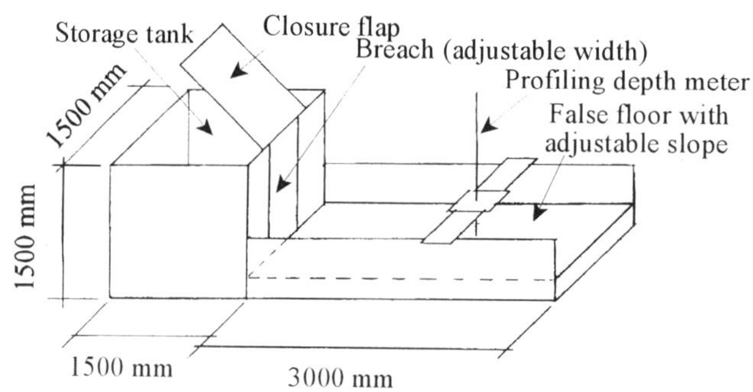

Figure 1 Big flume

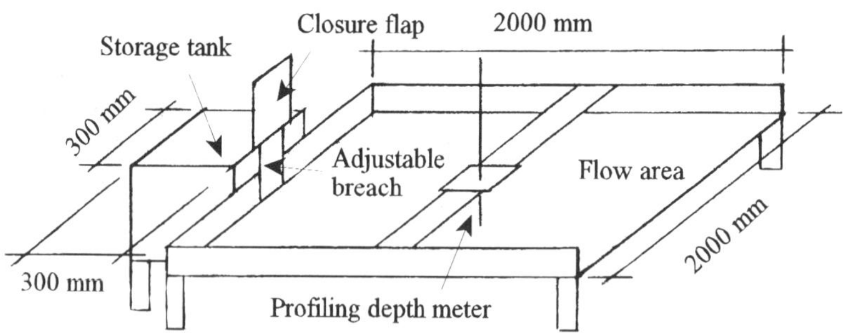

Figure 2 Small flume

Tailings slurry is initially mixed to a predetermined density in an industrial mixer. Using a rigid container of known volume a sample of the slurry is taken to determine its density. Water or air dried tailings is added to the batch mixture as necessary to obtain the required density. The slurry is placed into sealed plastic buckets and stored until the required amount of slurry has been produced. The buckets of slurry are re-mixed for a short time interval and poured into the storage section of the flume with the closure flap closed. During pouring of the slurry into the storage area, care is taken to reduce the drop height of the slurry to a minimum. At various time intervals during placement the slurry within the storage section of the flume is stirred so as to insure that a homogeneous mixture is obtained.

Upon completion of the placement of slurry, it is allowed to settle and consolidate with one-dimensional drainage. Once fully consolidated, the free surface water is removed, moisture content samples are taken and maximum shear strength determined, by means of a shear vane. The downstream conditions are set eg wet or dry surface, model trees or obstacles and the inclination of the ground surface. The breach width is set prior to the placement of the slurry.

The retaining flap is opened, and the material is allowed to flow. A video recording of the entire test is made for later examination and reference. With the aid of the trolley system spanning the width and length of the flume and a modified dip stick measuring device the runout profile of the flow failure can be measured.

3 MATERIAL USED IN FLUME TESTS

To date eight flume tests have been conducted; four small and four big. The material used in the small flume tests was gold tailings, referred to as gold 1 in figure 3. The material used in the big flume tests originates from the Merriespruit gold tailings dam (adjacent to the existing failure scar). The first two tests where performed with the material referred to as gold 2 and tests three and four with the material referred to as gold 3 in figure 3. The change in the Merriespruit gold tailings grade was as a result of requiring additional material and thus a second batch of sample material from the dam, which was of a finer grade. Figure 3 also shows the range of particle size distribution of material encountered at the Merriespruit tailings dam.

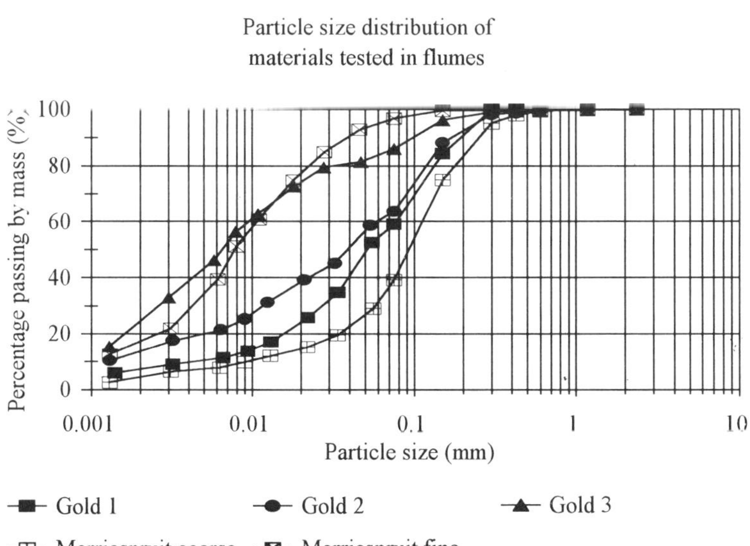

Figure 3 Particle size distribution of materials used in flume tests

4 FLUME TESTS

Table 1 tabulates the different flume tests performed to date.

Table 1 Flume tests

Tests	Flume	Material	w at placement (%)	Placement density (kg/m^3)	Height of slurry (mm)	Breach width (mm)	Downstream conditions
1	Small	Gold 1	55	1700	210	150	Level & wet
2	Small	Gold 1	85	1530	256	150	Level & wet
3	Small	Gold 1	85	1530	225	150	Level, wet & with trees
4	Small	Gold 1	40	1810	150	150	Level & wet
5	Big	Gold 2	40	1810	190	150	Level & wet
6	Big	Gold 2	40	1810	200	150	Level & wet
7	Big	Gold 3	40	1810	435	120	Level & wet
8	Big	Gold 3	40	1810	430	120	Level & dry

- Note that w is moisture content on a dry mass basis

Tests 1, 2 and 4 investigated the effect that placement moisture content/density has on the runout. Tests 2 and 3 investigated the effect downstream obstacles eg trees or houses has on the runout distance. For test 3, trees were simulated by 6 inch nails inserted into the flow board area, with steel wool for foliage. The scaling of the trees relative to the storage area was done by eye. The trees extended the entire width of the flow board, starting at 250 mm downstream of the face of the breach and extending to 550 mm. The positioning of the "trees" was random but densely concentrated.

Tests 5 to 8 form part of the investigation into the effects that initial height, ground surface conditions ie sloping, wet or dry and placement density have on the runout. As tests are currently being performed, only these initial tests results are currently available. Tests 5 and 6 tested the reproducibility of the tests. Tests 5, 6 and 7 investigated the effect an increase in dam height and a finer material has on the flow failure. Tests 7 and 8 investigated the effect ground surface conditions have on the runout length of flow slides.

5 SMALL FLUME TEST RESULTS

Table 2 tabulates various parameters measured for the resulting flume flow slides

Although pre-failure moisture contents for test 1 and 2 were the same, their runout lengths were different. The only variable which changed was the placement density/moisture content of the slurry, (refer to table 1). At this time it not understood why the failures behaved so differently, considering their pre-failure conditions were almost identical. It may be due to the higher moisture content at which test 2 was placed, so that the slurry segregated to a greater extent than test 1.

Settlement tests performed at different slurry water contents showed that a 40% moisture content slurry mix showed no segregation and a homogenous, uniform distribution throughout the consolidated slurry height. An 80% moisture content slurry showed a high degree of segregation with coarser particles at the bottom of the cylinder and a gradual transition to fine particles at the top This results in the coarser material at the bottom establishing a lower void ratio (higher density) than the finer material at the top, with a resultent variation in the void ratio/density profile of the two tests prior to failure. Upon breaching, the finer, looser material at the top of test 2 would thus have flowed further than that in test one.

The above is substantiated to some extent by the fact that test 4 which did not flow, was placed at a slurry moisture content of 40%, less than test 1. The pre-failure moisture content of the consolidated slurry was 32% compared to 33% for test 1, and a difference of 40 kg/m^3 in their pre failure density. Yet for such a small differences in conditions, test 4 did not flow. It is suggested that the placement moisture content/density of test 4 prevented segregation from occurring during consolidation. This produced a homogeneous material of uniform void ratio/density relative to test 1,2 & 3. Thus although the overall pre-failure moisture contents and densities for the four tests were similar, their density/void ratio profiles with depth were different. This resulted in different flow failure behaviour and runout distances.

The longitudinal profiles of tests 2 and 3 differ significantly from test 1 (refer to figure 4). The linear shape of the longitudinal slope of test1 and the concave shape of tests 2 and 3 suggest that these tests experienced greater segregation at failure than test 1. The steeper slope of tests 2 and 3 close to the breach indicate the deposition of coarser material with finer material being deposited further downstream at a flatter slope. Test 1, being more of a homogeneous material prior to failure would not show as much segregation during flow failure, thus developing a more constant profile. The overall steeper angle of the runout developed in test 1 compared to the flatter angle for test 2 and 3, may also result due to the viscosity/density of the runout material being different. This is reflected by the flow behaviour or rate of the flow slides, refer to table 2. The above points reinforce the idea that the pre-failure state of the material in the tests were different. In an attempt to verify this, a similar test is currently being performed in the big flume. Attempts are also being made to measure the void ratio/density and particle distribution of the consolidated slurries at different heights.

Comparing the longitudinal profiles of test 2 and 3 reveals the effect downstream obstacles have on the runout length of a flow slide, (figure 4). The longitudinal profile for the two tests at distance 0 to 550 mm, reveals that test 3 developed a steeper slope than test 2, whereas the slope for the two tests from distance 550 mm onwards is virtually the same. The trees which where placed at distances of 250 to 550 mm therefore produced a backwater effect to the flow slide of test 3, resulting in a steeper slope over this distance. This variation in slope at the start and end of the test 3 results in its longitudinal profile being more concave than test 2.

Table 2 Flume flow slide results

Tests	Runout length (mm)	Pre-failure material height (mm)	Pre-failure w (%)	Runout slope angle (°)	Pre-failure density (kg/m^3)	Flow speed and behaviour
1	540	170	33.9	5.4	1910	Rapid
2	1400	177	34.0	1.3	1910	Very rapid
3	1100	150	34.6	2.0	1900	Very rapid
4	0	140	33.0	NA	1940	None
5	1200	190	33.9	4.0	1880	Rapid
6	800	200	35.0	6.8	1870	Moderate, viscous
7	1450	395	38.2	7.6	1830	Moderate, viscous
8	1100	391	39.0	7.6	1825	Moderate, viscous

From the above tests and the big flume test discussed below it appears that the overall pre-failure condition of the tailings in terms of density and moisture content does not necessarily give an adequate description for the purpose of modelling (refer to table 2). This is supported by Spence & Guymer (1997) which state that *"A complete description of the variation of flow slide surface slope and runout distance with each experimental parameter could not be quantified because of difficulties with test sample preparation"*, although pre-failure measurements of density for tests showed a high degree of similarity.

6 BIG FLUME TEST RESULTS

Table 2 tabulates the various parameters measured for the resulting flume flow slides

Tests 5 and 6 were carried out to investigate the repeatability of tests. Although the pre-failure conditions for the two tests were almost identical, their runout distances, resting angle of runout and flow behaviour were quite different (refer to table 2). Examination of the video recordings revealed that test 5 displayed rapid flow upon initiating the failure and a fluid type behaviour. To the contrary, test 6 displayed a very gradual flow upon initiating the failure and a more viscous type behaviour. The duration of the two failures also differed with test 6 experiencing flow for 2 to 3 times longer than test 5. The steeper resting angle, of the runout material for test 6 to test 5, refer to figure 5, further supports the observation that test 6's flow behaviour was more viscous than test 5 and that their pre-failure conditions could not have been similar. On the basis of these two tests it appears that repeatability of tests is difficult to achieve. These tests further reinforce the argument put forward for the small flume tests that although the pre-failure conditions are similar, this does not necessarily indicate a similar state of the material (void ratio/density, material structure etc).

Test 7 examined the effect that a finer grade of material as well as an increase in consolidated material height has on the runout profile. Test 7 was therefore compared to both tests 5 and 6. In comparing tests 5 and 7, test 7 contained a finer grade of material and should therefore have displayed a relatively slower and more viscous flow behaviour, although both pre-failure conditions were similar. With this in mind test 7 still resulted in a longer runout distance than test 5. In addition, due to the viscous flow behaviour of test 7 the final runout angle was steeper (refer to figure 5).

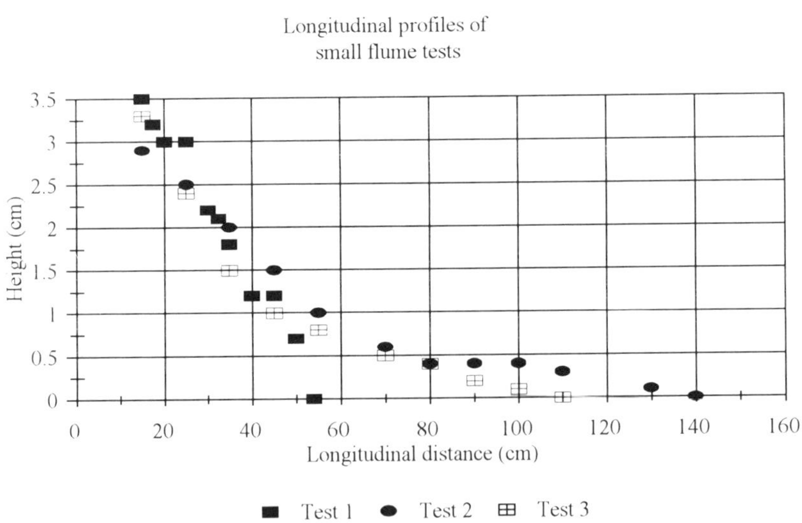

Figure 4 Longitudinal profiles of small flume tests

In terms of flow behaviour tests 7 and 6 displayed a greater similarity than test 5. Tests 7 and 6 displayed a slow viscous type of flow. This suggests to some extent that the pre-failure conditions could be similar for these tests, although test 7 was for a finer grade material. The outcome of these tests however again reveals that the runout distance of test 7 was greater than test 6. A surprising outcome however are the similar final runout angles of tests 6 and 7, which are steeper than for test 5. This reinforces the idea that tests 6 and 7's pre-failure conditions were similar but different to test 5, resulting in different flow behaviour and final resting angle of the runout material. Based on test 7 therefore an increase in storage material height produces a greater runout distance.

Test 8 investigated the effect the surface condition has on the runout distance. Blight (1997) has suggested that the runout distance of flow slides is affected by the surface conditions onto which they flow. He suggested that a wet surface results in a reduction in the shear strength at the interface of the surface and the runout material. It also results in the material directly intact with the ground surface becoming less viscous than the overlying material due to the additional water.

Comparing tests 7 and 8 definitely supports this idea. Although the pre-failure condition and the flow behaviour during failure for the tests were similar, test 8 produced a shorter runout distance. The final resting angles for the two tests was however identical. Surface conditions would therefore appear to have a significant effect on the runout distance of flow slides.

Comparing tests 8 and 6 (or even test 5) reveals that the runout distance of test 8 was greater than test 6 (slightly less than test 5). This suggests that height of storage material may a greater dominating factor to the run out distance than surface conditions. However one must bear in mind that the material used in test 5 was of a coarser grade than test 8. It is therefore planned to perform at test of a similar height to test 5and 6 using the finer grade material and a dry surface to firther investigate this point.

7 COMPARISON OF SMALL AND BIG FLUME TEST RESULTS

A comparison of test results obtained from the small and big flume is difficult. The main problem is the relatively small volume of material stored in the small flume as compared to the big flume. This resulted in the flow slides in the small flume terminating as a result of a depletion of available material to flow as compared to the big flume where the flow failure terminated as a result equilibrium conditions being attained.

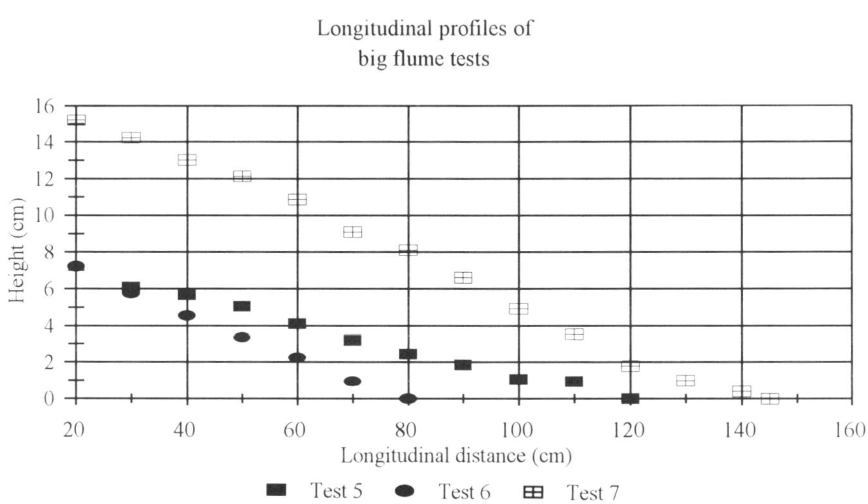

Figure 5 Longitudinal profiles of big flume tests

8 MASTER PROFILING OF FLOW SLIDES

The concept of a master profile for hydraulic fill beach profiles has been successfully applied by Blight (1994) and Fourie (1988) among others. This technique may therefore be applicable to the runout profiles of flow slides. Blight (1994) describes the master profiling technique and have presented profiles for different tailings and slurry densities. This technique was applied to the flume tests. Figures4 and 5 show the actual longitudinal profiles of the runout material for the flume tests. Figure 6 shows the longitudinal profiles of the runout material for the flume tests in dimension less height and length. With the exception of the tests 2 and 3 which were placed at a lower density, the profiles develop some unique "master" profile. Blight (1994) and Fourie (1988) have shown that the master profile changes as the slurry density is altered. This could thus explain the change in the master profile for tests 2 and 3 to some extent.

From this limited set of tests it seems the master profile can be applied to flow slides and the technique can be developed to estimate the runout distance of flow slides. It is however recognised that further tests and greater variation of conditions is required.

Comparison of the big flume tests and the Merriespruit flow slide were also made. Figures 5 and 6 show the runout profiles for the flume tests and the Merriespruit flow slide. Figure 7 shows the dimensionless profiles for the flume tests and the Merriespruit flow slide. In comparing the prototype and model runout profiles it should be noted that the prototype profile was obtained by aerial photogrammetry. It could also not be ascertained over what time period and at what time after the actual flow slide event the profiling was made. It should also be stated that surveying the profile of the runout material of a flow slide is not easy due to the loose and fluid like nature of the material.

Refering to figure 5 and 6 the resting angle of the runout material in the Merriespruit flow slide is much flatter than that obtained in the flume tests. It is believed that this is due to the viscosity, interface shear forces and capillary forces in the flume model being the controlling parameters compared to the potential and kinetic energy in the prototype. As a result of the scaling effect of the flume a direct correlation between flume and prototype flow slides cannot be made.

Refering to figure 7 there is a remarkable agreement in the master profiles of the prototype and flume flow slides. There is thus, some promise in applying the master profile technique to flow slides. As mentioned previously though, a more detailed investigation is required to substantiate this statement.

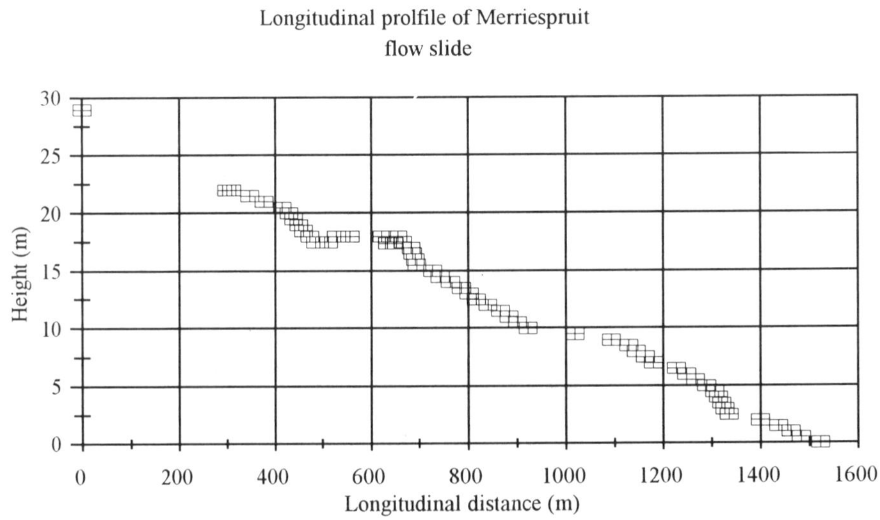

Figure 6 Longitudinal profile of Merriespruit flow slide

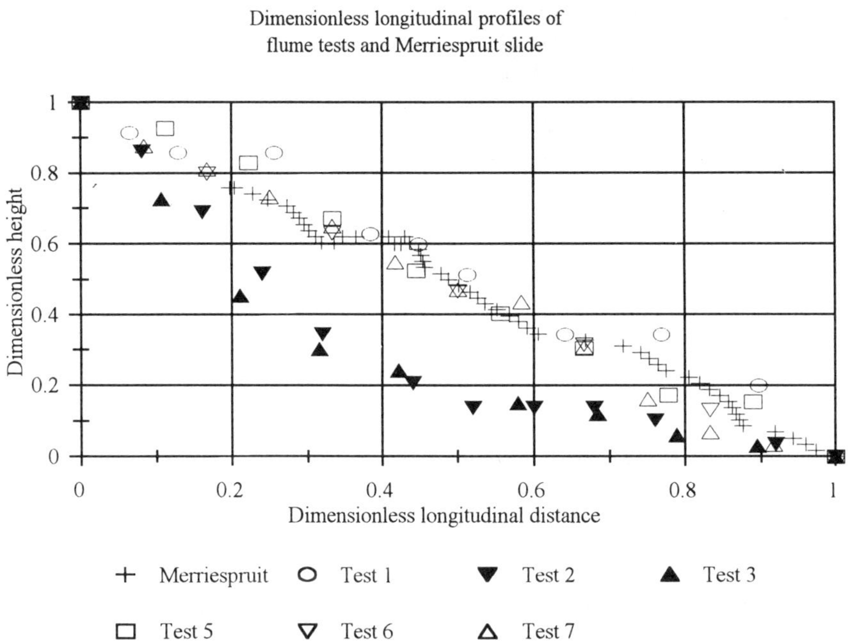

Figure 7 Dimensionless longitudinal profiles of flume tests and Merriespruit flow slide

9 TENTATIVE CONCLUSIONS

- The pre-failure moisture content and density of tailings slurry used in flume models does not necessarily give a good representation of the condition of the material
- Repeatability of the flume tests is extremely difficult. A better placement technique is therefore required to produce pre-failure material states/conditions that are similar.
- An increase in the storage material height increases the runout distance of a flow slide.
- The surface condition of the ground onto which the flow slide occurs affects the runout distance, a wet surface producing a greater runout distance than a dry one.
- The height of the storage appears to be have a more dominating effect on the runout distance than the surface condition.
- The flow behaviour during a flow slide is extremely complex, and goes through various phases during the entire flow slide.
- A direct correlation between flume models and prototype flow slides does not appear to exist.
- The application of "master" profiles to the runout profile of flow slides appears promising.
- Flume models provide an easy and convenient means of obtained a better understanding of the nature and behaviour of flow slides as well as which parameters affect flow slides and their runout distances.

10 REFERENCES

Blight G E (1997). Destructive mudflows as a consequence of tailings dyke failure. *Proceedings of the Institute of Civil Engineers, Geotechnical Engineering.* Vol 125 January : 9-18

Blight G E (1994). The master profile for hydraulic fill tailings beaches. *Proceedings of the Institute of Civil Engineers (London).* Vol 107 : 27-40.

Blight G E, Robinson M J & Diering J A C (1981). The flow of slurry from a breached tailings dam. *Journal of the South African Institute of Mining and Metallurgy*. January : 1-8

Blight G E, Thomson RR & Voster K (1985). Profiles of hydraulic-fill tailings beaches, and seepage through hydraulically sorted tailings. *Journal of the South African Institute of Mining and Metallurgy*. Vol 85 No 5 May : 157-161

Fourie A B (1988). Beaching and permeability properties of tailings. DJA van Zyl and SG Vick (eds). *Hydraulic Fill Structures, Geotechnical Special Publication No 21, 1988*. 142-154

Hungr O (1995). A model for the runout analysis of rapid flow slides, debris flow, and avalanches. *Canadian Geotechnical Journal*. Vol 32 : 610-623

Lucia P C, Duncan J M & Seed H B (1981). Summary of research on the case histories of flow failures of mine tailings impoundments. *Mine Waste Disposal Technology, Proceedings : Bureau of Mine Technology Transfer Workshop*, Denver, Colorado : 46-53

Spence K J and Guymer I (1997). Small-scale laboratory flowslides. *Geotechnique*. Vol 47 No 5 : 915-932

Wagner F, Howard C, Strydom K and Blight G (1997). The tailings dam flow failure at Merriespruit, South Africa. *Tailings and Mine Waste 1997* : 657-666

11 ACKNOWLEDGEMENTS

The above work could not have being performed without the financial assistance of the Foundation of Research and Development and the Merriespruit Fraser Alexander Tailings Bursary. Metago Environmental Engineers are thanked for allowing the use of the small flume flow slide data.

Tailings and Mine Waste '99 © 1999 Balkema, Rotterdam, ISBN 90 5809 025 6

Erosion-resistant crusting of slopes of fly ash dams

A. B. Fourie & G. E. Blight
University of the Witwatersrand, Johannesburg, South Africa

ABSTRACT: In South Africa, power station fly ash is disposed of in one of two ways; by dry dumping or by hydraulic deposition into ash dams. It has sometimes been observed that a marked hardening of the hydraulically deposited fly ash occurs, producing a very erosion-resistant surface. This paper discusses the characteristics of fly ash that enhance this hardening behaviour and whether atmospheric environmental factors influence the degree of hardening. Also considered is whether it is only a surface effect or whether the ash retains this added strength upon subsequent burial.

Laboratory specimens from two ash dams were prepared for testing by curing cylindrical specimens of ash in either a pure nitrogen or pure carbon dioxide environment The curing atmosphere had a significant effect on the strength gain of one of the ashes, but virtually no effect on the other ash. Dynamic cone penetrometer tests carried out on the slopes of the two ash dams confirmed the laboratory results.

1 INTRODUCTION

Thermal power stations contribute about 29000 MW to the South African national electricity grid at present. A typical power station burns 12 million tonnes of coal per year and the estimated mass of the resulting ash for all South African stations is about 21 million tonnes per year. Very little of this ash is used or is usable industrially and the vast majority (± 95%) must be disposed of on land. Until about 1984, all of the Electricity Supply Commission's (Eskom's) ash disposal facilities were dams into which ash was placed hydraulically (at a relatively high moisture content).

The vast majority of the power stations use coal from a dedicated (and adjacent) coal mine. The coal used by the different power stations is thus different, and the variation of geotechnical properties of different fly ash is the prime consideration of this paper.

2 POWER STATIONS INVESTIGATED AND BASIC GEOTECHNICAL PROPERTIES

Fly ash from two hydraulic fill ash dams (Duvha and Matla) was studied. Samples of ash (about 50kg from each station) were collected directly from the precipitators.

The particle size distribution curves for the two ashes are shown in Figure 1 and the results of basic geotechnical tests summarised in Table 1. Also shown on Figure 1 are grading curves for fly ash obtained from two dry ash dumps (designated as Lethabo and Kendal). The Duvha ash is seen to be significantly coarser than the other three, which are all very similar and indeed are similar to much of the data previously published in the literature (Heystee et al, 1991). The numbers in brackets after the specific gravity results for the Duvha and Matla samples are tests that were done on ash previously subjected to oedometer tests. There was no observed change in the specific gravity of the Matla specimens, but quite a large increase in the values for Duvha ash. The results are generally consistent with those reported by Brackley et al (1987), although they found higher values for the coefficient of consolidation (400m^2/year). The friction angles are all essentially the same and the cohesion intercept was consistently zero. As shown later in the paper, however, the development of significant cohesion is possibly due to the process of carbonation.

The primary difference between the Duvha ash and the other three ashes, aside from its coarser grading, is a greater compressibility. The value for the Duvha ash is up to ten times greater than that of the Matla ash, which may have some bearing on the storage capacity of the two impoundments, ie the Duvha ash will undergo greater settlement under a particular load, thus providing additional storage capacity. Aside from this difference, the basic geotechnical properties of fly ash appear to be well understood and well defined; it is the chemical properties and their significance that are as yet not fully appreciated, as discussed in the remainder of this paper.

3 BASIC CHEMICAL PROPERTIES

Chemical constituents were determined by X-ray fluorescence. The results for the two samples are summarised in Table 2. Once again the results are all fairly similar, although the iron content of

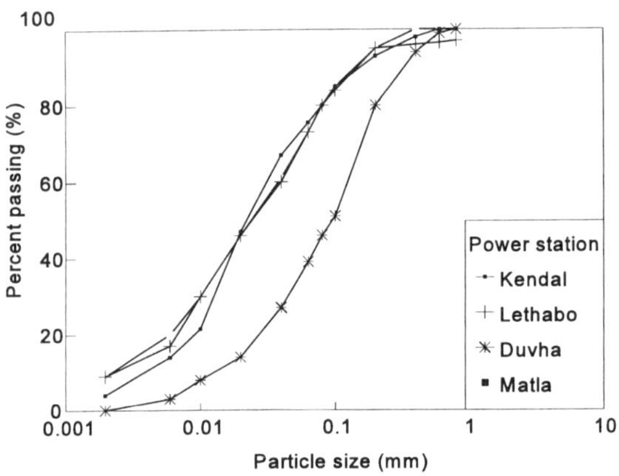

Figure 1: Particle size distribution curves for fly ash samples

Table 1: Basic geotechnical properties of four fly ashes

	Kendal	Lethabo	Duvha	Matla
Specific gravity	2.12	2.17	2.21 (2.24-2.79)	2.26 (2.26)
Compression index, Cc	0.25	0.13	0.34-0.4	0.02-0.16
Coefficient of consolidation c_v (m^2/year)	20-50	20-50		
d_{50} (μm)	22	22	100	30
Angle of friction (°)	36	34	35	39
Apparent cohesion (kPa)	0	0	0	0

Table 2: Basic chemical properties of fly ashes

Major elements (%)	Duvha	Matla
SiO_2	45.26	46.18
TiO_2	1.61	1.67
Al_2O_3	28.89	29.64
Fe_2O_3	8.98	5.2
MgO	1.84	2.5
CaO	8.35	9.75
P_2O_5	1.53	1.25
L.O.I.	2.25	1.38

Table 3: Chemical properties reported for fly ash from Duvha and Matla power stations (after Willis, 1987)

Mass (%)	SiO_2	Al_2O_3	Fe_2O_3	CaO	Quartz	Glass
Duvha	53	36	7.1	2.4	13	35
Matla	45	27	3.7	9.7	5.3	67.8

Duvha ash is high. Similar tests were carried out by Willis (1987) on samples of fly ash from Duvha and Matla power stations. His results are summarised in Table 3, where only selected parameters are tabulated.

These results are all for tests on precipitator ash and are the average of four replicate tests. The results are generally similar to those in Table 2, although for the Duvha ash the values of SiO_2 and Al_2O_3 have increased while the lime content has decreased quite substantially. An interesting observation is that in the Matla ash the quartz is present as glass in a much greater percentage than is the case for Duvha ash.

The likelihood of an ash exhibiting pozzolanic hardening properties has been correlated with the amount of (SiO_2+Al_2O_3+Fe_2O_3) present in the ash (ASTM Standard C 618-84). This standard suggests that if:

(SiO_2+Al_2O_3+Fe_2O_3) > 70% by mass of the sample

and LOI < 6% and SO_3 < 5%,

then the ash is likely to exhibit pozzolanic hardening. These requirements are satisfied by both the ashes tested and pozzolanic hardening would appear to be a possibility for these samples. However, as reported by Aitcin et al (1985), the above criteria are overly simplistic and account also needs to be taken of the amount of lime and sulphate present in the ash, since these two components enhance pozzolanic hardening. The presence of available water is another prerequisite for pozzolanic hardening.

4 STRENGTH GAIN DUE TO HARDENING

At sites where ash is disposed of hydraulically it has often been noticed that overtopping of the retaining embankment does not necessarily lead to the formation of erosion gullies, which would be the case for most conventional tailings dams. There thus appears to be the formation of a competent, erosion-resistant outer layer covering the surfaces of an ash dam. It is this observation that led to the investigation described in the remainder of this paper.

At a visit to Matla power station, a vertical section through the ash had been exposed in a trench. The side of the trench was sprayed with phenol phthalein, from the surface to a depth of 1m. Below about 10cm the ash changed to a pinkish colour, indicating carbonation was occurring above this depth. To check the possible strength gain due to carbonation, a series of laboratory strength tests was carried out, as described below.

Two air-tight 50ℓ polyethylene drums were equipped with inlet and outlet fittings through which gas could be circulated at a constant rate. One drum was continually flushed with nitrogen and the other with carbon dioxide. Specimens of Matla and Duvha ash were prepared as slurries and poured into cylindrical moulds 38mmϕ by 76mm long, and allowed to drain and consolidate. The moulds were removed and the specimens placed on trays suspended within the drums. They were then exposed to either a nitrogen or CO_2 atmosphere for different periods of time. Specimens were removed after 7, 14 and 28 days and subjected to unconfined, unconsolidated compression tests. The results are summarised in Figure 2 and each result is the average of two replicate tests. The Matla ash developed unconfined strengths that were more than three times greater than the Duvha ash specimens. These observations are consistent with the relatively greater amount of lime present in the Matla ash, which as mentioned earlier, is thought to enhance carbonation hardening by conversion of the calcium hydroxide content of the ash to calcium carbonate. Although there is a data point missing for the 28-day exposure Duvha specimen, it can be concluded that the increase in shear strength as a consequence of aging the specimens in a nitrogen environment is effectively zero for both Matla and Duvha specimens. Curing the specimens in a carbon dioxide environment, however, has a much more significant impact. The Matla ash undergoes an approximate two-fold increase in unconfined strength. There thus appear to be (at least) two factors contributing to the phenomenon of carbonation hardening. A greater amount of lime, as occurs in the Matla ash, provides some strength gain, which appears to be enhanced by exposure to a carbon dioxide environment. The reason that the Duvha ash experiences a decrease in strength from 7 to 28 days after an initial increase is more difficult to explain, particularly as the 28-day strength has decreased to a value almost equal to that of the ash exposed to nitrogen. This means that the process of carbonation has had no lasting effect on the strength for the Duvha ash, even though the calcium hydroxide content is relatively high.

When considering the implications of these findings it is important to remember that the Duvha ash is unrepresentative of typical ash (refer to Figure 1). The results of the Matla ash tests are therefore probably of more interest as they are more representative of a 'typical' ash. Since curing

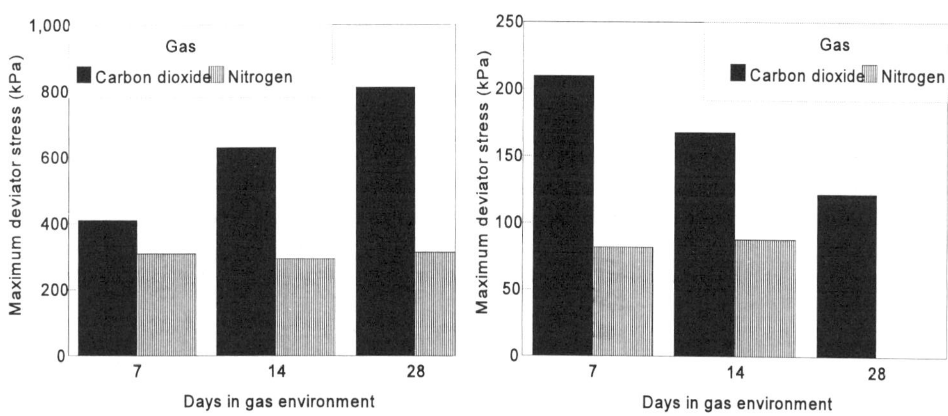

(a) (b)

Figure 2. Unconfined compressive strengths of (a) Matla and (b) Duvha specimens

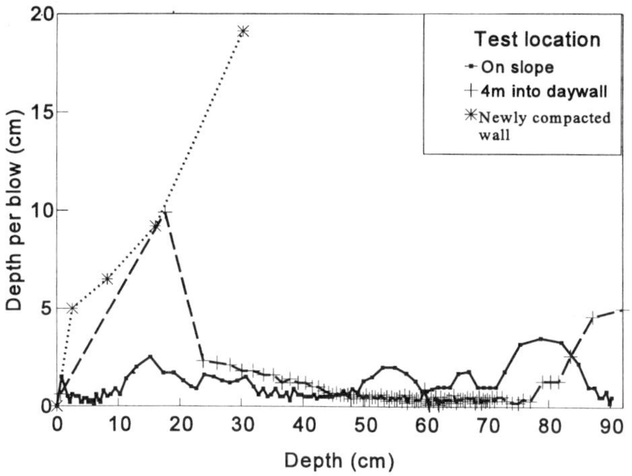

Figure 3: Penetration resistance of Matla fly ash.

took place under zero effective confining stress, the measured increase in strength due to hardening is likely representative (in terms of ambient stress conditions) of what occurs on the surface of hydraulically placed ash that is exposed to the atmosphere. It is clearly not representative of air quality, since the specimens were cured under a 100% CO_2 environment. The importance of knowing whether this effect may be relied upon, is that the erosion resistance (and thus the stability of the ash dams during overtopping) may be significantly increased by carbonation hardening.

5 FIELD MEASUREMENTS OF STRENGTH VARIATIONS

In order to determine whether the laboratory observed strength gains (particularly for the Matla ash) were realistic or whether they were primarily due to the saturated carbon dioxide atmosphere in which specimens were cured, field tests were carried out at both Matla and Duvha ash dams. The tests utilised a conventional dynamic cone penetrometer and the resistance profile with depth was determined at a number of locations on the outer slope of the dams, as well as within the inner, daywall paddock and on top of newly-constructed daywalls.

Results for Matla are summarised in Figure 3. The newly placed ash, despite having been compacted, has a very low resistance to penetration. The result illustrated for the slope of the dam was typical of eight separate tests and indeed at the base of the slope it was impossible to penetrate to a depth of more than 30cm below the surface. For the test on the slope, the penetration per blow was generally of the order of 0.5 to 1cm per blow. The penetration resistance of the ash 4m inside the daywall was also generally high, although there is evidence of a layering effect, eg at depths of 20cm and again at 90cm below surface, the resistance to penetration decreases.

Nevertheless, there is clear evidence that Matla ash left exposed to the atmosphere for a period of time undergoes a significant gain in strength and confirms the previously mentioned observation of an erosion-resistant slope. It is also important to note that the high strength ash is not restricted to the surface, but extends to a depth of at least 90cm and probably more. This is attributable to the method of deposition of the ash, whereby relatively thin layers of ash are exposed to the atmosphere and allowed to drain before placement of subsequent layers. The fly ash is thus exposed to carbon dioxide (which is present in air, albeit in a low concentration), in the presence of substantial free water. Subsequent burial by newly deposited ash does not affect this strength gain (at least for shallow depths).

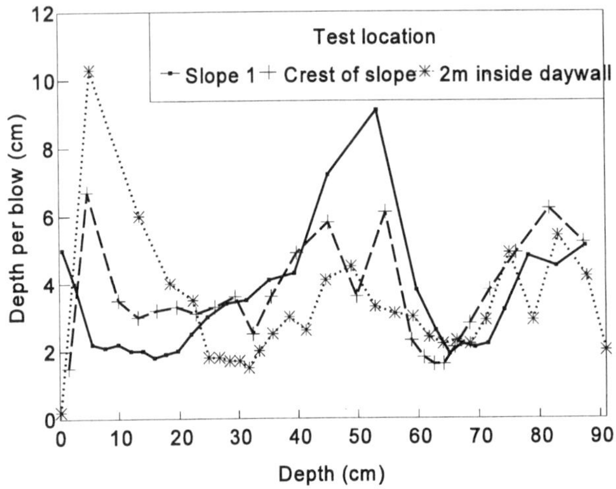

Figure 4: Penetration resistance of Duvha fly ash.

The results for the Duvha ash dam are shown in Figure 4. The resistance to penetration is generally much less than the Matla ash, with the greatest resistance corresponding to a depth per blow of 2cm. These results show even more pronounced evidence of layer formation, with all three tests showing very weak ash close to the surface, and at depths of about 50cm and 80cm below the surface. There is also a very good correlation between the depths at which the weaker and stronger layers occur at the three test locations, with the separation between two stronger (or two weaker) layers of about ½m corresponding to the thickness of an individual layer of hydraulically placed ash. Although there is thus some evidence of strength gain in the Duvha ash, it is not as significant as in the Matla ash and, as has been found on site, it is insufficient to prevent erosion of the slope of the ash dam, should localised overtopping occur.

Only two fly ash dams have been investigated in this study and it is thus not possible to generalise the findings. Previous work on South African fly ash dams by Stuart et al (1997) reported similarly erosion-resistant outer slopes to those encountered at the Matla ash dam. However, it does not at present appear possible to predict with a high degree of certainty how a particular fly ash will increase in strength with time. It does not appear to be only a matter of pozzolanic hardening, which is usually a fairly slow process, but to include some form of chemical hardening (termed carbonation in this paper) which requires the presence of carbon dioxide to facilitate the necessary reactions.

6 CONCLUSIONS

Laboratory tests on two fly ashes from hydraulic fill ash dams showed relatively little variation in basic geotechnical properties, despite differences in grain size distribution. However, in-situ the two dams were found to have significantly different strength properties, with the fly ash from Matla dam forming an erosion-resistant slope whereas the Duvha ash showed relatively little gain in strength. Laboratory tests on specimens of the two fly ashes showed that curing the ash in the presence of carbon dioxide increased the unconfined undrained shear strength of the Matla ash considerably, whilst having relatively little effect on the Duvha ash. Dynamic cone penetration tests in the field confirmed this general observation, with the Matla ash exhibiting extremely high penetration resistance and the Duvha ash relatively low resistance. In both cases, the penetration resistance was not constant with depth and indeed there seemed to be the development of alternating weak and strong layers. The thickness of these layers correlated quite well with the deposition thickness used at the dam, lending further credence to the conclusion that exposure to atmospheric carbon dioxide increased the degree of strength gain due to carbonation.

7 ACKNOWLEDGEMENTS

The financial assistance provided by the South African Electricity Supply Commission, Eskom, is gratefully acknowledged.

8 REFERENCES

Aitcin, P.C., Autefage, F., Carles-Gibergues, A. and Vaquier, A. (1985). Comparative study of the cementitious properties of different fly ashes. *Compositional Analysis*. Wiley and Sons, New York.

Brackley, I.J.A., Barker, M.J. and Smith, E. (1987). The stability of ash dumps and ash dams. *Proc International Conference on Mining and Industrial Waste Management*, Johannesburg 1987, pp 175-184.

Heystee, R.J., Johnston H.M. and Chan H.T. (1991). Coal fly ash properties and ash landfill design. *Proc. 1st Canadian Conference on Environmental Geotechnics*, Montreal, Canada, 65-71.

Stuart, R.J., Dorman, S.A., Geldenhuis, S.J.J. and Timm, R. (1997). Hydraulic ash disposal at Kriel thermal power station and its impact on the ground water regime. *Proc. 4th International Conference on Tailings and Mine Waste*, Fort Collins, USA, 1997, pp405-415

Willis, J.P. (1987). Variations in the composition of South African fly ash. *Proc. Conference on Ash - A valuable resource*, Pretoria 1987, 12pp.

Tailings and Mine Waste '99 © 1999 Balkema, Rotterdam, ISBN 90 5809 025 6

Improving the erosional stability of tailings dam slopes

G.E. Blight & F. Amponsah-Da Costa
Witwatersrand University, Johannesburg, South Africa

ABSTRACT: Preventing and repairing damage caused by the erosion of tailings dam slopes is costly and ongoing, but essential for protecting the surrounding environment from pollution by wind and water borne fugitive tailings. After identifying the major factors affecting tailings slope erosion, the concept of the "erosion rate surface" is introduced. The paper proceeds to examine the characteristics of erosion-resistant natural slopes with the object of re-creating natural erosion resistance on tailings slopes. The mechanics of wind erosion are examined and directions for future research described.

1. INTRODUCTION

In their quest to design ever more economically, geotechnical engineers tend usually to design the steepest slopes consistent with shear stability. It may be forgotten, however, that the surfaces of steep slopes can erode excessively and that the long-term costs of erosion repair may exceed the savings of using a steep slope in the first place. Rates of erosion from unprotected tailings slopes can be frighteningly large. Losses from unprotected slopes of gold tailings in South Africa of over 500 tons/ha/year (200 tons/acre/year) are quite common, while vegetated slopes can lose up to 200 tons/ha/year (Blight, 1989).

This is particularly the case when designing tailings dams, and even more particularly, gold tailings dams. The slopes of gold tailings dams in South Africa are traditionally built at 33°, and if this is judged too steep for shear safety, the slope is flattened by introducing berms or stepping back to reduce the overall slope angle. However, the 33° angle is retained between beams. As 33° is just about the pessimum angle for erosion, this results in severe and ongoing erosion problems, and makes it very difficult to rehabilitate the slopes without ongoing maintenance and repair. All sorts of excuses are made for adhering to this bad practice, none of which are valid when a long-term view is taken of the necessity for ongoing maintenance of the tailings dam after decommissioning. There is, however, a major problem: No one knows precisely how to design an erosion-free or even an erosion-resistant slope.

When considering the erosion of tailings slopes, both wind and water erosion need to be considered. Wind erosion is a particular problem in arid climates where it is difficult to establish vegetation on tailings, and water erosion is more of a problem in humid climates where vegetation can more easily be established. The presence of vegetation largely prevents wind erosion, but does less to prevent erosion by water. In semi-arid regions with well-defined wet and dry seasons, wind erosion is a problem in the dry season and water erosion in the wet season. Thus the effects of both agencies of erosion need to be considered.

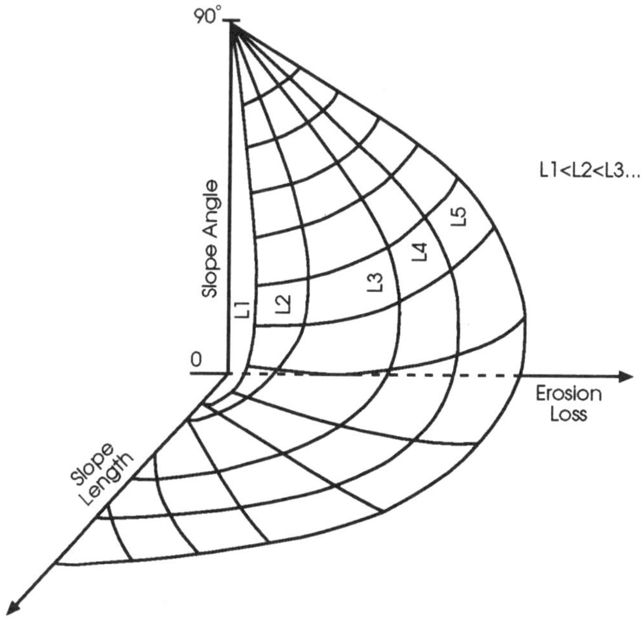

FIGURE 1 The erosion rate surface

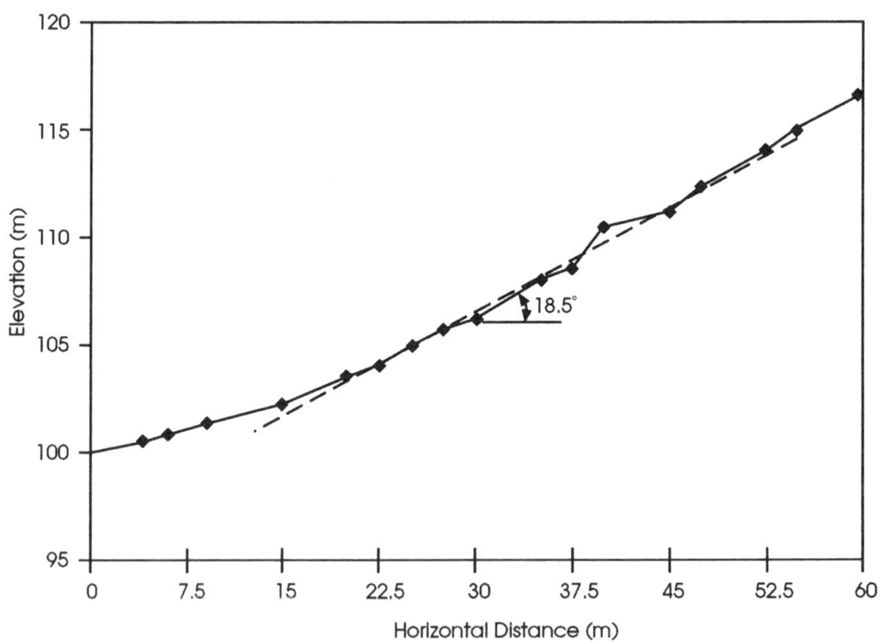

FIGURE 2 Profile of a natural erosion-resistant slope

2. FACTORS AFFECTING THE EROSION OF SLOPES

An equation, known as the Universal Soil Loss Equation (or USLE) (e.g. Evans and Kalkanis, 1977) has been used for many years to predict erosion losses from agricultural fields. The USLE identifies the most important parameters affecting erosion as:

- shear strength of the slope surface,
- slope length, and
- slope angle.

These have been confirmed as key parameters of erosion by observations and measurements of the erosion of the slopes of gold tailings dams in South Africa (Blight, 1989, 1991). These observations have shown that an "erosion rate surface" exists in a (slope angle-slope length-erosion loss) space, as shown in Figure 1. This surface shows that erosion rates increase with slope length, but are low both at very flat and very steep slope angles. The "belly" of the erosion rate "sail" at intermediate slope angles represents the range of slope angles usually used for tailings dams, which thus usually have the worst possible slopes for erosion losses.

The larger the surface strength of a slope, the lower the rate of erosion. Hence the effect of a varying surface strength would be like the wind on a spinaker sail, it would move the erosion rate surface in or out in the direction of the erosion loss axis. Alternatively, a separate erosion rate surface would apply for each value of surface shear strength. Protecting the surface of a slope, by armouring it or covering it with vegetation, has a similar effect to increasing the surface shear strength, and therefore moves the "sail" back towards the "mast".

Research has shown that obstructions that roughen a slope surface, such as a discontinuous layer of stone chips or rock fragments, also help to reduce erosion losses very considerably by dissipating the energy of water or wind moving over the slope (e.g. Roy and Jarrett, 1988). These are all pointers towards attaining an erosion-free slope.

3. THE NATURAL EROSION-RESISTANT SLOPE

When in doubt about the best way to achieve a geotechnical objective, it is often useful to study analogous situations of processes in nature (e.g. Smith et al, 1997). What are the characteristics of naturally erosion-resistant slopes? Figure 2 shows the profile of a natural slope of transported soil that has been derived from the weathering of greenstones (see Brink, 1979). The slope surface is covered by vegetation of various types (mainly grass and bushes) and is scattered with gravel and boulders. It appears to be very stable, with evidence that little erosion is occurring. Measurements of erosion on this slope (made by observing changes of the ground surface level against the tops of a series of steel pegs driven into the slope) showed on average accretion of less than 1 mm per year (13 tons/ha/year). This was produced by soil, eroded off the upper reaches of the slope, being deposited lower down. The slope surface is thus probably as stable as is possible under natural conditions. Other similar natural slopes in other climatic regions of South Africa have shown similar erosion/accretion figures.

What are the characteristics of this slope that render it erosion-resistant?

It is notable that the angle of the slope is 18.5°, not excessive, but also not unreasonably flat. (Incidentally, 18° is usually taken as the steepest angle on which agricultural machinery can work safely.) Figure 3 shows the variation of surface shear strength of the exposed soil along the slope profile. This has an average of about 27kPa. (The surface shear strength was measured by means of a Soiltest "Torvane"). To obtain a standard moisture condition for the shear strength measurements, the surface was doused with water, and when this had drained away, three shear

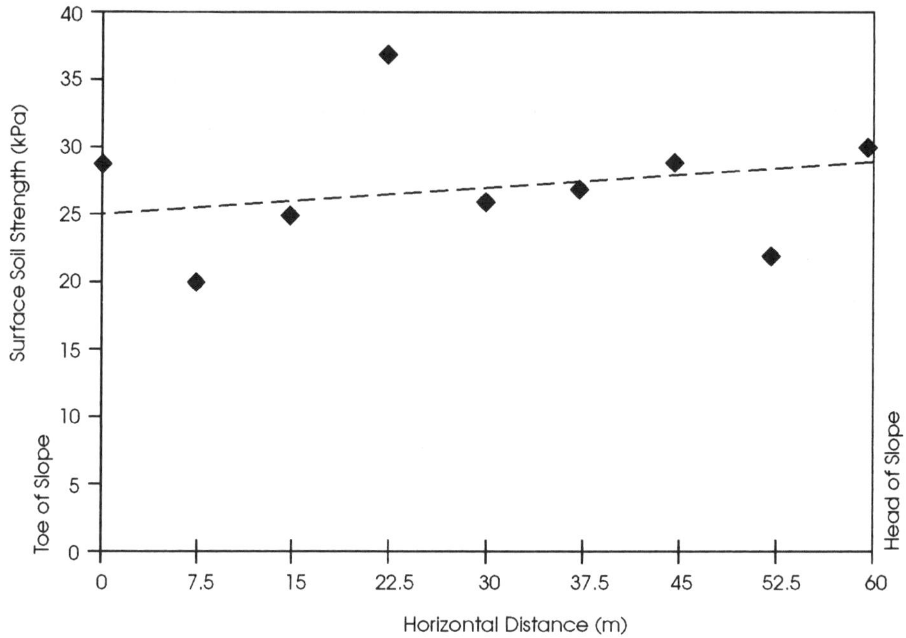

FIGURE 3 Surface soil strength of natural erosion-resistant slope of Figure 2

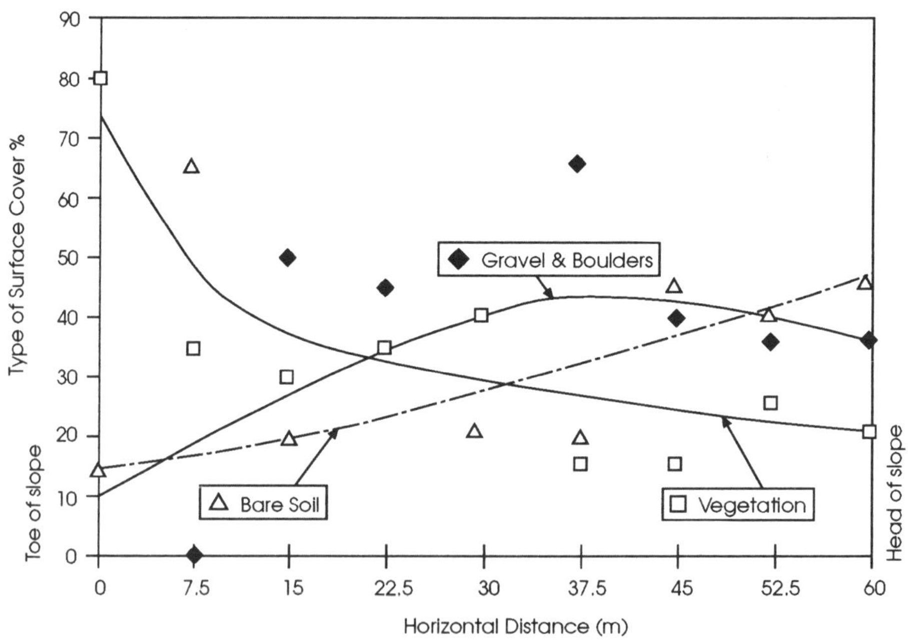

FIGURE 4 Surface cover of natural erosion-resistant slope of Figure 2

strength measurements were taken at each position and the average result reported. For a slope of unprotected gold tailings with a similar profile, length and particle size distribution to that of this soil slope, and in the same climatic region, one would expect an annual erosion loss of around 500 tons/ha/year (Blight, 1989). Hence the vegetation, and particularly the rock and gravel cover appear to be providing considerable and effective erosion resistance.

Figure 4 shows the surface cover of the slope, recorded as percentages of gravel and boulders, vegetation and bare soil surfaces. These were measured by tossing a square wooden frame measuring 0.5m x 0.5m at random onto the surface. The percentages of the three types of surface enclosed by the frame were then estimated independently by two observers who then compared and reconciled their observations. This observational technique has been checked by photographing the area enclosed by the frame and measuring the component areas accurately. Visual estimation and accurate area measurement agreed to the nearest 10%.
The percentage of vegetation tends to increase from top to bottom of the slope while the percentage of gravel and boulder cover decreases, as does the percentage of bare soil.

The measurements described above have been, and are being repeated on a number of other residual soil slopes originating from quartzites, sandstones and shales. Rates of erosion are also in the process of measurement as well as investigations of the shape of the slope profile and the components (soil, vegetation and rock) of the surface cover. In this way is hoped to identify the essential characteristics of an erosion-resistant slope.

4. EXPERIMENT TO TEST ALTERNATIVE PROTECTIONS FOR TAILINGS SLOPE SURFACES.

For the past 50 years, the slopes of gold tailings dams in South Africa have been protected against erosion by vegetation. The vegetation is seeded and established under irrigation, and initially flourishes. But because of the semi-arid climate in the gold-mining area and the steep 33° slopes on which it is growing, the vegetation inevitably declines in vigour once irrigation is stopped. After five years or so, most of the vegetation has usually been lost as a result of drought, fires in the dry season and erosion.

In recent years, the cost of establishing grass on slopes has escalated to the point where the mining industry is at last willing to consider alternative cheaper methods of slope protection. For this reason it was decided to undertake a large scale experiment into non-vegetative methods of protecting tailings slopes against erosion (Blight and Smith 1996). The experimental site is a south-facing slope with a slope angle of 16° over the lower two thirds of the slope length of 20m and 28° over the upper one third.. the slope has been divided into 11 panels each measuring 20m long (upslope) by 10m wide. Each panel is isolated from its neighbours by means of 0.5m high metal sheets partly dug into the tailings surface to form a low vertical wall. Each panel was originally equipped with a sprinkler irrigation system to simulate rain, 3 rain gauges and a set of 10 surface level pegs. The toe of each panel terminates in a catchment paddock to capture and hold solids removed from the slope by water erosion. Simulated rain was used to obtain initial results, but for the past 4 years the slopes have been exposed to natural weather. Figure 5 shows a typical plan of part of the experiment. The various surface treatments have been summarized in Table 1, which also summarises the results of the experiments to date.

Table 1 records the type of surface protection and its relative cost, actual erosion rates in tons/ha/year, relative erosion rates and a cost-effectiveness number represented by the product of relative cost (C) and relative erosion (E). It should be noted that the erosion rates were measured differently for phases 1 to 2 and 3 of the experiment. For phases 1 and 2 the material eroded off each panel was caught in the catchment paddocks at the toe of the slope, and this material was collected and weighed. During phase 3, an unusually heavy rainstorm caused tailings to be washed onto the test slopes from above, this rendering the origin of the mass of caught material

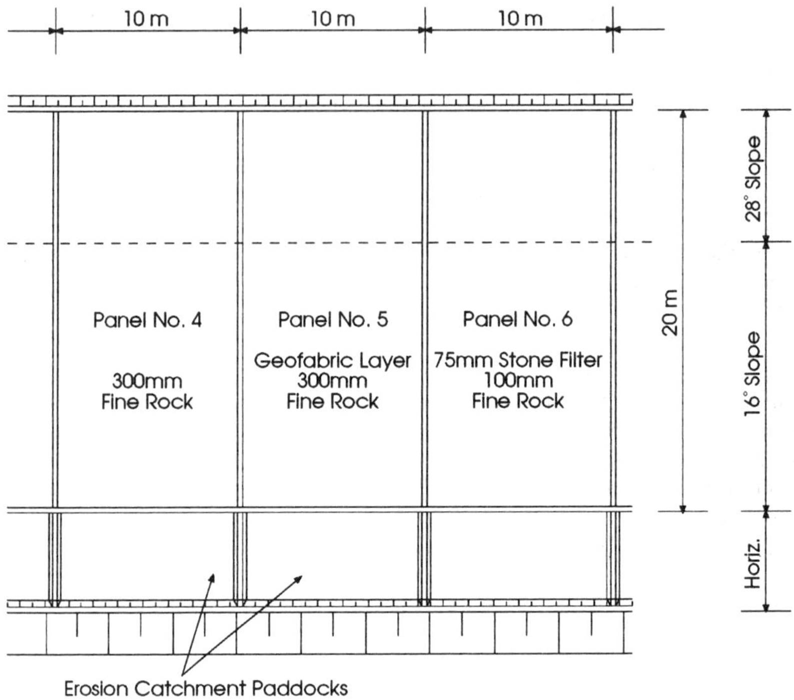

FIGURE 5 Part-layout of large-scale erosion experiment

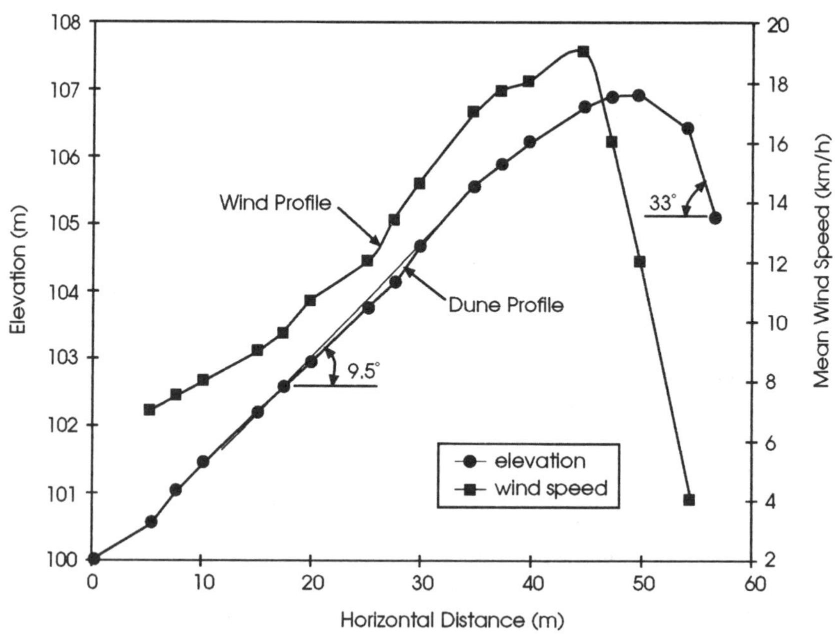

FIGURE 6 Physical and wind profiles for sand dune

questionable. For this reason, erosion for phase 3 was assessed by measuring the retreat of the slope surface against the surface level pegs. With hind - sight, this is, in any case a more reliable form of measurement, and also gives an indication of the distribution of the erosion loss from the slope. Unfortunately, the surface level pegs were removed from the conventionally grassed panel (Panel 1) when the grass was planted and there is thus no measurement for phase 3 for this panel.

On the basis of cost-effectiveness for phase 2, conventional grassing ranks just above no treatment at all, and for both phases 2 and 3, a soil layer covered with grass sods (Panel 10) rated top. However, the grass sods have visibly deteriorated with time and most of the grass has now died. Only the presence of the grass roots has maintained the effectiveness of the treatment. On Panel 9, which has a thicker soil layer, and was expected to outperform Panel 10, the grass has already died and the panel's rating has dropped from 1 to 8.

Table 1 shows that non-vegetative treatments occupy 6 of the first 7 places in the ranking, and should therefore be seriously considered for use in future. However, the study of natural slopes (Section 3) has shown that the slope surface probably does not have to be completely covered with rock or gravel, which would reduce the cost of the treatment with little or no loss in effectiveness. A simulated "natural slope" panel is being prepared at present to test this hypothesis. The simulated natural slope surface will be vegetated to the extent of about 33% cover and will have 33% covered with random sized rock and gravel particles. The remaining exposed area of tailings will be left unprotected.

5. WIND EROSION

Earlier measurements in the South Africa gold fields area (Blight, 1989, 1991) have shown that erosion losses from unprotected tailings surfaces are roughly evenly divided between water and wind erosion. Obviously, as a climate gets drier, wind erosion will dominate more and more, while in wetter climates it may be less important. This is, however, by no means certain. Because sand-dunes are shaped and formed by the wind, it is particularly interesting to study their shape, in relation to the varying velocity of the wind that has formed them. Figure 6 shows the profile of a wind-formed sand dune, with the windward side to the left. This windward slope of the dune (9.5°) is clearly related to the predominant speed of the wind, and its height to the available supply of sand. The leeward slope or slip face forms at the angle of repose of the sand, namely 33° in this case. The profile of the wind parallel to the slope surface and at a height of 200 mm above the sand has been shown superimposed on the elevational profile. This profile, typical of several measured on sand dunes, is similar in shape to the physical profile. The wind speed increases up the dune and then abruptly decreases shortly before the crest of the dune is reached. Once the crest starts to provide shelter from the wind, the wind speed falls considerably.

The wind profile was measured while a plume of sand was blowing from the crest of the dune, hence the threshold velocity for sand movement had been exceeded. As long ago as 1941, Bagnold showed that the rate of sand movement is proportional to the cube of the wind velocity. His equation is

$$q = K_t(V - V_t)^3 \tag{1}$$

where q is the rate of sand movement (units of MT^{-1})
V is the wind velocity
V_t is the threshold velocity and
K_t is a constant of proportionality.

Bagnold also showed from considerations of continuity that a dune can only maintain the same shape if the gradient of rate of movement of sand along its face (dq/dx) is proportional to the tangent of the angle of inclination (i) of the sand surface, i.e.

Table 1: Cost-effectiveness evaluation of slope protection methods

Panel no.	Treatment	Level & Compact	Level Only	Relative Cost/ha C (%)	Erosion Rate (tons/ha/yr)		Relative erosion E (%)		Cost-Effectiveness C x E (%)		Cost-Effectiveness Ranking	
					Phase 2	Phase 3	Phase 2	Phase 3	Phase 2	Phase 3	Phase 2	Phase 3
1	conventional grassing	✓	—	100	164	—	59	—	59	—	10	—
2	100 mm ballast (50 mm size)	✓	—	67	105	32	38	3	25	2	7	4
3	300 mm coarse rock	✓	—	62	170	12	61	1	38	1	8	1
4	300 mm fine rock	✓	—	62	38	96	14	9	9	6	3	7
5	geofabric + 300 mm fine rock	✓	—	120	22	15	8	1	10	1	4	1
6	75 mm of 6 mm stone + 100 mm fine rock	✓	—	66	42	82	15	8	10	5	4	6
7	300 mm fine rock	—	✓	54	118	75	43	8	23	4	6	5
8	250 mm open pit overburden	—	✓	64	203	161	74	16	47	10	9	9
9	250 mm soil + Ag Lime + grass sods	—	✓	120	19	72	7	7	8	8	1	8
10	100 mm soil + Ag Lime + grass sods	—	✓	96	21	15	8	1	8	1	1	1
11	zero-control (no treatment)	—	—	0	276	1029	100	100	0	0	11	10

Phase 2 - cumulative for 2 wet seasons (94/95 & 95/96)

Phase 3 - cumulative for 4 wet seasons (ending March 1998)

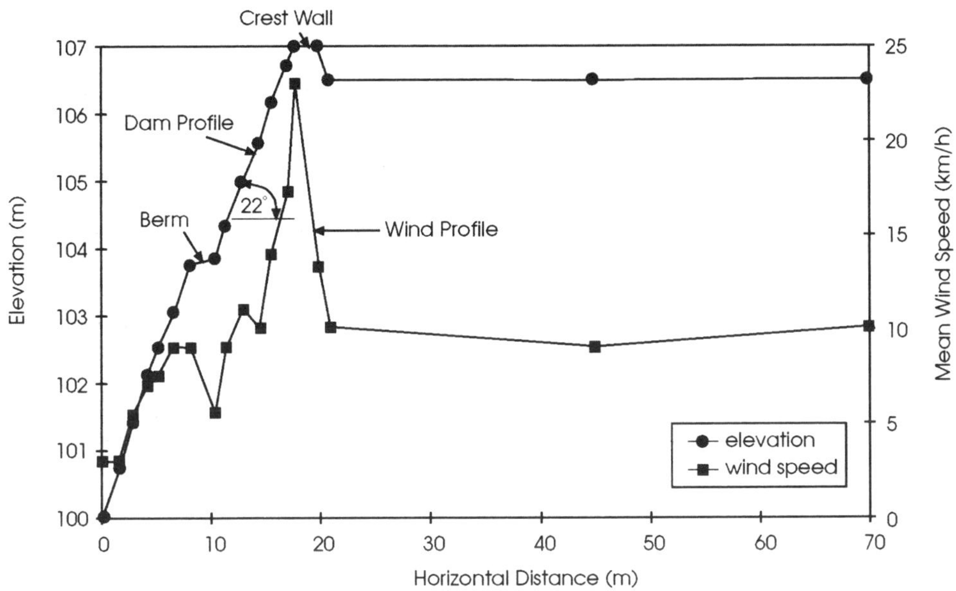

FIGURE 7 Physical and wind profiles for tailings dams

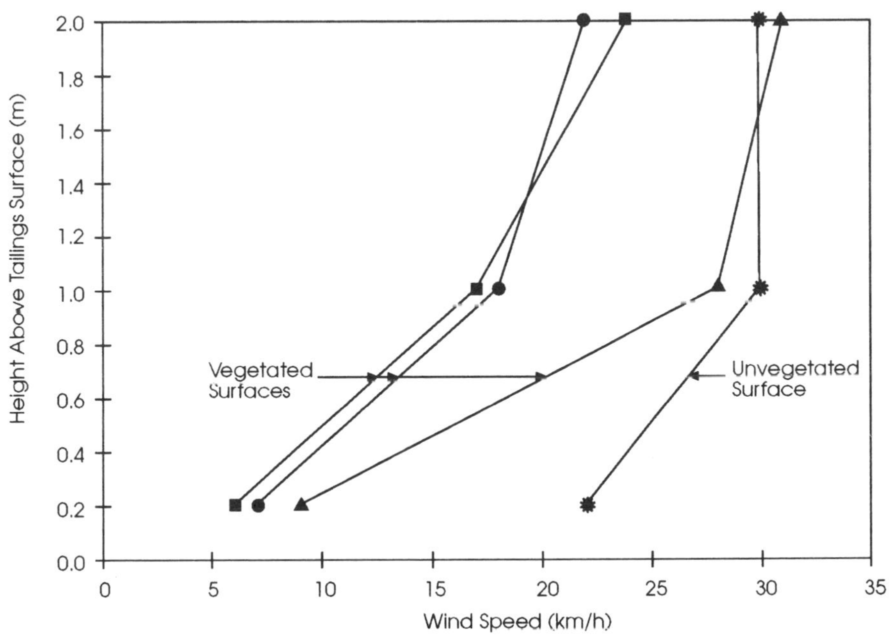

FIGURE 8 Effect of vegetation in moderating near-surface wind profile

$$dq/dx = K_T \tan i \quad (2)$$

These two requirements, however, are incompatible considering the form of the wind profile. There must be an additional regulatory mechanism to prevent the dune from being blown away.

Figure 7 shows similar physical and wind profiles for a tailings dam. The wind profile is again measured parallel to the surface at a height of 200 mm. The wind profile is also fairly similar to the physical profile of the tailings dam. It is interesting to see the localized drop in wind speed at the berm half-way up the height of the dam. Peak wind speed occurs at the leading edge of the crest wall, with a considerable drop in speed over the 2m width of the crest wall, and a further drop to a more or less constant wind speed over the flat surface of the dam.

From the wind profile one would expect the worst wind erosion to occur over the upper portion of the windward slope, with deposition occurring on the berm and just downwind of the crest wall. Because of the relatively low wind speed over the top of the dam, there should be little net erosion off the top. Figure 8 shows the effect of vegetation (80% cover of grass about 0.6m high) in moderating the near-ground wind speed over a tailings dam slope. Clearly, if the wind speed is so reduced at 0.2m above the surface, it will be close to zero right at the surface and the slope will be almost completely protected from wind erosion. A corresponding profile for an unvegetated surface also shows a reduction in wind speed close to the surface, but the reduction is much less than with vegetation.

6. FUTURE WORK

It is intended to continue the investigations of the erosion resistance characteristics of natural slopes and also the experiments on recreating naturally erosion-resistant surfaces on tailings dam slopes. Wind-tunnel tests are currently being started to investigate the effect of slope angle, shape, surface micro-roughness, etc on wind erosion, as well as re-examining the movement and deposition of sand on slopes of various inclinations and shapes.

It is hoped to tie this work in with field observations on a large tailings dam situated in a semi-desert environment. Use of this dam as an experimental site is being negotiated at present.

7. REFERENCES

Bagnold, R A (1941) *The physics of blown sand and desert dunes*. Methuen, London.

Blight, G E (1989) Erosion losses from the surfaces of gold-tailings dams. *Journal South African Inst. Mining and Metallurgy*, 89 (1), 23-29.

Blight, G E (1991) Erosion and anti-erosion measures for abandoned gold tailings dams. *Reclamation 2000: Technologies for Success*. 8th National Meeting, American Society for Surface Mining and Reclamation, Durango, Co, USA, 323-330.

Blight, G E and Smith, ME (1996) New ways of protecting tailings slopes against erosion. *Proc. Int. Symp. on Seismic and Environmental Aspects of Dams Design,* Santiago, Chile, 1, 207-211.

Brink, ABA (Ed)(1979). *Engineering Geology of Southern Africa, 1, The first 2000 Million Years of Geological Time*. Building Publications, Pretoria, South Africa.

Evans, W R and Kalkanis, G (1977) Use of the universal soil loss equation in California. *National Soil Erosion Conference*, Soil Conservation Society of America, 31-40.

Roy, B L and Jarrett, A R (1988) The role of coarse fragments and surface compaction in reducing erosion. *1988 International Summer Meeting*, American Society of Agricultural Engineers, Rapid City, S.D, Paper 88-2126.

Smith, G M, Waugh, W J and Kastens, M K (1997) Analog of the long term performance of vegetated rocky slopes for landfill covers. *Tailings and Mine Waste '97*, Balkema, Rotterdam, 291-300.

Tailings and Mine Waste'99 © 1999 Balkema, Rotterdam, ISBN 90 5809 025 6

Launching stone methods for stabilizing reclaimed waste sites

S.R.Abt, C.I.Thornton & C.M.Lipscomb
Department of Civil Engineering, Colorado State University, Fort Collins, Colo., USA
T.L.Johnson
NMSS, US Nuclear Regulatory Commission, Washington, D.C., USA

ABSTRACT: Launching stone (riprap) is a technique that has been utilized to protect migrating stream and river banks. Stone is sized and appropriately placed near the top of the bank at risk and then collapses as the river/stream migrates into the perched stone. The launching stone concept is proposed to protect tailings impoundments from incising gullies and intrusions. Stone is sized and placed on top of and partially down the slope of the impoundment cover. When a gully migrates up slope into the cover, the stone launches thereby armoring the leading edge of the gully. Preliminary results of the laboratory tests are presented.

1 INTRODUCTION

The reclamation of disturbed mined lands requires that the restored areas be contoured and protected against the forces of wind and water erosion. Perhaps one of the greatest threats to the long-term stability of a reclaimed uranium mill tailings and/or low-level waste impoundment is its susceptibility to gully incision. Gullies frequently observed on or near reclaimed impoundments are attributed to the mechanisms of a) rilling and gullying, derived from direct runoff of tributary areas on the top and sideslope areas of the embankment, and b) vertical and/or horizontal instability of adjacent drainage channels or streams that migrate by headcutting into the impoundment. Gullying not only influences the long-term impoundment stability, but also lends to the potential of releasing tailings or waste material to the environment which may be deemed hazardous to the health and safety of the public.

One erosion protection system used to assure the long-term stability of tailings and/or low-level waste impoundments places a protective rock riprap cover over the entire embankment. Although the rock cover is extremely effective in resisting the weathering and erosive forces of nature, it is expensive to implement. Therefore, a methodology is sought to provide a sufficient level of protection to assure the long-term stability of the impoundment while reducing the construction costs.

The concept of a launching rock riprap system was conceived to address the potential consequences of gully incision. The concept entails the placement of a riprap cover over the top slope and extending the rock down the sideslope from the embankment crest a portion of the slope length. When a gully or zone of erosive instability incises up the embankment slope to the rock cover, the sacrificial rock launches into the incision, armors the active eroding zone, and terminates further degradation. Key parameters entail the appropriate rock size, rock layer thickness, and distance the rock layer should extend downslope from the crest.

The objective of this study was to conduct a pilot testing program to evaluate the launching riprap cover system concept. A soil embankment was constructed and a riprap cover was placed on the top and on the upper slope near the crest. A runoff hydrograph was routed over the embankment and rock cover allowing a gully to develop and incise up the embankment slope. The rock characteristics and the findings of this pilot study are presented.

2 BACKGROUND

The concept of using launching stone to provide bank protection and stabilization to a migrating river channel was proposed and demonstrated by the U.S. Army Corps of Engineers (USACE) (1980, 1981). The Streambank Erosion Control Evaluation and Demonstration Act of 1974 permitted the USACE to test the launching stone concept at several locations along the Missouri River. At Vermillion, South Dakota (river mile 771), stone (riprap) was windrowed on the top bank of the outer river bend in 1979. During the next two years, the river slowly migrated thereby eroding the bank into preplaced windrowed revetment. As the bank migrated, the stone released down the bank face. The initial stones settled at the toe of the slope with subsequent releases keying into the toe stone and then stacking upslope at approximately 1-d_{50} thickness. Figure 1 illustrates the windrow and launching concept implemented at the Vermillion demonstration site. USACE (1980) indicated that the launching concept had promise and should be further developed.

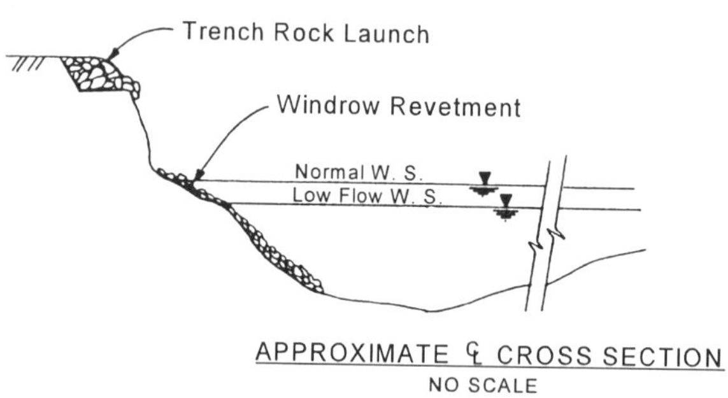

Figure 1. Channel bank launching concept.

The site was visited in 1996 to assess long-term bank stability. Figure 2 portrays the demonstration site along the left bank of the Missouri River. Figure 3 indicates how the stone stabilized the bank toe. Although the stone layer is dispersed and does not meet recommended revetment standards, the stone layer has stabilized the bank and mitigated river migration.

Figure 2. Vermillion, South Dakota demonstration site looking downstream in 1996.

Figure 3. Vermillion, South Dakota demonstration site left bank in 1996.

USACE (1984) further developed the launching concept by performing a model study to develop design criteria. Three types of windrow shapes were evaluated in the study; triangular, trapezoidal, and rectangular as presented in Figures 4a, 4b, and 4c, respectively. Variables that were considered that impact design criteria included the bed material, stone size, stone gradation, flow velocity, flow depth, bank height, potential toe scour depth, and windrow continuity. A series of expressions were developed for sizing the windrow volume. However, these findings have not been field tested nor has the criteria been officially adopted by the USACE.

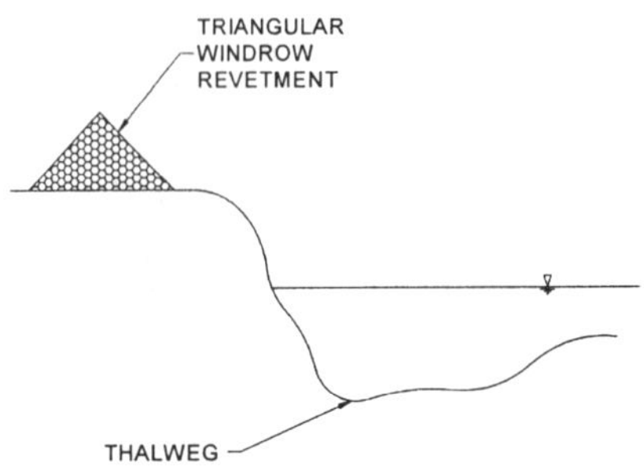

Figure 4a. Triangular windrow revetment.

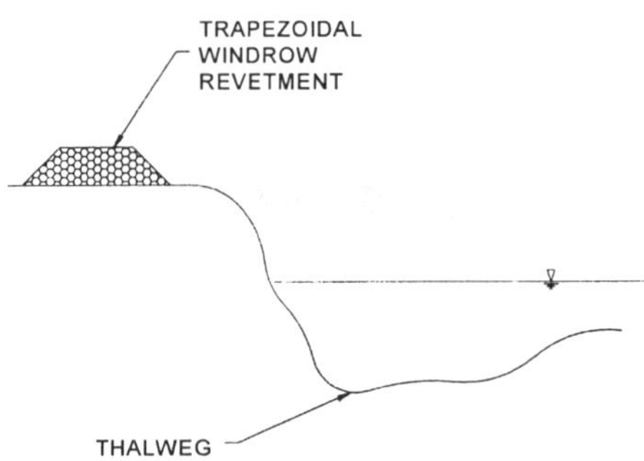

Figure 4b. Trapezoidal windrow revetment.

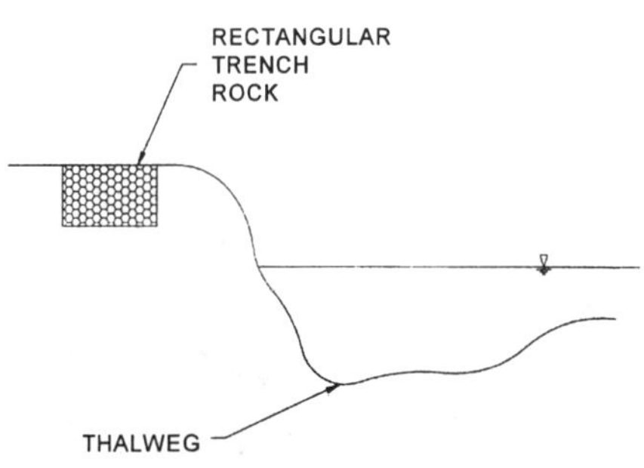

Figure 4c. Rectangular trench rock.

Skrebkov et al. (1991) presented a method for protecting underwater slopes with self-placement of stone. The essence of the procedure is that in the seasonal low-flow period of the river system, an additional shoulder of soil with a horizontal berm above the water level is filled to the design outline. The revetment stone intended for the slope is placed on the berm. During high flow or flood conditions, the stone collapses or launches as the slope and berm are eroded thereby forming the revetment. Skrebkov et al. (1991) conducted a series of laboratory tests. An expression was formulated for determining the volume of stone necessary for revetting 1 m of slope length:

$$V_{st} = h\left[\left(1 + m_2^2\right)^{1/2} + \left(m_1 - m_2\right)\right]\left[n_b n_{s1} / n_b - n_{s1}\right]\left[\left(d_{st} / k_s\right)k_{sf}\right] \tag{1}$$

where h is the calculated height of the slope; m_1 is the steepness at which movement of individual stones begins; m_2 is the steepness of the completely revetted slope; d_{st} is the average stone size; n_b is the number of stone layers on the berm; n_{s1} is the number of stone layers on the slope; k_s is the self-placement coefficient; and k_{sf} is the safety factor. The safety factor should not be less than a value of 1.5.

The launching stone concept has been applied by the USACE to stabilize riprap revetment placed on the outer bend of channel banks. Toe scour commonly occurs at the interface of the revetment base and the channel bed. As the bed scours, the riprap base drops thereby destabilizing the entire revetment mass. Knowing the maximum water depth in the bed, the designer can use a launchable riprap section placed at the toe on the existing bed (Maynord 1998). The before launch section thickness should be 2.5 to 4 times the thickness of the revetment if placed by mechanical means. The length of the before launch section depends on the required volume of rock. Riprap having d_{85}/d_{15} greater than 2 is recommended to prevent large voids. Specific design details are presented by the USACE (1994).

3 APPROACH

The pilot test program was formulated to evaluate the concept of the launching riprap cover system to protect a waste impoundment from the potential hazards of an incising gully or localized erosion phenomena. The approach used to evaluate the concept was to construct a simulated (model) embankment, place a riprap cover on the embankment top and upper side slope, and subject the cover system to an overtopping runoff hydrograph. The simulated embankment was prepared to allow a migrating gully to incise into the cover. The effectiveness of the launching riprap cover system can then be observed and evaluated.

3.1 *Test facility*

The testing program was performed at the Hydraulics Laboratory of Colorado State University. A flume 3.1 m wide, 18.3 m long and up to 3 m high was constructed as indicated in Figure 5. Water was conveyed from a sump via a 75-hp pump through a pipe network into a manifold and headbox. Flow filled the headbox and spilled over a headwall onto the test embankment overtopping onto the simulated cover system. Flow was then conveyed over the cover, down the embankment slope, through a material recovery basin, and out the wasteway. The pumping system had a capacity of 0.25 m^3/sec.

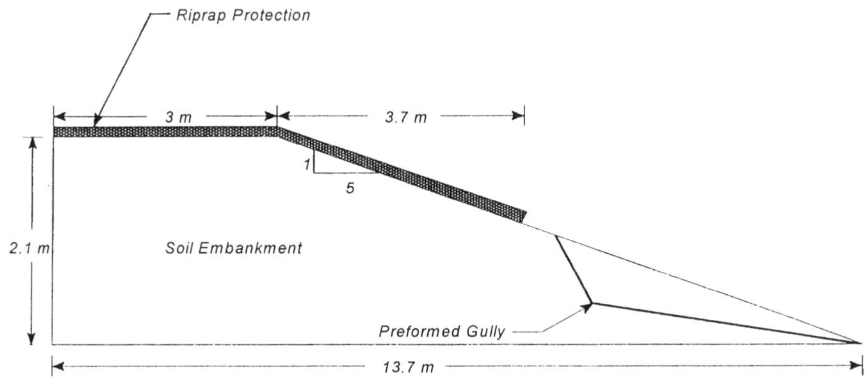

Figure 5. Schematic profile of test embankments.

The embankment was constructed to simulate a tailings/waste impoundment with rock cover. Each embankment consisted of a soil matrix with an 80:20 sand:clay mixture. Material was placed in 15 cm lifts and compacted to 90% of the maximum density, Standard Proctor. The embankments were 1.22 m or 2.1 m high, horizontal cap top length of 3.1 m, width of 3.1 m, and 1:5 side slope as schematically presented in Figure 5.

Rock riprap was placed on top of each embankment. A collapsing rock layer was placed on the surface of the upper slope extending 33% down the face length.

3.2 *Test program*

A series of six (6) tests was performed to evaluate the feasibility of a launching rock cover system. The primary variables influencing the mechanics of the launching rock cover system include the median rock diameter (d_{50}), rock layer thickness (expressed in the number of median diameters), and prospective fall distance to the bottom of the approaching gully expressed as the embankment height. A summary of test parameters is presented in Table 1.

Table 1. Test matrix.

Test No.	Riprap d_{50} Diameter in cm	Layer in cm	Preformed Gully	Height m	Stable Unit Discharge q - m^3/sec/m
1	6.4	12.8	No	2.1	0.020
2	8.9	17.8	No	2.1	0.037
3	8.9	26.7	Yes	2.1	0.042
4	8.9	17.8	Yes	2.1	0.037
5	3.3	13.2	Yes	1.22	0.030
6	12.7	27.4	Yes	1.22	0.053

3.3 *Test procedure*

In each test, the overtopping discharge was slowly started and incrementally increased until the rock layer was dispersed. The test procedure for each embankment configuration was identical. A summary of the procedure follows.

The test embankment was constructed and configured with the prescribed rock cover condition. The pipe network was charged and the head box was slowly filled. The test configuration was photographically documented. Video cameras were started and the entire event was recorded. When flow overtopped the embankment, flow was stabilized at the prescribed unit discharge (usually in units of 0.01 m^3/sec/m). Gully incision, flow concentrations and rock movement were documented. Flow was intermittently stopped to permit observation and documentation of the rock layer as deemed necessary. The discharge was incrementally increased until either the embankment and/or the rock protective layer was considered to not be effective. Upon conclusion of the test event, flow was terminated and the head box drained. The embankment was documented, the cover system removed, the embankment repaired, and the next cover system constructed.

4. FINDINGS

The objective of the sacrificial, riprap cover system was to launch stones into the intrusion as it incised up the embankment slope. The cover must contain sufficiently sized rock to resist the tributary flow and be sufficient in volume to adequately armor the intrusion and mitigate further incision. It is desirable that after the rock launches and the intrusion is stabilized, the upper slope not become exposed.

The test results of each cover system were observation based and linked to the unit discharge. Therefore, as each discharge was conveyed over the riprap cover system, the degree of rock movement was documented. The test continued until the maximum level of protection was provided prior to cover system and/or embankment deterioration. The maximum stable unit discharge relates to the unit discharge that provides maximum cover protection. For example, Figure 6a illustrates the pre-test condition for Test 3. Flows were incrementally conveyed over the embankment to a discharge of 0.028 m^3/sec/m as presented in Figure 6b. It is observed that the rock cover system launched into the intrusion thereby armoring the incision. Continued flow eroded the area adjacent to the incision until the rock formed a uniform protective barrier. Note that the embankment slope above the intrusion remained intact and that none of the upper slope is exposed. As the discharge increased, the rock was mobilized exposing the soil embankment in several locations. Therefore, the maximum stable unit discharge is 0.028 m^3/sec/m for this particular configuration.

Table 1 presents the maximum stable unit discharge for each of the 6 tests. It is observed that as the rock size in combination with the layer thickness increases (riprap volume), the maximum stable unit discharge increases. The determination of a sufficient volume of rock to launch into a migrating gully or other form of localized erosion to stabilize the disposal embankment is a critical component of the protective riprap cover system. A rock stability factor was developed as an indicator of minimum riprap stability. The stability factor (R_{sf}) is defined as:

Figure 6a. Test 3 pre-test embankment with riprap cover system.

Figure 6b. Test 3 after stone launching with stable unit discharge of 0.028 m^3/sec/m.

$$R_{sf} = d_{50} * Thickness \qquad (2)$$

where d_{50} is the median stone diameter in cm, and the thickness is the depth of the rock layer expressed in number of d_{50}s.

The stability factor was computed for each of the 6 test conditions and related to the maximum stable unit discharge as illustrated in Figure 7. An envelope expression was fit to the data indicating the minimum acceptable stability factor for a specified unit discharge.

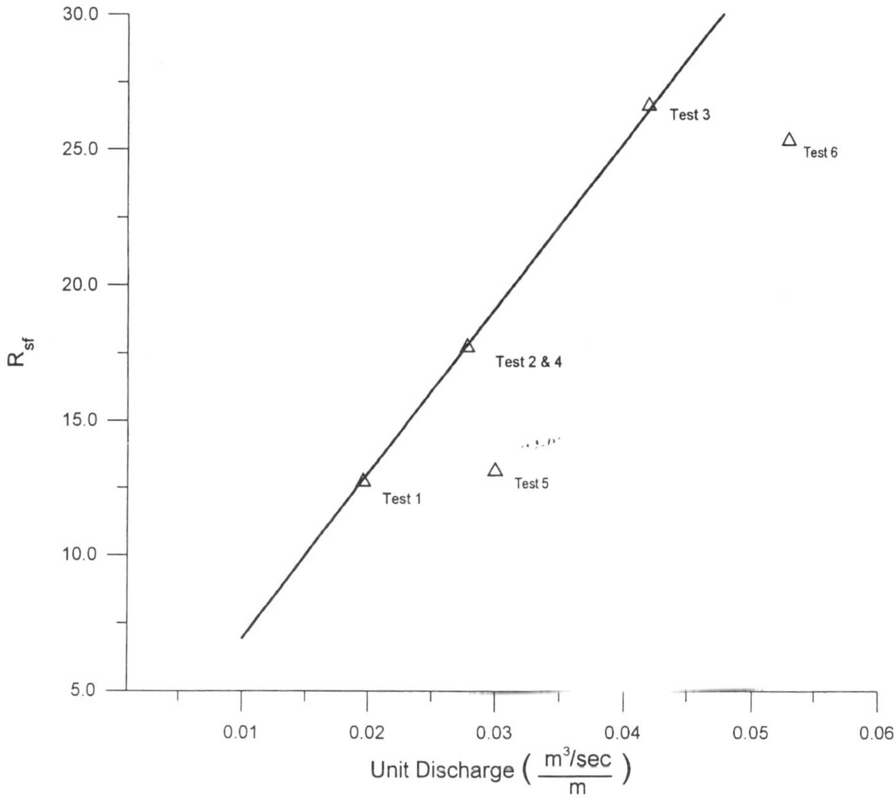

Figure 7. Stability factor versus unit discharge relation.

5 RECOMMENDATIONS

It appears that the sacrificial, launching stone protective system has potential field applications to mitigate intrusions into waste impoundments. Implementation of a launching protective system could potentially reduce the riprap cover needed to insure the long-term stability of buried waste materials. However, the data base must be expanded to validate the stability factor - unit discharge relation. Further, guidelines must be developed for system design and potential applications.

6 REFERENCES

Abt. S.R., Thornton, C.I., Batka, J.H. & Lipscomb, C.M. 1997. Investigation of gully stabilization methods with launching stone: pilot laboratory tests. Report prepared for the U.S. Nuclear Regulatory Commission, Washington DC , pp. 35, February.

Maynord, S.T. 1998. Corps of Engineers design for bank stabilization. Proceedings of the ASCE International Water Resources Engineering Conference, Memphis, TN, August 3-7, pp. 471-476.

Skrebkov, G.P., Afim'ina, K.A., Mukin, V.A. & Korotkov, V.E. 1991. Protection of underwater slopes by self-placement of stone. *Hydrotechnical Construction* 25(5):24-26, May.

Thornton, C.I., Abt, S.R. & Johnson, T.L. 1998. Gully stabilization methods utilizing launching stone. Proceedings of the ASCE International Water Resources Engineering Conference, Memphis, TN, August 3-7, pp. 748-753.

U.S. Army Corps of Engineers 1984. Laboratory tests to design windrow revetment for bank protection. MRD Hydraulic Laboratory Series Report No. 7, U.S. Army Engineer District, Omaha, NE.

U.S. Army Corps of Engineers 1981. Final report to Congress - The Streambank Erosion Control Evaluation and Demonstration Act of 1974, Missouri River Demonstration Projects. Appendix E, Volumes 1 and 2, Washington, DC, December.

U.S. Army Corps of Engineers 1980. Interim field evaluation of windrow revetment. Streambank Erosion Control Evaluation and Demonstration Program, U.S. Army Engineer District, Omaha, NE, March.

U.S. Army Corps of Engineers 1994. Hydraulic design of flood control channels. EM 1110-2-1601, Headquarters, U.S. Government Printing Office, Washington, DC.

Tailings and Mine Waste '99 © 1999 Balkema, Rotterdam, ISBN 90 5809 025 6

Tailings deposits with waste dams

C. Scognamillo & S. Barrera
Geotécnica Consultores S.A., Santiago, Chile

ABSTRACT: The exploitation of open pit mines generates a large quantity of waste material, which is usually disposed of in dumps located close to the pit. This material, which is basically granular and has a low fines content, in general presents a high permeability. The nearness of appropriate sites for depositing tailings, have made the use of waste material in the construction of deposit dams increasingly convenient.

In the last years, several large tailings deposits have been designed in Chile (two in operation for a long time and one just beginning) where waste has been used as construction material for the tailings contention walls. This paper presents the main characteristics of dams projected with waste, as far as their geometry, grain size distribution, compaction, draining, seepage controls, constructive aspects and interaction with the mine's operation, going deeply into the most relevant aspects considered in their design.

1 INTRODUCTION

Tailings produced by concentrator plants for copper or similar minerals have traditionally been disposed of in deposits, whose walls have been constructed using the coarse fraction of tailings (cycloned sand). Only in topographically favorable cases, such as where the presence of narrow gorges made it possible to contain large volumes of tailings with low-volume walls, or where the tailings' grain size distribution did not allow the production of enough sand for the construction of dams, were these deposits constructed with borrow material. The solution based on sand dams, is greatly advantageous in economic terms because among other things (1) it allows part of the tailings to be used in the contention walls for the deposit; (2) it reduces the volume of the impoundment itself by eliminating the coarse fraction of the tailings and, (3) it does not require an intensive use of borrow material for constructing the contention walls around the deposit.

In the last few decades, the progressive increase of open pit mines, as opposed to underground methods, has generated large amounts of waste material which is usually disposed of in dumps located close to the pit. This fact, besides certain topographical characteristics of the sites, have favored the inclusion of waste in studies for alternatives to provide construction material for the dams.

The study of the option to use waste material must consider, at the very least, the following aspects: (1) the additional cost of transporting material from the dumps to the tailings dam, (2) the grain size distribution and mineralogical characteristics of the material (possible acid generation), (3) compatibility between the waste extraction program and the program for constructing the tailings dams, (4) additional hauling equipment, (5) design considerations for the dams depending on the characteristics of hauling equipment used and the grain size distribution of the waste.

In the few last years, three deposits have been designed in Chile that have incorporated waste as construction material for contention walls. Two of them correspond to copper tailings and one to iron tailings. Fig. 1 shows the location of these deposits in the country. All of them are located in the north, where there is scarce rainfall (less than 150 mm/year on average).

The present paper provides a description of the main characteristics of the waste dams projected thus far, giving details about geometry, grain size distribution and compaction of waste, drainage systems, seepage controls and other details pertaining to the design and to the dams.

2 CASES OF TAILINGS DEPOSITS CONSTRUCTED WITH WASTE

2.1 *Candelaria Mine Tailings Deposit*

2.1.1 *General Background*

The site of the tailings deposit is located south of Copiapó, in a valley sited approximately 2 km north of the concentrator plant. The deposit's capacity is 366 million tons (Mtons) of tailings and the area that is finally covered will be some 450 hectares (Fig.2a). This deposit began operation in late 1994, and its second stage of construction is currently finished.

In this case the use of waste in the dam was the natural conclusion derived from the mine's waste disposal plan which sought to cover and close a creek adjoining the plant, which left an area that could potentially be used to deposit tailings. The analysis finally centered around the additional cost of anticipating the wall's filling, the over-hauling of waste (no further than 1 km) and the placement of waste in layers of 2 to 4 m requiring the use of a bulldozer to spread.

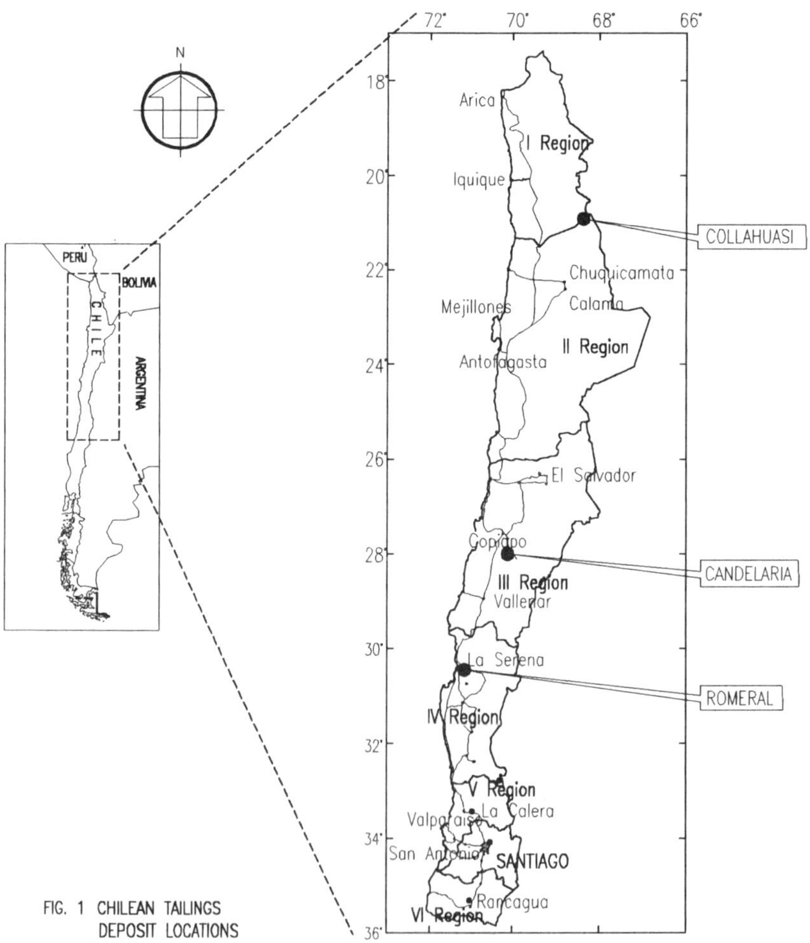

FIG. 1 CHILEAN TAILINGS DEPOSIT LOCATIONS

Tailings that result from the flotation process at a rate of 28,000 tpd are discharged by pumps through a pipe approximately 2000 m long, towards a feed box located at a saddle in the rear (south) end of the tailings deposit. From this point they are conducted by gravity through a pipe to the dam's crest, where they are discharged in multiple and variable positions. This is done to provide an adequate beach for the tailings, to control the position of the clear water pond and avoid the lifting of dust.

A large part of the water accumulated in the clear water pond is reclaimed using pumps and recirculated into the process. Water that seeps through the wall and the foundation soil is captured by the wall's drainage system and conducted to the cutoff trench, where it is also recirculated into the process.

The main dam is being constructed in stages, first with waste rock obtained by stripping and exploitation works from the mine's operation. Further on, two auxiliary dams, also made of rockfill, will be constructed to close two saddle points.

The initial dam design (and in particular the filters) considered waste material with a coarse grain size distribution, a significant percentage of particles larger than 50 cm, and a small amount of sandy and fine material. In practice this wasn't so because the waste was more a sandy gravel with a negligible fraction of sizes over 50 cm.

2.1.2 *Tailings Impoundment Dams*

The impoundment was built by the construction in stages of a main dam made of rockfill taken from the mine's excavation. This dam will reach a maximum height of 172 m with a final crest elevation of 792 m a.s.l. The upstream slope will be 1.8:1, the downstream slope will be 1.6:1 (H:V) and the crest width will be 20 m (see Table 1).

The starter dam, with an elevation of 680 m a.s.l., allows the storage of 15 Mtons of tailings with a 4 m freeboard. At the upstream slope of this dam there are two layers of granular material 3 m thick laid out horizontally to filter the tailings deposited and allow their drainage. After this wall's construction this aspect of the design was revised, mainly bearing in mind the waste's finer grain size distribution, which resulted in the replacement of the two layers of filters by a geotextile.

Table 1. Characteristics of the Main dam.

	Starter Dam	Final Dam
Crest elevation	680 m a.s.l.	792 m a.s.l.
Height (m)	50	172
Volume (Mm^3)	2.0	54.0
Upstream slope (H:V)	1.8:1	1.8:1
Downstream slope (H:V)	1.6:1	1.6:1
Material	Rockfill (*)	Rockfill (*)
Placement	In layers	In layers
Compaction	Passes with heavy equipment	Passes with heavy equipment

(*) Waste from the mine

The final cross section of the main dam (Fig. 2b) shows that it is made up of two parts and the difference corresponds to the different ways that the waste material was placed:

- Main sector, located upstream, where the rockfill is placed in layers 2.0 m thick, and compaction is done by bulldozers and by hauling trucks.
- Downstream sector, where rockfill will be placed in layers some 4.0 m thick and no special compaction is required.

The main characteristic of the disposal area is that it is formed from loose alluvial soils with paleoterraces on the valley floor and sound intrusive rocks on the slopes. According to this situation the treatment for the dam's foundation involved the removal of at least four meters of top alluvial soil, for the entire area between the wall's upstream line and the projection at the foundation level, on a 1:1 (H:V) inclination plane facing downstream passing by the downslope edge of the final crest. Where ancient terraced horizons are to be found, stripping 0.3 m deep should be done.

Once tailings deposited reach an elevation of 745 m a.s.l. it will be necessary to construct two auxiliary dams that will close two minor saddle points located on opposite sides of the main dam. The geometry and composition of these two dams will be similar to the main dam.

There is a base drainage system used to catch water from underground seepage caused by the consolidation of the tailings. It consists on finger drains or trenches with trapezoidal sections, filled with draining granular material. Above these finger drains there is a layer of filtering material and afterwards, rockfill from the dam's main body. Water that is caught this way is discharged into a channel of similar characteristics, which leads it down the center of the creek towards the seepage collection system.

2.1.3 *Seepage Collection System*

The seepage collection system is basically a cutoff trench some 27 m high and 130 m long, located 1200 m downstream of the main dam. Its purpose is to catch all seepage from the tailings impoundment.

This trench has a liner placed in zig-zag inside a low-permeability fill constructed with compacted material. These elements are placed inside a trench that is excavated until base rock is reached. This trench is trapezoidal in shape, with variable slopes 1:1 and 1:2.5 (H:V), 8 m wide at

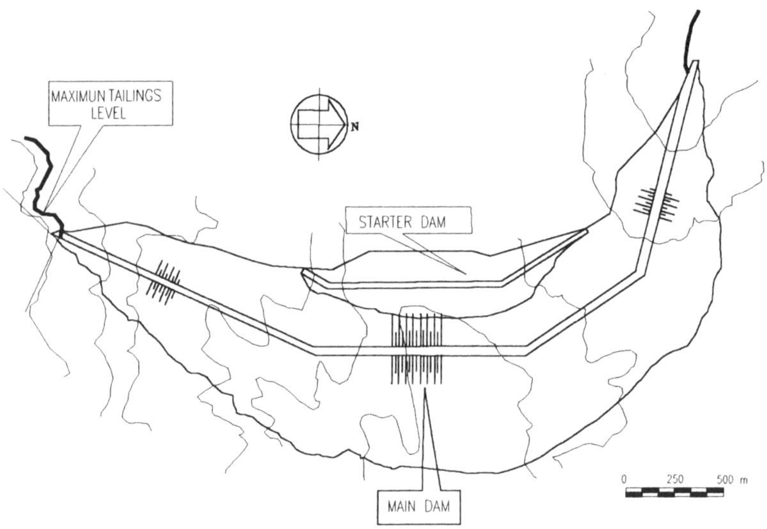

FIG 2a CANDELARIA TAILINGS DAM – PLAN

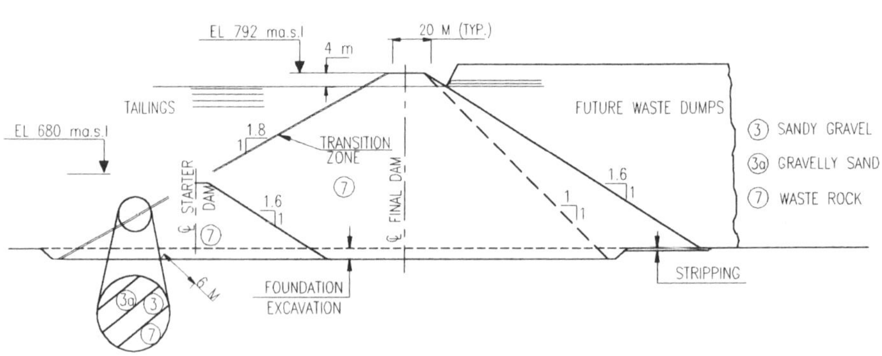

FIG 2b CANDELARIA TAILINGS DAM TYPICAL CROSS SECTION

the base and 27 m high maximum. At the bottom of the trench, on the upstream side, there is a drain made of 2 perforated steel pipes, with granular material for draining and filtering. This drain takes water towards the spillway channel. The liner is tightly fixed on the rock in its entire length. The rock must also be grouted to make it watertight, which further ensures the effectiveness of the cutoff trench.

The cutoff trench will be covered by the waste dump, so water that is caught is discharged into a 200 m long tunnel, and into an auxiliary vertical shaft located between the foot of the waste dump and the open pit. The tunnel has a longitudinal grade equal to 0.5% to ease the evacuation of water that enters it. The shaft is located approximately 40 m deep and at the bottom there is a pool from which water is taken out using pumps and driven to the plant for recirculation.

2.2 *El Romeral Mine Tailings Deposit*

2.2.1 *General Background*

The mineral deposit known as El Romeral is an iron mine located approximately 18 km north of La Serena. Tailings produced in the process (about 1,000 tpd) are taken to a deposit close to the mine.

This tailings deposit is located in an artificial "rinconada" (bend forged by two neighboring hills) , formed entirely by dumps of waste material, subproducts of the mining process(Fig.3a). The first dump, located on the left (east) side is 140 m high and is composed of waste material coming directly from the mine, and its grain size distribution varies from sand to blocks of a meter or more in diameter. The second dump, located on the right side of the deposit, is made up of low grade material, technically "material rejected from the plant" which is a subproduct of the processing plant. Its grain size distribution corresponds to a gravel with variable amounts of sand. The "rinconada" that is formed by these dumps is closed by a kinked axis dam which is constructed in stages using material rejected from the plant, and its height will reach 50 m towards the end of operation. For a height such as this and considering a 3 m freeboard, the deposit will have a total impoundment capacity of approximately 4.5 millions of m^3.

Tailings are discharged from the rear of the deposit, causing the formation of a small water pond in contact with the dam.

Similar to the Candelaria case, both the location of the tailings deposit and the definition of the wall's construction material was a natural consequence of how the mine and the processing plant manage materials. In this case, the tailings' fine grain size distribution did not make it possible to obtain sand in the quantity and quality needed for the construction of the dam.

2.2.2 *Contention Wall*

This wall is constructed using mainly material rejected from the plant, to form a trapezoidal wall approximately 50 m high at its highest (elevation 285 m a.s.l.) (Fig. 3b).

The wall is constructed in a total of 5 stages, and the its geometrical characteristics for each one of the stages are shown in Table 2 below:

Table 2. Geometrical characteristics of the contention wall.

Stage (1)	Crest			Slope	
	Elevation m.a.s.l.	Width (m)	Height (m)	Upstream H:V	Downstream H:V
1st Starter Dam	252	10	17	1.8:1	2.2:1
2nd	257	10	22	1.8:1	2.2:1
3rd (2)	266	10	31	1.8:1	1.8:1
4th	275	10	40	1.8:1	1.8:1
5th	285	10	50	1.8:1	1.8:1

(1) Stages 1 and 2 have already been constructed.

(2) Considering the construction material's excellent shear strength, the wall's stability was reviewed and starting from the 3rd stage the downstream slope was modified.

a) Construction material for the wall. The construction material for the contention wall corresponds to crushed material that is rejected by the plant. The grain size distribution varies from gravels to gravely sands, with a low fines content (less than 5% passing #200 sieve).

The particles that make up the material rejected by the plant are at the most 1" in size, angular in shape, with high individual shear strengths and good mechanical bonds.

The shear strength properties of this material were extensively researched during the project stage. A large number of direct shear tests (in large-size shear equipment) were carried out for different relative densities; compression and triaxial extension tests were also done.

Direct shear tests were done on materials with different grain size distributions (M1, M12, M4). The results were published by Bustamante (1989). Table 3 below shows a summary of these tests.

Table 3. Average internal friction values of material rejected by the plant.

Material	Internal Friction Angle (f)		
	Rd=20%	Rd=50%	Rd=85%
M1	40,3°	41,8°	43,5°
M12	43,4°	44,8°	45,5°
M4	41,7°	42,5°	43,2°
Average	41,8°	43,5°	44,1°

Rd: Relative Density.

As we can see internal friction angle values show a clear correlation with the material's level of compaction. As expected, however, they are only slightly sensitive to this parameter.

b) Placement of material. During the first stage of construction this material was placed in lifts and compacted with a heavy roller until a relative density of 75% is achieved, on average.

From the second stage onward the material is placed in layers 1.0 m thick which are not compacted in ways other than with construction or spreading equipment. This method results in relative densities above 50%, proven in rockfill tests made previous to the construction of the 2nd stage. The degrees of compaction obtained have been enough to guarantee the dam's stability and other important deformations caused by possible seismic movements.

2.2.3 *Cutoff Trench*

The cutoff trench is located at the upstream foot of the wall and here, an impermeable geomembrane is placed to intercept seepage that may percolate the more permeable stratum of the foundation soil. Its section is trapezoidal, with a base 0.8 m wide by 3.5 m deep and a lateral slope that varies between 1:1 and 1:5 (H:V) depending on depth. At the beginning of the operation some seepage was expected, but it was thought that it would quickly decrease once the bottom of the basin was covered with tailings. In practice, this is what happened.

2.2.4 *Drainage System*

The drainage system consists mainly in trenches laid out parallel to the wall's axis. The water that is caught by these trenches is conducted to a collecting drain that discharges it outside the wall, towards a seepage collection pool.

2.2.5 *Impermeabilization System*

The impermeabilization system is basically an impermeable geomembrane (PVC 1 mm thick) placed on the upstream slope of the dam and on the piles that make up the slopes adjacent to the dam. At the wall sector the geomembrane is placed above sand fill 2.0 m thick. This solution was adopted for the lack of fine impermeable materials in the area. When implemented in 1989, this became the first tailings deposit in Chile to apply this technology.

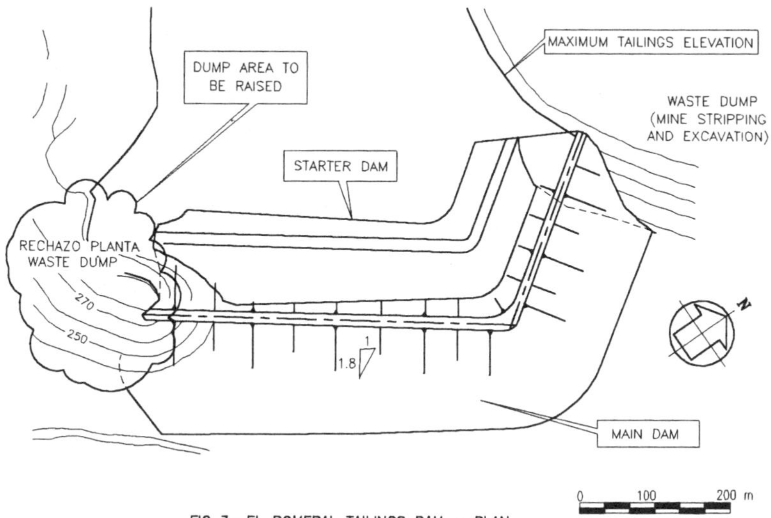

FIG 3a EL ROMERAL TAILINGS DAM - PLAN

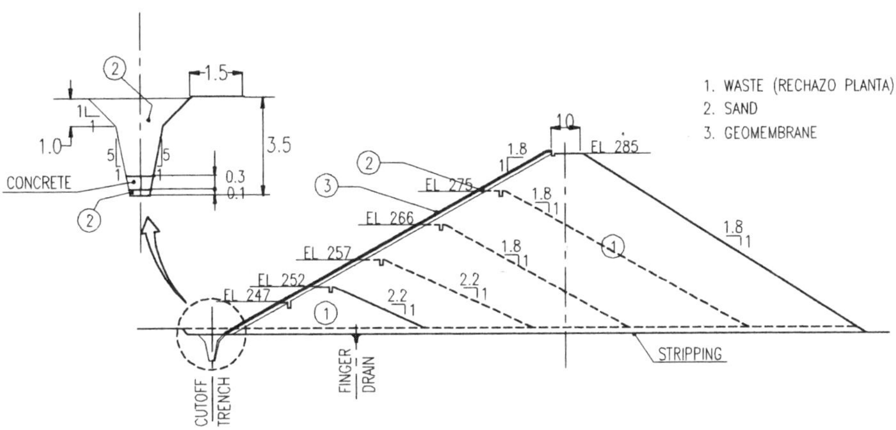

FIG 3b EL ROMERAL WASTE ROCK DAM TYPICAL CROSS SECTION

2.3 *Collahuasi Mine Tailings Deposit*

2.3.1 *General Aspects*

The Collahuasi Mine has two mineralized areas, Ujina and Rosario. It is located in the north of the country approximately 200 km southwest of Iquique, at an average elevation of 4500 m a.s.l. This mine will exploit an average 60,000 tpd of copper mineral; the tailings equivalent to over 95% of that amount will be sent to a tailings impoundment. The impoundment is in its final construction stage and starting operations.

The definition of the wall's cross section was the result of an extended analysis that considered the following aspects:

- Exploitation sequence of the mine, first Ujina for 6 years, then Rosario.
- Distance of the pits to the tailings impoundment. Ujina is 8 km away and Rosario 20 km away.
- Availability of tailings sand and cost of conveying and placing it on the wall.
- Availability and quality of waste material.

This analysis considered economic aspects: cost of overhauling and placing waste, additional waste hauling equipment and possible interference with the pit's exploitation, cost of placing the sand, stability aspects and associated operational risks; the conclusion was that a mixed solution was the most convenient with a starter dam made of borrow material, a dam made of waste for the first 6 years of operation and finally, a dam made of tailings sand.

2.3.2 *Contention Walls*

The Collahuasi tailings impoundment consists of three different types of dams (Fig.4a & 4b), which are: starter dam, waste dam and sand dam. A summary of general characteristics relative to these dams are presented in Table 4. Each one of them is briefly described below:

a) Starter Dam. The starter dam is constructed using borrow material, which is a gravely sand with maximum sizes between 2" and 3", and a low fines content. This material comes from the deposit's basin and it is placed in layers and compacted using rollers.

Below the starter dam, at the downstream end, a drain was constructed at a depth of at least 5.0 m and its direction is approximately parallel to the axis. The design also includes a perpendicular drain within, which conducts water to a reclaiming pond. Draining material is gravel or oversized particles wrapped in a geotextile to separate them from the natural soil.

b) Waste Rock Dam. The purpose of the waste rock dam is to allow the growth of the deposit and contain tailings until year 6 of operation. Waste rock will be hauled from the pit, placed in 2.0 m thick layers, and not submitted to compaction other than the one provided by equipment used to place and distribute the material. Material is hauled from the mine to the dam by the same trucks that handle the waste, which have a capacity for over 200 tons each.

The material from the mine's stripping deposited in waste dumps was submitted to grain size distribution tests so that its quality could be evaluated. The results and observations indicate that the material is a heterogeneous mixture with a matrix of silt, sand and gravel, and with rock sizes predominantly ranging up to 25 cm in maximum dimension, but with some larger rock sizes up to 90 cm in size. There was no evidence of segregation in larger sizes. The matrix material appears to provide complete infilling between larger sizes.

The waste material is not acid generating. That was verified by ABA test.

c) Sand Dam. Material for the construction of the sand dam will be obtained from the classification of tailings at the plant. Sands obtained in this fashion will contain between 15 to 20% fines content (smaller than 0.074 mm), which provides adequate permeability for the construction of a dam. The sand will be placed hydraulically in inclined lifts from the dam's crest. Its solid contents will be 65%. Water used in construction drains freely from the sand and is caught by the base drainage system so it can be returned to the plant.

After the sands are placed, they will be compacted by a bulldozer and vibratory roller in the downstream slope direction, until a minimum dry density equal to 95% in the Standard Proctor test is achieved.

All these dams have a small grade at the crest, rising towards the right abutment. This is intented to compensate for the estimated inclination of the tailings after disposal.

Table 4. Main Characteristics of the Dams.

Characteristics	Starter Dam	Waste Rock Dam	Sand Dam
Crest elevation (m.a.s.l.) (1)	4137	4151	4177
Maximum height (m)	21	35	64
Volume (m^3 x10^6)	1.3	6.0	25.1
Construction material	Borrow material	Mine waste	Cycloned sand
Disposal	Mechanical	Mechanical	Hydraulic
Compaction in layers	Horizontal	Horizontal	Inclined
Level of compaction (3)	95% MP	(2)	95% SP

(1) Crest elevation indicated is the minimum and corresponds to the left abutment (North - East). Towards the right abutment the elevation increases in a 0.5% gradient.

(2) The degree of compaction of the waste dam will be given by the movement of construction machinery and equipment for spreading material.

(3) MP: Modified Proctor, SP: Standard Proctor

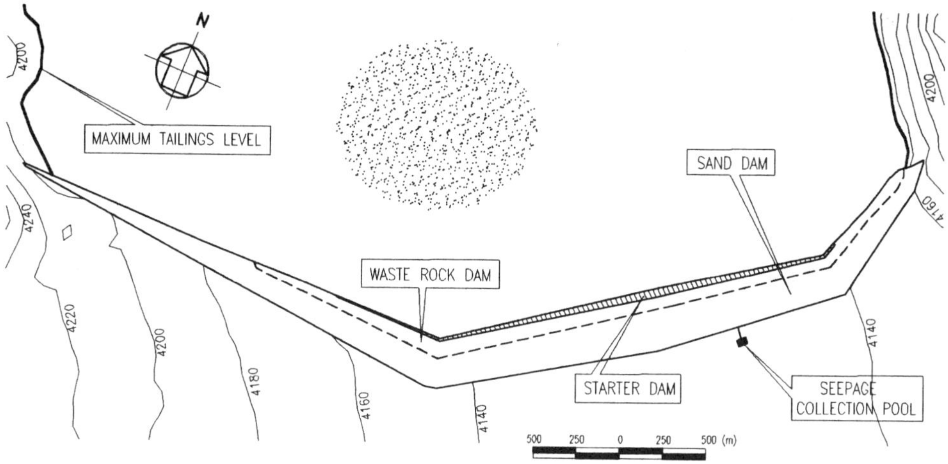

FIG 4a COLLAHUASI TAILINGS DAM – PLAN

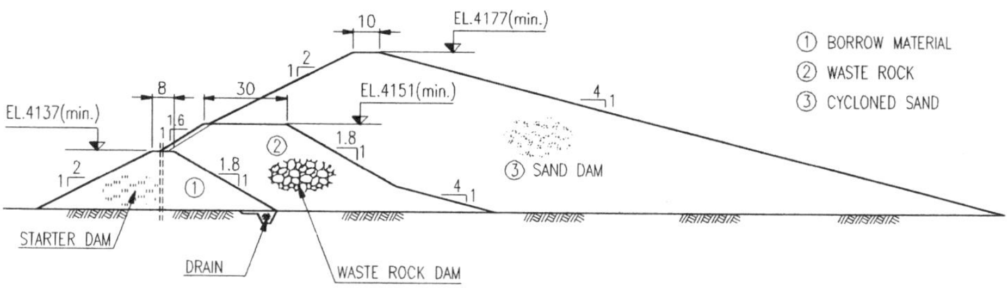

FIG 4b COLLAHUASI TAILINGS DAM – TYPICAL CROSS SECTION

These walls do not include an impermeabilization system in their upstream direction. During the first period seepage coming from the tailings will be caught by the drainage system of the starter dam. Afterwards, tailings disposed from the dam will impermeate the contact and will distance the clear water pond.

3 CONCLUSIONS AND COMMENTS

– The increase in the availability of waste material from the stripping of open pit mines has made this material an interesting alternative for the construction of impoundments for tailings. This depends on the characteristics of each case in particular, such as, the distance between the pit and the deposit, the cost of transporting this material, etc.

– The construction of dams using waste material has significantly simplified operational requirements when compared to the traditional system which was used in Chile, where cycloned sand is deposited using hydraulic methods. Requirements for the management of the clear water pond decrease and complex systems needed to deposit the sand are eliminated, such as: piping, slurry pumps, water for tailings dilution, etc.

– The waste materials studied in Chile and in particular those used in the cases presented in this paper, show excellent geotechnical characteristics (Valenzuela, 1989). Factors such as a high shear resistence, high permeabilities due to a low fines content, a varied distribution in the size of particles (well graded), together contribute to produce relatively high placement densities, even when there is limited compaction and the layers placed are thick. These details make waste structurally superior to tailings sand, because steeper slopes can be used and the dam's design is simplified. Besides, it reduces the need for complex drainage systems (usually expensive and hard to construct) and the need for more instrumentation, among other reasons.

– The use of waste material in dam construction requires an important coordination with mine operations, specifically operations related with waste disposal. This aspect may require increasing the quantity of hauling equipment or eventually modifying the operational shifts.

REFERENCES AND SOURCE OF INFORMATION

Bustamante B.,et al,1989. Obtención de Parámetros de Resistencia al Corte de Materiales Granulares Gruesos. Proceedings 2° Congreso Chileno de Ingeniería Geotécnica, Santiago de Chile, Abril 1989.

Valenzuela L.,1989. Estabilidad de Depósitos de Estéril. Proceedings 2° Congreso Chileno de Ingeniería Geotécnica, Santiago de Chile, Abril 1989.

Private reports prepared by Geotécnica Consultores for differents mines (Candelaria, El Romeral, Collahuasi).

Tailings and Mine Waste '99 © 1999 Balkema, Rotterdam, ISBN 90 5809 025 6

A test fill over a mine water treatment sludge impoundment

K.J. Dawson
CH2M Hill, Bellevue, Wash., USA

ABSTRACT: A case study of an instrumented test fill constructed over a sludge impoundment is presented. The impoundment is being closed as part of remedial actions at the Bunker Hill site in northern Idaho. The sludge is created by lime neutralization of acid mine drainage and consists mainly of oxides and hydroxides of iron, zinc, and manganese, with lesser amounts of gypsum and magnesium oxides and hydroxides. The purpose of the test fill was to determine the magnitude and rate of settlements that should be anticipated under closure loads and to determine constructibility constraints, such as bearing capacity and workability, of the sludge and soft sediments. Settlement and pore pressures created by placing up to 3.4 meters of fill over the sludge were monitored. A description of the mine water treatment process that produces the sludge, as well as the chemical composition and geotechnical engineering index properties of the sludge, is provided. Interpreted geotechnical engineering properties, such as the compression ratio ($C_{c\varepsilon}$), coefficient of secondary compression ($C_{\alpha\varepsilon}$), and coefficient of consolidation (c_v), back-calculated from monitoring of the test fill, are presented, as well as a summary of "lessons learned."

1 INTRODUCTION

A test fill was constructed over lime neutralization treatment sludge in one of two sludge ponds located on the Central Impoundment Area (CIA) for the Bunker Hill Mine in Kellogg, Idaho. Settlement and pore pressure were monitored during and after construction of the test fill. More complete data from the test fill can be viewed in the data report (CH2M HILL 1998a) and preliminary pond closure recommendations memorandum (CH2M HILL 1998b).

The Bunker Hill Mine tailings impoundment area, now known as the CIA, is roughly 18 meters (m) high and covers 80 hectares. Environmental cleanup of the site calls for placing approximately 1.3 million cubic meters of additional tailings from other locations, mostly within the Bunker Hill site, on top of the CIA, grading it to drain, and covering the area with a geomembrane and cover soil.

The sludge ponds are located on top of the CIA and will be closed as part of the remedial action. The ponds cover approximately 2.4 hectares. Grading to promote surface runoff will result in a maximum fill height of 3.4 to 4.9 m, corresponding to slopes of 2 to 5.5 percent, over the sludge. The ponds will be covered with the same geomembrane and cover soil as the rest of the CIA.

The test fill was constructed to determine special construction needs for placing the final fill and to estimate the magnitude and rate of settlement. When the test was devised, the sludge was too soft to support foot traffic. Information about the need for a reinforcing grid or geotextile and the minimum initial lift thickness could be gained from the test. No geotechnical information about the sludge settlement properties was available. The uncertainty in initial settlement estimates was roughly 2 m. The area was initially mandated for closure and capping with a geomembrane in one construction season. If the test fill had not been constructed, a costly preload would have been recommended and the level of risk would have been unknown.

The sludge is produced in the Central Treatment Plant (CTP), which started operation in 1974. The purpose of the CTP was to treat the combined effluents from the Bunker Hill Company mine, mill, lead plant, and zinc plant. The CTP, which is configured to use a lime high density sludge (HDS) process, currently treats a combination of water from the Bunker Hill Mine and surface water from other on-site areas disturbed by mining activities. This combined influent water has a pH of between 2 and 3 and contains dissolved metals primarily consisting of zinc, magnesium, manganese, and iron, and lesser amounts of aluminum, lead, and cadmium. The primary anion is sulfate.

The test fill was constructed in a corner of the second of two sludge ponds. Sludge at the bottom of the pond under the test fill was thought to be 10 to 15 years old, and sludge at the top of the pond was about 6 months old at the time of test fill construction.

The site elevation is approximately 700 meters. Average annual precipitation is 737 millimeters (mm), with only about 35 percent of this amount falling between the end of June and the end of November. The average summer temperature is 18°C, with a relative humidity just below 50 percent.

2 SLUDGE CHARACTERISTICS

The HDS process was developed by Bethlehem Steel in the 1960s in response to requirements for treatment of acid mine drainage (AMD) from coal mines. The key to the HDS lime neutralization process is that lime slurry is contacted with recirculated thickener underflow sludge in a sludge reaction and conditioning tank where the lime adheres to the recirculated sludge particles. The lime-treated underflow sludge overflows from this tank to a neutralization reactor where it is contacted with incoming AMD. In this process, precipitates form preferentially on the surface of the lime-coated sludge particles rather than as freely nucleating solids. Numerous recirculations of the sludge particles result in the growth of larger and denser particles, which enhances its settling rate and dewaterability compared to those formed by conventional lime neutralization.

Over the years since startup in 1974, the CTP has been operated using varying amounts of sludge recirculation. This has resulted in a range of sludge densities (percent solids by weight) being deposited in the sludge ponds. The specific deposited densities are unknown, but have varied between 1 and 30 percent solids. Sludge in the 1 to 10 percent solids range has the consistency of a viscous fluid, while sludge in the 20 to 30 percent solids range has the consistency of a highly thixotrophic, very soft clay. Corollary to this range in density is a range in size of sludge particles deposited and a range in the dewaterability of the deposited sludge. In general, higher deposited percent solids correlate with larger average particle size and better dewaterability.

A high density sludge production trial was conducted between March and July of 1997 (CH2M HILL, 1997). The trial was conducted to determine if sludge dewaterability could be increased and sludge volume decreased. Chemical composition and beginning and end solids content via filter press tests were conducted during the trial. Typical components of the dry sludge are listed in Table 1. Additional testing during the trial indicated that the sludge passes federal metals leaching criteria (TCLP), and thus is not a federally regulated hazardous waste. The sludge discharged by the CTP contains between 2 and 8 percent solids on a dry weight basis; on average the sludge is 3 to 5 percent solids.

Table 1. Typical sludge composition.

Major component	Approximate percentage by dry weight (%)
Iron oxides and hydroxides	25
Zinc oxides and hydroxides	25
Manganese oxides and hydroxides	22
Gypsum, $CaSO_4$+$2H_2O$	14
Magnesium oxides and hydroxides	12
Aluminum oxides and hydroxides	2

Filter press tests on the sludge currently produced at the CTP indicate a terminal density that is approximately 30 to 40 percent solids by dry weight (i.e., water content of 200 to 150 percent, where water content is the weight of water divided by the weight of solids) under loads of approximately 690 kilopascals (kPa). Additional water is not readily squeezed out of the sludge by increasingly larger stresses. Dewatering tends to come only through heating or chemical alteration.

Sludge produced for the past 10 years has probably been similar to that currently produced. Sludge produced between 10 and 20 years ago may have had a higher density, with a terminal percent solids of 50 to 60 percent (water content of 70 to 100 percent). Characteristics of the treatment plant indicate that the sludge was probably never of a lower terminal density than that currently produced.

2.1 Changes in sludge due to summer and fall drying

The test fill program was designed based on conditions observed in June of 1997. Large volumes of sludge were produced between March and June, the peak runoff period for the basin. In June, the surface of the sludge pond was wet. The sludge would support a person standing on a 0.6-m by 1.8-m plank, but would not support the weight of a person walking. Grab samples from the top 0.3 m were about 20 percent solids (water content of 400 percent), increasing to about 40 percent solids (water content of 150 percent) at a depth of 1.2 m. Presumably, the entire depth of sludge was saturated.

By the time the fill was actually constructed in December of 1997, the characteristics of the material at the surface of the impoundment had changed substantially. The change brought about by five months of drying is worthy of note. The surface of the sludge pond was moist, with a network of desiccation cracks approximately 13 mm deep. Foot traffic sank in less than 13 mm. A track-mounted Longyear 28 drill rig, weighing 3600 kg, was able to travel on the surface by leapfrogging 1.2-m by 2.4-m plywood sheets ahead of the tracks, so that each track was always supported by a 1.2-m-wide sheet (average bearing pressure of 6 kPa). There was no evidence of bearing capacity failure with this support method, suggesting an average shearing strength within the sludge of 1 kPa or more. The surface materials had the appearance of wet coffee grounds and low to medium plasticity based on field tests.

3 TEST FILL PROGRAM

3.1 Borings

Two weeks prior to test fill construction, hand auger and solid-stem, continuous flight auger borings were drilled through the sludge and decant sediments to visually characterize the materials, determine the depth of soft sediments, and install settlement plates at the bottom of the sludge layer. The locations of the borings are shown in Figure 1. The very soft, wet condition of the pond when the test was designed and estimates of 4.6 m for the sludge depth led to the initial use of a 102-mm hand auger. The hand auger was used for the first hole, SP1D, only. The firm surface conditions, deeper extent (eventually determined to average about 8 m) of the sludge, and equipment availability allowed a switch to a track-mounted Longyear 28 drill rig for the remainder of the sludge pond holes.

Hollow-stem auger drilling was attempted, but the material was so soft it pushed ahead of the split spoon sampler, resulting in no sample recovery. Instead, the holes were drilled with solid stem, continuous flight auger, moving the auger down at the same rate that the flights cut into the material. When the bottom of the boring was reached, as indicated by a sudden rise in drilling resistance, the augers were pulled out of the hole without spinning, the cuttings lodged in the flights of the augers were logged, and disturbed samples were collected for laboratory testing. In one of the borings, SP1P, the augers were pulled out at mid-depth of the sludge and 48-mm-diameter rings stacked inside of a split spoon sampler were pushed to collect intact samples. Other attempts at sampling were unsuccessful.

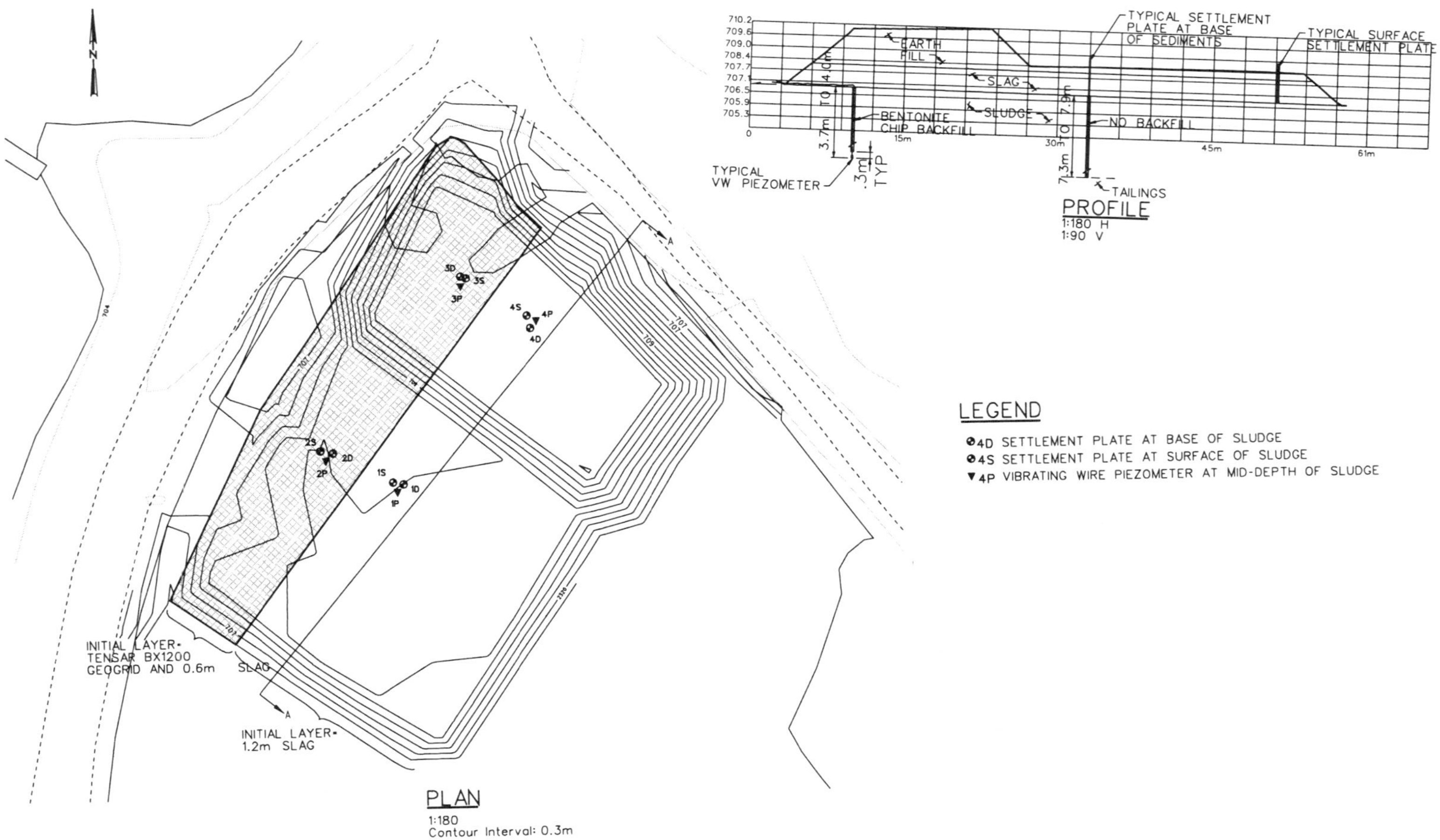

Figure 1. Test fill configuration.

3.2 Laboratory testing

Laboratory testing of disturbed samples consisted of 32 natural moisture contents, 5 Atterberg Limits, 3 specific gravity, and 1 percentage passing the number 200 sieve. These tests were performed primarily to determine the consistency of the material with depth and location, as well as to provide general classification. One consolidation and 3 unit weight tests were performed on the intact samples collected in the 48-mm inside-diameter brass rings. The results are summarized with the sludge descriptions in the Results section of this paper.

3.3 Instrumentation

Three types of instrumentation were installed between November 17 and 19, 1997:

1. Deep settlement plates, 64 mm in diameter, were installed through the borings to the bottom of the sludge to monitor the settlement of the underlying 9 to 12 m of mine tailings. Holes in the bottom plate could, potentially, allow groundwater levels to be monitored.
2. Surface settlement plates, 0.6 m square, were installed on the surface of the sludge to monitor settlement of these materials.
3. Vibrating wire, high-air entry, embankment-type piezometers were installed at mid-depth in the sediments to monitor pore pressure as the test fills were constructed and afterwards. Typically, a hole was drilled to within a foot of the desired tip location, then the piezometer was slowly pushed to the desired depth with an attachment on the end of the hand auger. The hole was backfilled to the surface with bentonite chips.

The settlement plates had inner risers that could be lengthened in sections as the fills were constructed. A casing, prevented from bearing directly on the settlement plate by a compressible styrofoam donut, isolated the inner riser from downdrag applied by the surrounding material.

3.4 Fill construction

Fill construction began on December 2, 1997. Two different initial lifts and two different total fill heights were constructed to evaluate fill constructibility. The initial lift conditions were: (1) 0.6 m of slag (mostly medium to coarse sand-sized material) over a Tensar BX 1200 geogrid, or (2) 1.2 m of slag without a geogrid.

The locations of the different initial lift treatments are shown in Figure 1. The slag, which provided drainage, was overlain by earth fill, a silty gravel with sand. Half of each fill was built to a final height of 1.7 m and the other half was built 3.4 m high, as shown in Figure 1. The different initial lift conditions were constructed to determine which surface treatment allowed for the easiest construction. Two different fill heights were represented to provide a range of likely loadings.

Initially, the slag was spread with a Komatsu D39P dozer. When no mudwaves or deformations were observed with the small dozer, a Caterpillar D5H LGP took over the spreading. Hauling trucks (Moxys) did not move onto the fill until approximately 1.2 m of earth fill were placed over the slag. The earth fill was placed in 0.3-m lifts, rolled for compaction, and periodically tested for density by a nuclear densometer. Compaction tests were also performed on the slag so that its unit weight could be accurately known.

Construction activities caused no visible deflections of the sludge or decant pond sediments. Even where the dozer pushed over as little as 1.2 m of unreinforced slag, there was no visible deformation.

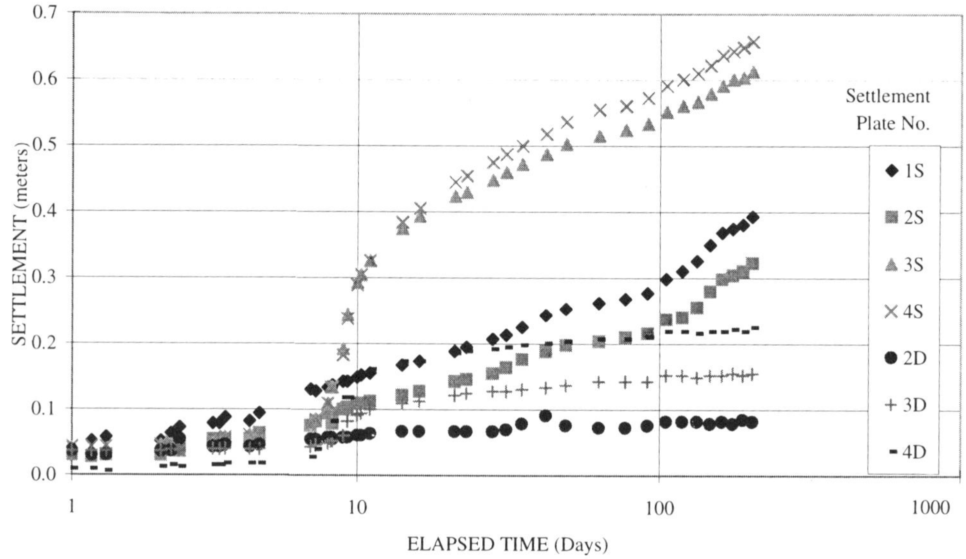

Figure 2. Settlement at sludge pond.

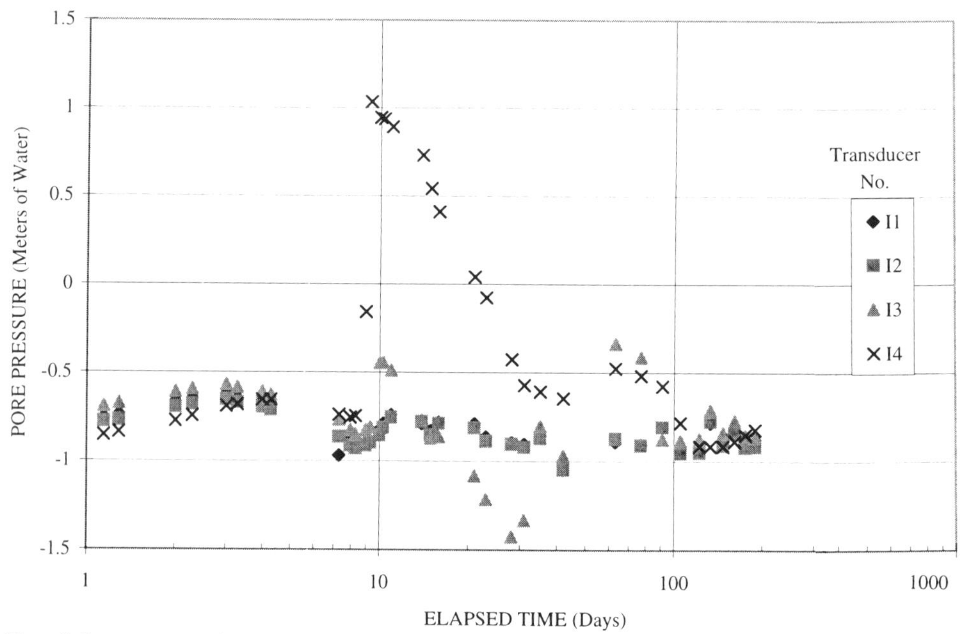

Figure 3. Pore pressure at sludge pond.

4 RESULTS

4.1 Subsurface conditions

Two different sludges and a silty sand interpreted to be mine tailings that underlie the pond were encountered in the borings. The upper sludge was about 5.3 m thick and was dark brown, moist, and very soft to soft. Partially hidden desiccation cracks extended approximately 0.3 m below the ground surface. The texture below the upper 0.3 m was generally smooth, but occasional zones of blocky (as in desiccated) material were also present. The material was non-dilatent and would roll to a thread somewhat larger than 3 mm in the field; however, laboratory tests indicated it was nonplastic. The moisture content tended to increase with depth, with a minimum of 147 percent, a maximum of 343 percent, and an average of 259 percent (corresponding to 41, 23, and 28 percent solids). The specific gravity was 3.5 and the moist unit weight averaged from 3 tests was 11 kilonewtons per meter cubed (kN/m^3).

The lower sludge ranged between 2.1 and 2.7 m thick, so that the bottom of the sludge pond was 7.3 to 8.2 m below the surface. The lower sludge was light gray-brown, moist, soft to firm, and contained conglomerations of light gray clay (visual classification) in a light brown matrix. The lower sludge graded to brown with a smooth texture, as in the upper sludge, in some layers. It was very sticky with no dilatancy and had an appearance and wet texture similar to hydrated granular bentonite. Laboratory tests indicated it was nonplastic, with a moisture content ranging from 55 to 343 percent, an average moisture content of 210 percent (corresponding to 65, 23, and 32 percent solids), and a specific gravity of 3.2.

The sludge was underlain by loose to medium dense, moist to wet silty sand. Free water was encountered in the sand at a depth of 7.9 m in one of the borings. There was no standing water within the sludge.

4.2 Settlement and pore pressure

Figures 2 and 3 show the settlement plate and piezometer readings, respectively, collected through the end of June 1998. Figure 4 shows the changes in total stress that occurred over the same time period as the fill was constructed. In the "elapsed time" plots shown, December 8, 1997, was set as time zero, 6 days after fill construction began and 9 days prior to when construction was complete. The selection of December 8 as the start time was somewhat arbitrary, but also selected as the date when most of the readings began to increase consistently.

In Figure 2, the upper two lines, for settlement plates 3S and 4S, show settlement of the surface of the sludge under 3.4 m of fill, at the loads shown in Figure 4. The third and fifth lines down, for settlement plates 1S and 2S, show the settlement of the surface of the sludge under 1.7 m of fill. The remaining lines, for plates 2D, 3D, and 4D, show the settlement of the sediments below the sludge. The actual settlement of the sludge alone can be computed by subtracting the settlement of the deep plate from the settlement of the corresponding upper plate.

The vibrating wire piezometer readings, shown in Figure 3, were adjusted for changes in temperature at each reading, but not barometric pressure. It was estimated that changes in barometric pressure of as much as plus or minus 51 mm of mercury could occur as fronts pass through the area. These changes could result in instrument changes as large as plus or minus 0.67 m of water.

The pore pressures started negatively, increasing to positive values after fill construction under the 3.4-m-high fill, and increasing to less negative values under the 1.7-m-high fill. The pore pressures returned to their pre-loading values 30 to 40 days after the beginning of fill construction (20 to 30 days after fill completion). The initial and final negative pore pressures reflect the unsaturated conditions observed during test hole drilling and indicated by water level readings (none measured) within the settlement plate risers during the test.

One of the laboratory consolidation test curves has been included as Figure 5. The rate of strain curves at other loads had similar shapes.

5 CONCLUSIONS

Based on settlement data and laboratory test results to date, the sludge does not appear to behave as commonly predicted by geotechnical theory. Typically in consolidation tests on soil, settlement proceeds at a relatively fast, linear rate when plotted on a logarithmic scale

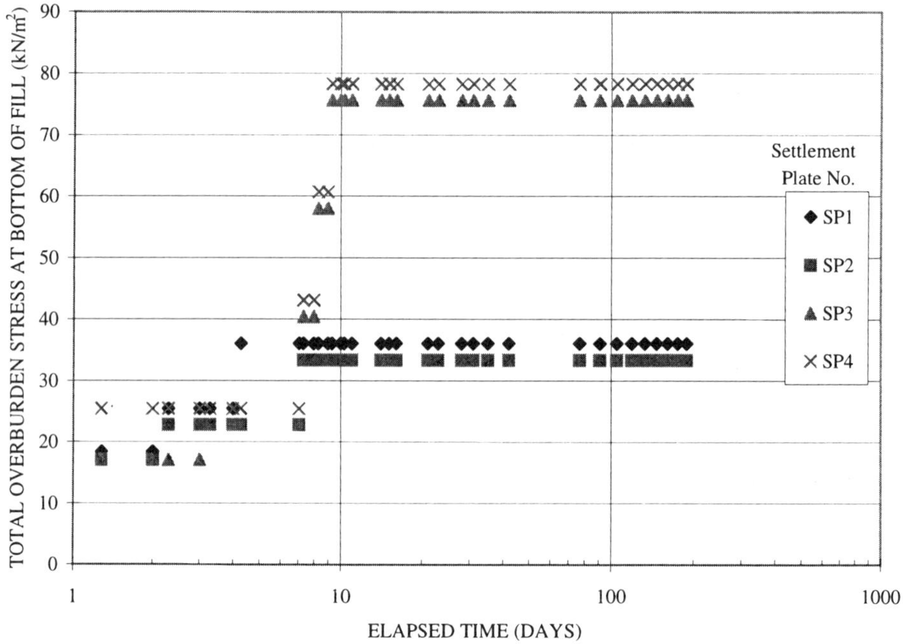

Figure 4. Added total stress vs. time.

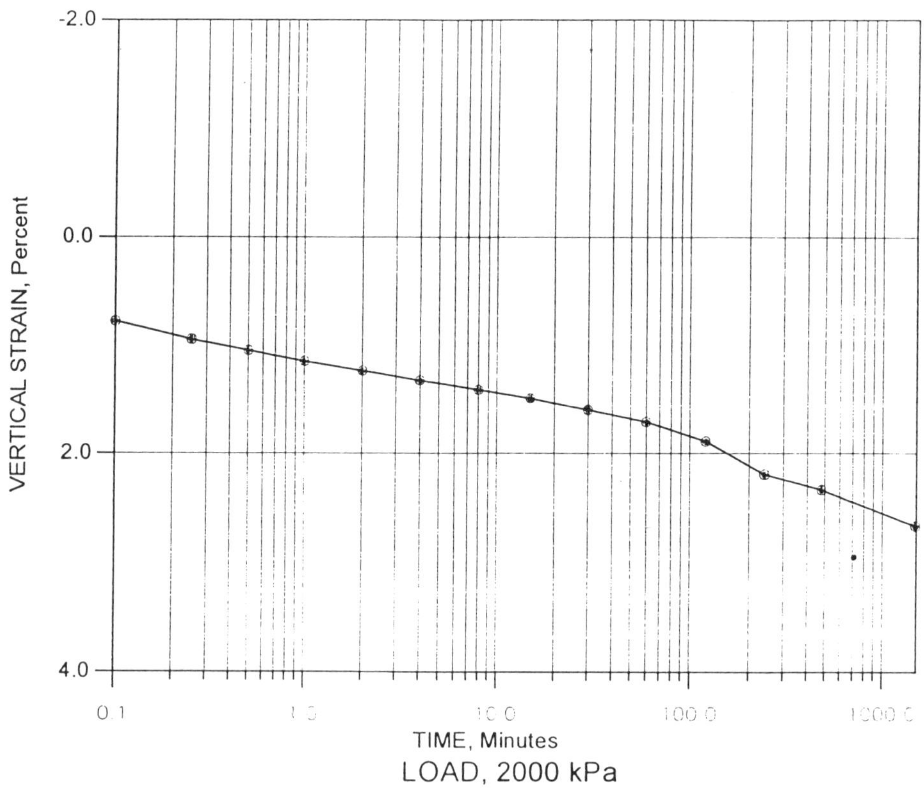

Figure 5. Typical laboratory rate of strain curve for sludge sample.

(primary settlement) and then gradually the rate of settlement decreases and continues at a lower linear rate (secondary consolidation). However, the test fill data show that the rate of settlement of the sludge is higher in the last 4 months than in the first 2 months following construction of the fill.

Based on the pore pressure data, primary consolidation of the sludge occurs relatively rapidly. Secondary consolidation is relatively large and will continue at an exponentially decreasing rate. The preliminary sludge properties estimated from the test fill monitoring program are listed below. There is considerable uncertainty in these properties, especially in the estimates for the compression ratio. The compression ratio was estimated by assuming a straight-line relationship (normally consolidated material) between the settlement at the end of pore pressure dissipation and logarithm of loading at two points, 1.7 and 3.4 m of load. The material is probably somewhat overconsolidated, due to periodic drying, and it is uncertain where the preconsolidation pressure lies in relation to these points. Monitoring of the test fill is continuing and these properties will be updated in the future.

Moist unit weight = 10.2 kN/m^3

Average water content = 220%

Compression ratio, $C_{c\varepsilon} = 0.2$

Coefficient of secondary compression, $C_{\alpha\varepsilon} = 0.03$

Coefficient of consolidation, $c_v = 0.3$ m^2/day

Provisions of the test fill program to examine various initial layer thickness and subgrade reinforcing options were not necessary because the surface of the sludge dried out and gained strength quickly over the 5 months with no use. It can be concluded that a 0.6-m initial layer, no reinforcing, and efficiencies allowed by using larger equipment can be achieved for similar sludges in similar climates if a few months of summer drying can occur before closure begins.

Other observations or lessons learned from the test fill include:

- Plan for redundant installations and use as many methods as possible to measure pore pressures. Electronic piezometers proved to be key in interpreting the results of this test, yet only the two piezometers under the highest fill showed any significant response to loading and the response was not consistent between the two. Experience on other projects suggests that there are numerous things that can and do go wrong with electronic piezometers.
- Install instrumentation that is easy to monitor over an extended time period. The vibrating wire piezometers and terminal box, though somewhat more expensive than pneumatic piezometers and a homemade protection system for the wire ends, were easily read by untrained personnel after initial setup and withstood the rigors of snow, heavy construction traffic, and temperature extremes.
- Establish a weather station to provide barometric pressure when using vibrating wire piezometers.
- Size test fills to match proposed construction conditions. Interpretation of settlement properties from a test fill will almost always be difficult because you cannot afford to collect settlement data at several different loads and still have multiple instruments at each load for redundancy. By sizing the heights of fill to match a likely range of fill heights that might be constructed, as we did on this project, the results of the test program still provide value even if you cannot precisely predict engineering properties.
- Consider laboratory triaxial consolidation testing of artificially prepared sludge samples for preliminary design. Sludge samples could be allowed to settle and consolidate to a water content approximating that of the in situ deposit in a supported triaxial cell. Consolidation testing of the resulting sample might provide estimates of consolidation properties within the normally consolidated range of the sludge.

ACKNOWLEDGEMENTS

The author would like to thank Don Anderson and Jim Stefanoff of CH2M HILL and Cami Grandinetti of the U.S. Environmental Protection Agency, Region X, for their assistance with this paper.

REFERENCES

CH2M HILL 1998a. *Bunker Hill Central Impoundment Area Test Fill Data Report - Draft - Bunker Hill Superfund Site, Kellogg, Idaho*. Prepared for U.S. Environmental Protection Agency Region X, July 8, 1998.

CH2M HILL 1998b. *Bunker Hill Superfund Site Preliminary CIA Sludge Pond Closure Recommendations*. Technical Memorandum to EPA, State of Idaho, and U.S. Army Corps of Engineers Management Team. May 22, 1998.

CH2M HILL 1997. *CTP High Density Trial Report Bunker Hill Superfund Site, Kellogg, Idaho*. Prepared for U.S. Environmental Protection Agency Region X, December 1997.

Tailings and Mine Waste '99 © *1999 Balkema, Rotterdam, ISBN 90 5809 025 6*

Predicting the deformation of tailings dams resulting from earthquake liquefaction

R.N.Kostaschuk & J.M.T.Wilkinson
Knight Piésold Limited, Vancouver, B.C., Canada

P.M.Byrne
University of British Columbia, Vancouver, B.C., Canada

ABSTRACT: During an earthquake loose soil materials such as mine tailings may liquefy, resulting in severe damage or failure of the tailings dam. In this paper, a mechanics-based total stress analysis approach is presented which uses the finite difference geomechanics modeling program FLAC to model the triggering of soil liquefaction and the resulting large displacement. In the approach, an earthquake motion is assigned to the bedrock surface under a dam and the induced cycles of shear stress are counted in each element of the model. When sufficient cycles have accumulated to trigger liquefaction in an element, the stress state in the element is set to that of a fluid and the strength and stiffness parameters are set to their post-liquefaction values. As earthquake shaking continues, additional elements may liquefy and deform in their attempt to mobilize resistance and return the system to dynamic equilibrium. In this way, the magnitude and pattern of displacements is predicted. This approach is used to back-analyze the response of two tailings dams at the Mochikoshi gold mine in Japan in 1978, when a magnitude 7.0 earthquake liquefied the tailings. The approach correctly models the failure of one dam during the main earthquake event, and also correctly simulates that a second dam deformed but did not fail during the main event. This result helps to verify the analysis approach, and lends confidence to applying the method for predicting earthquake response of an existing or planned tailings dam. A comparison of potential remediation measures to seismically upgrade a tailings dam is presented, where it is shown that a very significant construction cost saving can be realized if the proposed method is used rather than older simpler methods which make conservative assumptions.

1 INTRODUCTION

The damage that results when tailings liquefy behind a dam is caused by the very large reduction in stiffness and strength that occurs upon liquefaction. In assessing liquefaction response, it is necessary to determine whether liquefaction will be triggered in significantly large zones. If liquefaction is triggered, it is also necessary to compute displacements using post-liquefaction stiffness and strength parameters that reflect the large reduction which occurs upon liquefaction.

Many existing mine tailings dams are not adequate to resist seismic loading and require remediation which generally consists of densification, drainage or buttressing. Faced with the problem of providing remedial measures to improve the seismic stability of existing structures of dams in seismic areas, mine operators and engineers must design the remedial measures to be as cost-effective as possible. Using conventional, and possibly overly-conservative techniques, the cost of remediation of these structures may become excessive. The benefit of advanced, numerical modeling techniques include the potential to liquefy materials on an element-by-element basis, thereby more realistically predicting associated strength and stiffness losses and resulting displacements.

This paper presents a method using numerical modeling to assist Engineers in predicting soil liquefaction and the response of soil structures to earthquakes. The method is verified by modeling a case history of the seismic response of two similar Tailings Dams at the Mochikoshi Mine in Japan during a large earthquake in 1978. There were only subtle differences in the configurations of the two dams, but Dam No. 1 failed during the earthquake while Dam No. 2 did not fail until the following day after large aftershocks. The difference in response of the two similar dams to the initial main shock provides a rare opportunity to verify the modeling method with a high level of confidence.

This paper i) presents the Mochikoshi tailings pond case history, ii) provides a review of the current practice for seismic liquefaction prediction and displacement analysis, iii) describes the new numerical modeling approach using stress cycle counting to liquefy individual elements, iv) presents results of the new analysis method for both dams to verify the technique, and finally, v) compares the extent of buttressing required to seismically upgrade Dam No. 1 to prevent failure using both the current practice and the new approach.

2 THE MOCHIKOSHI TAILINGS POND CASE HISTORY

The tailings pond in operation at the Mochikoshi Gold Mine in 1978 was constructed in a bowl-shaped depression on top of a mountain by sealing off the perimeter with three dams as shown on Figure 1. The cross-sections of the two tailings dams that are studied here are shown on Figure 2. The dams were constructed using upstream methods. The upstream portion of the fill had a slope of 3 to 1 (horizontal to vertical) in both cases. The foundations of the dams comprised weathered, cobbly volcanic tuffs and the fill materials used to construct the dams were also of volcanic origin. There were no under-drainage systems beneath the tailings to collect consolidation seepage and relieve water pressures so the water table was high in both cases, located just below the fill placed in the upstream raises (Ishihara, 1984). The main difference between the two dams is that the crest of Dam No. 1 was 14 meters above the starter dam, while for Dam No. 2 it was only 10 meters.

On January 14, 1978, the Izu-Oshima-Kinkai earthquake with a Magnitude of 7.0 occurred 15 km off the coast of the Izu Peninsula, which is located 120 km south-west of Tokyo. The Mochikoshi Mine was only 36 km from the epicenter of the earthquake event, and experienced strong ground motion. The focal depth of the earthquake was presumed to be approximately 10 km (Ishihara, 1984).

The earthquake caused liquefaction of the tailings stored behind Dam No. 1 and failure of the dam during the event (Investigation Committee, 1979). An eye witness reported that

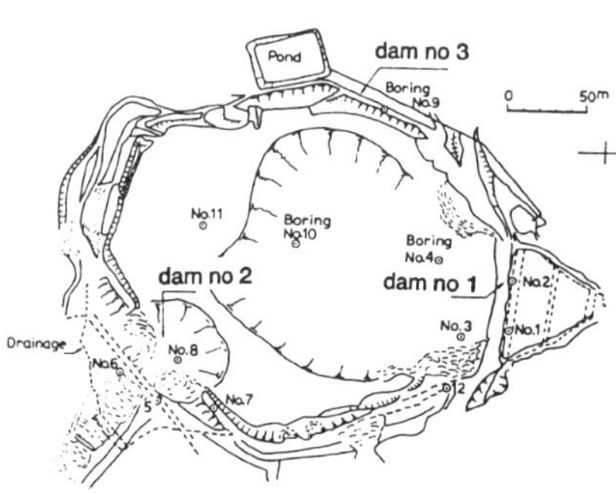

Figure 1. Plan view of the Mochikoshi tailings pond

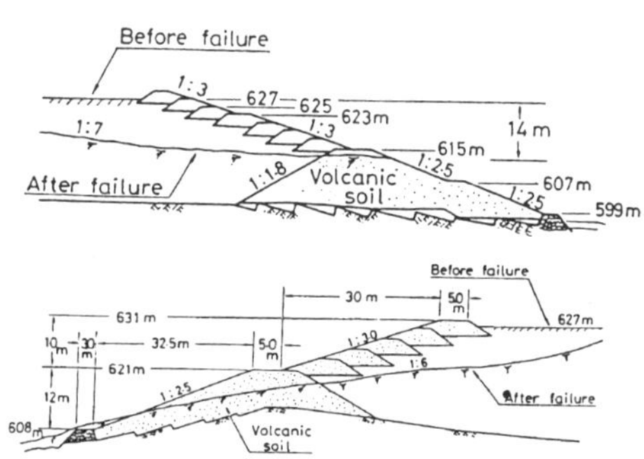

Figure 2. Cross-sections of the two dams studied
Dam No. 1 is the upper figure

within 10 seconds of the main shock, the face of the dam swelled out, the crest dropped, and a breach over the top of the dam occurred near the left abutment. Figure 2a shows the final surface after failure, and that the entire upstream fill portion of the dam failed. The slope of the tailings surface after failure was approximately 7 to 1 (horizontal to vertical).

Unlike Dam No. 1, Dam No. 2 did not fail during the main event. On the day following the main shock, two large aftershocks with Magnitudes of 5.8 and 5.4 occurred within 10 km of the mine. Dam No. 2 actually failed approximately 6 hours after these aftershocks, although significant deformation and cracking occurred earlier. It was postulated that the failure of Dam No. 2 was probably caused by re-distribution of pore pressure from the liquefied tailings into the upstream embankment fill (Ishihara, 1984). The slope of the tailings surface after failure was approximately 6 to 1. This paper only addresses the initial Magnitude 7.0 event and its effects, and the study of the delayed failure of Dam No. 2 is reserved for future work.

To perform real time modeling of the effects of the Magnitude 7.0 earthquake at the Mochikoshi tailings pond, it is necessary to know the input motion at the base of the dam. A key feature of the motion is the Peak horizontal Ground Acceleration (PGA). The analysis of earthquake liquefaction and subsequent deformation is sensitive to the PGA, and rational for the choice of values is presented here. There were no seismic recording stations near the mine, so all knowledge of the PGA at the site has to be derived from damage observations or from the use of earthquake attenuation relations and dynamic response analyses. A survey of overturned tombstones in cemeteries was conducted after the event, and it was estimated that the PGA at the mine would have been approximately 0.25 g (25% of gravity) at the ground surface.

Another method used by the Investigation Committee to assess the PGA at the site was dynamic response analysis. These analyses concluded that the peak horizontal acceleration at the surface level of the disposal pond was probably in the range of 0.15 g to 0.25 g. Ishihara (1984) summarizes the results of pseudo-static limit equilibrium stability analyses completed by the Investigation Committee. He concludes that a seismic coefficient of 0.17 was required to cause the factor of safety of Dam No. 1 to fall to unity (failure). Further, Dam No. 2 was more stable and had a factor of safety against failure of approximately 1.5 under the 0.17 seismic coefficient.

The PGA can also be assessed from attenuation relations which are based on statistical analyses of a number of recorded events, and give an estimate of the PGA from the magnitude of the earthquake and the distance from its focus to the site. For this study, two modern earthquake attenuation relations were used to estimate the PGA at the site resulting from a Magnitude 7.0 earthquake occurring at an epicentral distance of 36 km. The relation of Fukushima and Tanaka (1990) was used as the primary relation, being specific to strong earthquake ground motion in Japan. This method yields a value for the PGA of 0.185 g. The attenuation relation of Idriss (1993) was also used to estimate the PGA at the site. For design purposes, Idriss recommends using the mean plus one standard deviation values for the PGA, resulting in a value of 0.189 g.

Based on the above evidence, the Peak horizontal Ground Acceleration for the site is in the range from 0.17 g to 0.20 g.

For the real time numerical analyses, it is necessary to generate earthquake acceleration time histories, each with a PGA set within the above range. A Japanese earthquake record was selected for the analyses. The record has a PGA of 0.21 g and is from a Magnitude 7.4 event that occurred on June 12, 1978. The record has a peak ground acceleration of 0.21 g. A Fast Fourier Transform of the original record showed that there is little energy above 9.0 Hz, and all frequencies above this value were filtered out. The record was scaled to achieve the PGA values desired for the analyses and then integrated to a velocity time history that was applied to the base of the model.

Soil properties for the numerical models were based on work completed by the Investigation Committee after the failures. This work comprised Standard Penetration Tests (SPT's) and the recovery of disturbed and undisturbed samples for laboratory testing. The following table summarizes the index parameters for the fill and the tailings.

Table 1 - Results of Laboratory Testing of Tailings and Fill Materials

Soil Parameter	Tailings	Fill Material
Moisture content, w (%)	37	30 to 60
Specific gravity, G_s	2.73	
Void ratio, e	1.0	1.1 to 2.6
Bulk unit weight, γ_b (kN/m^3)	18.3	14 to 19 (15.7 used in models)
Percent passing No. 200 sieve (%)	60 to 95	40 to 90
Liquid limit (%)	27 to 31	
Plastic index (%)	0 to 10	
Unified soil classification	ML to CL	SM, ML and CL
Average effective friction angle, ϕ'	35°	35°
Average effective cohesion, c ' (kPa)	0	25 (0 used in model)
Cyclic resistance ratio, $\sigma_{dl}/2\sigma_o'$ (Note 1)	0.2	

Note 1. $\sigma_{dl}/2\sigma_o'$ is defined as the cyclic stress ratio to cause 5% double amplitude axial strain in 20 cycles in laboratory undrained cyclic triaxial compression tests

The soil parameters for the numerical modeling were based on normalized SPT blowcount $(N_1)_{60}$, and the fines content of the soil. Ishihara (1984) concludes that the pre-liquefaction SPT blow counts, N in the tailings ranged from 3 to 7 in the deeper parts of the tailings deposit that were not disturbed by the liquefaction and failures. This range of blowcounts normalizes to an $(N_1)_{60}$ of approximately 3 for the tailings. The fill for the starter dam and the upstream raises was a mixture of weathered volcanic tuffs and ashes from local borrows, which was lightly compacted. For the fill materials, the normalized $(N_1)_{60}$ blow count is approximately 5. For modeling purposes, the foundation was assumed to comprise material of similar grading as the fill materials, being comprised of the same weathered volcanic tuffs and ashes. However, the foundation had a higher blowcount, and a value for $(N_1)_{60}$ of 30 was used in the models.

3 ANALYSIS METHODOLOGY

3.1 *Analysis Procedure*

Liquefaction refers to a sudden large loss in stiffness and strength of granular soils under static or seismic loading. It is caused by the tending of loose granular soils to contract upon shearing. If such contraction is prevented or curtailed by the presence of water in the pores that cannot escape during the loading period, a large rise in pore water pressure can occur. If the pore pressure rises to equal the total stress, as it can under cyclic loading, the grain contact or effective stress drops to zero and the soil looses all its shearing resistance and readily deforms under applied shearing loads. However, as it deforms it will gradually dilate and develop some effective stress and strength and the strength at large strain is referred to as its residual strength. Soil prior to liquefaction can be quite stiff and strong. However, once triggered to liquefy it will have greatly reduced stiffness and strength.

In dealing with seismically induced liquefaction there are three concerns:

1. Will the design earthquake trigger significant zones of the dam to liquefy and if so,
2. Will a flow slide occur because of low residual strength, and if not,
3. What displacements will occur?

Current practice generally addresses these concerns with three separate analyses.

3.2 *Triggering Liquefaction - Current Practice*

The zones that may liquefy are identified by comparing the cyclic stress ratio (CSR) caused by the design or actual earthquake with the cyclic resistance ratio of the soil (CRR) through a factor of safety, F_{TRIG} = CRR/CSR. Zones having F_{TRIG} less than 1.0 to 1.3 are deemed to liquefy.

CRR can be obtained directly from cyclic loading tests on undisturbed samples, but is generally estimated indirectly from penetration resistance testing and field experience. CSR is commonly estimated from dynamic analyses using an appropriate time history of acceleration or velocity as base motion. An equivalent visco-elastic analysis is generally used in which the nonlinear behavior of the soil is incorporated through selection of shear moduli and damping values that are compatible with the level of computed shear strain. This involves an iterative process and computer programs such as SHAKE and FLUSH are commonly used to carry out such analyses. From these analyses, the peak horizontal dynamic shear stress τ_{max}, is identified for each element and CSR computed from CSR = $0.65\ \tau_{max}/\sigma'_o$ in which σ'_o = the vertical effective stress in the element prior to the earthquake. A simplified formula based on such analyses is sometimes used in place of a dynamic analysis, but requires an estimate of the surface ground motion A, which may be amplified relative to the base motion. Idriss (1990) provides some guidance on such amplification.

3.3 *Flow Slide - Current Practice*

Current practice is to specify post-liquefaction residual strengths in those elements predicted to liquefy, and assess stability from a standard limit equilibrium analysis. A computed factor of safety, $F_{flow} < 1$ would suggest large displacements and a flow slide, and remedial measures are probably required. On the other hand, $F_{flow} > 1$ does not insure that displacement will be tolerable, and displacement analyses are required to verify this.

3.4 *Displacement Analyses*

Current practice generally involves a Newmark type displacement analysis (Newmark, 1965). This involves idealizing a potential slide block identified from limit equilibrium analyses as a single-degree-of-freedom rigid block system on an inclined plane. The base plane is then shaken with the design earthquake, and the movement of the block down the plane computed. Newmark also suggested that the maximum displacement of the block, D, could be computed from a simple formula such as:

$$D = 6V^2/2gN$$

Where: V = the maximum velocity of the base
g = the acceleration of gravity
and N = the yield acceleration as a fraction of gravity required to initiate movement of the block down the slope. In practice, this is obtained from limit equilibrium analyses by finding the seismic coefficient to reduce the factor of safety to unity. Consideration should be given to the fact that the strength would change as liquefaction progresses.

3.5 *New Approach*

In this approach the triggering of liquefaction, the possibility of a flow slide and the displacement are evaluated from a single dynamic analysis. The time history of dynamic shear stress on the horizontal plane in each element is computed. As each half cycle of shear

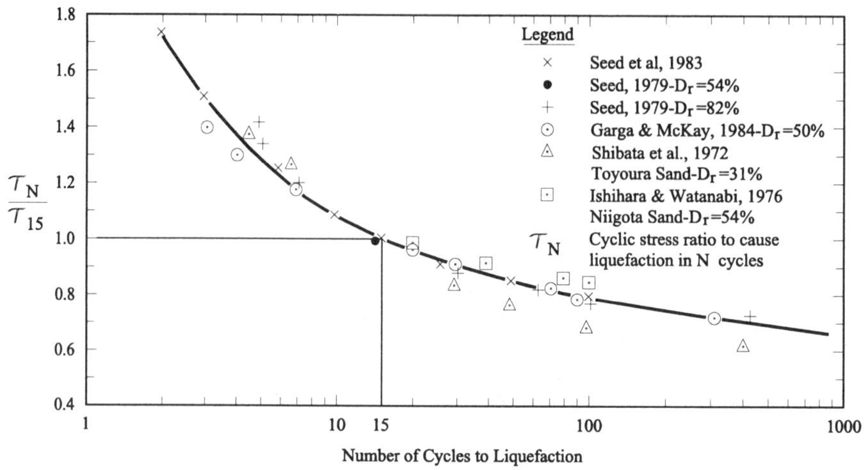

Figure 3. Chart for determining the equivalent number of τ_{15} cycles for triggering of liquefaction (Byrne, 1991)

stress is completed it is weighted in accordance with its value in comparison to τ_{15}, the value that would cause liquefaction in 15 cycles, and obtained from the Seed liquefaction chart (Youd and Idriss, 1997). The weighting used for each half cycle is based on laboratory testing as shown in Fig 3, which shows the ratio of τ_N/τ_{15} in which N is the number of cycles. 15 cycles of $\tau = \tau_{15}$ will cause liquefaction with a weighting factor of 1. Three cycles of $\tau = 1.5\ \tau_{15}$ will also cause liquefaction with a weighting factor of 5. Each stress pulse in each element is weighted as the shaking proceeds and when the count reaches 15, the element is triggered to liquefy.

Liquefaction is simulated by specifying a hydrostatic state of stress at the moment of triggering. This is achieved by setting the two horizontal stresses equal to the vertical stress and the shear stress equal to zero. In addition, very soft post-liquefaction stress strain and residual strength properties are specified. In this way each element will liquefy at a different time during the shaking and be prescribed a liquefied stress state and post-liquefaction properties while the shaking proceeds. Displacement, including the possibility of a flow slide, will be predicted if the residual strength is not sufficient for stability under the initial geometry. However, when geometry changes are considered then large displacements rather than a flow slide are predicted because as the slope flattens the driving stresses will reduce and a stable configuration becomes possible.

The computer program FLAC (Cundall, 1995) is ideally suited to perform such analyses. This is an explicit finite difference dynamic analysis program that solves the equations of motion in the time domain. Total rather than incremental equilibrium is satisfied at each time step so that specified stress changes associated with the triggering of liquefaction can be incorporated as well as the changed stress strain and strength relations. Because a plasticity approach is taken, the predicted stresses are curtailed so that they do not exceed the strength of the soil. Geometry changes can be accounted for in FLAC, but if deformations are very large the element grid may become unstable. The FLAC program was used to carry out the new liquefaction approach presented in this paper.

4 CASES ANALYZED AND ASSUMPTIONS

The geometry of the two dams studied is shown on Figure 4. The water table location and the $(N_1)_{60}$ values for the fill, foundation and the tailings are shown. The embankment fill zones are flagged as non-liquefiable in the analyses as they are above the water table. The fines content was set at 60 % for all the materials.

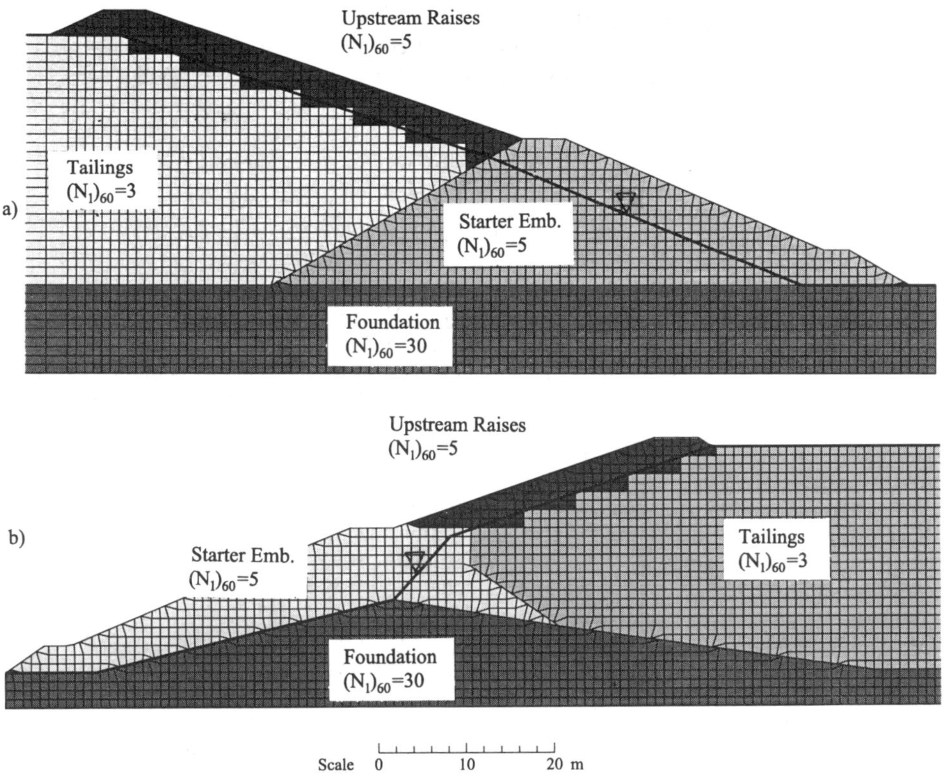

Figure 4 - Model geometry showing $(N_1)_{60}$ values and water table locations for a) Dam No. 1 and b) Dam No. 2

The analyses use a modification to the built-in Mohr-Coulomb model in FLAC. Both pre- and post-liquefaction strengths are set using cohesion and friction angle. Prior to liquefaction, the strength of all elements is set with a friction angle of 35° and zero cohesion. The shear modulus for the analysis is stress dependent, and is set from the $(N_1)_{60}$ value using the following equation (After Seed et al., 1986):

$$G_{max} = 0.6\,(20)\,(21.7)\,(N_1)_{60}^{1/3}\,P_A\,(\sigma'_m/P_A)^{1/2}$$

where: G_{max} = maximum (small strain) shear modulus
P_A = atmospheric pressure
σ'_m = mean effective confining stress
$G = 0.4\,G_{max}$ = average shear modulus to use in the numerical model

In addition to $(N_1)_{60}$ and the fines content, the following parameters are required for the dynamic analysis:

- Fraction of critical damping, D
- Post-liquefaction residual strength ratio, c_u/σ_{vo}'
- The strain required to mobilize the residual strength after liquefaction, γ_r

In the FLAC analyses, Rayleigh damping was used. The center of frequency of the Rayleigh damping was set at 2.5 Hz which corresponds to the predominant frequency of the earthquake input motion at the base of the model. The fraction of critical damping was set at 8 percent. The values for damping and the ratio of G / G_{max} were chosen because they produce agreement with SHAKE analyses for base accelerations in the range up to 0.25 g (Byrne et al, 1998).

If an element is triggered to liquefy its strength is reset to its residual value c_u, specified by a c_u/σ'_{vo} ratio where σ_{vo}' is the vertical effective stress prior to the earthquake. The c_u/σ'_{vo} ratio for the tailings was determined using the slope of the final failure surface. From infinite slope theory, it can be shown that:

$$c_u/\sigma_{vo}' = \frac{1}{SLOPE} \quad \frac{\text{total unit weight}}{\text{buoyant unit weight}}$$

where: SLOPE = the horizontal length of the failure surface for a vertical rise of unity

Note that the ratio of unit weights is approximately 1 for slip surfaces located near the water table and increases if the slip surface is deeper. The slope of the final failure surface for Dam No. 1 was 7, which produces a c_u/σ_{vo}' value of 0.143 for a slip surface just below the upstream fill of the dam. For Dam No. 2 the final slope was 6, resulting in a minimum c_u/σ_{vo}' value of 0.17 which increases with a deeper slip surface location. This range of values is consistent with work completed for the seismic upgrade of Duncan Dam (Byrne et al, 1994) and is also consistent with the experience of the authors on similarly graded silty tailings from gold mines.

The strain to mobilize the post-liquefaction residual strength, γ_r was set at 2.5 percent, and this together with the c_u value set the post-liquefaction shear modulus G_{liq}. This value of γ_r is based on a parametric study by Byrne et al (1999). In the study, static, post-liquefaction, one-dimensional FLAC simulations of shallowly sloping ground were performed. The results were compared to the empirical database and methods of Bartlett and Youd (1992). For post liquefaction residual strength ratios, c_u/σ_{vo}' ranging from 0.1 to 0.3, a value of γ_r of 2.5 %, was found to best reproduce the field-observed displacement values reported by Bartlett and Youd.

Using the parameters described in the foregoing, analyses were carried out for both dams over an range of PGA from 0.17 g to 0.20 g. The object is to study the predicted response of the two dams to the same earthquake event and compare with the observed Mochikoshi response reported by Ishihara (1984). It is emphasized that the only difference between the analyses for the two dams are the geometry (mainly Dam No. 1 is higher than No. 2) and the c_u/σ_{vo}' values which are slightly higher for Dam No. 2. The following section describes the results of the analyses.

5 RESULTS OF THE BACK-ANALYSES

A series of analyses, using 4 different values of Peak horizontal Ground Acceleration (PGA), were completed using the Dam No. 1 material parameters and geometry of Figure 5a. The PGA values used were 0.17 g, 0.18 g, 0.19 g and 0.20 g, resulting in limited deformations of 0.62 and 0.63 meters at the crest for the values of 0.17 g and 0.18 g respectively. For both of these cases, deformation was observed to accumulate and eventually stop during application of the earthquake time history. In contrast, results of the analyses using PGA values of 0.19 g and 0.20 g indicate the accumulation of deformations during application of the earthquake time history and continued large deformations after the strong ground motion has passed, eventually causing excessive deformations and failure. This occurs because the predicted extent of liquefaction increases with shaking level and results in a flow slide at shaking levels in excess of 0.19 g. The predicted model deformations match field observations of the failure of Dam No. 1 for a PGA in excess of 0.19 g, which is in good agreement with the estimated value of PGA at the site.

Figure 5 demonstrates the progressive development of a near-continuous zone of liquefied tailings at shallow depths of up to approximately 5 meters beneath the upstream fill raises of the dam and up to 17 meters below the dam crest. Limit equilibrium analysis of the Dam No. 1 geometry with fully liquefied tailings indicate a static factor of safety of only 0.55, confirming the FLAC prediction of failure under sufficiently liquefied conditions. Results of the FLAC and limit equilibrium analyses therefore suggest that dam stability and deformation are dependent on the PGA value.

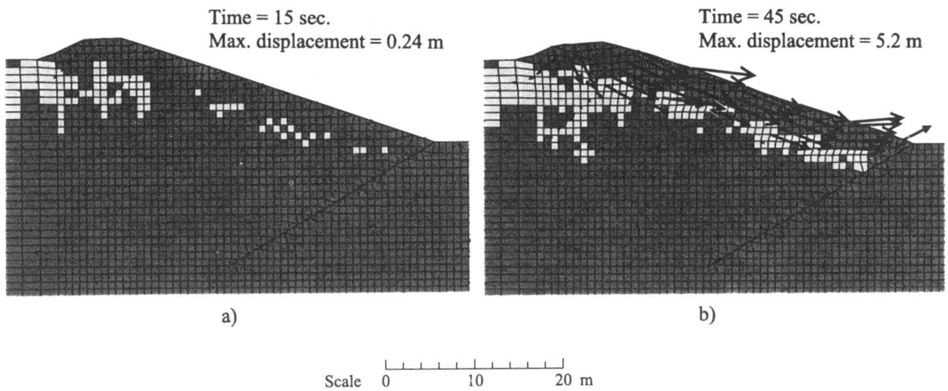

Figure 5. Deformed meshes for Dam No. 1 showing liquefied elements (in white) and displacement vectors

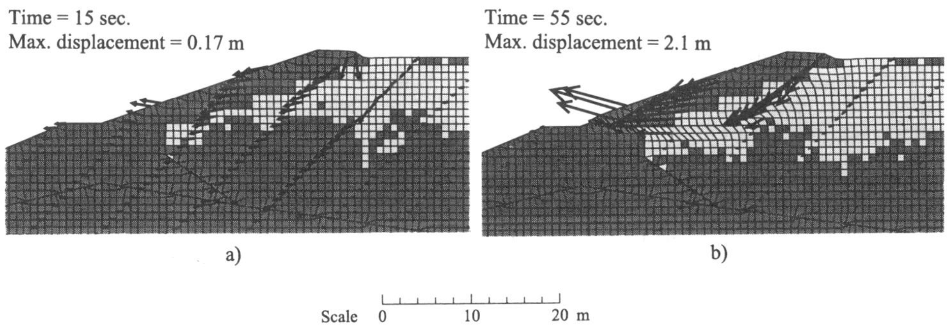

Figure 6. Deformed meshes for Dam No. 2 showing liquefied elements (in white) and displacement vectors

For the analysis of Dam No. 1 with a PGA of 0.19 g, a displacement vector of 1.4 meters is observed after 27.5 seconds of strong motion, and is observed to accumulate to 5.3 meters after 45 seconds of shaking. Additionally, the crest and toe of the of the upstream fill raises are observed to continue moving downslope during the seconds following the earthquake.

For Dam No. 2, the results of the model for the case using an input motion at the bottom of the mesh with a peak horizontal ground acceleration of 0.19 g are shown on Figure 6.

In this case, the tailings are predicted to liquefy to a depth of 13 to 15 m, which compares well to the depth of 15 m estimated from the drilling results after liquefaction (Ishihara, 1984). There is a continuous region of liquefied tailings under the upstream fill of the dam, but in this case failure is not predicted. Instead, equilibrium has been re-established after the earthquake by an overall flattening of the slope and all motion has stopped. The pattern of deformation of the model is similar to that seen by the eyewitness to the actual failure, where the face of the dam bulged out just above the starter dam. The maximum displacement vector length for this case is 2.1 meters, and the following presents some of the displacement and acceleration values at points in the model.

Location	Total displacement (m)		Peak horizontal
	Horizontal	Vertical	acceleration (g)
Downstream crest of the dam	1.7 downstream	0.9 down	0.22
Toe of upstream fill (in the bulge)	2.0 downstream	0.6 up	0.30
Top of the tailings			0.25

A limit equilibrium check indicates that the deformed geometry has a post-liquefaction static factor of safety near unity, which is consistent with the model result where equilibrium is re-established by deformation of the mesh.

An input motion with a peak ground acceleration of 0.17 g produced a maximum displacement vector of 0.3 m, but PGA values between 0.18 g and 0.21 g produced results ranging from 0.8 to 2.1 m. A value of PGA of 0.25 g for the input motion produces displacement vectors nearing 3 meters, but equilibrium is still re-established and motion stops.

In summary, the method of analysis produces model results that correctly simulate the fact that Dam No. 1 failed during the earthquake, where Dam No. 2 deformed but did not fail. The model also indicates that a value of PGA of 0.19 g or more is required to produce this result which is consistent with best estimates that can be made of the PGA that actually occurred at the Mochikoshi Mine on January 14, 1978.

6 REMEDIATION

Results of the current analyses of Dam Nos. 1 and 2 indicate that these dams would need remedial measures if exposed to seismic loading with a PGA in excess of 0.19 g. Amongst others, two cost-effective remedial measures would be considered in order to improve the seismic stability and reduce the predicted deformations of the dam structures: earthfill buttressing and drainage of the tailings mass. Determination of the size of the earthfill buttress or extent of drainage required to stabilize the dam cross section would traditionally be determined iteratively using the Newmark displacement analysis method described previously. Comparison of the required buttress sizes resulting from the use of the Newmark and FLAC methods will demonstrate the value of the proposed numerical modeling technique in terms of a reduced buttress cross-section to ensure seismic stability. Modeling results demonstrating the effectiveness of drainage within the tailings mass are also presented below.

Following the conventional Newmark method, two SHAKE columns were analyzed in the Dam No. 1 cross-section; one located at the dam crest and the other near the bottom of the upstream fill section. Both SHAKE columns intercepted the critical failure surface as determined using limit equilibrium methods. Results of the SHAKE analyses were used to determine peak accelerations and velocities at two locations along the critical failure surface due to the introduction of the earthquake time history (PGA = 0.19 g) at the base of the foundation. The critical pair of peak accelerations and velocities was used with the appropriate upper-bound Newmark equations to back-calculate a yield acceleration that would cause 1.0 meters of displacement along the failure surface. This yield acceleration was then used as a seismic coefficient in the limit equilibrium analyses to iteratively size the earthfill buttress that results in a predicted factor of safety of 1.0. The resulting buttress cross-section and critical failure surface are shown on Figure 7, for which 1.0 meters of displacement would be predicted using conventional techniques. The cross-sectional area of the buttress to provide this stability is 125 m^2. Over a crest length of approximately 500 meters, these remedial measures would require the placement of approximately 62,500 m^3 of compacted earthfill.

Using the proposed numerical modeling technique, the buttress shown in Figure 7 results in predicted (and acceptable) deformations of only 0.63 meters. In comparison to the cross-sectional buttress area of 125 m^2 prescribed by the conventional Newmark method, this buttress has an area of only 20 m^2, resulting in a reduction in fill volume of approximately 50,000 m^3. At a nominal unit rate for fill placement of \$5.00/m^3, this reduced volume would constitute a saving of \$250,000 in the cost of seismic remediation. The addition of a buttress to the toe of the upstream fill is known to have a three fold benefit in terms of stability: i) the buttress reduces the likelihood of liquefaction as confining stresses are increased in the tailings ii) if the tailings beneath the buttress do liquefy, the post-liquefaction strength is higher again due to the higher confining stress and iii) the buttress increases the length of the shear path through unliquefied embankment fill.

The effect of drainage of the tailings mass as a remedial measure was modeled by maintaining the water table as shown in Figure 4a, but by limiting all pore pressures below the water table to 20 kPa, thereby simulating the installation of horizontal or vertical drains (wick or other). The analysis results indicated the likelihood of shallow liquefaction immediately

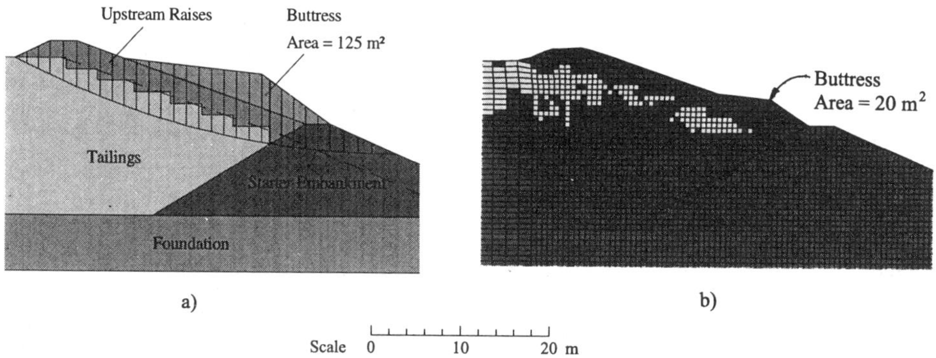

Figure 7. Buttress required to prevent the failure of Dam No. 1 using a) limit equilibrium and b) numerical modeling methods

beneath the upstream fill of the Dam No. 1 geometry and the resulting development of a failure surface at a shallower depth than in the original case with hydrostatic pore pressure. In this case the effectiveness of partial drainage of the tailings mass was therefore concluded to be less effective than earthfill buttressing, and continued to result in flow sliding within the dam although at a shallower depth.

7 SUMMARY

A new real-time numerical modeling method has been presented, wherein cycles of shear stress are counted in each element until enough have accumulated to trigger liquefaction. At liquefaction, the strength and stiffness parameters are reset to post-liquefaction values and dynamic analysis continues. In this way, the progression of liquefaction and the resulting displacements are more realistically modeled. The input parameters used for the model are essentially the same as used in conventional liquefaction assessments and analyses, and are readily available from field and laboratory testwork normally carried out for seismic design of dams.

The method is incorporated within the computer program FLAC and is verified by application to two dams at the Mochikoshi Mine in Japan, which was struck by a Magnitude 7.0 earthquake on January 14, 1978. The analysis produces model results that correctly predict that Dam No. 1 should fail during the main earthquake event, while Dam No. 2 should deform but not fail. These results help to verify the method of analysis for use in modeling other existing or proposed tailings dams.

To demonstrate the value of the new method, a modeling study is presented wherein the failed dam is hypothetically stabilized using a fill buttress. The required size of buttress is shown to be much smaller using this new procedure than if simpler, commonly-used analysis techniques are employed, resulting in a large reduction in fill placement volumes and cost. This result indicates that the use of more sophisticated techniques of modeling soil liquefaction and its effects can result in significant construction cost savings in the remediation of tailings dams to meet modern seismic standards.

REFERENCES

Bartlett, S.F. and Youd, T.L., 1992. Empirical analysis of horizontal ground displacement generated by liquefaction-induced lateral spread. *National Center for Earthquake Engineering Research, Technical Report NCEER-92-0021*

Byrne, P.M., 1991. Cyclic shear-volume coupling and pore pressure model for sand. *Second International Conference on Recent Advances in Geotechnical Earthquake Engineering and Soil Dynamics, St. Louis, MO, Paper No. 1.24, March 1991*

Byrne, P.M., Imrie, A.S. and Morgenstern, N.R., 1994. Results and implications of seismic performance studies for Duncan dam. *Canadian Geotechnical Journal, Vol. 31, No. 6, pp. 979-988.*

Byrne, P.M., Anderson, D.L. Naesgaard, E. and Atigh, E., 1998. Seismic behavior of box culverts in Richmond. *A Report for the City of Richmond, September 1998.*

Byrne, P.M., Beaty, M. and Dharmasotia, 1999. Predicting liquefaction induced displacements of earth structures with application to field experience. *Accepted for presentation at the Eighth Canadian Conference on Earthquake Engineering, June 1999.*

Cundall, P. 1995. FLAC, Version 3.3 - Fast Lagrangian Analysis of Continua. *Itasca Consulting Group Inc., Minneapolis Minnesota.*

Fukushima, Y. and Tanaka, T., 1990. A new attenuation relation for peak horizontal acceleration of strong earthquake ground motion in Japan, *Bulletin of the Seismological Society of America, 84, pp.757-783.*

Idriss, I.M. 1990. Response of soft soil sites during earthquakes. *H. Bolton Seed Memorial Symposium Proceedings, Vol. 2 pp. 273-289*

Idriss, I.M. 1993. Procedure for selecting earthquake ground motions at rock sites. *Report to the U.S. Department of Commerce, NIST GCR 93-625.*

Investigation Committee, 1979. Report on the incidents of Mochikoshi tailings disposal dams

Ishihara, K., 1984. Post-liquefaction failure of a tailings dam due to liquefaction of the pond deposit, *Proceedings of the First International Conference on Case Histories in Geotechnical Engineering, St. Louis Vol. 3, pp. 1129-1143*

Newmark, N.M. 1965. Effects of earthquakes on dams and embankments. *Geotechnique, Vol. 15, No. 2.*

Seed, H.B., Wong, R.T., Idriss, I.M. and Tokimatsu, K., 1986. Moduli and damping factors for dynamic analyses of cohesionless soils. *Journal of Geotechnical Engineering, ASCE, Vol. 112, No. 11.*

Youd, T.L. and Idriss, I.M., 1997. Proceeding of the NCEER workshop on evaluation of liquefaction resistance of soils. *Technical Report NCEER-97-0022, December 31, 1997.*

Tailings and Mine Waste '99 © 1999 Balkema, Rotterdam, ISBN 90 5809 025 6

Design, construction and performance monitoring of the submerged toe drain at Inco's rock dam

Kevin J. Semenoff
Klohn-Crippen Consultants Limited, Sudbury, Ont., Canada
Howard D. Plewes
Klohn-Crippen Consultants Limited, Richmond, B.C., Canada
Gary Remington
Inco Limited, Sudbury, Ont., Canada

ABSTRACT: The clogging of toe drains with chemical precipitates has been a historical problem at Inco's Central Area tailings dams. In a previous paper (Plewes and McDonald, 1996), it was reported that the precipitates formed when the iron-rich seepage water oxidizes as it emerges from the dams. Hence, submergence of drain materials to preclude oxygen is one means to prevent clogging.

Based on the findings of Plewes and MacDonald (1996), Inco decided to proceed with construction of a prototype submerged drain at their Rock Dam. Detailed design and construction was carried out in 1996. The main components of the submerged drain include a clay cutoff to bedrock to capture subsurface seepage, a pervious rockfill internal drain to collect the seepage, an internal high density polyethylene (HDPE) pipe to convey the seepage out of the drain, and a HDPE cover to prevent oxygen penetration. A secondary benefit of the HDPE cover is the separation of surface runoff from the contaminated seepage. Monitoring results in 1997 and 1998 show that the drain is anoxic (discharge water pH≈6.0) and therefore precipitate deposition inside the drain is expected to be minimal. The seepage flow rate from the drain is constant except during the spring freshet and following heavy rainfall events when the flow rate increases due to groundwater fluctuations in the dam and local infiltration. Long-term performance monitoring of the submerged drain is ongoing.

1. INTRODUCTION

Inco's Central Tailings Area near Sudbury, Ontario covers an area of about 5,500 acres (2225 hectares). Tailings have been continuously deposited in the impoundment since 1936 and future tailings deposition is planned for at least another 30 years. The present Central Tailings Area is comprised on six non-operating impoundments (termed the A, C-D, M, M1, P and Q Areas), and an operating impoundment (R Area).

The tailings deposited within the impoundments contain on average 6% by weight of total sulphides, predominantly as pyrrhotite. Oxidation of the tailings in the vadose zone of the impoundments results in acidic seepage conditions at the toe of the tailings dams. Seepage from the tailings dams is collected and pumped back into the tailings ponds (Plewes and MacDonald, 1996). It is predicted that sulphide oxidation reactions may persist for up to 400 years depending upon the thickness of the vadose zone formed in the impoundments and the sulphide content of the tailings (Coggan, 1992). Collection and treatment of seepage water may be required for a similar period under current operating conditions.

A concern related to the long-term performance of the Central Tailings Area tailings dams is the extensive formation of ferric hydroxide precipitates, which have lead to partial clogging of the toe drains within the dams. It is conceivable that growth of the precipitates over time could completely plug the toe drains and cause the phreatic levels in the dams to rise. This would reduce the overall stability of the dams and may lead to problems of erosion of the dam slopes.

In 1994, field and laboratory studies were carried out to investigate the cause of the chemical precipitates and to determine the extent of the drain clogging in various types of drain materials. The studies included sampling and geochemical testing of the pond and seepage waters to determine their properties and composition; test pits and drill holes to investigate the extent of the precipitate formations; and chemical testing of the precipitates to determine their composition. Based on this work, a model for the tailings seepage water geochemistry and precipitate formations was developed for the Inco site. The results of the 1994 studies along with a discussion of the findings and conclusions were presented in a previous paper (Plewes and MacDonald, 1996). One of the main conclusions contained in the paper was that at the Inco site, chemical precipitates do not form in saturated soil conditions below the water table. Hence, submergence of drain materials to preclude oxygen penetration could be an effective means to prevent clogging from chemical precipitates.

Based on the findings of Plewes and MacDonald (1996), Inco decided to proceed with construction of a prototype submerged drain at the Rock Dam located in the inactive P Area. The general concept of the submerged test drain, shown on Figure 1, is to allow uncontaminated runoff water from the surface of the dam to flow to the environment and to intercept internal seepage by means of a drainage trench at the dam toe. A seepage cutoff through the underlying foundation soils to bedrock would also be installed to restrict the migration of a contaminated groundwater plume passing below the bottom of the drainage trench. Maintaining the drains in a fully submerged condition to exclude oxygen would minimize the potential problem of drain clogging.

Design and construction of the prototype submerged drain at Rock Dam was carried out in 1996 to investigate the long-term effectiveness of submergence in preventing chemical clogging.

This paper presents the design and construction of the submerged drain, and the results of the ongoing monitoring program completed up to the middle of 1997.

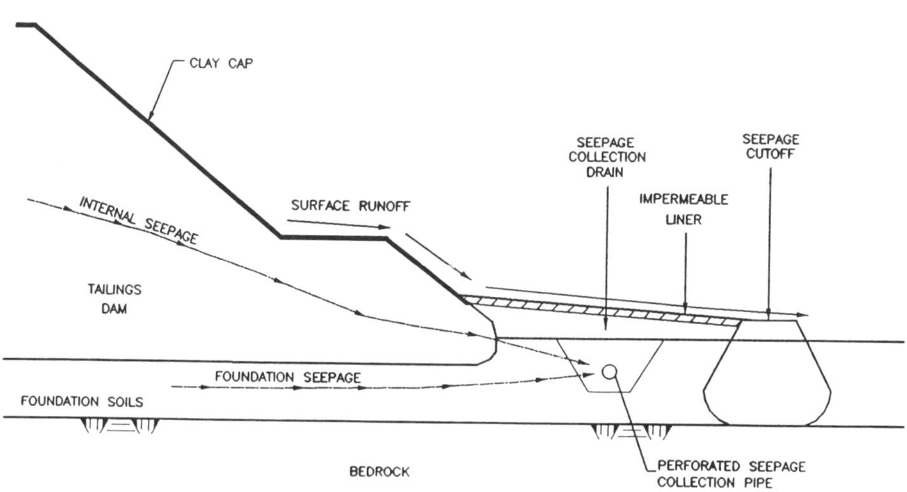

Figure 1 - Submerged drain concept

2. SITE DESCRIPTION

Rock Dam is typical of the dams in the inactive tailings area. These dams were constructed using the upstream construction method, beginning from starter dams. Tailings were spigotted from the crest of the starter dams and the dams were subsequently raised at 5H:1V slopes up to 160 ft (50 m) high. The starter dams in these older areas were constructed from locally borrowed soils, which ranged from impervious clay to semi-pervious silty sand and gravel. Most of these starter dams experienced erosion of the exterior slopes during the early years of operation from surface water runoff and seepage. The slopes were subsequently dressed with a 10 ft (3.0 m) to 20 ft (6.0 m) thick drainage blanket of smelter slag in an attempt to prevent further erosion.

Figure 2 shows a plan and typical section of Rock Dam with the submerged drain in place. The dam was started in the early 1960's and completed in 1984. After becoming an inactive dam, the surface of the tailings slope above the starter dam was capped with 1 ft (0.3 m) of compacted clay to minimize infiltration of surface water and to stabilize the slope against erosion. The surface of the cap has been reclaimed with vegetation principally comprised of grasses. Seepage from the dam is collected in a pond at the toe of the dam from where it is pumped to the P Area tailings pond.

Rock Dam was selected as the preferred test site for construction of the prototype submerged drain for the following reasons:

1. The dam is situated across a relatively narrow valley formed by bedrock at both abutments. The width of the dam is 1000 ft (300 m) at the crest and only 250 ft (75 m) at the starter dam.
2. The depth of overburden above bedrock was thought to be only about 10 ft (3 m), thereby minimizing costs for construction of the seepage cutoff to bedrock.
3. There is an appreciable estimated flow of about 70 USgpm (4.4 L/s) from the toe of the dam, which has remained relatively constant for the last 20 years. This known flow rate allows a meaningful comparison of seepage rates to be made before and after installation of the submerged drain.
4. The pH of the seepage water is uniformly between 6.5 and 6.7, and is high in Fe^{2+}. The high pH shows that little oxidation of the seepage water occurs before it exits the toe of the dam. Because pH and Fe^{2+} concentrations reduce quickly in the presence of oxygen, measurements of the changes in these parameters in the water exiting the submerged drain would give a good indication of the effectiveness of submergence in preventing oxidation of the water and, hence, deposition of precipitates.

3. DESIGN

The design of the submerged test drain at Rock Dam is indicated in Figures 3 and 4. The main features of the design are as follows.

- A clay-faced cutoff dam constructed on bedrock to intercept and capture subsurface seepage (Zones 1, 2 and 3).
- A coarse rockfill drain core to collect the internal seepage upstream of the cutoff dam (Zones 4 and 5).
- An HDPE pipe system inside the rock drain core to collect the seepage and convey it out of the submerged drain.
- An HDPE geomembrane liner to seal the top surface of the drain core to reduce oxygen penetration and to pass surface runoff water from the dam slopes over top of the drain. The liner is underlain and covered by granular bedding material (Zones 6 and 7).

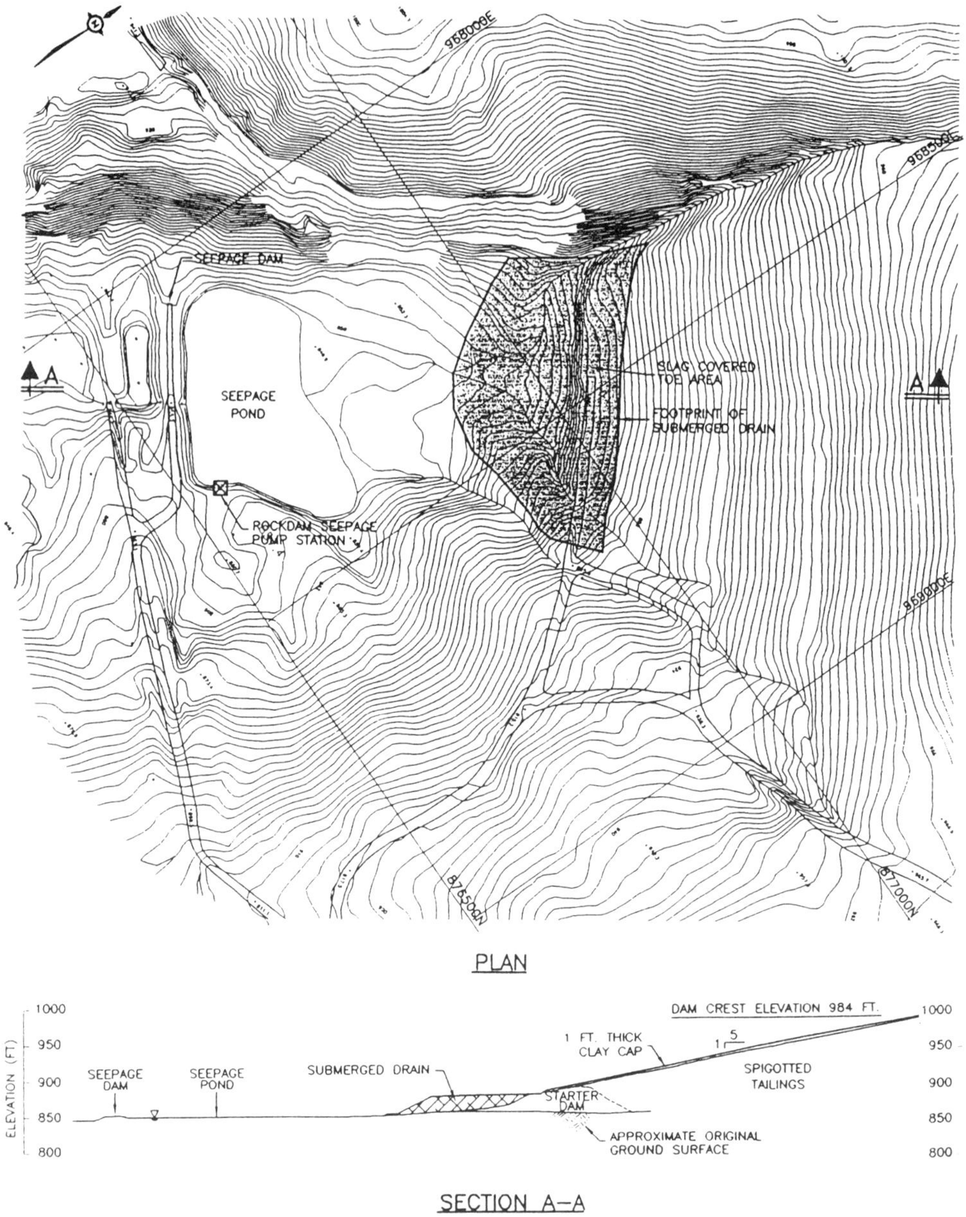

Figure 2 - Plan and typical sections of rock dam

- The provision of a small clay berm (Zone 10) and decant pipe at the crest of the HDPE geomembrane liner to permit water quality sampling and flow measurements of the surface runoff passing over the top of the drain.
- A rockfill cover for long-term erosion protection (Zone 8).

The outlet of the seepage conveyance pipe from the drain was terminated at the toe of the drain to permit flow measurements and water quality sampling. If the submerged drain is

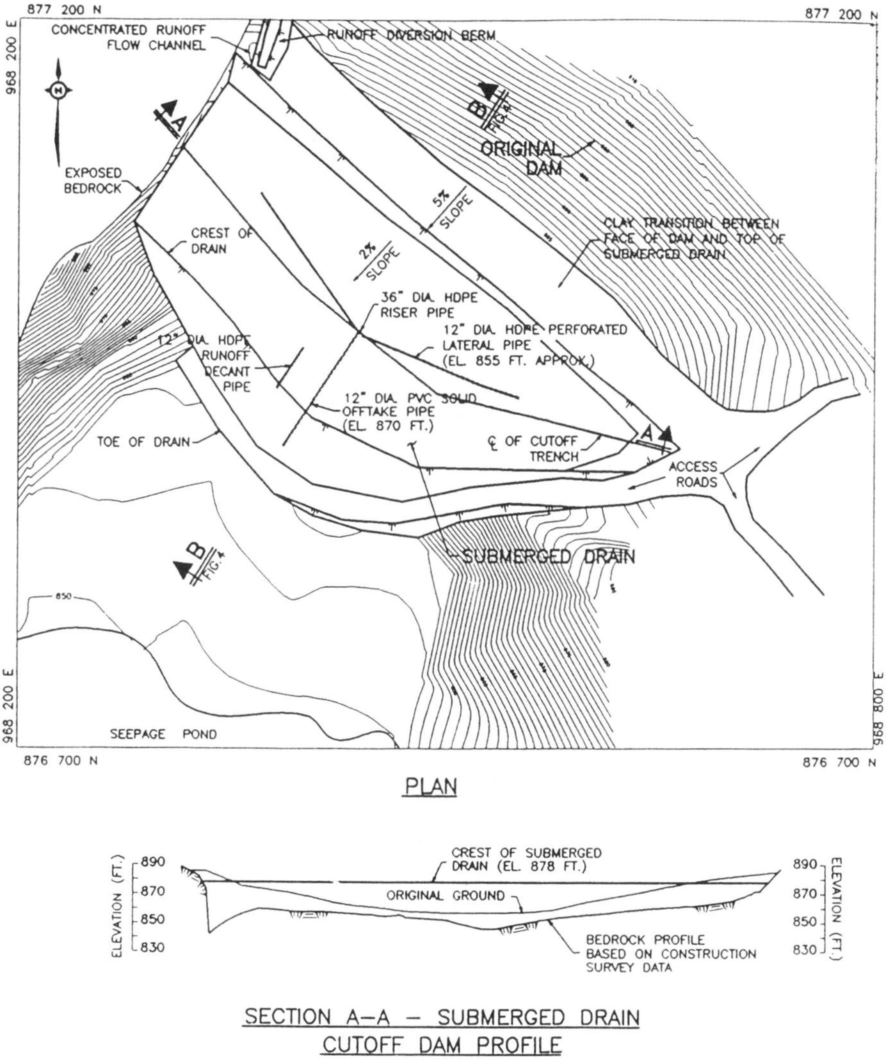

Figure 3 - Plan and profile of submerged drain

shown to be a viable long-term solution, the pipe could be extended to the existing reclaim pump station shown on Figure 2. The outlet of the pipe is fitted with two 90 degree elbows to keep the pipe flowing full at the point of discharge, thereby preventing the entry of oxygen into the conveyance pipe.

4. CONSTRUCTION

4.1 General

Construction of the Rock Dam submerged toe drain was carried out by a pre-qualified earthworks contractor between June and September, 1996. During construction, care was taken to ensure that all materials used in construction of the drain were inert. All rockfill materials

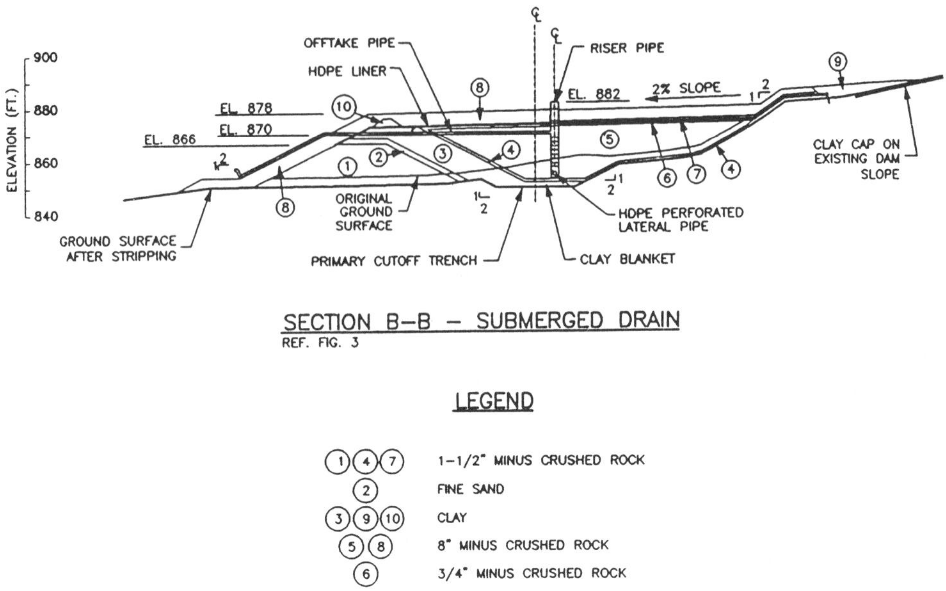

Figure 4 - Typical section of submerged drain

were produced by the contractor in a quarry developed within the active R4 tailings area. Acid Base Accounting (ABA) and ICP analysis tests on samples of rock from the quarry were conducted to establish that the rockfill materials from the quarry were not susceptible to sulphide oxidation or leaching. Filter sand and clay materials were imported from off-site borrow pits. An ICP scan was conducted on a sample of clay to confirm that the clay was not contaminated. HDPE geomembrane was supplied and installed by a reputable installation contractor who also carried out an extensive quality control program during the installation.

4.2 Execution

4.2.1 Seepage Cutoff Dam

Construction of the seepage cutoff dam started with overburden excavation to expose bedrock along the alignment of the cutoff dam. As the bedrock was exposed, labourers cleaned the bedrock surface using shovels and bars to remove compacted soil and pressure washed the surface to remove loose soil. An inspection of the cleaned bedrock surface showed the bedrock to be very competent with no visible fractures or joints. Placement of clay then proceeded in 1 ft (0.3 m) thick compacted lifts. The sand and crushed rockfill zones that also formed part of the cutoff dam were placed and compacted in sequence with the clay until the cutoff dam was raised to full height.

4.2.2 Drain Core

Construction of the drain core started once the cutoff dam was raised about 6 ft (1.8 m) above grade. To begin, about 2 ft (0.6 m) of the slag material placed over the original starter dam was removed since this material was largely contaminated with chemical precipitates. Next, a 1 ft (0.3 m) thick blanket of fine crushed rockfill was placed over the footprint area of the submerged drain between the cutoff dam and the original dam as a filter. This was followed by the placement of coarse rockfill to fill in the large bowl between the cutoff dam and the original dam. A perforated pipe collection system was installed within the coarse rockfill to ensure that collected seepage would flow freely to the centrally located riser pipe, which would then be decanted through an offtake pipe at a higher elevation. The top of the drain core rockfill was covered with two 0.5 ft (0.15 m) thick layers of fine crushed rockfill as a bedding material for the HDPE geomembrane.

4.2.3 Drain Cover

The geomembrane cover consisted of 60 mil HDPE sheets that were fushion welded together to form a continuous piece. All joints were vacuum tested for leaks and where leaks were detected, they were patched by extrusion welding and were tested again. Key trenches were excavated into the slope of the original dam and the clay zone of the cutoff dam to restrict the ability of surface runoff flows to penetrate the cover. Where the HDPE was in contact with the steep bedrock at the northeast abutment, a strip of clay was compacted against the exposed bedrock before the HDPE was placed. The HDPE was then placed in contact with the underlying clay strip and another strip of clay was placed that partially covered the edge of the HDPE and tied into the lower clay strip to form a continuous seal along the bedrock face. A layer of non-woven geotextile was then placed over the HDPE to prevent puncturing of the HDPE by the angular rockfill cover. Finally, a 1 ft (0.3 m) thick layer of fine crushed rockfill was placed over the geotextile followed by a 4 ft (1.2 m) thick layer of coarse crushed rockfill. The rockfill cover provides protection for the HDPE against UV degradation and freeze-thaw action, and prevents erosion of the surface by runoff from the slope of the original dam.

4.3 Quantities

A summary of the excavation and material quantities used to construct the submerged drain is contained in Table 1.

Table 1. Summary of submerged drain construction quantities.

Item	Quantity
Clearing and Grubbing	0.6 acres
Overburden Excavation	7,000 cu.yds.
Primary Trench Excavation	1,000 cu.yds.
8-inch Minus Rockfill (Type B)	23,800 cu.yds.
1.5-inch Minus Rockfill (Type C)	16,500 cu.yds.
Filter Sand (Type D)	600 cu.yds.
Clay (Type J)	9,400 cu.yds.
HDPE Geomembrane (60 mil)	64,700 sq.ft.
Geotextile (Non-Woven)	4,000 sq.ft.

5. MONITORING

Inco (Stuparyk and Carlaw, 1997) initiated a long-term monitoring program in 1996 after the submerged drain was completed to assess the effectiveness of the submerged drain. The program includes the following components:

- Monthly flow monitoring and water quality testing of discharge from the submerged drain under normal flow conditions (the flow monitoring frequency is increased to daily during the spring freshet).

➢ Sampling of surface runoff (rain plus snowmelt) from the slope of the existing dam above the submerged drain and from the runoff collection pipe at the crest of the submerged drain when such flows occur.

The results of the flow monitoring and water quality testing are presented below.

5.1 Flow

Figure 5 presents a plot of the measured flow rate from the submerged drain versus time for the period between October 1996 and June 1997. During normal flow conditions (i.e., outside of the spring freshet) the flow was found to be quite stable at approximately 50 USgpm (3.2 L/s). The recorded flow rate from the submerged drain under normal flow conditions agrees very closely with original seepage measurements from the tailings dam recorded in the 1970's and 1985. This suggests that the seepage cutoff dam is succeeding in capturing all of the seepage from the original dam.

During the peak runoff period of the spring freshet (April 1997), the flow rate from the submerged drain increased rapidly to more than three times the normal flow rate. This sudden increase in flow was initially attributed to infiltration of surface runoff into the submerged drain at the upper edge of the impervious cover.

In late 1997, remedial measures were implemented in an attempt to reduce the amount of infiltration into the submerged drain. The measures included construction of a diversion berm at the northwest abutment of the main dam and improvement of the clay transition at the upper edge of the impervious cover. During the spring freshet of 1998, the remedial measures were observed to be functioning well in routing runoff onto the surface of the submerged drain. However, the flow rate from the submerged drain again increased to a peak level similar to that recorded in 1997. This indicated that the remedial measures had no impact on reducing infiltration to the submerged drain. Therefore, it was concluded that the increase in outflow from the submerged drain during the spring freshet is due to seasonal groundwater fluctuations and to local surface water infiltration into the slope above the drain, probably at the right (northwest) abutment where surface runoff water is channeled against the exposed fractured bedrock.

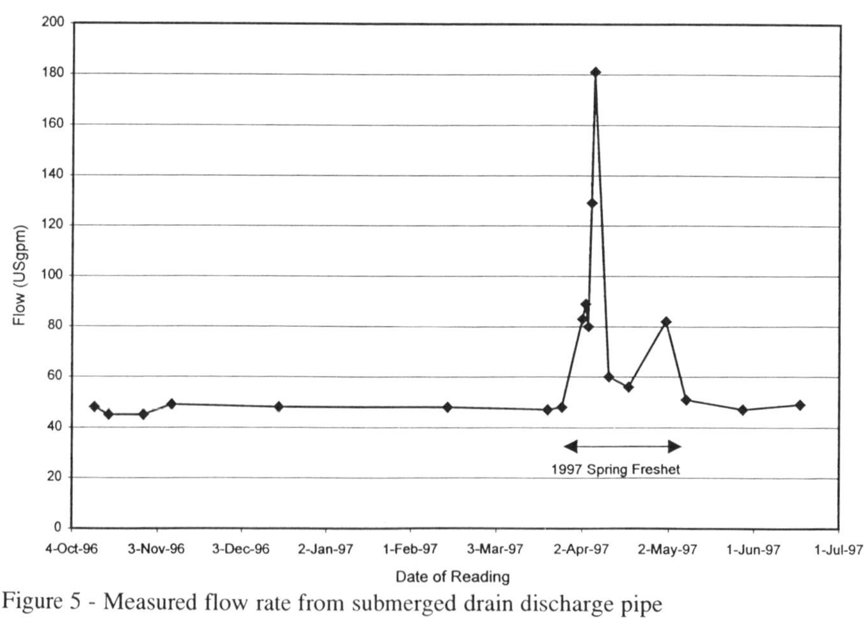

Figure 5 - Measured flow rate from submerged drain discharge pipe

5.2 *Water Quality*

During normal flow periods, discharge from the submerged drain is sampled and tested in the field for pH and Total Dissolved Solids (TDS). In addition, samples are submitted to a laboratory for Total Suspended Solids (TSS), SO_4 and total metals by ICP analysis. During the spring freshet, samples of snow and surface runoff are also collected for field and laboratory testing.

Table 2 summarizes the water quality data for tests conducted on samples of submerged drain outflow, snow and surface runoff for the period between October 1996 and June 1997. The test data on samples of submerged drain outflow have been split into two categories: a) sampled during spring freshet; and b) sampled outside of spring freshet.

The results clearly show that seepage outflow from the submerged drain is contaminated (Ni>15 mg/L) and requires treatment prior to release. The quality of the submerged drain outflow during the spring freshet showed only marginal improvement due to infiltration of cleaner runoff. The surface runoff is relatively clean (Ni<0.5 mg/L) and is comparable to the treatment plant effluent limits for nickel. However, the pH and TSS levels measured in the surface runoff have been shown to exceed effluent limits. Further monitoring is required before it can be concluded that surface runoff is suitable for untreated release.

Table 2. Summary of water quality test results from submerged drain monitoring program.

SAMPLE LOCATION	pH	TDS (g/L)	TSS	SO_4	Cu	Fe	Ni	Pb	Zn
Submerged Drain - Normal (1)	5.6 - 6.4	2.1 - 3.3	63 - 167	2950 - 3663	0.2 - 0.7	458 - 531	15.7 - 18.5	0.04 - 0.1	0.1 - [illegible]3
Submerged Drain - Spring Freshet(2)	5.8 - 6.3	1.5 - 3.0	5 - 29	1620 - 3070	0.2 - 1.0	204 - 525	9.5 - 16.4	0.05 - 0.09	0.09 - 0.8
Snow (3)	-	-	-	-	0.1 - 0.2	0.3 - 0.4	0.1 - 0.2	0.02 - 0.04	0.008 - 0.005
Surface Runoff (4)	4.8 - 7.5	0.05 - 75	23 - 4603	8 - 22	0.005 - 0.6	0.3 - 124	0.03 - 0.5	0.05 - 0.09	0.03 - 0.5

(1) Sampled October 1996 to June 1997 from submerged drain outlet pipe.
(2) Sampled April 2 to 18, 1997 from submerged drain outlet pipe.
(3) Sampled March 1997 from four locations around Rock Dam.
(4) Sampled April 6 to 16, 1997 from 3 locations around Rock Dam.
(5) Unless otherwise noted, concentrations are reported in mg/L.

Stuparyk and Carlaw (1997) also reported that the dissolved oxygen content of the outflow from the submerged drain was measured to be less than 1 mg/l and that the iron in the outflow was found to be entirely in the form of unoxidized ferrous iron (Fe^{2+}). These results would appear to indicate that the submerged drain is operating in an anoxic (i.e. oxygen-free) state. This is corroborated by the pH values of 5.6 to 6.4.

6. CONCLUSIONS

This paper summarizes the design and construction of a submerged drain at Rock Dam, which is located in Inco's Central Tailings Area. The initial results of an ongoing monitoring program are also presented. A study of the initial results from the monitoring program provides the following conclusions:

1. The Rock Dam submerged drain is effectively capturing seepage at the toe of the tailings dam. The measured flow rate from the submerged drain during normal flow conditions is in the order of 50 USgpm (3.2 L/s), similar to the historical flow rate recorded from the toe of the dam in the past. The drain discharge increases about threefold during the spring freshet. The increased flow rate is attributed to seasonal groundwater fluctuations and to local surface water infiltration into the slope above the drain.

2. Water quality of captured seepage from the submerged drain shows elevated metal concentrations, particularly nickel and iron, that are above treatment plant discharge limits. However, surface runoff separated from the seepage flow was found to be relatively clean and may be suitable for untreated release.
3. The dissolved oxygen content in the seepage from the submerged drain was measured to be less than 1 mg/L and the pH ranges between 5.6 and 6.4, which suggests that the drain is operating in an anoxic state. This observation is further supported by the fact that the iron contained in the seepage discharge is primarily in the form of unoxidized ferrous iron (Fe^{2+}). Therefore, the formation of precipitates and hence clogging of the submerged drain is not expected to occur.
4. Monitoring of the submerged drain is continuing to evaluate the long-term behaviour of the drain system.

REFERENCES

Coggan, C.J. 1992. Hydrogeology and geochemistry of the Inco Ltd., Copper Cliff, Ontario, Mine Tailings Impoundment. M.Sc. Thesis, University of Waterloo, Ontario, 159 pages.

Plewes, H.D. and T. MacDonald 1996. Investigation of chemical clogging of drains at Inco's Central Area tailings dams. In *Tailings and Mine Waste '96 Proceedings*:59-72. Rotterdam:Balkema.

Stuparyk, R. and S.M. Carlaw 1997. Rock Dam submerged seepage collection system – progress report. Unpublished internal document.

Tailings and Mine Waste '99 © 1999 Balkema, Rotterdam, ISBN 90 5809 025 6

The impact of elevated blanket drain installation on tailings dam operations

D.J.Venter & J.C.J.Boshoff
SRK Consulting, Johannesburg, South Africa

Y.Pather
Amplats Management Services, Johannesburg, South Africa

ABSTRACT: The tailings dam at Potgietersrust Platinum Mine in the Republic of South Africa is a large ring dyke impoundment with a maximum storage volume of 55 million m^3. Tailings is pumped via a 250 NB steel pipeline from the minerals process plant at a rate of 265 000 tpm and the tailings is deposited via a hybrid depositional system. This involves spiggotting during the day from the outer wall into paddocks and depositing via spray bars off the paddock wall into the basin at night over the main drain. This method of operation is employed in order to ensure coarse material is deposited over the main drain area, and the day wall paddock system ensures adequate freeboard is maintained during discharge of the serpentinitic tailings which is exceptionally fine and gives rise to a very flat beach profile. The main features and development of the elevated blanket drain, the purpose and benefits of this drainage method and the effect of this drainage method on the tailings deposition cycle period are considered and discussed.

1 INTRODUCTION

The Potgietersrust Platinum Mine is located approximately 30 km to the northwest of the town of Potgietersrust in the Northern Province of South Africa, Figure 1. The tailings dam complex is located approximately 1,5 km north of the plant and open pit. The dam is designed for a maximum safe height of 42m with an average outer slope angle of 20°, Figure 2. The dam was originally designed to accommodate 200 000 tpm for 30 years. The dam is currently accommodating 265 000 tpm.

The mining operation at Potgietersrust is by open pit mining and the ore body is characterised by serpentinised zones within the pyroxenite host rock. Because of the serpentinitic tailings which is exceptionally fine and gives rise to a very flat beach profile, alternative methods of drainage and deposition had to be investigated. This paper discusses the overall development of the elevated drainage system including main features, purpose and benefits.

Figure 1. Locality Map - Potgietersrust Platinums Limited.

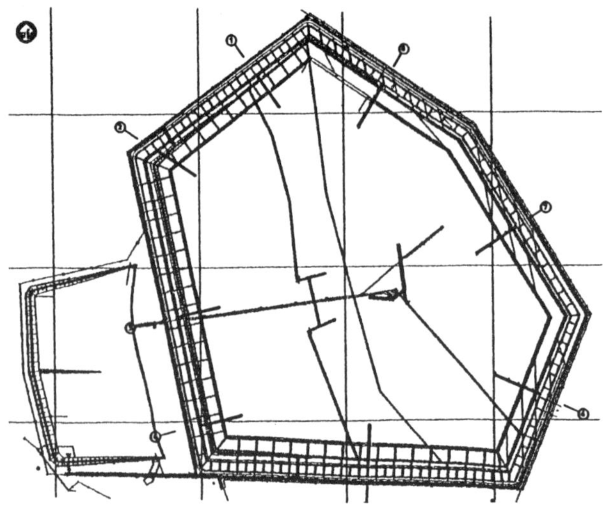

Figure 2. General arrangement of tailings dam complex.

2 BACKGROUND

The tailings dam at Potgietersrust Platinums Limited has been designed to be constructed in two phases. The first phase which was constructed in the lower areas of the site was designed with toe drains to control seepage regimes within the tailings around the outer parts of the tailings dam, Figure 3. This phase was also equipped with a main blanket drain to control seepage regimes within the dam and to ensure that pore pressures in the slope area will be minimised. The main blanket drain was constructed on top of concrete chimney intake structures which are located within wet placed tailings to reduce the probability of blinding of the drain by fines resulting from gravitational segregation of the tailings after discharge. It was envisaged that a conventional spigot method of deposition would be employed, notwithstanding prior knowledge that the product was likely to be of finer grind than found in most other mines. The second phase was constructed in the area of the site which was at a higher elevation and was also designed with an inner and outer toe drain to control seepage regimes within the tailings around the outer parts of the tailings dam. This phase was also equipped with a main blanket drain but without the concrete chimney inlet structures.

Following commissioning of the plant and tailings dam it was evident that very fine tailings had a tendency to accumulate in the plant circuit. These fines would periodically be flushed through the circuit and delivered to the tailings dam in slugs over periods ranging from a few hours up to two shifts (16 hours). The fines comprise soft talc-rich serpentinitic ores which are blended with other ore types of varying degrees of uniformity as a result of limitations in the efficiency of the blending process. At the same time other host rocks encountered in the ore body comprise pyroxenite and dolomite which contain minerals of a relatively high hardness index. As a result that while the percentage passing the 75 μm sieve size is 80% or higher, (indicating a fine grind), there is also a significant proportion of coarser particles of up to 0,25 mm, Figure 4. The net result is a variable tailings product that fluctuates from a fine, predominantly fine to medium silt to a coarse silt in medium sand which segregates more than is acceptable for conventional paddock operations. A further complicating factor is the presence of saponite within the serpentinite. The mineral is described as having a soapy feel as indicated by its name which is derived from sapo, soap and peotine, Greek for fat. The specific gravity of this mineral at 2,24 to 2,3 is low especially in comparison to the host rock which has a specific gravity in excess of 3,11. This causes severe clarity problems both on the dam and in the plant, and, on segregation results in the continuous deposition of extensive sheets of very fine silt to clay size particles of low permeability. The layers of fine material have the effect of an impermeable floor, while the layers of coarser materials with a higher permeability produce paths of intense seepage. Moreover, the talc and similar minerals result in a slurry of lower frictional resistance than is common on Merensky platinum tailings and therefore produce a shallow beaching angle. The results of the above factors are that :

- There is unlikely to be adequate beach freeboard generated during normal operations to provide the required stormwater holding capacity on the dam. This would imply a need for a method of operation which generates freeboard by formation of an outer wall. A paddock system permits the formation of this outer wall hydraulically.
- The fine slimes material is placed at a high moisture content as a result of its high specific area and tends to dry slowly and drain very reluctantly. This will influence cycle times if routine paddock methods of operation were to be employed.
- Layers of fine and coarse tailings are likely to form in the material deposited in the vicinity of the discharge. Such layering was found to be responsible for the Bafokeng disaster in 1974 and must be assiduously avoided.

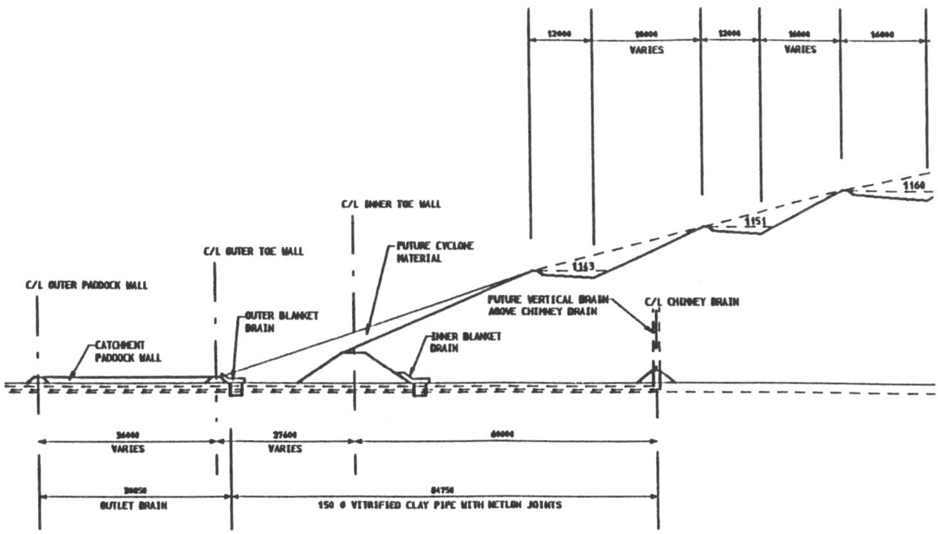

Figure 3. Cross section through the outer edge of the tailings dam showing toe drains and elevated drain.

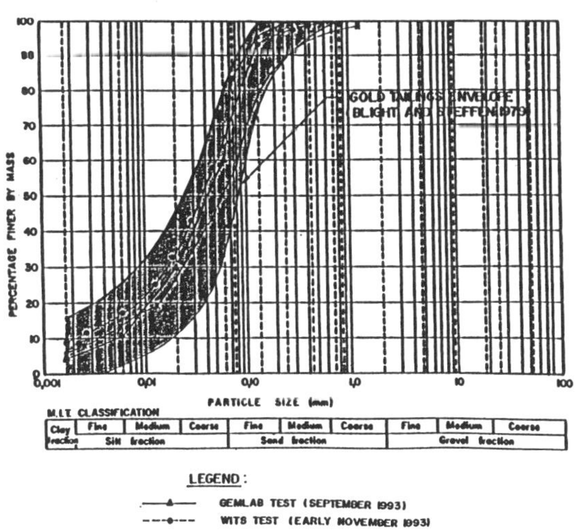

Figure 4. Particle size distributions of the coarse and fine products.

- Unless deposition of fine and coarse material can be controlled, there is a significant danger that under-drainage systems will be blinded and rendered ineffective in the long term which would have very severe effects on the slope stability of the dam and limit the maximum height of the dam.

In order to address each of the above concerns a method of operation which may best be described as a hybrid of the conventional paddock method used on gold tailings dams and the spigot deposition method has been designed. This method is described below:

3. DESCRIPTION OF HYBRID METHOD OF OPERATION

Figure 5 shows a typical cross section through the dam once the hybrid section was established. Noteworthy features of the section in Figure 5 are:-

- A relatively wide (50 m) daywall formed by beaching from spigot discharges directly across the daywall. At the inner wall of the daywall a number of "vent" pipe decants are used to draw supernatant water into the basin as well as some fines so as to encourage the deposition of coarser material at the outer wall and speed up drying times and improve the in-situ material density.
- To ensure that daywall formation is controlled and systematic, it was proposed that only a portion of the flow will be taken through the spigots ie. the spigots will be throttled down to restrict the flow per spigot.
- The use of paddock cross-walls to confine the flow to manageable areas and to facilitate control of the tendency of the tailings to flow laterally *along* the daywall as opposed to across the daywall.

4. PLACEMENT OF COARSE MATERIAL OVER THE ELEVATED BLANKET DRAIN.

To ensure that the operation of under-drainage systems is not hindered by layering over the main drain, it was proposed to employ a spray-bar pipe to be located at the inner wall of the daywall. This pipe would be equipped with 25 mm holes at 1 m centres and would discharge the tailings in the form of a jet. The jets would be angled downwards and would meet the beach just outside of the main drain area. This method of discharge would encourage the deposition of coarse material over the main drain and would apply in the period immediately preceding the formation of the main elevated blanket drain and would continue until a significant layer of coarse product has been formed over the drain, Figure 6.

It will however be necessary to ditch or rotovate the material above the main drain so as to mix the coarse material hydraulically deposited over the main drain. This is because while there will be a reduced tendency for layers to be formed as a result of flow meanders, those which will form as a result of variations in product will continue to occur.

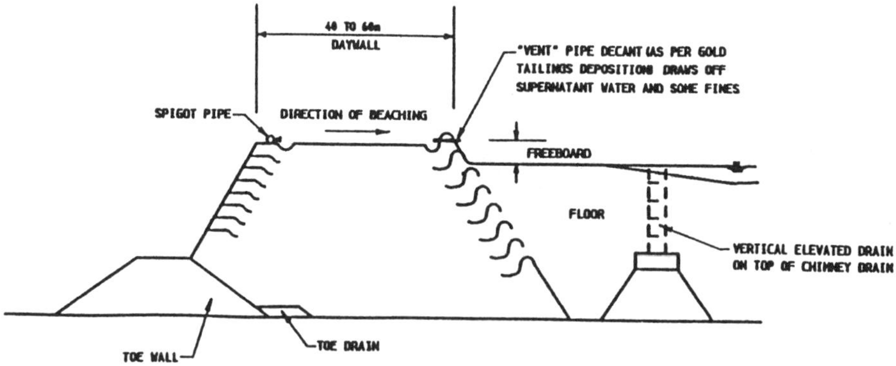

Figure 5. Typical section through the hybrid deposition system.

5. DEVELOPMENT OF THE ELEVATED BLANKET DRAIN

Precast concrete chimneys inlet structures filled with 6 mm stone were constructed approximately 50 m inside the inner toe wall on the northern, western and southern flanks of the tailings dam. The purpose of these chimney inlet structures is to facilitate the installation of the main drain behind the starter wall where the initial rapid rate of rise poses a problem due to the high moisture content of the placed tailings and the amount of fine tailings material in the vicinity of the main drain. These structures are spaced at 60m centres and will become operational following installation of the elevated sand blanket. Once the tailings has reached a level where the elevated sand drain is installed, this drain then plays a crucial role in maintaining a low phreatic surface and therefore maintaining an adequate margin of stability.
The elevated blanket drain also serves to prevent the phreatic surface from day lighting from the tailings slope. Seepage water is drained from each chimney structure by means of a 150mm diameter vitrified clay pipe with open netlon joints placed in a filter drain consisting of 19 mm stone, graded filter sand and geofabric to the outer toe wall.

5.1 Horizontal Elevated Blanket Drain

The elevated blanket drain initially consisted of a 300 mm thick horizontal blanket drain on top of the concrete chimney intake structures, Figure 7. This drain was designed when the effect of layering within the tailings dam was not evident and the horizontal blanket was designed wide enough to intercept any seepage that might occur and to prevent the phreatic surface from day lighting from the tailings dam slope.

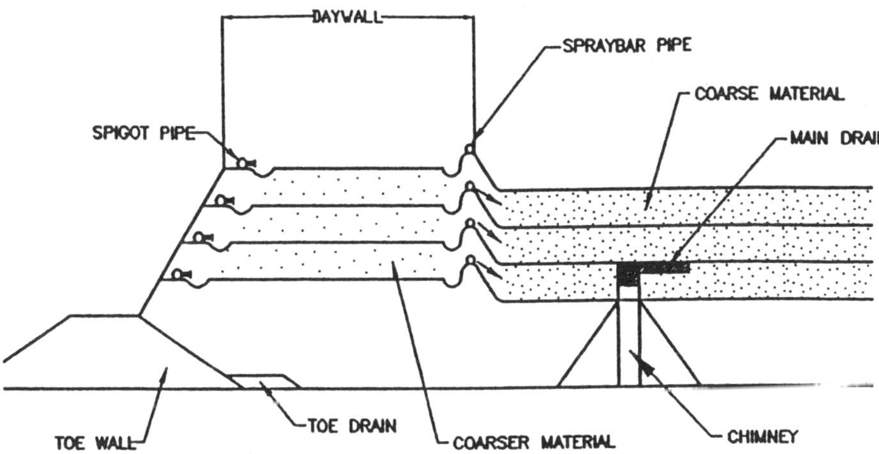

Figure 6. Typical section through hybrid system showing method of placement of coarse material over the elevated drain.

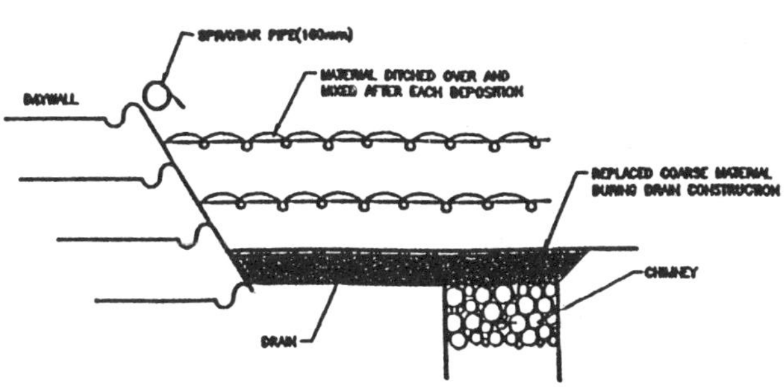

Figure 7. Initial horizontally orientated blanket drain.

5.2 Inclined Elevated Blanket Drain

As soon as the effect of layering and associated problems regarding operations of the tailings dam in terms of freeboard and beaching characteristics became evident, the horizontal elevated blanket drain was changed to an inclined elevated blanket drain, Figure 8. The inclined section of the drain was constructed once the height of the tailings had progressed to a level of approximately 4m above the concrete chimney intake structures and for the purpose of intercepting the layering and to allow for drainage of the coarse tailings between the layers. The drain comprised a 10 m wide graded filter drain of coarse tailings and filter sand inclined at 17° to the horizontal. However, once the height of the tailings had progressed to a level of approximately 5m above the construction level of the elevated drain, the tailings material had to be ditched or rotovated so as to mix the tailings material hydraulically deposited over the drains.

An area on the dam labelled for either ditching or the installation of an elevated blanket drain each time had to be taken out of operation for a period of sometimes up to six weeks. This allowed sufficient time for the tailings in the paddocks and above the main drain to consolidate and dry which enabled the tailings to accommodate equipment access onto the tailings dam. This also affected the tailings deposition cycle as tailings now had to be deposited on available and consolidated sections around the dam. The tailings deposition cycle therefore had to make allowance for an additional 3 to 4 weeks wherein the ditching had to take place. During each ditching exercise and drainage installation additional personnel and plant had to be established on site making it a very costly operational expenditure.

5.3 Vertical Elevated Blanket Drain

The ditching of tailings above the inclined elevated blanket drain and around the tailings dam had a negative impact on the tailings deposition cycle as the cycle became more complicated and had to be reviewed on a continuous basis in order to control the geometry and design levels of the rising tailings dam. After close evaluation and monitoring of the performance of the inclined elevated drain and resulting operational problems, a vertical elevated drain was designed to replace the inclined elevated blanket drain. The vertical drain comprises a 0,5 m wide graded filter drain of filter sand, Figure 9. The graded filter drain is installed by trenching and replacing the tailings with coarse to medium sized sand material which acts as both a filter and drain. The vertical drain is extended in stages and the depth of each lift depends on the depth the excavator can reach but is normally limited to the maximum free standing height permitted by safety regulations as well as the grain size and moisture content of the tailings material. To minimise the length of unsupported and exposed trench, backfilling with sand follows the excavator closely.

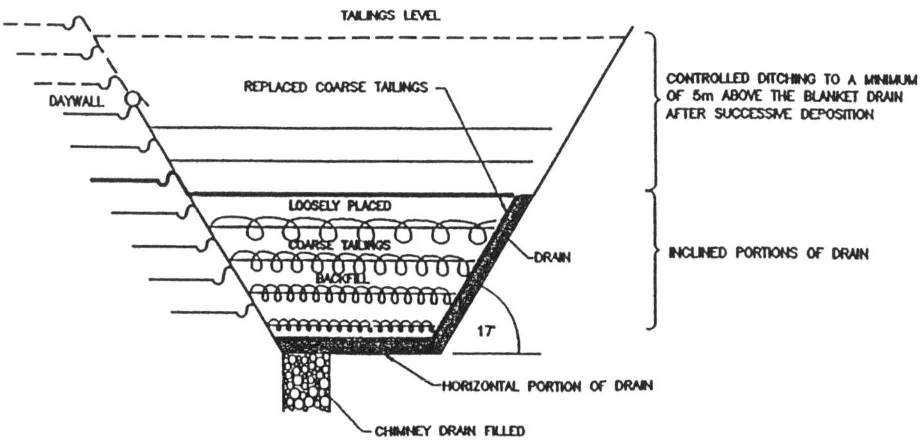

Figure 8. Inclined elevated blanket drain.

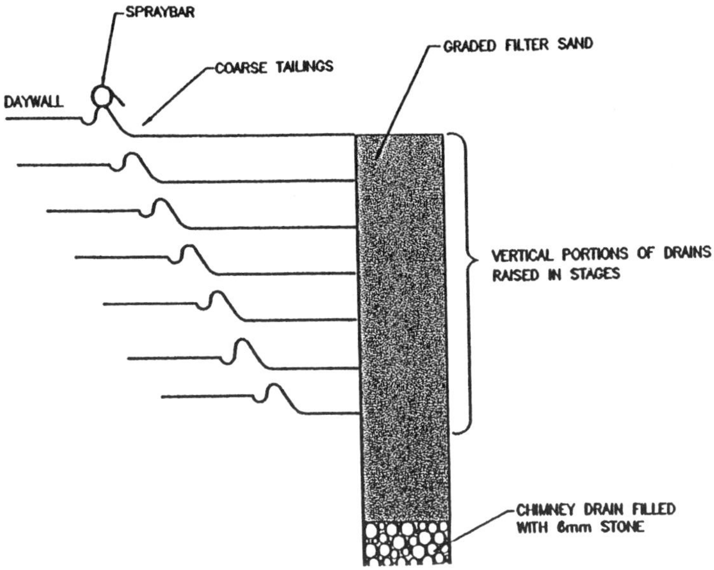

Figure 9. Vertical elevated drain.

6. BENEFITS OF THE VERTICAL ELEVATED DRAIN IN TERMS OF THE TAILINGS DEPOSITION CYCLE AND TAILINGS DAM OPERATION

The main advantage of the vertical elevated drain is the fact that it intercepts all seepage along horizontal paths. This will lower the phreatic line, decrease the hydrodynamic and pore pressure and accelerate consolidation over the main drain area. The drain is also constructed in stages taking account of pool location and dam height to ensure that the drain and outlets are always effective.

In terms of tailings dam operation, the vertical elevated drain can now be installed within the tailings deposition cycle. This is possible due to the fact that the vertical drain is extended with each completed 1.5 m lift of tailings deposited. The four to six weeks deposition cycle allows sufficient time for the tailings to dry, drain and consolidate. The vertical elevated drain can now also be installed at the same time that the day walls are raised in preparation for the next deposition cycle. Most of the material excavated to form the trench for the installation of filter sand for the vertical elevated drain, can now also be used to raise the inner wall of the day wall.

The simultaneous installation of the vertical elevated drain and the raising of the day wall now forms part of the routine procedure during each tailings deposition cycle around the dam. This action now forms part of the daily tasks of the tailings dam operation and no further large capital costs are incurred, neither is the tailings deposition cycle affected.

7. CONCLUSIONS

The spigot method of deposition remains the most cost effective method of disposal over the long term. It is however possible to revert back to this method of disposal over the majority of the life of the tailings dam once the elevated drain has been installed up to the first step-in level. Of all the factors affecting tailings dam stability, drainage is the one that can be implemented most easily, safely and economically throughout design, construction and operation.

Drainage of tailings dams, as well as their structure and construction, depend strongly on the physical characteristics of the tailings, especially on their coarseness and percentage and origin of the fines. Careful and correct drain construction during the early stages of dam development constitutes a cost effective assurance against future expensive remedial works.

8. REFERENCES

Blight, GE and Bentel, GM (1983). *The behaviour of mine tailings during deposition, Journal of South African Institution of Mining and Metallurgy, April,* 1983 : 73 - 86.

McPhail, GI (1995). *Prediction of the beaching characteristics of hydraulically placed tailings, PhD Thesis, University of the Witwatersrand, Johannesburg, South Africa.*

International Commission on Large Dams (ICOLD) 1994. *Tailings Dams - Design of Drainage - Review and Recommendations.* Bulletin 97, 107 pp.

Tailings and Mine Waste'99 © 1999 Balkema, Rotterdam, ISBN 90 5809 025 6

Laboratory simulation of consolidation in coal waste impoundments

R.A.Rohlf – *Kentucky Department for Surface Mining Reclamation and Enforcement, Frankfort, Ky., USA*
C.T.Justice – *Vulcan Materials, Knoxville, Tenn., USA*
R.J.Sweigard – *Department of Mining Engineering, University of Kentucky, Lexington, Ky., USA*
T.C.Hopkins – *Kentucky Transportation Center, University of Kentucky, Lexington, Ky., USA*

ABSTRACT: Three samples of coal slurry 305 cm in depth by 7.6 cm in diameter were consolidated under self-weight and incremental surcharge loading up to 138 kPa. Surface settlement and pore pressure at five different depths were measured during the consolidation tests. After full dissipation of pore pressure, samples of slurry were extruded from the consolidation tubes and subjected to consolidated undrained triaxial shear tests with pore pressure measurement. At the end of the consolidation tests, all samples of coal slurry remained at or above the liquid limit. Analysis of pore pressures generated during triaxial shearing showed that the pore pressure coefficient at failure ranged from 0.62 to 1.19 and increased as the water content of the slurry samples increased. Normalized effective shear strength of the slurry samples decreased as the water content increased.

1 INTRODUCTION

Coal waste is composed of rock, low quality coal, and other impurities extracted during mining and subsequently removed through the cleaning process at the preparation plant. The Kentucky coal mining industry is producing increasingly larger volumes of coal waste due to Clean Air Act requirements and the use of modern underground mining techniques which remove substantial amounts of material above and below the coal seam. These factors have resulted in a steady increase in the amount of coal waste produced per ton of coal mined. Increased rates of coal mine waste generation are causing not only larger disposal structures but also more rapid rates of construction.

Coal waste commonly occurs in two size fractions: coarse and fine. Coarse refuse is generally composed of sand, gravel, and larger sized particles while fine refuse is composed of silt and clay sized particles. Fine refuse usually has a high water content which results from the cleaning process. Fine refuse must either be dewatered prior to placement in a fill or be disposed of as a slurry in an impoundment.

Slurry impoundments are classified according to their method of construction: downstream, centerline, and upstream. The downstream method of impoundment construction begins with an initial embankment of coarse refuse with subsequent stages added on the downstream side of previous stages. The centerline method places subsequent stages of coarse refuse directly on top of the previous stage, while the upstream method places subsequent stages partially on impounded slurry. The upstream method of impoundment construction is the most critical since stability of the embankment is dependent on adequate consolidation and shear strength of the underlying slurry. Reclamation of slurry impoundments is often accomplished by grading down the embankment and covering the impounded tailings with approximately 6 m to 9 m of approved material. After reclamation, these structures are treated as fills.

The mechanical behavior of coal slurry, resulting from loading incurred by construction of an embankment by the upstream method or by covering slurry during reclamation, was simulated by consolidating three samples of coal slurry 305 cm in depth by 7.6 cm in diameter under self-weight and incremental surcharge loading up to 138 kPa. Surface settlement and pore pressure at five different depths were measured during the consolidation tests. After full dissipation of pore pressure, samples of slurry were extruded from the consolidation tubes and subjected to consolidated undrained triaxial shear tests with pore pressure measurement.

2 LABORATORY TESTING

2.1 *Consolidation columns*

Plexiglass tubes 325 cm in length by 7.6 cm in diameter were used for consolidating the slurry samples. The columns were constructed in two approximately equal length sections connected by flanges. Taps for water pressure gages were placed along the length of the columns at approximately 61 cm intervals, and taps for slurry water decantation were placed at 7.6 cm intervals near the top of the columns.

A surcharge load was applied to the slurry through a loading cylinder 6.4 cm in diameter. The loading cylinder contained a porous stone plug at the bottom to allow vertical drainage from the slurry during consolidation. A load container with a capacity of 68 kg of lead weight was mounted on top of the loading cylinder.

2.2 *Collection of slurry samples*

Three slurry samples were obtained from coal processing operations located in the Appalachian coal fields. The samples are designated by the number of the column into which they were placed for consolidation testing. Two slurry samples were obtained directly from an active refuse impoundment. The slurry placed in column #1 was taken from the pool area in the vicinity of the discharge point. The slurry placed in column #3 was collected from the discharge pipe by catching the slurry in a bucket as it discharged. The column #1 slurry had a lower water content and was coarser than the column #3 slurry. The slurry placed in column #2 was obtained directly from a coal preparation plant. Table 1 contains a summary of index properties for the three slurry samples.

2.3 *Consolidation tests*

Loading of slurry in refuse impoundments results from increasing levels of slurry or placement of coarse refuse on existing slurry, either during construction of an embankment stage for an upstream impoundment or during reclamation. These loading conditions were modeled in the laboratory by first allowing the slurry to consolidate under its self-weight and then subjecting the slurry to surcharge loads. Following the self-weight consolidation, the columns were subjected to surcharge pressures of 35., 69., 103., and 138. kPa. A surcharge pressure of 138. kPa represents a coarse refuse depth of approximately 9.1 m, assuming a coarse refuse density of 1520. kg/m^3.

Table 1. Slurry index properties.

Properties	Column #1	Column #2	Column #3
Collection point	pool area	preparation plant	discharge pipe
Water content (%)	23	86	64
Specific gravity of solids	2.55	1.87	2.19
Void ratio	0.59	1.61	1.40
Dry specific gravity	1.61	0.72	0.91
Liquid limit (%)	18	38	35
Plastic limit (%)	12	28	22
Fraction < .002 mm (%)	12	26	18
Fraction .002 - .05 (%)	16	51	40
Fraction > .05 (%)	72	23	42

2.4 *Extrusion of samples for triaxial testing and preparation of remolded samples*

After completion of the consolidation tests, the three columns were disassembled in preparation for extrusion of the samples. Samples were extruded from the columns by using the loading cylinder to carefully push a sample from the testing column. As the sample was being extruded, it was guided into a cardboard cylinder, which contained a latex membrane under vacuum. Once a sample of sufficient length had been guided into the cylinder, the sample was cut and the vacuum removed. The extruded sample, supported by the latex membrane, was then prepared in the triaxial chamber.

In addition to the samples extruded from the consolidation tubes, remolded samples were prepared from stored slurry material. To provide a uniform basis for comparison of triaxial test results between samples which had been consolidated in the slurry tubes and remolded samples, the remolded samples were prepared at an initial water content near the liquid limit.

Additional details on the equipment design and testing procedures can be found in Justice (1997).

3 RESULTS AND ANALYSIS

3.1 *Consolidation tests*

Incremental surcharge pressure and the resulting cumulative settlement for the three columns are presented in Table 2. Column #2 had the greatest cumulative settlement (62.5 cm) while column #1 had the lowest cumulative settlement (27.0 cm).

The average water content at the end of consolidation and liquid limit for each column are presented in Table 3. The average water content for each column was calculated using the relationships

$$e = e_o - (1 + e_o)\frac{\Delta h}{h_o} \tag{1}$$

and

$$w = e / G_s \tag{2}$$

where e is void ratio, e_o is the initial void ratio of the material placed in the column, h_o is the initial height of the slurry in the column (304.8 cm), Δh is the amount of settlement which occurred as a result of self-weight and surcharge loading, G_s is the specific gravity of solids, and w is water content (Whitlow 1990). The average water contents after consolidation were all equal to or greater than the liquid limit, and column #2 was significantly above the liquid limit (Table 3)

Reclamation of slurry impoundments is often accomplished by grading down the embankment and covering the impounded tailings with approximately 6 m to 9 m of approved material. After

Table 2. Load and cumulative settlement.

Surcharge (kPa)	Column #1 (cm)	Column #2 (cm)	Column #3 (cm)
self-weight	16.3	34.8	33.2
35.	20.1	45.5	41.6
69.	25.2	52.4	46.7
103.	26.0	57.2	52.0
138.	27.0	62.5	56.1

Table 3. Average water content at the end of consolidation.

	Column #1	Column #2	Column #3
w (%)	18	57	44
Liquid limit (%)	18	38	35

reclamation, these structures are treated as fills. The consolidation tests indicate that even with complete dissipation of pore pressure, coal slurry can remain at a water content above the liquid limit under loading conditions expected to exist under current construction and reclamation practices. If coal slurry remains above the liquid limit, failure of a slurry impoundment is likely to produce a flow of viscous material rather than a slide of earthen material as would be expected from a fill. Thus, continued integrity of the embankment may be vital in providing long-term stability of coal slurry impoundments.

3.2 *Estimated stress history for consolidation and triaxial tests*

Properties of consolidated and remolded samples used in triaxial testing including water content at the end of column consolidation, water content and void ratio after triaxial consolidation, and the triaxial confining pressure are presented in Table 4. All water contents at the end of column consolidation were above the liquid limit but all water contents after triaxial consolidation were below the liquid limit, regardless of the confining pressure.

Estimated stress-strain histories for the consolidated samples from the end of consolidation in the columns through consolidation in the triaxial cell prior to shearing are presented in Figure 1 in terms of the average void ratio and effective mean stress for each column. The points in Figure 1 include (1) the end of column consolidation; (2) the extruded sample placed in the latex membrane; and (3) consolidation of the sample prior to triaxial shearing at three different confining pressures.

Void ratio at the end of column consolidation was determined with Equation (1). The effective mean stress at the end of column consolidation was calculated using the relationships

$$\sigma'_v = \sigma'_s + (\gamma_{sat} - \gamma_w)\bar{h} \tag{3}$$

$$\sigma'_h = K_o \sigma'_v \tag{4}$$

$$K_o = 1 - \sin(\phi') \tag{5}$$

$$p' = (\sigma'_v + 2\sigma'_h)/3 \tag{6}$$

Table 4. Properties of consolidated and remolded samples used in triaxial testing.

Sample	Property	C1C(1)	C1R(2)	C2C	C2R	C3C	C3R
1	w_s (%)(3)	22.9	18.7	68.1	41.5	35.9	45.1
	w_o (%)(4)	14.6	15.5	37.4	29.8	26.8	23.5
	e_o(5)	0.372	0.395	0.699	0.557	0.587	0.515
	p_o (kPa)(6)	137.9	137.9	137.9	137.9	103.4	103.4
2	w_s (%)	21.4	18.3	65.3	40.2	35.5	41.7
	w_o (%)	13.3	15.0	33.5	27.4	25.0	24.3
	e_o	0.339	0.383	0.627	0.512	0.548	0.532
	p_o (kPa)	275.8	241.3	275.8	275.8	206.9	206.9
3	w_s (%)	19.6	17.2	43.8	39.6	35.8	31.7
	w_o (%)	13.3	14.0	31.2	26.7	24.2	25.4
	e_o	0.339	0.357	0.583	0.499	0.530	0.556
	p_o (kPa)	413.7	344.8	413.7	413.7	310.2	310.2
	ave w_s (%)(7)	21.3	18.1	59.1	40.4	35.7	39.5
	ave e_s (%)(8)	0.543	0.462	1.105	0.756	0.782	0.865
	Liquid limit (%)	18	18	38	38	35	35

(1) Column #1 consolidated samples.
(2) Column #1 remolded samples.
(3) Water content prior to triaxial consolidation.
(4) Water content after triaxial consolidation.
(5) Void ratio after triaxial consolidation.
(6) Consolidation pressure.
(7) Average water content prior to triaxial consolidation.
(8) Average void ratio prior to triaxial consolidation.

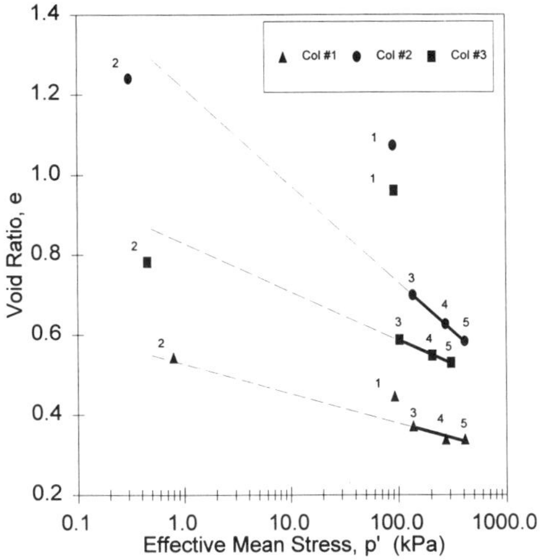

Figure 1. Estimated stress-strain history for column consolidated samples.

where σ'_v and σ'_h are the average effective vertical and horizontal stresses respectively, σ'_s is the effective stress due to surcharge loading, γ_{sat} is the saturated unit weight of the slurry at the end of consolidation, γ_w is the unit weight of water, $\bar{h}$ is the average height of the slurry in the column at the end of consolidation which was calculated as $(h_o - \Delta h) / 2$, K_o is the coefficient of earth pressure at rest, φ' is the effective friction angle, and p' is the effective mean stress. Equation (5) is Jaky's equation for determining K_o for normally consolidated soils (Whitlow 1990). The surcharge stress at the end of consolidation, assuming full dissipation of pore pressure, was 137.9 kPa for all columns. The void ratio and mean stress at the end of column consolidation are designated as point #1 in Figure 1.

Dismantling of the columns and extrusion of the samples into the latex membranes in preparation for triaxial testing produced a reduction in mean stress and change in void ratio. These points are designated as point #2 in Figure 1. Columns #1 and #2 showed an increase in void ratio while column #3 experienced a decrease in void ratio during the extrusion process. The void ratio for the extruded samples was the average e_s value taken from Table 4. The mean stress was estimated using Equations (3) - (6) for a sample 15.2 cm in length with no surcharge stress and $K_o = 1$ to account for overconsolidation (Whitlow 1990). Points #3, #4, and #5 in Figure 1 represent consolidation in the triaxial cell prior to shearing. Values of void ratio and mean stress for these points (e_o, p_o) are also contained in Table 4.

In all cases, the slurry samples showed a significant decrease in void ratio from the end of column consolidation to the initial triaxial confining pressure even though the mean stress is approximately equal for these two states (point #1 to point #3 in Figure 1). This indicates that the samples did not follow the recompression line during the unloading-reloading cycle generated by extrusion from the columns and consolidation for triaxial testing. Although much of the decrease in void ratio may have resulted from the normal hysteresis which occurs in cyclic loading, the decrease may also have been aggrevated by sample disturbance during extrusion or by the change in the consolidation stress state from K_o in the column to isotropic in the triaxial cell. Although great care was taken in extruding samples from the consolidation tubes, the decrease in void ratio prevented triaxial testing from being conducted at the higher water content which existed in the columns at the end of consolidation.

3.3 *Triaxial tests*

Results of the triaxial testing for consolidated and remolded samples are presented in Table 5. All cohesion values were zero as expected for normally consolidated samples. Values of the friction angle show no significant differences between consolidated and remolded samples, however the friction angle for the column #1 slurry was approximately 5° higher than the column #2 and #3 samples. This is likely due to the larger particle sizes in the column #1 slurry.

Friction angle versus plastic index is plotted in Figure 2 for consolidated and remolded samples from this study and for drained direct shear tests on fine coal refuse from Huang et al. (1987). The friction angle varies from approximately 27° to 38° for a narrow range in plastic index. Although scatter in the relationship is large, the trend and variation in φ' is within that for clays containing a range in compositions (Terzaghi et al. 1996)

Undrained shear strength versus liquidity index is presented in Figure 3 with the liquidity index given as

$$I_l = \frac{w - w_p}{w_l - w_p} = \frac{w - w_p}{I_p} \tag{7}$$

where w_l is the liquid limit, w_p is the plastic limit, and I_p is the plastic index. The data include results from unconsolidated-undrained triaxial tests from columns #1 and #2, unconfined compression tests conducted on fine refuse by Huang et al. (1987), and unconfined compression tests on slurry obtained from an impoundment (Zeng et al. 1998). The data are compared with undrained shear strength values for remolded clays from around the world (Terzaghi et al. 1996). The undrained shear strength of coal slurry presented in Figure 3 is considerably higher than remolded clays at high liquidity index. However for $0 \leq I_l \leq 1$, shear strength decreases rapidly as the liquidity index moves toward 1, thus indicating that the undrained shear strength of coal slurry decreases significantly as the water content approaches the liquid limit. The undrained shear strength for coal slurry is approximately 10 kPa for $I_l = 1$ (Figure 3).

The slurry samples experienced an increase in pore pressure and a decrease in effective stress during undrained shearing. This behavior is typical of normally consolidated and sensitive clays. When the confining stress, σ_3, is constant, the pore pressure coefficient, A, is given by

$$A = \frac{\Delta u}{\Delta \sigma_1} \tag{8}$$

for saturated soils where Δu is the change in pore pressure and $\Delta\sigma_1$ is the change in the major principal stress (Whitlow 1990). A summary of pore pressure coefficient values at failure is presented in Table 6. The pore pressure coefficient at failure, A_f, varied from 0.62 to 1.23 for the

Table 5. Strength parameters for consolidated and remolded samples.

	C1C	C1R	C2C	C2R	C3C	C3R
c'	0	0	0	0	0	0
φ' (°)	36.0	38.0	32.7	31.5	32.7	30.2

Table 6. Pore pressure parameter at failure for column consolidated and remolded samples.

Column	Sample	t'_f (kPa)	A_f	Column	Sample	t'_f (kPa)	A_f
C1C	1	46.	1.06	C1R	1	60.	0.83
	2	141.	0.62		2	110.	0.78
	3	196.	0.70		3	135.	0.89
	ave.		0.79		ave.		0.83
C2C	1	43.	1.16	C2R	1	51.	0.89
	2	84.	1.19		2	112.	0.72
	3	155.	0.90		3	151.	0.90
	ave.		1.08		ave.		0.84
C3C	1	35.	1.02	C3R	1	37.	0.88
	2	75.	0.93		2	67.	1.01
	3	98.	0.87		3	81.	1.23
	ave.		0.94		ave.		1.04

consolidated and remolded samples thus placing coal slurry generally in the category of a normally consolidated clay (Whitlow 1990). The average pore pressure coefficient for each set of triaxial tests, $\overline{A_f}$, is plotted versus average water content, $\overline{w_o}$, in Figure 4.

The effective stress point at failure, t'_f, can be normalized by dividing by the confining pressure, p_o where $t' = (\sigma_1 - \sigma_3) / 2$. The average normalized stress point at failure for each set of triaxial tests, $\overline{t'_f / p_o}$, is plotted versus average water content, $\overline{w_o}$, in Figure 5 for consolidated and remolded samples. Figures 4 and 5 indicate that the pore pressure at failure increases with increasing water content while the normalized effective shear stress at failure decreases.

Undrained instability typically develops when a soil structure yields under a condition of

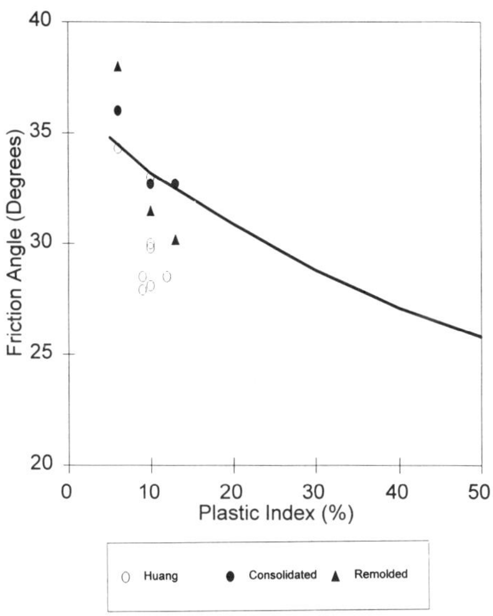

Figure 2. Friction angle versus plastic index.

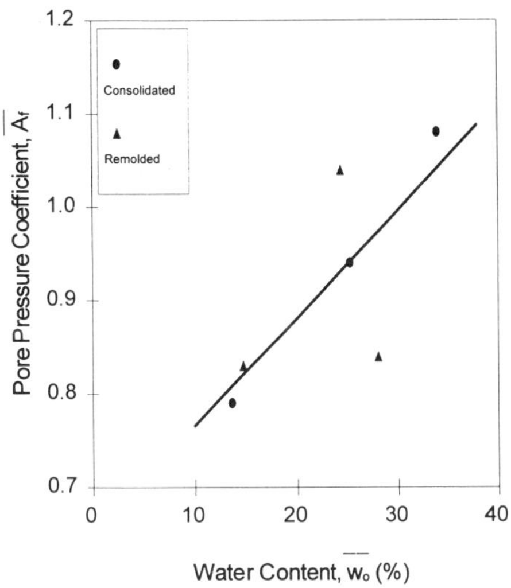

Figure 4. Average pore pressure coefficient at failure verses water content.

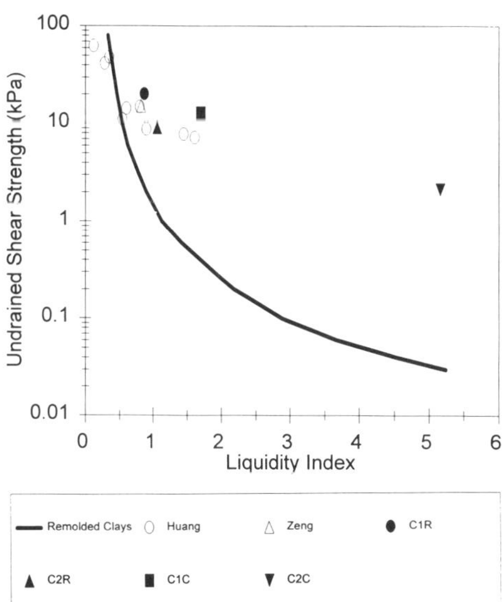

Figure 3. Undrained shear strength versus liquidity index.

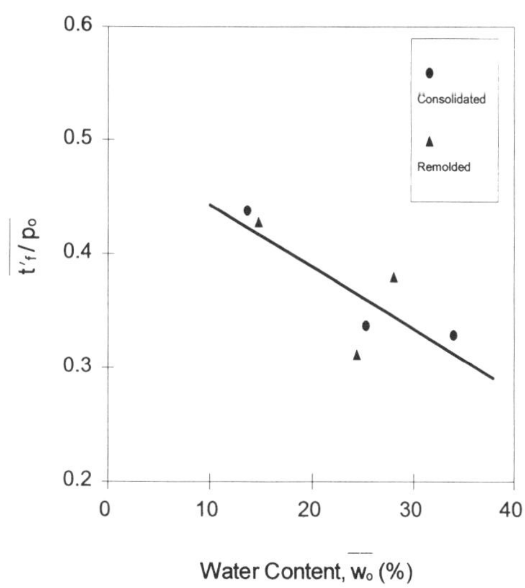

Figure 5. Average normalized shear stress at failure versus water content.

contractive shear (Terzaghi et al. 1996). Loading conditions generated during construction and reclamation of upstream slurry impoundments can be expected to produce normal consolidation and thus soil conditions conducive to the occurrence of contractive shear. Yielding within a slurry impoundment could be produced by a number of events including: (1) adding slurry or another embankment stage to an active upstream constructed slurry impoundment; (2) placing additional fill on top of a reclaimed impoundment either for mining or development purposes; (3) a long duration storm event or change in groundwater conditions which may increase the phreatic surface and seepage forces; (4) foundation movement due to collapse of underground mine works located below an existing slurry impoundment; (5) construction of buildings on a reclaimed slurry impoundment; and (6) vibration caused by an earthquake or nearby blasting.

The Mohr-Coulomb equation for undrained shear strength can be written

$$S_{uo} = c'_m + (\sigma - u_o - \Delta u)\tan\phi'_m \tag{9}$$

where S_{uo} is undrained shear strength; c'_m and φ'_m are the cohesion intercept and friction angle mobilized at yield, respectively; u_o is the hydrostatic or steady-seepage porewater pressure; and Δu is the shearing induced porewater pressure (Terzaghi et al. 1996). Although it is common to consider the hydrostatic or steady-seepage porewater pressure in stability analysis, including the shearing induced porewater pressure is much more difficult. Figure 4 shows that the pore pressure coefficient for coal slurry increases significantly as water content increases. If coal slurry in impoundments can remain at a high water content under current construction and reclamation practices, as the consolidation tests indicate, stability analyses which fail to consider shearing induced porewater pressure may over estimate the factor of safety.

4 SUMMARY AND CONCLUSIONS

Three samples of coal slurry 305 cm in depth by 7.6 cm in diameter were consolidated under self-weight and incremental surcharge loading up to 138 kPa. After full dissipation of pore pressure, all samples of coal slurry remained at or above the liquid limit. The consolidation tests indicate that coal slurry in impoundments can remain at a water content above the liquid limit under loading conditions expected to exist under current construction and reclamation practices. Since coal slurry can remain above the liquid limit, failure of an active or reclaimed slurry impoundment is likely to produce a flow of viscous material. Thus, continued integrity of the embankment may be vital in providing long-term stability of coal slurry impoundments.

After completion of the consolidation tests, samples of slurry were extruded from the columns and subjected to consolidated undrained triaxial shear tests with pore pressure measurement. Analysis of pore pressures generated during triaxial shearing showed that the pore pressure coefficient at failure increased as the water content of the slurry samples increased. Normalized effective shear strength of the slurry samples decreased as the water content increased. Although it is common to consider the hydrostatic or steady-seepage porewater pressure in stability analysis, including the shearing induced porewater pressure is much more difficult. However, if coal slurry in impoundments can remain at a high water content under current construction and reclamation practices, as the consolidation tests indicate, stability analyses which fail to consider shearing induced porewater pressure may over estimate the factor of safety.

Although the laboratory consolidation tests indicate that coal slurry can remain above the liquid limit under current construction and reclamation practices, field data collection is needed to verify this finding and to determine long-term groundwater flow conditions which exist in reclaimed slurry impoundments.

REFERENCES

Huang, H.H., J. Li & G. Weeratunga 1987. *Strength and consolidation characteristics of fine coal refuse*. Dept. of Civ. Engrg., Univ. of Kentucky, Lexington, Kentucky.

Justice, C.T. 1997. *Consolidation and shear strength of coal slurry*. M.S. Thesis, Dept. of Mining Engrg., Univ. of Kentucky, Lexington, KY.
Terzaghi, K., R.B. Peck & G. Mesri 1996. *Soil mechanics in engineering practice*, third edition. New York: John Wiley & Sons, Inc.
Whitlow, R. 1990. *Basic soil mechanics*, second edition. Hong Kong: Longman Group (FE) Ltd.
Zeng, X., J. Wu & R.A. Rohlf 1998. Seismic stability of coal-waste tailings dams. *ASCE Geotechnical Specialty Publication No. 75, Vol. II*, 950 - 961.

Tailings and Mine Waste '99 © 1999 Balkema, Rotterdam, ISBN 90 5809 025 6

Combined use of grouting and temporary groundwater control to stabilize and prevent erosion of gypsum stack embankments

P.C. Schmall & G.W. Lang
Moretrench American Corporation, Rockaway, N.J., USA

ABSTRACT: Due to nature of the deposition and material characteristics of gypsum stack embankments, they remain susceptible to erosion and unstable conditions due to internal water pressures and seepage forces. With an understanding of the foundation soils, a process combining grouting and temporary groundwater control can be implemented to cease the flow of boils that inherently develop at the embankment toes. The process has been developed and used successfully to completely seal off boils that have been as small as 1 gpm and as high as 1500 gpm.

INTRODUCTION

Central Florida is home to numerous phosphate mines, each leaving it's mark on the flat landscape with a mountainous stack of gypsum (calcium sulfate), a fine grained by-product of the fertilizer production process which behaves like a fine sand or silt. The gypsum is placed into containment stacks by hydraulic processes, in successive lifts, which in essence results in the creation of a dam embankment, which contains both the gypsum and the highly acidic process water, generally with a pH of approximately 2. The process water is typically deposited at the top of the stack, the gypsum settles out of the process water and the water is then recycled from a variety of sources. The height and breadth of the stacks reflect their age. Some stacks tower 300 ft above the landscape and cover many acres of land.

Due to the nature of the gypsum deposition process where water is continuously fed to the top of the embankment, high internal water levels are experienced within these high gypsum embankments as well as within the foundation soils below. Under normal conditions, the internal water pressures will dissipate under stable conditions as the process water flows down through the low permeability gypsum stack embankment and is collected by toe drains and/or trenches or pools at the toe of the embankment specifically for the purpose of intercepting seepage.

Boils occur at the embankment toe when the embankment toe seepage gradients exceed critical gradient and the material becomes "quick", i.e. when the water pressures near the surface exceed the weight of the overlying material. The piping of the foundation sands due to boiling condition at the toe often allows for flows to increase and subsequently erode more of the foundation soils, resulting in the embankment failure if not addressed.

Gypsum has the unique characteristics of developing piping paths because of the physical and chemical processes that take place as the stack height is increased. The stack filling procedures generally involve applying fresh "gyp" to seal or cover the piping paths. With some frequency, the original toe drainage design is not adequate and local or even extensive remedial procedures are necessary. Among these, slurry trenches, steel sheet piling, and permanent vacuum wellpoint systems have been employed.

The sites where these boils have been sealed have several features in common. The gypsum stack embankments were older, constructed before new regulations, and were sometimes constructed with minimal foundation improvement or any type of key-in to the native soils or toe drain prior to deposition of the first lift of gypsum. Most of the stack embankments were constructed as a gypsum deposited directly upon the native ground surface. Several geological features characterize the central Florida phosphate country geology where these projects have occurred; fine sandy soils of variable silt/clay content, and the presence of cemented sand or "hardpan" layers and clean sand strata at shallow depths. The natural bridging action of the cemented sands or "hardpan" often renders the soil immediately beneath the cemented layer looser than ordinary. With very low fines content, and of uniform gradation, these soils are highly susceptible to piping or movement with seepage. It is typically this loose, clean soil stratum immediately below a cemented sand layer where the piping path can develop, bringing the high pressure groundwater in the foundation soils below the central part of the stack to the embankment toe and exit as a boil.

Typical immediate response measures to stop or control the boils are not always adequate. Backfilling the boil with crushed stone to "weight" the area in an effort to cease the continued piping of sand is one typical response. Due to the vast differences in grain size between the stone and the flowing sand, the desired filtering effect is seldom achieved with a strong boil and the sand flows right through the pore spaces of the stone. Building a berm around the boil area to slow or stop the flowing water also results in limited success. Once the boil has "surfaced" and the high pressure source has in essence washed its way to the edge of the gypsum stack, the shallow cover permits the boil to easily find a new path of least resistance. We have seen boil locations move as much as 50 feet in only a few hours. Wellpoints or relief wells are effective as a preemptive measure only, and are of little benefit installed in undisturbed soils outside of the piping path.

A process has been developed and used successfully to completely seal off boils that have been as small as 1 gpm and as high as 1500 gpm. The objective of the process is not only to cease the boiling condition and movement of soils at the embankment toe, but also to restore groundwater pressures within and beneath the gypsum stack to safe levels. The process is very similar in methodology to that commonly used for chemically grouting cracks in concrete structures under flowing water conditions by diverting the flow of water through a temporary sealable outlet. Essential to the implementation of the process is a thorough examination of the foundation soil stratigraphy and an understanding the groundwater dynamics of the situation. With that understanding, delineating the piping path and sealing it off can be accomplished. One specific project (confidential client) is presented as a representative example.

PROJECT EXAMPLE, CENTRAL FLORIDA

Our example project is situated in central Florida. The gypsum stack is approximately 200 ft high, covering an area of more than 100 acres. A continuous flow of 50,000 gallons per day of process water carries the waste gypsum to the top surface of the stack. A 270 gpm boil developed immediately upon excavation of a shallow 40 ft by 40 ft area at the toe of the embankment for the construction of a new pumping station and moved approximately 30 cubic yards of soil into the excavation area. The boil moved several times across the excavated area within a couple of days of surfacing, perhaps due to backfilling with stone or attempting to isolate it within a small berm. With the fore-knowledge of the fine grained nature of the site sands, the amount of material deposited in the excavation area, the flow of water that was boiling up and the pressure that it was under, it was immediately evident that the water was coming through some kind of open void structure or piping path connected to the higher groundwater pressures under the middle of the stack.

Understanding the foundation soil stratigraphy

A series of probe holes and continuously sampled borings were advanced on the closest area of solid ground on the embankment side of the boil. Probe holes were drilled by direct rotary drilling using clean water so that wash return and fluid loss could be observed. The subsurface stratigraphy generally consisted of 10 feet of gypsum, a cemented sand or "hardpan" layer approximately 4 feet thick, over clean fine sand to the maximum depths probed. Several initial probes suggested loose conditions immediately below the hardpan and several probes experienced complete drilling fluid loss upon drilling through that zone indicating very high permeability or voidaceous conditions. Upon complete drilling fluid loss, a significant increase in turbidity was observed in the boil which confirmed that the piping path was encountered. Split spoon samples were taken in the first series of probe holes to provide confirmation of the nature of the soils and relative densities below the hardpan. The natural soils at that elevation were very clean, uniform fine sands as suspected.

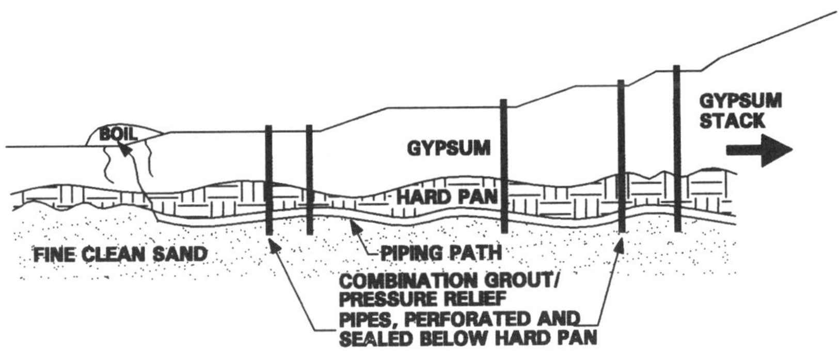

Sectional View through Piping Path

Installation and Testing of Pipes

Once it was determined that the zone of concern was situated immediately beneath the hardpan, every probe or soil boring advanced was completed with a perforated pipe which would allow either the introduction of grout or other materials to plug the flowpath, observation of water levels, or provide a means of relieving the pressure in the zone - which would be required at the time the boil is sealed off. Each pipe was constructed with 2" diameter PVC pipe, fitted with a borehole packer to seal the lower, perforated part of the pipe at or below the hardpan, and permit grouting of the annular space through and above the hardpan. Because high groundwater pressures would build up within the flow cavity once the boil is plugged, it was very important that each of the pipes was carefully grouted upon installation. If not carefully grouted, the improperly sealed annular space would become the new path of least resistance once the original boil is sealed, i.e. the ungrouted pipe would become the new boil location, further upslope and presenting greater danger to the embankment.

Several rows of pipes were installed so as to provide direct access to the flow path every 20 or 30 feet away from the boil. Numerous pipes were installed in each row, on a picket fence pattern, approximately 3 feet apart, with the intention of installing at least one pipe per row in direct communication with the piping path. The groundwater levels in each pipe reflected the relative proximity of the piping path.

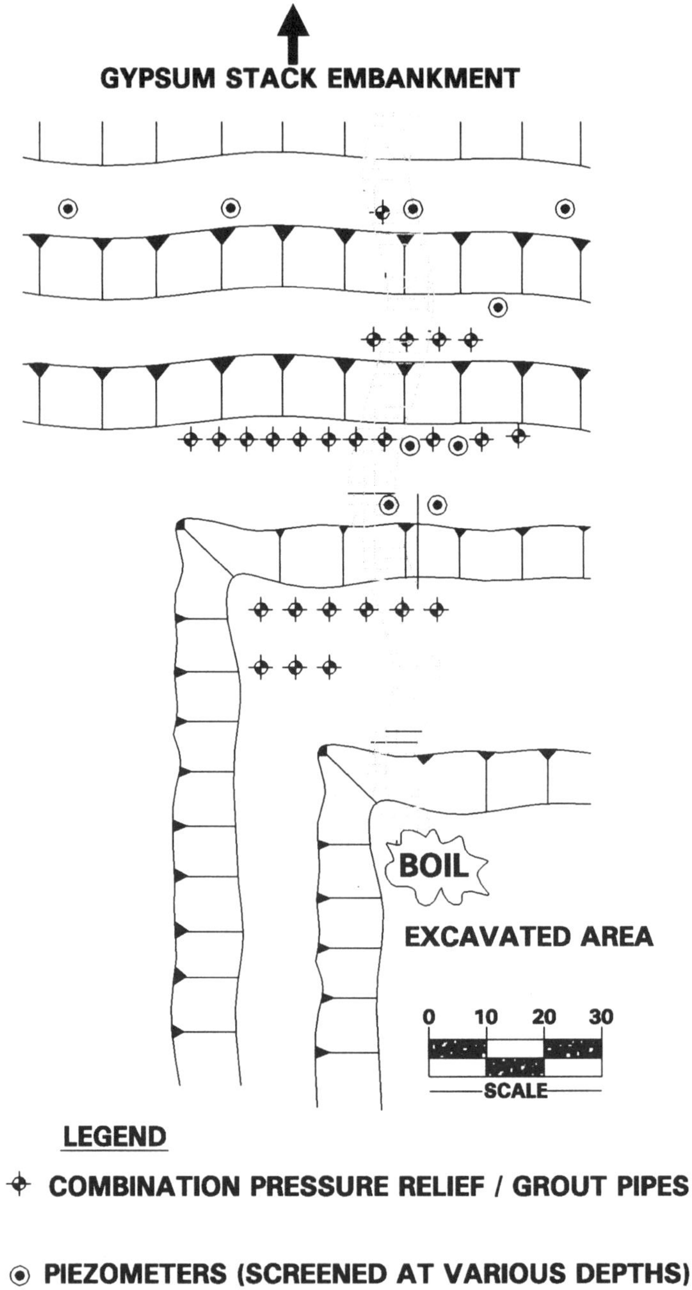

Plan View of Embankment Toe Area

Water levels were carefully monitored in the pipes at least three times per day during the pipe installation period which lasted approximately 5 days. The water levels observed in the pipes clearly reflected the high groundwater pressures associated with the boil. In those pipes where complete drilling fluid loss was experienced, the water levels were measured as high as

5 feet above ground surface, even as the boil flowed freely. Extended standpipes were fitted to each of these pipes so that they would not flow as an artesian well. The adjacent pipes which did not encounter voids, but only loose soils below the hardpan reflected the dissipation of the high groundwater pressure into the undisturbed or less disturbed soils. Over the 5 day pipe installation period, the piping path actually repositioned itself several times, almost like a river meandering slowly beneath the ground surface. These pipes which were clearly in the piping path, based on the groundwater elevation, would change from day to day. Sometimes it would be the disturbance of a probe hole penetrating into the piping path further upstream depositing a column of sand laden drilling fluid that would cause a sudden shift in the piping path location. At other times a shift would occur without any outside disturbance.

As mentioned previously, the pipes served several purposes; they would facilitate the introduction of grout or other materials to plug the boil, and they would provide a means of relieving the pressure in the piping path further upstream once the flowpath was plugged. On this particular project, the boil flowed at approximately 270 gpm. When some resistance was applied at the boil to decrease the flow, such as backfilling with stone, an increase in groundwater pressure corresponded with the decrease in boil flowrate. This is typical behavior of a confined, pressurized water source, i.e. an artesian aquifer. The concern in this particular case was a sudden increase in the groundwater pressure beneath the hardpan, and either a ground heave or the development of a new flowpath to the ground surface concurrent with plugging of the boil. It was the intent, once the boil was plugged off, to let the groundwater pressure continue to bleed off with the pipes flowing as artesian wells. Thus the total number of pipes installed in the piping path was capable of accommodating the total boil flow of 270 gpm.

Several tests were performed on the installed pipe array to confirm which pipes communicated directly with the boil, and the amount of groundwater which could be transmitted through those pipes in particular. Dye testing was performed to confirm communication between the pipes and the piping path, and water testing was performed to confirm flow capacities. Phosphorescent dye was pumped under slight pressure into each of the pipes which were suspected to be installed in or close to the piping path. The boil was observed with each injection. Several responses were recorded; the degree of coloration of the boil, the time required from introduction at the pipe to observation at the boil, and the length of time that the boil remained colored. As suspected, those holes which lost drilling fluid circulation upon installation and indicated high groundwater pressures communicated directly with the boil. A red phosphorescent dye pumped into the pipe appeared almost undiluted in the boil. The time delay between injection of dye and appearance in the boil varied directly with the distance the pipe was from the boil. Based on the flowrate at the boil and the time-distance relationship, it was estimated that the piping path had the water carrying capacity approximately equivalent to an 8 inch diameter pipe.

Water testing was conducted initially by pumping water into each pipe. Water take and pressure were recorded for each pipe. It was immediately apparent which pipes were set into undisturbed soil; water takes varied from 1.5 gpm to 4 gpm with pressures maintained below 8 psi so as not to blow out to the ground surface. Those pipes installed into the piping path took the maximum capacity of the test pump, 28 gpm, with no pressure. Each of the high flow pipes were also water tested by pumping with suction means and yielded a minimum of 20 gpm. It was believed that adequate pressure relief would be provided by the total flow capacity of the pipes in place and plugging of the boil could proceed.

Plugging of the Boil

Once an adequate array of pipes is in place, the next step in the process is to plug off the boil. The initial plug created at the boil will serve as a stopping block so that grout subsequently introduced into the piping path will flow under pressure back under the gypsum stack as far as possible with reasonable pressures.

Where voids were observed or suspected, a select mix of particulate materials (sands, fine gravels, etc.) are introduced into the pressure relief / grout pipes to create a graded filter plug at the throat of the boil, which typically has been loaded with gravel. The objective in introducing the particulates is to select or sequence the materials so that the plugging media is washed by the flowing water towards the boil and the plug is formed as close to the throat of the boil as possible. A concern is introducing a particulate material too coarse, un-moveable under the flowing water in the piping path, and subsequently forming a plug at the base of one of the pipes, thus leaving a length of piping path, or other tributary piping paths that may converge at the boil, separated from the other useable pipes further upstream and subsequent grout sealing. Good judgment and the ability to understand the signs and symptoms as they demonstrate themselves are necessary for the trial and error process that must occur.

The grout pipes installed and confirmed within the piping path itself are rather valuable entities. A typical boil seal project may involve the installation of 30 to 70 pipes to gain only a handful of key pipes in the piping path itself which will be used to do essentially all of the sealing work. On the typical project, there is generally a limited amount of real estate between the boil location and the steeper embankment slope between which numerous pipes must be installed and the whole operation must take place. Additional pipes installed from the sides of the embankment at greater distances from the boil are often highly impractical to install, and the likelihood of tapping the piping path is significantly diminished with the greater distance. Separating a length of piping path with good pressure relief / grout pipes by mistake would be a significant loss of valuable instruments, in addition to cutting off a void system from grout sealing.

In order to assure maximum travel distance of the particulate media, finer material is introduced first. The boil observed for media passing through so as to gage the effectiveness of the material and make changes as necessary. On this particular project in central Florida, bentonite-cement grout was introduced first into the closest open pipe to the boil. The grout flowed directly through the boil. Sawdust and wood shavings were then added to the grout, anticipating that the grout would flow at least as far as the boil, and the wood shavings would plug the coarse void spaces within the ¾ inch stone backfill in the boil. The grout was used to suspend the shavings and then immediately start grouting the piping path once the shavings took hold. Other plugging agents such as walnut shells, shredded cellophane, and even horse manure have been used in the past, typically for grouting openwork materials or leaks in cofferdams.

A 70 gallon batch of grout with wood shavings was pumped into the closest open pipe to the boil with immediate results; the boil slowed down from 270 gpm to about 45 gpm, and the next three upstream open pipes started flowing like artesian wells at about 40 gpm each. The pipes were allowed to "bleed" to relieve the groundwater pressures within the foundation soils so as to minimize the groundwater pressure build-up behind the plug which could have resulted in a blow-out of the plug or the creation of a secondary piping path. The 270 gpm total boil flow dropped to about 165 gpm from the boil and the 3 adjacent pipes, with a corresponding increase in groundwater levels, as displayed by the artesian conditions observed in the three flowing pipes. A second and third 70 gallon batch of grout was pumped; the boil slowed to less than 5 gpm, with a corresponding increase in pressure and flow from the upstream open pipes, but the plug blew out within minutes and the 270 gpm boil re-surfaced and the pipes stopped flowing. Sawdust and wood shavings were observed in the grout colored boil when it made its reappearance.

Approximately 2 cubic yards of a very uniform course sand was introduced next. As the sand was introduced over several hours and built up behind the gravel backfill, the boil slowed down again to about 45 gpm, and the three adjacent open pipes resumed flowing again at 40 gpm each. Grout with wood shavings was re-introduced and the boil was successfully slowed to a flow of less than 5 gpm which would permit cement grouting to take place.

Grout Sealing of the Piping Path

Once the boil was plugged, and the transfer of groundwater flow occurred from the boil to the pressure relief / grout pipes, grout was introduced into the piping path starting at the boil location, and working back into the stack. Pressure relief "bleeding" of the upgradient pipes was required concurrent with the grouting operation until the flowpath was grouted up a sufficient distance beneath the stack where there was adequate cover and groundwater pressures could be safely dissipated through the foundation soils.

Grout was pumped into each of the open pipes until a return of undiluted grout was observed in the adjacent pipe upstream. At the start of grouting, little or no pressure was required to pump the grout as it flowed into the open void space and the groundwater pressure was allowed to bleed from the pipes further upstream. When the return of grout was observed in the adjacent pipe upstream, the downstream pipe was capped and the injection moved to the upstream pipe which just started the artesian flow of grout. In this manner, the grouting of the void space was progressively worked back towards the gypsum stack.

As the grouting proceeded, and the open pipes yielded grout and were subsequently capped, fewer pipes remained to bleed off the groundwater pressure, and the groundwater pressures rose. Usually at this stage in the process, the open, ungrouted piping path is further towards the gypsum stack and there is greater soil cover and grout pressures can be safely increased to overcome the higher groundwater pressure. Accelerators were added to the cement grout to speed up the set time of the material so that higher grouting pressures needed further upslope were not transmitted to the piping path at shallower depths closer to the boil.

The central Florida project was well instrumented with piezometers which allowed us to understand what was occurring to the local groundwater regime during the project. It was apparent that the piping path brought groundwater under high pressure to the periphery of gypsum stack and elevated the water levels of the surrounding area. As the boil was plugged, and grouting proceeded back under the embankment, the further increase in groundwater pressure due to restricting the boil could be monitored and tracked. As the grouting proceeded back under the embankment, groundwater levels were observed to drop along the length of piping path that was grouted, and rise along the ungrouted upstream lengths.

SUMMARY

The basic conceptual steps of the process as conducted in the example project discussed herein and on several other projects can be outlined as follows:

- Understanding the foundation soil stratigraphy and conditions. The vulnerable or suspect strata must be determined.
- Delineating the approximate piping path(s) based on groundwater pressures at various depths within the foundation soils. An array of piezometers may be installed to accomplish this, at various distances from the boil, and possibly installed in several strata, at variable depths.
- Following the groundwater level indicators, a highly sensitive drilling process is utilized to locate the piping path(s) and install combination pressure relief / grout pipes within the piping path itself. The number and spacing of the pipes varies from site to site.
- A select mix of particulate materials are introduced into the pressure relief / grout pipes to create a graded filter plug at the throat of the boil. Particulate material is introduced until flow from the boil is slowed adequately to permit grouting.
- Concurrent with the plugging operation, the installed pressure relief / grout pipes are allowed to "bleed" and relieve the groundwater pressures within the foundation soils so as not to allow excessive pressure build-up behind the plug which may result in a blow-out of the plug or the creation of a secondary piping path and a new boil.

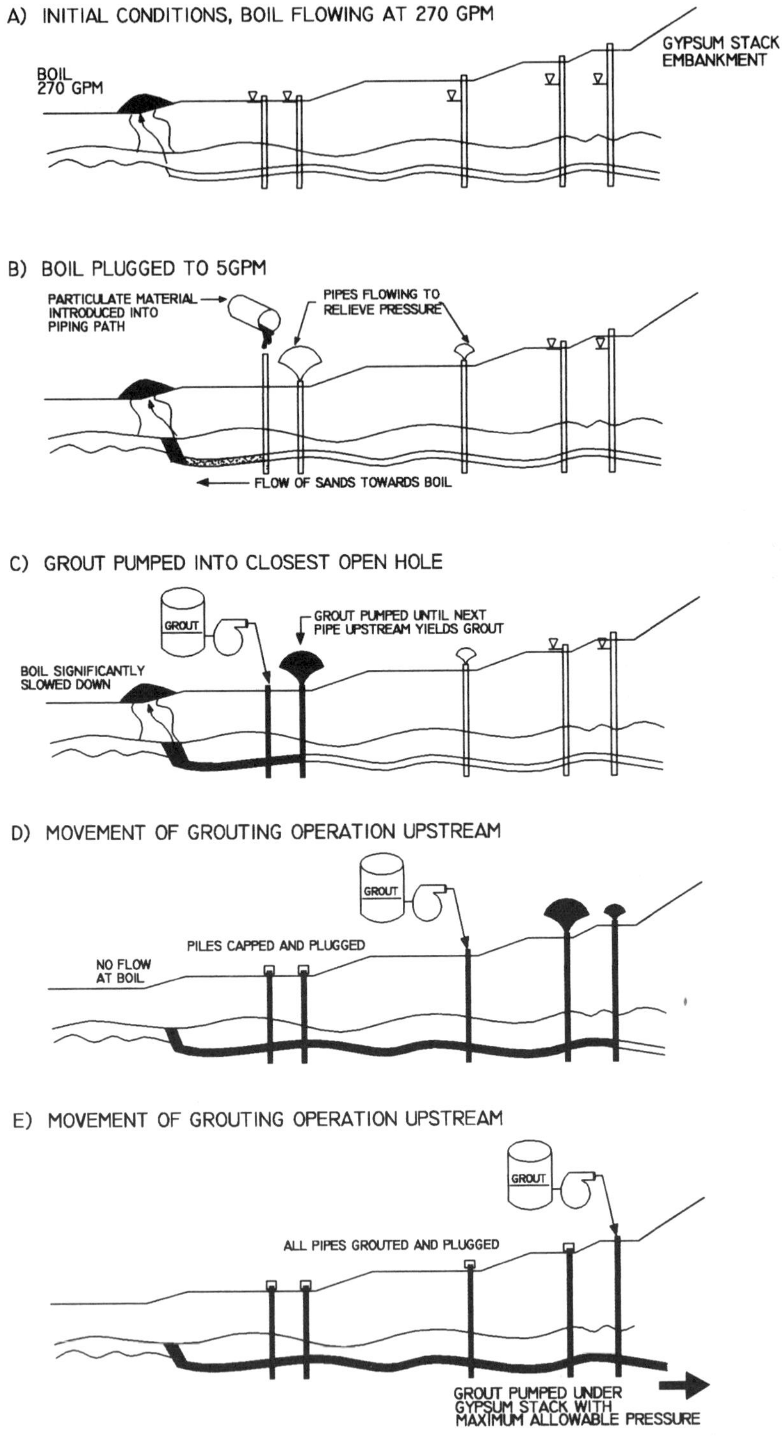

Typical Grouting Sequence

- Once the boil is plugged, and the transfer of flow has occurred from the boil to the pressure relief / grout pipes, grout is introduced into the piping path starting at the boil location, and working back into the stack. Pressure relief "bleeding" of the upgradient pipes is required concurrent with the grouting operation until the flowpath is grouted up a sufficient distance beneath the stack so that groundwater pressures may be adequately dissipated through the foundation soils. Accelerators are introduced into the grout to provide a quick set-time. Permanence of the grout material under the harsh chemical environment is considered in design of the grout mix.

CONCLUSIONS

It should be noted that although the approach to each project involves the same concepts, the conditions at each site are different and must be understood and addressed appropriately. Each project is essentially a new trial and error exercise involving a tremendous amount of observation, probing, testing, sampling, data interpretation, and educated guess work. The vast majority of the effort on any one of these projects is in delineating the piping path, and can only be done by understanding and recognizing the behavior of the groundwater. The actual steps followed to grout up the piping path are conceptually the same elementary techniques and methods that have been used repeatedly in the grouting of structures under flowing water conditions and in more specialized circumstances for grouting windows in cofferdams or subaqueous temporary structures.

Quality control and more importantly the use of good engineering judgment are essential to the safety of the operations. Perforating the landscape with drilling equipment, especially under artesian conditions, pumping water from open ended pipes, and intentionally building pressure within a flowing piping path are particular facets of this process that pose very real threats to already very sensitive and vulnerable site conditions if not conducted with the proper respect and more importantly the proper experience. Typically, each project is completed in fire drill mode to rectify the situation before the conditions worsen, and in the process a tremendous amount of quick decisions, observations, and interpretations of very subjective observations and data must be made in the trial and error process. An intuitive feel for what is occurring beneath the ground surface is absolutely essential. If at all possible, the work should be done in conjunction with an engineer knowledgeable in the process, and cognizant of the safety concerns, who can make the assessments of the conditions as they develop.

ACKNOWLEDGMENTS

Robert G. Lenz, P.E. and J. Patrick Powers, P.E. for their wealth of guidance and counsel.
Ardaman & Associates, Inc.

Tailings and Mine Waste '99 © 1999 Balkema, Rotterdam, ISBN 90 5809 025 6

Some considerations in the stability analysis of upstream tailings dams

T.E. Martin
AGRA Earth and Environmental Limited, Burnaby, B.C., Canada
E.C. McRoberts
AGRA Earth and Environmental Limited, Edmonton, Alb., Canada

ABSTRACT: Upstream constructed tailings dams represent a significant challenge to the geotechnical engineer in terms of analysis of their stability, in large part because the shear strength of the loose sands and fine grained or "slimes" components of such structures is open to considerable uncertainty. In particular, it is critical that the behavior of the tailings under shear (contractant or dilatant, failure under drained or undrained conditions) be understood, and that this understanding be incorporated into stability analyses. The continued development of critical state soil mechanics has served to focus attention on this issue, but this focus has generally been directed towards problems involving seismic liquefaction. However, contractant versus dilatant behavior in shear is equally important in terms of static stability, particularly for upstream tailings dams which so often are constructed of primarily contractant, potentially liquefiable materials.

This paper presents a review of drained versus undrained methods of static stability analysis of upstream tailings dams, and how these relate to dilatant versus contractant behavior. The authors support the view that for upstream dams constructed of contractant, potentially liquefiable tailings, both undrained strength analysis and steady state strength analysis should be considered, and that effective stress analysis for such structures can be fundamentally incorrect and unsafe.

1 BACKGROUND

The upstream method has been employed for many tailings dams. Two idealized typical sections of upstream tailings dams are shown in Figure 1. The past and continued attraction of the method is obvious, as it represents the most economical of all methods of tailings dam construction. Mill tailings are hydraulically separated - either by cycloning or spigotting on beaches or in cells - with the coarser sand sized fraction used to build a retention shell or dam and the fines collecting in a pond. In some cases when beaches are short, sands are placed subaqueously and as a result are very loose. As the dam raises the more stable subaerial beaches [which are often strengthened by desiccation] step out over loose subaqueous sands or fines (slimes.) As dams are usually raised against some sort of regional slope, or in a valley, water pools against the starter dam. This combination of circumstances usually occurs where the dam is highest. Thus as the dam is raised loose sands and weak slimes in a wide range of relative proportion and degree of intermixing are trapped in the downstream section, often with disastrous results.

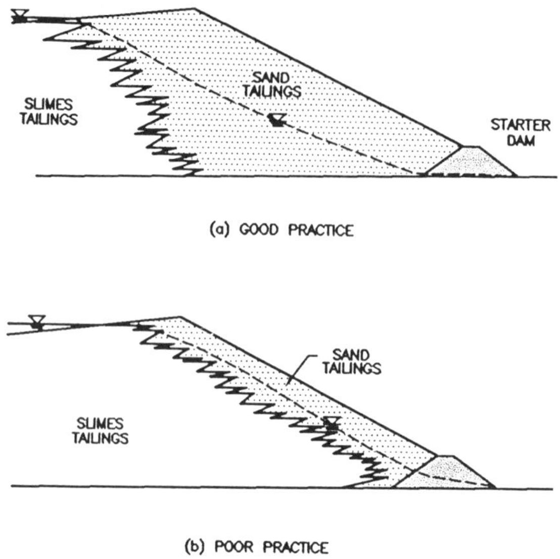

Figure 1. Idealized sections of upstream tailings dams.

Upstream tailings dams are unforgiving structures, and any one or combinations of improper design, construction and operation have resulted in a number of well-known, catastrophic failures, that have in some instances caused loss of life, such as the Stava failure in Italy (Berti et al., 1988, Chandler and Tosatti, 1995). The United States Committee on Large Dams (USCOLD, 1994) published a review of tailings dam failure records available to them in 1994. This review found that upstream-constructed tailings dams have recorded the largest share of documented failures.

The authors believe that many of these failures are due to the entrapment of loose sands and/or weak slimes in the downstream section. In many, and including recent designs reviewed by the authors, this factor is overlooked. This, as shall be described, is due to fundamental errors in the understanding of the operative strength of loose sands and weak slimes.

The susceptibility of upstream tailings dams to liquefaction and flow failures under seismic loading conditions is well known, and a number of case histories of such failures from the 1965 Chilean earthquake (Dobry and Alvarez, 1967), the 1978 Isu-Ohshima earthquake (Marcuson et al., 1979), and the 1985 Chilean earthquake (Castro and Troncoso, 1988) are well-documented. It is well understood that loose sands or weak slimes are contractant when sheared, and are therefore susceptible to this seismic triggering. What often appears to be not so well understood, and is the focus of this paper, is that undrained shear under static loading conditions can have the same consequences. As discussed in detail by McRoberts and Sladen (1992) there is little practical difference in the magnitude of the shear strength induced by seismic or undrained static loading.

The stability of tailings impoundments increases considerably with time once their operational life ends. This is particularly true for upstream tailings dams, and is due to the following factors:

1. Surface water is usually absent, particularly for impoundments regraded to shed runoff, allowing levels of saturation within the outer shell (and possibly the slimes) to gradually reduce.
2. Excess pore pressures induced by the raising of the impoundment will gradually dissipate, resulting in an increase in strength in the tailings slimes.
3. Capping of impoundments for closure reduces infiltration and allows for further reduction in saturation levels, particularly in dry climates.

4. Aging effects related to cementation and oxidation processes of tailings in the unsaturated zone, which may increase the liquefaction resistance of tailings by as much as 250% over 30 years (Troncoso, 1988, 1990).

5. Aging effects related to particle rearrangement resulting in macro-interlocking of particles and micro-interlocking of surface roughness (Joshi et al., 1995), a particularly significant mechanism given the angularity of tailings particles.

2 SAFE DESIGN AND OPERATION OF UPSTREAM TAILINGS DAMS

There is nothing fundamentally wrong with upstream tailings dams provided that key principles are adhered to in the design, construction, and operation of such dams. Upstream dams were originally designed in an empirical manner by mine operators, without the insights derived from geotechnical principles (Vick, 1992, Casagrande and MacIver, 1970). The vast majority of upstream tailings dams have performed satisfactorily. Based on experience, both successful and unsuccessful, this empirical design approach identified the importance of several key fundamentals for upstream tailings dams (Lenhart, 1950, Vick, 1992):

1. spigotting of a wide, sand (drained) tailings beach from the embankment crest;
2. avoiding situations whereby the dam slope is underlain by fine tailings (slimes) deposited within the water pond;
3. prevention of seepage emerging on the dam face; and
4. having a well-drained foundation.

The dam section shown in Figure 1(a) satisfies the criteria above, while the section shown in Figure 1(b) clearly does not.

These experienced-based principles remain valid, but do not directly address the criticality of operating and monitoring practices in maintenance of the safety of upstream dams. Nor do they directly address the issue of characterization of shear strengths in terms of drained or undrained behavior under shear. The authors therefore expanded the checklist above to the following eight fundamental rules for design, construction and operation of upstream tailings dams:

1. A sufficiently wide beach, relative to the ultimate height of the dam, must be maintained at all times, to achieve segregation of the coarser tailings sizes and to form a relatively strong, wide, drained (unsaturated), and/or dilatant (non-contractant during shear) outer shell. The dam slope must not be underlain by tailings slimes, unless the designer has satisfied Rule 4 below. The shell must be of sufficient width to retain the "bursting pressures" [see Casagrande and MacIvor] of the upstream contractant beach sands or slimes if they liquefy.

2. The rate of raising of the dam must be sufficiently slow such that there is a sufficient degree of dissipation of excess pore pressures in the outer shell and in the slimes, and such that excess pore pressure buildup does not occur in foundation materials.

3. There must be sufficient underdrainage (drainage blanket, finger drains) and/or a pervious foundation to maintain the sand shell in a relatively drained condition, and to prevent seepage from issuing from the face of the tailings dam.

4. Design analyses must include both undrained strength analysis (USA) and effective stress analysis (ESA), with design controlled by the analysis type giving the lowest factor of safety. A wide range of factors including material type, degree of consolidation and stress path must be assessed in assigning the appropriate USA.

5. A high degree of regular performance monitoring, reviews, and ongoing involvement by the designer is essential to check that design intent is being satisfied, to confirm design assumptions, and to identify any design changes that may be required.

6. Conventional upstream dams cannot be considered for areas of moderate to high seismicity. Improved upstream construction, involving a combination of compaction of the outer shell and good internal drainage, can be used in such areas.

7. The design must be consistent in terms of design requirements (e.g. minimum beach width) versus operational requirements (e.g. pond size required for clarification, storm storage and

freeboard). The geotechnical design of upstream tailings dams cannot be carried out in ignorance of operating constraints.

8. Seepage conditions within the dam must be well-defined, requiring a good understanding of pore pressure profiles and hydraulic gradients. The distinction between pore pressure measured at a given point, and saturation level, must be well understood and correctly applied in stability analyses, especially in instances where there is strong downward drainage.

The following sections of this paper are focused primarily on illustration and support of Rule 4 above. Rule 4 merits particular attention for the following reasons:

1. It becomes critical when Rule No. 1 is violated, as is frequently the case for upstream tailings dams;
2. A great many upstream tailings have been, and continue to be, designed and/or evaluated based on limit equilibrium slope stability analyses assuming only ESA parameters;
3. ESA analysis for upstream dams in many instances is based on fundamentally incorrect assumptions regarding pore pressures prior to and during shear failure, and can result in large overestimates of the factor of safety; and
4. USA analysis should be applied for staged construction, and upstream tailings represent a classic case of staged-construction.

Tailings deposited in upstream impoundments, and particularly tailings slimes, are generally loose to very loose, contractant during shear, and strain-softening (brittle). Exceptions occur where the outer shell of an upstream dam is compacted (e.g. Martin and Tissington, 1996), or where desiccation in arid climates leads to overconsolidated and unsaturated conditions within the tailings beach (Blight, 1988). Both of these exceptions are cases where the tailings are dilatant during shear (and/or unsaturated), and therefore the undrained strength is higher than the drained strength (negative pore pressures generated during undrained shear). For contractant tailings, the opposite is true, and the question then becomes whether shear will occur under drained (no shear-induced increase in pore pressure) or undrained (shearing induces increased pore pressure) conditions.

3 APPROPRIATE STRENGTH MODEL: USA VERSUS ESA

3.1 *Definitions of strength characterization*

ESA, following Ladd (1991) and Carrier (1991) is defined as limit equilibrium analysis that assumes effective stresses during shear are unchanged from those that existed immediately prior to the onset of shear. That is to say, measured insitu pore pressures describe the conditions at failure. In other words, ESA explicitly assumes that shear occurs slowly enough, and/or the material being sheared is sufficiently free-draining, that there are no shear-induced pore pressures. An ESA method of analysis is either correct if one is sure that no positive pore pressures are generated on shearing, or is conservative if it is known that shearing is dilatant. On the other hand if one did an ESA analysis using a method to predict the pore pressure response during shearing [i.e., at the moment of failure] then a correct answer would be obtained. What can be fundamentally wrong about ESA analysis is using existing as measured pore pressures in a dam to represent the conditions at failure. This simple fact is unfortunately not understood by many geotechnical engineers.

USA is an analytically economical way of accounting for the pore pressures generated by undrained shearing. USA is defined in this paper as limit equilibrium analysis that assumes shear occurs under undrained conditions. This type of analysis therefore accounts for positive shear-induced pore pressures during shear of contractant materials, and negative pore pressures in the case of dilatant materials. A convenient way to describe the USA strength is the undrained strength ratio (c_u/p'). The effective stress p' denotes the operative effective stress [or consolidation stress] that exists in a soil element at the moment that failure begins.

3.2 *Behavior of non-dilatant or contractant material during shear*

Traditional soil mechanics has and continues to develop strength models for two basic classes of materials, clays and sands. Many typical mill tailings streams produce relatively unique artificial materials that tend to be bounded by these more traditional models. However as the guidance offered by this rich literature database often seems to be ignored by tailings dam designers it is useful to briefly review some fundamental aspects of it.

The term USA-D is introduced, (D designates ductile) to denote in a simple way the response of normally consolidated clays during shearing. During shear excess pore pressures are generated but with straining the available strength does not fall off or reduce. Key references are for example the work of Ladd (1991) and Wroth (1984). This literature makes several key points:

1. The USA-D strength of normally consolidated clays as a first estimate is in the range of $c_u/p' = 0.20$ to 0.25.
2. The USA-D strength is dependent on stress path. For example see Mesri (1989) who indicates that for triaxial extension (TE) the c_u/p' can be from 50% to 70% of the triaxial compression (TC) strength and the strength in direct simple shear (DSS) is intermediate between these limits.
3. The USA-D is dependent on the method used to measure undrained strength. Different results are obtained by different methods (Wroth, 1984).

These issues are well known in geotechnical practice, the lesson being that the determination of USA-D strength is not straightforward. Specification of pore pressure response at failure in an ESA type approach is even less straightforward, as demonstrated by Carrier (1991).

The term USA-SS is used here to designate the SS or steady state of loose sands [also called the residual strength.] Following Casagrande (1975), Poulos (1988) and as discussed more recently by McRoberts and Sladen (1992) loose sand can exhibit an undrained strength response at high strain, and a response that is highly brittle. The strength of clean loose sands at high strain or at steady state can also be expressed in a normalized manner, see for example Wride et. al. (1998). These authors follow others in normalizing the back-calculated strength of several case records in which the residual or steady state strength can be obtained. The normalized strength from these case records is as low as 0.01 and ranges up to about 0.20.

A recent series of USA-SS laboratory test on loose clean sands has been reported by Yoshimine et.al. (1998) for TC, TE, and DSS modes. This work following on earlier studies by others indicates that at the relatively low strains measurable in the laboratory that - and entirely similar to USA-D behavior for clays - the mode of shear has a strong influence on the magnitude of the normalized c_u/p'.

3.3 *Collapse Surface Approach*

The collapse surface approach presented by Sladen et al. (1985), as an extension of critical state theory, provides a useful framework to illustrate the authors' contention that USA analysis is mandatory for upstream tailings dams where contractant materials are involved. The concept is particularly useful in this application because it ties together undrained behavior in shear, the need for triggering of undrained behavior, and the brittle, flowsliding nature so often associated with tailings dam failures.

The concept of a collapse surface is illustrated in Figure 2a, which shows that for a given void ratio (density), there is a unique condition of stress state at which collapse of the sand structure takes place and undrained failure (liquefaction) is initiated. This unique state of stress is termed the collapse surface, and it exists below the drained failure envelope for contractant tailings. Figure 2a illustrates that stress states lying on or above the collapse surface are highly unstable, as liquefaction can be triggered by even minor disturbance. Stress states below the collapse surface line are stable.

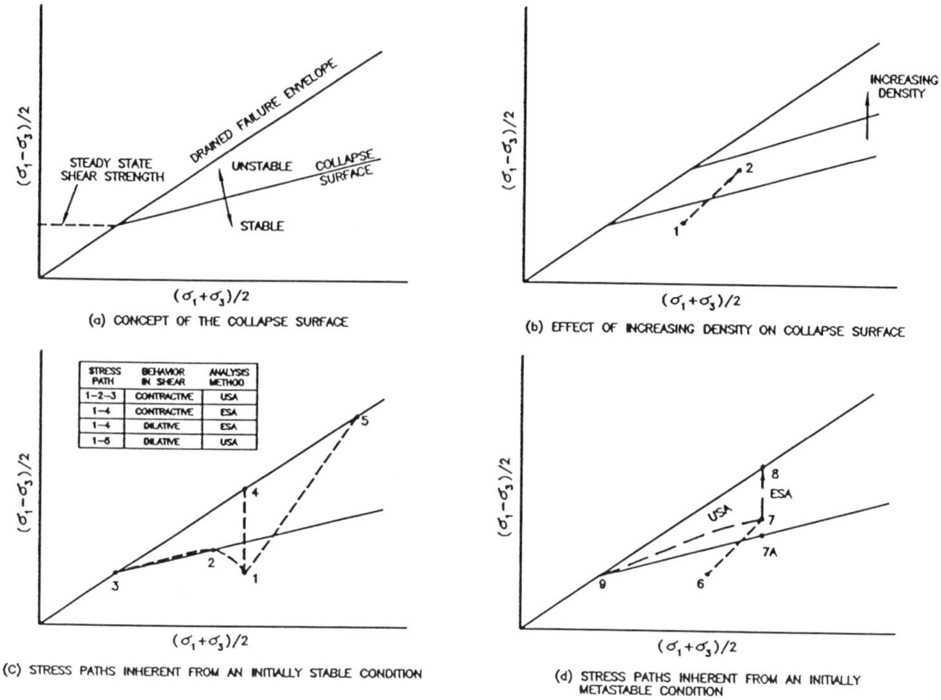

Figure 2. Collapse surface framework for comparison of USA and ESA approaches.

The collapse surface occurs in three-dimensional void ratio-shear stress-mean normal stress space. Figure 2b shows the effect of increasing density (decreasing void ratio) on the collapse surface, as projected onto two-dimensional shear stress - mean normal stress space. Also shown is a drained-loading stress path. Although point 2 lies above the collapse surface that corresponds to point 1, the fact that loading has taken placed under drained conditions, in concert with consolidation to a lower void ratio, has resulted in a shift upwards in the collapse surface line. Therefore, as long as undrained behavior is not triggered during the loading, a stable stress state can exist above the original collapse surface line, but only because the position of the collapse surface line has been shifted upwards during slow, drained loading.

Figure 2c presents stress paths for contractant and dilatant materials inherent to both USA and ESA limit equilibrium analysis. For a contractant material, USA assumes failure at point 2, with the mean normal effective stress at failure reduced from that at point 1 because of positive pore pressures generated during shear. USA also indicates a post-peak reduction in shear strength to point 3, which corresponds to the residual, or steady state strength. Stress path 1-4 represents that which corresponds to a conventional ESA, which by definition assumes slow shearing with complete dissipation of any shear-induced pore pressures during failure. Therefore, ESA assumes the mean normal effective stress at failure (point 4) to be the same as that at the initiation of shearing (point 1). It also assumes that failure occurs slowly. Figure 4c shows the shear stress at failure (and therefore the factor of safety) for ESA to be about twice that for USA, a typical result (Ladd, 1991) when comparing the two methods of analysis.

Assuming the material to be dilatant, then for undrained shear (assumed by USA), negative pore pressures developed by shearing would result in mean normal effective stress at failure (point 5) higher than that at the initiation of shear (point 1). Under fully drained shear assumed by ESA, failure would occur at point 4. Therefore, USA predicts a higher shear stress at failure (and therefore a higher factor of safety) than ESA for dilatant materials. In this case, ESA represents the more critical (and proper) method of analysis.

The differences between the stress paths assumed in the two types of analyses for contractant versus dilatant materials represent an important point. Advocates of ESA for upstream tailings dams, irrespective of whether saturated zones are contractant or dilatant, ignore completely the physical behavior of the materials under shear, surely one of the most important principles in soil mechanics. Furthermore, there is an inherent contradiction in accounting for contractant response when analyzing the seismic safety of upstream tailings dams, but failing to do so for static loading conditions. Finally, since ESA is the type of analysis applied for compacted embankments, it seems counterintuitive to blindly apply this type of analysis to uncompacted, loose tailings.

Figure 2d shows a case whereby loading from a point below (point 6) to above (point 7) the collapse surface occurs without failure, because undrained shear was not triggered during the loading. A stress state can therefore exist in the unstable zone, which in this case would be better referred to as the metastable zone, where spontaneous collapse (static liquefaction) can occur with even slight disturbance. USA for this circumstance would yield a factor of safety of less than 1 (failure at a shear stress corresponding to point 7A), and interpreted literally would suggest that such a stress state could not exist. Therefore, USA carried out in isolation as a standard limit equilibrium analysis ignores the need for some disturbance to trigger undrained behavior. Note again that ESA for such a metastable state ignores completely the potential for collapse of the soil structure, and gives a completely misleading impression as to the safety of the dam.

3.4 *Evidence Of USA Response In Upstream Tailings Deposits*

In a previous section we have reviewed the soil mechanics background and shown that for the typical range of soft normally consolidated clay soils to loose sands that a USA type strength response is often encountered. What then is the evidence from tailings deposits?

Probably the first reference on the normalized strength of copper tailings slimes is from Castro and Troncoso (1988) as summarized in Table 1.

Table 1. Data from Castro and Troncoso (1988)

Dam	Undrained Peak Strength Ratio USA-D	Undrained Steady State Ratio USA-SS	Comments
Cerro Negro CN4	0.27	0.07	Slimes, slightly plastic clayey silt PI of 5-20%
Veta deAgua VA1	0.21	0.11	Slimes, clayey silt.
El Cobre EC1	0.29	0.08	Slimes, no details.

Note: Tests at steady state done with rapid undrained vane tests.

This testing indicates both the USA-D type response as well as the substantially lower steady state or USA-SS mode.

A summary of other case records is given in Table 2.

Table 2. Summary of USA Response in Tailings

Case	Ratio c_u/p'	Reference
Aluminum red mud (residual or SS)	0.025	Poulos et. al. (1985)
Lead – Zinc non-plastic slimes	0.2 to 0.22	Vick (1990)
Copper tailings slimes , PI = 6%	0.26 to 0.33	Bromwell (1984)
Copper tailings slimes, PI = 10+/-3%	0.275	Ladd (1991)
Copper tailings slimes $N_{1,60}$ of from 1 to 5	>0.20	Vidic et. al. (1995)
South African slimes	0.25 CPT 0.17 to 0.4 Vane	Blight (1997)

Writing on possible mechanisms for the runout of tailings derived mudflows, Blight (1997) presents conflicting information on tailings strength from Merriespruit which clearly demonstrates the USA vs ESA issue. The author states that:

> "Tailings consist of sand and silt sized particles of milled rock. The tailings referred to in this paper contain hardly any clay-size particles and, when sheared, behave as frictional, cohesionless materials with angles of shearing resistance in the range of 29-35°."

Data from laboratory undrained dynamic shear tests report dilatant behavior, and Blight (1997) concludes that the low shear strengths that occurred in the field cannot be explained by postulating a form of undrained shear mobility of the tailings. However the author also presents the results of in situ vane and CPT tests (see Table 3 of Blight, 1997). The cone tip resistances are interpreted by Blight (1997) to give a shear strength gradient of about 2 kPa/m in the interior of the impoundment i.e., slimes or fine tailings. For an effective vertical stress gradient of about 8.0 kPa/m the USA-D ratio is therefore about 0.25. It is reasonable to think that these CPT probes are essentially undrained. The author also presents the results of vane testing. The vane peak tests, which are likely drained give a high strength/vertical effective stress ratio of 0.75 or more indicating dilatant response for drained shear. Remoulded vanes probably undrained give c_u/p' of 0.17 in the upper 15 m and about 0.40 in the lower 15-30 m of the deposit. It is not known if the laboratory tests were on undisturbed or remoulded samples. Experience indicates that obtaining either undisturbed samples of loose tailings or re-creating the in situ fabric by laboratory techniques is difficult. The field evidence in the form of vane and CPT probes clearly indicates undrained response, as does the flowslide nature of the failure itself. More exotic explanations of the failure hardly seem necessary.

The authors had occasion to investigate the recent failure of a tailings dam in South America designed in the early 1990's by an internationally known company experienced in tailings dam design. This dam was designed solely with an ESA framework. After the failure which was due to undrained loading and resulting shear failure a major site investigation was including SPT and CPT testing was undertaken by the designer. During this investigation slimes were encountered and described by the designer as follows:

> "Fine tailings (slimes) that induce porewater pressure response during cone penetration are interpreted by this classification as clayey silts to clay. This interpretation may be correct in interpreting predominant particle sizes, but suggests a degree of plasticity which is not present in non-plastic rock flour"

The designers went on to analyze the structure using an ESA approach and determined that it was safe, notwithstanding the recent failure, to raise the structure as the investigation had determined that the design parameters were consistent with the original design. Analysis of the CPT data using the same procedures as Vidic et. al. (1995) by the authors indicated in situ c_u/p' values of about 0.20. The use of this magnitude of strength readily explained the failure which occurred during a construction lift of the dam. In the authors' opinion this case record offers a classic example of the ESA versus USA issue and the absolute fallacy of assuming that "non plastic rock flour" can only have a so-called drained strength. In this particular case the designers recognized that the cone testing in slimes induced a pore pressure response but chose to rationalize the fact away by the "rock flour" model. The dam was condemned.

3.5 *Discussion*

Ladd (1991) discusses the differences between ESA and USA for staged construction, and argues convincingly that USA is the correct approach, because when failure of an upstream tailings dam does occur, it will be under rapid, undrained conditions if the tailings are contractant under shear, as is so often the case. Carrier (1991), in a paper that should be required reading for any engineer involved with upstream tailings dams, extends Ladd's argument specifically to the case of upstream tailings dams, for which he too convincingly advocates use of USA analysis. Carrier (1991) also pointed out the fallacy of justifying the assumption of drained shear in tailings on a high c_v value (i.e. assuming a relatively free-draining material would necessarily undergo shear under fully drained conditions). Comparing an upstream tailings dam to a stress-controlled triaxial test, Carrier pointed out that, once the peak deviator stress is reached, the strain rate to the steady state (residual) strength for an undrained, contractant sand can be about 170% per second (Been and Jefferies, 1985). At such a strain rate, even a highly free-draining material would undergo shear under undrained conditions.

For upstream dams composed of loose, contractant tailings, failure will therefore typically occur very rapidly and under undrained conditions, and this is borne out by the fact that upstream tailings dam failures so often take the form of massive flowslides. Unfortunately, such field behavior is often not duplicated in laboratory testing, where, out of testing convenience and equipment limitations, testing is carried out under strain-controlled rather than stress-controlled conditions.

Casagrande (1975) pointed out that rapid (i.e. undrained) collapse deformation is an important behavior related to flow liquefaction in the field that cannot be observed in strain-controlled tests in the laboratory. Zhang and Garga (1997), describing the results of triaxial tests on loose sands carried out under stress-controlled conditions with a rapid data collecting system, suggested that laboratory equipment limitations and procedures may inhibit collapse deformation behavior, supporting Casagrande's view. This work suggests some caution in the adoption of USA-SS strength characterization from laboratory testing, and especially of the strain-controlled variety.

The advocacy of USA for upstream tailings dams presented in this paper is therefore a well known approach, best supported by Carrier's (1991) back analysis of the well known Tyrone failure case history. The Tyrone failure provided a classic example of the misconceptions as to tailings strength. This seems to have its origin in the old soil mechanics precepts that sands were frictional and clays cohesive. The misconception is that if one has non-plastic rock flour such material is obviously frictional and an ESA analysis applies. The flawed corollary is that if slimes are not cohesive, then a USA is not required - irrespective of whether or not the slimes are contractant in shear, and potentially liquefiable. Given these misconceptions, it is not surprising that tailings dam failures continue to occur. These lessons, shorn of theoretical aspects, were well understood many years ago by the likes of Casagrande and MacIver (1970), Smith (1972), and Lenhart (1950), and yet still have not entirely permeated the practice of tailings dam design and analysis.

4 REQUIREMENT FOR TRIGGERING OF UNDRAINED SHEAR

ESA analysis typically overestimates the factor of safety by a factor of two relative to USA analysis. This in turn would suggest that a great many upstream tailings dams have USA factors of safety of less than one (since they are rarely designed using ESA to a factor of safety of 2 or more), and should have failed (based on limit equilibrium analysis), but have not. This in turn suggests that in many cases an undrained trigger has been absent.

The discussion above for Figure 2d addresses this apparent contradiction. Vick (1992), in a discussion of Ladd's Terzaghi lecture (Ladd, 1991) also addressed it, pointing out that for upstream dams constructed as shown on Figure 1a, which satisfy Rule No. 1 above, ESA had served well. Vick (1992) also discussed how the breakout of seepage on the dam slope could lead to rapid, undrained progressive failure even of relatively coarse (free-draining) but loose (contractant) sands. In this case, the seepage breakout triggered undrained shearing in a contractant material, even though drained conditions existed immediately prior to failure and the material was free-draining, a result consistent with the studies by Eckersley (1990). This reinforces the notion that some trigger is required to initiate undrained shear failure in upstream tailings dams.

For upstream tailings dams, unfortunately, potential triggers of undrained failure are numerous, and include those listed in Table 3.

Table 3. Triggering mechanisms for undrained failures of upstream tailings dams.

Mechanism	Trigger
Oversteepening at toe due to:	Erosion (intense storm runoff, pipeline break causing washout) localized, initially drained sloughing construction activities (excavation)
Overloading due to:	rapid rate of impoundment raising steepening at crest construction activities at crest
Changes in pore pressures due to:	seepage breakout on face of dam deterioration in performance of underdrainage measures inhibited volumetric creep concentrated tailings discharge from one location for extended period leakage/rupture of low level outlet accelerated rate of construction foundation and/or embankment movement intense rainstorms increased pond levels
Triggering collapse surface by reduction in mean effective stress (see Figure 3)	Consider an element of soil below the collapse surface with a low shear stress and high mean effective stress due to low or absent phreatic surface. Saturating the slope reduces mean effective stress, but leaves shear stress constant. The reducing mean stress results in contact with the collapse surface and liquefaction is triggered.
Overtopping due to:	severe storm runoff failure of diversion dams/ditches blockage and failure of spillways/decants seismic deformation and loss of freeboard
Acceleration/vibrations due to:	Earthquakes construction traffic Blasting

It is noteworthy that earthquakes represent perhaps the only of the above triggers for which analytical methods account for, and are indeed based on, the distinction between contractant versus dilatant material in shear.

Ladd (1991) observed that failure for contractant materials will occur under undrained conditions even if drained conditions prevailed immediately prior to the failure, an assertion supported by work by Eckersley (1990), in his modelling of flowslides. The most elegant demonstration of this, and the role of Sladen's collapse surface and drained triggering mechanisms can be found in a series of tests reported by Sasitharan et. al. (1993). Figure 3 reports on one of these tests. In this Figure the state boundary or collapse surface for a clean Ottawa sand is given at a void ratio of 0.809. In the lab a new sample of sand at a void ratio of 0.804 and at a stress state represented by a deviator stress of 100 kPa and mean consolidation stress of 300 kPa was prepared. This sample exists well below the state boundary surface and any minor excursions of shear stress or reduced or increased mean effective stress would result in a stable response to loading. However by appropriate manipulation of the loading conditions and back pressures, the mean effective stress on the sample was reduced but the deviator stress was kept constant. As the mean stress was reduced in a series of drained increments, the collapse surface was reached. At this time the void ratio had increased to 0.809, the collapse surface was reached and an undrained collapse of the sample was triggered. As shown on Figure 3 the deviator stress that could be supported by the sample reduced to about 20 kPa. The USA-SS strength for this sample is therefore about 20/300 or 0.067.

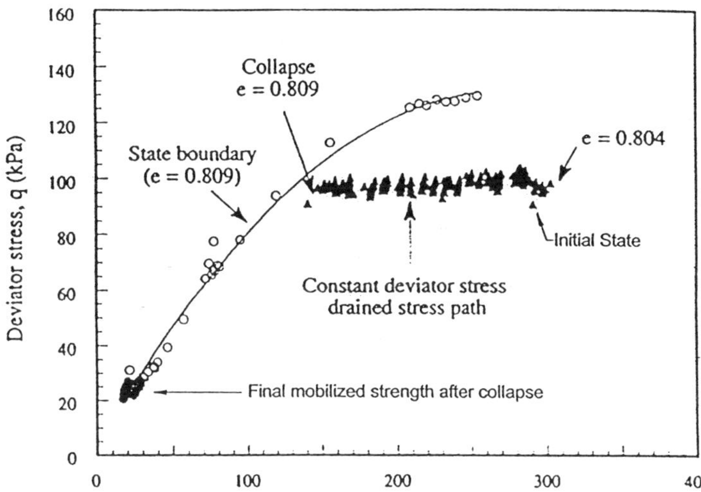

Figure 3. State boundary defined by void ratio 0/809 and constant deviator (q=100kPa) drained stress path. After Sasitharan et al (1993).

Consider then an element of soil in the relatively loose shell of a tailings dam. Assume a low phreatic surface and a resulting stress state below the collapse surface. This stress state has reasonable high shear stress and high mean effective stress due to low or absent phreatic surface. Assume further that the slope saturates due to heavy rainfall, lateral migration of the phreatic surface due to dam raising or some combination of these or other events. Saturating the slope reduces mean effective stress, but leaves shear stress constant. If the reducing mean stress results in contact with the collapse surface liquefaction is triggered. Right up to the initiation of collapse the soil elements have been drained. This is an example of drained loading triggering an undrained collapse.

5 RECOMMENDED APPROACH

The authors therefore strongly concur with Carrier's (1991) assertion that upstream dams should be assessed based on both USA and ESA analysis. However, in recognition of the metastable nature of upstream tailings dams, the USA analysis results must be interpreted in the context of the potential triggers of undrained shear listed in Table 1, rather than a limit equilibrium factor of safety in isolation. Unless considerable effort is expended in the design process and due consideration is given to all failure modes it is considered appropriate to assume that if a soil can liquefy it will.

Given that many of these triggers represent operational rather than geotechnical factors, it follows that it is essential that operational aspects (e.g. rate of raising) be integrated into any stability assessment and design of an upstream tailings dam. Furthermore, the margin of stability against shear failure of upstream tailings dams is better expressed jointly in terms of both the probability of undrained behavior being triggered, and a limit-equilibrium factor of safety.

The authors consider that, for a great many upstream tailings dams, especially those where the dam slope is underlain by slimes (i.e. violates Rule No. 1), ESA analysis has given the right answer (i.e. a stable dam), but for the wrong reasons. That is an undrained mechanism never existed. The fact that such dams have remained stable cannot be considered an endorsement of universal application of ESA analysis. Fortunate happenstance is no substitute for good design and analytical practice. Only where Rule No. 1 is satisfied, and the dam configuration is as shown on Figure 1a, can reliance solely on ESA be justified. Continued reliance on such an approach for upstream dams that violate Rule No. 1 ignores the fundamentals underlying triggering and pore pressure response during undrained shear in contractant materials, and is likely to lead to future failures of upstream tailings dams.

5.1 *Estimation of Drained Shear Strength Parameters for Analysis*

For assessments of the stability of a number of upstream tailings dams at base metal mines in which the authors were recently involved, little or no data was available with which to estimate the drained and undrained strength parameters for the tailings dams assessed. There is abundant case history experience to draw upon in terms of drained strength parameters. Effective friction angle (ϕ') values typically range between 25° and 35° for base metal tailings, with the lower portion of this range applying for silt/clay tailings slimes, and the upper portion applying for coarser tailings deposited on beaches. Table 2.8 of Vick (1990) gives a summary of typical values of ϕ' for various types of tailings.

5.2 *Estimates of Undrained Shear Strength Parameters*

The brief review presented earlier for normally consolidated clays and loose sands indicate two forms of USA behavior: ductile and brittle or steady state / residual modes. We have also reviewed a series of references where many examples of undrained strength ratios (c_u/p') reported in the literature, where a range of 0.2 to 0.3 appears to cover most of the cases for tailings at least for USA-D type response. This suggests that many tailings have an USA response very similar to normally consolidated clays.

However the test data presented by Castro and Troncoso (1988) introduces the considerable caution that a USA-SS response may also be present. Much of the data available and discussed above is based on insitu interpretations of CPT response.

What we have called USA-SS mode or strength response has been discussed in detail by Poulos (1988) who recommends an approach more conservative that the USA approach (which is based on peak undrained shear strength). For the USA-SS mode design analyses are carried out based on the steady state, or residual, undrained strength. This approach assumes that the worst case scenario (straining of the contractant tailings sufficient to cause undrained collapse of the soil structure) governs design, reducing shear strength from peak (USA analysis, point 2 on Figure 2c) to residual (Poulos method, point 3 on Figure 2c) levels. Carrier (1991) supports consideration of this approach. He emphasizes that it is necessary to consider the stress-strain characteristics of the tailings, with brittle, cohesionless tailings (strain to peak about 1% typically) more relevant and critical for a Poulos type analysis that clayey, ductile, cohesive slimes (strain to peak much larger than 1%).

The authors advocate that, for upstream dams, the Poulos approach should also be considered for both loose sands and well as slimes. This forces the stress-strain characteristics of the materials forming the dam to be considered in the analysis. The steady state approach is widely applied in analyzing the seismic stability of upstream tailings dams. Given that seismic loading is but one of many potential mechanisms of undrained loading (McRoberts and Sladen, 1992), it is inconsistent that the steady state approach not be at least considered under static loading conditions. Application of the method is not without its problems, particularly with respect to determination of the appropriate steady state strength, which has been the subject of much research and debate within the geotechnical profession for many years. Considering the difficulties of predicting the appropriate USA strength is however a step in the right direction; assuming the designer has at least abandoned a sole reliance on the ESA mode.

5.3 *Pore pressure conditions for stability analyses*

Use of both ESA and USA requires an estimation of the effective consolidation stresses at any given time, which in turn require a good understanding of the pore pressure conditions (hydrostatic versus downward drainage, normally consolidated versus excess pore pressures) within the dam. Pore pressure conditions within upstream tailings dams are very often complex, misunderstood, and improperly incorporated into stability analysis. Misinterpretation of piezometer data can easily occur if adequate piezometer coverage does not exist. Vick (1990) suggests that, for rates of impoundment rise of between 15 and 30 ft/year, excess pore pressures are usually assumed to dissipate as rapidly as the load is applied, and therefore a normally consolidated state (i.e. zero excess pore pressure) can be assumed. Mittal and Morgenstern (1976) also suggested this range as being sufficient to generate excess pore pressures in slimes.

The authors caution that these experience-based criteria on rate of rise can be safely applied only in cases with good underdrainage (permeable foundation relative to the tailings slimes), relatively smaller embankments (35 m in height or less) with relatively short drainage paths, and slimes free of significant clay content and plasticity. For example, the authors are aware of one large upstream dam in which very high excess pore pressures exist in the clayey slimes despite a rate of rise of only about 7 ft/year. Another example that emphasizes the need for caution is the Tyrone tailings dam, which failed under undrained conditions at a construction rate of 12 to 15 ft/year.

5.4 *Recommended approach*

To summarize, the authors provide the following general recommendations for any static stability analysis of an existing upstream tailings dam:

1. Determine whether or not the dam slope is comprised of materials that are contractant or dilatant under shear. Characterize pore pressure conditions within the dam, to properly determine effective stresses. If the materials are dilatant and/or fully drained (unsaturated), then only a ESA is required.
2. If the dam slope is fully or partially composed of contractant materials, then both ESA and USA should be carried out. The factor of safety from the USA will better represent the margin of safety of the dam. Considerable care must be exercised in the selection of the appropriate USA strength mode. Careful attention must be given to whether or not a USA-D or USA-SS mode might be triggered.
3. Review the operational and design factors necessary to assess the probability of undrained shear being triggered by the various mechanisms listed in Table 1. The probability of these mechanisms should be considered jointly with the USA factor of safety in evaluating the safety of the dam. Unless one is very sure that undrained triggers absolutely do not exist, it is considered prudent to assume the worst.
4. Evaluate the stress-strain behavior of the tailings comprising the dam. If the tailings are both contractant and brittle (low strain to peak strength, and significant post-peak reduction in strength), then a Poulos (steady state strength or USA-SS mode) analysis should be carried out.

The results of this analysis should be considered jointly with the assessed probabilities of undrained shear triggering mechanisms.

6 EXAMPLE RESULTS OF STABILITY ANALYSES

A dam configuration typical of several of those recently reviewed by the authors is shown on Figure 4. This dam was designed in the early 1990's and includes a compacted starter dam with drainage and filter zones as shown. The dam is being raised about 25 m above the starter dam crest, at a slope of 3H:1V, using spigotted tailings produced from a flotation circuit (base metals mine). The design specified that a minimum 30 m wide beach be maintained at all times. Fortunately, the operators were able to maintain somewhat wider beaches than this in all but one of the cases.

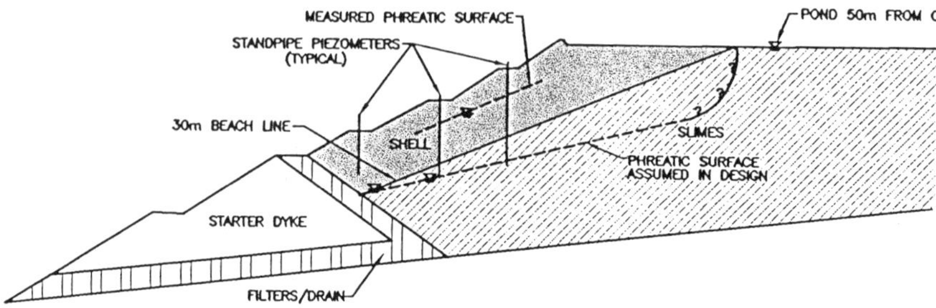

Figure 4. Typical design section of upstream dams reviewed.

The designer's stability analyses were of the ESA type, and assigned both the sand shell and the slimes an effective friction angle of 30°. Therefore, by assuming the same shear strength for both zones, the only effect of the beach width in terms of the design analyses was on the assumed location of the phreatic surface. The designer's analysis also assumed hydrostatic pore pressure conditions.

The stability analysis geometry and results obtained by the authors are illustrated graphically on Figures 5 and 6, for the design minimum beach width of 30 m. They clearly show that ESA provides an acceptable factor of safety to dam heights of up to 30 m (above the starter dam crest), while USA-D analysis indicates inadequate factors of safety. The USA analysis results shown on Figure 6 may well be slightly unconservative because they assume ESA can be applied for the outer sand shell. However, this material may also be contractant in shear, so undrained strengths could be more appropriate in the saturated portions of this zone. The analyses also do not account for the stress path dependence of USA-D strength discussed in Section 3.2. Moreover, depending on the ability of the tailings to absorb a degree of straining a lower USA-SS strength may be appropriate.

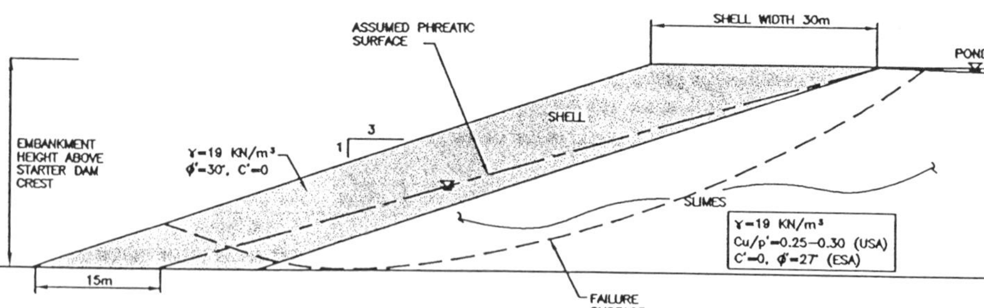

Figure 5. Stability analysis section.

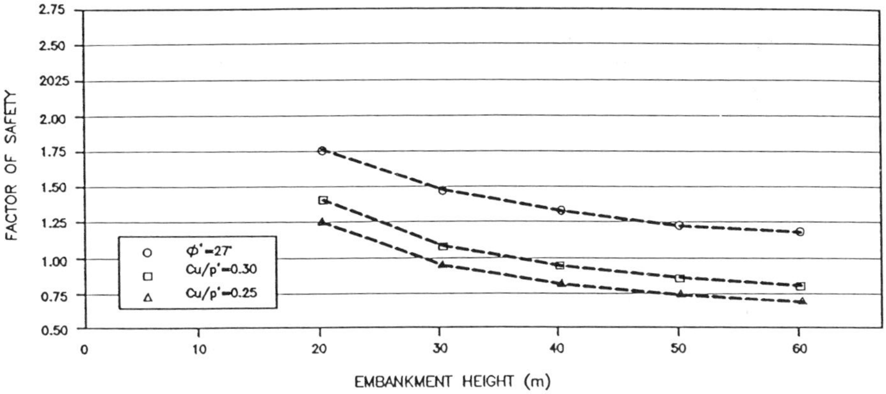

Figure 6. Results of preliminary stability analyses.

7 CONCLUSIONS

Upstream tailings dams are complex structures for which shear strength and pore pressure conditions are difficult to predict in advance. It is essential that stability analyses of these structures be predicated on a thorough understanding of the behavior of the tailings under shear, and an understanding of the likelihood of potential triggering mechanisms of undrained failure. Stability analysis of upstream tailings dams should be carried out using both drained strength analysis and undrained strength analysis. In some cases it may also be necessary to consider steady state strength analysis. This approach forces the designer to come to grips with the issue of contractant versus dilatant behavior in shear, and with the stress-strain characteristics of the material, two of the most fundamental precepts of soil mechanics that are ignored at the geotechnical engineer's peril.

REFERENCES

Been, K., & M.G. Jefferies 1985. A state parameter for sands. *Geotechnique*, Vol. 35, June, pp.99-112.

Berti, G., Villa, F., Dovera, D., Genevois, R., and J. Brauns 1988. The disaster of Stava/northern Italy. *Hydraulic Fill Structures*, ASCE Spec. Public. No. 21, pp. 492-510.

Blight, G.E. 1997 Destructive mudflows as a consequence of tailings dyke failures. *Pro. Inst. Civil Engrs. Geotech. Engng. 125, Jan., 9-18.*

Blight, G.E. 1988. Some less familiar aspects of hydraulic fill structures. *Hydraulic Fill Structures*, ASCE Spec. Public. No. 21, pp. 1000-1027.

Bromwell, L.G. 1984. *Stability investigation of Tailings Dam 1, Tyrone Mine, Tyrone, New Mexico*. Report to Phelps Dodge Corporation, Tyrone Branch, Tyrone, New Mexico, report prepared by Bromwell and Carrier, Inc.

Carrier, W.D. 1991. Stability of tailings dams. *XV Ciclo di Conferenze di Geotecnica di Torino*, Italy, November.

Casagrande, A. 1975. Liquefaction and cyclic deformation of sands: a critical review. *Harvard Soil Mech. Series No. 88.*

Casagrande, L., and MacIver, B.N. (1970). Design and construction of tailings dams. *Proceedings, Symposium on Stability in Open Pit Mining.*

Castro, G. and Troncoso, J., 1988 *Effects of 1985 Chilean earthquake on three tailings dams.* %th Chilean Conference on Seismology and Earthquake Engineering.

Chandler, R.J., and G. Tosatti, 1995. The Stava tailings dam failure, Italy, July 1985. *Proc. Inst. Of Civil Eng.*, Vol. 118, April, pp. 67-79.

Dobry, R., and L. Alvarez, L. 1967. Seismic failures of Chilean tailings dams. *Journ. Soil Mech. & Found. Eng. Div.*, ASCE, Vol. 93, No. SM6.

Eckersley, D. 1990. Instrumented laboratory flowslides. *Geotechnique*, 40(3), pp. 489-502.

Joshi, R.C., Achari, G., Kaniraj, S.R., and Wijeweera, H. (1995). Effect of aging on the penetration resistance of sands. *Canadian Geotech. Journ.*, Vol. 32(5), pp. 767-782.

Ladd, C.C. 1991. Stability evaluation during staged construction. The 22nd Karl Terzaghi Lecture, Boston, 1986, *Journ. of Geotech. Eng.*, ASCE, Vol. 117, April, pp. 537-615.

Lenhard, W. 1950. Control of tailings from washing plants. *Rock Products* (July), pp. 72-80.

Martin, T.E., and I. Tissington 1996. Design evolution of tailings dams at Inco Sudbury. Proceedings, *Tailings & Mine Waste '96*, January, pp. 49-57.

Poulos, S.J. (1988). Strength for static and dynamic stability analysis. *Hydraulic Fill Structures*, ASCE Spec. Public. No. 21, pp. 452-474.

Poulos, S., Robinsky, E., and Keller, T 1985 Liquefaction resistance of thickened tailings. *J.G.E., Vol. 111, No12 1380-1393.*

McRoberts, E.C, and Sladen, J., 1992 Observations on static and cyclic sand liquefaction methodologies. *Canadian Geotech. Journ.*, 29, 605-665.

Mesri, G. 1989 a revaluation of $s_{u(mob)} = 0.22\sigma'_p$ using laboratory shear tests. *Canadian Geotech. Journ.*, Vol. 26(1), 162-164.

Mittal, H.K. and N.R. Morgenstern 1976. Seepage control in tailings dams. *Canadian Geotech. Journ.*, Vol. 13(3), August, pp. 277-293.

Sasitharan , S, Robertson, P.K., Sego, D., and Morgenstern, N.R. 1993. Collapse behavior of sand. *Canadian Geotech Journ.*, Vol. 30, pp. 569-577.

Sladen, J.A., D'Hollander, R.D., and J. Krahn 1985. The liquefaction of sands, a collapse surface approach. *Canadian Geotech. Journ.*, Vol. 22, pp. 564-578.

Smith, E.S. 1972. Tailings disposal - failures and lessons. Proceedings, *1st International Tailing Symposium*, Tucson, Arizona, pp. 356-376.

Troconso, J. 1988. Evaluation of seismic behaviors of hydraulic fill structures. *Hydraulic Fill Structures*, ASCE Spec. Public. No. 21, pp. 475-491.

Troncoso, J. 1990. Failure risks of abandoned tailings dams. *Proc. Int. Symposium on Safety and Rehabilitation of Tailings Dams*, ICOLD.

USCOLD 1994. *Tailings dam incidents*. U.S. Committee on Large Dams, 82 p.

Vick, S. 1990. *Planning, design, and analysis of tailings dams*. Bitech Publishers, ISBN 0-921095-12-0.

Vick, S. 1992. Stability evaluation during staged construction - discussion. ASCE, *Journ. of Geotech,. Eng.*, Vol. No. 118, No. 8, August, pp. 1282-1289.

Vidic S, Mayne P., and Beckwith Geo., 1995 Profiling mine tailings with CPT *Proc. Int. Symp. On CPT, CPT'95, vol2 607-612.*

Wride, C.E., McRoberts, E.C., and Roberston, P.K. Reconsideration of Case Histories for Estimating Undrained Shear Strength in Sandy Soils. Manuscript submitted to the *Canadian Geotech. Journ.*, April 1998.

Wroth, C.P., 1984. The interpretation of insitu soil tests. *Geotechnique* 34, No. 4, 449-489.

Yoshimine, M., Robertson, P.K. and Wride, C.E. 1998 Undrained shear strength of clean sands. Manuscript submitted to the *Canadian Geotech. Journ.* April 1998.

Zhang, H. and V.K. Garga 1997. Quasi-steady state: a real behavior? *Canadian Geotech. Journ.*, Vol. 34(5), October, pp. 749-761.

Tailings and Mine Waste '99 © 1999 Balkema, Rotterdam, ISBN 90 5809 025 6

Characterization of pore pressure conditions in upstream tailings dams

T.E. Martin
AGRA Earth and Environmental Limited, Burnaby, B.C., Canada

ABSTRACT: Conventional upstream tailings dams are unique structures in that their shear strength and pore pressure conditions are often difficult to characterize, particularly during the design stage. Selection of shear strength models based on understanding contractant versus dilatant, and drained versus undrained, behaviour in shear was the subject of a companion paper. This paper addresses the complexity of proper characterization of pore pressure conditions in upstream tailings dams, and how this characterization is critical to selection of shear strength parameters and stability analysis methods. This paper reviews various pore pressure regimes that commonly exist within upstream tailings dams, and the significance of such conditions in terms the effective stress analysis (ESA) versus undrained strength analysis (USA) approaches. General recommendations are provided for the piezometric monitoring systems and geotechnical investigation techniques needed to properly characterize pore pressure conditions.

1 SIGNIFICANCE OF PORE PRESSURE CONDITIONS

Tailings dams constructed using the conventional upstream method (i.e. with no compaction of the spigotted tailings beaches that form the dam shell) are, in contrast to the simplicity of the construction method, highly complex structures requiring a high degree of ongoing monitoring and review to maintain their safety. Analysis of their stability is dependent on effective consolidation stresses, which in turn are controlled by pore pressure conditions. There is an important distinction to be made between pore pressures and saturation conditions that is often neglected by the convenient, simplifying and misleading assumption of hydrostatic conditions. Pore pressure and saturation conditions in large part govern whether the contractant tailings of which upstream tailings dams are often constructed will fail under drained (effective stress analysis, or ESA) or undrained (undrained strength analysis, or USA) conditions. As discussed in a companion paper (Martin and McRoberts, 1999), stability analysis of upstream tailings dams is only meaningful if the drained versus undrained behaviour distinction is understood and properly accounted for in the analysis.

Adequate monitoring and characterization of pore pressure conditions is also required to evaluate the adequacy of internal drainage facilities, a key component of upstream dams, and to give early warning signs of deteriorating performance. This is of particular importance because a gradual rise in saturation levels and pore pressures, in and of itself, can initiate static liquefaction (collapse) and flowsliding failures (Eckersley, 1990, Martin and McRoberts, 1999). It has been the author's experience that, more often than not, piezometric monitoring systems associated with upstream dams are inadequate, and in some cases nonexistent. Such inadequate monitoring systems can easily lead to misinterpretation of saturation and pore pressure conditions, with potentially disastrous results.

2 PREDICTION IN THE DESIGN STAGE

Prediction in the design stage of the pore pressure conditions that will exist within an upstream tailings dam is at best a difficult undertaking and is not possible in many instances. Pore pressure conditions are determined through complex interactions between a number of factors, including those listed below.

1. Boundary conditions: The underpinning for any seepage analysis problem, these must often be characterized as a 2-dimensional characterization of a 3-dimensional problem. In addition, boundary conditions will often vary significantly through the operating life of the facility, with transient rather than steady state conditions dominating.
2. Tailings properties: Factors to be considered include the void ratio versus permeability relationship, soil-water characteristic curve, the degree of hydraulic sorting on the tailings beach and the separation between the sand and slimes fraction, heterogeneity, and changes in ore/mill grind over time. These data are often not available during the design stage.
3. Sedimentation/consolidation conditions: Sedimentation and consolidation can often only be modeled, to a reasonable degree of accuracy, using finite strain consolidation methods, which require void ratio versus permeability relationships and a good understanding of often complex boundary conditions.
4. Deposition procedures: Tailings discharge procedures (i.e. single point discharge versus spigotting versus spray bars) can have significant affect on pore pressure regime. Here the designer is completely at the mercy of the operator.
5. Saturated versus unsaturated flow conditions: In arid climates, prediction and modeling of saturated versus unsaturated flow conditions is important, with the range of uncertainty in parameter selection often sufficient to make the difference between a saturated and unsaturated condition.

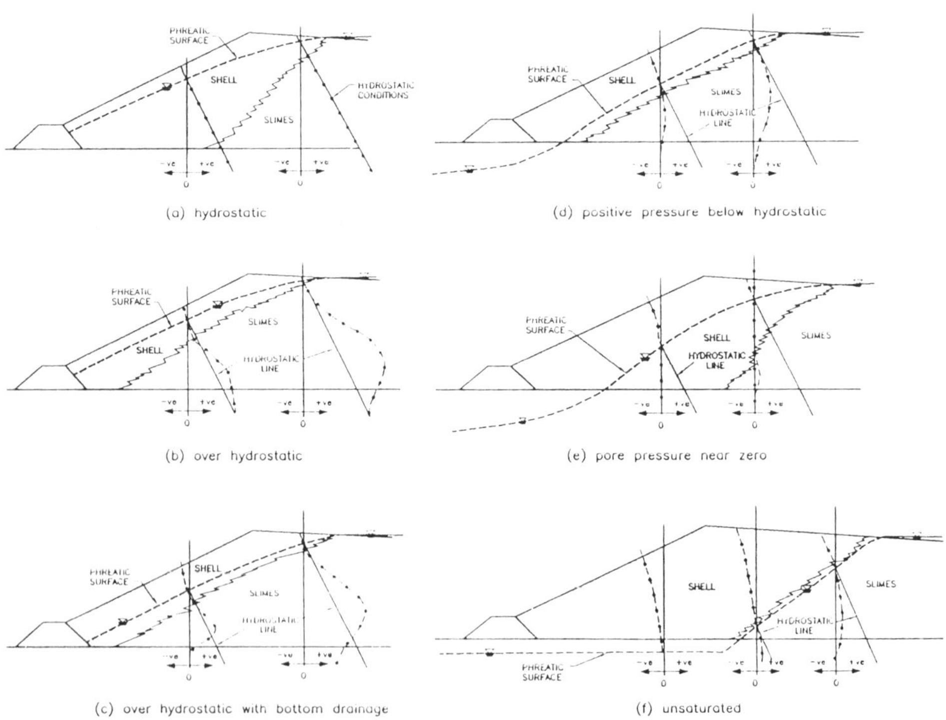

FIGURE 1
Pore pressure regimes

6. Chemical effects: In oxidizing tailings, chemical precipitate formation can result in significant reductions in permeability and plugging of internal drainage facilities.
7. Infiltration versus evaporation: Determination of infiltration versus evaporation losses for dams constructed in arid climates with systematic rotation of discharge points is a complex modeling exercise involving transient unsaturated flow analysis and detailed meteorology data.

The factors listed above, taken together with the difficulty in characterization of shear strength for conventional upstream tailings dams, and the many publicized failures of conventional upstream dams, make the author an ardent believer in Rule 1 as stated by Carrier (1991), who recommends "geotechnical engineers should avoid designing upstream tailings dams". However, because so many of these structures do exist, and so many geotechnical engineers have not and continue not to subscribe to Carrier's rule, geotechnical engineers do routinely become involve in the analysis and review of such dams designed by others. The remaining sections of this paper are therefore focused on the investigation and characterization of pore pressure conditions in existing upstream tailings dams.

3 PORE PRESSURE REGIMES

Despite the complexity of pore pressure conditions within upstream tailings dams, it is possible to define in general terms six idealized pore pressure regimes. These regimes are illustrated on Figure 1 and are listed in Table 1, which also lists the conditions under which these pore pressure regimes are most likely to exist. It is emphasized that the actual pore pressure conditions result from combinations of these and other factors not listed in Table 1, which represents at best a screening tool.

The rate of raising criterion is one for which some empirical guidelines have been reported in the literature, and is mentioned here as an example of the caution with which Table 1 should be considered. For example, Vick (1990) suggests that, for impoundments raised at between 4.6 m/year and 9.1 m/year, excess pore pressures usually dissipate as rapidly as load is applied, and a normally consolidated state can be assumed. Mittal and Morgenstern (1976) also suggest this range as generally being slow enough to prevent excess pore pressures in tailings slimes.

However, the author is familiar with a large upstream tailings dam for which high excess pore pressures exist within clayey tailings slimes underlying the dam slope, despite a rate of raising of only 2.1 m/year. In this particular case, the slow rate of raising and the relatively arid climate, which considered by themselves would tend to favor a fully consolidated condition, are dominated in terms of pore pressure conditions by a combination of the plasticity of the tailings slimes, foundation soils of about the same permeability as the tailings slimes, and the unfavorable drainage path conditions that result from the height and flat slope of the dam.

3.1 *Hydrostatic pore pressure conditions*

Figure 2 shows a dynamic pore pressure profile obtained in a piezocone sounding at the crest of an upstream tailings dam in eastern Canada. This dam is about 35 m in height, and is constructed of tailings rich in pyrite that are nearly 100% silt, with almost no sand and clay sizes. The tailings slimes are non-plastic. The dam is underlain by materials that are somewhat more pervious than the tailings, but the natural groundwater levels are close to surface. The rate of impoundment raising is only about 1 m per year. The results of pore pressure dissipation tests, carried out at 1 m depth intervals within the sounding, are also shown as discrete data points, and indicate a hydrostatic condition. The lower portion of the deposit indicates a primarily undrained response to cone penetration, requiring consideration of the USA approach for stability analysis.

Of additional interest at this site was the effect of the oxidation of the tailings on the pore pressure regime. The pervious starter dam, which serves as a toe drain, has in some areas been

Table 1. Summary of Pore Pressure Regimes

Regime	Fig. Ref.	Foundation Conditions: Hydrogeology	Foundation Conditions: k(foundation)	Plasticity of Tailings Slimes	Rate of Raising	Beach Width	Comments
hydrostatic	1a	natural groundwater table at or near ground surface	< or = k(tailings)	non to low plasticity	slow (< 6-8 m/year)	sufficiently wide such that slimes do not underlie slope	indicative of essentially horizontal seepage flow within deposit due to lack of underdrainage at base
over hydrostatic	1b	groundwater discharge area, or groundwater table near ground surface, and/or compressible foundation soils	usually < or = k(tailings)	low to high plasticity	rapid (> 8 m/year)	narrow, with slimes underlying slope	under-consolidated conditions within portion of tailings deposit and its foundation, with lack of underdrainage at base
over hydrostatic with bottom drainage	1c	groundwater recharge area	> k(tailings) by factor of 10 to 100	low to high plasticity	rapid (> 8 m/year)	narrow, with slimes underlying slope	under-consolidated conditions within portion of deposit, but foundation is fully consolidated and provides for double drainage conditions
positive pore pressures below hydrostatic	1d	low groundwater table (recharge area)	> k(tailings) by factor of 10 to 100	low to moderate plasticity	indeterminate	narrow, with slimes underlying slope	can indicate an under-consolidated condition where seepage is entirely vertical, or fully consolidated condition where flow has horizontal and vertical components
pore pressures near zero	1e	low groundwater table (recharge area)	> k(tailings) by factor of 10 to 100	non-plastic	slow (< 6-8 m/year)	wide beach, with slimes not underlying dam slope	Indicates a fully consolidated condition with downward seepage due to good underdrainage at base
unsaturated	1f	low groundwater table, arid climate	> k(tailings) by factor of 10 to 100	non-plastic	slow (<< 6m/year)	wide beach, with slimes not underlying dam slope	a condition that can probably only be achieved in operating impoundments in arid climates, with wide tailings beaches and systematic discharge and drying cycles.

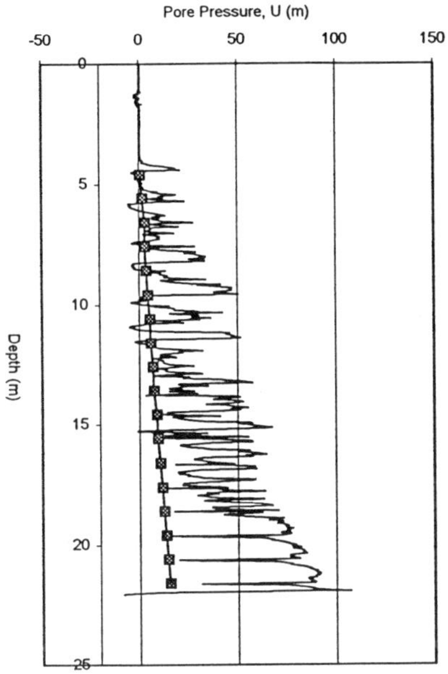

Figure 2. Piezocone pore pressure profile showing hydrostatic conditions

clogged with chemical precipitates due to acid mine drainage conditions, a condition noted at other sites as discussed by Plewes and McDonald (1996). These chemical changes have resulted in a reduction in toe drainage, causing seepage to discharge from the tailings slopes above the starter dams. This issue provides an example of how chemical aspects can affect pore pressure regimes.

Hydrostatic pore pressure conditions are commonly assumed within upstream tailings dams for the purposes of limit equilibrium stability analysis. This is often the result of one or more of: oversimplification for convenience of analysis, inadequate piezometric monitoring coverage, and/or misinterpretation of available piezometric data. Vertical seepage gradients cannot be identified without piezometers at multiple depths. Especially misleading are cases where standpipe piezometers are installed near the base of a tailings deposit in cases with downward drainage and a permeable foundation. This can lead to the incorrect and highly unconservative assumption in stability analysis that the phreatic surface is near the base of the tailings deposit.

The author was recently involved in a project where the designer assumed hydrostatic pore pressure conditions for stability analysis. Subsequent piezocone investigations (see Figure 3) showed that pore pressure conditions were in fact less than hydrostatic (and close to zero in the sand shell), due to downward drainage. The vertical effective stresses assumed by the designer in the stability analyses were therefore low (conservative), but this conservatism was eliminated by the designer's use of the ESA approach in the slimes zone, where USA would have been more appropriate. Carrier (1991) discusses a failure case history in which the designer was completely aware of a pore pressure condition less than hydrostatic, yet assumed a hydrostatic condition in ESA type stability analyses.

3.2 *Over hydrostatic without bottom drainage*

Figure 1b illustrates a pore pressure regime with pressures above hydrostatic due to an under-consolidated state. This condition typically requires at least one of: a rapid rate of raising, or

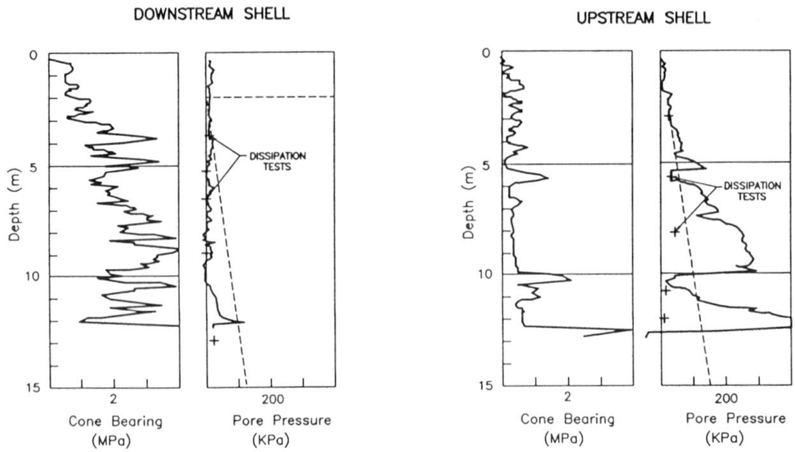

Figure 3. Piezocone profiles showing below hydrostatic conditions

some plasticity in the tailings slimes, together with foundation materials of similar or lower permeability than the tailings. This is the most unfavorable pore pressure regime in terms of stability, and it usually does not go unrecognized for the simple reason that it is hard to imagine such a dam being raised safely without the designer recognizing it. This condition highlights the importance of good underdrainage for upstream tailings impoundments, a principle understood in empirical terms many years ago by the likes of Lenhart (1950). This condition can be missed if piezometers are installed near the base of the deposit or at shallow depths, where pressures approach hydrostatic. It can also be missed if observation well standpipes (slotted along their entire length) are used, because the excess pressures at mid-depth in the deposit would probably be dissipated elsewhere within the standpipe.

For this condition, a USA type stability analysis is essential. If the tailings are in an under-consolidated (undrained) state, then if follows that during failure they will certainly undergo shear in undrained conditions (shear-induced pore pressures will not dissipate during failure). The use of ESA together with measured pore pressures does not recognize the presence of shear-induced pore pressures.

3.3 *Over hydrostatic with bottom drainage*

Figure 1c illustrates a condition of pore pressures above hydrostatic in a portion of the deposit due to an under-consolidated condition, but pore pressures less than hydrostatic nearer the base of the deposit due to some degree of underdrainage. The deposit is therefore consolidating under conditions of double drainage, and this is simply a variation of the condition illustrated on Figure 1b. The double drainage case is obviously much more favorable in terms of stability, although USA stability analysis is still required as described in Section 3.2.

This pore pressure regime can be missed with piezometers located only near the base of the deposit. Using such piezometers to assume hydrostatic conditions would greatly misrepresent the actual pore pressure regime. As an extreme example of this, if pore pressures approached zero near the base of the deposit, some might be tempted to think of the deposit as drained (unsaturated), by mistakenly equating pore pressure near the base of the deposit with the saturation level, and neglecting the under-consolidated state of much of the deposit.

3.4 *Positive pore pressures below hydrostatic*

Figure 1d indicates a commonly encountered condition where pore pressures are everywhere less than hydrostatic due to downward drainage, requiring foundation materials significantly more pervious than the tailings. This pore pressure regime is obviously more favorable than the three conditions discussed previously because of relatively higher consolidation stresses (and therefore shear strength) within the deposit. Some might be tempted to interpret pore pressures less than hydrostatic as indicating a fully consolidated condition within the deposit, which could lead to the further temptation of assuming a fully consolidated condition to be justification for sole reliance on an ESA approach for stability analysis. Carrier (1991) discusses a failure case history where the designer wrongly succumbed to both these temptations, with disastrous results.

It is important to remember that for vertical seepage through a slightly compressible, fully consolidated soil, pore pressures will be near zero (Lambe and Whitman, 1969, Carrier, 1991). Therefore, the pore pressure regime indicated on Figure 1d could well indicate an under-consolidated condition within the deposit. If so, since raising-induced pore pressures do not immediately dissipate, then neither would shear-induced pore pressures during failure. Therefore, the USA method of stability analysis is essential.

Pore pressures below may also simply indicate a horizontal seepage component (i.e. seepage is not entirely vertical) due to the coarser tailings forming the dam shell representing a drainage path. Therefore, other data besides pore pressures are required to determine if an under-consolidated state is likely to exist.

Figure 3 shows cone bearing and pore pressure data from two piezocone soundings carried out in an upstream dam in an arid region of South America, being raised at a rate of about 6 m/year. One sounding was carried out in the outer dam shell, with the other about 40 m upstream of the dam crest. Pore pressure dissipation tests were carried out at various depth intervals in both soundings to determine equilibrium pore pressures. The equilibrium pore pressures in the upstream sounding indicate a condition similar to that shown on Figure 1d. The dynamic pore pressures also clearly indicate a contractant response in the clayey-silt slimes to cone penetration.

When reviewing the data shown on Figure 3, it is evident how this pore pressure condition could be misinterpreted without pressure piezometers installed at multiple depths within the deposit. With a vertical seepage condition, pressure piezometers (of the pneumatic or electrical type) are essential, as standpipes can underestimate pore pressures. Conversely, observation well type standpipes (slotted along their entire depth) can be of great benefit since they indicate the level of saturation as opposed to pore pressure. In the case of a vertical seepage condition, pore pressure data may by itself not allow determination of saturation levels.

3.5 *Pore pressures near zero*

Figure 1e indicates a condition of essentially zero pore pressure within the tailings deposit, despite a relatively high saturation level. As discussed above, this condition can indicate a fully consolidated condition within the tailings deposit when vertical seepage conditions exist. Using only standpipe piezometers, and even pressure piezometers, it is possible to misinterpret this pore pressure regime as unsaturated, leading the designer to assume a drained condition that justifies sole reliance on ESA type stability analysis. Even if this condition is misinterpreted as unsaturated, the effective vertical stress would be calculated correctly for the purposes of stability analysis, since pore pressures are near zero despite the high level of saturation. However, a fully consolidated condition does not preclude contractant, undrained response under shear. For contractant, saturated tailings, a USA type stability analysis should be carried out. With this pore pressure regime, therefore, it is critical to recognize the difference between pore pressure and saturation.

The downstream piezocone sounding shown on Figure 3 indicated equilibrium pore pressures essentially at zero through the full depth of the sounding, which penetrated primarily silty sands as opposed to the upstream sounding. However, standpipe piezometers near this sounding do indicate a saturation level within 3 m of the surface of the deposit. Furthermore, the dynamic pore pressure profile from the sounding does indicate positive response to cone penetration. Standard penetration test (SPT) blowcounts (normalized to 60% energy) of 5 to 8 blows per foot were obtained in boreholes drilled adjacent to this sounding, indicating a loose condition. These data lead the author to conclude that the tailings in the dam shell as characterized by the downstream piezocone sounding are normally consolidated, saturated, loose, and contractant under shear. Therefore, for purposes of stability analysis, the shear strength of this material should be characterized using both the ESA and USA approaches.

Again, it is important to realize that pore pressures in the outer shell can, even with a very pervious underdrainage system, be well above zero due to the expulsion of porewater from the consolidating slimes to the upstream, since the dam itself represents a drainage path for the consolidating tailings deposit. Mittal and Morgenstern (1976) provide an insightful discussion of this issue.

3.6 *Unsaturated conditions*

Figure 1f indicates the shell portion of an upstream tailings impoundment in an unsaturated condition, with pore pressures being negative throughout the profile. The author believes this condition can probably only be achieved in arid climates (annual evaporation greatly exceeds annual precipitation) with systematic discharge of thin lifts and resting periods provided for air drying, and with good underdrainage. This is the only case for which, even with tailings that are contractant when sheared, sole reliance on ESA for stability analysis is supportable, albeit with the following cautions:

1. Section 3.5 discussed a pore pressure regime with zero pore pressures despite fully saturated conditions. Therefore, the designer must have sufficient data to distinguish between that condition (Figure 1e) and the unsaturated condition.
2. A degree of saturation of 100% is not required for undrained collapse behavior. For example, Yoshimi et al. (1989) presented laboratory test data where liquefaction was induced in materials with saturation levels as low as 80%. Therefore, saturation levels less than 100% (or even 95%) are, in and of themselves, not necessarily a guarantee of fully drained behavior under shear.
3. A determination of unsaturated conditions should be supported by moisture content data, and how this data relates to the soil-water characteristic curve. Sole reliance on zero pore pressures as indicated by pressure piezometers is not sufficient in all cases. Fine tailings slimes can remain tension-saturated due to capillary action.

Haile and Kerr (1989) describe the design and early operating experience of an upstream tailings impoundment at a base metals mine in a high rainfall area (net annual precipitation 2.4 m) on the west coast of Canada. The facility is located in a seismically active zone, and is founded on very pervious alluvium, which affords very effective underdrainage. The dam is being constructed of non to low plastic, primarily silt tailings, and is intended to function as a "fully drained system" using subaerial discharge methods. The design acknowledged "there is a possibility of liquefaction if saturated conditions exist", and assumed that "the rate of tailings rise (2.2 m/year) is sufficiently low to ensure full consolidation under two-way drainage during operations".

Haile and Kerr (1989) also described an extensive system of pneumatic and vibrating wire piezometers that showed negative or small positive pore pressures (range -1.5 m to +0.3 m equivalent head of water) at a time when the deposit was about 10 m thick (ultimate thickness to be about 25 m). Can this data be taken to confirm satisfactory operation, defined by Haile and

Kerr (1989) as "zero buildup of excess pore pressures throughout the structural zone of the tailings"? The author's experience is that pressure piezometer data alone is inconclusive in this regard, and can be equally representative of a below hydrostatic and fully consolidated condition (Figure 1e). The following methods are available to provide clarification in this regard.

1. Use of observation well type standpipes, screened for the full depth of the tailings, represent a very economical, simple, yet effective way to determine if saturated conditions exist. Stauffer and Obermeyer (1988) described how this sort of standpipe was used in a large upstream tailings dam to determine the location of the phreatic surface in a deposit with pore pressures less than hydrostatic due to downward drainage.
2. Piezocone soundings with pore pressure dissipation tests. If areas of positive dynamic pore pressure response are encountered, then this would indicate a condition at or very near saturation, even if the dissipations indicate essentially zero equilibrium pore pressure. Use of a resistivity module with the piezocone would also be of benefit, as the resistivity of a saturated soil is significantly less than that of a partially saturated soil.
3. A drilling and sampling program could be carried out to determine moisture content profiles. These profiles could be related to the soil-water characteristic curve(s) for the tailings to gain an appreciation of saturation conditions.

The various uncertainties, corrections, and potential sources of reading error associated with pressure piezometers is such that measured small negative pore pressures may in fact simply be zero readings. It is not clear that pressure piezometers can be used to accurately measure negative porewater pressures. Tensiometers probably offer a more reliable means of measurement of porewater tension (negative pressures).

In addition to the above, the following factors should be considered when evaluating pore pressure conditions within an upstream dam that is meant to be operated in a partially saturated condition.

1. As the deposit rises, the relative efficacy of underdrainage facilities could gradually diminish, particularly due to layering within the tailings deposit resulting in a change from unsaturated to saturated conditions.
2. Any significant, transient positive response of piezometers at depth in the deposit to rotation of discharge points and/or seasonal increases in precipitation would likely indicate a saturated condition. If the tailings remained in an unsaturated condition, then only the relatively shallow piezometers should show a response due to the moisture retention capacity of unsaturated soils.

The key point of the above discussions is that pore pressure conditions within a deposit that is to be operated as a drained facility must be confirmed as such, and piezometers alone, even pressure piezometers, may well be inadequate in this regard.

4 INVESTIGATION AND MONITORING OF PORE PRESSURE CONDITIONS

Piezometer monitoring systems appropriate for the various pore pressure regimes discussed in Section 3 vary depending on the regime and project specific conditions and requirements. Figure 4 presents an upstream dam section that illustrates in general terms what the author considers would be the ideal piezometer monitoring network. This network includes:

1. Observation well standpipes, slotted for their full length, extend through nearly the full depth of the tailings deposit, but not into the foundation soils. These standpipes will indicate the levels of saturation within the deposit. They must not extend into the foundation because if the foundation materials are highly pervious relative to the tailings, then this drainage may prevent water from collecting in the observation well.
2. Pressure piezometers, installed at various depths within the tailings deposit, are used to measure pore pressures and seepage gradients (horizontal and vertical components). These piezometers can also be used to assess the relative degrees of consolidation within the deposit

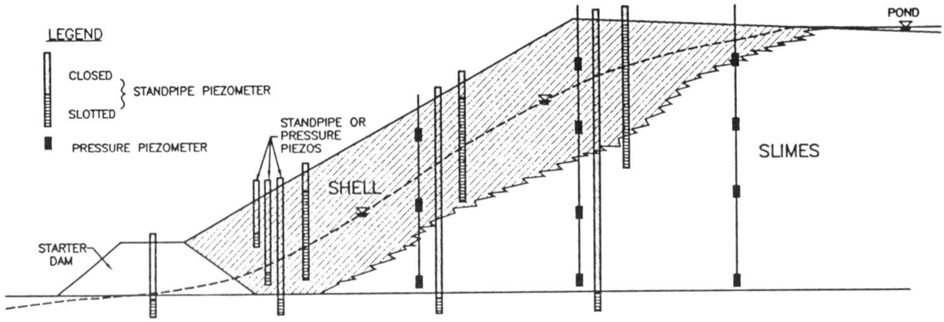

Figure 4. Section showing idealized monitoring scheme

profile, and to whether or not the deposit is responding in a drained or undrained manner to ongoing raising. This is a key consideration when considering ESA versus USA approaches to the stability analysis of the dam.

3. Standpipe piezometers can be installed instead of pressure piezometers in the dam shell if this is composed of relatively free-draining sands, although even in this case pressure piezometers are preferred since standpipes will underestimate pressures under vertical seepage conditions. Standpipe piezometers should also be installed in foundation materials (if these are relatively pervious relative to the tailings. However, for compressible, low permeability foundation materials (e.g. normally consolidated clays) of equal or lesser permeability, pressure piezometers are obviously required.

For field investigations of pore pressure conditions within upstream dams, the following general recommendations are made:

1. The piezocone test, with frequent pore pressure dissipation tests, represents a very rapid, economical, and effective means of pore pressure profiling, and of detecting contractant versus dilatant response in loose tailings. In deposits where partially saturated conditions are expected, or where vertical seepage conditions might make delineation of the phreatic surface problematic, addition of a resistivity module can be advantageous.

2. Sampled boreholes should be drilled adjacent to a number of the piezocone soundings for retrieval of representative samples for visual classification and laboratory testing. The laboratory testing should as a minimum include moisture content determinations and Atterberg limit testing for any materials with potential plasticity.

5 SUMMARY

Conventional upstream tailings dams are, in operational (if not safety) terms, the easiest and least expensive type of tailings dam to construct, and yet can be the most difficult to characterize in terms of shear strength parameters and pore pressure conditions. This paper has discussed a variety of idealized pore pressure conditions that can exist within upstream tailings dams, and has sought to present:

1. The importance of correct characterization pore pressures from the standpoint of selecting appropriate shear strength parameters and analysis methods;
2. The ease with which pore pressure conditions can be misinterpreted given inadequate piezometers systems and or a lack of understanding of saturation versus pore pressure issues; and
3. General recommendations for piezometer monitoring systems and investigation methods to ensure a correct characterization of the pore pressure regime.

REFERENCES

Eckersley, D. 1990. Instrumented laboratory flowslides. *Geotechnique* 40(3), pp. 489-502.

Haile, J.P. and T.F. Kerr 1989. Design and operation of the Myra Falls tailings disposal facility. *Geotechnical Aspects of Tailings Disposal and Acid Mine Drainage*, Proceedings, Vancouver Geotechnical Society, May.

Lenhart, W. 1950. Control of tailings from washing plants. *Rock Products* (July), pp. 72-80.

Martin, T.E. and E.C. McRoberts 1999. Some considerations it the stability analysis of upstream tailings dams. Proceedings, *Tailings and Mine Waste '99*, Fort Collins, Colorado.

Mittal, H.K. and N.R. Morgenstern 1976. Seepage control in tailings dams. *Canadian Geotech. Journ.*, Vol. 13(3), August, pp. 277-293.

Plewes, H.D. and T. McDonald 1996. Investigation of chemical clogging of drains at Inco's Central Area Tailings Dams. Proceedings, *Tailings and Mine Waste '96*, Fort Collins, Colorado, pp. 59-72.

Stauffer, P.A. and J.R. Obermeyer 1988. Pore water pressure conditions in tailings dams. *Hydraulic Fill Structures*, ASCE Spec. Publication No. 21, pp. 924-939.

Vick, S.G. 1990. Planning, design and analysis of tailings dams. Bitech, ISBN 0-921095-12-0.

Yoshimi, Y., Tanaka, K. and K. Tokimatsu 1989. Liquefaction resistance of a partially saturated sand. Soils and Foundations, Vol. 29(3), Sept., pp. 157-162.

Tailings and Mine Waste '99 © 1999 Balkema, Rotterdam, ISBN 90 5809 025 6

Plant dewatering and strengthening of tailings: Laboratory and computer simulations

M.J.Silva, K.W.Biggar & D.C.Sego
Department of Civil and Environmental Engineering, University of Alberta, Edmonton, Alb., Canada

D.S.Chanasyk & M.A.Naeth
Department of Renewable Resources, University of Alberta, Edmonton, Alb., Canada

ABSTRACT A greenhouse experiment was conducted to validate the model that simulates the strength enhancement mechanism of plants. Composite Tailings (CT) from the Alberta oil sands operated by Syncrude Canada Ltd. was used as the test material. CT having an initial solids content of approximately 65% solids was placed in lysimeters with no drainage at the bottom to prevent water loss other than via evapotranspiration. Five plant species were selected for testing. The solids content profiles and bearing capacities, which were measured in the greenhouse experiment, were compared with their respective values predicted with the model. The comparison shows a good agreement between simulated and measured values. The model is considered appropriate for predicting solids content profile and bearing capacity increase due to the plants for one growing season.

1 INTRODUCTION

The use of plants to dewater fine tailings has been identified as a mechanism that may economically enhance the surface stability of these weak deposits on a large scale. Plants growing in tailings accelerate the consolidation process and increase their bearing capacity in the root zone through evapotranspiration and root reinforcement.

Although plant dewatering of tailings has been recognized for a long time, it was not until recently that researchers showed interest in studying the mechanical effects that plants have on the tailings. Oil sands tailings from northeastern Alberta at 50% solids were planted with reed canary grass in trial pits. The tailings were dewatered to 80% solids in one growing season with a shear strength of about 120 kPa (Johnson et al. 1993). Stahl and Sego (1995) studied changes in surface stability of an 18.2 hectare coal tailings impoundment undergoing reclamation activities since reaching full capacity in 1989. Natural enhancement processes including dewatering through evaporation, evapotranspiration, and fiber reinforcement of plant root systems increased the shear strength of the surficial soils. Bearing capacity and surface stability of the coal tailings within the impoundment also increased. In some cases, the bearing pressure of the rooted tailings were 50-60% greater than that of the unrooted tailings at equivalent strain or relative settlement levels.

Prediction of the surface bearing capacity is essential in the management of tailings. Reclamation of these tailings cannot be initiated unless the deposit achieves a trafficable surface. A model capable of predicting the bearing capacity increase due to the strength enhancement mechanism of plants was theoretically described in Silva et al. (1998a). The model simulates one dimensional water flow in a tailings deposit, including the root zone, by solving a transient flow equation coupled with water extraction via the root systems. The model predicts the solids content profile and pore water pressure (matric suction) profiles, which are used to estimate the bearing capacity.

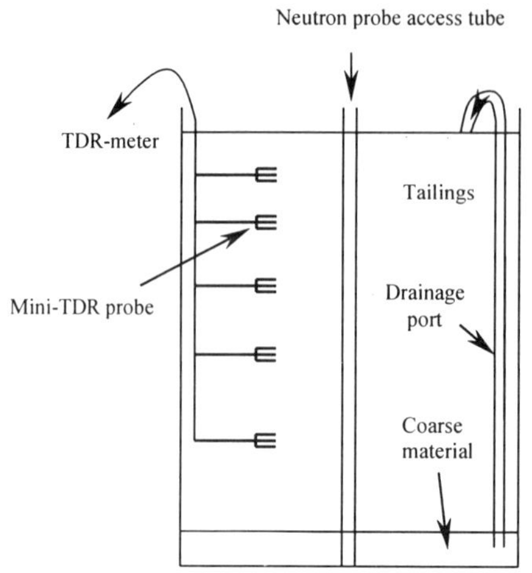

Figure 1. Instrumentation layout

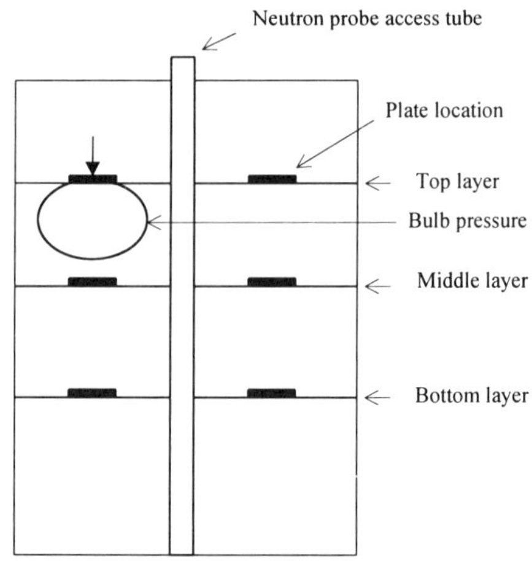

Figure 2. Bearing capacity tests

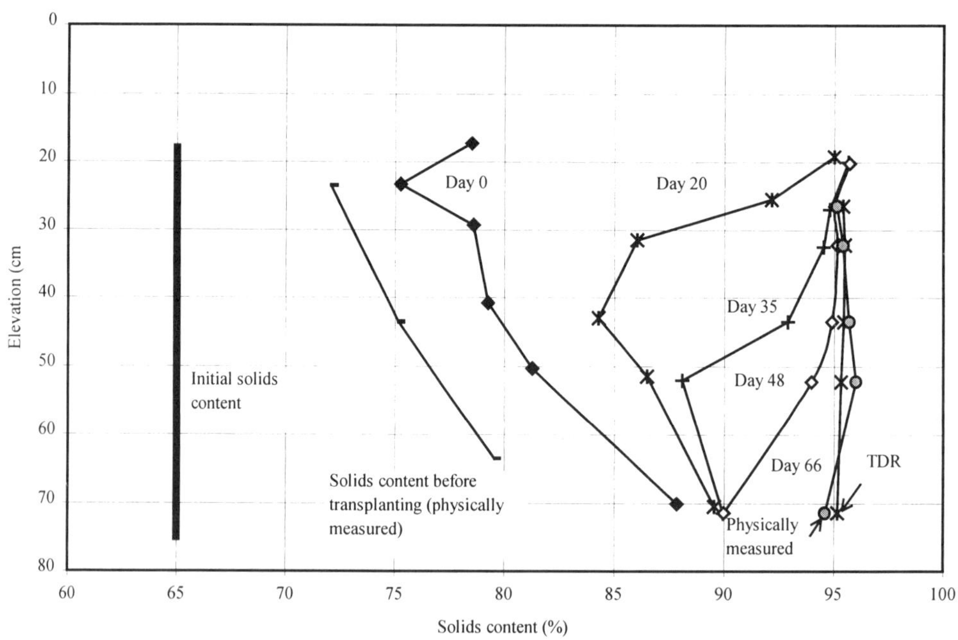

Figure 3. Solids content profile from TDR readings from 0 to 66 days compared to measured values at 66 days

This paper presents the experimental validation of the model. Values of solids content and bearing capacity profiles that were measured in a greenhouse experiment, were compared with their respective values predicted by the model. The analysis was conducted on the results of only one plant species (Reed canary grass). A complete analysis of the other species will be carried out as the experimental results become available.

2 LABORATORY PROGRAM

A greenhouse experimental program was conducted to evaluate the physical response of tailings to the plant dewatering mechanism as well as to provide a data set for the validation of the model. Oil sands Composite Tailings (CT) was used as the growth medium and five different plant species were selected for testing. These plant species proved the most viable in CT as a result of greenhouse experiments conducted to identify the most suitable plant species for dewatering CT (Silva et al., 1998b).

2.1 *Materials and method*

Composite Tailings (CT) was prepared by mixing tailings sand, Mature Fine Tailings (MFT), tailings pond water and gypsum, which were provided by Syncrude Canada Ltd. The amount of gypsum added was approximately 1200 ppm and the proportion of sand, MFT and water was such that the final CT mixture had 20% and 65% of MFT(fines) and solids content, respectively.

Forty five-gallon drums with a plastic liner were used as lysimeters. A saturated coarse material was placed at the bottom of each lysimeter to act as a filter. One end of a plastic tube, wrapped with geotextile, was inserted into the filter and the other end was located at the top of the lysimeter. This system created a double drainage, which accelerated the self-weight consolidation of the CT deposit. Aluminum tubes were installed in the center of each of the lysimeters to allow access for a neutron probe to monitor the solids content within the tailings. The lysimeters were filled with CT to a depth of about 0.8 m and self-weight consolidation was allowed to take place. Any expressed water was siphoned off and extra CT was added to restore the initial level. Six lysimeters were instrumented with five mini-TDR probes each built at the University of Alberta. The mini-TDR probes measured solids content in a smaller zone compared with that of the neutron probe. A layout of the instrumentation is presented in Figure 1.

The plant species used in the experiment were: Altai wildrye (*Elymus angustus*), Creeping foxtail (*Alopecurus arundinaceus*), Red top (*Agostis stolonifera*), Reed canarygrass *(Phalaris arundinacea)*, and Streambank wheatgrass (*Agopyron riparian*). Plants were started from seeds in root trainers and transplanted to the lysimeters after 4 weeks with the equivalent to 64 plants/m^2. Three replicates were used for each species and three replicates were left unplanted as a control. The lysimeters were placed in a growth room in a randomized complete block design. Air temperature and hours of light per day were set at 22 °C and 15 hours, respectively.

Water was added weekly to simulate average precipitation at Syncrude Canada's Mildred lake tailings pond from June through August.

2.2 *Measurements*

The climatic condition of the growth chamber was characterized by measuring the ambient relative humidity, minimum and maximum air temperature and potential evaporation. Surface temperatures were measured in the tailings and in the water of the evaporation pan.

Initial tailings elevations were recorded after transplanting. Thereafter, elevations were recorded in each lysimeter at regular intervals to determine the surface settlement. Solids content was measured every week using the TDR and neutron probes to obtain information about the water lost due to evapotranspiration in the planted lysimeters and evaporation in the

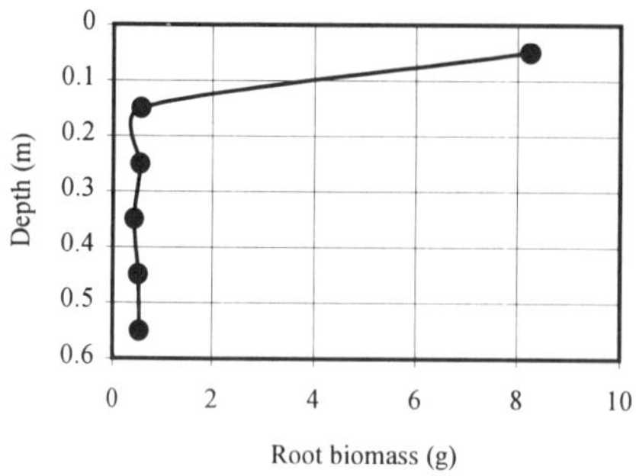

Figure 4. Root biomass profile

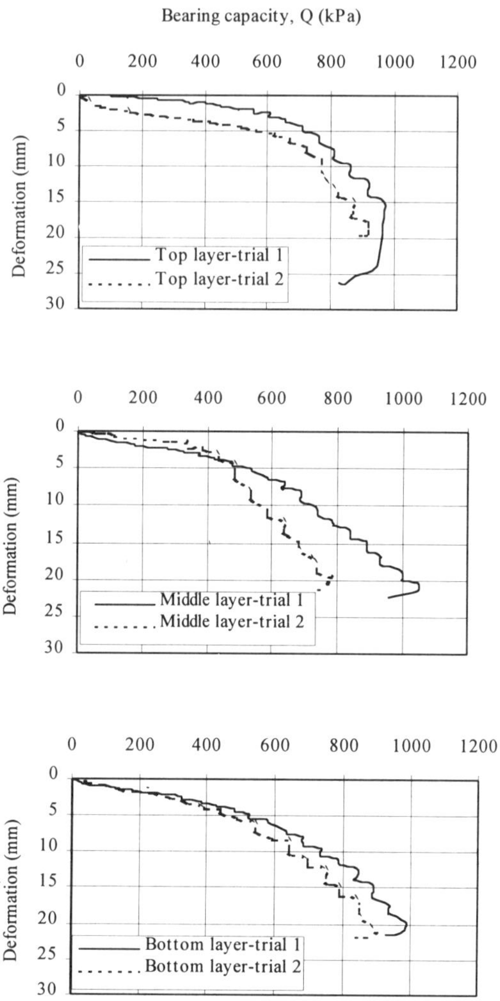

Figure 5. Bearing capacity tests

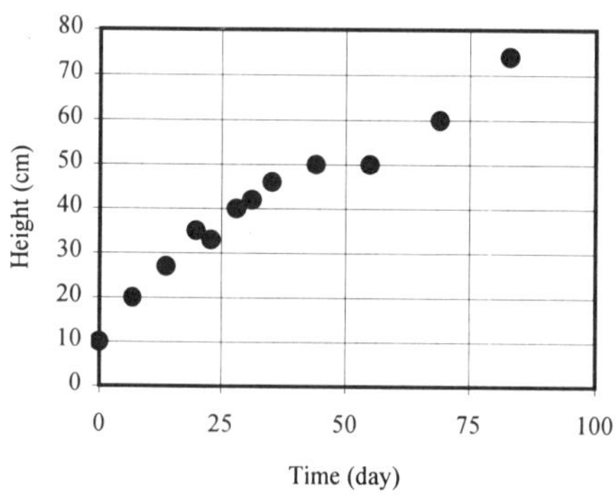

Figure 6. Plant height

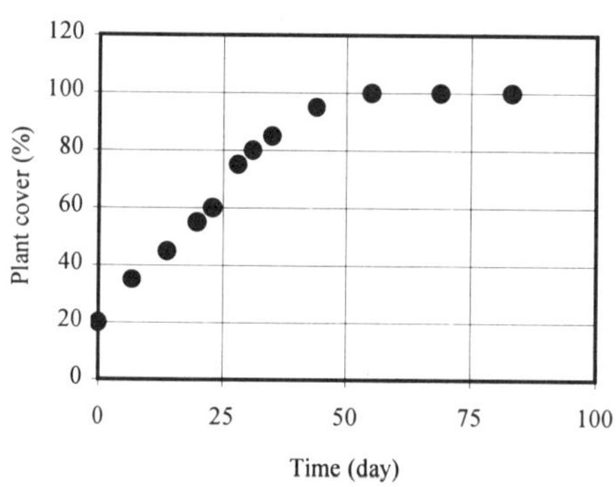

Figure 7. Plant cover

controls. Number of tillers, number of leaves, plant height, percent canopy cover, and leaf area were determined at regular time intervals through the experiment.

After ten weeks of plant growth, reed canary grass and creeping foxtail showed symptoms of water stress. Solids content measurements indicated that indeed the tailings deposit was near the wilting point, which is about 96% solids. This condition dictated the end of the test series. The plants were then cut to measure the dry shoot biomass.

Bearing capacity was measured in three different layers within each barrel by conducting plate load tests (Figure 2). The plate had a diameter of 80 mm. This size was chosen to eliminate any boundary effect during the test. Also, the vertical distance between two adjacent test layers was at least two times the plate diameter. This avoided overlapping of the zone of influence of the bearing tests.

Samples of tailings were taken at depths of every 10 cm to measure the solids content profile. Root density was measured by inserting a ring of known volume below every plate load test. The roots were washed and oven dried to measure the dry root biomass.

3 EXPERIMENT RESULTS

The measured solids content profiles on selected days are presented in Figure 3. The profiles indicate the influence of plant transpiration and root growth. At day 66 the solids contents were near the wilting point. Physical measurements of solids content at the end of the experiment matched very well with TDR measurements.

The root biomass profile is presented in Figure 4. The high concentration of roots near the tailings surface was perhaps the effect of the root trainers, which provided a constrained space for root development during the first four weeks. A second reason could be that, at the beginning, water was readily available at the surface causing a prolific root growth. When water was scarce the roots were forced to grow deeper to seek water. The root distribution does not follow a linear decreasing relationship as proposed in the theoretical model. Rather, the profile shows a uniform distribution.

Results of the plate load tests are shown in Figure 5. The bearing capacity test results are uniform in the tailings profile and are in accordance with the uniformity of the solids content profile shown in Figure 3 and the root biomass profile shown in Figure 4.

Plant height, percent ground cover, leaf area index, and potential evaporation are presented in Figure 6, 7, 8 and 9, respectively.

4 MODEL VALIDATION

The model was used to predict the plant dewatering process observed in the greenhouse experiment. Data obtained from the greenhouse experiment and from other sources were used as inputs to the model prediction.

The measured potential evaporation (E_0) was distributed into its evaporation and transpiration components using the model proposed by Ritchie (1972), who observed that the transpiration component was dependent on the leaf area index (LAI) according to the following expression

$$T_o = 0 \qquad LAI < 0.1$$

$$T_o = E_0(-0.21+0.7LAI^{1/2}) \qquad 0.1 \leq LAI \leq 2.7$$

$$T_o = E_0 \qquad LAI > 2.7$$

where T_o is the potential plant transpiration, E_o is the potential evaporation and LAI is the leaf area index, which is defined as the area of one side of leaves per unit area of soil surface (Jensen et al., 1990). According to this model after day 25 the water loss in the tailings was due solely to the plant transpiration, because after this day the LAI was equal or greater than 2.7 (Figure 8).

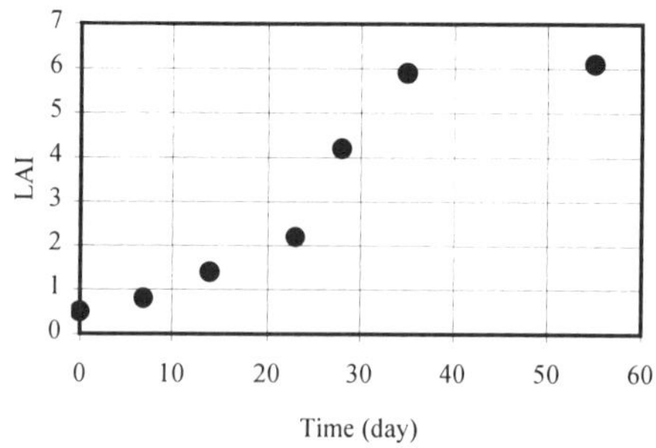

Figure 8. Leaf area index

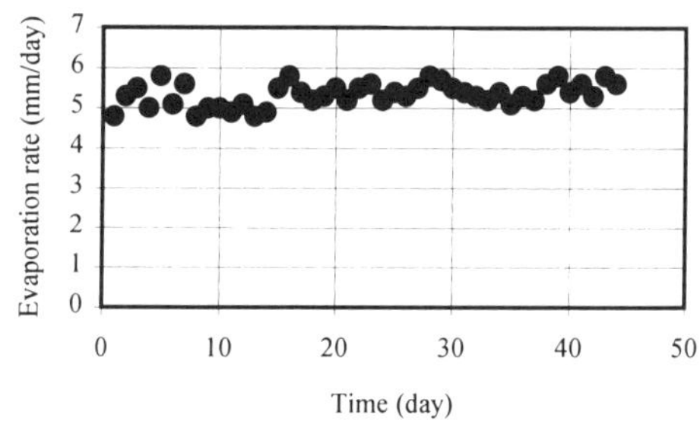

Figure 9. Potential evaporation (evaporation pan)

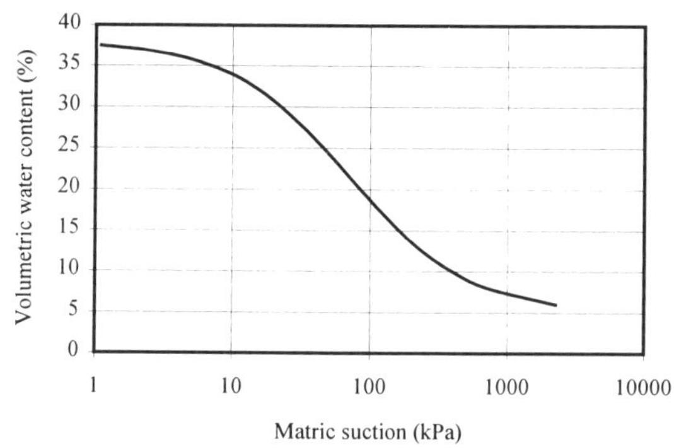

Figure 10. Soil water characteristic curve for CT

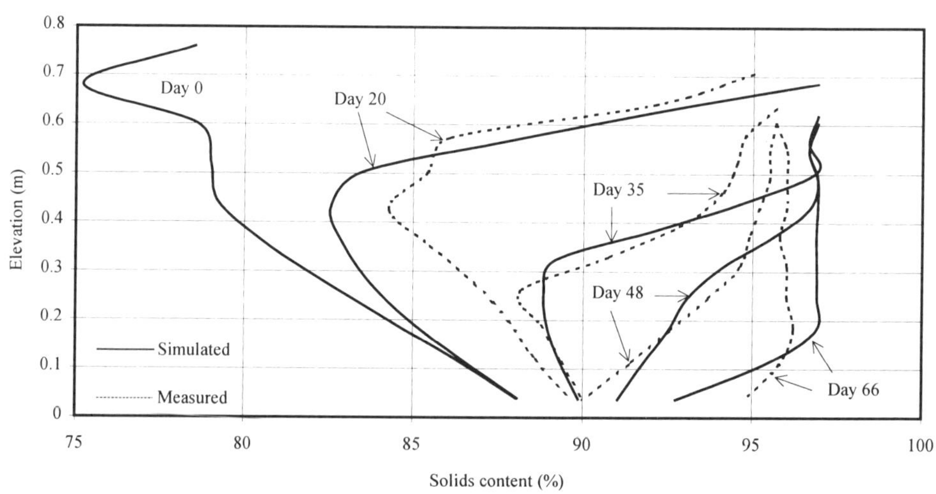

Figure 11. Measured and simulated solids content profiles

Table 1. Model parameters

Parameter	Qiu and Sego (1998)	Pollock (1988)
A (1/kPa)	0.9952	1.814×10^{-3}
B	- 0.1576	-0.09929
C (m/d)	5.18×10^{-4}	1.844×10^{-3}
D	1.3754	3.794
G_s	2.60	2.54

Specific gravity (G_s) and parameters of the saturated hydraulic conductivity and the effective stress versus void ratio relationship (Table 1) were obtained from Qiu and Sego (1998) and Pollock (1988). A power relation between the effective stress (σ') and the void ratio (e) was assumed $e=A(\sigma')^B$, where A and B are fitting parameters. For saturated conditions the hydraulic conductivity was assumed to be a function of the void ratio $K=Ce^D$, where K is the saturated hydraulic conductivity, and C and D are fitting parameters (Silva et al., 1998a).
The soil water characteristic curve for CT used as input in the model is presented in Figure 10.

The size of each time step was restricted to maintain a maximum variation in the volume of water of the compartments (Max $\Delta V_w = 0.1$ mm). Additionally, a maximum time step was prescribed (Max $\Delta t^{j+1} = 0.5$ day).

The measured and simulated solids content profiles are presented in Figure 11. The simulated values were obtained using Pollock's (1988) parameters as input in the model. Parameters from Qiu and Sego (1998) did not simulate very well the trends of the solids content. On day 0 the measured values were input in the model. The shape of this curve reflects the change in solids content due to the CT refilling of the lysimeters to restore the initial elevation. The model predictions for days 20 and 35 are close to the measured values. The laboratory trends are very well simulated. On days 48 and 66 the model predictions are in good agreement with the measured values.

The predicted matric suction profile for day 66 is presented in Figure 12. The tailings were at the wilting point and the matric suction was uniform through the deposit.

This model is unique because it includes the self-weight consolidation of the material. The soil is not assumed as a rigid structure, but as a deforming material. Figure 13 shows the dynamics of the void ratio changes in the tailings deposit on selected days. It is noted that on day 48 the deposit has reached the minimum void and no more deformation occurred. This agrees with measurements of the elevation of the tailings surface that after day 50 no more consolidation occurred.

The bearing capacity of the CT was predicted using the following parameters: $c = 3$ kPa, $\phi' = 30^{\circ}$, $c_R = 5$ kPa (c_R is the cohesion due to the root system), matric suction at the wilting point = 1500 kPa. The predicted bearing capacity was 1051.7 kPa, which is a little larger than the measured capacity (about 875 kPa). The contributions of each component to the total bearing capacity are 10.3% due to the saturated cohesion, 17.1% due to the root reinforcement, and 72.6% due to the matric suction. If the material has a moisture content equivalent to its field capacity (matric suction about 33 kPa) the predicted bearing capacity is 554.4 kPa, and the matric suction contribution will be 48%. If matric suction goes to zero after a heavy rainfall, the bearing capacity is predicted to be 288 kPa, and the roots will contribute 62.5% of this value.

5 BEARING CAPACITY ON CONTROL LYSIMETERS

Plate load tests conducted on the control lysimeters (no plants) resulted in low measured bearing capacities. The maximum bearing capacity at the surface of the tailings was 80 kPa. This bearing capacity was due to the contribution of only the effective cohesion (c = 3 kPa). Evaporation did not develop a significant matric suction at the surface of the tailings. At the middle layer the maximum bearing capacity was in the order of 20 kPa. At the bottom layer the material was so soft that it was not possible to conduct a test using the experimental apparatus.

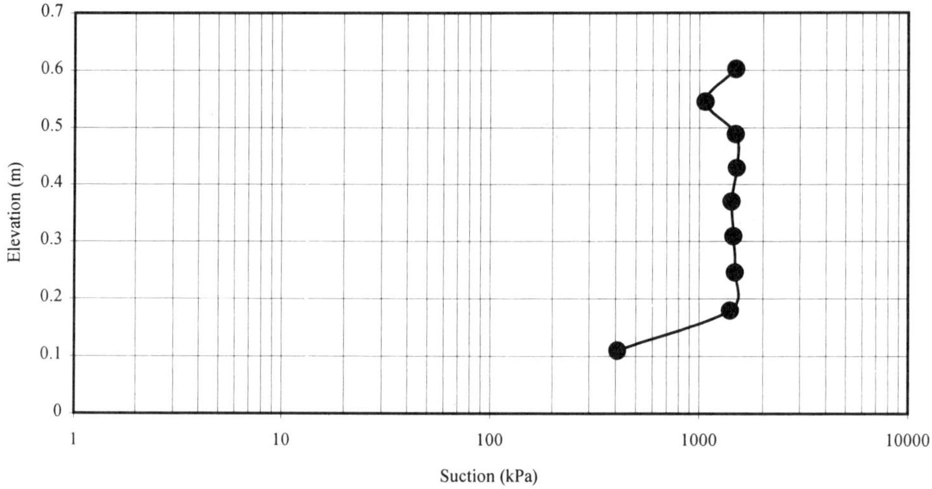

Figure 12. Matric suction profile

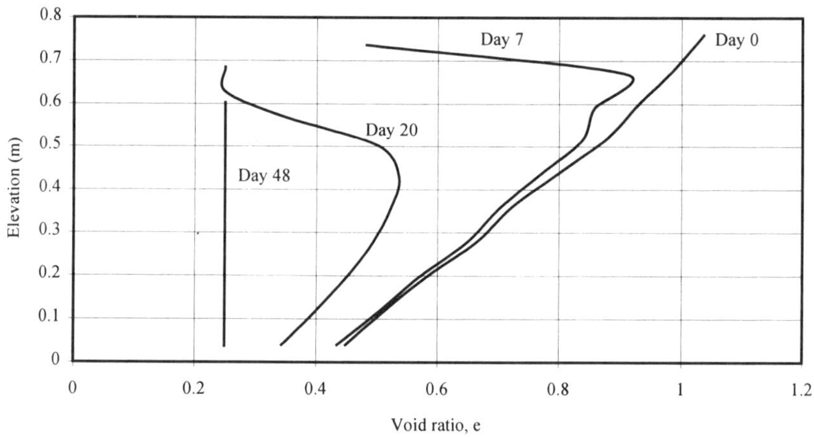

Figure 13. Void ratio profile

Comparing the bearing capacities of the planted lysimeters to the control lysimeters, the significance of the strength enhancement mechanism of the plants is evident.

6 CONCLUSIONS

A greenhouse experiment was conducted to evaluate the strength enhancement mechanism of plants growing on tailings. Measured values were compared with values predicted by the plant dewatering model. A good agreement was found between the measured and predicted solids content and bearing capacity profiles. The model is considered appropriate for predicting the strength enhancement of plants for one growing season.

The bearing capacity increase caused by the plant dewatering mechanism was beyond any expectation. The evaporation process, which was acting on the controls, did not cause any significant dewatering of the CT compared to the plants.

7 ACKNOWLEDGEMENT

The authors acknowledge the financial support of the National Science and Engineering Research Council of Canada and the first author thanks Luscar Ltd. for their Graduate Scholarship in Environmental Engineering. Special thanks is extended to Christine Hereygers, Steve Gamble and Gerry Cyre, whose technical assistance was extremely valuable.

REFERENCES

Jensen, M.E., Burman, R.D., and Allen, R.G., eds. 1990. *Evapotranspiration and irrigation water requirements*. Manual of practice No. 70, New York: ASCE.

Johnson, R.L., Bork, P., Allen, E.A.D., James, W.H., and Koverny, L 1993. *Oil sands sludge dewatering by freeze-thaw and evapotranspiration.* Alberta Conservation and Reclamation Council report No. RRTAC 93-8. ISBN 0-7732-6042-0.

Pollock, G.W. 1988. *Large strain consolidation of oil sand tailings sludge*. M.Sc. Thesis, University of Alberta, Canada.

Qiu, Y. and Sego, D.C. 1998. Engineering properties of mine tailings. *Proceedings of the 51th Canadian Geotechnical Conference*, Edmonton, Alberta.

Ritchie, J.T. 1972. Model for predicting evaporation from a row crop with incomplete cover. *Water Resources Research* 8(5):1204-1213.

Silva, M.J., Biggar, K.W., Sego, D.C., Chanasyk, D.S., and Naeth, M.A. 1998a. Plant dewatering of tailings: a theoretical model. *Proceedings of the 51th Canadian Geotechnical Conference*, Edmonton, Alberta.

Silva, M.J., Naeth, M.A., Biggar, K. W., Chanasyk, D.S., and Sego, D.C. 1998b. Plant selection for dewatering and reclamation of tailings. *Proceedings of annual meeting of the American Society for Surface Mining and Reclamation*. Pp. 107-117.

Stahl, R.P., and Sego, D.C. 1995. Influence of natural surface enhancement processes on bearing capacity and reclamation of fine coal tails. Paper submitted for review to the *Canadian Mining and Metallurgical Bulletin*, December 1995.

Tailings and Mine Waste '99 © 1999 Balkema, Rotterdam, ISBN 90 5809 025 6

Evaluation of mine tailings dams behavior – A case study

C. Priscu & H.S. Mitri
McGill University, Montreal, Que., Canada

H. Keira
AGRA-Monenco Quebec Incorporated, Montreal, Que., Canada

S.W. Jacobsz
Jones and Wagener, Rivonia, South Africa

ABSTRACT: A simple method integrating a 1-dimensional self-weight consolidation model for fine grained materials, together with a 2-dimensional stress/deformation finite element model is developed for stability evaluation of mine tailings dams. The scenarios of pond filling and dam construction are simulated, following the staged operation of the surface impoundment. A case study from an operating South African gold mine is used to perform the simulations and the parametric analyses for various tailings deposition rates. The structural behavior evaluation is made based on excess pore water pressure profiles and stress levels. The benefits of evaluating mine tailings dam behavior during various pond operating conditions are discussed in this paper. It is shown that adequate tailings deposition rates could be predicted in the future from such analyses for large operations.

1 INTRODUCTION

It is almost unanimously accepted that most incidents associated with mine tailings dams occur during mill/pond operation (U.S.C.O.L.D., 1994), when unpredictable changes from the normal conditions could take place. One of the engineering challenges that relate particularly to the upstream (u/s) method of construction, is the management and control of tailings discharges, especially in the case of large operations with high daily discharge rates of mill effluent.

In order to evaluate the behavior of such a structure, a method was proposed which makes use of classical numerical modelling tools to simulate the staged construction of tailings impoundments (Priscu et al., 1998). By combining self weight consolidation results into a stress-deformation analysis, excess pore pressures developed in soft under-consolidated slime zones underneath the subsequent raises of the daywall can be more realistically evaluated.

Here, the concept is applied to a case study of a South African gold mine, which is in operation for another 5 to 6 years (estimated), to reach a maximum height of 60 m over a total time period of more than 45 years. Although deposition rates are strictly kept within safe limits at this site, the existing in-situ data and laboratory tests allow for the simulation of various tailings deposition rates.

2 SITE DESCRIPTION

2.1 *Mining operation*

An approximately 50 m high tailings dam from a large South African gold mine was selected for the purpose of this study (this dam will be referred to as Dam A in the remainder of the paper). Tailings deposition at the dam site commenced in the late 1950s. The original construction consisted only of earth containment berms around the dam's perimeter, with a penstock and outlet pipe near the center. Dam A forms part of a tailings dam complex comprising eight compartments of which four are currently active. A typical cross section of the wall used in the analyses is shown in Figure 1, below. It is worth mentioning that at that time, the dam was not provided with an internal drainage system.

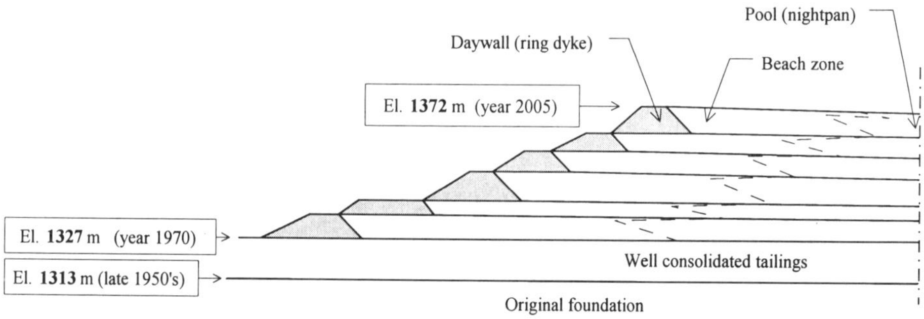

Figure 1. Typical cross section of Dam A (southern wall) and its operation sequence

A 14 m thick layer of tailings was deposited in the late 50's and in the 60', which have well consolidated ever since. From 1970, Dam A grew in height by approximately 1.3 m per year and is currently near the end of its operational life, which will terminate within the next 5 to 6 years when the dam reaches approximately 60 m in height.

Dam A was selected because of a reasonably detailed record of construction, and the availability of a number of piezocone (CPTU) and laboratory test results from past stability investigations, which were used to evaluate material parameters for the numerical analyses.

2.2 *Dam operation*

Dam A, like the majority of South African gold tailings dams, is constructed using the upstream semi-dry paddock method. Dams constructed by this method include two main components: the "daywall" and "nightpan".

The outer wall (dyke) or daywall of the dam is raised by constructing shallow, narrow paddocks along the outer perimeter of the dam wall. The tailings previously deposited are used to raise the paddock impounding walls with a tractor-plough combination or by labour intensive shovel packing. The paddock is then filled with approximately 150 mm to 200 mm of slurry from the delivery stations. The slurry flows from the delivery station into the daywall paddock and distributes by gravity in the daywall zone. It is then allowed to settle down before draining the excess water into the body of the dam through decant pipes located through the inner wall. Deposited material is left to dry and consolidate while deposition takes place at another delivery station. Consolidation and strength gain is achieved by desiccation. The nature of the deposition process results in a structure of very intense layering.

This wall-building operation takes place during the day when supervision is available, hence the name daywall. In order to achieve operational safety, discharge of tailings slurry at night takes place into the pond, hence the basin or the pond of the dam being known as the nightpan. The daywall is approximately 1 m to 2 m higher than the perimeter of the nightpan, and is generally raised mechanically by tractor-plough. Shovel packing is only used when conditions are too wet for the tractor to gain access. Water and slurry discharged into the nightpan runs down the concave beach surface of the dam towards the penstock, which is intended to be centred in the pool from where it drains accumulated water off the pond.

The dam is constructed through eight delivery stations at an average spacing of approximately 350 m. Under normal conditions, deposition cycles are irregular as the tailings dam complex (of which Dam A forms part) consists of four active compartments which are operated together with another independent adjoining tailings dam.

2.3 *Material characteristics*

Tailings deposited on the dam are of a quartzitic origin. Typical grading curves of tailings materials at various locations of Dam A are shown in Figure 2. Blight and Steffen (1979) remarked that gold tailings in South Africa have a remarkably uniform size distribution, the variation in grading from mill to mill and mine to mine being hardly any greater than from year to year in a single mill.

The strength parameters of the gold tailings comprising Dam A also show little variation. The solids in the slurry are a fine grained material with 71% passing the #200 mesh, characterized as a low plasticity sandy clayey silt (CL-ML in the Unified system).

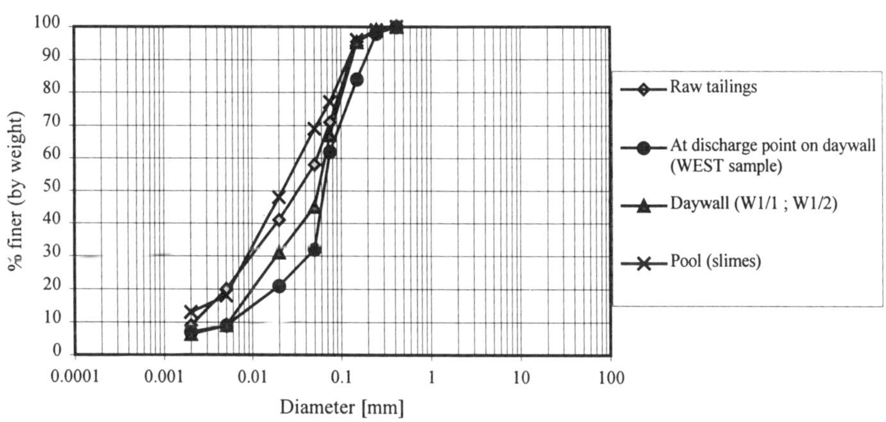

Figure 2. Grain size distributions of disposed materials at various locations on Dam A.

The effective friction angle and cohesion, determined from triaxial and shear box tests, amounts to averages of 37° and 0 kPa, respectively. This is in contrast with piezocone results which show large variation in dynamic pore pressures and penetration resistance, which correlates with a structure of intense layering, as mentioned above. The average in-situ dry density of gold tailings on Dam A amounts to approximately 1450 kg/m^3.

3 MODELLING APPROACH

3.1 *Construction simulation and models used*

The simulation of dam construction is achieved through incremental placement of layers of fill (tailings). The sequence should account for all major geometrical features and boundaries, while taking into account construction activities. The staged construction/operation simulation follows the time dependent increase in height of the pond. As presented in Figure 1, the actual operating life of this impoundment spans over 45 years, of which 35 are modeled in this analysis due to limited information on depositions previous to 1970's.

The typical structure of the Dam A tailings dam presented in Figure 1, and its three main structural components (the ring dike, the beach and the slimes in-and-around the pool) are modeled in different ways.

Self-weight consolidation of the fine fraction of tailings is modeled using the finite strain sedimentation/consolidation theory implemented in a finite difference (f.d.) code, following the methodology proposed by Somogyi (1980). This theory allows for nonlinear, time dependent simulation of slurry deposition and consolidation in-and-around the pool, which is located close to the center of the impoundment.

The ring dykes and the beach materials are modeled using the Modified Cam-Clay (MCC) model (Chang and Duncan, 1977), as implemented in CON2D f.e. code (D'Orazio et al., 1990), and later modified by Priscu and Keira (1996). The Modified Cam-clay model is an elasto-plastic cap-model, with stress-dependent stiffness and strength characteristics. The program treats the coupled problem of deformation and fluid flow, and can estimate movements and pore pressures in drained / undrained conditions.

The two different models are used together to simulate the construction of the impoundment. A more detailed description of the algorithm is provided elsewhere (Priscu et al., 1998).

3.2 *Methodology*

The approach follows closely the staged construction of an operating tailings dam. The main steps to follow in a typical analysis of an operating tailings pond are briefly outlined hereafter:

- Collect information on construction history and operation sequence (geometry, elevations and slurry discharges vs. time, major mill interruptions, changes in slurry parameters, etc.);
- Define loading (construction) stages for simulation based on the above data. Each loading stage may account for changes in material properties, geometry or slurry discharge rates;
- Based on pool elevations and cross-sectional layouts, evaluate (approximate) the distribution of solids within the structure, i.e. pool, beach and dam;
- Based on existing beach profiles, adjust and calibrate coefficients for the beach profile equation (Blight et. al, 1985; Blight, 1994), used to define intermediate boundaries between layers in order to generate the f.e. mesh, including future operating stages;
- Based on available in-situ and/or previously documented tests, determine particle segregation along the beach, identify material types and their geotechnical characteristics;
- Based on existing pool elevations, calibrate power curves coefficients to be used in self-weight consolidation of slimes. Perform simulations of the self-weight consolidation of slimes following pre-defined loading stages;
- Design and generate the f.e. mesh following pre-defined construction sequences and material distribution;
- Prepare modelling parameters for the f.e. analysis;
- Run the f.e. program that will include results from self-weight consolidation analysis.

Output results could be presented in terms of total or effective stresses, excess pore water pressures, deformations, and mobilized shear strength distribution within the structure.

4 MODELLING PARAMETERS

4.1 *Parameters for self-weight consolidation of slimes*

Tailings slurry, having a bulk unit weight of 12.25 kN/m^3 and a grading curve of solids as presented in Figure 2, is discharged at a pulp density of 31.8% (voids ratio of 5.8). The discharge history at the site is presented in Figure 4, in terms of pond elevation increase in m/year (or m^3/year/m^2 of pond surface). The period form 1970 to 1988 was estimated to have increased at an average rate of 1.3 m/year, limited data being available on this time period. From 1988 to 1997, the rate of rise varied from a low of 0.26 to a high of 2.43 m/year. It is projected that an average 1.3 m/year will be kept for the rest of the pond's useful life, i.e., the next 5 to 6 years.

The correlation curves used in the f.d. program to estimate the consolidation process of the slimes are presented in Figure 3. The correlations have been estimated form available in-situ tests, as well as from available data in published literature, relevant to gold tailings slurries (Caldwell et al., 1984; Gassner and Fourie, 1998), which were found within a reasonably close range. Calibration was performed based on available records of in-situ pond elevations vs. time.

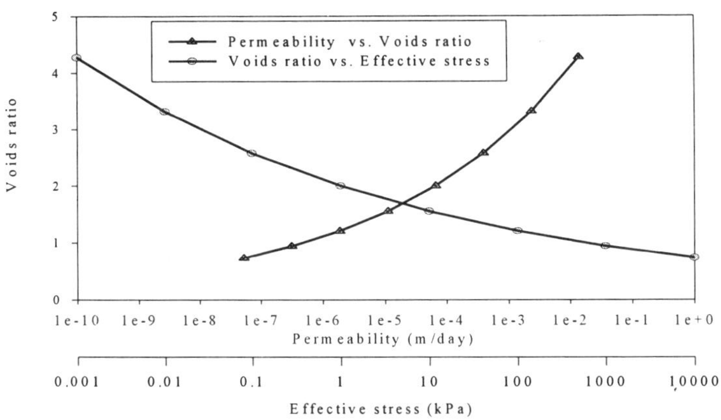

Figure 3. Permeability vs. Voids ratio vs. Effective stress; estimated correlations

Single drainage was considered as the impoundment was not underlain by any drainage system. Furthermore, the pond rests on almost 14 m thick layer of well consolidated tailings disposed in the late 50's and 60's. Given the low permeability of the old tailings layer, single upward drainage will take place during sedimentation and self-weight consolidation.

4.2 *F.e. modelling parameters*

The parameters used for modelling the ring dikes and beach consolidation were estimated from laboratory and in-situ tests results. Missing parameters were estimated from published data for materials with similar geotechnical properties. Table 1 presents the parameters used in the f.e. analysis for the four different structural material zones simulated in the case of actual deposition rate). Different parameters are used for beach and slime zones in simulations at

Table 1. Modelling parameters (MCC model) and geotechnical data for the f.e. analysis-actual rate of deposition

Parameter	Foundation	Daywall	Beach zone	Slimes zone
Total unit weight (kN/ m^3)	19.5	16.5	17	17
Slope of isotropic rebound curve, κ	0.0019	0.0055	0.008	0.009
Reference pressure (kPa)	50.	1.	1.	1.
Strength parameter	1.2	0.7	0.62	0.55
Horizontal permeability (m/s)	5.0E-07	5.0E-06	1.0E-05	2.2E-10
Vertical permeability (m/s)	1.0E-08	5.0E-07	8.0E-07	1.1E-10
Threshold degree of saturation	0.2	0.1	0.05	0.1
Initial voids ratio	0.6	1.05	1.25	1.39
Initial degree of saturation	0.9	0.7	1.0	1.0
Pressure at juncture (kPa)	100.	100.	100.	100.
Intercept after break (kPa)	100.	1.	1.	1.
Strength parameter after break	1.2	0.7	0.62	0.55
Pre-consolidation pressure (kPa)	0.	0.	0.	0.
Tensile strength (kPa)	1.	0.	0.	0.
Average stress "p_0" vs. Slope of virgin isotropic consolidation curve, λ	100. / 0.0023 500. / 0.07 990. / 0.025	318.5 / 0.0063 355. / 0.015 501. / 0.086	300. / 0.009 360. / 0.021 550. / 0.099	300. / 0.065 500. / 0.095 800. / 0.14
Mean effective stress "p" vs. Poisson ratio, ν	50. / 0.38 100. / 0.395 200. / 0.40	50. / 0.39 100. / 0.40 200. / 0.41	50. / 0.401 100. / 0.415 200. / 0.42	50. / 0.421 100. / 0.45 300. / 0.458

150% and 200% the actual rate. It is worth mentioning the differences between horizontal and vertical permeabilities in the structural zones, which reflect the strong structural heterogeneity of the dam. Parameters that describe characteristics of deposited slimes have the same initial values throughout each analysis only when deposition rates and slurry characteristics are kept constant. When slurry deposition or characteristics change from one stage to another, parameters vary and must be properly accounted for in the f.e. analysis, in order to realistically describe the initial consolidation state of slimes.

5 ANALYSES

5.1 *Self-weight consolidation of slimes*

Three different analyses were carried out at different tailings deposition rates, i.e., at 100%, 150% and 200% the projected rate. As presented in Figure 4, the results of the self-weight consolidation simulation are in reasonable agreement with the existing conditions at the site. It can also be observed that the maximum height of the impoundment is reached in the year 2005, 1995, and 1988 respectively. As expected, high percentage of fines in the slimes (77% passing the #200 mesh, see Figure 2), low permeability, and single drainage conditions result in low consolidation rates under self-weight.

Higher excess pore water pressures develop when deposition rates increase. Compared to the actual case (projected rate), excess pore water pressures increase at certain elevations by 15% and 35%, for the cases of 150% and 200% the projected rate, respectively. The resulting deposit is therefore more compressible, having a higher average voids ratio. From the above analyses, pond elevations and their corresponding distributions of pore water pressures, effective stresses and voids ratios are selected for each particular operating (loading) stage, to be further used in the f.e. program.

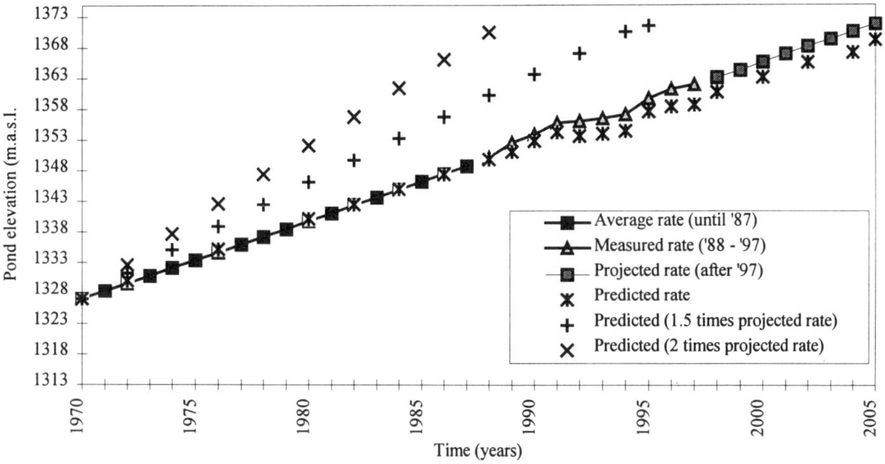

Figure 4. Measured and predicted values of pond elevations

5.2 *F.e. analysis of the tailings dam*

The cross section of Dam A was modeled from the toe of the South wall, up to the middle of the nightpan, where the supernatant pond is located, about 300 meters away. The f.e. mesh was automatically generated using GENCOM mesh generator (Mitri and Rizkalla, 1996), and comprises 358 4-node isoparametric elements, with a total of 423 nodes.

The distribution of simulated materials in the f.e. mesh is presented in Figure 5, and follows the zoning presented in Figure 1 for the daywall, the beach, the slimes and the underlying consolidated tailings. The latter, which forms the foundation is assumed to be fully consolidated, i.e., no pre-existing excess pore water pressure. This will allow for the analysis of the effect of deposition rate, alone.

The dam construction was simulated in six layers (loading stages). The mesh layout for each individual loading (construction) stage was approximated using the beach profile equation for gold tailings impoundments (Blight et al., 1985; Blight, 1994), calibrated on present beach geometry of Dam A.

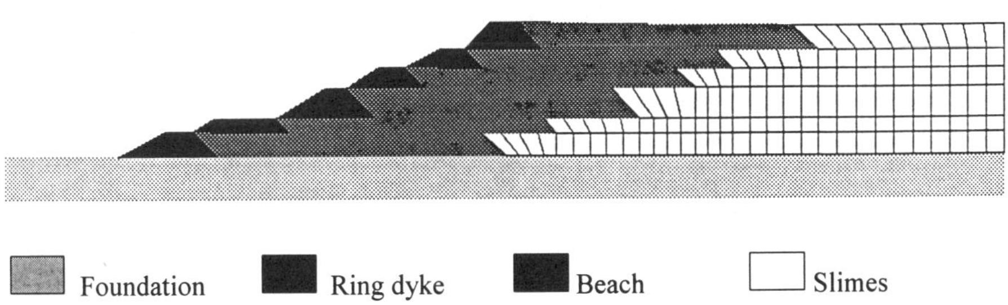

Figure 5. Materials distribution in the f.e. mesh

5.3 *Parametric study on tailings deposition rates*

The results of the f.e. analysis can be expressed in terms of stresses (total and effective), deformations, excess pore water pressures, deviatoric stress, mobilized, and total displacements. For the purpose of the present study, the distribution of excess pore water pressures and shear stresses are two of the main indicators of the structural behavior of the u/s tailings dam. In this paper, results from two analyses are presented: for the projected rate, and for the double rate of deposition.

Figure 6 shows the behavior of the structure at the end of its useful life, operated at the projected rate of deposition. It can be seen in Figure 6a that excess pore water pressures are concentrated in the center of the pond, with slimes having very low permeability coefficient, and single drainage conditions. This confirms the well known fact that excess pore pressures in tailings ponds dissipate extremely slow in fine grained slimes, even decades after the pond has been reclaimed.

Figure 6b shows the distribution of shear stresses, under the same operating conditions. Shear stresses are concentrated more in the dyke and the beach zones, extending into the foundation. This was also foreseen, as the dykes (which are relatively stiff materials compared to the beach and the slimes) in the u/s method of construction overlay the softer beach zones, concentrating stresses closer to the toe of the dam.

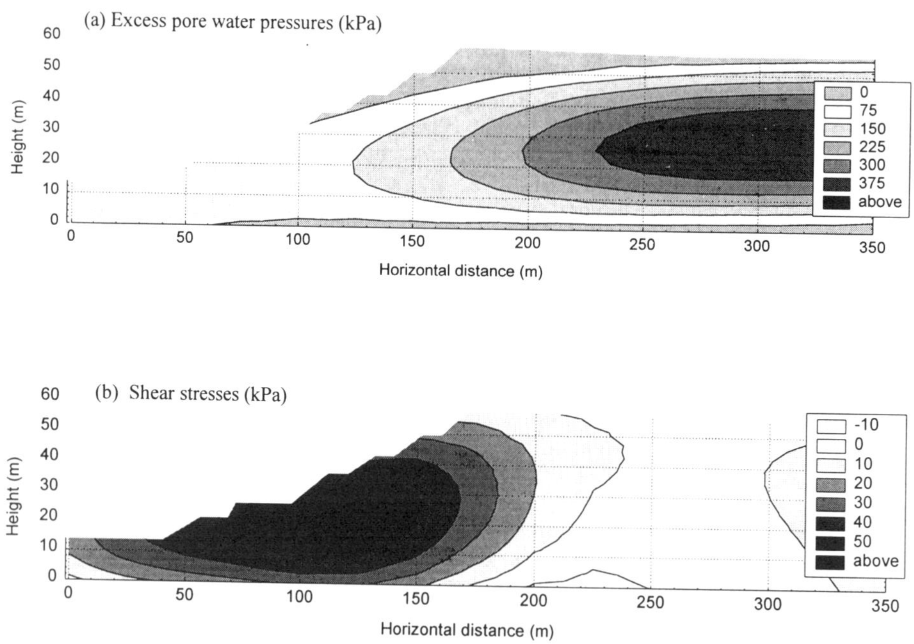

Figure 6. Results of consolidation analysis - actual rate of deposition, year 2005.

Figure 7 shows the dam behavior at the end of its operating life, if it were operated at double the deposition rate, that is approximately 2.6 m/year. It can easily be observed that excess pore water pressure increases in values both at mid-pond (25 to 50%) as well as in the beach zone underneath the dykes (50 to 100%). This increase in excess pore pressure is explained by the an increase in deposition rate, and implicitly a shorter period of time allowed for water to dissipate freely.

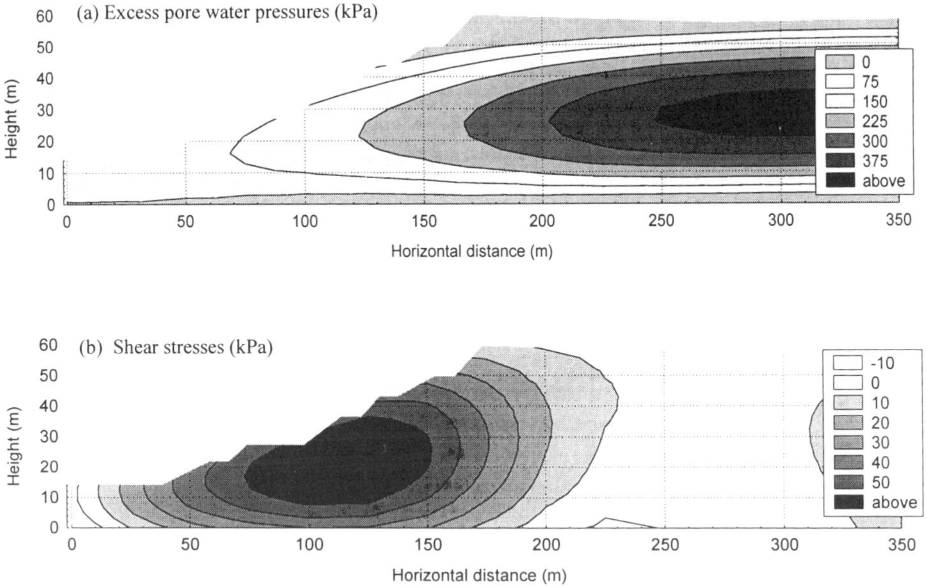

Figure 7. Results of consolidation analysis - twice the actual rate of deposition, year 1988.

Similarly, the shear stresses (Figure 7b) increase in value underneath the dykes (50 to 75%), and they propagate inwards, toward the pond. It can be observed that circular or curvilinear failures surfaces can, for high structures, develop not only in the beach zone but also at the beach/slimes interface.

At Dam A, the beach is kept wide enough so that a very small zone of deposited slimes is overlaid by the dyke and beach material. Furthermore, due to the method of operation which makes use of the dry climatic conditions in central South Africa, deposition rates are kept low at Dam A.

6 DISCUSSION

One of the limitations of the present numerical simulation is that average operating conditions were presented within each loading case, during which geotechnical parameters and characteristics of discharged slurry were considered constant throughout the life of Dam A.

For tailings pond operations with substantial changes in deposition rates or slurry characteristics throughout their useful life, such analyses may reveal weak structural zones from past tailings management, or they could predict future state of stress under various other operating conditions. Causes may be due to uncontrolled slurry discharge, change in slurry characteristics due to changes in ore deposit geology, increased mill capacity, pond receiving tailings from two different mills, delays/interruptions, etc.

7 CONCLUSIONS

A case study from an operating South African gold mine has been used to perform the numerical simulations and the parametric analyses for various tailings deposition rates. The scenarios of pond filling and dam construction have been simulated, following the staged operation of the surface impoundment. The behavior of the u/s tailings dam during various pond operating conditions has been assessed.

Preliminary results of the analyses have confirmed that, under high tailings deposition rates, tailings dams built by the u/s method may develop weak structural zones, by increasing excess pore water pressure levels, therefore decreasing shear stresses in the zones underneath the ring dykes. Presumably, safe tailings deposition rates for large tailings impoundments built by the u/s method of construction could be predicted at the design stage from such analyses.

ACKNOWLEDGEMENTS

The authors wish to acknowledge the kind support and feedback of Dr. F. Wagener, and of Jones and Wagener Consulting Engineers, Rivonia, South Africa, for sharing and allowing to publish the in situ and laboratory data.

REFERENCES

Blight, G.E. 1994. The master profile for hydraulic fill tailings beaches. *Proc. Instn. Civ. Engrs. Geotech. Engng*, 107, Jan, pp.27-40

Blight, G.E. & Steffen, O.K.H. 1979. Geotechnics of gold mining waste disposal. In Current geotechnical practice in mine waste disposal. Papers collected by the Committee on Embankment Dams and Slopes of the Geotechnical Engineering Division, ASCE, N.Y.

Blight, G.M., Thompson, R.R. & Vorster, K.1985. Profiles of hydraulic-fill tailings beaches and seepage through hydraulically sorted tailings. *J.S.Afr.Inst.Min.Met.* V85:5, pp.157-161.

Caldwell, J.A., Ferguson, K., Schiffman, R.L. & van Zyl, D. 1984. Application of finite strain consolidation theory for engineering design and environmental planning of mine tailings impoundments.In Yong and Townsend (eds): *Sedimentation / Consolidation Models-Predictions and Validations*, ASCE, San Francisco, CA, October 1, pp.581-606.

Chang, C.S. & Duncan, J.M. 1977. Analysis of consolidation of earth and rockfill dams. Report No. TE 77-3, Univ. of California Berkeley, Vol I, 127 p and Vol. II , 87 p.

D'Orazio, T.B., Sunami, S. & Duncan, J.M. 1991. CON2D90: A finite element computer program for analysis of consolidation. Virginia Polytechnic Institute, Blacksburg VA.

Gassner, F.W. and Fourie, A.B. 1998. Optimizing the allowable rate of deposition on tailings dams. Proc. of the 5th Int. Conf. On Tailings and Mine Waste '98, Ft. Collins, CO, Jan., pp. 241-247.

I.C.O.L.D.1989. *Tailings Dams Safety - Guidelines*. Bulletin no.74, Paris.

Jones & Wagener. 1996. Risk assessment of West Slimes Dam complex. J&W report no AA6/97/4934.

Mital, H. & Morgenstern, N.R. 1975. Parameters for the design of tailings dams. *Can. Geotech.J.* 12, pp. 235- 261.

Mital, H. & Morgenstern, N.R. 1976. Seepage control in tailings dams. *Can. Geotech. J.* 13, pp.277-293.

Mitri, H.S. & Rizkalla, M. 1996. GENCON/PRECON - a 2D finite element mesh generator/ graphical viewer. Dept. of Mining and Met. Eng., McGill University, Montreal, Quebec.

Priscu, C. & Keira, H. 1996. CON2D96 - A f.e. program for the analysis of stresses and deformations in earth structures. Groupe AXOR report no. 1029, Montreal.

Priscu, C., Mitri, H.S. & Keira, H. 1998. Behavior of mine tailings dams under high tailings deposition rates. Proc. of the 5th Int. Conf. on Tailings and Mine Waste '98, Ft. Collins, CO, Jan.,pp. 231-240.

Schiffman, R.L., Vick, S.G. & Gibson,R.E. 1988. Behavior and properties of hydraulic fills. In VanZyl and Vick (eds) *Hydraulic Fill Structures*, A.S.C.E. Geotechnical Special Publication No. 21, Ft.Collins,CO.

Somogyi, F. 1980. Large-Strain Consolidation of Fine-Grained Slurries. In Proc. of the Can. Soc. for Civil Eng. Annual Conference, Winnipeg, Manitoba, May 29-30.

U.S.C.O.L.D. 1994. *Tailings dams incidents*. Denver, CO.

Tailings and Mine Waste '99 © *1999 Balkema, Rotterdam, ISBN 90 5809 025 6*

Failure modes and effects analyses of potential ground movements at Golden Sunlight Mine

D.J. Miller
ESA Consultants Incorporated, Fort Collins, Colo., USA

ABSTRACT: Failure modes and effects analyses, or FMEA, is a structured, systematic process for evaluating risks to the environment by categorizing both the likelihood for an identified failure mode to occur (failure mode likelihood categories), and the associated potential environmental impacts (consequences categories). This paper describes an application of FMEA that was used to evaluate environmental consequences of potential ground movements at the Golden Sunlight Mine in Montana. The FMEA was done as part of an environmental impact statement (EIS). Prior to the EIS evaluations, the mine facilities (including plant structures, waste rock dumps, and one of two tailing impoundments) had been impacted by movements of massive, low-angle, planar landslides known as block slips. The block slips were subtle, ancient features that had not been identified prior to mine development. Mining activities remobilized movements of the primary slide block in April, 1994. Movements were successfully halted later that same year by offloading waste dump materials near the head of the slide, and placing a buttress of waste dump materials near the toe. Other ground movements at the site included shallow surficial landslides, structural foundation settlements, and waste dump slope failures. Some of these ground movement modes were associated with the block slip movements. The FMEA, in combination with comprehensive geotechnical analyses and professional judgement, provided a consistent methodology for categorizing the failure modes and evaluating risks associated with the complex ground movements and their interactions. The FMEA also provided a means to convey the risk analyses results in a concise tabular format, suitable for presentation in the EIS.

1 INTRODUCTION

Golden Sunlight Mine (GSM) is located approximately 30 miles east of Butte, Montana, near the town of Whitehall. GSM is an open pit gold mine that operated through the 1980's with a mill, one tailing impoundment, and a cyanide vat leach facility. A second tailing impoundment was authorized in 1990 to accommodate continued mining with expansion of the pit, and the east and west waste rock dumps. The pit is situated on the southern flank of Bull Mountain, targeting a 700 ft diameter breccia pipe within Precambrian sediments. The plant facilities, east waste rock dump, and tailing impoundments are situated east of the pit on valley fill sediments of Tertiary to Quaternary age. The general layout of the mine features is shown on Figure 1.

During its recent history, the mine's facilities have suffered a variety of complex, and sometimes interrelated, ground movements. These movements are categorized by several distinct mechanisms of instability, including (1) sliding of materials off Bull Mountain on steep, near-surface shear planes (landslides), (2) slip of massive earth blocks along deep, low-angle, pre-sheared bedding planes in the Tertiary valley-fill sediments (block slips), (3) sliding of fault-bounded blocks of bedrock adjacent to the pit, (4) instability of waste rock dump slopes, and (5) settlement of structural foundations caused by wetting of collapsible subsoils.

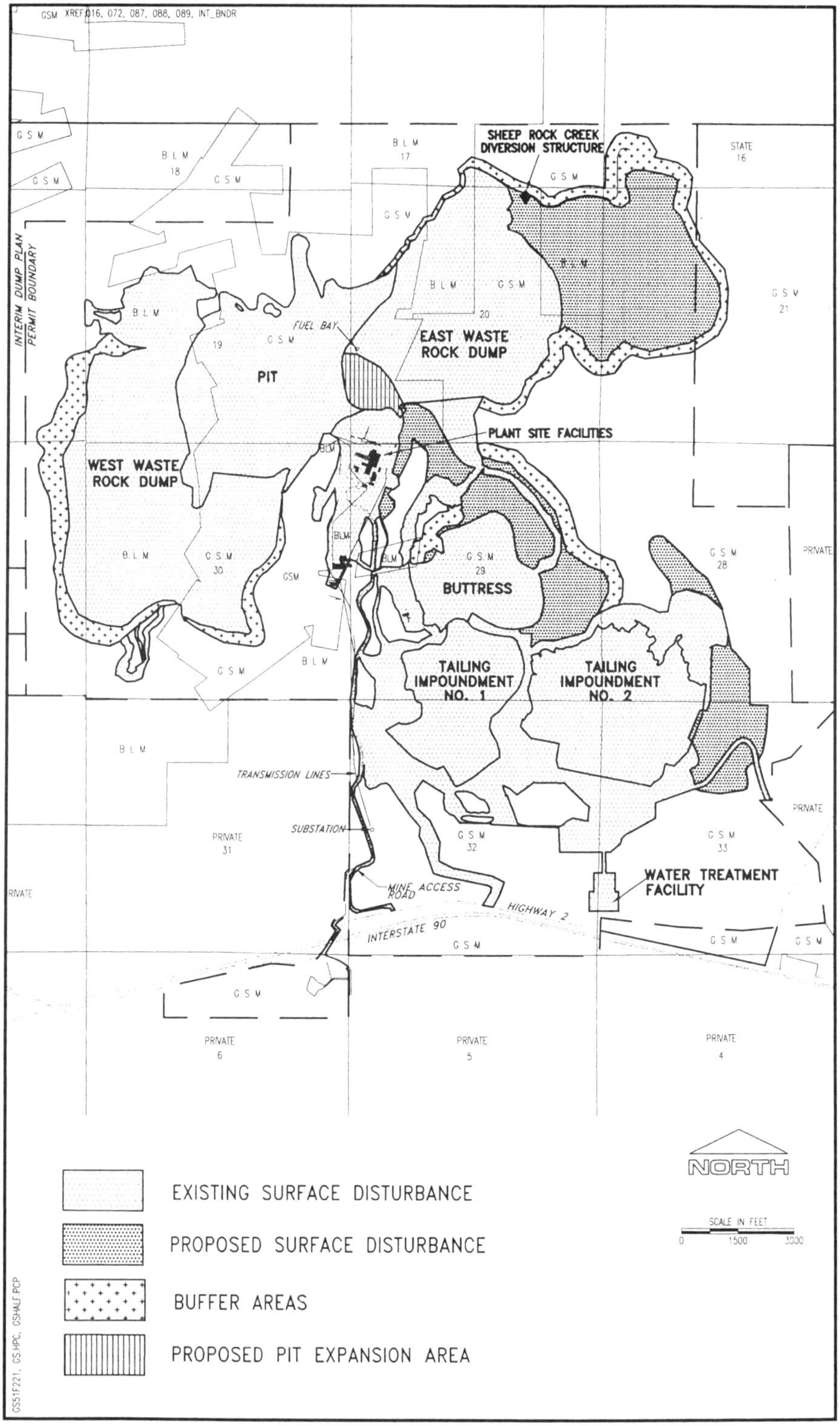

Figure 1. Golden Sunlight Mine general layout.

An evaluation of potential ground movements was done as part of an Environmental Impact Statement (EIS) that was submitted in April, 1998, to address potential environmental impacts of a proposed major expansion of the mine (GSM EIS, 1998). A failure modes and effects analyses (FMEA) was found to be a valuable tool for evaluating risks to the environment presented by the complex and interrelated ground movements at GSM. FMEA is a systematic method for categorizing both the likelihood for ground movement to occur (failure mode likelihood categories), and the associated environmental impacts (consequences categories). The FMEA resulted in a concise, tabular summary of environmental risk that is especially useful in a public document such as an EIS.

2 HISTORIC AND POTENTIAL FUTURE GROUND MOVEMENTS AT THE MINE

Geologic features relevant to ground movements at the Golden Sunlight Mine, including recent landslides and block slips, potential ancient landslide features, constructed and proposed landslide buttresses, and the outline of the proposed east dump expansion, are shown on Figure 2. Salient historic and future potential ground movements are described below.

2.1 *Midas Slump Landslide*

Geologists identified a surficial landslide northeast of the current mill facilities area during exploration of the site for its mineral resources. This natural feature was called the Midas Slump. The Midas Slump affected an area of approximately 86 acres and is outlined on Figure 2. The Midas Slump became an operational problem for the mine when waste rock was placed over the slump area. The waste rock embankment tended to settle excessively because the unstable foundation soils were continuously displacing out from under the dumped material. To stabilize this area, a waste rock dump was constructed around the down slope (south) end of the Midas Slump in late 1993. The intent was to provide a buttress, called the Midas Capture Dump, to contain the ground movement. This effort was successful and did arrest the Midas Slump ground movement. However, the placement of waste rock around the southern toe of the Midas Slump loaded an area that was a short time later discovered to be the head of an ancient, deep-seated block slip.

2.2 *Block Slips*

Ground movements were detected in portions of the pit, and beneath the plant site facilities and the east waste rock dump during the spring and summer of 1994. These movements were found to have been caused by mining-related reactivation of an ancient landslide feature called a block slip (Golder Associates Ltd. 1995a). Three distinct, but interrelated block slips were identified and given the names Sunlight Block, Rattlesnake Block, and Swimming Pool Block, as shown on Figure 2. The Sunlight and Rattlesnake Block slip movements occurred in valley fill sediments, in a geologic unit known as the Bozeman Group. The movements resulted from reactivation of ancient sliding that is believed to have begun during a period of active tectonism in the late Tertiary or early Quaternary. The sliding surfaces generally follow pre-sheared Tertiary bedding planes which dip down valley (southeast), at 3° to 6° from horizontal. The slip planes are very deep, typically about 150 to 300 feet below the ground surface, and occur in thin zones of weak, clayey materials within the Bozeman Group.

The sliding blocks evidently were stable in recent times, probably because of lower groundwater levels compared to conditions at the time the slides were initially activated. Movement of the Sunlight Block was initially reactivated by construction of the Midas Capture Dump portion of the east waste rock dump, which loaded the head of the block slip. Movement of the Sunlight Block is believed to have triggered movement of the adjacent Rattlesnake Block by drag forces acting along the Rattlesnake fault which separates the two sliding masses. Offloading of materials near the toe of the Rattlesnake Block where a borrow pit was previously located (in the vicinity of the Rattlesnake Block Buttress, as shown on Figure 2) may have also contributed to movement of that block slip. These movements, in

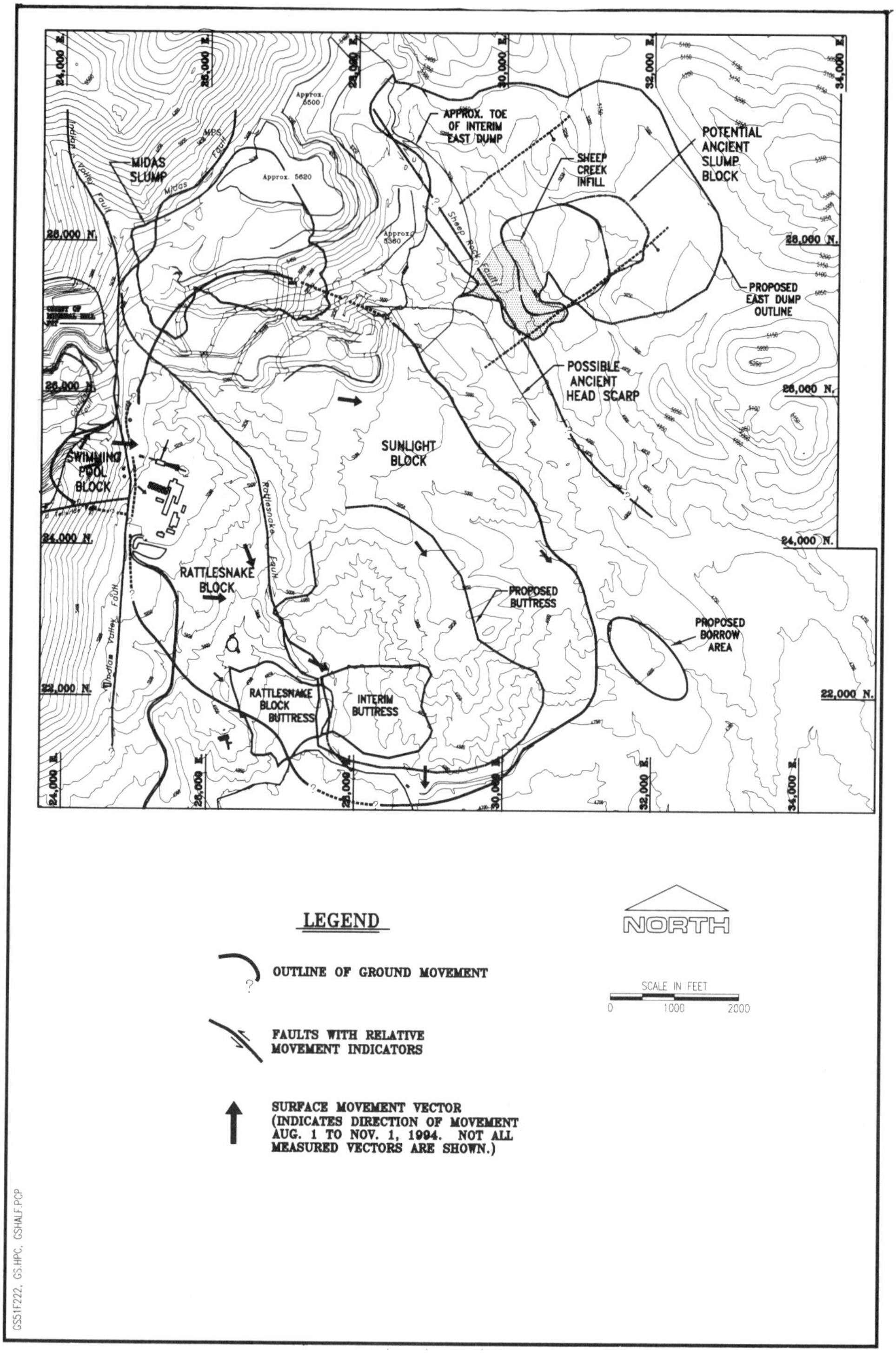

Figure 2. Principle ground movement features.

turn, are belived to have reduced lateral restraint along the east side of the smaller Swimming Pool block. The Swimming Pool Block, unlike the other two, was situated within Precambrian/Cretaceous bedrock, and was bounded on all side by faults. Reduced lateral stresses caused by pit excavation and movement of the Rattlesnake Block caused the generally eastward movement of the Swimming Pool Block.

Ground movement in the vicinity of the Swimming Pool Block initially was detected in April 1994. About that time, cracks were noticed in the concrete walls and floors within the plant area. Mill operations were suspended voluntarily by GSM on June 14, 1994. The full extent of the block movement was not understood until after August of that year. The area that was ultimately impacted by block slip movements was approximately 7,000 feet north-south by 3,800 feet east-west, and includes the mill/plant site, portions of the waste rock dumps, and the northern part of Tailing Impoundment No. 1.

Block slip movements were halted and stabilized by a combination of removal of waste rock from the head of the Sunlight Block, and placement of a stabilizing buttress at the toe of the Rattlesnake Block. The Rattlesnake Block Buttress was intended to fill in the North Borrow Area pit and halt the retrogressive failure at that location. A total of 15 million tons of waste rock were removed from the head of the slide area. Approximately 3.5 million tons of the off-loaded waste rock were placed as the Rattlesnake Block Buttress. The remaining 11.5 million tons were placed on stable ground northeast of the area of movement.

Geotechnical analyses estimate that waste rock dump unloading has increased the stability of the Sunlight Block by 11 percent relative to the condition prior to the onset of noticeable ground movement (i.e., prior to April, 1994). The safety factor for the failure area was estimated to be 1.22 prior to mining and modification of the topography (Golder Associates Ltd. 1995b). Additional buttressing was done during the interim permitting period, and more is scheduled as part of the proposed mine expansion plan to further improve the stability of the Sunlight Block relative to the pre-mining condition.

2.3 *Collapsible foundation soils*

The mill site buildings and related structures were constructed on a thick sequence of alluvial fan sediments that exist southeast of the pit. Evidence of foundation distress and structural damage was observed in the spring of 1994, and was attributed to foundation collapse settlement that occurred in conjunction with block slip movements. Geophysical mapping indicated that significant wetting had occurred near and beneath the southeast portion of the leach tanks (Golder Associates Ltd. 1995c). Tests revealed the presence of loose, granular soil layers within the alluvial fan deposits that underly the plant area. Upon wetting, rapid settlement or collapse of the loose zones occurred under the heavily loaded foundations. Total settlements up to 9 inches were estimated in the area. These settlements caused cracking of the floor slab around the tanks and opening of gaps in the structure. It was considered likely that wetting had occurred as a result of leakage from a buried fire line, which had been damaged by the east-west extensional ground movements across the mill site at the time of the block slips.

Remedial measures were implemented to improve the foundation conditions and enhance stability, including: compaction grouting, regrading the mill site to prevent ponding of water, installing a fiber reinforced PVC liner to provide a surface seal around the leach tanks, and sealing of open cracks and joints.

2.4 *Potential ground movement features relevant to planned mine expansion*

The proposed mine plan involved expanding the east waste rock dump towards the northeast, away from the identified head of the Sunlight Block. Geologic mapping and aerial photographic evidence indicated a possible Tertiary-age landslide (slump block) feature under the expanded dump area, and a possible head scarp feature east of the Sunlight Block scarp. These features are indicated on Figure

2. The possibility of mobilizing another unidentified block slip east of the Sunlight Block was investigated for this reason.

A geotechnical site investigation conducted by Golder in 1994-1995 (Golder Associates Ltd. 1995d) could find no conclusive evidence to connect the apparent surface manifestation of an ancient slump to a potential subsurface slip plane, and it was concluded that the aerial photo feature was not due to slumping. The detailed engineering geological site investigation identified important differences between the geologic conditions in the vicinity of the expanded east waste rock dump area and the conditions in the vicinity of the Sunlight Block. The expanded dump site was planned to be situated near the northern margin of the valley-fill deposits. The bedrock surface generally shallows beneath the proposed dump towards the northeast, and the northern margin of the proposed dump is founded on strong Precambrian bedrock. The orientation of the bedding planes in the sediments underlying the expanded dump was not conducive to a potential block slip movement direction in this vicinity, and also was an important distinction. These differing characteristics of the geology in the vicinity of the proposed expansion of the east waste rock dump indicate that there is only a moderate likelihood of initiating another block slip. However, the expanded dump was to cross over a drainage feature (Sheep Rock Creek) in which the weak Tertiary beds were partially daylighted. This foundation condition presented an increased likelihood of instability on the south and southeast slopes of the expanded east waste rock dump. Those slopes were therefore designed at shallower slope angles than was proposed for other dump slopes, with a toe buttress identified as the Sheep Creek Infill (Figure 2) proposed in the most critical areas. Further, the improved understanding of the geology east of the Sunlight Block revealed the potentially high risk of foundation instability for waste rock dump sites situated directly south of the proposed east dump expansion area. This knowledge affected one of the EIS alternatives which proposed using that area for waste rock placement.

2.5 *Earthquake-induced ground movement potential*

Golden Sunlight Mine is located within the Intermountain Seismic Belt (ISM), which is sharply defined by historic seismicity along about a 50 mile wide, northerly trending zone. Seismic analyses to assess earthquake stability of the waste rock dump slopes (including the buttress features), the tailing dams, and the Sunlight Block under a maximum credible earthquake (MCE) included both an evaluation of factors of safety against large scale slope failures, and an estimate of maximum displacements. Both results were considered in the FMEA.

3 GROUND MOVEMENT FAILURE MODES AND EFFECTS ANALYSES

The EIS involved evaluation of seven different alternatives. Failure modes and effects analyses proved to be a valuable tool, not only for performing the environmental impact evaluations, but also for presenting the results in the document.

The methodology followed in the ground movement FMEA was generally as outlined by Dushnisky and Vick (1996). The first step in the FMEA was to define the categories. Categories were defined for likelihood of ground movement (Table 1), and for the resulting environmental consequences (Table 2). A level of confidence was assigned both to the likelihood, and to the environmental consequence. Table 3 provides the definitions for level of confidence.

One of the challenges in the FMEA was to identify all the possible modes of failure. In addition to the ground movement features described in the previous section, the comprehensive FMEA included other ground movement failure modes, including potential pit wall instability, settlement of tailing impoundment covers, and instability of the tailing dams. A total of nine different primary modes of ground movement were identified and given two-letter symbols as follows:

BS = movement of Sunlight or Rattlesnake Block Slips
ED = slope failure on east waste rock dump
WD = slope failure on west waste rock dump
BB = slope failure on block buttresses
TC = settlement of tailing impoundment covers
TD = tailings dam slope failure
PW = pit wall instability
EB = movement of expanded east-dump-triggered block slip
DD = slope failure on alternative dump areas

Table 1. Likelihood Categories

Category	Likelihood of Occurrence	Chance of Occurrence
Negligible	$< 10^{-6}$	<1:1,000,000
Very low	10^{-6} to 10^{-4}	1:1,000,000 to 1:10,000
Low	10^{-4} to 10^{-2}	1:10,000 to 1:100
Moderate	10^{-2} to 10^{-1}	1:100 to 1:10
Significant	$> 10^{-1}$	> 1:10

Table 2. Environmental consequences categories

Category	Environmental Consequences
I	Negligible effect on water resources or the environment.
II	Failure will cause some environmental degradation in the vicinity of the mine by exposing limited volumes of ARD-generating materials, or releasing minor quantities of metals, cyanide, or sediments to the native soil and water resources systems.
III	Failure will degrade the environment in the vicinity of the mine by exposing large volumes of ARD-generating materials, or releasing major quantities of metals, cyanide, or sediments to the native soil and water resources systems. Long-term water quality and environmental degradation will occur if actionis not taken to remediate the failure.
IV	Failure will lead to major, long-term degradation of the natural environment and local water resources.

Table 3. Definitions for level of confidence

Likelihood/Consequence Confidence Category	Definition
Low	High level of uncertainty in selection of category or level; high variability possible because of lack of adequate basis for assessment.
Moderate or Medium	Some uncertainty in selection of category or level; moderate variability possible.
High	Confident estimate of category or level; adequate basis for assessment.

Table 4. Ground movement FMEA: proposed action alternative, movement of Sunlight or Rattlesnake Block Slips
Likelihood of block slip movement = moderate; confidence in likelihood = high

Symbol[1]	Stage[2]	Failure Mode	Likehood		Environmental Effects	Consequences		Potential Compensating Factors
			Category	Confidence		Level	Confidence	
SB1	O,C	East Dump slope failure or toe disruption	moderate	high	Exposure of ARD-generating materials to surface infiltration	II	high	Regular monitoring of block movements during mine life and after closure
SB2	O,C	Disruption of Midas Spring or North Borrow Spring subdrains	moderate	high	Exposure of ARD-generating materials to buried spring flows	III	medium	GSM commits to treating seeps
SB3	O,C	Disruption of reclamation covers on Block Buttresses, and Tailing Impoundment No. 1	moderate	high	Exposure of ARD-generating, and cyanide-bearing materials to surface infiltration	II	high	Regular monitoring of block movements during mine life and after closure
SB4	O	Settlement of plant-area structures due to: (a) spreading and subsidence; or (b)collapse settlement on wetting of foundation soils	(a) moderate (b) low	high	(a) None-minor structural distress (b) Cyanide solution spills through cracks in leach tanks	(a) I (b) II	medium	(b) Water lines are installed in a concrete corridor where rupture/leaks would be quickly detected and repaired.
SB5	O	Displacement or rupture of tailing slurry and return decant water pipelines	low to moderate	high	Spills of cyanide-bearing tailing slurry or reclaim water	II	medium	Existing pipes will be replaced in 1996 or 1997 by double-lined pipes
SB6	O,C	Pit wall instability near north side of Swimming Pool Block	low	medium	Greater exposure to water of ARD-generating materials	I	high	Pit water will be treated prior to discharge to groundwater system

[1]BS = Movement of Sunlight or Rattlesnake Block Slips;
[2] O = during operational life of mine and mill, C = after mine closure, reclamation activities completed, and treatment facilities installed and operating

Table 5. No action alternative summary of ground movement FMEA

		Likelihood Category				
		Significant	Moderate	Low	Very Low	Negligible
Consequence Category	I	TC2 PW1	BS4(a) PW2	BS6		
	II		BS1 BS3 ED1	BS4(b) BS5 WD1 BB1 BB3 BB4	TC1	
	III		BS2 ED2	WD2 BB2 PW3 ED3	TD3 TD4	
	IV				TD1 TD2	

Table 6. Proposed action alternative summary of ground movement FMEA

		Likelihood Category				
		Significant	Moderate	Low	Very Low	Negligible
Consequence Category	I	TC2 PW1	BS4(a) PW2	BS6		
	II		BS1 BS3 ED1 *EB1* *EB2*	BS4(b) BS5 WD1 BB1 BB3 BB4	TC1	
	III		BS2 ED2	WD2 BB2 PW3 *ED4*	TD3 TD4 *EB4*	*EB3*
	IV				TD1 TD2	

Separate FMEA tables were prepared for each primary mode of ground movement. Some tables were applicable to more than one EIS alternative. The complete ground movement FMEA was presented in the EIS (GSM EIS, 1998). One representative table (Table 4) is shown for the block slip (BS) primary failure mode to illustrate the process.

A specific, secondary failure mode was identified by a number following the two-letter ground movement symbol on Table 4. A brief description of each secondary failure mode was provided on the table. The likelihood of occurrence for each failure mode was categorized and assigned a level of confidence. A likelihood/confidence level also was assigned to each primary mode of ground movement. The likelihood/confidence level assignments for individual failure modes represent the overall evaluation for that failure mode, which already incorporated the likelihood for the the principle ground movement occuring. These likelihoods should not be interpreted as being additive or cumulative. The high confidence associated with the assignment of many of the likelihood and consequences categories in this FMEA would not have been possible without the extensive engineering geologic investigations, and rigorous geotechnical engineering analyses performed at GSM.

Results of the FMEA for the no action and proposed action alternatives are summarized on Tables 5 and 6. Entries on Table 6 (proposed action) that were different from Table 5 (no action alternative) are indicated by italics. Several conclusions could be drawn from the results displayed on Tables 5 and 6. The analyses indicated that most of the identified modes of failure had a low to very low probability of occurring. Moderately likely failure modes were primarily associated with potential block slip movements. The only failure modes with a significant probability of occurring were related to instability (ravelling) of the pit walls and consolidation settlements on the tailing impoundment covers. These potential ground movments were anticipated to have negligible environmental consequences. The failure modes with the most severe environmental consequences were tailings dam failures, whith very remote likelihoods of occurring.

A comparison between Table 6 and Table 5 revealed the only apparent difference between the no action and proposed action (mine expansion) alternatives with regard to potential ground movements related to the potential for initiating another block slip east of the Sheep Rock Creek drainage because of the expanded east waste rock dump. This action was interpreted to have a moderate likelihood of resulting in Level II environmental consequences, and a very low to negligible likelihood of resulting in Level III environmental consequences, as defined in Tables 2 and 3, respectively.

4 CONCLUSIONS

The complete FMEA for all seven EIS alternatives allowed comparisons to be made easily. The summary tables provided a succinct, visual display of the results that showed where most of the identified failure modes were placed with respect to likelihood and environmental consequences. The FMEA proved to be a valuable tool because it provided a rational, systematic method to break down a complex geologic problem into managable categories. The insight and understanding gained after performing the ground movement FMEA allowed a rational evaluation of environmental risk.

REFERENCES

Dushnisky, K. & S.G.Vick 1996. Evaluating risk to the environment from mining using failure modes and effects analyses. *Proceedings of Uncertainty '96, Uncertainty in the Geologic Environment*, American Society of Civil Engineers, Geotechnical Special Publication No. 58, Vol. 2: 848-865.

Golder Associates Ltd. 1995a. Investigation and evaluation of the earth block movements at the Golden Sunlight Mine. Report submitted to Golden Sunlight Mines, Inc., Whitehall, Montana, January 10.

Golder Associates Ltd. 1995b. Mitigation of the earth block movements at the Golden Sunlight Mine. Report submitted to Golden Sunlight Mines, Inc., Whitehall, Montana, August 1.

Golder Associates Ltd. 1995c. Geotechnical and geophysical investigation, foundation subsoil conditions mill site area,Golden Sunlight Mine. Report submitted to Golden Sunlight Mines, Inc., Whitehall, Montana, January 20.

Golder Associates Ltd. 1995d. Stability analysis, east dump. Report submitted to Golden Sunlight Mines, Inc., Whitehall, Montana, November 15.

GSM EIS 1998. Final environmental impact statement amending and adopting the draft environmental impact statement - Golden Sunlight Mine, U.S. Department of the Interior, Bureau of Land Management, Butte District and State of Montana, Department of Environmental Quality, April.

Tailings and Mine Waste '99 © 1999 Balkema, Rotterdam, ISBN 90 5809 025 6

Predicting the soil-water characteristics of mine soils

D.A.Swanson, G.Savci & G.Danziger
Savci Environmental Technologies, Golden, Colo., USA
R.N.Mohr & T.Weiskopf
Phelps Dodge Mining Company, Morenci, Ariz., USA

ABSTRACT: The soil-water characteristic curve for unsaturated mine soils (i.e., waste rock, and tailing) is a key parameter for predicting net infiltration in the design of soil caps, potential seepage from waste rock stockpiles, and the migration of potential seepage to groundwater. To expedite preliminary engineering analysis and design, computer programs are used to predict the soil-water characteristic curve based on particle size distribution. Observations of recent testing indicate that the theoretical models currently employed by these programs are accurate for uniform mine soils (i.e., tailing), but are not accurate for well-graded mine soils such as waste rock. Further research has shown that a combined theoretical/knowledge-based approach provides a more accurate prediction of the soil-water characteristic curve for mine soils such as waste rock.

1 INTRODUCTION

Hydrologic studies at mine sites frequently involve the simulation of water flow through unsaturated tailing and waste rock. For example, as illustrated on Figure 1, the components of a hydrologic study for a waste rock stockpile may include the analysis and design of soil caps, predicting potential seepage from the base of a stockpile and/or predicting the migration time of potential seepage to an underlying aquifer. The soil-water characteristic curve (SWCC), which defines the soil's ability to store and release water, plays a fundamental role in the hydrologic study of these components, and therefore, accurate and defensible quantification of the SWCC is important.

The duration for laboratory testing to measure the SWCC can range from several weeks to several months for typical mine soils. Feasibility studies and preliminary analysis and design projects at mine sites can be subject to deliverable times that therefore preclude laboratory measurement of the SWCC. Due to the lengthy testing times, computer programs based on theoretical models are sometimes used to predict the SWCC using the readily measured particle size curve (i.e., Gupta and Larson 1979). The most recent approach to predicting the SWCC combines theoretical models with a knowledge-based systems approach (Fredlund 1996). This combined approach allows the user to calibrate theoretical predictions to available laboratory SWCC data as a means of improving the accuracy of the theoretical prediction. This paper evaluates the theoretical and the combined theoretical/knowledge-based approach for predicting the SWCC of mine soils.

2 BACKGROUND

The soil-water characteristic curve (SWCC) defines a soil's ability to store and release water. A typical SWCC takes the form shown on Figure 2 and is illustrated by the relation between volumetric water content and soil matric suction. The volumetric water content at zero matric suction corresponds to the porosity of the soil, which represents the total volume of water that the soil can store. The change in water content divided by the change in soil matric suction (i.e., the slope of the curve) represents the storage potential. Therefore, the steeper curve over a specific range of soil matric suctions, the greater the water storage potential. The soil matric suction which corresponds to the initial draining of the soil pores (i.e., the first transition from flat to steep) is referred to as the air entry value. The water content at which the SWCC begins to flatten after the air entry value is called the residual water content. At zero water content the soil matric suction is approximately 1,000,000 kPa (Fredlund et al. 1994). This dry condition is achieved by oven drying the soil.

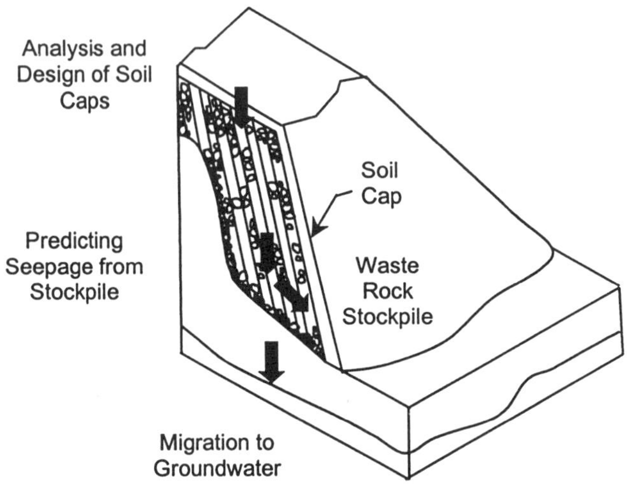

Figure 1. Schematic of a waste rock facility showing hydrologic conditions.

Methods to predict the SWCC using the particle size distribution have developed over the years. The earliest methods for predicting the SWCC from the particle size distribution were statistically based. The statistical models employ regression analyses on large data sets to predict the water content at specific matric suction values (i.e., Salter et al. 1966; Gupta and Larson 1979). Theoretical models were developed later and are based on the soil's pore-size distribution (which is determined from the particle size distribution). In this model, the particle size distribution is divided into a number of uniform particle sizes, and starting at the smallest particle size, a SWCC is estimated for each particle size and then summed to develop the

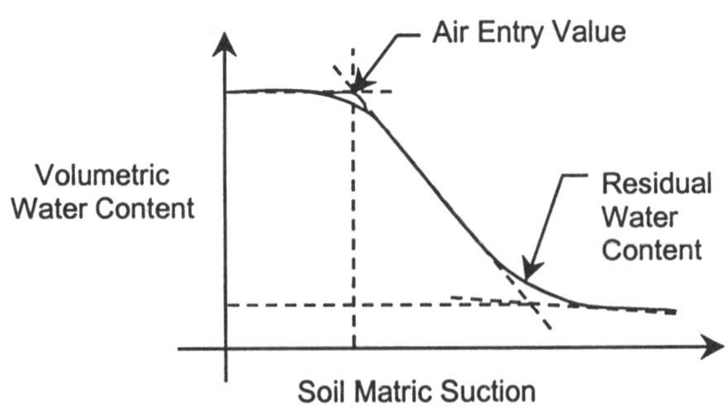

Figure 2. Components of a soil-water characteristic curve.

complete SWCC. Models that incorporate theoretical models include TheHyProS (Tarnawski and Wagner 1991), SOILPARA (Scientific Software Group 1998) and SOILVISION (SoilVision Systems Ltd. 1997).

The latest method for predicting the SWCC from the particle size distribution combines the theoretical model with a knowledge-based system (Fredlund 1996). The computer program SOILVISION (SoilVision Systems Ltd. 1997) combines the theoretical model with a knowledge-based system consisting of a data-base of soils (each with a measured particle size distribution and SWCC) and a quantitative means for calibrating theoretical predictions to a laboratory measured SWCC. Such versatility allows the accuracy of the prediction to be improved as more soils of a specific type are incorporated into the data-base for a given project.

3 PREDICTING SOIL-WATER CHARACTERISTIC CURVES OF MINE SOILS

The SWCC for three mine soils (i.e., two tailing samples and one waste rock sample) was predicted from the particle size distribution (Figure 3) using the theoretical method described in Section 2. The particle size distributions of the three mine soils are shown on Figure 3. The tailing samples have a relatively uniform particle size distribution (i.e., a narrow range of different particle sizes), whereas, the waste rock sample is well-graded (i.e., a wide range of particle sizes). The particle size distribution of this waste rock sample is slightly atypical due to its weathered condition; however, the well-graded particle size distribution is typical of many waste rock materials (Herasymuik 1996).

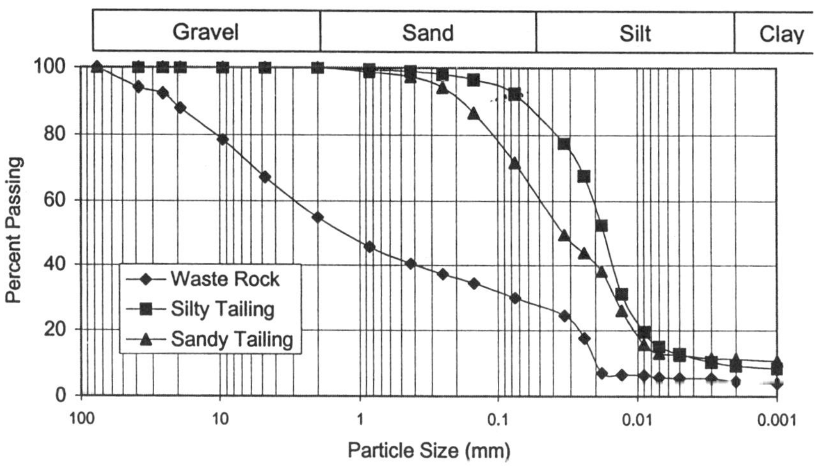

Figure 3. Particle size distribution of three mine soils.

The SWCC for each mine soil was predicted using the theoretical method employed in the SOILVISION model. The results of the predictions and the comparison to the laboratory measured SWCC for each mine soil is shown on Figure 4. As indicated on Figure 4, the predicted SWCC for the uniform tailing samples compared well to the measured SWCCs. In contrast, the SWCC for the waste rock was under-predicted.

The combined theoretical and knowledge-based approach capability of the SOILVISION model was used to improve the original prediction of the SWCC. A waste rock sample from the SOILVISION soil data-base having a similar particle size distribution to the waste rock sample shown on Figure 3 was identified by data query. The SWCC for this waste rock was already predicted and calibrated by adjusting the packing porosity parameter (theoretical prediction parameter) until the predicted SWCC matched the laboratory measured SWCC. The calibrated packing porosity for the data-base waste rock was used to predict the SWCC for the waste rock shown on Figure 3. This theoretical/knowledge-based prediction compared well to the laboratory measured SWCC (Figure 5), an obvious improvement from the prediction made using the theoretical prediction alone.

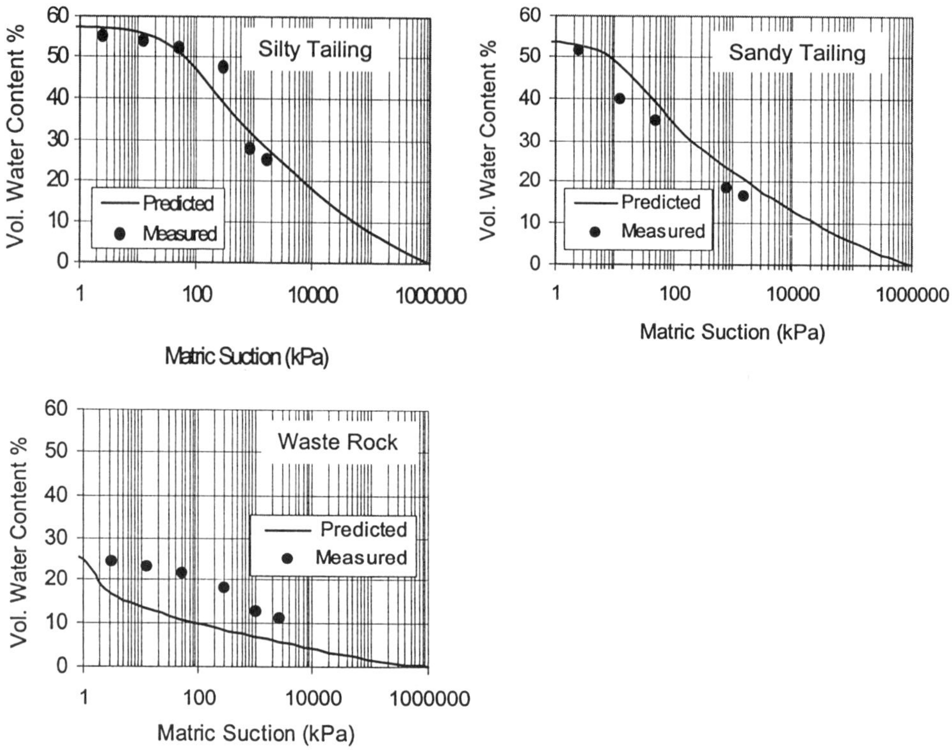

Figure 4. Predicted soil-water characteristic curves for three mine soils using the theoretical method.

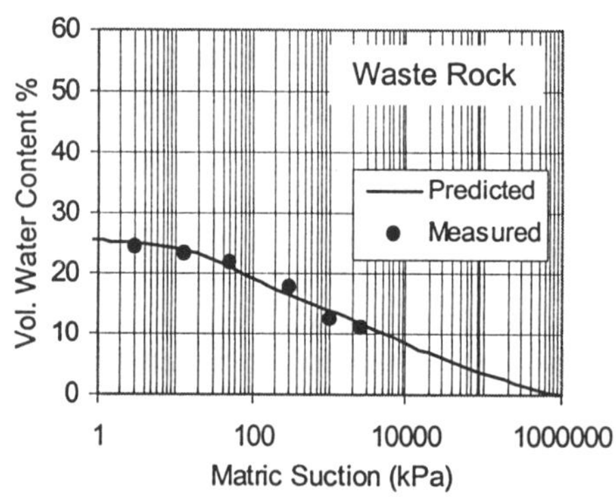

Figure 5. Predicted soil-water characteristic curve for the waste rock material using the combined theoretical and knowledge-based method.

4 DISCUSSION

The relatively poor comparison between predicted and laboratory measured SWCC for the waste rock sample and the good comparison for the tailing samples is due primarily to the shape of the particle size distribution. The tailing samples have uniform particle size distributions (i.e., predominantly sand or silt), whereas the waste rock sample has a very well-graded particle size distribution that includes clay, silt, sand and gravel size particles.

There are other factors (i.e., density) that should be taken into consideration when predicting the SWCC from the particle size distribution. Studies conducted by Veyera and Martin (1983) and Swanson and Barbour (1991) showed that density affects the shape of the SWCC for fine grained soils, while coarse grained soils such as sands are relatively unaffected by changes in density. Therefore, any mine soils that contain a fine grained fraction, such as all three mine soils shown in Figure 3, are susceptible to the effects of density. The combined theoretical/knowledge-based prediction demonstrated in Section 3 is particularly well suited to deal with these effects (provided that a sufficient data base of soils has been compiled to describe these effects).

5 CONCLUSIONS

The examples presented in this paper indicate that theoretical methods for predicting the SWCC using the particle size distribution are most accurate for mine soils exhibiting a uniform particle size distribution (i.e., tailing) and least accurate for mine soils exhibiting a well-graded particle size distribution (i.e., waste rock). Inaccurate SWCC data can translate into inaccurate hydrologic predictions such as prediction of net infiltration for the analysis and design of soil caps, predicting potential seepage from the base of stockpiles and predicting the migration time of potential seepage to an underlying aquifer.

Predictions for the SWCC of well-graded mine soils such as waste rock may be improved using a combined theoretical/knowledge-based approach. When SWCC data for similar soils is available to calibrate predictions, the combined theoretical/knowledge-based method can provide accurate predictions of the SWCC for mine soils for both well-graded and uniform mine soils for use in preliminary analysis and design.

REFERENCES

Fredlund, D.G., A. Xing & S Huang 1994. Predicting the permeability function for unsaturated soils using the soil-water characteristic curve. Canadian Geotechnical Journal. 31, 533-546.

Fredlund, M.D. 1996. Design of a knowledge-based system for unsaturated soil properties. M.Sc. Thesis. Department of Civil Engineering, University of Saskatchewan, Canada.

Gupta, S.C. & W.E. Larson 1979. Estimating soil water retention characteristics from particle size distribution, organic matter percent, and bulk density. Water Res Resch., Vol. 15, No. 6.

Herasymuik, G.M. 1996. Hydrogeology of a waste rock dump. M.Sc. Thesis. Department of Civil Engineering, University of Saskatchewan, Canada.

Scientific Software Group 1998. SOILPARA: Soil parameter estimator based on the statistical pore/particle size distribution model. Scientific Software Group.

SoilVision Systems Ltd. (1997). User's guide for a knowledge-based database program for estimating soil properties of unsaturated soils for use in geotechnical engineering. SoilVision Systems Ltd., Saskatoon, Saskatchewan, Canada.

Swanson, D.A. & S.L. Barbour 1991. The effects of loading on the moisture characteristic curve and permeability-suction relationship for unsaturated soils. Proceedings of the Canadian Society for Civil Engineering Annual Conference. Pages 194-203.

Tarnawski, V.R. & B. Wagner 1991. TheHyProS: Thermal and hydraulic properties of soils. Thermal Energy System, Computer Analysis and Design.

Veyera, G.E. and J.P. Martin 1983. Composition, Density and Fabric Effects on Bulk Waste Capillary Retention Characteristics. Role of the Unsaturated Zone in Radio-Active and Hazardous Waste Disposal. Ann Arbour Science.

Liners, covers and barriers

Tailings and Mine Waste '99 © 1999 Balkema, Rotterdam, ISBN 90 5809 025 6

Case studies on the use of geosynthetic clay liners in mining reclamation

Brad Miller
Colloid Environmental Technologies Company (CETCO), Mission Viejo, Calif., USA

ABSTRACT: Geosynthetic clay liners (GCLs) are manufactured barrier layers containing a high-quality sodium bentonite clay attached or adhered to geotextiles or a geomembrane. The GCL's low permeability and high strength make it an ideal replacement for low permeability soil or clay liners for mining reclamation projects.

This paper will detail a select number of case studies where the GCL has been utilized in mining reclamation projects in the United States. These case studies range from a simple GCL-bentonite slurry wall tie-in to a to a large 74 hectare Superfund reclamation cap. The case studies will summarize, where appropriate, project-specific design, regulatory, installation, schedule, and economic issues involved with each project.

1 INTRODUCTION

A GCL is a high quality, high-swelling sodium bentonite clay adhered or attached to manufactured geotextiles or flexible geomembranes. GCLs have been manufactured since the mid-1980's, initially used in double-lined landfill liner systems (Koerner, R.M., 1996). The structure of interlayer sodium cations makes sodium bentonite hydrophilic (water attracting). The sodium action clay structure readily accepts water molecules so that little free-water space is available in the clay voids, hence, maximum water permeabilities of 5×10^{-9} cm/sec are readily obtained using GCLs. GCLs contain clays that have high swell index characteristics of at least 24 ml per 2 grams of bentonite using ASTM Method D-5890, and maximum fluid loss values of 18 ml using ASTM Method D-5891. All GCLs contain an industry standard minimum of 3.65 kilograms of sodium bentonite clay per square meter of material as measured using ASTM Method D-5993.

There are unreinforced and reinforced GCLs available for use. Unreinforced products are manufactured using either adhesives and/or pressure. These products incorporate the use of a geotextile or flexible geomembrane to carry the bentonite layer. The unreinforced GCL is used on slopes gentler than 10:1, such as on the flat areas of reclamation caps, leaching facilities, and fluid containment basins. The reinforced GCLs are manufactured by needlepunching or stitch-bonding the top and bottom geotextiles together to encapsulate the sodium bentonite layer. The physical bonding of the geotextiles enhances the GCLs internal resistance to shearing and creep.

GCL is typically manufactured in rolls 4.2 to 4.7 meters wide and 45.7 meters long, and weigh upwards of 1,200 kilograms, so applicable offloading and installing equipment should be used that can withstand the heavy roll loads. The panels are overlapped typically 150 to 300 mm depending upon the application. If the GCL is not self-seaming (i.e., needlepunched products), then a thin layer of granular sodium bentonite is applied in the overlap seam at a typical rate of 375 grams per lineal meter. Self-seaming GCLs (i.e. Claymax) do not require the

Table 1 - Reclamation cap design issues

Design issues	Suggested design activities
Freeze-thaw	This design issue should not affect the performance of the GCL (see Kraus, et al, 1997, and Hewitt and Daniel, 1997), however, freeze-thaw could affect overlying drainage systems and vegetation growth.
Desiccation	This design issue should not affect the performance of the GCL (see Boardman and Daniel, 1996). If need arises to maximize GCL moisture retention, a design of a thicker than 300 mm cover soil layer is recommended.
Root penetration	Normal root growth has been shown to have minimal impacts on the performance of a GCL in a closure cover system. If deep rooted shrubs or trees are needed for reclamation plan, CETCO recommends digging a pilot hole first through the GCL, installing the vegetation, then backfilling with granular bentonite at the junction of the GCL and trunk.
Differential settlement	Most GCLs can withstand over 10 percent tensile strain before hydraulic conductivity increases (see LaGatta et al, 1997), therefore, this should not be a critical design issue for reclamation caps. If settlement is still a concern, increase overlap of panels to 12 to 18 inches to minimize panel overlap "pullout."
Subgrade preparation	Subgrade needs to be prepared so that protrusions greater than 13 millimeters are minimized. Additionally subgrade should contain minimal voids caused by too much coarse-grained soil material. If project dictates a geosynthetic cushion separation layer between GCL and subgrade, a 16 oz/sy nonwoven geotextile or 24 oz/sy Geocushion can be specified.
Liner stability	The critical interface in most cases is the GCL and veneer soil interface (if GCL-only barrier layer). Site-specific shear testing should be conducted at slopes 3:1 or steeper. May want to design a greater veneer stability by geogrid enhancement or pore pressure dissipation techniques. Pore water pressure dissipation can be designed using an earthen drainage layer or geosynthetic drainage blanket (see Richardson, 1997). Seismic design issues may also play a part if the site is in seismic impact zone. See Dombrowolski and Dellinger 1995 for more details designing GCL for steep sideslopes in seismic zones.
Chemical compatibility	Design using cover soils that do not have elevated leachable Ca and Mg. If elevated concentrations exist, conduct tier 1 and 2 compatibility tests on soil leachate. Soil leachate can be generated using simulated rainwater mixed with soil similar to EPA Toxic Characteristic Leaching Protocol (TCLP) test. If tier 2 tests indicate a higher than anticipated hydraulic conductivity, then design a GCL with a chemical resistant bentonite or use a composite laminate GCL.

additional sodium bentonite.

Field quality assurance is minimized when using GCLs because the GCL is a pre-manufactured product certified by the manufacturer to meet the design performance requirements. By contrast, field quality assurance is critical to the performance of a liner project when constructing a compacted clay liner (CCL).

Table 1 contains typical design issues for using GCLs in reclamation caps.

Table 2 - Selected project list

Project Name, Location	Application	Estimated size, m^2	Estimated Completion Date	Remarks
Arbiter Waste Repository, Butte, MT	Waste containment and cap	74,350	1992 & 1994	Hazardous waste repository and cap. GCL replaced 600 – 900 mm clay layer requirement.
Basin Creek Mine Cover, Basin, MT	Leach pad cap	12,490	1998	
Black Hawk Area Casino Mine Cover, Black Hawk, CO	Waste rock cap	6,000	Ongoing	Cover ARD-generating waste rock.
Brewer Gold Closure, South Carolina	Reclamation cap	191,500	1996 & 1997	Containment of ARD generating materials
Charter Oak Mine Cover, Avon, MT	Waste containment and cap	8,715	Sept. 1996	Mine repository w/GCL in base and cover liner.
Clark Tailings Repository, Butte, MT	Waste containment and cap	28,810	1997 & 1998	
Curlew Mine Tailings Repository, Missoula, MT	Waste containment and cap	17,425	Sept. 1996	Silver mine tailings repository. Project lined, filled, and capped within one month
Eagle Mine CTP Cap, Minturn, CO	Reclamation cap	47,400	Apr. 1996	Capping ARD potential tailings pile.
Joslyn Tailings Site, Helena, MT	Reclamation cap	7,900	1996	
Maxville Tailings Site, Maxville, MT	Reclamation cap	8,365	1997	State-managed mine clean up where remote tailings were consolidated under 0.8 hectares and capped.
Nancy Lee Mine Cover, Missoula, MT	Reclamation cap	41,820	1998	
Nellie Grant Mine Closure, Helena, MT	Reclamation cap	7,435	1998	
Oregon Gulch OU-10 Cover, Leadville, CO	Reclamation caps	58,375	Ongoing	GCL capping ARD-impacted tailings in ongoing Superfund project.
Rio Tinto Mine Mill Creek Channel, Rio Tinto, NV	Channel	5,110	1996	Diversion channel as part of Superfund remediation. Channel tied into 1,000 lf bentonite slurry wall.
Sharon Steel Tailings Cap, Midvale, UT	Reclamation cap	743,000	1996	Largest GCL cap ever constructed.
Summitville Cropsy Waste Pile Cap, Summitville, CO	Reclamation cap	46,500	Dec. 1995	Constructed in winter @10,000+ feet elevation.
Summitville HLP Cap, Summitville, CO	Leach pad cap	371,750	1997 and 1998	Special needlepunched GCL manufactured for capping project.

2 CASE STUDIES

Table 2 contains a selected list of reclamation mining projects where GCL has been utilized in the United States. Below is a discussion of some of these case studies listed in Table 2.

2.1 *Eagle Mine Consolidated Tailings Pile (CTP) Cover, Colorado*

The Eagle Mine, located near Minturn, Colorado, operated as a zinc and lead mine from the late 1800's through 1983. Over 7.25 billion kilograms of acid rock generating tailings were deposited in several areas. The Eagle Mine was placed on the Superfund National Priorities List (NPL) in 1984 because of the potential of the acid rock tailings threatening the watershed of the adjacent Eagle River. Additionally, children and employees at the Minturn Middle School were in danger of being exposed to windblown tailings.

A tailings reclamation plan was developed in 1988 that included consolidation and capping of approximately 283,000 m^2 using a multi-layered cap including a low permeability layer constructed from compacting the on-site fine tailings. It became apparent as construction progressed, that the fine tailings would need to be supplemented with another low permeability layer. The project engineer conducted a complete evaluation of available materials (compacted clay, FMLs, and GCLs) to cap the remaining 60,000 m2. Primary considerations were performance, cost, and constructability.

The GCLs was selected into a design that included 300 mm layer of protective material, 450 mm layer of 150 mm minus material for erosion and biointrusion protection, and a 300 mm layer of growth-sustaining cover soils.

GCLs were selected over compacted clay because of constructability and cost issues. The use of the GCL allowed the construction contractor to complete the project in a remote mountainous location with a short construction season. Approximately 1.3 truckloads of GCL was able to replace 150 truckloads of compacted clay for each acre of cap coverage.

2.2 *Sharon Steel OU1 Tailings Cover, Utah*

The Sharon Steel OU1 Tailings site, located in Midvale, Utah, was the former disposal site for residuals generated in a lead, copper, and zinc millings facility. Over 8.1 million cubic meters of contaminated material (soil and millings) was deposited from 1906 to 1971. Sharon Steel Tailings site was placed on the NPL in 1990 and a design for a composite low permeability cap was completed in 1995. The composite cap included GCL overlain by 30 mil PVC, geocomposite drainage net (used only on 2:1 sideslopes), 600 mm of cover soil, and 50-100 mm of top soil. The installed cost of the GCL (about $42,000 each hectare) was about 50 percent the cost estimated in the 1992 Record of Decision (ROD) for constructing a 600 mm layer of compacted clay liner.

The capping of over 74 hectares (743,000 m^2) began in Spring of 1996 and was completed by October 1996, six months ahead of schedule. The U.S. Bureau of Reclamation and State of Utah were the main oversight regulatory agencies. The installing contractor, Ogden Environmental, averaged approximately 19,400 m^2 each day of GCL. The installing contractor was able to rapidly deploy rolls of GCL because Ogden had specially designed and constructed hydraulically-driven roller GCL deployment assemblies.

2.3 *Brewer Gold Mine Reclamation Covers, South Carolina*

A GCL was recently used as the primary cap system for closure of the Brewer Gold Mine in north-central South Carolina. The Brewer Gold Mine closure included a specialized pit-backfilling approach because of increasing acid production from the heap leach pads, waste rock dump, and pit walls. The closure plan called for relocation of all heap leach and waste rock materials back into the site pits, and capping of the backfilled pits with a low-permeability liner system to limit oxygen and water contact with the acid generating materials. A GCL liner system was chosen following a cost/benefit evaluation that considered use of a GCL versus existing on-site clay (1 x 10-6 cm/sec permeability); or bentonite amendment of the on-site clay to achieve a 1 x 10-7 cm/sec permeability.

Construction operations for closure of the mine were performed sequentially, with each pit being capped immediately following backfill and final grading. Generally, the cap construction involved a) preparation of a suitable subgrade for installation of the GCL; b) installation of the GCL; c) placement and compaction of 450 mm of low-permeability soil; d) placement of 75 mm of loose top soil; and e) revegetation utilizing native flora.

The finished subgrade was prepared on the backfilled heap leach material (typically 20 mm-minus gradation) using a smooth drum roller to minimize the potential for protrusions in the subgrade. GCL deployment was initiated concurrent to completing subgrade preparation and subgrade acceptance by the on-site engineer. GCL was deployed using a tracked excavator with the spreader assembly attached to the bucket. All GCL panels were deployed parallel to the apparent direction of the slope with minimum 150 mm panel edge overlaps and minimum 600 mm panel end laps. All GCL panels were anchored at top of slope either using a 300 mm minimum panel run-out, or anchored into a 600 mm deep anchor trench. Completion of the cap required approximately 192,200 m^2. Daily GCL deployment rates varied because of many factors, however, a typical daily deployment rate 6,200 m^2 square feet on sideslope areas, to 13,500 m^2 on the flat areas.

Clay and topsoil placement operations varied with each pit because of varied constraints on access and borrow source locations. Most of the capped area was covered with a single 450 mm lift of clayey soil which was delivered to the cap with scrapers, spread with a dozer and compacted to 95 percent Standard Proctor with a pad foot roller. A typical 75 mm layer of topsoil was then spread over the clayey soil, and the topsoil revegetated.

Titan Environmental Corporation (now Schafer & Associates) was the design firm responsible for site closure design and construction management. Cap construction operations for the large pit was performed by Vecellio & Grogan, and the small pit by Amos Horton Grading.

2.4 *Summitville Cropsy Waste Pile Cover, Colorado*

A high priority ARD-generating waste pile at the Summitville Mine was capped using approximately 46,500 m^2 of needlepunched reinforced GCL under winter conditions in November and December of 1995. The Summitville Mine is an NPL site located in the Rocky Mountains at about 3,560 meters. The waste rock and heap leach materials at the site contain high ARD-potential. The US Bureau of Reclamation (USBR) designed a barrier cap in 1995 covering the Cropsy Waste Pile to minimize surface infiltration of runoff and snow melt and subsequent leaching of low pH, metal-laden water into the already-impacted groundwater. A compacted clay liner was not designed because of poor constructability, high costs, and freeze-thaw issues. GCL was considered to be the best solution for the project because of its economic advantages and short construction time.

Environmental Chemical Corporation (ECC) began installing the GCL in late November and completed the project by mid-December, 1995. High winds were encountered during installation and daytime temperatures were just above freezing. Fortunately, no heavy snowstorms were encountered during the installation. ECC deployed about 6,000 m^2 of GCL each day, and covered the GCL with about 450 mm of soil and top soil.

2.5 *Black Hawk Area Casino Waste Rock Cover, Colorado*

About 6,000 m^2 of ARD-generating waste rock was capped using a reinforced GCL as a part of a future hotel and casino development in Black Hawk, Colorado. The GCL was placed using normal construction equipment at approximately a 3 percent grade. A wedge of up to 31 meters of compacted rock was placed on top of the GCL to final grade where a paved access road will be placed. The needlepunched reinforced GCL used in the project met the peak shear strength design requirements of 25 degrees and an apparent cohesion of 12 kPa. The project was approved by the Colorado Department of Public Health and the Environment.

Stewart Environmental Consultants of Fort Collins, Colorado was the Environmental design engineer, D.H. Blattner of Avon, Minnesota prepared the subgrade, installed the GCL, and covered the GCL. The project began in August of 1998, and is still ongoing.

3 SUMMARY

Over 1.7 million m^2 of GCL has been installed in mining reclamation applications because it is easy to install, cost effective, meets leakage performance requirements, and is resistant to many physical and climatic issues that effects the performance of compacted soil or clay liners. Additionally, GCLs are more readily accepted by Federal and State Regulatory Agencies for covering tailings and waste rock features.

4 REFERENCES

Bisch, S.D., and Quinones, B.E. 1998 "Reclamation of the Londonderry Mine/Maxville Tailings Site near Maxville, Montana," Proceedings of the *Fifth International Conference on Tailings and Mine Waste '98*, Fort Collins, Colorado, January 26-28, 1998, pp 779-787.

Boardman, B.T. and Daniel, D.E. 1996 "Hydraulic Conductivity of Desiccated Geosynthetic Clay Liners" *Journal of Geotechnical Engineering*, Vol. 122, No. 3, pp 204-208.

Breitenbach, A. 1998 "Geomembrane Pad Liner Failures Under High Heap Fill Loads" *Geotechnical Fabrics Report*, January/February 1998, Vol. 16, No. 1, pp 38-40.

Bronson, J. et. al. 1997 "Control of Acid Rock Drainage by Pit Backfilling," Proceedings in *Fourth International Conference on Acid Rock Drainage*, Vancouver B.C. Canada, May 31-June 6, 1997, Volume III.

Dobrowolski, J.G., and Dellinger, A.S. 1995 "Innovative Cover Design for 2:1 Slopes Complying with California Closure Criteria" in *ASCE Geotechnical Special Publication No. 53*, Edited by Dunn R.J., and Singh, U.P., pp 35-50.

Hewitt R.D., and Daniel D.E. 1997 "Hydraulic Conductivity of Geosynthetic Clay Liners After Freeze-Thaw" *Journal of Geotechnical and Geoenvironmental Engineering*, Vol.123, No. 4, April 1997, pp 305-313.

Koerner, R. M, PhD 1996 "Geosynthetic Clay Liners, Part One: an Overview" *Geotechnical Fabrics Report*, May 1996, Vol. 14, No. 3, pp 22-25.

Kraus, J.F., et. al 1997 "Freeze-Thaw Cycling and Hydraulic Conductivity of Bentonitic Barriers" *Journal of Geotechnical Geoenvironmental Engineering*, Vol. 123, No. 3, March 1997.

LaGatta, M.D., et al 1997 "Geosynthetic Clay Liners Subjected to Differential Settlement" *Journal of Geotechnical Geoenvironmental Engineering*, Vol. 123, No. 5, May 1997.

Richardson, G.N. 1997 "Designing GCLs for Sideslopes" Proceedings of *Geosynthetics 97*, Long Beach, California, Volume III.

Ruhl, J.L, and Daniel, D.E. 1997 "Geosynthetic Clay Liners Permeated with Chemical Solutions and Leachates" *Journal of Geotechnical Geoenvironmental Engineering*, Vol. 123, No. 4, April 1997.

5 ACKNOWLEDGMENTS

The author thanks Joe Kaul with Ted D. Miller & Associates, Jon Bronson with Schafer & Associates, and Tom Norman with Stewart Environmental for helping in review the case studies. In addition, thanks go to Chuck Hornaday with CETCO for his support with the manuscript review.

Tailings and Mine Waste '99 © 1999 Balkema, Rotterdam, ISBN 90 5809 025 6

Construction quality assurance of a geosynthetic clay liner at high elevation

William Brewer, Donald Reimer & Christopher Wienecke
Advanced Terra Testing Incorporated, Lakewood, Colo., USA
Becky Morfitt
US Bureau of Reclamation, Denver, Colo., USA
Ronald Frobel
R.K. Frobel and Associates, Lakewood, Colo., USA

ABSTRACT: A geosynthetic clay liner (GCL) was used to cap the 178,000 m^2 heap leach pad at the Summitville Mine Superfund Site in Colorado. Due to the remote location, elevation and site conditions, unique problems were encountered for the construction of the heap leach pad cap system and subsequently for the construction quality assurance for the project. This paper addresses some of these unique problems encountered to allow the readers to consider them in regard to projects with similar site conditions.

1 SITE LOCATION AND HISTORY

The Summitville Mine Superfund Site is an abandoned gold mine in southwestern Colorado. It is located within the San Juan Mountain Range of the Rocky Mountains, approximately two miles east of the continental divide, at an average elevation of 3,500 m. The actual mine site covers approximately 2,800,000 m^2.

In addition to the high altitude the site has significant weather impacts. The site receives over 760 cm of snow annually. Snow first starts accumulating in October and does not melt until mid May to June. Between May and October the site receives and average of 46 cm of rain with 40 to 80 days of significant rainfall. High winds are common at the site.

Mining operations at the site began in 1875, and continued through the 1990's with some intermittent interruptions. In 1985, Summitville Consolidated Mining Company, Inc. (SCMCI) commenced an open pit gold mine. Gold was extracted by percolating cyanide solution over crushed ore placed on a lined heap leach pad. In December 1992, after SCMCI had vacated the site, the U.S. Environmental Protection Agency (EPA) initiated emergency response actions to prevent catastrophic release of cyanide solutions from the heap leach pad. Water treatment was initiated to treat leachate from the heap leach pad and acid mine drainage under the emergency response. The site was added to the National Priority List in June 1994.

2 HEAP LEACH PAD DESCRIPTION

The heap leach pad is a dish-shaped structure constructed within the Cropsy Creek drainage. The Cropsy Creek was diverted to avoid surface water infiltration into the pad. A composite liner of HDPE geomembrane and compacted clay underlies the pad and is suspected to have been damaged during construction. There were four well cans that extended through the pad, which were used to pump the leachate. A network of gravel lined trenches and perforated pipes underlie the liner and intercept ground water flows and leakage. There are approximately 5,100,000 m^3 of crushed ore within the pad.

The material within the pad is composed of crushed latite ore which was leached with a sodium cyanide solution to remove gold and silver. These materials consist of angular gravel, with sand and clay-size material. Chemical analysis of the leachate and pad materials showed the presence of cyanide and the potential for hazardous working conditions.

The first phase of the cleanup was pumping and treating of the contaminated water (leachate) within the pad. This was to be followed with biotreatment to inoculate the pad with cyanide-destroying bacteria. The treatment of the leachate began in December 1992 and the treated water was recirculated through the pad. In the summer of 1995, a drilling program was initiated to characterize the material within the pad. Based on the results of the data from the drilling program, it was decided that the rinsing operation had sufficiently reduced the cyanide level and biotreatment was not necessary.

3 BACKGROUND INFORMATION ABOUT CAPING THE HEAP LEACH PAD

The heap leach pad cap was designed to provide a hydraulic barrier against infiltration of precipitation into the pad. The cap is a long term remediation that will prevent the pile from becoming saturated, allowing water to overflow into the down slope dike, in turn causing instability of the dike and the leach pad. A secondary benefit is the reduction in the amount of potentially contaminated leachate to be treated.

The cap design combines low saturated hydraulic conductivity materials with lateral drainage considerations. Past construction experience has shown that the ability to construct an acceptable compacted clay liner at this site is very difficult. Weather conditions at the site make construction difficult, the elevation reduces equipment efficiency, and the closest potential borrow source of uncontaminated clay material is approximately eighty kilometers from the site. Therefore, it was decided to construct the cap with geosynthetic materials.

The pad configuration, prior to remediation, required reshaping to allow proper installation of the liner due to its irregular shape, steep slopes, holes, and flat areas. The reshaping involved adding approximately 650,000 m^3 of waste rock material excavated from adjacent areas and moving approximately 270,000 m^3 of material within the leach pad footprint. The reshaping configuration was selected to achieve a balance between cap performance, resistance to surface erosion, waste rock disposal quantity, and constructability. The original design incorporated 4H:1V side slopes with benches every 56 m of slope length. The benches were required to facilitate construction of the liner system based on the anticipated geosynthetic roll lengths, and minimize soil erosion of the final grade. The contractor requested that the shape be reevaluated to accommodate the moving of the materials. The slopes were flattened to 5H:1V and the benches were removed. Near the end of the project the amount of material within the footprint was reevaluated and the eastern side of the pad was reshaped to accommodate the amount of material present, with slopes shallower than 5H:1V.

4 DOUBLE GCL LINER SYSTEM

The cap for this project consists of a geosynthetic clay liner (GCL) placed directly on the recontoured heap leach pad. Overlying the GCL is a geocomposite drain, then another layer of GCL. The GCL consists of, from top to bottom, a nonwoven needlepunched 270 g/m^2 geotextile, a layer of sodium bentonite, and a composite of a light woven geotextile needlepunched to a 200 g/m^2 nonwoven geotextile. The 270 g/m^2 nonwoven geotextile has a dense mat of coarse and fine fibers. During the GCL composite manufacturing process, closely spaced barbed needles drive the fibers from the 270 g/m^2 geotextile through the composite into the lower geotextile. Fibers protruding through the woven side are then heat set. The effect of this needle punching and heat setting is to supply internal shear strength to the GCL. The manufactured roll dimensions of the GCL were 45 m long by 4.7 m wide.

A drainage layer was placed between the layers of GCL to collect and discharge any water that infiltrated through the top layer. A natural material such as sand would need to be purchased from a commercial processor and hauled over eighty kilometers or processed from rock on site. This would not have been cost effective, therefore, a geosynthetic drain was used. The geocomposite drain was comprised of a HDPE geonet with nonwoven geotextile heat bonded to both sides. The geotextile is required to increase the friction angle between the GCL and the geonet and to prevent the GCL from blocking the geonet as it swelled.

The leading edge of the existing heap leach pad bottom liner is buried from 0.5 to 6 m below the surface and excavating to this surface could cause excessive damage to the liner. Connecting the cap to the original bottom liner was not critical on three sides of the pad, as groundwater flows were already cut off by other means. In these areas a cutoff wall of GCL extends a minimum of 1.5 m beyond the leading edge of the bottom liner. On the south side, the cap extends vertically into a trench to a point below the original bottom liner. Additionally, the cap material lines the drainage ditch encircling the bottom of pad as further protection against water entering the pad.

The cap surface is covered by a 1.2 m protective layer of crushed ore (approximately 210,000 m^3) amended with lime or crushed limestone, 15 cm of topsoil, and is seeded with a mixture of native vegetation.

5 REMOTE LOCATION OF THE SITE

The Summitville Mine Superfund Site is located within the Rio Grande National Forest approximately 29 km off the nearest paved road. Due to the lack of usable native materials within the site's boundaries and the fact that the site is located within a National Forest, all materials needed to be trucked into the site. Therefore, the use of geosynthetic materials for lining the heap leach pad was more cost effective than using native materials.

During the course of the lining project, approximately 170 semi-tractor trailer loads of geosynthetic materials were brought to the site. The road to the site is a narrow, winding, dirt road that rises in elevation from ~2,700 to ~3,500 m with 100 m dropoffs in some locations. Due to the extreme weather conditions at the site, fresh snow covered the road when the first truck load of liner was delivered in June 1997. By the end of the 1997 construction season, the contractor was pulling the semi-tractor trailers up the last 10 km of the road with the use of a D-5 bulldozer, due to the amount of snow on the road. Between these events, rain storms or dust clouds made the road difficult to navigate.

Some of the truck loads of materials arrived with rolls that were damaged during transport. This damage consisted of broken cores due to the rolls shifting during transport. It is not known if this damage was caused during the trip up the National Forest road or during the initial transport from the factory.

6 WEATHER CONDITIONS

As mentioned above, weather conditions also affected the project's timing and quality. The last snowfall for the 1996/1997 winter season fell on the 16th of June 1997 with the first snowfall of the 1997/1998 winter season falling on the 21st of September 1997. In the days between these snow falls, installation was delayed or stopped on several occasions due to rain or wet subgrade conditions.

6.1 *Rain*

Precipitation, mainly in the form of rain, affects the installation of all geomembranes. This is typically due to the welding or seaming operations required for the majority of geomembranes. If the panel to be welded becomes wet, the welding or seaming method is affected and the seam quality is diminished. GCL does not require this method of bonding the panels together, however the bentonite within the GCL should be maintained near its received moisture condition until it is covered in order to function as intended. For this project, the GCL material was to be maintained below twenty percent moisture content prior to the cover soil placement to be considered acceptable. The material for this project was delivered at a moisture content between 8 and 12 percent. Rain also damaged the subgrade for the capping project and is discussed in the subgrade section of this paper.

6.2 *Wind*

Strong winds were also a problem at the site on occasion. One the benefits of using a GCL product is that it is more resistant to the wind than other liner materials due to its weight. The GCL used for this project weighed approximately 5000 g/m^2. On several occasions the edges of the GCL and GNGC were flipped by the wind and needed to be corrected before deployment or cover soil placement could continue. On one occasion five panels of GCL were displaced 2 to 5 meters from their deployed positions by strong winds. After these panels were returned to their positions deployment was halted for the day due to the wind.

6.3 *Snow and freezing conditions*

Snow and cold weather were also problems for this project and caused the eventual shut down of work for the 1997 construction season. When snow first started falling in September, it did not affect the project any differently than the rain had. The main concern was the moisture content of the GCL and the subgrade conditions. Work would be halted or stopped until these events passed and the subgrade conditions could be corrected, then work would resume.

One of the main concerns with snow which was not a concern with rain was the possibility of creating void spaces under the liner system. After large snow events the contractor needed to clean the surface of the subgrade with bulldozers. This was to ensure the liner system was deployed directly onto the subgrade and not over snow that could later melt out and cause voids under the liner. These voids would allow the liner system to settle unsupported and potentially could rupture the liner system.

The lining project was ultimately stopped for the 1997 construction season due to frozen subgrade conditions on November 17, 1997. From the 9th until the 17th of November only 370 m^2 of liner system had been installed due to almost continuous snow fall. Prior to this period of time 1,800 - 4,000 m^2 per day was typical.

On the 17th, after four days without any deployment, the sky was clear, however, the subgrade was frozen solid and subgrade preparation was not possible. Upon arrival at the site and seeing that the weather was clear and the subgrade was still exposed after three days of work by the contractor, lining work was proposed to start. However, when the contractor started excavating the needed anchor trench the backhoe could not break the frozen surface. A D-10 bulldozer with a rip hook was brought in to break the surface of the subgrade and after several tries succeeded in starting the anchor trench. The backhoe was then used to clean out the trench and work continued toward starting the deployment of liner. However, after the first two panels of GCL were deployed it was noted that the frozen subgrade was puncturing the liner system as equipment traveled over the liner during deployment. It was decided that this was unacceptable. The GCL that had been deployed was repaired and the remaining layers of geosynthetics were deployed to complete the area started. Once the area was completed the lining project for the 1997 construction season was stopped.

The liner system was originally planned to be completed in one construction season. Accordingly, there was no provision for having the leading edges of the liner system exposed over the winter. The specifications called for the GCL to be kept dry until it was under load from the cover soil. As it was unlikely that the GCL could be kept dry over the seven months before lining could continue it needed to be placed under load.

The CQA team, contractor and liner installer devised a plan for winterization of the leading edge and it was accepted by the Bureau of Reclamation. Winterization for the liner system used the geonet composite on site as a sacrificial layer at the edges of the liner system. This protective layer was then covered with 0.6-1.2 m of cover soil in order to keep the GCL under load during the likely hydration that would occur during spring runoff.

In June of 1998, when the leading edges of the liner system were excavated it showed that the winterization had served its purpose. The GCL was relatively dry due to the loading, with little damage due to the sacrificial layer of geonet composite protecting the GCL.

7 SUBGRADE

Typically GCL manufacturers call for the subgrade to contain minus 5 cm rounded rocks supported in a matrix of finer materials with no material in excess of 1.3 cm loose on the surface. This is to ensure the GCL is not damaged during placement and will function as specified. The subgrade for the liner system at the site was a minus 8 cm crushed latite ore with some fines due to the leaching operation.

The fines within the latite ore were easily washed away during rain events and did not function well as a bonding matrix for the crushed ore. Therefore, loose material on the surface of the subgrade was a continuous problem. This loose material was an extra concern due to the method of deployment for the geosynthetics. Each roll of geosynthetics used weighed roughly 1,150 kg and could not be handled by hand. Therefore, low ground pressure equipment was used to deploy the materials. Even though the equipment used exerted low ground pressures, if the equipment turned while passing over loose material under the liner damage occurred. Overall, the contractor maintained the subgrade in a manner that reduced the amount of loose materials. However, in certain areas which were referred to as "Rock Pockets" there were little or no fines present. The subgrade would appear to be prepared properly until the equipment passed over and kicked up numerous rocks. Upon deploying the next layer of geosynthetics the lower layer would become damaged. After recognizing the problem with "Rock Pockets" the repair methods to the subgrade were adjusted to reduce the problem. Several solutions were used to correct theses areas including: covering the area with material containing more fines; excavating the area and replacing the material with more suitable fill; or covering the problem area with scrap pieces of GCL. All of these repair methods reduced the damage to the liner system.

Another problem was wet/saturated subgrade that could not be deployed on. This problem occurred mainly at the base of the slope due to the concentration of fines washed down from the slopes. However, the problem did occur on the side slopes as well. The subgrade material would become so saturated that even the low ground pressure equipment would leave ruts after it passed over the area. The effects of precipitation on the subgrade varied throughout the site. In some areas a rain event would saturate the subgrade for a few days in other areas the subgrade could be deployed on the same day as a rain event. Little could be done to correct the subgrade that was over saturated other than wait for the area to dry out and hope another area could be prepared for deployment. Saturated subgrade was the main cause for delays and down days during the construction of the liner system other than the rain events themselves.

In addition waste rock from other areas around the site was deposited into the heap leach pad. Some of the waste rock buried within the pad was in access of one meter in diameter. On several occasions this waste rock was at the surface of the prepared subgrade and needed to be excavated and removed or covered with fill material to make an acceptable subgrade for the liner system.

8 GEOSYNTHETIC MATERIAL TESTING

The Bureau of Reclamation required that the geosynthetics be sampled and tested after they arrived on site. The remote location of the site, combined with the need for quick turn around times for the tested materials, required that a laboratory be set up on site during the 1997 construction season. Due to the lower volume of geosynthetic materials delivered to the site during the 1998 construction season the geosynthetics were sampled on site, then taken to an off site laboratory for testing.

Both the geosynthetic clay liner and geonet geocomposite were required to be tested at a frequency of one sample every 4,600 m^2 of material to ensure there acceptance for the project. The GCL was tested for peel strength and tensile strength to determine the textile strengths of the GCL after manufacture. The bentonite within the GCL was tested for fluid loss, free swell, and moisture content. The GCL was also tested for mass per unit area to ensure the proper amount of bentonite was present within the GCL, and for Flux to ensure its hydraulic barrier potential. The GNGC was tested for ply adhesion and mass per unit area.

8.1 *Unacceptable testing results*

During the course of the project only one area of the testing was found to have unacceptable results. This was in part due to the fact that there was no established ASTM procedure for the test at the time of the project, and partly due to a single group of samples received at the site.

The lack of an ASTM procedure occurs frequently in the geosynthetics field as manufacturers produce new materials faster than committees can write new procedures. Therefore, manufacturers, institutions and engineers tend to develop testing methods prior to official ASTM methods being issued. In this case the method used by the BOR and the manufacturer differed. Both test methods were run in essentially the same manner, however, the method of interpreting the final data varied significantly. The BOR method called for recording the average of the

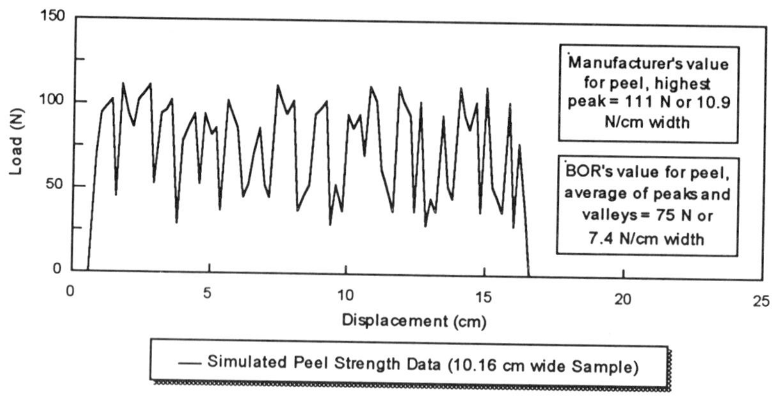

Figure 1. Comparison between BOR method of interpretation and the manufacturers method of interpretation for the results from a peel strength test.

peaks and valleys of the graphical test result as the test value, While the manufacturer recorded the highest peak as the test value (Figure 1).

This caused some concern at the beginning of the project when approximately 30% of the materials received at the site were not meeting the specification value, based on the BOR method. Once the difference in testing methods was discovered, both methods were employed on all of the remaining samples. The BOR had based the design on a GCL meeting their test method for peel, therefore, that method was required to accept roll goods for the 5H:1V slopes. If a sample did not meet the BOR method for acceptance but passed the manufacturers method, the roll goods were accepted for use on the 20H:1V slopes at the top of the heap leach pad.

A group of forty four rolls was received at the site that did not meet the BOR method or the manufacturer's method. Manufacturer representatives verified that the field testing was being performed properly with equipment that was in calibration and had the results verified by their own laboratory. Once the results were verified the manufacturer removed the group of rolls and replaced them with acceptable rolls.

All of the other testing performed for the GCL and the GNGC resulted in acceptance of the roll goods for the project.

8.2 *Description of testing performed*

The following is a brief description of the testing performed on the geosynthetic materials used for the capping of the heap leach pad.

Mass per unit area of the GCL was determined by ASTM D 5993. For each sample, five individual specimens were subsampled, weighed, and measured. The minimum value for acceptance was 3,900 g/m^2.

The peel strength of the GCL by the BOR's method was determined following ASTM D 413, modified for GCLs. This method utilized a grip separation of 7.62 cm and a grip width of 10.16 cm, testing a 10.16 cm wide by 20.32 cm long specimen. For each sample, five individual specimens in the direction of manufacture (machine direction) were tested. The graphical output was then analyzed calculating an average of the maximum and minimum values of each specimen. The minimum acceptance value was 831.9 N/m, based upon original design conformance test data performed. The Bureau of Reclamation later changed the acceptance value to 656.7 N/m based on additional internal shear test data.

The peel strength of the GCL, by the manufacturer's method, was determined following ASTM D 4632, modified for GCLs. This method utilized a grip separation of 5.08 cm and a grip width of 10.16 cm, testing a 10.16 cm wide by 20.32 cm long specimen. For each sample, five individual specimens in the direction of manufacture (machine direction) were tested. The graphical output was then analyzed by picking the single highest peak value of each specimen. The minimum acceptance value was 656.7 N/m.

Wide width strip tensile testing was performed following ASTM D 4595, modified for a 10.16 cm width. For each sample, five individual specimens in the direction of manufacture (machine direction) were tested. The graphical output was then analyzed by picking the peak value of each specimen, and analyzed for peak strength and the percent strain at break. The minimum values for acceptance were 1751.3 N/m peak strength, and 9.0 percent strain at failure.

The flux of a GCL material was determined following ASTM D 5887. Each test specimen was subsampled from the sample roll and shipped to an off site laboratory testing. Once in the laboratory, it was further subsampled as a 10.16 cm diameter specimen cut from the field sample. This 10.16 cm diameter specimen was then placed into a triaxial chamber and saturated. Following the test, the acquired test data was entered into a computer spreadsheet, and a permeability and an index flux were calculated. The minimum value required for acceptance was 1.0 x 10^{-7} $m^3/m^2/sec$ for index flux. This test was originally required on one out of every four samples, however due to consistently acceptable flux test results, the BOR requested a reduction in frequency, such that flux was tested only if other bentonite tests failed to reach their required values.

The bentonite component of the GCL was also required to be tested in the CQA testing program. This bentonite was split down to an approximate fifty gram sample, and pulverized completely so as to pass a number 100 standard sieve, with at least 65 percent passing a number 200 standard sieve. This ground sample was then separated into specimens for the free swell, fluid loss and moisture content tests.

The moisture content of bentonite is determined following ASTM D 2216. A portion of the sample, approximately 10 grams, was weighed and placed into an oven maintained at 105 degrees centigrade for drying overnight. After the sample was thoroughly dried, it was weighed, and the data entered into a computer spreadsheet, resulting in the moisture content calculation. The maximum value allowed was 12 percent moisture.

The free swell of the bentonite was determined following ASTM D 5890. A portion of the ground sample, approximately 10 grams, was placed into an oven maintained at 105 degrees centigrade for drying overnight. Upon removal from the oven, a 2.00 gram portion was removed for the free swell test. A one hundred milliliter graduated cylinder was filled to the ninety milliliter mark with distilled water, and the 2.00 grams of ground, oven dried bentonite was slowly placed into the cylinder in increments of 0.1 gram. These 0.1 gram increments were tapped into the cylinder over a one minute period, with at least ten minutes between each increment. Once all of the bentonite was in the cylinder, distilled water was added to raise the level to the 100 milliliter mark, and an initial temperature was measured. At least sixteen hours later, the temperature and the amount of swell were recorded. The minimum value required was 24 milliliters of swell per 2 grams of bentonite.

The fluid loss of bentonite is determined following ASTM D 5891. The test specimen is composed of 22.5 grams of the ground bentonite and 350 milliliters of distilled water, mixed for a period of twenty minutes. Following this mixing period, the initial temperature was measured, and the sample was allowed to cure for a period of at least sixteen hours. After the curing period, the temperature was again measured, and the sample was again mixed for five minutes to break the gel, after which it was placed into a filter press under 689.5 kPa of air pressure. The volume of filtrate pushed out of the filter press was collected and measured. The filter cake resulting from the test was observed for thickness and visual description. The maximum value for fluid loss was 18 milliliters.

Mass per unit area of the GNGC was determined by ASTM D 5261. For each sample, five individual specimens were subsampled, weighed, and measured. These mass measurements were then entered into a computer spreadsheet, and an average value for the mass per unit area was calculated. The minimum value for acceptance was 1200 g/m^2.

The ply adhesion of the GNGC was determined by ASTM D 413, modified for a 10.16 cm wide sample. Five specimens were tested by peeling the top geotextile off the geonet, while the other five specimens tested the adhesion of the bottom geotextile to the geonet. The graphical output was then analyzed by picking the peak value of each specimen. The minimum acceptance value was 656.4 N/m.

9 SUMMARY AND CONCLUSIONS

The final liner system for the project was constructed in a quality manner within typical industry standards. Weather related problems were the major source of concern for the quality of the final project. However, due to its location these problems needed to be worked around in order to complete the project in a timely fashion. The subgrade was a continual problem due to the nature of the subgrade materials and the contractor worked to solve these problems in an acceptable fashion. Upon discovery of unacceptable geosynthetic materials being delivered to the site, the manufacturer replaced the materials in question.

Tailings and Mine Waste'99 © *1999 Balkema, Rotterdam, ISBN 90 5809 025 6*

Industry activities for improving material and installation specifications for geomembranes

R.W.Thomas
TRI/Environmental Incorporated, Austin, Tex., USA

ABSTRACT: The last five years have seen a focused effort to improve the quality of geomembrane specifications. The activities for Manufacturing Quality Control include the withdrawal of Standard 54 by the National Sanitation Foundation, the developmental efforts of standard specifications for polyethylene geomembranes and the new standard specification for poly (vinyl chloride). Recent Manufacturing Quality Assurance activities include corporate specifications for large companies, in-plant sampling, and the accreditation of independent laboratories. Installation activities include certification of third party construction quality assurance technicians and the formation of an association of geosynthetic installers. There is also a new destructive seam testing protocol that may reduce the number of destructive seam tests, and thus the number of holes cut into geomembranes. All these efforts have a common goal to improve the quality of geomembrane installations while simplifying the specifications required to ensure high quality.

1 INTRODUCTION

Specifications are a necessary and important part of any geomembrane installation. They cover everything from the earthwork to the final inspection of the completed installation. Good specifications reduce conflicts, help the construction process proceed more smoothly, and keep the project costs under control. Poor specifications can have the opposite effect. They can cause extensive delays, cost overruns, and are often contributors to litigation. The importance of good specifications cannot be overstated.

One of the problems seen in the geosynthetics industry is that the engineers writing specifications are often not familiar with the products they are specifying. So, they have little choice but to copy a specification written in the past, or to create one to the best of their ability with their limited knowledge. Both of these scenarios can be disastrous. The problem with old specifications is that the geosynthetics industry is a growing and dynamic one. There is more learned each year about what properties are important, which tests should be used to measure the properties, and how the tests should be performed. One must be an active participant in standards organizations like the American Society for Testing and Materials (ASTM), and technical societies like the North American Geosynthetics Society (NAGS) to possibly keep up with the state of the practice. Without direct knowledge of the most recent advances in test methods, one cannot possibly write an appropriate specification dealing with geomembranes.

The purpose of this paper is to discuss the most recent activities concerning the practice of specifying geomembranes. Three specific areas will be addressed. They are Manufacturing Quality Control (MQC), Manufacturing Quality Assurance (MQA), and Installation.

2 MANUFACTURING QUALITY CONTROL

The purpose of manufacturing quality control (MQC) is for the manufacturer to make sure the product being made meets all the desired properties. A program needs to be in place to measure certain properties at certain frequencies. This information allows the manufacturer to make process changes that will result in the target properties being met more consistently. When this is accomplished, the process runs more smoothly and the cost of manufacturing is reduced. The key to a good QC program is to perform the minimum number of tests that ensure a quality product. Too few tests will result in an inconsistent product and too many tests is simply a waste of time and money.

Obviously, the manufacturers need to take the lead in the development of a standardized MQC program. Their input is critical because they know best what should be required to make a consistent product. On the other hand, there is also a tendency by manufacturers to allow certain standards to become relaxed, if they are more difficult to meet consistently. Non-conforming product costs time and money. Therefore, the best standardized specification will come from manufacturers and others working together with the goal of providing the best product to the industry.

2.1 National Sanitation Foundation Standard ANSI/NSF 54 - 1993

This standard, titled "Flexible Membrane Liners", was first published in 1983 and was most recently revised in 1993 (NSF,1993). The standard was written by a team that had representatives from across the industry including regulators, design engineers, consultants, and manufacturers. It covered fifteen different types of geomembranes and had tables of pertinent properties, minimum values for the properties, and recommended testing frequencies for quality control. It also contained an annex (Annex A) with specific information concerning how certain tests were to be performed. Most of the language in the annex made general ASTM standards more specific for geomembranes. One excellent example is the tensile properties for HDPE geomembranes. If one were to just consult the appropriate ASTM standard (D638), there would be no guidance concerning the specimen dimensions, the rate of extension, or what gage lengths to use for calculating elongation-at-yield and elongation-at-break. NSF 54 provided this information. In this case, its a Type IV dumbbell specimen, the rate is 2 inches per minute, the gage for yield is 1.3 inches and the gage for break is 2.5 inches. This kind of information was only found in NSF 54 and has become standard industry practice.

Overall, this standard was acceptable, but it never was fully embraced by the industry. Design engineers still had a tendency to create their own requirements regardless of the recommendations in the standard. In 1996, NSF announced that it would no longer revise or publish this specification. On the one hand, this will have little effect on the industry because the standard was not widely used. On the other hand, it was the most appropriate standardized specification that was available. Standardized specifications are extremely important. They provide guidance for selecting the correct tests and performing them at appropriate frequencies. If written by consensus of all the interested parties, a standardized specification gives the message that the "technical experts" have agreed that following the specification is a technically appropriate thing to do.

2.2 Geosynthetics Research Institute Standard GM-13

A new standard specification has been issued by the Geosynthetics Research Institute (GRI) at Drexel University (GRI,1998a). The title of this specification is "Test Properties, Testing Frequency and Recommended Warrant for High Density Polyethylene (HDPE) Smooth and Textured Geomembranes". It was published partly in response to the withdrawal of NSF 54 and also as a way to promote standard tests and test frequencies for HDPE geomembranes. The idea of a standard specification has wide acceptance, but there are some controversial issues associated with this particular standard. First, it has to be recognized that this document was written mainly by the manufacturers. The owner, design engineering and testing laboratory communities contributed little to the document. Therefore, there are disagreements about some of the content of the standard. It is important that these technical issues be resolved because some of them affect the long-term performance of the geomembranes. It also will help the standard become accepted if it truly represents the best efforts from all the interested parties.

2.3 PVC Geomembrane Institute Standard

The PVC Geomembrane Institute (PGI) is a trade association that promotes the proper use of PVC geomembranes in many applications, including mining. It is composed of eight member companies and is currently managed from the Department of Civil Engineering at the University of Illinois. This group has published two documents to assist in the specification process. The first is a Construction Quality Control Manual and the other is a standard specification. The manual contains a large amount of technical information about PVC and how it should be properly installed. The specification (PGI 1197) provides a list of appropriate properties, the test methods used to evaluate the properties, and recommended minimum values for the properties (PGI, 1997). The PGI has essentially made a promise to the industry that its member companies will provide products exceeding all of the minimum values listed in their standard. Although this standard was written by a group of manufacturers, it is essentially identical to the one previously published by NSF. That fact and the fact that some of the minimum requirements have been raised gives credibility to the document. However, one important omission is the appropriate test frequencies. There is no guidance on this issue in PGI 1197.

3 MANUFACTURING QUALITY ASSURANCE

The purpose of Manufacturing Quality Assurance (MQA) is to make sure that the material that arrives on a job site meets all the requirements in the specification. Some MQA plans contain more testing than the manufacturer performs and others contain less. Ideally, the MQA plan should be a subset of the MQC plan. There should be a certain level of trust that the manufacturer will do whatever he can to provide a quality product. However, since the manufacturer does not test every roll of material, it is reasonable to double check some of the more important properties. It is also valuable for an independent party to perform the tests.

3.1 Corporate Specifications with Annual Review

One of the ways large corporations are addressing the MQA issue is to publish a standard specification to be used by all its employees. This is now commonly done by the large waste management firms. These companies might have more than a dozen major projects each year in different parts of the country. Additionally, their MQA requirements can be distributed to geomembrane manufacturers, so all the interested parties know at the bid stage what is expected. And, they know that there will be consistency from job-to-job. This has worked very successfully over the past few years. Another aspect of the corporate plans is annual review by geomembrane experts. This way, the specifications can stay current with accepted industry practice. One of the difficulties encountered though, is replacing outdated test methods. Often times, outdated test methods live longer than they should because they are contained in permits issued by state regulators. The owners can either apply for a permit modification or provide results from the outdated test methods. It is often more cost effective to simply run the old test, even though the result has less meaning technically.

3.2 *In-Plant Inspections*

There is a new and improved way in which construction materials are being monitored. In the past, samples from the rolls that were delivered to the construction site were collected in the field and sent to a third party laboratory for product conformance testing. The field personnel waited for a few days until the testing was completed before they could begin the installation. If there were any difficulties with the material, further delays occurred while the problems were settled. Then, if it was decided that the material truly was non-compliant, it would have to be returned to the factory and replacement material sent from the factory. This entire process could cost thousands of dollars and days of delay.

This entire process can now be done independently in the manufacturing plant. Independent test laboratories have hired local personnel in the cities where manufacturing plants are located and trained them to monitor the production of geomembranes and collect samples from the rolls as they are manufactured. This is one of the most successful new developments in the industry. Both the buyer and the seller win. The buyer has an advocate in the plant monitoring the production event. This person can easily track resin batches, collect resin certificates, obtain MQC results, record production problems, and deliver samples to the laboratory before the material is shipped. The manufacturer wins because there is no risk that any material will be sent back from the field. The material receives final acceptance before it is shipped. Non-compliant material can simply be diverted and either downgraded in quality or sold to a buyer with less restrictive specifications.

3.3 *Laboratory Accreditation*

In 1995, the Geosynthetics Accreditation Institute (GAI) was formed at Drexel University as part of the Geosynthetics Institute (GSI). The Geosynthetics Research Institute (GRI) is also part of GSI. The purpose of GAI was to develop and implement an accreditation program for laboratories. The resulting program is known as the GAI-Laboratory Accreditation Program (GAI-LAP). The program is intended to insure that specific laboratories are capable of properly rendering the tests that they contract to perform. The focus of the program is on equipment, documentation, and testing protocols. The desired result is to give credibility to those laboratories that are properly equipped and prepared to perform specific tests. Additionally, the laboratories are required to have a quality manual, written standard operating procedures, and to prepare specific test reports for each test.

There are 26 laboratories that have received accreditation. They each are accredited to perform a different number of tests. Therefore, just because a lab is accredited does not necessarily mean that they have been accredited to perform any specific test. The only way to know which labs can perform which tests is to obtain a complete listing from GAI. Of the 26 laboratories, 14 are third party, independent contract laboratories. These are the laboratories that should be used to perform MQA testing.

4 INSTALLATION

Installation specifications are another area where industry activities are being focused. The installer community has claimed for many years that they continue to be put into positions where the plan as offered by the design engineer cannot or should not be built at the site. The problem seems to be that the specifications are more prescriptive and not based on the final performance of the installation. As with many construction projects, a geomembrane cannot always be installed exactly as the plans describe. Especially when it comes to corner areas and areas around sumps and penetrations. Another problem expressed by the installers is the lack of training and experience of the third party inspectors.

4.1 International Association of Geosynthetic Installers

The installers have responded to these issues by forming a trade association. Formed in 1995, the International Association of Geosynthetic Installers (IAGI) now has over 140 member companies located in 15 countries. The main purpose of this organization is to finds ways to get the knowledge of the installers incorporated into specifications written by engineers. One way they are doing this is to write their own product specific specifications for the installation of geomembranes. There are currently two committees working on specifications for HDPE and PVC. Besides installers, the committees are composed of manufacturers, designers, and consultants. Their desire is to publish their recommendations in a form that will be used by engineers. This will help engineers write specifications that result in projects that will run more smoothly and are more easily built.

Other issues that IAGI members are addressing include a certification program for installers and ways to reduce the number of destructive seams that are required. The certification is an issue that effects Construction Quality Control (CQC). The installers are responsible for their own QC and one way to improve in this area is to require that installers are certified by some independent body. This could be company certification or individual certification. The program is in the earliest phase of development, but there will probably be some kind of independent audit of an installers equipment, procedures, training program and personnel. This type of program will assure all the interested parties that the installation company will provide qualified installers.

4.2 NICET Certification for CQA

The National Institute for Certification in Engineering Technologies (NICET) began certifying Construction Quality Assurance (CQA) personnel in 1992. The purpose of CQA is to monitor the installation process to ensure that the installation is done according to the specification. The impetus for the certification program came from the USEPA, which concluded in the mid-1980s that the persons performing independent CQA were often unqualified. The program covers a variety of inspection activities including material acceptance, field storage, deployment, field seaming, and seam testing. It also contains specific elements covering four types of geomembranes, geotextiles, geonets, geogrids, geosynthetic clay liners, and geosynthetic appurtenances. The certification is awarded based on the number of work elements passed by written exams. As one completes more tests, they can be certified at four levels. The most qualified technicians are NICET Level IV. As of May, 1996, there were over 300 certified technicians for CQA (Klarr, 1996). The tests are offered four times a year at more than 120 cites across the country. Requiring NICET certified technicians for CQA is another way specifiers can improve the quality of their installation projects.

4.3 Destructive Seam Testing

This is one of the major issues concerning geomembrane installations. Although this discussion is appropriate for all geomembranes, it seems to be most important for HDPE. The standard practice in the industry is to cut a sample for destructive testing every 500 linear feet. Most seams are made by the automatic double track fusion welders. When the three feet length of seam is cut out, it is replaced with an extrusion seam. If one assumes six inches of overlap on all sides, the hole left after cutting out the sample is patched with a seam 12 feet in length. So, 3 feet of fusion seam are replaced with at least 12 feet of extrusion seam. This is important because extrusion seams are more difficult to prepare, are less consistent, and are more susceptible to environmental stress cracking (Thomas, 1998). There is a general agreement that less holes should be cut for destructive testing, but no one is sure what an appropriate solution would be. Several different options are being explored.

In Germany, a process has been developed that eliminates the practice of cutting holes in seams. It involves data acquisition welders and protective tape on the edges of the geomembranes. The welders track all the important welding variables like sheet temperature, ambient temperature, wedge temperature, speed, and pressure. These data are collected so it can be verified that the conditions stayed constant along a length of seam. The protective tape is pulled off right in front of the welder to protect the sheet from surface contamination. Then, destructive samples are taken at the beginning and the end of a seam length. If these pass, and the conditions were constant, it is assumed that all the seam is acceptable.

There also has been some developmental work on a non-destructive method for testing seams with infrared thermography (Peggs, 1995). This technique seems to work well for breeches in seams, but does not yet have the required sensitivity to determine if a seam might peel or not.

A new method for sampling has also been recently introduced (GRI, 1998b). Its a guide called "Selecting Variable Intervals for Taking Geomembrane Destructive Seam Samples Using the Method of Attributes". The premise of the method is the frequency of sampling depends on the recent results. If good seams are consistently made, the frequency of sampling is decreased. When failures occur, the frequency in increased. Its a great idea because it rewards success by decreasing the number of required samples. The main concerns raised about the method is that its fairly complicated, involves continual calculations, and works best on large projects. It needs to be evaluated in the field to determine its appropriateness.

5 CONCLUSION

Specifications have a large impact on the ultimate success of any construction project, including the installation of geomembranes. There is a large amount of information available, but it has often been difficult for design engineers to access it. The geosynthetics industry has recently focused its efforts on finding ways to disseminate the knowledge gained concerning specifications over the last ten to fifteen years. The results of these efforts have been standardized specifications, in-plant product acceptance, laboratory accreditation, certification of field personnel, and the formation of a trade association of installers. All these activities have made it easier for design engineers to locate the information they need to write better specifications. The ultimate result of these activities will be higher quality geosynthetics projects.

6 REFERENCES

GRI, 1998a. Test Properties, Testing Frequency and Recommended Warrant for High Density Polyethylene (HDPE) Smooth and Textured Geomembranes. GRI GM-13, Geosynthetics Research Institute, Philadelphia, PA.

GRI, 1998b. Selecting Variable Intervals for Taking Geomembrane Destructive Seam Samples Using the Method of Attributes. GRI GM-14, Geosynthetics Research Institute, Philadelphia, PA.

Klarr, R., 1996. A Brief History of NICET. *Geotechnical Fabrics Report*, September, pp.34-36.

NSF, 1993. Flexible Membrane Liners. Standard 54-1993, NSF International, Ann Arbor, MI.

Peggs, I., 1995. Assessing Geomembrane Seam Strength - Nondestructively. *Geotechnical Fabrics Report*, June/July, pp.16-20.

PGI, 1997. Specification for PVC Geomembrane., PGI-1197, PVC Geomembrane Institute, Urbana, IL.

Thomas, R.W., 1998. Evaluating the Stress Crack Resistance of HDPE Seams. *Proceedings of the Sixth International Conference on Geosynthetics*, Atlanta, GA, pp.349-352.

Tailings and Mine Waste '99 © 1999 Balkema, Rotterdam, ISBN 90 5809 025 6

Hydraulic conductivity of compacted mine rock backfill

Todd H. Hill
Golder Associates Inc., Irvine, Calif., USA
Craig H. Benson
University of Wisconsin-Madison, Wis., USA

ABSTRACT: Field hydraulic conductivity testing was conducted on compacted mine waste rock used to backfill a copper mine pit. Testing was conducted using two large-scale methods commonly employed for compacted clay that were modified to account for the higher hydraulic conductivity and larger particles in the mine rock. The tests were conducted on a test pad and on waste rock placed in the reclaimed pit. Similar hydraulic conductivities were obtained using the various test methods, although hydraulic conductivities measured with larger test methods were less variable than those measured with smaller test methods. Hydraulic conductivities in the pit were also very similar to those measured on the test pad prior to backfilling. Hydraulic conductivities were also estimated using empirical correlations based on gradation. The empirical methods over-predicted the field hydraulic conductivity of the waste rock.

1 INTRODUCTION

Mine waste and abandoned mines have come into focus in recent years as an environmental problem in connection with acid rock drainage (ARD). ARD is caused by the combination of oxygen, water, and earthen materials containing sulfide compounds. Combination of these three components induces oxidation of iron and the release of protons, resulting in a low pH condition in the pore fluid (Eriksson and Desounti 1997, Jambor 1994). The low pH results in dissolution and leaching of heavy metals, which can be toxic.

The hydraulic conductivity of the mine waste backfill governs the rate at which groundwater will re-flood a reclaimed mine pit. Understanding the time required for re-flooding is important because it defines the period during which the waste rock will be subjected to significant oxidation. The objectives of the study described in this paper were (1) to measure the field hydraulic conductivity of the mine waste rock used to backfill a surface mine located in northern Wisconsin and (2) to compare the field hydraulic conductivities to those estimated using particle size characteristics. Field tests were conducted in two phases during a 1-year period between June 1996 and April 1997. Testing was initially conducted on a test pad prior to backfilling. Confirmation tests were then conducted in the mine pit during backfilling. Testing was conducted using two-stage borehole permeameters (TSBs) and sealed double ring infiltrometers (SDRIs).

2 MATERIALS AND METHODS

2.1 *Mine Pit and Waste Rock*

The mine pit was 61-m deep, 180-m wide, and 700 m long. Waste rock used to backfill the pit was defined as rock having a metals concentration < 3%. The waste rock was further designated as Type I or Type II, corresponding to rock containing <1% sulfur or >1% sulfur, respectively. Testing showed that Type I waste rock does not generate ARD, whereas Type II waste rock could generate ARD. Type II waste rock was mixed with approximately 1 to 5 kg of crushed limestone per 1000 kg of rock prior to backfilling to maintain a pore water pH of 6.5 and prevent ARD while the mine re-floods . Limestone was not added to the Type I waste rock.

The Type II waste rock was a sericite schist or an andalusite biotite schist (hereafter referred to as sericite and andalusite, respectively). Sericite rock could be distinguished by its blue-gray color, as opposed to the reddish color of the andalusite rock. Large bulk samples (180 to 360 kg) of both rock types were collected from the test pad and analyzed to determine their particle size distribution, Atterberg limits, bulk specific gravity, and absorption. Particle size distribution curves were developed according to ASTM D 422 (mechanical sieve and hydrometer) on materials passing the U.S. No. 4 sieve. Particles retained on the U.S. No. 4 sieve were separated by their nominal dimension into the following size fractions: (1) > 279 mm; (2) 279 mm – 76 mm; (3) 76 mm – 19 mm; (4) 19 mm – 4.75 mm (U.S. No. 4 sieve); (5) < 4.75 mm.

Particle size distributions for both materials are shown in Fig. 1. Both rock materials are coarse-grained soils according to the Unified Soil Classification System (USCS) (ASTM D 2487) and are well graded. The sericite and andalusite materials had d_{50} of 50 and 13 mm and a fines content of 12% and 16%, respectively. The clay size fraction (particles < 2 μm) was < 2% for both materials. Atterberg limits testing on the P40 fraction resulted in a non-plastic designation. Both materials classify as GW-GM in the Unified Soil Classification System.

The bulk specific gravity and absorption of the fraction > 38 mm were determined according to ASTM C 127. The bulk specific gravities of the sericite and andalusite rock were 3.05 and 2.47, and the absorptions were 3.0 and 3.8%, respectively.

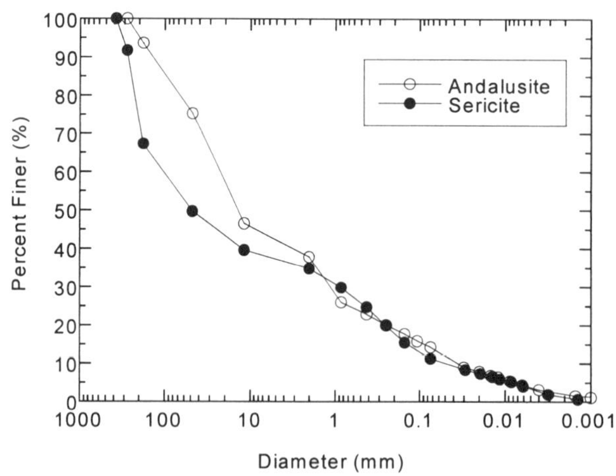

Fig. 1. Particle Size Distributions for the Waste Rock.

2.2 *Field Test Methods*

The test pad was constructed in 1996 using three methods of compaction: vibratory compaction, static padfoot compaction, and a combination of dump truck and dozer compaction. Analysis of data from the test pad showed no significant difference between the dry unit weight or hydraulic conductivity for the different compaction methods (Civil Engineering Consultants 1996). Thus, the hydraulic conductivity data from the test pad were pooled into two groups corresponding to sericite or andalusite rock.

Dump trucks and dozers were used to compact the pit backfill tested in 1997 based on the results from the test pad. Details of the field compaction procedures and the hydraulic conductivity tests can be found in Civil Engineering Consultants (1996), Benson and Hill (1997), and Hill (1998). The hydraulic conductivity of the sericite and andalusite rock exhibited similar characteristics, and thus only the andalusite is described in the remainder of this paper. However, the sericite rock was approximately an order of magnitude more permeable than the andalusite rock.

2.2.1 Sealed Double Ring Infiltrometer (SDRI) Tests

Large-scale measurements of vertical hydraulic conductivity were made with sealed double ring infiltrometers (SDRIs) using the methods described in ASTM D 5093. A schematic of a typical SDRI is shown in Fig. 2. Infiltration rate is obtained by measuring the rate of flow into the waste rock beneath the inner ring (Daniel 1989). Hydraulic conductivity was calculated from infiltration rate assuming a hydraulic gradient of 1. A unit gradient was assumed because soil suctions measured during the tests were zero, indicating that near saturated conditions existed beneath the SDRI for considerable depth (Hill 1998).

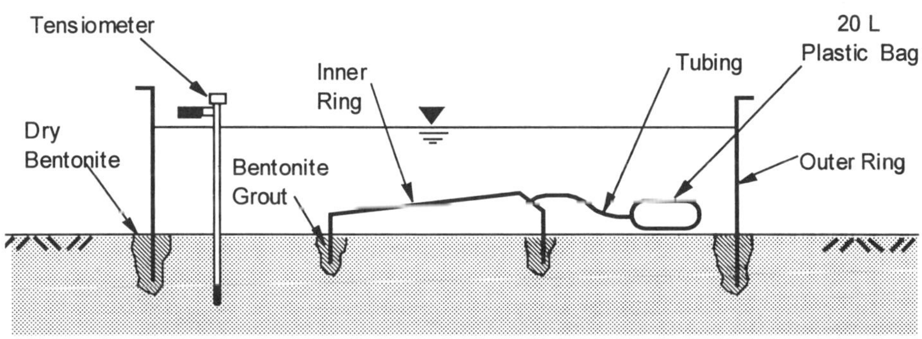

Fig. 2. Schematic of Sealed Double-Ring Infiltrometer (SDRI).

2.2.2 Two Stage Borehole (TSB) Tests

TSB tests were conducted on the test pad and in the backfilled pit in accordance with procedures described in the most recent ASTM draft standard (April 1997). The TSB test is a two-stage falling-head infiltration test conducted in a cased borehole (Fig. 3). Stage 1 is performed with the bottom of the casing flush with the base of the borehole. A standpipe is attached to the casing and then both are filled with water. After Stage 1 is finished, the borehole is extended an additional 1 to 1.5 casing diameters and then testing is repeated again. Each stage yields an apparent conductivity (K_1 and K_2, respectively), which are then converted to the horizontal and vertical hydraulic conductivities as described by Daniel (1989) and Trautwein and Boutwell (1994).

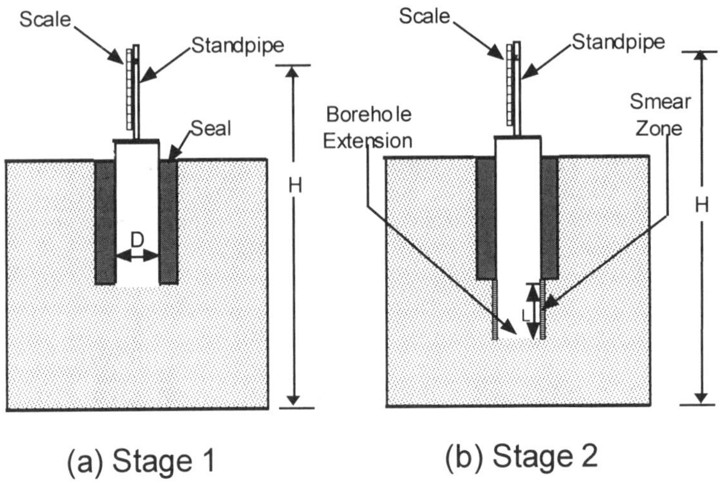

Fig. 3. Schematic of Two-Stage Borehole (TSB) Test Set-up: (a) Stage I and (b) Stage II .

The casing diameter was increased to accommodate the large particle sizes and high hydraulic conductivity typical of mine waste rock, and to assess whether the hydraulic conductivity was scale dependent. Casing diameters of 100 mm, 200 mm, and 300 mm were used. The diameter of the standpipe was also varied between 100 mm and 300 mm to increase testing times, and allow more precise head measurements to be made. The time between head measurements generally ranged from 10 seconds to 20 minutes depending on the infiltration rate and standpipe diameter.

3 RESULTS

3.1 *Field Hydraulic Conductivities*

Field hydraulic conductivities of compacted andalusite rock in the test pad and the mine pit are shown in Fig. 4 with box plots. The center of the box corresponds to the geometric mean vertical hydraulic conductivity (k_{gv}). The upper and lower boundaries of the box represent the geometric mean hydraulic conductivity ± 1-log standard deviation (1-lnσ), respectively.

Tests conducted on the test pad using 100-mm-diameter TSBs yielded a k_{gv} of 7.3×10^{-6} cm/s with a 1-lnσ range between 3.5×10^{-4} cm/s and 1.5×10^{-7} cm/s. The 200-mm-diameter TSBs installed in the test pad yielded a k_{gv} of 2.0×10^{-5} cm/s (1.5×10^{-5} cm/s to 2.6×10^{-5} cm/s), and the 300-mm-diameter TSBs a k_{gv} of 7.6×10^{-6} cm/s (5.2×10^{-5} cm/s to 1.1×10^{-6} cm/s). One SDRI was installed in the portion of the test pad constructed with andalusite rock. This SDRI yielded a vertical hydraulic conductivity (k_v) of 3.2×10^{-6} cm/s.

On average, each of the test methods yielded similar hydraulic conductivity (Fig. 4). However, greater variability was obtained with the 100-mm TSBs. In some cases a large piece of rock was directly beneath the casing, whereas in others crushed rock or pieces of rock forming large macropores existed beneath the casing. These features had less effect on the larger TSBs because they comprised a smaller portion of the area of flow beneath the casing.

Experience gained during field-testing showed that the 300-mm TSBs were also easier to install than the 100-mm or 200-mm TSBs. Less disturbance occurred during installation, and smear along the sidewalls and base was easier to remove.

Based on the results from the test pad, two TSB tests were conducted on the andalusite rock in the mine pit using only 300-mm-diameter casing. An SDRI test was also conducted for comparison. Results of these tests are also shown in Fig. 4 and are noted as "Pit". The k_{gv} of the andalusite rock placed in the mine pit was found to be 8.5×10^{-6} cm/s using the TSBs, and fell within a narrow 1-lnσ range (8.8×10^{-6} cm/s to 8.2×10^{-6} cm/s). Consequently, the box for the TSBs appears as a thick line. The SDRI yielded k_v of 5.3×10^{-6} cm/s, which is similar to the k_{gv} obtained from the TSBs.

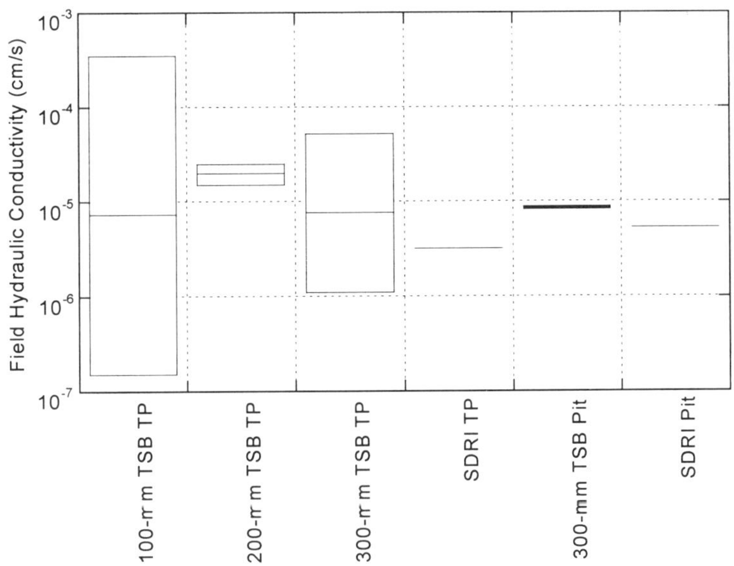

Fig. 4. Hydraulic Conductivities Measured Using TSBs and SDRIs on Test Pad and in Pit.

The similarity of the hydraulic conductivities measured on the test pad and in the pit suggested that the material placed in the pit was hydraulically comparable to that in the test pad. Because of this similarity, additional hydraulic conductivity testing was not warranted and the remainder of the pit was backfilled using unit weight measurements as the primary basis to control compaction.

3.2 *Hydraulic Conductivity Estimates Based on Empirical Equations Using Particle Size Data*

Although little experimental data exist regarding the hydraulic conductivity of compacted mine waste rock, data do exist corresponding to granular soils used for earth dams and sand filters.

Empirical equations have been developed based on the data to estimate the hydraulic conductivity (k) of granular materials. Two such equations are known as "Hazen's equation" and "Kenney's equation" (Hazen 1892, Kenney et al. 1984).

Hazen's (1892) equation is:

$$k = 1.0 D_{10}^2 \quad (1)$$

where D_{10}, the "effective size," is in mm and k is in cm/s. The empirical coefficient (1.0 in Eq. 1) accounts for the shape of the pore channels in the direction of flow and the total volume of pores, which are a function of grain shape and gradation (Terzaghi et al. 1996). The D_{10} of the andalusite rock is 0.025 mm, which yields $k = 6.3 \times 10^{-4}$ cm/s.

Kenney's equation is similar to Hazen's equation, and has the form:

$$k = \beta \gamma_w D_5^2 / \mu \quad (2)$$

where D_5 is the diameter (in mm) corresponding to 5% finer, γ_w is the unit weight of the pore fluid, μ is the dynamic viscosity (μ), and β is an empirical parameter ranging between 0.5 and 1.0. As in Hazen's equation, k is in cm/s. For an average β of 0.75 and $D_5 = 0.008$ mm for andalusite rock, $k = 4.7 \times 10^{-5}$ cm/s.

Both Hazen's and Kenney's equations over-predict the field hydraulic conductivity (k_f) of the compacted mine waste rock. Hazen's equation over-predicts k_f by about two orders of magnitude, whereas Kenney's equation over-predicts k_f by about one order of magnitude. These over-predictions should be expected, because both Hazen's and Kenney's equations were developed for clean sands and the andalusite mine rock contains 16% fines. The prediction using Kenney's equation is better because it was developed for compacted sands used in dams, whereas Hazen's equation was developed for sand filters which typically are looser and thus have higher hydraulic conductivity.

Kenney's equation could be adapted for use with the andalusite waste rock by changing the coefficient β to 0.1. However, additional particle size analysis and hydraulic conductivity testing would be necessary to verify that the equation could be used with confidence.

4 SUMMARY AND CONCLUSIONS

Tests were conducted on compacted waste mine rock to determine its field hydraulic conductivity. The waste rock was being used to backfill a mine pit during reclamation. The tests were conducted on a test pad prior to reclamation and during backfilling for confirmation. Sealed double-ring infiltrometers (SDRIs) and two-stage borehole permeameters (TSBs) having casings with diameters of 100, 200, or 300 mm were used to measure the hydraulic conductivity.

Comparison of the hydraulic conductivity measurements conducted on the test pad showed that all of the test methods yielded similar hydraulic conductivity, on average. Thus, the hydraulic conductivity of the compacted rock was not scale-dependent. However, the hydraulic conductivities measured with the small 100-mm-diameter TSBs exhibited substantial variability because the base of the casing permeated a relatively small cross-sectional area. As a result, flow from the casing was often dominated by local heterogeneities (e.g., a large piece of rock, fragments of crushed rock, or large macropores between smaller pieces of rock).

To minimize the effects of local heterogeneities, the large 300-mm-diameter TSBs and an SDRI were used to measure the hydraulic conductivity of the backfill in the pit. The large TSBs were also easier to install with limited disturbance. Both methods yielded similar hydraulic conductivities. The hydraulic conductivities measured in the pit also were similar to those measured on the test pad, indicating that the backfill was being placed as intended.

Hydraulic conductivity of the compacted rock was also estimated using Hazen's equation and Kenney's equation. Both are empirical equations based on particle size characteristics. Hydraulic conductivities estimated with both equations were substantially higher than the actual hydraulic conductivity of the compacted rock primarily because the rock contained significant fines.

5 ACKNOWLEDGEMENT

The Flambeau Mining Company (FMC) of Ladysmith, Wisconsin provided financial support for this study. However, the opinions and conclusions expressed in this paper are those of the authors, and are not necessarily consistent with the opinions or policies of FMC.

6 REFERENCES

Benson, C. and Hill, T., (1997), "Results of Field Hydraulic Conductivity Tests Conducted on Mine Backfill: Flambeau Mine," Environmental Geotechnics Report 97-4, Department of Civil and Environmental Engineering, University of Wisconsin-Madison, Madison, Wisconsin.

Civil Engineering Consultants (1996), "Results of Field Hydraulic Conductivity Tests Conducted on a Compacted Mine Rock Test Pad," Verona, Wisconsin.

Daniel, D., (1989), "In Situ Hydraulic Conductivity Tests for Compacted Clay," *Journal of Geotechnical Engineering*, ASCE, Vol. 115, No. 9, 1205-1226.

Eriksson, N. and Desounti, G., (1997), "Combined Effects of Dissolution Kinetics, Secondary Mineral Precipitation, and Preferential Flow on Copper Leaching from Mining Waste Rock," *Water Resources Research*, American Geophysical Union, Vol. 33, No. 3, 471-483.

Hazen, A., (1892), "Physical Properties of Sands and Gravels with Reference to Use in Filtration," *Report to Mass. State Board of Health*, 539.

Hill, T., (1998), "Field and Laboratory Measurements of the Hydraulic Conductivity of Compacted Mine Waste Rock," Master of Science Thesis, University of Wisconsin-Madison, Madison, Wisconsin.

Jambor, J., (1994), "Mineralogy of Sulfide-rich Tailings and Their Oxidation Products," *Short Course Handbook on Environmental Geochemistry of Sulfide Mine-Wastes*, Waterloo, Ontario, Canada, 59-102.

Kenney, T., Lau, D., and Ofoegbu, G., (1984), "Permeability of Compacted Granular Materials," *Canadian Geotechnical Journal*, Vol. 21, 726-729.

Terzaghi, K., Peck, R., and Mesri, G., (1996), *Soil Mech. in Engineering Practice*, Wiley Interscience Publishers, Third Edition.

Tailings and Mine Waste '99 © 1999 Balkema, Rotterdam, ISBN 90 5809 025 6

Solute transport in unsaturated soil beneath two abandoned evaporation ponds, Gas Hills, Wyoming: 1. Characterization of soils

Jorge C.Troncoso & Charles D.Shackelford
Department of Civil Engineering, Colorado State University, Fort Collins, Colo., USA

ABSTRACT: Laboratory and field tests were performed to determine the properties of the soils necessary to perform radionuclide transport modeling for two evaporation ponds at UMETCO's Gas Hills uranium mine in Wyoming. Soil properties were determined for a compacted clay liner and underlying waste rock, including saturated hydraulic conductivities, soil-water retention curves, exchangeable cations, and an adsorption isotherm for uranium. The results of these tests will be used subsequently to model the unsaturated migration of uranium from the evaporation ponds.

1 INTRODUCTION

This paper details the characterization of the soil properties necessary to perform unsaturated contaminant transport simulations at two evaporation ponds at UMETCO's Gas Hills uranium mine in Wyoming shown in Fig. 1. During the operating life of the mine, the ponds were used for storage and evaporation of mill waste solutions and tailings liquor. The ponds were built over existing waste rock piles. Containment of the solutions in the ponds was provided by a 91.4-cm (3-ft)-thick compacted clay liner. A typical schematic profile of the ponds is shown in Fig. 2.

The soil properties determined in this study are summarized in Table 1. Testing was performed on samples of: (1) the borrow source (BS) used for the clay liner, (2) the clay liner (CL) recovered from excavated pits and boreholes in the existing liner, and (3) the waste rock (WR) recovered from various locations in the waste rock piles, some of which were subjected to infiltration from the ponds (contaminated waste rock).

2 PHYSICAL PROPERTIES

2.1 *Index Properties*

The specific gravities of the waste rock and clay liner were measured according to ASTM D 854 as 2.63 and 2.66, respectively. Grain-size analyses and plasticity tests were performed on the clay liner material and waste rock. The grain-size analyses and plasticity tests were performed according to ASTM D 422 and ASTM D 4318, respectively. The results, presented in Figs. 3 and 4 indicate that the clay liner classifies as a low plasticity clay (CL) according to the Unified Soils Classification

Table 1 Property category and purpose

Type of property	Soil[1]	Location of measurement	Classification of soils
Index	BS	Lab	General characterization of soil
	CL	Lab	
	WR	Lab	
Density	BS	Lab	Density and porosity characterization
	CL	Lab & Field	
	WR	Lab & Field	
Chemical	BS	Lab	Compare soils
	CL	Lab	
	WR	Lab	
Saturated hydraulic conductivity	BS	Lab	Establish upper bound on hydraulic conductivity
	CL	Field	
	WR	Lab & Field	
Soil-water retention characteristics	BS	NA	Characterize unsaturated flow
	CL	Lab	
	WR	Lab	
Adsorption	BS	Lab	Characterize uranium adsorption
	CL	NA	
	WR	NA	

(1) BS = borrow source; CL = clay liner; WR = waste rock

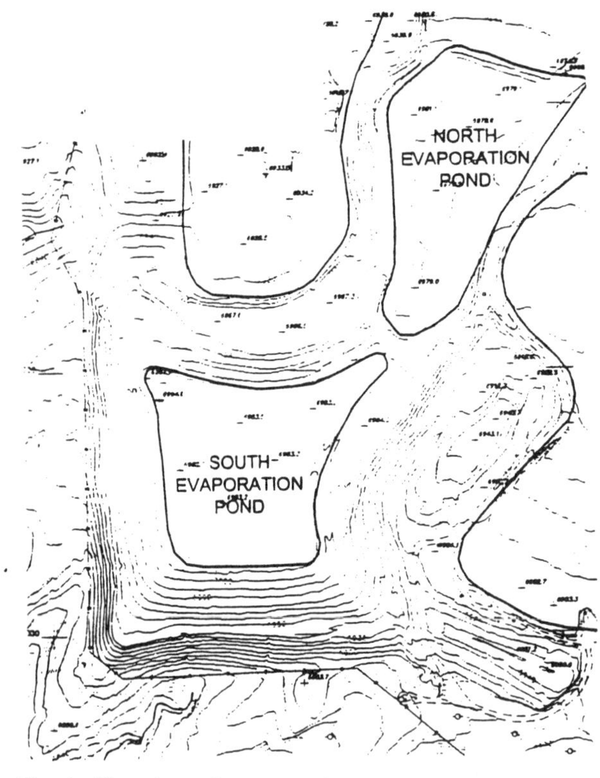

Fig. 1 Plan view of evaporation ponds

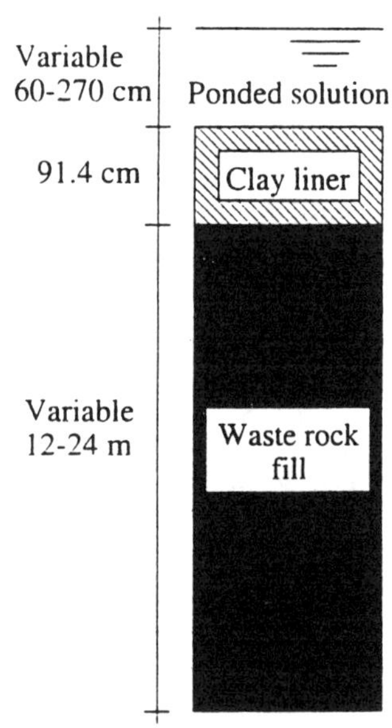

Fig. 2 Schematic cross-section of soil profile

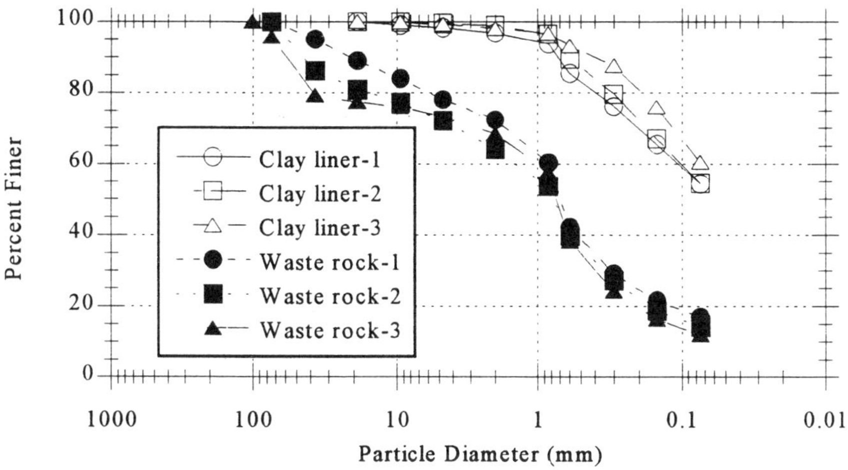

Fig. 3 Particle-size distributions for clay liner and waste rock.

System (USCS), whereas the waste rock classifies as a silty sand (SM). The plasticity results indicate an increase in the plasticity of the soils that were exposed to contamination (clay liner versus borrow source, and contaminated waste rock versus uncontaminated waste rock). Lambe and Whitman (1969) indicate that the plasticity of medium to high activity clay minerals, such as illite and montmorillonite, increases when the dominant exchange cation is monovalent (Na^+, K^+) versus di- or tri-valent (Ca^{2+}, Fe^{3+}). This observation is consistent with the decrease in exchangeable calcium of the clay liner relative to the borrow source (see Table 3).

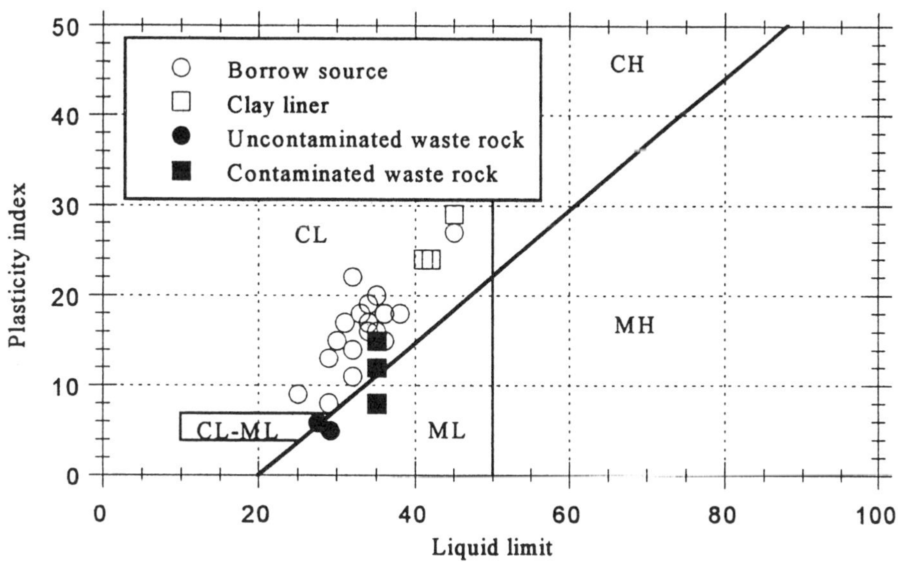

Fig. 4 Classification of borrow source material and clay liner, and fine fraction of waste rock.

Table 2 Density and water content data.

Method	Parameter	Soil(3)	
		CL	WR
Nuclear densometer(1)	ρ_d (kg/m^3)	1630	NM
	w (%)	17.9	NM
	θ_i	0.29	NA
Shelby tube(2)	ρ_d (kg/m^3)	1720	1628
	w (%)	16.9	10
	θ_i	NA	0.16

NM: not measured; NA: not applicable

(1) ρ_d, w_i, θ_i = average dry density, gravimetric water content, and volumetric water content, respectively, based on 7 measurements.

(2) ρ_d, w_i, θ_i = average dry density, gravimetric water content, and volumetric water content, respectively, based on 14 measurements for the clay liner and 45 measurements for the waste rock.

(3) CL = clay liner; WR = waste rock

2.2 *Densities*

"In-situ" densities and gravimetric water contents were determined for the clay liner and waste rock. The average densities and water contents are indicated in Table 2. For the Shelby tube results, dry densities were obtained from the wet densities calculated as the total weight of soil in the known volume of the Shelby tube. For the clay liner, the dry densities obtained from the Shelby tubes are slightly higher than those obtained from the nuclear densometer. This difference in density is consistent with compression and densification of the soil resulting from side friction between the soil and the Shelby tube as it is pushed into the ground.

Compaction tests also were performed on the different materials. Modified Proctor tests (ASTM D 1557) resulted in an average ρ_{dmax}=2040 kg/m^3 and w_{opt}=8.8 % for the clay borrow source, and ρ_{dmax}=1990 kg/m^3 and w_{opt}=8.5 % for the waste rock material. Standard Proctor tests (ASTM D 698) resulted in an average ρ_{dmax}=1750 kg/m^3 and w_{opt}=16.2 %, and ρ_{dmax}=1860 kg/m^3 and w_{opt}=12.3 % for the existing clay liner and waste rock materials.

3 CHEMICAL PROPERTIES

The results of the chemical analyses performed on the borrow source material and the clay liner are presented in Table 3. The results show that the measured cation exchange capacity (CEC) is similar for the clay material obtained from the borrow source and the existing clay liner material. This indicates that, in principal, the soil sampled as the borrow source clay has similar adsorption properties as the soil that was originally placed as liner. Therefore, the adsorption isotherm for the liner before it was subjected to infiltration from the evaporation ponds can be calculated based on tests performed on the source clay material.

Table 3 Chemical properties of soils.

Property	Borrow source[1]		Clay liner[2]	
	Avg.	Range	Avg.	Std. Dev.
Cation exchange capacity, CEC (meq/100 g)	28.1	23.9-31.0	25.3	12.27
Exchangeable cations (meq/100 g):				
Calcium	4.3	1.4-7.5	1.6	0.79
Magnesium	0.5	0.2-0.7	1.1	0.63
Potassium	0.1	0-0.1	0.0	0.03
Sodium	0.1	0-0.2	0.2	0.17
Sum:	5.0		2.9	
Soluble salts (meq/L):				
Boron	0.0	0.0-0.1	0.0	0.0
Calcium	19.7	10.9-28.1	33.1	5.57
Magnesium	3.7	2.4-5.4	52.4	21.23
Potassium	0.0	0.0	0.0	0.0
Sodium	1.4	1.2-1.8	7.0	4.17
Chloride	1.1	1.0-1.3	5.0	3.78
Nitrate (as N)	0.0	0.0-0.0	0.0	0.0
Sulfate	120.7	67.2-169.0	300.3	143.09
Specific conductance @ 25 C (μS/cm)	1773	1220-2310	2489	678
pH	6.9	5.4-7.7	3.9	0.72

(1)Average based on 3 measurements.
(2)Average based on 10 measurements.

The results also show that the sums of the common exchangeable cations are much less than the cation exchange capacity in the source and existing clay liner. These differences imply that the source clay also has been subjected to replacement of a significant portion of the exchangeable cations commonly present in clays. The source clay, which was taken from strata very near or in the mineralized area of the mine pit, probably was subjected to natural leaching of solutions containing dissolved cations of higher replacing power than the typical exchangeable cations found in soils (Mitchell, 1993).

The soluble salts in the existing clay liner are higher than the borrow source material as expected since the liner has been subjected to infiltration of concentrated salt solutions (see Table 4). The increase in specific conductance also reflects this infiltration.

The decrease in pH also is indicative of some infiltration from the highly acidic waters contained in the evaporation ponds. The pH values of 6.2 and 4.3 were measured in uncontaminated and contaminated waste rock, respectively, immediately below the clay liner. This decrease in pH reflects the occurrence of some infiltration and percolation through the clay liner and into the underlying waste rock.

Table 4 Chemical characteristics of the solutions in the evaporation ponds[(1)](UMETCO 1998).

Property	North Pond	South Pond
Metals (mg/L):		
Calcium	632	1126
Iron	20190	7851
Potassium	0	86
Magnesium	9898	3562
Sodium	2353	5127
Non-metals (mg/L):		
Chloride	469	6950
Sulfate	122886	63289
Uranium (pCi/L)	205459	112574
pH	2	1.9

[(1)]Average values measured in the ponds from 12/8/87 to 6/24/93.

4 SATURATED HYDRAULIC CONDUCTIVITY

Double-ring infiltration tests performed (ASTM D 3385) were performed to determine the saturated hydraulic conductivities of the existing clay liner. The results are presented in Fig. 5. The saturated hydraulic conductivities were obtained by applying the Green-Ampt model for infiltration (Green and Ampt 1911) to the data obtained from the field infiltration tests as follows:

$$t = \frac{(\theta_s - \theta_i)}{k}\left\{L - (d + \psi_e)\ln\left(1 + \frac{L}{d + \psi_e}\right)\right\} \tag{1}$$

where θ_s = saturated volumetric water content or porosity, θ_i = initial volumetric water content, k = saturated hydraulic conductivity, t = time for the wetting front to arrive to a depth L, d = depth of ponded water, and ψ_e = effective suction at the wetting front.

The effective suction at the wetting front was calculated as one-half of the air entry suction of the soil obtained from the soil-water retention curves that are presented in the following section in accordance with Bouwer (1966). The initial volumetric water content and the saturated volumetric water content (porosity) were obtained from the "in-situ" density and moisture content measurements (Section 2). Values of ψ_e=85 cm, θ_s=0.38 and θ_i=0.3 were used in the calculations.

The average hydraulic conductivity, k, was calculated as the mean of the individually calculated k values from each data point in each test. The results from the infiltration tests can be compared with results of laboratory constant-head tests performed by UMETCO on compacted source clay liner before the construction of the ponds. These tests resulted in an average saturated hydraulic conductivity of 9 x 10^{-8} cm/s with a maximum of 3.9 x 10^{-6} cm/s and a minimum of 7.4 x 10^{-8} cm/s. Thus, the hydraulic conductivity values measured in the field are higher than those measured in the laboratory. This is consistent with a scale effect between laboratory- and field-measured hydraulic conductivities of compacted clay liners (Daniel 1984, Benson et. al. 1994). However, the observed difference in k values also may be attributed, in part, to desiccation of the clay liner.

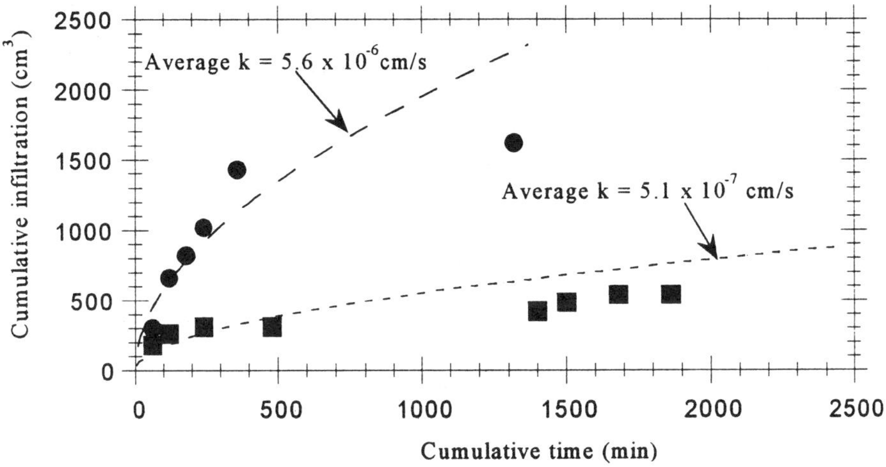

Fig. 5 Infiltration versus time data for field double-ring infiltrometer tests.

Borehole packer tests were performed by UMETCO on the "in-situ" waste rock before the evaporation ponds were built. These results indicated an average hydraulic conductivity of 1.72×10^{-4} cm/s (UMETCO 1998).

5 SOIL-WATER CHARACTERISTICS

Soil water-retention curves were measured for the clay liner and the waste rock based on the results of tests performed according to ASTM D 2325. The discrete data obtained in the tests was regressed with the continuous function developed by van Genuchten (1980) as follows:

$$\theta = \theta_r + \frac{\theta_s - \theta_r}{[1+(\alpha h)^n]^m} \tag{2}$$

where

$$m = 1 - 1/n \tag{3}$$

and h is the matric suction, θ_r is the residual volumetric water content, and α and n are the fitting parameters. The results of the tests and curves fitted to the test data are shown in Fig. 6. The fitting parameters obtained are summarized in Table 5.

The data obtained from the clay liner samples that had a similar porosity (CL-2 and CL-3) produced similar fitted curves. The tests performed on sample CL-1, with a higher dry density and lower porosity produced different results, indicating the need to adequately reproduce in the laboratory the conditions in the field. As indicated in Section 2, the "in-situ" dry density of the clay liner was on the order of 1630 kg/m^3, therefore, the soil-water retention curve obtained from CL-1 was considered representative of the conditions encountered in the field. The air entry suction of 170 cm of water obtained from this curve was used to calculate the saturated hydraulic conductivity from the infiltration data.

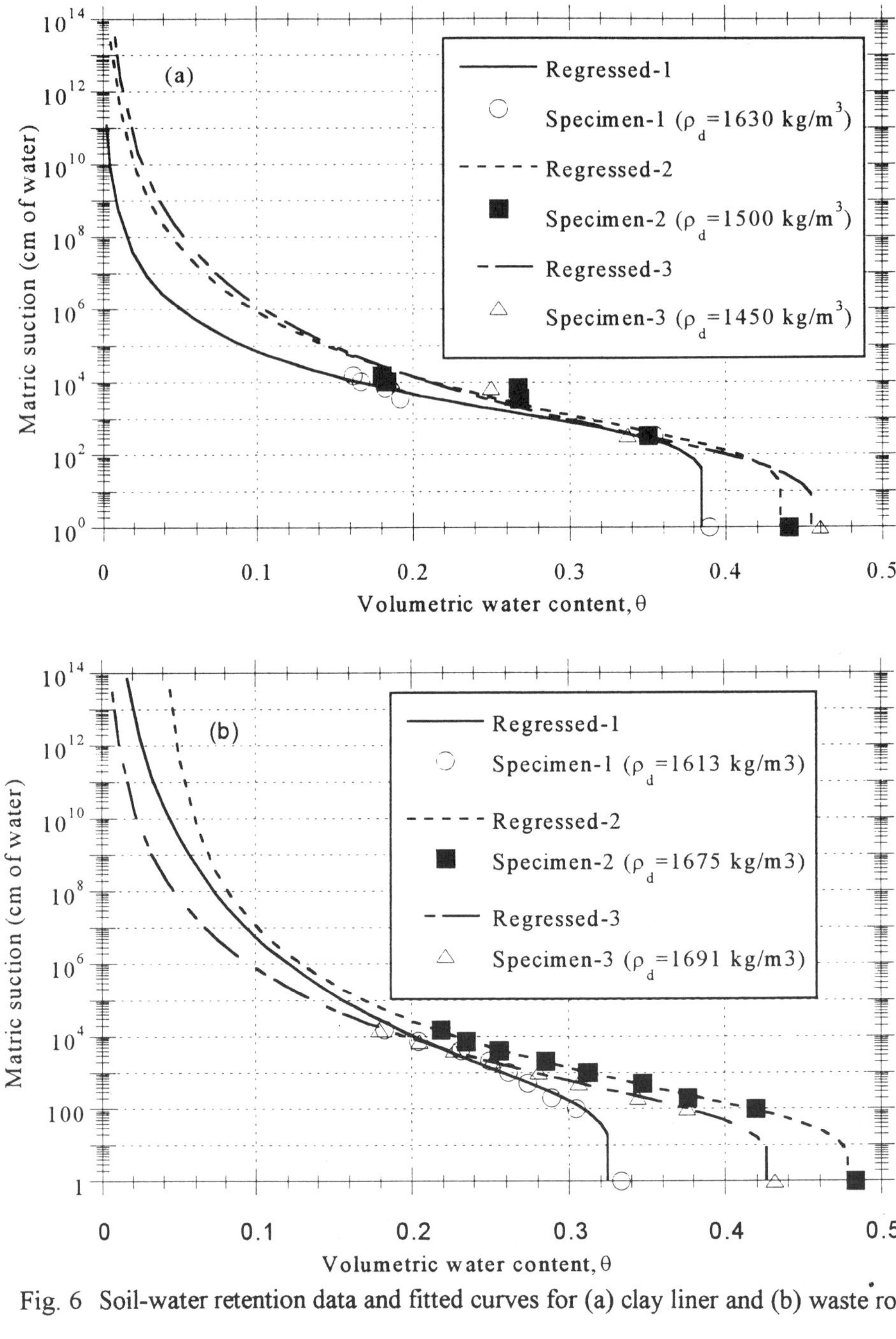

Fig. 6 Soil-water retention data and fitted curves for (a) clay liner and (b) waste rock.

6 ADSORPTION

Twenty-four hour batch equilibrium tests (Griffin et. al., 1986) were performed to determine the adsorption isotherm for uranium of the original clay liner. Five batch tests were performed with contact solutions containing varying concentrations of

Table 5 –Summary of fitted values for soil-water characteristics.

Sample	Saturated volumetric water content, θ_s	Residual volumetric water content, θ_r	Fitting parameters [1] α (1/cm)	n
CL-1	0.39	0	0.003	1.26
CL-2	0.44	0	0.007	1.17
CL-3	0.46	0	0.017	1.15
WR-1	0.33	0	0.008	1.11
WR-2	0.48	0.04	0.020	1.16
WR-3	0.43	0	0.016	1.16

[1]Based on van Genuchten (1991)

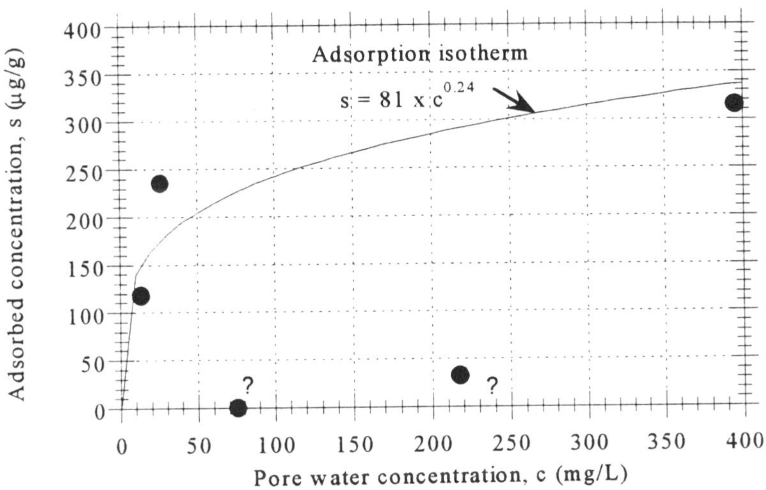

Fig. 7 24-hr batch equilibrium tests results (ASTM ES 10-85)

uranium so as to determine the distribution coefficients for the range of concentrations expected in the field. Field geochemical conditions were simulated by using source clay material and a contact solution made with the most significant constituents at concentrations similar to those observed in the ponded waters. The five contact solutions contained 0.9 mg/L of calcium, 2.5 mg/L of iron, 1.9 mg/L of magnesium, 13.7 mg/L of potassium, 10.3 mg/L of sodium, 28.2 mg/L of chloride, 8.3 mg/L of nitrate, 10 mg/L of sulfate, 104 pCi/L of radium-226, 1254 pCi/L of thorium-230, 51.7 pCi/L of lead, and activities of uranium varying from 280,000 pCi/L to 14,000 pCi/L. The soil-solution ratio (w/w) was maintained at 1:20. The pH was maintained at approximately 1.5 corresponding to the lower limit of the pH measured in the ponds. The results are presented in Fig. 7. The Freundlich type adsorption isotherm regressed to the experimental data also is shown in Fig. 7.

The secant distribution coefficients (K_d) calculated from tests performed with pore water concentrations of 13.1 mg/L and 26.2 mg/L were both 9.01 mL/g indicating linearity over this lower range of pore water concentrations. Thus, the test results show a linear trend for low pore-water concentrations (c < 26.2 mg/L) of uranium. However, at the pore-water concentration of 395.2 mg/L, the secant K_d is approximately 0.8 mL/g, indicating strong non-linearity in the adsorption of uranium at higher concentrations.

7 CONCLUSIONS

Laboratory and field tests were performed to determine the properties of soils for unsaturated radionuclide transport modeling. The conclusions are as follows.

- The hydraulic conductivity values of the clay liner based on short-term, double-ring infiltration tests performed in the field are approximately one to two orders of magnitude higher than those previously reported on the basis of laboratory measurements. This difference in hydraulic conductivity is consistent with a scale effect and may be attributed, in part, to desiccation of the clay liner.
- The soil-water retention curves varied depending on the initial density of the soil specimen. Thus, careful control on the specimen density is required when measuring soil-water retention curves.
- Twenty-four hour batch equilibrium tests were performed to estimate the adsorption of uranium to the clay liner. The adsorption was linear at relatively low pore-water concentrations ($c < 26.2$ mg/L) but became highly non-linear at higher pore-water concentrations.
- The clay liner and underlying waste rock has been subjected to infiltration from the ponded waters as is evidenced by the decrease in pH, increase in soluble salts and increase in plasticity of the fine fraction of both soils.

8 ACKNOWLEDGEMENTS

Financial support for this study is being provided by both UMETCO and the National Science Foundation under the grant MSS-9257305. In particular, the support of Curt Sealy of UMETCO as well as the field assistance provided by UMETCO personnel at their Gas Hills site in Wyoming is appreciated. Finally, the cooperation of Shepherd Miller, Inc., is appreciated.

REFERENCES

Benson et. al. 1994. Representative Specimen Size for Hydraulic Conductivity Assessment of Compacted Soil Liners. *Hydraulic Conductivity and Waste Contaminant Transport in Soil,* ASTM STP 1142, D.E. Daniel and S.J. Trautwein, Eds., ASTM 3-29.

Bouwer, H 1996. Rapid Field Measurement of Air Entry Value and Hydraulic Conductivity of Soil as Significant Parameters in Flow System Analysis. *Water Resources Research*. 2:729-738.

Daniel, D.E. 1984. Predicting Hydraulic Conductivity of Clay Liners. *J. of Geotechnical Engineering, ASCE.* 110:285-300.

Green and Ampt 1911. Studies on Soil Physics. Part I: the flow of air and water through soils. *J. Agr. Res.* 4:1-24.

Griffin et. al. 1986. Batch Type 24-h Distribution Ratio for Contaminant Adsorption by Soil Materials. *American Society for Testing and Materials.* 6:390-408.
Mitchell, J.K. 1993. *Fundamentals of Soil Behavior*. Second Edition, Wiley.

Umetco 1998. *Site Characterization Report East Gas Hills Reclamation Project.* Volume 1.

van Genuchten, M.Th. 1980. A Closed-Form Equation for Predicting the Hydraulic Conductivity of Unsaturated Soils. *Soils Sci. Soc. Am. J.*, 44:892-898.

van Genuchten et. al. 1991. *The RETC Code for Quantifying the Hydraulic Functions of Unsaturated Soils.* EPA/600/2-91/065.

Tailings and Mine Waste '99 © *1999 Balkema, Rotterdam, ISBN 90 5809 025 6*

Soil covers for sloped surfaces of mine waste rock and tailings

B. Boldt-Leppin
Saskatoon, Sask., Canada

M. O'Kane
O'Kane Consultants Incorporated, Saskatoon, Sask., Canada

M. D. Haug
M.D.Haug and Associates Limited, Saskatoon, Sask., Canada

ABSTRACT: A literature search of soil cover design principles and applications demonstrated that the design and monitoring of these cover systems was completed with respect to the horizontal or top of the waste rock piles. The influence of the sloped surfaces on hydraulic performance of the cover system was either ignored or the slopes were covered with impermeable geo-membranes in order to avoid infiltration or exfiltration across the protective soil cover. This paper summarizes the literature search and presents key design principles. The analogy to hillslope hydrology is proven. Case studies are presented to demonstrate current state-of-the-art technology.

1 INTRODUCTION

Considerable fundamental and applied research has been devoted to investigate the performance of soil cover systems for mine waste rock and tailings. Soil cover systems are geotechnically designed and their, at site, water balance is numerically modelled to predict long term performance (Filipek *et al.,* 1996). Soil cover systems are used to prevent or limit leachate or gas production in waste containment systems. A review of existing literature indicated that the majority of soil cover designs dealt with vertical flow of heat and moisture in horizontal layers, using numerical models as well as laboratory and/or field scale physical models. Field performance monitoring typically focused on one-dimensional performance. In reality, mine waste rock dumps have steep slopes as a result of construction methods and placement configurations. A typical configuration for closure of a waste rock pile is shown in Figure 1. The hydraulic performance of a soil cover system placed on these slopes and its ability to function as an oxygen ingress and water infiltration control system will be different than that predicted by idealized one-dimensional numerical models.

The application of one-dimensional modelling software does not take into consideration the two-dimensional nature of a waste rock heap. The hydraulic behavior of real covers, hence, would be idealized by one-dimensional models. The objective of this literature research was to determine the influence of the slope of a waste rock heap on the hydraulic behavior of a soil cover system, and to evaluate the design, construction and monitoring criteria for the slopes. The literature research formed the basis for a two-dimensional field performance study that included two-dimensional saturated-unsaturated numerical modelling followed by construction and monitoring of a field test plot on a sloped waste rock surface.

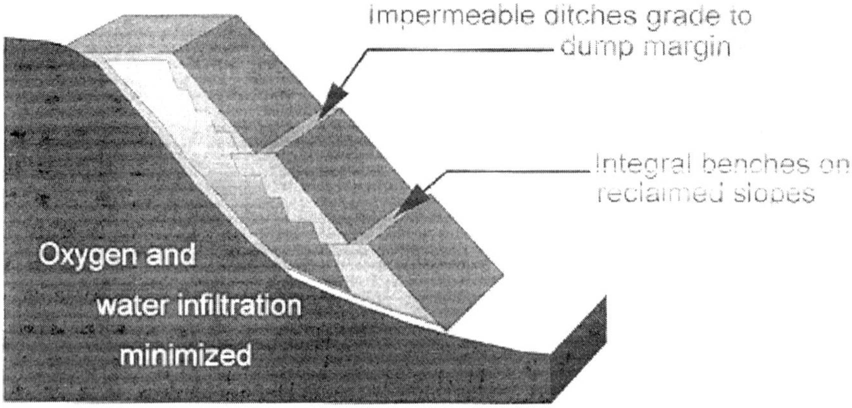

Figure 1. Waste rock pile build in lifts at hill slope (after Filipek *et al.* 1996).

2 ANALOGY TO NATURE

The topographically controlled groundwater flow system developed in fine grained soils in waste heaps is analogue to natural groundwater flow on hillslopes. The distribution of the hydraulic heads is controlled by the topography (Freeze and Cherry, 1979), raising the water table to conform with the topography. The flow system is mainly influenced by low hydraulic gradients controlled by the slope, low hydraulic conductivities, high capillary forces due to the fine grained soils, and spatial homogeneities.

2.1 *Hillslope hydrology*

Weeks *et al.* (1992) identified six components in hillslope hydrology, with flow occurring either singly or in combination: infiltration-excess overland flow, saturation overland flow, return flow, saturated subsurface flow, unsaturated subsurface flow, and groundwater flow. The fundamental hydrological principles studied at natural hillslopes (catchments) are valid at any slope exposed to meteorological conditions. The components of the hillslope hydrological cycle are shown in Figure 2.

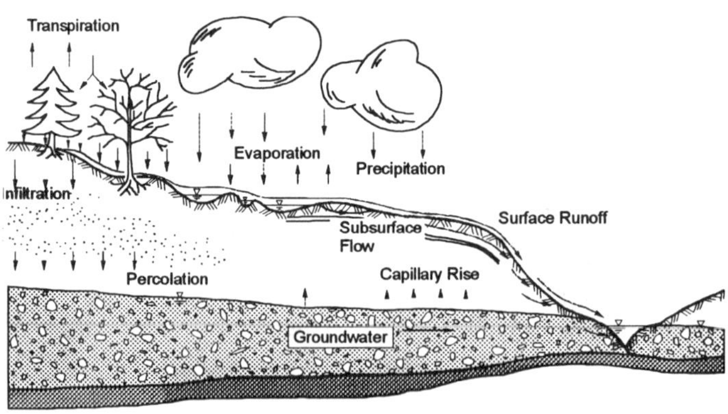

Figure 2. Components of the hillslope hydrological cycle (redrawn after Bronstert, 1994).

2.2 *3-D modelling of hillslope hydrology*

Physically based models to simulate catchment hydrology are computationally very demanding. The high computational effort for two- and three-dimensional numerical solutions their application to non-academic, practical problems. Most 2-D and 3-D numerical models incorporate simplifications to the simulations in order to reduce computational time. Bárdossy *et al.* (1995) suggested a simple and efficient method to solve Darcy's law for unsaturated flow by employing a fuzzy rule system. The method approximates the solution of the partial differential equation with the help of fuzzy rules. The error of replacing the exact numerical solution of the partial differential equation with a fuzzy solution is small in comparison with other uncertainties such as estimation of material properties, and the solution algorithm is orders of magnitude faster.

The method using fuzzy rules has been implemented in the distributed and physically based hydrological modelling software HILLFLOW (Bronstert and Plate, 1997). HILLFLOW simulates the relevant hydrological processes of interception, evapotranspiration, infiltration into the soil matrix and possibly macro pores, moisture movement in the soil matrix, surface runoff, and subsurface storm flow. Three versions of HILLFLOW were developed. The one-dimensional version is designed to simulate vertical, column type soil water balances. In the two-dimensional version lateral flow is added, enabling the simulation of situations typical in hillslope hydrology, such as subsurface storm flow, lateral matrix water movement due to soil layering and anisotropy, surface runoff from the slope, or lateral saturation excess at the hillslope bottom. The three-dimensional version expands to the second direction for lateral flow, enabling the simulation of hydrological processes in an entire catchment with landscapes of arbitrary shape.

3 STATE-OF-THE-ART SOIL COVER DESIGNS

The knowledge derived from the study of water flow in natural systems can be directly applied to the design of waste deposits and their soil cover systems. The main hydrological systems inside the refuse pile are surface flow, percolation in the waste material, flow across the cover system waste material interface, and groundwater flow. Precipitation is the major input flux to the water balance of a cover system and is balanced by percolation and evapotranspiration. The lateral inflow and outflow into the individual layers and on the surface should be taken into account when the cover system is on a sloped surface.

The engineered design for surface cover systems in principle consist of layers of high and low permeability materials (Günther, 1988). The sequence of the layers determines the hydrological behavior of the cover systems and separates the systems into saturated or unsaturated systems. The design principle for saturated systems is described by Bonaparte and Giroud (1996). The US Environmental Protection Agency (US-EPA, 1989) recommends a multi-layer design with a uniform slope of at least 3% but not more than 5%. The design principles for an unsaturated system on a slope is described by Ross (1990), such that the sequence of soil layers (fine grained material over lying coarse material) utilizes the capillary barrier concept.

3.1 *Design principles of capillary barrier system*

The physical principles of capillary barriers used in the design of soil covers, was investigated by Frind *et al.* (1976), Sailie (1985), Rasmuson and Eriksson (1986), Nicholson *et al.* (1989,1991), and Ross (1990). The capillary barrier concept is based on the principle of unsaturated flow between soils of different textures. In an initially dry two layered system, with fine textured soil overlying coarse textured soil, surface infiltrating water will not move into the coarse textured soil, until the fine textured soil is at or near saturation (the pressure head is zero). Capillary flow to the coarse layer can only occur when the pore size just filled in the fine material matches the smallest pore size in the coarse material. If during an infiltration event the fine textured material can drain and maintain negative pressure at the interface with the coarse material, the coarse material would remain dry. This requirement can be fulfilled by constructing a slope, sufficient to establish a lateral gradient, at the interface of the two materials.

Ross (1990) calculates the relative permeability of a two layered, dipping isotropic system under steady state flow conditions. Infiltrating water, not able to enter the lower coarse soil must flow laterally above the contact. Near the up-dip boundary the lateral flow will be small, and the soil will remain relatively dry. Farther down-dip the lateral flow increases. To accommodate the increased flow, the hydraulic conductivity must increase, the pressure potential becomes less negative, and the degree of saturation increases. At points far enough down-dip, the downward flow might reach the rate of infiltration into the coarse material. At this point, reaching the diversion capacity of the system, the capillary barrier will be ineffective. The diversion capacity and maximum effective width of the fine grained soil layer is dependent on the dip angle and saturated hydraulic conductivity of the fine grained material. An anisotropic fine grained layer with a greater hydraulic conductivity parallel to the contact will extend the performance of a capillary barrier (Stormont, 1995).

3.2 *Hydrologic modelling of protective barriers*

The suitability of cover designs should be either numerically evaluated before construction of the cover, or *in situ* through test designs and field performance monitoring of the test plots. Field performance monitoring must account for two-dimensional hydraulic performance.

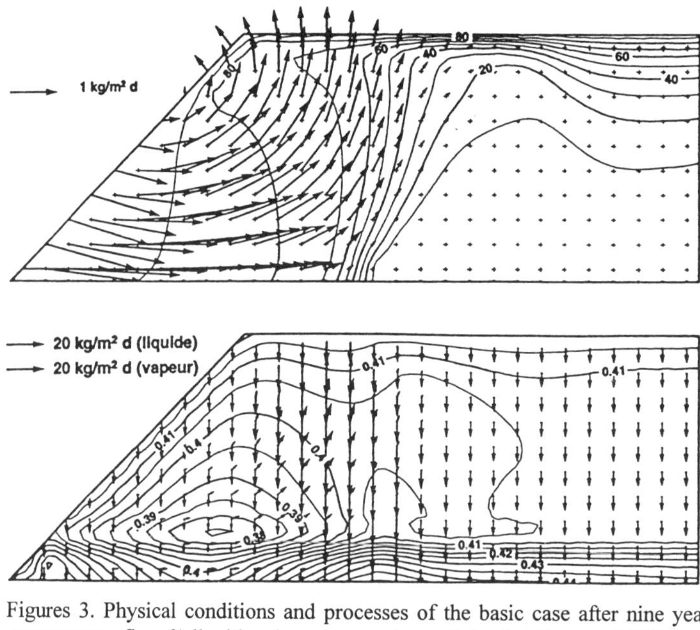

Figures 3. Physical conditions and processes of the basic case after nine years of AMD production; *a*) oxygen mass flux, *b*) liquid and vapor mass fluxes (after Lefebvre and Gélinas, 1995).

Computer modelling techniques present an effective way to simulate the performance of the design before construction since every site chosen for construction of a waste deposit is different with respect to climate and availability of appropriate construction materials. Mathematical modelling techniques can be either based on numerical solutions or analytical solutions of the governing equations. Analytical solutions are limited by the condition of flow in a homogeneous layer of finite vertical thickness and infinite lateral extension. Numerical solutions based either on hydrological concepts, chemical reactions to predict the acid mine drainage (AMD) process, or a combination of both are limited by computer power.

The importance of numerical simulations of hydrological and geochemical processes in waste rock heaps, and especially on their slopes, before construction of a cover is illustrated in Figure 3. Lefebvre and Gélinas (1995) applied the numerical model, TOUGH AMD, developed to represent the main processes involved in AMD: hydrology, gas transfer, heat transfer, geochemistry, and mass transport to a basic case. The influence of the slope on the transport of oxygen into the waste heap through convection is illustrated in Figure 3*a*. The water circulation through the waste rock heap is shown in Figure 3*b*. Liquid water moves downward from the surface during infiltration in the unsaturated zone, however, water vapor is transported laterally into the dump by the relatively cold air entering mainly at the toe of the slope. Powers and O'Neill (1986) simulated natural convection in a two-dimensional laterally confined, saturated porous medium. The simulations showed distinct cell patterns and heat transfer characteristics when the medium was inclined relatively to the horizontal. These changes in the isotherms of convection cells with inclination could directly effect the movement of water vapor in the unsaturated cover of reactive waste rock dumps.

3.3 *1-D numerical modelling*

SoilCover (1997) is a one-dimensional finite element model for saturated-unsaturated flow, and coupled heat and mass transfer (liquid and vapor). The physically based method is coupled to the atmosphere to allow for prediction of actual evapotranspiration and surface flux boundary conditions on the basis of atmospheric conditions, vegetative cover, and soil properties. The predicted surface flux boundary conditions can be used in conjunction with two-dimensional saturated-unsaturated seepage modelling programs.

The computer program SoilCover was applied by Wilson *et al.* (1995) to predict the performance of waste rock covers under two different climatic conditions (humid and dry). The authors demonstrated that SoilCover could be applied to model the infiltration and exfiltration of barrier, as well as capillary systems. Infiltration and exfiltration, however, was only predicted for one dimensional flow. Lateral flow was not considered in the one-dimensional modelling program.

3.4 *2-D numerical modelling*

The program Seep/W (GeoSlope 1991) was applied by Yanful and Aubé (1993) to model the moisture retention of a soil cover consisting of clay placed between an upper fine sand layer and lower coarse sand layer. The simulated moisture contents and hydraulic heads at various depths were compared with those measured in a column laboratory test. Long term predictions for evaporation and no evaporation show that the clay layer would remain saturated even under prolonged dry periods assuming one-dimensional flow.

The combination of SoilCover and Seep/W in predictive modelling was applied by O'Kane *et al.* (1996) to evaluate closure alternatives to the Dona Lake Mine tailings impoundment. The measured and predicted water contents were compared for models with different water table conditions. The changes in water content were calculated in response to extreme climatic conditions.

The suitability of the HELP computer model for simulation of the water balance of landfill cover systems was studied by Berger *et al.* (1996) at the Georgswerder landfill in Hamburg, Germany. HELP is a quasi two-dimensional program, coupling several one-dimensional sub-models for vertical and lateral flow one vertical profile at a time. HELP models only gravitational forces eliminating its application for capillary barriers. The authors found that the model predicted the measured lateral drainage. However, predicted percolation through the barrier system did not agree with field measurements.

The finite difference numerical model FIDHELM was applied to study the gas transport mechanism related to the generation of acidic drainage at a waste rock pile at the Heath Steele mine in New Brunswick (Bennett *et al.*, 1995). FIDHELM simulates the gas, water and heat flow through potentially acid generating porous media in two spatial dimensions. In an example of a waste rock heap it was used to represent a truncated cone and predicted results were compared with the measured pore gas oxygen concentration. The discrepancy between the calculated and measured results was attributed to the assumed homogeneity of the waste heap in the simulation. The influence of the sloped surface on the oxygen production rate was depicted. The simulation showed that for higher air permeability values the pore gas oxygen mass fraction ISO-terms were retarded at the lower part of the slope.

3.5 *3-D numerical modelling*

Numerical modelling of water and oxygen transport in layered soil covers for deposits of pyrite mine tailings was carried out by Collin and Rasmuson (1990), using the integrated finite difference computer code TRUST, developed by Narasimhan (1975). The code allows the calculation of saturated and unsaturated water transport in three dimensions, applying the Darcy and Richard equations. The calculations were carried out for a homogeneous and isotropic medium, reducing the hydraulic conductivity to a simple function of water content. In the unsaturated case, only desaturation of the pores is accounted for. The input data for the computer code requires the geometrical description and hydraulic material data in the form of capillary pressure and hydraulic conductivity versus moisture content. The water supply at the top surface was modelled using strongly simplified precipitation and evapotranspiration data. The conceptual cover design had a sloped surface (5%) and sloped interlayers parallel to the surface, allowing runoff and lateral water transport. The evaluation of the cover system was carried out in three steps: water transport, diffusivity calculations, and oxygen transport reactions.

A two-phase (liquid-gas) migration/multi-migrant transport model (TENSIC) was developed by Pollmer and Eckart (1997) to model the pollutant load in the unsaturated zone of mine dumps. The model was applied to forecast the development and the composition of the effluent at the base of the Beerwalde mine dump, deposited 25 years ago and not covered until recently. The differential equation describing the two-phase flow system was transformed into a space- and time discretized form. The solutions for the gaseous- and aqueous phase were transferred into a contaminant transport model.

4 CASE STUDIES

Aubertin *et al.* (1997) used Seep/W for two-dimensional modelling of unsaturated inclined layers (silt placed between upper and lower sand layers). The authors investigated the moisture distribution along a theoretical slope to ensure a high degree of saturation in a cover system in order to prevent gas transport across the cover. The effects of a 4% slope on desaturation of the cover layers after 60 days of drought were demonstrated. The saturation profile in the two sand layers was similar to the one-dimensional simulation, however, the profile in the fine grained

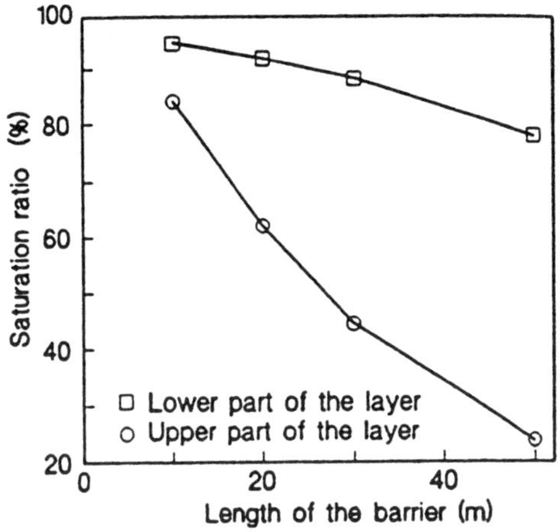

Figure 4. Effect of the slope length on saturation ratio near the lower and upper part of the silt layer (from Aubertin *et al.,* 1997).

silt layer showed a marked desaturation, particularly in the upper part of the layer. The elevation difference in this example was 2m, which was slightly below the air entry value of the silt. The boundary conditions in the simulation together with the hydrological regime created conditions where the moisture retention in the silt layer was not sufficient to maintain saturation near the top of the slope. Figure 4 illustrates the saturation ratio in the fine grained material layer at the end of a 60 day drought at the top and bottom of the layer, as a function of slope length. The results show that for the materials used in this example, a 4% slope is sufficient to desaturate the top portion of a relatively short slope length.

A composite soil cover for an experimental waste rock pile at the Heath Steele mine in Newcastle, New Brunswick, was studied by Yanful *et al.* (1993). The experimental waste rock pile, constructed by end load dumping of waste rock onto an impermeable synthetic membrane, had a surface area of 2100m^2, and maximum height of 5m. The pile was contoured with a maximum slope of 33%. The cover design was evaluated using laboratory column studies and one-dimensional saturated-unsaturated flow modelling.

The water balance of a lysimeter constructed on the slope of the cover of a municipal landfill, Karlsruhe, Germany, was simulated by Bronstert and Plate (1997) using the two-dimensional computer model HILLFLOW and compared with the *in situ* measurements. The model was able to calculate vertical (interception, infiltration, evaporation) and lateral flow. The simulations indicated desaturation of the clay barrier at the top of the slope after prolonged dryness. In the coarse sand, directly overlying the clay, however, an increase in moisture was especially pronounced at the toe of the slope. The results were compared with the moisture measurements of a neutron probe and are shown in Figure 5. The adaptation of a computational model, designed to calculate large scale hill flow hydrological processes, to a small slope on a landfill cover is clearly demonstrated.

A multi-dimensional subsurface moisture flow simulation was conducted to evaluate a low-level radioactive disposal facility design (Nichols and Meyer, 1996). The design featured a sloping sand/gravel capillary barrier designed to route natural infiltration around a concrete vault. Three barrier slopes (2:5, 1:10, and 1:5) and three sand/gravel combinations were analyzed. The porous flow code used for the simulation was a three-dimensional, finite-element code, VAM3DCG (Huyakorn and Panday, 1993). The code uses the Galerkin finite-element method to solve the variably saturated flow equation using non-orthogonal grid discretizations and robust efficient numerical techniques.

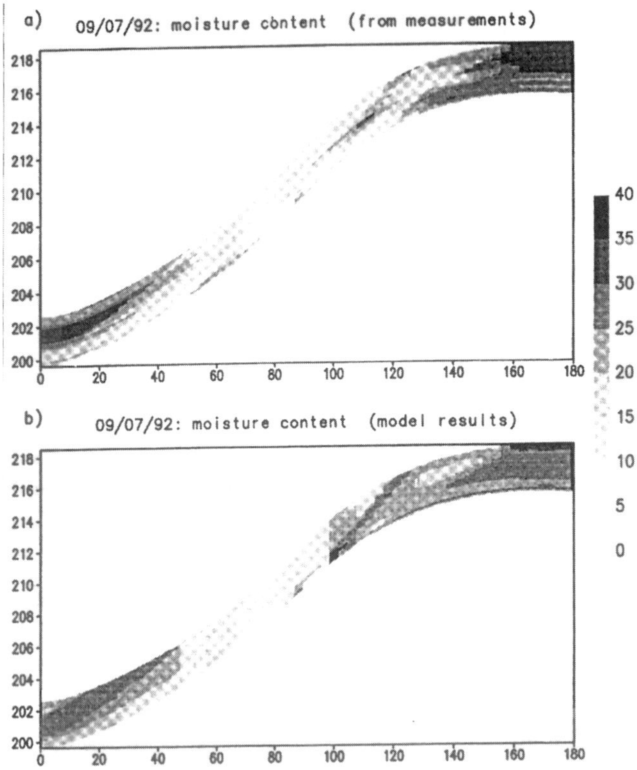

Figure 5. Measured and modeled soil moisture in the first and second half of the simulation period (after Bronstert and Plate 1997).

The three-dimensional problem was conceptualized as a capillary barrier surrounding the concrete vault, with the capillary barrier being the dominant flow mechanism. The design was analyzed in two- and three-dimensional modes. Comparison with two-dimensional performance monitoring demonstrated that the two-dimensional analysis resulted in underestimating the barrier effectiveness. Cross-slope flow as a result of pressure gradients produced in the third, cross-slope, dimension was observed. The barrier effectiveness was found to be highly sensitive to the sand/gravel material properties. The influence of the slope of the barrier was found to be less important when high quality materials (i.e. more moisture retention capability) were used in the simulation.

5 CONCLUSIONS

Environmental regulations in the industrialized countries require the safe and long term disposal of hazardous waste. Long term solutions to prevent contamination of the surface water and groundwater are investigated by scientists and engineers around the world. Engineered multi-layer soil cover systems, designed to prevent the infiltration of water and oxygen into the hazardous waste disposal systems, are studied intensely.

Acid mine drainage (AMD) is one of the largest environmental liabilities facing the mining industry. Sulphidic waste rock piles and tailings are the most significant sources for acid generation on a mine site. The research strategies are focused on prediction, prevention, treatment and control. The application of soil cover systems on waste rock heaps and tailings deposits is researched intensely in Australia, Canada, Sweden, and the United States.

The design of a soil cover requires the knowledge of the micro-climate at the site of the waste heap and geotechnical/soil mechanical information on the materials used for construction. Numerical modelling is applied to calculate the optimum layer design, using the available data, in order to reduce the infiltration of water and oxygen into the mine spoil.

The literature search showed that numerical designs for the soil covers are typically carried out assuming horizontal waste rock surfaces. The importance of a slight slope (1-2%) of the drainage barrier in order to accommodate lateral drainage is mentioned in most publications; however, in the one-dimensional numerical calculations this slope is not represented. A one-dimensional numerical analysis of the two-dimensional flow problem is valid if the slope is of sufficient length and the point of evaluation is sufficiently far from the beginning or the end of the slope. Hydraulic performance of the two-dimensional system is a function of surface flux boundary conditions, slope length, slope angle, cover material properties, and underlying waste material properties. Hence, each case is unique and requires the application of comprehensive predictive tools and monitoring methodologies to ensure the chosen soil cover alternative meets closure requirements. In the researched case studies the special hydrological condition in the slopes of soil covered waste deposits is sometimes acknowledged by covering the steep side slopes with impermeable geomembranes in order to avoid infiltration or exfiltration through the protective cover on the slopes.

The side slopes of a waste deposit cover system are typically short and steep. The infiltration rates, runoff and the hydrogeological conditions of the cover system, are thus different for an almost horizontal surface (1-2% slope) as compared to the side slopes (10-50%). The hydrological conditions of such steep slopes covered with multi-layer soil systems can be compared to natural hill slopes. Bronstert and Plate (1997) applied two- and three-dimensional hill flow hydrology modelling to demonstrate the lateral flow in multi-layered natural and engineered systems. The simulated conditions are compared with the measured conditions and good agreement was found.

The hydrologic conditions of the horizontal surface of a soil cover system can be well described and modeled by available one-dimensional numerical models. The hydrologic conditions of the steep slopes; however, represent a two-dimensional system and should be modelled and monitored accordingly.

REFERENCES

Aubertin, M., Chapuis, R., Bouchentouf, A. and Bussière , B. 1997. Unsaturated flow modelling of inclined layers for the analysis of covers. *Proceedings, Fourth International Conference on Acid Rock Drainage*, Vancouver, BC, May 31-June 6, Vol. 2, pp. 731-746.

Bárdossy, A., Bronstert, A. and Merz, B. 1995. 1-, 2- and 3-dimensional modeling of water movement in the unsaturated soil matrix using a fuzzy approach. *Advances in Water Resources*, **18**(4):237-251.

Bennett, J.W., Garvie, A.M., Pantelis, G., Ritchie, A.I.M., Bell, A.V. and Noël, M. 1995. Comparison of measured and predicted transport processes controlling oxidation in the waste rock piles at the Heath Steel mine site. *Proceedings, Sudbury '95, Conference on Mining and the Environment*, Sudbury, Ontario, May 28-June 1, pp. 1017-1026.

Berger, K., Melchior, S. and Miehlich, G. 1996. Suitability of Hyrologic Evaluation of Landfill Performance (HELP) model of the US Environmental Protection Agency for the simulation of the water balance of landfill cover systems. *Environmental Geology*, **28**(4): 181-189.

Bonaparte, R. and Giroud, J.P. 1996. Waste containment systems for pollution control: Part II - Hydraulic design and performance. In *Advances in groundwater pollution control and remediation.* M.M. Aral (ed.), Kluwer Academic Publishers, Dordrecht, The Netherlands, pp. 437-477.

Bronstert, A. 1994. *Modellierung der Abflussbedingung und der Bodenwasserdynamik von Hängen.* Ph.D. thesis, Institut für Hydrologie ind Wasserwirtschaft, Universität Karlsruhe (TH), Germany.

Bronstert, A. and Plate, E.J. 1997. Modelling of runoff generation and soil moisture dynamics for hillslopes and micro-catchments. *Journal of Hydrology*, **198**:177-195.

Collin M. and Rasmuson, A. 1990. Mathematical modelling of water and oxygen transport in layered soil covers for deposits of pyrite mine tailings. *Proceedings, GCA-MAC Joint Annual Meeting*, Vancouver, BC. J.W Gadsby, J.A. Malick and S.J. Day (eds.), BiTech Publishers, Vancouver, BC, May 16-18, pp. 311-333.

Filipek, L., Kirk, A. and Schafer, W. 1996. Control technologies for ARD. *Mining Environmental Management*, December: 4-8.

Freeze, R.A. and Cherry, J.A. 1979. *Groundwater.* Prentice-Hall Inc., Englewood Cliffs, NJ.

Frind, E.O., Gillham, R.W. and Pickens, J.F. 1976. Application of unsaturated flow properties in the design of geologic environments for radioactive waste storage facilities. In *Finite Elements in Water Resources.* Proceedings, First International Conference on Finite Elements in Water Resources, Princeton University, USA, July. W.G. Gray, G.F. Pinder and C.A. Brebbia (eds.), Pentech Press, London, pp. 3.133-3.163.

Geo-Slope. 1991. *Seep/W User Manual.* Geo-Slope International Calgary, Alberta.

Günther, K. 1988. Surface coverings for tips and contaminated sites. *Proceedings, Contaminated Soil '88, 2nd International TNO.BMFT Conference on Contaminated Soil,* Dordrecht, Boston, April 11-15, Hamburg, BRD. K. Wolf, J. van den Brink and F.J. Colon (eds.), Kluver Academic Publishers, pp. 561-577.

Huyakorn, P.S. and Panday, S. 1993. *VAM3DCG-Variably saturated analysis model in three-dimensions with preconditioned conjugate gradient matrix solvers: Documentation and user's guide version 1.1.* HydroGeoLogic, Inc. Herndon, VA.

Lefebvre, R. and Gélinas, P.J. 1995. Numerical modeling of AMD production in waste rock dumps. *Proceedings, Sudbury '95, Conference on Mining and the Environment,* Sudbury, Ontario, May 28-June 1, pp. 869-878.

MEND. 1992. *Heath Steel waste rock study phases 1 to 3. MEND Project 2.31.1(a).* Energy, Mines and Resources Canada, CANMET.

MEND. 1994. *Engineering design and construction report; Phase IV - Composite soil cover acid waste rock study Heath Steele mines. MEND Project 2.31.1(b).* Energy, Mines and Resources Canada, CANMET.

MEND. 1996*a*. *Monitoring report; Phase IV - Composite soil cover acid waste rock study Heath Steele mines. MEND Project 2.31.1(b).* Energy, Mines and Resources Canada, CANMET.

MEND. 1996*b*. *SoilCover user manual. Version 2.0. MEND Project 1.25.1.* Energy, Mines and Resources Canada, CANMET.

Narasimhan, T.N. 1975. A unified numerical model for saturated-unsaturated groundwater flow; Ph.D. thesis. Lawrence Berkley Laboratories, University of California.

Nichols, W.E. and Meyer, P.D. 1996. Multidimensional water flow in a low-level waste isolation barrier. *Ground Water*, **34**(4):659-665.

Nicholson, R.V., Akindunni, F.F., Sydor, R.C. and Gillham, R.W. 1991. Saturated tailings covers above the water table: The physics and criteria for design. *Proceeding, Second International Conference on the Abatement of Acidic Drainage*, Montréal, Sept, 16-18, Vol.1, pp. 443-460.

Nicholson, R.V., Gillham, R.W., Cherry, J.A. and Reardon, E.J. 1989. Reduction of acid generation in mine tailings through the use of moisture-retaining cover layers as oxygen barriers. *Canadian Geotechnical Journal*, **16**(1):1-8.

O'Kane, M. 1995. *Instrumentation and monitoring of an engineered soil cover system for acid generating mine waste.* M.Sc. thesis, University of Saskatchewan, Saskatoon, SK, Canada.

O'Kane, M., Wilson, G.W., Barbour, S.L. and Swanson, D.A. 1995. Aspects of the performance of the till cover system at Equity Silver Mine Ltd. *Proceedings, Sudbury '95, Conference on Mining and the Environment*, Sudbury, Ontario, May 28-June 1, pp. 565-573.

O'Kane, M., Bews, B.E., Wilson, G.W. and Ferguson, K.D. 1996. Field measurements, field response modelling, and predictive modelling of the tailings impoundment at Dona Lake Mine. *Proceedings, 98th General Meeting of CIM*, April 28-May 2, Edmonton, Alberta.

Pollmer, K. and Eckart, M. 1997. Description of reaction- and transport processes in the zone of aeration of mine dumps. *Proceedings, Fourth International Conference on Acid Rock Drainage*, Vancouver, BC, May 31-June 6, Vol. 4, pp.1499-1514.

Powers, D.J. and O'Neill, K. 1986. Natural convection in sloping porous layers. *Proceedings, VI International Conference on Finite Elements in Water Resources*, Lisboa, Portugal, June, 1986. Sá da

Costa, A., Melo Baptista, A., Gray, W.G., Brebbia, C.A. and Pinder, G.F. (eds.), Springer Verlag Berlin.

Rasmuson, A. and Eriksson, J.-Ch. 1986. *Capillary barriers in covers for mine tailing dumps*. Internal Report, Royal Institut of Technology, S-100 44 Stockholm, Schweden.

Ross, B. 1990. The diversion of capillary barriers. *Water Resource Research*, **26**(10):2625-2629.

Sailie, L. 1985. Some aspects of sandwiched soil cover design for toxic waste repositories. *Engineering Geology*, **21**:321-326.

SoilCover. 1997. *The design of soilcover systems. SoilCover Version 4.0 User Manual*. Unsaturated Soils Group, Department of Civil Engineering, University of Saskatchewan, Saskatoon.

Stormont, J.C. 1995. The effect of constant anisotropy on capillary barriers. *Water Resource Research*, **31**(3): 783-785.

US-EPA, 1989. *Final covers on hazardous waste landfills and surface impoundments*. Technical Guidance Document EPA/530-SW-899-047.

Weeks, O.L., Mansell, R.S. and McCallister, S.W. 1992. Evaluation of soil top-cover systems to minimize infiltration into a sanitary landfill: A case study. *Environmental Geology and Water Sciences*, **20**(2):139-151.

Wilson, G.W., Barbour, S.L., Swanson, D. and O'Kane M. 1995. Instrumentation and modelling for saturated/unsaturated performance of soil covers on acid generating waste rock. *Hydrogéologie*, **4**:99-108.

Yanful, E.K. and Aubé, B.C. 1993. Modelling moisture-retaining soil covers. *Proceedings, Joint CSCE-ASCE National Conference on Environmental Engineering*., Montréal, QUE, July 12-14, Vol. 1, pp. 273-288.

Yanful, E.K., Bell, A.V. and Woyshner, M.R. 1993. Design of a composite soil cover for an experimental waste rock pile near Newcastle, New Brunswick, Canada. *Canadian Geotechnical Journal*, **30**: 578-587.

Tailings and Mine Waste '99 © 1999 Balkema, Rotterdam, ISBN 90 5809 025 6

Design construction and monitoring of a waste rock cover using pulp and paper residues

Alexandre Cabral, Fabien Burnotte & Guy Lefebvre
Department of Civil Engineering, University of Sherbrooke, Que., Canada
Gail Amyot
Cambior Incorporated
Gaëtan Lacasse
Groupe S.M. Incorporated

ABSTRACT: A waste rock fill potentially generating acid mine drainage (AMD) near Woburn, Quebec, Canada, was capped by a cover system including a capillary barrier constituted of P&P (deinking) residues. The results of tests conducted during construction as well as the first results pertaining to the monitoring program are presented. These results seem to indicate that the capillary barrier is performing as required: the deinking residue layer is keeping a high degree of saturation and O_2 is not reaching the bottom of the barrier (thus AMD is not being produced).

1. INTRODUCTION

A waste rock fill located near Woburn, Quebec, Canada, known as the Clinton site, was designed to be capped by a cover system including a capillary barrier constituted of deinking residues. The former Clinton mine is located in a limited access zone presently being used by the forestry industry. A mine shaft with a surface of 2 000 m^2 and depth of up to 20 m was backfilled with potentially AMD generating waste rock gathered all around the site. The purpose of this barrier is to prevent gaseous oxygen from reaching the rock fill, thus to curb the oxidation of sulfide rich minerals and the resulting generation of acid mine drainage (AMD). The relatively low hydraulic conductivity of the compacted deinking residues barrier also helps reducing infiltration into the rock fill.

2. MATERIALS AND METHODS

2.1 *Cover design*

The cell capping covers a surface of 5 000 m^2. It consists of a cover system composed of the following layers from top down (see Figure 1): a thin vegetative cover, a granular layer, a capillary/hydraulic barrier constituted of deinking residues, and a layer of coarse waste rock.

The top layer is for the support of vegetation, the protection of the cover against erosion and the rehabilitation of the natural environment. The granular material acts as a confinement layer; the applied load helps improving the barrier's desired properties. This layer also helps preventing rodents and other small animals from digging holes into the barrier.

The deinking residue layer acts as an O_2 barrier in two ways: 1) due to its high water retention capacity (capillary properties), it keeps a high degree of saturation (Sr%), thus reducing atmospheric O_2 migration by four orders of magnitude (Yanful, 1993). 2) Due to the high organic content of the deinking residues and to microbial activity, the barrier consumes atmospheric oxygen. As it will be discussed further in the text, 100% efficiency is obtained in the prevention of O_2 migration. Due to the relatively low hydraulic conductivities achieved, the deinking residues also acts as a hydraulic barrier, helping reduce infiltration into the waste fill.

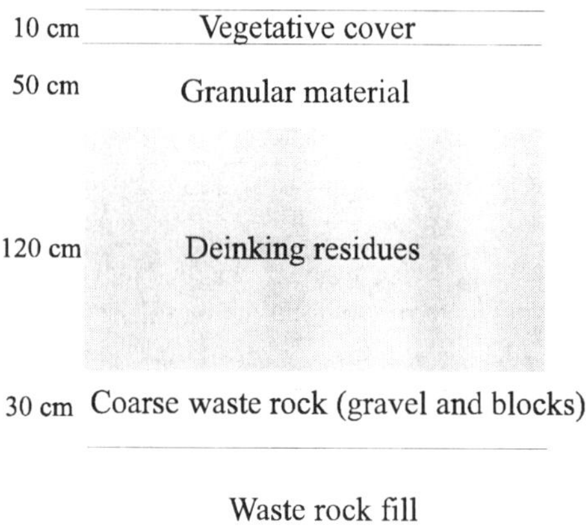

Figure 1. Profile of the cover system constructed at Clinton site

The bottom most layer is constituted of coarse waste rock. The difference in capillary properties between this layer and the deinking residue layer creates a hydraulic contrast - or capillary barrier effect -, which insures that the deinking residue layer will not loose its water to the underlying layer, unless a minimum gradient of 1 is originated by the accumulation of a film of water on the surface. This effect is well explained by Nicholson et al. (1989), Shackelford et al. (1994), etc.

2.2 Instrumentation

Site instrumentation allowing for the monitoring of the evolution of several indexes and properties of the capillary barrier, as well as the monitoring of environmental quality indicators is constituted by the following:

- 3 lysimeters;
- 20 Time Domain Reflectometry (TDR) probes;
- 18 thermocouples;
- 3 infiltration wells for permeability measurements
- 3 shallow wells (installed in the granular material covering the P&P layer in order to monitor the eventual presence of a subsurface - or perched - water table),
- 3 settlement plates.

A plan view of the cell is given in Figure 2, with the location of three Sites (A, B, C), each equipped with an instrumented profile as depicted in Figure 3. Also, several piezometers were installed in the mine residue fill for monitoring of water quality. Surface water samples are also collected in ditches around the site.

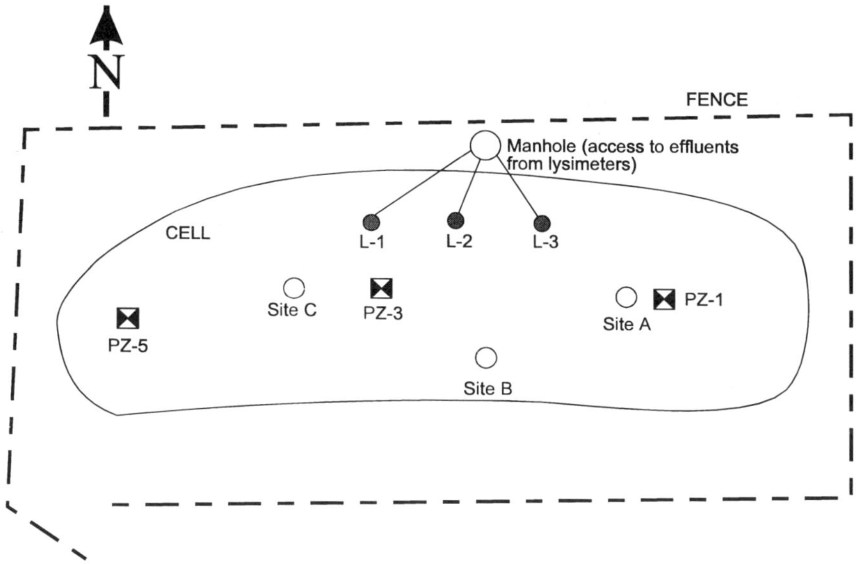

Figure 2. Plan view of the Clinton cell

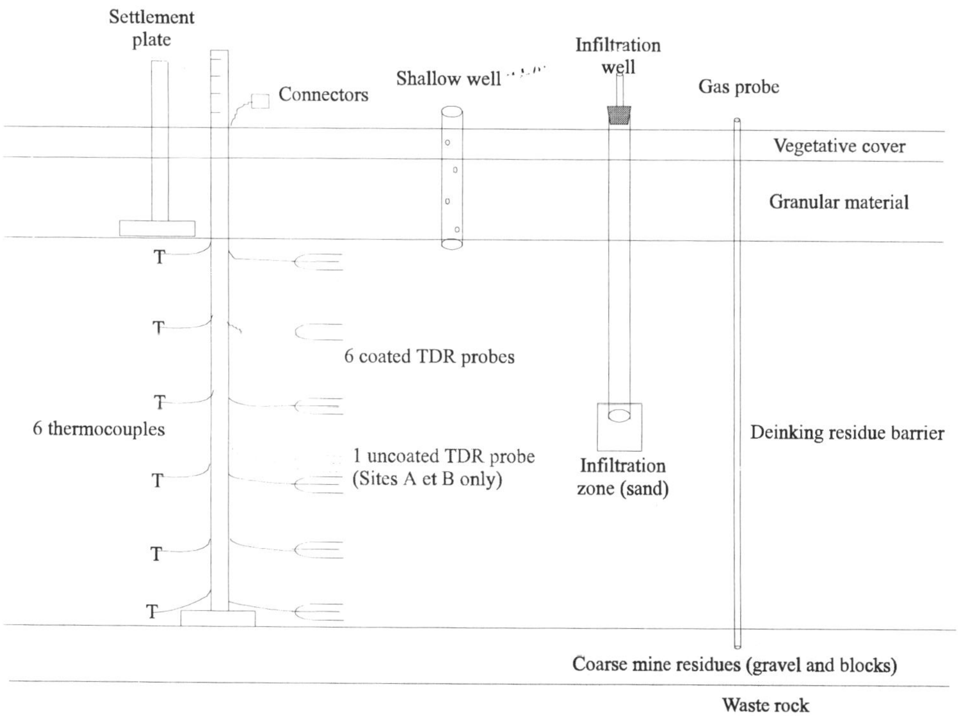

Figure 3. Schematic of an instrumented profile

2.3 Properties of deinking residues

For the construction of the capillary barrier, deinking residues from two paper mills were used. Their general characteristics, presented in Table 1 were obtained at the Geoenvironmental laboratory of Sherbrooke University.

Table 1. General characteristics of deinking residues used at Clinton Site

Index or property	Paper Mill	
	KF	PC
w_{opt} (%)	≈ 90	≈ 80
w (%)	120 - 150	150 - 180
% Above optimum	30 - 60	70 - 100
G_S	1.95 - 1.98	2.06
% Organic	≈ 60	≈ 50

Optimum water contents (w_{opt} %) of around 80% to 90% were obtained by Standard Proctor compaction tests. Lower values (w_{opt} = 50 to 60 %) have been reported in the literature for other P&P residues (e.g. Cabral et al., 1998; Kraus et al., 1997; Moo-Young and Zimmie, 1996; Zimmie et al., 1993). The as-received water content varied between 120 and 180%. The density of solids (G_S) was around 2.0, which is slightly higher than the values reported by Zimmie et al. (1993) for residues having organic matter contents between 50 to 60%.

Extensive laboratory and field tests have shown that the effect of compaction on the desired properties is optimized when the residues are compacted at water contents at least 50% above w_{opt}. At this level, lower values of hydraulic conductivity are obtained and the water retention properties are enhanced. It was expected that mixing the two types of residues KF (slightly too dry) and PC (wetter) would result in an acceptable barrier.

2.4 Construction

The Clinton site was capped in September 1997. Compaction of the ~1.2 m thick deinking residue layer was made in three lifts of approximately 0.4 m each. Each layer was spread by a bulldozer and compacted by three passes of a loaded ten-wheel truck (Figure 4). Several sand cone tests were performed on each layer to verify the in-situ density and placement water content.

3. RESULTS

3.1 Water content

The water content profiles were obtained by two different methods: the gravimetric method (samples taken from boreholes) and the TDR method (TDR probes). Results for site A are given in Figure 5a. A mean value of approximately 145% was found for the gravimetric water contents measured at the end of construction (24-09-97) at Site A. Local variations of more than 10% were observed (142% < w % < 155%) along the profile. This shows that a certain time is needed for equilibrium of the water content to be reached inside the barrier. Mean

Figure 4. Construction of the deinking residue layer

values of 135% (within the range found for KF residues) and 165% (within the range found for PC residues) were obtained at Sites B and C, respectively. Given the differences in the mean water contents between the sites, it can be said that a perfect mix of the two types of residues (KF and PC) was not perfectly achieved. More precise construction quality instructions have to be formulated in order to avoid a "patchwork"-like mixing. However, the desired optimum plus 50% was achieved nearly everywhere.

The TDR water contents for Site A are given in Figure 5b. Readings taken right after the placement of the probes (09-09-97 for the 3 deeper probes and 24-08-97 for the 3 superior probes) give significantly lower values than those obtained afterwards. The lower initial readings probably are related to air gaps that were formed around the wires of probes. With placement and compaction of the remaining sublayers, these gaps were closed. In general, the values obtained using TDR probes show that there is a tendency towards equilibrium of the water content along the profile. In order to plot the water contents obtained with the TDR (volumetric water content - θ) as gravimetric water contents (Figure 5), a conversion had to be made using the dry density of the material. The latter varied with time, as the material consolidated. TDR water content profiles for Sites B and C show a similar behavior.

3.2 Organic content and temperature profiles

The organic content profiles are given in Figure 6a. Organic contents vary from 49% to 58%, corresponding to the two types of residues. The average organic content of Site C corresponds more closely to that of residues produced at the PC mill. At Site B it is highly possible that more KF residues were employed. At Site A a mix of the two seem to have been placed.

The temperature profile for Site A is given in Figure 6b. First, an increase in temperature in the first few weeks can be due to aerobic degradation after placement, the mixing of residue causing aeration of it. Then, a decrease in temperature is related with atmospheric conditions (Winter), and a low increase in the Spring (same behavior for Sites B and C).

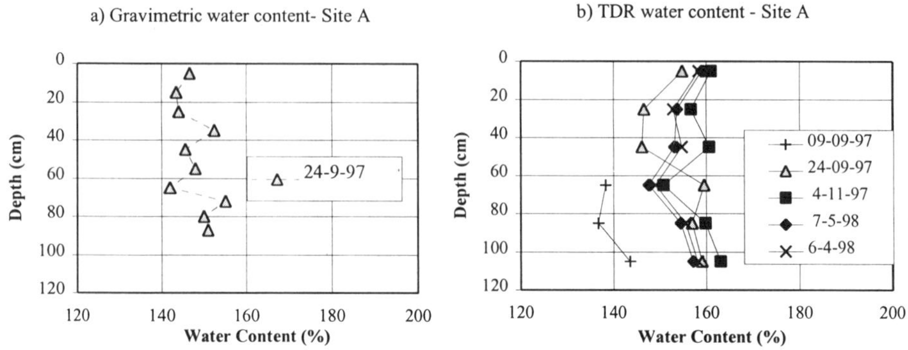

Figure 5. Water content profiles at the Clinton site

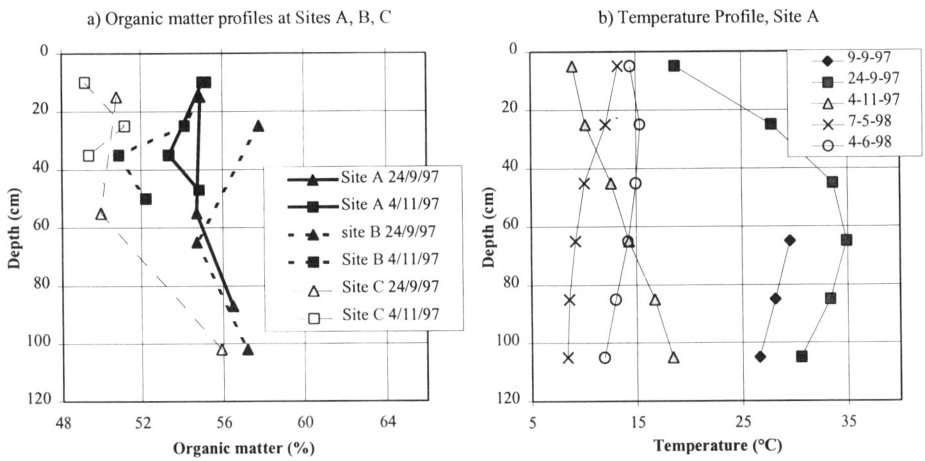

Figure 6 - Organic content and temperature profiles

3.3 Gas profiles

Gas profiles for Site A are presented in Figure 7. During construction (09-09-97), the O_2 is still present due to recent mixing, although at very low concentrations (less than 2% at Site A). Already, CO_2 and CH_4 are present at significant concentrations; a clear indication that biological activity is taking place in the deinking residue barrier. At the end of construction (24-09-97), gas measurement was only possible at the drained bottom of the layer. This is due to the high water saturation of the material (water enters the gas probe when it is inserted into the deinking residue layer). It can be observed that the O_2 concentrations at the end of construction was already nil. CO_2 and CH_4 concentrations increased from after-placement levels. The very high CH_4 concentration found at the bottom indicates that anaerobic conditions prevail within the barrier after a while. Nine months later (4-6-98), CH_4 concentrations increased, while the CO_2 concentration decreased to reach levels corresponding to that of a biogas (Gandolla et al. 1995). Sites B and C show the same tendency.

Despite the production of gas, odor has not been a problem at various sites being monitored by the Geoenvironmental group. In fact, it has been observed in core samples taken from a 10 year-old dedicated P&P-residue landfill that the material keeps the same visual and odor characteristics as fresh deinking residues. The study of gas production rates has not been completed yet. However, recent tests performed in the Geoenvironmental laboratory at the University of Sherbrooke have shown that gas production decreased dramatically under temperatures below 20 °C.

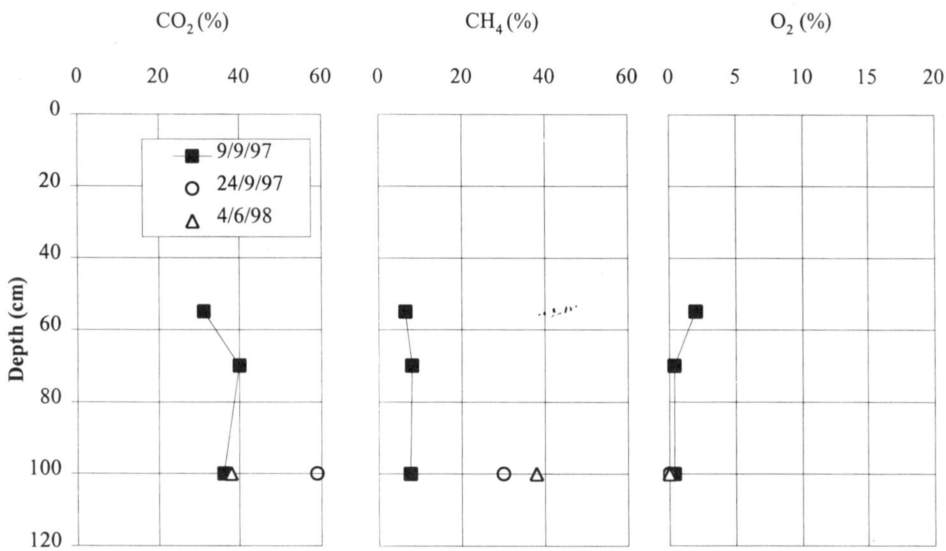

Figure 7 - Gas profiles

3.4 Water Quality

Data on water collected before (1995) and after (1998) construction is presented in Table 2. It can be seen that pH inside the mine shaft was acceptable initially, but that the water leaving the site was becoming increasingly acid due to oxidation of the sulfide rich mine wastes. After construction, the pH measured in piezometers located inside the cell (former mine shaft) is practically neutral. It is important to note here that during the backfilling of the shaft the oxidized rocks were mixed with lime (buffer material). This was aimed at creating an environment where *Thiobacillus ferooxidans* can no longer thrive (pH above 3.5 to 4.5); in this case, the acidification process would require a significant input of oxygen. Given that oxygen can no longer reach the waste rock fill, it is likely that AMD should no longer be a problem at the Clinton site.

Copper concentrations are well below the criteria. Although the concentration of zinc dropped significantly, it is still above the criteria. Given the neutral pH observed, it is expected that it will be under the criteria in the near future. This will be confirmed by further monitoring and analyses.

Table 2. Water quality before and after (1995 and 1998) construction of the cell.

	pH		Cu (ppm)		Zn (ppm)	
	1995	1998	1995	1998	1995	1998
In the mine shaft	6.6 - 6.9	6.6-6.8	0.4 - 3.5	0.01-0.1	0.7 - 1.7	0.4-2.0
Exfiltration from the site	4.3 - 5.6	7.2	2 - 30	<0,005	5 - 50	1
Criteria [1]	6.5 - 9.5		0.30		0.50	

[1] Directive 019 from the Ministry of the Environment of Quebec.

3.5 Density and consolidation

The sand cone tests performed during compaction (on each lift) showed that the degree of compaction was lower than that obtained with the Standard Proctor. The water content measured with sand cone test samples were below the water contents obtained in the profiles measured at end of construction. This is probably due to drying of the testing surface before the test was actually performed. The corrected initial densities are indicated as points A, B, and C in Figure 8. For the correction, the water content was adjusted to the levels measured in the profiles obtained at the end of construction, keeping the same energy of compaction.

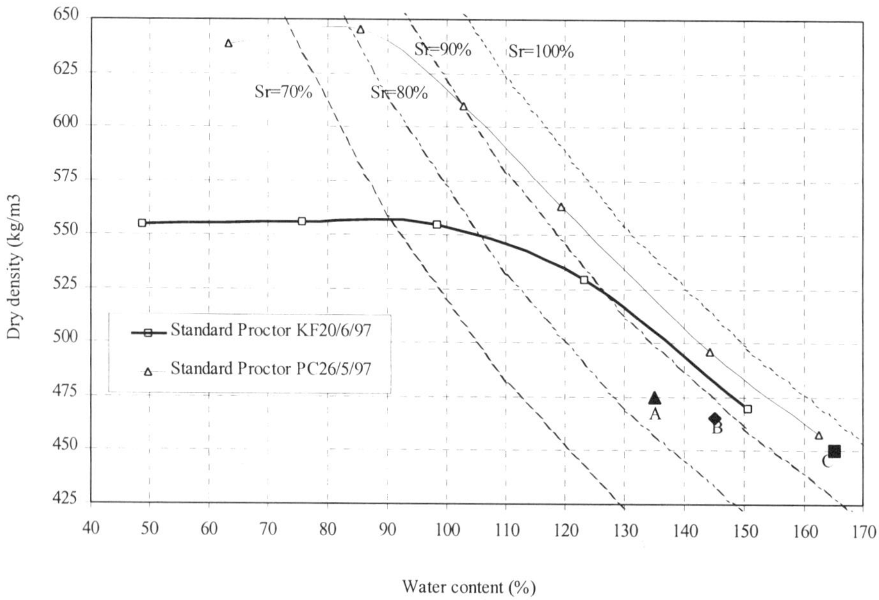

Figure 8. Standard Proctor compaction curves and field densities at the Clinton site

The three settlement plates installed have provided preliminary information on the consolidation behavior of the deinking residue layer under 18 kPa (average stress exerted by the granular material layer, plus the organic soil layer and half of the thickness of the deinking residue layer). Settlement data is presented in Table 3. The data shows that settlement varies from 4,2 to 12,3%, depending on the initial water content, degree of compaction and type of residue. The increase in the degree of saturation observed at each site further reduces the chances of O_2 diffusing through the barrier.

Table 3. Data on consolidation of the deinking residue layer

Index or property	Site A	Site B	Site C
Initial thickness (m) (24-09-97)	1	1,2	1,14
Final thickness (m) (07-05-98)	0,94	1,15	1
Settlement (%)	6,0	4,2	12,3
w (%) initial	145	135	165
w (%) final	140	130	148
Initial dry density (after compaction) (kg/m^3)	465	475	450
Final dry density (after 8 months of consolidation (kg/m^3)	493	494	505
Sr i (%)	88	84	96
Sr f (%)	92	85	100

3.6 Hydraulic conductivity and infiltration

In-situ permeability tests were performed using the TSB (Two Stage Borehole) method (Trautwein and Boutwell, 1994). Only the second stage was done and it was assumed that the horizontal and vertical hydraulic conductivities (k_h, k_v) were equal. A scheme of the infiltration wells installed at the Clinton site for permeability measurements is given in Figure 9. The first tests, performed at the end of construction (24-09-97), were difficult to interpret due to the difficulty of evaluating the water level. The latter was probably influenced by the dissipation of interstitial water pressure after compaction. The results of the tests performed eighth months (04-06-98) after the cap was finished are presented in Table 4.

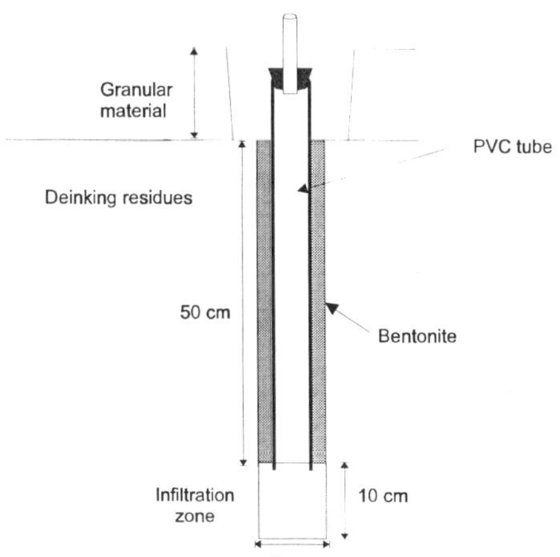

Figure 9. Scheme of the wells installed at Clinton site.

Table 4. Field hydraulic conductivity of the deinking residue layer, Clinton site

	Site A	Site B	Site C
k(cm/s) (04-06-98)	4×10^{-6}	1×10^{-6}	2×10^{-6}

The hydraulic conductivities measured are all in the 10^{-6} cm/s range, which is similar to the hydraulic conductivity of a fine silty material. Due to the relatively low hydraulic conductivity of the deinking residue barrier, it is expected that infiltration through the cover will be highly minimized during the wet season or following snow melt. During the dry season, there is absence of a hydraulic gradient. Given the capillary barrier effect, infiltration through the cover should be virtually nil.

Permeability tests were performed in the laboratory with the same material compacted under the Standard Proctor energy and tested under several loads. For the 18 kPa load, the k-value obtained ($k = 5 \times 10^{-7}$ cm/s) is two times to one order of magnitude lower than the field values. Particularities relating to the methods of permeability measurements and interpretation, as well as differences between samples in the field and in the laboratory can explain the gap between field and laboratory k-values.

It is expected that infiltration through the cover has not stabilized yet and is influenced by the consolidation of the cover. As a consequence, lysimeter data should be used with care. Very little water (less than 50 ml) has been collected from lysimeters L2 and L3. In the case of lysimeter L1, 1.5 liters of water were collected between 07-05-98 and 19-08-98. Based on this preliminary value, the operational hydraulic conductivity (k_{op}) would equal 7×10^{-8} cm/s. If a wet condition on the surface is considered to occur ten percent of the time (it rains - in average - one day out of ten in the region) a value of k (not the k_{op}) = 7×10^{-7} cm/s is obtained. This is quite similar to the k measured in the field, at Site B.

The shallow wells have been installed only lately. As a consequence, no data is yet available concerning the accumulation of water in the granular material layer placed over the deinking residues.

4. FINAL COMMENTS

The Clinton site was capped by a cover system including a capillary barrier constituted of deinking residues. Field instrumentation included 3 lysimeters, 20 TDR probes, 18 thermocouples, 3 infiltration wells, 3 shallow wells, and 3 settlement plates. Also, several piezometers were installed in the mine residues fill.

During construction the density of the compacted deinking residues was continuously measured by means of sand cone tests. The first results of the monitoring program are reported in this paper. Water content, temperature and gas profiles were obtained for three different zones. The quantities of the infiltration water were monitored by sampling the lysimeters. The quality of these waters will be reported once the monitoring program for the Fall 98 is completed. Samples of surface waters collected from a ditch contouring the site were analyzed for some chemical elements and compounds. Only Zn is still slightly above provincial criteria.

Gas (O_2, CO_2, and CH_4) profiles are also obtained at each of the three zones. O_2 is not found at the bottom of the barrier, whereas high concentrations of CO_2, and CH_4 were identified. The latter confirms that microbial activity is taking place inside the barrier. As a consequence, a few centimeters below the surface of the deinking layer the oxygen concentration is negligible; when it can be measured. In fact, due to the high degree of saturation of the barrier, gas sampling is not possible just a few centimeters below the surface.

Field hydraulic conductivity (Two Stage Borehole tests) were performed and yielded k-values in the 10^{-6} cm/s range. Laboratory permeability tests yielded k-values from two times to one order of magnitude lower. This gap is attributed to imperfections in the testing and interpretation methodology, as well as to differences in the conditions prevailing in the laboratory and in the field. A field testing program, presently under way, aim at improving the final properties of a deinking residue layer through - among others - better compaction procedures.

Based on the above, it can be said that the capillary barrier is performing as required: the deinking residue layer is keeping a high degree of saturation and zero O_2 concentrations were measured at the base of the P&P barrier interface, which means that O_2 is not diffusing, thus AMD will no longer be an environmental threat at the Clinton site.

5. ACKNOWLEDGEMENTS

The authors would like to thank the encouragement given by several members of the team working on the deinking residues project. Their comments and suggestions were greatly appreciated. The support of Claude Audet, from Cascades Industries is also greatly appreciated.

6. REFERENCES

Cabral, A.R., Lefebvre, G., Burnotte, F., Proulx, M-F. and Racine, I. 1998. Developments in the use of deinking residues in cover systems for acid generating mine tailings. Proc. Tailings and Mine Wastes, Ft. Collins, 1998. pp. 379-388.

Gandolla, M., Acaia, C., Fischer, C. 1995. Techniques, exploitation et gestion des décharges contrôlées. École Polytechnique Fédérale de Lausanne, Suisse, avril 1995.

Kraus, J.F. Benson, C.H., Van Maltby, C., Wang, X, 1997. Laboratory and field hydraulic conductivity of three compacted paper mill sludges. *J. of Geot. and Geoenv. Eng.*, Vol. 123, No. 7, pp.: 654-662.

Moo-Young, H.K. and Zimmie, T.F. 1996. Effects of organic decomposition on paper mill sludges used as landfill cover material. *Proc. 2nd Intern. Conf. Environemental Geotechnics*. Osaka, Japan, November 1996. Ed. M. Kamon, Balkema, 827-832.

Nicholson, R.V., Gilham, R.W., Cherry, J.A., Reardon, E.J. 1989. Reduction of acid generation in mine tailings through the use of moisture-retaining cover layers as oxygen barriers. Canadian Geotechnical Journal, 26: 1-8.

Schackelford, C.D., Chang, C.K. and Chiu, T.-F. 1994. The capillary barrier effect in unsaturated flow through soil barriers. Proc. 1st Intern. Conf. Env. Geotechnics. Edmonton, 1994. pp. 789-793

Trautwein, S.J. and Boutwell, G.P. 1994. In situ conductivity tests for compacted soil liners and caps. ASTM STP 1142. Daniel, D.E. and Trautwein, S.J., eds. Philadelphia, 1994. pp. 184-223.

Yanful, E.K. 1993. Oxygen diffusion through soil covers on sulphidic mine tailings. *ASCE Journal of Geotechnical Engineering*, 119(8): 1207-1228.

Tailings and Mine Waste '99 © 1999 Balkema, Rotterdam, ISBN 90 5809 025 6

Alligator Ridge tailings impoundment closure: A 100-yr performance of a soil cover

L.L.Newman & G.W.Wilson
Unsaturated Soils Group, University of Saskatchewan, Sask., Canada
M.F.Reed
Alligator Ridge Project, Bald Mountain Mine, Placer Dome N.A., Ely, Nev., USA

ABSTRACT: Predicting the flow of water between the ground surface and the atmosphere is a critical issue in the design of soil covers. Both short-term and long-term net infiltration rates are required to ensure that effluent stored within the tailings as well as precipitating water does not release constituents into the surrounding environment. The soil cover at Alligator Ridge was designed by characterizing the tailings and cover materials and by using the numerical model SoilCover. The results indicate that a net upward flux develops within the soil cover profile due to a moisture deficit in the system. The result of a 100- year simulation showed that after approximately 15 to 20 years of successive evaporation, the reduced water contents in the profile affected the ability of the tailings to transport moisture to the surface for evaporation. Subsequent movement of water within the tailings was primarily a result of vapour flow.

1 INTRODUCTION

Soil cover systems are commonly constructed over tailings and waste rock as one component of an overall decommissioning strategy. A soil cover is usually designed to limit the ingress of oxygen and/or water into the underlying waste material and is used to provide a suitable growing medium for vegetation. At many sites, the primary objective of a soil cover is to ensure that effluent stored within the tailings as well as precipitating water does not release constituents to the surrounding environment at levels that could degrade the local and regional water system. The ability to predict the flow of water between the soil surface and the atmosphere is a critical issue in the design of soil covers. Both short-term and long-term net infiltration rates are required to accurately predict the performance of the soil cover system into perpetuity.

In February 1997, the Unsaturated Soils Group at the University of Saskatchewan became involved the decommissioning of the Tailings Impoundment (TI) facility using a soil cover at the Bald Mountain Mine - Alligator Ridge Project. The research program had three main objectives: To hydraulically characterize the available soil cover material and the tailings; to determine an optimum cover thickness that would support vegetation and act as a store and release cover system to minimize infiltration into the tailings; and to determine the long-term performance of the optimum soil cover design.

2 BACKGROUND INFORMATION

The Alligator Ridge gold mine is a Placer Dome North America - Bald Mountain Mine property located in east central Nevada at an elevation of 2,012m above sea level (Figure 1). The climate

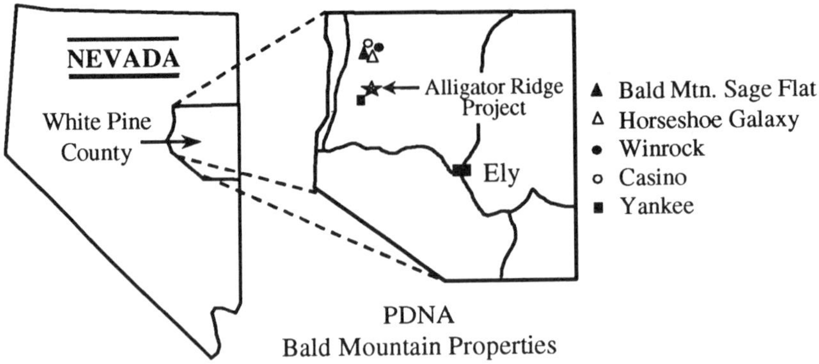

Figure 1. Location of the Alligator Ridge mine site.

is classified as semi-arid with an annual precipitation of 200 to 300mm. Annual pan evaporation from nearby Ruby Lake has been measured at 1542mm.

The tailings impoundment (TI) was constructed in 1987 and is located in a shallow valley that drains west to east with an upstream catchment area of approximately 29 km^2. The subsoil conditions at the site consist of alluvial deposits which are typically less than 3m thick at the edges of the valley but increase to deposit depths of 9 to 12m in the center of the valley (SRK 1987). The TI became inactive in 1990 after approximately 500,000 tons of material had been deposited into the facility.

A primary feature of the Alligator Ridge TI is the presence of a compacted soil liner and drainage system. A coarse drainage layer was placed over a 30cm compacted soil liner and pipes installed in the drainage layer were attached to a small sump that was used to remove seepage from the base of the tailings. This system was operational for several years during filling of the impoundment, but has not experienced drainage for many years. It is believed that the coarse layer is currently drained and is acting as a capillary break, inhibiting the downward flow of tailings pore-fluid.

Flow of surface water in the area is intermittent. A drilling program initiated during the site investigation revealed the phreatic surface to be located at a depth of approximately 150 to 180m below the ground surface. No perched water tables or seepage springs were found to be located within an 8km radius of the impoundment area.

A series of test pits were excavated in August of 1996 to determine the nature of the tailings within the impoundment. The test program revealed that the tailings consisted of a 30cm layer of brown, oxidized tailings underlain by 480cm to 900cm of black, unoxidized tailings. The tailings were found to be fairly moist at a depth of 120cm and were of a homogeneous consistency throughout the excavated profile (Dunlap 1996).

The proposed cover design for the Alligator Ridge TI consisted of placing 0.6 to 0.9 m of loose colluvium material, which was native to the area directly on top of the existing tailings. The purpose of a loose cover system such as the one proposed, is to prevent infiltrating water from entering the tailings. The infiltration is limited through a store and release mechanism. This type of cover system performs well in arid climates and its success has been demonstrated at the Golden Sunlight Mine in Montana and the Kidston Gold Mine in Australia (USG, 1997a). The store and release cover system relies on a cover material which possesses a significant storage potential and a climate which maintains a high evaporation rate to extract water stored within the cover and therefore prevent infiltration. This approach is an alternative to incorporating a material with a lower hydraulic conductivity such as a compacted clay barrier. Compacted clay barriers are known to fail in arid climates as a result of drying and cracking.

3 MATERIAL CHARACTERIZATION

Two material samples were sent to the University of Saskatchewan from Alligator Ridge for material characterization. The samples consisted of a full gradation tailings and the colluvium material intended for use as a soil cover for the existing TI. The laboratory program included grain-size analyses (both mechanical sieve and hydrometer analyses), measurement of the soil-water characteristic curve and determination of the saturated hydraulic conductivity for each sample.

3.1 *Grain-size analyses*

The tailings material was determined to be a uniform and fine-grained material with 87% finer than 0.075mm (#200 sieve) and 60% finer than 0.037mm. The colluvium material is a coarse, well-graded material with 40% finer than 4.75 mm (#4 sieve) and 15% finer than 0.075 mm (#200 sieve).

3.2 *Soil-water characteristic curves*

Figure 2 presents the measured soil-water characteristic curves for both the tailings and colluvium materials.

The soil-water characteristic curve for the fine-grained tailings was established using a small, modified pressure plate (Fredlund & Rahardjo 1993). Preliminary tests revealed that the tailings had a tendency to experience a significant volume change during desaturation. As a result, the tailings sample was pre-consolidated to a pressure of 120 kPa before it was placed in the modified pressure plate. The matric suctions applied to the tailings sample ranged from 0 to 500 kPa. The colluvium material contained a significant fraction of large-sized particles (60% of the particles were larger than 4.75 mm). As a result, the soil-water characteristic curve was

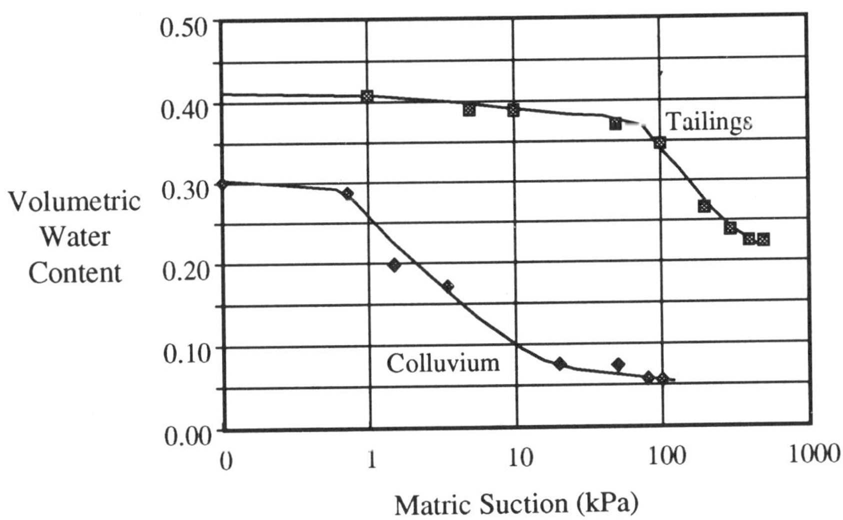

Figure 2. Soil-water characteristic curves for the Alligator Ridge tailings and colluvium material.

established using a large-scale, modified pressure plate, which enabled the entire sample to be used in the characterization (Herasymuik, 1996). The matric suctions applied to the colluvium sample ranged from 0 to 100 kPa.

The air entry value (AEV) of the tailings was measured to be approximately 80 kPa or 8 m of negative water pressure head. Once the AEV is exceeded, the tailings, due to their uniform nature, desaturate relatively quickly. The AEV of the colluvium sample is 0.7 kPa or 0.07 m of negative water pressure head. The drying curve for the colluvium is not as steep as the drying curve for the tailings due to the well-graded nature of the colluvium. However, the colluvium begins to stabilize at a residual water content at approximately 80 to 100 kPa.

3.3 *Saturated Hydraulic Conductivity*

The saturated hydraulic conductivity of the tailings was determined using a modified oedometer apparatus, which measures both volume change and saturated hydraulic conductivity. The consolidation pressures applied to the tailings sample ranged from 5.8 kPa to 717 kPa resulting in void ratios of 3.13 to 0.38 respectively. Over the tested void ratio range, the saturated hydraulic conductivity of the tailings varied almost two orders of magnitude from 9.6×10^{-6} cm/s to 1.6×10^{-9} cm/s.

A falling head permeameter test was used to determine the saturated hydraulic conductivity of the colluvium material. Over a series of four tests, the average hydraulic conductivity of the colluvium was determined to be 3.0×10^{-4} cm/s.

4 NUMERICAL ANALYSIS

The numerical modelling program was carried out using SoilCover (USG, 1997). SoilCover is a transient, one-dimensional, finite element, soil-atmosphere flux model that links the sub-surface saturated/unsaturated groundwater system and the atmosphere above the soil to represent the soil-atmosphere continuum.

The model uses a physically based method for predicting the exchange of water and energy between the atmosphere and a soil surface. SoilCover predicts the evaporative flux from a saturated or an unsaturated soil surface on the basis of atmospheric conditions, vegetation cover, soil properties and antecedent soil conditions. A modified Penman formulation (Wilson, 1990) is used to compute the actual rate of evaporation from the soil/atmosphere boundary. The theory is based on the well-known principles of Darcy's and Fick's Laws which describe the flow of liquid water and water vapour, and Fourier's Law to describe conductive heat flow in the soil profile below the soil/atmosphere interface.

4.1 *SoilCover input parameters and boundary conditions*

The modelling program was designed to identify the optimum thickness of a loose cover that would minimize infiltration and potential long-term seepage from the TI. Three different soil cover depths (60cm, 75cm and 90cm) were used in the modelling program. All three profiles included a tailings depth of 900cm.

Site specific climatic data which included daily max/min temperatures, relative humidity values, average daily wind speeds and daily precipitation were used as input for SoilCover. The data was obtained from the Bureau of Land Management's remote access weather station located 0.8 km from the TI. As neither pan evaporation or net radiation values were available from the site, measured monthly pan evaporation values from nearby Ruby Lake were used.

The weather data used in the modelling program covered the period from January through December, 1994. The total precipitation recorded during that year was 203mm. To simulate the effectiveness of the store and release soil cover mechanism during a "wetter" year, additional simulations were completed for each of the three profiles using 305mm/year. The additional 102mm was distributed evenly over the existing precipitation events to be consistent with the other meteorological conditions reported for those days. The total annual pan evaporation for 1994 was measured at Ruby Lake to be 1542mm.

The presence of the compacted soil liner and drainage system existing under the TI provided the basis for the initial boundary conditions used in the modelling program. It is believed that the coarse drainage layer is currently drained and is acting as a capillary break inhibiting the downward flow of tailings pore-fluid. The test pit excavations conducted within the TI in 1996 revealed an oxidized layer approximately 30cm thick underlain by 9m of black, unoxidized tailings. The air entry value of the tailings was determined during the laboratory program to be approximately 80kPa or 8m of water pressure head, which would indicate that a significant portion of the tailings were tension-saturated within the TI when the drainage initially subsided. Evaporation would have since decreased the water contents near the surface of the tailings.

The initial boundary condition used in the model, consisted of a slightly negative suction at the base of the tailings to simulate the capillary break between the drainage layer and the tailings material. The suctions increased linearly to a maximum suction of 90kPa at the top of the 9m tailings profile. The initial suction profile through the cover material was 90 kPa throughout, which corresponds to the residual moisture content of the colluvium material.

5 SOILCOVER RESULTS

The output generated by SoilCover includes daily fluxes and net annual cumulative fluxes of potential and actual evaporation, potential and actual evapotranspiration (if applicable), net infiltration and surface runoff. Potential evaporation is calculated throughout the simulated day using net radiation (or pan evaporation as was used in this simulation), air temperature, relative humidity and wind speed. Actual evaporation is also computed during the each simulated day based on the moisture available in the soil for evaporation and the climatic conditions imposed at the soil/atmosphere interface. Net infiltration for each day is calculated as follows:

$$NI = P - AE - R \qquad (1)$$

where NI = net infiltration (mm/day), P = precipitation (mm/day), AE = actual evaporation (mm/day) and R = runoff (mm/day).

In addition, SoilCover cumulates net fluxes throughout the simulation at a user-defined node. For all of the Alligator Ridge simulations, the user-defined node was selected to be at the interface between the soil cover and the tailings. The SoilCover generated output values are summarized in Table 1. Negative net infiltration fluxes indicate a net loss of water from the profile across the surface node. Positive fluxes at the user-defined node indicate an upward movement of pore-fluid from the tailings into the soil cover.

The primary objective of the soil cover at Alligator Ridge was to ensure that infiltrating water and water stored within the tailings would not leach out of the impoundment and degrade the waters of the state of Nevada. The purpose of the store and release cover was to take in the water from the precipitation events and then release the moisture by an evaporative process later.

The key parameter used in determining the optimum soil cover thickness at Alligator Ridge was the net flux calculated at the base of the cover. A negative net flux would represent percolation of surface water through the soil cover into the underlying tailings, indicating that the soil cover is not functioning as a store and release system. For all the simulations modelled, the net flux calculated at the base of the cover was positive indicating a net upward movement of tailings pore-fluid into the soil cover. This indicates that there was actually a moisture deficit within the cover system. There was not enough moisture applied during the course of the year to satisfy the evaporative demand and so a net upward gradient formed. On the basis of the net flux values calculated at the base of the cover, a 60cm soil cover is sufficient to limit a positive infiltrative flux into the tailings.

Table 1. SoilCover generated output.

Simulation	Cover Thickness cm	Measured Annual Precipitation mm	Measured Potential Evaporation mm/yr	Calculated Actual Evaporation mm/yr	Calculated Net Infiltration mm/yr	Calculated Flux at Base of Cover mm/yr
Case 1	60	203	1542.0	215.3	-11.6	11.9
Case 2	60	305	1542.0	317.2	-11.8	11.2
Case 3	75	203	1542.0	215.2	-11.5	11.7
Case 4	75	305	1542.0	316.3	-10.9	11.2
Case 5	90	203	1542.0	215.6	-11.9	12.1
Case 6	90	305	1542.0	317.1	-11.7	11.2

The effect of increasing the amount of annual precipitation for all three profiles can be evaluated by comparing the actual evaporation values. The actual evaporation fluxes increased by approximately 102mm/yr which is equal to the additional precipitation added to the system. The extra water applied to the system only partially satisfied the water deficit within the system. Increased precipitation meant more water was available for evaporation and so actual evaporation values increased over the year (see Table 1). The demand for water from the underlying tailings was slightly reduced resulting in slightly less negative flux values at the base of the cover.

Figures 3 and 4 show the degree of saturation profiles for the 60cm and 90cm soil cover profiles.

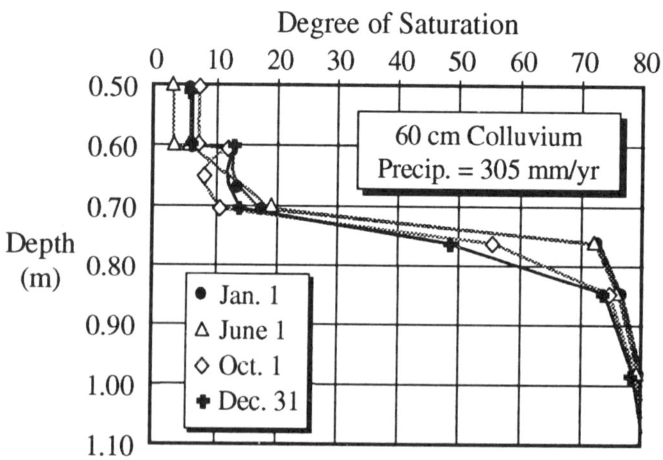

Figure 3. Degree of saturation profile for a 60 cm loose colluvium cover with 305 mm/yr precipitation.

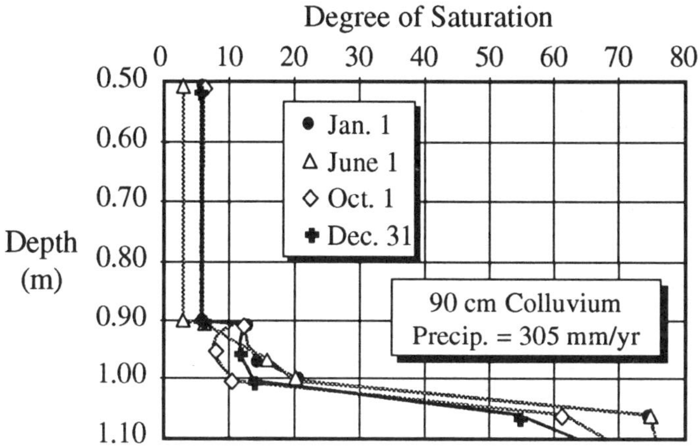

Figure 4. Degree of saturation profile for a 90cm loose colluvium cover with 305 mm/yr precipitation.

It can be seen in the two degree of saturation figures that there is very little difference in moisture redistribution between the two simulations. Throughout the year, the degree of saturation in both cover profiles wet and dried, but remained below 10% saturation for the entire year. The tailings gradually desaturated as a result of the pore-fluid moving upward towards the cover/tailings interface.

5.1 *A 100-year numerical simulation to evaluate long-term performance*

The results from the numerical modelling program to determine the optimum cover thickness showed that 60cm of loose colluvium successfully limited the downward movement of water into the tailings. The results also showed that a net upward gradient developed at the tailings / soil cover interface, even under wet year conditions. The formation of this upward flux raised the question of how long the mine can expect the significant migration of constituents from the underlying tailings pore-fluid into the soil cover.

The 100-year simulation investigated the performance of a 60 cm, loose colluvium cover and an applied precipitation rate of 305mm/yr. The purpose of the long-term simulation was to determine the extent of tailings desaturation following placement of the soil cover. Plant transpiration was simulated by defining the presence of a "poor" quality grass with a maximum root depth of 30cm. The growing season was identified as being from April 1 to October 1 of every year.

The initial degree of saturation profile and the saturation profile after 10, 40 and 100 years are presented in Figure 5. The water contents within the cover layer fluctuated over any given year (see Figure 3), but the net yearly change in moisture within the soil cover was negligible. The tailings profile, however, showed a gradual and steady drying trend that was driven by the net yearly evaporative demand at the surface of the cover.

Figure 6 shows the net upward flux from the tailings into the soil cover for ten year intervals during the long-term simulation. The initial rate of water loss from the tailings can be attributed to higher water contents within the tailings profile. At higher water contents, the ability of the tailings to transport a greater quantity of water to the surface and plant roots is enhanced due to a greater unsaturated hydraulic conductivity. After approximately 15 to 20 years of excessive evaporation, the reduced water contents and subsequent decrease in unsaturated hydraulic conductivity, affect the ability of the tailings to transport moisture upwards for evaporation. It should be noted that SoilCover takes into account moisture distribution by vapour movement as

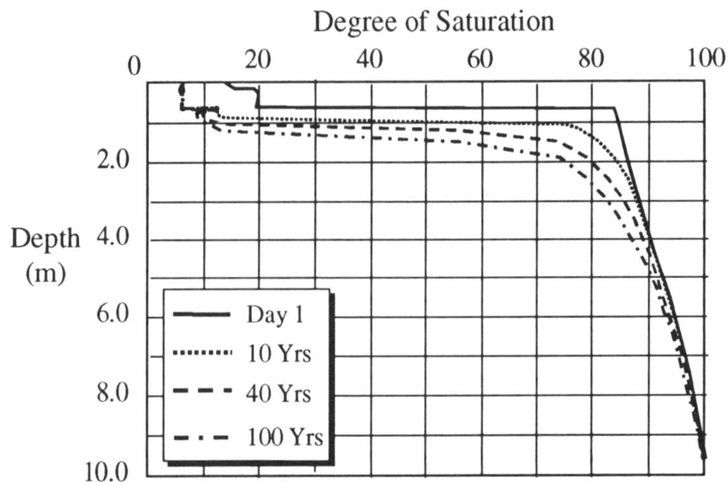

Figure 5. 100-yr degree of saturation profile for a 60cm cover with an applied precipitation of 305 mm/yr.

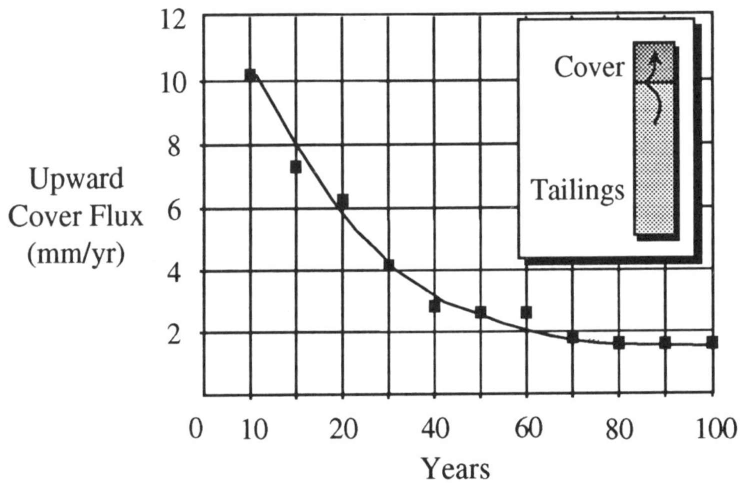

Figure 6. Net upward flux determined at the base of the cover over 100 years.

well as liquid flow. Moisture movement from the tailings into the soil cover tapered off over the 100 year simulation to a steady-state rate of approximately 1.6 mm/yr (Figure 6).

6 SUMMARY AND CONCLUSIONS

The optimum thickness of soil cover required over the existing tailings impoundment at Alligator Ridge was determined to be 60cm of native colluvium. A 100-year numerical simulation was completed to determine how long a significant flux of tailings pore-fluid can be expected to migrate upward into the soil cover. The modelling results showed that the initial vertical moisture movement is relatively large at 10 mm/yr, but drops off rapidly to a steady – state rate of 1.6 mm/yr after 70 to 80 years. The numerical modelling program shows that the regional waters of the state will not be affected by the tailings impoundment at Alligator Ridge.

Both the state and federal agencies (Nevada Division of Environmental Protection and the Bureau of Land Management) have reviewed the numerical modelling program described in this paper and have accepted the design of the soil cover at Alligator Ridge and the concept of using SoilCover as a predictive modelling tool for the design of soil covers.

REFERENCES

Dunlap, Shannon W. 1996. ARM Tailings Impoundment Characterization Sampling Plan. *Report submitted to the Bureau of Mining Regulation and Reclamation*, Carson City, NV.

Fredlund, D.G., Xing, Z. & Huang, S.Y. 1994. Predicting the permeability function for unsaturated soils using the soil-water characteristic curve. *Canadian Geotechnical Journal*, Vol. 31, No. 4: 533-546

Fredlund, D.G. & Rahardjo, H., 1993. *Soil Mechanics for Unsaturated Soils.* New York: John Wiley & Sons

Herasymuik, G. M., 1996. *Hydrogeologic characterization of an existing waste rock pile.* M. Sc. Thesis, University of Saskatchewan, Saskatoon, Saskatchewan, Canada.

SRK, 1987. Design of the Tailings Impoundment for the Alligator Ridge Mine. *Internal Report from Steffen Robertson and Kirsten (Colorado) for Amselco Minerals, Inc.* Report No. 03004/1.

USG, 1997a. *A compilation of papers on recent work on mine reclamation and decommissioning.* Unsaturated Soils Group. Department of Civil Engineering, University of Saskatchewan, Saskatoon, Canada.

USG, 1997b. *SoilCover V.4.0 User's Manual,* Unsaturated Soils Group. Department of Civil Engineering, University of Saskatchewan, Saskatoon, Canada.

Wilson, G.W. 1990. *Soil evaporative fluxes for geotechnical engineering problems.* Ph.D. Thesis, University of Saskatchewan, Saskatoon, Saskatchewan, Canada.

Tailings and Mine Waste '99 © 1999 Balkema, Rotterdam, ISBN 90 5809 025 6

Comparative reclamation soil cover modeling in an arid environment

P.E. Kowalewski
SRK Consulting, Lakewood, Colo., USA

ABSTRACT: Estimates of post-closure percolation through a reclamation soil cover over tailings at a proposed mine in southern New Mexico were completed in 1996 using the Hydrologic Evaluation of Landfill Performance (HELP) model version 3.04. Recent work by Khire et al. (1997) and Meyer et al. (1996) has shown that the HELP model may overestimate percolation rates in arid and semi-arid environments. These recent findings led to the hypothesis that the percolation rates previously determined at the Copper Flat site may be high. To support this hypothesis, the HELP model and SoilCover, a finite element program, were run in one year transient simulations under various conditions. The results obtained using the models were then compared to see if the HELP model estimated higher rates of percolation through the modeled soil profiles and to assess whether additional modeling at the Copper Flat site using the SoilCover model is warranted.

1 BACKGROUND

In 1994, Alta Gold Company Inc. (Alta) purchased the Copper Flat property from Gold Express Corporation. The Copper Flat site was initially developed and operated for a 3-month period in 1982 by the Copper Flat Partnership, Ltd. The area was reclaimed in 1986 (BLM, 1996). Alta proposes to reinitiate mining activities at the historic Copper Flat site near Hillsboro, New Mexico. Copper and molybdenum would be extracted from ore through standard milling procedures. The site is located approximately 3 miles northwest of Hillsboro and 30 miles southwest of Truth or Consequences, New Mexico. Mining activities at the site would occur for approximately 10 years, followed by approximately 3 to 4 years of reclamation.

The site is located in an arid environment. On an average annual basis, the site receives approximately 13 inches (330 mm) of precipitation while pan evaporation approaches 85 inches (2150 mm).

In order to fulfill the requirements of the New Mexico Mining Act, Alta had to present a reclamation plan for the site. The reclamation plan included the reclamation of the waste rock dump and tailings impoundment through the construction of soil covers over the tailings impoundment and waste rock dump surfaces to minimize infiltration of meteoric waters through the waste material. As part of the permit requirements, estimates of infiltration through the reclamation soil cover had to be presented for the purpose of assessing potential impacts to regional groundwater. The Hydrologic Evaluation of Landfill Performance (HELP) model version 3.04 (Schroeder et al., 1995) was used to estimate infiltration through the reclamation soil covers constructed over the waste rock facilities and the tailings impoundment.

Recent work by Meyer et al (1996) and Khire et al (1997) involving field tests using instrumented cover systems has shown that the HELP model may significantly overestimate infiltration through soil covers constructed at sites located in semi-arid or arid climates.

Several finite element computer codes (e.g. UNSAT, UNSAT-H, TRACER3D, HYDRUS-1D, HYDRUS -2D, and SoilCover) have been developed for determining the near-surface water balance in landfills and soil cover systems. The finite element codes present more mathematically rigorous representations and solutions of the equations that govern flow through the near-surface system.

2 INTRODUCTION

These recent findings have led to the development of the hypothesis that the HELP model may have overestimated the amount of infiltration passing through the soil cover, through the waste material, and impacting the groundwater at the Copper Flat site. To test this hypothesis, transient simulations for a soil profile consisting of the reclamation soil cover and tailings slimes was analyzed using the HELP model, version 3.06 (Schroeder et al., 1996), and a finite element code, SoilCover, version 4.0 (Geo-Analysis 2000, 1997).

The results obtained from SoilCover were then compared to results obtained using the HELP model. Significant variables compared from each model included: infiltration into the soil profile, evapotranspiration, and percolation through the soil column.

The goal of this report is not to present estimates of long-term percolation of meteoric water through the soil profile after reclamation is completed, but rather to assess whether further analyses utilizing an alternative modeling approach to the HELP model is warranted. This assessment is important because long-term percolation of meteoric waters through the tailings could impact regional groundwater quality. The potential impact involves the elevation of sulfate (SO_4^{2-}) levels in the groundwater due to leaching of sulfate from the waste materials in the tailings impoundment. As such, accurate estimates of long-term percolation rates are important with respect to the determination of any potential future impacts to the regional groundwater quality.

3 GEOTECHNICAL TESTING OF COVER SOILS

During reclamation, a soil cover, consisting of a minimum of 12 inches (30 cm) of locally derived alluvium will be placed over the tailings to minimize infiltration of meteoric water into the underlying tailings. The alluvium will be stripped from the tailings impoundment footprint area during impoundment construction and stockpiled for use during reclamation. Based on grain size analyses of the alluvium at the site, the soil is classified in the range of a clayey sand with gravel (SC) to a clayey gravel with sand (GC). The alluvium will be revegetated with native grasses to provide resistance to erosion and to aid in the removal of water from the soil profile through evapotranspirative processes. The alluvium may need to be amended with fertilizers or organic material to promote growth. The ability to revegetate the alluvium is not addressed in this report.

Bulk samples of alluvium and tailings ("slimes") material were obtained from the Copper Flat site and tested at Steffen Robertson and Kirsten's (SRK's) Denver geotechnical laboratory. Two samples of alluvium (Samples A and B) were obtained from different locations within the tailings impoundment area to compare physical properties. The following geotechnical tests were performed:

- Particle Size Distribution (ASTM D2217-85)
- Saturated Hydraulic Conductivity (ASTM D5084-90 or D2434-68)
- Moisture-Density Relationship (ASTM D698)
- Capillary Moisture Retention Test (ASTM D3152-72)

A summary of the geotechnical testing completed on the alluvium and tailings are included in Table 1.

Table 1 – Geotechnical Testing Summary

Soil	Grain Size Distribution (% passing Standard Sieve) 2"	1"	#4	#200	Maximum Dry Density (pcf)[1]	Optimum Moisture Content (wt %)[2]	Saturated Hydraulic Conductivity (cm/sec)
Tailings	100	100	100	63	---	---	$8.3x10^{-6}$
Alluvium A	100	96	60	20	118.5	12.7	$1.0x10^{-3}$
Alluvium B	100	93	59	13	121.6	11.8	$1.3x10^{-4}$

Capillary Moisture Retention Testing (moisture content (vol/vol) at given soil suction)

Soil	0 bar[3] 0 kPa	0.3 bar 30 kPa	0.5 bar 50 kPa	1.0 bar 100kPa	2.0 bar 200 kPa	5.0 bar 500 kPa	10.0 bar 1000 kPa
Tailings	0.3874	0.2998	0.2981	0.2884	0.2841	0.2689	0.2015
Tailings (R)	0.4145	0.3027	0.3009	0.2891	0.2838	0.2673	0.2020
Alluvium A[4]	0.5356	0.3819	0.3801	0.3746	0.3706	0.3621	0.3107
Alluvium A (R)[4]	0.4716	0.3695	0.3677	0.3593	0.3550	0.3403	0.2869
Alluvium B[4]	0.4886	0.2904	0.2892	0.2835	0.2793	0.2711	0.2348
Alluvium B (R)[4]	0.4917	0.2880	0.2864	0.2760	0.2700	0.2492	0.2038

Notes: (1) pcf – pounds per square foot
(2) moisture content in weight percentage; moisture content equals weight of water divided by dry weight of soil ($w = W_W/W_S$)
(3) 1 bar = 1019.7 centimeters (cm) of water
1 kilopascal (kPa) = 10.2 cm of water
(4) Laboratory results for these samples need to be adjusted using the coarse particle correction factor presented by Brakensiek et al., 1986

The results of the laboratory hydraulic conductivity testing of the tailings slimes do not necessarily reflect the hydraulic conductivity that will be achieved in the field. The tailings at the Copper Flat site will be deposited using a subaerial deposition technique. The subaerial deposition technique results in a stratified tailings mass, which introduces anisotropy in the tailings mass. Studies have shown the degree of anisotropy (k_H/k_V) of in-situ subaerially deposited tailings can range between 2 and 10 (Vick, 1983). The laboratory testing conducted on the tailings was on disturbed samples, which would destroy any stratification within the tailings sample. Without the stratification, the vertical permeability calculated using the flexible wall permeameter could be an order of magnitude higher than the in-situ permeability of the subaerially deposited tailings.

The two samples of alluvium are similar with respect to grain size distribution, hydraulic conductivity, and moisture retention properties. Therefore, an average of the values obtained through laboratory testing can be used to represent the properties of the alluvium in the modeling exercises.

4 SOIL COVER MODELING

4.1 Model Input Parameters

Both programs require soil properties to be input for each soil in the profile to be modeled, and both require climate data to be input in a daily format. The specific formats required for each program are somewhat varied. The following is a summary of the soil and climatic parameters used as input for the HELP and SoilCover models.

4.1.1 Soil Parameters

The HELP model requires the input of soil porosity, field capacity, wilting point, and saturated hydraulic conductivity for each soil type, while the SoilCover model requires at least three points on the soil water characteristic curve (SWCC) for each soil as well as the saturated hydraulic conductivity. Tables 2 and 3 provide a summary of the soil parameters used in the HELP and SoilCover models, respectively.

Table 2 – Soil Values Used in the HELP Model

Soil	Porosity (vol/vol)	Field Capacity (vol/vol)	Wilt Point (vol/vol)	Saturated Hydraulic Conductivity (cm/sec)
Alluvium[1]	0.288	0.19	0.15	$6x10^{-4}$
Tailings	0.401	0.30	0.20	$1x10^{-5}$
				$1x10^{-6}$ [2]

Notes: (1) Values are different from values presented in Table 1 due to application of coarse particle correction factor as per Brakensiek et al., 1986

(2) Simulations were run using both values of permeability due to the uncertainty associated with the measurement of the saturated hydraulic conductivity using a disturbed sample in the laboratory (see discussion in Section 3)

4.1.2 Climatic Parameters

Two different simulations were run for each soil profile: a "normal" year and a "wet" year. The "normal" year consisted of 365 days, with an average annual precipitation close to that expected at the Copper Flat site (approximately 13 inches). The "wet" year consisted of 365 days, with an average annual precipitation equal to approximately 125% of normal precipitation totals. The 125% value was considered to be representative of a typical wet season at the site.

Two years of daily climatic data were prepared for input into the HELP and SoilCover models. The only variable that was changed between the two years was the total depth of precipitation and the precipitation distribution. Actual daily precipitation distributions from the Hillsboro station were used in the simulations as opposed to simply multiplying the daily values in the normal year distribution by 125%.

All rainfall events were assumed to have a duration of 24 hours. This was done to maintain consistency between the two models. The HELP model assumes a 24 hour distribution, while the SoilCover model allows the duration for each precipitation event to be specified in the input file. The assumption that the precipitation occurs over 24 hours reduces the amount of runoff, which represents the worst case condition with respect to infiltration of meteoric water into the modeled soil profile.

A monthly summary of the precipitation data used in the models is presented in Table 4.

4.2 Modeling Methodology

The base cases used in the analyses consisted of two different soil profiles, both of which were used in transient simulations with a duration of one year. All simulations were run from January 1st through December 31st (365 day duration). The two profiles consisted of: 1•foot of alluvium over 20 feet of tailings (tailings $k_{SAT} = 1x10^{-5}$ cm/sec), and 1 foot of alluvium over 20 feet of tailings (tailings $k_{SAT} = 1x10^{-6}$ cm/sec).

The HELP model does not allow boundary conditions to be applied to the lower surface of the profile, whereas the SoilCover model requires a boundary condition to be specified. As a result, two different bottom boundary conditions were specified in the SoilCover runs: a zero pressure boundary (equivalent to saturated condition at the base of the column) and a moisture content equal to the soil's field capacity.

Table 3 – Soil Values Used in the SoilCover Model

Parameter	Soil Type: Alluvium	Soil Type: Tailings
k_{SAT} (cm/sec)	$6x10^{-4}$	$1x10^{-5}$ $1x10^{-6}$ (1)
M_V (1/kPa)	$5x10^{-3}$	$2x10^{-5}$
Weighted Quartz Percentage in Soil	0.8(2)	0.9(2)
Mass Specific Heat (J/m^3C)	800(2)	850(2)

Moisture Retention Data

Suction (kPa)	Moisture Content (vol/vol)	Suction (kPa)	Moisture Content (vol/vol)	Suction (kPa)	Moisture Content (vol/vol)
0.10	0.288	0.10	0.401	0.10	0.387
29.98	0.192	1.0	0.400	0.29	0.376
49.97	0.191	10.0	0.375	0.98	0.368
99.93	0.187	29.98	0.301	3.14	0.328
199.86	0.181	49.97	0.300	9.80	0.143
999.33	0.150	99.93	0.289	19.60	0.106
		199.86	0.284	61.84	0.068
		499.66	0.268	246.27	0.041
		999.33	0.202	1553.30	0.039

Notes: (1) Simulations were run using both values of permeability due to the uncertainty associated with the measurement of the saturated hydraulic conductivity using a disturbed sample in the laboratory (see discussion in Section 3)
(2) Default values for till and mine tailings presented within SoilCover documentation used for alluvium and tailings, respectively.

Table 4 - Monthly Summary of Precipitation Data Used in Models

Month	Normal Year (in)	Normal Year (mm)[1]	Wet Year (in)	Wet Year (mm)[1]
January	0.60	15.24	0.75	19.05
February	0.50	12.70	0.59	14.99
March	0.35	8.89	0:43	10.92
April	0.30	7.62	0.39	9.91
May	0.50	12.70	0.64	16.26
June	0.65	16.51	0.84	21.34
July	2.50	63.50	3.10	78.74
August	2.52	64.01	3.40	86.36
September	2.37	60.20	2.51	63.75
October	1.25	31.75	1.43	36.32
November	0.59	14.99	0.82	20.83
December	0.98	24.89	1.25	31.75
Annual Total	13.11	332.99	16.15	410.21

Note: [1] SoilCover requires data input in SI units

In addition, SoilCover allows an observation point to be specified, which allows drainage to be observed at two points in the soil profile. In the SoilCover analyses, an observation point was used at a depth of 6 feet below the ground surface.

5 MODELING RESULTS

There are three important variables presented in the output from each model: percolation through the soil column (out of the bottom of the profile), infiltration into the soil profile (at the soil-atmosphere boundary), and the total evapotranspiration. The HELP model presents annual totals for precipitation, runoff, evapotranspiration, percolation, and a combined change in water storage term in its output. The SoilCover model calculates a total water balance error in terms of a percentage of the total precipitation. Tables 5 and 6 contain a summary of the simulations completed using the HELP and SoilCover models, respectively.

Table 5 – Summary of HELP Model Results

Run Description	Precipitation (Norm/Wet)	Evaporative Depth (in)	HELP Model Output: Infiltration (in)	ET[1] (in)	Percolation (in)
20' Tailings (10^{-5})	Norm	18	13.11	13.35	0.2591
20' Tailings (10^{-5})	Wet	18	16.15	15.94	0.2490
20' Tailings (10^{-5})	Norm	24	13.11	14.25	0.2375
20' Tailings (10^{-5})	Wet	24	16.15	17.54	0.2555
20' Tailings (10^{-5})	Norm	30	13.11	14.86	0.2701
20' Tailings (10^{-5})	Wet	30	16.15	18.04	0.2958
20' Tailings (10^{-5})	Norm	36	13.11	15.48	0.2074
20' Tailings (10^{-5})	Wet	36	16.15	18.71	0.2720
20' Tailings (10^{-6})	Norm	18	13.11	13.37	0.0151
20' Tailings (10^{-6})	Wet	18	16.15	16.23	0.0254
20' Tailings (10^{-6})	Norm	24	13.11	14.25	0.0317
20' Tailings (10^{-6})	Wet	24	16.15	17.64	0.0317
20' Tailings (10^{-6})	Norm	30	13.11	14.84	0.0317
20' Tailings (10^{-6})	Wet	30	16.15	17.92	0.0309
20' Tailings (10^{-6})	Norm	36	13.11	15.23	0.0261
20' Tailings (10^{-6})	Wet	36	16.15	18.46	0.0267

Notes: [1] Evapotranspiration; the sum of evaporation and transpiration losses in the soil profile

6 DISCUSSION OF RESULTS

The HELP and SoilCover models produced different estimates of percolation through the soil profiles, while producing similar estimates of infiltration into the soil profile and evapotranspirative losses from the soil profile. This indicates the two models are producing different estimates of the amount of water going into, or being released from, storage within the soil column. In both of the base case simulations the HELP model estimated higher amounts of percolation through the soil column.

6.1 Alluvium over Tailings ($k_{sat} = 1x10^{-5}$ cm/sec)

A total of 8 HELP model simulations and 4 SoilCover simulations involving the alluvium over tailings (k_{sat} = $1x10^{-5}$ cm/sec) were completed. Comparing the results from the HELP and SoilCover base cases under normal (average) precipitation conditions (HELP model evaporative depth = 18 inches, SoilCover bottom boundary condition equal to field capacity of soil), the HELP model predicted approximately three and a half times as much percolation through the soil column as the SoilCover model (0.2591 inches versus 0.074 inches).

Table 6 – Summary of SoilCover Results

Run Description	Precipitation (Norm/Wet)	Bottom Boundary Condition[2]	SoilCover Model Output: Infiltration (mm)	ET (mm)	Percolation (mm)[3]
20' Tailings (10^{-5})	Norm	P = 0 kPa	333.01	341.24	-128.54
20' Tailings (10^{-5})	Wet	P = 0 kPa	409.96	400.69	-128.54
20' Tailings (10^{-5})	Norm	M.C. = F.C.	333.01	340.67	1.88
20' Tailings (10^{-5})	Wet	M.C. = F.C.	409.87	400.70	1.88
20' Tailings (10^{-6})	Norm	P = 0 kPa	333.01	332.64	-65.21
20' Tailings (10^{-6})	Wet	P = 0 kPa	409.86	394.25	-64.86
20' Tailings (10^{-6})	Norm	M.C. = F.C.	333.01	332.64	0.21
20' Tailings (10^{-6})	Wet	M.C. = F.C.	409.80	394.26	0.21

Notes:
1 In all of the simulations a maximum rooting depth of 18 inches was used to be consistent with the HELP model analyses.
2 Two different bottom boundary conditions were modeled, a matric suction equal to zero kPa (P = 0 kPa) [zone of saturation at base of soil profile] and a moisture content at the base of the profile equivalent to the soil's field capacity (M.C. = F.C.) [drained soil column]
3 Negative (-) values indicate upward flow at the base of the soil profile; water flowing into the profile at the base.

Under average precipitation conditions, calculation of the net flux across the soil-atmosphere boundary in both the HELP and SoilCover models shows that there is a negative flux across the boundary. The negative flux indicates more moisture is removed from the soil profile through evapotranspirative processes than is replenished through precipitation infiltrating into the soil column. This leads to the observation that the percolation occurring through the soil profile is a result of water being released from storage within the column and not meteoric water passing through the soil profile. The fact that percolation is not reduced by increasing the evaporative depth further supports this observation.

Under "wet" (125% of average annual precipitation) conditions, the HELP model predicted approximately three and a half times as much percolation through the soil column as the SoilCover model (0.2490 inches versus 0.074 inches). Increasing the evaporative depth to 36 inches in the HELP model did not reduce the amount of percolation through the soil column.

In the HELP model wet-year run, 16.15 inches of precipitation infiltrated into the column, while 15.94 inches was removed from the soil column through evapotranspirative processes, resulting in a gain of 0.21 inches of water in the soil profile. This value is less than the amount of deep percolation predicted in the model, which leads to the observation that water is draining from storage within the profile. Once again, this observation is supported by the fact that percolation was relatively unaffected by increasing the evaporative depth.

It should be noted that the amount of percolation predicted by SoilCover remained the same for both the normal and wet-year simulations, while the amount of percolation predicted by the HELP model slightly decreased in the wet-year simulation (from 0.2591 inches to 0.2490 inches). As discussed above, the values of percolation may be indicating water percolating through the soil profile is due to release of water from storage as opposed to the passing of meteoric water through the column.

The SoilCover analyses performed using a pressure equal to zero kilopascals (P = 0 kPa) bottom boundary condition showed an upward gradient at the base of the soil profile. In both the normal (average precipitation) and wet-year (125% of average annual precipitation) simulations, approximately 5 inches of water flowed into the soil profile across the bottom boundary.

An additional set of HELP model simulations were run in an attempt to imitate the initial conditions (saturation at the base of the column) used in the SoilCover simulations. The saturated conditions were imitated by saturating the bottom 1 foot of the soil profile. The results obtained from the SoilCover model could not be duplicated using the HELP model due to the fact that the HELP model does not model upward flow (capillary rise) outside of the evaporative zone.

6.2 Alluvium over Tailings ($k_{sat} = 1x10^{-6}$ cm/sec)

A total of 8 HELP model simulations and 4 SoilCover simulations involving the alluvium over tailings ($k_{sat} = 1x10^{-6}$ cm/sec) were completed. Comparing the results from the HELP and SoilCover base cases under normal (average) precipitation conditions (HELP model evaporative depth = 18 inches, SoilCover bottom boundary condition equal to field capacity of soil), the HELP model predicted approximately twice as much percolation through the soil column as the SoilCover model (0.0151 inches versus 0.0083 inches). Increasing the evaporative depth to 36 inches did not reduce the amount of percolation through the soil column.

Under normal precipitation conditions, application of the continuity equation across the soil profile shows that in the SoilCover model, a small amount of infiltrating water goes into storage, while in the HELP model, the source of the percolation is the release of water from storage within the profile. This is similar to the observation made for the previous case involving the alluvium over the tailings ($k_{SAT} = 1x10^{-5}$ cm/sec).

Under "wet" conditions, the HELP model predicts approximately three times as much percolation through the soil column as the SoilCover model (0.0254 inches versus 0.0083 inches). Increasing the evaporative depth to 36 inches did not reduce the amount of percolation through the soil column.

It should be noted that the amount of percolation predicted by SoilCover remained the same for both the normal and wet-year simulations, while the amount of percolation predicted by the HELP model increased in the wet-year simulation (from 0.0151 inches to 0.0254 inches).

In the HELP model wet-year run, 16.15 inches of precipitation infiltrated the column, while 16.23 inches was removed from the soil column through evapotranspirative processes, resulting in a net loss of 0.08 inches of precipitation. This value supports the observation that the percolation through the soil column is due to the release of moisture from storage as opposed to due to the percolation of meteoric water.

The SoilCover analyses performed using a pressure equal to zero kilopascals (P = 0 kPa) bottom boundary condition showed an upward gradient at the base of the soil profile. In both the normal and wet year simulations, approximately 2.6 inches of water flowed into the soil profile across the bottom boundary.

As in the previous case, an additional set of HELP model simulations were run in an attempt to imitate the initial conditions (saturation at the base of the column) used in the SoilCover simulations. Once again, the results obtained from the SoilCover model could not be duplicated using the HELP model due to the models inability to model upward flow outside of the evaporative zone.

7 CONCLUSIONS

The results of the HELP and SoilCover modeling show that the HELP model estimated between two and three times more percolation through the soil profile than the SoilCover model. It should be noted that these percolation values are not long-term estimates of percolation at the Copper Flat site, but were used as a basis of comparison between the HELP and SoilCover models.

Each of the two water balance models used in the analyses have advantages and disadvantages when compared to the other model used in the analyses. In the following section, the merits and drawbacks of each program are highlighted.

7.1 The Hydrologic Evaluation of Landfill Performance (HELP) Model

The main advantage of using the HELP model is the ability to model a large number of alternative designs in a relatively short period of time. The HELP model contains several soil and climatic databases, which allows the user to perform simulations for durations up to 100 years even if the user does not possess all of the required data from the project site. In these analyses, the HELP model took approximately 30 seconds to execute a one year simulation using a computer with a 133-Mhz Pentium processor and 16 Mb of RAM. Longer simulations, on the order of 100 years in duration, may take up to 5 minutes using a similar computer.

There are several shortcomings of the HELP model when compared to other models that predict the water balance in soil profiles. For example, the HELP model does not allow boundary conditions to be assigned to the soil profile and the model does not account for capillary rise (upward flow in the soil profile) outside of the user defined evaporative depth.

The HELP model does not allow a bottom boundary condition to be specified for the soil profile in the model. In addition, the model assumes the permeability at the base of the profile is infinite (i.e. the profile can drain freely). Therefore, careful attention must be given in the event that the soil profile drains into a relatively closed system. If the materials below the soil profile cannot transmit flow as rapidly as the modeled soil profile, estimates of percolation predicted by the HELP model may be unduly high.

The HELP model does not account for capillary rise in the soil profile outside of the user-defined evaporative depth, which has a significant effect in arid and semi-arid climates. In arid and semi-arid climates, significant upward gradients can form in soil profiles due to high evaporative losses from the soil surface. These effects are more significant when modeling soil profiles containing fine-grained soils.

The HELP model represents the soil water characteristic curve (SWCC) of each soil used in each simulation using 4 moisture values. Three of the moisture values (porosity, field capacity, and wilt point) are input by the user, while the fourth value (residual moisture content) is determined by the model using an empirical relationship. The SWCC is then generated using a function developed by Brooks and Corey. In coarse-grained soils, the slope of the SWCC is steep. The method the HELP model uses to estimate the SWCC is not able to model this curve steepness effectively. This will lead to errors in the calculation of the unsaturated hydraulic conductivity of the soils and the resultant percolation values.

7.2 SoilCover Model

The SoilCover model is a relatively rigorous program which requires detailed data to be input to evaluate a soil cover. The benefits of using the program include the representation of physical processes through complex mathematical functions. Where the HELP model makes a significant amount of simplifying assumptions, the SoilCover model very accurately simulates the actual processes occurring. Two examples of this are the calculation of the soil water characteristic curve (SWCC) and the requirement of boundary conditions in the SoilCover model.

The SoilCover model calculates the SWCC for each soil using up to 20 points input by the user. This allows the user to accurately represent each soil's SWCC, including steep curves generally produced when modeling coarse-grained soils. This representation of the SWCC also leads to more accurate estimates of the unsaturated hydraulic conductivity of the soils, which in turn leads to accurate estimates of percolation.

In the analyses presented in this paper, two different bottom boundary conditions were used: a pressure equal to 0 kilopascals (P = 0 kPa) and a moisture content equal to the field capacity of the soil. These two boundary conditions were selected to represent the range of bottom boundary conditions that could exist in the tailings impoundment. If the water table remains relatively close to the surface, the bottom boundary condition will resemble the P = 0 kPa boundary condition, while if the tailings drain completely, the bottom boundary condition will resemble the moisture content equal to field capacity (gravity-drained moisture content) boundary condition.

As can be seen from the analyses, the selection of the bottom boundary condition in the SoilCover model has a considerable effect on the estimate of percolation values. Using the P = 0 kPa bottom boundary condition shows upward flow in all of the soil profiles ranging from 2.5 to 5 inches per year. Using the moisture content equal to field capacity boundary condition shows percolation rates ranging from 0.0083 to 0.074 inches per year. These two boundary conditions obviously show considerable differences in results. Therefore, careful thought must be given to the bottom boundary condition assigned to the soil profile in the SoilCover model.

The drawbacks to the SoilCover model include intensive data requirements and the amount of time required for execution of the model. The simulations included in these analyses took approximately 24 hours to execute a one year simulation using a computer with a 133-Mhz Pentium processor and 16 Mb of RAM. In addition, the model can only be run for a period of 365 days. If a simulation greater than 365 days is desired, the input file must be updated and the program restarted, as opposed to the HELP model, which allows up to 100 years of simulation to be calculated in one model run.

The SoilCover model requires daily climatic data to be input for each day of simulation. Daily data required by the program include temperature (maximum and minimum), relative humidity (maximum and minimum), pan evaporation, and precipitation. If an initial estimate of cover performance is desired before significant data collection has been performed at the site, data collected at other locations must be used. The validity of using the data collected at other sites can be questioned in some situations, depending upon the proximity of the station collecting data to the site, the similarity in location with respect to elevation, orientation of the site, site exposure, and proximity to mountains or other topographic factors affecting the local climate.

7.3 Summary

A definite advantage can be realized through the use of an integrated modeling approach using both the HELP and SoilCover models to assess percolation through reclamation soil covers. The use of the HELP model as an initial screening tool to compare cover alternatives, followed by a more in-depth analysis on the preferred alternative using the SoilCover model appears to be the most cost–effective approach.

8 ACKNOWLEDGEMENTS

Robert N. Pratt and James S. Goff of Alta Gold Company, Inc. are acknowledged for granting permission to use data generated during reclamation cover studies at the Copper Flat site.

9 REFERENCES

Bureau of Land Management (1996). "Draft Environmental Impact Statement Copper Flat Project;" United States Department of the Interior, Bureau of Land Management, Las Cruces, NM.

Khire, M.V., Benson, C.H., and Bosscher, P.J. (1997). "Water Balance Modeling of Earthen Final Covers," American Society of Civil Engineers Journal of Geotechnical and Geoenvironmental Engineering, August 1997, pp. 744-754.

Kowalewski, P.E. (1997). "Comparative Water Balance Modeling of a Reclamation Soil Cover in an Arid Environment;" Master's Degree Engineering Report, Colorado School of Mines. Golden, CO. Submitted/Defended November 1997.

Meyer, P.D., Rockhold, M.L., Nichols, W.E., and Gee, G.W. (1996). "Hydrologic Evaluation of Methodology for Estimating Water Movement Through the Unsaturated Zone at Commercial Low-Level Radioactive Waste Disposal Sites," NUREG/CR-6346 PNL-10843 Pacific Northwest Laboratory, Richland, WA.

Tailings and Mine Waste '99 © 1999 Balkema, Rotterdam, ISBN 90 5809 025 6

Permeability testing of slurry trench cutoff walls at a mine tailings facility

Howard D. Plewes
Klohn-Crippen Consultants Limited, Richmond, B.C., Canada
Kevin Semenoff
Klohn-Crippen Consultants Limited, Sudbury, Ont., Canada

ABSTRACT: A series of soil-bentonite (SB) and cement-bentonite (CB) slurry trench cutoff walls were installed along the perimeter of a Northern Ontario tailings impoundment to reduce environmental impacts on the local groundwater system. Permeability of the walls was measured by laboratory testing of the backfill materials during construction, rising and falling head response tests in standpipes installed in the walls, construction of an enclosed 20 ft by 20 ft cell (CB only), and construction of a 10 ft by 10 ft mass block with a standpipe installed in the centre (CB only). The in-situ permeability of the SB wall was found to be in the order of 3×10^{-8} cm/s while the laboratory tests on small samples yielded permeability values of 6×10^{-7} cm/s to 2×10^{-7} cm/s. The in-situ permeability of the CB backfill was determined to be about 5×10^{-6} cm/s which was at the upper bound of the laboratory measurements which ranged from 4×10^{-6} cm/s to 1×10^{-8} cm/s. The apparent scale effect of the permeability of the CB walls was a key finding from the in-situ test work.

1 INTRODUCTION

As part of the development of the tailings storage facility for a large base metals mine in Northern Ontario, a series of slurry trench cutoff walls were installed in 1996 along the perimeter of the tailings impoundment to reduce environmental impacts on the local groundwater system. Prior to constructing the critical cutoff walls, an evaluation and comparison of both soil-bentonite (SB) and cement-bentonite (CB) wall types was carried out by construction of full-scale test walls. These walls provided data to evaluate the two construction methods for the site and to measure in-situ permeability characteristics of the walls. Eventually, CB walls were selected because of issues of trench stability.

Of particular interest in the comparison work was the measurement of the in-situ mass permeability of the two wall types. Permeability of the walls was measured by laboratory testing of the backfill materials during construction, rising and falling head response tests in standpipes installed in the walls, construction of an enclosed 20 ft by 20 ft cell (CB only), and construction of a 10 ft by 10 ft mass block with a standpipe installed in the centre (CB only). Extensive in-situ permeability testing was considered warranted for the project because of the general lack of published in-situ permeability test data on slurry trench walls. In general, most publications report the results of laboratory tests on small samples of the wall materials while true in-situ performance test data is rarely reported. The mine owner also viewed the work as an investment for future possible applications of slurry trench cutoff walls at their facilities.

This paper reviews the construction and installation of the test walls, and the results of the in-situ permeability measurements.

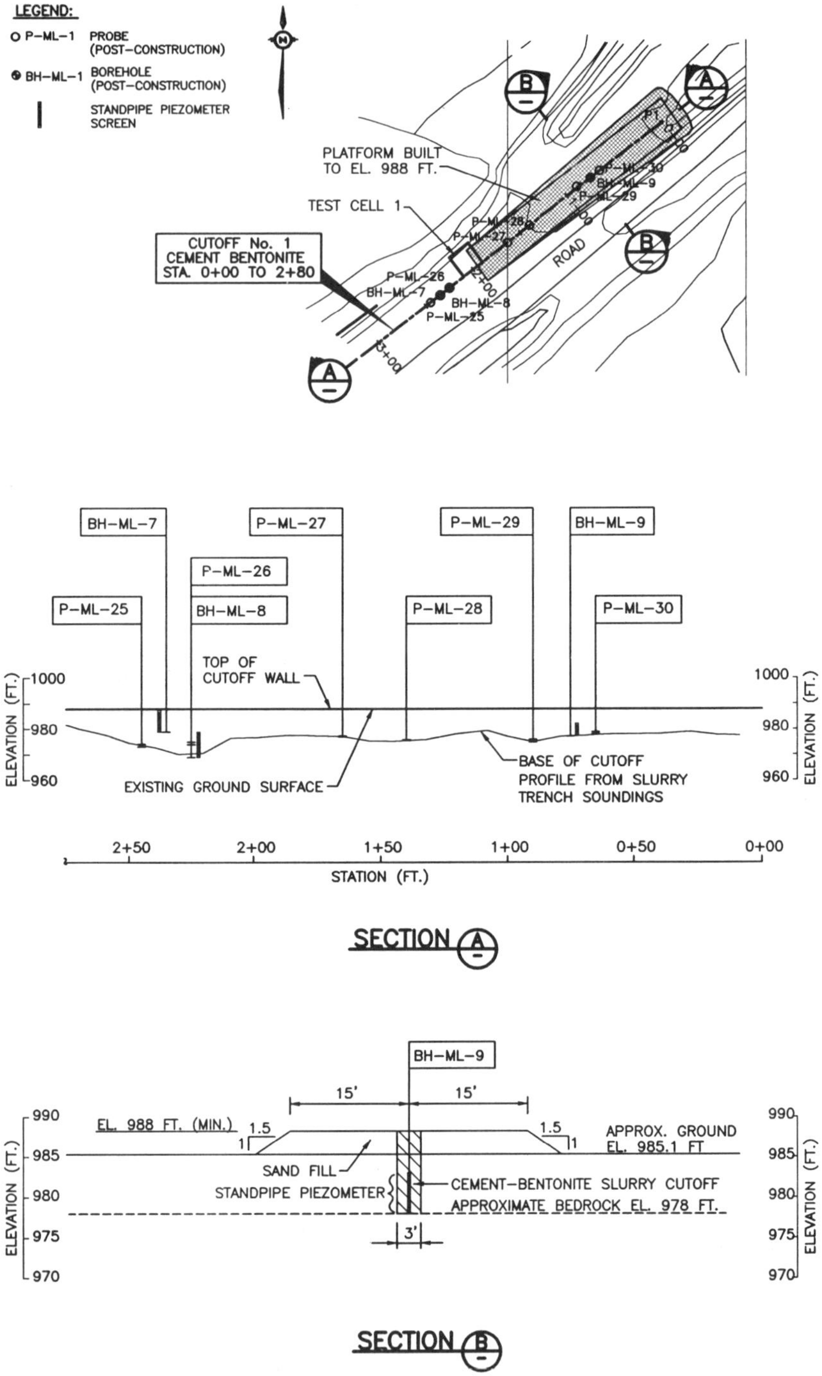

Figure 1 – Plan and sections of Site A – Cutoff No. 1

2 SLURRY TRENCH WALL CONSTRUCTION

The tailings area is impounded by a series of perimeter dams that are being constructed of spigotted tailings using the upstream construction method. The perimeter dams are located in shallow valleys between topographic highs in the regional bedrock surface. These valleys feature irregular terrain with bedrock exposed or thinly mantled by glacial till in higher areas; and sand and silt, lacustrine clays and silts, and muskegs in the base of the valleys.

Seepage passing through the dams will be collected in a toe seepage ditch. The water will be conveyed in the ditch to a central pump station where it will be reclaimed to the tailings impoundment. The seepage ditch is excavated through bedrock at higher elevations, but is founded in the overburden soils within the low points of the valleys. Generally, the bedrock is of low permeability and the ditch in these sections is unlined. The overburden soils in the base of the valleys range between a very low permeability clay to pervious coarse soils and till deposits. In order to reduce the quantity of seepage from the dams bypassing under the seepage collection ditch, a series of slurry trench cutoff walls were installed through the overburden soils to the underlying bedrock. These seepage cutoff walls were installed downstream of the seepage collection ditch to direct all seepage from the dams into the collection ditch.

A total of seven separate cutoffs were installed at the tailings facility. Five of the cutoffs were installed in an area referred to here as "Site A" and these were composed of shallow cutoffs to depths of 10 ft (3 m) to 33 ft (10 m). Figure 1 shows the plan and sections of Site A - Cutoff No. 1. The remaining cutoffs were installed at locations referred to here as "Site B" and "Site C", and ranged in depth up to 56 ft (17 m). Table 1 summarizes the details of the wall installations.

Prior to constructing the deeper and more critical cutoffs at Site B and Site C, it was decided that a comparison of both the soil-bentonite (SB) and concrete-bentonite (CB) wall types would be carried out by construction of full-scale test walls at Site A. As given in Table 1, three SB cutoffs were constructed using standard side of the trench backfill mixing and remote batch mixing of the soil-bentonite backfill. In both cases, the backfill consisted of an imported fine sand (15-30% fines content) mixed with 4% bentonite. Two CB walls were installed with concrete/water (c/w) ratios varying between 0.2 and 0.3. In the case of Site B and Site C, the bedrock was grouted prior to installation of the slurry cutoff walls. All walls were excavated 3 ft (0.9 m) in width.

Table 1. Summary of slurry trench cutoff walls

Location	Type	Length (m)	Area (m^2)	Maximum depth (m)	Comments
Site A					
Cutoff No. 1	CB	85	289	6.2	c/w = 0.2 and 0.3
Cutoff No. 2	SB	178	773	5.5	Batch Backfill Mixing
Cutoff No. 3	CB	75	295	5.9	c/w = 0.2
Cutoff No. 4	SB	74	284	7.0	Side of Trench Backfill Mixing
Cutoff No. 5	SB	51	244	7.0	Side of Trench Backfill Mixing
Site B	CB	157	1352	16.5	c/w = 0.3
Site C	CB	131	1051	17.1	c/w = 0.3

A detailed evaluation and comparison of the various wall types was made over a wide range of criteria. These included:

- continuity of the wall types;
- contact with the underlying bedrock;
- wall permeability;
- trench wall stability;
- construction problems;
- effectiveness of QA/QC procedures; and
- cost.

It is beyond the scope of this paper to review all the findings and conclusions from the evaluation of the two wall types. However, the key factor in selecting the CB wall type for the

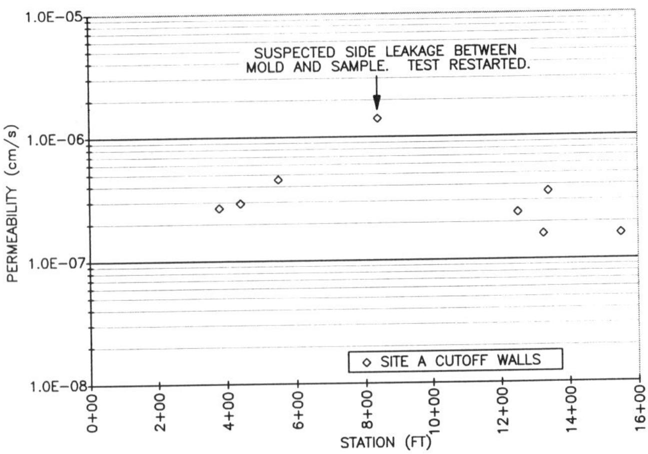

Figure 2 – Laboratory permeability test results for soil–bentonite backfill

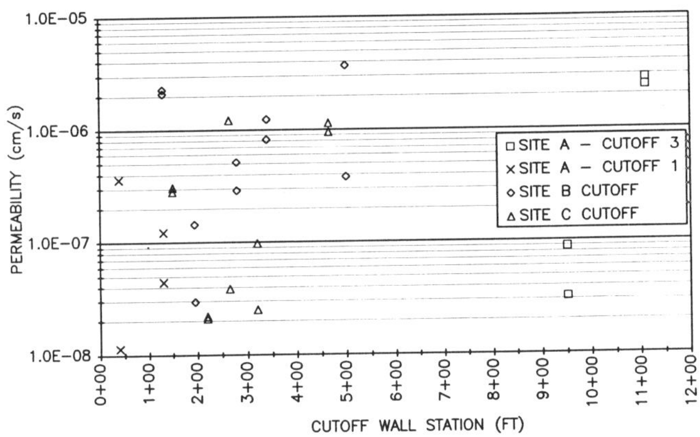

Figure 3 – Laboratory permeability resullts for cement–bentonite backfill

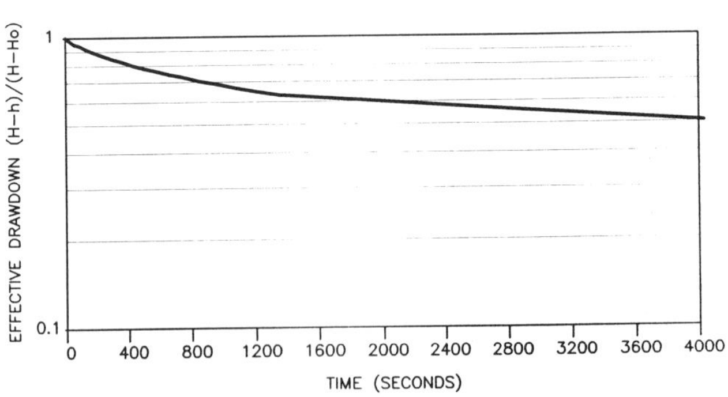

Figure 4 – Typical falling head standpipe response test result

deeper cutoffs at Site B and Site C was the better control of the trench wall stability in the soft clays and loose, saturated sands that were present at these dam sites.

3 LABORATORY PERMEABILITY TEST RESULTS

Samples of the SB and CB backfill were recovered from the trenches during construction. In the case of the SB, the samples were preserved in sealed containers for transportation to the laboratory. The SB samples were poured into triaxial test apparatus and the permeability determined at a consolidation pressure of 25 kPa and gradient of 20. The samples of CB were cast in the field in 75 mm diameter by 150 mm tall plastic moulds and then cured in a moisture room. Samples were batched in sets of four to allow strength and permeability determination at 7 days and 28 days. The permeability was determined in a triaxial cell apparatus at a confining stress of 25 kPa and gradient of 20.

Figures 2 and 3 present the results of the permeability tests for the SB and CB walls, respectively. The permeability of the SB backfill ranged narrowly from 6×10^{-7} cm/s to 2×10^{-7} cm/s. The range of data for the CB backfill was higher, varying from 4×10^{-6} cm/s to 1×10^{-8} cm/s.

4 IN-SITU PERMEABILITY TEST RESULTS

4.1 *Standpipes in Wall*

Rising and falling head response tests were conducted in standpipe piezometers installed inside the cutoff walls to determine the in-situ permeability characteristics of the backfill materials. For these tests, a slug equivalent to a 2 ft to 5 ft change in the static water level in the standpipe was introduced with water level readings then taken at regular intervals to monitor the rate of rise and fall of the water levels back to the static level. Figure 1 shows a typical standpipe installation. Figure 4 presents the falling head response test data for this standpipe.

Many of the piezometers installed in the Site A cutoff walls were difficult to test because the low static water level in this area meant that part of the screened section of the piezometers was often above the static water level. Analyses of these tests included the effect of the sand pack around the standpipes. None of the difficulties associated with test analysis of the Site A piezometers were encountered at Site B and Site C because care was taken during the installation of piezometers at the latter sites to ensure that all screened sections were installed below the static water table outside the walls

The field test data were analyzed using both a computer program (Bouwer-Rice Method) and a computer spreadsheet (Hvorslev method). The Bouwer-Rice solution, which takes into account the effect of the unsaturated sand pack, was determined to be most appropriate for the Site A cutoff walls. For Site B and Site C, the Hvorslev solution was determined to be adequate. The permeabilities calculated for the tests are summarized in Table 2. The in-situ permeability values were reduced by a factor of 2 to account for the thin wall geometry as compared to the infinite thickness assumed by standard analytical solutions. The factor of 2 was determined by finite element seepage analyses.

For the SB walls, the single test value of 3×10^{-8} cm/s was lower than the laboratory triaxial tests (6×10^{-7} to 2×10^{-7} cm/s) on SB samples collected during construction. The very low field permeability of the SB backfill was corroborated by the very soft condition of the material more than 1 month after the completion of construction. There is insufficient data to draw any definite conclusions between the field and laboratory measurements.

For the CB walls, the range of permeabilities from the field tests (2×10^{-5} to 1×10^{-6} cm/s) are at the upper end of the range of permeabilities from the laboratory triaxial tests (4×10^{-6} to 1×10^{-8} cm/s). This result suggests that scale effect is a significant factor in determining the permeability of CB cutoff walls.

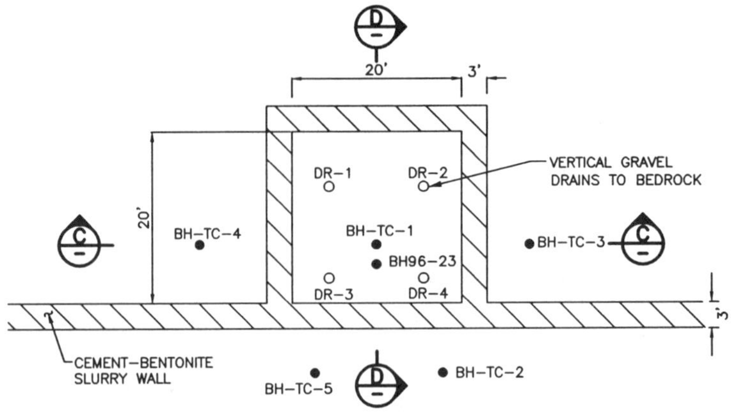

Plan – Layout of cement–bentonite test cell 1

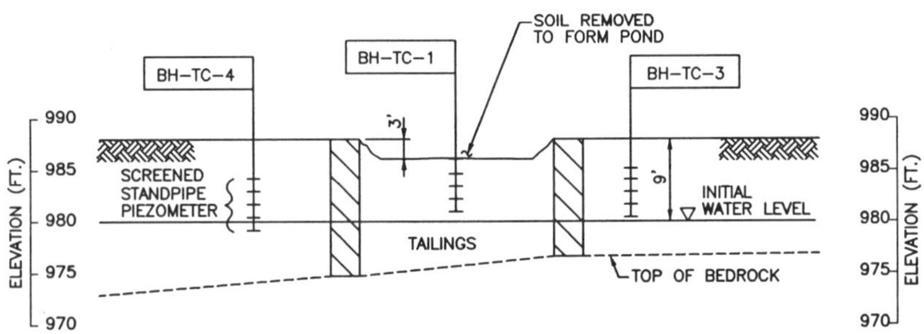

SECTION C

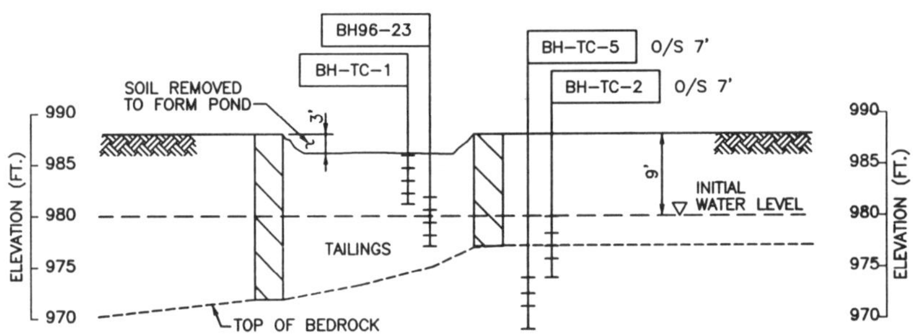

SECTION D

Figure 5 – Plan and sections of cement–bentonite test cell

Table 2. Summary of in-situ wall permeabilities measured in standpipe response tests

Location	Type	Station (ft)	Permeability (cm/s) Falling Head Test	Permeability (cm/s) Rising Head Test	Comments
Site A					
Cutoff No. 1	CB	0 + 75	9 x 10^{-6}	1 x 10^{-6}	c/w = 0.2
		2 + 35	9 x 10^{-6}	3 x 10^{-6}	c/w = 0.3
Cutoff No. 3	CB	10 + 20	5 x 10^{-6}	3 x 10^{-6}	c/w = 0.2
Cutoff No. 5	SB	15 + 20	3 x 10^{-8}	-	Side of Trench Backfill Mixing
Site B	CB	1 + 80	4 x 10^{-6}	-	c/w = 0.3
		3 + 30	1 x 10^{-5}	2 x 10^{-5}	c/w = 0.3

4.2 *Test Cell*

During construction of Site A - Cutoff No. 1 (CB), a square test cell measuring 20 ft by 20 ft was constructed with the main cutoff wall forming one of the side walls of the test cell. This test cell was used to conduct a full-scale permeability test to assess the mass permeability of the cutoff wall, base contact, and bedrock. The location of the test cell is shown on Figure 1. Detailed plan and sections of the test cell are shown on Figure 5.

Prior to conducting the test, piezometers were installed inside and outside of the test cell to allow careful monitoring of groundwater levels during the test. Standpipe piezometers were installed at two depths in the overburden soils inside the test cell; and in the overburden, bedrock and overburden-bedrock contact outside of the test cell. In addition, four vertical sand and gravel drains were installed through the overburden to bedrock inside the test cell to ensure full saturation of the overburden when the test was conducted. At the test location, the overburden soils consisted mainly of tailings which had been previously deposited in the area.

After the piezometers were installed, a hydraulic excavator was used to carefully remove about 3 ft of overburden from the inside of the test cell. The excavation was surveyed to accurately calculate the excavated volume and then the excavation was filled with water to the top of the wall to saturate the overburden within the test cell. Water level readings in the pond and piezometers were taken every few hours for the first day and twice daily thereafter. After each reading, the water level was topped up to maintain maximum head. After ten days, the water level readings indicated that the overburden had become fully saturated and the water levels in the various standpipes had stabilized at close to steady state flow conditions. The final test was then started.

The final test was conducted by filling the inner pond in the evening, measuring the piezometer and pond water levels immediately after filling, and recording time. The following morning, the same water levels were again measured and the time was recorded. This procedure was repeated four times in dry weather conditions. Figure 6 graphically presents the water level readings taken during the saturation and final test periods. The average seepage loss from the cell was found to be 10.5 cm^3/s (0.16 Usgpm).

A simple approach to estimate the equivalent wall permeability is to assume that all seepage passes through the wall (i.e. no bedrock seepage) and all head loss occurs within the wall. For the wall area of 91 m^2, and head loss of 9 ft (2.7 m) across the 3 ft (0.9 m) wide wall, the equivalent wall permeability is then calculated to be:

$$k = \frac{10.5 \text{ cm}^3/\text{s}}{91 \text{ m}^2 \times (100)^2 \times {}^{2.7}/_{0.9}} = 4 \times 10^{-6} \text{ cm/s} \quad (1)$$

A second approach was to conduct a two-dimensional finite element seepage analysis of a representative section through the test cell using a computer seepage analysis program. A best match fit of the seepage rate and observed water level rises in the tailings and bedrock yielded

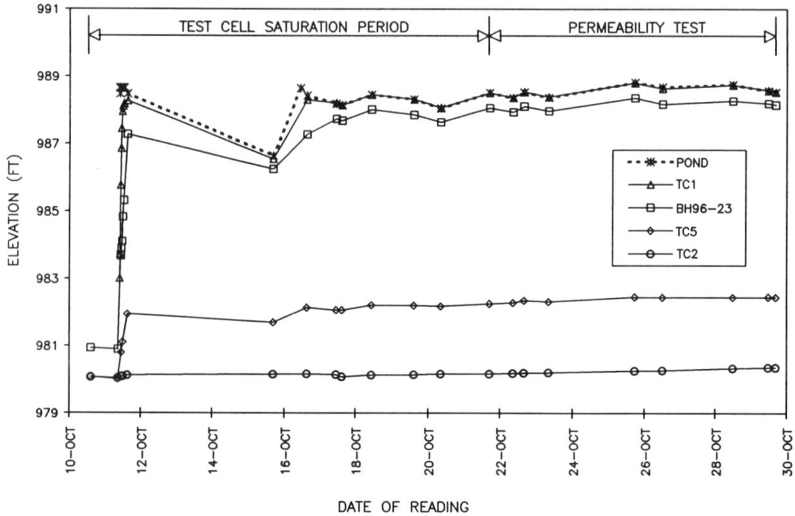

Figure 6 – Water level reading for cement–bentonite test cell

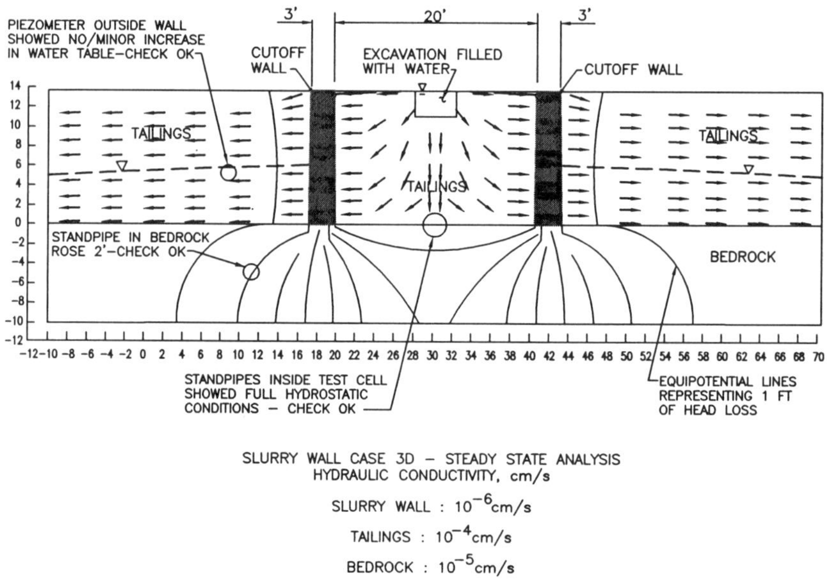

Figure 7 – Finite element seepage analysis of test cell

the following set of best case permeability parameters:

- k_{wall} = 3×10^{-6} to 5×10^{-6} cm/s
- $k_{tailings}$ = 10^{-3} to 10^{-4} cm/s
- $k_{bedrock}$ = $<10^{-5}$ cm/s

A sample output from the computer generated solution is presented on Figure 7. The range of wall permeability of 3 x 10^{-6} cm/s to 5 x 10^{-6} cm/s compares quite favorably with the wall permeability estimated in the first approach. The average estimated wall permeability of 4 x 10^{-6} cm/s is slightly lower than the average permeability of 7 x 10^{-6} cm/s measured by the standpipes in the CB walls (Table 2).

4.3 *Test Block*

A 10 ft wide by 10 ft long test block was excavated to a depth of 43 ft within the final cutoff wall alignment at Site C. A single standpipe was installed in the centre of the wall and the mass permeability of the CB determined by falling and rising head response tests. The results are given in Table 3 below. Because of concerns of the quality of the seal of the standpipe in the test block, the results for the rising head test are considered more applicable. The permeability of the test block was therefore taken to be 3 x 10^{-6} cm/s. This is very close to the results from test cell.

Table 3. Standpipe response test results for large scale test block

TEST METHOD	PERMEABILITY (cm/s)
Falling Head Test	9 x 10^{-6}
Rising Head Test	3 x 10^{-6}

5 CONCLUSIONS

1. The in-situ field permeability of the CB wall was determined by three completely different methods. The results are as follows:
   - Standpipes in CB walls: 1 x 10^{-6} to 2 x 10^{-5} cm/s
   - Full Scale Test Cell: 3 x 10^{-6} to 5 x 10^{-6} cm/s
   - Standpipe in CB test block: 3 x 10^{-6} cm/s

   The results indicate that the average mass permeability of the CB walls is about 5 x 10^{-6} cm/s. This permeability value is consistent with other data reported in the literature for CB walls.

2. The in-situ permeability of the CB walls of 5 x 10^{-6} cm/s determined by field tests is at the upper range of the permeabilities determined by laboratory testing (4 x 10^{-6} cm/s to 1 x 10^{-8} cm/s). This result suggests that scale effect on permeability testing of CB backfill is a significant factor and reliance only on laboratory testing may be suspect.

3. The single in-situ field permeability of 3 x 10^{-8} cm/s for the SB walls was less than the range of laboratory permeability values (6 x 10^{-7} cm/s to 2 x 10^{-7} cm/s). While the basis for comparison is very limited, it is our opinion that the laboratory test methods provided a realistic assessment of the in-situ permeability and the results may in fact be conservative.

Groundwater and geochemistry

Tailings and Mine Waste '99 © 1999 Balkema, Rotterdam, ISBN 90 5809 025 6

Determination of migratory parameters with the help of mass transfer inverse problem solution

E.D.Pavlik & G.N.Pavlik
Rostov State University, Rostov-on-Don, Russia

ABSTRACT: Poor prediction ability of pollution migration in grounds and aquifers of the rocks is connected with a high complexity extent of a hydrogeochemical system. The methods suggested have permited to compensate a traditional deficit of regime observation data at the particular object. The migratory parameters obtained with their help have ensured a reliable solution of the direct boundary problem of mass transfer and allowed to make the prediction calculations for the change of ecological situation near a pollution source.

1. INTRODUCTION

The reliability of calculations and the results of mathematical simulation of migration are determined by the parameters of the real system "polluted waters - ground medium". Determination of these parameters for a lot of pollutants is difficult and, up to now, insolved problem. Hydrogeological data necessary for their determination can be established by the way of the vertical holes. Interpolation between vertical lines is a rather difficult task since there are no data about the continuity of the layers between holes and invariability of their thickness. Under such conditions the assurance of prediction using mathematical simulation increases greatly if the coefficients are found with the help of inverse problem solution. Though the solution of inverse problems is connected with great mathematical difficulties due to their poor attributes and ill-definition of statement. However the model parameters found in such a way integrally include spatially-temporal changeability of properties determining water-mass trasfer in nature, inaccuracy in the description of chemical interactions and primary direction of solution movement through micropores and cracks.

In the present article an attempt to use the method for the determination of migratory characteristics of rocks with the help of mass transfer invers problem solution has been made. This direction in the investigations appeared in the connection with the necessity of scientifically valid choice of engineering measures for the protection from pollution Saburovo Water Intake supplying the town Voskresensk with drinking water. As it was described in papers (Pavlik 1996,1997) the Intake is situated in a zone of effect of the main pollution source - in the dump of phosphogypsum and pyrite cinders stored on the bank of the river Moscow. The mistakes in choice of parameters when mathematical simulation of migratory processes may lead to mistaken predictions and as a result - or to the extremely high expenditures for the water-protecting measures or, on the contrary, to the situation when even taking concrete actions the pollution of the Intake will take place not in the periods and not by those components which are predicted.

2. STATEMENT AND SOLUTION OF INVERSE PROBLEM

Traditional description of water-mass transfer processes in unsatureted zone is based on a classical Richard's equation for unsaturated water flow and Fick's equation for convective-dispersive transfer of salts (Galitskaya & others 1997).
If the change of solution density is small, and processes of transverse dispersion are allowed to be ignore then the analysis of the main regularities of mass transfer is reasonably to limit by the examination of elementary one-dimensional migratory flows

$$R_D \frac{\partial c}{\partial t} = \frac{\partial}{\partial x}\left(\frac{D \partial c}{\partial x}\right) - V \frac{\partial c}{\partial x} - \lambda \cdot c \qquad (1)$$

at the constant concentration in the entrance border

$$c(0,t) = c_* \qquad (2)$$

and uniform initial condition

$$c(x,0) = 0 \qquad (3)$$

where

R_D - coefficient of deceleration corresponding to conventionally linear adsorption ;

D - coefficient of hydrodispersion;

V - filtration velocity of pore water;

λ - coefficient of degradation whose physical meaning is described in (Shakelford 1997).

To solve the initial-boundary value problem the magnitudes R_D , λ, D should be known. They are determined according to the results of laboratory and field hydrogeological · investigations, experimentally-migratory studies, observations of underground waters and changes in hydrogeological situation at the areas of revealed pollutions.
It is mostly difficult to determine parameter D characterizing total effect of molecular diffusion and convective trasport of a substance in ground medium. In consequence of combination of diffusive and convective factors in coefficient D its magnitudes vary in a wide range depending on rock type, porosity, cracking.
Under such conditions the assurance of prediction using mathematical simulation considerably increases if parameter D is found from the solution of inverse problems.
Below it is presented one of the possible statements of inverse problem for the determination of coefficient of hydrodispersion D according to the measurement data of substance concentration at some distance from the pollution source.
We shall consider a particular case of the problem (1-3) at the constant movement velocity of pore water and absence of (bio)-degradation (λ=0).
The formulated problem is related to the category of inverse coefficient problems for the equations of parabolic type. This problem is ill-defined one and when numerical processing the regularization is required (Tihonov & Arsenin 1990).
Let us consider linearized statement of the main problem (1-3). let us represent $D(x)$ in the form of small-parameter ε expausion:

$$D(x) = D_0 + \varepsilon \cdot D_1(x) + \varepsilon^2 \cdot D_2(x) + \ldots \qquad (4)$$

Solution (1-3) we also seek in the form of a series:

$$c = c_0(x,t) + \varepsilon \cdot c_1(x,t) + \varepsilon^2 \cdot c_2(x,t) + \ldots \tag{5}$$

Substituting it into the main equation (1) and equating terms at the equal power we shall obtain two direct problems relatively $c_0, c_1, c_2 \ldots$ and so on. If we limit ourselves to the examination of two first terms of the series we shall obtain direct problems for c_0, c_1.

Problem I

$$\begin{aligned} R_D \frac{\partial c_0}{\partial t} &= D_0 \frac{\partial^2 c_0}{\partial x^2} - V \frac{\partial c_0}{\partial x} \\ c_0(0,t) &= c_* \\ c_0(x,0) &= 0 \end{aligned} \tag{6}$$

Problem II

$$\begin{aligned} R_D \frac{\partial c_1}{\partial t} &= D_0 \frac{\partial^2 c_1}{\partial x^2} + \frac{\partial}{\partial x}\left(D_1(x) \frac{\partial c_0}{\partial x}\right) \\ c_1(0,t) &= 0 \\ c_1(x,0) &= 0 \end{aligned} \tag{7}$$

2.1. CONSTRUCTION OF PROBLEM I SOLUTION

We shall seek the solution of the initial-boundary value problem (6) in the form:

$$c_0(x,t) = \exp(\lambda \cdot x + \mu \cdot t) F_0(x,t) \tag{8}$$

The function $F(x,t)$ satisfies an equation:

$$\frac{\partial F_0}{\partial t} = a^2 \frac{\partial^2 F}{\partial x^2} \tag{9}$$

with the following boundary and initil conditions:

$$\begin{aligned} F(0,t) &= c_* \\ F(x,0) &= 0 \end{aligned} \tag{10}$$

where

$$a^2 = \frac{D_0}{R_D}$$

$$\lambda = \frac{V}{2D_0}$$

$$\mu = -\frac{V^2}{4D_0 R_D}$$

Problem (9-10) has the known solution:

$$F(x,t)=\frac{c_*}{2}\exp(-\mu\cdot t)\left[\exp(-i\sqrt{\alpha\cdot\mu}Erfc\left(\frac{1}{2}\sqrt{\frac{\alpha}{t}}-i\sqrt{\mu\cdot t}\right)-\right.$$

$$\left.-\exp(i\sqrt{\alpha\cdot\mu}Erfc\left(\frac{1}{2}\sqrt{\frac{\alpha}{t}}+i\sqrt{\mu\cdot t}\right)\right] \tag{11}$$

where

$Erfc(z)=\frac{2}{\sqrt{\pi}}\int_z^{\infty}\exp(-t^2)dt$ - calculus of probability integral;

$$\alpha=\frac{x}{a}$$

Then we shall use an asymptotic expansion of function $Erfc\sqrt{z}=1-\frac{e^{-z}}{\sqrt{\pi\cdot z}}, z>>1$ and then the solution of the initial-boundary value problem (6) has a form:

$$c_0(x,t)=\frac{c_*}{2}\exp(\lambda\cdot x)[\exp(-i\sqrt{\alpha\cdot\mu})\frac{\exp(\frac{1}{2}\sqrt{\frac{\alpha}{t}})}{\sqrt{\pi}(\frac{1}{2}\sqrt{\frac{\alpha}{t}})^{\frac{1}{2}}}+$$

$$\exp(i\sqrt{\alpha\cdot\mu})\frac{\exp(\frac{1}{2}\sqrt{\frac{\alpha}{t}})}{\sqrt{\pi}(\frac{1}{2}\sqrt{\frac{\alpha}{t}})^{\frac{1}{2}}}] \tag{12}$$

Formula (12) can be used for the datermination of the main term of expansion (4) D_0. Let be known the magnitudes of polluting component concentration in a certain point (hole) situated at a distance d from the pollution source for two moments of time t_1, t_2. Let us construct the equation for the determination a as follows:

$$\frac{c_0(d,t_1)}{c_0(d,t_2)}=\sqrt{\frac{t_1}{t_2}}\exp[(-\frac{1}{2}\sqrt{\alpha})(\frac{1}{\sqrt{t_1}}-\frac{1}{\sqrt{t_2}})] \tag{13}$$

Limiting ourselves to the first term of expansion (14) and taking the logarithm of both parts of equality (13) we shall obtain:

$$\ln\frac{c_0(d,t_1)}{c_0(d,t_2)}=\frac{1}{2}\ln\sqrt{\frac{t_1}{t_2}}+\frac{1}{2}\sqrt{\alpha}(\frac{1}{\sqrt{t_2}}-\frac{1}{\sqrt{t_1}}) \tag{14}$$

Let all degrees for the magnitudes of function $c_{01}, c_{02}, c_{03}...c_{0N}$ in the point with coordinate d be done with the equal accuracy.

Denote $A=d/a$ and find A according to the method of the least squares. The assessment of peramcter A is made such that deviation square sum of the measured magnitudes c_{0k} from the calculated ones, that is a value:

$$S=\sum_{k=1}^{N}[\ln\frac{c_0(d,t_{1k})}{c_0(d,t_{2k})}-\frac{1}{4}\ln\frac{t_{1k}}{t_{2k}}-\frac{1}{2}A\left(\frac{1}{\sqrt{t_{2k}}}-\frac{1}{\sqrt{t_{1k}}}\right)]^2 \tag{15}$$

takes on the least value.

Minimization of function $S=S(A)$ amounts to solving of an equation:

$$\frac{\partial S}{\partial A}=0 \tag{16}$$

From equation (16) we obtain the basic formula for the determination of A:

$$A=\frac{\sum_{k=1}^{N}\left(\ln\frac{c_0(d,t_{1k})}{c_0(d,t_{2k})}-\frac{1}{2}\ln\sqrt{\frac{t_1}{t_2}}\right)}{\frac{1}{2}\sum_{k=1}^{N}\left(\frac{1}{\sqrt{t_{2k}}}-\frac{1}{\sqrt{t_{1k}}}\right)} \tag{17}$$

2.2. CONSTRUCTION OF PROBLEM II SOLUTION

The solution of problem II we seek in the form:

$$c_1(x,t)=\exp(\lambda\cdot x+\mu\cdot t)F_1(x,t)$$
$$c_0(x,t)=\exp(\lambda\cdot x+\mu\cdot t)F_0(x,t) \tag{18}$$

Substituting (18) in (7) we have obtained an equation in partial derivatives relatively to function F_1 with zero initial and boundary conditions:

$$R_D\frac{\partial F_1}{\partial t}=D_0\frac{\partial^2 F_1}{\partial x^2}+\frac{\partial D_1}{\partial x}(\frac{\partial F_0}{\partial x}+\lambda\cdot F_0)+D_1(\frac{\partial^2 F_0}{\partial x^2}+2\lambda\frac{\partial F_0}{\partial x}+\lambda^2\cdot F_0) \tag{19}$$

Applying to the equation (19) Laplace transform and missing intermediate calculations we shall obtain the solution of problem II in the forms:

$$c_1(p,x)=-\frac{c_*}{p}\exp(\gamma\cdot x)\int_0^x D^*(\xi)\left[\frac{\gamma}{\gamma_1 D_0}\left(\frac{V}{D_0}-\gamma\right)\exp(\gamma_1(x-\xi))-\frac{\gamma^2}{D_0\gamma_1}\right]d\xi \tag{20}$$

where

$$\gamma=\frac{V}{2D_0}-\frac{1}{2}\gamma_1$$

$$\gamma_1=2\sqrt{\frac{V^2}{4D_0^2}+\frac{pR_D}{D_0}}$$

$$D^*(\xi)=\frac{D_1(\xi)}{D_0}$$

p - parameter of Laplace transform.

Applying to the equation (20) an inverse Laplace transform with allowance for boundary condition

$$c(d,t) = c_* f(t) \tag{21}$$

we shall obtain an integral equation of Fredgolm of I kind to find function $D_1(x)$:

$$\int_0^d D^*(\xi)K(\xi,t)d\xi = f_1(t) \tag{22}$$

where kernel $K(\xi,t)$ is present in the form:

$$K(\xi,t) = \frac{1}{2\pi \cdot i}\int_{c-i\cdot\infty}^{c+i\cdot\infty} K_1(d,\xi,p)\frac{\exp(\gamma\cdot d)}{p}\exp(pt)dp$$

$$K_1(x,\xi,p) = \frac{\gamma}{\gamma_1}((\gamma+\gamma_1)\exp(\gamma_1(x-\xi)) - \gamma) \tag{23}$$

and functions $f_1(t), f(t)$ is an inverse Laplace transform from the form

$$\bar{f}_1(p) = -\bar{f}(p) + \frac{\exp(\gamma\cdot d)}{p}$$

$$\bar{f}(p) = \lambda\frac{\partial D_1}{\partial x} + \lambda^2 D_1 - \frac{\sqrt{p}}{a}\left(\frac{\partial D_1}{\partial x} + 2\lambda D_1\right) + \frac{p}{a^2}D_1$$

As it is known (Tihonov & Arsenin 1990) a construction of integral equation (22) solution is an ill-defined problem and requires a regularization when constructing an approximate solution.

The approximate solution of equation (22) can be made on the basis of 3 approaches.

1. Digitization of assumed equation by the way of integral substituon with integral sum according to some quadrature formula is made. As a result a system of linear algebraic equations with degenerate or poorly attriduted matrix is obtained.
2. Digitization of smoothing functional is made and then a problem of function minimization of many variables with following determination of regularization parameter is solved.
3. Digitization of boundary problem is made and then an obtained system of linear algebraic equations is solved.

3. SPECIFIC APPLICATION OF THE METHODS

To illustrate the method presented we limit ourselves to expansion (4) by dominant term of series .

Coefficient of hydrodispersion integrally reflecting the properties of hydrogeological medium is determined by formula (17) using the results of the sample analysis obtained from the holes in a vicinity of phosphogypsum dump.

Researched region in valley of in the river Moscow in geological meaning is represented by Quarternary deposits, late Carboniferous limestones, middle Carboniferous chalky clays and dolomites (Pavlik 1997). The quartenary deposits consist of filling and aluvial formations.

The filling soils are represented by phosphogypsum and pyrite cinders; the alluvial soils are represented by sandy loams, loam soils and sands of different granular with the inclusion of gravel and pebble on a border with the original rocks. Alluvial deposits are water-bearing rocks for water-bearing horizon, that has a direct hydraulic connection with surface waters.
There is Podolsk-Myachkov exploited aquifer in the reseached region that lays on depth 35-80 m from the surface.
The main task of mathematical simulation is to determine the extent and the area of pollution spreading entering this horizon from phosphogysum dump. The most toxic components of phosphogypsum, a by-product of ferfilizer production, are fluorides, iron, phosphates, sulphates and some heavy metals. Their polluting capacity is determined in the aggregate with the total toxicity and quantitative content in wastes. Table.1 shows the data on the content of sulphates (mg/l) sampled from the holes.

Table 1.

Hole №	1989 year	1989 year	1990 year
6T	878	5621	9679
5K	333	3325	9629
84	1734	5279	5999

Using formula (17) we take 1960 - a year of phosphogypsum dump disposal - as a beginning of count $(t=0)$, d - hole depth, $d=20$ m, $R_D=1$

Then for hole № 84 we obtain:

A=936 D=0.00117 m^2/24-hours

for hole № 5K

A=2685.6 D=0.00041 m^2/24-hours

for hole № 6T

A=1909.6 D=0.000573 m^2/24-hours

CONCLUSIONS

In our opinion the new and rather simple methods allowing to find in explicit form the main term of hydrodispersion coefficient expansion are presented.
In such a case the results of standard experimental-migratory investigations are used that simplifies a procedure of informational supply of mathematical models considerably.

REFERENCES

Galitskaya I.V., Golubeva G.A., Chesalov S.M 1997. Studying the role of rocks in formation of human-induced pollution of ground waters in zones affected by the industrial - waste dumps (with special reference to Voskresenskyi industrial zone). *Geoelоliya, RSA*, № 1, pp. 58-69.

Pavlik G.N 1997. Application of mathematical simulation techniques to prediction of diffusion of pollutants from a phos-phogypsum dump. *Proceeding of the V International Conference on Talings and Mine Waste'97*, Colorado,USA, Balkema, pp. 393-399.

Pavlik G.N., Babayan V.R 1996. Mathematics modelling of migration process and distribution of contaminations from phosphogypsum wastes. *Foundation and Soil Mechanics*, Stroyizdat, № 4, pp. 13-17.

Shackelford C.D 1996. Modeling and analysis in environmental geotechnics: An overview of practical applications. *Proceedings of 2-nd Int. Congress on Environmental Geotechnics*, Osaka, Japan, Balkema, Rotterdam, pp.1375-1404.

Tihonov A.M., Arsenin A.A 1990. Methods of solution of uncorrect problems Moscow, Nauka, p.p. 37-84.

Tailings and Mine Waste '99 © 1999 Balkema, Rotterdam, ISBN 90 5809 025 6

Induced hardpan formation in a historic tailings impoundment, Røros, Norway

D.C. Ettner
Geocare AS, Oslo, Norway

G. Braastad
Norwegian Pollution Control Authority, Oslo, Norway

ABSTRACT: Historic sulfide mines at Røros, Norway, are protected from surficial disturbance by a conservation plan. Therefore, traditional remediation methods to reduce acid rock drainage (ARD) at these mines would alter the landscape, and can not be implemented. An alternative remediation method to provide a subsurface chemical barrier by inducing hardpan formation with limestone/lime amendments has been tested at the Storwartz tailings impoundment. After one year, hardpan layers composed of sulfates and iron oxyhydroxides were observed. The hardpan layers significantly reduced the saturated hydraulic conductivity of the tailings and pore gas studies showed lower oxygen concentrations in shallow tailings. This field test showed that an induced hardpan can be formed at shallow depths in deeply oxidized tailings, and under extreme climatic conditions. The hardpan would function as a chemical barrier to reduce ARD and would allow the historical landscape to be kept intact.

1 INTRODUCTION

The historic Røros mining district is located approximately 150 km SE of the city of Trondheim in the county of Sør-Trøndelag, Norway (Figure 1). During 333 years of continuous mining activity, ending in 1977, the district produced copper, zinc and sulfur from several massive sulfide deposits. The town of Røros including nearby slag piles are presently included in UNESCO's World Heritage List. The UN's International Council on Monuments and Sites and the Norwegian Directory for Cultural Heritage are discussing plans to increase the area of preservation to include some of the district's mines. In spite of the harsh winter climate, the sulfidic mine wastes generate acid rock drainage (ARD) which severely impacts nearby water courses. Efforts for site remediation in the area are in conflict with efforts for historic preservation, especially when sources for ARD are considered a part of the historical landscape (Grøndalen 1997).

Of particular conflict is the preservation of the Storwartz mine area, where copper ore was first discovered and mined. This area includes Old Storwartz mine (1645-1700), New Storwartz Mine (1708-1920's), Hestkletten Mine (1690-1756), and a flotation plant/tailings impoundment (1947-1970) which also contains sulfidic waste from Olavs Mine (1937-1972). Yearly runoff of ARD from sulfidic rock dumps and the tailings impoundment is estimated at 1900 kg Cu, 4200 kg Zn and 8.7 kg Cd (Arnesen 1996). The waters from Storwartz drain into Prestbekken Creek which is devoid of aquatic life along its 3 km stretch. Prestbekken Creek contributes water to Djupsjøen Lake and the Hitter River, which flows into the Glomma River, Norway's longest river (Figure 1). Studies show that aquatic life in Djupsjøen Lake, Hitter River and Glomma River is negatively impacted by ARD (Kjellberg 1991; Grande et al. 1996). The Norwegian Pollution Control Authority (SFT) set a goal to reduce the total release of copper and zinc 60 to 90% from ten of the largest abandoned and active mines combined, by 1995 (SFT 1992). This goal was achieved, although remediation plans at Storwartz were halted because of concerns from the Norwegian Directory for Cultural Heritage.

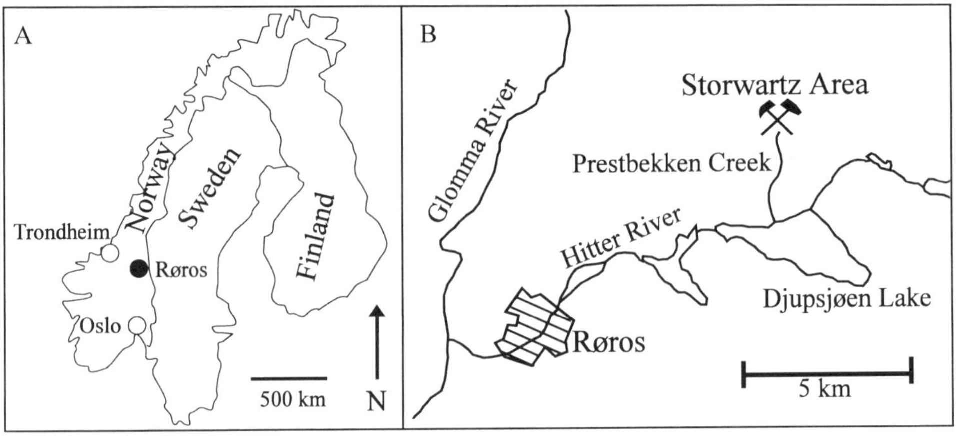

Figure 1 General map of Scandinavia with location of Røros (A) and location map of Røros area (B).

At other similar Norwegian sulfide mines remediation techniques have included capping with fine-grained moraine material, with or without synthetic membranes, and flooding with water (Steffen et al. 1991; Ese 1995). However, at Storwartz these measures would cause significant landscape alterations of the area and are not considered acceptable with regards to the historical conservation plan.

The induced development of a shallow, subsurface hardpan in the tailings impoundment was considered to be a potential strategy to control ARD from the tailings impoundment at Storwartz, without altering the landscape or the reddish oxidized colors. Laboratory tests by Chermak and Runnells (1997) showed that hardpan layers could be induced to form by applying limestone or lime directly over sulfidic material. The observed precipitation of gypsum ($CaSO_4 \cdot 2H_2O$) and iron oxyhydroxides ($FeO(OH)$) significantly reduced the permeability of the test columns. Chermak and Runnells (1997) concluded that hardpan layers could provide self-healing chemical barriers which would reduce the overall permeabilities of mine waste, thereby limiting ARD generation by reducing oxygen and water infiltration. Naturally formed hardpans have been observed in tailings impoundments in Canada and Sweden (Boorman & Watson 1976; McSweeney & Madison 1988; Blowes et al. 1991; Lin & Qvarfort 1996; Tassé et al. 1997), in or near rock waste piles in Sweden and Colorado, USA (Whitney et al. 1995; Herbert 1995), and were also observed at Storwartz during this study.

With the agreement of the Norwegian Directory for Cultural Heritage and Røros Museum a field test at the Storwartz tailings impoundment was started in 1997. The goal of this test was to induce gypsum/oxyhydroxide iron hardpan formation and to measure the effect on water and oxygen infiltration. In contrast to Chermak and Runnells' (1997) study with fresh sulfidic waste, this attempt to induced hardpan formation was conducted within a deeply oxidized layer of tailings waste.

2 TEST CONSTRUCTION

In July, 1997, four 4 m^2 test plots were chosen for testing at Storwartz. Narrow trenches were dug to 1 m depth and PVC walls surrounding each plot were inserted. Observations during trenching showed a general oxidation profile of a surfical red-brown oxidized layer, a deeper brown semi-oxidized layer with large lenses of dark gray unoxidized tailings, and deep unoxidized tailings below one meter. A discontinuous hardpan layer was observed at the bottom of the oxidized zone during cell construction. Meteoric water mobility procedure tests (NDED 1990) were conducted on each tailings profile to determine the amounts of available acid and metals in the materials (Table 1).

In three of the four test cells the top 5 cm of oxidized tailings were removed and replaced with the following:

Cell I)	3 cm of fine and coarse crushed limestone covered by 2 cm of oxidized tailings,
Cell II)	3 cm of fine and coarse crushed limestone covered by 2 cm of unoxidized tailings,
Cell III)	Reference Cell. No tailings were removed and no limestone or lime were added, and
Cell IV)	3 cm of a quick lime and crushed limestone mixture covered by 2 cm of oxidized tailings.

In each cell water collection wells, declined at 45°, were constructed so that water in the unsaturated zone may be sampled. The wells were designed to sample at depths of 25 to 50 cm and 75 to 100 cm.

Table 1 Selected results from meteoric water mobility procedure leach tests

Sample		Top oxidized layer	Semi-oxidized layer	Unoxidized lens	Semi-oxidized layer
Depth		0 - 9 cm	9 - 52 cm	52 - 75 cm	75 - 100 cm
Parameter					
pH		2.6	5.1	6.3	3.6
Ca	(mg/L)	36	600	580	520
Fe	(mg/L)	7.0	28	0.34	930
K	(mg/L)	10	15	18	15
S	(mg/L)	170	830	520	1600
Cu	(mg/L)	7.1	0.15	0.03	0.34
Zn	(mg/L)	9.6	450	36	1100
Cd	(mg/L)	0.021	2.2	0.047	1.9

2.1 Topographic and climatic variables

The Storwartz tailings impoundment is located 850 m above sea level on the gentle slopes of Bergfjellet Mountain. The area is above tree line and especially prone to wind erosion (Myran & Ulvang 1996).

During the one year test period the total precipitation measured at Røros was 599 mm, with a total of 221 days of precipitation (DNMI 1998). Average monthly temperatures at Røros over this period ranged from 14.1°C to -7.4°C (DNMI 1998). Approximately 200 mm of this precipitation fell as snow, with runoff occurring in April and May.

2.2 Monitoring methods

The test cells were monitored in September of 1997 followed by a second monitoring round during July and August of 1998. Monitoring of the test cells included infiltration testing, penetration testing, pore gas testing, and paste pH testing. Water infiltration testing was conducted with a double ring infiltrometer. Both a "pocket penetrometer" and a Bush recording soil penetrometer were used for hardness testing. The pore gases were analyzed with a PID field analyzer attached to a pump and probe which was inserted into the tailings. The collection wells placed in the unsaturated zone, intended for collecting pore waters, did not collect enough water for analysis.

Samples of natural and induced hardpan were collected. These samples were investigated and analyzed with a SEM at the Institute for Geology, University of Oslo.

3 TESTING RESULTS

Induced hardpan layers were not observed after 2 months, however hardpans were found after one year. The residual limestone layer and deeper hardpan layers were observed in all three of the manipulated cells. The surface of the test cells did not show any sign of physical change or discoloration due to the testing. In Cell II the unoxidized tailings that had been applied to the surface were oxidized and showed no color differences with respect to the other cells. Dark red to brown hardpans were observed during trenching to a depth of 24 cm in each cell (Table 2). A poorly formed natural hardpan was also observed in Cell III.

During SEM analyses amorphous and crystalline iron oxyhydroxide (FeO(OH)), gypsum ($CaSO_4 \cdot 2H_2O$) and jarosite ($KFe_3(SO_4)_2(OH)_6$) were observed. In samples from Cells II and III the amorphous iron oxyhydroxide occurred as widespread, thin, spalling grain coatings (Figure 2). In the Cell II sample the iron coating was observed in places to be layered with bladed to platy crystals growing between layers (Figure 3). In all the samples, minute clumps of amorphous iron oxyhydroxides were observed. In samples from Cells II and IV, ball-shaped clusters and single crystals of jarosite were observed (Figure 3). The crystal clumps were up to 20 μm in diameter with individual crystals around 0.5 to 1 μm in size. In the Cell II sample, the jarosite clumps were observed on iron oxyhydroxide spalled-edges (Figure 3), indicating that they formed after the iron coatings. Bladed gypsum crystals were observed rarely in Cells II and IV samples (Figure 4). The large majority of the intragranular spaces observed in the hardpan samples appeared to be open.

Table 2. Observed hardpan layer formation depth and mineralogy

	Depth to hardpan	Mineralogy
Cell I	10-12 cm 17-24 cm	amorphous iron hydroxide clumps
Cell II	9-10 cm 16-18 cm 20-24 cm	jarosite, gypsum, amorphous iron hydroxide clumps and coatings
Cell III (Reference Cell)	19-24 cm	amorphous iron hydroxide coatings
Cell IV	15 - 20 cm	jarosite, gypsum, amorphous iron hydroxide clumps

Figure 2 SEM photomicrograph of spalling amorphous iron oxyhydroxides coating grains. Space bar is 10 μm.

Penetrometer measurements of the cells showed that Cells I, II and IV were generally harder than the reference cell (Figure 5). Of the two methods, the pocket penetrometer correlated best with the observed hardpan layers.

Paste pH analysis of 3 cm intervals, were conducted to a depth of 24 cm in each cell (Figure 6). The reference cell displayed a constant paste pH of around 2.6 for all the samples. The limestone layers in the manipulated cells were clearly distinguishable, showing high pH values.

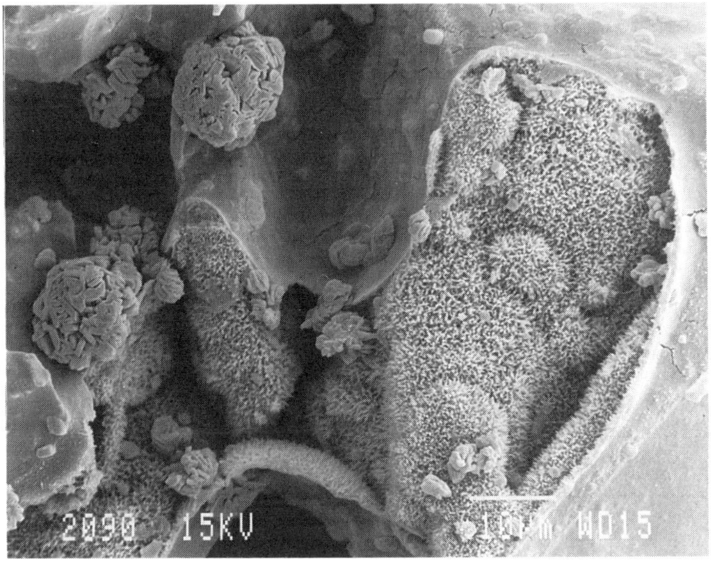

Figure 3 SEM photomicrograph of ball-shaped jarosite crystal clumps sitting on layers of platy iron oxyhydroxides. Space bar is 10 µm.

Figure 4 SEM photomicrograph of larger bladed gypsum crystals. Space bar is 100 µm.

Under the limestone layers the paste pH values were variable, but higher than those found in the reference cell. The observed hardpan layers correlated roughly with changes in the paste pH profiles.

The saturated double ring infiltration tests on the surface of each test cell showed that infiltration rates in the treated cells were significantly reduced, with respect to the reference cell. The reference cell had a saturated hydraulic conductivity of 1.1×10^{-5} m/s. Cell IV showed the best results with a total hydraulic conductivity of 6.6×10^{-6} m/s, whereas Cells I and II showed results of 8.8×10^{-6} m/s and 8.9×10^{-6} m/s, respectively. These results showed that the cells with induced hardpan layers created barriers with reduced permeability.

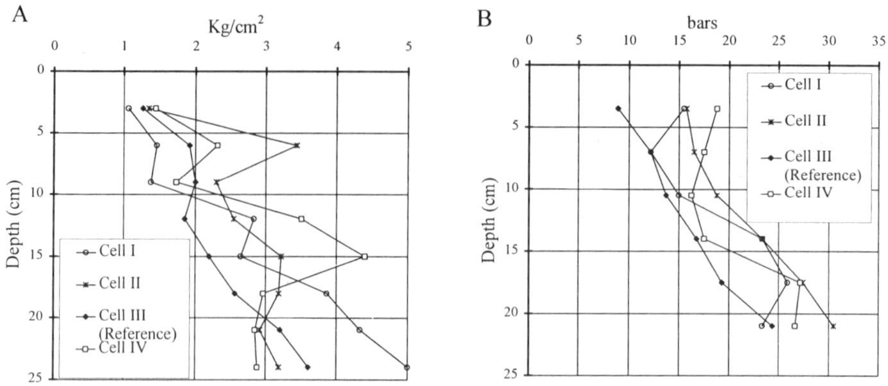

Figure 5 Penetration measurements conducted on test cells with a "pocket penetrometer" (A) and with a "Bush recording soil penetrometer" (B).

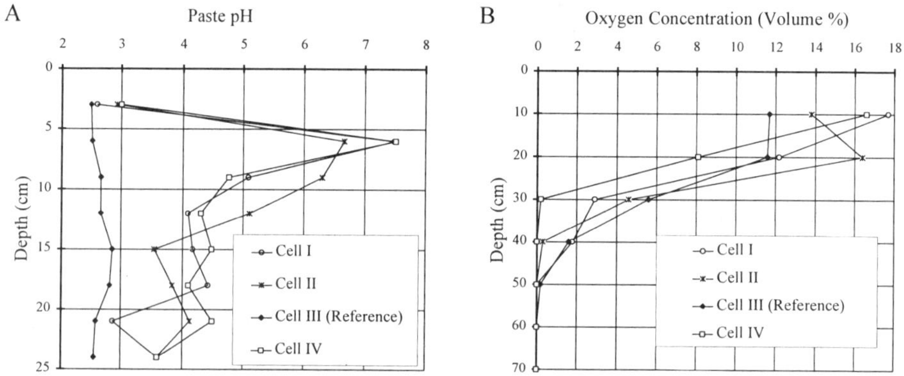

Figure 6 Paste pH (A) and volume percent oxygen (B) measurements in the test cells.

Pore gas measurements showed that relatively high concentrations of oxygen were present in the top 20 cm of tailings, although oxygen concentrations in the underlying tailings were reduced in the manipulated cells (Figure 6). In Cell IV the oxygen percent dropped to 0.2 volume % at 30 cm and was not detectable in deeper measurements, and in Cell II the oxygen volume percent was reduced to 0.3 volume % at 40 cm and not detectable at 50 cm. For Cell I, the oxygen concentrations were not detectable at 50 cm, and in the reference cell oxygen concentrations reached 0.2 volume % at 50 cm and not detectable at 60 cm.

4 DISCUSSION

The data collected showed that sulfate and iron oxyhydroxide crystallization had been induced to varying degrees in the oxidized layer of the test cells by the addition of limestone and lime. Penetration tests indicated that the secondary minerals resulted in the cementation of the tailings so that they were generally harder than the reference cell tailings. The formation of several discrete hardpan layers were also induced.

In each cell, an unique mineral assemblages was found in hardpans at different depths. The mineralogical assemblages and depths of the hardpans were probably dependent on the degree of limestone and lime dissolution, the Eh/pH boundaries and the availability of dissolved calcium, Fe^{+2}/Fe^{+3} iron and sulfate.

Paste pH measurements showed that limestone and lime were dissolved and HCO_3^- was carried downward into the underlying tailings. Although no acid-base accounting (ABA) has yet been conducted to determine the acid generating potential over long periods, the meteoric water mobility procedure (MWMP) testing showed short-term availability of acids for limestone dissolution. Where fresh tailings were allowed to oxidize over the limestone layer (Cell II), the dissolution and transport of HCO_3^- into underlying layers appeared to be increased. Dependent on the amount of HCO_3^- in solution, and the degree and depth of the overlying oxidized zone, different pH boundaries apparently formed. As the pH of the pore fluids increased, residual ferrous iron was oxidized and iron oxyhydroxides precipitated in the pore spaces. At depths where abundant acidic pore fluids dominated, such as near the bottom of the oxidized zone, jarosite crystals precipitated. Gypsum may have formed as diffuse crystallization, rather than in discrete hardpan layers, thus contributing to the general hardening of the manipulated cells. The general reactions to form the observed hardpan minerals are as followed. For iron oxidation and oxyhydroxide precipitation:

$$4Fe^{2+} + O_2 + 4H^+ \rightarrow 4Fe^{3+} + 2H_2O \qquad (1)$$

$$Fe^{3+} + 2H_2O \rightarrow FeO(OH) + 3\,H^+ \qquad (2)$$

For carbonate dissolution and gypsum precipitation:

$$SO_4^{2-} + CaCO_3 + 2H^+ + H_2O \rightarrow CaSO_4{\cdot}2H_2O + CO_2 \qquad (3)$$

For jarosite precipitation:

$$K^+ + 3Fe^{3+} + 2SO_4^{2-} + 6\,H_2O \rightarrow KFe_3(SO_4)_2(OH)_6 + 6H^+ \qquad (4)$$

Saturated hydraulic conductivity and pore gas testing indicated that the secondary mineral crystallization in the pore spaces reduced the permeability of the tailings. Observations of the mineral precipitates with the SEM, showed that the majority of pore spaces were still open and that further mineral crystallization would have space to proceed if favorable conditions persist. Although the particular layers which reduced the saturated hydraulic conductivity could not be identified, oxygen barriers exist between 20 and 50 cm depth in the cells.

5 CONCLUSIONS

This work showed that the addition of small amounts of crushed limestone and lime just below the surface of the tailings resulted in the formation of hardpan layers within one year and under extreme climatic conditions. The hardpan layers are formed by iron oxyhydroxide, jarosite and gypsum which coat particles and fill pore spaces. This filling of the pore spaces appears to provide a barrier which reduces the infiltration of water and oxygen in the shallow layers. The oxygen barriers would tend to reduce the oxidation of the sulfides in the tailings, whereas the reduction of water would limit the rinsing-out of available metals and acid from oxidized tailings. This combined effect of providing a barrier for oxygen and water is expected to reduce ARD generation. After one year the best induced hardpan with respect to reduced oxygen and water infiltration, was formed in Cell IV with a limestone and lime mixture. The technique could be used where limestone and lime is applied directly to sulfidic waste, as shown by Chermak and Runnells (1997), or applied to the surface of deeply oxidized waste, as found in this study.

This induced hardpan technique may be used for the remediation of ARD at Røros and other mine sites where mine waste is included in a historical conservation plan. In addition, hardpan chemical barriers may provide a cost-effective solution at other mine sites where limestone is easily available.

Further work is needed to determine the continued development of hardpan over a longer period of time, to determine mineralogy with XRD analyses, and to determine the chemistry of the water which seeps through the hardpan layers.

6 ACKNOWLEDGMENTS

This research has been financed by the Norwegian Pollution Control Authority, and partially by the Norwegian Research Council (NFR). Thanks is given to the Norwegian Directory for Cultural Heritage, Røros Museum, and Røros Community for their permission to test this method at the historical Storwartz Mine. Limestone for testing was generously donated by Eithun A/S quarry at Meldal. We also thank John Chermak and Donald Runnells at Shepherd Miller Inc. and Christian Zwach for many useful discussions and suggestions, and Berit Løken Berg at the University of Oslo for her assistance with the SEM analyses.

REFERENCES

Arnesen, R.T. 1996. *The Storwartz Project, Documentation of the impact of mining activity on the environment, Part I: Water chemistry investigations.* Norwegian Institute for Water Research (NIVA) Rep. LNR 3476-96. (in Norwegian)

Blowes, D.W., E.J. Reardon, J.L. Jambor & J.A. Cherry 1991. The formation and potential importance of cemented layers in inactive sulfide mine tailings. *Geochim. Cosmochim. Acta*, 55:965-978.

Boorman, R.S. & D.M. Watson 1976. Chemical Processes in abandoned sulfide tailings dumps and environmental implication for northeastern New Brunswick. *Canadian Inst. Mining Metallurgy Bull.* 69: 86-96.

Chermak, J. & D. Runnells 1997. Development of chemical caps. *Mining Engineering*, technical papers, June.

Den Norwegian Meteorological Institute (DNMI) 1998: *Climatic data summary*. (in Norwegian)

Ese, H. 1995. Storage of mining waste - Examples of remediation methods used in Norway. In *Proceedings of Contaminated Site Remediation Methods Conference, Gol, Norway, 24-26 April 1995*: Norwegian Geotechnical Association (NGF). (in Norwegian)

Grande, M., S. Andersen, P. Brettum, K. Hylland, & J.E. Løvik 1996. *The Storwartz Project, Documentation of the impact of mining activity on the environment, Part II: Biological investigations of the Hitter River Watercourse.* Norwegian Institute for Water Research (NIVA) Rep. LNR 3473-96. (in Norwegian)

Grøndalen, R. 1997. Permanent conservation of a pollution source. *Bellona Magasin* 4,3-6. (in Norwegian)

Herbert, R.B. 1995. Precipitation of Fe oxyhydroxides and jarosite from acidic groundwater. *Geologiska Föreningens i Stockholm Förhandlingar* 117: 81-85.

Kjellberg, G. 1991. *Remediation oriented monitoring of the upper Glåma River in 1990.* Norwegian Institute for Water Research (NIVA) Rep. 800212. (in Norwegian)

Lin, Z. & U. Qvarfort 1996. A study of the Lilla Bresjön tailings impoundment in mid-Sweden - A comparison of observations with RATAP model simulations. *Appl. Geochem.* 11:293-298.

McSweeney, K. & F.W. Madison 1988. Formation of a cemented subsurface horizon in sulfidic minewaste. *J. Environ. Qual.* 17:256-262.

Myran, T. & R. Ulvang 1996. *Dust and wind-blown sand from the Storwartz tailings impoundment.* SINTEF Rep. SFT22 A96015. (in Norwegian)

Nevada Division of Environmental Protection (NDED) 1990. *Meteoric water mobility procedure.* Nevada Bureau of Mining Regulation and Reclamation, 9/19/90.

Statens Forurensningstilsyn (SFT) 1992. *Sites with hazardous waste, contaminated soils and polluted sediments: Plan of action for remediation.* Norwegian Pollution Control Authority Rep. 92:32. (in Norwegian)

Steffen, Robertson & Kirsten (B.C.) Inc. 1991. *Acid rock drainage remediation measures at four Norwegian Mines*, British Colombia Acid Mine Drainage Task Force Report.

Tassé, N., D. Germain, C. Dufour & R. Tremblay 1997. Hard-pan formation in the Canadian Malartic Mine tailings: implication for the reclamation of the abandoned impoundment. *Fourth International Conference on Acid Rock Drainage, Vancouver, B.C., Canada, May 31 - June 6, 1997.* Proceedings, Vol. IV: 1797-1812.

Whitney, G., K.J. Esposito & K.N. Sweeney 1995. Mineral reactions in a Colorado mine dump: Implications for remediation in arid and semi-arid environments. In G.E. Schuman & G.F. Vance (eds) *Decades Later: A time for reassessment, Proceedings of the 12th Annual National Meeting of the American Society for Surface Mining and Reclamation, Gillett, WY, June 3-8*, 557-586.

Tailings and Mine Waste '99 © 1999 Balkema, Rotterdam, ISBN 90 5809 025 6

Antfarms: Mine dump simulation studies in the laboratory

Robert B. Vaughn, Robert J. Horton, Mark R. Stanton, David Fey, Steven Sutley & Peter Theodorakos
US Geological Survey, MS 973, Denver Federal Center, Colo., USA

ABSTRACT: The simulator systems ("Antfarms") we have been developing are an attempt to build practical analogs of mine dumps that can be monitored over more realistic sampling intervals and multiple wet/evaporate/dry (WED) cycles in or near a home laboratory. The simulators are plumbed and instrumented for detailed geochemical, geophysical, and microbiological data collection; and are sized and shaped to address some limitations of standard laboratory apparatus and protocols. Design and materials details are given along with preliminary data from the first year of a prototype experiment on baseline geochemistry and geophysics using dump material from a quartz/base-metal vein mine. The prototype tank has been taken through two WED cycles from August 1997 to August 1998 and is showing some geochemical and mineralogical parallels with the parent mine dump. Low pH effluent with high total dissolved solids (TDS) is produced within 30 minutes of addition of deionized (DI) water, and evolution of Fe^{2+}, Fe^{3+}, and dissolved oxygen (DO) suggest interesting chemistry and perhaps microbiology occur within the simulator. Preliminary interpretation of geophysical data collected during WED cycles suggest that self-potential (SP) anomalies are associated with oxidation-reduction reactions occurring near the boundary between the saturated and unsaturated zones.

1 INTRODUCTION

Most studies of mine dumps take place during one or a few visits to a remote site to collect data, observations, and samples. Further work is then usually conducted in the laboratory with instrumental and visual analyses, and standardized simulations such as leach protocols. Our simulators and experiments are designed to address the size, time, scale, and instrumentation limitations of some of the standardized laboratory protocols such as bulk chemical analyses, leach tests, and small (<50 kg) size simulations without spending unrealistic time on the mine dump site(s).

In our large-scale simulators (see below), we have attempted to address the problems of scale, particle size, ion-exchange paths, migrating reaction fronts, precipitates, natural diurnal and seasonal cycles, evaporation, O_2 fugacity, pile surface chemistry, and longer (>6 months) time effects. As with most attempts to address weaknesses in standard analytic techniques; our simulators have their own limitations and compromises (see table 1), and should be viewed as complementary to, and not a substitute for, the standard methods. Nevertheless; by using our approach, it is hoped that more of the complexity of a mine dump system can be modeled over longer time spans and larger size scales inside the laboratory than is possible with conventional techniques.

The study is being conducted in parallel with a multi-disciplinary field investigation of the source mine waste dump (Vaughn, et al., this volume) to test the usefulness and accuracy of these simulators in the modeling of mine waste dump behavior in situ. This paper describes some initial results from a prototype system and some modifications and directions for future work.

TABLE 1: Limitations and strengths of different investigative methods

	Bulk Chemistry	**Most Standard Leach Tests**	**Column Simulators**	**"Antfarm" Simulators**
Representative particle size	----	---	-/+	+++
Multiple cycles	----	---/+	+	++++
Detects changes over mid to long-term	----	---/+	-/+	+++
Examine evaporative and capillary effects	----	----	---/+	++++
Allows examination of air/surface phenomena	----	----	--	++++
Addresses chromatographic effects	----	----	-/+	+++
Oxic vs anoxic reactions simultaneously	----	----	--	+++
Evolution and examination of internal structure	----	----	--/+	++
Investigation of biogenic contributions	----	----.	+++	+++
Rapid results	+++	+	--	----
Relatively low cost	+++	--	---	----
Results easily comparable to other methods	+++	+++	+	-/+
Meets regulatory requirements	+++	+++	**N/A**	**N/A**

+ = well suited - = poorly suited -/+ = variable, depending on specifics

2 SIMULATOR DESCRIPTION

2.1 Tank design

The prototype tank (Figure 1) is constructed from 6.4 mm acrylic sheet with outside dimensions of 196 cm x 84 cm x 19 cm. It is supported by a heavy angle-iron frame raising the tank approximately 20 cm above the floor. It has three plastic ball valves at each end of the tank (total of six) for introducing or sampling fluids at different depths within the mine waste. There is a drain with a plastic ball valve on the bottom of the tank and a glass tube manometer was added before the beginning of the second WED cycle to determine the approximate saturated fluid level within the simulator. At this point the simulator looks like a giant antfarm; hence the name. The tank was externally wrapped in heavy aluminum foil and 14 mm extruded polystyrene closed-cell foam (R-Value: 3.1) for light-proofing and insulation.

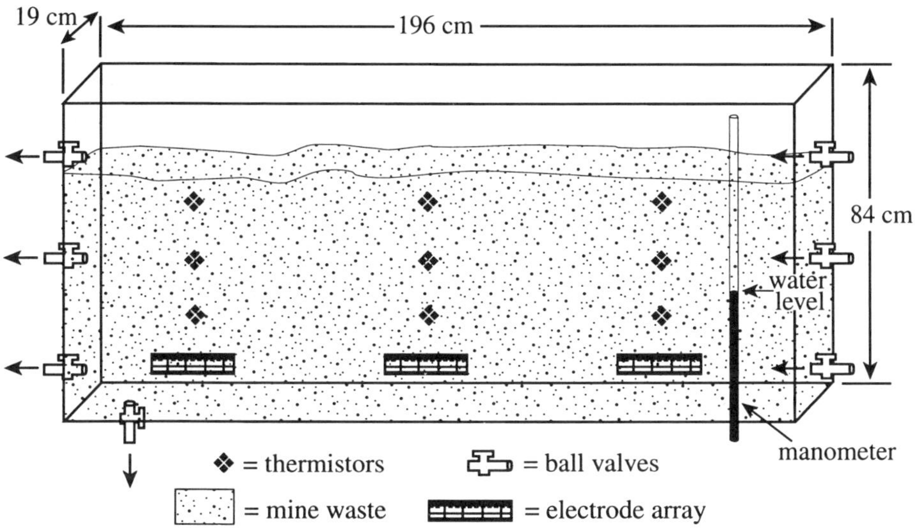

Figure 1: Schematic of prototype mine waste simulator ("antfarm")

2.2 *Instrumentation*

The tank's geophysical instrumentation consists of 9 thermal sensors and 3 electrode arrays. The thermal sensors provide temperature data from different depths and positions within the tank. The electrode arrays allow resistivity and self-potential (SP) data to be collected.

The thermal sensors are precision thermistors, with a resolution of 0.2 ^{0}C., connected to teflon coated wire leads. To protect the thermal sensors from acidic conditions within the tank, the thermistors were encapsulated in epoxy resin. The thermal sensors are located at 3 depths (top, middle, and bottom), and 3 lateral positions (input end, center, and output end) across the tank (see figure 1).

The electrode arrays consist of 4 platinum wire electrodes on a solid teflon rod 2.5 cm in diameter and 10.5 cm long. The electrodes are bright platinum wire (0.8 mm diameter) looped around the rod, spaced 2.5 cm apart. The electrodes are connected to teflon coated wire leads. The electrode arrays were buried 15 cm from the bottom the tank and located at the input end, center, and output end of the tank (see figure 1).

3. EXPERIMENTAL METHODS

The tank was filled (initially 10 cm from the top of the tank, after settling, 15 cm from top of tank) with 390 kg of mine waste material from a polymetallic vein mine located (see Vaughn, et al., this volume) in the Leadville mining district a few miles east of Leadville, Colorado. For the first WED cycle, the simulator was left dry for 11 days while a series of baseline geophysical measurements was made. On day 12, 65 liters of (DI) water were poured onto the top surface of the mine waste over a period of 30 minutes and allowed to percolate into the material. After 20 days the tank was drained; producing only 3 liters of low pH (2.3) water. The rest of the water had either evaporated, was chemically bound in hydrous mineral species, trapped in damp colloidal material, or retained as hygroscopic water. The tank fill was allowed to dry for 87 days, then another 65 liters of DI water was added, beginning the second WED cycle.

3.1 *Geochemistry*

During the second WED cycle, aqueous samples were collected from the bottom valve on the output end of the tank. The fluid was collected in a clean glass beaker and a 120 ml aliquot of unfiltered raw sample was taken for microbiological isolations. The rest of the samples were filtered through 0.45μm cellulose-acetate filters mounted on 50 ml plastic syringes. The filters were saved for x-ray diffraction (XRD) and microscopic analyses of the collected solids.

All samples were placed into acid and DI water rinsed amber bottles to inhibit potential light-activated reactions. Three aliquots were taken: (1) 60 ml of filtered, unacidified sample for anions (F^-, Cl^-, PO_4^-, NO_3^-, SO_4^{2-}) by ion chromatography analysis (Arbogast, 1996, p 149-153); (2) 60 ml of filtered, acidified sample for 40-element (Arbogast, 1996, p 83-89) inductively-coupled-plasma atomic emission spectroscopy (ICP-AES); (3) 30 ml of filtered, unacidified sample for immediate Fe^{2+} and $Fe_{(tot)}$ spectrophotometric analysis. The iron analyses were completed within 30 minutes of collection of the raw sample using a spectrophotometric method (Hach*, 1997) to quantify Fe^{2+} and $Fe_{(tot)}$ ions. Fe^{3+} ion content was determined by difference ($Fe_{(tot)} - Fe^{2+} = Fe^{3+}$).

3.2 *Geophysics*

Temperature, SP, and DC resistivity measurements were made 2 to 3 times per week. Data were recorded on log sheets and later keyed into a PC computer for data manipulation, display, and interpretation. Temperature data was collected by measuring the resistance of the thermistors using a digital multi-meter. Resistance measurements were converted to degrees Centigrade using a formula derived from laboratory calibration curves. The laboratory air temperature was also monitored using a thermal sensor identical to those buried in the tank.

SP data was collected by measuring the spontaneous voltage generated across the electrode array using a digital multi-meter. For each array, 4 SP voltage readings were made between the different electrodes. SP measurements were always made before the DC resistivity measurements, to avoid induced polarization effects.

DC resistivity measurements were made on each array by transmitting DC current into the two outer electrodes on that array and measuring the resulting voltage across the two inner (potential) electrodes. The transmitter was a constant current DC power supply, the receiver was a digital multi-meter. The resistivity was calculated using the ratio of the received voltage to the applied current, multiplied by an array geometry constant. To correct received voltages for SP effects, the SP across the potential electrodes was measured prior to the DC resistivity measurements. For each resistivity measurement, the voltage was recorded at 3 different current levels (2.00, 1.75, and 1.50 milliamps). To compensate for induced polarization effects, the polarity of the transmitter was reversed and the measurements were repeated.

4 RESULTS AND DISCUSSION

4.1 *Geochemistry*

No geochemical analyses were performed in WED cycle #1 because this run was used for geophysics baseline data. Figures 2 , 3 and 4 show concentrations of dissolved species during WED cycle #2. Fe and S geochemistry dominates the simulator, similar to results observed in the actual waste pile.

Figure 2 illustrates the changes in Fe species (Fe^{2+}, Fe^{3+}, and $Fe_{(tot)}$), S (as SO_4^{2-}), dissolved oxygen (DO), and physical parameters such as temperature (°C) over a period of 660 hours. The greatest changes in aqueous concentrations and physical parameters take place within approximately the first 130 hours, after which a near "steady state" appears to be reached.

30 minutes after the beginning of the run, total iron consists of >95% Fe^{3+} and <5% Fe^{2+}. After this time; however, the amount of Fe^{2+} (~3000 ppm) that is solubilized up to about 130 hours exceeds that of Fe^{3+} (~1000 ppm), suggesting that the initial loading of Fe^{3+} to the system is the result of dissolution of water (acid)-soluble Fe^{3+}-bearing solids, perhaps Fe-oxide and "jarosite-type" phases.

At 133 hours, there is a marked change in Fe chemistry when Fe^{2+} becomes the dominant iron species and continues to increase relative to Fe^{3+}. During the entire reaction period, DO slowly decreases from >5 ppm to <1 ppm, indicating that O_2 is being consumed by oxidative weathering reactions.

The change from Fe^{3+} to Fe^{2+} dominance in the total iron pool indicates that although the "new" reduced Fe^{2+} is being released into the system, it is not undergoing oxidation to Fe^{3+}. The rate of Fe^{2+} oxidation in highly-acid (pH <2) systems is quite slow unless mediated by microbes, with half-times on the order of 1000 days (Stumm and Morgan, 1981). The increasing dominance of Fe^{2+} and decrease in DO suggest O_2-depletion contributes to a reduction in the rate of Fe^{3+} formation. The decrease in Fe^{3+} also suggests that iron and sulphur oxidizing microbes, which were previously-isolated from the tailings material (Stanton, et al., 1997), are unable to oxidize the available Fe^{2+} faster than it is formed as a result of the depletion of O_2. Nearly all Fe oxidizers (e.g. *Thiobacillus sp.*) are obligately aerobic. The reduced O_2 appears to inhibit or retard the ability of the microbial species present to oxidize Fe^{2+}. However, the specific microbial role in Fe redox chemistry in the simulators will be elucidated by future work.

The apparent lack of Fe^{2+} oxidation may be representative of isolated microenvironments in the source pile (but perhaps not of the pile as a whole) where diffusion of O_2 may become a limiting factor. That is, the actual pile probably has sufficient exchange with the atmosphere that does not lead to total O_2 depletion, whereas the tank system permits O_2 depletion and inhibits replenishment of DO. The O_2 conditions in the tank, with respect to microbial action, are probably best-described as microaerophilic (lowered but non-zero DO and gaseous O_2 concentrations), rather than anaerobic (no H_2S odor was detected in the tank).

Figures 3 and 4 depict changes in major element and trace metal aqueous concentrations over the course of the 2nd WED cycle. Major elements, defined here as Fe, S, Al, Si, Ca, Mg, and Mn, show a sharp increase in concentration within 24 hours. The sharp rise is followed by a gentler but generally increasing concentration until 660 hours (the end of the run). Mg; however, is an exception to the general trend because its concentration decreases after the initial sharp rise in the first 24 hours.

The trace metals such as Cu, Pb, and Zn show changing concentrations similar to major elements during the run. That is, a steep rise in aqueous concentration within 24 hours, followed by a gentler increase for the rest of the run. Pb, however, is an exception. Pb decreases after reaching its maximum aqueous concentration at 95 hours, suggesting there is a sink for this metal in the system. Geochemical modeling (Allison, et al., 1991) predicts that anglesite ($PbSO_4$) is saturated and should precipitate from solution. However, anglesite has not yet been identified in the solids, and it is unknown where the precipitation of this mineral, if any, is physically occurring in the tank. Given the complex mixture of aqueous ions and brine-like composition of the fluid, it is unlikely that a "pure" $PbSO_4$ is forming. Rather, mixed-metal phase(s) such as plumbojarosite ($PbFe_6(SO_4)_4(OH)_{12}$) may account for part of the decrease in aqueous Pb concentrations.

The redox chemistry is likely dominated by the oxidation of pyrite (FeS_2). Generation of acidity (H^+) by oxidation of Fe^{2+} to Fe^{3+} and hydrolysis of Fe^{3+} to precipitate Fe^{3+}-bearing solids is represented by equations 1 thru 3, respectively. This FeS_2 oxidation cycle could, in principle, continue until FeS_2 is exhausted and all Fe is in "Fe^{3+}-oxide" solids. However, the net gain of Fe^{2+}(aq) over Fe^{3+}(aq) in the fluids from the simulator (figure 2) suggests the cycle is being interrupted or slowed dramatically

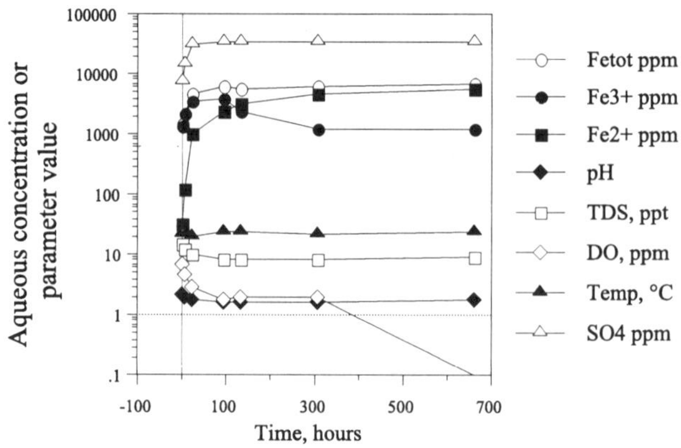

Figure 2. Aqueous iron, sulfate, dissolved oxygen (DO), total dissolved solids (TDS), pH, and temperature during WED cycle #2.

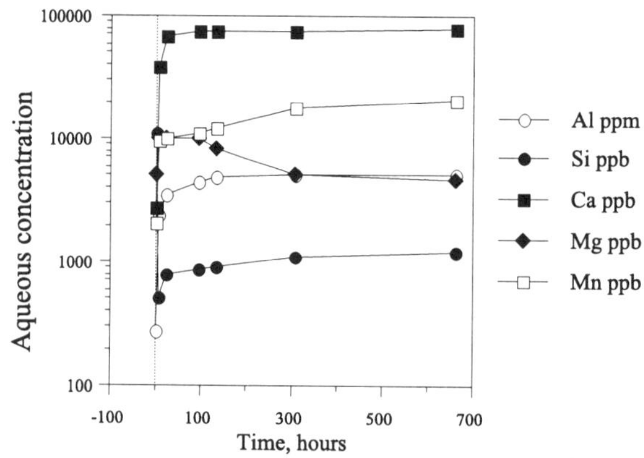

Figure 3. Aqueous major element concentrations during WED cycle #2.

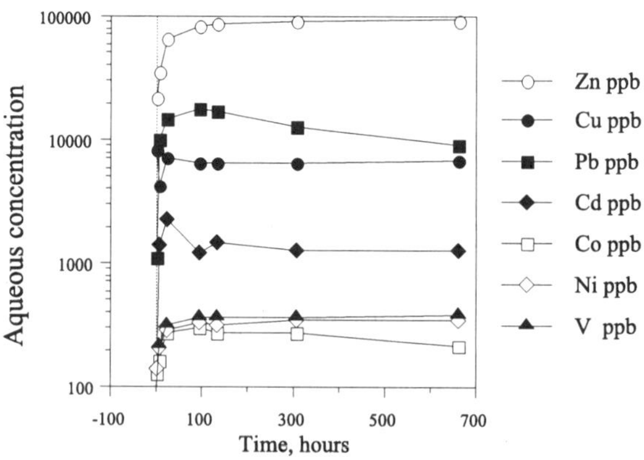

Figure 4. Aqueous trace metal concentrations during WED cycle #2.

at the Fe^{2+} oxidation step. The observed decrease in DO from samples of bottom fluids and the known slow kinetics of abiogenic Fe^{2+} oxidation at acid pH appear to permit a net accumulation of Fe^{2+}.

$$FeS_2 + 3.5\ O_2 + H_2O \rightarrow Fe^{2+} + 2SO_4^{2-} + 2H^+ \qquad (1)$$

$$Fe^{2+} + 0.25\ O_2 + H^+ \rightarrow Fe^{3+} + 0.5\ H_2O \qquad (2)$$

$$Fe^{3+} + 3\ H_2O \rightarrow Fe(OH)3(s) + 3H^+ \qquad (3)$$

$$FeS_2 + 14\ Fe^{3+} + 8\ H_2O \rightarrow 15\ Fe^{2+} + 2SO_4^{2-} + 16\ H^+ \qquad (4)$$

Equation 4 shows a mechanism also exists to reduce the concentration of Fe^{3+}(aq) "released" by the initial addition of water to the simulator. In this reaction; Fe^{3+} is consumed in oxidizing sulphur in FeS_2 to SO_4^{2-}, and the continued reduction in DO (equations 1 and 2) results in an increase of Fe^{2+}. Although DO decreases, it is not completely depleted even at the termination of data acquisition (660 hrs.), suggesting there is sufficient diffusion of O_2 into the deeper simulator water to support lower (but positive) DO levels.

4.2 *Geophysics*

Absolute measurements and the relative differences between thermal sensors were used to interpret the temperature data. Two significant observations were made regarding the temperature data. First, major changes in the tank's temperature are directly related to changes in room air temperature. Second, a thermal gradient was identified within the tank, the top of the tank is cooler than the bottom. The thermal gradient became less pronounced as the water level in the tank decreases with time (primarily due to evaporation). A 1° C thermal gradient, between the bottom and top sensors (approx. 60 cm), was observed in the first two weeks following the addition of water. After one month, this gradient had decreased to about 0.5° C, and after three months it had decreased to about 0.25° C.

The initial relatively large 1° C gradient may be primarily due to evaporative cooling at the surface. The relatively small gradient observed late in the drying portion of the cycle may represent a combination of surface cooling due to evaporation of residual water and possibly exothermic reactions deeper in the tank.

The SP values in the dry tank were very stable, with magnitudes on the order of 0 to 20 millivolts (mV) for the 3 electrode arrays. The addition of high-purity DI water caused the magnitude to increase slightly, on the order of 10 mV. The SP remained at relatively constant values until the tank was drained. Upon draining, the magnitude of the SP began to increase on all arrays. Ten to fifteen days after draining, SP magnitudes began to peak, with the center array measuring over 200 mV. As the tank continued to dry out, all SP magnitudes decreased and all readings changed polarity (positive to negative) at approximately the same time (figure 5).

The observed SP voltages are thought to be produced by oxidation-reduction reactions (see reactions 1 through 4 above). Oxidation in one area of the tank produces electrons that migrate to reducing areas. Within the tank the position of the water table may coincide with a redox boundary. The large SP anomaly (figure 5), observed after the tank was drained, is thought to represent the passing of the redox boundary across the arrays. The observed change in polarity may represent a

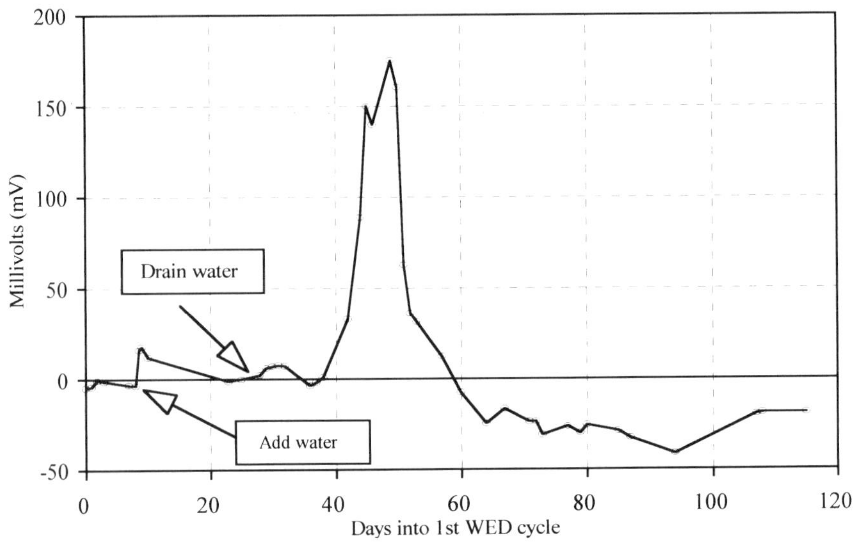

Figure 5. Antfarm self-potential (SP) data (center electrode array)

change from reducing to oxidizing conditions, or vise versa, near the bottom of the tank where the electrode arrays are located.

Resistivity measurements of the dry mine waste were about 320 ohm-m. Immediately after the DI water was added the resistivity dropped to about 10 ohm-m. Over the next 48 hours the resistivity slowly decreased to about 5 ohm-m. The resistivity remained fairly constant, at about 4.8 ohm-m, until the tank was nearly dry. As drying continued, the resistivity increased slightly to about 5.8 ohm-m. 87 days after the tank was drained, the resistivity had only increased to 6.1 ohm-m.

The resistivity of the dry mine waste was relatively high due to the lack of water. When DI water was added to the tank, the soluble mineral phases dissolved, resulting in a resistivity decrease due to ionic conduction through the high total dissolved solids (TDS) in the water. The continued decrease, from 10 to 5 ohm-m, represents a further increase in TDS due to acidic dissolution of the mine waste. This decrease in resistivity is correlated with the increase in major and trace element concentrations seen in the geochemical analyses (figures 2, 3, and 4). The relatively constant 4.8 ohm-m resistivity suggests the water was at a steady TDS concentration. Upon drying, the slight resistivity increase results from the loss of conductive pore fluids. High concentrations of hydrated ions in hygroscopic water, water absorbed to mineral grain surfaces, and the presence of hydrous mineral phases results in the relatively low-resistivity of the damp mine waste.

5 SUMMARY AND CONCLUSIONS

5.1 *Geochemistry*

Dissolved iron and sulphur species dominate the geochemistry of the simulator; other major elements such as Al and Mn are also readily dissolved in the simulator waters, but in lesser quantities than Fe^{3+}, Fe^{2+}, and SO_4^{2-}. In the tank fluids the most abundant aqueous trace-metal ions are Pb, Zn, and Cu, with lesser amounts of Cd, Co, Ni, and V. Decreases in Pb, Cd, and Co ion concentrations during later samplings indicate a sink exists for these ions within the system. The changing concentration of Fe^{2+} relative to Fe^{3+} with SO_4^{2-} remaining essentially constant and DO decreasing, indicates active FeS_2 oxidation, but a slower rate (abiotic plus biotic) of Fe^{2+} oxidation. Geochemical and microbial results from the simulator suggest that the "real" mine dump pile consists of numerous microenvironments where the geochemical processes and reaction rates are temperature, moisture, and O_2 controlled.

5.2 *Geophysics*

Temperature, SP, and DC resistivity measurements were made to observe changes as a function of time and changing water conditions. Thermal gradients observed in the simulator are attributed to evaporative cooling at the surface and possibly exothermic reactions deeper in the tank. Observed SP anomalies are thought to produced by oxidation-reduction reactions. The resistivity of the dry mine waste decreased an order of magnitude when water was added to the tank. Draining the tank resulted in only a slight resistivity increase. The persistent low-resistivity is attributed to high TDS concentrations in hygroscopic water and in water absorbed to mineral grain surfaces.

5.3 *Modifications*

Lessons learned and questions raised from the prototype tank have produced some design changes and instrumentation modifications and additions to later tanks and experiments. The next two full-size tanks have the same external dimensions but are constructed of 9.5 mm acrylic sheet for additional strength. The steel angle frames have been modified to allow for fine adjustment for added support of the plastic tank. The insulation/ light-proofing material has been changed to an 8 mm aluminized " bubble wrap"construction insulation (R-Value: 9.8). Acrylic lids have been fabricated so that different gases may be introduced or sampled, and several air-tight ports have been added for instrumentation cables and fluid/gas sampling tubes. Two smaller (76 cm x 84 cm x 19 cm) tanks with the same modifications have been constructed for investigations into biogenic contributions to geochemical reaction processes and rates and other geomicrobiological questions. Screened sampling tubes (3 mm id.) have been added at different depths for fluid, gas, and micro-organism sampling.

Geophysics instrumentation has been modified in the next generation of tanks; the electrode arrays have been lengthened and are positioned vertically to make SP and resistivity measurements as a function of depth. The electrode arrays now consist of 28 platinum wire electrodes on a 13 mm diameter teflon rod, 76 cm long.

6 ACKNOWLEDGMENTS

We would like to acknowledge and thank the landowners who have graciously allowed us to work on their property. Thanks to Norrie Robbins whose unceasing enthusiasm and encouragement helped move these investigations along and to Murray Beasley for all of his patience, cooperation, and active help while sharing the "antfarm" lab.

REFERENCES

Allison, J.D., Brown, D.S., and Novo-Gradac, K.J. 1991. *MINTEQA2/PRODEFA2, a geochemical assessment model for environmental systems, ver 3.0 user's manual*: EPA/600/3-91/021: U.S. Environmental Protection Agency.

Arbogast, B.F. (ed.) 1996. *Analytical methods manual for the Mineral Resource Surveys Program*, U.S. Geological Survey: U.S. Geological Survey Open-File Report 96-525.

Hach Chemical Company*, 1997. *User's manual.*

Stanton, M.R, Vaughn, R.B., Swayze, G.A., and Robbins, E.I., 1997. *Mineral crusts or microbial mats?: alteration of surficial mine tailings, Leadville District, CO.*: Abstract, 4th International Symposium on Environmental Geochemistry.

Stumm, W. and Morgan, J.J. 1981. *Aquatic chemistry, 2nd ed.*: New York: John Wiley and Sons.

Vaughn, et al. 1999. *A year in the life of a mine dump: a diachronic case study.* this volume.

* NOTE: Any use of trade or product names is for informational use only and does not imply endorsement by the U.S. Government.

Tailings and Mine Waste '99 © 1999 Balkema, Rotterdam, ISBN 90 5809 025 6

A year in the life of a mine dump: A diachronic case study

Robert B.Vaughn, Mark R.Stanton & Robert J.Horton
US Geological Survey, MS 973, Denver Federal Center, Colo., USA

ABSTRACT: Studies of the geochemistry, geophysics, and local climate of a mine dump were initiated in July 1997 and continue at the time of this writing (September 1998). The dump is located at 3660 m. on semi-arid alpine tundra to the east of Leadville, Colorado. The site is of particular interest because of the presence of a large iron-stained area devoid of vegetation (kill-zone) running down slope from the base of the pile. The site has four instrumented wells with thermistors for ground temperature profiles. Detailed micro-climate observations and temperature profiles at well sites have been kept for over a year, at weekly to monthly sampling intervals. Preliminary data and observations suggest acid runoff occurs only during a brief period in late spring during normal weather years, causing the kill-zone; and acidity and total dissolved solids (TDS) content of runoff waters is much reduced within a few hundred meters down slope of the mine dump pile.

1 INTRODUCTION

Most studies of abandoned mine waste sites in the rural western U S are conducted in one or a few visits because of time, distance, and or climate constraints. While this approach often produces good scientific data of the current state of the site; the picture it paints, is by its nature, a synchronic one. This type of study produces a snapshot of the systems' interactions and processes that are occurring at the moment the observations and samples are taken. Or in the case of analyses and experiments conducted after returning to the laboratory, such studies often produce an altered version of the real story because of simplified parameters within the laboratory setting. This paper describes some initial results from a portion of an ongoing diachronic study of a mine waste site near Leadville, Colorado with a large "kill-zone" barren of vegetation originating from sulfide-bearing mine wastes and extending 300 meters down slope. From the geometry of the zone, the plant distribution and species, and the mineral species present in the waste material, it is suspected that acid water runoff was being generated by the mine waste and causing the die-off of plants below the pile. The study is investigating the overall geochemical, mineralogical, climatological, biological, microbiological, and geophysical baselines of this site in conjunction with a larger U.S. Geological Survey (USGS) interdisciplinary team project to investigate multiple mine waste sites in different locations. The field investigations are being conducted in parallel with laboratory studies (Vaughn, et al., this volume) over two or more annual cycles. The data and conclusions in this paper are preliminary and the study is planned to continue at least through summer of 1999.

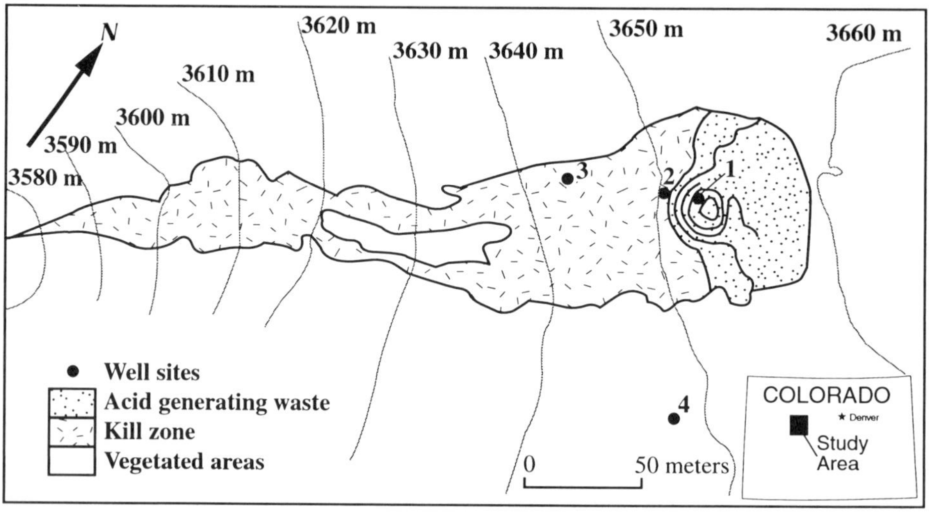

Figure 1: Simplified site map; unlabeled contours are at 2 meter intervals.

2 SITE DESCRIPTION

2.1 *Geology*

The site is located 5 km east-southeast of Leadville, Colorado, within the historic Leadville mining district (Figure 1). The two source shafts for the mine waste material were in an altered monzonite porphyry host rock with quartz/base-metal vein deposits. The major sulfides are pyrite, galena, and sphalerite with trace copper sulfides (Emmons et al., 1923, p 322). The mine was worked from 1901 through 1935, mainly for gold (Behre, 1953, p 164-165). The elevated waste pile dimensions are approximately 35 m x 25 m x 10 m for the undisturbed portion of the pile, and about the half the original pile has been recently (estimated late 1980's) spread around upslope of the remaining pile by bulldozer (covering the original shafts for hazard abatement) to an average cover depth of 1 to 2 m. Estimated total sulphide-rich mine waste site (disturbed and undisturbed) is approximately 5,500 m^3 to 9,000 m^3 (1.25x 10^4 to 2.2x 10^{4}t)

The major minerals present in the bulk composite surface sample (< 0.15 mm fraction) of the waste pile were determined by X-ray diffraction analysis(XRD). Dominant minerals are quartz, jarosite, mica, and kaolinite. The heavy mineral concentrate also shows smaller amounts of cerrusite, barite, pyrite, and ferberite (Desborough, pers com.,1998).

2.2 *Climate*

Temperature extremes measured at the site for the period 7/15/97 through 8/1/98 were 30° C to -30° C with a more usual range of 17° C to -13° C during daylight hours. Daytime highs are seldom above 0° C from mid December through early April.

The area around the site is in the rain shadow of the Sawatch mountain range and receives 25 cm to 32 cm of precipitation during an average year based on interpolations of climate data from four National Weather Service (NWS) sites (NOAA, 1998) which bracket the waste pile in location and elevation, and our own observations from 6/1/97 to 8/24/98 (continuing). Rain and hail are the primary form of precipitation from June through August, but snow and "snail" (soft hail) can occur at any time during the year. Between early October and mid May, precipitation is mostly snow, with about 130 cm to 180 cm (liquid water equivalent = 12 cm to 18 cm) accumulating in the form of seasonally permanent snow cover. Very little snow melt (liquid water detectable at the ground surface) occurs through the middle of May. Based both on NWS data (NOAA, 1998) going back to 1949 and our own observations for two springs (1997 and 1998), most of the melting occurs within a period of less than 2 weeks during late May to late June (depending on whether late spring is unusually warm or cold). Nearly constant high winds occur at the site, with average wind velocities of >9 m/sec (>20 mph) in the summer and >13 m/sec (>30 mph) in the winter.

Cloud cover varies depending on the season, but most (> 75%) days from early May through mid July have sun in the morning and early afternoon. Because of high altitude (>3600 m), insolation is high (25% more visible light and twice the UV flux of sea level) (Zwinger and Willard, 1972). A combination of high isolation, the nearly constant winds, a low relative humidity, and high altitude lead to high evapo-transpiration rates.

2.3 *Vegetation*

One of the most striking features of the area surrounding the mine waste pile is a large (~300 m long, ≤100 m wide) iron/manganese-stained area (figure 1) of rock and gravel pavement with almost no living plants visible inside its boundaries (see below). This area runs down slope from the base of the waste dump pile to a small dry gully at the major terminus. The baseline plant community surrounding this "kill-zone" is a moderately impoverished subalpine/ lower alpine shrub, forb, and grass community with lichens and mosses present in moderate amounts depending on micro-topography and drainage. A few subalpine margin trees (*Picea engelmannii*, *Abies lasiocarpa*, and *Pinus flexilis*) were present in pre-mining times.

Bilberry (*Vaccinium cespitosum*), an acid tolerant low (<10 cm high) woody shrub, is the only vascular plant even sporadically present within the kill-zone, and surviving plants are stunted or dying within the area. The only other macro flora observed within the zone proper are several species of lichens growing several centimeters above the local average ground level on the top surfaces of larger (>20 cm) rocks. These colonies are usually surrounded by dead or dying colonies of lichens lower on the rock surfaces. There is one large vegetated island and a few scattered fragmental plant communities on ground slightly higher (several cm) than the average local topography within the kill-zone.

3 METHODS

The waste pile, kill-zone, and adjacent terrain were surveyed with high precision GPS, laser theodolite, and survey chain in July of 1997, and a digital topographic base map was constructed for geophysical and geochemical investigations. Grids for geophysical surveys, geochemical sampling locations, and sites for biogeochemical investigations were located to within a relative accuracy of ± 4 cm horizontally and vertically, and absolute accuracy of ± 15 m (horizontal) and ± 5 m (vertical) (Hein and Fitterman, 1998, in press). A one-time detailed geochemical sampling of the minus 20 mesh surface material (upper 5 cm) from the waste pile was done in August 1997 by a team of geochemists, geologists, analytical chemists, and geomicrobiologists, and will be the subject of future reports. Approximately one ton of waste material (unsized) was sampled from a single area on the east side of the pile for long-term studies in controlled laboratory conditions.

Four slotted-PVC cased wells (4 cm id.) were drilled with a stainless-steel hand auger at: (1) top of waste dump pile, (2) downhill foot of pile, (3) ~50 m downhill in the kill-zone, and (4) on undisturbed, vegetated tundra ~75 m east of pile (figure 1). The maximum depths of the well casings are: (1) 85 cm, (2) 83 cm, (3) 52 cm, (4) 54 cm. Thermistor depths at the well locations are: (1) 2 cm, 85 cm, 191 cm; (2) 2 cm, 47 cm, 83 cm; (3) 2 cm, 85 cm, 107 cm, and (4) 1 cm. Temperatures at well locations have been recorded on a pseudo-random basis (site visits are affected by weather, e.g. blizzards and high winds) since 9/25/97. Sampling intervals range from approximately twice a week to once a month (Figures 2 & 3). Snow depths/ground coverage, snow water content, air temperature, approximate wind speed, cloud conditions, and other weather conditions are recorded; and photographs of the waste pile, kill-zone, and surrounding area are taken during each site visit.

3.1 *Geochemistry*

Solid samples ($\leq$0.5 kg) were collected from the surface and up to 20 cm deep into the pile. These included fine-grained surface crusts, scrapings from weathered surfaces of rocks, FeS_2-bearing solids, and red- (iron-oxide) and yellow- (jarosite?) stained, weathered dump material. Examination and analyses by optical microscopy, X-ray diffraction analysis (XRD), X-ray fluorescence analysis (XRF), ion microprobe, and scanning electron microscopy (SEM) are in progress.

All surface runoff samples were collected into acid and deionized water (DI) rinsed amber bottles to inhibit potential light-induced reactions. Three aliquots were taken: (1) 60 ml of filtered (0.45μm), unacidified sample for anions (F^-, Cl^-, PO_4^{3-}, NO_3^-, SO_4^{2-}) by ion chromatography analysis (Arbogast, 1996, p 149-153); (2) 60 ml of filtered (0.45μm), acidified sample for 40-element inductively-coupled-plasma atomic emission spectroscopy (ICP-AES) (Arbogast, 1996, p 83-89); (3) 30 ml of filtered (0.45μm) acidified (HCl) sample for Fe^{2+} and $Fe_{(tot)}$ analysis (Hach*, 1997). The samples were filtered through 0.45μm cellulose acetate filters mounted on 50 ml plastic syringes. Temperature, pH, and conductivity of flowing surface waters were recorded simultaneously at each sampling site with portable meters. Only pH and conductivity results for summer 1998 were available at the time of this writing. The detailed geochemistry work on surface runoff and shallow groundwater is still in progress and data will be published in 1999.

3.2 *Geophysics*

The site was surveyed using a variety of techniques to characterize the geophysical properties of the waste and surrounding host rocks, and to identify which geophysical techniques were best able to observe seasonal effects of acid water generation. Initial site characterization included a magnetic survey and a ground conductivity survey.

Based on the site characterization surveys, and preliminary results of laboratory experiments (Vaughn, et al., this volume) ground conductivity and self-potential (SP) methods were chosen for the diachronic study. Ground conductivity measurements respond to ground water quantity and quality, and the data can be collected rapidly, an important consideration for surveys requiring a large number of measurements. The SP method measures naturally occurring potentials generated by pressure, temperature and chemical gradients in the subsurface. Laboratory experiments (see Vaughn et al., this volume) indicated that large SP anomalies may be generated by oxidation-reduction reactions in the waste pile.

3.3 *Geomicrobiology*

Bulk samples (see 3.1 above) were used to determine water content. Sieved (<230 mesh) 1.0 g aliquots of solids from samples taken at different depths and locations in the waste pile were used to characterize populations of *Thiobacillus ferrooxidans* and *Th. thiooxidans*. The 1 g samples were inoculated into serial dilutions (10^1 to 10^{-6}) of media selective for *Thiobacillus* species. These media dilutions were checked for growth during a one month incubation at 25^0 C. Table 1 shows the results of the 10^{-3} series. Isolates were examined by visible light microscopy and showed numerous rods and bacilli present.

TABLE 1: Results of 1000X dilution series for *Thiobacillus* species.

Sample No.	Location/Type	Depth (cm)	Th. ferrooxidans	Th. thiooxidans	H_2O, wt %
Sun-2B	EF crust	Surface	++	++	3.0
Sun-3B	NF jarosite	Surface	---	++	ND
Sun-4B	NF 'porous'	Surface	---	+	6.1
Sun-6B	EF	1 cm	+	++	1.2
Sun-7B	EF, Fe_2O_3	Surface	+	++	0.9
Sun-10	NF 'scraping'	Surface	---	---	ND
S-3	Surface fines	Surface	++	+	ND
SU2-A	NF-top	15 cm	++	+++	8.4
SU2-B	NF-middle	16 cm	++	+++	9.0
SU2-C	NF-middle	20 cm	++	+++	NA
SU2-D	NF-top	15 cm	++	---	8.3
Control	---	---	---	---	

EF=east face; **NF**= north face; **+** = very slight positive; **++** = slight positive; +++ = positive; **---** = no growth; **ND** = not detected; **NA** = not analyzed.

4 RESULTS AND DISCUSSION

Observations on site from May 1997 through August 1998 indicate flowing surface water is present for only a short period(s) of time (~ 2 weeks) in late spring between mid-May and late June during average weather years. The exact timing of liquid water flow on the surface is influenced by the amount of snowfall or rain and temperature variations in May and early June. From the end of melt-off through October no flowing water was seen even during heavy rainstorms; either on the waste pile, the barren kill-zone, or adjacent vegetated tundra. From November through March, little or no liquid water is present on the surface.

During the peak melt-off event (May 25 through June 5 in 1998), sheet flow water was seen on nearly all ground surfaces including under remaining snowpack, bare kill-zone, and snow-free vegetated areas. Small scale solifluction was observed on the faces of most mine waste piles between 3200 m and 3700 m as well as at the down slope termini of the kill-zone, and gully cutting was observed at the major terminus of the kill-zone. Nearly all surface flow is produced by rapidly melting snowpack and (based on two summers' observations and NWS data from nearby weather data sites) any contribution from a normal year's rainfall is insignificant. During melt-off, all soils are still frozen (Figure 3) at depths ranging from 0 cm under the insulating snow pack to <5 cm in highly insolated, wind-sheltered areas. The melting snowpack water is concentrated into sheet flow across the surfaces down slope from the waste pile. Under snow fields, water is protected from evaporation; and from freezing until at least several hours after sunset.

By late June to early July, with the disappearance of snow fields, most exposed surfaces have thawed to depths of 50 cm to 100 cm; although the subsurface remains frozen at greater depths (Figures 2& 3). At this time the highly permeable soil and fractured subsoil rock absorb any rainfall nearly instantly and little or no long distance surface flow occurs in this area under normal weather conditions.

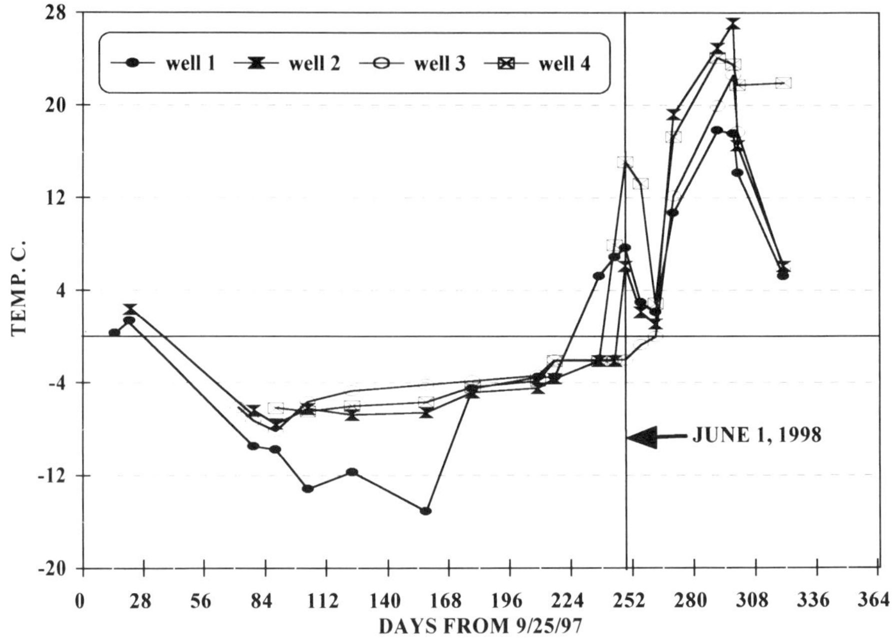

Figure 2: Temperature at ground surface; well 1 = top of waste pile, well 2 = foot of pile, well 3 = kill-zone, well 4 = vegetated ground

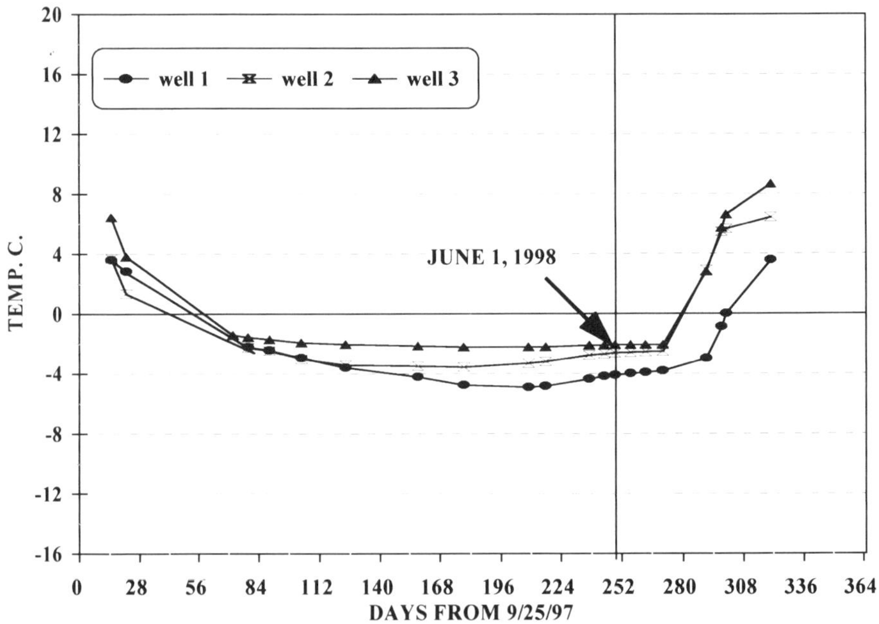

Figure 3: Temperature at depth; well 1 = 190 cm, well 2 = 84 cm, well 3 = 107 cm

4.1 *Geochemistry*

During peak run-off (June 1,1998), 200 to 400 liters per minute of water were flowing off of the main kill-zone terminus 300 m down slope from the waste pile (see figure 1) at 1:00 PM MDT. This water had a pH of 4.0, a conductivity of 0.1 mS (milli-Siemens), and a temperature of 0° C. This compares with clean snow melt: pH 5.5, 0.0 mS, 0° C; flowing water on vegetated tundra (near well 4): pH 5.4, 0.0 mS, 0° C; flowing water at down slope base of waste pile (near well 2): pH 2.9, 1.2 mS, 1.2 °C, and flowing water in the kill-zone (~ 25 m to ~100 m down slope from pile): pH 3.3 to 3.7, 0.1 to 0.4 mS, 0° C to 1.4 °C. Obviously, the mine waste material at the top of the kill-zone is producing low pH waters, but pH rises and conductivity (which is closely correlated with total dissolved solids) tends to decrease going downhill in the kill-zone.

4.2 *Geophysics*

The SP method measures naturally occurring electrical potentials generated from electrochemical, electrokinetic, and thermoelectric processes. Electrochemical potentials can be generated by a number of processes including, but not limited to, ion diffusion due to chemical gradients and electron transfer associated with oxidation-reduction reactions. Electrokinetic potentials may be generated due to fluid flow through fractured or porous media. Thermoelectric potentials may be generated by diffusion of ions and electrons due to thermal gradients.

At the site, chemical gradients produced by dissolution of waste rock and by oxidation-reduction reactions involving the sulfide minerals could produce electrochemical potentials. Runoff and flowing ground water through the waste rock or fractures could produce electrokinetic potentials. Exothermic chemical reactions (sulfide oxidation?) could generate a thermoelectric potential.

The SP data shows a broad, low-magnitude response of about 25 mV associated with the waste rock pile. This may be an electrochemical potential produced by oxidation-reduction reactions within the pile. A high amplitude SP anomaly was observed in a laboratory mine dump simulator experiment (Vaughn, et al., this volume), and is attributed to oxidation-reduction reactions associated with the vadose-saturated zone boundary.

Comparison of late spring and early summer SP data collected along the same profile, shows two distinct differences. First, a 50 mV anomaly associated with the position of a snow field was observed in the spring data, but not in the early summer data. This anomaly is thought to be electrokinetic in nature, produced by melt water flowing under and out of the snow field.

Secondly, the summer SP data appears to have more variability between individual readings than the spring data. This variability or "noise" in the summer SP data may be due to differential melting of the frozen alpine tundra, which would affect local ground water conditions. The variable summer SP values might also be the result of localized plant root ion exchange activity, which can produce bioelectric potentials. Neither the differential melting nor bioelectric mechanisms would have be active in the spring because the ground was still frozen (figures 2& 3).

The ground conductivity methods measure the electrical conductivity of the subsurface. The electrical conductivity of the subsurface is governed primarily by the amount of water present, the TDS (total dissolved solids) of that water, and the mineralogy of the subsurface. Saturated, high-porosity ground is more conductive than unsaturated, low-porosity ground. Water high in TDS is more conductive than low TDS water. Conductive minerals, such as clays (e.g. montmorillonite) or sulfides (e.g. pyrite) can significantly increase the electrical conductivity of the ground if present in relatively high concentrations.

The instruments used for the ground conductivity survey are electromagnetic induction instruments that measure the conductivity of the subsurface to shallow depths of about 4 meters. The ground

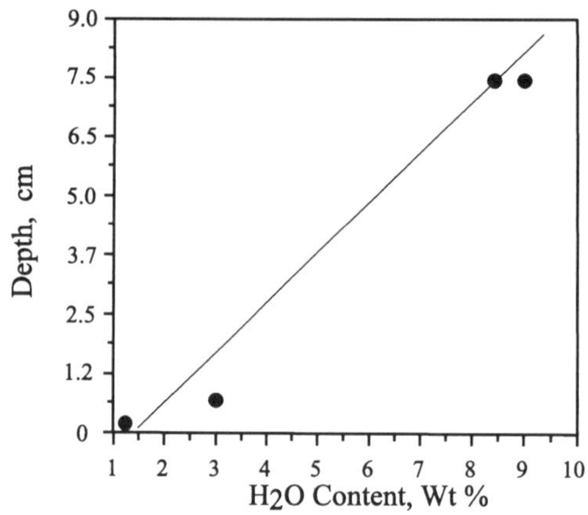

Figure 4. Water content (weight %) in solid samples vs depth into mine waste pile.

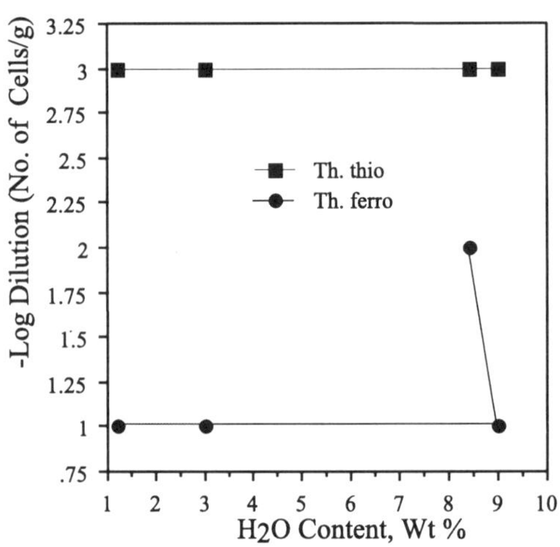

Figure 5. Abundances of *Thiobacillus thiooxidans* and *Thiobacillus ferrooxidans* vs water content in solid samples from mine waste pile.

conductivity survey was conducted along east-west grid lines that crossed the waste pile and the kill-zone (figure 1). Conductivity measurements on the waste pile ranged from about 20 to 55 millimhos per meter (mmhos/m). Immediately below the waste pile, on the downhill side, the conductivities ranged from 10 to 18 mmhos/m. Crossing the kill-zone, conductivities are 3.3 to 6.5 mmhos/m; whereas the surrounding undisturbed vegetated areas ranged from 1.4 to 2.0 mmhos/m.

4.3 *Geomicrobiology*

Thiobacillus ferrooxidans and *Thiobacillus thiooxidans* are chemoautotrophic bacteria that obtain energy by oxidation of Fe^{2+} (*Th. ferrooxidans*) or S^{2-} (*Th. thiooxidans*). Under some conditions (which we hope to determine more precisely in natural mine waste systems), these bacteria may be important in influencing the rates and preferred pathways of several important acid-generating geochemical reactions decomposing sulfides within mine waste piles. *Th. ferrooxidans* and *Th. thiooxidans* were not isolated at dilutions $>10^{-3}$ (Table 1), which suggests relatively low cell numbers in the pile. Dry, surficial crust samples were the least productive for thiobacilli; deeper, wetter (H_2O >8 wt%) samples generally showed a greater number of positive samples for bacterial growth. In most samples, *Th. thiooxidans* seemed more abundant relative to *Th. ferrooxidans* (Figure 5). Although unidentified fungi were present in nearly all inoculated tubes, there was no visible growth of fungi in the pile itself.

5 SUMMARY AND CONCLUSIONS

Preliminary data appears to support the conclusion that in ordinary weather years, no low pH, flowing, surface water is generated except during the short (1-2 week) period at peak snowmelt, when the ground is frozen to within a few cm of the surface and any liquid water must move as sheet or gully flow. Even at peak melt-off, pH increases from 2.9 to 4.0 and TDS decreases substantially within a few hundred meters down slope from the waste pile. The primary source of H^+ in acid water runoff is thought to be pyrite oxidation with minor contributions from other metal sulfides, sulfo salts, and perhaps soluble sulfates. From the pH of spring runoff, the presence of acidophilic bilberry, the shape of the kill-zone governed by waters coming from upslope pyrite-rich waste material, and the survival of lichens and flora communities at local micro high-spots within the barren area our initial conclusion is that the kill-zone is caused by acid-runoff from the waste pile during a brief period each spring. Further work on heavy metal content of runoff waters will help confirm or modify this hypothesis.

The present study emphasizes the importance of both detailed local climate and topographic knowledge in evaluating a mine waste dump's potential for amount, pH, and rate of acid generation. This knowledge along with detailed laboratory analyses helps more accurately forecast the real impact, both local and down drainage of a specific site.

5.1 *Geophysics*

Seasonal changes in the sites SP data were observed. Some of the SP anomalies are electrokinetic in nature, produced by spring surface runoff. SP variations attributed to electrochemical mechanisms have not yet been observed. However, as summer proceeds the ground continues to thaw, and precipitation and melt can infiltrate the interior of the pile, providing water for redox reactions. The fall SP data may show electrochemical activity.

The ground conductivity survey indicates the waste rock is more conductive than the surrounding undisturbed areas. The high conductivity of the waste rock is attributed to its high porosity relative to the host rock, high TDS ground water, and high concentrations of conductive minerals including sulfide minerals and clay. Conductivity measurements also indicate the kill-zone has a lower conductivity that the surrounding undisturbed areas, attributed to high TDS from the waste pile.

5.2 *Geomicrobiology*

The heterogeneity of grain sizes, variable moisture content, and differences in elemental chemistry (particularly iron and sulphur) in the pile are reflected by variable but generally low populations of thiobacilli. It appears that the often windy, dry, sunny, hot, surficial environment is quite harsh and not conducive to the survival and proliferation of these bacteria at the surface. Deeper within the pile, where water content is higher and micro-climate conditions are more stable and less harsh, greater numbers of thiobacilli were obtained from solid samples. Thus, if these microbes are active in decomposition of the iron and sulphur bearing solids, it probably occurs to the greatest extent in the interior of the pile. Overall, the cell numbers ($\leq 10^3$ cells/g) are quite low compared to commercial heap-leach systems or perennial streams affected by mine drainage as in the eastern U.S., where cell numbers may exceed 10^6/g.

5.3 *Future work*

Repeat geophysical measurements are planned for spring, early summer, late summer and fall through the remainder of 1998 till summer of 1999. The geophysical properties of the subsurface are expected to change throughout the seasonal freeze-thaw cycles and as liquid water infiltrates the waste pile and surrounding ground.

Future work will include studies of Fe and S oxidation reaction rates that may be mediated by the microbial isolates, and examination of the pile materials for other types of microbes that may play as yet unknown roles in alteration of the pile rocks and the preferred reaction paths of waste pile chemical processes.

6 ACKNOWLEDGMENTS

We would like to acknowledge and thank the landowners who have graciously allowed us to work on their property. Thanks to Norrie Robbins whose unceasing enthusiasm and encouragement helped move these investigations along.

REFERENCES

Arbogast, B.F. (ed.) 1996. *Analytical methods manual for the Mineral Resource Surveys Program, U.S. Geological Survey*: U.S. Geological Survey Open-File Report 96-525.

Behre, C.H. 1953. *Geology and ore deposits of the west slope of the Mosquito Range*: U.S. Geological Survey Professional Paper 235.

Desborough G., 1998. pers. comm.

Emmons, S.F., Irving, J.D., Loughlin, G.F. 1923. *Geology and ore deposits of the Leadville Mining District, Colorado*: U.S. Geological Survey Professional Paper 148.

Hach Chemical Company*, 1997. *User's manual.*

Hein A.S. and Fitterman D.V. 1998 (in press). *Topographic surveys of selected mine dumps near Silverton and Leadville, Colorado*: U.S. Geological Survey Open-File Report 98-???.

National Oceanic and Atmospheric Administration (NOAA), *1998. Climatologic data from Leadville, Leadville 2 SW, Climax, and Twin Lakes weather stations, Colorado 1948 -1998.*

Vaughn, et al. 1999. *Ant farms: mine dump simulation studies in the laboratory.* this volume

Zwinger A.H. and Willard B.E. 1972. *Land above the trees*. New York: Harper & Row

* NOTE: Any use of trade or product names is for informational use only and does not imply endorsement by the U.S. Government.

Tailings and Mine Waste'99 © 1999 Balkema, Rotterdam, ISBN 90 5809 025 6

Wet closure of exposed tailings using surficial lime slurry application

Andrew Evans, Jr
Department of Integrated Science and Technology, James Madison University, Va., USA

ABSTRACT: The wet closure is a remediation technology that has been used to reduce heavy metal contaminant mobility and toxicity. Exposed shoreline tailings comprising 40.5 acres (16.4 ha) located on the southern perimeter of Pond 2 in the Warm Springs Operable Unit were to be wet closed, with Pond 2 discharge being required to meet Interim Water Quality Standards. Field studies were undertaken to evaluate the suitability of direct surficial lime additions to exposed tailings. Results indicated that lime slurry application rate of 8.96 Mg ha^{-1} (4.0 ton/acre) would be optimal in reducing metal release into the overlying water column. Metal concentration in Pond 2 discharge during closure showed slightly higher dissolved Cu concentrations (0.055 - 0.105 mg/L) than predicted from column studies (0.050 mg/L). Limnological processes within Pond 2 related to the spring algal bloom and lake turnover was believed responsible for the observed differences.

1 INTRODUCTION

The Warm Springs Ponds (WSP) is a large settling pond facility that was historically used to treat Silver Bow stream inflows for suspended loads originating from upstream mining and ore processing operations. The Anaconda Copper Company made the first attempt at controlling the sediment load carried into the Clark Fork River from Silver Bow Creek during the period 1918 - 1920 by building a 6.06 m (20 foot) high tailings dam on Silver Bow Creek near the town of Warm Springs, MT, effectively creating Pond 1 (Figure 1). The need for increased detention time within the Pond system necessitated the construction of additional upstream dams resulting in the formation of Pond 2 during the 1920's and Pond 3 during the period 1956 - 1959.

In 1990, an Interim Record of Decision was issued by the U.S. EPA (U.S. EPA, 1990) which outlined the preferred alternatives for remediation within the active areas of the pond system (Ponds 2 and 3). Among the selected alternatives was the wet closure of exposed tailing impoundment areas within the ponds. It is estimated that approximately 14,527,400 m^3 (19,000,000 CY) of tailings, pond bottom sediments and associated soils, are contained within the Warm Springs Ponds Operable Unit of the Silver Bow Creek/Butte Area NPL Site (Anaconda, MT)

The exposed surface tailings containing high concentrations of readily soluble copper and zinc salts were located along the Pond 2 southern shoreline and occupied approximately 16.4 ha (40.5 acres). These tailings were to be wet closed, permanently covered with water by raising the Pond 2 water level, as part of the remedial action proposed for the site. The flooding of these tailings posed several problems which included the feasibility of surficial lime application to highly compressible tailings; possible adverse impact on aquatic life in the Pond and the Clark Fork River due to dissolved metal concentrations; and the need to meet Interim Water Quality Discharge Standards during the initial Pond raise.

To address these problems, soil column studies were undertaken to evaluate the suitability of direct surficial lime slurry application to the exposed tailings and to determine the optimal lime application rate. Supplemental field studies were undertaken to evaluate the implementation of this proposed remediation and to understand the interaction of various chemical - physical - biological processes associated with the flooding process.

2 MATERIALS AND METHODS

Soil column samples were collected within the main Pond 2 area from the exposed tailings located near the southern portion of the Pond 2 (Figure 1). Intact soil columns were collected by driving 20.32-cm diameter x 91.44-cm length of PVC piping into exposed tailings to a depth of 30.48 cm. The lower edges of the soil column were leveled with the bottom edge of the piping using a stainless steel spatula and capped with a PVC end plate. Paired soil columns were collected a nine sampling locations for a total of 27 soil columns. Prior to filling the soil columns with Pond 3 water, water collected from Pond 3 was analyzed for alkalinity; pH; total and dissolved Cu, Zn, Fe, Ca, Mg, Na, and K, using ICP; anions analyzed included Cl^-, PO_4^{3-}, SO_4^{2-} and NO_3^- and were analyzed colorimetrically using an Alpchem Autoanalyzer. Nine of the soil columns were filled with 14.8 liters of Pond 3 water collected near the Pond 3-east outlet tower. Pond 3 water was added slowly down the side of the PVC column using a Tygon™ tubing in order to minimize re-suspension of the tailings. Aliquots of the flooding water were collected every fifteen minutes and analyzed for the above parameters for the first six hours of the run and then sampled once hourly for a total run time of 12 hours.

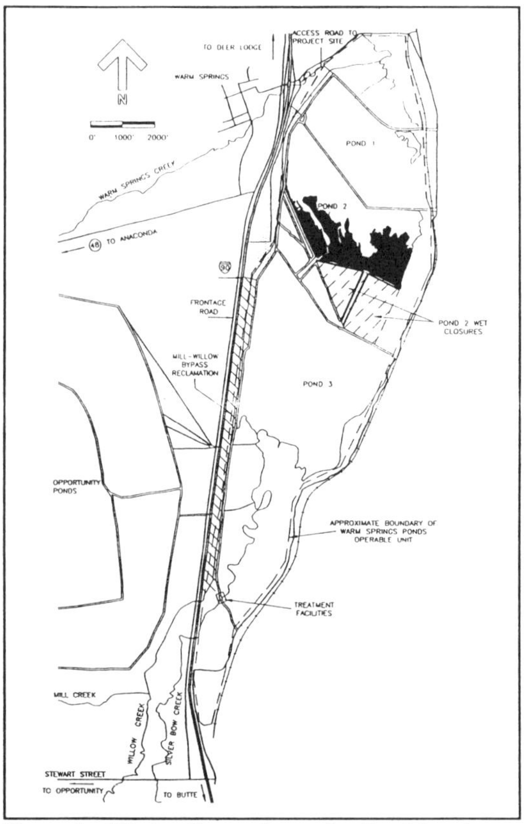

Figure 1. Site location and map for Pond 2 wet-closure study, Warm Springs Ponds. Shaded area represents exposed tailings that were wet closed.

A second set of nine soil columns was flooded as described above and the water was allowed to equilibrate 5 minutes (0.083 hours), after which time the solution pH was measured. The solution pH was adjusted to 9.3 by the addition of lime slurry using technical grade $Ca(OH)_2$ (hydrated lime). The selection of the pH 9.3 supernatant value corresponded to the optimal operating pH for the pond treatment facility. Lime was added to 100 ml of column water, stirred for one minute, and then added back to the water column as slurry. The water column/lime slurry mixture was stirred for 30 seconds using a paddle to insure uniform mixing. After two hours, the pH of the water column was measured and additional lime slurry was added to bring the pH up to 9.3. Solution pH was measured at two-hour time intervals for a period of 12 hours and then approximately every 24 hours, for a total period of 10 days. On days 3 and 10, additional lime was added as needed to adjust the solution pH to 9.3. The sum of these lime additions was considered to be the total lime requirement for the flooded tailings and was used in the following surfical lime addition study.

The remaining soil columns received a surficial lime slurry addition prior to flooding with 14.8 l of Pond 3 water. Lime slurry was prepared by mixing 29.1-g technical grade $Ca(OH)_2$ with 100 ml Pond 3 water and surface applying the mixture to each soil column. After application, the slurry was allowed to equilibrate one hour, at which time the soil column was flooded with Pond 3 water. The sampling scheme for ions and pH in the water column was similar to that used for the no lime treatment described previously

During the actual flooding of Pond 2 during the following spring, composite water samples were collected weekly from Ponds 2 and 3 discharge and analyzed for Cu, Zn, Fe, pH, SO_4^{2-}, Chl *a* and Total Suspended Solids (TSS). The TSS were sequentially extracted (fractionated) to determine the Cu partitioning between various suspended solid phases (Tessier 1979).

3 RESULTS AND DISCUSSION

The flooding of soil columns with Pond 3 water without lime addition resulted in the rapid release of Cu, Zn and Fe into the water column due to the dissolution of surface salts (Figure 2). Maximum dissolved metal concentrations were observed during the first four hours of flooding. After an initial

Fe concentration increase, dissolved Fe concentration decreased rapidly for study times > 0.083 hr and continued to decrease during the study period. The water columns were observed to be over saturated with respect to amorphous $Fe(OH)_3$ and hematite. Saturation indexes for these minerals were determined using MINTEQA2 (Allison et al. 1993). Based on reaction kinetics and mineral solubility, amorphous $Fe(OH)_3$ was initially precipitated and controlled Fe^{3+} activity during the course of these studies. The transformation of $Fe(OH)_3$ to hematite (Fe_2O_3) is a relatively slow process and would not be expected to be a factor controlling Fe^{3+} activity during this study. Both dissolved Cu and Zn concentrations were observed to decrease slightly with time. These

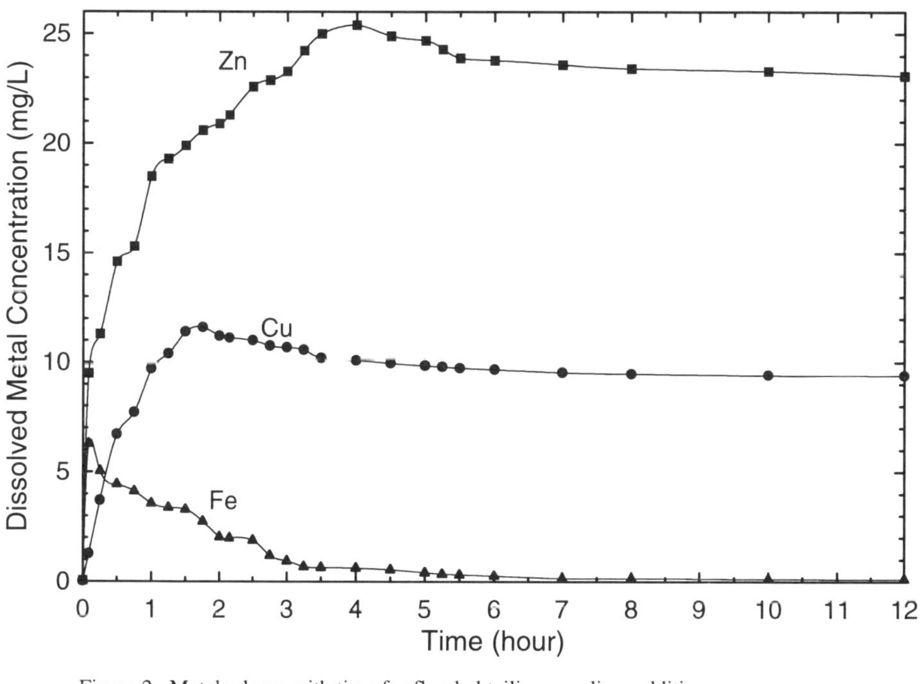

Figure 2. Metal release with time for flooded tailings, no lime addition.

decreases, however, were relatively small 1 - 2 mg/L for Cu and Zn, compared to the quantity of dissolved metals that were initially released. Removal of dissolved Cu from solution, for study times < 1.75 hr, was attributable to the co-precipitation/adsorption of Cu by amorphous $Fe(OH)_3$ (Leckie et al. 1980). For reaction times > 1.75 hours, precipitation of Cu minerals was unlikely due to relatively low solution pH and solution under saturation with respect to Cu solid phases. However, scavenging of Cu by freshly precipitated amorphous $Fe(OH)_3$ was possible, since the water column was saturated with respect to $Fe(OH)_3$ throughout the study. However, the scavenging efficiency of amorphous $Fe(OH)_3$ was reduced due to lower solution pH (Kinniburgh et al.1976). The flooded soil columns were not saturated with respect to common Zn solid phases, such as $Zn(OH)_2$ and $ZnCO_3$. Some Zn removal from the water column was possible due to adsorption onto amorphous $Fe(OH)_3$ (Kinniburgh and Jackson 1982). Zinc retention via this process would probable be minimal due to the selective adsorption of Cu over Zn by $Fe(OH)_3$. Zinc removal from solution would also be minimized by possible exchange reactions between Cu and Zn on the surfaces of freshly formed colloidal material such as amorphous $Fe(OH)_3$ and between Ca and (Cu, Zn) on soil solids (Bolt 1982). Dissolution processes were accelerated by the rapid decrease in solution pH. The rapid increase in sulfate concentration from 81.5 (Pond 3 background) to 310 mg/L was observed and was attributed to the dissolution of sulfate salts (Figure 3). A gradual increase in SO_4 concentration was observed during the study period, reaching a maximum of 400 mg/L after 12 hours.

The effect of surficial lime application prior to column flooding can be seen in Figure 4. The initial concentration of dissolved Fe, Cu and Zn present in solution was substantially lower compared to the unlimed treatment (Fig 4). At time = 0.083 hr, the water columns were saturated with respect to amorphous $Fe(OH)_3$ (S.I. = 3.703), copper ($Cu(OH)_2$ (S.I. = 0.023) and $Zn(OH)_2$ zinc hydroxide (S.I. = 0.438). Dissolved metals concentrations were observed to decrease rapidly for time > 0.083 hr, with a minor fluctuations being observed in the dissolved Zn concentration (see inset Figure 4). Calculated saturation indexes for the flooded soil columns for time >0.083 hr. indicated that the solutions were saturated with respect to amorphous $Fe(OH)_3$ and hematite and undersaturated with respect to Cu and Zn hydroxides and oxides. At low dissolved metal concentrations, the activity of dissolved Cu and Zn has been shown to be controlled by amorphous $Fe(OH)_3$/hematite and various colloidal components such as aragonite and calcite present in fresh

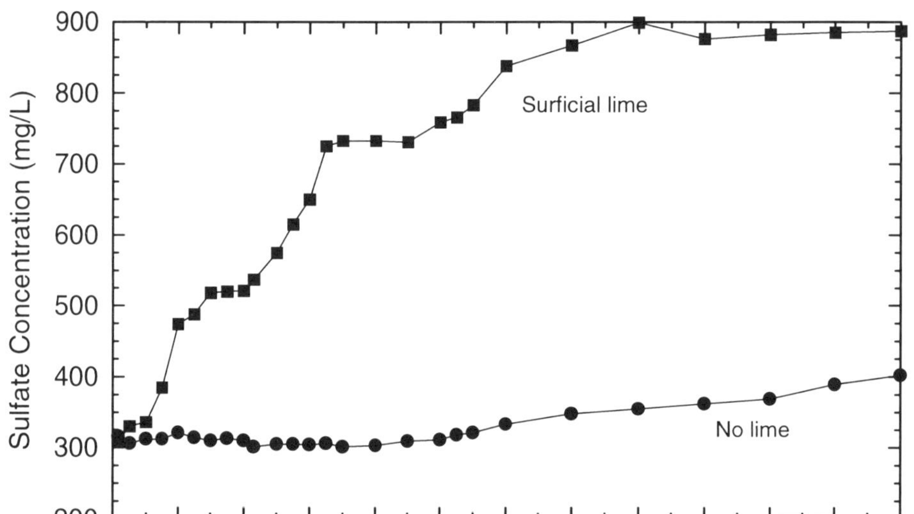

Figure 3. Comparison of surficial lime addition and no lime treatment on sulfate release into the water column.

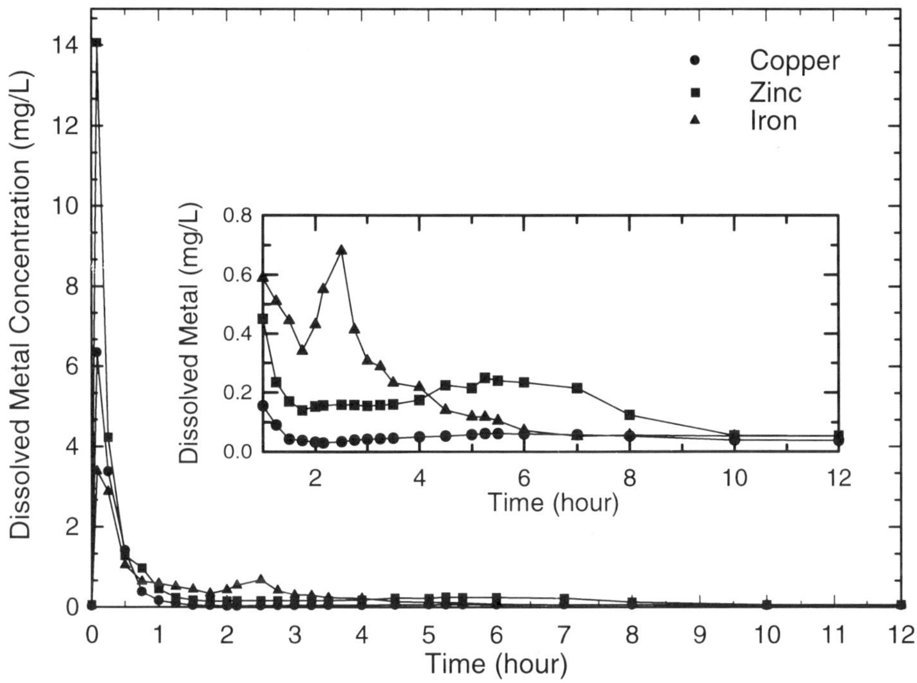

Figure 4. Metal release into the water column with time after surficial lime application.

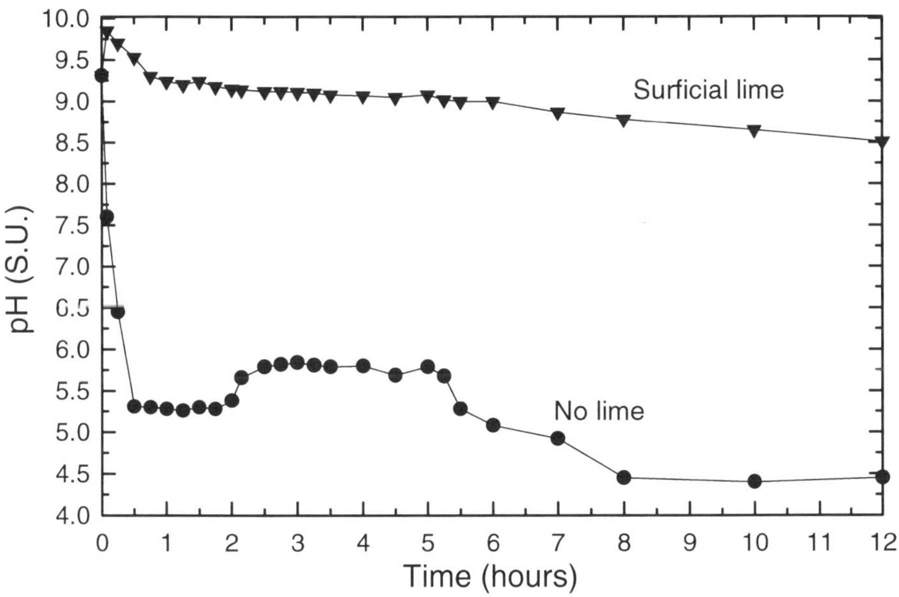

Figure 5. Water column pH with time for no lime and surficial lime treatments.

water (Sigg 1987). From Figure 4, it can be seen that the flooded soils, which received surficial lime application, appear to have equilibrated at approximately 0.050 mg/L dissolved Cu with minor fluctuations being observed. Dissolved copper concentration within the flooded columns at the end of the study period were in compliance with the Tier I Daily Standards established by EPA for Pond 2 discharge (U. S. EPA 1991).

Soil columns receiving surficial lime application exhibited a slight increase in the water column pH from 9.3 to 9.7 during the initial flooding (Figure 5). The solution pH of the water columns decreased gradually during the study period from 9.7 to 8.0. Surficial lime additions resulted in a significant increase in dissolved sulfate concentration with time (Figure 3) from 81.5 mg/L to 887 mg/L after 12 hours. The increase in sulfate concentration above 310 mg/L was attributed to anion exchange process between OH^- and exchangeable SO_4^{2-} present on soil solids (Radian 1979) and has been observed previously in both laboratory and field studies (Chao 1964; Parfait and Smart 1978). The step like release of sulfate into solution suggests that diffusion may be a limiting process in the release of sulfate into the water column.

A comparison of dissolved Cu concentration from the surfical lime addition study (Figure 4) with field measurements made at the Pond 2 discharge after the lime slurry addition and during the raise of Pond 2 elevation (Figure 6), indicated slightly higher than predicted dissolved Cu concentrations in the range 0.055 - 0.105 mg/L. Several additional environmental factors were operating during this period. Limnologically Pond 2 was undergoing ice-off and thermal turnover, which was subsequently followed by a large spring algal bloom (Williamson and Evans 1994; Evans and Williamson 1994). Biological activity due primarily to the algal bloom is believed responsible for a shift in equilibria between various Cu fractions associated with the Total Suspended Solids (TSS) component. TSS samples collected at the Pond 3 and 2 discharge stations indicate a large shift in Cu associated with the exchangeable and carbonate fractions of the TSS (Figure 7). During the peak of the algal bloom, March 17 - April 20, 1993, it can be seen that more Cu was associated with the readily available biological fractions, the exchangeable and carbonate fractions, in Pond 2 compared to Pond 3, than with the iron oxide and organic matter fractions. This shift is important in that Cu associated with these solid fractions can potentially be complexed either directly to algal cell walls or by organic exudates associated with algal biological activity. Dissolved Cu complexed with organic exudates are not readily precipitated as oxides or hydroxides. Consequently these Cu complexes could pass through the Pond treatment system, resulting in Cu exceedances in the allowable discharge. To determine the Cu complexation potential for both the

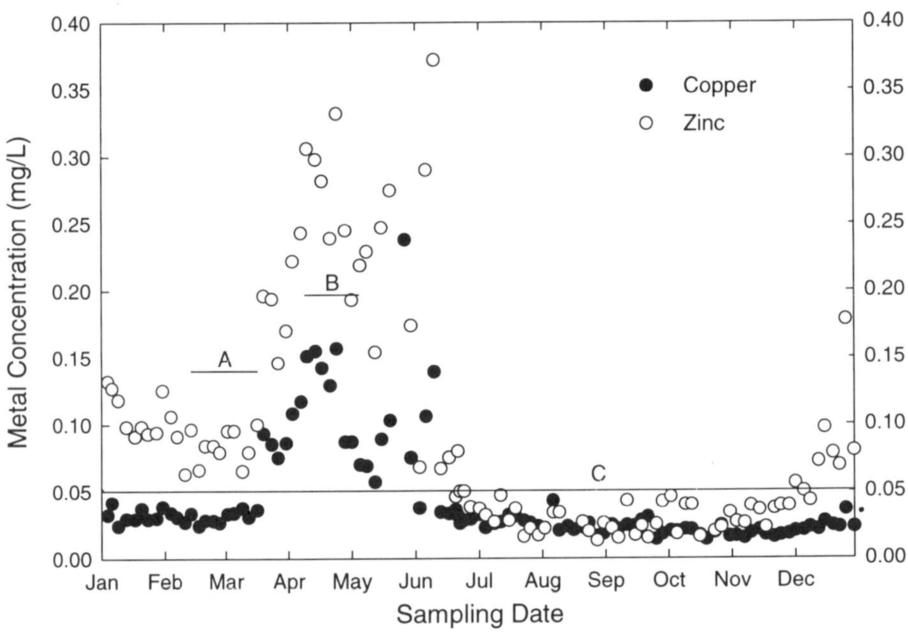

Figure 6. Dissolved metal concentration in Pond 2 discharge. a) Period of Pond 2 raise/exposed tailings flooding. b) Period of Pond 2 bypass, no discharge from Pond 2. c) Predicted copper metal concentration from column study.

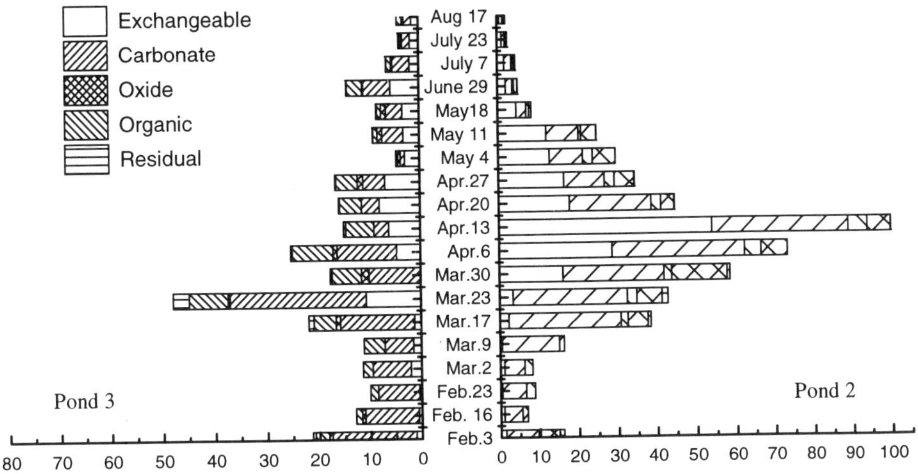

Figure 7. Sequential extraction of copper from filterable solids collected at the discharge of Ponds 2 and 3 during flooding of exposed mine tailings.

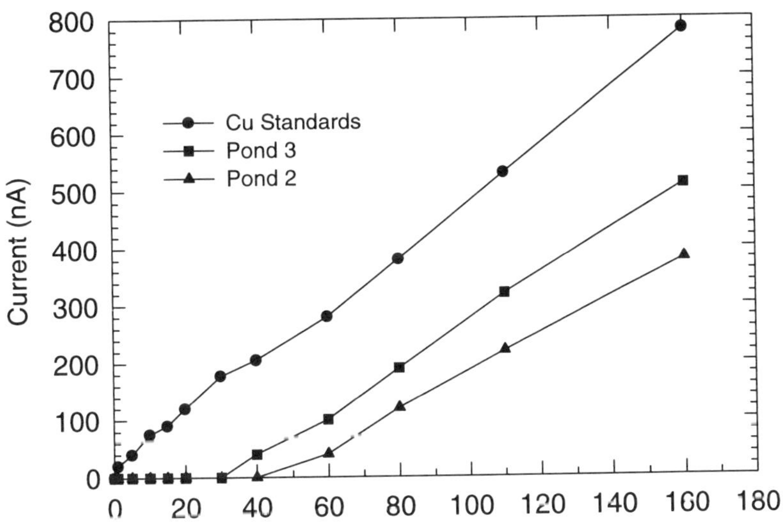

Figure 8. Complexation titration for Pond 2 and Pond 3 water. Deviation from x-axis indicates saturation of complexing agents with copper.

Pond 2 and Pond 3 discharge water during the algal bloom water samples were collected and a complexation titration was conducted. The results of this titration are shown in Figure 8. Increases in measured solution current corresponded to the saturation of the complexing agent with Cu and the measurement of free Cu in solution. The complexation potential was estimated for Pond 2 to be 0.047 mg/L, which was in good agreement with the difference observed between that predicted from the column studies and that observed in Pond 2 discharge. The Cu complexation potential in Pond 2 would be expected to vary seasonally and be a function of algal concentration (Williamson and Evans 1994; McKnight and Morel 1979).

REFERENCES

Allison, J.D., D.S. Brown and K.J. Novo-Gradac. MINTEQA2/PRODEFA2, A Geochemical Assessment Model for Environmental Systems. Environmental Research Laboratory, U.S.EPA, Athens, Georgia.

Bolt, G.H. Soil Chemistry B: Physico-Chemical Models. pp. 205-249. Elsevier Publishing, New York.

Chao, T.T. 1964. Anionic effects on sulfate adsorption by soils. Soil Sci. Soc. Am. Proc. 28:581-583.

Evans, A. Jr. and R.L. Williamson. 1994. Geochemical processes within the Warm Springs Ponds, Anaconda, Montana. Proceedings of the First International Conference on Tailings and Mine Waste 1994. Fort Collins, CO, USA. 19-21 January, 1994. pp.137-147. Balkema Publishers, Rotterdam, Netherlands.

Kinniburgh, D.G., M.L. Jackson and J.K. Syers. 1976. Adsorption of alkaline metals, transition, and heavy metal cations by hydrous oxide gels of iron and aluminum. Soil Sci. Am. J. 40:796.

Kinniburgh, D.G. and M.L. Jackson. 1982. Concentration and pH dependence of calcium and zinc adsorption by iron hydrous oxide gel. Soil Sci. Soc. Am J. 46:56.

Leckie, J.O., M.M. Benjamin, K. Hayes, G. Kaufman and S. Altman. 1980.Adsorption coprecipitation of trace metals from water with iron hydroxide. EPRI Document CS-1513.

McKnight, D.M. and F.M.M. Morel. 1979. Release of weak and strong copper-complexing agents by algae. Limnol. Oceanogr. 24(5):823-837.

Parfitt, R.L. and S.C. Smart. 1977. The mechanism of sulfate adsorption on iron oxides. Soil Sci.Soc. Am J. 42:48-50.

Rajan, S.S.S. 1979. Adsorption of sulfate and charge relationships in Allophanic clays. Soil Sci. Soc. J. 43:65-69.

Sigg, L. 1987. Surface Chemical Aspects of the Distribution and Fate of Metal Ions in Lakes. In W. Stumm (Ed.), Aquatic Surface Chemistry, pp. 319-349. John Wiley and Sons, New York.

Tessier, A., P.G.C. Campell and M. Bission. 1979. Sequential extraction procedure for the speciation of particulate trace metals. Anal. Chem. 51:844-851.

U.S. EPA. 1990. Record of Decision. Silver Bow Creek/Butte Area NPL Site, Warm Springs Ponds Operable Unit, Upper Clark Fork River Basin, Montana.

U.S. EPA. 1991. Unilateral Administrative Order for the Silver Bow Creek/Butte Area, Superfund Site, Warm Springs Ponds Active Area. Docket No. CERCLA-VIII-91-25.

Williamson, R.L. and A. Evans, Jr. 1994. The effects of limnological parameters on lime slurry treated water in large detention ponds. Proceedings of the First International Conference on Tailings and Mine Waste 1994. Fort Collins, CO, USA. 19-21 January, 1994. pp.137-147. Balkema Publishers, Rotterdam, Netherlands.

Tailings and Mine Waste '99 © 1999 Balkema, Rotterdam, ISBN 90 5809 025 6

Bi-phasic neutralization of evapo-concentrated acid mine drainage

L. Pucik Ericksen & W.J. Walker – *Geomega, Sacramento, Calif., USA*
C.K. Maslanka – *USA*
R.G. Burau & R.J. Zasoski – *Department of Land, Air and Water Resources, University of California at Davis, Calif., USA*
C.G. Schreier – *Prima Environmental*

ABSTRACT: During the bi-phasic neutralization process described here for evapo-concentrated acid mine drainage (AMD) containing high levels of ferric iron, arsenic is removed through adsorption to iron precipitates before other less hazardous constituents are removed. Addition of lime to maintain iron hydroxide precipitation at the 70% removal equivalence point results in optimal levels of arsenic removal and mass precipitation during Phase I. After separation of Phase I sludge, a traditional neutralization/precipiation process on the remaining AMD is performed (Phase II). Application of this process to AMD of pH 2.2 and containing primarily 1,600 mg/L iron (over 90% ferric), 30 mg/L arsenic, 740 mg/L aluminum, and 6,900 mg/L sulfate in a field pilot test generated a low volume of Phase I sludge containing 2,300 mg/kg arsenic and 8,300 mg/kg iron. Phase II generated a large volume of sludge containing mostly iron and aluminum hydroxides and gypsum, but containing no detectable arsenic.

1 INTRODUCTION

In a project sponsored by the Lahontan Regional Water Quality Control Board (LRWQCB), a bi-phasic neutralization process developed by researchers in the Department of Land, Air and Water Resources at the University of California at Davis (UCD) was applied to evapo-conentrated acid mine drainage (AMD) containing elevated arsenic, iron, acidity, and other metals at Leviathan Mine in Alpine County, CA. The treatment process is designed to remove metals and neutralize pH using a two-step process. The first phase removes arsenic from the AMD and generates a separate and relatively small arsenic-laden sludge. In the second phase, other metals are removed and large amounts of sludge are generated. Results of pilot scale implementation and testing of this process on-site are described here.

2 BACKGROUND

2.1 *The site*

The Leviathan Mine site is an abandoned mining operation located approximately 20 miles southeast of Lake Tahoe on the eastern slope of the Sierra Nevada Mountains at an elevation of 7,000 feet. Underground mining of copper and copper sulfate at the site began in the 1860's and was followed by open-pit mining of sulfur in the 1930's. Activity at the site ceased in 1962. In 1985, as part of a pollution abatement project, five lined evaporation ponds were constructed to collect combined acid mine drainage from the pit underdrain system, the adit, the springs, and the seeps. Annual overflow from these ponds into nearby Leviathan Creek has created the need for an on-site water treatment system capable of bringing this discharge into compliance with water quality standards.

Characterization of AMD in the ponds by the UCD group (Burau et al., 1995) showed that it is of low pH (2.3) and high redox potential (Eh ~ 0.87 V). Concentrations of primary constituents of concern vary seasonally: sulfate (2150 - 15300 mg/L), iron (900 - 4790 mg/L), aluminum (560 - 3440 mg/L), arsenic (18 - 113 mg/L), nickel (11 - 48 mg/L), and copper (5 - 27 mg/L).

2.2 *Treatment process chemistry*

Common treatment for this type of discharge is neutralization, in which cationic metals are removed from solution via precipitation (Mine Waste Study Team of UC Berkeley, 1988). Precipitation is induced by raising solution pH, which induces the formation of solid metal hydroxides, oxides, and oxyhydroxides. Because different metal species exhibit different pH-solubility relationships, the choice and control of the pH endpoint in this process are critical for ensuring that the target metals remain insoluble.

Because AMD at Leviathan is characterized by high concentrations of iron and aluminum, the pH-solubility relationships of these two compounds drive control of the neutralization process. A pH endpoint in the 6.5 to 8.5 range provides minimum solubility for these elements (Burau et al., 1995). Above this range, aluminum in particular becomes increasingly soluble. Below this range, sufficient removal by precipitation may not occur.

The choice of endpoint range becomes more complex when removal of the other trace constituents must also be considered. Cationic metal trace constituents may be removed during the neutralization process by precipitation, co–precipitation, or adsorption onto iron oxide surfaces. Nickel and copper are best sorbed in pH ranges 3.5-6 (McKenzie, 1980), which are below the optimal iron/aluminum removal range.

It is the removal of arsenic, however, that becomes the focus of optimizing competing neutralization processes (Burau et al., 1995). Arsenic content in the neutralization process waste stream determines the toxic classification of the sludge generated by treatment. Sludge that fails the Total Threshold Limit Concentration (TTLC) test for arsenic is considered a hazardous waste and therefore requires more rigorous handling and disposal from an environmental standpoint. Furthermore, because it is present in site AMD as arsenate, an anion, it behaves differently than the cationic trace constituents of concern.

Soluble arsenic concentrations are usually controlled by redox conditions, pH, biological activity, and physical processes (sorption) rather than by solubility equilibria (Bhumbla and Keefer, 1994). Sorption is the mechanism most likely to remove soluble arsenic due to the soluble nature of most arsenic salts. Adsorption of arsenic is controlled by a variety of factors. Arsenate is sorbed strongly by iron and aluminum oxyhydroxides (Fuller et al., 1993; Anderson et al., 1976). The sorption process is affected by pH, with strongest sorption of arsenate usually occurring under acidic conditions.

Arsenic removal, then, is best achieved in acidic pH ranges through adsorption processes, while removal of the other metal species present in the AMD requires a neutral pH range to induce complete precipitation. Treatment thus involves accurately controlling pH conditions to induce processes that will ensure removal of both the cationic and anionic species. Kinetic control of the system must also be maintained to ensure that precipitation and adsorption processes, which have different rates of reaction, are given sufficient time to effect removal.

2.3 *Bi-phasic neutralization*

It is with these constraints in mind that the UCD group developed a biphasic neutralization approach for treating AMD at the site (Jacobson, 1997). Using a two-step process in which pH is sequentially raised with lime, they were able to remove all constituents of concern while generating relatively small amounts of arsenic-laden sludge in bench-scale laboratory tests. During this process, arsenic is first removed by adsorption to iron precipitates. This is achieved by maintaining iron removal at the 56% equivalence point, which occurs in the iron hydroxide buffering zone. The buffering zone occurs when addition of extra base is utilized by iron hydroxide precipitation:

$$Fe(III) + 3OH^{-} \rightarrow Fe(OH)_{3(s)} \tag{1}$$

Solution pH remains nearly constant in this zone as long as excess soluble iron is available to buffer the addition of base. Given enough reaction time, it is in this zone at the 56% iron removal equivalence point that maximum arsenic removal occurs. Because the pH at which this occurs (~2.8) is lower than the precipitation points of the other metals present, a small volume of sludge containing mostly iron and arsenic is generated. This sludge is then separated from the AMD before the pH is raised again in the second treatment step. In this second step, pH is raised to the neutral range in order to remove all other constituents of concern. A large amount of sludge with low arsenic content is generated and can be disposed of separately.

The project described here was a pilot-scale test of this biphasic neutralization treatment approach. The goal of the project was to demonstrate the potential use of this process for on-site treatment of large volumes of AMD. Issues considered included the feasibility of successfully controlling pH and reaction kinetics in a scaled-up system. The methods employed and the results obtained during this project are described below.

3 METHODOLOGY

Lime slurry treatment reagent was first prepared by mixing 50 lbs of $Ca(OH)_2$ with 50 gallons of fresh water, obtained from Leviathan Creek upstream of the AMD ponds.

Four Phase I tests were run. Test 1 was a uncomplete startup trial and is not discussed here. AMD to be treated was pumped at approximately 2.9 GPM into the mixing line, where it was combined with the lime slurry flow at approximately 0.02 GPM. This combined flow was then pumped into the 2,000 gallon reactor tank, where it was continuously mixed with an aerator for the duration of the reaction. Filling of the tank continued for 4 hours, resulting in a volume of approximately 700 gallons. Influent to the tank was then shut off and the solution was allowed to continue mixing for another 2 to 2.5 hours. Samples of the reaction fluid were taken from a sampling port at the bottom of the tank every half hour during this process and analyzed for pH and temperature. Samples were then passed through a 0.45 μm filter and analyzed for total iron, ferrous iron, and sulfate. These analyses were performed using a field spectrophotometer and Hach brand colorimetric and turbidimetric reagents (FerroZine for total iron, 1,10 Phenanthroline for ferrous iron, and SulfaVer4 for sulfate). Every hour, a sample to be saved was collected, 0.45 μm filtered and preserved with nitric acid. Solid material from the filtration process was saved as a sludge timepoint sample. At the end of the mixing period, the solution was pumped into the 1,200 gallon clarifying tank and allowed to stand overnight. The next morning, the reaction solution was pumped through a filter press and the effluent was returned to the cleaned reaction tank for Phase II processing. Sludge from the press was collected and stabilized by adding $CaCO_3$ at a rate of 10% (w/v). Collected aqueous samples were analyzed by EPA method 6010 for iron and arsenic and sludge was analyzed by EPA methods 6010/7000 for various metals.

To determine the amount of lime slurry required to neutralize the Phase II solution, a 3 gallon aliquot was first titrated with lime slurry in a bucket. This dosing rate was used to determine the total amount of lime slurry needed to titrate the 650 gallons of solution in the reactor tank. The lime slurry flow rate was then set to deliver this dose over a two hour period to the continuously

Table 1. Results of Pond 1, Treatment 2, Phase I.

Lime flow	0.0225 GPM	(1lb/gallon slurry)
AMD flow	2.88 GPM	
Stop filling	2:48 PM	

		Reaction solution								Sludge		
Timepoints	Time	pH	Fe total	Fe total	Fe(II)	EC	Temp.	As	Al	Fe	As	Al
		field	field	lab	field	field	field	lab	lab	lab	lab	lab
[min]			[mg/L]	[mg/L]	[mg/L]	[mS/cm]	[C]	[mg/L]	[mg/L]	[mg/kg]	[mg/kg]	[mg/kg]
0	9:48 AM	2.40	942	1500	250	>20	11.0	29		15	29	
60	10:48 AM	2.53	725	660	250	>20	11.3	9.1		120000	3400	
120	11:48 AM	2.41	419	560	69		12.0	6.5		110000	2800	
180	12:48 PM	2.46	374	510	61	>20	16.0	5.3		110000	2800	
210	1:18 PM	2.61				>20	14.3					
240	1:48 PM	2.44	177	500	57	7.2	16.4	5.1		110000	2800	
270	2:18 PM	2.56					12.7					
300	2:48 PM	2.42	260	470	54	6	12.1	4.6	690	110000	2700	820

mixing reactor tank. After this amount had been delivered, the lime slurry influent line was shut off and the system was allowed to mix for another 1.5 hours. During this process, samples were collected and analyzed every 15 minutes as described in the Phase I section above. Samples for preservation were collected every hour. The solution was then pumped into the clarifying tank and allowed to settle. After settling overnight, the volume of sludge produced was recorded and the supernatant was analyzed. The sludge was then filtered through the filter press. Effluent was routed into Pond 1 and sludge samples were collected and stabilized as during Phase I.

4 RESULTS

4.1 *Phase I*

Results of Pond 1, Treatments 2-4, Phase I are shown in Tables 1 through 3. With a few exceptions, total iron field analyses were generally within 15% of laboratory results.

The maximum pH reached during the lime addition period was approximately 2.6 in all cases. For treatments 2 and 4, pH eventually dropped down to approximately 2.4 during the mixing period. These values are somewhat lower than expected for the iron hydroxide buffering range under standard conditions (pH ~ 2.8), although this depression may be due to the lower ambient temperatures experienced in the field. Sulfate concentrations fluctuate slightly but appear to stay relatively constant in all of the tests.

Arsenic removal, which is the goal of Phase I, was 89% in Treatment 3 and 79% in Treatment 4. Table 4 shows a comparison of arsenic and iron removals versus lime dose. Lime doses are expressed in a ratio of milliequivalents lime to milliequivalents total iron. Better arsenic removal was achieved with the higher lime dose (0.88 versus 0.81) and corresponded with the 71% iron removal versus the 57% removal. In both tests, arsenic removal at two hours after starting neutraliztion was approximately 92% of the maximum removal obtained by the end of tests. Thus, although arsenic concentrations continued to steadily decline for the duration of the 6 - 7 hour tests, the bulk of removal was achieved within two hours.

Filtered sludge collected during the Phase I tests confirms this. Sludge arsenic concentrations essentially reached their maximum within the first two hours in both Treatments 3 and 4. As expected based on the greater soluble arsenic removal obtained in Treatment 3, average arsenic content in this sludge (3,000 mg/kg) was higher than in the Treatment 4 sludge (2,300 mg/kg). Iron contents were also higher (113,000 in Treatment 3 versus 83,000 in Treatment 4). The ratio of arsenic to iron in sludge from both tests is approximately 0.027.

Other constituents including calcium and toxic cadmium, chromium, and vanadium were removed from solution during this phase, as shown by their concentrations in final Phase I sludge (Table 5). Processing of this sludge through the filter press was somewhat difficult because the material retained a fair amount of moisture.

4.2 *Phase II*

Filtered effluent from Treatment 4, Phase I was subjected to Phase II processing. Results of this test (Pond 1, Treatment 4, Phase II) are shown in Table 6.

Lime was added for 37 minutes into the 650 gallons of Phase I effluent to increase pH in the reaction mixture. This dose (approximately 470 meq lime per gallon of effluent) was based on a field titration of the effluent. After the lime was shut off, there was a 30 minute lag time in which pH continued to rise as the reaction mixture was stirred. After the pH stabilized at 6.4, a second dose of lime (approximately 63 meq per gallon of reaction mixture) was added over a 5 minute period in order to reach the target pH endpoint of 7.5. Due to the low buffering capacity

Table 2. Results of Pond 1, Treatment 3, Phase I.

Lime flow 0.02 GPM (1lb/gallon slurry)

AMD flow 2.94 GPM

Stop filling 12:37 PM

		Reaction solution								Sludge	
Timepoints	Time	pH field	Fe total field	Fe total lab	Fe(II) field	SO_4^{2-} field	EC field	Temp. field	As lab	Fe lab	As lab
[min]			[mg/L]	[mg/L]	[mg/L]	[mg/L]	[mS/cm]	[C]	[mg/L]	[mg/kg]	[mg/kg]
0	8:37 AM	2.46	858	1400	990	7400	>20	11	27	5600	160
60	9:37 AM	2.62	814	890	410	7400	6.33	11.2	16	120000	3300
90	10:07 AM	2.69					>20	10.6			
120	10:37 AM	2.61	560	500	480	6900	6.23	9.3	5.7	120000	3200
150	11:07 AM	2.67					6.09	11.3			
180	11:37 AM	2.65	351	450	127	6300	6.12	11.9	6.1	100000	2600
210	12:07 PM	2.67					6.22	10.8			
240	12:37 PM	2.70	444	430	80	7000	5.93	13.4	5.6	110000	2900
270	1:07 PM	2.63					6.17	11.1			
300	1:37 PM	2.59	380	410	77	6500	6.09	12.2	3.9	110000	3000
330	2:07 PM	2.60					5.87	12.1			
360	2:37 PM	2.63	372	400	72	5500	5.96	13.1	2.9	120000	3200

Table 3. Results of Pond 1, Treatment 4, Phase I.

Lime flow	0.02 GPM	(1lb/gallon slurry)
AMD flow	2.8 GPM	
Stop filling	12:38 PM	

		Reaction solution								Sludge		
Timepoints [min]	Time	pH field	Fe total field [mg/L]	Fe total lab [mg/L]	Fe(II) field [mg/L]	SO_4^{2-} field [mg/L]	Temp. field [C]	As lab [mg/L]	Al lab [mg/L]	Fe lab [mg/kg]	As lab [mg/kg]	Al lab [mg/kg]
original AMD*	8:10 AM	2.21	1317	1600	96	6900	6.3	30	740			
0	8:07 AM	2.73	1371		91	8000	5.1			17000	260	730
31	8:38 AM	2.66					5.0					
60	9:07 AM	2.64	807	830	158	6800	9.4	10	680	73000	2100	640
90	9:37 AM	2.71					11.8					
120	10:07 AM	2.65	884	780	124	7200	13.1	9.5	680	82000	2300	710
149	10:36 AM	2.53					8.8					
180	11:07 AM	2.58	588	620	103	6400	11.7	5.4	660	81000	2200	650
214	11:41 AM	2.51					12.1					
240	12:07 PM	2.59	697	700	101	7000	10.3	7.9	670	90000	2500	650
270	12:37 PM	2.56	691		122	6000	11.5					
300	1:07 PM	2.65	789	650	124	6800	11.8	7	640	84000	2300	740
330	1:37 PM	2.48					11.4					
360	2:07 PM	2.45	600	680	114	6400	11.4	6.7	690	77000	2100	650
390	2:37 PM	2.5	624				11.4					
420	3:07 PM	2.45	623	710	125	7200	11.6	6.3	740	95000	2600	710
1418	7:45 AM	2.4	563		113	6400	7.3					

* Original AMD was from pond. Timepoint 0 was from influent mixture to reactor.

Table 4. Iron and arsenic removal versus lime dose for Pond 1, Phase I.

Treatment	Lime dose [meq lime/meq Fe]	Fe removal [%]	As removal [%]
3	0.88	71	89
4	0.81	57	79

in this region, however, the additional dose overshot this endpoint and a final pH of 8.7 was obtained.

Aluminum and iron were completely removed from solution during this process, although some re-dissolution of aluminum is observed after the second dose of lime. The residual soluble arsenic remaining after Phase I treatment was also completely removed during Phase II. Finally, 70% of the sulfate present was removed during this stage.

Analysis of final Phase II sludge showed arsenic to be non-detect (<5 mg/kg), suggesting that the residual arsenic left in solution after Phase I does not significantly affect Phase II sludge quality. Estimates of arsenic mass balance based on previous laboratory tests by the UCD team, however, suggest that arsenic content in the Phase II sludge must be higher. Further study is needed. Calcium concentration (170 mg/kg) in the sludge showed that additional calcium was also removed during this phase.

Inspection of the clarifying tank after an overnight settling period revealed that the sludge accounted for approximately 40% of the mixture volume. Filtration of this material through the filter press yielded well formed, relatively dry, thick cakes of sludge.

5 SUMMARY AND CONCLUSIONS

The use of bi-phasic neutralization for treatment of evaporation pond AMD at Leviathan Mine was successfully demonstrated in pilot scale plant tests performed at the site. This process generated a low-volume of sludge in Phase I containing 2,300 mg/kg arsenic and 8,300 mg/kg iron. Trace toxic metals including chromium, cadmium, and vanadium were also removed during this phase. Phase II generated a large volume of sludge containing mostly iron and aluminum. It contained no detectable arsenic or other trace toxic metals.

It appears that the optimal lime dose for Phase I was 0.88 meq lime per meq iron in the Pond 1 AMD. At this dose, approximately 71% of the iron and 89% of the arsenic were removed. The majority of this removal (92%) was achieved within two hours of the start of the test. Arsenic removal is quite sensitive to a change in lime dose: reducing the dose by 8% reduced iron removal to 57% and arsenic removal to 79%. This lower removal, however, was still capable of providing non-detect levels of arsenic in the subsequent Phase II sludge. Because this dosing range is so compact and because pond water chemistry fluctuates during the year, laboratory bench-scale batch testing was the fastest and most efficient method for determining an effective dose during the field trials.

Similarly, bench scale titration of Phase I effluent for Phase II dosing was the most efficient means of dose determination during the field test. Because the reaction has an approximately 30 minute lag time and the mixture is relatively unbuffered in the neutral range, achieving the target endpoint in Phase II requires precise dosing and reaction-time control. Overdosing will result in redissolution of aluminum into the aqueous phase.

Table 5. Concentrations of metals in Pond 1, Treatment 4, Phase I sludge.

Metal	Calcium	Cadmium	Chromium	Vanadium
Concentration [mg/kg]	8600	12	67	130

Table 6. Results of Pond 1, Treatment 4, Phase II. Treatment of 650 gallons.

Lime flow	0.67 GPM	(1lb/gallon slurry)							
Lime off	1:17 PM								
Lime on	2:41 PM								
Lime off	2:46 PM								

Timepoints [min]	Time	pH field	Fe total field [mg/L]	Fe total lab [mg/L]	Fe(II) field [mg/L]	SO_4^{2-} field [mg/L]	Temp. field [C]	As lab [mg/L]	Al lab [mg/L]
0	12:40 PM	2.44	769	590	134	6600	10.1	3.3	680
17	12:57 PM	4.13					9.8		
30	1:10 PM	4.50	280		22	5200	9.9		
45	1:25 PM	5.21					10.8		
60	1:40 PM	6.09	5.45	0.45	1.43	3500	10.6	3.7	0.62
75	1:55 PM	6.23					10.6		
90	2:10 PM	6.41	0.436		0.09	3000	10.4		
105	2:25 PM	6.36					10.2		
120	2:40 PM	6.36	0.398	<0.1	0.08	2400	10.2	<0.002	0.11
135	2:55 PM	8.73					10.2		
150	3:10 PM	8.74	0.572	<0.1		2000	10.4	0.005	1.8

The results of this study show that on-site biphasic neutralization is a feasible treatment process for evapo-concentrated AMD containing high levels of ferric iron. The process offers a waste-handling cost advantage in that the bulk of the sludge generated does not contain arsenic and can be classified as non-hazardous. In studies currently underway, the stability of Phase II sludge and its potential for on-site disposal are being investigated.

6 REFERENCES

Anderson, M.A., Ferguson, J.F. & Gavis, J. 1976. Arsenate adsorption on amorphous aluminum hydroxide. *Journal of Colloid and Interface Science* 54(3):391-399.

Bhumbla, D.K. & Keefer, R.F. 1994. In J.O. Nriagu (ed.), *Arsenic in the environment*. New York: Wiley.

Burau, R.G., Jacobson, A.R., Zasoski, R.J. & Walker, W.J. 1995. Final report to California Regional Water Quality Control Board Lahontan Region: Bench scale studies to treat acid mine drainage at the Leviathan Mine Alpine County, California. Interagency agreement 2-703-160-0.

Fuller, C.C, Davis, J.A. & Waychunas, GA. 1993. Surface chemistry of ferrihydrite: Part 2. Kinetics of arsenate adsorption and coprecipitation. *Geochimica et Cosmochimica Acta* 57:2271-2282.

Jacobson, A.R. 1997. The biphasic neutralization of acid mine drainage from the Leviathan Mine in Alpine County, California. M.S. thesis in Soil Science, U.C. Davis.

Mining Waste Study Team of UC Berkeley 1988. Mining waste study - Final report. State of CA Water Resources Control Board, California Department of Health Services, California Department of Conservation.

McKenzie, R.M. 1980. The adsorption of lead and other heavy metals on oxides of manganese and iron. *Australian Journal of Soil Research* 18:61-73.

Tailings and Mine Waste '99 © 1999 Balkema, Rotterdam, ISBN 90 5809 025 6

Predicting the arsenic source-term from buried uranium mill tailings

Donald Langmuir
Hydrochem Systems Corporation, Golden, Colo., USA
John Mahoney & Anjali MacDonald
Hydrologic Consultants Incorporated, Lakewood, Colo., USA
John Rowson
COGEMA Resources, Saskatoon, Sask., Canada

ABSTRACT: The proposed JEB Tailings Management Facility (TMF) to be emplaced below the groundwater table in northern Saskatchewan, Canada, will contain mill tailings from McClean Lake, Midwest and Cigar Lake uranium ore bodies. Laboratory tests and geochemical modeling were performed to examine ways to reduce arsenic and nickel concentrations in TMF porewaters to minimize contamination of surrounding groundwaters.

Based on the laboratory tests and modeling, COGEMA Resources proposes to remove As and Ni from mill effluent waters by adding ferric sulfate and slaked lime to produce a raffinate Fe/As molar ratio of at least 3 to 5 and a nominal pH of 8 to 9. Statistical, thermodynamic and kinetic analyses of the aging-test data indicate that arsenic and nickel source term concentrations of 1 and 2 mg/L or less, respectively, are probable in porewaters of the TMF at 4°C for at least 10,000 years after tailings disposal. During this period, the arsenic will occur chiefly in ferric and nickel arsenate minerals that have been precipitated from the raffinate, with smaller amounts adsorbed by hydrous ferric oxides, aluminum oxides and clays present in the tailings.

1 INTRODUCTION

Uranium deposits of the Athabasca basin in northern Saskatchewan Canada, produce about one-third of the world's uranium (National Resources Canada, 1998). These deposits, which include those of Cigar Lake, Midwest, McClean Lake, Rabbit Lake, McArthur River, and Key Lake among others, are relatively high-grade, with average ore grades that often exceed 1% and may be as high as 15-32% U (Cramer and Smellie, 1994; Delaney et al., 1998). Unlike most other deposits, they may also be high in nickel (5%) and arsenic (10%). Crushing, acid leaching and oxidizing the ore in the mill to dissolve and extract the uranium may produce high arsenic concentrations in resultant acid raffinate solutions and in the porewaters of limed tailings emplaced in a tailings management facility (TMF).

Donahue et al. (1998) discuss historic disposal practices for the tailings of the Athabasca basin. The favored current approach which is used at Rabbit Lake and Key Lake mine sites, is disposal of the fine-grained tailings in mined-out pits below the water table. This method has the advantage that the release rate of a contaminant such as arsenic from a TMF to surrounding groundwaters is limited by its diffusion rate. The treatment of tailings has historically involved lime addition to attain a pH greater than 10 before tailings disposal. This approach removes metals such as Ni, but can lead to porewater As concentrations as high as 5-15 mg/L (Delaney et al., 1998; Donahue et al., 1998). To reduce arsenic concentrations to 1 mg/L or less in TMF porewaters in the JEB pit, COGEMA Resources proposes adding ferric sulfate to acid tailings raffinates that have Fe/As values less than 3-5, prior to lime addition for pH control. This modified approach leads to precipitation of the arsenic in ferric arsenate (scorodite). Lime would then be added to attain a nominal pH of 8 to 9. This paper discusses the scientific basis for this approach.

2 GEOCHEMICAL CONTROLS ON ARSENIC CONCENTRATIONS

2.1 *Adsorption and mineral equilibria*

Dissolved arsenic occurs in natural waters chiefly as arsenate (As[V]) species, and less often as arsenite (As[III]) or methyl (organic) arsenate species. Arsenate forms such as $HAsO_4^{2-}$ dominate in oxidized (high Eh) waters, whereas arsenite species (e.g., $H_3AsO_3^{o}$) may occur in low Eh waters, which are oxygen-deficient or anaerobic. In tailings and raffinates and in TMF porewaters, however, As(III) is likely to be a minor As form relative to As(V). For example, after liming for pH control in the mill, typical tailings porewaters have pH values above 7-8, are slightly oxidizing, and contain chiefly $HAsO_4^{2-}$ (As[V]).

Under oxidizing to slightly reducing conditions, removal of dissolved arsenic from water occurs by its adsorption onto mineral surfaces, and in high arsenic waters by precipitation of arsenate minerals. Under more reducing conditions dissolved arsenic may precipitate as insoluble sulfide and arsenide minerals. Such minerals are the chief mode of As occurrence in association with the uranium ores of the Athabasca Basin (Watkinson et al., 1975).

Arsenic is most strongly adsorbed by oxides and hydroxides of iron(III) and aluminum, and to a lesser extent by clay minerals. In general, As(V) species are more strongly adsorbed than arsenite (As[III]) species. The strongest sorbent for arsenic is freshly precipitated iron(III) oxide, which is also termed ferrihydrite, or hydrous ferric oxide (HFO). Fuller et al. (1993) have found that freshly precipitated HFO can adsorb or coprecipitate up to 0.7 moles of As(V) per mole of iron. This corresponds to a minimum Fe(III)/As weight ratio of about 1.1/1. A considerable literature exists on the measurement and modeling of As(V) and/or As(III) adsorption by HFO (Pierce and Moore, 1982; Goldberg, 1986; Robins et al., 1988; Fuller et al., 1993; Bowell, 1994; Langmuir and Mahoney, 1997). As(V) and/or As(III) adsorption by clays has been less measured or modeled (Frost and Griffin, 1977; Goldberg and Glaubig, 1988; Manning and Goldberg, 1997). Some authors have described arsenic removal by HFO as an adsorption process, regardless of the arsenic concentration. However, there is strong evidence that when high concentrations of Fe(III) and As(V) are present in acid solutions, and the pH is gradually raised, ferric arsenate solids are precipitated before HFO can attain saturation and precipitate (Harris and Krause, 1993; Demopoulos et al., 1995).

In raffinate solutions, which are high in both As(V) and cations including Ni, ferric sulfate addition and neutralization lead to the precipitation of poorly crystalline arsenates that include scorodite ($FeAsO_4 \cdot 2H_2O$), and as the pH is increased cabrerite-annabergite $[(Ni,Fe,Mg)_3(AsO_4)_2 \cdot 8H_2O]$[1]. Scorodite can precipitate under acid, oxidizing conditions, and is unstable relative to HFO at pH values that can range from 3 to 7, depending on the iron(III) and arsenic concentrations. In iron-rich raffinates, scorodite precipitation at low pH may so reduce dissolved As(V) concentrations that the nickel arsenate cabrerite-annabergite remains undersaturated as the pH is increased.

---

[1]Annabergite is the pure nickel end-member.

3 LARGE-SCALE AGING TESTS

Large-scale tests performed under the direction of Drs. Michel Roche and Isabelle Picquet in the COGEMA/SEPA laboratories in Bessines, France, were designed to evaluate the impact of iron(III) additions on dissolved arsenic concentrations in tailings raffinates of the McClean Lake, Midwest and Cigar Lake ores, and in porewaters within tailings to be emplaced in the TMF. Procedure 1 tests were designed to approximate the reactions that take place in the plant tailings neutralization circuit. These short-term tests, which involved the addition of slaked lime and ferric sulfate to raffinate solutions and tailings, were intended to define optimum conditions for removal of dissolved arsenic from the raffinates during processing of the tailings. All procedure 1 tests were performed at 25°C in contact with atmospheric oxygen (air), and were run for 3 hrs, the expected residence time of the tailings in the neutralization circuit. Procedure 2 tests, which were continued for times that ranged from 5-6 days to as long as 40-50 days, were designed to help predict dissolved arsenic concentrations in porewaters within the TMF for years after tailings disposal. These tests were performed under oxygen-free (anaerobic) conditions as expected in the TMF. Most procedure 2 tests were carried out at 25 and 50°C, with a few also run at 4°C, the probable temperature after burial in the TMF. Example compositions of tailings raffinates chosen for study are given in Table 1.

Table 1. Chemical analyses of untreated raffinates prior to ferric sulfate addition in tailings proposed for disposal in the JEB TMF. Also shown are chemical analyses of tailings solutions after ferric sulfate and $Ca(OH)_2$ addition and aging. All tailings solutions listed were aged at 25°C.

	Raffinates			Neutralized and Aged Solutions		
	JEB	McClean Lake	Cigar Lake	JEB 120 hrs	McClean Lake 144 hrs	Cigar Lake 128 hrs
Fe/As (molar)	2.56	4.64	7.54	5.0	5.0	7.5
*pH	(0.85)	0.75	(0.40)	7.32	7.53	7.31
Eh (mV)	576	623	588	437	369	301
Alkalinity (mM/L)				0.43	0.64	0.37
Al	503	393	4280	<0.02	<0.02	0.03
Σ As	271	275	580	0.79	1.7	0.36
As^{5+}	218.1	222.5	454	<0.2	1.21	<0.2
As^{3+}	52.9	52.5	126	<0.2	0.49	0.36
Ca	441	126	724	471	482	444
Cl	35.2	30.1	81	34.2	34.8	34.3
Σ Fe	517	950	3260	<0.02	<0.02	0.05
Fe^{2+}	413	419	2670	<0.02	<0.02	<0.02
Fe^{3+}	104	531	590	<0.02	<0.02	<0.02
*H^+	363	(179)	403	--	--	--
K	219	174	388	95.6	89.5	151
Mg	303	61.7	2330	221	77.9	1470
Na	25.8	31.7	129	546	609	478
Ni	144	164	453	0.52	0.2	0.36
NO_3	53.6	35.9	40.5	1.2	2.2	3.7
P	69	23	156	0.69	1.13	1.13
Ra (in Bq/L)	43	1.18	82.71	66.9	4.78	169.2
SO_4^{2-}	26900	26100	67200	3487	2862	9204
U	0.07	<0.05	1.64	0.08	<0.05	<0.25

* Values of H^+ concentration in parentheses have been estimated from measured pH values. Values of pH in parentheses have been estimated from measured H^+ concentrations.

Concentrations are in mg/L unless otherwise indicated.

3.1 *Aging test results*

Previous work and results of the large-scale tests indicated that for a given pH, higher Fe/As ratios in tailings raffinates lead to more complete arsenic removal from solution. Published studies show that arsenic removal is most complete in the pH range from about 3 to 5. During early test work involving tailings, it was observed that liming to pH values between 7 and 8 generally resulted in adequate arsenic removal without causing unacceptable increases in the dissolved nickel. Most long-term experiments were therefore performed at nominal pH 8 or 9 with the result that final pH values were usually between pH 7 and 8. Statistical tests showed that for tailings being considered for TMF disposal, the pH generally varied independent of time. Aging test results for example tailings solutions are given in Table 1.

At nominal pH 8, arsenic concentrations did not change with time in the 4°C aging tests, and only changed with time in three of seven tests run at 25°C. Examination of the test results for 25 and 50°C at 120 and 144 hrs, indicates that most of the increase in arsenic concentrations occurs prior to 72 hrs. The data collected at 72 hrs and thereafter was assumed to have most nearly approached steady state conditions, and so was treated as a single population for each run. Accordingly, the arsenic concentrations reported after 72 hrs for runs at 4, 25 and 50°C between pH 7 and 8 were averaged for each run. These averages are plotted versus the molar Fe/As ratio of each raffinate in Fig. 1. The data have been fit with curved regression lines, which have r^2 values that range from about 0.8 to 0.9. The plot shows that arsenic removal increases with increasing Fe/As ratio. Arsenic removal also increases with decreasing temperature and is greatest at 4°C, the expected temperature of TMF porewaters. The Fe/As ratios of tailings that would be emplaced in the TMF range from 3.0 to 7.7. The regression curve for 4°C, shows average As concentrations less than about 1 mg/L when Fe/As ratios equal or exceed 3.0.

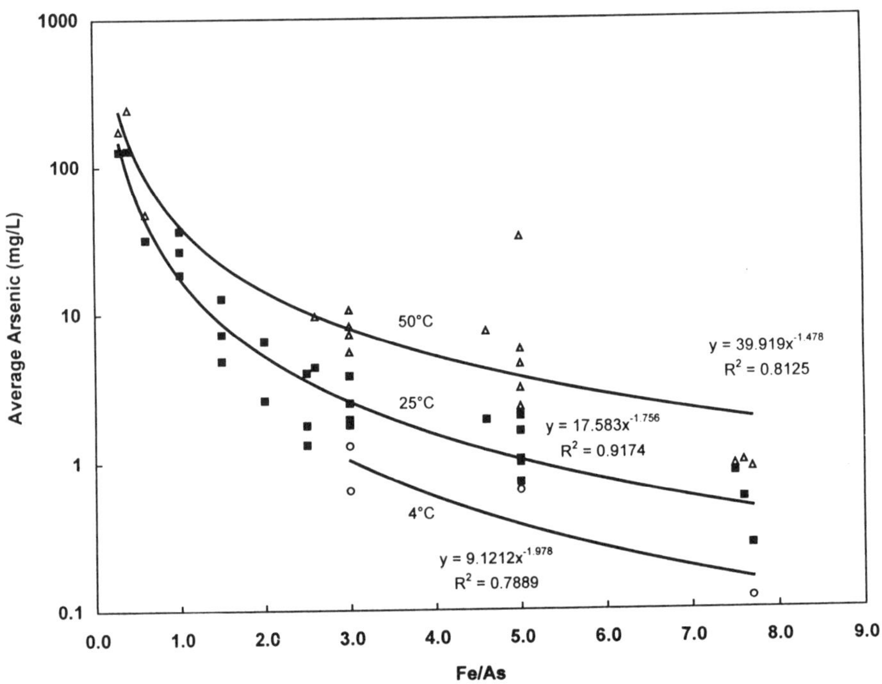

Figure 1. Average arsenic concentrations in procedure 2 experiments with pH values between 7 and 8 for times after 72 hrs as a function of tailings raffinate Fe/As ratio and temperature.

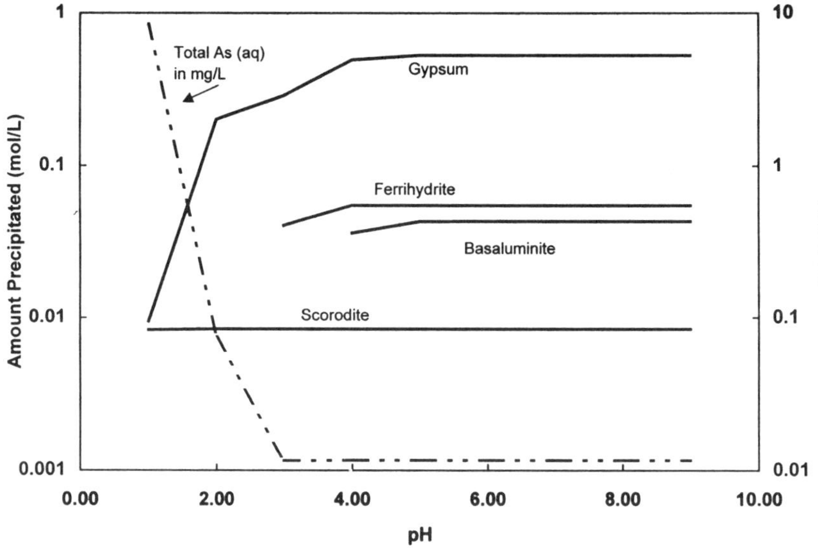

Figure 2. Precipitation of secondary phases as a result of adding $Ca(OH)_2$ to Cigar Lake raffinate (Fe/As = 7.5) as predicted with the geochemical code MINTEQA2. Also shown is the predicted concentration of As(V) as a function of pH.

3.2 *Composition and mineralogy of the tailings after treatment*

Several approaches were taken to determine the composition of the phases precipitated from the acid raffinate solutions as a consequence of ferric sulfate and calcium hydroxide addition. First, final compositions of solutions after aging times of 120 hrs or later were subtracted from original raffinate compositions to determine amounts of each oxide incorporated in secondary solids. Mole percent Ca and SO_4 were roughly equal in all of the processed tailings, consistent with the occurrence of gypsum ($CaSO_4 \cdot 2H_2O$), which constitutes most of the precipitated solids in all of the tailings.

As a second method of identifying the precipitated phases, liming of the raffinates was simulated using computer models MINTEQA2 (Allison et al., 1991) and PHREEQC (Parkhurst, 1995). Among the phases that were usually at saturation and are capable of precipitating within the short times of the procedure 1 experiments (minutes to a few hours) were hydrobasaluminite ($Al_4(SO_4)(OH)_{10} \cdot 15H_2O$), scorodite, and ferrihydrite (Fig. 2).

Separately, the composition and mineralogy of eight tailings samples were determined using a variety of techniques[2]. The tailings were chemically analyzed for their content of major and minor elements and element oxidation states. Analyses also included x-ray diffraction, scanning electron microscopy, electron microprobe, and petrographic work. All of the arsenic in the tailings was as As(V), with none found as As(III). Average weight percents in four tailings samples proposed for disposal in the TMF[3] in decreasing amounts, were illite clay (26.1%), quartz (25.1%), gypsum (22.0%), hydrobasaluminite (13.1%),

[2]Analyses and interpretation by Dr. Maynard Slaughter, Crystal Research Laboratories, Golden, CO. Some results provide by COGEMA's SEPA Laboratory, Bessines, France.

[3]These four tailings had been limed but no ferric sulfate added. Scorodite and cabrerite-annabergite amounts were as high as 6.3 and 14.9%, respectively in non-TMF tailings.

kaolinite (3.4%), Mg-chlorite (3.1%), cabrerite-annabergite (2.2%), and goethite (2.0%; amorphous and crystalline forms with the latter probably from the rock). Phases averaging less than 1% included coffinite (0.9%), rutile (0.6%), scorodite (0.4%), galena (0.3%), iron (II, III) oxyhydroxide (0.2%), hydrous phosphate (0.2%), chalcopyrite (0.1%) jarosite (0.1%) and pyrite (0.01%).

Among these phases, the secondary phases, which are those precipitated entirely from the raffinate, included gypsum, hydrobasaluminite, iron (II, III) oxyhydroxide, all of the amorphous 'goethite', cabrerite-annabergite, scorodite, jarosite, and hydrous (Ca) phosphate. Most of the arsenic in the tailings is present in the cabrerite-annabergite and scorodite, with the rest adsorbed or coprecipitated with relatively amorphous Al and Fe(III) oxyhydroxide phases and adsorbed by the clays. The nickel is present in the cabrerite-annabergite, with smaller amounts also adsorbed or coprecipitated with the oxyhydroxides and adsorbed by the clays.

4 CHEMICAL EQUILIBRIUM AND KINETIC MODELING OF AGING TEST RESULTS

4.1 *The dissolved arsenic versus Fe/As relationship*

As Fe/As ratios are increased for a given total arsenic concentration, scorodite (Fe/As = 1) apparently precipitates before significant arsenic is available for adsorption. The curves in Fig. 1 reflect scorodite and cabrerite-annabergite precipitation for Fe/As ratios below about 1. At greater Fe/As values, the remaining dissolved arsenic is controlled by adsorption. To confirm this explanation of arsenic behavior, some of the 25°C data in Fig. 1 was reproduced with the geochemical model PHREEQC (Parkhurst, 1995). The modeling was performed in two steps with increasing Fe/As ratios. The first step allowed for mineral precipitation. The second added reactions to model arsenic adsorption by HFO. Concentrations of species in the raffinates at their lowest Fe/As ratios, were the initial values for the modeling. Ferric sulfate was added to each raffinate in steps to duplicate the Fe/As ratios used in the experiments. The pH was then raised and maintained at pH 7.5 by $Ca(OH)_2$ addition. In the first step, scorodite, annabergite, gypsum and HFO were allowed to precipitate if they became supersaturated. If HFO was precipitated in step 1, its amount was used in a calculation of arsenic adsorption. Adsorption by HFO was computed with the diffuse layer model (Dzombak and Morel, 1990) assuming a surface area of 600 m^2/g, and a site density of 0.2 moles of sites per mole of FeOOH (the formula for HFO used in the model). It was necessary to vary the solubility product of scorodite by ±0.3 log units (≈1%) among the different tailings in order to optimize the model fit to the data in Fig. 1.

4.2 *Effect of temperature on arsenic concentrations*

Immediately following the short-term tests at 25°C, temperatures of some of the processed tailings solutions were raised to 50°C or lowered to 4°C for the long-term tests. Arsenic values increased rapidly at 50°C and dropped at 4°C to values near those shown in Fig. 1. The change was too rapid to have been caused by a readjustment in arsenate mineral/solution equilibria, and probably resulted instead from a readjustment of arsenate adsorption reactions. Such behavior is consistent with the negative adsorption enthalpy for anions reported by Machesky (1990). A generalized adsorption reaction for arsenate onto tailings solids may be written:

$$SOH + H_2AsO_4^- + H^+ = SH_2AsO_4 + H_2O \quad (1)$$

where SOH and SH_2AsO_4 are sorption sites occupied by hydroxyl and arsenate. The regression equations in Fig. 1 may be solved for Fe/As = 5 to give arsenic concentrations of 0.38, 1.1 and 3.7 mg/L at 4, 25 and 50°C. Examination of the generalized adsorption reaction (Eq. 1) shows that the adsorption K_{eq} value is proportional to $1/[H_2AsO_4^-]$ at constant pH. The van't Hoff equation, which describes the effect of a change in temperature (T) on K_{eq} (Langmuir, 1997) may be used to calculate the heat of the adsorption reaction, ΔH_r^o, from the concentration data at 4, 25 and 50°C. This gives -8.3 and -9.3 kcal/mol between 4 and 25°C, and between 25 and 50°C, respectively. The average arsenate adsorption enthalpy is thus ΔH_r^o = -8.8±0.7 kcal/mol.

4.3 *Reactions controlling the arsenic release rate*

The rate of dissolution of the arsenate minerals scorodite and cabrerite-annabergite probably controls the changes in arsenic concentrations with time in long-term tests. Above pH 7, the breakdown reaction of scorodite to form HFO may be written:

$$FeAsO_4 \cdot 2H_2O(s) + H_2O \rightarrow Fe(OH)_3(s) + HAsO_4^{2-} + 2H^+ \quad (2)$$

The breakdown of a cabrerite-annabergite that contains 33% of Mg substituting for Ni, may be written:

$$(Ni_{2.00}Mg_{1.00})(AsO_4)_2 \cdot 8H_2O(s) \rightarrow 2Ni(OH)_2(s) + Mg^{2+} + 2H^+ + 2HAsO_4^{2-} + 4H_2O \quad (3)$$

where it is assumed that the nickel precipitates as nickel hydroxide, which is most likely as pH values approach 8. At lower pH values Ni^{2+} concentrations will be controlled instead by adsorption reactions.

Given the chemical complexity of the tailings, it is not possible to identify the elementary rate equations that control arsenic concentrations. However, the overall equations describing the breakdown of scorodite and cabrerite-annabergite provide useful insights into reaction controls. Based on the breakdown reaction for scorodite, the rate of increase in arsenic concentrations as a function of time can be described by[4]

$$\frac{d(As(V))}{dt} = k_+ \frac{A_w}{M}(scorodite) - k_- \frac{A_w}{M}(HFO)[HAsO_4^{2-}]^n[H^+]^m \quad (4)$$

In this expression k_+ and k_- are the forward and reverse rate constants. A_w and M are the wetted surface area of solid scorodite or HFO exposed to solution, and the mass of water, respectively. The exact dependence of the rate on dissolved arsenic and hydrogen ion concentrations is not known. This uncertainty is expressed by the fact that exponents n and m for these species have not been defined. A similar expression may be written for the breakdown of cabrerite-annabergite.

The scorodite breakdown equation provides insights that are consistent with observation. These include that the rate of breakdown of scorodite is slower in tailings with higher Fe/As

[4]See also Demopoulos et al. (1994)

ratios, consistent with increasing amounts of HFO causing back reaction towards scorodite. Secondly, breakdown rates increase with increasing pH, also consistent with the rate equation.

Breakdown rates of the arsenates should also be proportional to their degree of undersaturation in the tailings, with arsenic concentrations increasing most rapidly in the most undersaturated porewaters. The saturation index of scorodite computed for all runs at 25°C which had measurable iron (>0.02 mg/L) is plotted against the pH in Fig. 3. The plot shows that crystalline scorodite is at or above saturation in the raffinates and probably also in porewaters up to about pH 6-7, but becomes increasingly undersaturated relative to HFO as the pH further increases. Scorodite SI values in Fig. 3 range from about -0.1 to +0.9. The MINTEQA2 saturation calculation has assumed -log K_{sp} = 25.87 for scorodite at 25°C. It seems likely that SI (scorodite) = +0.9 represents equilibrium with a relatively amorphous scorodite. If so, its -log K_{sp} is about 24.1.

The computed saturation state of annabergite shows similar pH-dependence in most cases, although the mineral is undersaturated in the raffinates. The neutralized raffinates were not sampled at pH values below about 6-7, however, computer modeling suggests that annabergite attains saturation in many of the tailings at some pH between 5 and 6.

An important question is how long into the future the proposed reaction mechanism of arsenic release is likely to dominate release rates. To answer this question, the regression equations for arsenic releases from JEB and Sue A+B tailings derived from the 25°C aging test data for nominal pH 8, were used to estimate arsenic porewater concentrations after 10,000 yrs and a million years. Predicted arsenic concentrations were 1.2 and 1.9 mg/L at 10,000 yrs, and 1.3 and 2.2 mg/L at 10^6 yrs, respectively. JEB tailings contains 0.74% scorodite, and Sue A+B tailings 1.21% scorodite. To attain these predicted porewater concentrations assuming the sole arsenic source is scorodite dissolution, it is necessary to dissolve about 0.01% of the scorodite in JEB and Sue A+B tailings. This indicates that the wetted surface areas of scorodite and HFO (the A_w terms in Eq. 4) will be nearly constant with time and that a change in the reaction mechanism is unlikely.

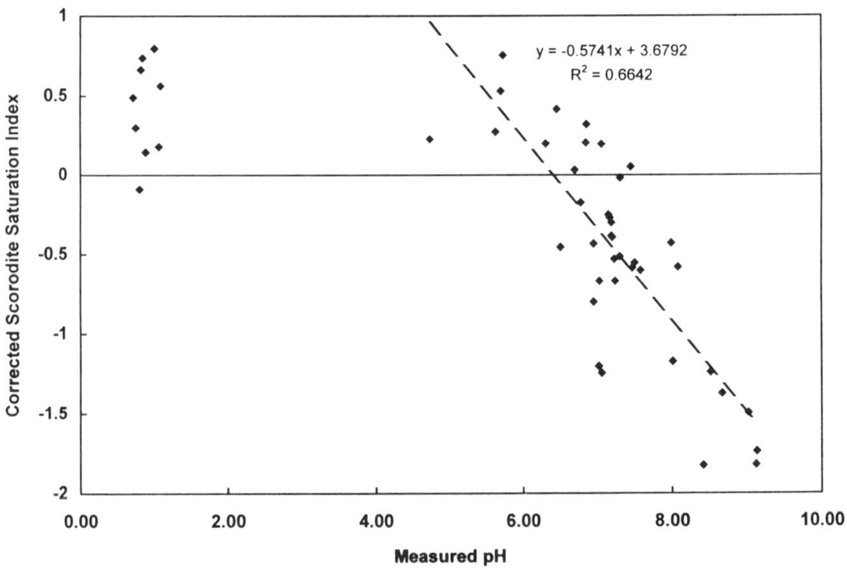

Figure 3. Scorodite saturation index (SI) values versus measured pH at 25°C for all tailings, times, and pH values. SI (corrected) = 0.5 log($[Fe^{3+}][AsO_4^{3-}]/K_{sp}$).

4.4 *The activation energy of arsenate mineral dissolution reactions*

Examination of arsenic concentration versus log time plots shows that when there are significant positive trends to the data, the slopes of the regression lines are steeper at 50 than at 25°C, and steeper at 25 than at 4°C. The slopes of the trend lines are proportional to the rate of the reaction. In this case the reaction is in fact at least two reactions that include the very slow breakdown of both scorodite and cabrerite-annabergite. We cannot quantify the separate rates of these reactions, but through the regression slopes can compare their combined overall rates at 4, 25 and 50°C. This can be accomplished through the Arrhenius equation:

$$\log \frac{k_1}{k_2} = \frac{E_a}{2.303\ R}\left[\frac{1}{T_2} - \frac{1}{T_1}\right] \qquad (5)$$

(Langmuir, 1997), in which the log quotient is the difference between the log of the reaction rate constants k_1 and k_2 at temperatures T_1 and T_2 (°K), respectively. E_a and R are the activation energy of the reaction and the gas constant. Because slopes of the regression lines are proportional to the rate constants, we can substitute values of the slopes for the rate constants in the Arrhenius expression and solve for the activation energy. The resultant activation energy is constant, independent of time, even though the instantaneous reaction rates decrease with time at all temperatures.

Regression lines with significant positive slopes at 50 and 25°C are available for 10 tailings runs with specific Fe/As ratios. Comparison of the slopes leads to an average E_a

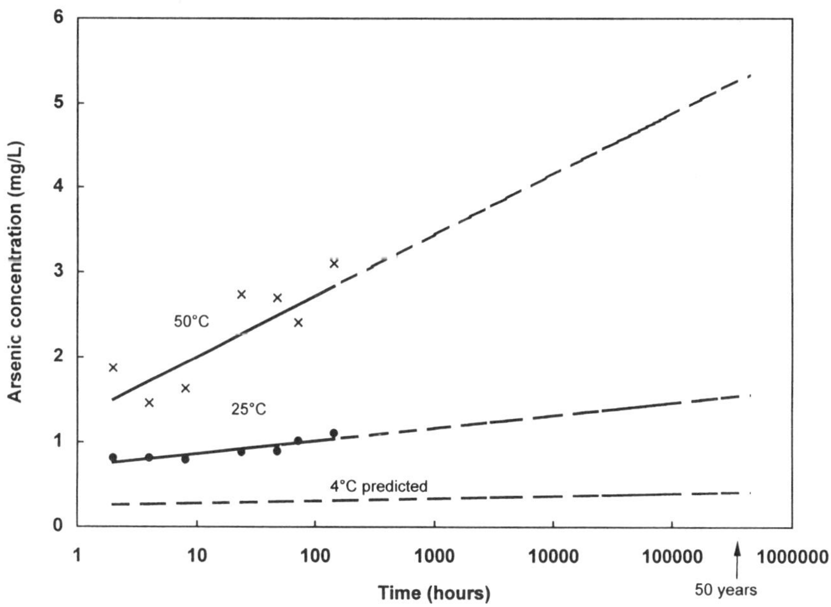

Figure 4. Projected arsenic concentrations in Sue A+B tailings (Fe/As = 3.0) at different temperatures. Solid lines are regression fits of the plotted empirical data at 50 and 25°C with R^2 value of 0.737 and 0.783, respectively. Dashed lines are predicted. The arsenic concentration at 2 hrs and 4°C has been predicted from the regression curve for 4°C in Fig. 1. The slope of the 4°C line is based on the slopes of the 25 and 50°C lines and the Arrhenius equation.

Table 2. Predicted arsenic concentrations in tailings pore waters at nominal pH 8 and 4°C for tailings proposed for burial in the TMF.

Tailings	Fe/As (molar)	Data Trend	Source of Trend Line	As (mg/L) predicted 50 yrs	10,000 yrs
JEB	5.0	no trend	data regression	0.68 ± 0.13	0.68 ± 0.13
JEB	5.0	y=0.0132x+0.2114	Arrhenius equation	0.29 ± 0.03	0.32 ± 0.05
Sue C	7.7	no trend	data regression	0.10 ± 0.05	0.10 ± 0.05
McClean Lake*	5.0	no trend	data regression	0.43 ± 0.30	0.43 ± 0.30
McClean Lake	5.0	y=0.0546x+0.0806	Arrhenius equation	0.39 ± 0.22	0.51 ± 0.30
Sue A+B	3.0	y=0.0269x+0.2920	Arrhenius equation	0.44 ± 0.04	0.51 ± 0.07
Cigar Lake*	7.5	no trend	data regression	0.16 ± 0.11	0.16 ± 0.11
Blend 1*	7.6	no trend	data regression	0.13 ± 0.08	0.13 ± 0.08
Blend 2	3.0	no trend	data regression	0.54 ± 0.3	0.54 ± 0.3
Blend 2	3.0	y=0.0697x+0.1918	Arrhenius equation	0.58 ± 0.26	0.75 ± 0.37

Uncertainty in predicted values based on the Arrhenius equation reflect the standard deviation of the average activation energy. Uncertainty in predicted values which are averages of the data showing no trend, are standard deviations of the averages.

* Data for 25°C which showed no trend adjusted to 4°C using regression relationships in Figure 1.

Blend 1 = (75% Sue C) + (25% Cigar Lake)

Blend 2 = (55% Midwest) + (15% Sue A+B) + (30% Cigar Lake)

value of 13.4±4.2 kcal/mol. This value may be used to predict long-term arsenic concentrations in the TMF at 4°C. For the calculation, an initial arsenic concentration is estimated from the Fe/As ratio of a particular tailings using the equation of the curved regression line for the 4°C data in Fig. 1. The arsenic concentration at 4°C computed from this equation is conservatively assumed to lie on the linear regression line at 2 hrs. The slope of the linear regression line for the 4°C data is then calculated from the slope of the lines at 25 or 50°C through the Arrhenius equation, assuming E_a = 13.4 kcal/mol. Aging tests were not run on Sue A+B (Fe/As = 3.0) at 4°C. The average E_a and regression lines for the 50 and 25°C aging data have been used to estimate the arsenic release rate at 4°C with the result shown in Fig. 4. Equations that describe arsenic concentrations as a function of time for the different tailings at 4°C derived using this approach,, are given in Table 2. Also shown are predicted arsenic concentrations in the TMF after 50 and 10,000 yrs, based on extrapolation of the 4°C regression equations. The table suggests that arsenic concentrations in the TMF at 4°C and nominal pH 8 may not exceed 1 mg/L after 50 to 10,000 yrs.

It is important to note that reaction rates have been accelerated in the aging tests by placing the reaction vessels on a shaker table which creates mixing and convective flow of solutions. In TMF porewaters, however, dissolved reactants will move by diffusion only. Diffusion rates will be extremely low in the bulk of the tailings because of the near absence of arsenic concentration gradients except at the outer edge of the TMF.

5 SUMMARY AND CONCLUSIONS

The acid raffinates of tailings proposed for burial in the JEB Tailings Management Facility (TMF) have molar Fe/As ratios that range from 0.3 to 7.7. In order to control arsenic concentrations (the arsenic source term) in porewaters of the tailings after their burial in the TMF, COGEMA Resources proposes adding ferric sulfate to the raffinates to increase the Fe/As ratios of Sue A+B and Blend 2 tailings to 3, and the JEB and McClean Lake tailings to 5. Tailings/raffinate pH values will then be adjusted to nominal pH 8 to 9 by slaked lime addition, to result in a pH of 7-8 in TMF porewaters.

Large-scale laboratory tests of the McClean Lake, Midwest and Cigar Lake ores were performed to study the effects of adding different amounts of ferric sulfate and slaked lime to the tailings/raffinates for arsenic and nickel control. Short term (3 hr) tests run at 25°C were designed to duplicate conditions during tailings processing in the JEB mill. Aging tests, which used products of the short-term tests as their starting materials, were run for times that ranged from 5 to 50 days, and temperatures of 4, 25 and 50°C. The aging tests were designed to predict arsenic and nickel source term concentrations in the TMF at 4°C for times up to 10,000 yrs.

Mineralogical analysis of the processed tailings and geochemical modeling of the test results indicate that most of the arsenic and nickel precipitate in the minerals scorodite and cabrerite-annabergite as the pH is increased by slaked lime addition. In tailings raffinates with sufficiently high Fe/As ratios, HFO is subsequently precipitated and adsorbs most of the remaining arsenic. A kinetic model is presented that describes scorodite breakdown with time to form HFO and release dissolved arsenic. The model explains the observation that arsenic concentrations and the arsenic release rate are reduced at high Fe/As ratios and in nominal pH 8 porewaters relative to those of nominal pH 9.

Statistical and kinetic analyses of the long-term test data have been used to predict arsenic and nickel concentrations at 4°C in TMF porewaters after 50 and 10,000 yrs. Where neither approach was applicable for a particular tailings, but the test data for 25°C showed no trend, arsenic concentrations at 4°C have been computed from the average concentration at 25°C, using the regression equations in Fig. 1. For arsenic, the results of these several calculations are given in Table 2. Predictions based on the three approaches give consistent results, and show that arsenic and nickel concentrations in porewaters are unlikely to exceed 1 and 2 mg/L, respectively, even after 10,000 yrs of burial in the TMF.

REFERENCES

Allison, J.D., D.S. Brown, and K.J. Novo-Gradac 1991. *MINTEQA2/PRODEFA2*, a geochemical assessment model for environmental systems, Version 3.0 User's Manual. U.S. Environmental Protection Agency, EPA/600/3-91-021.

Bowell, R.J. 1994. Sorption of arsenic by iron oxides and oxyhydroxides in soils. *Applied Geochemistry*, v. 9, p. 279-286.

Cramer, J.J. and J.A.T. Smellie 1994. Final report of the AECL/SKB Cigar Lake analog study. *Whiteshell Laboratories*, Pinawa, Manitoba, Canada, ROE 1L0, Report AECL-10851.

Delany, T.A., D.E. Hockley, J.T. Chapman and N.C. Holl 1998. Geochemical characterization of tailings at the McArthur River Mine, Saskatchewan. *Proceedings from Tailings and Mine Waste '98 conference*, p. 571-578.

Demopoulos, G.P., D.J. Droppert, and G. Van Weert 1994. Options for the immobilization of arsenic as crystalline scorodite. In B. Harris and E. Krause (eds), *Impurity control and disposal in hydrometallurgical processes*, Montreal Quebec: CIM Publication.

Demopoulos, G.P., D.J. Droppert, and G. Van Weert 1995. Precipitation of crystalline scorodite ($FeAsO_4$ $2H_2O$) from chloride solutions. *Hydrometallurgy,* v. 38, p. 245-261.

Donahue, R., M.J. Hendry, and P. Landine 1998. Distribution of arsenic and nickel in uranium mill tailings, Rabbit Lake, Saskatchewan, Canada. *Submitted to Journal of Applied Geochemistry.*

Dzombak, D.A. and F.M.M. Morel 1990. Surface complexation modeling - hydrous ferric oxide. New York: *Wiley Interscience.*

Frost, R.R. and R.A. Griffin 1977. Effect of Ph on adsorption of arsenic and selenium from landfill leachate by clay minerals. *Soil Sci. Soc. Am. J.*, v. 41, p. 53-57.

Fuller, C.C., J.A. Davis, and G.A. Waychunas 1993. Surface chemistry of ferrihydrite – Part 2, kinetics of arsenate adsorption and coprecipitation. *Geochem. et Cos. Acta*, v. 57, p. 2271-2282.

Goldberg, S. 1986. Chemical modeling of arsenate adsorption on aluminum and iron oxide minerals. *Soil Sci. Soc. Am. J.*, v. 50, p. 1154-1157.

Goldberg, S. and Glaubig, R.A. 1988. Anion sorption on a calcareous, montmorillonitic soil-arsenic. *Soil Sci. Soc. Am. J.*, v. 52, p. 1297-1300.

Harris, G.B. and Krause, E. 1993. The disposal of arsenic from metallurgical processes - its status regarding ferric arsenate. Paper presented at *The Paul E. Queneau International Symposium on "Extractive Metallurgy of Nickel, Cobalt, and Associated Metals", TMS Annual Meeting*, Denver, Colorado, USA, February 21-25, 1993.

Langmuir, D. 1997. Aqueous environmental geochemistry. New Jersey: *Prentice Hall.*

Langmuir, D., and J.J. Mahoney 1997. Geochemical controls on arsenic levels in some mine tailings and groundwaters associated with saturated zone tailings disposal: *paper presented at International Symposium on Geology and the Environment GeoEnv '97*, Istanbul, Turkey, September 1-5.

Machesky, M.L. 1990. Influence of temperature on ion adsorption by hydrous metal oxides. In D.C. Melchior and R.L. Bassett (eds), *Chemical modeling of aqueous systems II*, ACS Symposium series 416, 1990, p. 282-292.

Manning, B.A. and S. Goldberg 1997. Adsorption and stability of arsenic (III) at the clay mineral-water interface. *Environ. Sci. & Techn.*, v. 31, no. 7, p. 2005-2011.

Natural Resources Canada 1998. Canadian uranium statistics fact sheet. *Natural Resources Canada, Uranium and Radioactive Waste Division*, Energy Resources Branch.

Parkhurst, D.L. 1995. *PHREEQC* – A computer program for speciation, reaction-path, advective-transport, and inverse geochemical calculations. *U.S. Geological Survey Water-Resources Investigations* 95-4227.

Pierce, M.L. and C.B. Moore 1982. Adsorption of arsenite and arsenate on amorphous iron hydroxide. *Water Res.*, v. 16, p. 1247-1253.

Robins, R.G., J.C.Y. Huang, T. Nishimura, and G.H. Khoe 1988. The adsorption of arsenate ion by ferric hydroxide. In *Arsenic metallurgy fundamentals and applications, TMS Sym. Proc.*, p. 99-112.

Watkinson, D.H., J.B. Heslop, and W.D. Ewert 1975. Nickel sulphide-arsenide assemblages associated with uranium mineralization, Zimmer Lake area, Northern Saskatchewan. *Canadian Mineralogist,* vol. 13, p. 198-204.

Tailings and Mine Waste '99 © 1999 Balkema, Rotterdam, ISBN 90 5809 025 6

Adsorption capabilities of clinoptilolite relating to mine drainage remediation

J.S.Gilbert, P.M.O'Meara & T.R.Wildeman
Department of Geochemistry, Colorado School of Mines, Colo., USA
J.G.Crock
Mineral Resources Program, United States Geological Survey, Colo., USA
G.A.Desborough
Mineral Resources Program, United States Geological Survey, Colo., USA

ABSTRACT: This study concerns the possible application of 11 natural zeolites of the clinoptilolite variety to the remediation of acid/metal mine drainage. Laboratory experiments consisted of mixing each zeolite with the following single element solutions: Ag, Al, As, Ca, Cd, Cs, Cu, Fe^{2+}, Fe^{3+}, K, Mg, Mn, Pb, Rb, SO_4, and U. Also, distilled water and two mine drainage samples were mixed with each zeolite. Metal adsorption on a per gram zeolite basis was calculated using ICP – MS and ICP – AES analytical data, and shows that some of these natural zeolites are extremely effective at sequestering metals. However, in response to adsorption, there is substantial desorption of Na, Al, Ca by the zeolites causing possible concern. Zeolite composition obtained by XRF seems to show a correlation between the element desorbed and abundance in the zeolite. Important to note is that the various zeolites behaved differently in terms of metal adsorption/desorption, permitting selection of a few that would be best suited for use in the field.

1 INTRODUCTION

Mining operations and undisturbed mineral occurrences throughout the world are responsible for contaminating watersheds with potentially toxic metals; pollution which, if not contained, is able to migrate downstream and affect other localities. The source of much of this contamination is various metal sulfide minerals, particularly pyrite (FeS_2). When exposed to air and water, these sulfide minerals undergo reactions that generate significant amounts of acidity and soluble heavy-metals. The resulting solution, being highly acidic – a pH around 3 is common – is very reactive and thus is able to dissolve or leach more metals from rock it comes in contact with. Also, the acidic nature of the solution allows the metals to be transported in their most soluble form.

Many studies have demonstrated that natural clinoptilolite-rich rocks (CRRs) with their ion-exchange capabilities are very effective at removing metals from solution (Desborough 1994; Bremner and Schultze 1993; Zamzow and Schultze 1993). Once adsorbed, the degree of metal retention or exchangeability seems to be a function of the pH of the initial, metal rich water; the pH and concentration of the exchange medium; the CRR to exchange solution ratio; and the CRR (Desborough 1995). The focus of this study is the adsorption/desorption characteristics of 11 clinoptilolite-rich rocks (CRRs) from 10 deposits in the Western U. S. and 1 deposit in Canada (Table 1) in order to assess heavy metal mine drainage remediation during storm runoff situations. Specifically, laboratory experiments were designed to determine CRR total compositions and the effects of pre- and post-CRR exposure to acidic, single element stock solutions within a sulfate matrix. Also, two mine drainage samples were treated with the CRRs in order to determine the effect of multi-ion competition since it is well noted that Ca^{2+} and other non-metallic cations, commonly present in large concentrations in mine drainage, can significantly interfere with metal uptake (Desborough 1994; Zamzow and Schultze 1993).

The goals of the these experiments and those in the future are to determine: 1) the total amount of metal able to be adsorbed by a given CRR; 2) the retention of the metal(s) by the CRR and the major controlling factor(s); and 3) the rate of metal uptake, i.e., the time necessary to achieve equilibrium. The performance of the CRRs was quite variable and seemed to be a function of their compositions, especially in terms of the adsorption capacity and dominant cation(s) desorbed by a particular CRR. These variations in effectiveness permit selection of a few of the most promising CRRs for future lab/field studies.

Table 1. Abbreviations, deposit locations, and diluent minerals in raw clinoptilolite-rich rock (CRR) samples used in this study. Minerals, in decreasing order of abundance, determined by X-ray diffraction by G.A. Desborough and Richard Sheppard, U. S. Geological Survey [clay signifies layer silicates with a 10-angstrom spacing; tr.-trace] (Adapted from Desborough 1996 and 1994).

Locality and CRR abbreviation	Diluent minerals
British Columbia	
Princenton-PBC	Opal, feldspar, mordenite, quartz (tr.)
Colorado	
Creede-CCO	Opal, K-feldspar, plagioclase, quartz, clay
Idaho	
Crisman Hill-CHI	Opal, K-feldspar
Preston-PWI	Opal
Montana	
Grasshopper Creek-GCM	Opal, K-feldspar, plagioclase
Nevada	
Fish Creek-FCWN	K-feldspar, quartz, plagioclase
New Mexico	
Winston-SNM	Opal, clay (tr.)
South Dakota	
Rocky Ford-SDH	Plagioclase, calcite, quartz, opal, K-feldspar, clay (tr.)
Texas	
Alamito Creek-ACT	Opal, plagioclase, clay
Tilden-ZEO	Opal, clay, calcite
Wyoming	
Ft. LaClede-FLW	Plagioclase

2 ION-EXCHANGE AND COMPOSITION OF CLINOPTILOLITE-RICH ROCKS

Ion-exchange behavior of zeolites is primarily due to a fixed, negative surface charge which attracts and holds certain cations, exchanging them when other cations more preferred by the zeolite are present. This fixed or constant charge is the result of a deficiency in positive charge from structural replacement of cations with lower valence states or presence of crystal imperfections (Zelazny and others 1996). The strong affinities or selectivities of clinoptilolite for certain cations have been reported by many researchers – Zamzow and Schultze (1993); Bremner and Schultze (1993); and Boylan (1993). The relative affinities of different clinoptilolites for different cations become important when considering the specific application of the CRR. For example, if a particular CRR had a high affinity for Ca, use of this CRR in the treatment of acid mine drainage (AMD) with large amounts of Ca and other heavy metals might not be advantageous because of the greater exchange capacity required to capture the heavy-metals, in addition to the Ca. The same logic would apply to other cations such as Al, Fe, and Mg which are known to compete with heavy metals for exchange sites (Bremner and Schultze 1993). Important to note is that affinities reported for clinoptilolite should be considered general in the sense that it is possible for others to have different affinities presumably due to their specific composition, i.e., major exchangeable cations.

The CRRs vary in clinoptilolite content, and the composition of clinoptilolite is also variable primarily due to differences in the major exchangeable cations (Ca, K, Na) and trace elements present. Total elemental composition was determined using a wavelength dispersive X-ray fluorescence (XRF) method (Table 2) and inductively coupled plasma – atomic emission spectroscopy (ICP –AES) (Gilbert and others 1998) on powdered and digested samples (Arbogast 1996). The results from the two different analytical techniques seem to compare reasonably well. Major cation content proved to be variable in the samples; some CRRs were enriched in either Na, K, or Ca while others contained a mixture of 2 or 3 of these elements. Desborough (1996 and 1994) also reports compositions of some of the CRRs used in this study.

3 LABORATORY METHODS

Sized samples of the CRRs were obtained by screening crushed material through a minus 10-mesh (2 mm) sieve. Five grams of this < 2 mm fraction were then mixed (in 50 mL Nalgene™ centrifuge tubes) with 25 mL of the various single element stock solutions (Table 3), discussed below, and two natural mine drainage samples from Summit county, Colorado (Burleigh and McClelland Tunnels). The tubes containing the 5 g of CRR and 25 mL of solution were then horizontally shaken (80 cycles per minute) for 4 hours, a duration long enough for the exchange reaction to equilibrate as determined by preliminary studies not shown here and by other researchers (Zamzow and Schultze 1993). After the 4-hour time period, aliquots of the exchange solution were centrifuged at 5000 rpm for 15 minutes, filtered through a 0.45 μm filter, acidified with nitric acid, and then analyzed by inductively coupled plasma – mass spectroscopy (ICP – MS) or ICP – AES depending on the element of interest. Some of the aliquots analyzed by ICP – MS had to be diluted to prevent saturation of the instrument's detector. The specific solutions diluted and the extent of dilution was element specific. The pre- and post- exchange stock solution compositions in raw form are all located in the report of Gilbert and others, 1998. This data was used to calculate micrograms of metal adsorbed per gram of CRR in order to evaluate the performance of each CRR relative to one another.

The concentrations of the elements in the solutions were chosen to simulate a worst case scenario in terms of values normally seen in AMD. All stock solutions (except mine drainage

Table 2. Composition of clinoptilolite-rich rocks determined by x-ray fluorescence (XRF).

Sample	SiO_2 (%)	Al_2O_3 (%)	Fe_2O_3 (%)	MgO (%)	CaO (%)	Na_2O (%)	K_2O (%)
FLW	63.3	12.9	1.28	0.59	1.49	4.45	1.60
ACT	63.5	12.7	2.82	0.89	4.25	1.77	2.42
CHI	67.4	11.0	2.27	0.23	1.23	0.94	4.85
CCO	67.4	12.5	1.35	0.72	2.42	1.18	3.85
GCM	66.4	12.5	0.76	0.78	2.39	1.64	3.26
FCWN	66.9	13.9	0.72	0.46	2.10	2.94	3.93
ZEO	65.4	12.1	1.11	0.94	4.39	1.11	1.18
PWI	67.4	10.6	1.70	0.45	2.23	0.59	4.19
SNM	65.7	12.0	1.28	1.27	3.08	0.60	2.27
SDH	63.9	11.9	2.00	0.81	2.20	2.97	4.01
PBC	66.3	12.2	2.42	0.76	2.21	2.54	2.51
Average	65.8	12.2	1.61	0.72	2.54	1.88	3.10

Sample	TiO_2 (%)	P_2O_5 (%)	MnO (%)	Loss on Ignition-925 C	Total (%)
FLW	0.20	0.15	0.01	13.7	99.67
ACT	0.36	0.15	0.05	10.1	99.01
CHI	0.28	0.24	0.03	10.1	98.57
CCO	0.23	0.13	0.01	9.04	98.83
GCM	0.08	0.11	0.03	11.0	98.95
FCWN	0.24	0.09	0.01	7.71	99.00
ZEO	0.19	0.09	0.10	12.2	98.81
PWI	0.27	0.10	< 0.01	11.4	98.93
SNM	0.22	0.12	0.06	12.3	98.90
SDH	0.27	0.20	0.05	10.7	99.01
PBC	0.23	0.11	0.04	9.27	98.59
Average	0.23	0.14	0.039	10.7	98.94

Table 3. Stock solutions, concentrations (μg/L), and pH.

Solution:	Ag	Al	As	Cd	Cu	Fe^{2+}	Fe^{3+}	Mg	Pb	SO_4	Zn	Cs & Rb	U (VI)	Ca	K	Mn
Concentration	7.2	190	18	15	360	820	180	280	2.5	5200	430	1.3	2.6	430	22	36
pH[a]	2.5	2.5	3.0	3.0	3.0	2.0	2.5	2.5	3.0	2.5	2.5	3.0	2.5	2.0	2.5	2.0

[a]A matrix of 5000 μg/L sulfate was created using concentrated sulfuric acid; pH adjusted to 2-3 with NaOH (aq).

samples) contained a 5,000-ppm sulfate matrix from the addition of concentrated sulfuric acid and the pH of each was adjusted to 2-3 with NaOH (aq). Different amounts of NaOH were needed to achieve the desired pH; therefore, the Na concentration in the individual solutions is variable; a point that may seem trivial but turns out to be quite important in terms of Na desorption. Measurements of pH were taken prior to exchange and are shown in Table 3.

Table 4. Summary of CRR adsorbtion/desorption[a] for various stock solution elements.

CRR	Ag	Al	As	Ca	Cd	Cs	Cu	Fe^{2+}	Fe^{3+}
FLW	-36	-400	-82	-670	-75	-8.0	-890	-2700	-740
ACT	-35	-950	-83	4800	-76	-8.0	-1700	-3900	-910
CHI	-35	-400	-84	1700	-71	-8.0	-180	-1250	-850
CCO	-35	-150	-83	1000	-71	-7.9	-180	-1350	-630
GCM	-36	-250	-78	1300	-71	-8.0	120	-500	-570
FCWN	-36	-250	-82	900	-75	-8.0	-230	-600	-520
ZEO	-36	-950	-84	1300	-76	-8.0	-1700	-3985	-910
PWI	-35	0	-82	1200	-69	-8.0	-380	-1400	-520
SNM	-36	-250	-82	1800	-70	-8.0	120	-900	-570
SDH	-36	-950	-83	780	-76	-8.0	-1100	-3350	-910
PBC	-36	-300	-85	1500	-75	-8.0	-330	-1600	-470

CRR	K	Mg	Mn	Pb	Rb	SO_4	U	Zn
FLW	-60	-620	-180	-12	-4.7	-23000	-12	-1700
ACT	22	-570	-170	-12	-4.7	-23000	-12	-1100
CHI	1700	-370	-180	-12	-4.2	-23000	-12	-900
CCO	540	-220	-170	-12	-4.5	-24000	-12	-750
GCM	270	-17	-170	-12	-4.6	-23000	-13	-500
FCWN	310	-170	-170	-12	-4.4	-23000	-13	-450
ZEO	840	-820	-180	-12	-4.5	-22000	-12	-1400
PWI	1300	-170	-170	-12	-4.5	-24000	-12	-750
SNM	160	-17	-140	-12	-4.7	-23000	-12	-500
SDH	240	-520	-170	-12	-4.7	-23000	-12	-1400
PBC	12	-320	-180	-12	-4.7	-23000	-13	-700

[a]μg element adsorbed (-) or desorbed (+) per gram of CRR.

4 RESULTS AND DISCUSSION

4.1 *Ion Exchange in Single Element Stock Solutions*

As mentioned above, metal adsorptions were calculated on a microgram metal adsorbed per gram of CRR basis. Theses values are reported (Gilbert and others 1998) for a suite of elements in each stock solution ion exchange in order to see what was desorbed from the CRRs as a consequence of adsorption of the element of interest. Adsorption/desorption was calculated with respect to the solution; therefore, a negative value signifies removal from solution, which was assumed to be accomplished by CRR adsorption. However, direct precipitation of metal bearing phase could also account for metal removal from solution.

Table 4 is a summary of the adsorption/desorption values for the element of interest in each stock solution, e.g., Zn capture from the 430 ppm Zn stock solution, Cu capture from the 360 ppm stock solution, etc. The data in this table clearly shows that some of the CRRs, namely FLW, ACT, ZEO, SDH, have adsorption capacities which are much greater than the rest. For example, the amount of Zn adsorbed by FLW is almost 4 times larger than what was removed by GCM. ZEO and SDH showed almost 3 times and ACT around 2 times larger amounts adsorbed relative to GCM. The same pattern holds true for Cu, Fe(II), Fe(III), and Al and Mg to some extent. Results for Pb, Cs, Rb, U, Ag, and As are not as useful in making comparisons between the CRRs because the values are all very similar. Presumably, the concentration used in each of these solutions was not large enough to show differences in removal of these specific metals because each CRR removed about the same amount which happened to be all that was in solution.

To begin to understand the variations in adsorption properties, consider FLW, which is Na-rich; ACT and ZEO, which are Ca-rich; and SDH which is Ca-/K-rich. These compositional differences relative to the other CRRs seem to account for the wide variation in metal removal seen between the CRRs. More specifically, the Na- and Ca- rich appear to capture more metals than the K-rich CRRs. Also, intimately related to this observed composition – adsorption interdependence is the relationship between composition of CRR and desorption.

For a Na-rich CRR, one would expect most of the exchangeable cations to be Na; therefore, in an ion-exchange reaction, when a cation preferred over Na is present, the CRR would be expected to desorb Na and adsorb the more preferred cation. This is exactly what is seen for FLW, a Na-rich CRR. For example, in the Fe(II) stock solution, FLW adsorbed 2700 μg of Fe(II) per gram of CRR. In keeping with the laws of solutions and their mandatory charge balance constraint, there must be a release of cations by FLW. Table 5 shows that a large amount of Na, 8600 μg/g, was indeed desorbed by FLW; thus, most Fe(II) displaced Na from FLW because the amounts of Ca (Table 6), Al (Table 7), and K (Gilbert and others, 1998) desorbed were substantially smaller than the value for Na. Figure 1 displays these values graphically and shows the relationships very well. The same pattern holds true for ACT, a Ca-rich CRR, which desorbs substantial Ca in response to exchange with large amounts Fe(II) (Figure 2).

Important to note is that the values in Table 4 do not necessarily represent maximum metal capture abilities of the CRRs. The results do show that some of the CRRs have greater affinities for certain metals than others, e.g., Mg, Al, Zn, and in some cases, the results are such that one can assume a CRR has reached adsorption capacity of a certain metal because it removed << 50 % of the element of interest from solution. This is demonstrated by GCM adsorption of Mg which is approximately 50 times less than the value for ZEO (Table 4). These data allow comparisons to be made between the CRRs permitting selection of those with the best performance. Those selected will then be studied more thoroughly in the future.

Table 5. Summary of CRR Na adsorption/desorption[a] in various stock solutions.

CRR	Na - Ag	Na - Al	Na - As	Na - Ca	Na - Cd	Na - Cu	Na - Fe^{2+}	Na - Fe^{3+}	Na - K
mg/L Na[b]:	n/a[c]	2.3	n/a	2.3	n/a	2.6	0.17	0.92	2.2
FLW	n/a	1500	n/a	3200	n/a	5000	8600	2400	1800
ACT	-5800	-2000	n/a	-1800	n/a	-1500	830	-1200	-1200
CHI	-5800	-2000	n/a	-1800	n/a	1000	1900	-720	-1200
CCO	-5800	-3000	n/a	-2300	n/a	-2000	830	-1700	5800
GCM	n/a	-1500	n/a	-1300	n/a	1000	2600	-220	2800
FCWN	n/a	-1000	n/a	-830	n/a	2000	2500	-270	-170
ZEO	n/a	1000	n/a	670	n/a	3000	3800	1400	7300
PWI	-5900	-4500	n/a	-3800	n/a	-3000	530	-2100	2800
SNM	-5800	-2500	n/a	-1800	n/a	-500	180	-1500	-1200
SDH	n/a	1500	n/a	2200	n/a	5000	6100	3400	1300
PBC	n/a	-500	n/a	-330	n/a	2000	2900	880	-170

CRR	Na - Mg	Na - Mn	Na - Pb	Na - SO4	Na - U	Na - Zn	Na - DI[d]	Na - Burl.[e]	Na - McC.[f]
mg/L Na:	2.6	BDL	n/a	n/a	n/a	2.7	0.05	11	12
FLW	500	600	n/a	n/a	n/a	1000	24	1400	1500
ACT	-3000	320	n/a	n/a	n/a	-3000	14	440	440
CHI	-2500	330	n/a	n/a	n/a	-3000	3.5	650	690
CCO	-3500	270	n/a	n/a	n/a	-4000	6.0	390	440
GCM	-2000	360	n/a	n/a	n/a	-2500	15	700	740
FCWN	-2000	500	n/a	n/a	n/a	-1500	22	800	940
ZEO	-1000	750	n/a	n/a	n/a	-500	45	1200	1300
PWI	-5500	200	n/a	n/a	n/a	-5000	2.2	320	350
SNM	-2500	110	n/a	n/a	n/a	-2500	6.5	150	160
SDH	500	750	n/a	n/a	n/a	500	55	1400	1500
PBC	-1500	440	n/a	n/a	n/a	-1000	20	1000	1100

[a]μg element adsorbed (-) or desorbed (+) per gram of CRR. [b]Na concentration in stock solution prior to ion exchange. [c]n/a - Na not analyzed for. [d]Distilled water control. [e]Burleigh tunnel mine drainage, 11 ppm Na. [f]McClelland tunnel mine drainage, 12 ppm Na.

Table 6. Summary of CRR Ca Adsorption/Desorption[a] in various stock solutions

CRR	Ca - Ag	Ca - Al	Ca - As	Ca	Ca - Cd	Ca - Cu	Ca - Fe^{2+}	Ca - Fe^{3+}	Ca - K
FLW	50	700	50	-670	31	1100	700	310	650
ACT	500	3400	280	4800	240	2900	2300	2800	3500
CHI	220	2500	160	1700	130	2800	2100	1600	1900
CCO	290	3800	310	1000	260	3300	2600	2300	6000
GCM	190	2500	190	1300	150	3400	2500	1600	2900
FCWN	190	2100	130	930	110	2900	2300	1600	1700
ZEO	270	3100	95	1300	75	3000	2700	2300	2800
PWI	310	4500	360	1200	280	3000	2800	2800	7500
SNM	230	2700	200	1800	160	3700	2900	1900	2400
SDH	200	2100	45	780	32	1500	2000	1700	900
PBC	160	2200	150	1500	130	2900	1800	1500	1800

CRR	Ca - Mg	Ca - Mn	Ca - Pb	Ca - SO_4	Ca - U	Ca - Zn	Ca – DI[b]	Ca - Burl.[c]	Ca - McC.[d]
FLW	650	0	60	55	40	750	0.50	-400	-600
ACT	3500	320	300	390	280	4700	4.4	140	140
CHI	2000	31	160	180	140	2200	3.1	-260	-370
CCO	3900	190	300	210	270	3600	1.0	27	-7.5
GCM	2100	60	180	220	160	2300	1.0	-180	-230
FCWN	1400	50	150	190	131	2300	1.0	-240	-350
ZEO	1500	30	120	200	120	2400	6.0	-310	-470
PWI	4300	200	350	190	320	4200	0.40	42	-7.5
SNM	2200	255	190	260	170	2600	1.5	140	140
SDH	750	11	70	100	60	1200	2.0	-390	-580
PBC	1800	29	150	190	130	2100	1.5	-250	-360

[a]µg element ad. (-) or desorbed (+) per gram of CRR. [b]Ca concentration in stock solution prior to ion exchange. [c]Burleigh tunnel mine drainage, 82 ppm Ca. [d]McClelland tunnel mine drainage, 120 ppm Ca.

Table 7. Summary of CRR Al adsorption/desorption[a] in various stock solutions

CRR	Al - Ag	Al	Al - As	Al - Ca	Al - Cd	Al - Cs, Rb	Al - Cu	Al - Fe^{2+}	Al - Fe^{3+}
FLW	78	-400	0.035	200	-1.9	2.5	28	120	270
ACT	-1.7	-950	0.0029	0	-2.1	-0.20	0	0	0
CHI	53	-400	0.014	200	-2.0	3.4	55	150	190
CCO	93	-150	0.0024	300	-2.1	7.8	70	210	340
GCM	68	-250	-0.0061	200	-2.1	2.0	43	190	250
FCWN	58	-250	0.017	110	-2.1	-0.15	0	45	180
ZEO	-1.7	-950	-0.0059	0	-2.0	-0.12	0	0	0
PWI	100	0	-0.0053	380	-1.2	15.3	180	320	400
SNM	83	-250	0.0034	260	-2.1	5.3	50	160	260
SDH	-1.7	-950	0.042	0	-2.1	-0.19	0	0	0
PBC	73	-300	0.013	190	-2.1	0.040	4.0	130	290

CRR	Al - K	Al - Mg	Al - Mn	Al - Pb	Al - SO_4	Al - U	Al - Zn	Al - DI[b]	Al - Burl.[c]
FLW	8.0	0	11	0.70	-10	0.44	60	6.5	4.0
ACT	0	0	0	0	-13	-0.69	0	6.0	-0.14
CHI	24	4.5	1.5	1.8	-6.0	2.0	90	14	-0.25
CCO	65	1.0	0	1.8	-1.5	4.2	160	2.5	-0.18
GCM	9.5	0	0	0.23	-8.0	-0.28	95	2.0	-0.23
FCWN	0	0	0	0.037	-13	-0.70	5.0	0.87	-0.23
ZEO	12	0	85	0.30	-13	-0.65	0	18	-0.16
PWI	180	27	10	10	12	12	220	1.1	-0.18
SNM	16	0	0	0.70	-11	1.7	95	0.67	0.24
SDH	0	0	190	0.020	-13	-0.68	0	18	3.1
PBC	0	0	14	0.0031	-12	-0.66	60	6.0	-0.19

[a]µg element adsorbed (-) or desorbed (+) per gram of CRR. [b]Distilled water control. [c]Burleigh tunnel mine drainage, 51 ppb Al.

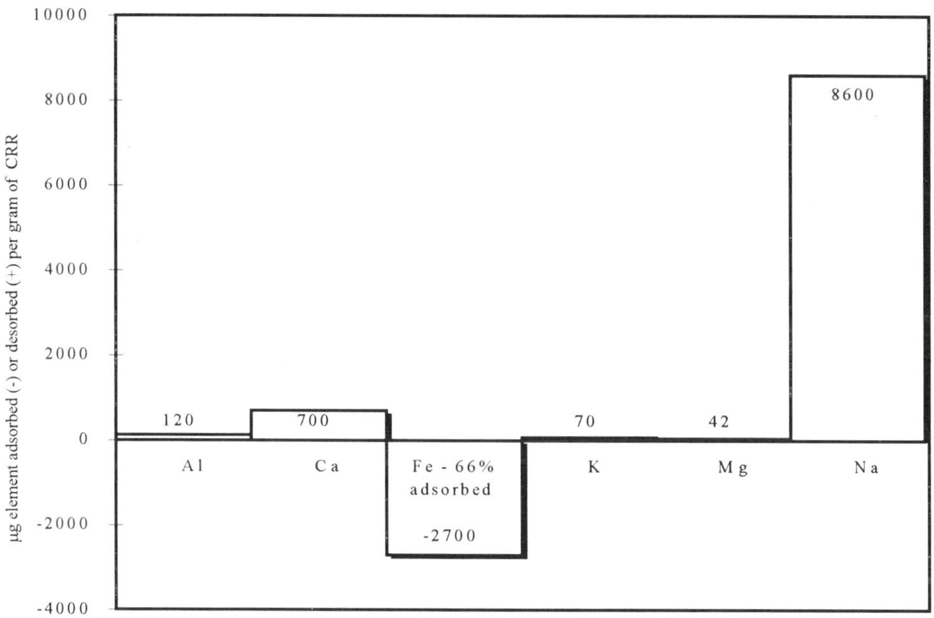

Figure 1. FLW adsorption/desorption in 820 ppm Fe^{2+} stock solution. Note the large amount of Na desorbed.

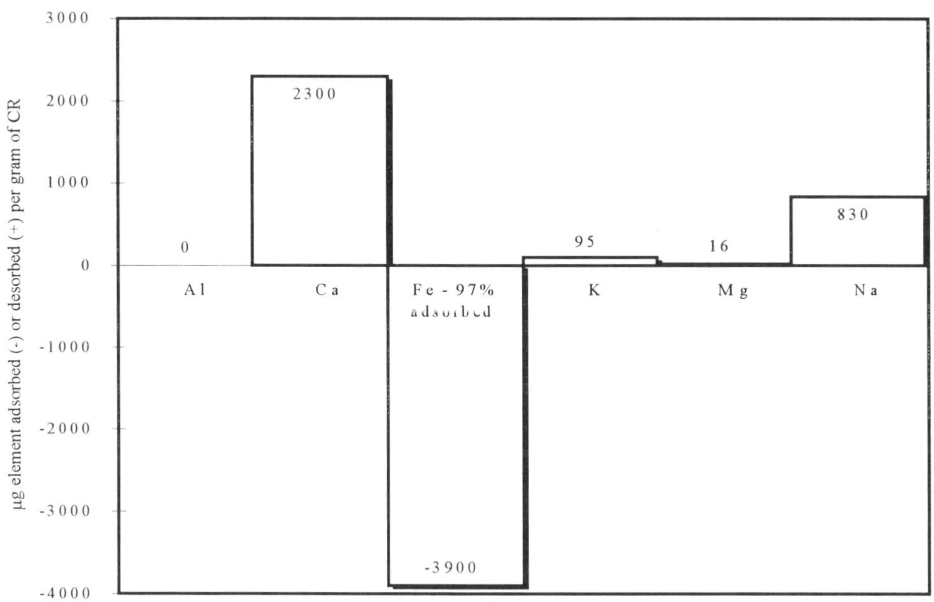

Figure 2. ACT adsorption/desorption in 820 ppm Fe^{2+} stock solution. Note the large amounts of Na and Ca desorbed consistent with its composition.

Some elements of principal concern in AMD are Cu, Zn, Ag, Pb, Cd, Hg, and As. For the most part, the CRRs that adsorbed the largest amounts of these metals were FLW, ACT, ZEO, and SDH. These CRRs in turn also desorbed the largest amounts of cations such as Al, Na, and Ca. Al and Na desorption causes some concern because of fish and plant toxicity (Drever 1997; Kesler 1994; Driscoll and others 1980). Of these two, Al would be considered more toxic but less mobile (Kesler 1994) while Na would be less toxic but more mobile. Increased mobility or solubility of Al occurs at low pH values. Ca desorption is not of great concern for two reasons. First, Ca desorption increases water hardness which increases the tolerance of trout for heavy metals. Second, trout need at least 15-25 mg/L Ca, in the absence of heavy metals, to survive (Desborough 1994). With this in mind, it is extremely important to pay close attention to those CRRs that readily adsorb Ca. In this study, FLW was the only CRR to remove Ca from the Ca stock solution, and similar behavior, of this specific CRR, was seen by Desborough (1994).

Table 8. Selected elemental concentrations (mg/L) in McClelland and Burleigh mine drainage.

	Na	Mg	Al	SO4	K	Ca	Mn	Fe	Zn	Cd	Pb	U	Ni	Rb	Sr	Cu
McClelland	12	46	5	450	1.3	120	18	1.5	7.4	0.045	0.024	0.032	0.17	0.01	1.5	1.3
Burleigh	11	37	0.051	350	1.3	82	3.6	0.26	70	0.15	0.023	0.016	0.05	0.009	1.7	0.0003

4.2 *Ion Exchange in Mine Drainage Samples*

The compositions of the Burleigh and McClelland mine drainage solutions are shown in Table 8. Both contain large amounts of sulfate but the McClelland water is enriched in Ca, Al, Fe, Mn, and Cu relative to Burleigh which is enriched in Zn.

The performance of the CRRs in McClelland drainage was quite variable (Table 9). All adsorbed approximately the same amount of Al, Fe, Pb, Cd, and Cu while they varied widely in the quantity of Mg, SO_4, Ca, Mn, Zn, and Ni captured. The largest removals of Mg, Mn, Zn, and Ni were accomplished by FLW, ZEO, and SDH. ACT also removed large amounts of these metals, especially Zn. Sulfate was adsorbed in large amounts by all CRRs except FLW, the only sample to adsorb K. The main elements released in response to adsorption were Na and Ca. K was also desorbed; however, not in very large quantities, in fact, the most K-deficient CRR desorbed the most K. If a particular CRR desorbed a large amount of Na, then it seemed to adsorb Ca, and this was the case for FLW, CHI, GCM, FCWN, ZEO, SDH, AND PBC. Generally, the more Na desorbed, the more Ca adsorbed. If Na was not released in very large quantities, then Ca was desorbed. This was seen in ACT, CCO, PWI, and SNM. Of all the CRRs, FLW, ACT, SDH, and ZEO desorbed the most Na and adsorbed the most Ca, and of these 3, FLW showed the largest values with respect to Na and Ca. Important to note is that these 4 CRRs performed the best in terms of metal uptake showing that there is a definite relationship between the amount of metal adsorbed by and desorbed from the CRR. In summary, the more metal adsorbed from the AMD, the more metal desorbed.

The results of CRR exposure to Burleigh mine drainage (Table 10) were very similar to those for McClelland. The desorption patterns for Na, Ca, and K mentioned above are identical to those seen here. Again, sulfate was removed in large quantities by all CRRs except FLW. Zn removals were substantial, and were one order of magnitude greater than that seen in McClelland which had one order of magnitude less Zn than Burleigh. This shows that in most cases, the amount of a certain metal removed is proportional to its concentration in solution simply because there is more available. This relationship breaks down if a CRR shows high affinity for a specific metal, no matter what its concentration. For example, the concentration of sulfate was 5 times greater than that of Zn; however, the amount of Zn removed per gram of CRR was greater than sulfate. This could be because sulfate is an anion while Zn is a cation. In another example, the concentration of Ca was also higher than that of Zn, but some CRRs adsorbed while others desorbed Ca. The same 4 CRRs – FLW, ACT, ZEO, SDH – showed the best performance in terms of amount of metals removed from solution.

Table 9. Summary of CRR adsorption (-) and desorption (+) in McClelland tunnel water. Values are μg element ads./des. per gram of CRR

CRR	Na (+/-)	Mg (+/-)	Al (+/-)	SO_4 (+/-)	K (+/-)	Ca (+/-)	Mn (+/-)	Fe (+/-)
FLW	1600	-220	-23	-35	-2.1	-590	-88	-7.3
ACT	490	-170	-25	-240	25	140	-75	-6.2
CHI	790	-180	-25	-190	400	-340	-76	-7.3
CCO	440	-130	-25	-190	63	42	-64	-6.6
GCM	740	-97	-25	-140	38	-190	-59	-7.3
FCWN	840	-68	-25	-240	58	-310	-62	-7.3
ZEO	1300	-210	-25	-240	110	-470	-85	-7.3
PWI	350	-120	-25	-240	160	43	-65	-6.5
SNM	180	-43	-25	-190	29	140	-48	-6.1
SDH	1500	-220	-24	-240	43	-580	-87	-7.3
PBC	1000	-160	-25	-240	18	-340	-78	-7.3

CRR	Zn (+/-)	Cd (+/-)	Pb (+/-)	U (+/-)	Ni (+/-)	Rb (+/-)	Sr (+/-)	Cu (+/-)
FLW	-36	-0.22	-0.12	-0.13	-0.78	-0.025	-7.3	-5.9
ACT	-36	-0.22	-0.12	-0.087	-0.70	0.075	-6.1	-6.2
CHI	-34	-0.20	-0.12	-0.16	-0.68	0.90	-6.7	-6.1
CCO	-32	-0.19	-0.12	-0.16	-0.58	0.30	-5.5	-5.9
GCM	-33	-0.19	-0.12	-0.10	-0.56	0.065	-5.9	-5.8
FCWN	-35	-0.21	-0.12	-0.11	-0.57	0.30	-6.7	-5.9
ZEO	-37	-0.22	-0.12	-0.12	-0.80	0.24	-5.9	-6.3
PWI	-32	-0.18	-0.11	-0.16	-0.45	0.34	-6.8	-5.9
SNM	-29	-0.16	-0.12	-0.14	-0.47	0.050	0.17	-6.1
SDH	-37	-0.22	-0.12	-0.082	-0.81	0.015	-7.2	-6.2
PBC	-36	-0.22	-0.12	-0.082	-0.77	-0.0064	-6.7	-6.2

Table 10. Summary of CRR adsorption (-) and desorption (+) in Burleigh tunnel water. Values are μg element ads./des. per gram of CRR.

CRR	Na (+/-)	Mg (+/-)	Al (+/-)	SO4 (+/-)	K (+/-)	Ca (+/-)	Mn (+/-)	Fe (+/-)
FLW	1400	-180	7.2	-75	-2.3	-380	-18	-1.3
ACT	500	-150	-0.15	-280	19	140	-15	-0.74
CHI	700	-160	-0.09	-280	330	-240	-16	-1.3
CCO	420	-120	-0.25	-280	54	82	-14	-1.3
GCM	700	-95	-0.25	-230	36	-140	-13	-1.3
FCWN	750	-65	-0.25	-230	49	-210	-13	-1.3
ZEO	1200	-180	-0.20	-230	99	-330	-17	-1.3
PWI	310	-110	-0.11	-280	140	92	-14	-0.78
SNM	170	-45	-0.25	-280	26	140	-11	-0.79
SDH	1400	-180	7.7	-230	41	-390	-18	2.7
PBC	1000	-140	-0.26	-280	16	-240	-13	-1.3

CRR	Zn (+/-)	Cd (+/-)	Pb (+/-)	U (+/-)	Ni (+/-)	Rb (+/-)	Sr (+/-)
FLW	-350	-0.77	-0.10	0.0025	-0.22	-0.025	-8.3
ACT	-330	-0.75	-0.11	-0.0075	-0.19	0.065	-7.3
CHI	-330	-0.73	-0.11	-0.067	-0.22	0.70	-7.8
CCO	-310	-0.69	-0.11	-0.052	-0.17	0.25	-6.9
GCM	-300	-0.68	-0.11	-0.023	-0.18	0.060	-7.2
FCWN	-310	-0.72	-0.12	-0.013	-0.17	0.24	-7.8
ZEO	-350	-0.76	-0.11	-0.032	-0.14	0.20	-7.5
PWI	-310	-0.67	-0.11	-0.012	-0.16	0.25	-7.9
SNM	-260	-0.58	-0.12	-0.028	-0.15	0.050	-1.8
SDH	-350	-0.77	-0.11	0.0075	-0.24	0.020	-8.2
PBC	-340	-0.76	-0.11	0.033	-0.22	-0.0065	-7.9

5 SUMMARY AND CONSLUSIONS

The results of the experiments proved to be very useful in making comparisons of 11 different CRRs. All of the samples showed varied ion-exchange behaviors in single element stock and mine drainage solutions. The CRRs that seemed to adsorb more metals from solution were FLW, ACT, ZEO, and SDH, and these samples also desorbed the most Na and Ca. This shows that the metals are predominantly exchanging with Na and Ca, that Na- and Ca-rich CRRs are more effective at metal removal, and that substantial adsorption results in substantial desorption. Another important finding is the correlation between CRR composition, in terms of Na, Ca, and K, and the dominant cation desorbed. For example, FLW, the most Na-rich CRR, consistently released the most Na in the exchange reactions.

Therefore, further tests will be done on FLW, ACT ZEO, and SDH in order to determine which of these CRRs is best suited for use in a treatment system. The CRRs PWI and CHI will also be tested in the future because of their behavior in mine drainage solutions and their compositions (K-rich). These 2 CRRs demonstrated relatively good adsorption capabilities without the release of large amounts of Na.

REFERENCES

Arbogast, B.F. 1996. Analytical methods for the mineral resource surveys program. *U.S. Geological Survey Open-File Report* 96-525.

Boylan, J.A. Jr. 1995. Decreased permeability of clinoptilolite-rich tuffs resulting from cation exchange and associated precipitation reactions. In D.W. Ming & F.A. Mumpton (eds), *Natural Zeolites '93: Occurrence, Properties, Use*: 325-337. New York: International Committee on Natural Zeolites.

Bremner, P.R. & L.E. Schultze 1995. Ability of clinoptilolite-rich tuffs to remove metal cations commonly found in acidic drainage. In D.W. Ming & F.A. Mumpton (eds), *Natural Zeolites '93: Occurrence, Properties, Use*: 397-403. New York: International Committee on Natural Zeolites.

Desborough, G.A. 1994. Efficacy of heavy-metal capture by clinoptilolite-rich rocks from heavy-metal-polluted water in five drainages in Colorado: *U.S. Geological Survey Open-File Report* 94-140.

Desborough, G.A. 1995. Extraction of metals from raw clinoptilolite-rich rocks exposed to water in heavy-metal polluted drainages: *U.S. Geological Survey Open-File Report* 95-56.

Desborough, G.A. 1996. Potential use of clinoptilolite-rich rocks for capture and retention of soluble lead in aqueous systems such as soils, contaminated drainages, and waste water: *U.S. Geological Survey Open-File Report* 96-715.

Drever, J.I. 1997. *The geochemistry of natural waters*. New Jersey: Prentice Hall.

Driscoll, C.T., J.P. Baker, J.J. Bisogni Jr., & C.L. Schofield 1980. Effect of aluminum speciation on fish in dilute acidified water. *Nature*. 284: 161-164.

Gilbert, J.G., P.M. O'Meara, J.G. Crock, T.R. Wildeman, & G.A. Desborough 1998. The adsorption capabilities of clinoptilolite-rich rocks as it relates to mine drainage remediation. *U.S. Geological Survey Open-File Report*, in press.

Kesler, S.J. 1994. *Mineral resources, economics and the environment*. NewYork: Macmillan.

Zamzow, M.J. & L.E. Schultze 1995. Treatment of acid mine drainage using natural zeolites. In D.W. Ming & F.A. Mumpton (eds), *Natural Zeolites '93: Occurrence, Properties, Use*: 405-413. New York: International Committee on Natural Zeolites

Zelazny, L.W., H. Liming, & A. Vanwormhoudt 1996. Charge analysis of soils and anion exchange. In D.L. Sparks (ed.), *Methods of Soil Analysis: Part 3 – Chemical Methods*:1231-1253. Madison, Wisconsin: Soil Society of America, Inc.

Tailings and Mine Waste '99 © 1999 Balkema, Rotterdam, ISBN 90 5809 025 6

Hydrogeologic uncertainty and the design of a remedial well field

M. Kautsky
MACTEC-ERS, LLC, US Department of Energy, Grand Junction, Colo., USA

ABSTRACT: Uncertainty concerning the nature of hydrogeologic conditions becomes apparent during engineering design of a remediation well field. Difficulties arise because subsurface conditions are seldom, if ever, adequately defined to ascertain plume geometry, aquifer heterogeneity, geochemical conditions, and contaminant sources. The effects of hydrogeologic uncertainty on the design of a remediation well field are evaluated for a uranium mill tailings disposal cell near Tuba City, Arizona. Important parameters affecting the fate and transport of contaminants in the groundwater are the aquifer hydraulic conductivity, total and effective porosity, and sorption onto the solid phase. Site characterization data indicate that the performance of a remediation well field should be predictable using groundwater flow modeling and particle tracking in combination with a phased implementation strategy; however, the eventual performance of the well field could be hampered by a large disparity between total and effective porosity.

1 INTRODUCTION

Remediation-well-field design is complicated by subsurface conditions that are mostly unknown. Field programs designed to determine subsurface conditions provide glimpses of what those conditions might be. However, regardless of the care taken in the planning and execution of a site characterization program, measurement and sampling error is assimilated into the data. Often "rules of thumb" are used to describe measurement imprecision. For example, hydraulic conductivity measurements are often regarded as "good to within an order-of-magnitude."

During design, predictions of groundwater flow and plume behavior take on new consequences because project planning and budgeting decisions must be made. During this phase of a project, typical site characterization data that are viewed by scientists as estimates of the physical world take on important economic consequences. Predictions based on groundwater-flow and solute-transport models, which are even further removed from the truth than the original data, must be viewed with particular skepticism.

This paper explores the uncertainty of field measurements and its effect on planning a well field. It also evaluates the methods for numerically modeling a field site to design a well field. Examples are taken from the Uranium Mill Tailings Remedial Action (UMTRA) groundwater project site near Tuba City, Arizona (Figure 1). Characterization data for the site suggest that conditions are favorable for active remediation, but the desired objectives may nevertheless be difficult to achieve.

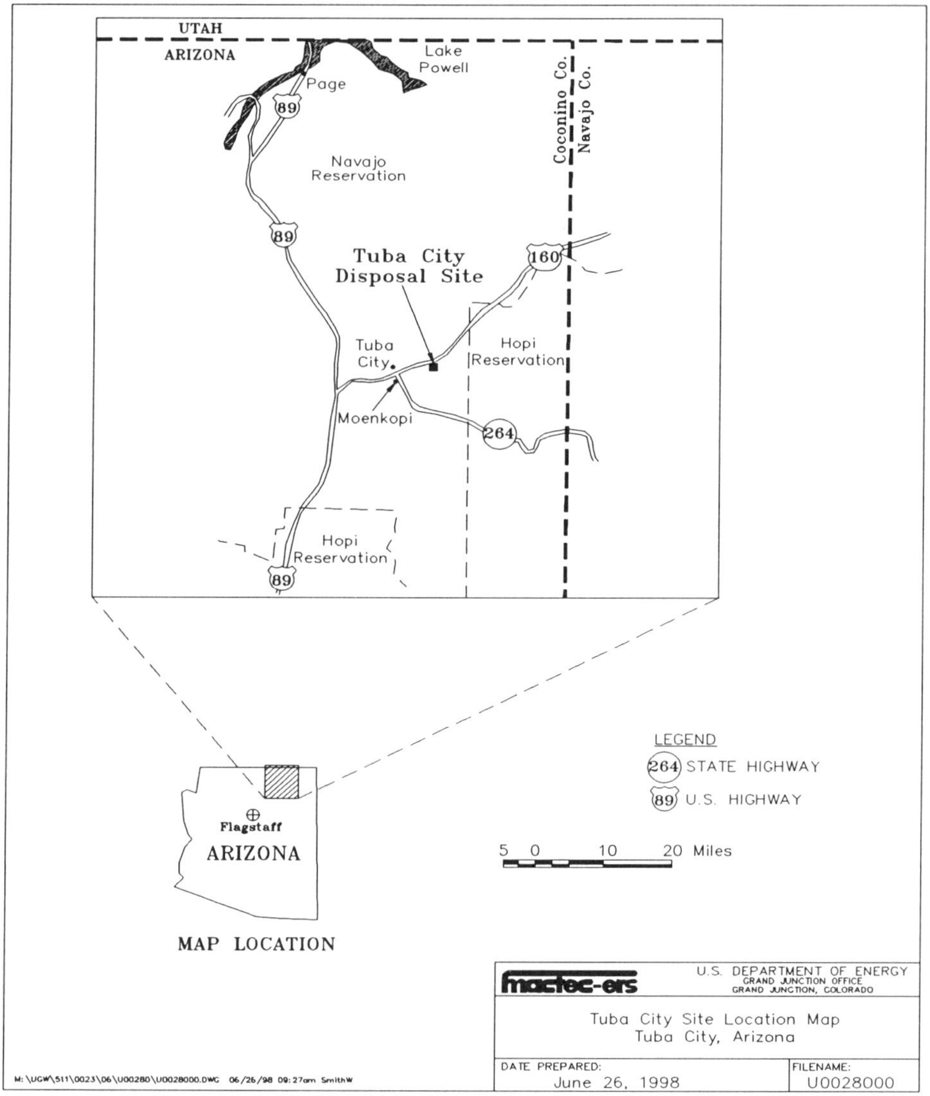

Figure 1. Location of the UMTRA site near Tuba City, Arizona

2 DESIGN CRITERIA

Numerous programmatic issues are evaluated during the engineering phase of a project. Evaluations are made to determine if the regulatory framework allows the contamination to be addressed using the most economically conservative solutions first. These solutions might consist of either natural flushing or supplemental standards as a basis for achieving compliance. If active remediation using pump and treat is the only remaining alternative, the following questions become important:

- What is the volume of the plume?
- How many withdrawal and injection wells are required and where should they be located?
- What pumping rates can the wells sustain?
- How many pore volumes are needed to reduce contaminant concentrations to target levels?
- How will the pumping be implemented to keep pace with projected funding levels?

3 CHARACTERIZATION DATA

Considerable resources are expended to characterize the physical and chemical properties of the aquifer. Characterization may last several years before active remediation begins. Some of the important characterization information is described below.

3.1 *Physical properties*

Hydraulic conductivity (K) of the formation is perhaps the most basic and important of all aquifer parameters and one that is most often evaluated during a site investigation. Site-scale hydraulic conductivity is normally assessed using either piezometer tests or pumping tests, or both. Questions about which type of test to rely on have been discussed throughout the past 15 years. Freeze and Cherry (1978) advocated piezometer tests for contamination projects because of their simplicity and low cost. They further argued that piezometer tests can provide adequate data in many cases where pumping tests are not justified.

However, recent experience has shown that pumping tests need to be an essential part of aquifer characterization because the resulting parameter values are averaged over a larger and more representative aquifer volume. Butler et al. (1996) cite numerous examples of investigations where the measurements of hydraulic conductivity with piezometer tests are biased on the low side relative to pumping tests because in many cases there appears to be a scale dependence in K. The basic hypothesis is that pumping test data are strongly affected by conduits of relatively high K, while slug tests rarely sample such heterogeneities.

Another advantage of pumping tests is that two or more components of hydraulic conductivity may be assessed from pumping test data provided that the data lend themselves to such analyses. These analyses are possible because pumping tests sample a larger portion of the aquifer. By contrast, slug test data are essentially point measurements.

Since the response of the aquifer to pumping stresses should be known when a remedial well field is designed, the importance of having pumping test data to evaluate hydraulics is inescapable. Another often overlooked advantage of a pumping test is that it provides a unique opportunity to monitor contaminant concentrations with time and space during the test.

Characterization for the UMTRA surface project included slug tests that were performed during the mid-1980s. A series of short-term and long-term pumping tests have been performed by the UMTRA groundwater project during the past three years (MACTEC-ERS 1998). Both series of tests provided hydrologic parameters for the aquifer. The mean horizontal hydraulic conductivity at the site appears to be about 0.2 meters/day, and the mean vertical hydraulic conductivity is approximately 0.009 meters/day. Vertical anisotropy ratios ranging from 1 to 1000 are common in modeling applications (Anderson and Woessner 1992). Vertical hydraulic conductivity at the site may be controlled by interdune deposits consisting of silt, and occasionally limestone beds up to 1 km^2 in area, that are randomly distributed areally and vertically.

Long-term pumping tests at the Tuba City site have provided information about hydrologic parameters as well as contaminant concentrations that help to evaluate the potential effectiveness of pumping on mass removal. The duration of the long-term tests was approximately 3 months. Figure 2 presents the results of uranium concentration-versus-time measurements for selected monitoring wells during the test. Since the concentrations did not drop appreciably during the long-term test, it appears that pumping may be an effective method to remove contaminants from the aquifer.

Total porosity and its relationship to effective porosity is also important to consider when the flow takes place in a fine-textured medium where adhesion (i.e., the attraction to the solid surface of the porous matrix of the molecules of fluid adjacent to it) is important (Bear 1979). The definition of effective porosity, however, is not always clearly understood. The classic perception of effective porosity is the measurable volume of "dead-end" pores that do not contribute to fluid transport. In a larger sense, effective porosity refers to the connectivity of pores through the solid medium. In this sense, the effective porosity is more closely related to permeability than to total porosity (Domenico and Schwartz 1990).

Table 1 indicates that the aquifer at Tuba City has a narrow range of particle sizes and the particles themselves are very fine grained. On a large scale, the total porosity of the formation is about 30 percent. The effective porosity as measured during a tracer test (MSE 1997) may be one tenth of the total porosity. This suggests that adhesion forces could impede aquifer remediation, but the degree to which this may occur and how it will manifest itself, is uncertain. Additional laboratory tests are underway to measure effective porosity according to the method of Corey (1986).

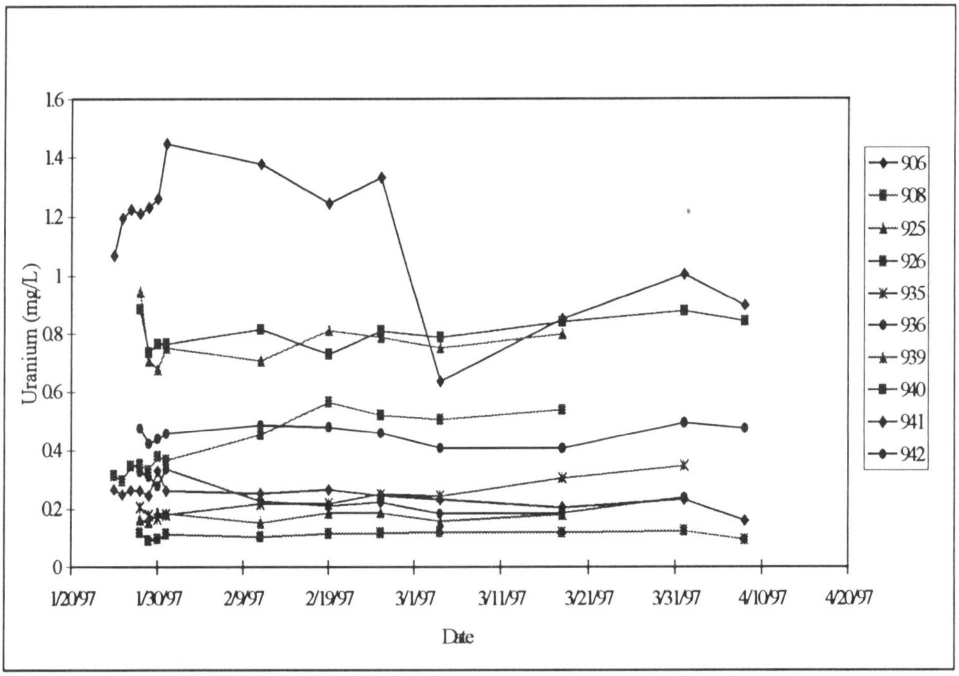

Figure 2. Uranium concentration versus time for selected wells during long-term pumping tests.

Table 1. Physical properties of selected Tuba City core samples

Sample ID	Depth (m)	Grain size distribution (gravel: sand: fines)	Total porosity (percent)
MW 930	9.5	0:60:40	15.9
MW 930	13.7	0:88:12	30.6
MW 930	16.5	0:90:10	
MW 930	17.1	0:93:7	29.0
MW 937	11.6	0:92:8	30.8
MW 937	13.1		32.2
MW 937	16.2	0:95:5	30.4
MW 937	16.5	0:95:5	30.8
MW 941	14.0	0:95:5	31.3
MW 941	16.7	0:93:7	28.6
MW 941	17.4	0:95:5	25.8

3.2 *Chemical properties*

Contamination at the Tuba City site consists predominantly of uranium, nitrate, and sulfate. The uranyl (+VI) form of uranium appears to be the predominant uranium species at the site (MACTEC-ERS 1996). It forms strong aqueous complexes with carbonate and uranyl dicarbonate $[UO_2(CO_3)_2]^{2-}$ (Krauskopf 1967).

Adsorption of uranyl to mineral grains in the Navajo sandstone (mostly quartz and feldspar) is likely to be insignificant. Analysis of rock core collected from the aquifer indicates that the distribution coefficient is very near zero (MSE 1997). However, uranyl is known to adsorb to ferric oxyhydroxide in relatively high concentrations (Morrison et al. 1995). It is likely that adsorption to ferric or manganese minerals is the principal mode of uranium-transport retardation at the site. The high concentration of carbonate in the groundwater near the disposal cell favors the partitioning of uranium to the dissolved phase. In distal portions of the plume, where dissolved carbonate concentrations are lower, adsorption of uranyl to oxide or oxyhydroxide minerals may be a dominant process.

High concentrations of dissolved uranyl together with high concentrations of dissolved nitrate and sulfate imply that these constituents are relatively conservative and that significant flushing should occur after just one pore volume. Consequently, it is believed that pumping and treatment may be an acceptable strategy for aquifer cleanup at the Tuba City site. This would also imply that conventional groundwater flow modeling combined with particle tracking would provide enough numerical precision to forecast the aquifer response to pump and treat.

The distribution of chemicals in the aquifer defines the shape of the groundwater plume. The ideal way to define the shape of the plume is to treat it as if it were an ore body. However, the economic incentive to define the plume with high precision may be absent during plume characterization. If the plume moves slowly and there are no immediate downgradient receptors, the plume can be defined adequately for planning the remediation program, and the shape of the plume can be defined more precisely using the observational approach during remediation.

Each site has its own unique conditions that make plume definition problematic. At the Tuba City site, the entire aquifer is about 150 meters thick, but the contamination is only present in the upper one-fifth of it. Contaminant concentrations are at background levels in six monitoring wells screened at depths in excess of 25 meters below the water table. On the basis of these data, the lower four-fifths of the aquifer is believed to be uncontaminated. However, the rate at which concentrations decrease with depth is unknown.

The observation that concentrations may decrease with depth prompted a study that used vertical-profiling techniques and a diffuse multi-layer sampler. The sampler relies on chemical studies of groundwater obtained from a series of dialysis cells suspended in packed intervals of the well screen. Chemical concentrations are evaluated for each sampled interval and may be plotted as a function of depth. Figure 3 illustrates the data for the Tuba City site. The data indicate that vertical changes in concentration are nondetectable across the screened portion of each well. A major uncertainty associated with these measurements is whether the duration of the sample-equilibration period, that is, the length of time between emplacing the sampler and collecting the sample, was sufficient to allow complete cessation of the turbulence created by inserting the tool into the hole.

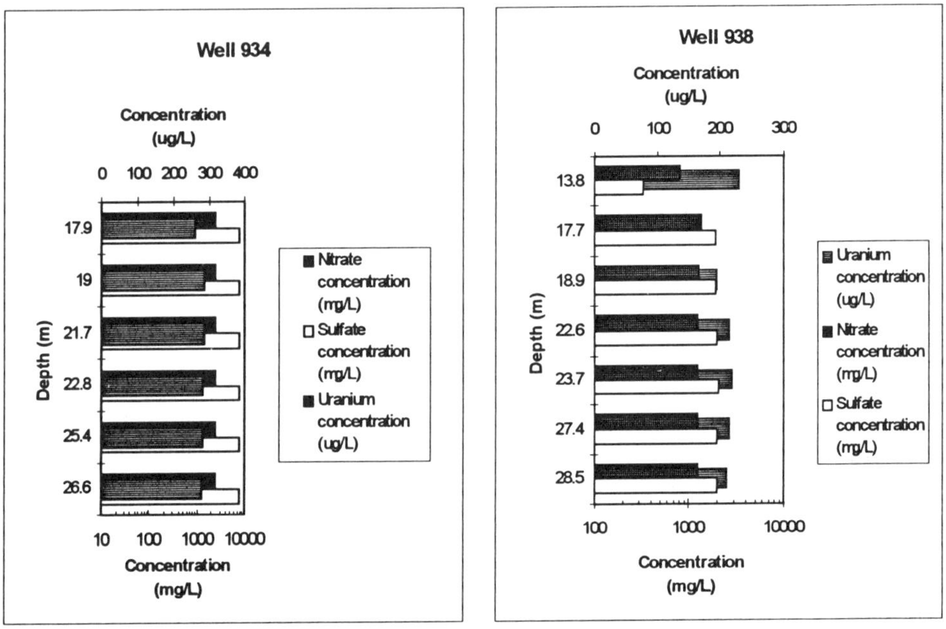

Figure 3. Depth-concentration profiles for wells 934 and 938.

4 CONCEPTUAL MODEL

A site conceptual model is normally developed for each UMTRA site before initiation of numerical modeling. The site conceptual model is a compilation of hydrologic processes expressed in terms of a water balance. Because of uncertainty in the computed magnitude of each component of the water balance, it is best to consider the water balance components as approximations that may change as more is learned about the flow system. Development of the conceptual model enables the water balance obtained from numerical modeling to be examined in terms of the field water balance and also makes possible an evaluation of any sources of errors.
Table 2 presents the site conceptual model for the Tuba City UMTRA site. The table shows that most of the groundwater entering the site is due to natural groundwater inflow from the north. Most of the outflow from the site is due to evapotranspiration and discharge to Moenkopi Wash

about 1800 meters southeast of the site. Other minor components of the site conceptual model are also listed in Table 2.

The mineral gypsum precipitates in the wellbore in some wells along the downgradient edge of the disposal cell where the hydraulic conductivity is lower. This area has elevated hydraulic head conditions, and chemical saturation indices indicate oversaturation with respect to both gypsum and calcite. Consequently, the lower measured hydraulic conductivities are interpreted as real features that formed as acidic solutions percolating through the tailings dam dissolved calcite cement and transported the minerals downgradient where they are now precipitating.

Table 2. Water balance based on field data for the UMTRA repository near Tuba City.

Flow component	Computed flux (cm/yr)	Surface area (m^2)	Inflow (m^3/day)	Outflow (m^3/day)
North-northwestern boundary	110	4.5×10^5	1.34×10^3	0
West-southwestern boundary	NA	NA	0	0
East-northeastern boundary	NA	NA	0	0
Evapotranspiration between middle terrace and cliff bands (ET_1)	5.0[a]	4.9×10^6	0	6.74×10^2
Evapotranspiration at greasewood area (ET_2)	16	1.1×10^5	0	4.87×10^1
Evapotranspiration along cliff bands (ET_3)	110	1.56×10^5	0	4.70×10^2
Discharge to Moenkopi Wash	110	1.23×10^5	0	3.68×10^2
Diffuse recharge	1.6	5.2×10^6	2.28×10^2	0
Maximum transient drainage rate: thin slimes	1.2	8.9×10^4	2.92×10^0 (at t = 40 yr)	0
Maximum transient drainage rate: thick slimes	0.42	4.65×10^4	5.35×10^{-1} (at t = 70 yr)	0
Totals			1.57×10^3	1.56×10^3

[a] Estimated from literature values (Nichols 1993)

5 NUMERICAL MODELING STUDIES

Numerical modeling simulations are used to predict the effect of proposed management alternatives on hydraulic conditions within the aquifer. Numerical modeling begins with model calibration. For the design of a remediation well field, model calibration should consist of qualitative and quantitative techniques such as matching simulated heads to observed heads, making sure that the simulated groundwater flows in the same directions as the actual flow, watching that the field water balance terms agree reasonably with the site conceptual model, and minimizing the spatial bias within the model domain. Detailed recommendations for the practice of modeling are available in Anderson and Woessner (1992) and ASTM D5447 (04.09), D5718 (04.09), D5611 (04.09), D5490 (04.09). With a structured approach, the modeler may effectively defend the uniqueness of the calibrated solution and help silence the critics who would assert that models can show whatever one desires.

5.1 *Calibration of the numerical flow model*

Modeling for the engineering design phase of an UMTRA Project well field follows the same general procedures that are recommended for the modeling industry as a whole. The basic procedure involves selecting a model domain to match the site conceptual model, assigning a sufficient number of finite difference grid cells to produce piecewise continuous solutions to the flow problem, choosing realistic boundary conditions, and assigning parameter values and calibration targets and determining suitable calibration–acceptance criteria. In the UMTRA Project the general practice is to involve stakeholders in the initial stages of the modeling project, so that fewer disagreements arise during the later stages of the modeling study. The overall philosophy for the modeling is to honor all valid data in the numerical simulation.

Both quantitative and qualitative model calibration targets are used. An important quantitative calibration criterion is to express the calibration in terms of the residual standard deviation divided by the total range in head. This ratio, expressed as a percent, should normally be within 10 percent, but the modeler should strive to achieve less than 5 percent error (Rumbaugh 1998). In the Tuba City model, this error term achieved a minimum of 0.6 percent.

Trial and error simulations can be used initially during the simulations, but it is best to combine the trial and error simulations with sensitivity analysis to evaluate the effect that ranges in parameter uncertainty may have on the simulation. By identifying the parameters to which the model is sensitive, the calibration effort can be accelerated greatly. Figure 4 illustrates the use of a sensitivity analysis where the error terms are expressed as a function of parameter multipliers applied to zones of hydraulic conductivity, recharge, and boundary conditions. The sensitivity analysis can be expanded during a later phase of the project to improve the calibration even further using inverse modeling techniques.

The use of Windows-driven pre– and post–processors has greatly enhanced the ease of conducting simulations and the rate of calibrating a model. Literally hundreds of simulations may be run in an 8-hour day with modern Pentium-based machines. Consequently, model calibration is now a relatively short-lived portion of the overall project.

Precautions were taken during the modeling to predict the groundwater flow velocities on the basis of the total porosity, since the entire void volume contains the groundwater and the dissolved constituents. A sensitivity analysis approach was taken to evaluate the effect that porosity values have on the duration required to flush one pore volume.

5.2 *Predictive Simulations*

Predictive simulations are used to forecast potential response of the aquifer to pumping stresses. At tailings sites where geochemical conditions favor the contaminants remaining in solution as pumping progresses, the use of flow modeling with particle tracking appears to yield useful

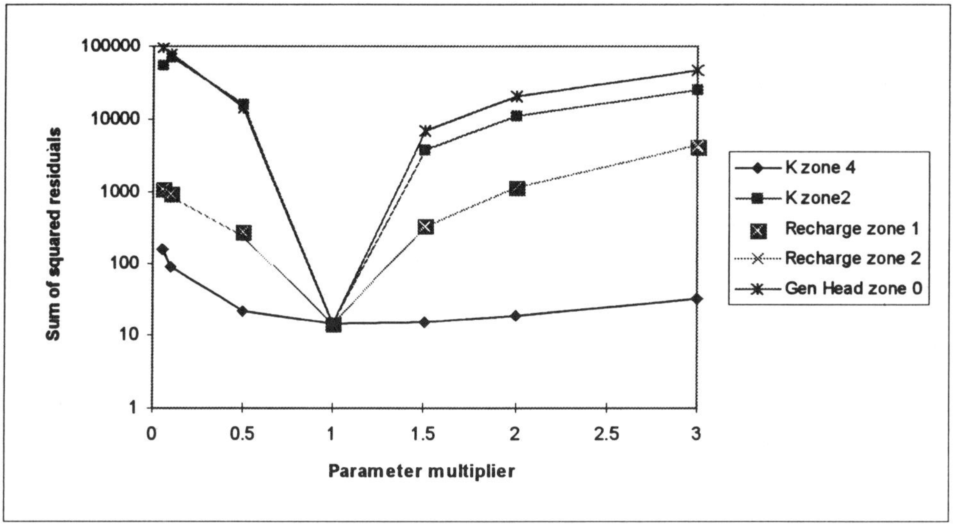

Figure 4. An example of sensitivity analysis applied to selected zones of a numerical model.

results. This approach is relatively straightforward and expedient because the additional rigor and concomitant uncertainty resulting from transport modeling can be avoided.

Particle tracking simulations are also used to evaluate the relative effectiveness of alternate deployment strategies for the well field. These simulations help the design team to evaluate proposed well locations and the relative effectiveness of alternate ramp-up strategies for the well field. They also enable the team to direct expenses to portions of the plume that will require the most time to achieve compliance with the groundwater standards.

Particle tracking simulations for the Tuba City site were used in this manner. The simulations resulted in a design with extraction wells along the downgradient margin of the disposal cell, discharge of treated groundwater to an infiltration trench along the upgradient margin of the disposal cell to increase the hydraulic gradient and enhance flushing, localized five-spot pumping arrays in the centroid of the plume, and discharge of treated groundwater along the downgradient margin of the plume to control contaminant migration. Deployment of the final 200 gallon-per-minute well field will be accomplished with a phased escalation of pumping in combination with performance-assessment monitoring, analysis of parameter uncertainty, revised calibration of the flow model, and particle tracking analyses. Stochastic simulations that evaluate the probability of plume capture and incorporate parameter uncertainty are also underway.

6 CONCLUSIONS

The structured technical approach is described as a basis for accomplishing the design of a remediation well field. The basis for this approach is that parameter uncertainties are managed through a phased escalation of pumping capacity. Because the contaminants are oxidized and most constituents will migrate as conservative species, geochemical conditions at the site appear to favor significant mass reduction through pump and treat. This conclusion is supported by laboratory tests of chemical sorption and desorption. The main impediment to removing the contaminants from the ground appear to be the large disparity between total and effective porosity. This disparity may favor the retention of contaminants on the solid phase due to adhesion of the groundwater to the solid phase. Cyclical or pulsed pumping periods may be the only effective way to remove this contamination.

7 REFERENCES

Anderson, M. P., and W.W. Woessner 1992. *Applied Groundwater Modeling: Simulation of Flow and Advective Transport.* San Diego: Academic Press, Inc.

Bear, J. 1979. *Hydraulics of Groundwater: Water Resources and Environmental Engineering.* New York: McGraw-Hill, Inc.

Butler, J. J. Jr., C.D. McElwee, and W. Liu 1996. Improving the Quality of Parameter Estimates Obtained from Slug Tests. *Ground Water* 34(3): 385–576.

Corey, A. T. 1986. *Mechanics of Immiscible Fluids in Porous Media*: Englewood: Water Resources Publications, 230 p.

Domenico, P. A., and F. W. Schwartz 1990. *Physical and Chemical Hydrogeology.* New York: John Wiley and Sons, Inc.

Freeze, R. A., and J. A. Cherry 1979. *Groundwater*: Englewood Cliffs: Prentice-Hall, 604 p.

Krauskopf, K. B. 1967. *Introduction to Geochemistry: International Series in the Earth and Planetary Sciences.* New York: McGraw Hill, Inc.

MACTEC-ERS 1996. *Groundwater Geochemistry of the Tuba City Site.* U.S. Department of Energy Grand Junction Office: Calculation No. U00034AA.

MACTEC-ERS 1998. *Tuba City Interim Action.* U.S. Department of Energy Grand Junction Office: Calculation No. U00178AA.

Morrison, S., J. R. Spangler, and V.J. Tripathi 1995. Adsorption of Uranium on Amorphous Ferric Oxyhydroxide at High Concentrations of Dissolved Carbon and Sulfur: *Journal of Contaminant Hydrology.* (17) 333-346.

MSE Technology Applications 1997. *Report on Injection-Extraction Pilot Test at Tuba City, Arizona.* Butte, MT. 97MSE-1558.

Nichols, W. D. 1993. Estimating Discharge of Shallow Groundwater by Transpiration from Greasewood in the Northern Great Basin: *Water Resources Research*, v. 29, p. 2771-2778.

Rumbaugh, James, III. 1998. personal communication with the author of Groundwater Vistas.

Tailings and Mine Waste '99 © 1999 Balkema, Rotterdam, ISBN 90 5809 025 6

Aquatic selenium at Cominco's Red Dog Mine: Sources, speciation, distribution, and control

M. Edwards – *Cominco Research, Trail, B.C., Canada*
J.E. Kulas & J.O. Weakley – *Cominco Alaska Incorporated, Kotzebue, Alaska, USA*
W.J. Kuit – *Cominco Limited, Vancouver, B.C., Canada*
N.S. Bloom & D. Wallschläger – *Frontier Geosciences, Seattle, Wash., USA*

ABSTRACT: Cominco's Red Dog Mine has undertaken an extensive study of aquatic selenium as a result of recent changes to its water discharge permit. The new limit for selenium is 4.9 μg/L, while treated effluent contains about 9 μg/L. This study included, in part, a review of analytical methods, the development of a selenium speciation technique, and an evaluation of treatment methods. Selenate and selenocyanate were identified as the two important species appearing in the process. A fairly constant background of selenate was present throughout the water system. Selenium levels increased 5-10 fold with the addition of sodium cyanide prior to the rougher flotation circuit. All of the additional selenium was in the form of selenocyanate, $SeCN^-$. This species decomposes or precipitates in the tailings impoundment via an unknown mechanism. The water entering and discharged from the effluent water treatment plant contains only selenate. As this species is not readily removed with conventional treatment methods efforts are now focused on the control and/or removal of selenocyanate.

1 INTRODUCTION

The Red Dog Mine, situated in northwest Alaska, is the world's largest producer of zinc concentrates (Figure 1). The mine is operated by Cominco Alaska Incorporated (Cominco Alaska) and the deposit is owned by the NANA Regional Corporation. In the mine's concentrator plant, zinc- and lead-containing minerals are recovered from the ore using conventional froth flotation technology. Combined production of zinc and lead concentrates is expected to exceed 1 million tonnes in 1998.

To minimize the impact of the mine's activities on the surrounding aquatic environment, an extensive water management system is used. Tailings from the concentrator, area run-off, and water that has passed through disturbed ground in the open-pit mine and plant site, are collected in a tailings impoundment which is located in the South Fork Red Dog Creek. The impoundment water contains dissolved metals/metalloids, primarily zinc with lesser amounts of iron, lead, cadmium, and others. Water reclaimed from the tailings impoundment is treated in one of two water treatment plants (WTP) to remove the metals by precipitation, clarification, and polishing sand filtration. The treated water is either recycled for use in the concentrator or, in order to maintain a reasonable water level in the tailings impoundment, is discharged during the spring and summer months to Middle Fork Red Dog Creek. This creek flows into Ikalukrok Creek, which in turn flows into the Wulik River and then into the Chukchi Sea. During the discharge season up to 6.6 billion litres (2.4×10^9 gal) of treated water are discharged at a flow rate of about 38,000 L/min (10,000 gpm). Metals concentrations in the receiving waters have decreased relative to premining conditions resulting in an improved environment for aquatic life. Studies by the Alaska Department of Fish and Game indicate an increase in the utilization of Red Dog Creek by chum salmon and Arctic Grayling.

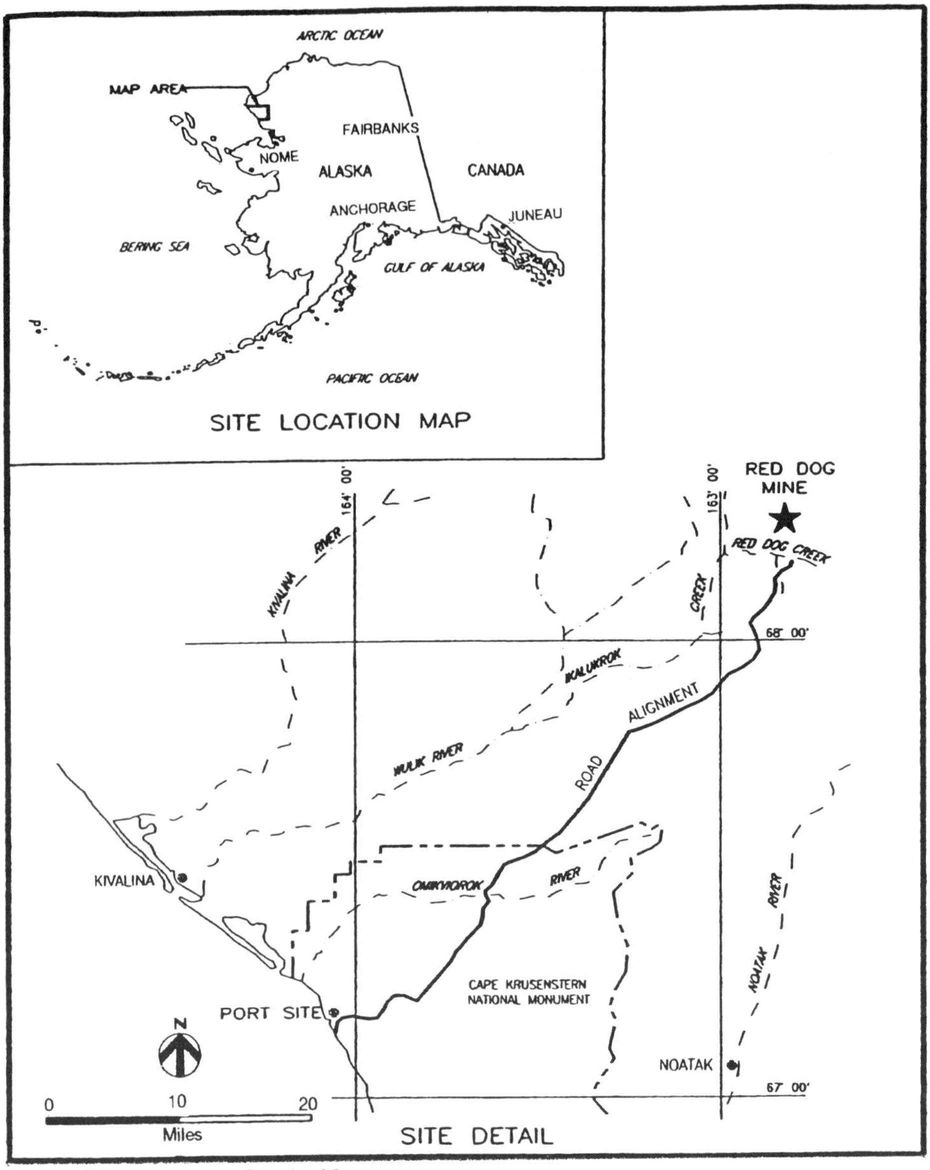

Figure 1. Red Dog Mine Site Location Map.

The quantity and quality of the discharged water are subject to various federal and state permit regulations. The most recent permit, issued in July 1998, resulted in a significant reduction in the concentration limits for many metals and other analytes (Table 1). These reductions reflect a shift from technology-based to water quality-based permit criteria. Modifications to the effluent water treatment plant, WTP #2, have enabled the operation to meet the new limits for all metals/metalloids but selenium, whose limit is now 4.9 μg/L. This limit for selenium is based upon the U.S. EPA freshwater chronic criterion of 5 μg/L, established in 1987. Selenium is an essential nutrient but is toxic/teratogenic at high concentrations [Frankenberger & Engberg 1998]. Consequently, Cominco Ltd. initiated an extensive study of selenium at the Red Dog Mine. This paper describes the current findings of this study with an emphasis on the sources, speciation, distribution, and control of selenium at the mine.

Table 1. Effluent Quality Limitations and Recent Discharge Quality for Red Dog Outfall 001 (Monthly Avg.).

Analyte	1985 NPDES Permit	1998 NPDES Permit	Aug. 1998 Performance
Cadmium, Total, μg/L	50	2.0	1.0
Copper, Total, μg/L	150	15.1	<5
Lead, Total, μg/L	300	8.1	1.2
Selenium, Total, μg/L	-	4.9	8.7
Zinc, Total, μg/L	750	119.6	59

2 SOURCES OF SELENIUM

In an effort to determine the routes of selenium entry into various waters at Red Dog, extensive characterization of selenium occurrences in the ore, reagents and grinding media was completed. In addition, comprehensive mill sampling campaigns were conducted to establish trends in total dissolved selenium concentrations through each of the major process steps. These efforts were directed towards identifying routes of selenium introduction into process solutions and identifying methods to reduce selenium in discharged water through source control.

A lack of reliable and accurate analytical methods for the measurement of low-level selenium in complex matrices, i.e. <50 μg/L, initially compounded the problem of identifying sources. An interlaboratory performance study, conducted as part of this work, was used to identify reliable laboratories for selenium analysis [Möller et al. 1998]. This study indicated that the analysis of selenium at low levels could be problematic even for reputable laboratories. Therefore, Red Dog currently relies on a very limited number of selected laboratories for total selenium analysis.

The analysis of 227 ore, host rock, and regional rock samples indicated that selenium is ubiquitous throughout the Red Dog geology (Table 2) [Coulter 1997]. Median selenium concentrations were typically 5 to 10 ppm with the highest concentrations observed being 150 ppm. While no discrete selenium mineral has been identified, it is likely that it is present in a wide range of minerals in which it displaces small amounts of sulfur. The chemical properties of selenium are very similar to those of sulfur. No significant amount of selenium was found in the chemical reagents used in the concentrator circuit and the water treatment plant, the grinding media, or the blasting agents used in the pit. The only significant source of selenium in the Red Dog water circuit was concluded to be ore. An estimated 50 kg/d of selenium enters the mill with the ore. However, only a small fraction is solubilized in the mill circuit.

Table 2. Selenium Concentrations in Red Dog Ore, Host Rock and Regional Rock Samples.

Material	Median Se Concentration (mg/kg)	Range of Se Concentrations (mg/kg)	Standard Deviation in Se Analyses (mg/kg)
Ore	5.0	0.0 to 108	17
Host Rock	7.6	0.0 to 146	29
Regional Rock	9.9	0.0 to 55	14

Solutions in the concentrator circuit were surveyed to generate a profile of dissolved selenium concentrations through each unit operation. Figure 2 illustrates the typical dissolved selenium profile for the Red Dog circuit. The treated reclaim water used in the grinding and flotation circuits contains about 10 μg/L selenium. During grinding the concentration increases slightly to 15-20 μg/L. Following the initial addition of reagents prior to lead flotation there is a sharp increase in dissolved selenium levels to about 40-50 μg/L. The levels remain relatively high through the lead and zinc rougher circuits. The final tails stream exiting the mill has a selenium concentration of about 20-30 μg/L, though this concentration decreases to about 10 μg/L in the tailings impoundment, likely as a result of dilution with other waters and possibly through chemical precipitation. Therefore, the tailings impoundment, which typically contains about 15 billion litres (4 x 10^9 gallons) of water, contains a total of about 150 kg of dissolved selenium. There is no significant decrease in the selenium concentration as reclaim water is treated in the effluent water treatment plant. At a discharge water flow rate of 38,000 L/min some 0.5 kg of total selenium is discharged to Red Dog Creek each day.

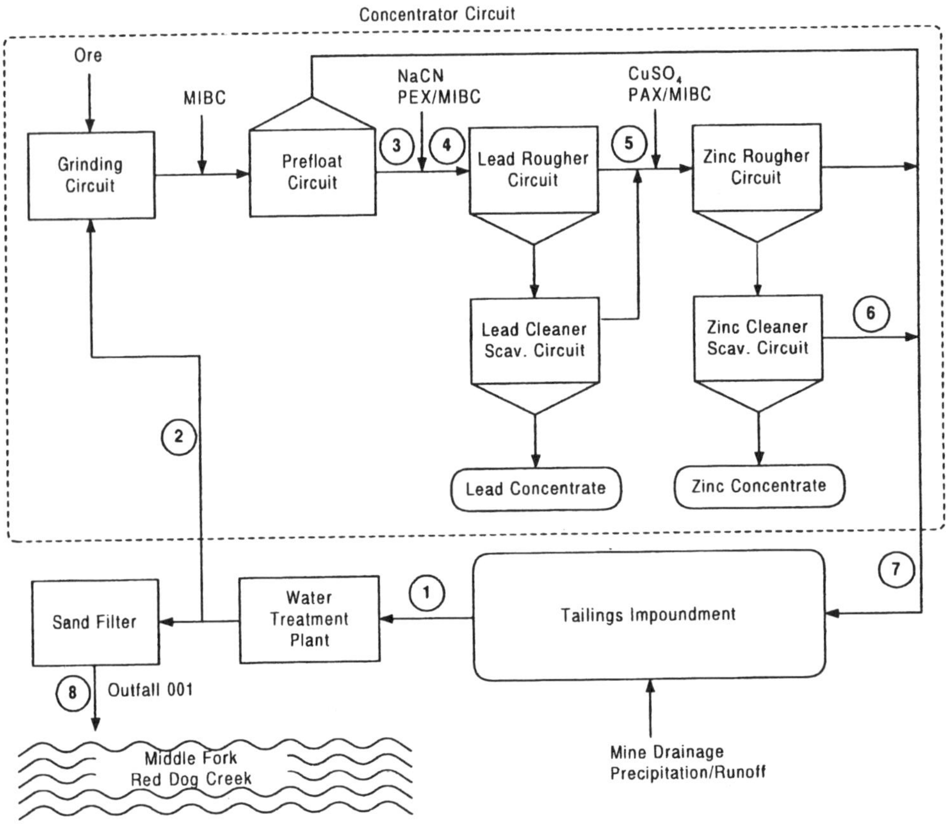

Sampling Point	Se Concentration (μg/L) as SeO_3^{2-}	as SeO_4^{2-}	as $SeCN^-$	Σ Species[a]	Se/Total[b]
1. Reclaim Water	<2	8.6	<2	8.6	9.5
2. Process Water	<2	7.6	<2	7.6	9.2
3. Flotation Feed	<3	5.0	4	9.0	15
4. Lead Rougher Feed	<4	7.8	50	58	48
5. Lead Rougher Tails	<5	9.0	51	60	56
6. Zinc Scav. Tails	<6	6.6	<2	6.6	10
7. Final Tails	<7	6,4	17	24	20
8. Final Effluent	<8	8.6	<2	8.6	9.5

a. Arithmetic sum of Se in selenite, seleante, and selenocyanate.

b. Dedicated total selenium analysis (HG-AFS).

Figure 2. Dissolved Selenium Concentrations in the Red Dog Mill and Water Circuit

3 SELENIUM SPECIATION AND DISTRIBUTION

Selenium exhibits a complex solution chemistry very similar to that of sulfur [Greenwood & Earnshaw 1984]. A listing of several important selenium species is presented in Table 3. In addition to these species many organoselenium compounds are known. The identification of the selenium species present in aquatic systems is critical for the development of selenium control and treatment strategies. However, low-level selenium speciation analyses are not generally available through commercial laboratories. In addition, the commonly used Cutter speciation method is often unsuitable for some selenium species such as selenocyanate and for some complex matrices [Cutter 1978; Wallschläger & Bloom 1998, 1999]. Therefore Cominco Alaska and Kenncott Utah Copper retained Frontier Geosciences of Seattle, Washington to develop a new selenium speciation method. The method relies on species separation via ion chromatography followed by hydride generation and finally detection using atomic fluorescence spectroscopy (IC-HG-AFS). Full details of this method are described elsewhere [Wallschläger & Bloom 1998, 1999]. Although the method is still under development it has already provided valuable insight into the sources and subsequent aquatic behavior of selenium at Red Dog. This new selenium speciation technique has been used for the analysis of plant, laboratory, and environmental samples.

Table 3. Summary of Selenium Species.

Species	Name	Selenium Valence
Se^{2-}	Selenide	-2
$SeCN^-$	Selenocyanate	-2
Se^0	Elemental Selenium	0
SeO_3^{2-}	Selenite	+4
SeO_4^{2-}	Selenate	+6

Results from a selenium speciation survey of the concentrator circuit are provided in Figure 2. The key findings were:

- Selenate was present in all samples at levels from 5-10 μg/L and was the only species present in reclaim water and final effluent.
- Selenocyanate was the predominant species (70%-85%) in the lead and zinc rougher circuits and in the final tails stream.
- Selenocyanate was discharged from the mill into the tailings impoundment but was not detected in the water reclaimed from the pond.
- Selenite was not detected in any of the samples.
- There was no evidence for the presence of organoselenium compounds.

As a result of these findings it became apparent that the increase in selenium concentrations in the flotation circuit was primarily related to reagent additions, and to the use of sodium cyanide in particular. Sodium cyanide is a commonly used flotation reagent that acts to depress the flotation of pyrite and sphalerite. Cyanide is known to react with many sulfur species and minerals to produce the thiocyanate anion, SCN^- [Wang & Forssberg 1996]. Consequently, thiocyanate is often present in sulfide flotation circuits. Cyanide reacts with selenium species/minerals in an analogous manner to yield the selenocyanate anion [Jennings & Yannopoulos 1974]. The reactions of cyanide with sulfide and selenide are shown below. Krishnan & Zingaro [1974] postulated that 80% of the selenium in precious metals tailings pond water is present as selenocyanate and the remainder is selenate. However, others have not substantiated this, possibly due to the inadequacies of existing selenium speciation techniques.

Recently selenocyanate was found to be the predominant form of selenium discharged in effluent from some petroleum refineries in the San Francisco Bay area [WSPA 1995]. The use of specialized selenium speciation techniques played a critical role in this finding.

$$(x-1)CN^- + S_x^{2-} \rightarrow (x-1)SCN^- + S^{2-} \quad (1)$$

$$(x-1)CN^- + Se_x^{2-} \rightarrow (x-1)SeCN^- + Se^{2-} \quad (2)$$

The absence of selenocyanate in the water reclaimed from the tailings impoundment indicates that this species decomposes or precipitates in the pond. The analogous sulfur species, thiocyanate, is chemically and biologically degradable, with the decomposition products being ammonium, carbonate, and sulfate [Mudder & Smith 1991]. This suggests that selenate would be the predominant end product of selenocyanate oxidation. However, it has been shown that under certain conditions selenite is formed, along with lesser amounts of selenate [WSPA 1995]. The degradation of selenocyanate in Red Dog tailings water is currently under investigation.

To refine our understanding of the relationship between selenium dissolution and the use of cyanide, laboratory-scale flotation tests were conducted and solution samples were subjected to selenium speciation analysis. The key findings of this test work were:

- A small amount of selenate, ca. 5 µg/L, was present in solution after grinding. This 'background' level remained roughly constant through flotation regardless of the test conditions.
- A 75% reduction in the cyanide dose only reduced the selenocyanate concentration by 30%.
- The other flotation reagents in use at Red Dog did not appear to have a significant effect on selenium dissolution.
- Premixing some of the key flotation reagents did not reduce selenium concentrations significantly.
- No selenite was detected in any of the samples.

Since the water used in these tests was free of selenium the selenate detected after grinding must have come from either the dissolution of selenate salts present in the ore or from the oxidation of selenide minerals during milling. It is not known what proportion of the selenate present in the discharge water is accounted for by this mill-generated selenate. A mass-balance model for aquatic selenium is under development.

A limited number of water samples from the receiving environment have also been subjected to speciation analysis and to date only selenate has been detected. Upstream of the effluent discharge point in Red Dog Creek, the total selenium level is about 2-4 µg/L, while just below the discharge point the concentration increases to about 5-8 µg/L. Levels in the Ikalukrok Creek are about 1-2 µg/L.

A preliminary study of fish distribution and life history profiles identified the Ikalukrok Creek as the habitat of concern for possible selenium exposure [EVS 1998]. Of the four fish species that reside in the Ikalukrok the potential for maternal transfer of selenium to developing offspring appears to be significant for only the Arctic grayling and slimy sculpin. For both species gonadal development appears to coincide with feeding periods in the receiving environment. The two other resident species, chum salmon and Dolly Varden char, are unlikely to be affected by selenium as the pathway for exposure via the food web is not complete. Chum salmon and Dolly Varden char are the predominant species in Ikalukrok Creek and are of primary subsistence and recreational value. To date no adverse aquatic life impacts have been identified. An extensive biological monitoring program is conducted annually.

4 SELENIUM CONTROL

Both treatment and preventative measures have been investigated as means of controlling aquatic selenium at the Red Dog Mine. While many technologies have been identified for the removal of selenium from waste waters very few have been proven effective on a commercial scale [Chamberlin 1996]. An extensive study of selenium removal technologies, conducted in 1997 by Cominco Ltd. and external consultants, concluded that the selenium present in Red Dog waters is not readily removed by conventional means such as iron co-precipitation, hydroxide precipitation, sulfide precipitation, adsorption, oxidation, reduction, or through the use of proprietary reagents [Mudder & Botz 1997]. However, as this work was conducted prior to the development of the selenium speciation method described above, it was assumed that selenate and possibly selenite were the only species of concern. Red Dog's selenium control studies are now focused on the role played by selenocyanate.

Eliminating the use of cyanide in the flotation circuit may appear to be a potential source control strategy. However, cyanide is an essential reagent for the depression of pyrite and sphalerite and cannot be replaced. To limit the amount of selenium that may be leached by cyanide and to ensure that residual cyanide is not present in effluent, Red Dog has restricted its use to only that absolutely essential to maintain suitable operations. Further reductions in cyanide use may be possible with the use of proprietary reagents [Bulatovic & Wyslouzil 1995].

Our research efforts are currently directed towards determining the fate of the selenocyanate discharged in the mill tailings stream. Laboratory-scale tests are underway to determine the decomposition mechanism and products for this selenium species. This work will also provide insight to the factors that affect its decomposition. In addition to continuing on-site work Cominco Alaska has just initiated a 2- to 3-year study of selenium at the University of Idaho. The ultimate aim of this project is to establish a means of controlling the quantity of selenium discharged in treated effluent. This may include source control in the mill, treatment of the tailings stream, control of the tailings impoundment chemistry, or treatment immediately prior to discharge.

The Western States Petroleum Association (WSPA) conducted a study of technologies for the removal of selenocyanate from refinery waste water [WSPA 1995]. Iron co-precipitation (with an oxidation step), ion exchange, and anoxic biological treatment showed significant promise for selenocyanate removal. While the selenium concentrations in the WSPA study were much higher (i.e. mg/L) than at Red Dog, the first two removal methods are worthy of evaluation.

5 SUMMARY AND CONCLUSIONS

Cominco's Red Dog Mine faces the unprecedented challenge of reliably and economically reducing the selenium concentration of its effluent to less than the permit limitation of 4.9 μg/L. The concentration in effluent is currently about 9 μg/L. A lack of suitable methods for the analysis and speciation of low-level selenium has contributed to this challenge. The problem of analytical reliability was overcome with the identification of reliable chemical analysis techniques and laboratories. In addition, the development of an advanced selenium speciation technique has provided considerable insight into the chemical behavior of selenium in the Red Dog water circuit.

The only significant source of dissolved selenium in the Red Dog water circuit was determined to be ore. Two selenium species were identified in the concentrator circuit. A relatively small amount of selenate was present in the front of the circuit while selenocyanate dominated in flotation. The selenate is believed to have come from either the dissolution of selenate salts present in the ore or from the oxidation of selenide minerals during grinding. Selenocyanate formation was the result of the addition of sodium cyanide, a critical flotation reagent. This species decomposes or precipitates in the tailings impoundment via an unknown mechanism.

The selenium present in feed to the effluent water treatment plant was exclusively selenate. As this species is not amenable to removal by conventional means current selenium control efforts are now focused on selenocyanate. The removal of selenocyanate from the mill tailings stream or the control of its decomposition/precipitation in the tailings impoundment are two approaches to selenium control that will be investigated. A fundamental study of the chemistry of selenocyanate will also be conducted to facilitate the development of removal methods.

REFERENCES

Bulatovic, S. & D.M. Wyslouzil 1995. Selection and evaluation of different depressant systems for flotation of complex sulfide ores. *Minerals Eng.* 8(1/2): 63-76.

Chamberlin, P.D. 1996. Selenium removal from waste waters – an update. *Randol Gold Forum '96*. 119-127.

Coulter, G.A. 1997. Selenium distribution in Red Dog rocks. Cominco Alaska internal memorandum.

Cutter, G. 1978. Species determination of selenium in natural waters. *Anal. Chim. Acta*. 98(59).

EVS Consultants 1998. Species profiles and food webs related to selenium exposure in the Upper Wulik River near the Red Dog Mine. Report to Cominco Alaska.

Frankenberger, W.T.Jr. & R.A. Engberg 1998. *Environmental chemistry of selenium*. New York: Marcel Dekker, Inc.

Greenwood, N.N. & A. Earnshaw 1984. *Chemistry of the elements*. Oxford: Pergamon Press.

Jennings, P.H. & J.C. Yannopoulos 1974. Recovery and refining of selenium. In Zingaro & Cooper (ed.), *Selenium*. New York: Van Nostrand Reinhold.

Krishnan, V. & R.A. Zingaro 1974. Coordination compounds in which selenium functions as the donor atom. In Zingaro & Cooper (ed.), *Selenium:* New York: Van Nostrand Reinhold. 337-407.

Möller, G. , Smith, B.W., Steinhoff, P.J., and D.W. Warner 1998. Interlaboratory performance analysis in the determination of total selenium in water. Report to Cominco Alaska.

Mudder, T. & A. Smith 1991. *The chemistry and treatment of cyanidation wastes*. London: Mining Journal Books.

Mudder, T. & M. Botz 1997. Evaluation of selenium control technologies for the Red Dog Mine. Report to Cominco Alaska.

Wallschläger, D. & N.S. Bloom 1998, Simultaneous determination of selenite, selenate and selenocyanate in waters by IC-HG-AFS. *6th Int. Symposium On the Uses of Selenium and Tellurium, Scottsdale, 10-13 May*.

Wallschläger, D. & N.S. Bloom 1999. Selenium speciation in mining process waters by IC-HG-AFS. Paper to be presented at *Proceedings of Tailings and Mine Waste '99, Fort Collins, 24-27 January 1999*.

Wang, X.H. & K.S.E. Forssberg 1996. The solution electrochemistry of sulfide-xanthate-cyanide systems in sulfide mineral flotation. *Minerals Eng.* 9(5): 527-546.

WSPA 1995. Selenium removal technology study – final report.

Tailings and Mine Waste '99 © 1999 Balkema, Rotterdam, ISBN 90 5809 025 6

Geochemistry of acid mine waters in an eastern anthracite strip mine

C.M.Loop, W.B.White & B.E.Scheetz
Materials Research Laboratory, The Pennsylvania State University, University Park, Pa., USA
R.Hornberger
Pennsylvania Department of Environmental Protection, District Mining Operations, Pottsville, Pa., USA

ABSTRACT: This paper reports on the progress in geochemical modeling of the 135,000,000 gallon pool contained within an abandoned strip mine in eastern Pennsylvania that is the first segment of a three part PA Department of Environmental Protection demonstration project. In this demonstration, fly ash from a cogeneration facility is being placed "dry" into standing acid mine water in a large stripping pit known locally as the "Big Gorilla." The original, pre-World War I mining of this site occurred in the Mammoth vein (Llewellyn Formation, Pennsylvanian) in a relatively small synclinal valley known as the Silverbrook syncline. The valley's total water balance is estimated at 435 million gallons of water, approximately 25% of which are contained within the stripping pit, with the remainder present in underlying deep mines. The outflow of the syncline lies at a level 30 feet below the surface of the water in the Big Gorilla. Pre-demonstration water analyses suggest that the stripping pit and the remaining waters in the syncline are in chemical balance with each other. With the placement of the fly ash into the Big Gorilla, the acid mine water changed from a pre-placement pH of 3.6 to a current value of 11.6. Some 110,000 tons of ash have been placed at the site with only approximately one third of this tonnage actually in contact with the water. Analyses of the water chemistry of the site have shown that the ash placement has resulted in direct, significant, and homogeneous change in the entire pool. Metals of concern include Zn, Al, Mn, and Fe which have lowered concentrations in the treated water.

1 BACKGROUND

The Pennsylvania Department of Environmental Protection's Regional Mining Office in Pottsville, in conjunction with the Wilkes-Barre Regional Office, has issued three demonstration permits for the placement of co-generation derived fly and bottom ash into pre-act stripping pits in the middle anthracite coal basin in eastern Pennsylvania. The permits allow placement of a) fly ash slurry into standing water [Shen Penn Project]; b) fly ash slurry into a dry stripping pit [Nickerbocker Project]; and c) dry ash into standing water [Big Gorilla Project].

The objective of these demonstration projects is to show that co-generation derived fly ash can be placed into standing acid mine water without detrimental environmental effects. Preliminary studies have explored methods for ash placement that would minimize turbidity and maximize consolidation. The slurry placement into a dry hole completes the wet/dry options and will allow an evaluation of mobility of water-born fly ash through fractures into the underlying mine pool.

2 SITE HISTORY

The first project takes place at the present day site of Northeastern Power Company (NEPCO), whose main operations are production and sales of electricity by burning culm as fuel and

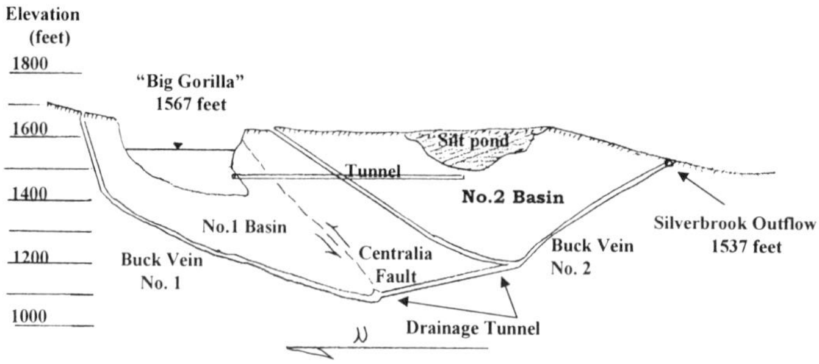

Figure 1. Cross-section of the site showing the acid mine pool, the deep mine connection and the outflow from the Silverbrook basin.

secondarily providing excess energy for the production of hothouse plants. Historically, the location has been actively strip mined and deep mined since the 1800s. At one point, the strip mined pit known as the Big Gorilla was filled with coal silt that was later reclaimed for fuel use. A cross-section of the deep mines, the infilled Big Gorilla, and the interconnection between the mines is shown in Figure 1. Due to extensive mining in the area, acid mine drainage is prevalent throughout the basin.

Before the Big Gorilla Project was begun, NEPCO had been distributing its fly and bottom ash onsite to reclaim previously very acidic soils. The ash was placed in 6 foot layers in an attempt to approximate the pre-mining grade. These areas were seeded with vetch, native grasses, hardwoods, and conifers. Wild game began to populate reclaimed areas. However, land available for onsite reclamation became more scarce, and the acidic waters of the Big Gorilla remained both a hazard and an eyesore. The placement of dry ash in the Big Gorilla signaled another beneficial use of fly and bottom ash that might otherwise have been destined for a landfill.

3 COAL GEOLOGY

This site is in the Eastern Middle field of the Pennsylvania anthracite region. The major coal beds in this region, including the Buck Mountain vein and the Mammoth vein, are all contained in the Llewellyn Formation (Inners, 1988). The Llewellyn Formation consists of alluvial plain sediments and contains, in addition to the anthracite, interbedded, dark-gray, carbonaceous sandstones, conglomerates, siltstones, claystones, and shale, in fining-upward cycles (Inners, 1988). The Llewellyn also contains an abundance of pyrite and siderite throughout the sandstones, within the coal beds, and as concretions in the claystones and siltstones (Inners, 1988).

4 BASIN HYDROLOGY

The Silverbrook Basin is approximately 1 mile wide at its broadest point and approximately 5 miles long (A.W. Martin, 1973). The basin is a syncline with the Centralia thrust fault breaking the surface as the northern face of the Big Gorilla. Whereas the Big Gorilla is the result of surficial mining of the Mammoth vein, the fault allowed the Buck Mountain vein to be mined at two locations beneath the Big Gorilla. Figure 1 shows the two portions of the Buck Mountain vein connected by the Gordon Slope, as well as the Number 1 Tunnel connecting the Big Gorilla to the southern Buck Mountain vein.

Similar in a way to conduit flow in karst, the abandoned mine workings direct water preferentially through the subsurface. This complicates the definition of the water table. It is estimated that 300 million gallons of water are present in the subsurface mines and approximately 135 million gallons of water are in the Big Gorilla. The Big Gorilla is approximately 400 feet wide, 1400 feet long, and is up to 85 feet deep when the water level is at its normal elevation of 1567 feet above sea level.

The outflow for the basin occurs at 1537 feet above sea level, where the southern Buck Mountain vein intersects the land surface. This location, the Silverbrook outflow, forms the headwaters for the Little Schuylkill River. The river receives approximately 16,400 lbs. of acid daily from various sources in the Little Schuylkill River Basin (A.W. Martin, 1973). It supports no fish and supplies no domestic or industrial water (A.W. Martin, 1973). Although the report citing these facts is over twenty-five years old, it is thought to be accurate in that conditions in the area have changed little.

A primary question in this study is the influence of ash deposition in the Big Gorilla on the chemistry of the Silverbrook basin, which directly affects the Little Schuylkill River. The connection of the Mammoth and Buck Mountain veins is established, yet the permeability of the silt filling the tunnels and slopes is unknown. The extent to which the Big Gorilla is draining laterally into the mine workings is also yet to be determined. Ash deposition into the Big Gorilla has taken place since August of 1997, but chemical change in the Silverbrook outfall has only been detected in the most recent round of sampling.

5 GEOCHEMICAL ANALYSES

Both the ash and the water have been analyzed in an attempt to predict long term effects of ash deposition in the Big Gorilla. The ash has been characterized by x-ray diffraction, TGA and wet chemical analyses. The pre- and post-deposition chemistry of the Big Gorilla and the Silverbrook outflow has been determined based on the requirements of Module 25, which is the chemical criteria for ash placement in mine reclamation, as specified by the Pennsylvania Department of Environmental Resources (PA DER, 1993).

5.1 *Ash Chemistry*

The mineralogy of the fly ash has been characterized by bulk qualitative and quantitative x-ray diffraction, thermal gravimetric analysis and wet chemical analyses. These analytical methods were chosen to complement the weakness of each of the respective approaches as a means of getting at the true quantitative proportions of the mineral phases. These results are summarized in Table 1.

Table 1. Mineralogical Analysis of Culm-derived Fly Ash Used in the "Big Gorilla" Reclamation Project

Mineral	*Formula*	*Analytical method*	*Wt. %*
Quartz	SiO_2	QXRD, wet chem	20.0
Mullite	$Al_6Si_2O_{13}$	QXRD	7.0
Hematite	Fe_2O_3	QXRD	2.0
Portlandite	$Ca(OH)_2$	DTA, QXRD, wet chem	6.5
Calcite	$CaCO_3$	QXRD, wet chem	0.4
Gypsum	$CaSO_4 2H_2O$	QXRD	0.5
Soluble silicate	Meta-clays	QXRD	63.6
Total			100

Using these proportions and nominal mineralogical compositions for the phases identified, the bulk chemical composition of the fly ash was calculated (Table.2). This calculated analysis is compared to the average from five analyses of culm-derived ash drawn from the geographic area of the Big Gorilla. As can be seen, the two analyses are very close suggesting that our identification of the constituents is reasonably good.

The chemical make-up of the ash is thought to be relatively constant, because approximately the same amount of lime is added to the boiler each day, and the culm is brought from either on-site or locally. The temperature of the boiler is consistent at 800ºC. Thus, similar minerals should form every day.

Table 2. Comparison of Calculated Bulk Chemical Analysis vs. Average Composition of Five Culm-derived Ash Analyses

Oxide	*Calculated ash composition [wt %]*	**Average composition of culm-derived ash analyses***
SiO_2	64.47	60.69
Al_2O_3	18.82	23.11
Fe_2O_3	5.74	5.08
CaO	5.45	5.44
MgO	1.48	1.15
K_2O	3.77	2.90
SO_3	0.26	1.60

* n = 5

5.2 *Pre-ash Placement Water Chemistry*

The Big Gorilla was sampled twice in 1993 for the parameters included in the Module 25. It mimicked the Silverbrook in pH, acidity, sulfate, zinc, and aluminum concentrations of 3.2, 38.2, 55 mg/L, 0.22 mg/L, and 3.5 mg/L, respectively. The acid mine pool was initially somewhat chemically stratified. For example, calcium concentrations ranged from 3.95 mg/L at the surface to 5.25 mg/L at depth. The stratification has been lost since the beginning of ash injection, possibly due to the mixing caused by the ash placement into the standing water. The fine-grained ash appears to flow down the 90 foot slope to the bottom of the mined pit in a mannor similar to turbidity currents that occur in the natural environment in present dav delta formation. This slurry drags the overlying water in contact with it to the bottom. The data from surface water samples of the Big Gorilla in July 1993 are shown in Table 3.

The Silverbrook has been monitored monthly since 1989 by the Pennsylvania Department of Environmental Protection (PA DEP). Measurements of pH appear cyclical on a variable time scale, with measurements ranging from 3.2 to 4.9. Alkalinity was always found to be below detection, and aluminum ranged from 0.95 mg/L to 12 mg/L.

Three monitoring wells are sampled on a quarterly basis, also in accordance with Mod 25. The wells do not appear to be directly connected to the Big Gorilla, the Silverbrook, or each other. This conclusion is based on consistently higher pH values than any of the other sampling locations (well 2) and much higher aluminum concentrations (well 9). Well 3 was constructed only a year before ash deposition in the Big Gorilla. Although it is less than 150 feet away from the Big Gorilla, its chemistry is much closer to that of the Silverbrook, however, it has been so variable that it is difficult to state with conviction that the two are related.

5.3 *Post-ash Placement Water Chemistry*

Since ash deposition began in August 1997, the chemistry of the Big Gorilla has been highly variable, in contrast to that of the Silverbrook. The pH in the Big Gorilla has ranged from 6.2 to 11.6. Sulfate values are between 180 and 369 mg/L, and alkalinity values have ranged from 7 to 244 mg/L as $CaCO_3$. Concentrations of dissolved aluminum have ranged from 0.131 to 1.35 mg/L. These order of magnitude changes in the chemical analyses indicate a system in transition. Surprisingly, the Big Gorilla is more homogeneous than before the ash placement.

Current measurements from the Big Gorilla's water are found in Table 3.

The monitoring wells have shown little change due to ash deposition, and until recently, the same was true for the Silverbrook outflow. Analyses from the last sampling event show an alkalinity value above 1 mg/L of $CaCO_3$ for the first time ever. This is interpreted as preliminary evidence that the Silverbrook outfall and the Big Gorilla are hydrologically connected. As mentioned previously, the question of transport through the deeper mine workings is as of yet unsolved.

Based on both the primary and secondary drinking water standards, there was no health risk associated with the pre-ash water of the Big Gorilla, and there is no health risk associated with the post-ash water of the Big Gorilla and Silverbrook outflow, other than their high alkalinities and acidities.

6 GEOCHEMICAL INTERPRETAION AND MODELING

6.1 pH Trends

The variability of the chemistry in the Big Gorilla is of interest in determining the long term impact of ash placement in the acid mine pool. The most direct controls appear to be the timing and quantity of ash deposition. In Figure 2, the mass of ash deposited is compared with the pH of the water. There is a hiatus in placement during the winter of 1997-1998. This is because the weather made conditions unsafe for both the trucks that dump the ash and the bulldozers that push it into the pit. The most interesting aspect of this time is the drop in pH (to 6.2) found on the March sampling date. After the deposition resumed in April, the pH rose steadily from 9.6 in June to 11.6 in August 1998. Thus, the timing and extent of ash placement seem to be the greatest controls in the geochemistry. This is thought to be due to the CaO content of the ash.

It is possible to use the estimated volume of the Big Gorilla pool and the added tonnage of ash to rationalize the observed chemistry of the water. The volume of 1.35×10^8 gallons is 5.11×10^8 liters. 110,000 tons of ash containing 5.45% chemically active CaO (Table 2) calculates to 9.75×10^7 moles of CaO (or $Ca(OH)_2$ after reaction with water). The measured pH of the mine pool is taken as 11.4 (Table 3). This pH, assumed 2.51×10^{-3} moles/liter of OH^-. Providing this alkalinity for the entire mine pool requires 6.42×10^5moles of CaO which is only 0.66% of the CaO available in the ash. For comparison, consider the Ca^{2+} in solution. Before treatment, the pool contained only 3.95 mg/L of Ca^{2+} (Table 3). After treatment, the calcium

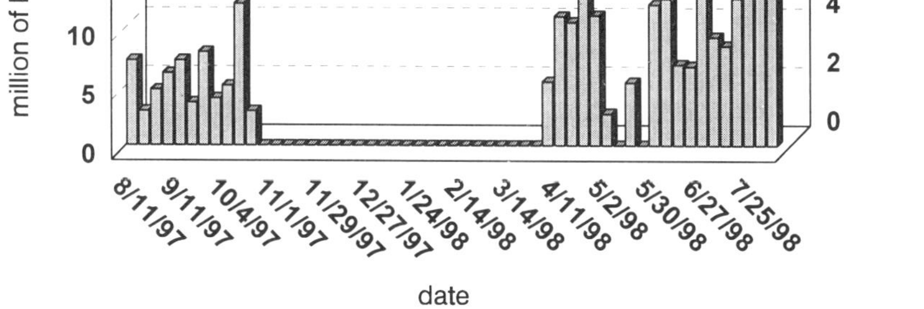

Figure 2. Relationship between ash deposition and pH

concentration represents 2.48 x 10^6 moles of CaO in the entire mine pool which is only 2.54% of the available CaO.

6.2 PHREEQC Modeling

PHREEQC is the updated version of the U.S. Geological Survey's computer program for speciation, reaction-path, advective-transport, and inverse geochemical calculations (Parkhurst, 1995). It was applied to speciating the initial solutions, then performing reaction calculations to determine chemical equilibrium. This was done for three waters: a) the Big Gorilla before ash deposition (July 1993), b) the Big Gorilla after ash deposition (August 1998), and c) the Silverbrook outfall (August 1998). The data for entry to PHREEQC was taken from PA DEP analyses (Table 3). The charge balance for each simulation was within limits of error in the analyses. It is thought that all the elements of interest have been included.

The PHREEQC program uses the observed concentrations of various dissolved species to calculate saturation indices for all minerals that contain those species. The waters are in equilibrium with a few minerals, supersatureated with respect to a few others and highly undersaturated with respect to still others. The saturation index is a log function, so the range of saturation for the entire suite of possible minerals spans tens of orders of magnitude. The degree of saturation, as expressed by the saturation index, does provide a useful way of

Table 3. Analytical data for pre-ash placement in the Big Gorilla and current data for both the Big Gorilla and the Silverbrook outflow.

Analyte	*Big Gorilla 1993*	*Big Gorilla 1998*	*Silverbrook 1998*
pH	3.2	11.4	4.3
Conductivity (umhos/cm)	210	1273	443
Alkalinity (mg/L $CaCO_3$)	<1	172	6.2
Acidity (mg/L)	37.8	<1	78
Turbidity (NTU)	0.24	1.6	<1
Dissolved solids (mg/L)	98	311	268
Chloride (mg/L)	6	3	9
Sulfate (mg/L)	59	369	193
Aluminum (mg/L)	4.2	1.32	6.81
Arsenic (mg/L)	<0.001	<0.004	<0.004
Barium (mg/L)	<0.1	0.0013	0.028
Cadmium (mg/L)	0.0007	<0.01	<0.01
Calcium (mg/L)	3.95	195	27
Chromium (mg/L)	<0.001	0.051	<0.05
Copper (mg/L)	0.02	0.012	0.023
Iron (mg/L)	0.4	<0.02	17.7
Lead (mg/L)	0.001	<0.001	<0.001
Magnesium (mg/L)	4.5	0.193	9.6
Manganese (mg/L)	0.72	<0.01	1.64
Mercury (mg/L)	<0.0002	0.0014	<0.001
Nickel (mg/L)	<0.04	<0.05	0.091
Nitrogen (mg/L)	<0.5	0.09	<0.04
Potassium (mg/L)	1	10.4	1.84
Selenium (mg/L)	<0.002	<0.007	<0.007
Silica (mg/L)	na	19.11	15.96
Sodium (mg/L)	1.15	9.87	9.85
Zinc (mg/L)	0.2	<0.01	0.295

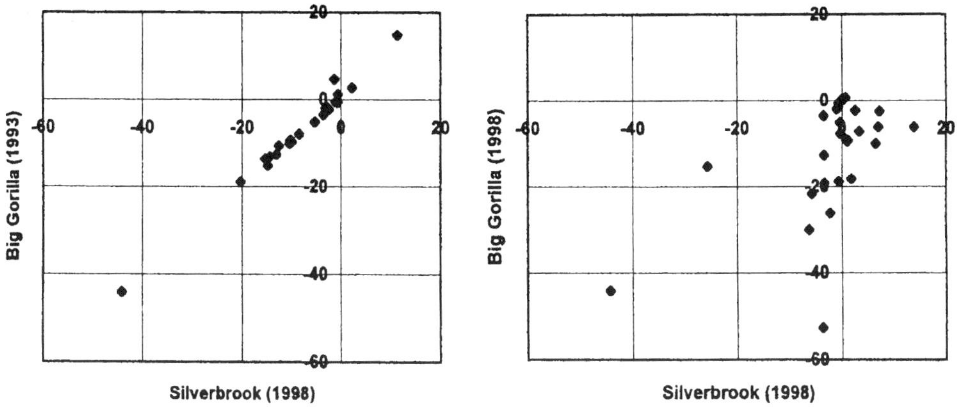

(A) (B)

Figure 3. Plots of saturation indices for 18 different mineral species calculated by PHREEQC for A) the pre-ash placement waters of the Big Gorilla vs. the Silverbrook outflow and B) the post-ash placement waters of the Big Gorilla vs. the Silverbrook outflow .

comparing the water in the Big Gorilla with the water in the Silverbrook outflow. If the saturation indices of the two waters are equal for all minerals, the water can be said to be chemically equivalent.

As can be seen in Figure 3(A), the pre-ash placement water in the Big Gorilla was chemically equivalent to the Silverbrook outflow (the Silverbrook has not significantly changed in chemistry since the beginning of measurements in 1989). As mentioned previously, the ash deposition in the Big Gorilla has altered the chemistry dramatically. That this water is no longer chemically equivalent with the Silverbrook outflow water is evident in Figure 3(B). A key question in this ongoing study concerns the time it will take for these two waters to re-establish equivalence, and what species will be present at that point.

6.3 Heavy metals

Even in this early stage of treatment, it is apparent that raising the alkalinity of the mine pool is suppressing the concentrations of heavy metals. Those that appeared in significant concentrations in the untreated pool and in the Silverbrook outflow, Al, Fe, Mn and Zn, have been reduced: Al by a factor of 4, and the others taken below limits of detection. Mostly, the reductions can be accounted for simply by the reduced solubility in alkaline water. The solubility of aluminum reaches a minimum in the near-neutral regime and rises again at higher pH.

7 SUMMARY AND CONCLUSIONS

Addition of fly ash to the acid mine pool has resulted in the following effects:

(1) A rapid rise in pH from pre-treatment values in thre range of 3.2 to post-treatment values in the range of 11.4. The pH rise was rapid and engulfed the entire pool.
(2) The increased pH and increased calcium concentration in the pool can be accounted for with only a few percent of the active CaO available in the fly ash. A large reserve of latent alkalinity remains sequestered in the fly ash.
(3) Saturation indices for various minerals as calculated by the PHREEQC computer program show that the pre-treatment water in the Big Gorilla mine pool and the water in the

Silverbrook outflow were chemically equivalent. Fly ash treatment has upset this equivalence.

(4) Heavy metal concentrations are significantly reduced by the increased alkalinity in the mine pool.

8 REFERENCES

A.W. Martin Associates, Inc. 1973. Little Schuylkill River Basin Mine Drainage Study, Project WCE SL-167. Prepared for the Pennsylvania Department of Environmental Resources.

Inners, J.D. 1988. The Eastern Middle Anthracite Field, *Proceedings of the 53rd Annual Field Conference of Pennsylvania Geologists*. October 1988:32-39. PA Geologic Survey.

PA Department of Environmental Resources 1993. Module 25A - Ash Generation and Ash Quality Information. ER-MR-311. Bureau of Mining and Reclamation, Harrisburg, PA.

Parkhurst, D.L. 1995. User's Guide to PHREEQC. WRI Report 95-4227, U.S. Geological Survey, Lakewood, CO.

9 ACKNOWLEDGMENTS

The authors would like to extend their gratitude to Northeastern Power Company for their financial support and especially to Mr. James Wetzel for his technical support and enthusiasm for this project.

Tailings and Mine Waste '99 © 1999 Balkema, Rotterdam, ISBN 90 5809 025 6

Soil pyrite determination using indirect Fe-based method with ICP-AES

L.C. Dharmasri & W.H. Hudnall
Louisiana State University Agricultural Center, Louisiana Agricultural Experiment Station, Agronomy Department, Baton Rouge, La., USA

ABSTRACT: Pyrite determination techniques are time consuming and complicated. They indirectly determine pyrite content either based on pyritic-Fe or pyritic-S. Some techniques need calibration for the soil type either because of the digestion method or they are based on empirical relationships. The Fe-based estimation involves separation of pyritic and non-pyritic Fe. Schneider and Schneider (1990) suggested a method that indirectly estimates pyrite based upon pyritic-Fe. We modified their method to be used with Inductively Coupled Plasma-Atomic Emission Spectroscopy (ICP-AES) to quickly estimate pyrite content in soil samples. The methodology is presented in detail with pyrite data for a coastal organic soil in Louisiana, USA and an acid sulfate soil in Matara, Sri Lanka. Using digestion tubes and a digestion block, 45 samples can be analyzed in two days. This method also enables one to determine more than 40 elements simultaneously from both digestions by using the ICP-AES.

1 INTRODUCTION

A suitable method is required for pyrite determination to assess pyrite accumulations. A few different techniques are available for pyrite determination. The qualitative determination methods include X-ray diffraction and Scanning Electron Microscopy. The X-ray diffraction technique can also be used as a semi-quantitative method to estimate pyrite content. The available quantitative determination methods indirectly estimate pyrite content either from pyritic Fe or pyritic S. Willett and Beech (1987) proposed an empirical relationship to estimate pyrite content based on carbon estimates for acid sulfate soils. Iron based sequential extraction and differential measurement techniques are used to estimate pyrite. Sulfur based techniques fractionate S into organic and inorganic S. The inorganic fraction is separated into four groups as HCl-soluble S (mainly sulfate and H_2S), acid volatile S (AVS) (monosulfides and dissolved sulfides), elemental-S and pyritic-S (Begheijn et al., 1978; Krairapanond et al., 1991a, 1991b, and 1992;).

The Fe based estimation involves separation of pyritic and non-pyritic Fe. This is directly related to the estimation of the rate of pyritization (Leventhal and Taylor, 1990). More detailed Fe based schemes are used in acid sulfate soil analysis, where pyrite, as well as, jarosite fractions are estimated separately (Lord, 1982; Raiswell et al., 1994). Considering the need to estimate pyritic material, a quick and easy method for separating pyritic and non- pyritic Fe is required. The S-based techniques may be complicated because of the presence of S in many different forms, especially since gaseous forms of S may be easily lost. Lord (1982) suggested a selective extraction method for pyrite determination. The $HF-H_3BO_3$ solution used for silicate dissolution dissolved pyrite and required a correction factor to adjust for pyrite content. The indirect estimate of pyrite based upon the separation of pyritic and non-pyritic Fe produced promising results (Schneider and Schneider, 1990). This method is based on selective extraction of total non-pyritic and pyritic-Fe in 2N HNO_3 acid and the extraction of non-pyritic Fe in 5N HCl acid, followed by the differential calculation of pyritic-Fe and pyrite content. Komadel (1990) suggested the use of atomic absorption spectroscopy to avoid several steps needed for the spectrophotometric method or complexometric titration for Fe determination suggested by Schneider and Schneider (1990a and 1990b). More than 40 elements can be determined simultaneously by using ICP-AES. Our objective was to modify this method to analyze a larger number of samples quickly using ICP-AES for soil pyrite determination.

2 MATERIALS AND METHODS

2.1 *Sample Description and Preparation*

Organic soil samples were collected from a saline marsh in the Barataria Bay marsh in Louisiana, USA and mineral soils were collected from an acid sulfate soil in Matara, Sri Lanka. Samples from different horizons of the soil profiles were used in this study. Soil horizons were designated according to Keys to soil taxonomy (Soil Survey Staff, 1996). We described the organic soil profiles. For the acid sulfate soils, profile descriptions given by Deturck (1994) were used. Soils were stored in a cold room until analysis and analyzed as soon as possible in the wet state to avoid pyrite oxidation.

2.2 *Analytical Methods*

The method used for pyrite estimation in coal and ceramics by Schneider and Schneider (1990) was modified for the analysis of extractions for Fe by the ICP-AES. The recipe was adjusted based on the bulk density of the organic soils. Six scoops (1/8 tsp = 0.6 mL) from each soil sample were measured into six digestion tubes that were calibrated to 75-mL volume. Tubes were separated into two batches of three sub-samples to represent three replicates for each sample; one was for non-pyritic (NP) and the other set was for pyritic and non-pyritic (NP+P) Fe digestion. The soil sample was placed at the bottom of the tube. Six mL of concentrated HCl were added to each tube with soil under a fume hood, and the tube was covered with a small glass (“reflux”) funnel. The soil was stirred within the acid to make a suspension and left overnight, covered with "reflux" funnels. The next day, 23 mL of deionized (DI) water were added into each tube. The digestion procedure was different for NP and NP+P sets.

2.2.1 *Digestion for Non-pyritic Fe*

To digest the NP set a batch of 45 tubes with soil and acid mixture were placed at the same time onto the digestion block, and the block heater was turned on. It took approximately 15 to 20 min to start boiling. The samples were boiled for 1 h and cooled. The tubes were filled to the 75-mL level with 1:10 HCl, then mixed well and filtered through #42 Whatman filter paper. The filtrate was highly acidic; therefore, it was diluted to reduce the acidity and for better detection by the ICP-AES.

2.2.2 *Digestion for Non-pyritic and Pyritic-Fe*

For the NP+P set, after adding 23 mL of DI water, the batch of 45 tubes with soil and acid mixture were boiled for 30 min, and then 25 mL of 3N HNO_3 were added to each tube while they were on the digestion block. Acid was slowly added using a funnel. It took another 10 to 15 min for the mixture to start boiling. Samples were boiled for another 30 min and cooled. The tubes were filled to the 75-mL level with 2N HNO_3, mixed, and filtered through #42 Whatman filter paper. This filtrate was also diluted before analysis by ICP-AES.

2.3 *Estimation of Pyrite Content*

The pyritic-Fe content was determined based on the difference between Fe content of NP+P and NP digestions. The pyrite (FeS_2) content was estimated from pyritic-Fe content based on the molar ratios given in the following conversion equation using the mean of three sub samples.

$$\text{Pyritic-Fe} * 2.14828 = FeS_2$$

Because of the organic soils used in the study, soil volume was used and pyrite content is expressed on volume basis. Based on soil bulk density, pyrite content can be expressed on weight basis or as a percentage for the organic soils. However, for the mineral soils weighed samples can be used for analysis as in the method described by Schneider and Schneider (1990).

3 RESULTS

3.1 *Pyrite Estimates for Coastal Organic Soils*

Pyrite contents for the organic soil horizons are presented on volume basis because of the low bulk density of these soils. Bulk density varied from 0.20 to 0.31 Mg M^{-3} for these soils. Table 1 presents the pyrite estimates for streamside of the saline marsh. Values represent averages for three sampling sites within the streamside. High coefficient of variation (CV) may represent spatial variability of pyrite accumulation in the field or may be due to non-homogeneous samples used for analysis. Our samples were not well mixed as Schneider and Schneider (1990) proposed for their samples, because we wanted to avoid pyrite oxidation due to drying.

Table 1. Pyrite content for different soil horizons at streamside of the saline marsh.

Soil Horizon	Mean Pyrite Content (mg L^{-1})	Coefficient of Variation (%)
Oa1	1518.3	65.4
Oa2	381.7	24.0
Oa3	702.0	15.8
Oa4	1660.3	41.1

3.2 *Pyrite Estimates for Acid Sulfate Soils*

Table 2 presents pyrite estimates for mineral acid sulfate soil horizons. The bulk density for these soils ranged from 0.678 to 1.307 Mg M^{-3}. High CV is also observed with these samples and the source of variation may be similar to that of the organic soils.

Table 2. Pyrite content for different soil horizons of the acid sulfate soil.

Soil Horizon	Mean Pyrite Content		Coefficient of Variation
	(g L^{-1})	(g Kg^{-1})	(%)
Ap	24.571	36.240	24.6
AB	27.207	22.408	37.6
Bj1	167.721	128.301	25.0
Ap*	38. 374	44.457	27.2

* from a different location.

Our pyrite estimates ranged from 3.62% to 12.83% for the acid sulfate soils in Matara, Sri Lanka and seem comparable to the earlier reports. Pyrite was found in the subsurface layer (i.e.,

Bj1 horizon) of these soils in concentrations ranging from 1 to 8% by Deturck (1994) using the method of Begheijn et al. (1978).

4 CONCLUSIONS

When an ICP-AES is available, the method proposed by Schneider and Schneider (1990) can be used to quickly estimate pyrite content in soil samples. Final steps involving preparation for spectrophotometric method or complexometric titration for Fe determination can be skipped. Using digestion tubes with the digestion block, 45 samples can be analyzed in two days. The use of homogeneous samples may help to obtain reproducible results (small CV) but pyrite can be oxidized during sample preparation (drying, grinding and sieving) underestimating the pyrite content. This method enables one to directly estimate the pyritization rate, as well as, to determine more than 40 elements simultaneously from both digestions by using ICP-AES.

5 ACKNOWLEDGMENTS

This research was supported by NASA under cooperative agreement NCCW-0059 and the Louisiana Board of Regents/LEQSF under the contract NASA/LEQSF (1994-97)-IMP-03. We thank Sri Lanka Department of Agriculture and the staff at the Agriculture Research Station, Thelijjawila, Matara, Sri Lanka.

6 REFERENCES

Begheijn, L. Th., N. van Breemen, and E. J. Velthorst. 1978. Analysis of sulfur compounds in acid sulfate soils and other recent marine soils. *Commun. Soil Sci. Plant Anal.* 9(9):873-882.

Deturck, P. 1994. *Iron toxicity to rainfed lowland rice in Sri Lanka.* Dissertation De Agricultura, Katholleke Universiteit te Leuven, Belgium. 162p.

Komadel, P. 1990. Comment on indirect method for the determination of pyrite in clays and shales after selective extraction with acid solutions. *Ceramic Bulletin* 69(10):1711.

Krairapanond, N., R. D. DeLaune, and W. H. Patrick, Jr. 1991a. Sulfur dynamics in Louisiana coastal freshwater marsh soils. *Soil Sci.* 151(4):261-273.

Krairapanond, N., R. D. DeLaune, and W. H. Patrick, Jr. 1991b. Seasonal distribution of sulfur fractions in Louisiana salt marsh soils. *Estuaries* 14:17-28.

Krairapanond, N., R. D. DeLaune, and W. H. Patrick, Jr. 1992. Distribution of organic and reduced sulfur forms in marsh soils of coastal Louisiana. *Org. Geochem.* 18:489-500.

Leventhal, J., and C. Taylor. 1990. Comparison of methods to determine degree of pyritization. *Geochimica et Cosmochimica Acta* 54:2621-2625.

Lord III, C. J. 1982. A selective and precise method for pyrite determination in sedimentary materials. *J. Sediment. Petrol.* 52:664-666.

Raiswell, R., and D. E. Canfield, and R. A. Berner. 1994. A comparison of iron extraction for determination of degree of pyritisation and the recognition of iron-limited pyrite formation. *Chemical Geology* 111:101-110.

Schneider, J. W., and K. Schneider. 1990a. Indirect method for the determination of pyrite in clays and shales after selective extraction with acid solutions. *Ceramic Bulletin* 69(1):107-109.

Schneider, J. W., and K. Schneider. 1990b. Reply to comment on indirect method for the determination of pyrite in clays and shales after selective extraction with acid solutions. *Ceramic Bulletin* 69(10):1711.

Soil Survey Staff. 1996. *Keys to Soil Taxonomy.* Seventh edition. USDA-NRCS, Washington, D.C.

Willett, I. R., and T. A. Beech. 1987. Determination of organic carbon in pyritic and acid sulfate soils. *Commun. in Soil Sci. Plant Anal.* 18(7):715-724.

Tailings and Mine Waste '99 © 1999 Balkema, Rotterdam, ISBN 90 5809 025 6

The use of sequential extractions for the chemical speciation of mine wastes

R.W. Leinz, S.J. Sutley & P.H. Briggs
US Geological Survey, Denver, Colo., USA

ABSTRACT: Sequential extractions are applied to the chemical speciation of samples of mine waste composited from waste piles at two inactive mine sites. The results show that metals are more favorably partitioned into the sulfide phase in one composite and into the crystalline Fe-oxide phase of the other composite. Normalizing the results of the extractions of five individual size-fractions by the weight percent of the fraction in the composite, shows that the distribution of metals in the extracted phases is independent of particle size, but the amount of each phase present in the fraction may vary. Differences between total extractable metals determined by total digestion of a separate split of the composites and by summation of the results of the extractions is attributed to pulverizing the sample. Dissolution of jarosite, important because of its common association with pyritic mine wastes, in the sequential extraction steps is still uncertain but will determined in the near future.

1 INTRODUCTION

Phase-specific partial extractions, used for some time to chemically characterize soils and sediments, have been advocated by geochemists for use in mineral exploration (Chao and Theobald, 1976; Chao, 1984) on the premise that metals migrating from a buried ore deposit to the earth's surface are partitioned into mineral phases, the analysis of which can help locate the deposit. Tessier and others (1979) succinctly state that sequential extractions provide important information about the "origin, mode of occurrence, biological, and physicochemical availability, mobilization, and transport of trace metals." Griffioen and Broers (1993) argue for the use of sequential extractions for studying the fate of pollutants in the subsurface. From these and other reports in the literature it can be surmised that sequential extractions can provide information on the partitioning, or speciation, of metals in various mineral phases in mine wastes, which bears directly on assessing the availability of the metals to the environment.

This paper presents the results of sequential extractions of composite samples collected from waste dumps at two inactive polymetallic mines in Colorado, one located in the Animas basin near Silverton (composite 1) and the other located in the Leadville district (composite 2) (Figure 1). Extractions were conducted on five size-fractions derived from the < 2mm fraction using an extraction scheme based largely on the scheme of Griffioen and Broers (1993). The applied scheme is designed for the sequential release into solution of metals residing in the following, operationally defined phases: (1) water-soluble phase, (2) Mn and amorphous Fe oxides, (3) crystalline Fe oxides, (4) silicates, and (5) sulfides.

2 METHODS

At each site (Figure 1), the waste dumps were divided into 30 cells of about equal size and a minimum of thirty, 50-100g samples were randomly collected from each cell. Individual samples were collected from the surface of the piles to a depth of about 5 to 10 cm. The samples were combined and allowed to air. The dry composites were placed into a v-blender and tumbled for five minutes to break up aggregate material, after which they were sieved through a 2mm stainless steel screen. A split of the <2mm material of each composite was further reduced into five size-fractions (Table 1).

The five size-fractions of each composite were sequentially extracted by a scheme including steps for dissolving the crystalline Fe oxide phase, the silicate phase, and the sulfide phase taken from the scheme recommended in Griffioen and Broers (1993). The step for the dissolution of the Mn and amorphous Fe oxide phase was taken from Chao (1984). A sub-sample of the <2mm bulk material, pulverized to 0.074mm, was analyzed for multiple elements by inductively-coupled plasma atomic emission spectrometry (ICP-AES) and for mineralogy by X-ray diffraction (XRD). The applied extraction scheme is described as follows. The weights and extraction volumes for each step are given in Table 2.

Step 1, water-soluble metals. The sample was extracted with deionized water in a reciprocating shaker for 2 hours. Separation of residue from the solution for this and the subsequent steps was by centrifugation and (or) filtration of the supernatant fluid through a 0.45μ membrane filter.

Step 2, metals associated with Mn and amorphous Fe oxides. The residue from step 1 was extracted with 0.25M $NH_2OH \cdot HCl$ in 0.25M HCl for 30 minutes in a water bath at 50°C.

Step 3, metals associated with crystalline iron oxides. The residue from step 2 was extracted with a combined solution of 0.2M ammonium acetate and 0.2M oxalic acid in 0.1M ascorbic acid in a water bath at 94°C for 2 hours.

Step 4, metals residing in the silicate fraction. The residue from step 3 was shaken with 10M HF for 8 hours after which 5 g of H_3BO_3 was added per 30mL of HFand the mixture was shaken for another 8 hours.

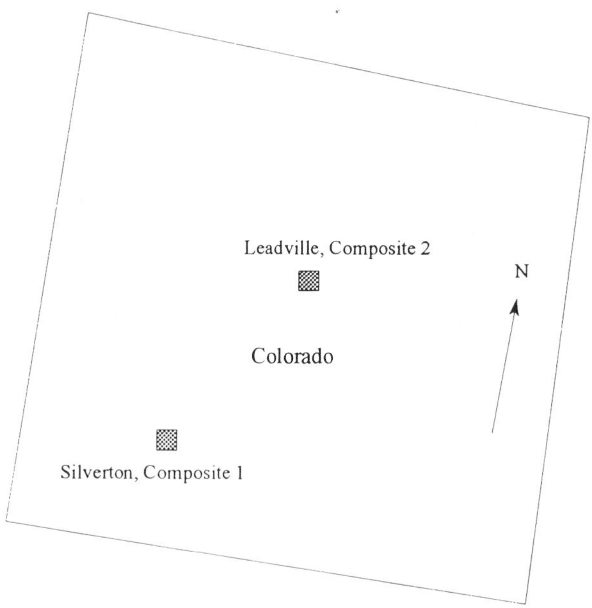

Figure 1. Locations where composites 1 and 2 were collected from waste piles at inactive mines.

Table 1. Range of particle sizes in each of 5 fractions, obtained from the <2mm fraction of composites 1 and 2, used in sequential extraction.

Fraction	Range of particle size	Weight % in composite	
		Composite 1	Composite 2
1	<2mm to >0.425mm	55.40	55.08
2	<0.425mm to >0.125mm	27.31	25.74
3	<0,125mm to >0.075mm	8.95	12.42
4	<0.075mm to >0.043mm	7.57	6.44
5	<0.043mm	0.77	0.33

Table 2. Sample weights and extraction volumes for the 5-step sequential extraction of mine waste composites 1 and 2.

	Fraction 1, weight 17 g	Fractions 2-5, weight 1.0 g
Step	Volume (mL)	Volume (mL)
1	340	20
2	340	20
3	340	40
4	225	30
5	10	40

Table 3. Analyses of composites 1 and 2 for mineralogy by XRD and total concentrations of Fe, Pb, Zn, and Cu by ICP-AES following total digestion (I) , and by summation of their concentrations in each phase in each size fraction (II), determined by AAS. XRD and ICP-AES analyses were on a pulverized split of the <2mm bulk material.

XRD analysis (%)			ICP-AES (I) and AAS (II) analyses (μg/g)		
Component	Composite 1	Composite 2	Element	Composite 1	Composite 2
Quartz	28	17	Fe I	96000	67000
Mica	15	22	Fe II	56000	43000
Jarosite	7	5	Pb I	2300	1800
K-spar	<1	8	Pb II	1800	1600
Kaolinite	----	6	Zn I	1600	220
Clinochlor	----	1	Zn II	2600	300
Pyrite	6	1	Cu I	1100	270
Amorphous	40	40	Cu II	1200	300

Step 5, metals associated with sulfide phases. The residue from step 4 was shaken for 2 hours with concentrated HNO_3, the solution was diluted with 1.5 volumes of deionized water.

An aliquot of the water extract was immediately acidified with HNO_3 and the acidified extract, and the extracts from the other steps, were analyzed for metal content by atomic absorption spectrophotometry (AAS).

3 RESULTS AND DISCUSSION

Table 3 shows the total concentration of Fe, Pb, Zn and Cu in the composites determined in the pulverized split of the <2mm fraction following total digestion (with $HF/HNO_3/HClO_4$). Figures 2 and 3 depict the concentrations of these metals in each phase of the composites as the sum of their concentration in the five size-fractions, adjusted for the amount of each fraction in the composite. Summation of the concentration of metals in each phase, thus determined, is presented as another form of total concentration in Table 3. Figure 2 shows the highest concentration of Fe to be in the sulfide phase of composite 1 whereas Fe predominates in the crystalline Fe-oxide phase of composite 2

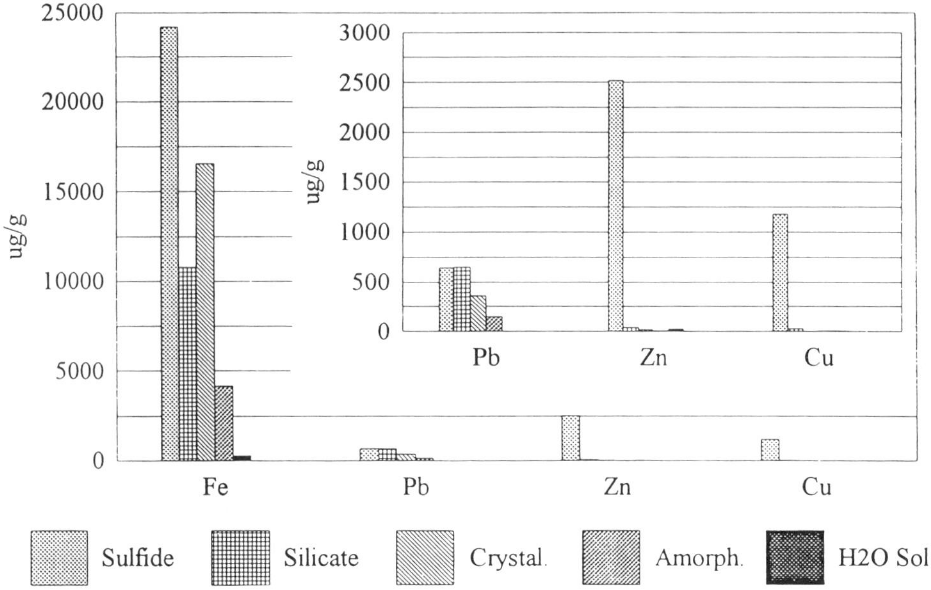

Figure 2. The total concentration of metals in each of 5 phases of composite 1 represented as the sum of their concentrations in the 5 size-fractions.

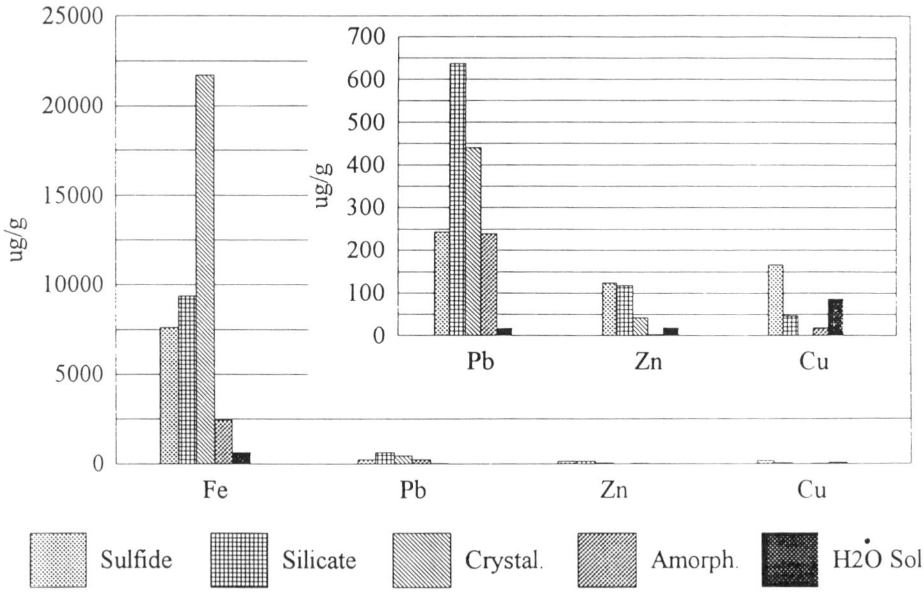

Figure 3. The total concentration of metals in each of 5 phases of composite 2 represented as the sum of their concentrations in the 5 size-fractions.

(Figure 3). Zinc and Cu occur mostly in the sulfide phase of composite 1, and Pb in the silicate and amorphous Fe-oxide phases (Figure 2). Composite 2 contains more Pb than Zn and Cu, and the Pb is seen in Figure 3 to occur mostly in the silicate and the crystalline phases.

The predominance of Fe, Zn and Cu in the sulfide phase of composite 1 compared to composite 2 agrees with the results of mineralogical analysis by XRD (Table 3) which shows composite 1 to contain 6% by weight pyrite compared to 1% pyrite in composite 2. The results are an indication of more extensive oxidation of pyrite at the Leadville site than at the Silverton site. Table 3 shows that of the quantified minerals present, jarosite is a major iron-bearing mineral in the composites, with mostly hydronium jarosite [$(H_3O^+)Fe_3(SO_4)_2(OH)_6$] occurring in composite 1 and the potassium end member [$KFe_3(SO_4)_2(OH)_6$] predominating in composite 2. The extent of dissolution of jarosite in the steps of the sequential extraction is uncertain, but , because it is important for the complete and accurate speciation of the composites, will be evaluated in the near future. Because of its common association with pyritic mine wastes, understanding the behavior of jarosite is of further importance. The difference between total concentrations of Fe determined in the <2mm split following total digestion and by summation of its concentrations in the size fractions (seen in Table 3) is most likely an effect of pulverizing the <2mm material to 0.074mm.

The data in Figures 2 and 3 were obtained by normalizing the concentration of metals in the five size- fractions by the weight percent of each fraction in the sample, and summing the results. The resulting data is thus presented as μg/g of the composite, not of the size fraction. Figure 4 shows the contribution of Fe in each size-fraction to the total iron in composite 2 obtained by summation and the weight percent of each fraction in the composite. With the exception of the sulfide phase, the contribution of Fe in each phase to the total Fe in composite 2 corresponds well with the weight percent of each size-fraction in the composite. The data for Fe in Figure 4 are, for the most part, representative of the data for Fe in composite 1 and for Pb, Zn, and Cu in both composites. This

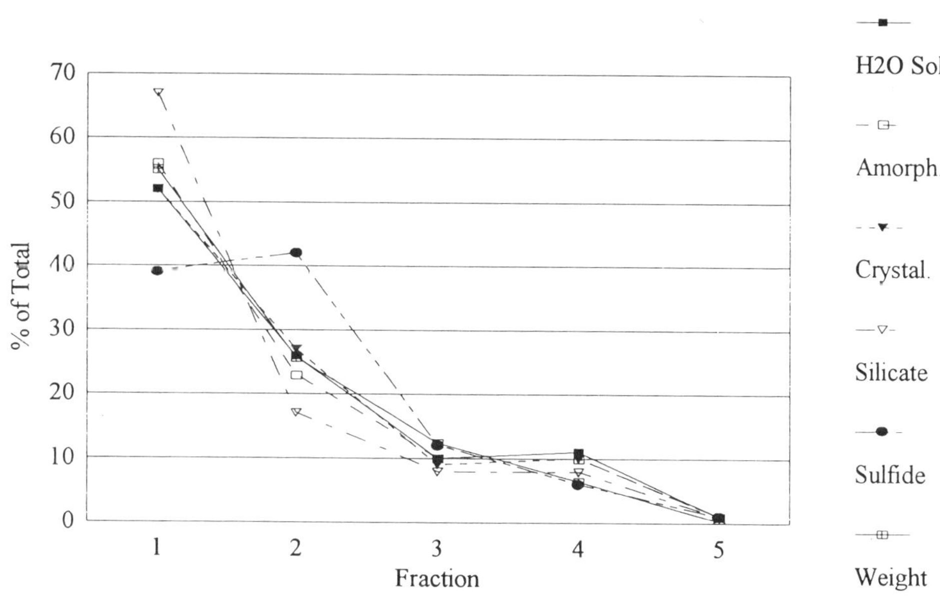

Figure 4. The concentration of Fe in each of 5 phases in the 5 size-fractions of composite 2, given in percent of the total Fe in the phase, and the weight percent of each fraction in the composite.

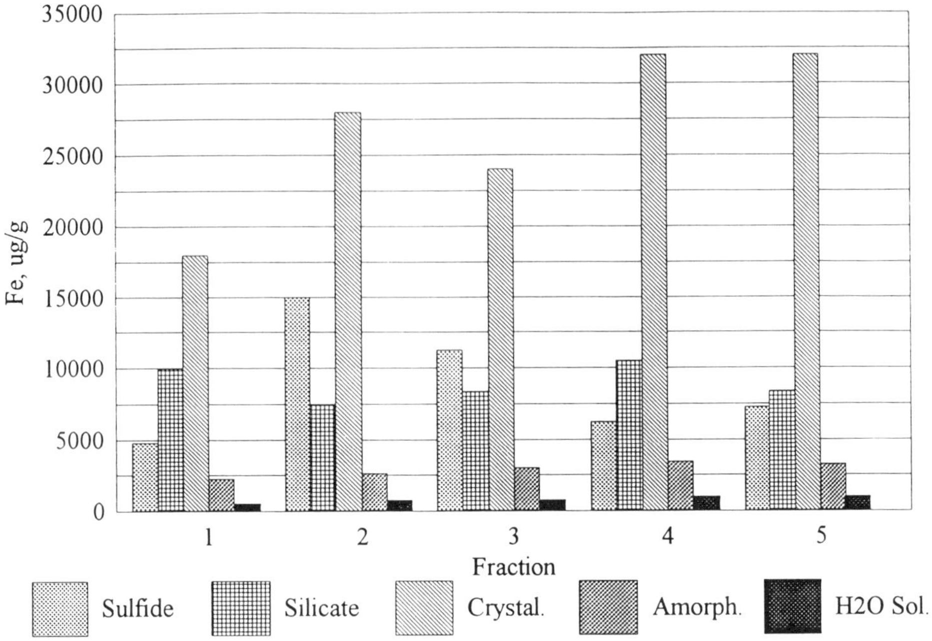

Figure 5. The concentration of Fe in each of 5 phases in the 5 size-fractions of composite 2.

indicates that the distribution of Fe and the other metals in each phase is independent of particle size. A nearly equal amount of Fe occurs in fractions 1 and 2 although in composite 2 there is about twice as much fraction 1 as fraction 2. The data in Figure 4 should not be confused with the actual concentration of Fe in each fraction, which is shown in Figure 5, where it is seen that the distribution of Fe in the water-soluble, amorphous and silicate phases are remarkably similar throughout the fractions. Iron in the sulfide phase is seen to decrease from fraction 2 to fraction 4, where it is nearly equal to that found in fraction 5. The distribution of sulfide-Fe between fractions in Figure 4 is thought to reflect more rapid weathering of small sulfide grains, which is also evidenced in the overall increase of Fe in the crystalline phase from the coarser to finer fractions.

4 CONCLUSION

Sequential extractions applied to composite samples collected from two inactive polymetallic mines show that metals are more favorably partitioned into the sulfide phase at one of the sites and into the crystalline Fe-oxide phase at the other site. In this respect, the data is corroborated by XRD mineralogical analysis. The results show that distribution of metals in the extracted phases is independent of particle size, but the amount of each phase present in each fraction may vary. Analysis of the composite following total digestion of a pulverized split gives total element concentrations which differ from the total concentrations obtained by summing the results of the extractions. The behavior of jarosite, commonly associated with pyritic mine wastes needs to be evaluated for more meaningful speciation of mine wastes.

REFERENCES

Chao, T.T. & Theobald, P.K., Jr., 1976. The significance of secondary iron and manganese oxides in geochemical exploration. *Econ. Geol.* 71: 1560-1569.

Chao, T.T. 1984. Use of partial dissolution techniques in geochemical exploration. *J. Geochem. Expl.* 20: 101-135.

Griffioen, J. & Broers, H.P. 1993. Characterization of sediment activity: the feasibility of sequential extraction techniques. *Report OS 93-65A, TNO Environmental and Energy Research, TNO Institute of Applied Geoscience*: Delft, the Netherlands.

Tessier, A., Campbell, P.G.C., & Bisson, M. 1979. Sequential extractions for the speciation of particulate trace metals. *Anal. Chem.* 51: 844-851.

Tailings and Mine Waste '99 © 1999 Balkema, Rotterdam, ISBN 90 5809 025 6

Selenium speciation in mining process waters by IC-HG-AFS

D.Wallschläger & N.S.Bloom
Frontier Geosciences Incorporated, Seattle, Wash., USA

ABSTRACT: We present a method for the speciation of inorganic selenium in waters. Selenite, selenate and selenocyanate are separated by ion chromatography. Selenate is then prereduced to selenite on-line with HBr at 100 °C. Afterwards, all species are converted to selenium hydride with potassium borohydride, purged from the eluent, and detected by atomic fluorescence spectrometry. The method's performance is characterized in pure water and in real world aqueous samples. Detection limits range from 10 to 200 ng/L. Selenium speciation is determined in different kinds of environmental and process waters. The method's performance and the obtained speciation results are compared to a method for operationally-defined selenium speciation, and advantages and disadvantages of both approaches are discussed. We also present conclusive evidence of selenocyanate in mining process environments, suggesting that this compound may have relevance for mobilization, distribution and remediation processes of selenium in mining operations involving the use of cyanide.

1 INTRODUCTION

Selenium is a trace element exhibiting ambiguous (eco)toxicological properties. As an essential micronutrient, if selenium is present in insufficient quantities, severe deficiency symptoms occur (Keshan Disease Research Group 1979). Moderate uptake has been shown to affect human health positively, e.g. by reducing cancer risk (Salonen et al. 1984), and in some cases, elevated uptake may even be used for acute medication of serious diseases (Schrauzer 1998). On the other hand, excess chronic uptake results in severe negative effects on wildlife (Skorupa 1998). Those environmental concerns may result from natural processes, but are usually enhanced by human activities. Prominent examples of processes where Se can become an important environmental concern include Se enrichment in aquatic systems as a result of irrigation practices (Seiler 1998), release of Se associated with petroleum production and processing (WSPA 1995), and mobilization of Se as a side-effect of various types of mining activities (Adams & Pickett 1998; Naftz & Rice 1989; Edwards et al. 1999).

In all of the above biogeochemical processes, the binding state ("speciation") of selenium is one of the most important factors determining the magnitude of the process and the resulting effects. Inorganic selenium usually occurs in two environmentally relevant oxidation states (+4 and +6), which show significant differences with respect to both toxicity and environmental fate. Additionally, there's elemental Se (oxidation state 0) and a variety of organic Se species (oxidation state -2), such as selenoaminoacids or volatile organic Se compounds, which also have totally different geochemical and ecotoxicological properties. It is therefore necessary to distinguish between these different species in each investigated system in order to fully understand what is happening, and why.

Unfortunately, the analytical techniques necessary for investigating those processes are presently not nearly as developed as the questions they should help to answer. This is especially true for solid matrices, such as animal and plant tissues, soils and sediments, but even in aquatic systems, problems are sometimes encountered when Se speciation is to be determined. Up to date, operationally-defined wet chemistry is most commonly used for this speciation analysis, in combination with hydride generation - atomic absorption spectrometry (HG-AAS) as a detection method (Cutter 1978). Briefly, this approach (referred to as the "Cutter-method" in general and throughout this paper) makes use of the fact that only Se(IV) forms SeH_2 when reacted with borohydride ("hydride generation", HG). So the sample is first measured directly to determine Se(IV), and then other species are derived from additional measurements after chemical treatments. In a second step, Se(VI) is reduced to Se(IV), and total inorganic Se (TISe) is measured. Se(VI) is then calculated as the difference between TISe and Se(IV), Finally, the whole sample is oxidized, then reduced to Se(IV), and total Se (TSe) is measured. Organic Se is likewise estimated as TSe minus TISe. The same "selective reduction" approach can be used with different detection techniques, e.g. atomic fluorescence spectrometry (AFS) (Bloom et al. 1997).

Besides requiring three separate analyses and two sample preparations, this technique has the fundamental shortcoming that it detects only Se(IV), so important species like Se(VI) are inferred from indirect measurements and defined operationally, relying on the functioning of the chemistry underlying the chemical pretreatment steps. Both the hydride generation reaction for Se(IV) and the prereduction for TISe determination are prone to severe interferences from real world matrix components (Dedina and Tsalev 1995), potentially causing serious errors in the derived speciation information. Thus, it is desirable to develop selenium speciation techniques that a) can positively measure Se(VI), as well as other relevant Se species, and b) separate the Se species of interest from interfering matrix compounds prior to the analytical determination. This can be achieved by means of hyphenated speciation techniques, in which a chromatographic separation is coupled on-line to an element-specific detector.

If we narrow down the speciation task to inorganic Se species in waters, ion chromatography (IC) is an ideal separation method, because both important oxidation states of selenium can be expected to occur in anionic form under natural conditions. Se(IV) and Se(VI) have been speciated before using IC with detection by HG-AFS (Pitts et al. 1995) or by inductively-coupled plasma-mass spectrometry (Guérin et al. 1997). In addition, IC is also capable of separating selenocyanate (SeCN), which is an important Se species in mining processes (Edwards et al. 1999) and petroleum refining (WSPA 1995), from the other two anions. AFS represents a cheap, selective and ultra-sensitive detection system for Se after hydride generation (Bloom et al. 1997). In this paper, we present the coupling of IC with HG-AFS for the simultaneous determination of selenite, selenate and selenocyanate in waters at the sub-µg/L-level, compare its performance to the "Cutter-method", and apply it to the determination of selenium speciation in a variety of real samples, particularly in waters related to mining processes.

2 METHOD

The developed method (fig. 1) consists of four individual steps, which are coupled together on-line:

1) ion chromatographic separation
2) prereduction of Se(VI)
3) hydride generation
4) detection by atomic fluorescence spectrometry

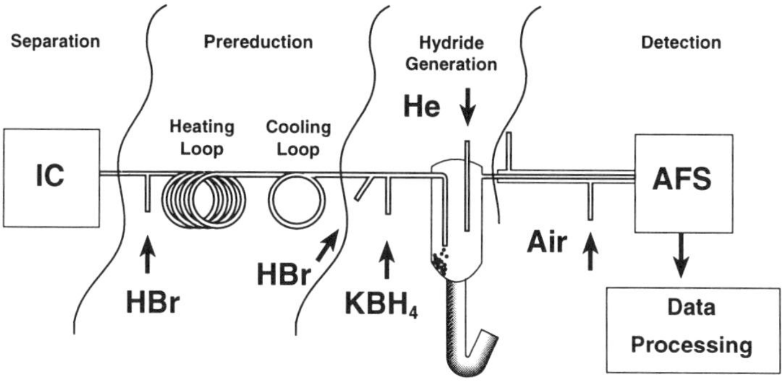

Figure 1. Schematic setup of the developed method for selenium speciation by IC-HG-AFS.

The three anions are separated on a low-capacity anion-exchange column (IonPac AS4S-SC, 4mm, Dionex, Sunnyvale, CA). Perchlorate is used as the eluent anion (Janos & Aczel 1996) in a gradient elution program with a run time of 15 min. Selenate is then quantitatively prereduced to selenite on-line by reaction with concentrated HBr (D'Ulivo 1989) at 100 °C in 70 seconds. A short cooling loop was introduced to reduce the evolution of water vapor in the following hydride generation step. Next, Se(IV) is converted to SeH_2 by reaction with KBH_4 in concentrated HBr medium. The formed SeH_2 is purged from the liquid phase by a combination of the H_2 produced from the HG reaction and a smaller additional He stream, that is introduced in the gas-liquid separator. To avoid quenching of the AFS-signal, most of the water vapor is removed from the gas stream by a semipermeable dryer-membrane (Corns et al. 1992) before the analyte enters the detector. Here, the AF signal is recorded at 196.0 nm after atomization in a hydrogen diffusion flame and excitation by a boosted Se hollow cathode lamp, and processed electronically.

The three selenium species elute in the order Se(IV), Se(VI), SeCN (fig. 2) and are baseline-separated in 10 min. Under the current chromatographic conditions, separation between Se(IV)/Se(VI) and Se(VI)/SeCN is each on the order of 3 min. This allows identification and determination of small amounts of one species in the presence of large amounts of another one. All three peaks show comparable peak shape with only very slight tailing, even for the very strongly retained SeCN, which is classified as a "lipophilic" anion.

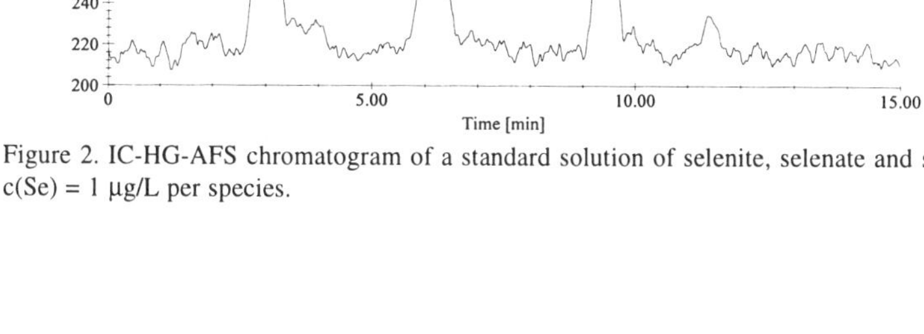

Figure 2. IC-HG-AFS chromatogram of a standard solution of selenite, selenate and selenocyanate; c(Se) = 1 µg/L per species.

3 RESULTS

After optimizing most of the instrumental and chemical parameters involved in the method, we conducted a method detection limit study according to USEPA protocol (USEPA 1986) and a linear range study, and then applied the method to the analysis of most of the real samples depicted in fig. 5. The derived analytical figures of merit (table 1) represent the current state of development, but additional studies are underway to lead to further improvement, especially with regard to even lower detection limits and better reproducibility.

Table 1. Current analytical figures of merit for the developed method for Se speciation by IC-HG-AFS.

parameter	Se(IV)	Se(VI)	SeCN
estimated detection limit [μg/L]	0.08	0.17	0.08
linear range [μg/L]	0.1-100	0.2-100	0.1-500
reproducibility [%]	< 10	< 20	< 10
matrix spike recovery [%]	51-73	64-122	59-148

The shown detection limits are conservative estimates based on the reproducibility of low level spikes in deionized water. However, realistic instrumental detection limits, based on the signal to noise ratio, are approximately one order of magnitude lower (around 10 ng/L Se per species). The worse detection limit for Se (VI) in comparison to the other two species is entirely caused by the poorer reproducibility, probably due to temperature instability during the prereduction process. Within the analytical error margins, all three Se species are determined with the same sensitivity. This proves that Se(VI) is quantitatively converted to Se(IV) during the on-line prereduction step. It also eliminates concerns that selenocyanate may decompose during the prereduction step, despite the compound's instability in highly acidic media. The method shows a linear range spanning at least three orders of magnitude, from the detection limits up to Se concentrations of several hundred μg/L. Even at lower concentrations, we observe no deviation from linearity, because the method is not limited by substantial blanks. The reproducibility from real samples is not worse than from standard solutions, and is generally significantly better than the maximal uncertainties indicated in table 1. The recovery of matrix spikes appears to be incomplete for Se(IV) and too variable for the other two species, but we expect this to improve as the method development continues.

When the method was applied to the analysis of real samples, the method's current detection limits were too high to permit the determination of Se speciation in rainwater collected at an unpolluted background location. Analysis of a Se-polluted coastal seawater delivered a Se(VI) peak corresponding to about half of the total Se content determined by HG-AFS, but also showed an equally big peak eluting at the dead volume of the chromatographic system. Additional experiments showed that this effect is caused by the excess of chloride anions in seawater competing with the selenium anions on the IC column, leading to the "exclusion" of Se(IV), and maybe also Se(VI), from the separation process. Matrix spikes confirmed that SeCN is not impacted by this process. This exclusion effect has since been replicated for other common anions (nitrate, sulfate, hydroxide), and thus currently limits the determination of Se speciation in high salinity matrices, unless the Se concentrations are high enough to overcome the problem by dilution.

In contrast to the poor performance in those two matrices, the developed IC-HG-AFS method performs well in some more difficult matrices (samples are the same that were used

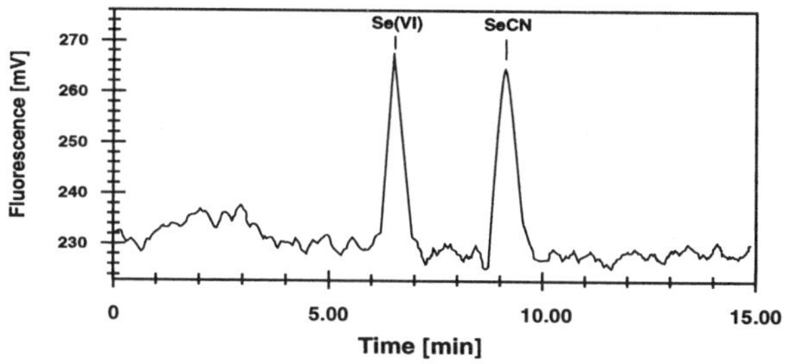

Figure 3. Selenium speciation in the tailings pond of a zinc/lead mining process.

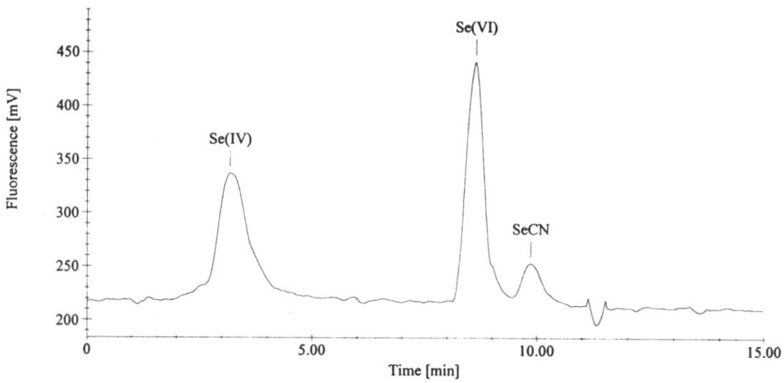

Figure 4. Selenium speciation in a waste water from a noble metal extraction process.

for evaluation of the "Cutter-method" in fig. 5). Analysis of the fly ash pond waters shows that Se(IV) is the only Se species present at detectable concentration levels. The measured Se(IV) concentrations were, again, lower than the determined total Se concentrations. Part of this discrepancy may be due to particulate Se in the samples, which is not analyzable by the chromatographic technique. In the two samples from a zinc and lead mining process, the IC-HG-AFS technique showed unequivocally that Se(IV) is not present in detectable amounts. Determination of Se(VI) was not affected by the matrix in both samples, and spike recoveries were good. In the reclaim water from the tailings pond, Se(VI) was the only Se species present, but in the mine tailings pond itself, we were additionally able to identify SeCN (fig. 3). In that sample, approximately one third of the total Se was present as Se(VI), and two thirds as SeCN. In both mining samples, the sum of the individual determined Se species agreed well with the total Se measured independently by HG-AFS.

The method was also applied to waste water samples from a noble metal extraction process. In this process, Se(IV) was identified as an additional species, and SeCN was again found to contribute significantly to the Se speciation (fig. 4). In contrast to the previous mining waste waters, these samples were acidified to pH 2 prior to analysis, so it is possible that the shown chromatogram does not represent the original Se species distribution, as some of the original SeCN may have decomposed during storage. The sum of the determined species agreed well with the independently determined total Se content, but minor degradation of SeCN may still have occurred. Alternatively, some SeCN may have been converted to one of the other species. We will conduct stability studies of individual Se species under different physical and chemical conditions in the future to address this issue.

4 DISCUSSION

The developed method for Se speciation by IC-HG-AFS allows for the determination of the major inorganic Se species in most environmental and industrial process waters down to the sub-µg/L concentration level, which corresponds to absolute detection limits of only a few picogram Se. Thus, it is sufficiently sensitive for monitoring Se speciation well below both the current freshwater chronic criterion for total Se of 5 µg/L (USEPA 1987) and the proposed criterion maximum concentration for Great Lakes waters, which would be between 13 and 186 µg/L, depending on the actual Se speciation in the respective water (USEPA 1996). Still, it is either not sensitive enough or too perturbed by matrix interferences to allow Se speciation in certain important environmental waters, such as rain water or sea water. Thus, further improvements on the method's sensitivity are necessary to permit Se speciation even in the most uncontaminated samples, or to overcome matrix interferences by dilution.

In order to assess the accuracy of the new method and demonstrate some of the problems associated with the "Cutter-method", we compared the results discussed above with the performance of operationally-defined Se speciation (modified after Cutter 1978) (fig. 5). In this diagram, a wide variety of environmental and industrial waste water samples, spanning over four orders of magnitude in their total Se concentrations (indicated below the x-axis), are sorted in order of increasing matrix complexity ("difficulty") from left to right. To facilitate comparison, total Se in all samples has been normalized to 100 %, and Se(IV) and total inorganic Se are expressed as fractions thereof.

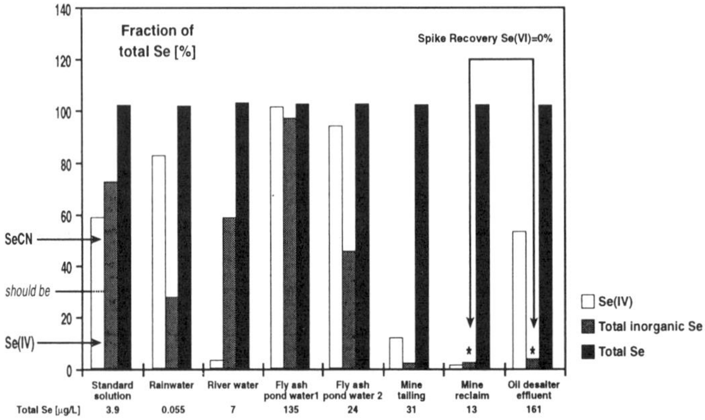

Figure 5. Performance of operationally-defined selenium speciation (Cutter 1978) in a variety of aqueous matrices.

Two characteristic features can be seen in fig. 5. In accordance with many previous results in our lab, total inorganic Se in many samples (rain water, fly ash pond water 2, mine tailing and oil desalter effluent) is considerably lower than Se(IV). This is not only theoretically impossible, because TISe comprises Se(IV), but it also leads to the calculation of negative Se(VI) concentrations. In samples like the rain water (close to the detection limits) or fly ash pond water 1, this may still be explained by analytical uncertainty. In the other samples, however, this indicates not only incomplete conversion of Se(VI) to Se(IV) during prereduction, but also suggests some losses of Se(IV) during the same treatment (possibly via reduction to elemental Se). In the last two samples, this effect goes even further, in that Se(VI) matrix spikes are quantitatively not recovered. This strongly suggests that some matrix components alter the chemistry of the prereduction reaction, thereby yielding much too low TISe concentrations and, consequently, prohibiting the determination of Se(VI) in those samples. Trials in our lab showed that humic substances inhibit the reduction of Se(VI), and we suspect the phenols act similarly.

The example of the mine tailings pond water (fig. 3) demonstrates clearly that the presented IC-HG-AFS method is far superior for assessing Se speciation in those cases where the "Cutter-method" delivers poor results. By this technique, both the determination of Se(VI) and the recovery of matrix spikes is unproblematic, even though the dilution of the matrix prior to IC-HG-AFS analysis (due to the higher detection limits) is less than for the "Cutter-method". This suggests that the interfering unidentified matrix compounds are separated chromatographically from the Se species, and thus do not affect the HG reaction any longer. Moreover, IC-HG-AFS also permits identification and quantification of SeCN, which proves to be the dominant Se species in that particular sample. This is inherently impossible to do with the "Cutter-method", because there is no separate chemical treatment step that would selectively convert SeCN to SeH_2. In fig. 5, the standard solution on the left contained approximately 1 ppb each of Se(IV), Se(VI), SeCN and selenomethionine (as an example of an organic Se compound). Within the analytical errors, it seems that SeCN in this case is mostly appearing as Se(IV) in the resulting speciation pattern, but we've also seen it behave like Se(VI) in standard solutions. Although SeCN is technically a Se(IV) compound, SeCN would be wrongly interpreted as selenite by this result of the "Cutter-method". In normal cases, we can expect SeCN to distribute to some extent between all three fractions of the "Cutter-method", thereby leading to misinterpretations for all three operationally-defined species Se(IV), Se(VI) and organic Se. This is likely the reason why the "Cutter-method" detected some Se(IV) in the two zinc/lead mining samples, whereas the IC-HG-AFS method proved its absence. Finally, in those mining samples, the "Cutter-method" suggests that all Se in the samples is organic, which is simply proven wrong by the IC-HG-AFS technique.

Similar misinterpretations occur for the waste water sample from the noble metal extraction process (fig. 4). Again, the "Cutter-method" gives a low result for TISe, which leads to the false conclusion that there is no Se(VI) present, but that half of the total Se are organo-Se species. By comparison, IC-HG-AFS proves the absence of other Se species besides Se(IV), Se(VI) and SeCN. Also, Se(IV) determined with the "Cutter-method" appears to constitute half of the total Se, whereas that fraction is only a quarter when Se(IV) is measured by IC-HG-AFS. According to the IC-HG-AFS analysis, SeCN is present in the sample at a concentration comparable to Se(IV), which suggests that the Se(IV) determined by the "Cutter-method" is actually the sum of Se(IV) and SeCN.

In conclusion, while the "Cutter-method" usually yields good Se(IV)/Se(VI) speciation results in fairly clean water samples, such as river waters and, especially, marine waters, the chemistry underlying its operationally-defined speciation concept may be seriously impacted by complex matrices. This leads to either senseless or seriously wrong speciation patterns, which do not allow quantification of certain Se species and/or cause severe misinterpretation of the Se speciation in the sample. The same happens when other selenium species than selenite, selenate and organic Se compounds are present as major species, because their distribution between the three fractions gives wrong results for all three operationally-defined Se species. As the "Cutter-method" does not have any indicator for the presence of other species and/or malfunction of the underlying chemistry, these interferences lead to complete misinterpretation of the present Se speciation. "Direct" speciation methods, combining chromatographic separation and specific detection, such as the presented IC-HG-AFS technique, have to be employed in such cases to obtain accurate Se speciation data. However, it has to be emphasized that these problems are not universal, and that the "Cutter-method" probably works well in most common environmental applications. Some additional characteristic features of the two approaches to Se speciation are compared in table 2.

The discovery of selenocyanate in mining process waters represents one of the first occasions that this species has been unequivocally determined in natural or industrial environments. Little is known about this compound in general, and its environmental behavior in particular. Its formation pathway is currently under investigation, but textbook

Table 2. Advantages and disadvantages of the developed method for Se speciation by IC-HG-AFS in comparison to operationally-defined Se speciation (Cutter 1978).

IC-HG-AFS	"Cutter-method"
advantages	disadvantages
no sample preparation	requires two chemical pretreatment steps
one analysis	three separate analyses
direct measurement of Se(VI)	Se(VI) is only calculated by difference, increasing uncertainty
positive signal for each Se species	Se species detected in groups (organic Se) or based on operational definition
good performance for Se(VI) in difficult matrices	often yields "negative" Se(VI) concentrations in difficult matrices
allows determination of SeCN	SeCN is distributed between fractions and cannot be identified or quantified
unknown major Se species are detected and can be identified later	unknown major Se species remain undetected and yield wrong speciation information
disadvantages	advantages
currently not very reproducible	very reproducible, except for Se(VI)
currently insufficient detection limits for low Se waters	very low detection limits
separation performance currently impaired in high salinity matrices	works extremely well in seawater

inorganic chemistry suggests that it could be formed by reaction of cyanide with elemental Se or selenite. Additional studies have since reconfirmed the existence of this species in various other mining environments, and seem to suggest that selenocyanate can be a major or even the dominant Se species in different kinds of mining environments that employ cyanide in their processes. In other mining processes that do not employ cyanide, we did not detect SeCN. Consequently, we propose that the determination of Se speciation, especially of selenocyanate, holds the key for understanding the biogeochemical processes that Se undergoes during the mining processes. This knowledge will help to comprehend its mobilization and fixation and might even prove useful for designing remediation processes.

5 CONCLUSION

The presented method for Se speciation in waters by IC-HG-AFS has been shown to be a valuable alternative to the "Cutter-method" when unusual Se species or interfering matrix compounds are present in the sample. In comparison to the "Cutter-method", the technique performed better for Se(VI) determination in difficult matrices. Hyphenated methods like these are the only option when selenocyanate is a compound of interest. Some problems are still associated with the measurement of Se speciation in high salinity matrices and at extremely low concentrations, but further development should overcome those issues, making the method a promising candidate to replace the "Cutter-method" in all applications in the long run. We have pointed out the importance of SeCN for Se speciation in mining process waters using cyanide, and similar findings have been reported for refinery waters, particularly so-called "sour waters" (WSPA 1995). The formation mechanism and fate of selenocyanate in these environments should be investigated in the future in order to understand the causes of environmentally relevant Se problems occurring during these technical processes, and find solutions for their remediation.

ACKNOWLEDGMENTS

This work was performed as part of a research project funded jointly by Cominco Ltd., Kennecott Utah Copper Co. and Frontier Geosciences, Inc. We are grateful for technical support received from Dionex Corp. and P.S. Analytical, Ltd.

REFERENCES

Adams, D.J. & T.M. Pickett 1998. Microbial and cell-free selenium bioreduction in mining waters. In W.T. Frankenberger & R.A. Engberg (eds.). *Environmental chemistry of selenium*: 479-499. New York: Marcel Dekker Inc.

Bloom, N.S, M.L. Gauthier & D. Wallschläger 1997. Ultra-trace speciation of selenium by automated hydride generation - flame atomic fluorescence spectrometry (HGAFS), In *Proc. 3rd Int. Conf. Anal. Geol. Environ. Mat., Vail, CO, 1-5 June 1997.*

Corns, W.T., L. Ebdon, S.J. Hill, & P.B. Stockwell 1992. Effects of moisture on the cold vapour determination of mercury and its removal by use of membrane dryer tubes. *Analyst* 117: 717-720.

Cutter, G.A. 1978. Species determination of selenium in natural water. *Anal. Chim. Acta* 98: 59-66.

Dedina, J. & D.L. Tsalev 1995. *Hydride generation atomic absorption spectrometry*, Chichester: John Wiley & Sons.

D'Ulivo, A. 1989. Studies on the determination of selenium by hydride generation non-dispersive atomic fluorescence spectrometry using hydrobromic acid-based reaction matrices. *J. Anal Atomic Spectrom.* 4: 67 70.

Edwards, M., W. Kuit, J.E. Kulas, J.O. Weakley, D. Wallschläger & N.S Bloom 1999. Aquatic selenium at Cominco's Red Dog mine: sources, speciation and distribution. In *Proc. of the Tailings and Mine Waste '99 Conf., Fort Collins, CO, 24-27 January 1999.*

Guérin, T., A. Astruc, M. Astruc, A. Batel & M. Borsier 1997. Chromatographic ion-exchange simultaneous separation of arsenic and selenium species with inductively coupled plasma-mass spectrometry on-line detection. *J. Chromatogr. Sci.* 35: 213-220.

Janos, P. & P. Aczel 1996. Ion chromatographic separation of selenite and selenate using a polyanionic eluent. *J. Chromatogr.* 749A: 115-122.

Keshan Disease Research Group 1979. Epidemiologic studies on the etiologic relationship of selenium and Keshan disease. *Chin. Med. J.* 92: 477-482.

Naftz, D.L. & J. Rice 1989. Geochemical processes controlling selenium in groundwater after mining, Powder River Basin, Wyoming, U.S.A. *Appl. Geochem.* 4: 565-576.

Pitts, L., A. Fisher, P. Worsfold & S.J. Hill 1995. Selenium speciation using high-performance liquid chromatography - hydride generation atomic fluorescence with on-line microwave reduction. *J. Anal. Atomic Spectrom.* 10: 519-520.

Salonen, J.T., G. Alfthan, J.K. Huttunen & P. Puska 1984. Association between serum selenium and the risk of cancer. *Am. J. Epidemiol.* 120: 342-349.

Schrauzer, G.N. 1998. Selenium: present and future applications in medicine. In *Proc. 6th Int. Symp. on the Uses of Selenium and Tellurium, Scottsdale, AZ, 10-12 May 1998.*
Seiler, R.L. 1998. Prediction of lands susceptible to irrigation-induced selenium contamination of water. In W.T. Frankenberger & R.A. Engberg (eds.). *Environmental chemistry of selenium*: 397-418. New York: Marcel Dekker Inc.
Skorupa, J.P. 1998. Selenium poisoning of fish and wildlife in nature: lessons from twelve real-world examples. In W.T. Frankenberger & R.A. Engberg (eds.). *Environmental chemistry of selenium*: 315-354. New York: Marcel Dekker Inc.
USEPA 1986. Definition and procedure for the determination of the method detection limit. *Fed. Reg.* 51: 23703.
USEPA 1987. *Ambient water quality criteria for selenium.* USEPA report EPA 440/5-87-008. Springfield, VA: National Technical Information Service.
USEPA 1996. Proposed selenium criterion maximum concentration for the water quality guidance for the Great Lakes system. *Fed. Reg.* 61: 58443-58449.
Western States Petroleum Association (WSPA) 1995. *Selenium removal technology study: project final report.* Concord, CA: WSPA.

Tailings and Mine Waste '99 © 1999 Balkema, Rotterdam, ISBN 90 5809 025 6

Efflorescent sulfate salts – Chemistry, mineralogy, and effects on ARD streams

D.C. Keith, D.D. Runnells, K.J. Esposito, J.A. Chermak & S.R. Hannula
Shepherd Miller Incorporated, Fort Collins, Colo., USA

ABSTRACT: During the dry summer months significant amounts of efflorescent salts accumulate around intermittent seeps and surface streams at Iron Mountain, California as a result of evaporation of acid rock drainage. Those salts include: rhomboclase $[(H_3O)Fe^{3+}(SO_4)_2 \cdot 3H_2O]$, ferricopiapite $[Fe^{3+}{}_5(SO_4)_6O(OH) \cdot 20H_2O]$, bilinite $[Fe^{2+}Fe^{3+}(SO_4)_4 \cdot 22H_2O]$, alunogen $[Al_2(SO_4)_3 \cdot 17H_2O]$ kalinite $[KAl(SO_4)_2 \cdot 11H_2O]$, gypsum $[CaSO_4 \cdot 2H_2O]$, and hexahydrite $[MgSO_4 \cdot 6H_2O]$.

The dissolution of these salt accumulations during the first major storm of the wet season produces a characteristic and strong hydrogeochemical response in surface waters, but the response is subdued in later storms. On a weight basis, Fe-sulfate salts are capable of producing more acidity than Al- or Mg-sulfate salts. The acidity released from Fe-bearing salts is due to the hydrolysis of Fe^{3+} and the precipitation of poorly crystalline oxyhydroxides or oxyhydroxysulfates. In addition to lowering the pH of the streams, these salts appear to be a significant source of dissolved metals.

INTRODUCTION

Some of the most perplexing issues facing environmental scientists and regulators involve estimating the impacts of sources of contamination to groundwater and surface streams in areas that have been disturbed by mining. This investigation focuses on distinguishing the relative contributions of dissolved efflorescent sulfate salts, acidic groundwater, and surface runoff to metal loads in Boulder Creek at the Iron Mountain Mine in Shasta County, California.

Boulder Creek is one of three perennial streams at Iron Mountain, which ultimately drain into the Sacramento River, and thence into San Francisco Bay (Figure 1). The pH at the mouth of Boulder Creek, where it enters Spring Creek (Figure 1), is typically around 3 during dry summer base flow conditions and rises to above 4 during storms in the late wet season. These acidic conditions maintain elevated concentrations of dissolved trace metals compared to other natural streams.

The primary, or ultimate source, of acidity to ground water and surface streams in the Boulder Creek is from the oxidation of pyrite and other sulfide minerals in mineralized areas of the watershed. However, as shown in this investigation, efflorescent sulfate salts are an important intermediary source of acidity and metals. The investigations presented below were designed to gain an understanding of these intermediate sources of acidity and metal loads in Boulder Creek during baseflow and storm conditions. The findings of these investigations should be relevant to all sites impacted by ARD.

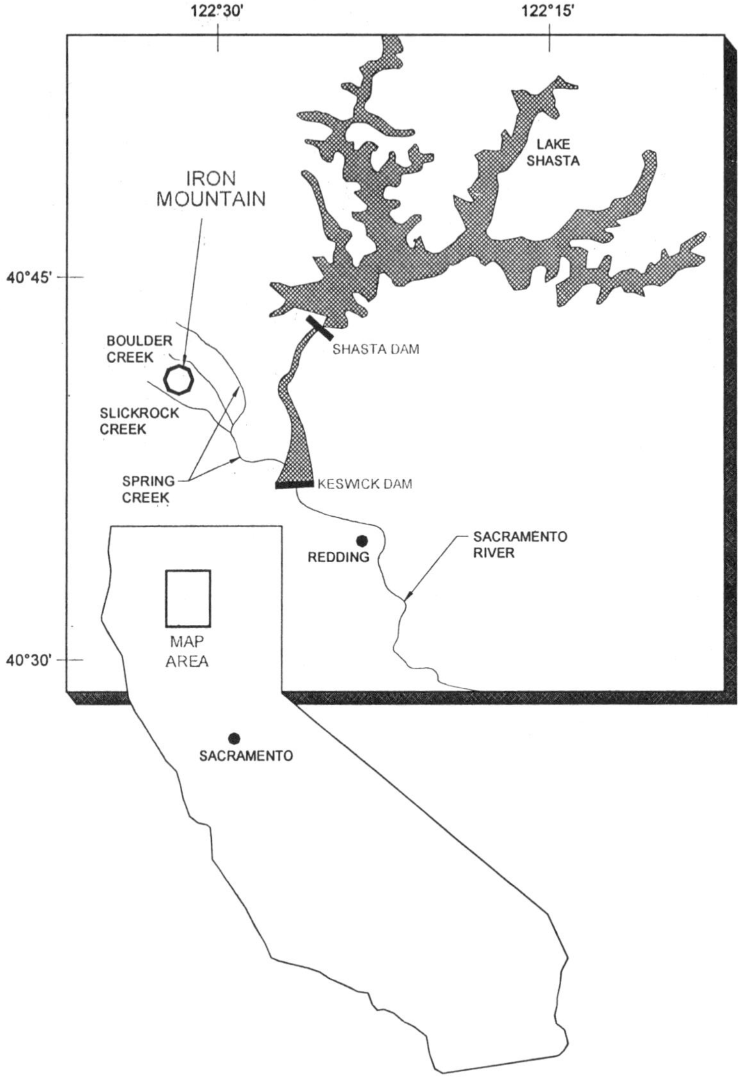

Figure 1 Location map of study site

HYDROGEOCHEMICAL SETTING

The Boulder Creek drainage basin is located approximately 16 km west of Redding, California (Fig. 1), draining the northeastern side of the Iron Mountain. Iron Mountain contains extensive sulfide deposits and is part of the West Shasta Mining District. The ore deposits consist of massive pyrite with minor amounts of copper and zinc sulfides. The ore host rocks in this district are hydrothermally altered Devonian volcanic rocks, including the keratophyric Balaklala rhyolite and the spilitized Copley Greenstone (Kinkel *et al.*, 1956). There are two major sulfide deposits along Boulder Creek, the Richmond and Hornet deposits, that were originally a continuous lens of massive sulfide cut later by normal faults (Kinkel *et al.*, 1956). The Hornet deposit lies upstream of the Richmond deposit, and the first significant appearance of ARD in Boulder Creek is a short distance downgradient from the original outcrop of the Hornet ore body and the historic Hornet Mine workings.

The area of the Boulder Creek drainage basin is approximately 700 hectares. Headwaters originate at an elevation of approximately 1060 meters (m), and at its discharge point into Spring Creek the elevation of Boulder Creek is approximately 430 m. The elevation drop of 630 m occurs along a 6100 m run, which is equivalent to a 10 percent grade. The mean annual precipitation of the area varies from 140 cm (55 inches) at lower elevations to 200 cm (80 inches) at higher elevations (USEPA, 1992). The mean annual precipitation based on the long-term data from a weather station at Shasta Dam is approximately 150 cm (60 inches) and is representative of precipitation that can be expected for lower elevations of the Boulder Creek watershed (USEPA, 1992). About 85 percent of the annual precipitation falls between November and April. Snow is common above elevations of 610 m between November and March. Flows in Boulder Creek range from less than 34 m^3/hr (150 gpm) in the dry summer and fall months to over 79,500 m^3/hr (350,000 gpm) during flash floods which occur in the winter wet season.

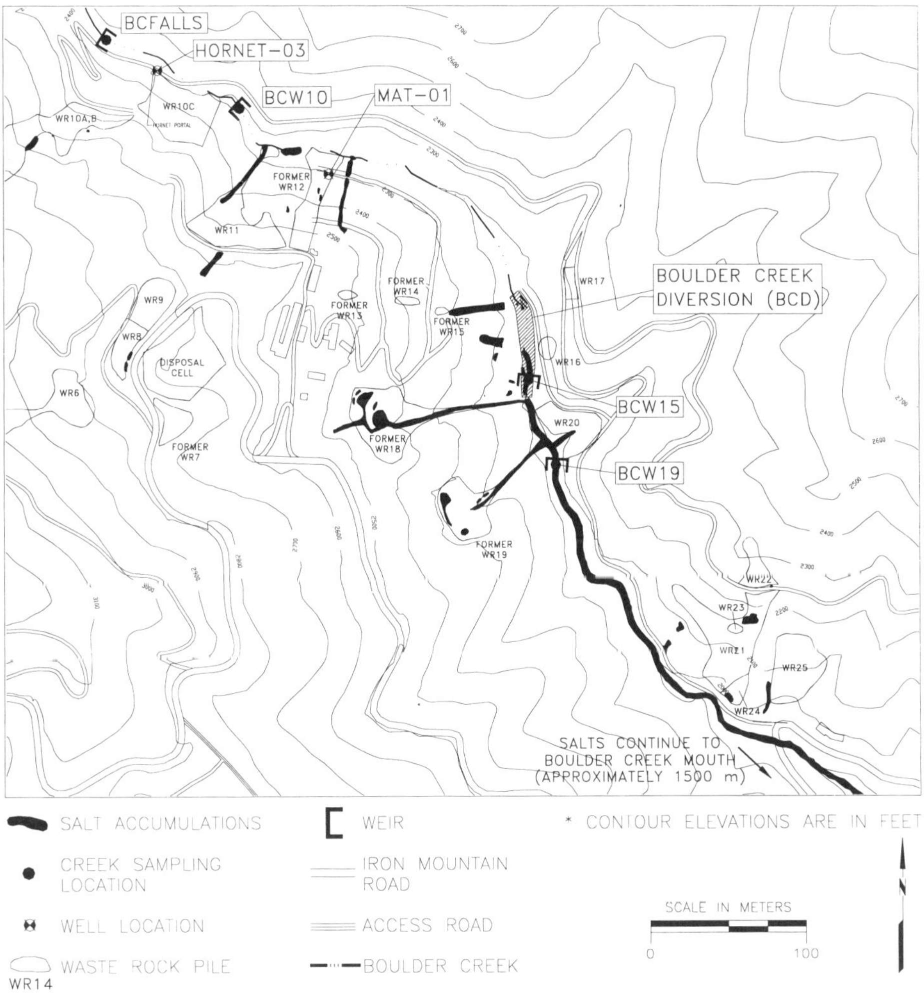

Figure 2 Location of primary stream monitoring stations and other relevant surface features within the Boulder Creek watershed

Table 1 Chemical Analysis Of Salts Associated With Gossan At Iron Mountain

	Gossan Salt 1	Gossan Salt 2	Gossan Salt 3	Gossan Salt 4	Gossan Salt 5	Gossan Salt 6	Gossan Salt 7	Gossan Salt 8	Gossan Salt 9
Cu (mg/kg)	2920	70.2	574	51.5	33	137	301	346	95.3
Zn (mg/kg)	3780	22.9	27.8	24	28.2	35.2	23.2	43.1	33.3
% Moisture	16.2	6.8	5.9	5.8	6.5	4.3	5.2	6.6	7.1
Minerals Identified by XRD	Gypsum Hexahydrite Kalinite	Ferricopiapite Rhomboclase	Ferricopiapite Rhomboclase	Ferricopiapite Rhomboclase	Hexahydrite Rhomboclase	Ferricopiapite Rhomboclase	Rhomboclase	Ferricopiapite Rhomboclase	Ferricopiapite Rhomboclase

Table 2 Coloro, Mineralogy, and Chemical Analysis Results of Salts in Boulder Creek Watershed

Sample I.D.	Color Description	Mineralogy (1)	Copper (mg/kg)	Zinc (mg/kg)	Iron (mg/kg)	Aluminum (mg/kg)
Landslide between WR-11 & WR-12	White, trace yellow	K, G	1830	5070	6200	58900
WR-11 (#2)	Reddish yellow	Fc, Rh, Al, H	1430	7060	22500	49100
WR-12M	White	Al, G, K	1600	4970	1770	43900
Boulder Creek @ LSMP-8	White to olive yellow	Al, K, G, H	3360	6500	11700	43600
FracSeep	White to yellow	Fc, Rh, H	4090	17200	4500	43500
WR-11 (#1)	White to pale yellow	Fc, Rc, G, K	2090	9930	42600	41700
Boulder Creek @ BCCP	White to yellow	Fc, Rh, Al, K	1130	12200	11700	40900
BGSeep	White	Al, K, H	3140	10300	9280	39100
WR-11 (#3)	White	K, Bi	1040	6450	45100	37700
WR-25 B	White to Pale Yellow	Rh, K, H, G	878	574	33700	37000
Slickrock	White to pale yellow green	Fc, Rh, G, K	3660	4350	10800	35400
WR-21 B	White to reddish yellow	K, G, H, Al	1610	7900	4560	35200
Seepbed @ WR-19T	White, trace pale yellow	Al	2370	9250	537	34400
WR-12TS	White	Al, K, H	2450	8070	472	27500
Boulder Creek between WR-11 & WR-12	White to olive yellow	Fc, Al, H	3400	17000	59500	26500
WR-19 B,M	White	K, G	845	6680	657	25500
Iron Volcano	White	Rh, Al, K, H	293	13000	3700	25000
WR-18B	Yellow	Fc, Rh, Al, K	1760	4920	89700	23000
Secondary Slide	White to olive yellow	Fc, Al, Rh, G, H	5320	5450	91100	18900
WR-18T	White to yellow	Fc, Rh, Al	872	3150	131000	14900
WR-23 B	Yellow	Fc, Rh, G, K	234	660	142000	7730
WR-22 B	Yellow	Fc, Rh, G	125	280	105000	6720
Hornet-yellow	Yellow	Fc, Rh	781	52	126000	6630
WR-23 M	Yellow	Fc, Rh, G	491	1580	16000	5690
Hornet-white	White	Fc, Rh	662	10	132000	2490
WR-10A,B (#1)	White to yellow	Fc, Rh	221	124	160000	2040
WR-10A,B (#2)	White to yellow	Fc, Rh	505	53	178000	1870
Tramway	White to yellow	Fc, Rh	159	420	167000	1670
BFP-1	Yellow	Fc, Rh	95	26	163000	831
WR-19T	White to	K, G, H, Al	0	3	19	211
Average Values	******************	************	1550	5440	59000	24600

(1) Al = alunogen; Bi=bilinite; Fc = ferricopiapite; G=gypsum; H=hexahydrite; K=kalinite; Rh=rhomboclase

Methods

For this investigation, a network of stream monitoring stations and wells along and within Boulder Creek was utilized to obtain periodic flow measurements and water samples. The locations of these stations and other relevant surface features within the study area are shown in Figure 2. The primary stream monitoring stations are located at weirs within Boulder Creek and are designated using the prefix BCW. In addition to the permanent stations shown in Figure 2, Boulder Creek water quality was monitored using mini-piezometers and grab samples taken throughout the monitored reach.

Water quality parameters that were measured in the field included: pH, electrical conductivity, ferrousiron, temperature, and oxidation/reduction potential (Eh). Laboratory measured constituents include dissolved Al, As, Ca, Cd, Cu, Fe, K, Mg, Mn, Na, Pb, SiO_2, Zn, SO_4, and total dissolved solids. All samples collected for laboratory analysis of dissolved metals were filtered (<0.45 μm) and preserved with nitric acid in the field.

The mineralogic character of 31 samples of salt collected at different locations throughout the watershed was determined by XRD, and chemical analysis (including determination of Al, Cu, Fe, and Zn concentrations) of each sample was determined by ICP. In addition, the distribution of surface salt accumulations at Iron Mountain was determined in a mapping program conducted in October of 1995. Notable salt accumulations identified in the field reconnaissance and mapping within the Boulder Creek watershed are shown on Figure 2.

Results

During the first major storm of the wet seasons of 1995 and 1996, high metal loads (concentration multiplied by discharge) were observed in Boulder Creek. In those storms, the pH of the creek remained low (approximately 3), and metal loads were significantly higher than loads that were predicted considering simple dilution of the Creek by rainwater. Data from the first significant storm of the 1995/1996 wet season show that increasing concentrations and loads of Fe, Al, and Cu measured at the mouth of Boulder Creek were coincident with increasing discharge. By comparison, the increase in metals load in later wet season storms was small relative to metals loads in the early season storm.

MINERALOGY

The salt mapping and characterization program showed that salt accumulations in the Boulder Creek watershed generally occur as thin crusts on rocks and soil, in locations where intermittent seeps and surface streams exist. Salt accumulations also occurred in areas where surface materials contained significant concentrations of sulfide minerals, even where obvious seeps were not present. The most notable examples of salt accumulations in the absence of groundwater seeps were on residual natural sulfide lenses and nodules within the massive gossan outcrops at the top of Iron Mountain.

The results of XRD examination and the Cu and Zn contents of nine samples of salts from natural gossan outcrops at Iron Mountain are shown in Table 1. These salts, which formed on residual sulfide lenses and nodules within the gossan, were primarily composed of ferricopiapite and rhomboclase, and had higher concentrations of Cu than Zn, presumably as solid-state substitution in the crystal lattice.

The results of XRD and chemical analyses of 31 salt samples collected from seeps, intermittent surfaces streams, and residual waste rock piles, are presented in Table 2. These results show that Fe-sulfates, including rhomboclase ($(H_3O)Fe^{3+}(SO_4)_2{\cdot}3H_2O$), ferricopiapite ($Fe^{3+}{}_5(SO_4)_6O(OH){\cdot}20H_2O$), bilinite ($Fe^{2+}Fe^{3+}(SO_4)_4{\cdot}22H_2O$), and the Al-sulfates, including alunogen ($Al_2(SO_4)_3 \cdot 17H_2O$) and kalinite ($KAl(SO_4)_2{\cdot}11H_2O$), were the dominant phases in the crusts. Accessory minerals included gypsum ($CaSO_4{\cdot}2H_2O$) and hexahydrite

($MgSO_4 \cdot 6H_2O$). The bulk concentrations of Fe, Al, Cu, and Zn of each sample are also listed in Table 1. These results clearly show that higher concentrations of Cu and Zn are associated with Al-sulfates than with Fe-sulfates from the same area.

DISCUSSION

The stream monitoring results indicate that there is a step-like increase in metals and decrease in pH in Boulder Creek as a result of the addition of groundwater from three identifiable reaches along the southwestern (mineralized) bank of the creek. The addition of ground water within each reach exerts an incremental impact on the composite water quality at the mouth of Boulder Creek, which is modified by storm runoff in the wet season.

The effect of dilution by rain on the water quality of Boulder Creek was assessed using the chemistry of Boulder Creek at baseflow as a starting point in a geochemical model. The base flow prior to the initial storm of the 1995/1996 wet season was approximately 59 m^3/hr (260 gpm) and there was approximately 29,500 m^3/hr (130,000 gpm) of flow at BCMO during the peak of the storm (a 500X dilution).

A reaction path simulation of dilution,in EQ6 (Wolery and Daveler, 1992) using 500 parts of clean water (represented by diluting water from BCFalls, Figure 2) into water below the reaches where groundwater impacts are the most significant shows that simple dilution would produce pH values of approximately 5.1. This pH is a full 2 units higher than the actual stream pH values observed in the initial storm of the 1995/1996 wet season. These results indicate that simple dilution does not account for the observed changes in the pH of Boulder Creek in early wet season storms.

Effect of Sulfate Salt Dissolution

Chemical responses to storms are almost instantaneous in Boulder Creek. There are no obvious surface sources of metals and acidity that could respond so quickly to increases in discharge in Boulder Creek, as is observed, other than soluble sulfate salts. The results of EQ6 (Wolery and Daveler, 1992) calculations, which simulated the titration of sulfate salts into diluted Boulder

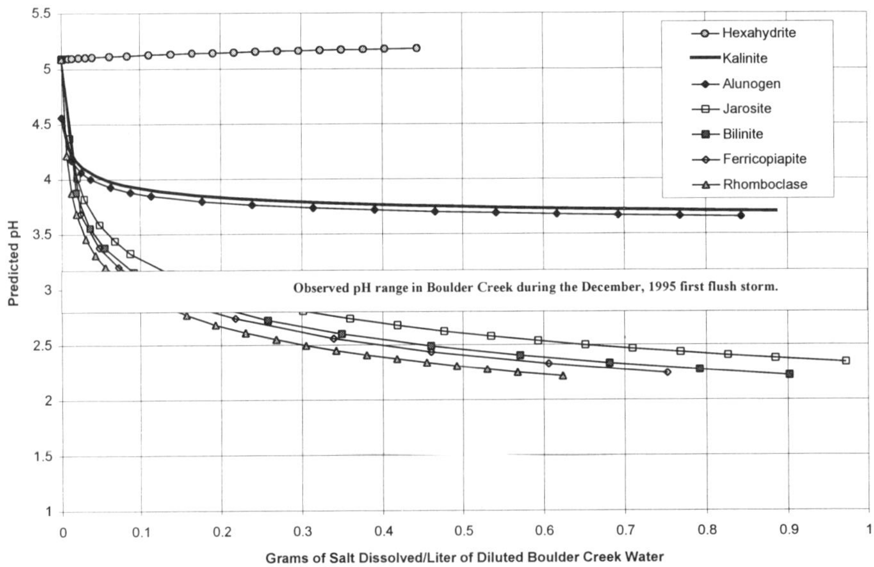

Figure 3 Predicted change in pH as sulfate salts are titrated into diluted Boulder Creek water

Creek water, show that the addition of small amounts of Fe-bearing salts can lower the pH to below 3.0, as observed in the creek. The influence of these minerals on the chemistry of storm runoff was previously discussed by Bayless and Olyphant (1993). The modeling results presented here additionally show that rhomboclase dissolution causes the most rapid decrease in pH per gram of salt dissolved, followed by ferricopiapite, bilinite, and jarosite (Figure 3). The additional acidity due to rhomboclase dissolution results from the release of the hydronium ion (H_3O^+) within its crystal structure. In contrast, it is not possible to lower the pH to the observed stream values with any amount of dissolution of the Al- or Mg-bearing minerals hexahydrite, alunogen, kalinite, or gypsum.

CONCLUSIONS

Studies of mine drainage have historically emphasized the oxidation of sulfide minerals as the predominant source of acidic solutions in ARD impacted areas. Our results show that efflorescent sulfate minerals, resulting from the oxidation of pyrite, are a potent intermediary source of acid and metals. Over the long term, weathering of metal-sulfide minerals is certainly the ultimate source of ARD; however, this long-term effect is transmitted to water by the precipitation and dissolution of soluble by-products, such as hydrated sulfate minerals. These minerals are generally most abundant on the ground surface and are concentrated by evaporation of sulfate-rich waters. The chemistry of Boulder Creek during storms following prolonged dry seasons at Iron Mountain indicates that these salts add significant acidity and dissolved metals to surface runoff upon dissolution.

ACKNOWLEDGMENTS

The authors are grateful to Stauffer Management Company for their financial support of this research and for their permission to publish these results. We also would like to thank SVL Laboratories, Kellogg, Idaho for contributing chemical analyses of the Boulder Creek stream precipitates.

REFERENCES CITED

Bayless E.R. and Olyphant G.A. (1993) Acid-generating salts and their relationship to the chemistry of ground water and storm runoff at an abandoned mine site in southwestern Indiana, U.S.A. *Journal of Contaminant Hydrology* 12, 313 - 328

Kinkel A.R., Hall W.E. and Albers J.P. (1956) Geology and base metal deposits of the West Shasta copper-zinc district, Shasta County, California. *U.S. Geol. Survey Professional Paper* 285, 156 p.

U.S. Environmental Protection Agency (EPA), 1992. Remedial Investigation Report Boulder Creek Operable Unit Iron Mountain Mine Redding, California, May.

Wolery, T.J. and Daveler, S.A. (1992) *EQ6: A Computer Program for Reaction Path Modeling of Aqueous Systems: User's Guide and Documentation.* Lawrence Livermore National Laboratory, University of California, Livermore, CA, 255 pages.

Tailings and Mine Waste '99 © 1999 Balkema, Rotterdam, ISBN 90 5809 025 6

Estimating groundwater discharge by hydrograph separation, Iron Mountain, CA

S.R.Hannula, D.C.Keith, K.J.Esposito, J.A.Chermak & D.D.Runnells
Shepherd Miller Incorporated, Fort Collins, Colo., USA

ABSTRACT: Iron Mountain is located in the West Shasta Mining District in California. An investigation of the generation of acid rock drainage and metals loading to Boulder Creek at Iron Mountain was conducted. As part of that investigation, a hydrograph separation technique was used to determine the contribution of groundwater to total flow in Boulder Creek.

During high-flow storm events in the winter months, peak flow in Boulder Creek can exceed 360,000 gpm, and is comprised of surface runoff, interflow, and groundwater discharge. Total groundwater discharge to the creek, as estimated from the hydrograph separation analysis, approaches 5,000 gpm during the high-flow season. Using drainage basin apportionment and other studies, the groundwater contribution from the mineralized area of Boulder Creek was determined as a proportion of the total groundwater discharge.

INTRODUCTION

One of the many hydrologic and geochemical investigations at Iron Mountain near Redding, California is the characterization of groundwater and its role as a transport mechanism of metals and acidity to Boulder Creek. Quantifying the contribution of acidity and metals from groundwater to Boulder Creek is important in establishing appropriate remediation goals for the site.

This paper focuses on estimating groundwater discharge as a proportion of the total discharge in Boulder Creek during the wet season. The total discharge in Boulder Creek is comprised of groundwater, surface runoff, and interflow. Each component is a fluctuating contribution, depending on the season and intensity of storm events. The groundwater discharge component is estimated using a hydrograph analysis method described by Moore (1992). The resulting groundwater discharge values are then apportioned to the mineralized area along Boulder Creek and used in a metals loading analysis. The primary purpose of this paper is to estimate the groundwater contribution to Boulder Creek; however, a short discussion of the metals loading due to groundwater is also included.

SITE BACKGROUND

Iron Mountain is located within the historic West Shasta Copper-Zinc Mining District approximately 9 miles west of Redding, California (Figure 1). The district was mined for copper, silver, gold, zinc, and pyrite. Mining at Iron Mountain began in the 1860s and continued until 1963. The mineralized areas drain either to Boulder Creek to the north or Slickrock Creek to the south. These two drainage basins are tributaries to Spring Creek, which ultimately discharges into the Sacramento River.

In 1982, the Iron Mountain Mine was placed on the Environmental Protection Agency's National Priorities List for superfund restoration activities. Historically, and since being listed as a superfund site, various remedial efforts have greatly reduced metals loading from Iron

Mountain. This investigation is part of the current effort to better understand the hydrogeology of the site and to help establish a basis for evaluating future remediation alternatives.

HYDROGEOLOGY OF BOULDER CREEK

Boulder Creek is a perennial stream with yearly behavior that can be divided into two regimes: a dry season and a wet season. During the dry season (from May to October), the area receives very little precipitation and Boulder Creek discharge can recede to a low of 100 gpm. During the wet season (from November to April) intense rainfall events have produced discharges in Boulder Creek in excess of 360,000 gpm.

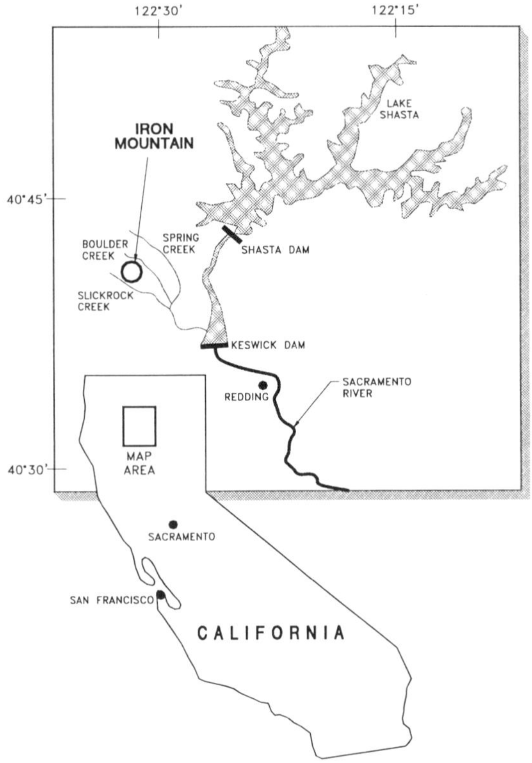

Figure 1 Location map

The Boulder Creek drainage basin is approximately 1,732 acres. The headwaters of Boulder Creek originate at an elevation of approximately 3,470 feet and the mouth, at the confluence with Spring Creek, is at an elevation of approximately 1,400 feet. The average grade is 10 percent. The rapid response of the flow to storms and the steep grade make Boulder Creek a very dynamic stream.

Mining-related activities primarily occurred on approximately 106 acres west of Boulder Creek, between weir stations BCW12 and BCW19. Figure 2 shows the potentiometric surface and the location of most monitoring wells. A prominent geologic feature within the area of Figure 2, is a landslide that extends from above the Richmond Mill Buildings to Boulder Creek. The landslide is comprised of young slide debris and fragments of older underlying Balaklala Rhyolite and Copley Greenstone bedrock (Kinkel et al., 1956). The slide debris is comprised primarily of large blocks and cobbles of massive and disseminated gossan, with a matrix of a fine sand to gravel with substantial clayey material. The thickness of the landslide ranges from zero to over 60 feet, and averages near 40 feet (Cooksley, 1995). The presence of small slope failures in the form of debris flows at several locations within the site indicates periodically over-saturated conditions in the subsurface (Chorely et al., 1985). Evidence of fresh cohesionless material is common, particularly after a heavy rain.

The contact between the colluvium (landslide) and the underlying bedrock is a zone of fractured and weathered bedrock. This zone is considered to be the bottom of the near-surface aquifer. From geophysical and drilling data, the bedrock beneath the colluvial layer appears to be competent with low permeability. The movement of groundwater within the bedrock is generally considered to be slow relative to the near-surface groundwater system.

The natural leaching of the heavily mineralized gossan and underlying bedrock contributes a significant metals load to Boulder Creek. This naturally occurring metals load is transmitted to Boulder Creek via groundwater and interflow, hence understanding the groundwater characteristics is essential to understanding the metals loading in Boulder Creek.

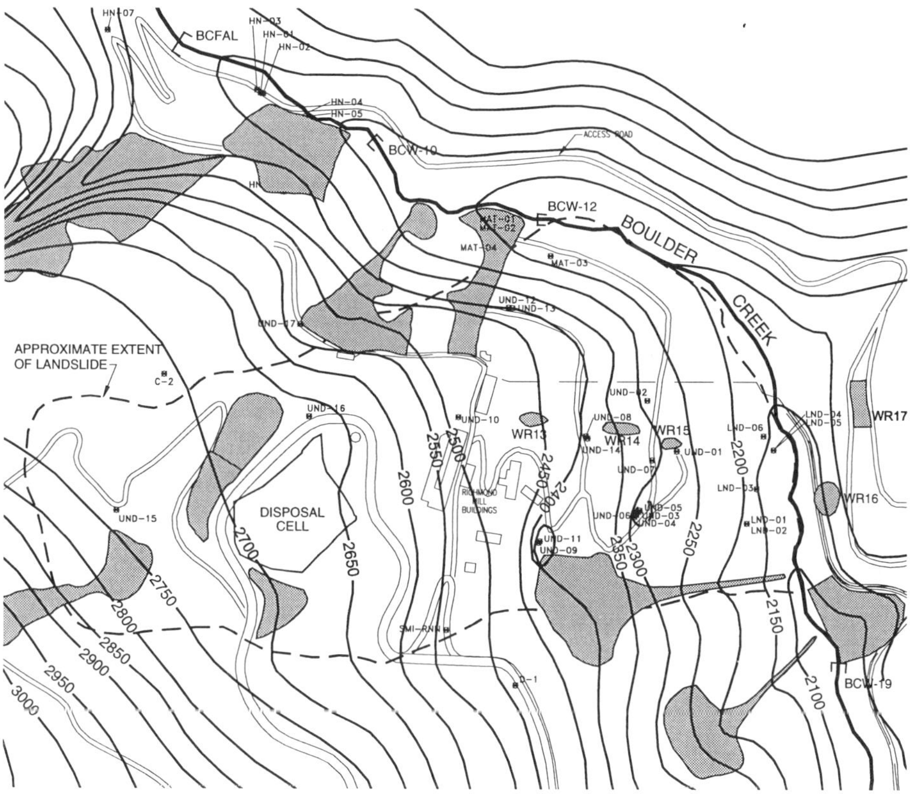

Figure 2 Boulder Creek mineralized area potentiometric map and monitoring well locations

ESTIMATE OF GROUNDWATER FLOW FROM THE ENTIRE BOULDER CREEK DRAINAGE BASIN

The potentiometric surface shown in Figure 2 generally follows the land surface topography. The large groundwater underdrain toward the northwest is caused by the underground mine workings. The gradient of the groundwater table does not change significantly throughout the year. However, the water level in individual monitoring wells is variable and can change as much as 17 feet in response to a storm or as little as 1 foot over a season. The changes in water level of the monitoring wells are not significant compared to the potentiometric relief across the site, therefore there is little change in groundwater gradient. The changes in water level alter the saturated thickness of the aquifer, which proportionally alters the transmissivity.

To distinguish the groundwater flow component from the other flow mechanisms (interflow and surface flow), a hydrograph separation technique described by Moore (1992) was used.

The reader is referred to the original paper by Moore (1992) for the detailed development of the hydrograph separation technique. Essentially, this technique determines the groundwater contributions to the total discharge in a creek. Figures 3 through 5 show the hydrograph of Boulder Creek, measured at the Boulder Creek mouth (BCMO), between April 1995 and March 1996. As shown in Figure 4, the lowest groundwater discharge during the summer months of 1995 averaged near 150 gpm. Peak flow of approximately 360,000 gpm occurred on February 17, 1996 (Figure 5). The oscillations in the hydrograph in the summer months were caused by diurnal effects, probably when evapotranspiration was significant relative to the total discharge.

The hydrograph separation analysis begins by constructing a semi-log plot of stream discharge at BCMO versus time. The resulting curve is analyzed for straight line portions (Figure 3). In the absence of rain, the recession of streamflow should plot as a straight line on a logarithmic graph because the release of groundwater from storage can be described by the law of exponential decay (Chow, 1964). The straight line(s) have slopes characteristic of the rate of drainage from the dominant groundwater reservoir.

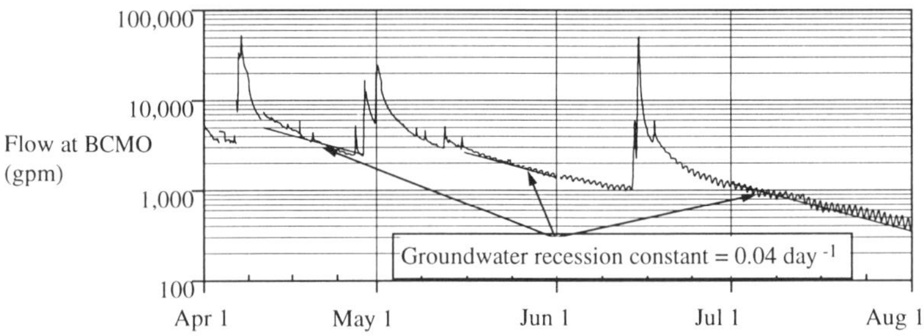

Figure 3 Boulder Creek hydrograph for April through July 1995

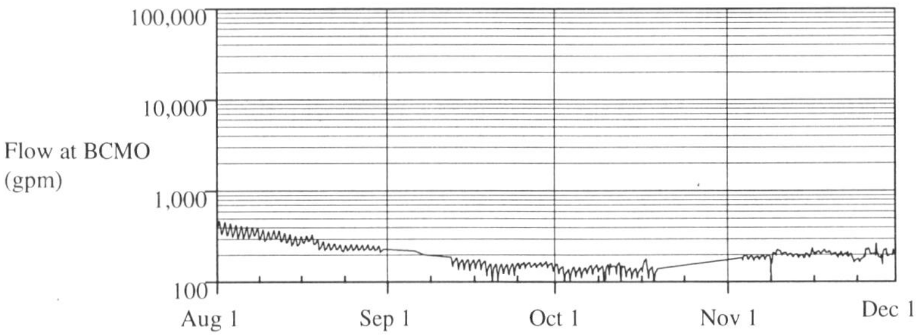

Figure 4 Boulder Creek hydrograph for August through November 1995

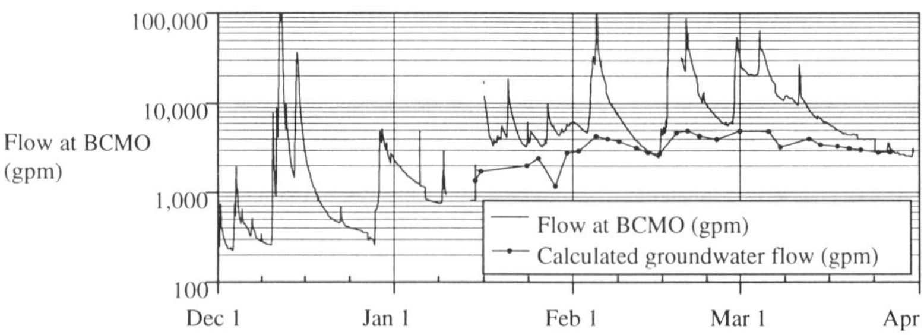

Figure 5 Boulder Creek hydrograph and estimated groundwater discharge for December 1995 through March 1996

Recession constant

Figure 3 shows the recession limb of the hydrograph forming a straight line three times from April to August 1995 with similar slopes in each instance. The slope of the line is the groundwater recession constant (K_r), which is determined from the time required for stream discharge to decline one log cycle (Moore, 1992; Chow, 1964; Mau and Winter, 1997); as shown by equation 1:

$$K_r = \ln(Y_1 / Y_2) / (t_2 - t_1) = \ln(Q_1 / Q_2) / (t_2 - t_1) \tag{1}$$

where Y_1 is groundwater stage and Q_1 is streamflow at time t_1 and Y_2 is groundwater stage and Q_2 is streamflow at time t_2.

The resulting recession rate of groundwater is 0.04 ± 0.005/day, in which the error is based on a mean-square propagation of individual errors (estimated) through the calculations. The recession constant is assumed to be constant throughout the year, except for the following circumstance; as groundwater levels recede, the conductive flow paths probably empty and the remaining groundwater will then drain through deeper, less conductive pathways. These changes in flow paths alter the rate at which groundwater drains, thus changing the groundwater recession constant (i.e. where a straight line no longer fits the curve). As shown in Figure 4, the flow in Boulder Creek does not continue to decline to zero, but levels off at an average of 150 gpm, with a recession rate that approaches zero.

Average Specific Yield and Average Transmissivity

The hydrograph separation technique described by Moore (1992) can also be used to estimate the average specific yield, and average peak transmissivity, which can be used to determine groundwater discharge during the wet season.

To calculate the average specific yield of the basin, the following equation (Moore (1992) was used.

$$Q_t / K_r = AS_yY_t \tag{2}$$

Rearranging equation 2 for specific yield gives the following:

$$S_y = Q_t / K_rAY_t \tag{3}$$

The parameters necessary to calculate specific yield (S_y) are groundwater discharge (Q_t), the recession constant (K_r), drainage area (A), and average groundwater stage (Y_t). On May 13, 1995, Boulder Creek discharge (Q_t) was approximately 3,000 gpm and was considered to be composed of 100 percent groundwater discharge (Figure 3). As mentioned previously, the drainage basin covers 1,732 acres (A), and the recession constant (K_r) is 0.04 ± 0.005/day. The average groundwater stage (Y_t) is considered to be the height of groundwater relative to some baseline level. The average groundwater stage (Y_t) chosen for the calculation is the height of groundwater above the levels recorded on December 6, 1995, because the levels on December 6, 1995 represent the lowest level to which groundwater dropped during the year. Water levels during the remainder of the year should be higher, yielding a positive stage. Table 1 shows the calculation of average groundwater stage for May 13, 1995, (Y_t = 2.90 ± 0.14 ft). The calculated stage (Y_t) is assumed to be an average groundwater stage, because heterogeneities across the site make no single well representative of the actual groundwater stage. Using these parameters in equation (3) yields an average specific yield of 0.066 ± 0.009 (unitless), calculated as follows:

$$S_y = \frac{3{,}000\ \frac{\text{gal}}{\text{min}}}{0.04\ \text{day}^{-1} * 1{,}732\ \text{acres} * 2.90\ \text{ft}} * \frac{1\ \text{acre}}{43{,}560\ \text{ft}^2} * \frac{\text{ft}^3}{7.481\ \text{gal}} * \frac{1{,}440\ \text{min}}{\text{day}}$$

$$= 0.066 \pm 0.009\ \text{(unitless)} \tag{4}$$

The estimated error of ± 0.009 is based on a mean-square propagation of individual estimated errors through the calculations.

Table 1 Specific yield calculation, May 13, 1995 data.

Well Identifier	Groundwater elevation (season low) (12-6-95) (feet)	Groundwater elevation (5-13-95) (feet)	Groundwater stage, Y_t (5-13-95) (feet)
HN-02	2330.33	2331.09	0.76
HN-04	2322.05	2322.28	0.23
HN-05	2321.85	2321.96	0.11
LND-02	2168.93	2168.85	-0.08
LND-03	2147.84	2150.63	2.79
LND-04	2141.46	2141.39	-0.07
RNN-01	2532.55	2534.51	1.96
SMI-01	3148.2	3155.97	7.77
SMI-10	2986.06	2993.22	7.16
SMI-11	3240.04	3246.35	6.31
UND-02	2289.76	2294.68	4.92
UND-03	2333.49	2330.69	
Average groundwater stage, Y_t, (ft)			2.90
Measured groundwater discharge, Q, (gpm)			3,000
Average specific yield, Sy			0.066

Table 2 Input data for groundwater discharge example calculation.

Well Identifier	Lowest groundwater elevation 12-6-95 (feet)	Groundwater elevation 2-14-96 (feet)	Stage, Y_t 2-14-96 (feet)	Groundwater elevation 2-21-96 (feet)	Stage, Y_t 2-21-96 (feet)
HN-02	2330.33	2330.66	0.33	2331.7	1.37
HN-04	2322.05	2321.80	-0.25	2322.43	0.38
HN-05	2321.85	2321.56	-0.29	2321.72	-0.13
LND-01	2157.57	2158.93	1.36	2159.54	1.97
LND-02	2168.93	2169.35	0.42	2169.33	0.4
LND-03	2147.84	2148.65	0.81	2149.28	1.44
LND-04	2141.46	2142.28	0.82		
RNN-01	2532.55	2534.18	1.63	2536.25	3.7
SMI-01	3148.2	3152.29	4.09	3160.66	12.46
SMI-10	2986.06	2991.43	5.37	2995.12	9.06
SMI-11	3240.04	3245.05	5.01	3246.1	6.06
UND-01	2286.71	2291.89	5.18	2296.84	10.13
UND-02	2289.76	2297.23	7.47	2303.12	13.36
UND-03	2333.49	2334.42	0.93	2334.77	1.28
UND-05	2281.84	2281.79	-0.05	2281.76	-0.08
UND-06	2308.35	2308.23	-0.12	2308.62	0.27
UND-07	2334.31	2336.48	2.17	2337.55	3.24
UND-08	2384.88	2394.85	9.97	2401.17	16.29
UND-09	2384.38	2385.47	1.09	2385.53	1.15
UND-11	2436.1	2442.28	6.18	2443.16	7.06
UND-12	2395.08	2396.92	1.84	2398.07	2.99
UND-16	2656.78	2664.47	7.69	2666.01	9.23
E-1	3216.73	3217.91	1.18	3219.41	2.68
Average Stage, Y_t (ft)			2.73		4.74
Calculated groundwater discharge, Q (gpm)			2,826		4,905
Average specific yield, Sy (unitless) (from equation 4)			0.066		0.066

The average peak transmissivity can be calculated from the following equation (Moore, 1992).

$$T_0 = S_y Y_0 K_r a_0 / I \tag{5}$$

The subscript "0" indicates time of peak discharge and the value of the parameters should be utilized from that time. Several variables must be defined to calculate average peak transmissivity. Average path length (a_0) is described as the average distance between the drainage divide and the creek, and is approximately 1,885 feet. The average hydraulic gradient (I) is estimated from the potentiometric surface (Figure 2) to be approximately 0.38 with an estimated error of ± 0.1. The average groundwater stage (Y_0) during peak flow conditions (February 21, 1996) is approximately 4.74 ± 0.14 feet (Table 2). Table 2 shows the data used to calculate groundwater stage on February 14, 1996, and February 21, 1996, along with the calculated groundwater discharge using a specific yield value of 0.066.

Average peak transmissivity for the basin is therefore calculated as follows:

$$T_0 = \frac{0.066 * 4.74 \text{ feet} * 0.04 \text{ day}^{-1} * 1{,}885 \text{ feet}}{0.38} = 62 \pm 20 \ \frac{\text{ft}^2}{\text{day}} \tag{6}$$

The results from the method used here describe the entire basin and are not specific to one particular location. Equation (5) can also be used to calculate the average transmissivity of the basin during non-peak flows.

Groundwater contribution to Boulder Creek

This technique (Moore, 1992) can also be used to estimate the groundwater discharge into Boulder Creek under any flow condition. To do this, equation 2 is rearranged to solve for groundwater discharge (Q_t). Several assumptions have to be made to use equation 2:

1. The area (A) that is actively draining following a storm is assumed constant. During the wet season, when the groundwater levels are above base level, the entire drainage basin can be assumed to be draining (A = 1,732 acres).
2. The soil moisture conditions are assumed to be constant during the wet season. During the first storm of the season, the soil moisture content is low and begins to increase in response to infiltrating water. For the remainder of the wet season, the soil moisture content and hence, the infiltration rate, is assumed to remain relatively constant. Furthermore, the average specific yield (S_y) and the groundwater stage (Y_t) are assumed to have a more consistent relation when soil conditions are constant.
3. The groundwater recession constant is assumed to remain constant (K_r = 0.04 ± 0.005/day).

As an example of how these assumptions are used, the data in Table 2 can be used to calculate groundwater stage (Y_t) and the corresponding groundwater contribution (Q_t) on both February 14, 1996, and February 21, 1996. Peak groundwater discharge (February 21, 1996) is calculated by using equation 7 and as shown in detail by equation 8. For this example, the required parameters are the average specific yield of the basin (0.066 ± 0.009) and stage of groundwater on February 21, 1996, (4.74 ± 0.14 ft) (Table 2), as follows:

$$Q_{peak} = A S_y Y_{peak} K_r \tag{7}$$

$$Q_{peak} = 1{,}732 \text{ acres} * 0.066 * 4.74 \text{ feet} * \frac{0.04}{\text{day}} * \frac{43{,}560 \text{ ft}^2}{\text{acre}} * \frac{7.481 \text{ gal}}{\text{ft}^3} * \frac{\text{day}}{1{,}440 \text{ min}} \tag{8}$$

$$= 4{,}905 \pm 900 \frac{\text{gal}}{\text{min}}$$

The groundwater stage used in equation 8 is the highest average level observed during the 1995\1996 wet season. Therefore, the peak groundwater discharge is estimated to be 4,905 gpm.

Using the same approach as was used in equation 8, estimates of groundwater contribution to Boulder Creek were made for each date for which a respective estimate of groundwater stage could be calculated for the 1995/1996 wet season. The results show groundwater discharge increases as groundwater stage increases. Figure 5 shows the actual hydrograph along with the estimated groundwater contributions.

Groundwater discharge estimates were not made with this procedure for the transition period from the dry season to the wet season (December to mid-January) because the relationship between groundwater stage and discharge is dominated by changes in storage and soil moisture.

APPORTIONMENT TO THE MINERALIZED AREA

Figures 3 through 5 illustrate the discharge hydrograph of Boulder Creek as measured at BCMO from April 1995 through March 1996 (SMC, 1996). Included in Figure 5 is the estimate of the total groundwater discharge for the basin. However, the metals loading does not originate from the entire basin. The principal area of interest is between stations BCW12 and BCW19 on the west side of Boulder Creek within and near the landslide area (Figure 2). To calculate the component of groundwater discharge originating from the west side of the creek between stations BCW12 and BCW19, the hydrograph (Figures 3 through 5) was apportioned based on several studies briefly summarized below.

The estimate of the minimum contribution of groundwater from the west side of Boulder Creek between BCW12 and BCW19 is based on the area of drainage into that reach. The predominantly mineralized area on the west side on Boulder Creek between stations BCW12 and BCW19 covers 6 percent of the total drainage basin area; therefore the minimum proportion of groundwater discharging from the mineralized area is estimated to be 6 percent of the total groundwater discharge.

The estimate of the maximum groundwater discharge from the mineralized area is based on drainage basin apportionment, a lithium tracer test, and local geology. The total sub-basin drainage area between BCW12 and BCW19 is approximately 16 percent of the total drainage basin area. Lithium tracer tests indicate that between 20 and 27 percent of the total flow in Boulder Creek originated from between BCW12 and BCW19. Field observations of the local geology show that the east side of the creek has insignificant colluvial thickness, and therefore does not transmit as much groundwater to Boulder Creek as does the landslide. Together, these studies suggest that the maximum groundwater contribution from the mineralized area is 20 percent of the total groundwater flow during high flow conditions.

Using the peak groundwater flow of 4,905 gpm calculated above, the estimated groundwater discharge from the mineralized area ranges from a minimum of 300 gpm (6%) to a maximum of 980 gpm (20%) during the wet season. These estimates of groundwater discharge from the mineralized area are important in determining future remedial options and effectiveness.

USING GROUNDWATER DISCHARGE VALUES TO CALCULATE METALS LOAD

Groundwater discharge estimates such as these can be used to estimate metals loading from groundwater to streams in mineralized areas. For example, the potential metals loading from groundwater was estimated using an average metal concentration in the groundwater (i.e. from monitoring wells) and multiplying it by the groundwater discharge rate estimated for the mineralized area.

Groundwater characteristics and metals loading due to groundwater were investigated at Iron Mountain during the dry season by diverting Boulder Creek around the base of the landslide through a temporary culvert and collecting the water that entered the isolated stream channel (Hall and Ekoniak, 1997). The average copper and zinc concentrations measured during that study were approximately 6 and 11 milligrams per liter (mg/L), respectively, and the flow rate from the isolated channel was approximately 120 gpm. Wet season groundwater loading estimates can be made using these concentrations and the wet season groundwater discharge rates (300 to 980 gpm). This calculation yields a potential copper loading rate between 0.9 and 3 lbs/hr and a potential zinc loading rate between 1.7 and 5.5 lbs/hr during the wet season.

CONCLUSIONS

A hydrograph separation technique was used with weir and water table measurements to determine the groundwater discharge from Boulder Creek drainage basin during the wet season of 1995/1996. A peak groundwater discharge of 4,905 gpm was calculated. Basic drainage basin apportionment and other field studies suggest that between 6 and 20 percent of the groundwater discharge originates from the mineralized area between stations BCW12 and BCW19. A simple metals loading analysis for the same reach of Boulder Creek suggests that the potential copper loading rate from groundwater during the wet season is between 0.9 and 3 lbs/hr and between 1.7 and 5.5 lbs/hr for zinc.

REFERENCES

Chorley, R.J. Schumm, S.A. & Sugden, D.E. 1985. *Geomorphology*. Methuen & Co. New York, New York.

Chow, V.T. 1964. *Handbook of Applied Hydrology*. McGraw Hill. New York.

Cooksley Geophysics, Inc. 1995. Refraction seismic survey of landslide area at Boulder Creek at the Iron Mountain Mine, Shasta County, California. May.

Hall L.E. & Ekoniak P.E. 1997. Determination of natural background and metal loading to Boulder Creek at baseflow conditions, Iron Mountain Mines, Redding, California. *Geological Society of America Annual Meeting Abstracts*. 1997. Salt Lake City, Utah. p. 126.

Kinkel, A.R., Hall W.E. & Albers J.P. 1956. Geology and base-metal deposits of west Shasta copper-zinc district Shasta county, California. Professional Paper 285. U.S. Geological Survey.

Mau, D.P., & Winter T.C. 1997. Estimating ground-water recharge from streamflow hydrographs for a small mountain watershed in temperate humid climate, New Hampshire, USA. *Ground Water*. v. 35. no. 2. pp 291 - 304.

Moore, G.K. 1992. Hydrograph analysis in a fractured rock terrane. *Ground Water*. v. 30. no. 3. pp. 390-395.

Stauffer Management Company (SMC). 1996. Iron Mountain Summary Report, Stream Monitoring Program, Rainfall and Metals Loading Data. Wilmington, Delaware.

Tailings and Mine Waste '99 © 1999 Balkema, Rotterdam, ISBN 90 5809 025 6

Uncertainty: Geochemical water quality predictions of an open pit mine

Walter L. Niccoli & Jim B. Finley
Shepherd Miller Incorporated, Fort Collins, Colo., USA

ABSTRACT: Following cessation of active mining in an open pit, water discharging from multiple groundwater sources began filling the pit. One of the major environmental questions regarding permitting of pit-lakes concerns the long-term evolution of pit-lake chemistry. Shepherd Miller, Inc. (SMI) was commissioned to predict the future concentrations of constituents in the pit lake. This paper considers the application of the Latin Hypercube (LHS) sampling technique for estimating the potential range in predicted sulfate concentrations due to the variability in sulfate concentrations in the various sources of water discharging to the pit-lake. Based on the uncertainty analysis, SMI concluded that there was a 60% probability that the future concentration of sulfate in the pit-lake would exceed the concentration predicted using average sulfate concentrations as model input. The LHS approach provided an efficient and robust method for incorporating uncertainty analysis in pit-lake modeling, especially when compared with the computational burden of Monte Carlo methods.

INTRODUCTION

As part of the mine closure, an assessment of long-term water quality in the pit lake was required. Using a state of the art water balance model and known hydrologic input parameters, the steady-state water level was predicted to develop in approximately 200 years. Although the pit-lake will be through-flowing, evaporation dominates the flow of water from the pit-lake. SMI subsequently predicted, using average influent concentrations, that the sulfate concentrations in the lake would be approximately 1,800 mg/L 1,000 years after cessation of mining and would ultimately reach an equilibrium concentration of approximately 1,838 mg/L. As part of the overall modeling effort, an uncertainty analysis for the concentrations of constituents was also incorporated into the combined water and chemical mass balance model. This paper discusses the analysis of uncertainty as applied to the concentration of sulfate in the pit-lake.

APPROACH

In order to describe the uncertainty analysis, a general explanation of the pit lake geochemical model for sulfate is in order. Figure 1 shows the conceptual sulfate mass balance developed for the pit lake. Basically, five hydrologic flow components contribute sulfate mass to the pit lake where evaporation concentrates the sulfate. The five main flow components contributing sulfate mass to the pit lake include: (1) non-impacted ground water flow from the Upper Formation, (2) ground water flow from the unaffected Lower Formation, (3) impacted ground water flow from the Upper Formation, (4) impacted flow from the Lower Formation, and (5) precipitation. Yearly average concentration values for each of the five hydrologic flow components were quantified using: (1) data collected from a series of monitoring wells that were completed in each of the aquifers, and (2) regional meteorological information. Sulfate concentrations associated with these five hydrologic flow components were then evaluated using the uncertainty analysis as described in the following section.

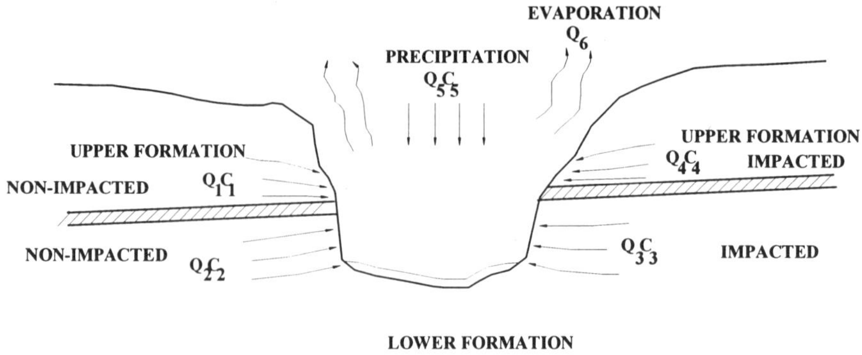

Figure 1. Conceptual sulfate mass balance model

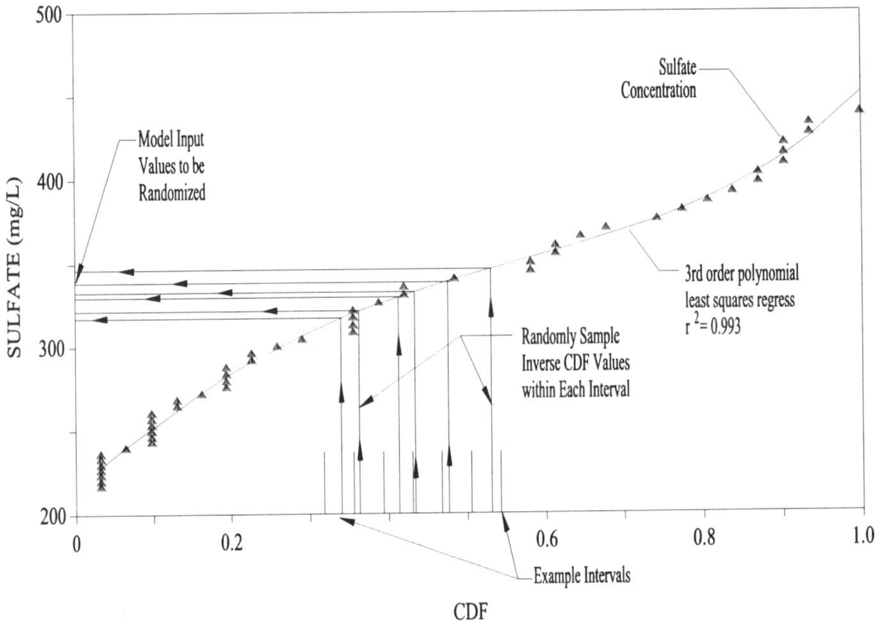

Figure 2. Example of the LHS technique

UNCERTAINTY ANALYSIS

To address the uncertainty in selecting values of sulfate input concentrations, SMI applied the Latin Hypercube Sampling (LHS) technique as described by McKay et al. (1979). LHS is similar in concept to the Monte Carlo approach for quantifying uncertainty. The Monte Carlo technique is based on simple random sampling (SRS) of the input variables and generally requires a large number of realizations (simulations) to provide statistically meaningful results. In contrast, the LHS combines SRS and stratified sampling techniques to provide statistically significant results with substantially fewer simulations. Compared to the Monte Carlo approach, LHS produces an unbiased estimate of the mean and a probability distribution function of the model output, while creating a smaller variance. The variance reduction of LHS translates to fewer model simulations necessary to obtain the same degree of precision as SRS without data stratification. To generate probability distributions of final pit-lake sulfate concentrations, the LHS was performed using the following steps:

1. Assign an inverse cumulative distribution function (cdf) for the sulfate concentrations in each hydrologic flow component. The inverse cdf is needed to back calculate the parameter value for each randomly chosen cdf value.

2. Choose the number of simulations (N) to be performed.

3. Divide the cdf for each variable into N equi-probable intervals.

4. For each interval, choose a random sample, x, from the inverse cdf and develop a data set for each parameter (this data set will reproduce the cdf of each parameter).

5. Randomly select sulfate concentrations from each data set (one for each of the hydrologic flows) to create model input sets.

6. Use an the coupled hydrologic and geochemical model to conduct a simulation for each model input set. Based on the results of N model runs, develop a cumulative probability function for the model simulations.

These steps are illustrated by the hypothetical example shown on Figure 2.

RESULTS AND DISCUSSION

The inverse cdfs based on the five sulfate inflow concentration data sets are shown on Figures 3 through 7. Polynomial and linear relationships were developed to represent the inverse cdfs in order to automate the LHS process. LHS was then performed on each of these cdfs, and an input data set was generated for each source of hydrologic flow, and N model input sets were created by randomly selecting from each data set. A geochemical dynamic systems model was then used to conduct a simulation for each model input set. The results of the uncertainty analysis generated N sets of output showing the time-evolution of sulfate concentration in the pit-lake (Figure 8).

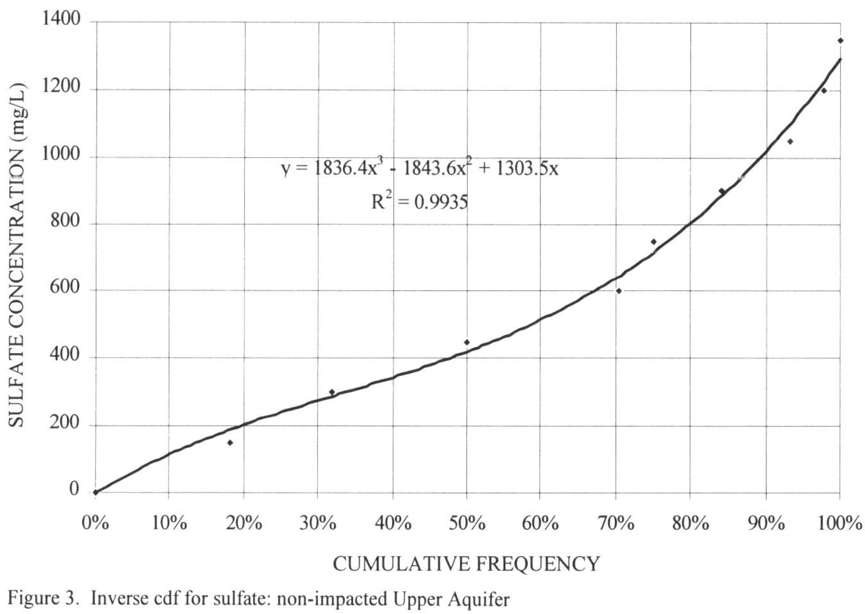

Figure 3. Inverse cdf for sulfate: non-impacted Upper Aquifer

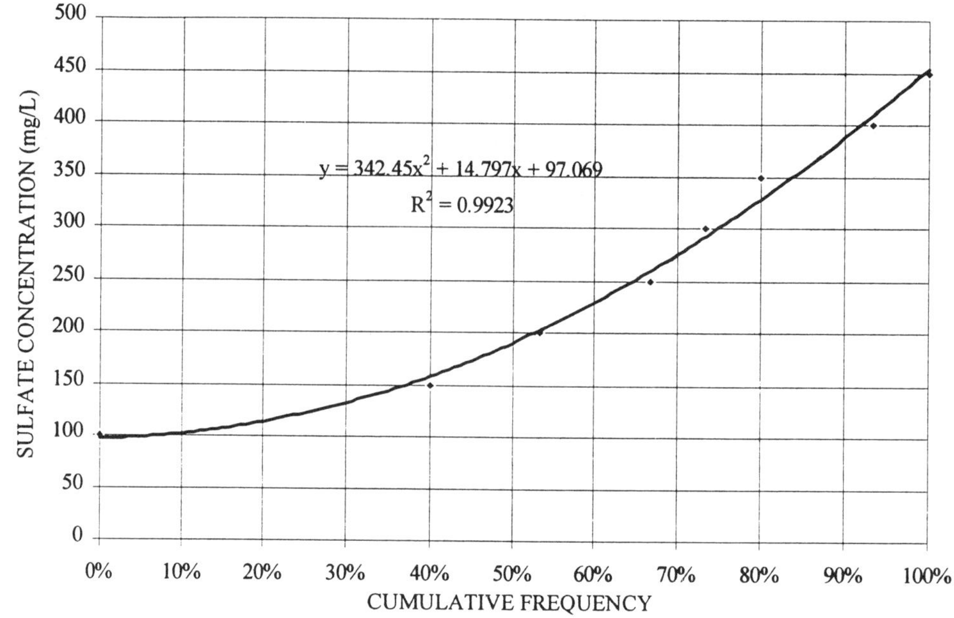

Figure 4. Inverse cdf for sulfate: non-impacted Lower Aquifer

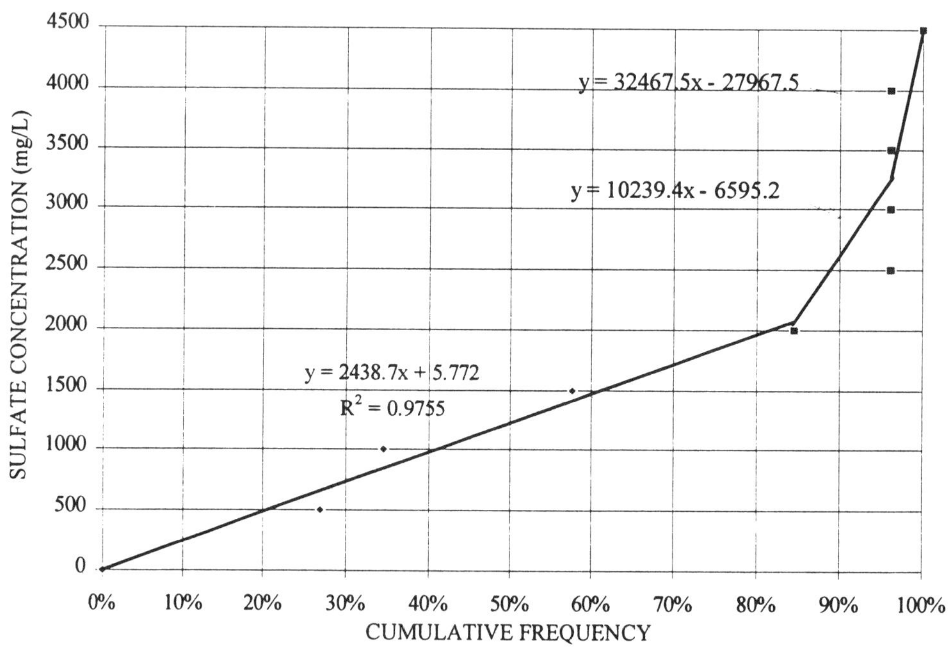

Figure 5. Inverse cdf for sulfate: impacted Lower Aquifer

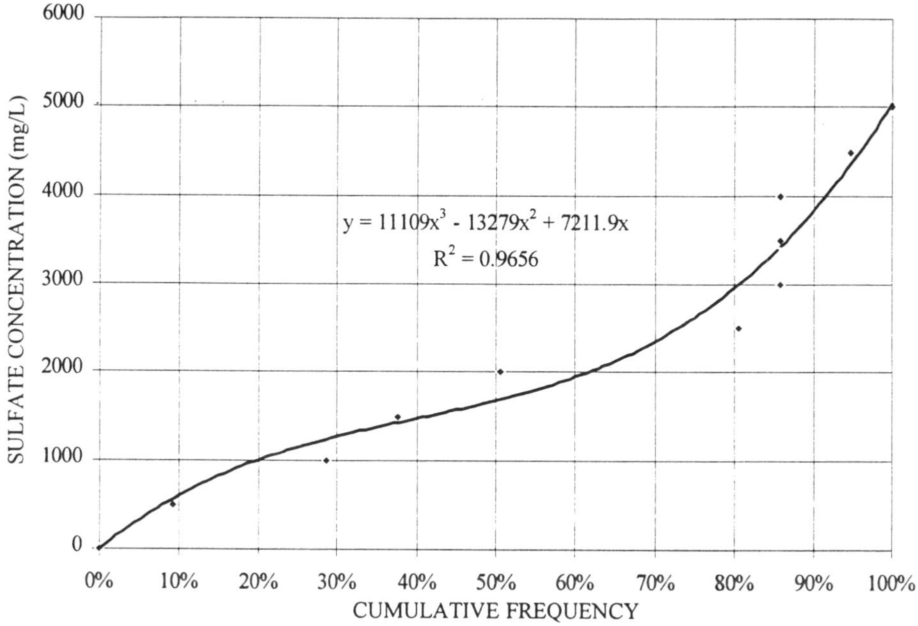

Figure 6. Inverse cdf for sulfate: impacted Upper Aquifer

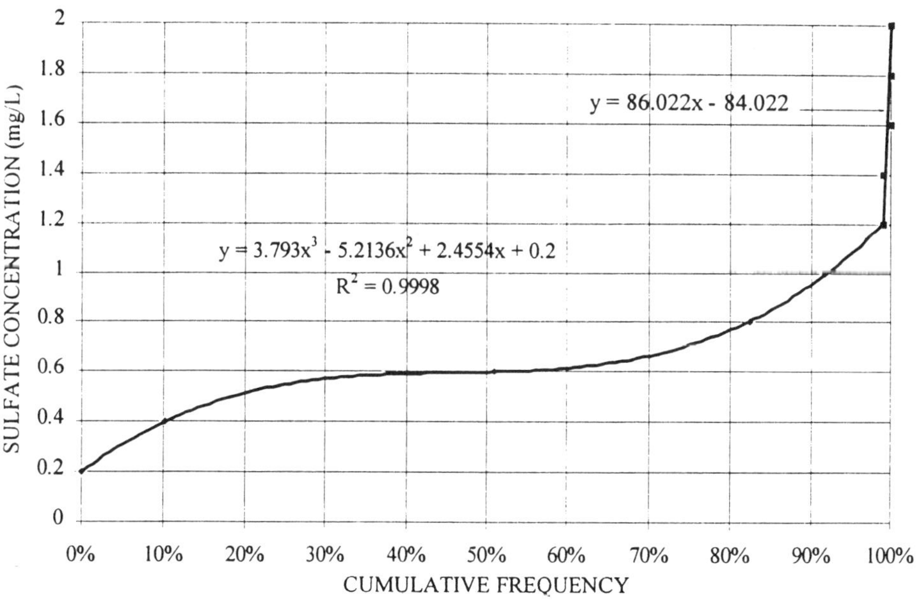

Figure 7. Inverse cdf for sulfate: precipitation

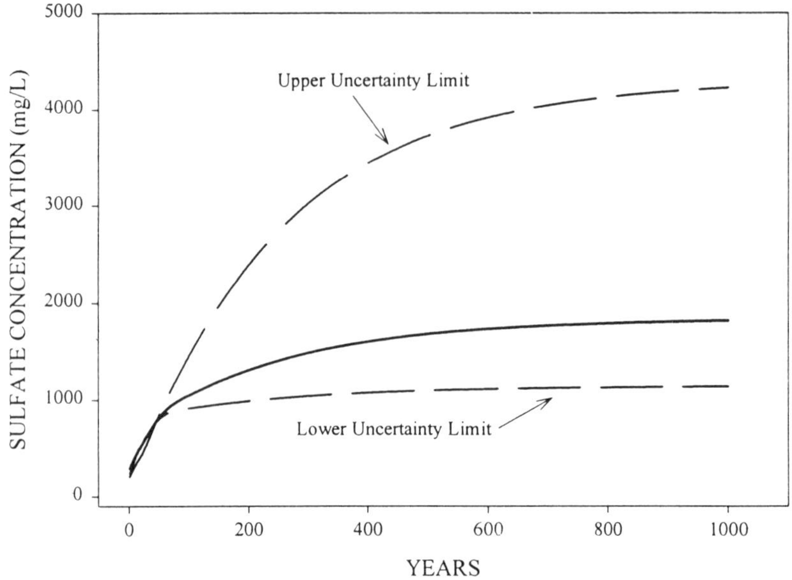

Figure 8. Sulfate concentration in the pit-lake as a function of time. Heavy line is the concentration of sulfate using average concentrations for the hydrologic flows

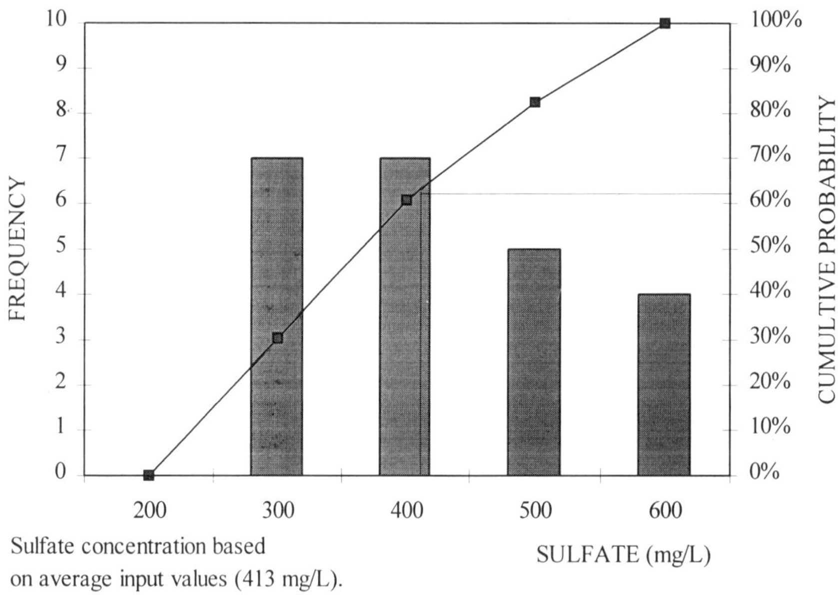

Figure 9. Probability of sulfate concentrations in the pit lake after 10 years of filling

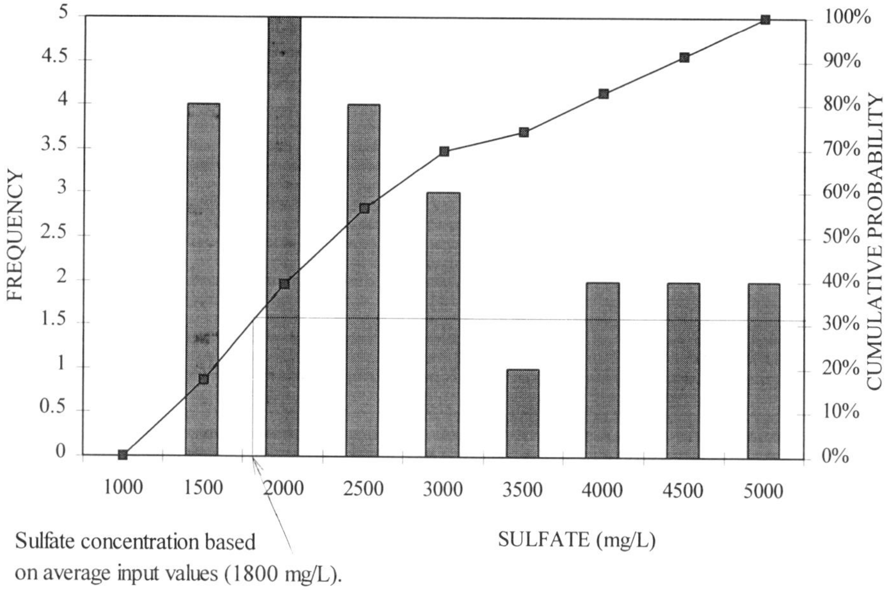

Figure 10. Probability of sulfate concentrations in the pit lake after 1000 years of filling

Based on 20 realizations, the cumulative probability for sulfate concentrations in the pit lake at year 10 is presented on Figure 9. As shown, there is essentially a 100% certainty that after 10 years of filling, the sulfate concentrations in the pit lake will be below 575 mg/L. Based on the cdf shown on Figure 9, there will be a 40% chance that sulfate concentrations will exceed the value predicted using average input concentrations. The probability of exceeding the predicted value increases with time and by year 1000 there is a 70% chance that the predicted value will be exceeded. Two possible reasons for this increasing trend are: (1) the sulfate concentration distributions for the five hydrologic flows do not follow a normal distribution, and (2) the relative proportions of the 5 hydrologic flows change with time.

CONCLUSIONS

Through the use of LHS, the uncertainty associated with the pit-lake sulfate concentration predictions was quantified. The most significant finding is that the probability of exceeding the predicted sulfate concentration from average input values increases with time. The LHS method provided an efficient and statistically robust method to estimate the uncertainty in sulfate concentrations with a fraction of the computational effort required to perform a standard Monte Carlo uncertainty analysis. Additionally, the LHS method provided an estimate of the most statistically likely predicted sulfate concentration in the pit lake.

REFERENCES

McKay, M.D., R.J. Bechman, and W.J. Conover, 1979. "A Comparison of Three Methods for Selecting Values of Input Variables in the Analysis of Output from A computer Code," Technometics. Vol. 21, No. 2, May, pp. 239-245.

Surface water quality

Tailings and Mine Waste '99 © *1999 Balkema, Rotterdam, ISBN 90 5809 025 6*

A multistakeholder approach to update Canadian mine effluent regulations

W.B. Blakeman
Mine Environmental Consulting Associates, Ottawa, Ont., Canada

P.G. Finlay
Minerals and Metals Division, Environment Canada, Ottawa, Ont., Canada

C.G. Dumaresq
National Environment Effects Monitoring Office, Environment Canada, Ottawa, Ont., Canada

ABSTRACT: The Federal *Fisheries Act* prohibits the discharge of substances harmful to fish and their habitat in all Canadian fish bearing waters except as authorized by federal regulations. The Metal Mining Liquid Effluent Regulations (1977) set authorized limits for metals and suspended solids in, and the radioactivity and pH of, effluents discharged from all base metal, iron ore, and uranium mines which commenced operation after 25 February 1977. Federal mine effluent guidelines apply to mines that were in operation prior to February 1977. Currently 85 mines are subject to the regulations or the guidelines.

In 1993, Environment Canada initiated a multistakeholder process, Assessment of Aquatic Effects of Mining, AQUAMIN, to identify and assess the effects of mining on the aquatic environment and to determine the efficacy of the MMLER in protecting fish/fish habitat prior to amending this regulation.

This paper discusses the AQUAMIN process carried out by scientific and technical representatives of the mining industry, federal and provincial agencies, environmental organizations and aboriginal communities. As well, it outlines the AQUAMIN recommendations and the objectives and activities of Environment Canada's multistakeholder path forward to implement these recommendations.

1 INTRODUCTION

In 1990, the Canadian government issued an Environmental Policy Document, which called for the "updating and strengthening" of the federal *Fisheries Act* regulations that applied to the pulp and paper and the metal mining sectors. These regulations were developed and promulgated in the mid 1970's.

This paper describes the initial and current multistakeholder processes undertaken by Environment Canada in responding to the 1990 Environmental Policy Document.

2 BACKGROUND

The metal mining sector generates approximately $12 billion in annual sales and accounts for 2-4% of Canada's Gross Domestic Product. The major metal commodities are: copper, nickel, silver, lead, iron, uranium, zinc, gold and molybdenum. Currently approximately 85 metal mines are subject to the Metal Mining Liquid Effluent Regulations and Guidelines (1977), and in the order of 50 gold mines are subject to provincial or Territorial Water Board effluent quality requirements.

As shown in Figure 1, the greatest numbers of base metal, iron ore and gold mines are located in the Precambrian Shield from Newfoundland to Quebec, Ontario, Manitoba and the Northwest Territories. The Appalachian region of New Brunswick and Quebec also contains

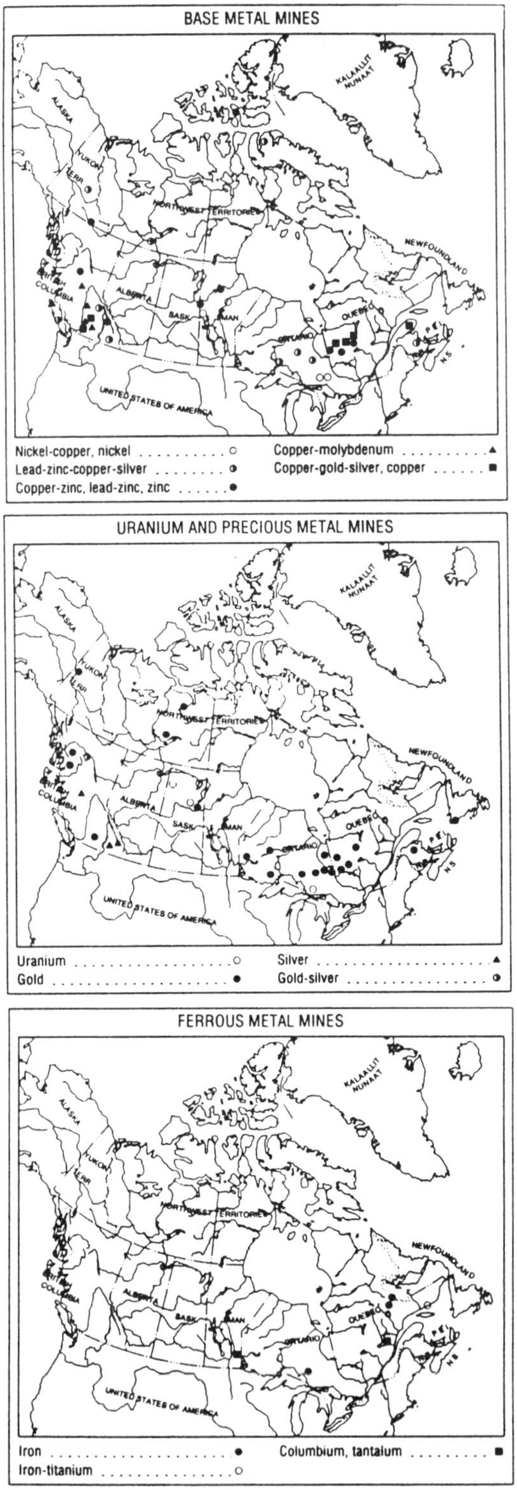

Figure 1.
Mining Association of Canada. 1997. *Mining in Canada--Facts and Figures*, 1996.

base metal, primarily lead-zinc, mines. As well, the Western Cordillera of British Columbia and the Yukon support base metal, gold and molybdenum mining operations. In addition, two lead-zinc mines are in operation on islands in the high Arctic.

3 THE USE OF THE *FISHERIES ACT* IN CONTROLLING THE QUALITIES OF INDUSTRIAL EFFLUENTS

The water pollution provisions of the *Fisheries Act* are the strongest federal environmental tools because they absolutely prohibit the discharge of harmful substances to all Canadian fish bearing waters! The principal provisions of the Fisheries Act that apply to industrial effluents prohibit the discharge of harmful substances into fish bearing waters anywhere in Canada except where authorized by regulations under the Act.

The federal Metal Mining Liquid Effluent Regulations (SOR/77-178) were promulgated under section 36(2) of the *Fisheries Act,* on 25 February 1977. These regulations apply to all base metal, iron ore and uranium mines coming into production or expanding after 25 February 1977 and prescribe authorized limits for total arsenic, copper, lead, nickel, zinc, suspended matter and the radioactivity of radium 226 in, and the pH of, effluents discharged to fish bearing waters.

The regulations define "metal mine" as any mine or mill producing or processing ores containing arsenic, copper, lead, nickel, and zinc plus such other metals as molybdenum, cadmium, titanium, selenium, antimony, tungsten, mercury, etc.

The Metal Mining Liquid Effluent Regulations do not apply to gold mines using the cyanidation process to recover gold that accounts for 50% or more of the value of the mine's production. In 1997 approximately 50 gold mines were using cyanide and were not subject to MMLER.

The intent of these regulations is to limit the discharge of the above identified deleterious substances from base metal, uranium and iron ore mines. New, expanded and reopened mines are subject to Regulations, whereas mines existing in 1977 are covered by the federal Metal Mining Liquid Effluent Guidelines which contain effluent quality objectives that are numerically equal to the MMLER effluent quality limits. A mine effluent that meets the Guideline objectives is considered to be meeting the spirit of the law.

Table 1. Authorized Levels of pH

Parameter	Min. Monthly Arithmetic Mean	Min. pH in a Composite Sample	Min. pH in a Grab Sample
pH	6.0	5.5	5.0

Table 2. Authorized Levels of Substances (Total concentrations except for Ra 226)

Item Substance	Max. Authorized Monthly Arithmetic Mean Concentration (mg/L)	Max. Authorized Concentration in a Composite Sample (mg/L)	Max. Authorized Concentration in a Grab Sample (mg/L)
1. Arsenic	0.5	0.75	1.0
2. Copper	0.3	0.45	0.6
3. Lead	0.2	0.30	0.4
4. Nickel	0.5	0.75	1.0
5. Zinc	0.5	0.75	1.0
6. Total Suspended Matter (TSM)	25.0	37.5	50.0
7. Radium 226	10.0 pCi/L	20.0 pCi/L	30.0 pCi/L

The MMLER apply uniformly as national baseline standards to provide the necessary protection to fish and other aquatic life. However, in certain locations where they fail to protect the fishery resources, more stringent requirements may be imposed by the federal government through a site-specific regulation. Where provincial requirements are more stringent than the federal regulations, the more stringent requirements prevail.

The MMLER effluent quality limits are based upon and are achievable through the application of Best Practicable Technology (B. P. T.) available in the mid 1970's, which in this case, was pH control and physical settling.

The numerical limits in the federal regulations are comparable to standards in other Canadian jurisdictions. For example, Ontario's effluent limits are identical to those in the MMLER for arsenic, copper, lead, nickel and zinc. However they set a more stringent limit for total suspended matter as well as a limit for total cyanide and an upper pH limit.

The MMLER authorized limits are presented in Tables 1 and 2.

4 RATIONAL FOR EXCLUDING GOLD MINE & MILL EFFLUENTS FROM THE MMLER

Commercially proven treatment methods for the control of cyanide bearing liquid and solid wastes from gold mines were not available in the mid 1970's. Untreated cyanide bearing gold mine effluents are extremely toxic to fish. In 1977 Environment Canada excluded gold mines using cyanidation from the application of the MMLER. Instead, a gold mining working group was formed to encourage and promote the development of technologies to reduce the levels of cyanide in gold mill effluents. During the late 1970's and early 1980's, Canada developed and implemented several cyanide treatment technologies, including alkaline chlorination, the Inco SO_2/air oxidation process, hydrogen peroxide oxidation, cyanide regeneration and the optimization of natural degradation in tailings impoundments. In some instances, effluents from natural degradation systems are subsequently subjected to one or more of the aforementioned "treatment" technologies in order to meet standards prescribed by other regulatory agencies.

All of the approximately 50 Canadian gold mine effluents are now subject to limits for total and weak acid dissociable cyanide specified in provincial regulations, permits, and territorial Water Board licenses.

5 THE INDUSTRY'S MMLER COMPLIANCE PERFORMANCE

Compliance is based upon the monthly mean average of weekly samples of effluents discharged to natural receiving waters from a final control point. If all 8 parameters meet the MMLER (or MMLEG) limits for a given month, the mine is "in compliance" for that month. Annual compliance is based upon the percentage of operating months that the mines' effluents met all prescribed limits.

The industry's compliance performance has improved significantly over time as shown in Table 3. This improvement is due to the installation of effective effluent treatment systems, a

Table 3. Compliance Performance--Continuous Improvements

Year	Regulated Number of Mines	%	Guideline Number of Mines	%
1982	20	85	113	63
1986	30	60	81	65
1990/91	20	87	51	82
1994	35	95	38	95

cooperative relationship between the Mining Association of Canada and Environment Canada, as well as the Mining Association's 1987 commitment for its member companies to achieve 100% compliance.

6 ISSUES IDENTIFICATION WORKSHOP

In November, 1992, Environment Canada held a multistakeholder workshop to determine how the provisions of the existing 1977 Metal Mining Liquid Effluent Regulations could be "updated and strengthened" as required by the 1990 Environmental Policy Document. This workshop was attended by representatives of the mining industry, other federal government departments, provincial agencies, the Canadian Environmental Network and Aboriginal groups (First Nations). A major objective of this workshop was to determine if a mandatory Environmental Effects Monitoring Program (EEM) for metal mines should be required by the amended MMLER.

The most significant recommendations arising from this work shop were that:

• the effects of Canadian mining on the aquatic environment be identified and assessed through a review of an extensive body of existing reports prior to amending the regulations;

• the assessment process be carried out in a multi-disciplinary/multistakeholder basis; and

• the assessment be based upon existing information rather than being dependent on new (lengthy) research programs.

In view of the concerns pertaining to mandatory EEM that were included in the amended Pulp and Paper Regulations (1992), members of the mining sector took the position that if Environment Canada proposed to require EEM for metal mine effluents, members of the industry's scientific and technical staffs should participate in determining if there was a need for a national mining EEM program. If an EEM Program was found to be necessary, the mining sector wished to participate in the development of the relevant guidance documents and protocols.

7 THE AQUAMIN PROGRAM

The Assessment of the Aquatic Effects of Mining in Canada (AQUAMIN) was initiated in June, 1993, in response to the forgoing recommendations arising from the Issues Identification Workshop. AQUAMIN was not to be a research program, but rather it was to review and assess existing scientific reports pertaining to the effects of mining on fish, water and sediment quality, fish habitats and other aquatic biota in the receiving environment.

This program was directed by a senior level steering group, co-chaired by a representative of Environment Canada and the Mining Association of Canada (MAC). The other members represented federal and provincial environmental and natural resources agencies, industry, environmental and aboriginal organizations.

Five regional multistakeholder multidisciplinary working groups (engineers, geologists, fisheries scientists, hydrologists, biologists, mining and environmental consultants, etc.) were organized to review and assess reports on the effects of mining operations on the aquatic environments in the Precambrian Shield, the Appalachian Region, the British Columbia and Yukon Cordillera, the Prairies and the Arctic Islands.

The activities of the regional working groups were coordinated by a Secretariat at Environment Canada and supported by a multistakeholder technical working group. Members of the mining industry, academic institutions, environmental agencies, environmental and Aboriginal organizations, and federal and provincial environmental, mining and natural resources agencies submitted a total of approximately 800 reports to AQUAMIN pertaining to the effects of mining on the aquatic environment. These reports were screened by a consultant to ensure that they met prescribed criteria pertaining to:

• currency of the information (post 1985),
• base line conditions and the use of "control" sites,
• appropriate sampling, analytical and statistical procedures, and
• quality assurance/quality control measures, etc.

Over 700 reports that met the screening criteria became the "AQUAMIN Database" and were distributed among the five regional working groups. The members of each regional working group reviewed the reports from their respective areas of expertise, and collectively produced a synthesis working group report of their findings. In addition, based upon the availability of scientific information, the five working groups prepared a total of 18 site specific case studies. Upon completion of the program, the AQUAMIN Database was donated to the Mining Environment Database at Laurentian University in Sudbury, Ontario.

In summary, AQUAMIN achieved its objectives of identifying and assessing the biological and chemical effects of metal mining effluents on fish, water, sediments and benthos in Canadian rivers, streams and lakes. The review of the existing EEM information indicated that mining effluents are affecting the receiving environments, but that there are wide variations in the magnitude and extent of these effects. The results of AQUAMIN suggest that the effects of effluents discharged from the newer (post 1977) mines are less pronounced than those caused by the older mines. This is likely a function of improved quality of the effluents being achieved by the waste water treatment systems and improved waste water management programs at the post 1977 mines. AQUAMIN also indicated that it is often difficult to identify specific causes of observed effects in the receiving environment, and that commonly there are wide variations in the methods and procedures employed by different companies, consultants or regulatory agencies in carrying out the various EEM programs.

The regional reports and case studies were summarized and consolidated in a final consensus report that was completed in April, 1996, and released to federal and provincial agencies and the general public in October, 1997. This report contains recommendations to Environment Canada and Fisheries and Ocean pertaining to:

- amendments to the MMLER,
- the development of an EEM program for mines,
- the development of an Environmental Code of Practice for Mines, and
- areas requiring future research.

The implementation of the AQUAMIN recommendations will result in a national environmental protection framework consisting of three components.

a. A federal regulation to ensure a consistent, national minimum quality of effluents being discharged to aquatic ecosystems, i.e., the Metal Mining Liquid Effluent Regulations (MMLER);

b. Site specific requirements where necessary to ensure adequate protection of some aquatic ecosystems; and

c. A national environmental effects monitoring (EEM) program that would be implemented to provide information to industry and government decision-makers and the public regarding the effectiveness of both environmental protection measures and long-term regulatory strategies.

The Report advanced a total of 51 recommendations under the above three framework components, and 3 recommendations regarding the development of an up-dated Environmental Code of Practice for Mines.

8 GENERAL RECOMMENDATIONS FROM THE AQUAMIN PROGRAM

It is recommended that:

- the MMLER be amended to improve the quality of effluents discharged to aquatic ecosystems;
- Environment Canada establish a process to review the MMLER in light of future environmental effects monitoring results;
- a National Environmental Effects Monitoring Program be required under the amended MMLER;
- the revised effluent quality limits in the MMLER be based upon the Best Available Technology, economically achievable, and that the results from EEM and toxicity identification evaluation/ toxicity reduction evaluation be utilized to encourage continuous improvement in environmental protection. Additional key recommendation are presented in the Appendix.

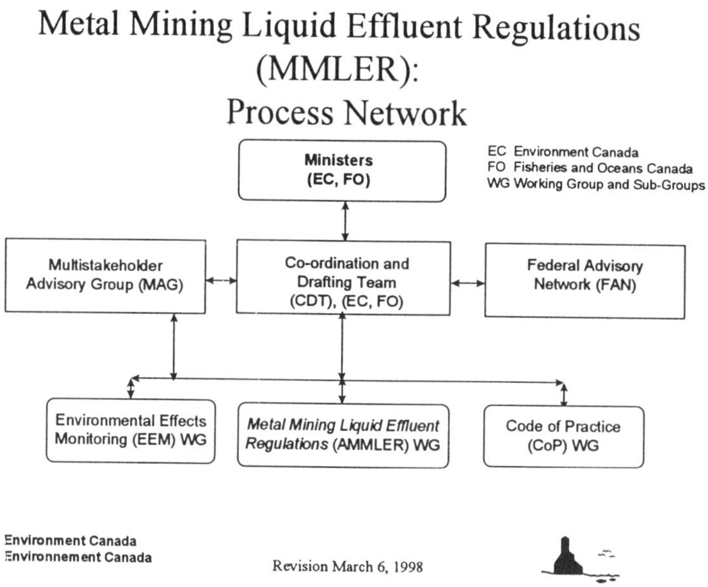

Figure 2.

9 PATH FORWARD

Environment Canada is currently (1997-98) implementing the full set of AQUAMIN recommendations through a multistakeholder, multidisciplinary process involving other federal government departments, such as Fisheries and Oceans, Natural Resources Canada, Indian Affairs and Northern Development, and the Department of Justice as well as provincial agencies, the Mining Association of Canada, and Aboriginal and Environmental Groups. The implementation process is being lead by Environment Canada in consultation with a Multistakeholder Advisory Group, a Federal Advisory Network, and three Scientific/ Technical Working Groups that will draft the modernized MMLER in accordance with the AQUAMIN recommendations, develop the EEM Program, and develop an Environmental Code of Practice for Mines. (Figure 2)

First priorities are being devoted to addressing the AQUAMIN recommendations to draft the modernized MMLER, and also to develop the EEM Program requirements and the Technical Guidance Documents. The drafting working group is reviewing international mine effluent standards and identifying modern, economically achievable treatment technologies. This will ensure that the effluent standards in the amended MMLER can be achieved through the implementation of BAT. The EEM documents are being developed by five multistakeholder scientific working groups, consisting of several former AQUAMIN participants. These working groups will focus on the roles of fish, benthic invertebrates, water and sediment quality and sub lethal toxicity as tools for assessing the effects of a mine's effluents on site specific receiving environments over time.

As the MMLER are promulgated under the *Fisheries Act*, the Minister of Fisheries and Oceans will submit the draft Amended Metal Mining Liquid Effluent Regulation and an accompanying Regulatory Impact Analysis Statement (RIAS) to the Cabinet Committee on Regulations prior to its official publication in the Canada Gazette.

10 LESSONS LEARNED

Over 100 individuals representing industry, environmental and Aboriginal groups, provincial agencies, consultants, other federal government departments and Environment Canada were able to produce a consensus report containing 51 recommendations that have been accepted as a complete package by Environment Canada. This was achieved as a result of:

- a well designed, open and transparent multistakeholder process;
- a commitment to respect the opinions of all participants;
- financial and "in kind" support totaling $140 000 over 3 years by the federal government and the Mining Association of Canada;
- a realistic time frame;
- the dedication and expertise of industry, government and non-government organizations' professional and support staffs; as well as
- the freedom to carry out AQUAMIN as a "low key" scientific program having the full support of government, industry, environmental and aboriginal stakeholders.

In conclusion, multistakeholder consultative scientific review processes can work provided that all parties "buy into" the process and that it has a realistic time table, sufficient funding, and is based upon scientific methods.

REFERENCES

Blakeman, W. B., J. E. MacLatchy and M. J. Stefanski. 1995. *The Development and Review of the Metal Mining Liquid Effluent Regulations* presented at the Canadian Institute of Mining and Metallurgy, AGM, Halifax, Nova Scotia.

Environment Canada. 1977. *Metal Mining Liquid Effluent Regulations Report,* EPS, 1-WP-77-1.

Environment Canada. 1998. *Assessment of the Aquatic Effects of Mining in Canada*, AQUAMIN Final Report.

Mining Association of Canada. 1997. *Mining in Canada--Facts and Figures,* 1996.

APPENDIX

MAJOR SPECIFIC RECOMMENDATIONS

With respect to the APPLICATION of the Regulations, it is recommended that:

- the current Metal Mining Liquid Effluent Guidelines be rescinded;
- the amended MMLER apply to all metal mines, including gold mines (excluding placer operations);
- there be a transition period to ensure that any mines that are not under the existing MMLER or not in compliance have a reasonable period to improve their control systems;
- the amended MMLER apply when the average effluent flow rate exceeds a minimum level, specified in the regulation, over a specified period of time; and
- the amended MMLER apply after commercial production ends, during the period when the closure plan for an operation is being implemented and that they cease to apply when the closure plan for an operation has been fully implemented to the satisfaction of the relevant regulatory agencies.

With respect to TOXICITY, it is recommended that;

- the amended MMLER include a requirement that all metal mine operations conduct and report periodic testing of effluent for acute lethality using current standard test methods;
- if effluents are consistently non-acutely lethal, the frequency of testing should be reduced;
- a failed test should trigger increased testing frequency; and
- if an effluent is consistently acutely lethal, the operator should be required to conduct toxicity identification evaluation/ toxicity reduction evaluation in an attempt to determine the cause of toxicity and to report the results. Results should be considered in establishing site-specific requirements ; and

as soon as an adequate body of data on the results of acute lethality testing and toxicity identification evaluation/ toxicity reduction evaluation is available, a review be conducted to determine if the MMLER should be reviewed to require non-acutely lethal effluent.

With respect to ENVIRONMENTAL EFFECTS MONITORING, it is recommended that:

• the amended MMLER include a requirement that mine operators develop, conduct, and report on a site-specific EEM program that monitors key components of aquatic ecosystems (e.g., water, sediments, fish, benthic invertebrates) and uses tools and methods that are appropriate to site conditions;

• an environmental effects monitoring program be developed for all metal mines in Canada;

• an EEM program take a progressive, phased approach to determine any effects in the aquatic environment in accordance with standard scientific methods;

• an EEM program address site-specific questions and that practitioners (stakeholders) be encouraged to design the most appropriate study for each site;

• the EEM program be effective and efficient, that all information collected have a clearly identifiable purpose, and that maximum use be made of historical data and relevant regional data;

• the EEM program ensure that study designs are scientifically defensible and that quality assurance is given high priority. Study designs must be placed in the context of multiple samplings before (where possible) and after development, and at both exposed and reference locations;

• guidelines for the process of determining what is to be considered an acceptable or unacceptable effect should be developed by Environment Canada in consultation with the Department of Fisheries and Oceans and other stakeholders;

• Environment Canada establish a central registry of reports and data generated through EEM in the mining industry; and

• Environment Canada institute an independent and periodic peer review of the EEM program to provide recommendations on revisions to the environmental protection framework.

With respect to the ENVIRONMENTAL CODE OF PRACTICE FOR MINES, it is recommended that:

• the current code be updated and additional sections be included to address environmental management and monitoring; and that

• the updated Environmental Code of Practice for Mines continue to have a significant focus on issues related to water management and the prevention of water pollution.

Tailings and Mine Waste '99 © 1999 Balkema, Rotterdam, ISBN 90 5809 025 6

Characterization of suspended sediments and colloids in Clear Creek, Colorado

Katherine A. Morris, James F. Ranville & Donald L. Macalady
Department of Chemistry and Geochemistry, Colorado School of Mines, Wheatridge, Colo., USA
Kathleen S. Smith
US Geological Survey, Denver, Colo., USA

ABSTRACT: Clear Creek is a small watershed located west of Golden, Colorado. As a consequence of past mining activities, Clear Creek water contains elevated levels of metals, including Al, As, Cd, Cu, Fe, Mn, and Zn. Some of these metals are sorbed and transported on solid phases. Because sediment transport and surface chemical reactions (such as sorption) depend on particle size, suspended sediments in water samples from five sites along Clear Creek and North Clear Creek were size fractionated for analysis. Size fractions for chemical analysis were <250μm, <25μm, <5μm, <0.45μm, and less than 30,000 Daltons. Size fractions for sediment characterization and mass analysis were 250μm-25μm, 25μm-5μm, 5μm-0.45μm, and 0.45μm-30,000 Daltons. In Clear Creek the bulk of the sediment mass (56-79%) was present in the largest fraction (250μm-25μm), whereas in North Clear Creek the bulk of the mass (57%) was in the 25-5μm fraction. A smaller, but significant amount (5-27%) was present in fractions smaller than 5μm. Given the much higher specific surface area of the fine particles, this fraction of the sediments may be the most important for sediment-transported metals in Clear Creek.

1 INTRODUCTION

The Clear Creek watershed, located in Clear Creek and Jefferson counties in Colorado, has played host to numerous historic mining operations, nearly all of which have left mine waste and underground drifts and shafts. As surface water and groundwater infiltrate these sites, the water chemistry can be considerably changed, often becoming more acidic and liberating metals from the surrounding rock. This metal-rich, acidic water enters Clear Creek as surface runoff or groundwater. The potential exists for sorption of metals onto suspended sediments (>1μm) and colloids (<1μm) downstream from mine sites (Smith et al, 1998). This is of interest in part because the speciation of metals influences toxicity. Metals that sorb onto fine particles and colloids have the potential to travel extreme distances in a river. This may pose a latent threat because the particles and colloids may reach new water chemistry conditions that do not support metal sorption. Discovering if certain metals preferentially bind to specific size fractions of the suspended sediments and colloids in the river is one more step toward understanding how to effectively treat mine and tailings waste to prevent metal contamination in water. This paper presents the initial results of a study of metal transport by various sediment size fractions in Clear Creek, and focuses on the amount and size distribution of the suspended sediment and colloids. Future publications will present the results of metal analysis of the suspended sediments and colloids.

2 METHODS

Water samples containing suspended sediments from five sites along Clear Creek and North Clear Creek, both impacted by runoff from historic mining operations, were collected and separated into five fractions based on the size of suspended sediments. Separations were achieved using sequential sieving and tangential flow filtration, with all of the particles and portions of each filtrate retained for chemical and mass analysis and particle counting. Recovered particles were centrifuged and then dried to obtain masses. Particle counts were obtained by diluting size-fractionated filtrate samples and analyzing them with a single particle optical sensor.

2.1 STUDY SITES

Study sites were chosen to investigate the impact of mining activities in Idaho Springs and along North Clear Creek on the water chemistry of Clear Creek. Study sites required the presence of a bridge accessible to foot traffic for sample collection. Sites included a footbridge at Cheyenne Street in Golden, a bridge over North Clear Creek 0.2 miles northwest of the junction of US 6 and Route 119, a bridge on County Road 314 located off I-70 exit 243, the Gilson Street bridge in Idaho Springs, and the 23rd Street bridge in Idaho Springs. Comparison of the water chemistry at the 23rd and Gilson Street bridges allowed examination of the impact of the Argo tunnel treatment plant on Clear Creek because Argo tunnel effluent enters Clear Creek between the two sampling locations. The inclusion of sites both above and below North Clear Creek (CR 314 and Golden, respectively) allowed examination of the impact of North Clear Creek on the main stem of Clear Creek.

2.2 FIELD PROCEDURES

Field measurements of alkalinity were made with a field analysis kit (Hach Chemical Company, Loveland, CO), that uses the bromocresol green-methyl red endpoint. Temperature, pH, and conductivity were measured with a multi-function field meter (Horiba U-10). The pH was calibrated with 4 and 10 buffers, and the conductivity was calibrated at three concentrations of potassium chloride. Small volume water samples were processed through a 250μm sieve, with portions retained for metal analysis. An additional portion was further processed by filtration through a 0.45μm disposable filter for ion chromatography and metal analysis. This was done to allow comparison of the "dissolved" chemistry obtained by the tangential-flow filtration process to standard filtration methods that define dissolved as < 0.45um.

2.3 SAMPLE COLLECTION

Sample collection dates were May 28, 1998, near the peak of the spring runoff, and July 16, 1998 during the trailing limb of the hydrograph. A partial sample set was collected from North Clear Creek and Golden immediately following a region-wide storm event on August 1, 1998. Samples were collected using a modified integrated sampling method designed to obtain a water sample that represents a cross-section of the creek water. Integrated samples are especially important when considering suspended sediment and colloid transport because these particles are not equally distributed throughout a river cross section. When flows were high, and the creek could not be waded, a suspended sampling device (DH-76) was lowered from a bridge. When the creek could be waded, a handheld sampler (DH-81) was used. The methods employed with both samplers were similar and involved measuring the width of the creek and dividing to obtain five evenly spaced collection sites. The sampler was then introduced into the water at a uniform rate, allowed briefly to touch the creek bed, and then retrieved at the same uniform rate. A similar rate was applied in all five collections. To ensure equitable distribution of sample volume from the five collection segments the sample collection bottle was just filled with one trip across the river. The total volume collected at each sampling location was approximately ten liters. The Golden storm sample was the only non-horizontally integrated sample and is thus termed a "grab." This sample was, however, depth integrated over the single collection location.

2.4 SIEVING AND TANGENTIAL FLOW FILTRATION

Separations were accomplished by serially sieving and filtering the entire ten-liter volume of water. At each filtration stage a portion of the water and sediment was removed for chemical analysis. The first step in the process was to pass the sample through a 250μm sieve and to collect a small portion for chemical analysis. The remaining <250μm filtrate was then poured through a 25μm sieve, and <25μm filtrate samples were taken. The particles retained on the sieve were collected for the 250μm-25μm sediment fraction. This process was repeated to obtain the remaining size fractions for both suspended sediment in water and concentrated sediment samples. The 5μm separation was achieved using a nylon mesh sieve in a Millipore Minitan-S tangential flow filter (TFF) apparatus (see below). And the 0.45μm and 30,000Daltons separations were obtained using Minitan durapore and regenerated cellulose filters respectively in Minitan TFF.

Tangential flow filtration (TFF) was used in an effort to ensure that particles collected were indeed representative of a given size fraction (Ranville & Schmiermund, 1998). This method of filtration is designed to filter larger volumes while minimizing plugging of the pore spaces in the filter. In this technique, filtrate flows through the filter membrane as in traditional filtration. However, there is also a tangential flow of "retentate" which passes over the filter membrane perpendicular to the direction of the filtrate and at a velocity several times that of the filtrate. This tangential flow is recirculated until the desired volume of filtrate is obtained. The purpose of the tangential flow is to introduce a shear velocity across the surface of the filter membrane sufficient to detach any large particles that have accumulated on the surface of the membrane, thus permitting particles smaller than the designated pore size to continue to pass through the membrane. After processing, some of the particles greater than the rated pore size are concentrated in the retentate. A portion of these particles also coats the surface of the membrane. These particles are recovered by placing the membrane in a plastic bag, adding filtrate, and gently rubbing the filter from outside the bag to dislodge the particles. The concentrated retentate is then combined with the filtrate used to clean the filters, and the entire volume is centrifuged to collect the particles.

2.5 SINGLE PARTICLE OPTICAL SENSING

In addition to reserving sediment and aqueous samples for mass and chemical analyses, samples of each suspended sediment water fraction above 0.45μm were reserved for single particle optical sensing (SPOS) using an AccuSizer 770 (Beckett, 1986). With this method, a thin stream of dilute sample is passed through a sensor that employs laser diffraction and extinction to measure both the size of particles and count the number of each size particle. As a particle passes through the sensor, the light reaching the detector is reduced. The reduction in light is directly proportional to the size of the particle. By calibrating the instrument with monodisperse polystyrene spheres, the size of each particle in the sample is measured, one particle at a time. In this way a particle size distribution can be obtained for each size fraction.

2.6 ION CHROMATOGRAPHY

Samples for ion chromatography (IC) were obtained both in the field with 0.45μm disposable filters, and again in the lab after 0.45μm TFF. Samples were analyzed on a Dionex ion chromatograph for fluoride, chloride, nitrate, phosphate and sulfate. Prior to analysis, samples were refrigerated to minimize bacterial degradation, but no preservatives were used.

3 RESULTS

Only the results from the May 28 sample are presented in this paper. Table 1 shows field measurements of water chemistry. In-stream measurements of pH (Table 1) are unexpectedly high for rivers receiving acid mine drainage. Similarly, alkalinity values are low, especially at North Clear Creek, suggesting that the buffering capacity of the creeks may be compromised.

Table 1. Field measurements of water chemistry from Clear Creek, Colorado.

Site	Time	T (deg C)	Conductivity (mS/cm)	pH (in field)	Alkalinity (mg/L)
Golden	9:05am	9.2	0.122	7.5	22
North CC	11:10am	9.2	0.126	7.1	14
CR 314	12:15pm	9.2	0.113	7.3	20
Gilson St	1:30pm	10.6	0.104	7.3	20
23rd St	3:19pm	11.4	0.100	6.7	20

Ion chromatography results are given in Table 2. Based on the Golden field replicate, the uncertainty in the IC analysis is <20%. Phosphate is not listed because concentrations reported for all samples were zero. Sulfate sorption onto iron oxides can reduce surface charge and cause flocculation of iron colloids. Hence, a change in sulfate concentrations may affect particle sizes. Sulfate concentrations are highest, as expected, in North Clear Creek, because that branch of the river currently has the greatest input of untreated mine drainage. There is no appreciable increase in sulfate concentration between 23rd St and Gilson St due to the Argo tunnel, which is consistent with the reduction of acid input by the Argo treatment facility.

Table 2. Ion chromatography results for 0.45μm filtered water samples collected from Clear Creek, Colorado. Note: The Golden Duplicate is an analytical duplicate.

Site	Fluoride (mg/L)	Chloride (mg/L)	Nitrate (mg/L)	Sulfate (mg/L)
Golden	0.3	4.5	0.4	22.8
Golden Duplicate	0.3	4.8	0.6	19.4
North CC	0.2	4.2	0.4	30.2
CR 314	0.3	3.3	0.5	19.7
Gilson St	0.4	3.2	0.4	20.3
23rd St	0.3	3.0	0.4	18.2
DI blank	0.1	0.0	0.0	1.6

Table 3 shows the raw mass distribution data from which the percent mass distribution in Table 4 was derived. Though actual mg/L concentrations are not presented, masses obtained for all sediment samples are derived from nearly equivalent volumes of filtered water.

Table 3. Original masses of suspended sediment from Clear Creek, Colorado.

Site	250 - 25 μm (g)	25 - 5 μm (g)	5 - 0.45 μm (g)	0.45 μm - 30000 Daltons (g)	Total (g)
Golden	0.55585	0.11368	0.01976	0.01674	0.70603
N. Clear Creek I	0.09072	0.15488	0.02765	0.00745	0.28070
N. Clear Creek II	0.07751	0.16009	0.02183	0.01359	0.27302
CR 314	0.17553	0.03743	0.00774	0.00677	0.22747
Gilson St	0.15294	0.03935	0.00501	0.01068	0.20798
23nd St	0.13539	0.03883	0.05859	0.00805	0.24086

Percent mass distributions of suspended sediments are presented in Table 4. Based on the North Clear Creek field replicate, the uncertainty in the percent mass distribution is <25%. The suspended sediment load is generally dominated by the largest size fraction (250µm-25µm). However, the two field replicates at North Clear Creek indicate that in that sample the 25µm - 5µm fraction contains the majority of the sediment load. Another interesting distribution is in the unusually high sediment load in the 5µm - 0.45µm fraction for 23rd St, a site which receives the least input from mine drainage. Although it is generally the least abundant size fraction, a significant amount of mass (2-5%) is present in the 0.45µm-30000 Dalton fraction.

Table 4. Percent mass distribution of suspended sediment from Clear Creek, Colorado. Note: The labels N. Clear Creek I and II indicate field duplicate samples.

Site	250 - 25 µm	25 - 5 µm	5 - 0.45 µm	0.45 µm – 30000 Daltons
Golden	79	16	3	2
N. Clear Creek I	32	55	10	3
N. Clear Creek II	28	59	8	5
CR 314	77	16	3	3
Gilson St	74	19	2	5
23rd St	56	16	24	3

Finally, particle size analyses have been conducted for all sites for the <25µm and <5µm fractions. Figures 1 and 2 are included to illustrate the types of distributions obtained using SPOS. The two data sets represented are from the <25µm suspended sediment water fractions. The curves in the population distribution (Figure 1) represent the total number of particles counted of a given diameter versus diameter. The fact that the curve tapers to zero near 25µm suggests that a good separation was achieved in the size fractionation. The difference in height of the two curves is not truly representative of the concentrations of the two water samples. The Golden sample was actually slightly more than three times more concentrated than the North Clear Creek sample.

The volume-weight difference curves of Figure 2 are obtained by computing the spherical volume of a particle of a given dimension, multiplying by the number of particles of that size and dividing by the combined particle volumes. The results are strongly biased toward the larger particles because it is in these particles that the most mass resides. In some cases, the single peaks at the right of the plot represent only one particle each. The fact that the Golden curve displays more of the massive particles is in agreement with the differences between the Golden and North Clear Creek samples seen in the percent mass distribution.

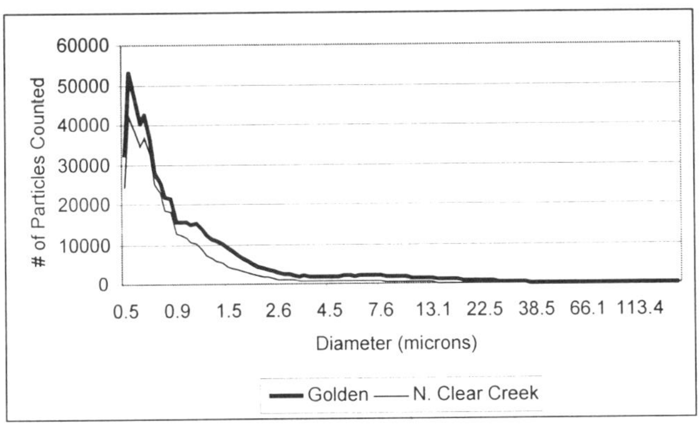

Figure 1: Population distribution of <25µm suspended sediment at Golden and North Clear Creek from Clear Creek, Colorado.

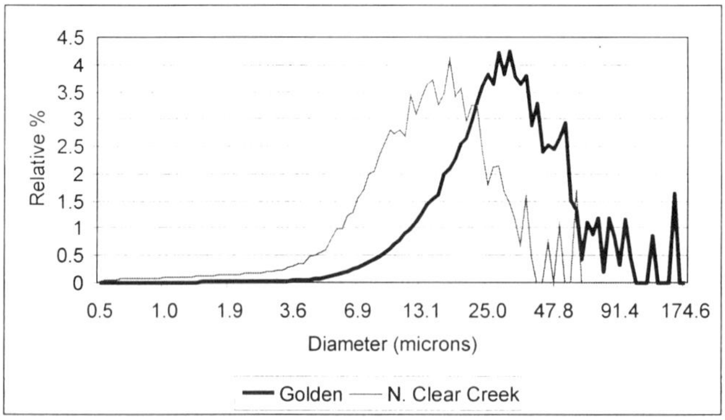

Figure 2: Volume-weight difference distribution of <25μm suspended sediment at Golden and North Clear Creek from Clear Creek, Colorado.

4 SUMMARY AND CONCLUSIONS

Much of the suspended sediment mass is found to reside in the largest size fraction (250μm - 25μm), and metals associated with this fraction may take longer than smaller fractions to be transported downstream. However, substantial masses of fine particles and colloids (<5μm) exist in the creek loads, and metals associated with this fraction may be more reactive because of the much greater specific surface area of the particles. The greatest amount of <5μm material occurs in North Clear Creek, with the exception of the anomalously high concentration at the 23rd Street site. In view of these results, the potential for metal sorption and transport on fine sediment and colloids from mine drainage up North Clear Creek is high. Future work will involve metal analysis and geochemical modeling of the fate and transport of these particles and any associated metals.

REFERENCES

Beckett, Ronald, "The Composition and Surface Properties of Suspended Particulate Matter." Hart, Barry T., ed. 1986. *Water Quality Management: The Role of Particulate Matter in the Transport and Fate of Pollutants.* Melbourne: Water Studies Centre, Chisolm Institute of Technology.

Ranville, James F., and Schmiermund, Ronald L. "An Overview of Environmental Colloids." Macalady, Donald L., ed. 1998. *Perspectives in Environmental Geochemistry.* New York: Oxford University Press. 512p.

Smith, Kathleen S., Ranville, James F., Plumlee, G.S., and Macalady, Donald L. "Predictive Double-Layer Modeling of Metal Sorption in Mine-Drainage Systems," Jenne, A. E., ed. 1998. *Adsorption of Metals by Geomedia.* San Diego: Academic Press. 583p.

ACKNOWLEDGMENTS

Funding was provided by the Edna Bailey Sussman Foundation. Thanks to Dan Schultz, Kevin Norton, and Brian Marlow for field assistance, and to Rich Harnish for IC analysis. We also wish to thank Katherine Walton-Day, James Martinez, and Bruce Honeyman for the use of additional equipment and Randall Carver for his exceptional formatting skills.

Tailings and Mine Waste '99 © 1999 Balkema, Rotterdam, ISBN 90 5809 025 6

Trace element dispersion from tin deposits of the western Seward Peninsula, Alaska

Cynthia C. Parnow, Richard J. Goldfarb, Karen D. Kelley & Allen L. Meier
US Geological Survey, Denver Federal Center, Colo., USA
Geoff York
US Geological Survey, Anchorage, Alaska, USA

ABSTRACT: A study of the dispersion patterns of trace elements (As, Be, Cu, Li, Pb, Sn, W, and Zn) typically enriched in tin deposits and occurrences in the western Seward Peninsula indicates significant downstream accumulation of arsenic in stream sediments, whereas surface waters consistently contain low concentrations of all dissolved trace elements. Sediments collected in the headwaters of streams draining the slate-hosted Potato Mountain tin occurrences contained as much as 530 ppm As and levels remained as high as 78 ppm as far as 7 km downstream. Samples collected directly below the Lost River mine, hosted in carbonate rocks, contained as much as 1400 ppm As in stream sediments, with concentrations as high as 260 ppm As characterizing stream sediments 8 km downstream at the mouth of Lost River. None of the surface waters collected from either area contained more than 9 ppb As and pH's were consistently near neutral even directly below mine tailings.

INTRODUCTION

A geochemical study to determine the dispersion patterns and possibility of accumulation of metals and arsenic in stream sediments and waters was carried out on the western Seward Peninsula, Alaska in areas of mined tin deposits and unmined tin occurrences (fig. 1). This part of the Seward Peninsula was targeted for detailed study based upon an assessment of regional element distribution maps, which were compiled from stream- and lake-sediment geochemical data collected during the National Uranium Resource Evaluation (NURE) program (Hoffmann and Buttleman, 1996). Interpretation of the regional NURE data indicated that stream sediments collected in parts of the western Seward Peninsula contained as much as 635 ppm As and 144 ppm Sn. These concentrations are considered anomalous when compared to regional background concentrations of 5-35 ppm As (n=919) and 7-12 ppm Sn, (n=1405) (Parnow and others, in press).

Tin mineralization in the western Seward Peninsula occurred during granite emplacement into carbonate and clastic sedimentary rocks between 80-70 Ma (Sainsbury, 1969). Tin is mainly found in cassiterite [SnO_2], and associated accessory minerals include pyrite [FeS_2], arsenopyrite [FeAsS], galena [PbS], sphalerite [ZnS], stannite [Cu_2FeSnS_4], scheelite [$CaWO_4$], chalcopyrite [$CuFeS_2$], tourmaline [$(Na,Ca)(Li)(Al)_6(BO)_6(Si_6O_{18})(OH)_4$], fluorite[$CaF_2$], beryl [$Be_3Al_2(Si_6O_{18})$], and wolframite [$(Fe,Mn)WO_4$]. The tin deposits occur as skarn and greisen lodes hosted in altered carbonate rocks and granite, respectively (e.g., the Lost River mine), or as placer accumulations derived from veins and replacement zones in slates (e.g., the Potato Mountain area). Over the last hundred years slightly more than 3000 metric tonnes (t) of tin has been mined from the area (Hudson and Reed, 1997). The focus of this paper is to define, compare, and contrast differences in drainage geochemistry and element dispersion patterns that result from the erosion of two contrasting styles (skarn, greisen and vein versus replacement, vein and placer) of mineralization and in chemically different host rocks (carbonates versus

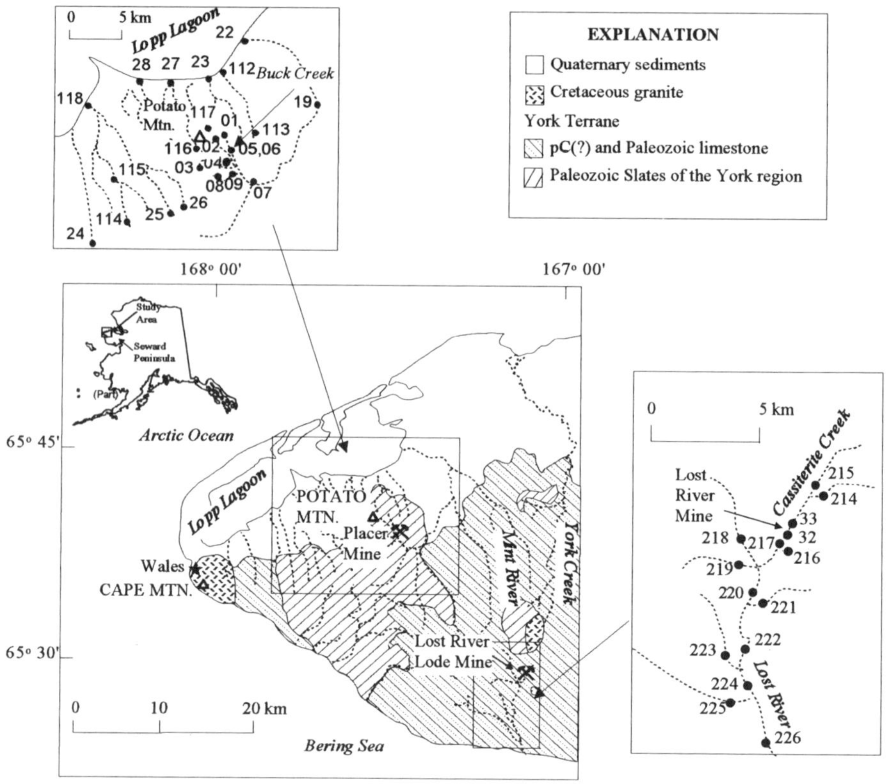

Figure 1. Sample locations and general geologic map for the western Seward Peninsula tin deposits (geology after Sainsbury, 1972).

slates). Special attention is given to arsenic because it potentially could have a significant effect on the near shore marine environment.

REGIONAL GEOLOGY

The western Seward Peninsula consists mainly of Precambrian(?) and lower Paleozoic argillaceous and dolomitic limestones and slates (fig. 1), which are together referred to as the York terrane (Sainsbury, 1972). Much of the York terrane was mapped by Sainsbury (1972) as Precambrian in age, but is lithologically similar to what is now recognized to be lower Ordovician rocks found elsewhere on the Seward Peninsula (Till and Dumoulin, 1994). The York terrane is typically subdivided into two major units. A slate unit, informally termed the Slate of the York region, consisting of graphitic, slaty, schistose or phyllitic rocks that were originally comprised of carbonaceous siltite, with lesser laminated to thinly-bedded mudstone, siltstone, fine-grained sandstone, calcareous sandstone, and carbonaceous limestone (Sainsbury, 1972). Higher iron content (5.0-6.9 %) in many samples of the slate possibly indicates the presence of diagenetic pyrite (Potter and others, 1980). The second major unit is a thick sequence of fossiliferous limestone, dolomitic limestone, and argillaceous limestone. Both the slate- and carbonate-dominant units were intruded by granites and altered by tin-rich hydrothermal fluids in the Late Cretaceous. The northern portion of the study area is covered in a thick sequence of Quaternary sediments.

TIN DEPOSITS

The tin lode deposits at Lost River occur as skarn, greisen and breccia (Dobson, 1982). In addition to cassiterite, small veins and pods with minerals enriched in As, Be, Cu, Li, Pb, W, and Zn were deposited over a wide area in the Lost River watershed during hydrothermal activity. All the tin mined at Potato Mountain was from small placer deposits, although geophysical evidence and a large zone of hornfels suggest a buried pluton, indicative of a lode occurrence much larger than the few small recognized veins and replacement zones (Hudson and Reed, 1997). The cassiterite in the recognized lodes is often associated with pyrite and arsenopyrite (Mulligan, 1965; Hudson and Reed, 1997).

Much of the 3,000 t of tin production from the western Seward Peninsula has come from the placer deposits at Cape Mountain and Potato Mountain (fig. 1). Placer production from the Potato Mountain region was about 1000 t, with the majority of the production (700 t) occurring between 1911-1919 from Buck Creek (fig 1) using a dredge. Our field investigations in 1997 showed little remaining evidence of the past mining operations in the Potato Mountain area. A few abandoned buildings, some old prospect pits and mounds, and the large dredge in Buck Creek are the only remnants of past mining activity.

The Lost River mine operated intermittently between 1904-1955. The majority of the production was from the lode skarn deposit, which yielded more than 300 t of tin from calc-silicate-altered rocks between 1952-1955. Another 85 t of tin was mined from associated placer deposits, which were active into the late 1960's. The workings at the lode mine, located along Cassiterite Creek (fig.1), consisted of one main adit and three smaller ones, and a small mill. All of these workings have now either collapsed or are flooded. The mineralized zone has also been extensively explored by pits, trenches, diamond drill holes and several collapsed adits, which surround the main workings. Mine waste and tailings that surround the workings are piled along the east bank of Cassiterite Creek, from stream level upward for about 50 m. Small seeps drain parts of these mine waste piles and discharge into Cassiterite Creek. Sainsbury (1969) estimated that there may be as much as 200,000 t of tin remaining in the abandoned lode deposit.

METHODS

A total of 38 stream sediments and 39 waters were collected from sites in areas of historic mining operations and in their surrounding drainage basins in summers of 1997 and 1998 (fig. 1). Three 60-mL water samples were collected in high-density polyethylene (HDPE) bottles at each site. Two samples, one filtered and one unfiltered, were taken for major, minor and trace element analysis. The filtered sample was passed through a 0.45-mm-filter and analyzed for dissolved element concentrations. The unfiltered sample was analyzed for dissolved plus suspended element concentrations. Both samples were acidified with ultra-pure concentrated nitric acid to a pH of less than 2. A third filtered, unacidified sample was collected for sulfate and alkalinity analysis. This sample was refrigerated until analysis. Water samples collected for cation analysis were analyzed for a broad suite of elements by inductively-coupled argon plasma-mass spectroscopy (ICP-MS). Sulfate was analyzed by ion-chromatography (IC), and total alkalinity was measured by titration as bicarbonate (HCO_3^-). Field measurements included pH, conductivity, and temperature.

Stream sediments were composited from active parts of the streambed and were sieved in the field to minus-10-mesh (2 mm). The sediments were decomposed using a three acid total decomposition method. The resulting solution was analyzed for 40 major, minor and trace elements by inductively coupled plasma-atomic emission spectrometry (ICP-AES). For more detailed descriptions of sample collection and references for analytical methods see Parnow and others (in press).

Table 1. Selected stream sediment and water data for Potato Mountain.

Filtered Water Samples											Stream Sediments							
Field No.	pH	SO4 ppm	Alkalinity ppm	As ppb	Be ppb	Cu ppb	Li ppb	Pb ppb	W ppb	Zn ppb	As ppm	Be ppm	Cu ppm	Li ppm	Pb ppm	Sn ppm	W ppm	Zn ppm
Samples collected from mineralized areas on Potato Mountain																		
01	4.9	39	3	< .2	.05	2	9	<.05	<.05	10	53	< 4	30	20	30	< 20	<40	71
02	5.7	7	4	< .2	.05	<.5	.5	<.05	<.05	<.5	510	1	42	94	29	62	5	10
03	6.1	24	27	< .2	<.05	<.5	1.2	<.05	<.05	<.5	43	< 1	27	79	32	8	6	120
04	7.3	52	34	.6	<.05	<.5	1.6	<.05	<.05	<.5	150	< 1	24	76	26	13	7	120
05	6.5	45	19	.2	<.05	<.5	2.7	<.05	<.05	<.5	130	< 1	29	80	31	7	17	120
06	7.0	53	69	.3	<.05	<.5	1.2	<.05	<.05	<.5	73	< 1	26	96	31	9	7	94
07	6.9	43	41	< .2	<.05	<.5	1.8	<.05	<.05	<.5	65	< 1	23	77	27	8	7	110
08	6.5	9	26	2	<.05	<.5	.9	<.05	<.05	<.5	85	< 1	26	74	29	8	7	120
09	7.0	38	52	< .2	<.05	<.5	1.9	<.05	<.05	<.5	16	< 1	21	64	28	5	6	110
113	7.7	27	93	< .2	<.05	<.5	1.2	<.05	<.05	<.5	60	< 1	17	130	30	13	6	96
114	7.6	22	38	< .2	.05	<.5	.2	<.05	<.05	1	95	< 1	23	80	29	6	12	110
116	7.3	75	17	< .2	<.05	<.5	.2	<.05	<.05	.6	100	< 1	25	78	30	15	8	130
117	4.7	54	1	.7	.2	6	3.7	<.05	<.05	6	530	2	59	108	30	36	13	72
Samples collected at mouths of streams feeding Lopp Lagoon																		
23	7.5	24	41	.3	<.05	<.5	2	<.05	<.05	<.5	78	< 1	16	79	25	14	8	120
27	5.5	25	3	.3	<.05	.8	2.8	<.05	<.05	8	78	< 1	9	51	20	14	8	60
28	6.7	45	8	< .2	.05	<.5	1.4	<.05	<.05	1	32	< 1	13	70	23	8	10	100
112	7.3	36	130	< .2	<.05	<.5	1.2	<.05	<.05	<.5	25	< 1	15	60	24	10	6	110
118	8.0	28	17	< .2	<.05	<.5	.6	<.05	<.05	<.5	24	< 1	14	62	21	7	7	98
Samples distal to mineralized areas																		
19	8.1	22	101	< .2	.09	<.5	1.3	<.05	<.05	1	35	< 1	29	56	25	6	6	102
24	7.9	92	73	< .2	<.05	<.5	.8	<.05	<.05	<.5	63	< 1	42	52	28	16	7	150
25	7.6	36	62	< .2	<.05	<.5	1	<.05	<.05	<.5	33	< 1	26	79	29	10	6	113
26	7.2	58	148	< .2	<.05	<.5	1.8	<.05	<.05	<.5	40	< 1	20	71	28	6	8	91
113	7.7	27	93	< .2	<.05	<.5	1.2	<.05	<.05	<.5	60	< 1	17	130	30	13	6	96
114	7.6	22	38	< .2	.05	<.5	.2	<.05	<.05	1	95	< 1	23	80	29	6	12	105
115	7.4	35	18	< .2	<.05	<.5	.8	<.05	<.05	<.5	23	< 1	23	74	28	6	8	120

Table 2. Selected stream sediment and water data for Lost River area

Filtered Water Samples											Stream Sediments							
Field No.	pH	SO4 ppm	Alkalinity ppm	As ppb	Be ppb	Cu ppb	Li ppb	Pb ppb	W ppb	Zn ppb	As ppm	Be ppm	Cu ppm	Li ppm	Pb ppm	Sn ppm	W ppm	Zn ppm
Samples collected from mineralized areas																		
32	8.3	2.0	74	7.6	< 0.05	< 0.5	0.4	0.4	0.2	1.0	1400	130	290	770	770	140	210	5200
33	8.0	3.3	80	180	< .05	.7	.6	< .05	.4	10	no data	no data	no data	no data	no data	no data	no data	no data
216	8.3	< .08	66	2.9	< .05	< .5	< 1	.2	.07	1	150	86	13	190	110	31	570	170
217	8.0	2	71	8.9	< .05	< .5	< 1	.2	.04	2	450	12	30	78	230	90	1400	220
220	8.2	< .08	66	1.8	< .05	< .5	< 1	< .05	< .02	2	52	14	16	52	91	19	<20	200
221	8.0	2	75	2.4	< .05	< .5	< 1	.2	< .02	< .5	200	24	39	120	140	49	390	310
222	8.2	2	75	4.6	< .05	< .5	< 1	.1	< .02	2	280	26	79	170	120	66	350	450
223	8.1	3	85	1.5	< .05	< .5	< 1	.06	< .02	< .5	70	45	19	130	69	20	90	110
224	8.0	2	77	4.3	< .05	< .5	< 1	< .05	< .02	1	480	32	90	260	130	80	240	580
226	8.2	2	77	2.8	< .05	< .5	< 1	.08	< .02	< .5	260	15	51	120	83	44	220	250
Samples distal to mineralized areas																		
214	8.3	1	66	< .3	< .05	< .5	< 1	< .05	< .02	< .5	< 10	< 1	2	8	< 4	9	<20	24
215	8.2	< .08	99	< .3	< .05	< .5	< 1	.06	< .02	2	< 10	< 1	6	10	< 4	5	<10	18
218	8.3	2	68	< .3	< .05	< .5	< 1	.1	< .02	< .5	< 10	2	4	18	20	11	30	33
219	7.8	2	68	< .3	< .05	< .5	< 1	< .05	< .02	< .5	< 10	1	15	42	14	6	<10	57
225	8.1	.8	78	< .3	< .05	< .5	< 1	.69	< .02	< .5	< 10	5	10	18	32	6	<20	47

GEOCHEMISTRY DISTAL TO KNOWN TIN OCCURRENCES

Stream sediments collected in drainages underlain by the Slate of the York region that are distant from any known tin occurrences are characterized by concentrations of 23-95 ppm As, 6-16 ppm Sn, 20-42 ppm Cu, 52-130 ppm Li, 25-30 ppm Pb, 6-12 ppm W and 91-150 ppm Zn (table 1). Many of the concentrations are extremely high for typical stream sediments solely derived from clastic rocks. These concentrations are believed to indicate the widespread distribution of other small tin lode occurrences rather than simply background concentrations from the slates. Sediment samples collected in the drainages underlain by carbonate units in the Lost River area, upstream from any known tin occurrences, are characterized by lower As (<10 ppm), Sn (5 to 11 ppm), Li (8-42 ppm), and Zn (18-57 ppm) concentrations, but higher W (<10-30 ppm), with similar background values for Pb (<4-32 ppm) (table 2).

Waters from areas distal to known mineral occurrences and in areas underlain by either slate or carbonate have extremely low concentrations of dissolved trace elements. Concentrations for As, Be, Cu, Pb, Sn, W, and Zn are at or below detection limits[1] for most of the waters collected. The pH values of the waters draining the slates distal to Potato Mountain are mainly slightly alkaline and have dissolved sulfate concentrations of 22-92 ppm (table 1), likely reflecting weathering of the abundant diagenetic pyrite. The waters collected from the drainages underlain by carbonates are alkaline and have very low dissolved sulfate concentrations (<0.8-2 ppm) (table 2).

GEOCHEMISTRY AND MINERALOGY OF MINERALIZED AREAS

The predominant tin-bearing mineral at Potato Mountain is cassiterite, occurring in quartz veins and disseminated in the slates, from which the placer deposits are derived. Accessory minerals in the lodes are arsenopyrite, pyrite, and scheelite. Stream sediments collected in drainages containing these small lodes contain as much as 530 ppm As. A few samples also have as much as 36-62 ppm Sn. Other elements that are normally enriched in the tin lode deposits, such as Be, Cu, Li, Pb, W and Zn, are typically only slightly enriched when compared with sediments derived from the slates distal to Potato Mountain. Arsenic concentrations are similar throughout the Potato Mountain drainages regardless of whether the stream had been mined for placer deposits or not.

Filtered waters collected from drainages below Potato Mountain were generally acidic to neutral (pH 4.7-7.3), and have low dissolved arsenic concentrations ($\leq$ 2 ppb) (table 1). The extremely low pH's of 4.7 and 4.9 were from samples collected at the highest elevations sampled (sites 01 and 117), where patches of snow were contributing to the stream flow. The acidic water at site 01 is believed to reflect a large contribution of the flow from snowmelt, whereas the acidic water at site 117 may be due to water-rock interaction. Although dissolved Cu, Li, and Zn were elevated relative to the rest of the waters sampled (table 1), the absolute abundance of trace elements are still extremely low. Sulfate and alkalinity concentrations were highly variable, ranging from 7 ppm-75 ppm and 1-69 ppm, respectively. The broad range in these parameters probably reflects the sporadic occurrence of carbonate interbeds within the slates.

The mineralogy of the Lost River deposit is more complex due to the higher reactivity of the carbonate host rocks, and possibly, a more voluminous hydrothermal system. Tin is found as stannite [Cu_2FeSnS_4] in addition to cassiterite, and the main tungsten-bearing mineral is wolframite. In addition to the tin and tungsten mineralization, Lost River contains significant beryllium resources (Sainsbury, 1969). Sulfide mineralization is more extensive at Lost River, and accessory minerals include chalcopyrite, sphalerite, and galena as well as pyrite and arsenopyrite. A sediment sample collected from a stream draining the mine dump and tailings a the Lost River mine (site 32) contained 1,400 ppm As, 130 ppm Be, 290 ppm Cu, 770 ppm Li, 770 ppm Pb, 140 ppm Sn, 210 ppm W, and 5,200 ppm Zn.

[1] 1) Detection Limits in filtered waters for selected elements: < 0.2 or <0.3 ppb As; <0.05 Be; <0.5 ppb Cu; <1 ppb Li, <0.05 Pb, <0.6 ppb Sn, <0.02 ppb W, <0.5 ppb Zn

A stream sediment collected immediately downstream from the main mine workings along the bed of Cassiterite Creek was enriched in As (450 ppm), Li (78 ppm), Pb (230 ppm), Sn (90 ppm), and W (1400 ppm) (site217). Very high Be (86 ppm), Li (190) ppm, and W (570 ppm) concentrations in sediment from an adjacent stream (site 216) reflect small undeveloped Be-rich lode occurrences.

A water sample (site 33) collected from a seep draining the tailings pile contained dissolved concentrations of 180 ppb As, <0.05 ppb Be, 0.7 ppb Cu, 28 ppb Fe, <0.05 ppb Pb, <0.5 ppb Sn, 0.4 ppb W, and 10 ppm Zn. Carbonate rocks buffered the pH of the seep to 8.0 which restricted solubilities to low levels for most trace elements despite enrichments of many metals in the tailings. The relatively small discharge passing through the mine tailings is rapidly diluted by Cassiterite Creek. Surface waters of the creek, collected at sites a few hundred meters downstream from the mine runoff (sites 32 and 217) contained 7.6-8.9 ppb As, 1-2 ppm Zn, and background amounts of other trace elements. The pH and sulfate concentrations of the waters collected from these same sites did not differ from the levels for waters draining the host carbonates distal to the mine workings.

DISPERSION OF METALS AND ARSENIC

Table 1 lists selected stream sediment and water data for samples extending as far as 7 km downstream from the Potato Mountain lode tin occurrences and placer workings. Stream sediments collected at the mouths of the streams along Lopp Lagoon (sites 23 and 27), whose headwaters begin at Potato Mountain, contained values of 78 ppm As and 14 ppm Sn. This reflects transport of grains from the lode occurrences to the near-shore environment. Filtered waters collected from sites 23 and 27 contained dissolved As concentrations of 0.3 ppb, which further shows little solution transport from metalliferous occurrences in the headwaters of these streams. Dissolved sulfate levels were less than 50 ppm in this area of Quaternary sediments. The sample collected at site 27 was relatively acidic (pH 5.5), and had slight enrichments in Cu (0.8 ppb) and Zn (8 ppb), but this may be attributable to local production of significant amounts of organic acids from the more boggy tundra at this location.

The sediment samples collected along Lost River were elevated in all ore-related metals and arsenic for the entire 8-km-long channel length below the junction with Cassiterite Creek (fig.1). As, Be, Cu, Fe, Li, Pb, Sn, W, and Zn concentrations remained relatively high along the entire length of Lost River, below the junction with Cassiterite Creek (table 2), approximately 8 km. Beryllium concentrations were highest in tributaries to the Lost River (45 ppm at site 223 and 86 ppm at site 216), possibly reflecting the presence of unmined vein occurrences distal to the main Lost River skarn deposit, as previously mentioned by Sainsbury (1969).

Similar to the waters draining the Potato Mountain tin deposits, dissolved trace elements were near or below detection limits in the water samples collected from the Lost River. Dissolved arsenic in samples collected along Cassiterite Creek increased from less than 0.3 ppb above the mine workings to 8.9 ppb below the mine discharge (table 2). Along the Lost River, dissolved arsenic levels also increased, from <0.3 ppb upstream from any occurrences, to 2.4-4.6 ppb below influxes from Cassiterite Creek and other unmined tin-bearing skarn occurrences. Dissolved sulfate remained very low (≤ 3 ppm), and alkalinity (68-85 ppm) was relatively high throughout the watershed dominated by carbonate rocks. The carbonates consistently buffered pH values to 8.0-8.3, even along Cassiterite Creek immediately below the mine workings.

ENVIRONMENTAL IMPACT OF TIN DEPOSITS

Acid mine drainage normally associated with heavy metal mining is caused principally by the weathering and oxidation of pyrite. Tin deposits, whether they are skarn, greisen or vein, generally do not pose significant environmental impact due to the low sulfide mineral content of these deposits (Hammarstrom, et. al, 1995, Elliot, et. al, 1995). In this study, concentrations of dissolved cations are low due to the small amount of pyrite and other sulfide minerals in the tin deposits, so there is little potential for significant acid water generation. In addition, any acidic waters that are generated are quickly buffered by the surrounding carbonate host rocks. Furthermore, arsenic, a potentially hazardous element in the tin ores is relatively immobile and stable as arsenopyrite over the pH ranges found in the study area. The EPA safe drinking water limit for arsenic is 50 ppb and, except for the single sample taken from a seep draining the tailings pile at the Lost River mine, all of the waters sampled are well below this level.

Figure 2 shows the compositions, in terms of sulfate, alkalinity, and iron of surface waters downstream from the Potato Mountain and Lost River deposits. Waters collected below the Lost River deposit have little variation in composition, whereas those waters collected from the Potato Mountain area are more variable. Previous work (Mulligan, 1965, Sainsbury, 1969, and Dobson, 1982) indicates that mineralization at the Lost River mine is more sulfide-rich than Potato Mountain. But the low concentrations of sulfate and dissolved metals in downstream surface waters along Lost River indicate minimal oxidation of these accessory sulfide minerals and buffering due to the carbonate host rock. Greater variability in the alkalinity and sulfate concentrations in waters from the Potato Mountain area partly reflects the oxidation of diagenetic pyrite within the clastic rocks and variable amounts of carbonate units interbedded in the slates. Low metal concentrations in these same waters again is a result of natural buffering of the water from interbedded carbonates within the slates.

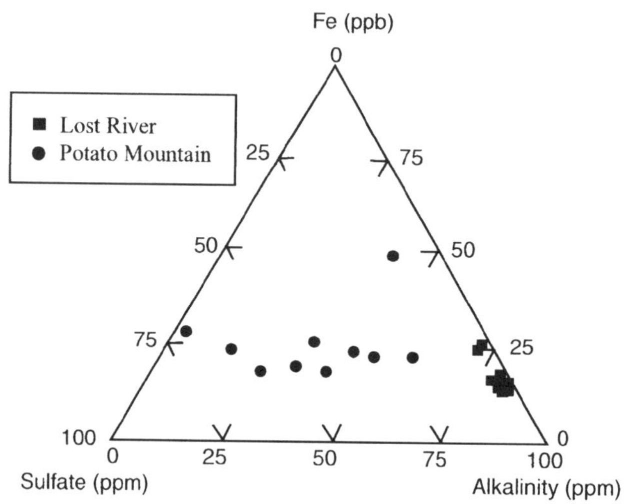

Figure 2. Ternary diagram of iron-sulfate-alkalinity in filtered water samples for selected sites from the Lost River and the Potato Mountain tin deposits.

CONCLUSIONS

Our data indicate significant downstream sediment transport of As, Be, Cu, Li, Pb, Sn, W, and Zn from tin lodes at Lost River and to a lesser extent from those at Potato Mountain. Arsenic is the only element found in sediments with significantly elevated concentrations that might possibly pose an environmental hazard. However, neither arsenic nor any metals are being transported in significant concentrations in surface waters. A comparison of the concentrations of arsenic in waters versus sediments (figs. 3A and B), clearly shows that although anomalous concentrations of arsenic are being transported as far as 8 kilometers downstream from both deposits in the sediment loads, surface waters are unaffected by historic mining activities.

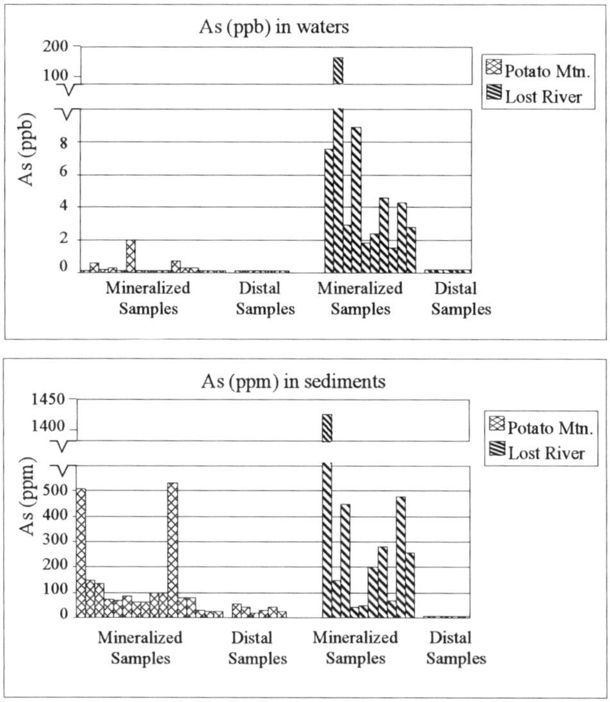

Figure 3: Comparison between dissolved arsenic in waters A) and arsenic in sediments B) for selected samples from the Lost River and Potato Mountain tin deposits.

This study found no indication of serious present-day negative impacts from the historic production of tin from the western Seward Peninsula. Exposed material from the mining activity itself probably did not cause a significant increase in surface water metal values compared to what would have been present prior to human activity. There is evidence that the lode mining at Lost River has caused some downstream transport of arsenic and metals in stream sediments, but the impact is negligible downstream from the placer mining activities at Potato Mountain. Future study should investigate dispersion of metals and/or arsenic from these deposits into the near-shore marine environment. Anomalous arsenic and tin levels in the stream sediments at the mouths of streams above Lopp Lagoon and at Lost River suggest that metals may be entering the marine environment, but their impact, if any, on marine organisms remains uncertain.

ACKNOWLEDMENTS—We would like to thank the following people for their support: Toby Anungazuk of the Wales Native Corporation, Wales, Alaska, Barrett Cieutat and Laurie Balistrieri for their help with the field work, Paul Briggs, Pete Theodorakos and Rich Wanty of the USGS for their help with the geochemical analyses, and Bob Eppinger, John Gray, and Cliff Taylor, Jr., also of the USGS for their insightful reviews.

REFERENCES

Dobson, D.C. 1982. Geology and alteration of the Lost River tin-tungsten-fluorine deposit, Alaska. *Economic Geology*. 77: 1033-1052.

Elliot, J.E., Kamilli, R.J., Miller, W.R., and Livo, K.E. 1995. Vein and greisen Sn and W deposits. In du Bray, E.A., (ed.), *Preliminary compilation of descriptive geoenvironmental mineral deposit models*. U.S. Geological Survey Open-File Report 95-831: 61-69.

Hammarstrom, J.A., Elliot, J.E., Kotlyar, B.B., Theodore, T.G., Nash, J.T., John, D.A., Hoover, D.B., and Knepper, D.H. 1995. Sn and (or) W skarn and replacement deposits. In du Bray, E.A., (ed.), *Preliminary compilation of descriptive geoenvironmental mineral deposit models*: U.S. Geological Survey Open-file Report 95-831: 54-61.

Hoffmann, J.D. and Buttleman, K. 1996. National Geochemical Database: 1. National Uranium Resource Evaluation (NURE) hydrogeochemical and stream sediment reconnaissance (HSSR) data for Alaska, U.S. Geological Survey Digital Data Series DDS-18-B.

Hudson, T. L. and Reed, B.L. 1997. Tin Deposits in Alaska. *Economic Geology Monograph* 9:450-465.

Mulligan, J.J. 1965. Tin lode investigations, Potato mountain area, Seward Peninsula, Alaska. *U.S. Bureau of mines Report of Investigations 6587*: 85 p.

Parnow, C.C., Goldfarb, R.J., and Kelley, K.D., Meier, A.L, Sutley, S.J, and Theodorakos, P.M. (in press). Analytical results of stream sediment, heavy-mineral concentrates, and water data collected from northwestern Seward Peninsula, Alaska. *U.S. Geological Survey Open-File Report.*

Potter, P., E., Maynard, J. B., Pryor, W., A. 1980. *Sedimentology of shale; study guide and reference source.* New York: Springer-Verlag

Sainsbury, C.L. 1969. Geology and ore deposits of the central York Mountains, western Seward Peninsula, Alaska. *U.S. Geological Survey Bulletin 1287*: 101 p.

______, 1972, Geologic Map of The Teller Quadrangle, western Seward Peninsula, Alaska: *U.S. Geological Survey Map I-685*. scale 1:250,000.

Till, A.B., and Dumoulin, J.A. 1994. Geology of the Seward Peninsula and Saint Lawrence Island. In Plafker, G., and Berg, H.C. (eds), *The Geology of Alaska: Boulder, Colorado Geological Society of America, The Geology of North America. G-1*: 141-152.

Tailings and Mine Waste '99 © 1999 Balkema, Rotterdam, ISBN 90 5809 025 6

Metal leaching through a fluvial tailings deposit along the upper Arkansas River, Colorado

Kathleen S. Smith, Paul J. Lamothe, Allen L. Meier & Katherine Walton-Day
US Geological Survey, Denver, Colo., USA
James F. Ranville
Department of Chemistry and Geochemistry, Colorado School of Mines, Golden, Colo., USA

ABSTRACT: Fluvial deposits of mine tailings occur along the upper Arkansas River south of Leadville, Colorado, as a result of historical mining activities. We excavated an intact 8-inch diameter core from a small fluvial tailings deposit (approximately 0.1 km^2) located about 13 km downstream from Leadville. To simulate leaching of metals from the fluvial tailings deposit to shallow ground water, we performed column-leaching experiments and examined the amount of metals leaching from the core. We applied deionized water to the top of the core and collected leachates from ports at the base of the core. Unsaturated, partially saturated, and fully saturated conditions were simulated. Leachates were analyzed for pH, metal concentrations, and specific conductance. For approximately the first 100 hours of unsaturated flow through the core (12 L of deionized water), conductivity and metal concentrations rose steadily and pH fell to values below 3. Metal concentrations peaked at about 170 mg/L Zn, 110 mg/L Al, 35 mg/L Fe, 8.7 mg/L Mn, 4.0 mg/L Cd, 1.9 mg/L Cu, and 0.1 mg/L Pb. The results demonstrate that large amounts of acidity and metals can be readily leached from these fluvial tailings deposits. The pH remained fairly stable under partially and fully saturated conditions, and Zn, Al, Mn, Cd, and Cu concentrations declined. Iron concentrations showed a slight increase during the partially and fully saturated conditions. Lead exhibited unique behavior throughout the leaching experiment and spiked when partially and fully saturated conditions were achieved. This behavior indicates that there is a readily available reservoir of Pb throughout the core. In light of these data, flooding of the fluvial tailings deposits along the upper Arkansas River may result in increased Pb concentrations locally in the shallow ground water. However, since Pb is fairly immobile in iron-rich systems at pH > 4, this situation probably does not represent a threat to the Arkansas River.

1 INTRODUCTION

The floodplain of the upper Arkansas River, south of Leadville, Colorado, contains several fluvially deposited mine tailings from historic mining operations in the Leadville area. These deposits are a possible nonpoint source of acid and metal contamination to surface and ground water. Studies at other fluvial tailings sites document that stormwater runoff from tailings can be a source of acidity and metals to surface and ground water (Nimick & Moore 1991; Bayless & Olyphant 1993).

Our study site is in the upper Arkansas River basin approximately 13 km south of Leadville, Colorado (Figure 1). The size of the site is about 0.1 km^2 and it is predominantly barren of living vegetation. Recently the land use in the surrounding area has been cattle grazing. There is a distributary channel that flows through the study site and the tailings mainly are deposited between the Arkansas River channel and this distributary channel.

The fluvial tailings deposits at the site are generally fine-grained, overbank deposits containing mixtures of tailings and other sediment. The tailings themselves are composed of mine waste washed downstream from the Leadville Mining District, which contains gold, silver, lead, and

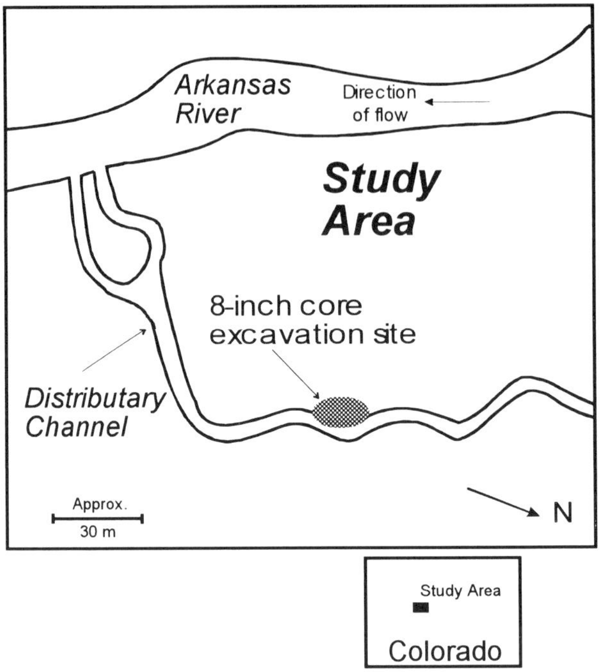

Figure 1. Schematic of the study area showing location of core excavation site.

zinc ores mined over the last 100+ years. Cored material from the fluvial tailings deposit is very heterogeneous. Generally the top of the deposit consists of a fine-grained pyrite-rich layer, the middle portion of the deposit is clay-rich with sand and silt lenses, and the bottom is organic-rich underlain by a sand and gravel shallow aquifer. The dominant minerals are quartz, feldspar, and mica (Jerz 1998; Smith et al. 1998b).

In this study we examined the leaching behavior of an intact 8-inch core excavated from the fluvial tailings deposit. We investigated the leaching behavior under unsaturated, partially saturated, and fully saturated conditions to simulate leaching behavior of the fluvial tailings deposit under a variety of naturally occurring and hypothetical hydrologic states. Previous work at this site shows that metals and acidity can be readily leached from the fluvial tailings material (Walton-Day et al. 1996; Jerz 1998; Smith et al. 1998b).

2 METHODS

An 8-inch core was excavated intact from the bank of the distributary channel that cuts through the tailings deposit. A clear polymethylmetharcrylate tube was placed on top of the bank. The fluvial tailings around the tube were slowly excavated and the tube pushed down to encase the remaining material. This process was repeated until the shallow aquifer material was reached (approximately 60 cm of overlying material). The bottom of the tube was fitted with a polyvinylchloride cap and the joint sealed with silicone cement. The cap contained drainage ports that were designed to separate water draining along the interface between the cored material and the inner edge of the tube (side-wall flow; edge port) from the water draining through the center of the cored material (center port). A polypropylene cloth prevented fine material from passing from the cored material into the cap. The cap was filled with quartz sand to minimize dead volume.

Deionized water was applied to the top of the core at a rate of 2 mL/min using a peristaltic pump and was allowed to gravity drain through the core. This resulted in unsaturated flow conditions. Leachate samples were collected from ports at the bottom of the core (in this paper we discuss only samples collected from the center port). Leachate was collected at varying times and sample weight, pH, and specific conductance were measured. Subsamples of the leachate were taken and saved for chemical analyses. Results presented here are for unfiltered samples. Filtered samples (0.45 μm) were also collected, but little difference was noted between filtered and unfiltered results.

After approximately 20 days, the application rate was reduced to 1 mL/min for 3 days. About 23 days after the beginning of the experiment the core was allowed to partially fill from the bottom to about half full. This partially saturated state is intended to simulate rising of the underlying shallow ground water table into the fluvial tailings deposit. Partial saturation was maintained for about 29 days with continual application at 1 mL/min. After that time, the core was allowed to completely saturate and saturation was maintained for 9 days with continual application of deionized water at the same rate. During saturation, ferrous iron and dissolved oxygen were monitored in addition to pH and specific conductance. Acidified samples were analyzed by inductively coupled argon plasma - mass spectroscopy (ICP-MS).

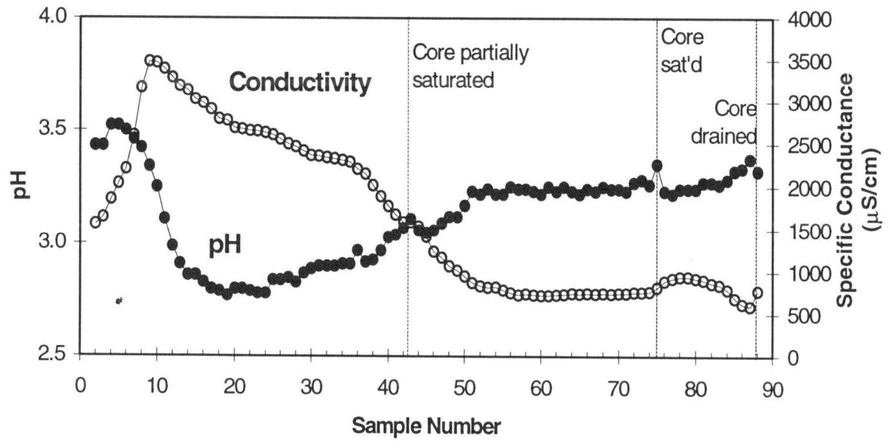

Figure 2. Conductivity and pH values of fluvial tailings core leachate samples collected under unsaturated, partially saturated, and fully saturated conditions.

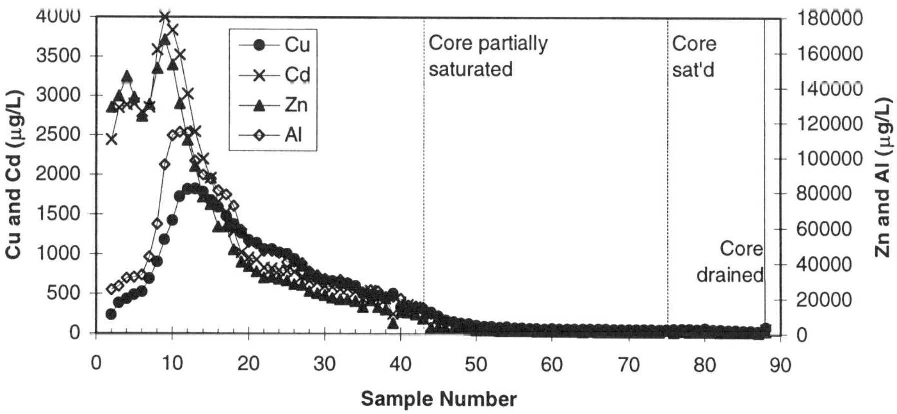

Figure 3. Copper, cadmium, zinc, and aluminum concentrations in fluvial tailings core leachate samples collected under unsaturated, partially saturated, and fully saturated conditions.

3 RESULTS AND DISCUSSION

3.1 *Unsaturated leaching conditions*

The near-surface area of the core contained obvious deposits of efflorescent salts that are readily soluble. We assume that the majority of the acidity and dissolved metals in the leachates originate from these salts at and near the top of the core. The salts were visibly depleted over the course of the leaching experiment. Initial pH and specific conductance measurements (Figure 2) and Cu, Cd, Zn, Al, Mn, Fe, and Pb concentrations (Figures 3-5) show that there is a reservoir of readily leachable material in the core that can be removed under unsaturated conditions.

The pH of the leachate (Figure 2) drops quickly over the first 100 hours to less than pH 3, remains low, and then slowly rises to pH > 3. The low pH corresponds to the slow tailing off of Fe (Figure 4) and probably is related to acid produced by the precipitation of hydrous Fe oxide minerals. After a rapid increase over the first 100 hours, the specific conductance slowly decreases over the course of the leaching experiment (Figure 2).

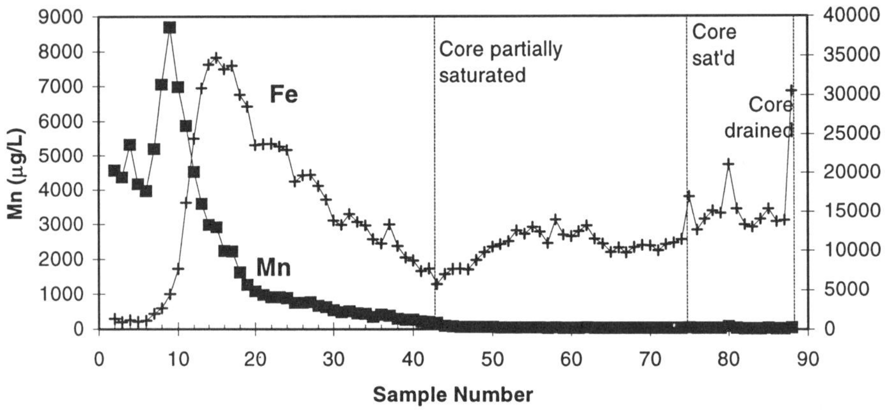

Figure 4. Iron and manganese concentrations in fluvial tailings core leachate samples collected under unsaturated, partially saturated, and fully saturated conditions.

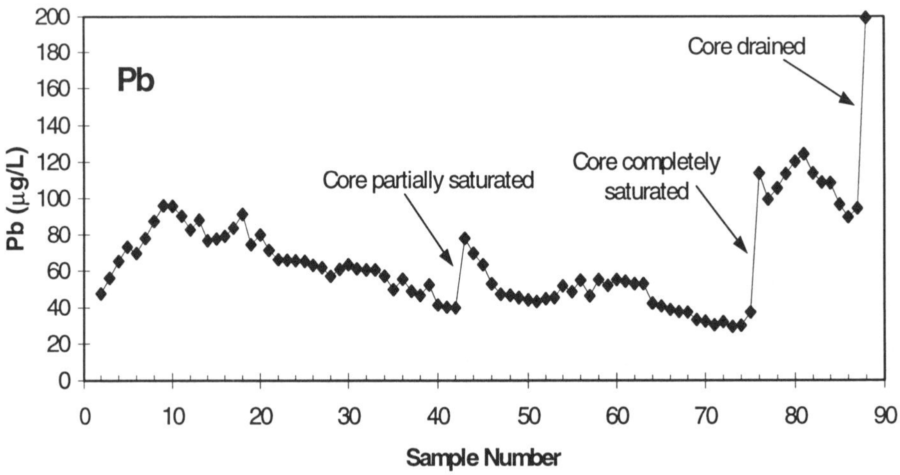

Figure 5. Lead concentrations in fluvial tailings core leachate samples collected under unsaturated, partially saturated, and fully saturated conditions.

Concentrations of all the metals, except Pb, increased to a maximum after about 100 hours (around sample number 10; 12 L of deionized water) since the beginning of the leaching experiment (Figures 3 and 4). Peak concentrations of some of the metals are slightly offset from each other and have different features before and after the peak (as illustrated in Figures 3 and 4). Most notable are Cu, Al, and Fe, whose migration through the core seems to be somewhat retarded compared with the other metals in Figures 3 and 4. It is likely that Fe is precipitating in the core as hydrous Fe oxides. The behavior of Pb (Figure 5) is different from the other metals in that there is a less obvious peak in concentration and a very broad tail. This may indicate that Pb is being retained in the core by mechanisms such as precipitation of anglesite ($PbSO_4$) or sorption onto hydrous iron oxide minerals in the core.

3.2 *Partially and fully saturated leaching conditions*

About 23 days after the beginning of the leaching experiment the core was allowed to partially fill from the bottom up to about half full by raising the exit tube from the center port. The fill line was well below the upper pyrite-rich layer of the core and below any zone of obvious salt formation. This partially saturated state is intended to simulate rising of the underlying shallow ground water table into the fluvial tailings deposit as might take place during the spring. Partial saturation was maintained for about 29 days with continual application at 1 mL/min. After that time, the core was allowed to completely fill and complete saturation was maintained for 9 days with continual application at 1 mL/min.

During partial saturation, pH slowly increased from about 3.1 to 3.3, and specific conductance slowly decreased and leveled off (Figure 2). There was a slight jump in pH and in specific conductance when the core was completely filled. Specific conductance also jumped slightly when the core was drained, and pH decreased slightly (Figure 2). On the whole, partial and complete saturation of the core did not have a large effect on pH or specific conductance.

Partial and complete saturation also did not have much of an effect on Cu, Cd, Zn, Al, and Mn (Figures 3 and 4). After the initial peak of these metals under unsaturated conditions, they steadily decreased throughout the remainder of the leaching experiment as their source was depleted.

Iron concentrations slightly increased during partial and complete saturation of the core (Figure 4). Shortly after partial saturation of the core, dissolved oxygen dropped to near zero and ferrous iron rose from near zero to roughly 100% of the total iron. The increase in total Fe concentration may be due to dissolution of hydrous iron oxide minerals present in the core.

Lead concentrations spiked when the core was partially and completely saturated, and again when the core was drained (Figure 5). These spikes in concentration represent the presence of a readily available reservoir of Pb throughout the core. This Pb reservoir may be a soluble mineral phase present throughout the core or it may be lead sorbed onto hydrous iron oxides present in the core. In light of these data, flooding of fluvial tailings deposits along the upper Arkansas River may result in increased Pb concentrations locally in the shallow ground water. However, since Pb is fairly immobile in iron-rich systems at pH > 4 (Smith et al. 1998a), Pb will likely be immobilized prior to reaching the Arkansas River.

4 SUMMARY AND CONCLUSIONS

We performed column-leaching experiments with an intact 8-inch core excavated from a fluvial deposit of mine tailings along the upper Arkansas River south of Leadville, Colorado. For approximately the first 100 hours (12 L of deionized water) of unsaturated flow through the core, conductivity and metal concentrations rose steadily and pH fell to below 3. We attribute these peaks in metal concentrations and acidity to the leaching of readily soluble salts present at and near the top of the core. The pH slowly rose and leveled off under partially and fully saturated conditions, and Zn, Al, Mn, Cd, and Cu concentrations declined as the source material was depleted. Iron concentrations showed a slight increase during the partially and fully saturated conditions possibly due to the dissolution of hydrous iron oxide mineral phases under anoxic conditions. Lead exhibited unique behavior throughout the leaching experiment and spiked when partially and fully saturated conditions were achieved. This Pb behavior suggests that there is a readily available reservoir of Pb throughout the core, which could either be a soluble mineral phase or Pb sorbed onto hydrous iron oxide minerals in the core.

5 ACKNOWLEDGMENTS

We would like to acknowledge the landowner who has graciously allowed us to work on his property. Thanks to Phil Hageman and Mike Singleton for their help in the field collecting the core. We are grateful for the help Jeane Jerz gave us in setting up the experiments and to Don Macalady for the use of his laboratory at the Colorado School of Mines. Jim Crock and Reini Leinz provided helpful reviews of this paper. Analytical chemistry was performed at the U. S. Geological Survey Mineral Resources Program Analytical Laboratories, Denver, Colorado. We acknowledge Monique Adams for her help with the ICP-MS analyses. This work was funded through the U. S. Geological Survey Toxic Substances Hydrology Program.

REFERENCES

Bayless, E.R. & G.A. Olyphant 1993. Acid-generating salts and their relationship to the chemistry of groundwater and storm runoff at an abandoned mine site in southwestern Indiana, U.S.A. *J. Contaminant Hydrology* 12:313-328.

Jerz, J.K. 1998. *Mechanisms of acid and metal release from a fluvial tailings deposit*. M.S. Thesis, Colorado School of Mines, Golden, Colorado.

Nimick, D.A. & J.N. Moore 1991. Prediction of water-soluble metal concentrations in fluvially deposited tailings sediments, Upper Clark Fork Valley, Montana, U.S.A. *Applied Geochemistry* 6:635-646.

Smith, K.S., J.F. Ranville, G.S. Plumlee, & D.L. Macalady 1998a. Predictive double-layer modeling of metal sorption in mine-drainage systems. In E.A. Jenne (ed.), *Adsorption of metals by geomedia--variables, mechanisms, and model applications*:521-547. San Diego: Academic Press.

Smith, K.S., S.J. Sutley, P.H. Briggs, A.L. Meier, K. Walton-Day, J.F. Ranville, & J.K. Jerz 1998b. Trends in water-leachable lead from a fluvial tailings deposit along the upper Arkansas River, Colorado. In *Proceedings of the Fifth International Conference on Tailings and Mine Waste '98, Fort Collins, Colorado, January 26-29, 1998*:763-768. Rotterdam: Balkema.

Walton-Day, K., J.K. Jerz, J.F. Ranville, J.B. Evans, & K.S. Smith 1996. Effects of fluvial tailings deposits on the quality of surface and ground water at a site in the upper Arkansas River basin, Colorado. In *Abstracts with Programs, Geological Society of America Annual Meeting, Denver, Colorado, October 28-31, 1996*:A-466.

Tailings and Mine Waste '99 © 1999 Balkema, Rotterdam, ISBN 90 5809 025 6

Using flood-analysis techniques to estimate dissolved-zinc concentrations

K.J. Leib & W.G. Wright
US Geological Survey, Durango, Colo., USA

M.A. Mast
US Geological Survey, MS 415, Denver, Colo., USA

ABSTRACT: Recurrence, probability, and duration of dissolved-zinc concentrations are needed to describe fish exposure to different zinc-concentration ranges and flow regimes in the mining-impacted rivers of the Upper Animas Watershed. The recurrence interval (in days), probability (in percent), and duration (in days) of dissolved-zinc concentrations for the Animas River downstream from Silverton, Colorado, were estimated using frequency analysis techniques. Samples collected during water year 1997 were used to develop a correlation between dissolved-zinc concentrations and streamflow by using high-flow and low-flow regression models. A log-Pearson type III distribution was fitted to the estimated dissolved-zinc data to estimate recurrence intervals and exceedance probabilities for concentrations that range from 200 to 800 μg/L (micrograms per liter). Durations for dissolved-zinc concentrations were estimated using the duration-curve method in conjunction with the high-flow and low-flow models. Recurrence intervals for dissolved zinc ranged from 1.1 days for concentrations of 200 μg/L to about 8 days for concentrations of 800 μg/L. Daily exceedance probabilities for dissolved zinc ranged from 13 percent for 800 μg/L to 95 percent for 200 μg/L. For dissolved-zinc durations, 200 μg/L was equaled or exceeded 365 days per year, and 800 μg/L was equaled or exceeded 37 days per year.

1 INTRODUCTION

Trace-metal concentrations in rivers draining hard-rock mining districts show trends that correlate with seasonal flow regimes. The concentrations of most dissolved constituents in streams decline during periods of high streamflow (Anderson et al. 1997). Of particular interest to scientists are periods when trace metal concentrations reach levels of acute and chronic toxicity for fish. The frequency and duration of toxic metal concentrations are generally undefined for high-elevation rivers and streams because of difficult access and limited resources. A method of evaluating the frequency for when acute and chronic toxicity levels are exceeded seasonally would therefore be useful when assessing aquatic habitat and remediation.

This report describes methods used for frequency analysis of dissolved-zinc concentrations in micrograms per liter (μg/L) in terms of recurrence interval (in days), probability (in percent), and duration (in days) in the mining-impacted Animas River downstream from Silverton, Colorado. The sampling site (A72) is part of the U.S. Geological Survey Abandoned Mine Lands Initiative (AML) monitoring network and is located at the outflow of the study area. The AML is part of a larger interagency agreement by Federal land-management agencies and Department of the Interior science bureaus to remediate pollution from abandoned mine lands using the watershed approach.

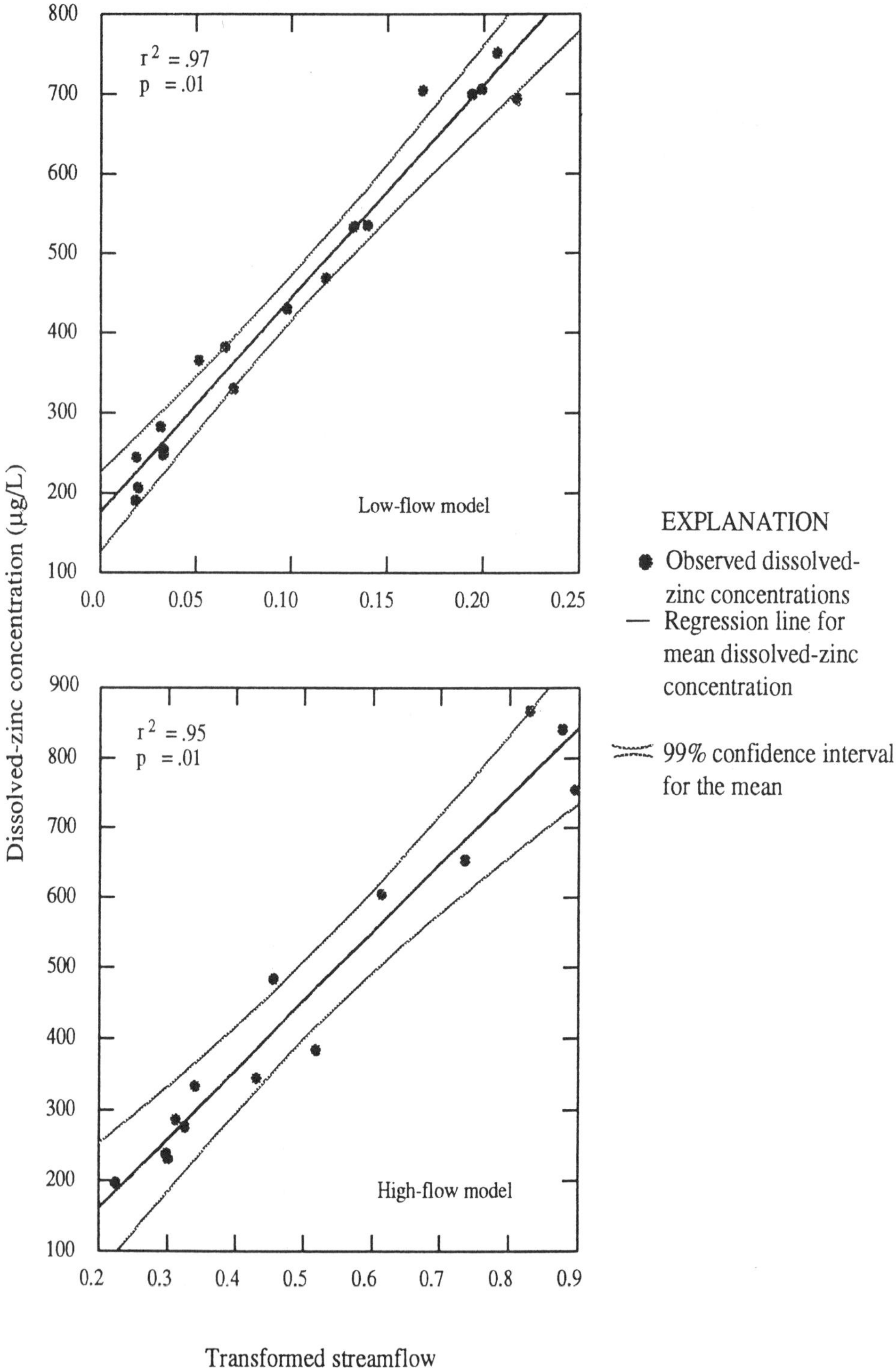

Figure 1. Regression relation of dissolved-zinc concentrations and hyperbolically transformed streamflow at A72.

2 CORRELATION BETWEEN DISSOLVED-ZINC CONCENTRATION AND STREAMFLOW

Mean daily streamflow and water-quality data collected by the U.S. Geological Survey and Sunnyside Gold Corporation in water year 1997 at A72 were used to determine two simple linear regression models. The high-flow model includes the effects of flushing of constituents during the rise of the snowmelt hydrograph. The low-flow model includes the effects of limited constituent supply during the recession of the snowmelt hydrograph and lack of surface-water inflow during winter months. The relation between dissolved-zinc concentration and streamflow (Figure 1) is described by equations 1 and 2:
Low-flow model (July 1- February 28):

$$Y_l = 1{,}000[0.1767 + 2.6786(x_l)] \tag{1}$$

where Y_l = estimated dissolved-zinc concentration (μg/L), x_l = transformed streamflow = 1/{1+b[Q(35.31)]}, b = 0.065, and Q = mean daily streamflow (m^3/sec),
High-flow model (March 1-June 30):

$$Y_h = 1{,}000[-0.0331 + 0.9742(x_h)] \tag{2}$$

where Y_h = estimated dissolved-zinc concentration (μg/L), x_h = transformed streamflow = 1/{1+b[Q(35.31)]}, b = 0.0017, and Q = mean-daily streamflow (m^3/sec). Transformed streamflow (x_n) is a hyperbolic transformation that better approximates a normal distribution for the dissolved-zinc data and improves mean daily streamflow (Q) as a predicting variable. Transforming the streamflow variable decreased curvature and heteroscedasticity in residual versus predicted value plots for both the high-flow and low-flow models. The models were checked for serial correlations by plotting the residuals over their respective time periods. No evidence of a serial correlation was observed for the models at this phase of data collection for site A72.

Streamflow data from A72 are used in the two-part model to estimate daily dissolved-zinc concentrations ranging from 200 to 800 μg/L. Estimated dissolved-zinc concentrations are used in the frequency analysis. Because no major diversions or impoundments exist upstream from A72, all mean daily streamflow record (1992-97) was used in conjunction with equations 1 and 2 to estimate dissolved-zinc concentrations. Estimated dissolved-zinc concentrations are then used in the frequency analysis. These dissolved-zinc concentrations generated from the 1992-97 streamflow record can be used to quantify the effects of remediation in the watershed when compared to future frequency analysis developed using post-remediation models.

3 FREQUENCY ANALYSIS OF DISSOLVED-ZINC CONCENTRATIONS

3.1 *Recurrence interval and exceedance probability*

Dissolved-zinc concentrations were predicted by using equations 1 and 2 and combined with flood-frequency analysis techniques to estimate recurrence intervals, exceedance probabilities, and annual durations of dissolved-zinc concentrations. Recurrence interval (T) is the average length of time within which a given event will be equaled or exceeded (Linsley et al. 1975). Exceedance probability (P) describes the likelihood of dissolved-zinc concentrations equaling or exceeding a given value in terms of percentages and is the reciprocal of T. Flood frequency analysis is traditionally used in conjunction with streamflow data for predicting design floods for structures and levees. For example, annual maximum streamflow data for a period of 50 years would provide statistically useful information when predicting the recurrence interval and probability of a 100-year design flood. Forecasting beyond the total number of years in the data set is possible because frequency analysis normalizes the distribution of flow to what a complete period of record would represent.

This report uses the same frequency analysis techniques, but applies them to the predicted dissolved-zinc concentrations. This application is possible because the concentrations are dependent on mean daily streamflow. Each water year is assumed to be a separate period of record 365 days long, where mean-daily-zinc concentrations are treated like annual maximum streamflow values in a traditional flood analysis. Based on this assumption, the estimated dissolved-zinc concentrations for each water year can be normalized by applying a log-Pearson type III distribution. The log-Pearson type III distribution technique is recommended for flood frequency analysis by the Water Resources Council (Interagency Advisory Committee on Water Data 1982). Although the true distribution of dissolved-zinc concentrations for any given water year is not known, distributions using the log-Pearson type III frequency analysis technique provided T and P values comparable to those of observed mean daily streamflow. Therefore, if the distribution of mean daily streamflow is normalized for any given water year, it is assumed that dissolved-zinc concentrations are normalized on the basis of relations defined by equations 1 and 2.

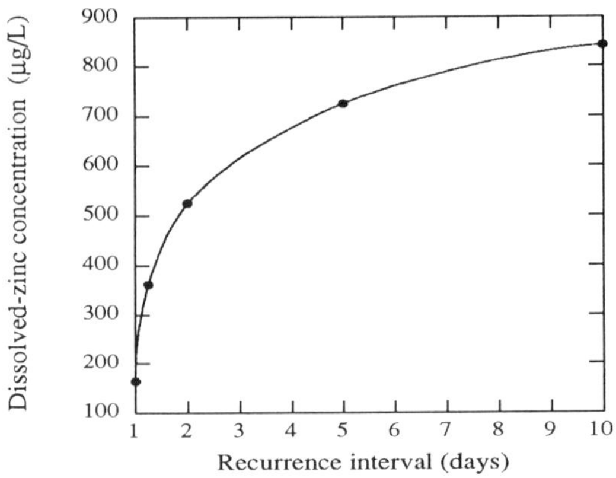

Figure 2. Average number of days in which a given concentration of dissolved zinc will be equaled or exceeded.

The log-Pearson type III frequency analysis was performed on each of the generated dissolved-zinc data sets from water years 1992-97. The results were then averaged to obtain one final data set. Recurrence-interval values for dissolved-zinc concentrations ranged from 1.1 days for concentrations of 200 µg/L to about 8 days for concentrations of 800 µg/L (Figure 2). Daily exceedance probabilities for dissolved zinc ranged from 13 percent for 800 µg/L to 95 percent for 200 µg/L (Figure 3).

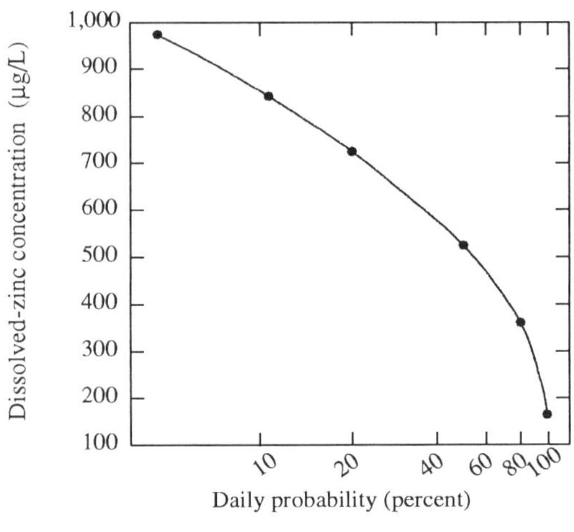

Figure 3. Probability that a given dissolved-zinc concentration will be equaled or exceeded during a day.

3.2 *Durations for dissolved-zinc concentrations*

Durations for zinc concentrations were determined using a frequency-analysis technique known as the calendar-year flow-duration-curve method. The quality of surface streams is commonly shown by duration curves of sediment, turbidity, hardness, or some other characteristic of surface-water quality (Searcy 1959). The duration curve used for this analysis (Figure 4) was constructed using the calendar-year method, which averages mean-daily streamflows for each day of the year for all years of record. Traditional flow-duration curves are cumulative frequency

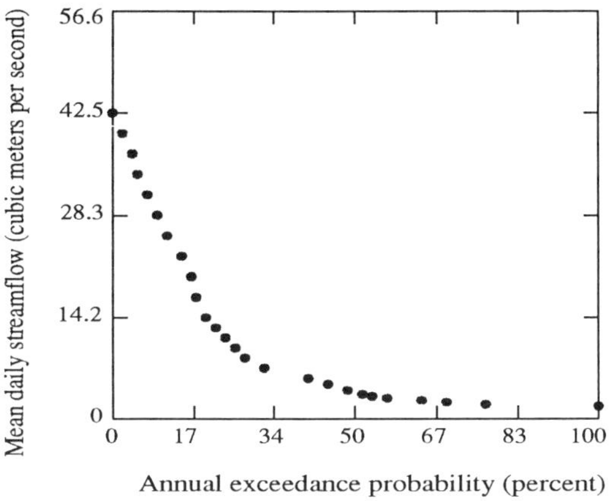

Figure 4. Calendar-year flow-duration curve for water years 1992-97.

curves that show the percentage of time specified streamflows were equaled or exceeded. The theory and construction of duration curves is outlined by Searcy (1959). Durations in this report are defined as the number of days that dissolved-zinc concentrations will equal or exceed a given value in a 365-day period and are referred to as zinc-duration days. This definition differs from traditional duration definitions in two ways: (1) the percentage of time values are equaled or exceeded is converted to days per year by applying that percentage to 365, and (2) streamflow values are converted to dissolved-zinc concentrations by using equations 1 and 2. By using this approach, durations can be reported in terms of annual percentages and number of days per year. Table 1 lists zinc-duration days calculated from the low-flow and high-flow models. Note that zinc-duration days may not occur consecutively due to concentration increases from rainfall events.

4 DISCUSSION

Results from the method described herein can determine the magnitude of water-quality degradation and, for existing conditions, help assess the frequency of acute and chronic levels of toxicity for aquatic life as well as the feasibility of proposed water-quality standards. Levels of acute and chronic toxicity for aquatic life, during water year 1997 high and low flow, were calculated at A72 for dissolved zinc using hardness-corrected toxicity equations provided by the Colorado Water Quality Control Commission (CWQCC 1995) and data collected as part of AML. During high flow (March 1-June 30), calculated acute toxicity thresholds ranged from 27.0 to 93.7 μg/L with a median value of 33.6 μg/L and calculated chronic toxicity thresholds ranged from 24.5 to 84.8 μg/L with a median value of 30.4 μg/L. During low flow (July 1-February 28), calculated acute toxicity thresholds ranged from 31.6 to 137 μg/L with a median value of 70 μg/L and calculated chronic toxicity thresholds ranged from 28.6 to 124 μg/L with a median value of 63.4 μg/L. Based on the results from the frequency analysis, dissolved-zinc concentrations at A72 are all above the calculated toxicity levels. The State of Colorado Water Quality Control Commission has proposed a standard of 225 μg/L for dissolved zinc at A72 (CWQCC 1995). The frequency analysis indicates that dissolved-zinc values equal to or less than 225 μg/L occur approximately 1-2 days during high flow and approximately 4-8 days during low flow (table 1). The frequency analysis suggests the proposed standard level is not attainable at this time.

Remediation in the watershed may temporarily increase levels of dissolved-zinc concentrations due to the disturbance of channels and exposure of waste rock during relocation. Additional sampling through the year 2000 by the U.S. Geological Survey as part of AML will help address water-quality trends resulting from remediation by refining the low-flow and high-flow models used in frequency analysis.

Frequency analysis would benefit from characterization of storm-runoff hydrographs. The lack of data for this aspect of the seasonal hydrograph may affect the applicability of the regression models. Without the incorporation of a storm-runoff model, dissolved-zinc concentrations may be underestimated.

Table 1. Zinc-duration days for water year 1997 in the Animas River downstream from Silverton, Colorado [µg/L, micrograms per liter]

High-flow model (March 1-June 30)		Low-flow model (July 1-February 28)	
Zinc-duration days	Zinc µg/L	Zinc-duration days	Zinc µg/L
1	851	1	723
9	837	74	659
16	824	96	609
29	812	102	568
37	800	121	534
38	788	129	505
44	776	132	481
48	754	138	442
50	723	152	399
53	694	162	368
58	651	191	332
64	612	200	307
65	578	206	290
68	547	211	276
72	519	213	265
72	493	221	257
73	449	227	244
73	412	229	234
78	380	234	227
82	352	240	222
88	328	241	217
95	306	242	214
101	287	243	211
120	229		
122	217		

REFERENCES

Anderson, S.P., Dietrich, W.E., Torres, R., Montgomery, D.R., & K. Loague 1997. Concentration-discharge relationships in runoff from a steep, unchanneled catchment. *Water Resources Research*: 211-225.

Interagency Advisory Commitee on Water Data 1992. Guidelines for determining flood flow frequency. *Bulletin 17B of the Hydrology Committee*, USGS, Office of Water Data Coordination: Reston, Virginia.

Linsley, R.K., Jr., Kohler, M.A., & J. H. Paulhus 1975. *Hydrology for engineers.* New York: McGraw-Hill.

Searcy, J.K. 1959. *Flow-Duration Curves*: U.S. Geological Survey: Water-Supply Paper 1542-A.

Colorado Water Quality Control Commission (CWQCC) 1995. *Stream classifications and water quality standards,* Water Quality Control Commission triennial review: Denver, Colorado.

Tailings and Mine Waste '99 © *1999 Balkema, Rotterdam, ISBN 90 5809 025 6*

The measurement of biologically-available metals – A new approach

J. Jay McNee & Alan J. Martin
Lorax Environmental Services Limited, Vancouver, B.C., Canada

ABSTRACT: Environmental scientists have experienced several problems associated with the collection of discrete, uncontaminated water samples for trace metal analysis. The main challenge has been to preserve sample integrity from the perspectives of contamination and metal speciation. Compounding the problem is the fact that dissolved metal concentrations in natural waters (particularly sea water) are often below the analytical detection limits of even the most sensitive analytical equipment. Another major shortcoming of existing measurement techniques is that although they can define total and dissolved trace metals in water samples, they cannot quantify the biologically-available (toxic) fraction. Since it is the latter fraction that controls metal uptake and toxicity in aquatic biota, total or dissolved contaminant concentrations are not the best tools for defining water quality or effluent discharge criteria. An in situ water sampling device has been developed to measure biologically-available or "labile" metals in natural waters and effluents. The active mechanism involves the use of a polyacrylamide gel layer to quantitatively control metal transfer to a trace-metal-adsorbing, gel-imbedded resin. Because the resin only adsorbs free metal ions, and since metal bioavailability is related to the free-ion activity, the device provides a direct measure of the biological-availability of metals in the environment sampled. The passive nature of this device affords it the potential to address many of the limitations presented by conventional sampling methods such as sample contamination introduced during collection or filtration and changes in trace metal speciation prior to analysis. Laboratory studies and field trials indicate that the sampler behaves as anticipated and illustrate that biologically-available metal concentrations in many environments are a small fraction of the total concentrations and often significantly less than the dissolved concentration. The implications of this on regulatory criteria are discussed in addition to presentation of select field data and water quality sampling strategies.

1 INTRODUCTION

The collective understanding of trace metals cycles in the environment has been advanced significantly since the early 1980's with the advent of "ultra-clean" trace metal protocols for sample collection and handling (Bruland, 1979; Nriagu *et al.* 1993). Despite an improved understanding of trace metal geochemistry, the regulatory framework for the environmental assessment of water quality (*i.e.*, the reliance on total metal concentrations) has remained virtually unchanged in the face of the obvious limitations associated with such an approach. For example, it is acknowledged that total metal concentrations are a poor proxy for metal bioavailability particularly where high suspended sediment loads exist. Nonetheless, the use of dissolved metal concentrations has been discouraged from use by regulators for several reasons, not the least of which is the potential to introduce contamination during the process of sample filtration. The problem is exacerbated by the fact that receiving water guidelines have been progressively lowered by regulatory agencies in well-meaning but often misguided efforts to prevent environmental impacts. These criteria incorporate significant "safety factors" designed to minimize perceived risks and are in many cases, more restrictive than levels found naturally in many aquatic systems where biological communities thrive. Despite the recognition that total metal concentrations are not an appropriate measure for impact assessment, the mining industry has reached an impasse in which there is an acknowledged deficiency of tools available to address these issues. This problem is best illustrated by the willingness of regulators to acknowledge natural mitigation processes such as organic complexation of trace metals yet there is an inability to incorporate such considerations into site-specific criteria.

2 BACKGROUND AND THEORY

Recent data suggest that the toxicity of a given trace metal is mitigated by complexation with various aquatic agents; most notably, dissolved and particulate organic matter are known to offset toxicological impacts of many trace metals for a given concentration. The reason for this phenomenon relates to the manner in which trace metals are taken up by aquatic organisms. It is now well documented that the most effective modes of metal uptake occur only for free, uncomplexed, trace metal ions (Campbell, 1995); toxicity or bio-availability is directly related to the free-ion-activity of the metals in solution. Complexation of trace metals reduces their solution activity *in situ* and diminishes their respective toxicity or biological availability.

Until recently, polarographic analyses were the only techniques available for the measurement of metals species in natural waters (Hirose and Sigimura, 1985; Bruland, 1989; 1992). Here, metals are plated out of the stirred sample solution onto a mercury electrode. Speciation is inferred from the understanding that only free, ionic metals and weak metal complexes with rapid dissociation kinetics are reduced onto the electrode forming the amalgam; particulate or organically complexed metals are not plated and consequently are not detected by the technique. Samples for metal speciation are collected in the field by conventional water sampling devices (*e.g.* Go-Flo bottle), maintained as close to in situ conditions as possible (*i.e.* refrigeration or storage on ice) and transported back to a laboratory for analysis as soon as possible. This approach constitutes the principal weakness in determinations of speciation; the time between sample collection and analysis opens an opportunity for speciation to be altered, particularly given that ambient conditions are not well preserved (Byrne *et al.* 1988). In addition, conventional sampling techniques are still sensitive to the potential to introduce contamination artifacts. The gel sampler approach not only measures free metal activity but also addresses many of the short-comings of existing techniques.

The gel sampler consists of a layer of polyacrylamide gel which separates the solution to be sampled from a gel-bound ion exchange resin (Figure 1). Since approximately 90% of the polyacrylamide gel is water, trace-metal transport through the gel is controlled by molecular diffusion (Davison and Zhang, 1994; Krom et al., 1994), and is therefore well-defined. Metal ions are rapidly and quantitatively bound by the resin contained within the inner gel layer (Paulson, 1986; Pai, 1990), creating a steep concentration gradient in the outer gel layer (Figure 1). The amount of labile trace metal that accumulates on the resin can be related directly to the concentration in the solution, the exposure time and the diffusion coefficient of the labile species. The latter two variables are known or can be determined empirically, allowing for the determination of the labile metal concentration; the details are presented below.

The polyacrylamide gel technique is based on measurement of a trace metal flux through the gel. The flux of trace metals may be quantified by applying Fick's first law of diffusion as expressed by the following equation:

$$F = \frac{DC_b}{\Delta g} \quad (1)$$

where F is the flux of trace metal ion diffusing from the water column to the resin layer, C_b is the concentration of trace metal in the bulk solution, Δg is the thickness of the gel layer and D is the molecular diffusion coefficient of a free metal ion in the gel (*i.e.* Li and Gregory, 1974; Krom and Berner, 1980; Krom *et al.* 1994). To determine the concentration of metal in the solution, equation (1) can be rearranged and written as:

$$C_b = \frac{C_r \Delta g \Delta r}{Dt} \quad (2)$$

where C_r is the concentration of trace metal accumulated on the resin after a given time, t, and Δr is the thickness of the resin layer.

The metal ion concentration measured using this technique is a time averaged mean concentration in the solution; longer deployment times result in greater quantities of metal

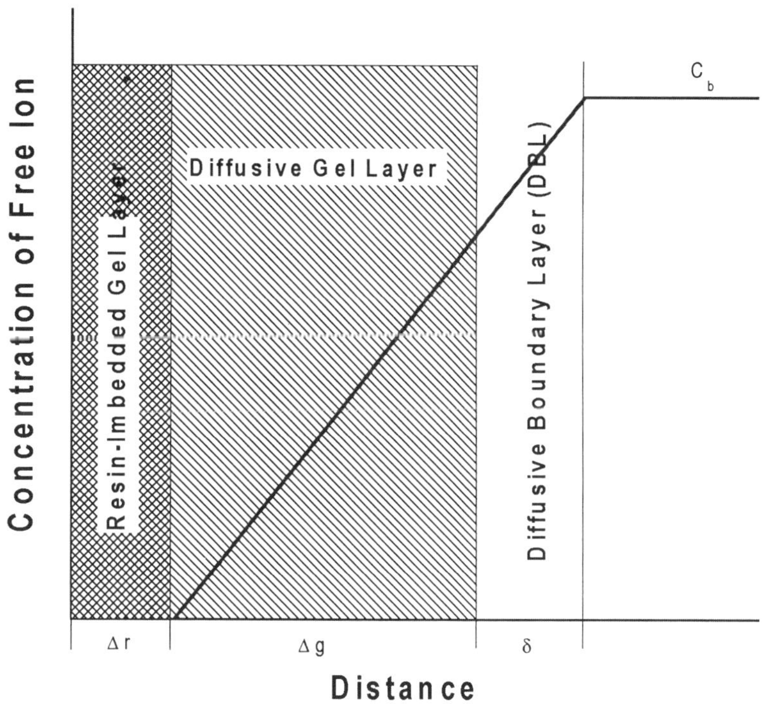

Figure 1 - Schematic diagram of a polyacrylamide gel sampler and the representation of the free metal concentration from the natural water through to the resin layer.

accumulation on the resin. This in situ preconcentration enables the measurements of ultra-trace concentrations of labile metals in natural waters, and also overcomes the contamination problems which plague conventional techniques. For example, longer deployment times increase the mass of metal on the resin and thus improve the effective detection limit. Accordingly, longer deployment times and larger masses of analyte on the resin progressively decreases the probability of introducing significant contamination.

The gel sampler methodology is a kinetic speciation and size-exclusion technique. Thus, large (*i.e.* >0.45 μm particle diameter) will not pass through the sampler to the resin layer and only free ions or weakly associated complexes will be available for uptake by the resin. In other words, a tightly bound organic-metal complex such as EDTA-Cu or NTA-Cu (both complexes have low biological-availability) are not readily detected by the sampler. However, a weak complex (perhaps a humic acid-Cu interaction) would have a larger equilibrium component of free metal in solution and would thus be reflected as a higher concentration in the sampler. Because it is the free ionic activity of the metals which are retained by the sampler, increasingly strong complexing agents will retard accumulation of metal on the sampler in the same way that they diminish bioavailability or toxicity of metals in aqueous solutions. The gel sampler in essence measures a representation of the labile, biologically-available fraction of potentially toxic trace metals.

3 RESULTS AND DISCUSSIONS

A suite of laboratory and field experiments have been performed to characterize the response of the sampler to different environmental condition. Sampler response to parameters such as temperature, turbulence, pH and organic complexation have been assessed and incorporated into the overall sampler performance and calibration. A synopsis of some of the findings in addition to some of the possible applications of the technology are presented below.

All work has been performed in Class 100 clean room. Polyacrylamide gels and resin layers were prepared with trace-metal-cleaned reagents and then acid-washed prior to use to lower blank values and improve precision and accuracy. The gel sampler is comprised of a resin layer overlain by a polyacrylamide gel layer and filter constrained within an acid-washed plexiglas holder. The assembled sampler is stored wet in distilled, deionized water until it is deployed in the solution to be measured. Each batch of samplers are calibrated with metal stock, uncomplexed metal solutions of known activity in order to compensate for potential variability in successive polyacrylamide castings. The "calibrated" sampler assemblies were then subjected to variations in parameters such as temperature, pH, organic complexation, and turbulence to determine the response of the device under the conditions to which it might be subjected. Typical calibration responses to laboratory solutions are presented in Figure 2.

Figure 2 illustrates how the gel sampler continues to accumulate metal on the resin in a linear fashion during its deployment. As predicted by the laws of diffusion, the rate of accumulation of metal on the resin is governed by its flux through the outer gel layer. Because flux is determined by the concentration gradient across the gel layer, the primary response of the sampler to different metal concentrations is manifest as a change in the slope of accumulated metal with time (Figure 2). Using these calibration data, and knowing the duration of sampler deployment and the *in situ* temperature of the environment sampled allows for back calculation of the labile metal concentration in solution using the equations outlined in the previous sections.

Two experiments were run to assess the response of the sampler to various environmental conditions. The first set of experiments explored sampler response to organic complexation while the second set examined the influence of effluent dilution into seawater.

Laboratory experiments were run to assess the influence of dissolve organic matter complexation on trace metal bio-availability as determined by the gel sampler. Stock

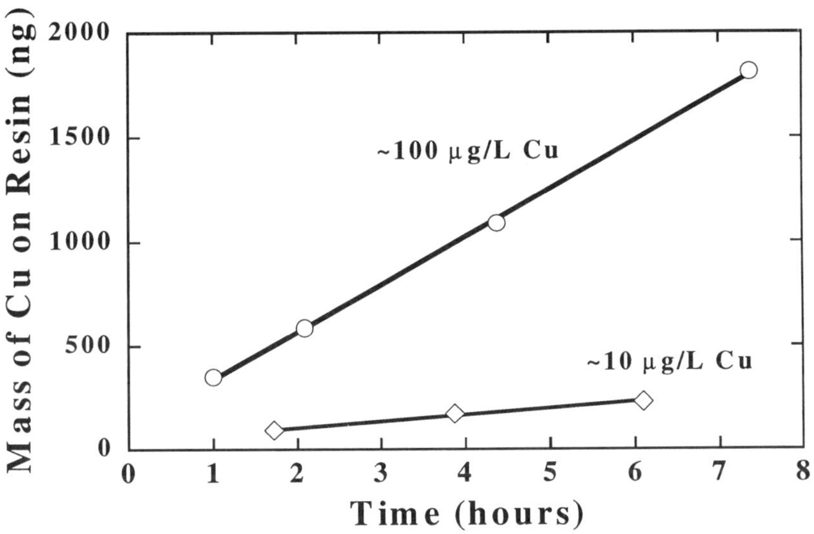

Figure 2 - A calibration experiment (mass of Cu on the resin vs. time) in which replicate samplers were exposed to a ~100 μg/L and ~10 μg/L copper solutions for various lengths of time. The different concentrations of metals are manifest as different accumulation rates of metal on the resin layer (*i.e.* slope of the line).

copper solutions of 20.8 ppb had one of the organic ligands ethylenediamine tetra acetic acid (EDTA, a strong organic complexing agent) or citric acid (a comparatively weak complexing agent) added to it. The gel sampler was deployed in each solution including an uncomplexed solution for 4 hours. The callibration data and triplicate sampler data are presented in Figure 3.

As illustrated in Figure 3, the gel sampler appropriately represents the bio-availability associated with the various metal-organic complexes. The uncomplexed metal stock was measured as 100% available while the organically complexed solutions proved to be available inversely related to the strength of the organic-metal complex. EDTA, the strongest of the ligands, resulted in the lowest available concentration with only 2.3 of the 20.8 ppb Cu being detected by the sampler. With the weaker ligand citric acid, 7.4 ppb of the 20.8 ppb Cu was detected (Figure 3). The implication of these data is that the gel sampler is a far better proxy for bio-available metals than the dissolved concentration particularly where trace-metal toxicity is mitigated by organic matter complexation.

In order to test the response of the sampler to effluent mixing in seawater, samples were collected from Britania Creek (an acid contaminated stream) which drains into Howe Sound (a coastal seawater fjord adjacent to Vancouver, B.C.). Filtered (0.45 μm) end-members (*i.e.* river water and seawater) were mixed to simulate estuarine mixing from 0 to 9 parts per thousand. Sub-samples were collected and gel samplers were deployed at several salinities to determine the dissolved and labile Cu concentrations (Figure 4).

As illustrated in Figure 4, the dissolved and labile Cu behave in a similar fashion; both fractions exhibit non-conservative removal from solution with increasing salinity as indicated by the concave-upward profiles beneath the ideal mixing line. Such behaviour is consistent with the geochemistry of copper in estuarine environments where copper is known to interact with metal oxyhydroxides and organic matter through the transitional environment presented by the mixing zone (Sholkovitz and Copland, 1981; Paulson, 1988; 1994). However, there is a marked increase

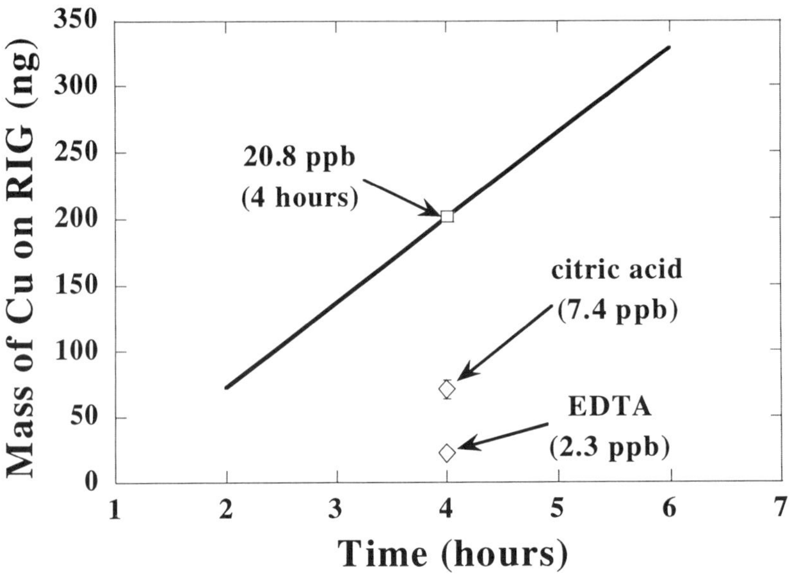

Figure 3 – The callibration data (line) and gel sampler representations of the stock solution, and complexed solution with additions of the organic complexing agents EDTA and citric acid. Error bars of triplicate analyses fall within the open symbols.

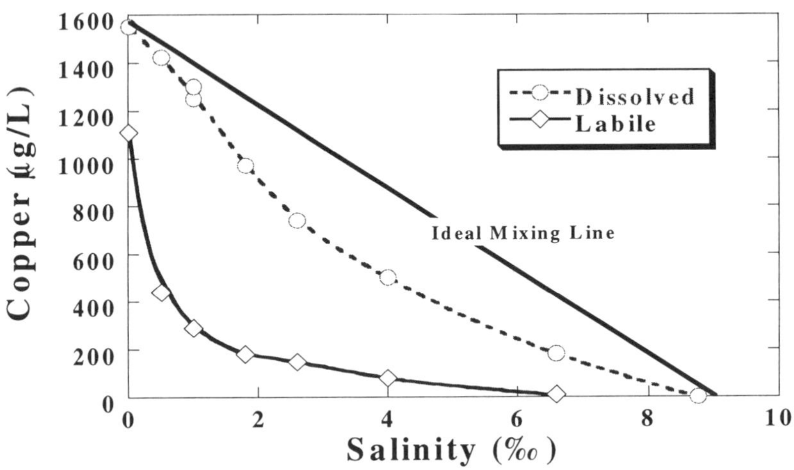

Figure 4 - Dissolved and labile copper concentrations in various dilutions of AMD-contaminated river water and seawater. The straight line represents ideal mixing between the two end-members with no geochemical or complexation interactions.

in the reactivity of labile compared to the dissolved copper with a correspondingly lower biological availability. In other words, bioavailable copper in this case, is only a small fraction of the dissolved copper and the bioavailable component is removed from solution (mitigated naturally) at a greater rate than either dilution or geochemical precipitation of the dissolved fraction (Figure 4).

It is anticipated that similar behaviour occurs in many effluent discharge scenarios where metal-laden effluent is released to a receiving water body. In such cases, even when organic complexing capacity exists, or where geochemical reactivity is anticipated to naturally mitigate deleterious impacts, the conservative regulatory approach allows only for considerations of dilution. It is believed that the use of the gel sampler as a tool in such cases would allow for more realistic estimates of bioavailability, which would benefit all stakeholders.

4 CONCLUSIONS

Present "main-stream" sampling technologies are incapable of addressing many of the issues identified by the present regulatory framework. The development of the polyacrylamide-based gel sampler has resulted in a method which has the potential to answer many of the questions regarding metal bioavailability which have plagued the mining industry for the last two decades. The main advantages of the gel technique over existing approaches for determining metal speciation include:

- the technique is designed for in situ use, which minimizes the risk of contamination during sample handling and transport to the laboratory;
- there is no need for sample filtration since particulate metals are excluded by the sampler;
- the volume of samples to be shipped to an analytical laboratory is reduced;
- metals are fixed *in situ* thereby avoiding changes in speciation during transit to the laboratory;
- the ability to easily archive samples for long-term storage thereby offsetting analytical costs;
- metals are concentrated in the resin and therefore ultra-low analytical detection limits are not required;
- the method is insensitive to variations in salinity; and
- there are no matrix interferences in contaminated environments.

The problems typically encountered with existing sampling techniques include:

- the potential for trace metal contamination during sample handling;
- a lack of preservation of speciation during sampling and storage;
- matrix interferences during chemical analyses (especially in sea water or heavily contaminated fresh water);
- analytical limitations imposed by low concentrations of metals (especially in sea water) in the absence of preconcentration; and
- an inability to quantify the labile or bioavailable fraction of the trace metal inventory.

Collectively, there are many advantages of the gel sampler over more conventional sampling techniques. Field trials in concert with use of the sampler in bioassays are presently underway in order to move the sampler technology from the laboratory to practical application in the field and are the subject of ensuing papers.

REFERENCES

Boyle, E. A., Huested, S. S. And Jones, S. P. 1981. On the distribution of copper nickel and cadmium in the surface waters of the North Atlantic and North Pacific Ocean. *Jour. Geophys. Res.*, 86: 8048-8066.

Bruland, K. W. 1989. Complexation of zinc by natural organic ligands in the central North Pacific. *Limnol. Oceanogr.*, 34: 269-285.

Bruland, K. W. 1992. Complexation of cadmium by natural organic ligands in the central North Pacific. *Limnol. Oceanogr.*, 37: 1008-1017.

Bruland, K. W., Donat, J. R. And Hutchins, D. A. 1991. Interactive influences of bioactive trace metals on biological production in oceanic water. *Limnol. Oceanogr.*, 36: 1555-1577.

Bruland, K. W., Franks, R. P., Knauer, G. A. And Martin, J. H. 1979. Sampling and analytical methods for the determination of copper, cadmium, zinc and nickel at the nanogram per liter level in sea water. *Alan. Chim. Acta*, 105: 233-245.

Campbell, P. G. C. 1995. Interactions between trace metals and aquatic ogranisms: a critique of the free-ion activity model. In: *Metal Speciation and Bioavailability in Aquatic Systems.* John Wiley and Sons.

Coale, K. H. And Bruland, K. W. 1990. Spatial and temporal variability in copper complexation in the North Pacific. *Deep-Sea Res.*, 37: 317-336.

Davison, W. And Zhang, H. 1994. In situ speciation measurements of trace components in natural waters using thin-film gels. *Nature*, 367: 545-548.

Hirose, K. And Sigimura, Y. 1985. Role of metal-organic complexes in the marine environment - A comparison of the copper and ligand titration methods. *Mar. Chem.*, 16: 239-247.

Krom, M. D. And Berner, R. A. 1980. The diffusion coefficients of sulphate, ammonium and phosphate ions in anoxic marine sediments. *Limnol. Oceanogr.*, 25: 327-337.

Krom, M. D., Davison, P., Zhang, H. And Davison, W. 1994. High resolution pore water sampling with a gel sampler. *Limnol. Oceanogr.*, 39: 1967-1972.

Leland, H. V. And Kuwabara, J. J. 1985. Trace Metals. In: *Fundamentals of Aquatic Toxicology: Methods and Applications*. Hemisphere Publishing Corp., New York.

Li, Y.-H. And Gregory, S. 1974. Diffusion of ions in sea water and in deep-sea sediments. *Geochim. Cosmochim. Acta*, 38: 703-714.

Nriagu, J. O., Lawson, G., Wong, H. K. T. And Azcue, J. M. 1993. A protocol for minimizing contamination in the analysis of trace metals in Great Lakes waters. *J. Great Lakes Res.*, 19: 175-182.

Pai, S.-C., Whung, P.-Y. And Lai, R.-L. 1988. Pre-concentration efficiency of Chelex-100 resin for heavy metals in seawater. Part 2: Distribution of heavy metals on a Chelex-100 column and optimization of the column efficiency by a plate simulation method. *Anal. Chim. Acta*, 211: 271 - 280.

Pai, S. C., Fang, T. H., Chen, C. T. A. And Jeng, K. L. 1990. A low contamination Chelex-100 technique for shipboard pre-concentration of heavy metals in seawater. *Mar. Chem.*, 29: 295-306.

Paulson, A. J. 1986. Effects of flow rate and pre-treatment on the analysis of trace metals in estuarine and coastal seawater by Chelex 100. *Anal. Chem.*, 58: 183-187.

Paulson, A. J., H C Curl, J. And Gendron, J. F. 1994. Partitioning of Cu in estuarine waters, I. Partitioning in a poisoned system. *Mar. Chem.*, 45: 67 - 80.

Rainbow, P. S. 1990. Heavy metal levels in marine invertabrates. In: *Heavy Metals in the Marine Environment*. CRC Press, New York.

Sholkovitz, E. R. And Copland, D. 1981. The coagulation, solubility and adsorption properties of Fe, Mn, Cu, Ni, Cd, Co and humic acids in river water. *Geochim. Cosmochim. Acta*, 45: 181-189.

Tailings and Mine Waste'99 © 1999 Balkema, Rotterdam, ISBN 90 5809 025 6

Hydrogeologic controls on the geochemical function of the Forest Queen wetland, Colorado

M.C.Pavlik, K.E.Kolm, S.Winkler & J.Emerick
Division of Environmental Science and Engineering, Colorado School of Mines, Golden, Colo., USA

T.R.Wildeman
Department of Chemistry and Geochemistry, Colorado School of Mines, Golden, Colo., USA

R.Robinson
Bureau of Land Management, Denver, Colo., USA

ABSTRACT: The Forest Queen wetland receives acid mine drainage from the collapsed adit of the Forest Queen mine. This study was performed to determine how the wetland functions in terms of its ability to serve as a source or sink for metal contamination. The geochemical function of the wetland is closely tied to the hydrogeologic function, which fluctuates seasonally.

1 INTRODUCTION

This study was performed on a natural wetland below the collapsed adit of the Forest Queen mine, northeast of Silverton, Colorado. The Forest Queen wetland is located in the San Juan Mountains of Colorado. The elevation of the wetland is 9800 feet. It is approximately one acre. The wetland receives acid mine drainage from the collapsed adit. The purpose of this study was to quantitatively characterize the Forest Queen wetland as a source or sink for metals.

The amount of metals entering and leaving the wetland are closely tied to seasonal fluctuations of the water table and runoff. The hydrologic regime was studied to quantify the amount of metals retained in the wetland. The hydrologic patterns of the wetland were characterized in 1997 and 1998 for the relatively dry season in October, pre-snowmelt in April, snowmelt in June, and monsoon season in August. This paper provides a preliminary analysis of how the hydrology of the wetland changes with season. Based on this preliminary assessment, an estimate is made on whether Fe, Al, and Zn issuing from the mine adit are retained within the wetland.

2 DESCRIPTION

The wetland can be separated into two hydrogeologic units: an upper peat layer and a lower unconsolidated layer composed of sand, gravel and boulders. The peat layer, which has a saturated thickness up to 10 feet, serves as a low hydraulic conductivity unit. The direction of groundwater flow is predominantly vertical based on seasonal hydrology and local head distribution. The unconsolidated layer located beneath and around the peat, serves as a high hydraulic conductivity unit. The direction of flow is predominantly horizontal in the unconsolidated layer based on head distribution relative to the Animas River and surrounding slope aquifers.

The direction of surface water flow within the wetland is overwhelmingly horizontal. Surface water consistently enters the wetland from the collapsed adit by way of a small pond, which overflows into the wetland. Figure 1 shows the location of the collapsed adit, ponds, and other features referred to later in the paper. During the high runoff in the snowmelt period, surface water also enters the wetland through beaver channels to the north and springs at the base of the slope to the east. A pond dammed by an abandoned railway grade and beaver dams bound the west side of the

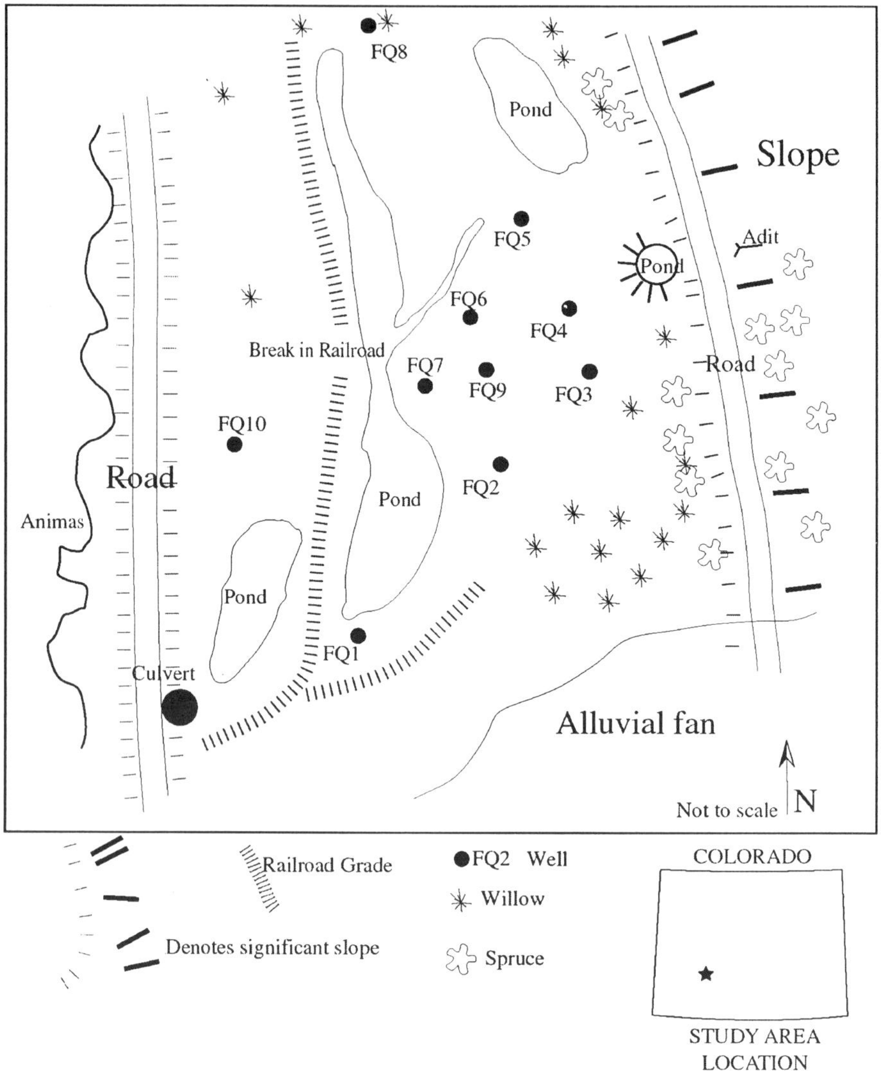

Figure 1. Site Map

wetland. The surface water typically discharges through a break in the railroad grade, then flows over tailings deposited by a flood, and under the road through a culvert into a braided reach of the Animas River. The peat hydrogeologic unit can be separated into three hydrogeochemical layers: an oxidized surface layer, which comprises the top 8-12 inches, a middle reduced zone, and a lower oxidized zone. The surface of the peat has visible Fe and Al precipitates.

The vegetation cover of the wetland is primarily sedges and mosses. The plant community is primarily composed of *Carex aquatilis*, *C. utriculata*, and *C. praegracilis*, *Hordeum jubatum*, *Calamagrostis canadensis* and *C. stricta*, *Deschampsia caespitosa*, *Juncus balticus*, and *Sphagnum* moss. Willows bound the perimeter along the north and south sides. Blue spruce and Douglas fir grow up the slope on the east side of the wetland. The west side past the railroad grade is mostly devoid of vegetation due to the tailings.

3 METHODS

To determine hydrology and geochemistry of the peat layers, fifteen wells were installed with screening at deep, middle, or shallow depths. The wells were drilled using a power auger, for up to a depth of four feet, and/or a hand auger for depths deeper than four feet. The screened section of casing was lined with silica sand, sealed off with bentonite, and capped with concrete. Ten of the wells are nested at five locations such that in a group of two wells, one is shallow and the other deep. Wells were installed in October. Ten of the shallow wells were installed the first week and water levels were measured after the wells recharged overnight. The deep wells were completed October 18th. The water levels measured on October 1997 reflect the water levels after the wells were allowed to recharge overnight. The other water levels, shown on Figure 2, were measured upon arrival to the site in April, June, and August.

The wells, FQ1, FQ10 and FQ8 are screened in alluvium. The wells FQ7D and FQ2D reach the bottom of the peat, but do not extend into the colluvium. The remaining wells are solely screened in the peat.

The wetland was characterized chemically by field measurements and samples analyzed in the laboratory. Eh, pH, ionic conductivity, temperature, and alkalinity were measured in the field. Additionally, groundwater samples were tested for dissolved oxygen using a Hach spectrophotometer kit. Dissolved concentrations of cations, in particular Fe, Al, and Zn, were measured in the laboratory using a Perkin Elmer Inductively Coupled Plasma-Atomic Emission Spectrometer (ICP-AES). The ICP-AES analyses were performed on .45 μm filtered water samples, which had been preserved with nitric acid. A Dionex Ion Chromati-ograph (IC) was used to measure concentrations of flouride, chloride, bromide, nitrate, phosphate, and sulfate. The samples were filtered with a .45 μm filter and were not acidified; however, they were frozen to significantly reduce bacterial degradation.

Peat samples were collected in October 1997 at varying depths. We assumed that metal concentrations in the peat would not change significantly with the seasons. The metals content of the peat was analyzed using a strong acid digestion. A filtered sample of each extract was tested for metals on the ICP-AES.

4 HYDROLOGICAL SYSTEM

The wetland occupies a low area of topographic relief. The collapsed adit is located on the adjacent steep slope below an avalanche chute. Acid mine drainage discharges from the collapsed adit throughout the year. Surface water from the wetland discharges through the break in the railroad. The main road lies parallel to the railroad grade approximately 50-100 feet beyond the railroad grade, forming another dam between a braided reach of the Animas River and the surface discharge zone of the wetland.

Surface water levels in the wetland vary seasonally on the order of almost two feet. Snow melt in April causes water to collect on the surface of the wetland, raising water levels. Nested wells near the slope also show an upward movement of water in the peat at this time. During the high runoff period in June, water level in the ponds on the west end rises almost two feet, and the topographically higher, east side of the wetland is inundated.

The surface water in the wetland also rises in conjunction with the water level in the Animas River. As the water level of the Animas River rises commencing in April, water seeps under the road into typically dry abandoned river channels and beaver channels. This surface water flows over tailings and enters the north side of the wetland. The water is stored in ponds within the wetland and flows out through the break in the railroad grade.

The unconsolidated material in October was not saturated. When wells FQ2D and FQ7D were installed, the holes were augured to the bottom of the peat. Water that had been present throughout the auguring process immediately drained through the bottom of the borehole. During the high runoff period in the spring, the water level in the unconsolidated material under the peat rises. In June, the high water table in the unconsolidated material forces groundwater infiltrating through the slope into and around the peat. Slope groundwater appears as springs near the slope break at and above the wetland boundary. One spring discharged an estimated 30 gallons per minute in June. However, this spring did not flow in August or during any other visit to the site. In June, the water level in nested well FQ3D was higher than

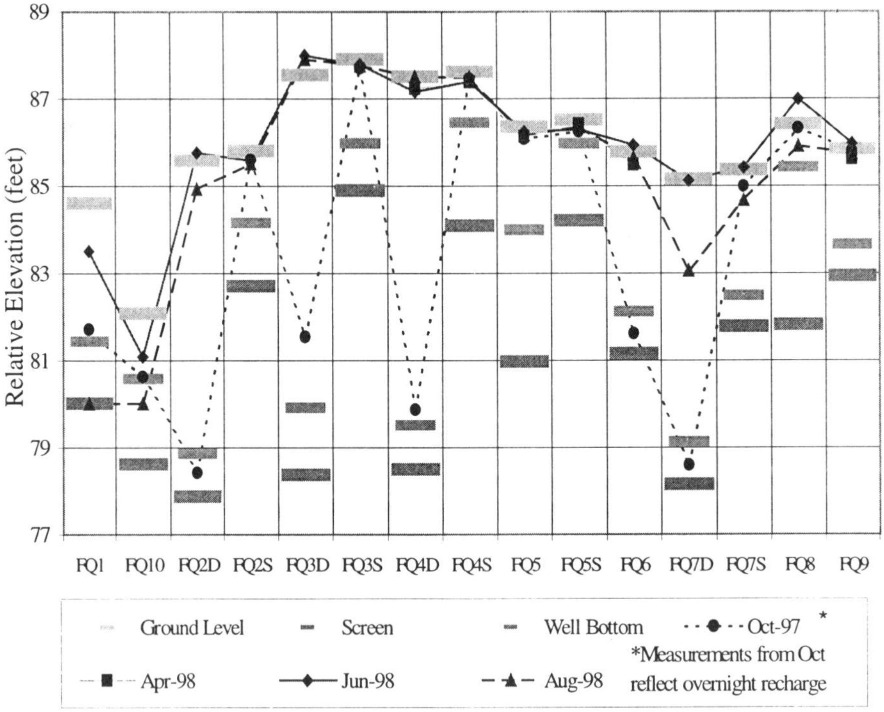

Figure 2. Water Levels

the level in FQ3S, indicating upward vertical movement of water. Figure 2 shows relative water elevations, ground elevation, and screening elevation for all of the wells. However, it seems the rest of the year, the vertical movement of the peat groundwater is downward.

The surface water level of the wetland ponds and the Animas River dropped substantially during the monsoon season (July/August), and continued to drop throughout the fall, which is typically dry. Surface water from the north and the major spring on the east no longer visibly flowed into the wetland. Between June and August, water in the western pond dropped about 1.5 feet. Surface water in August and October enters the wetland exclusively through the collapsed adit and precipitation. The surface water is lost through the break in the railroad grade, evaporation, and evapotranspiration. Flows measured in October 1997 with a flume at the exclusive surface water source and discharge point of the wetland were 17.0 gallons per minute (gpm) and 9.3 gpm respectively.

Groundwater levels in the shallow peat wells do not change substantially throughout the year. Water levels in wells FQ2S, FQ3S, FQ4S, FQ5, FQ5S, and FQ9 vary within about one half foot seasonally (Figure 2). Water levels in wells around the perimeter of the wetland, but still within the wetland vary more, on the order of 1 to 4 feet. This implies that the water moves so slowly through the peat that, between the flow from the adit, precipitation, and minor amount from the slope groundwater, the wetland can maintain saturated or close to saturated conditions in the upper, peat zone throughout the year.

5 GEOCHEMICAL SYSTEM

The hydrogeochemical environment in the wetland is highly variable from location to location. Metal concentrations fluctuate a limited amount seasonally at any particular location. Locations that typically have high concentrations of a particular metal maintain a relatively high concentration annually.

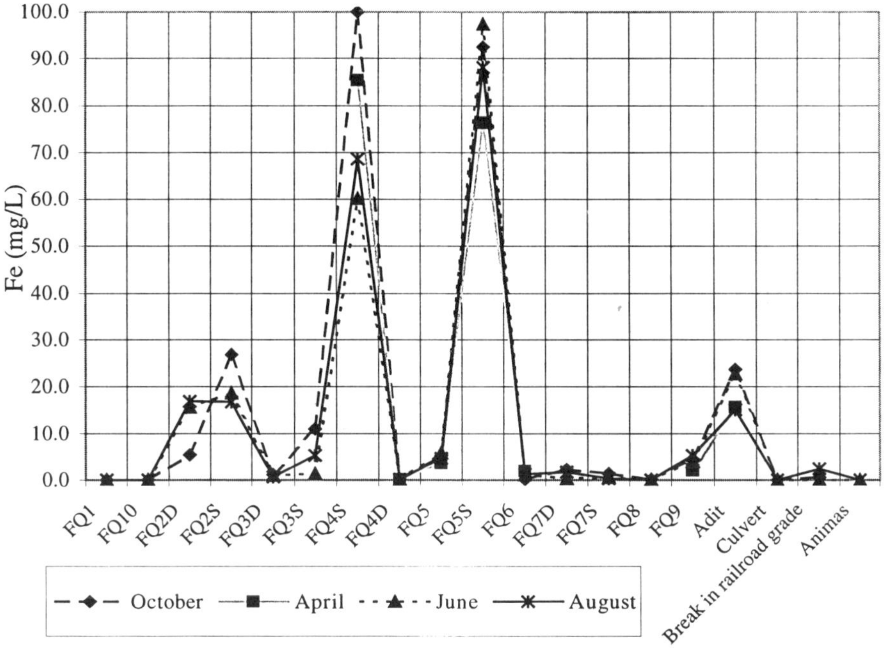

Figure 3. Seasonal Changes in Fe Concentrations

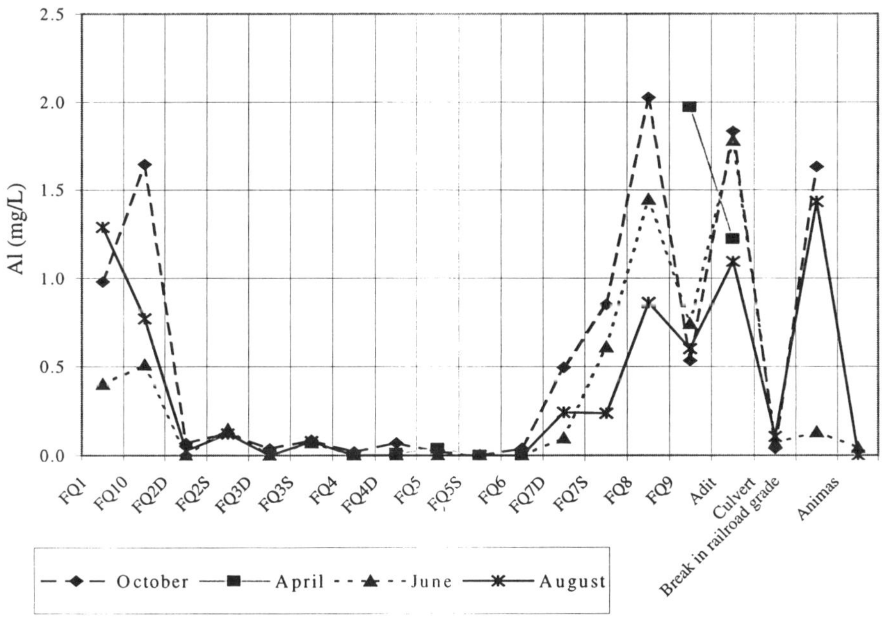

Figure 4. Seasonal Changes in Al Concentrations

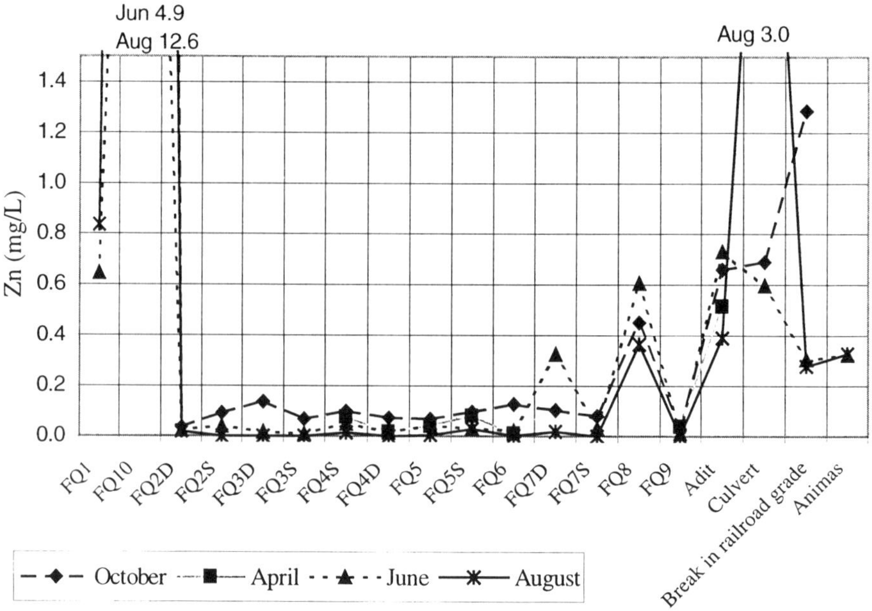

Figure 5. Seasonal Changes in Zn Concentrations

The pH of wells in the middle of the wetland typically ranges between 5.75 and 7.25. The pH of wells around the boundaries, the adit, and surface water outside of the wetland area range from as low as 3.5 at the adit and 6.75 in the Animas River. The pH changes a minor amount seasonally. Alkalinity is present in the wetland, varying from 3.5 to 630 mg/L $CaCO_3$, and fluctuates seasonally. Little to no Alkalinity was measured along the boundary and outside of the wetland year round.

There are at least two sources of contamination into the wetland. The consistent, perennial point source is from the collapsed adit. The primary metals carried in the adit effluent are Fe and Al, at average concentrations of 19.3 and 1.5 mg/L, respectively. The contribution of Zn from the adit is relatively small at 0.57 mg/L. The main non-point source of metal pollution is from the tailings that were washed to the north and west boundary of the wetland. In June, the concentration of Zn flowing into the wetland from the north was 0.205 mg/L. The concentration of Al was .049 mg/L and Fe was below detectable limits. We have measured significant amounts of Zn in the tailings. Additionally, some seeps along the slope break have high concentrations, 0.8 to 2 mg/L, of Al.

The surface of the wetland has Fe and Al precipitates on it. The precipitates and metal concentrations have a well-defined annual spatial pattern of concentrations. Orange Fe precipitates decrease with distance from the adit. White Al precipitates appear starting where the orange precipitates decrease (Wildeman, et al., 1997).

The spatial pattern of Fe and Al precipitates on the surface of the wetland is similar to the concentrations measured in wells. The highest Fe concentrations, from wells FQ4S and FQ5S, shown of Figure 3, were found in the shallow wells closest to the source. The concentration of Fe in these wells neared 100 mg/L. The screening for these wells (Figure 2) allows capture of water in the reduced zone just below the oxidized layer. At this point ferric oxyhydroxides precipitated in the surface sediment can be reduced and the iron mobilized as Fe(II). In general, Fe concentrations in wells diminish farther from the adit. This is because surface sediments no longer contain appreciable Fe. Also, water in the corresponding deep wells has Fe concentrations below 6 mg/L. This implies that vertical flow through the peat is very slow.

Aluminum concentrations are the highest farther away from the adit. At FQ7S and FQ9 the Al concentrations were as high as 2.0 mg/L as shown on Figure 4. However, a significant

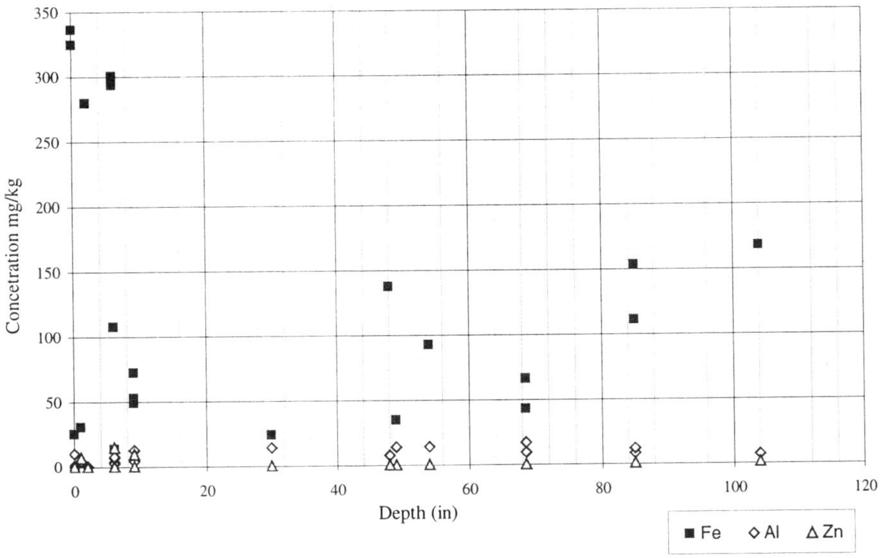

Figure 6. Metal abundances in sediments at varying depths

amount of Al is added to the wetlands during the spring runoff. High Al concentrations in the slope groundwater add additional Al to the system.

Zinc concentrations, shown on Figure 5, in general do not fluctuate as a function of distance from the adit or with depth in the wetland. Fluctuations of Zn concentrations occur in June when the additional Zn source is added from the tailings north of the wetland. High concentrations of Zn are also located along the path the surface water takes , across the lower part of the wetland and can be seen in wells FQ8 and FQ10. Wells FQ1 and FQ7D show abnormally high concentrations of Zn. Although these wells do not collect surface water, hydrologically, they connect with the groundwater in the unconsolidated material that is impacted by the tailings.

Abundances of metals in the peat soil are shown on Figure 6. The abundances of Al, and Zn are fairly constant throughout the thickness of the peat. Iron abundance are high in the oxidized portion of the peat, low in the middle, reduced section of the peat, and increase slightly at depth. The changes in iron with depth are consistent with mobilization of Fe in the middle, reduced zone.

6 DISCUSSION

With the seasons, the hydrologic regime of the Animas River and the groundwater discharge from the slope outside of the wetland fluctuate a high degree. Water levels in the beaver ponds contained within the wetland fluctuate seasonally almost 2 feet. Water levels on the surface of the wetland fluctuate only 6 inches at the most, seasonally. Water movement and storage in peat soils has proved to be difficult to quantify (Baird, 1997). Peat may have a porosity greater than 0.9, and the variability of the hydraulic conductivity of peat may be between 7.74×10^{-6} to 2.85×10^{-8} ms^{-2} (Baird, 1997). Water recharge to the Forest Queen wetland is from the surface water entering via the adit, springs and slope groundwater discharge, precipitation, and the surge of surface water during the spring runoff. The hydraulic conductivity of the peat is low enough to maintain saturated or near-saturated conditions from the spring recharge, adit discharge, and precipitation for the whole year. Additional water is stored in the beaver ponds within the wetland. Water storage in the ponds may allow enough time for water to percolate into the peat and help keep it saturated.

Iron is thought to enter the wetland system almost exclusively from the mine adit. The hydraulic conductivity of the peat is low enough to inhibit vertical movement of metal laden surface water. Most of the horizontal water movement in a peatland occurs near the surface of the peat layer (Wieder, 1986, Boelter, 1965, Korpijaakko and Radforth, 1972). The surface portion of the wetland is where we found the highest concentrations of Fe in the peat soils and water. The deep wells, except FQ7D, maintain low concentrations of Fe annually. In October, when dilution effects are minimal, Fe concentrations and flow leaving the wetland are substantially less than the Fe concentrations and flow entering the wetland. Aluminum is retained to a much lesser degree than Fe. Iron and Aluminum are retained as visible oxyhydroxide precipitates in the surface sediments. Wieder and Lang (1985), and Wieder et al. (1986) showed oxide formation to be an important mechanism for metal removal from solution and retention in peat in natural and constructed wetlands for coal mine drainage. Oxyhydroxide formation appears to be the dominant process in the Forest Queen wetland. However, Al retention is complicated because water from the slope is a significant source of Al.

Zinc is not retained within the wetland. Generally, the surface of the wetland maintains a pH below 6, and thus, $ZnCO_3$ and $Zn(OH)_2$ are incapable of precipitating. Within the groundwater system, the retention of Zn is complicated by the fact that Animas River has a Zn concentration that is higher than the mine drainage and the tailings have a higher concentration of Zn that contaminates the water that flows through them.

7 CONCLUSION

Across the surface of the wetland, Fe and Al from the mine adit water are retained in the surface sediments and Zn is not retained. The high concentration of Fe in the upper part of the peatland implies that treatment of the mine drainage occurs on the surface of the wetland. It appears Fe is retained to varying degrees year round. Aluminum appears to be retained by the natural wetland system. However, the slope water is a significant source of Al, and it is not clear that there is a sink for this Al.

Zinc in the mine effluent is not retained on the wetland surface, and at times of high runoff, a greater load of Zn is entering the wetland from water that has flowed over and through the tailings that are north and west of the wetland. For this stretch of the Animas River, these multiple sources of Zn make treatment more difficult.

When natural wetlands are compared to constructed treatment systems, it was always assumed that the competing aerobic and anaerobic processes may limit the treatment capability of a natural system (Wildeman, Brodie, and Gusek, 1993). However, this study has shown that the multiple sources that deliver water to the natural wetland can be more of a problem in making the assessment of whether contaminants are retained by the natural system.

8 REFERENCES

Baird, A.J., 1997. Continuity in hydrological systems. In Robert L. Wilby (ed.), *Contemporary Hydrology*, Chichester: John Wiley and Sons Ltd.

Boelter, D.H., 1965. Soil Science, 100: 227.

Korpijaakko, M. and Radforth, N.W., 1972. Studies on the hydraulic Conductivity of Peat. *Proceedings of the Fourth International Peat Congress*, Helsinki, Finland.

Wieder, R.K., Lang, G.E., and Whitehouse, A.E., 1985. Metal removal in *Sphagnum*-dominated Wetlands: Experience with a Man-made wetland system. In R.P. Brooks (ed.). *Wetlands and Water Management on Mined Lands*, The Pennsylvania State University: University Park.

Wieder, R.K., and Lang, G.E., 1986. Fe, Al, Mn, and S chemistry of *Sphagnum* peat in four peatlands with different metal and sulfur input. *Water, Air, and Soil Pollution* 29: 309-320.

Wildeman, T.R., G. Brodie, and J.J. Gusek, 1993. *Wetland Design for Mining Operations*, Richmond, B.C:BiTech Publishers.

Wildeman, T.R., J.W. Dinkel, R.M. Smith, and M.L. McCallister, 1997. Field assessment of Fe(III), Al, and dissolved O_2 for passive treatment options. In *Proceedings of the Fourth International Conference on Acid Rock Drainage*, 4: 1659-1672, Ottawa: Mend.

Tailings and Mine Waste '99 © 1999 Balkema, Rotterdam, ISBN 90 5809 025 6

Mining-related impacts to Lago Junin, Perú

Alan J. Martin & J. Jay McNee
Lorax Environmental Services Limited, Vancouver, B.C., Canada

ABSTRACT: Field sampling was conducted at Lago Junin, Peru, in May/June, 1997 and February/March, 1998, representing the dry and wet seasons, respectively, in order to characterize the physical and geochemical limnology of the lake system. The sampling strategy was designed to assess the magnitude and extent of mining-related impacts. Low concentrations of trace metals measured in surface waters during the dry season contrasted greatly from the high values observed during the wet period. The seasonal variability can be attributed to increased loadings associated with the river discharge and changes in lake circulation resulting from hydroelectric dam operations. Such mechanisms are responsible for the periodic degradation in water quality in the inner and main basins, and the continual widespread accumulation of metal-rich sediments. High-resolution pore water profiles of dissolved Fe, Mn, SO_4^{2-} and H_2S illustrate that the organic-rich sediments (20-25 wt.% organic carbon) in the main basin become anoxic within a several millimetres of the sediment-water interface. The pore water profiles of Cu, Zn and Pb exhibit consumption profiles at depths consistent with sulphate reduction zones, suggesting they are sequestered as authigenic sulphide minerals. Diffusive influxes calculated for Cu, Zn and Pb suggest that diffusion mechanisms contribute insignificantly to the accumulation rates of these metals. The efflux of dissolved Zn associated with the remobilization of labile particulates in the interfacial horizons could contribute significantly to the water column inventory of dissolved Zn at this site. The predominantly mining-related deposits in Upamayo are characterized by large remobilizations of copper and zinc in the deep acidic (pH < 4) pore waters, resulting in minor releases of these metals to the overlying water column. The acid-generating properties are believed to stem from sulphide oxidation reactions which are accelerated by periodic wetting and drying episodes associated with the annual cycle of lake drawdown and recharge. Assuming the input of mining-related contaminants into the inner and main basins of Lago Junin can be restricted, via either river clean-up or modifications to dam operations, a passive approach to lake restoration is recommended for these areas.

1 INTRODUCTION

In recent times, the major research objectives on mining-contaminated systems have progressed from initial surveys of sources and degree of impact, toward more detailed investigations of the mechanisms controlling the distribution, mobility and bioavailability of different metal species. Only by generating an understanding of the processes governing the behaviour of mining-related contaminants can recovery predictions and remediation strategies be adequately assessed.

As part of the project for the "Control of Mining-Metallurgical Environmental Contamination in the Mantaro Basin" a limnological assessment of Lago Junin, Perú, was conducted in order to investigate the extent and magnitude of mining-related impacts to the system. Specifically, the objectives were four-fold: 1) To characterize the inputs of mining-related contaminants to the lake system; 2) To determine the biogeochemical processes governing the behaviour and mobility of trace metals; 3) To determine the direction and significance of benthic fluxes of trace metals; and 4) To assess how the chemistries of the water column, underlying sediments and associated porewaters change seasonally. In addition to providing insight into the behaviour of metal contaminants in polluted lake-systems, the project was ultimately designed to provide a basis from which to recommend remediation strategies.

2 ENVIRONMENTAL SETTING AND BACKGROUND

Lago Junin occupies a large, shallow basin at 4,200 m above sea level in the Peruvian altiplano, located approximately 300 km northeast of Lima (Figure 1). The lake system has been receiving mining-related inputs via the Rio San Juan for several decades. Contaminant loadings reflect the activities of several base-metal mining operations 10-20 km upstream which have contributed acid rock drainage (ARD), tailings and tailings pond overflow to the river drainage. Large inventories of mining-derived material rich in Zn, Cu and Pb exist along the riverbanks. The Rio Yahuar, which enters Upamayo, drains agricultural lands to the west of Lago Junin. A hydroelectric dam operation situated at the downstream end of Upamayo supplies power to the surrounding industry (Figure 2). The lake and surrounding drainages are characterized by relatively-high alkalinity and pH (pH from 7.5 to 8.0). The hard waters stem from the high concentrations of magnesium and carbonate salts, which are in turn derived from the weathering and transport of abundant carbonate rocks in the drainage system.

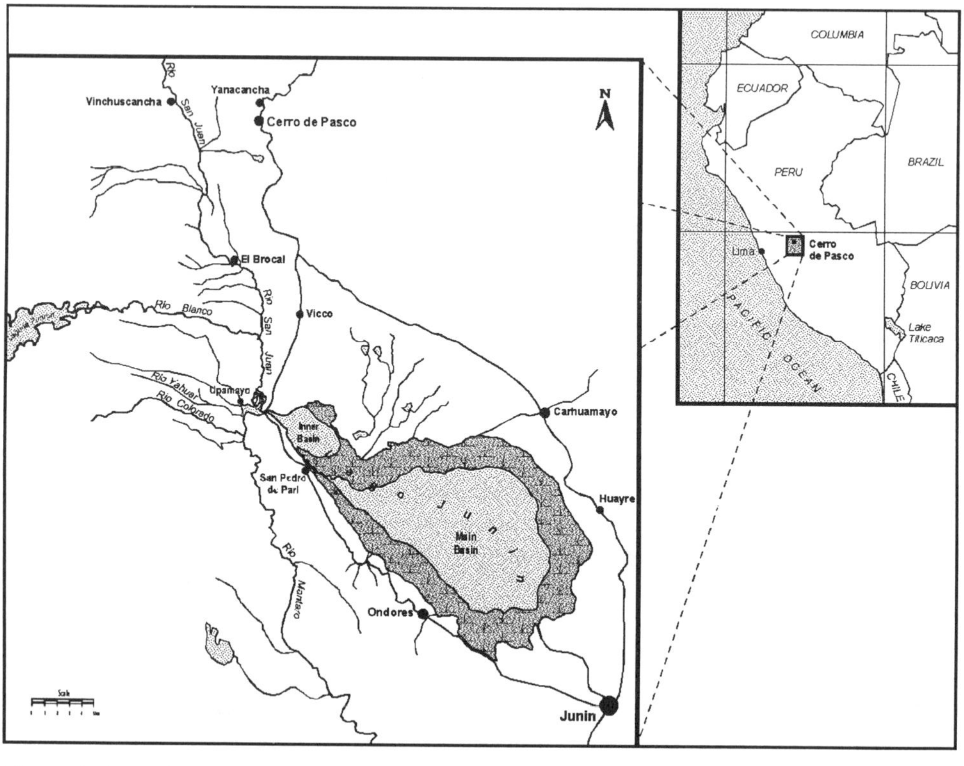

Figure 1. Location map of Lago Junin, Perú.

Lago Junin is naturally divided into three basins: 1) Upamayo and lower reaches of the Rio San Juan; 2) a shallow inner basin which is connect to Upamayo by a narrow channel; and 3) the main basin of Lago Junin which is connected to the inner basin via a system of narrow channels (Figure 2). Upamayo comprises a small (~2 km long) shallow basin (~ 1 m in depth) which borders large areas of exposed and partially submerged deltaic deposits, predominantly derived from mining-related sources. The inner basin of Lago Junin occupies a large, shallow-water region (1 to 2 m in depth) which is bordered by extensive vegetation (totora). The main basin of the lake (~20 x 10 km) is essentially enclosed by up to 3 km of totora which restricts circulation to and from the lake. The organic-rich sediments in the lake (20-25 wt.% organic carbon) are reflective of high productivity in the relatively-deep water column (4 to 10 m).

Although the Junin region exhibits little annual temperature variability, precipitation varies substantially. Monthly average precipitation ranges from 150 mm per month in February to less than 10 mm per month in June (ECSA 1996). The resulting seasonal fluctuations in discharge from the Rio San Juan dictate dam operations associated with lake drawdown and recharge. Typically, during the wet months from December to May, the flow of water exiting Upamayo is restricted which in turn results in a lake level rise. The resulting reservoir of water is used to drive the hydroelectric power station when the dam gates are open in the dry season from June to November. Dam operations foster a seasonal lake-level amplitude of up to 2 m. The seasonal hydrological shift has important ramifications with respect to the migration of river-derived contaminants.

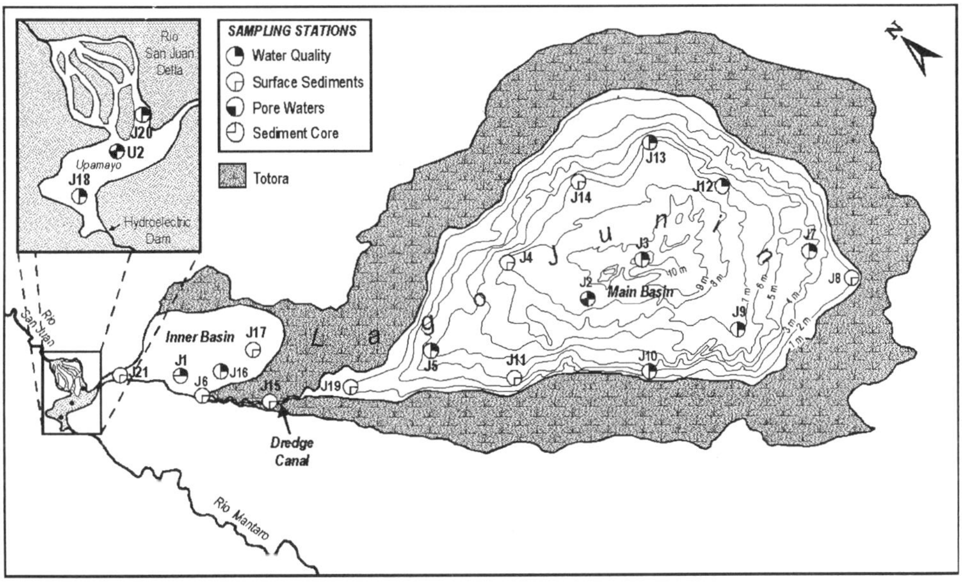

Figure 2. Locations of water column, surface sediment, sediment core and pore water sampling stations, Lago Junin, Perú.

3 SAMPLING AND METHODS

Surface water samples were collected during both the dry and wet periods at five sites along a transect from the Rio San Juan to the southeastern extremity of Lago Junin (Figure 2). Depth profiles were collected at some sites in order to assess vertical heterogeneity. Samples were obtained using a 5 L Teflon-lined Go-Flo sampling bottle weighted with a 10 kg plastic-coated ballast. The Go-Flo was deployed on a nylon line and triggered using a Teflon messenger. Subsamples to be analyzed for total metals were collected in acid-washed polyethylene bottles and acidified to pH 2 with ultra-pure nitric acid (Seastar Chemicals, Sidney, British Columbia). Subsamples to be analyzed for dissolved metals were filtered through a 0.45 μm polysulfone membrane and then acidified. A YSI probe (Model 50B) was used for *in situ* measurement of temperature and dissolved oxygen in the water column.

The post-depositional cycling of metals and nutrients in the lake sediments and their exchange with the overlying water column was assessed by high resolution sampling of the pore waters and bottom water using dialysis arrays (peepers). Peepers were deployed at two sites: one in the central basin of the lake proper (Station J2) and the other in predominantly mining-related sediments in Upamayo (Station U2) (Figure 2). The samplers consist of a Plexiglas plate (120 x 20 x 2.3 cm) with two vertical rows of 6.5 long x 0.9 cm wide and 1.0 cm deep horizontal chambers spaced at 1.25 cm intervals centre to centre, covered with a Gelman HT-450 membrane. To avoid oxidation artifacts, the samplers were de-oxygenated prior to deployment as per the protocols outlined by Carignan and Gächter (1994). Peeper deployments were performed via diver insertion and allowed to approach diffusive equilibrium with the adjacent pore waters and bottom waters for a period of 14 days. Once retrieved, the pore waters were subsampled in a nitrogen environment for dissolved metals, sulphate and hydrogen sulphide. Metals samples were preserved by acidification to pH 2 with nitric acid. Sulphate samples were frozen and samples to be analyzed for hydrogen sulphide were fixed with 2 M Zn-acetate.

Undisturbed sediment cores typically a few decimetres long were collected at both peeper sites via diver retrieval. Surface sediments were collected at 19 sites to assess the horizontal distributions of heavy metals (Figure 2). Sampling was carried out using an Ekman stainless-steel grab which sampled a sediment area of 225 cm^2 to a depth of approximately 10 cm. The top 1.5 cm was collected for surficial sediment analysis. All sediment subsamples were frozen until analysis for total metals, carbonate carbon and total C, N and S.

Determinations of metal concentrations in the water column and pore waters were performed using inductively-coupled plasma mass spectrophotometry (ICP-MS). Pure (99.998%) ICP-grade argon was used as the plasma and carrier gas. Operating conditions were optimized daily to obtain maximum sensitivity. Concentrations of the respective metals were calculated using a linear calibration curve derived from certified metal standards. An internal standard (In^{115}) was applied to all samples, standards and blanks to correct for sensitivity variations during analysis. Iron was measured by graphite furnace atomic adsorption spectroscopy (GFAAS) using a Varian Spectra AA300 with Zeeman background correction. Sulphate concentrations were measured by ion chromatography using a chemically suppressed conductivity detector (Dionex). Precision (1σ rsd) was estimated to be better than ± 0.5%. Total sulphide (ΣH_2S) was determined in pore waters using a diamine reagent and measured spectrophotometrically using a Bausch and Lomb Spectronic 2000 analyzer.

Metal concentrations were determined in the solid phase by X-ray fluorescence spectrometry using pressed powders and glass-discs for minor and major elements, respectively. An aliquot of each freeze-dried sediment sample was ground in a tungsten-carbide mill prior to preparation for XRF measurements. Organic carbon, nitrogen and sulphur concentrations were determined by combustion/gas chromatography using a Carlo-Erba NA-1500 elemental analyzer (1σ rsd = 0.6, 3.5 and 3.5%, respectively). Carbonate carbon was determined by coulometry with a precision of 1.9%. Organic carbon was determined from the difference between the total and carbonate carbon values (combined precision of 1.9%). Accuracy of the elemental analyses was monitored by including aliquots of the standard sediment sediment references MESS-1, BCSS-1 and PACS-1 with each batch of samples. Reagent grade $CaCO_3$ was used to assess routinely the accuracy of the coulometer.

4 RESULTS AND DISCUSSION

The Rio San Juan presents the major pathway of mining-related inputs to the Lago Junin system. River waters exhibit elevated levels of several trace metals in comparison to local background values, particularly with respect to Cu, Pb, Hg and Zn (Table 1). The levels of these latter elements, which are hosted primarily in particulate phases, are in exceedence of aquatic life thresholds. Due to a combination of sewage and blasting reagent inputs to the river system, concentrations of ammonium and nitrate are also elevated. Increased river flow associated with abundant precipitation in the wet season results in greater loadings of metals to the lake system during this period (Table 1). An inspection of metal partitioning suggests that the higher riverine concentrations of total metals largely reflect increases in particulate metal phases. Widespread accumulations of metal-rich deposits distributed along the course of the Rio San Juan represent the source of erodible sediments.

Table 1. Comparison of concentrations of mining-derived contaminants in the Rio San Juan (Station J20) between the dry and wet season surveys.

	[1]Background		Dry season		Wet season	
Parameter	Total (μg/L)	Dissolved (μg/L)	Total (μg/L)	Dissolved (μg/L)	Total (μg/L)	Dissolved (μg/L)
Arsenic	6.3	4.5	4.2	2.7	12.7	1.7
Cadmium	[2]<0.05	<0.05	1.0	0.9	2.7	1.4
Chromium	<0.5	<0.5	<0.5	<0.5	<0.5	<0.5
Copper	0.9	0.5	45	15	69	18
Iron	260	<30	220	<30	2,700	<30
Lead	0.4	0.1	10	0.7	121	0.8
Mercury	<0.05	<0.05	0.25	0.05	1.05	<0.01
Nickel	<0.1	<0.1	2.1	2.0	3.5	2.9
Selenium	<1.0	<1.0	<1.0	<1	<1.0	<1.0
Zinc	<0.5	<0.5	292	230	970	360
Sulphate	[3]N/A	5	N/A	87	N/A	111
Nitrate	N/A	<5	N/A	86	N/A	176
Ammonium	N/A	<5	N/A	282	N/A	28

[1]Samples collected in the Rio San Juan 15 km upstream of mining influences.
[2]Value represents "less than" the indicated detection limit value.
[3]Denotes "not applicable".

Invariant profiles of temperature and dissolved oxygen in the water column of Lago Junin during both the wet and dry seasons indicate that well-mixed conditions persist throughout the year (Figure 3). Dissolved oxygen concentrations remained at or near saturation values at all depths. The homogeneous vertical distributions of trace metals and major ions also indicate that periodic wind-mixing of the shallow water column inhibits the development of any chemical stratification associated with the transfer of dissolved constituents across the sediment-water interface. The vertical thermal homogeneity can be attributed to the minimal variability in seasonal air temperatures.

Pronounced seasonal variability in trace metal concentrations was evident between the dry and wet seasons. During the dry period, high metal levels were largely localized to the discharge zone of the Rio San Juan. With respect to Zn, Cu and Pb, the river signature of these elements was largely indiscernible 3 km to the east at Station J1 (Figure 4). Conversely, during the wet period, surface waters of the inner basin (Station J1) and northwest portions of the main basin (Station J5) were characterized by significantly elevated concentrations of trace elements. The concentrations of total Zn, Cu and Pb at J5 of 387, 52 and 101 μg/L, respectively, represent respective seasonal increase factors of 14, 10 and 400. The inter-seasonal similarity in metal concentrations at stations J2, J3 and J7 indicates that the extent of the "contaminant front" was limited to the northwest areas of the main basin at the time of sampling.

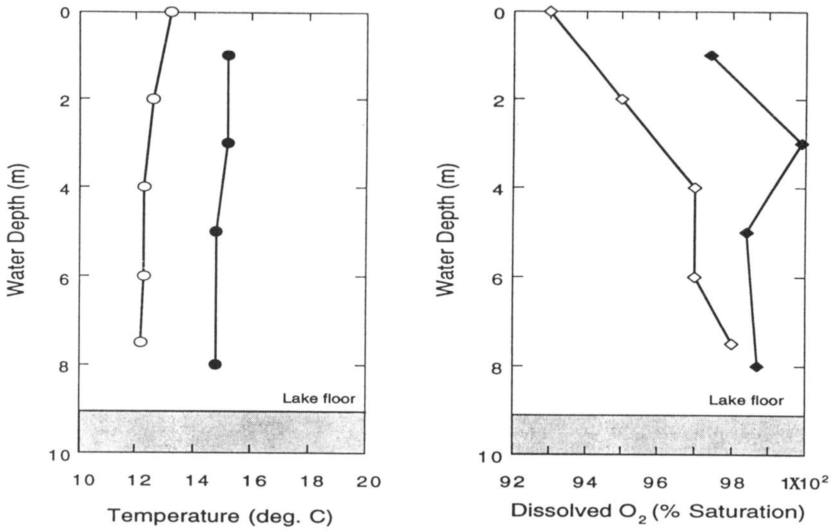

Figure 3. Water column profiles of temperature and dissolved oxygen for the dry (open symbols) and wet (closed symbols) season at Station J2, Lago Junin.

The seasonal trace metal data suggest that changes in the hydrological regime strongly influence the migration of river inputs. Specifically, the increased transport of metal-rich waters into the inner and main basins of Lago Junin during the wet season is predicted to be a function of both: 1) greater loadings of mining-related contaminants associated with increased river discharge; and 2) the change in lake circulation resulting from the hydroelectric dam operations. During periods of lake drawdown (*i.e.*, when the flood gates are open), the circulation in Upamayo is predicted to be predominantly uni-directional, with river inputs draining directly through the dam into the Rio Mantaro (Figure 2). Therefore, river inputs to the inner and main basins are predicted to be minor during this period. Water exchanges with the main basin are further restricted during low-water periods due to the large expanses of emergent totora which act as both a physical and chemical filter. Conversely, during periods of lake recharge (*i.e.*, when the dam gates are closed) loadings from the Rio San Juan are directed into the inner and main basins of Lago Junin. The transport of river-borne material is further enhanced during the wet period due to the concomitant increase in river discharge and associated metal loadings.

The surface sediments in Lago Junin exhibit widespread enrichments of heavy metals. The concentrations of solid-phase Zn, Cu and Pb observed are much greater than typical crustal values for these elements of ~80, 60 and 120 mg/kg, respectively (Turekian and Wedepohl 1961). The distributions of Zn, Cu and Pb exhibit considerable inter-site variability, which does appear to be solely related to distance from the Rio San Juan. Zinc, for example, is characterized by lake-wide enrichments with maximum values occurring at sites located in the distal lake regions (J7, J9, J13 and J14). The lateral distributions of Cu and Pb exhibit a similar degree of spatial variability; however, surface concentrations generally decrease with distance away from the river source (Figure 5).

Enrichments of metals in sediments can occur through post-depositional processes (i.e., precipitation of authigenic minerals) and via the accumulation of metal-rich particulates. However, the pore water data demonstrate that diagenetic mechanisms could not produce enrichments of the magnitudes observed in Lago Junin sediments (see below). This implies that accumulations of metal-rich particulates derived from the Rio San Juan are responsible for the observed metal distributions. Although at the time of sampling in the wet season the extent of the "contaminant front" was limited to a zone northwest of Station J2, the widespread enrichments of metals in the surface sediments suggest that the front periodically extends to the distal areas of Lago Junin. Such events most likely coincide with periods of

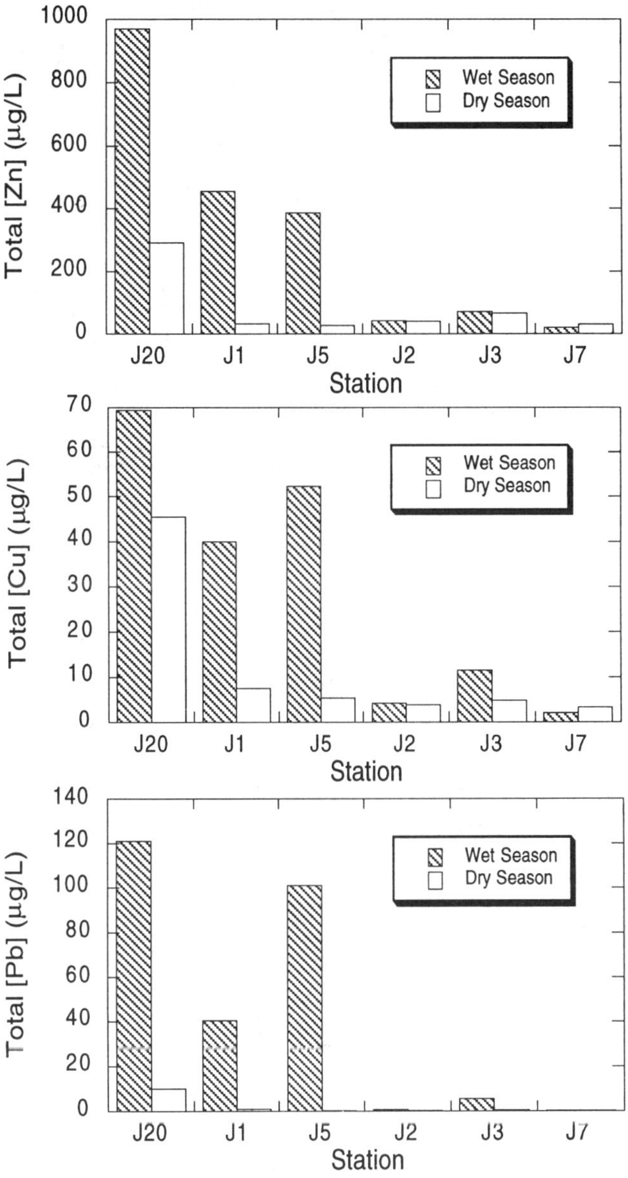

Figure 4. Comparison of total Zn, Cu, Pb concentrations in surface waters between the wet and dry seasons along a transect between the Rio San Juan (J20) and the distal regions of Lago Junin (J7). Station locations are shown in Figure 2.

high river-discharge and high lake-level. The large distances over which river-borne material must migrate also suggest that Zn, Cu and Pb are preferentially associated with finer grain-size fractions.

High-resolution sampling of the bottom and interstitial waters was conducted in order to assess the biogeochemical controls governing the mobility of trace metals in Lago Junin sediments. At station J2, profiles of dissolved Fe, Mn, SO_4^{2-} and H_2S indicate that the organic-

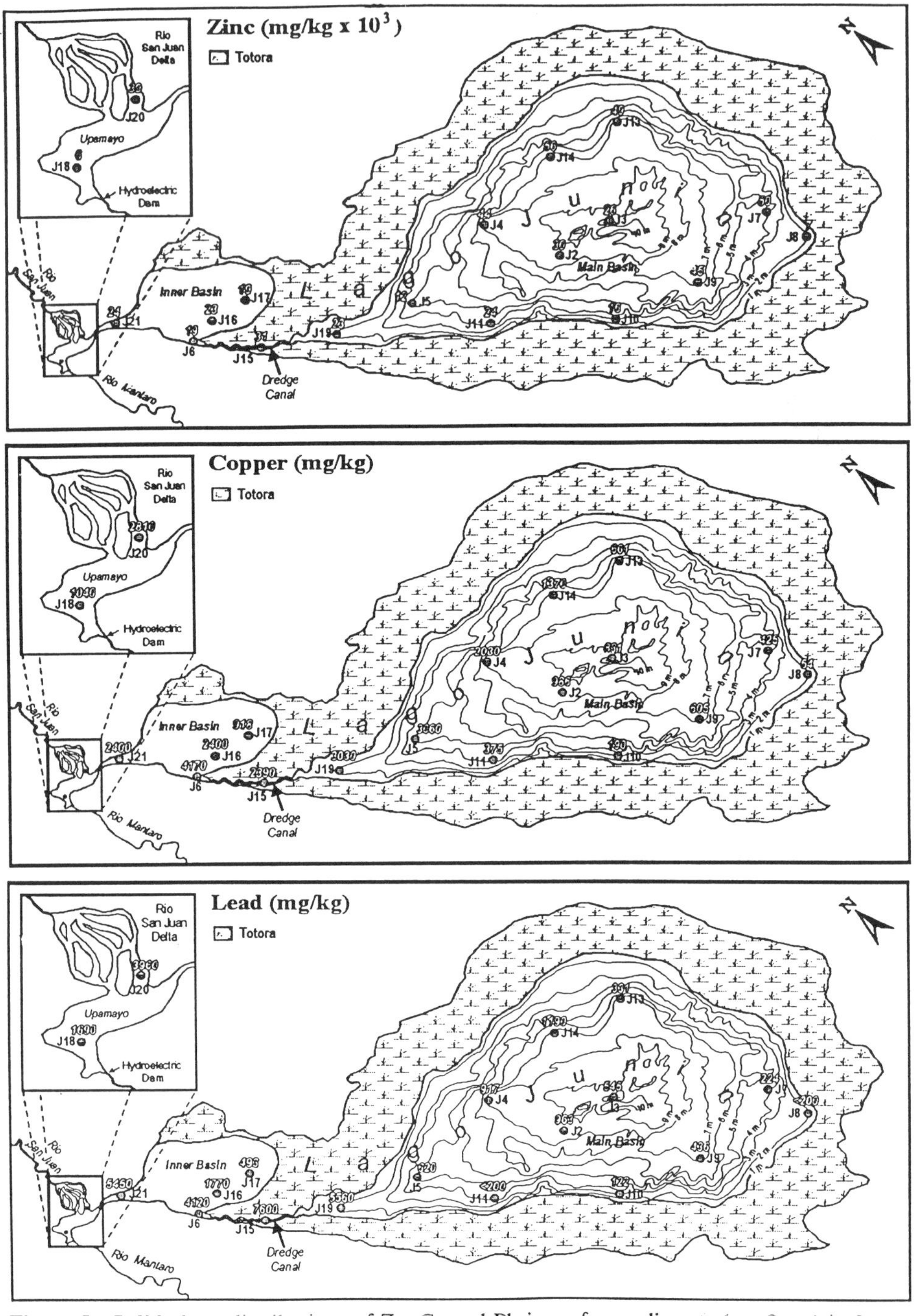

Figure 5. Solid-phase distributions of Zn, Cu and Pb in surface sediments (top 2 cm) in Lago Junin.

rich (20-25 wt.% organic carbon) sediments become anoxic at very shallow sediment depths. The reductive dissolution of manganese- and iron-bearing solid phases liberates their dissolved counterparts to the pore waters in the upper few centimetres of the sediment column (Figure 6). Since ferrous iron can not exist in significant concentrations in the presence of molecular oxygen, the abundance of the former in the upper horizons is indicative of sub-oxic conditions (Stumm and Morgan, 1981). Further support for sediment anoxia is provided by the sulphate and hydrogen sulphide profiles (Figure 6). The reduction of sulphate results in the precipitous decline in sulphate concentrations from over 60 mg/L at the sediment-water interface to less than 5 mg/L at a depth of 10 cm. Hydrogen sulphide, which is an end product of sulphate reduction, appears in the pore waters in the upper two centimetres. The data, which are consistent with established redox zonation principles (*e.g.*, Froelich *et al.* 1979; Postma and Jakobsen 1996), collectively imply that fully anoxic conditions are present in close proximity to the sediment-water interface.

Trace metal profiles at Station J2 demonstrate that dramatic changes in chemistry occur across the interfacial sediment horizons. Maxima of dissolved Zn and Ni evident at the sediment-water interface indicate that these elements are simultaneously diffusing to deeper sediment depths and into the overlying water column (Figure 7). The magnitude of Zn enrichment is particularly large; the pore water maximum is ~250 μg/L greater than background water column concentrations. The remobilization depths of Zn and Ni occur 2 to 4 cm above the predicted horizon of Mn(IV) and Fe(III) reduction suggesting that these pore water enrichments are not the result of reductive processes. Rather, the interfacial maxima likely reflect the oxidative dissolution of labile organic-particulates at the sediment-water interface. The cycling of organic matter has been invoked to control the distributions of Zn and Ni due to their nutrient-like distributions in the oceans (Bruland 1980) and lakes (Hamilton-Taylor et al. 1996). Several authors have offered interfacial oxidation mechanisms to account for near-surface pore water maxima of Ni and Zn (Martin 1996; Westerlund et al 1986; Klinkhammer 1980; Morfett et al 1988). However, reports of pore water enrichments in excess of 50 μg/L are rare. It is proposed that the periodically high concentrations of Zn in the water column facilitates its sorption onto and/or incorporation with labile organics in concentrations great enough to support interfacial

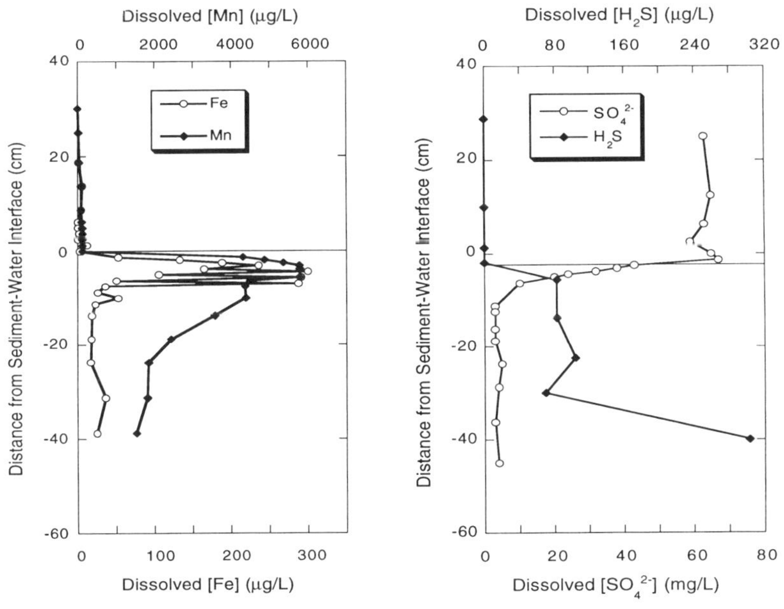

Figure 6. Profiles of dissolved Mn, Fe, H_2S and SO_4^{2-} across the sediment-water interface at Station J2, Lago Junin.

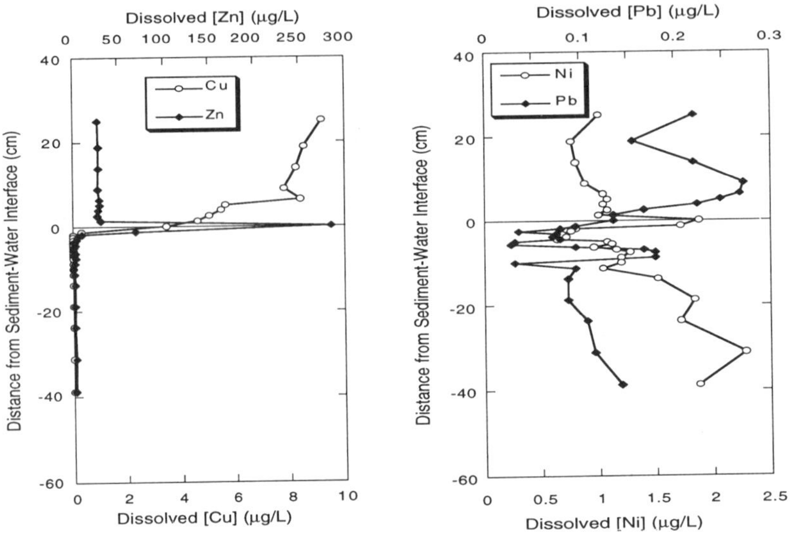

Figure 7. Profiles of dissolved Zn, Cu, Ni and Pb across the sediment-water interface at Station J2, Lago Junin.

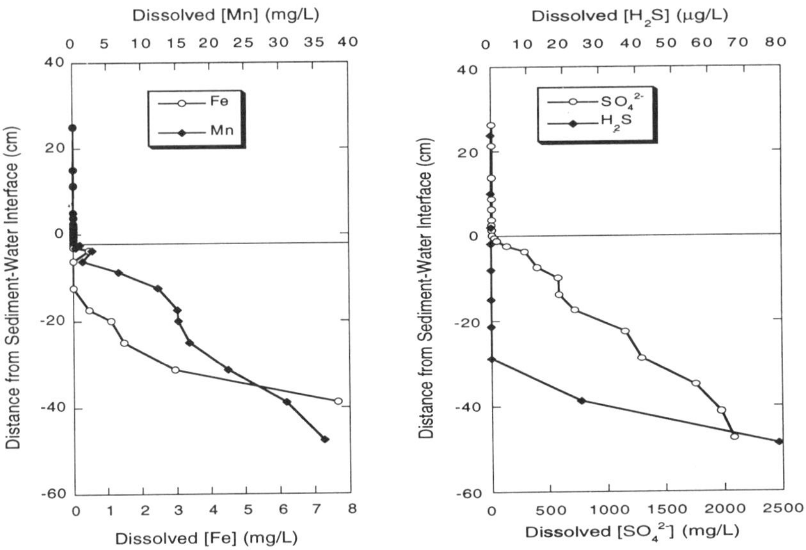

Figure 8. Profiles of dissolved Mn, Fe, H_2S and SO_4^{2-} across the sediment-water interface at Station U2, Upamayo region, Lago Junin.

releases of such magnitude. Indeed, the surface sediments in the main basin are severely contaminated with respect to trace metals, with Zn concentrations at J2 of 30,000 mg/kg (see below). A similar hypothesis was put forth by Martin (1996) who observed comparable magnitudes of Zn enrichment in the near-surface pore waters in a mining-contaminated lake in northwestern Ontario.

Sedimentary effluxes of trace metals from the sediments to the overlying waters were calculated in order to obtain an estimate of the potential impact of metal release to the water column. Water quality impact calculations were based on a conservative residence time of Lago Junin of 2.5 years and an average water column depth of 6 m. The diffusive efflux rate for Zn suggests that inputs from the lake sediments to overlying waters could contribute significantly to the water column concentration of 30 μg/L measured at this site. Although significant in terms of the water column inventory, Zn effluxes do not result in water column concentrations which exceed aquatic life thresholds. Release rates were negligible for Ni.

Profiles of trace metals at Station J2 respond to changes in the redox environment with depth. Steep decreases in the concentrations of Zn and Cu reflect the removal of these metals from pore solution (Figure 7). Copper concentrations decrease from ~9 μg/L in the water column to below detection limits (0.1 μg/L) at approximately 2 cm below the sediment-water interface. Similarly, Zn values decrease from ~275 μg/L at the interface to ~2 μg/L below 2 to 3 cm depth. The consumption zones, which occur in the upper few centimetres of the sediment column, are consistent with sulphide production horizons, suggesting Zn and Cu are precipitated as authigenic sulphides. Sulphide precipitation has been proposed to limit the concentrations of these elements in the pore waters of both pristine lakes (Hamilton-Taylor and Davison 1994; Williams 1992) and mining-impacted systems (Pedersen 1983; Pedersen et al. 1993; Martin 1996). Minor removal can also be observed for Ni and Pb (Figure 7); however, the magnitudes of the diffusive influxes for these elements are considerably lower. Collectively, the data imply that the sediments in the main basin of Lago Junin are serving as a year-round sink for Zn and Cu, and to a minor extent, Pb and Ni. Diffusive influx calculations for Zn, Cu, Ni and Pb indicate that the precipitation of authigenic phases contributes insignificantly to the accumulation rates of these elements.

The diagenetic state of sediments in Upamayo (Station U2) was assessed via examination of various pore water constituents. The reductive dissolution of Mn and Fe oxides results in increases in the concentrations of dissolved Mn(II) and Fe(II) the shallow pore waters (Figure 8). The abundance of dissolved iron in the pore waters indicates suboxic conditions are achieved by approximately 20 cm. Pore water sulphate concentrations also increase in the top 5 cm in the sediment column. The pore water maxima of over 1,500 mg/L represents a two order of magnitude increase over bottom water values (Figure 8). Since the magnitude of sulphate remobilization is so large, sulphate removal via reduction reactions is not discernable, and as a result, such profiles offer little with respect to the diagenetic state of the deposits. However, the hydrogen sulphide data demonstrate the presence of fully anoxic conditions at a depth of 30 cm (Figure 8).

The pore waters at U2 are characterized by a pronounced decrease in pH. Values drop across the sediment-water interface to a minimum of ~pH 3.2 at a depth of 40 cm (Figure 9). The decrease in pH in the pore waters likely reflects the result of acid-generating reactions accelerated by periodic wetting and drying associated with dam operations. During the period 1988 to 1997, annual changes in lake level averaged over 1.7 m and ranged from 0.85 to 2.5 m (Centromin, 1998, personal communication). During the wet season survey, the lake level was near its seasonal maximum. Since the water depth at station U2 was only 70 cm, it is likely that the sediments at U2 would be exposed to the atmosphere for a considerable period of time each year. In an average year, unsaturated conditions would likely extend to a sediment depth of approximately 1 m. This annual periodicity of unsaturated and saturated conditions likely fosters conditions which encourage sulphide oxidation in the pyrite-rich (2-4 wt.% S) deposits at this site. To contrast, pore water pH at Station J2 remains invariant with depth (Figure 9).

The pH depression in the pore waters at Station U2 has a significant effect on metal remobilization. The remobilization of Mn and Fe, for example, likely reflects a combination of both changes in redox potential and pH with depth (Figure 8). Concentrations of dissolved Cu, Zn, Ni and Pb also exhibit pronounced increases with depth, and reach near-maxima at a depth of 20 cm. The magnitude of remobilization for Cu and Zn is particularly large, with levels reaching 30 and 130 mg/L, respectively (Figure 10). The steepest slopes of the metal profiles, which occur in the upper 20 cm, coincide with the steepest slope observed for pH. The data suggest that the acid-generating environment promotes the remobilization of metal-rich solid phases at depth. Metals are predicted to be hosted as both sulphide minerals and secondary oxidation products in the Upamayo deposits. The increasing metal concentrations with depth fosters the upward diffusion of dissolved Cu, Zn, Pb and Ni towards the sediment-water interface. A considerable proportion of the upward migrating ions appears to precipitate in the oxic surface sediments, presumably with iron and manganese oxides. However, a small portion escapes into the overlying water column. Lead profiles are also characterized by a second zone of release at the sediment-water interface (Figure 10). Such remobilization, possibly derived from the oxidation of labile particulates at the sediment surface, imposes a negligible impact on water quality.

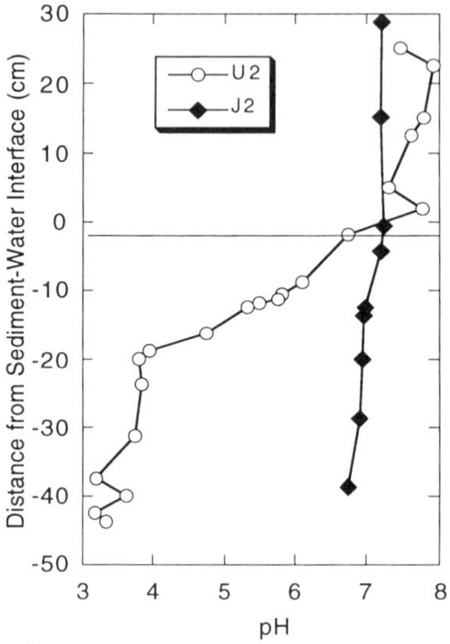

Figure 9. Comparison of pH across the sediment-water interface at Stations J2 and U2, Lago Junin.

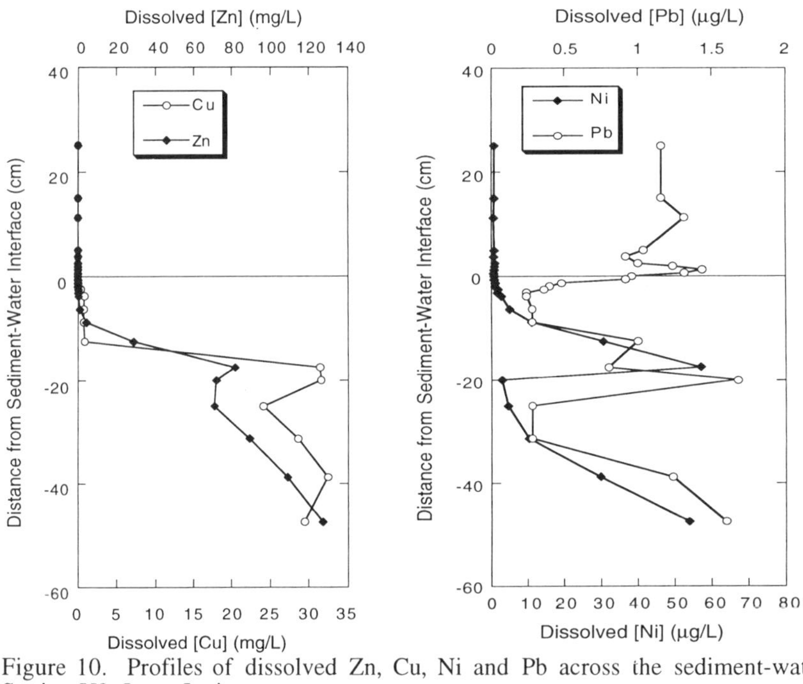

Figure 10. Profiles of dissolved Zn, Cu, Ni and Pb across the sediment-water interface at Station U2, Lago Junin.

The water level fluctuations in the shallow Upamayo region, and the resulting periodic exposure of the sulphur-rich deposits, will result in continued acid-generation and metal remobilization. Although fluxes to the overlying water column were minor at the time of sampling (late wet season), release rates may be considerably higher at the onset of flooding in the early wet season. A pulse of dissolved metals to the water column, for example, may result from the dissolution of oxidation products generated over the unsaturated period. Studies which have examined the reactivity of unoxidized, sulphide-rich tailings deposits have demonstrated that rates of sulphide oxidation are highly reduced in subaqueous environments (Pedersen et al. 1993; Pedersen 1985; Lorax 1998). Specifically, rates of acid generation and metal remobilization resulting from sulphide oxidation have been shown to impact negligibly the chemistry of overlying waters. From a geochemical perspective, therefore, permanent flooding of the deposits in Upamayo presents a suitable remediation strategy to minimize sediment reactivity.

5 CONCLUSIONS

The migration of mining-related contaminants to the Lago Junin system is largely dictated by annual changes in lake hydrology. The circulation shift, which reflects the influences of hydroelectric damming operations and seasonal fluctuations in river discharge, results in increased transport of river-borne material to the inner and main basins of Lago Junin. This mechanism is responsible for the periodic degradation in water quality and the widespread accumulation of metal-rich sediments.

In terms of lake recovery, inputs of contaminant loadings to the inner and main basins must be restricted in order to maintain an acceptable level of water quality and permit the accumulation of natural sediments. The progressive burial of the contaminated deposits with natural sediments will prevent future releases of metals. Since the dominant transport mechanism of dissolved metals in sediments is diffusion, and because mass transport by diffusion is effective only over short distances (i.e., a few centimetres), accumulation of a relatively thin layer of natural sediments over the contaminated deposits is sufficient to isolate them chemically from the water column. In addition, such a passive approach to lake restoration will ensure the development and health of natural benthic communities.

As long as an erodible source of contaminant sediments exists along the Rio San Juan, and the dam continues to operate on its current schedule, a pulse of particulate and dissolved metals into Lago Junin can be expected to occur on an annual basis. Therefore, remedial options include removing the source of contamination upstream, or modifying dam operations to ensure that river flow is continually directed through the dam.

REFERENCES

Bruland, K. W. 1980. Oceanographic distributions of cadmium, zinc, nickel and copper in the North Pacific. *Earth Planet. Sci. Lett.* 47: 176-198.

Carignan R., St.-Pierre, S. & R. Gächter 1994. Use of diffusion samplers in oligotrophic lake sediments: effects of free oxygen in sampler material. *Limnol. Oceanogr.* 39(2): 468-474.

ECSA 1996. Estudio para la recuperacion del Lago Junin: Cerro de Pasco. Empresa Minera del Centro Del Peru S.A.

Environment Canada 1995. Canadian Water Guidelines.

Froelich, P.N., Klinkhammer, G. P., Bender, M. L., Luedtke, N. A., Heath, G. R., Cullen, D., Dauphin, P., Hammond, D. & B. Hartman 1979. Early oxidation of organic matter in pelagic sediments of the eastern equatorial Atlantic: suboxic diagenesis. *Geochim. Cosmochim. Acta* 43: 1075-1090.

Hamilton-Taylor, J. & W. Davison 1994. Redox-driven cycling of trace elements in lakes. In *Lakes II:* 129-152.

Hamilton-Taylor, J., Davison, W. & K. Morfett 1996. The biogeochemical cycling of Zn, Cu, Fe, Mn and dissolved organic C in a seasonally anoxic lake. *Limnol. Oceanogr.* 41: 408-418.

Klinkhammer, G. P. 1980. Early diagenesis in sediments from the Eastern Equatorial Pacific, II. Pore water metal results. *Earth Planet. Sci. Lett.* 49: 81-101.

Lorax Environmental Services Ltd. 1998. Geochemical assessment of pyrite concentrate in a tailings pond. April, 1998.

Martin, A. J. 1996. The geochemistry of Fe, Mn, Ni, Cu, Zn and as in the water column, sediments and pore waters in a seasonally anoxic lake. M.Sc. Thesis, University of British Columbia, Vancouver, Canada.

Morfett, K., Davison, W. & J. Hamilton-Taylor 1988. Trace metal dynamics in a seasonally anoxic lake. *Environ. Geol. Water Sci.* 11(1): 107-114.

Pedersen, T. F. 1985. Early diagenesis of copper and molybdenum in mine tailings and natural sediments in Rupert and Holberg Inlets, British Columbia. *Can. J. Earth Sci.* 22: 1474-1484.

Pedersen, T. F. 1983. Dissolved heavy metals in a lacustrine mine tailings deposit- Buttle Lake, British Columbia. *Mar. Poll. Bull.* 14: 249-254.

Pedersen, T. F., Mueller, B., McNee, J. J. & C. A. Pelletier 1993. The early diagenesis of submerged sulphide-rich mine tailings in Anderson Lake, Manitoba. *Can. J. Earth Sci.* 30: 1099-1109.

Postma, D. & R. Jakobsen 1996. Redox zonation: Equilibrium constraints on the Fe(III)/SO_4-reduction interface. *Geochim. Cosmochim. Acta* 60(17): 3169-3175.

Stumm W. and J. J. Morgan 1981. *Aquatic Chemistry.* New York: Wiley-Interscience.

Turekian, K. K & K. Wedepohl 1961. Distribution of the elements in some major units of the earth's crust. *Geol. Soc. Am. Bull.* 72: 175-192.

Westerlund, S. F. G., Anderson, L. G., Hall, P. O. J. & A. Iverfeldt 1986. Benthic fluxes of cadmium, copper, nickel, zinc and lead in the coastal environment. *Geochim. Cosmochim. Acta.* 50: 1289-1296.

Williams, T. M. 1992. Diagenetic metal profiles in recent sediments of a Scottish freshwater loch. *Environ. Geol. Water Sci.* 20(2): 117-123.

ACKNOWLEDGEMENTS

This project was financed through a joint agreement between the Peruvian Ministry of Energy and Mines (MEM) and the Interamerican Development Bank (IDB). The authors wish to thank Luis Alberto Sanchez and José Mogrovejo (MEM), and David Tither (IDB) for their support and continued interest throughout the study.

Tailings and Mine Waste '99 © 1999 Balkema, Rotterdam, ISBN 90 5809 025 6

Oxygen isotopes of dissolved sulfate as a tool to distinguish natural and mining-related dissolved constituents

W.G.Wright
US Geological Survey, Water Resources Division, Durango, Colo., USA
D.K.Nordstrom
US Geological Survey, Water Resources Division, Boulder, Colo., USA

ABSTRACT: Natural and mining-related dissolved-constituent concentrations need to be distinguished in a watershed affected by abandoned mines to prioritize subbasins for remediation and to assist with the establishment of water-quality standards. The oxygen isotopes of dissolved sulfate can be used to distinguish between natural and mining-related sources of dissolved constituents. Several methods employing the oxygen isotopes of dissolved sulfate can be used to determine the relative amounts of natural and mining-related dissolved constituents in water: (1) the isotope-dilution equation for simple mixing zones (two sources and one receiving stream); (2) the isotope mass-balance equation for streams receiving dissolved sulfate from multiple geologic sources; and (3) graphical relations and the mathematical solution of simultaneous equations in a watershed approach. Using the different methods for data collected during low flow, about 71 to 75 percent of the dissolved-constituent concentrations are from natural sources in selected subbasins of the Upper Animas Watershed.

1 NATURAL AND MINING-RELATED CONCENTRATIONS OF DISSOLVED CONSTITUENTS

Natural and mining-related concentrations of dissolved constituents need to be distinguished in watersheds affected by abandoned mines in order to prioritize subbasins for remediation and to assist with the establishment of water-quality standards. The typical method for distinguishing between natural and mining-related sources of dissolved constituents is a mass-balance approach, in which all mines and natural streams are sampled synoptically (a snapshot in time), and a mass balance is obtained for a conservative constituent (such as dissolved sulfate or zinc). However, in the mountainous Upper Animas Watershed (Figure 1), this can be a monumental task subject to errors, and theoretically conservative constituents might not be truly conservative. Oxygen isotopes of dissolved sulfate can be used to distinguish natural and mining-related sources on a watershed basis, and the results can be used to verify mass-balance calculations of natural and mining-related dissolved constituents. When used in mass-balance calculations, oxygen isotopes of dissolved sulfate can provide very accurate estimates of natural and mining-related dissolved constituents.

2 THEORETICAL DESCRIPTION OF THE OXYGEN ISOTOPES OF DISSOLVED SULFATE

Data for the oxygen isotopes of dissolved sulfate (symbol $\delta^{18}O_{SO_4^{2-}}$) can provide insight into the processes that formed the sulfate. In some geologic settings, $\delta^{18}O_{SO_4^{2-}}$ data can reflect the mineralogy of the dissolved-sulfate source. The $\delta^{18}O_{SO_4^{2-}}$ data also can determine whether sulfate reduction has occurred in the water. The analytical measurement of $\delta^{18}O_{SO_4^{2-}}$ is more precise than other hydrologic parameters (such as discharge and analytical determination of inorganic constituents), and could be useful for quantitative calculations.

In sulfide-mineralized geologic regions that have been mined, such as the Upper Animas Watershed (Figure 1), the oxidation of sulfide minerals produces dissolved sulfate (symbol SO_4^{2-}) in natural springs and in mine drainage. These oxidation processes involve oxygen that has an isotope composition indicative of the reaction mechanism. Oxygen in the dissolved sulfate has two possible sources--dissolved oxygen (symbol $O_{2(aq)}$ for oxygen in the aqueous phase) and oxygen in the water molecule (symbol H_2O). Oxygen isotopes of dissolved sulfate are expressed in per mil (or parts per thousand) relative to the Vienna Standard Mean Ocean Water (VSMOW) on a scale that is normalized such that the $\delta^{18}O$ of the Standard Light Arctic Precipitation (SLAP) water is -55.5 per mil exactly. Precision of $\delta^{18}O_{SO_4^{2-}}$ measurements is $\pm$0.1 per mil.

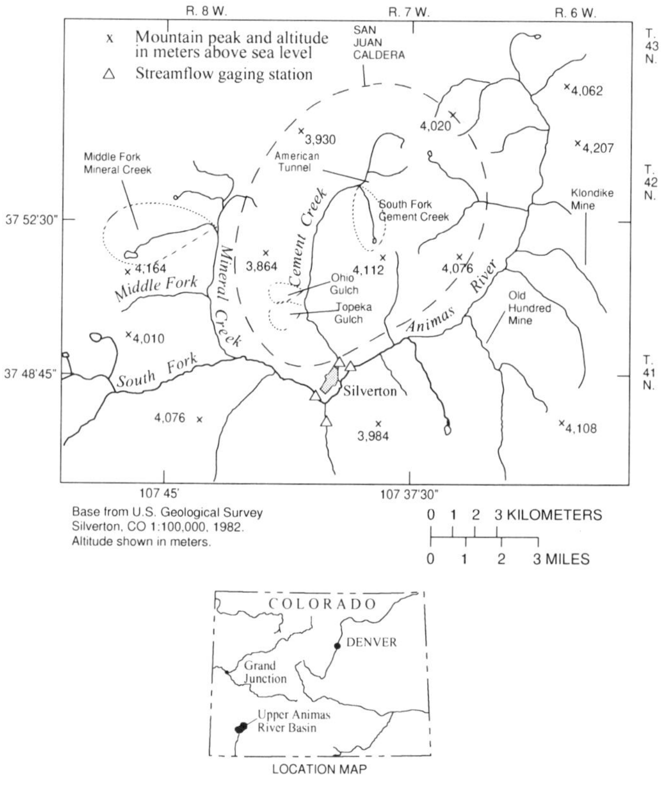

Figure 1. Upper Animas Watershed and locations of study areas.

Numerous studies of sulfide-mineral oxidation have been conducted in the last two decades. Of particular interest are those studies conducted under conditions similar to environments producing acid mine drainage. The following reactions commonly are used to represent the overall oxidation processes of pyrite (symbol FeS_2):

$$FeS_2 + 7/2\,O_{2(aq)} + H_2O \rightarrow Fe^{2+} + 2SO_4^{2-} + 2H^+ \quad (1)$$

$$FeS_2 + 14Fe^{3+} + 8H_2O \rightarrow 15Fe^{2+} + 2SO_4^{2-} + 16H^+ \quad (2)$$

Reaction (1) is limited by the availability of dissolved oxygen and reaction (2) is limited by the oxidation rate of Fe^{2+} to Fe^{3+}. The ferrous iron (Fe^{2+}) produced from reactions 1 and 2 is subsequently oxidized further to produce ferric iron (symbol Fe^{3+}), which can precipitate as iron oxyhydroxide:

$$Fe^{2+} + 1/4O_{2(aq)} + H^{+} \rightarrow Fe^{3+}{}_{(solid)} + 1/2H_2O \qquad (3)$$

Sulfur oxyanions, such as thiosulfate ($S_2O_3^{2-}$), polythionate ($S_nO_6^{2-}$), and sulfite (SO_3^{2-}) are sulfur compounds intermediate to the oxidation pathway of FeS_2 to SO_4^{2-}. The oxidation of sulfur oxyanions can occur quickly, and the resulting SO_4^{2-} may acquire oxygen via exchange with H_2O (Van Stempvoort & Krouse 1994). Oxidation of sulfur and sulfur oxyanions may proceed as follows:

$$S_2^{2-}{}_{\text{in pyrite}} + OH^{-} \rightarrow S_2O_3^{2-} \qquad \text{(Moses et al. 1987)} \qquad (4)$$

$$S_2O_3^{2-} + OH^{-} \rightarrow S_4O_6^{2-} \rightarrow SO_4^{2-} \qquad \text{(Xu and Schoonen 1995)} \qquad (5)$$

$$S_2O_3^{2-} + OH^{-} \rightarrow SO_4^{2-} \qquad \text{(Xu and Schoonen 1995)} \qquad (6)$$

The oxygen molecules (whether in $O_{2(aq)}$ or in H_2O) that are involved in reactions (1)-(6) can, except during sulfate reduction, retain the original isotopic composition of the source of the oxygen. The oxygen isotope composition of H_2O (symbol $\delta^{18}O_{H_2O}$) varies with the altitude of the precipitation that deposited the water on the Earth's surface and with other processes such as evaporation, transpiration, and latitude. For the Upper Animas Watershed, $\delta^{18}O_{H_2O}$ averages about -15.9 per mil. The oxygen-isotope composition of atmospheric oxygen that dissolves in water (symbol $\delta^{18}O_{O_{2(aq)}}$) has the value of +23 per mil (Horibe et al. 1973). Graphically relating $\delta^{18}O_{SO_4^{2-}}$ data to dissolved SO_4^{2-} data can indicate the geochemical mechanisms involved with the formation of the dissolved SO_4^{2-}.

The isotopic composition of $\delta^{18}O_{SO_4^{2-}}$ that results from the reactions can indicate whether the dissolved sulfate is from natural or mining-related sources. If reaction (1) is the predominant reaction pathway, the resulting $\delta^{18}O_{SO_4^{2-}}$ might consist of a greater percentage of the $\delta^{18}O_{O_{2(aq)}}$ composition, as compared with the $\delta^{18}O_{H_2O}$ composition. If reaction (2) is the predominant oxidation pathway, aqueous ferric iron ($Fe^{3+}_{(aq)}$), rather than $O_{2(aq)}$, is the primary agent responsible for oxidation of pyrite in acid conditions. All of the reactions (1)-(6) are mediated by bacteria in the environment; however, mining accelerates the weathering processes through both exposure of fresh minerals to oxygen and increased populations of sulfide-mineral oxidizing bacteria (Taylor et al. 1984). In accelerated weathering environments, reactions (4), (5), and (6) might occur very quickly; consequently, the resulting $\delta^{18}O_{SO_4^{2-}}$ might consist of a greater percentage of the $\delta^{18}O_{H_2O}$ composition, as compared to the $\delta^{18}O_{O_{2(aq)}}$ composition, because of oxygen exchange between H_2O and SO_4^{2-}.

Effects on $\delta^{18}O_{SO_4^{2-}}$ values might be caused by the relative distribution of different sulfur minerals in different rock types. The dissolution of gypsum, barite, and alunite affect the $\delta^{18}O_{SO_4^{2-}}$ compositions. Two samples of vein gypsum collected from a mine dump pile in the Cement Creek subbasin of the Upper Animas Watershed had $\delta^{18}O_{SO_4^{2-}}$ values of +9 and +10 per mil. In a geochemical environment dominated by pyrite oxidation, dissolution of gypsum imparts a mixture of the two processes on the resulting $\delta^{18}O_{SO_4^{2-}}$ compositions.

Microbially-mediated reduction of SO_4^{2-} to sulfide (symbol H_2S, which can be present in the dissolved or gaseous phases) can result in substantial enrichments of sulfur isotopes (symbol $\delta^{34}S_{SO_4^{2-}}$) and $\delta^{18}O_{SO_4^{2-}}$ in residual SO_4^{2-}. The lighter ^{32}S molecule in SO_4^{2-} is preferentially utilized over the ^{34}S molecule by microbes, which enriches (or makes more positive) the $\delta^{34}S_{SO_4^{2-}}$ composition of the remaining SO_4^{2-}. Likewise, the lighter ^{16}O molecule is preferentially utilized over the ^{18}O molecule by microbes, which enriches (or makes more positive) the resulting $\delta^{18}O_{SO_4^{2-}}$. In addition, in geochemically-reducing conditions, the oxygen molecule in $O_{2(aq)}$ and in H_2O become parallel electron acceptors during dissimilatory reduction of SO_4^{2-}; hence both $\delta^{18}O_{O_{2(aq)}}$ and $\delta^{18}O_{H_2O}$ may become enriched isotopically, whereby further FeS_2 oxidation by these isotopically-enriched oxygen molecules produces isotopically enriched $\delta^{18}O_{SO_4^{2-}}$ (C. Kendall, U.S. Geological Survey, written commun. 1997). Several collapsed mines in the Upper Animas Watershed show indications of SO_4^{2-} reduction: zero $O_{2(aq)}$, smell of H_2S gas, and isotopically heavy $\delta^{34}S_{SO_4^{2-}}$ and $\delta^{18}O_{SO_4^{2-}}$ compositions. There is, however, some debate concerning the electron donors available for SO_4^{2-} reduction. Mold on wood timbers in the abandoned mines may provide sufficiently labile organic matter for SO_4^{2-} reduction; however, this hypothesis has not been tested.

3 OXYGEN ISOTOPES OF DISSOLVED-SULFATE DATA FROM THE UPPER ANIMAS WATERSHED, SOUTHWESTERN COLORADO

The Upper Animas Watershed has complex geology related to the approximately 28 million-year-old San Juan Caldera (Figure 1) and placement of intracaldera lavas. Regional propylitic alteration introduced sulfur as sulfide minerals which are present in veins and disseminated throughout the rocks in varying proportions. Late-stage hydrothermal mineralization is typified by quartz vein filling accompanied by chalcopyrite, pyrite, galena, gold, rhodonite, silver, and sphalerite. Vein gypsum and barite also were introduced by late-stage hydrothermal venting. Copper-molybdenum porphyry complexes and acid-sulfate hydrothermal alteration systems also are present in the region.

Water-quality samples were collected during low flow from natural springs, draining mines, and streams in the Middle Fork Mineral Creek, South Fork Cement Creek, Ohio Gulch, and Topeka Gulch subbasins (Figure 1), as well as from the Klondike Mine, the Old Hundred Mine, and the American Tunnel (Figure 1). Graphically relating the $\delta^{18}O_{SO_4^{2-}}$ data to dissolved SO_4^{2-} shows a considerable spread in the data, probably caused by gypsum dissolution. When relating all $\delta^{18}O_{SO_4^{2-}}$ data to dissolved SO_4^{2-} and dissolved-calcium concentrations, the importance of gypsum to $\delta^{18}O_{SO_4^{2-}}$ compositions is evident (Figure 2). Data points with lower dissolved-calcium concentrations and lighter $\delta^{18}O_{SO_4^{2-}}$ compositions represent the sulfide-oxidation region (Figure 2); data points with lower dissolved-calcium concentrations and heavier $\delta^{18}O_{SO_4^{2-}}$ compositions represent the sulfate-reduction region (sulfate reduction that occurs in collapsed, anoxic mines) (Figure 2); and data points with higher dissolved-calcium concentrations represent a mixture of sulfide oxidation and gypsum dissolution (Figure 2).

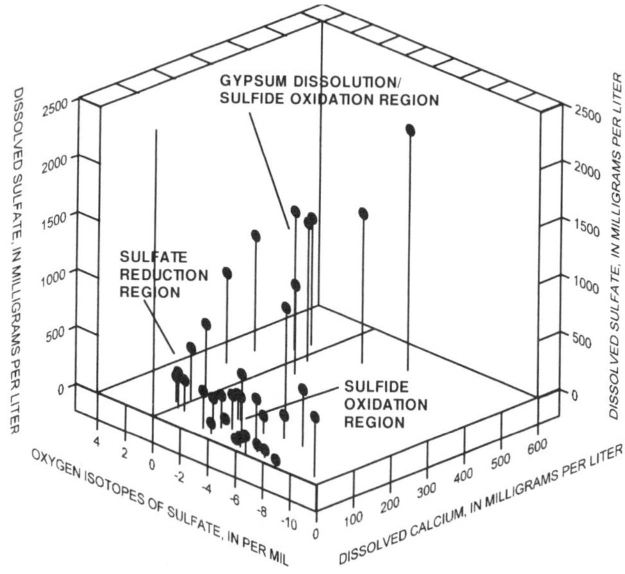

Figure 2. Oxygen isotopes of sulfate related to dissolved sulfate and dissolved calcium in water from ground-water sites (springs and mines, excluding streams) analyzed as of 1998 in the Upper Animas Watershed, during low-flow conditions.

Graphically relating $\delta^{18}O_{SO_4^{2-}}$ to SO_4^{2-} can distinguish natural from mining-related dissolved sulfate. Figure 3 shows data from ground-water samples collected during low flow in the Cement Creek subbasin (rock types from the center of the caldera). The isotopically heavier $\delta^{18}O_{SO_4^{2-}}$ compositions probably are affected by gypsum dissolution or SO_4^{2-} reduction, or both. When comparing natural springs to mine drainage in Figure 3, there is an isotopic shift to the left (decrease in isotope values) and a sulfate shift up (increase in sulfate concentrations) on the graph from natural springs to mine drainage. This might be explained by the incorporation of a greater percentage of $\delta^{18}O_{H_2O}$ in $\delta^{18}O_{SO_4^{2-}}$ and greater dissolved sulfate concentrations possibly because of bacterially-mediated sulfide oxidation (the "kinetic shift"). To illustrate the kinetic shift, three pairs of spring/mine combinations (a-a', b-b', and c-c') are shown on Figure 3 where the springs and mines are situated in similar rock types. There is a shift to the left and up on Figure 3 from the springs to the mines. A t-test of the paired data indicated that the kinetic shift is significant ($p = 0.3$, $r^2=0.78$). The kinetic shift is a result of the mining-related acceleration of weathering processes. The data point for Cement Creek at the mouth of the subbasin, an amalgamation of all water types in the Cement Creek subbasin, plots in the middle of all data points.

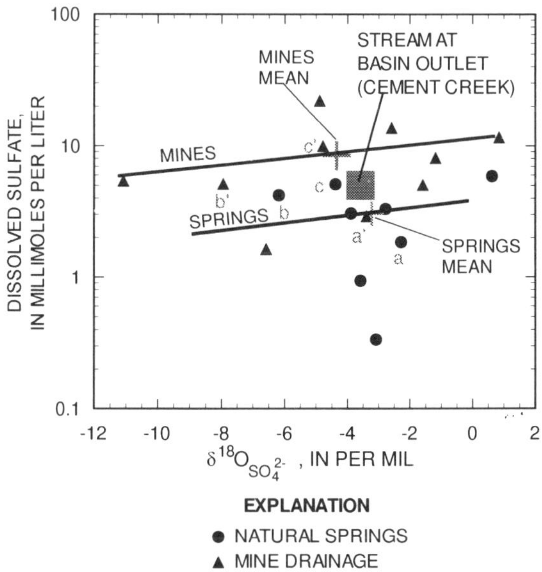

Figure 3. Oxygen isotopes of sulfate related to dissolved sulfate in water from springs and mines in the Cement Creek subbasin during low flow.

4 USE OF OXYGEN ISOTOPES OF DISSOLVED SULFATE

A typical method for distinguishing between natural and mining-related dissolved constituents in a subbasin is to collect water-quality samples, measure discharges, and perform loading mass-balance calculations, in which all mines and natural sources are accounted for. This effort can be a monumental task. The method described in this paper uses the oxygen isotopes of dissolved sulfate in a watershed approach so that all water sources (natural and mining-related) do not need to be sampled, only a representative sampling of the different site types is needed. Current work in the Upper Animas Watershed is, however, using the traditional mass-balance approach to distinguish natural and mining-related dissolved constituents, and the isotope method will be used to verify the mass-balance approach.

The oxygen isotopes of dissolved sulfate also can be used to perform mass-balance calculations. The error associated with analysis of water samples for the $\delta^{18}O_{SO_4^{2-}}$ isotopes is

±0.1 per mil or ±1 percent; therefore, the mass-balance calculations will have great accuracy using isotopes. Also, in a geologically heterogeneous subbasin, it is possible to determine the source of the dissolved sulfate using $\delta^{18}O_{SO_4^{2-}}$ data.

The following example describes the use of oxygen isotopes of dissolved sulfate in a simple mixing problem to determine how much sulfate came from a particular mine. The sites and conditions of a location in Topeka Gulch are described as follows:

Topeka Gulch:

TOPEKA8 = Natural alpine stream upstream from mine
$\delta^{18}O_{SO_4^{2-}}$ = -6.8 per mil
TOPEKA9 = Large draining mine (80 percent of the flow in Topeka Gulch)
$\delta^{18}O_{SO_4^{2-}}$ = 0.83 per mil
TOPEKA10 = Mining-affected stream downstream from mine
$\delta^{18}O_{SO_4^{2-}}$ = 0.3 per mil.

To determine the percent of dissolved sulfate that comes from the TOPEKA9 mine, the isotope-dilution equation can be used (Shearer & Kohl 1993):

Isotope-dilution equation:

$$\text{Percent TOPEKA9} = \frac{(\delta^{18}O_{TOPEKA8} - \delta^{18}O_{TOPEKA10}) \times 100}{(\delta^{18}O_{TOPEKA8} - \delta^{18}O_{TOPEKA9})} \quad (7)$$

$$= [(-6.8) - (0.3)/(-6.8) - (0.83)] \times 100$$

= 93 percent of the dissolved sulfate came from the mine.

In this simple example, the isotope-dilution equation calculated that 93 percent of the dissolved sulfate came from the mine at TOPEKA9. Using dissolved sulfate in a typical mass-balance loading calculation, 95 percent of the sulfate came from the mine. When a water-balance approach was used, 80 percent of the water came from the mine. The discharge of the stream and the mine are needed to perform the sulfate mass balance, and errors are inherent in the discharge and sulfate measurements. When the isotope-dilution equation was used, discharges were not needed, which eliminated the errors in streamflow-discharge measurements and in sulfate analyses.

An example using oxygen isotopes of dissolved sulfate in multiple geologic settings can be applied in the Middle Fork Mineral Creek subbasin (Figure 1). About one-half of the subbasin consists of chlorite-epidote-calcite dominated intracaldera lavas, and one-half of the subbasin consists of rocks altered by a copper-molybdenum porphyry complex. An isotope mass balance is obtained so that natural and mining-related sources balance at the mouth of the subbasin.

Isotope mass-balance equation:

$$\delta QC_{natural} + \delta QC_{mining\text{-}related} = \delta QC_{Middle\ Fork\ at\ mouth} \quad (8)$$

where: δ = $\delta^{18}O_{SO_4^{2-}}$, in per mil
Q = discharge at the site, in cubic meters per second
C = dissolved-sulfate concentration, in milligrams per liter

The known value is $\delta QC_{Middle\ Fork\ at\ mouth}$, and $\delta_{Middle\ Fork\ at\ mouth} = 0.0$ per mil.

All of the mines in the Middle Fork Mineral Creek subbasin were sampled during low-flow conditions. Values of $\delta^{18}O_{SO_4^{2-}}$ for the mines ranged from -5.8 to +0.2. Using isotope mass balance, the flow-weighted average $\delta^{18}O_{SO_4^{2-}}$ composition for the mining-related water in the subbasin was calculated as: $\delta_{mining\text{-}related}$ = -0.76 per mil, which is an amalgamation of

$\delta^{18}O_{SO_4^{2-}}$ compositions from mining-related sources.

Because two of the three values in equation (8) are known, the remaining unknown can be calculated: $\delta_{natural} = +0.32$ per mil, which is an amalgamation of $\delta^{18}O_{SO_4^{2-}}$ compositions from natural sources in the subbasin. Because this composition is weighted towards the positive side, and there is gypsum and barite in the copper-molybdenum porphyry complex (contributing heavy $\delta^{18}O_{SO_4^{2-}}$ compositions), much of the SO_4^{2-} concentration at the mouth of the Middle Fork Mineral Creek subbasin probably comes from the porphyry complex. The isotope-dilution equation now can be used to determine the percent of mining-related dissolved sulfate that discharges from the mouth of the Middle Fork Mineral Creek: Mining-related sulfate = 29 percent; therefore, natural sulfate = 71 percent.

The percent of natural and mining-related dissolved zinc also can be determined using the $\delta^{18}O_{SO_4^{2-}}$ data and the solution of simultaneous equations. Data for $\delta^{18}O_{SO_4^{2-}}$ are graphically related to dissolved zinc (Figure 4) for ground-water sources only (to separate the effects of stream mixing processes), and the Cement Creek sample plots in the middle of the data. By determining the geometric mean of the two groups, the result is two end members and a receiving stream, and the proportion of the two end members in the Cement Creek sample can be solved using simultaneous equations. For the data shown in Figure 4, solution of the simultaneous equations indicated that, during low-flow conditions, 75 percent of the dissolved zinc in Cement Creek comes from natural springs and 25 percent of the dissolved zinc comes from mine drainage. This estimate is for ground-water sources only and does not include the seepage of dissolved zinc from mine dump piles or other mine workings. This estimate might vary during snowmelt-runoff periods.

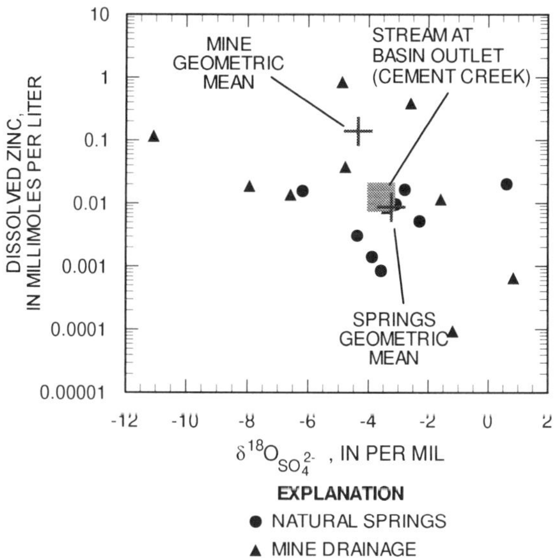

Figure 4. Oxygen isotopes of sulfate related to dissolved zinc in water from springs and mines in the Cement Creek subbasin during low flow.

A t-test of the $\delta^{18}O_{SO_4^{2-}}$ means shown in Figure 4 indicated that the differences in the means are significant to the $p \leq 0.4$ level. A t-test of the dissolved-zinc means shown in Figure 4 indicated that the means are significant to the $p \leq 0.3$ level. Hence, there is a 60 to 70 percent confidence that the means are significantly different, and this confidence must be considered in the interpretation of the simultaneous equations results.

5 SUMMARY AND CONCLUSIONS

The oxygen isotopes of dissolved sulfate ($\delta^{18}O_{SO_4^{2-}}$) are a useful diagnostic tool to describe the geochemical processes that incorporate dissolved sulfate into water from sulfide-mineral oxidation and from gypsum dissolution. The percentage of natural and mining-related sources of dissolved constituents can be determined using $\delta^{18}O_{SO_4^{2-}}$ data in a watershed approach. Using the isotope dilution equation in simple mixing zones near a mine site, accurate estimates of the percentage of natural and mining-related dissolved sulfate can be determined, and discharge measurements are not needed for the calculation. Using the isotope mass-balance method in a heterogeneous geologic setting, $\delta^{18}O_{SO_4^{2-}}$ data indicated that 71 percent of the dissolved sulfate in the Middle Fork Mineral Creek subbasin came from natural sources and 29 percent came from mining-related sources. Using the graphical method and solution of simultaneous equations in a watershed approach, $\delta^{18}O_{SO_4^{2-}}$ data indicated that 75 percent of the dissolved zinc in the Cement Creek subbasin came from natural sources.

REFERENCES

Horibe, Y., Shigehara, K. & Y. Takakuwa 1973. Isotope separation factor of carbon dioxide-water system and isotopic composition of atmospheric oxygen. *J. Geophys. Res.* 78(15): 2625-2629.

Moses, C.O., Nordstrom, D.K., Herman, J.S. & A.L. Mills 1987. Aqueous pyrite oxidation by dissolved oxygen and by ferric iron. *Geochim. et Cosmochim. Acta* 51: 1561-1571.

Shearer, G. & D.H. Kohl 1993. Natural abundance of nitrogen-15 isotopes: Fractional contribution of two sources to a common sink and use of isotope discrimination. In Knowles, R., & T.H. Blackburn (eds), *Nitrogen isotope techniques:* 89-125. San Diego, Calif., Academic Press, Inc.

Taylor, B.E., Wheeler, M.C. & D.K. Nordstrom 1984. Stable isotope geochemistry of acid mine drainage--Experimental oxidation of pyrite. *Geochim. et Cosmochim. Acta.* 48: 2669-2678.

Van Stempvoort, D.R. & H.R. Krouse 1994. Controls of $\delta^{18}O$ in sulfate--Review of experimental data and application to specific environments. In Alpers, C.N. & D.W. Blowes (eds), *Environmental geochemistry of sulfide oxidation:* 446-480. Washington, D.C., American Chemical Society, Symposium Series 550.

Xu, Y. & M.A.A. Schoonen 1995. The stability of thiosulfate in the presence of pyrite in low-temperature aqueous solutions. *Geochim. et Cosmochim. Acta* 59: 4605-4623.

Remediation and reclamation

Tailings and Mine Waste '99 © 1999 Balkema, Rotterdam, ISBN 90 5809 025 6

Use of risk assessment in developing rational clean-up strategies

J.A.Johnson & J.D.Nelson
Shepherd Miller Incorporated & Colorado State University, Fort Collins, Colo., USA

C.O.Sealy
Umetco Minerals Corporation, Grand Junction, Colo., USA

ABSTRACT: Uranium mining and milling operations in the western United States have left some areas with naturally occurring radionuclide concentrations in soils and sediments elevated above regional background values. Existing regulations governing the clean-up of these areas are generally based on acceptable residual radionuclide concentrations in soil and do not take into account the characteristics of the region, potential risks to humans, and the real risks to the environment. Clean-up operations can increase overall human health risks and damage valuable ecological and cultural resources. A risk-based approach to decommissioning can reduce overall human health and ecological risks. The potential human health risks associated with a hypothetical site were calculated assuming a Ra-226 concentration of 100 pCi/g in stream bed sediment. The analysis showed a maximum exposure to an individual of less than 15 mrem per year, based on reasonable land use and exposure assumptions. This calculated annual dose is well within the variability of regional background doses.

1.0 INTRODUCTION

Decommissioning of sites where radioactive materials have been mined or processed generally involves cleaning up large areas to residual radionuclide levels mandated by regulatory authorities. These levels may not be appropriate for sites or portions of sites in sparsely populated or ecologically sensitive areas. Regulatory standards require extensive clean-up that would have serious ecological consequences with little or no reduction in risk to members of the public. A risk-based approach to decommissioning, balancing the risks and costs of various clean-up options, would reduce overall ecological and human health risks and result in cost-effective clean-up strategies. Such options include (1) clean-up to meet existing regulatory standards, (2) partial removal of affected soils and sediments to achieve the greatest effect in risk reduction, and (3) no action. Stakeholder involvement in the decision-making process, and risk communication are important aspects of successful adoption and implementation of cost-effective clean-up strategies.

2.0 WESTERN U.S. URANIUM MILL ENVIRONMENT

Uranium mining and processing areas in the western United States, such as the Gas Hills Uranium District in Wyoming or Ambrosia Lake in New Mexico may cover from a few square kilometers up to 40 or 50 square kilometers. The climate in these areas is typically harsh with strong winds and extreme temperatures. The mean monthly wind speeds can range up to 8 meters per second in the winter and spring. Winter temperatures may reach as low as -40 degrees celsius. Summer temperatures may reach 38 degrees celsius.

Population in these areas is generally sparse. There may be isolated ranches or individual houses within a few miles. Typically the nearest permanent town is several miles away, but in some cases a "company town" is located within or near the mill site. Many of these "company towns" are no longer in use and have been demolished. On the other hand, the areas support abundant wildlife including deer, antelope, rabbits and other small mammals, and reptiles.

3.0 HISTORIC RADIONUCLIDE RELEASES

At many sites, radionuclides were released intermittently during the operational period. Thorium-230 releases may have occurred as routine discharges in the form of overflow from the tailings impoundment during the early years of uranium milling activities. Tailings solids with elevated radium-226 concentrations may have been released during tailings dam breaches or other incidents.

4.0 PRELIMINARY HUMAN HEALTH RISK ASSESSMENT

The human health hazards associated with current and potential future use of a typical site and its immediate environs were evaluated in a generic preliminary human health risk assessment for a relatively narrow drainage adjacent to a large site. The purpose of the assessment was to demonstrate the advantages of using a risk-based approach in developing clean-up strategies. This discrete area is assumed to have been impacted by historic mill operations.

At the generic site, as with many actual mill sites, the only reasonable land use for the surrounding area is cattle and sheep grazing and recreational hunting. The maximally exposed individual was assumed to be a sheepherder who might spend as much as two weeks per year in the area. For this example, grazing permits for sheep on public land limit the amount of time the herd can remain in one area.

Pathways of exposure considered in the assessment of potential dose and risk to the sheepherder include the following:

- Direct gamma radiation
- Incidental ingestion of surface water
- Incidental ingestion of soil and sediment
- Ingestion of meat from animals consuming surface water, soil, and vegetation in the vicinity of the site
- Inhalation of radon emanating from the tailings solids
- Inhalation of resuspended sediments or soils from the streambed or banks.

The stockman was assumed to spend 14 days per year camped in the affected area with 50 percent of his time spent in the drainage in which the releases were contained.

RESRAD, a computer code developed by the U. S. Department of Energy (Yu, 1993) for determination of residual radioactive material guidelines, was used to estimate potential doses from the following pathways:

- Inhalation of radon and radionuclides in airborne particulate matter
- Ingestion of soil
- External gamma radiation.

Groundwater pathways were excluded from the analysis, assuming that it would be highly improbable that wells for domestic water use would be developed in the affected area. In addition, the area covered by the impacted sediments and soils was assumed to be relatively small, thus would have little effect on ground water quality.

RESRAD assumes an infinite plane for the calculation of external gamma radiation doses. A shape factor was applied to the calculated dose to account for the fact that the impacted area was not an infinite plane but more closely resembled a narrow ribbon of impacted surface materials (i.e., a creek or stream bed).

Table 1 Summary of Results of the Maximum Dose Estimate (mrem/year)

Pathway	Sheepherder
External gamma (percent of total)	14 mrem/year (90%)
Inhalation of dust (percent of total)	0.8 mrem/year (5%)
Ingestion of soil (percent of total)	0.6 mrem/year (4%)
Ingestion of meat (percent of total)	Not Applicable
Inhalation of radon and its short-lived daughters (percent of total)	0.1 mrem/year (1%)
TOTAL	15 mrem/year

Table 2 Distributions Used in the Monte Carlo Assessment

Variable	Units	Distribution	Parameter	Value	Reference
Concentration	pCi/g	Uniform	minimum, maximum	Th-230 (72, 273) Ra-226 (10.5, 101) U-234 (5.1, 22) U-235 (0.23, 1.0) U-238 (5.1, 22)	All concentrations are from hypothetical distribution estimates.
Occupancy Factor	unitless	Triangular	minimum, most likely, maximum	0.0016, 0.004, 0.019	The minimum value is estimated based on 1 hour/day exposure for 14 days; the most likely is for 2.5 hour/day for 14 days; maximum is for 12 hour/day for 14 days. All are professional judgment.
Ingestion Rate	grams/year	Triangular	minimum, most likely, maximum	10.95, 18.25, 36.5	USEPA, 1996
Mass Loading	grams/m^3	Triangular	minimum, most likely, maximum	9.0E-06, 1.0E-04, 2.5E-04	Gilbert et al., 1983
Inhalation Rate	m^3/year	Truncated LogNormal	mean, standard deviation, minimum, maximum	7358, 4730, 3639, 11038	USEPA, 1996
Shape Factor	unitless	Uniform	minimum, maximum	0.41, 0.673	Calculated for an individual next to and in a drainageway.
Dust Fraction	unitless	Triangular	minimum, most likely, maximum	0, 0.25, 1.0	Best professional judgment.
Exposure Duration	year	Triangular	minimum, most likely, maximum	1, 7, 30	Most likely value is for a farm worker from USEPA (1996); maximum is from USEPA (1996) and represents the upper end of the occupational mobility tenure.

Doses from the meat and surface water ingestion pathways were calculated manually. RESRAD does not provide for consumption of impacted surface water by humans or animals. Dose conversion factors used for calculating dose from radionuclide intake by ingestion were taken from U.S. Environmental Protection Agency Federal Guidance #11 (USEPA, 1988).

The concentrations of Ra-226, Th-230, and U-nat used in the RESRAD analysis were 100 pCi/g, 275 pCi/g, and 45 pCi/g respectively. The total maximum estimated annual dose to a sheepherder was 15 mrem per year. Over 90 percent of the dose was attributed to direct gamma radiation. The maximum reasonable dose estimates for the generic site under assumed existing conditions are given in Table 1.

A Monte Carlo simulation was performed for the dose to sheepherder. Since nearly all of the dose was due to external gamma radiation, the simulation was simplified to include only that pathway. The estimated ranges and distributions of numerical values projected for each of the parameters used in determining the external gamma dose are shown in Table 2. The distribution showed a most likely value of approximately 2 mrem per year and a 95% probability that the annual dose to the hypothetical sheepherder would be less than 11 mrem per year. These calculated doses are within the range of variability in background doses.

5.0 CONCLUSION: RISK-BASED DECISION-MAKING

The decision as to what course of action to take during clean-up of any site should be made on the basis of potential human health risk and ecological consequences. The balance between maintaining a valuable habitat and avoiding a potential radiation dose to a member of the public that is in the range of background variability in the United States should be an integral part of developing clean-up criteria. The potential risk to workers excavating generally inaccessible areas and potential harm to cultural and environmental resources, as well as the cost of remediation activities must be considered in the decision process. The benefits to society of spending large amounts of money to mitigate negligible potential risks must be weighed against benefits which might be realized from other use of these resources.

Risk-based decision-making with regard to remediation of uranium mining and milling facilities can reduce the overall risk to human health and the environment and preserve valuable resources for the future.

6.0 REFERENCES

Gilbert, T.L. Et al. 1983. *Pathways Analysis and Radiation Dose Estimates for Radioactive Residues at Formerly Utilized MED/AEC Sites,* ORO-832, prepared by Argonne National Laboratory, Argonne, IL, for U.S. Department of Energy, Oak Ridge Operations.

U.S. Environmental Protection Agency (USEPA). 1988. *Limiting Values of Radionuclide Intake and Air Concentration and Dose conversion Factors for Inhalation, Submersion, and Ingestion.* EPA-520/1-88-02.

U.S. Environmental Protection Agency (USEPA). 1996. *Exposure Factors Handbook.* EPA 600/P-95/002Babc. Draft. August.

Yu, C., A.J. Zielen, J.J. Cheng, Y.C. Yuan, L.G. Jones, D.J. LePoire, Y.Y. Wang, C.O. Loureiro, E. Gnanapragasam, E. Faillace, A. Wallo III, W.A. Williams, and H. Peterson. 1993. *Manual for Implementing Residual Radioactive Material Guidelines Using RESRAD Version 5.0.* Environmental Assessment Division. Argonne National Laboratory. Argonne, IL. September.

Tailings and Mine Waste '99 © 1999 Balkema, Rotterdam, ISBN 90 5809 025 6

Reclamation management practices of waste spoil dumps in the Australian coal industry

E.K.H.Goh & T.O.Aspinall
Department of Mining, Minerals and Materials Engineering, University of Queensland, Brisbane, Qld, Australia
J.S.Kuszmaul
Department of Geology and Geological Engineering, University of Mississippi, Miss., USA

ABSTRACT: The environmental impact of mining and evolving environmental legislation has received increased attention worldwide in the last two decades (Bradfield, Shultz & Stone 1996). The potential impacts associated with unstable spoil dumps from mine operations is the focus of concern both by the mining industry, environmental legislative agencies and members of the public. Engineered slopes of mine spoils may be stable at the end of construction, but they can deteriorate over time. There is thus the need to increase the base of knowledge on the existing practices of spoil dump design and reclamation. Information concluded from the analysis of the industrial survey carried out on Australian spoil dump management practices are presented in this paper.

1 INTRODUCTION

Open-cut coal mining operations in the Bowen Basin of Central Queensland cost \$15,000 to \$25,000 per hectare for regrading and reshaping of waste spoil dumps to achieve stable conditions (Roe, Mulligan and Bell 1996). Owing to the costs involved in the construction of stable slopes and the regulatory requirements aimed at minimising off-site effects, the efficient design of spoil dumps has emerged as a significant engineering task (Williams 1996). Erosional and mass stability are important criteria to be considered in the design of spoil dumps (Hannan & Bell 1993). Efficient engineering design and erosion management of spoil dumps for mine reclamation assists in maintaining the expected standards of environmental compliance at optimum costs. Environmental management of spoil dumps should thus be considered an integral phase throughout the life of an engineering project.

Owing to the scarcity of published information on spoil dump design and reshaping management practices for Australian mines, an industrial survey was carried out by the principal author to expand this knowledge base. Questionnaires were distributed to mining companies to obtain a representative sample of the mining industry within Australia involved in the design, construction and reclamation of overburden spoil dumps. The questionnaire was structured to obtain data on the following aspects of spoil dump design and reclamation management systems:

- physical properties of spoil dumps and their environment,
- spoil dump design and the reshaping process to meet legislative requirements,
- spoil dump management systems,
- industrial erosion mitigation measures for spoil dumps, and
- operational costs for reclamation of spoil dumps.

Questionnaires were distributed to mining companies to depict a representative sample of the mining industry within Australia involved in the design, construction and reclamation of overburden spoil dumps. Figure 1 illustrates the sites of coal mining operations in Australia. The selection of the operations involved were based on the location and climatic environment. Care was taken in this industrial survey to prevent sampling biases (Williams 1978). Sampling bias is due to the occurrence of a high percentage of non-responses during the industrial survey which run the risk of under-representing a particular distinctive feature in the overall survey results. Twenty-seven questionnaires from the coal industry were answered and used in this analysis. The total number of surface coal operations specifically involved in overburden spoil dump production and reclamation management is 35 (Roe, Mulligan & Bell 1996, Hannan & Gordan 1996, Minerals Council of Australia 1995). The response rate of the coal industry to the spoil dump survey amounts to 27 out of 35 coal operations. The very high response rate of 77% clearly indicates that the overall survey results represent a vast majority of the Australian coal mining industry involved in systematic spoil dump management and reclamation. The information derived from this spoil dump design and reclamation industrial survey is summarised and presented in this paper.

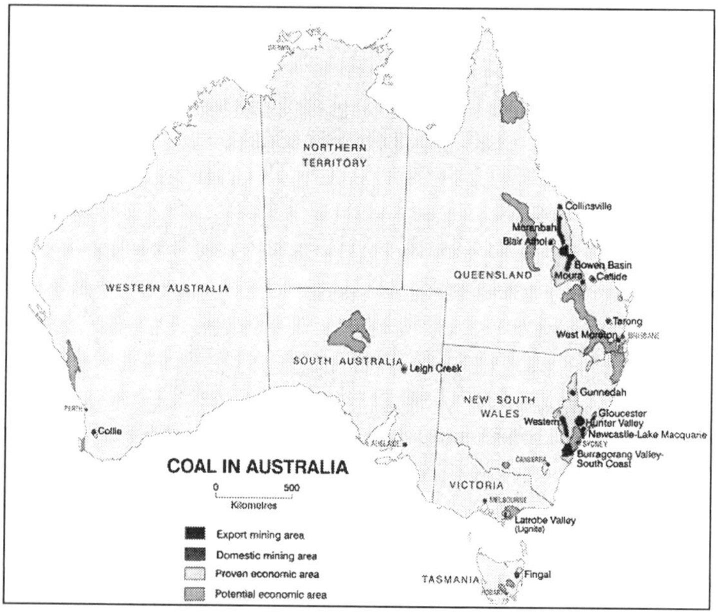

Figure 1. Coal producing sites in Australia (Australian Dept. of Primary Industries and Energy 1998)

2 OPERATIONAL SITE CHARACTERISTICS

Of the fully-completed questionnaires received from the coal mining industry, it was observed that in terms of mining methods, all those twenty-seven sites which responded of carrying systematic spoil dump design and reclamation management are surface mine operations. Table 1 outlines the operational size of the coal mines surveyed and the site characteristics of overburden spoil dumps. Of all the industrial responses received from the survey, 63% were from the semi-arid (dry hot/warm) with seasonal rainstorms climatic and 37% were from the temperate (mild humid) regions.

The distribution of the general topography from the areas surrounding the surveyed mine sites indicates that the majority of the sites surveyed, amounting to 77%, have an undulating topographic profile.

Table 1. Operational size and site characteristics of spoil dumps at mines surveyed

Type of mineral extracted	Coal
No. of mines surveyed	27
Lease Area (hectares)	1300 - 27132
Area of current mine (hectares)	340 - 5500
Annual overburden spoil production (* $10^6 m^3$)	4 - 130
Present area (hectares)	80 - 4000
Expected area at mine closure (hectares)	180 - 10000
Typical size of individual dump sites (hectares)	1 - 2065

3 CHARACTERISTICS OF OVERBURDEN SPOIL DUMPS

Figure 2 summarises the distribution of the soil type texture, based on the <2mm factions, of overburden spoils from the surveyed mine sites. 87% of the overburden spoil types surveyed are that of sandy clay, silty clay loam, medium clay and sandy clay loam. The majority of the responded survey (75%) indicated that the swell factor of their spoil dump material lies in the range of 15 - 25%.

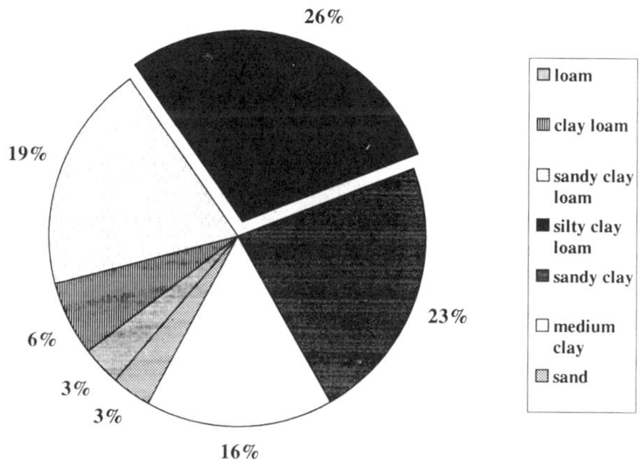

Figure 2. Distribution of the types of overburden spoils soil texture of at surveyed sites

4 SPOIL DUMP DESIGN AND MANAGEMENT

The design process of spoil dumps at the mine site is a critical phase in mine planning since the design must conform to the requirements of evolving legislation, the need for cost optimisation and the necessity for erosion minimisation. An efficient engineering design of spoil dumps aims to ensure the long term stability of the dumps at the end of construction. It also seeks to reduce the erosion potential through correct selection of an initial stable landform design. Implementation of effective planning and operating procedures, through recontouring and grading, aids in the reshaping of spoil dumps to a safe, stable and non-erodible environment. From the spoil dump industrial survey undertaken, 88% of the responses indicated that a combination of engineering experience using deterministic and probabilistic computer modelling techniques are the normal practices for spoil dump engineering design and site management.

From statistical analysis of the survey, it was noted that the factors used in the design of stable spoil dumps are: climate, rate of erosion, height of dumps, slope gradient, slope length, equipment, soil characteristics, legislation and costs involved. The importance of the various design variables is illustrated by the percentages of the companies basing their spoil dump designs on the respective design factors (Figure 3). Figure 3 shows that the main variables most emphasised upon by the industry in the spoil dump design are slope gradient, height of

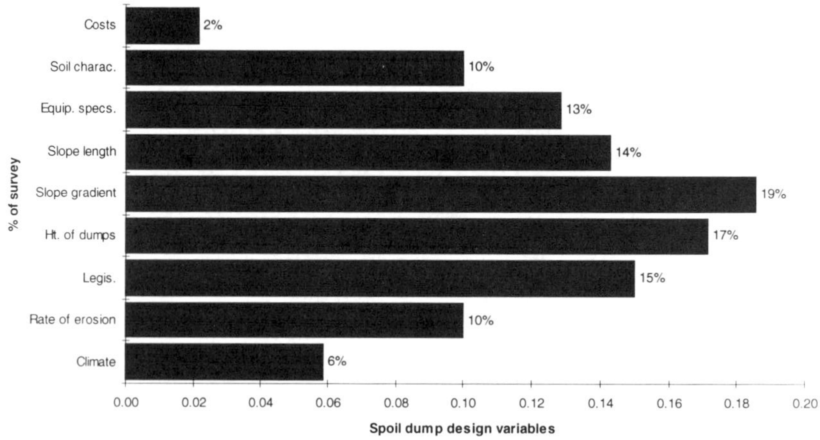

Figure 3. Industrial responses of variables used for spoil dump design

dumps, legislation and slope length. These variables are also used in the determination of the rate of erosion of the spoil dumps at the respective mine sites.

The findings of the industrial survey indicate that the principal modes of overburden removal currently used for spoil dump management are that of draglines, truck/shovels systems and bucket wheel excavators with draglines being the most widely-used technique. From this coal industrial survey, it was noted that the ratio of dragline operations to that of the truck/shovel system is 8:1. The majority of the companies surveyed, 95%, had a designed angle of repose for their pre-reshaped overburden spoil dumps set at a slope gradient of 37° (75%).

Statistical analysis of typical dragline spoil dump profile characteristics based on the responses from the coal mining industry indicate that:

- 94% of the sites have heights of designed dumps ranging from 25-125 m,
- 75% having distance between dump crests in the range of 45-75 m,
- 77% of those surveyed have widths of spoil dumps in the 50-100 m range, and
- 64% of the responses have lengths of spoil dumps ranging from 1 - 2 km.

5 SPOIL DUMP RECLAMATION

The mode (66%) of the annual rate of overburden spoil production for the Australian coal industry is 10-50 million m^3. The majority of the companies surveyed (70%) had an annual spoil dump reclamation rate of less than 90 hectares/year. Results of the survey analysis indicate that all the spoil dumps in the industry were either reshaped with benches, terraces or reduced to that of the surrounding undulating landscape prior to revegetation for improved erosional and mass stability.

The majority of the responded survey, 67%, indicated that the track-type bulldozer is the most suitable equipment for the reshaping of spoil dumps to legislative requirements. The survey results indicate that the 574 kW (770 hp) powered track-type bulldozer is the most sought after equipment by the mining industry for spoil reshaping with the U-blade as the most frequently used dump reshaping attachment (Figure 4).

Data analysis of the industrial survey indicated that the 574 kW bulldozer models most commonly used in the industry for spoil dump regrading are the Caterpillar's D11R and the Komatsu's D475. The industrial survey statistical performance analysis of the 574 kW bulldozer-type at mines using single units for part-time weekly spoil dump reshaping jobs indicate that the average material haul distance for spoil dump reshaping is 55 metres and that the average daily amount of material dozed is 0.48 hectares per day (Table 2).

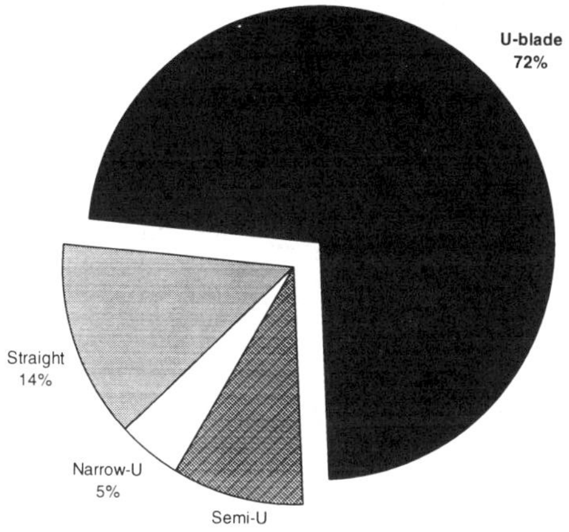

Figure 4. Response of industry on bulldozer earthmoving attachments for reshaping spoils

Table 2. Statistical spoil dump performance analysis of 574 kW bulldozer-type

	Average haul distance per trip (m)	Daily performance (ha/day)	Hourly performance (m^3/hr)	Part-time weekly equipment usage (days/wk)	Annual performance (ha/yr)
Mean	55	0.48	753	5	41
Std. Deviation	8	0.16	235	2	15
Coefficient of variation (%)	14.3	32.7	31.2	40.2	37.2

The time frame for the dozing cycle depends on the distance that the material is transported per trip in the reshaping of the dumps. From statistical analysis it was noted that the average ratios of the loading, transfer and return times for each complete bulldozer cycle for spoil dump reshaping is 1.6: 6: 2.7 (Table 3).

Table 3. Statistical analysis of bulldozer reshaping cycle for spoil dumps

Reshaping cycle of bulldozer	Cycle time (s)	
	Mean	Std. Deviation
Loading	16	13.0
Transfer	60	36.9
Return	27	18.0

Statistical analysis of the bulldozer spoil dump reshaping process indicate that:

- 69% of the sites surveyed have a average dozing distance of less than 100 m,
- 64% of the responses having a spoil dump slope gradient for dozing activity of less than 20%,
- 66% of those surveyed from the industry have a job efficiency of 45 - 50 min/hr.

The majority of the companies surveyed, 92%, indicated that equipment for spoil dump reshaping is assigned on a part-time basis. 62% of the companies which responded indicated that 5-7 days per week were allocated on a part-time basis for the usage of earth-moving equipment, primarily bulldozers, for spoil dump reshaping (Table 4). The mode for part-time weekly dozing equipment usage for spoil dump reshaping at sites reviewed is 5 days.

The response of the coal operations investigated, namely 78%, uses the practice of greater then 10 cm topsoil thickness for placement on the reshaped spoil dumps at the respective sites prior to revegetation. The recommended depth of replaced topsoil for already reshaped spoil dump reconstruction is in the range of 15-30 cm (Bell 1996). This means that the majority of the industry surveyed do conform to the recommended guidelines for topsoil placement. It was also noted that the popular practice in the coal industry for topsoil being stockpiled before reuse is in the 1 - 2 years storage time frame (75%).

Table 4. Percentage industrial responses on days per week equipment used for spoil dump reshaping

Part - time						Full-time
1 day/wk	3 days/wk	5 days/wk	7 days/wk	Variable	Weekends	7 days/wk
8 %	4 %	35 %	27 %	15 %	4 %	7 %

All the mine sites surveyed indicated that vegetative cover, namely that of trees, shrubs, grass and legumes, are finally established in the revegetation of the spoil dumps. Of all the industrial responses received from the survey, 68% of the companies reviewed indicated that grass and trees are the most popular vegetative types for spoil dump revegetation.

6 EROSION MITIGATION MEASURES

Construction of erosion mitigation measures aids in the long term stability of the rehabilitated spoil dump. Analysis of the industrial questionnaire survey shows that all the organisations surveyed indicated that erosion prevention techniques of one type or another are utilised in the stabilisation of spoil dumps. The erosion mitigation measures popularly used in the industry include surface ripping, contour banks, sediment traps, diversion banks, runoff velocity control, lined waterways, grassed waterways and hydromulching. The graphical statistical findings, in Figure 5, indicate that the more popular erosion minimisation techniques for improving spoil dump stability is that of ripping and the construction of contour banks. Rippers attached to bulldozers assist in reducing the compaction resulting from heavy machinery traffic during spoil dump reshaping operations. The ripping process aids in minimising erosion and increasing dump stability by increasing water infiltration and reducing runoffs on the spoil dump surfaces.

7 OPERATION COSTS FOR SPOIL DUMP RECLAMATION MANAGEMENT

Optimisation of costs for the efficient design and construction of spoil dumps assists in determining the final viability of any mining operation. Cost optimisation in rehabilitation

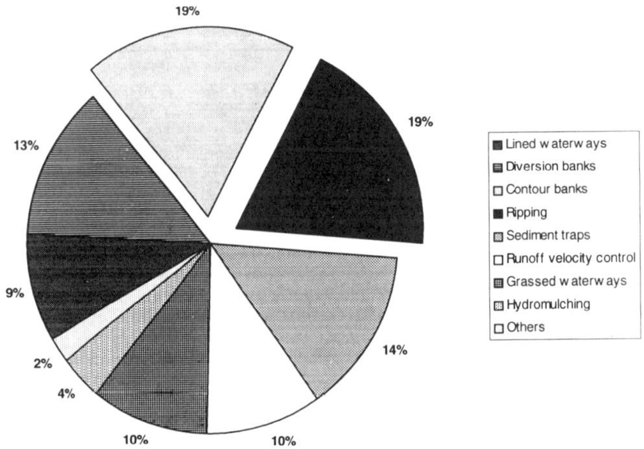

Figure 5. Industrial response on the application of erosion mitigation measures

Table 5. Statistical cost analysis of spoil dump reclamation phases

Phases for dump reclamation	Mode (%)	Mean (%)	Std. Deviation (%)
Reshaping/ recontouring costs	70	60	20
Drainage construction costs	5	8	6
Topsoil application costs	5	17	13
Ripping costs	5	4	3
Revegetation costs	4	7	10
Supplementary costs (eg. research)	5	4	21

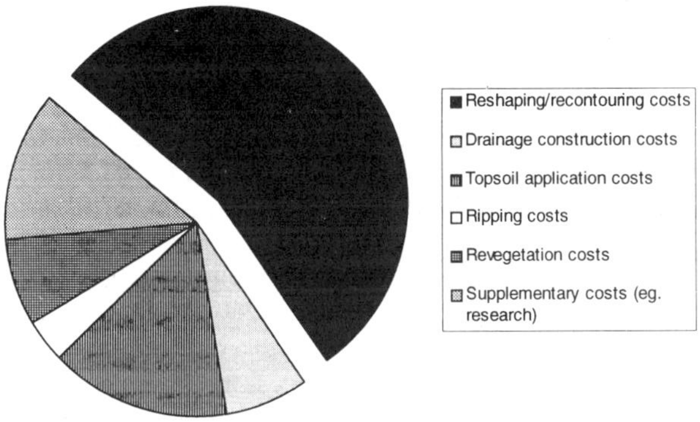

Figure 6. Proportional costs for the different components of spoil dump reclamation

operations seeks to minimise the expenditure for overburden handling through systematic engineering design by avoiding the double rehandling of spoil. Hoare (1985) observed that in some instances the rehabilitation costs can be as high as $100,000 per hectare if additional environmental repairs are required for disturbed sites. Such occurrences provide an incentive to ensure that rehabilitation works are done once and done well to avoid these excessive supplementary high costs. Careful planning and optimal design of spoil dumps can assist in saving thousands of dollars during the construction and reclamation of spoil dumps. Table 5, from survey analysis, indicates the division of the costs associated with the different phases of spoil dump reclamation to overall reclamation costs to meet legislative requirements.

Spoil dump reshaping costs accounts for the highest percentage of overall reclamation costs from the site studies carried out (Figure 6). 88% of the coal companies surveyed indicated that the amount spent on the annual reclamation of spoil dumps amounts to less than 10% of the total annual mine operating budget.

8 CONCLUSIONS

Twenty-seven complete questionnaires were returned for this spoil dump survey denoting a industrial response of 77%. The industrial responses to the survey questionnaire is considered a good representative sample of the Australian coal mining industry involved in the systematic management and reclamation of spoil dumps since the total number of surface coal operations specifically involved in overburden spoil dump production and management is 35. Information presented in this paper could be viewed as a quantitative guide to the spoil dump design and reclamation management practices in the Australian coal mining industry.

Efficient spoil dump design and reclamation practices are acknowledged as important phases in engineering mine design and planning. Assessing the risks attached to any selected spoil dump design is the aim of the surface coal mining industry worldwide involved in overburden spoils handling. Though the industry is in the phase of altering the methodology of designing spoil dumps, very little information is presently available to assist in the design process for a wide range of expected field conditions. This industrial survey is an integral phase of the research study on spoil dump designs. The results of this survey will be beneficial to the coal industry in assisting to optimise the costs in spoil dump design and reclamation practices.

REFERENCES

Australian Dept. of Primary Industries and Energy 1998. *Coal Maps*. Australian Department of Primary Industries and Energy: Canberra.

Bell, L. C. 1996. Reclamation of disturbed land. In D. Mulligan (ed), *Environmental management in the Australian minerals and energy industries - principles and practices*: 227-261. Sydney: UNSW Press.

Bradfield, P.J., C.E. Schulz & M.J. Stone 1996. Regulatory approaches to environmental management. In D. Mulligan (ed), *Environmental management in the Australian minerals and energy industries - principles and practices*: 46-73. Sydney: UNSW Press.

Caterpillar. 1997. *Caterpillar Performance Handbook - Edition 28*: 1.35 - 1.56. Illinois: Caterpillar Inc.

Hannan, J. C. & L.C. Bell 1993. Surface reclamation. In A.J. Hargraves & C.H. Martin (eds), *Australasian coal mining practice - Monograph series No.12:* 260-280. Parkville: Australasian Institute of Mining and Metallurgy.

Hannan, J. C. & R.M. Gordan 1996. Environmental management of coal mines in the Hunter Valley, NSW. In D. Mulligan (ed), *Environmental management in the Australian minerals and energy industries - principles and practices*: 227-261. Sydney: UNSW Press.

Komatsu. 1997. *Specifications and Application Handbook - Edition 18*: 1B.1 - 1B.37. Tokyo: Komatsu Ltd.

Louthean, R. 1997. *The Australian Mines handbook 1997/98 edition*: 207-300. West Perth: Louthean Publishing and Minmet Australia.

Minerals Council of Australia 1995. *Australian Mining and Mineral Operations map*. Canberra: Minerals Council of Australia.

Roe, P.A., D.R. Mulligan & L.C. Bell 1996. Environmental management of coal mines in the Bowen Basin, Central Queensland. In D. Mulligan (ed), *Environmental management in the Australian minerals and energy industries - Principles and practices*: 290-315. Sydney: UNSW Press.

Williams, W.H. 1978. *A sampler on sampling*: 231-232. New York: John Wiley & Sons Inc.

Tailings and Mine Waste '99 © 1999 Balkema, Rotterdam, ISBN 90 5809 025 6

Rehabilitation of abandoned deposits of mineral residues

J.H.Troncoso
Catholic University of Chile, Chile

D.A.Troncoso
Catholic University of Maule, Chile

ABSTRACT: Rehabilitation of abandoned deposits of mineral residues by agricultural means may be effectively accomplished to adapt the landforms for production and recreation, to protect the slopes and the plateaus against erosion and to decrease potential for acid generation. Long term stability of the deposits has to be realistically predicted before investments and permitting are decided. The analyses should consider verifications of increased shear strength and stiffness, related to changes in chemical composition, as shown in this paper.

1 INTRODUCTION

Abandonment of deposits of mineral residues is a subject which presently poses critical problems for the mining industry due to the magnitude of their permanent environmental impacts and to the eventual stability risks they represent for the neighbor areas .

Many countries like Chile, where mining activities are concentrated close to populated areas, to valuable agricultural lands or to important water courses, need to apply with urgency rehabilitation programs. Otherwise, thousands of hectares covered by mineral residues will be lost for production and many others will be contaminated forever.

Furthermore, vulnerability of deposits to the effects of natural disasters create unacceptable risks for the population downstream. Rehabilitation procedures have to include, therefore, stabilization of the soil structures of the deposits to make them safe against maximum foreseeable events such as earthquakes and floods.

Rehabilitation programs require interdisciplinary work of agricultural and environmental sciences and geotechnical engineering to face specific problems of erosion, acid generation, transport of contaminants, slopes stability, flow failures potential and flooding scour. First, joint efforts should help to optimize solutions for the control of negative impacts and to minimize risks of catastrophic failures; second, they should aim to develop safe and economic reclamation plans.

This paper presents first results of interdisciplinary research efforts devoted to investigate aging effects in the stability of tailings deposits. The project aims to obtain basic properties of soils structures as functions of time oriented to support more realistic predictions of long term behavior of the soil structures. Knowledge of these properties should permit to evaluate the risks of investments in diverse scenarios of rehabilitation from minimum reinforcements to large agricultural developments for production of timber, food and/or recreational purposes.

2 INFLUENCE OF FINE FRACTION IN SOIL PROPERTIES

The fine fraction of mineral residues, composed of silts and clays, plays important roles in the properties of the deposits. Rehabilitation projects have to consider specially the effects of the fine fraction in the stiffness and cyclic strength of the retaining structures, in the chemical reactions which lead to acid generation and in the interaction with fluids, vital for growth of vegetation.

The following results of previous research on the influence of fines in soil properties measured in compacted specimens, in laboratory (Troncoso, 1992; Troncoso and Verdugo, 1985), are useful to reveal the importance of silts in the mechanic and dynamic behavior of tailing dams.

2.1 *Shear modulus*

Shear modulus, G, decreases as shear strain, γ, increases in all tailings soils. The shape of the curve $G = G(\gamma)$ is very much dependent of the structural characteristics of each soil including fabrics and interparticle bonds.

Fine content and void ratio are the most important parameters which determine the value of shear modulus. For same void ratio, e, shear modulus decreases as fine content increases, from clean sand to 30% silty sand, as observed in Figure 1. Decrease in stiffness may be explained, in tailings soils produced by crushing parent rocks, due to decrease of interlocking resistance as finer particles fill in voids between coarser grains. On the other hand, shear modulus increases with the square-root of the effective mean normal stress.

2.2 *Cyclic strength*

Cyclic strength ratio, $\sigma_d/2\ \sigma_0'$, decreases as fine content increases, as shown in Figure 2. This unfavorable influence of fines is related to decrease in stiffness and in permeability. More deformable and less permeable structures dynamically loaded under undrained conditions naturally undergo larger pore pressures and eventually fail in less number of cycles due to progressive deterioration. In this respect it has to be assumed that each new cycle of loading acts over a structure which effective confining stress has decreased as pore pressure has increased and therefore, shear modulus has decreased. Lower modulus lead to higher increments of pore pressure and to lower confining stresses, and so on, until failure occurs.

2.3 *Damping ratio*

Damping ratio, D, increases as shear strain, γ, increases as shown in Fig. 3, which summarizes results of cyclic triaxial tests of chilean tailings soils. Void ratio, or relative density, does not present a clear influential tendency in energy dissipation capabilities.

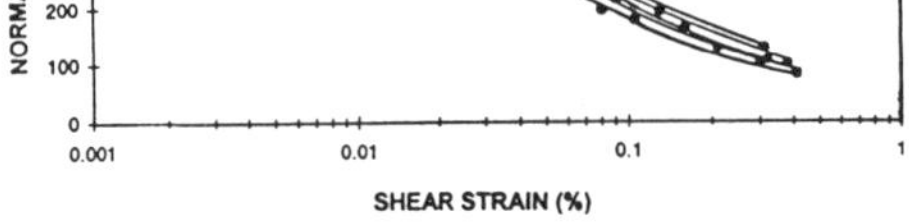

Figure 1. Shear Modulus of silty sands in cyclic triaxial compression.

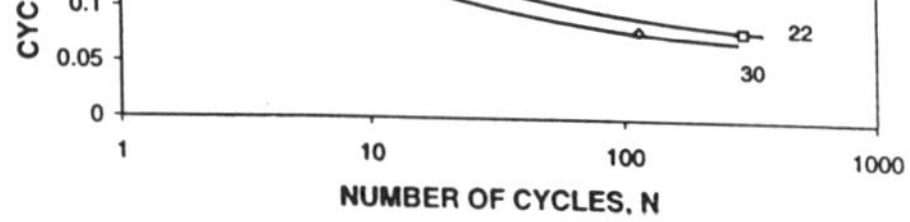

Figure 2. Cyclic strength of silty sands

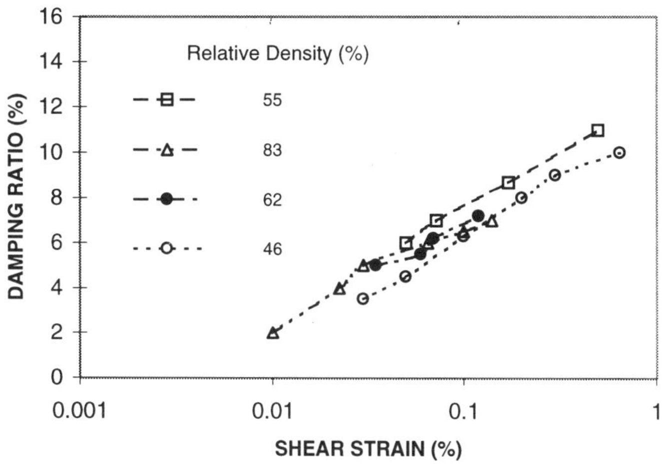

Figure 3. Damping ratio of silty sands in cyclic triaxial compression

2.4 Chemical composition

Chemical composition of mineral residues is very important for acid generation analyses and for feasibility studies of agricultural use. It is also interesting for interpretation of aging effects through double layer theory. Table 1 includes results of chemical tests in typical shallow samples of sandy silts from an abandoned Chilean deposit of copper tailings. Results are compared with Canadian tailings abandoned in similar way, without isolation from oxygen, and with isolation under water, as published by Garga and Troncoso (1991). These results indicate the usefulness of effective barriers for oxygen to prevent acid generation.

Table 1. Chemical composition of tailings soils in abandoned deposits.

Constituent	Canadian mine (1)			Chilean mine
(mg/Kg)	Underwater	Shore	Dry	Dry
Copper	0.25	6.23	797	500
Zinc	15.50	44	218	100
Lead	0.35	1.05	1.27	200
Calcium	145	2,590	2,360	24,600
Magnesium	18	121	577	9,800
Sulfate	240	7,340	35,400	5,800

(1) Source: Hamilton and Fraser (1976) quoted by Garga and Troncoso (1991)

3 EVOLUTION OF SOIL PROPERTIES

Mechanic properties of tailings soils change with time under effects of consolidation, desiccation and aging. These changes affect stability as they increase the strength and the stiffness of the soil structures.

Different methods of subsurface exploration have been used in Chile to monitor the evolution of properties in deposits of tailings of different ages. The aim has been to determine time functions which could represent the evolution of properties of soils with age

Weaker materials in a tailings deposit are the sandy silts impounded in the reservoir, however, shear strength of these soils also increases with age as evidenced by cone penetration resistance measured in deposits of different ages. Troncoso (1998) has proposed the following equation derived from point resistance, q_c:

$$q_{ct}/q_{co} = (t / t_0)^a \qquad (1)$$

where: q_{ct} = point resistance at age t (yr), q_{co} = point resistance at age t_o (yr), with: a = 0.29 for $2 \leq t \leq 14$ year.

Fine content is an important characteristic for the development of bonds which are responsible of aging effects in soil structures. This is because chemical bonds created between solid particles and adsorbed double layers are relatively stronger for smaller grain sizes. Chemical composition of grains and fluids which form the mine tailings is therefore interesting for the investigation of evolution because initial bonds may change with time as chemical reactions proceed and as seepage of fluids charged with acids and salts transport them from shallow to deeper layers.

Results of in-depth electric resistivity and downhole shear wave velocity measurements are presented in Figure 4 together with SPT and index properties of soils obtained in an investigation performed in an old tailings dam 20 year after abandonment. Electric resistivities are relatively high in comparison with those obtained in natural sedimentary deposits of sandy silts. Shear wave velocities also reveal stiffer structures than most lacustrine silts. Damping ratios, calculated from particle velocity attenuation records, in the wave propagation tests, are also included in the profiles. These results indicate that the tailings structures of fine soils, such as silts, which normally form weak structures, evolve favorably with age. It is therefore concluded that it is possible to count with positive increments of strength and stiffness in the soil structures which form the deposits of fine mineral residues as result of aging effects. These gains help to balance the larger scale loadings which correspond to maximum probable events.

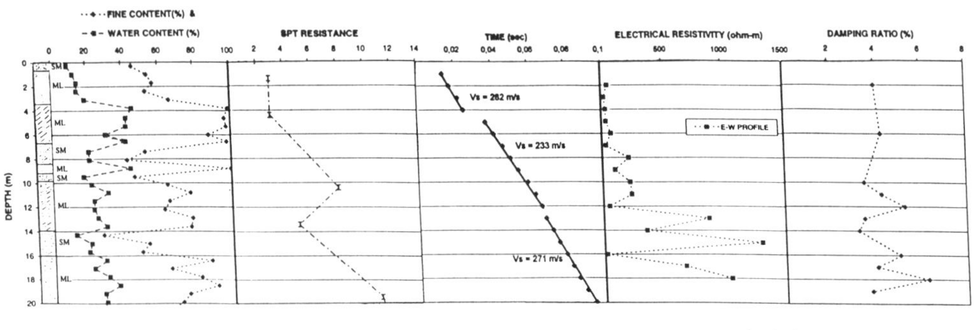

Figure 4. Index properties, SPT resistance, electric resistivity and down-hole shear wave velocity and damping ratio measured in 20 year-old tailings dam.

4 EVALUATION OF STABILITY OF ABANDONED DEPOSITS

4.1 Need to evaluate stability

Abandoned deposits of tailings soils remain saturated for long periods of time due to the high capacity of fine sandy silts to retain adsorbed water. Therefore, pore water pressure may increase and shear strength may decrease during undrained loading. Consequences for slope stability may be large displacements and even flow failures.

Studies of rehabilitation of an abandoned deposit have to begin with detailed evaluation of the stability of the soil structures which form the retaining dams and the contained sediments in order to quantify the possible risks.

Knowledge of risks shall permit:

1) To design reinforcement works as necessary to satisfy safety requirements of the structures of the deposit and to protect the population and the properties located downstream.
2) To analyze the most appropriate ways to rehabilitate considering the dangers for persons and for capital investments.
3) To define the limitations to be imposed to the construction works to avoid weakening of the critical structures.

Although most of the failures of tailings dams which have caused losses of human lives and catastrophic environmental damages have occurred in operating tailings dams, there have been a few cases of large size failures of abandoned deposits. The main differences have been the presence of surface waters which have eroded and breached a dam after cracks have opened or the increase of pore pressures below the water table which have led to liquefaction of flow failures. Water is, therefore, an important factor to be considered in stability during and after rehabilitation.

4.2 Stability analyses

The analyses of stability of abandoned deposits have to be performed using the most realistic models available in order to obtain useful results. This condition is imposed by the need to predict behavior under the worst foreseeable loadings proper of extreme events such as maximum credible earthquake and maximum probable flood. The extreme loadings should cause failures which could not be economically avoided but consequences have to be calculated in order to prevent unacceptable losses.

Finite element models are presently the most complete tools to analyze stability of abandoned deposits as they permit to calculate actual states of stress and deformations under static or seismic conditions. Results of finite elements analyses permit to determine critical zones inside dams, to calculate liquefaction and flow failures potentials and to estimate residual displacements. However, results of these analyses are fully dependent of the parameters introduced in the models to represent stress-strain and hydraulic conductivity relationships.

The evolution of the properties of the soils which form the abandoned deposits affect the stiffness as well as the strength of the materials. While strength increases, as presented in the previous section, are favourable for stability of dams, stiffness increases may be unfavourable for some seismic events. This last effect is related to the possibility of resonance in response of dynamic structures which natural period coincides with the predominant period of excitation.

In conclusion, the stability of abandoned deposits may be evaluated realistically by means of mathematical models which include the effects of evolution of soil properties with age. Once this evaluation is completed it is necessary to design auxiliary works devoted to reinforce the structures, if necessary, and to preserve safe water balances in the deposit.

4.3 Safe water balance

Rehabilitation of an abandoned deposit may need to maintain shallow ponds of water covering the reservoir and/or layers of soils supporting vegetation, for prevention of acid generation or for commercial or recreational objectives. On addition, restoration of natural water courses may need to habilitate a river bed over the deposit as alternative to diversion tunnels or buried conduits (Troncoso, 1998).

Evaluation of stability needs to include effects of maximum probable flood and corresponding water balances in the long term behavior of abandoned deposits. Maximum hydraulic heads in the deposit have to be limited by conservative design of spillways including safe discharge over protected areas of the dam slopes as second line of defense to prevent uncontrolled overtopping.

5 STABILIZATION OF TAILINGS WITH VEGETATION

Stabilization of tailings is fundamental not only to prevent failures and damages to downstream zones but also to mitigate environmental impacts and to recover land for wildlife and/or productive use through agriculture and forestry.

Stabilizing the impoundment against long-term wind and water erosion is often a difficult task to achieve. In addition, it is crucial for avoiding environmental contamination and for returning the area to productive uses.

Daniel and Koerner (1993) have described several cover systems. Their main objectives for mine wastes are to separate tailings from surface environment and to restrict infiltration of water into the deposit. As tailings extend over large areas, cover systems must be inexpensive and should require minimal or none maintenance. To promote vegetative settlement over tailings accomplish objectives of protection and minimize wind and water erosion and may return the affected area to its original appearance and land use.

Surface soil layer of tailings is permanently affected by incoming water from rain and snowfall and occasional water streams formed during intensive rains and snowmelt periods. Usually, this water is drained from the tailings by basal ducts, relief wells, spillways and water diversion systems (Spichiger *et al.*, 1996a,b). These structures demand high investments and continuos maintenance. On the contrary, vegetation cover returns water to the atmosphere via evapotranspiration and requires low maintenance and infrastructure.

Establishment of vegetation on tailings presents advantages over other cover systems. Nevertheless, special conditions are required to allow plant development along different stages of growth. Normally, tailing substrate has high heavy metal and low macroelements (e.g. N, P and K) concentration, nil organic matter content, low pH and lack of adequate texture and structure for plant growth. In addition, drought and wind restrict vegetative establishment. Although heavy metals in proper balance, such as zinc and copper, are essential micronutrients, in excess may be toxic to living organisms.

The success to establish a vegetative cover on tailings substrate depends primarily on the knowledge of tailings conditions, weather and economic factors. These vary according with the extraction process of the mineral, location of the deposit and management decisions of the mining company. Substrate analyses for typical Chilean copper tailings deposit show concentrations above 500 mg/Kg for copper, considered elevated and may be toxic for plants (Russell, 1988; Bohn *et al.*, 1985). However, other toxic elements, as arsenic, iron and zinc, remain in concentrations below levels of risk, for plant and animal performance. Acid generation and heavy metal precipitation may be controlled by adequate management of discharge of tailings and maintenance of shallow water ponds over the tailings for long periods of time, during operation, and fast cover with vegetation after closure.

Copper tailings are feasible for developing vegetation (Figure 5), as shown by Lobos and Silva (1990), who have reported successful growth of shrub and tree species of *Eucalyptus globulus, Eucalyptus camaldulensis, Acacia cyanophylla, Acacia melanoxylon, Acacia dealbata and Pinus radiata*, without any correction for fertility or pH. However, plants had a depressed growth, notorious N and P deficiencies and low bud activity during spring. After amendments to increase levels of N, P, K and pH, trees, shrubs and herbaceous plants, as *Phalaris tuberosa*, were established successfully. In addition, species like *Erodium sp., Escholtzia californica, Polygonum sp. Baccharis solisi* and *Lolium sp.* invaded naturally. This vegetation establishment controlled effectively wind erosion and allowed domestic and native animal grazing. Other experiences have concluded that bush and herbaceous species as *Atriplex nummularia, Galenia secunda, Acacia cianophila, Acacia semperflorens* and *Tamarix articulata,* and the tree species *Eucalyptus camaldulensis, Eucalyptus globulus, Pinus radiata* and *Quillaga saponaria*, could be successfully grown over copper tailings, requiring none or only irrigation at establishment (Cohen and Moenne, 1991). Vegetation has upgraded the stability of the deposit, improved slope stability, prevented wind and water erosion, and improved surroundings.

Development of techniques to promote vegetation establishment, as the Air-Earth Interface model (Dixon, 1997) and fertirrigation are fundamental to accelerate natural plant succession, to enhance productivity of tailings and to reduce costs of revegetation.

Figure 5. Rehabilitation of abandoned deposit of tailings with pines and eucaliptus

6 CONCLUSIONS

Rehabilitation of deposits of mineral residues is necessary to prepare them for the worst loading events proper of the long term periods which correspond to abandonment.

Reinforcement of the soil structures and restoration of the natural water courses are minimum actions to be taken towards rehabilitation. Vegetation of areas of slopes and impoundments are possible ways to further prevent negative environmental impacts such as eolic and rain erosion, acid generation and scenery degradation.

Positive increments of cone penetration resistance, cyclic strength and shear wave velocity have been detected in abandoned deposits of copper tailings in Chile. Knowledge of these aging effects in the mechanic properties of soil materials is useful to optimize the designs for abandonment. This is so because gains in shear strength and in stiffness of soils of mineral residues increase with time and these gains are significative to counterbalance maximum loads probable to occur in long-term periods in the life of the deposits.

Geophysical methods of subsurface exploration such as electric resistivity and shear wave velocity measurements are useful tools to verify structural changes associated to variations in geochemical properties and to characterize the soils in conjunction with index properties.

Establishment of vegetation over copper tailings is limited by conditions as drought, low fertility and soil porosity, rather than high concentration of toxic elements. Agriculture experiences of different types of vegetation established in old tailings deposits have permitted to identify some species which grow better according with location and climate. Future research should determine plant species adaptation to natural tailing conditions, with the objective to improve stability, restore natural ecosystem, reduce erosion and contamination, and overall return of land to productivity through safe animal and human use.

7 ACKNOWLEDGEMENTS.

The authors appreciate the financial support of the Chilean National Fund for Scientific and Technological Research, FONDECYT, to Project Nr. 1971259 on Aging Effects in Dynamic Properties of Soils, which has permitted to perform in-situ and laboratory tests in abandoned deposits of tailings. The continuous support of mining companies to research works on tailings deposits is greatly appreciated.

8 REFERENCES

Bohn, H.L., B.L. McNeal & G.A. O´Connor 1985. Soil chemistry. 2nd edition. John Wiley & Sons.

Cohen, M. & G. Moenne 1991. Tailings deposits at the El Soldado Mine Compañía Minera Disputada de Las Condes. *Proc. IX Panamerican Conference on Soil Mechanics and Foundation Engineering,, Viña del Mar, Chile.*

Daniel, D.E. & R.M. Koerner 1993. Cover systems. In D.E. Daniel (ed.), *Geotechnical practice for waste disposal*: 455-496. London: Chapman & Hall.

Dixon, R.M. 1997. Accelerating revegetation through infiltration control: Principles and practices. *Proc. Fourth International Conference on Tailings and Mine Waste, CSU, U.S.A.*

Garga, V. & J.H. Troncoso 1991. Design and evaluation of behavior of tailings dams. General Report. *Proc. IX Panamerican Conference on Soil Mechanics and Foundation Engineering, Viña del Mar, Chile.*

Lobos, C. & R. Silva 1990. Vegetation growth over an abandoned deposit of copper tailings (in spanish). Publication of Codelco Chile – Visión El Teniente.

Rusell, E.W. 1988. Russell's soil conditions and plant growth. 11th ed. London: Longman Group UK

Spichiger, A., R. Edwards & T. López 1996a. Piuquenes tailings dam rehabilitation and site closure. *Proc. International Symposium on Seismic and Environmental Aspects of Dams Design: Earth, Concrete and Tailings Dams (Vol. 1), Santiago, Chile.*

Spichiger, A., R. Feliú & A. Edwards 1996b. Los Leones tailings dam site closure project stage. *Proc. International Symposium on Seismic and Environmental Aspects of Dams Design: Earth, Concrete and Tailings Dams (Vol. 1), Santiago, Chile.*

Troncoso, J.H. 1998. New perspectives in management of operating and abandoned deposits of mineral residues. *Third International Congress on Environmental Geotechnics, Lisboa, Portugal.*

Troncoso, J.H. 1992. Fundamentals of earthquake geothecnical engineering. (in spanish). Catholic University of Chile (2nd Ed.)

Troncoso, J.H. & R. Verdugo 1985. Silt content and dynamic behavior of tailings sands. *Proc. XI International Conference on Soil Mechanics and Foundation Engineering, San Francisco, USA.*

Tailings and Mine Waste '99 © 1999 Balkema, Rotterdam, ISBN 90 5809 025 6

Open pit reclamation at the Kennecott Ridgeway gold mine

R. Dorey & M. Willow
SRK Consulting, Lakewood, Colo., USA

R. Duckett
Kennecott Ridgeway Mining Company, Columbia, S.C., USA

ABSTRACT: The Kennecott Ridgeway Gold Mine, located outside Ridgeway, South Carolina, is scheduled to complete mining and ore processing in late 1999. The project completed mining in its South Pit in 1996. As part of the overall reclamation planning for the project, methods to manage sulfide bearing waste rock and exposed sulfides in the pit walls were developed to allow reclamation of the pits as fresh water lakes. Through detailed waste classification and amendment addition to both waste rock and pit water, the pit is being backfilled using subqueous disposal to alleviate sulfide oxidation while maintaining satisfactory quality pit water. Development of methods to measure stored acidity in the waste rock and monitoring of pit water quality were required. Design of a liquid base amendment addition system and the detailed sequencing of rock placement and water management were intrinsic to the success of this approach to pit reclamation.

Since 1996, the monitoring programs have shown that a high quality water is being maintained in the South Pit.

1 INTRODUCTION

1.1 *Project Description*

Kennecott Ridgeway Mining Company (KRMC) operates a 15,000 tons-per-day open-pit gold mine located approximately 5 miles east of the town of Ridgeway, and 25 miles north of Columbia, South Carolina. The ore bodies are hosted in a sequence of metamorphosed sediments and volcanic rocks with gold mineralization associated with hydrothermal alteration zones. In general, the upper portions of the ore bodies are oxidized with the rocks occurring as a saprolite. With depth oxidation decreases and sulfide mineralization, mostly as pyrite, occurs. Since December 1988, the mine has extracted gold and silver from two open-pit deposits, the South Pit and the North Pit. The pits extend to between 300 and 400 feet in depth. In total, approximately 60 million tons of ore have been mined at a stripping ratio averaging 1:1, ore to waste. The South Pit has been mined to design depth and is currently being reclaimed. The North Pit is currently in production, with enough reserves to last through 1999.

1.2 *Reclamation Plan Overview*

Based on planned and permitted reclamation activities and the terms and conditions of the mining permit and reclamation plan, the Ridgeway mine open pits will be reclaimed as deep water lakes. The primary objective, therefore, of pit closure is to create and maintain water quality conditions that are consistent with this proposed post-closure land use.

Establishment and maintenance of good water quality conditions will primarily be focused on the control of acid generation and acid rock drainage (ARD) in the backfill and pit walls.

Acceptable water quality conditions include a neutral to alkaline system with low concentrations of dissolved metals. In the short term, control of acid generation will be accomplished by treatment of pit influent, amendment of waste rock backfill and, as necessary, treatment of pit walls, and by sequencing the backfill operations in such a way that subaqueous conditions are rapidly established. In the long term, the mitigation of acidic conditions and maintenance of water quality will be accomplished by establishing an anoxic, subaqueous environment to control sulfide oxidation and acid generation from backfill and pit walls. This concept for reclamation of the open pits has been approved by the South Carolina Department of Environmental Control (DHEC).

The pre-mining water table in the pit areas was generally within 30 to 40 feet of ground surface. Modeling of the post mining reclaimed pits shows that permanent lakes will be established with lake levels controlled by outlet structures. By routing surface runoff from the mine area to the pits it is calculated that filling of the South Pit will require approximately seven years followed by some nine years for the North Pit. Management of the water entering and filling the pits can be considered in two phases: during active mining and following reclamation of the remainder of the mine area.

During the active mining and backfilling of the pits treatment of both the waste rock and influent water will be performed. After reclamation, the management of the influent and in pit water will be performed.

2 WASTE MANAGEMENT

Backfilling of the pit with waste rock commenced when mining was completed in the South Pit and will extend through the completion of mining in the North Pit. Waste rock backfill will be subjected to field testing to determine available acidity concurrently with mining and placement in the South Pit, and will be neutralized by base amendment such that when placed in the pit water, no net impact to pit water quality will be realized. Water quality monitoring and modification of water chemistry will be conducted continuously during this period to guide backfill operations and maintain water quality.

Backfill placement and water management will be conducted in a manner that results in rapid establishment of subaqueous conditions in order to mitigate ongoing acid generation. A water balance and pit backfill schedule have been developed to support these activities.

Initial mining of the North Pit involved the development of a waste rock management area adjacent to the pit. Long term management of this waste rock area involves its remining and placement in the North Pit. In addition, waste rock used to develop haul roads and ramps which are to be removed at reclamation will be remined and used in the North Pit backfill.

At the end of reclamation, no waste rock will be left on surface with the exception of the waste rock used in the construction of the engineered fill for the projects' tailings impoundment.

2.1 *Determination of Waste Rock Acidity*

Waste derived from the pits varies in both its acid generating potential and the degree of oxidation the acid generating material has undergone. Unlike the conventional approach to defining acid generating materials based upon static, or acid base accounting, and kinetic test programs, the backfilling and inundation of the waste rock requires a measurement of the leachable acidity in the materials. That is, upon placement under water, the acidity stored within the rock that will become available to the pit water, requires definition.

By maintaining an alkaline buffered pit water chemistry and using base amendments to neutralize the acidity stored in the waste rock, water quality can thereby be maintained.

A simple field test was developed to determine base addition requirements for the waste rock placed in the South Pit concurrent with backfill . By correlating acidity measured in field tests to the results of waste rock neutralization tests, the quantity of base additive required to neutralize available acidity in the backfill was determined. The field test procedure involves:

- Placing a waste rock sample in a plastic container and submerging the sample in deionized water;
- Allowing the submerged sample to achieve an equilibrium pH;
- Decanting or draining a fixed quantity of the pore water (+/- 1 liter) after equilibration; and
- Titration of the pore water sample with a standard base solution to neutrality (pH 7.0) to determine available acidity in the waste rock.

This test is conducted on blast hole chip samples prior to blasting and excavation.

Leach tests to determine available acidity contained in rock samples derived from the North Pit walls were performed. The leaching tests indicate available acidity ranging from 0.03 to 4902 mg/kg. In general, materials in the upper levels of the pit exhibit higher available acidity than materials from lower portions of the pit.

2.2 *Estimation of Base Addition Requirements*

In order to estimate the addition requirements for various alternative base additives, a laboratory test program was performed. Locally and commercially available base additives and raw materials considered for testing included:

- Slaked lime;
- Dolomitic limestone;
- Crushed limestone;
- Lime kiln dust;
- Agricultural lime;
- Phosphate rock;
- Sodium hydroxide; and
- Trisodium phosphate dodecahydrate (TSP).

For additives such as lime or sodium hydroxide (NaOH), neutralizing reactions occur rapidly. For materials such as dolomite, limestone and phosphate rock, neutralizing reactions occur more slowly as a result of the limited solubility of the additive. The purpose of the testing was to determine minimum base addition rates that would be required for short-term waste rock neutralization.

Conditioning of the backfill only requires sufficient base to maintain alkaline to neutral conditions during backfill rather than attempts to provide a neutralizing potential equal to or greater than the acid generating potential. Once inundated, the establishment of subaqueous conditions control long-term sulfide oxidation. The test procedure for each base additive was as follows:

- A measured amount of base additive was mixed with a waste rock sample;
- The sample was submerged in deionized or distilled water. Once submerged, additional mixing did not occur so as to simulate field disposal conditions;
- The pH and conductivity was monitored until equilibrium conditions were attained;
- The amount of base additive applied and the pH and conductivity as a function of time were recorded;
- The test repeated by either increasing or decreasing the base addition such that the desired pH endpoint was achieved.

When the pH endpoint was achieved (pH 7.0 or greater), the pore water was analyzed for pH, alkalinity/acidity, sulfate, arsenic, aluminum, copper, iron, lead, manganese, and zinc.

The purpose of testing each of the these additives was to provide flexibility in the choice of waste rock amendments. Based upon reaction time, neutralizing ability and cost, the dolomitic and crushed limestone, agricultural lime and phosphate rock were rejected as amendments. The relationship of available acidity to amendment rates were then developed for slaked lime, sodium hydroxide (NaOH), kiln dust and trisodium phosphate (TSP).

2.3 *Waste Placement*

A pit filling schedule for the South Pit was developed such that inundation of waste rock backfill would occur within a period of less than four months. This is accomplished by end dumping waste rock into the pit lake from selected elevations. When the combined waste rock and pit water volume approaches the active placement level, the site of waste placement is moved to a higher elevation within the pit.

After placing approximately 50 acre-feet of water in the pit, dumping of waste rock commenced in a horizontal lift at the base of the pit. In approximately one month, the pit level reached this level and dumping operations were relocated to a higher level. The placement of waste rock will not result in a "struck level" backfill surface. Rather, the pit, at the completion of end dumping at each level, will contain a pond or lake of variable depth. As end dumping is commenced at a higher level, backfill material is placed in the pit lake and then extended over the previously placed fill.

3 WATER MANAGEMENT

Water quality during pit backfilling is controlled by monitoring and adjusting base additions to the backfill and adjustment of the pit water inflow quality. Initially, the pit water was amended to achieve an alkalinity in the range of 100 mg/L as measured as calcium carbonate. This level of alkalinity was chosen to provide some confidence that the inundation of the backfill, and runoff from areas tributary to the pit during storm events, would not result in an unacceptable decrease in pH. This is accomplished through the addition of milk-of-lime directly into the feed line supplying water to the pit. The amendment circuit is a modification of the existing NPDES water treatment circuit which formerly collected and discharged site surface waters.

To allow pit water to be amended during dry periods when runoff from the site is insufficient to allow pit filling, a recycle pump and pipeline is operated.

Following the completion of waste rock backfilling, the water pumped to the pit will continue to be amended to maintain excess alkalinity in the pit water. Periodically during pit filling, the introduction of organic matter will be performed. The general reclamation of the pit includes regrading of the pit rim. Vegetation stripped from these areas and selected harvesting of trees from around the site provides a source of the organic matter. Through natural decay of the organic matter, a reduced dissolved oxygen level can be achieved.

Oxygen scavenging compounds such as sodium sulfite may be added from time-to-time to diverted inflows at the existing water treatment facilities. These compounds will be used to reduce the oxygen concentration in the pit water, and facilitate the development of reducing conditions in the subsurface.

During inundation, some ARD will report to the pit water from the leaching of the pit walls. The period following the completion of mining and backfill, however, may require possible pit wall treatment to alleviate ARD. As a contingency measure, in-situ water treatment to maintain adequate pH and water quality may also be required.

3.1 *Base Amendment System*

Components for the modified base amendment circuit are located adjacent to the existing NPDES treatment facility. The principle component for the system is a liquid amendment agitator tank in which the milk-of-lime product is stored. A 10-inch diameter pipe transports the amended water to the South Pit. During filling operations, an 8-inch line is used to transport the water to the bottom of the pit. This minimizes re-aeration of the water as it cascades down the pit walls.

3.2 *Lake Bed Preparation*

At the completion of waste rock backfill placement, an organic rich sediment layer will be slurried into the pit. This material will be derived from the various sediment control structures constructed within the mine area. These ponds and dams have naturally developed organic sediments and in shallow areas support a variety of wetland vegetation species.

Following the settling of the sediment in the pit, the rapid development of an anoxic condition will occur in the material above and within the backfill. Continued inflow of sediment carrying runoff during and following reclamation of the areas tributary to the pit will continue to enhance the generation of the lakebed.

4 NORTH PIT RECLAMATION

Reclamation of the North Pit will involve excavation and placement of material from the North Pit Waste Rock Disposal Facility into the North Pit. This activity is scheduled to begin after the completion of mining in late 1999. Material from the ramps and surficial reclamation activities, such as the excavation of roads and the foundations of the low grade ore stockpile, will also be placed in the pit bottom.

The backfill will fill the lowest levels of the pit with approximately 150 feet of material. Material will be placed in horizontal lifts and compacted by haul traffic.

Inundation of the North Pit will commence in 2004, leaving four years between the time of backfill placement and the onset of flooding.

4.1 *Wet Cover System*

The wet cover system involves the reclamation of the surface of the backfill to create a wet cover consisting of several soil layers. The cover would be composed of approximately 6 inches of organic matter above the backfill, overlain by two to three feet of compacted, low-permeability saprolite. A one-foot thick layer of sediment and organic-rich topsoil would be placed on top of the saprolite. As for the South Pit lakebed generation, the sediment and topsoil would be obtained from the reclamation of the projects' sediment ponds. The underlying saprolite layer would be compacted to decrease permeability to allow standing water to be maintained and the composite cover would then be inundated with a shallow layer of water. A small berm at the east end of the pit bottom, adjacent to an overflow collection sump, would regulate the water depth. Vegetation would rapidly become established in a manner similar to other areas of the site. The source for these plants would be plants currently in the sediment ponds, which have already proven tolerant to the site conditions and would reestablish themselves above the cover. The effect of the wet cover would be to reduce the influx of oxygen to the underlying waste.

Sources of water inflow to the pit include precipitation, some of which would fall directly onto the wet cover and some onto the pit walls, and groundwater inflow. During precipitation events, excess water would overtop the berm and flow into the sump, the floor of which would be composed of inert, fine grained waste rock allowing percolation of water into the backfill. Flow to the sump and groundwater inflow would be allowed to collect in the bottom of the pit and begin to flood the backfill. The water level in the waste would be monitored to provide that it does not reach the saprolite layer. Evaluation of the water tributary to the pit indicates that sufficient void space will exist in the backfill to absorb the water during the period prior to flooding of the pit. Should the water level become too high, excess water would be pumped to the South Pit through the current base amendment treatment system.

In addition to controlling sulfide oxidation within the waste, the constructed cover and wetland is intended to form an organic layer that will produce an anaerobic lakebed in which reducing conditions will help control potential oxidation products leaching from the waste material and entering the water column once the pit is flooded.

The third objective of the wet cover is to provide some degree of treatment of pit wall runoff that will collect above the saprolite cover. Due to the reactive nature of portions of the pit walls, this runoff water is expected to exhibit a depressed pH and contain elevated dissolved metals concentrations.

5 POST RECLAMATION WATER MANAGEMENT

As discussed above, runoff from the mine area is anticipated to fill the South Pit to approximately its post reclamation level by year 2004. To limit the potential of a release from the pit, and to allow the pit lake chemistry to equilibriate, a level ten feet below the outlet will be maintained. From 2004 through 2013, surface runoff from the then reclaimed mine area will be routed to the North Pit. Through active pit water management, the North Pit lake will be maintained in an alkaline condition.

Once full, the North Pit will be allowed to release water through a series of wetland areas to the South Pit. Surface runoff would then also be redirected to the South Pit lake. After approximately one year, the South Pit lake would reach its outlet level and release of water to the receiving stream would occur. The water balance and water chemistry were modeled using a mass balance approach with an annual analysis of equilibrium water chemistry using the PHREEQ-C model. In this model, the flow rate and water chemistry of water sources in the mine area were mixed in both the South and North Pit lakes. Finally, the release from the North Pit lake was included in the South Pit model and the ultimate water quality of the water released to the receiving stream modeled.

Evaluation of the post reclamation pit lake limnology indicate that a stable stratified, meromictic condition will exist.

The results of the post reclamation model show the ultimate water to be of extremely high quality. Elevated calcium and sulfate are predicted while heavy metals are predicted to be significantly below anticipated aquatic criteria for the receiving stream.

6 OPERATIONAL MONITORING

Since 1996, both the waste rock backfill and the South Pit water quality have been extensively monitored. Initial monitoring of stored acidity in the waste rock indicated that insufficient oxidation was occurring in the freshly mined material from the lower levels of the North Pit to warrant amendment of that material. Localized area of waste from exposed pit walls did indicate amendment was warranted. The waste monitoring program was accordingly modified to reduce testing of all but exposed pit wall materials.

The alkalinity of the pit water has been maintained throughout the filling period. Several extreme storm events have occurred at the site with Hurricane Earl producing approximately eight inches of rain in September, 1998. These events do depress the alkalinity and occasionally the pH to neutral levels, however, the restoration of acceptable levels of alkalinity has been relatively easy to achieve.

Water quality has been maintained at better than drinking water standards with only minor occurrences of zinc, copper and iron at parts per billion levels.

In the summer of 1998, the emergence of algae in the pit water was observed suggesting the continued development of a viable fresh water lake environment.

Tailings and Mine Waste '99 © 1999 Balkema, Rotterdam, ISBN 90 5809 025 6

Demonstration of metals-fixation technologies on the Coeur d'Alene River, Idaho

B.C.Williams & R.E.Nimmer
Idaho Water Resources Research Institute, University of Idaho, Moscow, Idaho, USA
S.L.McGeehan
Soil Science Division, University of Idaho, Moscow, Idaho, USA

ABSTRACT: A field demonstration program is ongoing to compare *in situ* metals-fixation and metals-removal technologies on the Coeur d'Alene River. The river banks and upland materials consist of fluvially-deposited mine wastes and natural sediments. Metal loading to the river results from erosion of the banks as well as seep discharge of pore waters from the river banks. The demonstration technologies solicited for this project are directed to address near-bank ground water that discharges to the river. The technologies demonstrated in the field include encapsulation with alumino-silicates, fixation by apatite mineralization, and phytobioremediation. The field test plots will be evaluated according to several performance factors, including (a) reduction in metals leachability, (b) soil fertility, (c) chemical stability of the final phase, (d) cost per volume or area, (e) operation and maintenance requirements, (f) performance or stability in freezing weather, and (g) potential risk to riparian ecosystems.

1 INTRODUCTION

The U.S. Environmental Protection Agency and the Idaho Water Resources Research Institute (IWRRI) invited vendors and investigators to participate in field demonstration of metals-fixation and metals-removal technologies in sediments mixed with tailings along the Lower Coeur d'Alene River, Idaho. Technologies were invited tó be *in situ* or *ex situ*, and were required to be compatible with riparian ecosystems. The emphasis of the demonstration program is for participants to perform field-scale demonstrations on site; however the program was structured so those participants could perform bench tests using site materials prior to mounting their field-scale demonstration. Potential entrants were informed that no financial support would be provided by EPA, but that there were no restrictions on how participants funded their work from other sources.

IWRRI staff identified and characterized a location within the Coeur d'Alene River Basin where representative mine waste materials and trace-element enriched sediments were located. An invitation to participate was circulated to more than 170 people, and a RFP was provided to interested parties. Interestingly, the majority of the respondents were teams that learned of the program through word of mouth. Respondents to the RFP were asked to provide information including description of process, the metals fixed or stabilized, the long-term stability of the treated phase, whether the process requires excavation, a letter of reference if process was proprietary, cost per area or volume, and the largest scale at which the technology had previously been applied. Respondents were also required to inform IWRRI of all chemicals used so that they could be checked for risk to the environment, and in the case of pytoremediation, plant lists so that county agents could review them in terms of weed propagation.

Composite samples were collected by IWRRI for demonstrators to bench-test their technology using site-specific materials. The composite samples treated by demonstrators were analyzed by IWRRI for leachability using the Toxic Characteristic Leaching Procedure (TCLP, EPA, Method 1311) with attention to lead, cadmium, zinc and arsenic; as well as review of fertility with respect to the parameters of pH, total nitrogen, phosphorus, potassium and organic matter. The constituents lead, zinc and cadmium are target metals in the Idaho Division of Environmental Quality assessment of Coeur d'Alene River for Total Maximum Daily Loads (TMDL's). The Coeur d'Alene River has designated uses in the Idaho water quality standards (IDAPA 16.01.02110,01.ee) of agricultural water supply, cold water biota, primary and secondary contact recreation with salmonid spawning protected for the future (IDEQ, 1998).

At the time of this writing (September, 1998) demonstrators are launching their field demonstrations. After they have applied their treatment or technology, reporting and field assessment will be provided by IWRRI and also by the demonstrators, according to the protocols described in this paper.

2 PHYSICAL SETTING

In the late 1800's and early 1900's, the Lower Coeur d'Alene River received tens of millions of metric tons of mine wastes that are now intermixed with sediments. The mine wastes contain zinc, lead, cadmium, and other metals from mine waste and tailings that were produced by numerous mines in the Silver Valley. Until the 1920's, sediments containing mine waste were discharged directly into the South Fork of the Coeur d'Alene River and its tributaries. Below the confluence of the North and South Forks, the gradient flattens significantly, which has caused contaminated sediments to be deposited in the bed and banks of the river in this area.

As described by IDEQ (1998), contamination to the Coeur d'Alene River continues by erosion of the river bank and floodplain, and/or metals release from sediments in the river bed, bank, and floodplain. There are three main mechanisms contributing to this problem:

1. Erosion has accelerated since the late 1960's due to changes in river energy. Prior to the 1920's mine tailings were dumped into the South Fork of the Coeur d'Alene River which increased the sediment load. The energy of the river was sufficient to carry the bed load; hence, the sediments were deposited in the bed, along the bank, and on the floodplain. Since the late 1960's point source controls for sediment discharge were implemented. The reduced bedload has caused an increase in the river's energy that causes greater erosion rates. Based on aerial photographs, an estimation of the river bank erosion rate is six inches per year.

2. Water level management and boat traffic increase erosion rates of the river bank; therefore, increasing the metals loading to the river. The water level of the river is lowered after September. The bank now above the low water mark has little to no vegetation and a small growing season window. Boat wakes erode this zone more rapidly causing the bank to collapse into the river. Boats have also become larger, faster, and more numerous in the river.

3. Metals-contaminated ground water seeps out of the river bank and into the river. When the water level is low (fall and winter months) this ground water comes in contact with the oxygenated air and some of the metals precipitate along the bank. During the high water period (spring and summer) these precipitates are then washed into the river, delivering a delayed colloidal load to the river system.

The Coeur d'Alene River flows from the confluence of the North and South Forks of the Coeur d'Alene River near Enaville, Idaho, westward to its discharge to Coeur d'Alene Lake near Harrison, Idaho. The reach of interest for the demonstration projects is the low-gradient reach from Kingston to Harrison. The river bank is subject to water level fluctuations of 8 feet or more each year due to seasonal changes in runoff, and management of the Post Falls Dam. The river banks consist of fluvially-deposited mine wastes and natural sediments. Metal loading to the river results from erosion of the banks, as well as seep discharge of pore waters from the banks as the water level lowers. The demonstration projects solicited for this program are focused on the second of these mechanisms, that is, bank storage water that discharges to the river.

The site that was selected for these demonstration projects is on the Cataldo Mission Flats, on land owned by the Mine Owners' Association just west of the Cataldo Mission State Park on Interstate-90 between Cataldo and Rose Lake, Idaho. In addition to hosting fluvially-deposited materials, this land also received dredge spoils removed from the river by the Mine Owners' Association from 1929 through the 1960's to prevent flooding of the lower river. The study area is in upland floodplain materials, approximately 20-30 feet back from the high water mark. Soils at the site are level to nearly level, very deep and poorly to somewhat poorly drained. A high water table is present in the Spring. Slickens, Pywell, and Fluvaquentic Arents (SCS, 1994) are the most extensive soils at the site. The Slickens soils are derived from coarse to medium-textured ground rock that has commonly undergone chemical treatment during the ore-milling process. Significant areas of the Slickens soils are devoid of vegetation due to phytotoxic metal concentrations. The Slickens soils are susceptible to severe water erosion and channelization due to occasional flooding. The Pywell series formed in organic materials and occupy the lower landscape positions. Flooding hazard is frequent for long periods of time with the potential for channelization. The Fluvaquentic Arents are fine textured soils that have been disturbed by mining activities. These soils are found on stream terraces. This landscape position results in a low flooding hazard but a high potential for wind erosion.

3 TECHNOLOGIES BEING DEMONSTRATED

At the time of this writing, four demonstrators (Table 1) have fulfilled the requirements to perform field tests this field season. The technologies being demonstrated fall into two general categories (a) excavation, mixing of reagents for chemical stabilization followed by re-emplacement; and (b) phytoremediation. The Frutchey/UFA team is employing a mixture of the two approaches by tilling the apatite treatment into the soil and then revegetating with redtop grass (*Agrostis alba*) which has proven resistant to metal toxicity in the Coeur d'Alene River valley.

Table 1. List of demonstrators that are performing field tests during the 1998 field season.

Demonstrator(s)	Description of Technology
KEECO, Inc., Lynwood, WA with, Metcalf & Eddy, Santa Barbara, CA	Chemical microencapsulation using calcium silicate
Frutchey, Cataldo, ID with, UFA Ventures, Inc., Richland, WA	Revegetation by redtop grass with phosphate-induced metal stabilization (PIMS) using natural apatite
ICF Kaiser, Seattle, WA with, Forrester Environmental, Hampton, NH	Metals fixation by nucleation and coprecipitation during *in situ* apatite crystallization
Dr. Valerie Chamberlain, Professor, Geology University of Idaho, Moscow, Idaho with, Bill Owens, Graduate Student, Geography University of Idaho, Moscow, Idaho	Phytobioremediation using selected plant mixtures

4 BENCH TESTS

IWRRI prepared composite samples for demonstrators so that each participant could develop and refine site-specific metals-fixation strategies prior to the field phase of the project. Subsamples, collected at 9 random locations across the demonstration site, were thoroughly mixed to form the composite. Samples for Composite #1 were collected from the 0-12" depth while Composite #2 represents the 12-24" depth. Composite samples were assessed for As, Cd, Pb, and Zn by IWRRI using a modified[1] Toxic Characteristic Leaching Procedure (TCLP, EPA Method 1311) prior to and after bench treatment by each investigator.

Composite Sample #1 exhibited low metals concentrations in the untreated sample reflecting low contamination in the surface 12 inches (Table 2). Each demonstrator providing a treated sample achieved a reduction in TCLP-extractable Cd and Pb relative to the untreated sample. Variable results were obtained for Zn. The Chamberlain/Owens team has not yet returned a treated composite sample due to the time required for phytoremediation. Arsenic data are not reported because all concentrations were below instrument detection limits.

Initial metal concentrations in Composite Sample #2, collected from the 12 to 24 in. depth, were much higher with Pb exceeding the TCLP regulatory level. Each participant was again able to decrease the extractable metal concentrations with the most significant changes observed for extractable Pb and Zn. In particular, final Pb values were below the TCLP regulatory level in two of the three remediation processes.

Soil pH did not change significantly for either of the apatite-based technologies (Table 2). The aluminosilicate method increased pH from 6.8 to 11.8 - a value that could present problems for plant growth once the site is revegetated. Neither the apatite- nor the aluminosilicate-based technologies adversely influenced other soil fertility parameters such as macronutrient availability and organic matter content (data not shown).

Table 2. TCLP-extractable metals and pH values for bench-tested composite samples.

	Composite Sample #1 (0-12 in.)				Composite Sample #2 (12-24")			
	TCLP-Extractable (mg/L)				TCLP-Extractable (mg/L)			
	Cd	Pb	Zn	pH	Cd	Pb	Zn	pH
Untreated	0.14	0.74	6.9	6.8	0.33	43.8	51.8	6.8
TCLP levels	1.0	5.0	NA	NA	1.0	5.0	NA	NA
KEECO/M&E	0.13	0.53	3.7	9.8	0.27	13.6	22.0	11.8
Frutchey/UFA	0.11	0.49	7.4	7.0	0.19	1.6	21.9	7.0
ICF Kaiser/FE	NA	NA	NA	NA	0.17	0.3	24.8	7.5

5 SPATIAL VARIABILITY OF METALS IN FIELD PLOTS

Because extreme spatial variability of metals in the fluvially-deposited sediments was expected at many scales, it was deemed imperative that IWRRI document pre-treatment variability for each plot. Documentation of the variability allowed demonstrators to plan for the full range of metals concentrations and provided plot-specific baselines for post-treatment performance assessment of each technology.

Samples were collected at 9 points within each 250 ft^2 plot (Figure 1). Each point was sampled to a depth of 24 inches in 6 inch increments (providing 24 to 36 samples per plot). Each sample was analyzed separately for modified TCLP-extractable Cd, Pb, and Zn.

---

[1] The modified TCLP extracts were obtained using EPA procedure 1311 with the following exceptions: a 5 g/100 mL soil solution ratio was used and the analytes are As, Pb, Zn, and Cd. The modified-TCLP tests were performed by the UI Soil Science Division Lab (non-EPA certified).

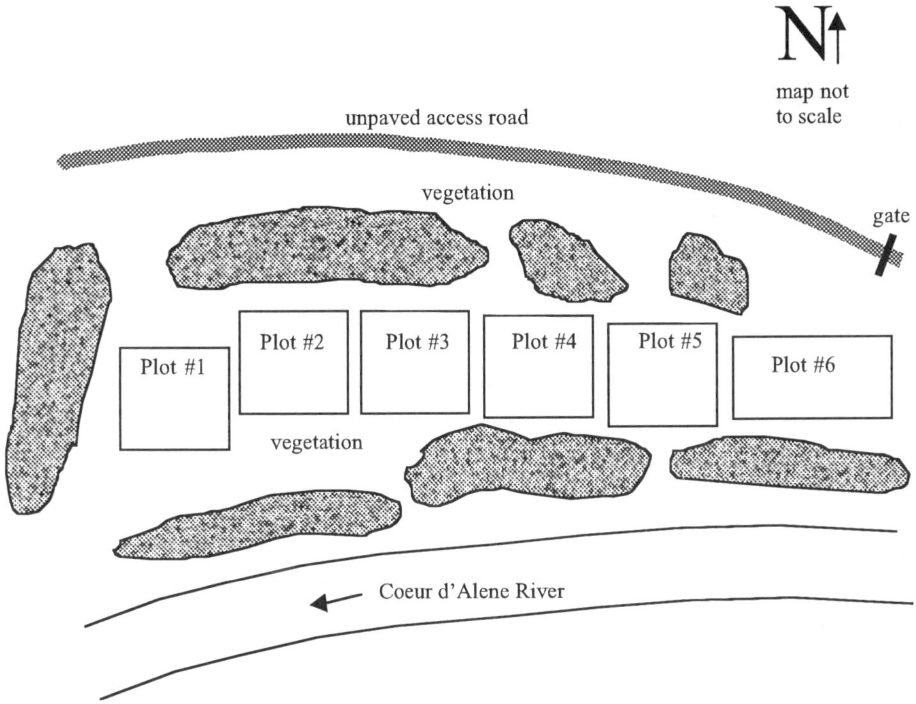

Figure 1.

Representative spatial variability data for Plot #1 and #6 are summarized in Figures 2 and 3. Significant variability in TCLP-extractable Pb and Zn was observed both within and among the plots. Cadmium concentrations were generally low and less variable (data not shown). Lead concentrations are much higher in Plot #1 than in Plot #6 at each depth. Within Plot #1, Pb concentrations increased with depth with maximum concentrations (30-50 mg/L) observed in the 18-24 in. samples (Figure 2). Zinc also increased with depth in Plot #1 exceeding 55 mg/L at the 18-24 in. depth. Overall, Plot #1 exhibited a wider range in Pb and Zn concentrations at each depth. A somewhat different distribution of Pb and Zn was observed in Plot #6. Zinc concentrations were highest in the surface 6 in. and decreased significantly with depth, while Pb concentrations were relatively low throughout the depth profile. These data clearly show the need to collect detailed spatial variability information on a small scale prior to developing metals remediation strategies for sites subject to fluvial deposition.

6 PROTOCOL FOR FINAL ASSESSMENT OF FIELD PERFORMANCE

The performance standards against which the field demonstrations of the technologies will be evaluated fall into two categories, those to be assessed by IWRRI, and those to be reported by the demonstrators.

6.1 IWRRI Review

Standard TCLP (EPA-certified) tests will be obtained for two composite samples corresponding to the first and second foot of the treated plot. In addition, a modified TCLP will be performed on the treated area at the same sample density used to assess the pre-treatment plot variability. The Synthetic Precipitation Leaching Procedure (SPLP,

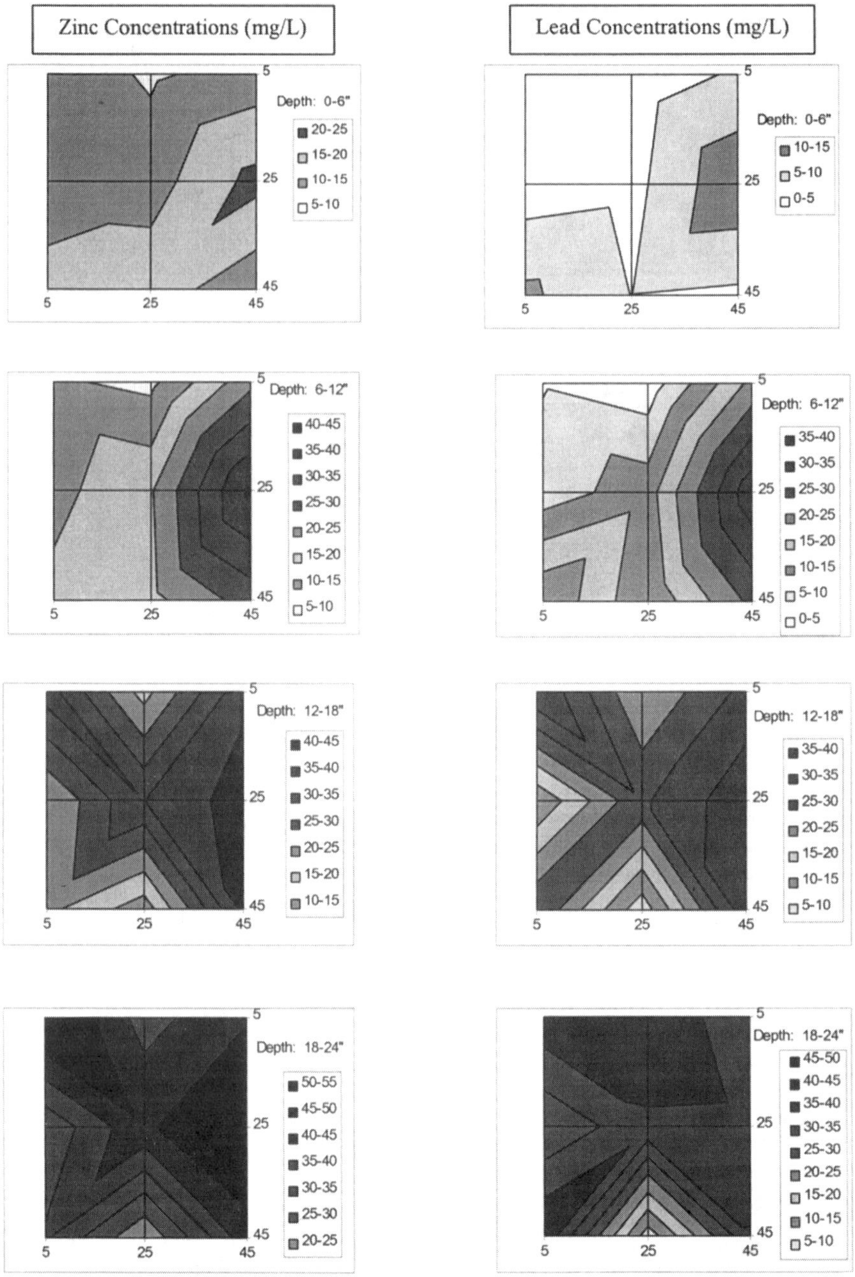

Figure 2. Plot 1 spatial distribution of metals concentrations from modified-TCLP extracts (distances in feet).

EPA, Method 1312) will be employed also because it uses a neutral extraction solution. Comparison of the pre- and post-treatment leachable metals concentration will constitute one measure of performance. A second measure will be assessment of fertility including pH, total nitrogen, phosphorus, potassium and organic matter. Results will be reported with respect to the pre-treatment range of metals found in each plot.

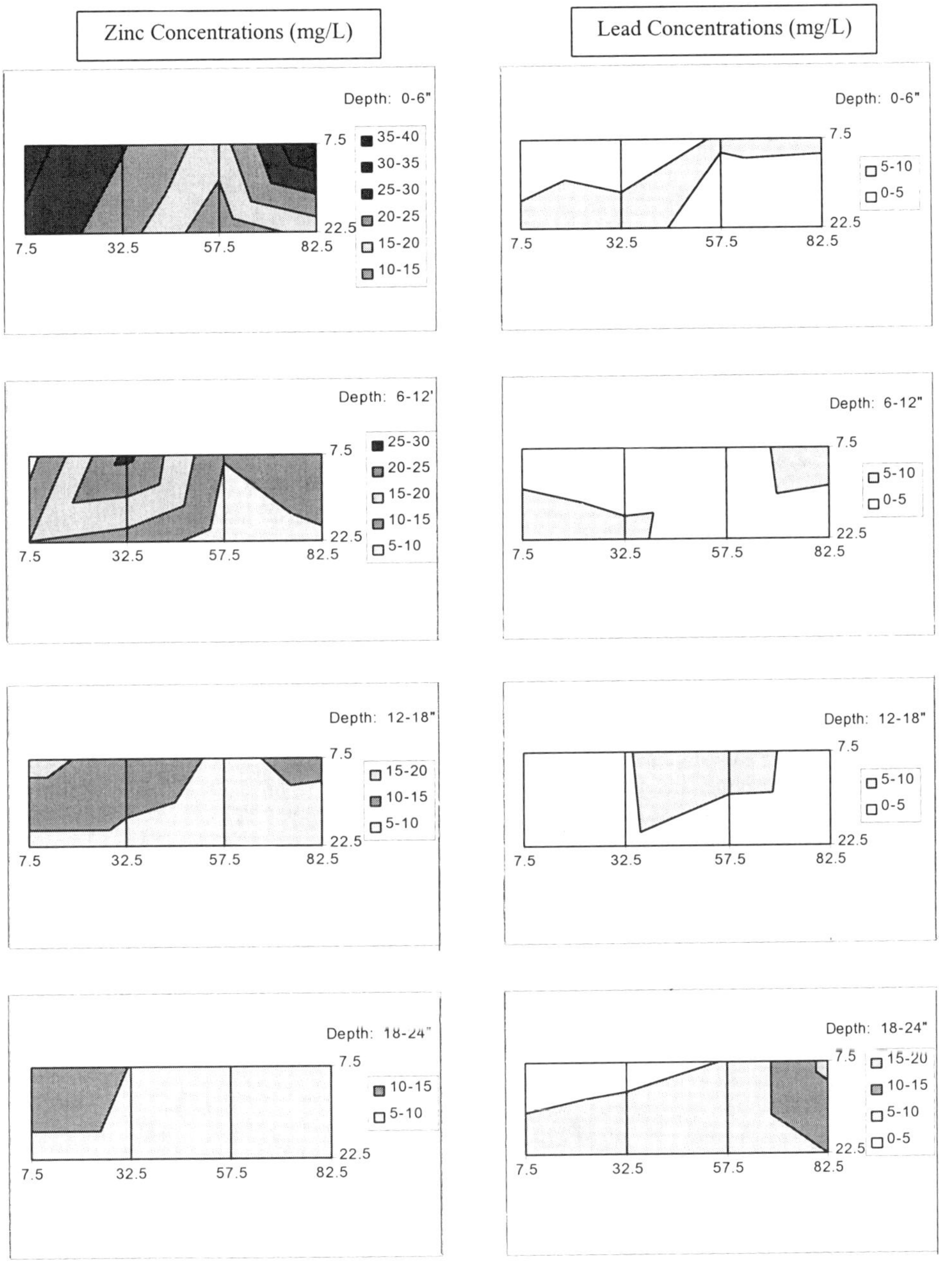

Figure 3. Plot 6 spatial distribution of metals concentrations from modified-TCLP extracts (distances in feet).

6.2 Demonstrator report

Demonstrators will be required to provide a final report that includes the following information:

- Discussion of any potential long-term degradation products from chemicals or biological components applied during treatment (beneficial, neutral, or potentially harmful).

- Cost per volume or per area on a test plot basis as well as projections for cost scaling for larger projects.
- Operation and maintenance requirements.

7 SUMMARY AND CONCLUSIONS

Demonstrators are in the process of performing field tests of chemical fixation and phytoremediation on tailings mixed with sediments in near-bank floodplain materials on the Lower Coeur d'Alene River. On the basis of IWRRI's analysis of bench tests treatments, all contributing participants were able to reduce levels of TCLP leachable cadmium, lead and zinc for Composite Sample #2 as compared to the untreated composite sample. In terms of fertility, soil pH was not changed by either of the apatite-based technologies. The aluminosilicate method increased pH from 6.8 to 11.8 - a value that could present problems for plant growth once the site is revegetated. Neither the apatite- nor the aluminosilicate-based technologies adversely influenced other soil fertility parameters such as macronutrient availability and organic matter content.

Within-plot variability data have been provided to demonstrators for their use during plot treatment. Between-plot variability, although within one order of magnitude, is sufficient such that it will require site-specific consideration of baseline concentrations when reporting individual technology performance.

DISCLAIMER

The information in this document has been funded in part by the U.S. Environmental Protection Agency under cooperative agreement number X-990964. Mention of certain technologies, trade names, or commercial products does not constitute endorsement or recommendation for use.

ACKNOWLEDGEMENTS

Financial support for administration of this project was provided by the Environmental Protection Agency. Input received from Earl Liverman and Nick Ceto of EPA, Region 10, during the inception and duration of this project is greatly appreciated. The work of Roy Mink, Dan Kealy, Kathy Canfield-Davis and Peggy Hammel of IWRRI was also vital. In addition, we thank the Mine Owners' Association for providing the land required for the demonstration projects. Finally, we would like to acknowledge the efforts of the demonstrators themselves, who provided their time and technology, and helped us to refine the program design.

REFERENCES

IDEQ, 1998. *Coeur d'Alene River water quality assessment and total maximum daily toad to address trace (heavy) metals criteria exceedences.* Idaho Division of Environmental Quality, Northern Idaho Regional Office, Coeur d'Alene, ID.

SCS, 1994. *Coeur d'Alene River cooperative river basin study.* U.S. Department of Agriculture – Soil Conservation Service, Boise, ID.

Tailings and Mine Waste '99 © 1999 Balkema, Rotterdam, ISBN 90 5809 025 6

Mine waste and water management at the Upper Blackfoot Mining Complex

R.C.Anderson
Hydrometrics Incorporated, Helena, Mont., USA

B.G.Hansen
McCulley, Frick and Gilman Incorporated, Missoula, Mont., USA

ABSTRACT: The Upper Blackfoot Mining Complex (UBMC) is located at the headwaters of the Blackfoot River in Lewis & Clark County, Montana. Underground lead, zinc, and silver mining activities occurred at several mines within the UBMC, beginning near the turn of the century and terminating in the 1950s. Drainage from adits and erosion of waste rock piles within the UBMC were shown to contribute to water quality degradation in the Upper Blackfoot River. In 1993, private mining companies began implementing a systematic remediation program in the UBMC under a proactive agreement with the Montana Department of Environmental Quality (MDEQ). The program includes relocation of waste rock to a series of three engineered repositories and passive treatment of waters draining from the Mike Horse Mine. This paper describes the mine waste relocation efforts, repository design and construction, the treatment system design, and the performance of the treatment system in terms of metal removal.

1.0 INTRODUCTION

The UBMC, also known as the Heddleston District, is located in Lewis & Clark County approximately 15 miles northeast of Lincoln, Montana (Figure 1). The UBMC is located at the headwaters of the Blackfoot River, which flows to the west and enters the Clark Fork River near Missoula, Montana. The Mike Horse Mine is situated on Mike Horse Creek, a tributary of Beartrap Creek. The confluence of Beartrap Creek and Anaconda Creek comprises the beginning of the Blackfoot River. The Anaconda Mine is located adjacent to this confluence (Figure 1). The UBMC includes several distinct mining properties (e.g., the Mike Horse, Anaconda, Paymaster, Edith, and Mary P mines) that were operated at various times between the late 1800s and the early 1950s. The mining focused on silver-, lead-, and zinc-bearing ores, though minor amounts of gold were produced during the district's early years. All mining at the UBMC was underground. Parties currently implementing remediation work at the UBMC are ASARCO Incorporated (Asarco) and ARCO.

1.1 *Remediation Overview*

In 1990, environmental conditions within the UBMC were typical of abandoned, underground hard-rock mining sites in the western United States. Accumulations of waste rock and mill tailings

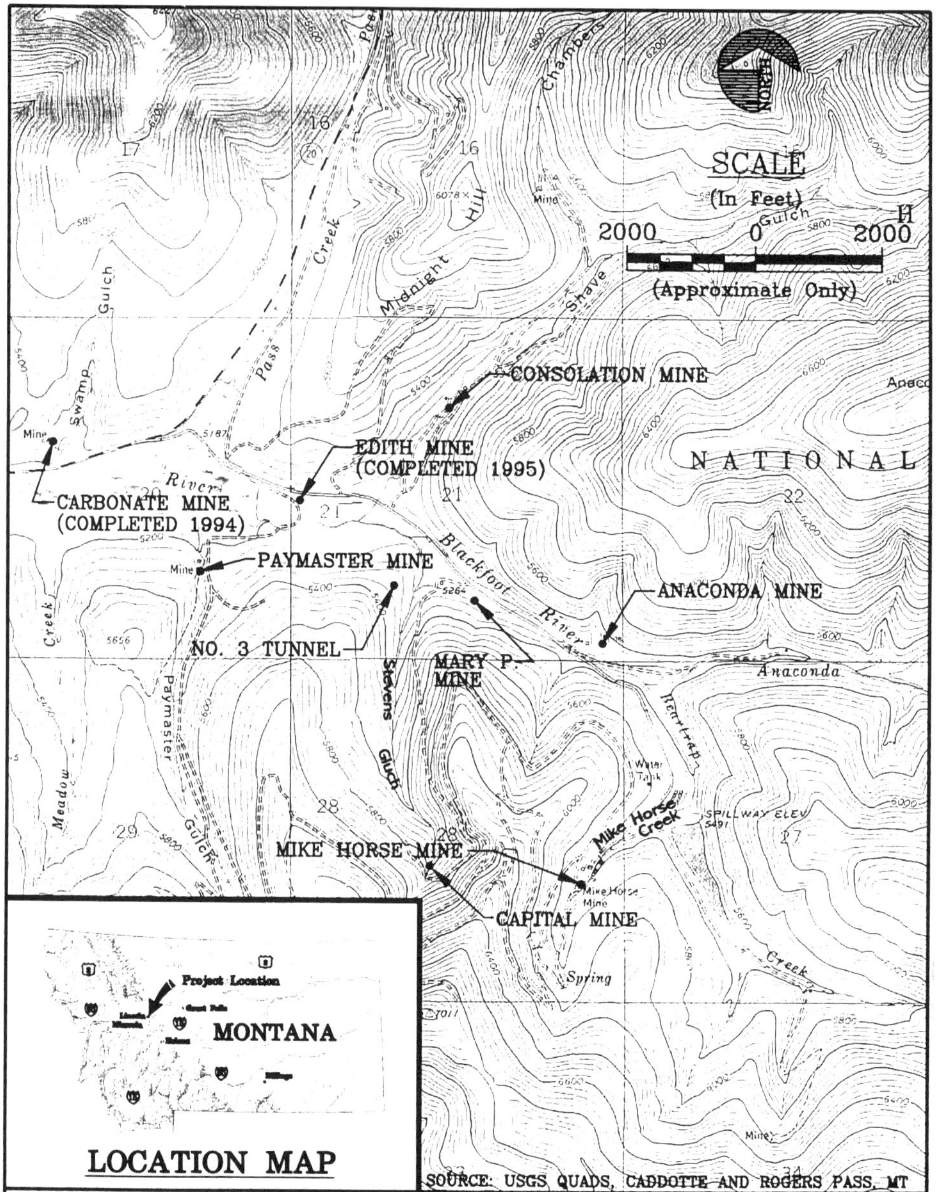

UPPER BLACKFOOT MINING COMPLEX
PROJECT LOCATION AND VICINITY MAP

FIGURE 1

UPDATE TIME: 1:00PM
013\ 0018\065\0127\HEL\090998\I:\STORAGE\01898A14.DWG

were located in stream bottoms where active erosion and dissolution of metals occurred. Acid rock drainage from several collapsed mine adits entered the Blackfoot River and its tributaries. These mining-related sources have affected the water quality of the upper Blackfoot River as described in reports by Spence (1975), Moore (1990), Hydrometrics (1992), McCulley, Frick & Gilman, Inc. (1993), PTI (1993, 1994), and Menges (1997).

In 1993, Asarco and ARCO began a comprehensive remedial program to address environmental issues within the UBMC. These activities have been conducted proactively and in cooperation with the MDEQ. Acid rock drainage from the 300-level adit of the Mike Horse Mine was identified as the single most dominant source of metal loading to the Blackfoot River. Discharge from this single source accounted for more than 90 percent of the mining-related loading of metals to the headwaters of the river. Thus, the remedial efforts have focused on decreasing metal loadings from this source. Lesser sources of metal loading to the Blackfoot River included (1) drainage from the Anaconda Mine adit and shaft; (2) leaching from mine wastes at the Carbonate, Anaconda, Edith, Paymaster, No. 3 Tunnel, and Mike Horse mines; and (3) drainage from the Paymaster Mine (see locations on Figure 1).

The remedial activities commenced in 1993 with the removal of waste rock and tailings from Swamp Gulch at the Carbonate Mine site, adjacent to State Highway 200. Construction of a passive treatment system for the Mike Horse Mine drainage commenced in 1994 and was substantially completed by 1996. Removal of waste rock from the Anaconda and Edith mine areas was implemented concurrently with construction of the treatment system. Removal of waste rock from the Paymaster Mine site was implemented in 1996.

2.0 MINE WASTE MANAGEMENT

Large volumes of mine waste rock and tailings are present within the UBMC. Waste rock piles associated with the Carbonate, Anaconda, Edith, and Paymaster mines were removed and placed in engineered repositories to limit leaching and subsequent mobilization of metals. An accumulation of mill tailings was also removed at the Carbonate Mine.

2.1 *Carbonate Mine*

The Carbonate Mine was selected for initial remedial actions at the UBMC because of its visible location adjacent to State Highway 200 (see Figure 1) and because waste rock and mill tailings were in direct contact with water in Swamp Gulch. The remedial approach entailed removal of approximately 11,500 m^3 of waste rock and tailings from the drainage and placement of these materials in a repository located a few hundred feet uphill at the upper Carbonate waste rock pile.

Quicklime was used to stabilize the relocated materials, which were saturated and non-compactable. The repository design includes a top cover consisting of a geosynthetic clay liner (GCL) with a 30-cm-thick vegetated soil cover. The repository sides are sloped at 2:1 and include a 15-cm-thick gravel drainage layer covered with 30 cm of soil. The final repository covers an area of approximately 1 ha. A portion of the repository covers a mine shaft, which was closed using a concrete plug prior to repository construction. Runon control ditches with riprap-lined outfalls protect the repository from hillside runoff flows.

The lower Carbonate area was backfilled using borrow gravel, and the area was regraded to re-establish a wetland within Swamp Gulch. The backfilled area was planted with Carex (spp.) and willows, and a new culvert extension was installed to facilitate drainage under Highway 200 to the Blackfoot River.

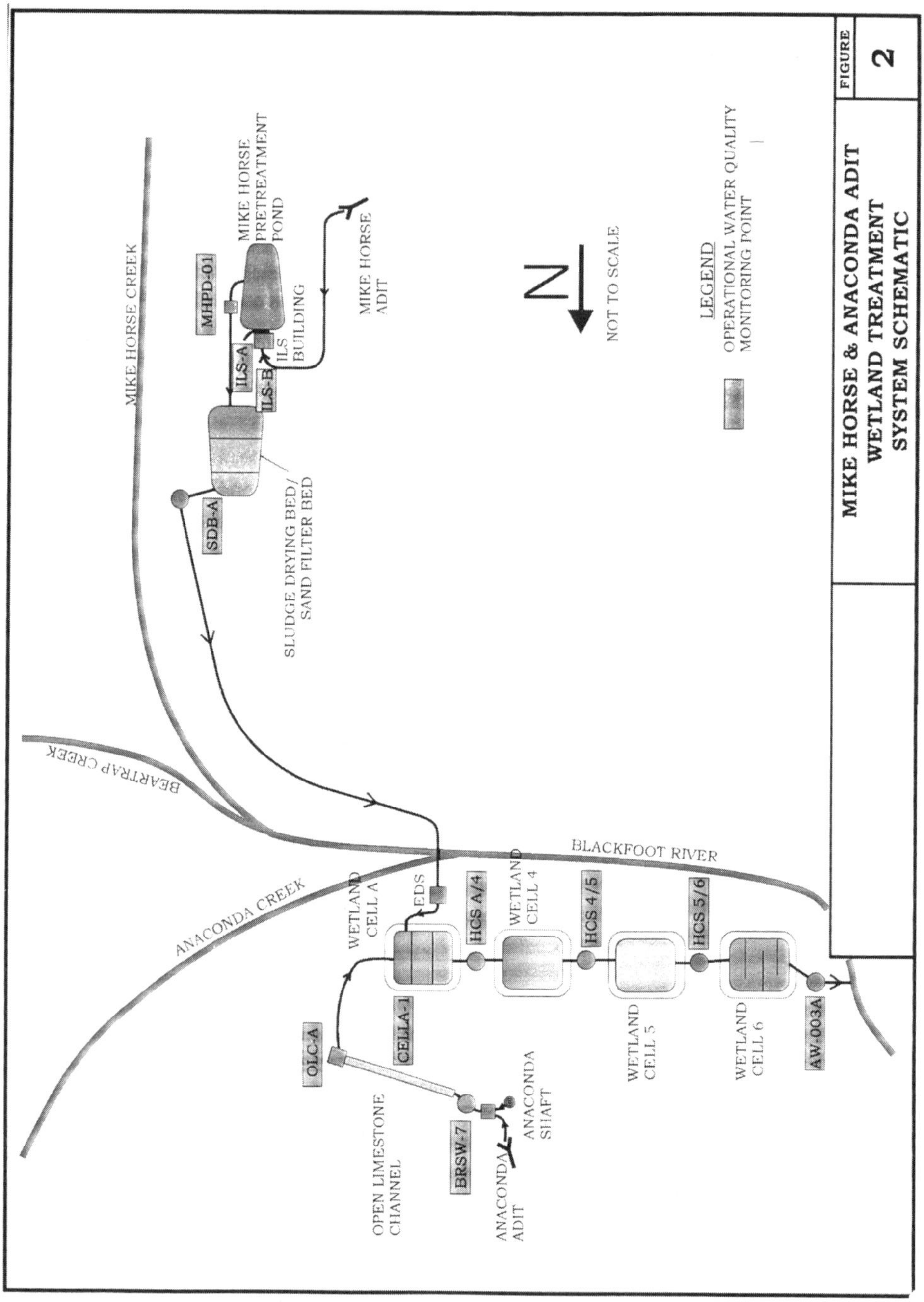

UPDATE TIME 2:40 PM
080\0018\0037\HEL\090998\I:\STORAGE\COREL\01898A04.CDR

2.2 *Anaconda and Edith Mines*

Waste rock was removed from the Anaconda and Edith mine areas and placed in an engineered repository adjacent to the Mike Horse Mine 300-level adit. Removal of waste rock from the Anaconda Mine area provided the associated benefit of rendering a relatively large area of flat land usable for construction of a wetland treatment system to treat discharge from the Anaconda and polish pretreated drainage from the Mike Horse Mine. Waste rock at the Edith Mine area actually comprised bulk ore samples collected from the No. 3 Tunnel by the Anaconda Company in the 1960s. Overall, approximately 29,700 m^3 of waste material was placed in the Mike Horse repository during 1994 and 1995. The majority of this material (approximately 25,700 m^3) originated from the upper and lower waste rock piles at the Anaconda Mine.

Waste material was compacted in lifts as it was placed in the Mike Horse repository. Design components of the repository include: (1) a limestone drainage layer at the repository base; (2) a subsurface water collection and drainage system; (3) a 45-cm sublayer of mine waste amended with cement kiln dust to limit acid production; (4) a 30-cm growth medium cap with runoff control berms and vegetated surfaces; and 5) a synthetic liner on the upper, flat surface. The Mike Horse repository was designed and constructed to accommodate the placement of a sludge dewatering/sand filtration bed system on its top surface. This bed comprises a component of the pretreatment system for the Mike Horse Mine drainage.

2.3 *Paymaster Mine and No. 3 Tunnel*

Waste rock removal was implemented at the Paymaster Mine and No. 3 Tunnel areas in 1996. This material was placed in a third engineered repository located near the Paymaster Mine. Three distinct waste rock piles, totaling approximately 6,200 m^3, were present in Paymaster Gulch and corresponded to adits associated with the Paymaster mine.

Per an arrangement with MDEQ, all material placed in the Paymaster repository was fully amended with lime kiln dust or cement kiln dust at rates required to eliminate long-term acid production. Full amendment of the material was implemented to obviate the need for access restrictions and water quality monitoring at the repository. Overall, the repository was built to allow for flexibility in terms of accommodating other waste rock accumulations originating within the UBMC.

3.0 PASSIVE WATER TREATMENT SYSTEM

As previously noted, drainage from the Mike Horse Mine was identified as the dominant source of metal loading to the Blackfoot River within the UBMC. An innovative passive treatment system was designed and constructed to address the commingled drainage from the Mike Horse and Anaconda mines. This system was substantially completed in 1996 and went on-line in October of that year. Key considerations in the design and operation of this system are: (1) the lack of infrastructure (e.g., electrical power) within the UBMC and (2) maintenance of the rural and undeveloped nature of the area. The characteristics of the Mike Horse Mine drainage are discussed in Section 4.

A schematic of the Mike Horse/Anaconda treatment system is presented in Figure 2. The system design focuses on initial pretreatment of the Mike Horse drainage to increase pH and remove iron followed by commingling with the Anaconda drainage and polishing in subsurface flow, anaerobic constructed wetlands to polish trace metals (e.g. zinc) in the drainage. Discharge from the treatment system enters the Blackfoot River and is permitted under the Montana Pollutant Discharge Elimination System (MPDES).

Pretreatment of the mine drainage begins with an anoxic limestone drain (ALD) emplaced within the Mike Horse 300-level adit. The ALD consists of approximately 120 m^3 of limestone (15 to 30 cm fragments) through which the drainage must flow prior to exiting the mine. A concrete mine adit plug with flow-through pipe was installed downstream of the ALD near the adit portal to impound waters within the mine and allow for attenuation of spring peak flows. On exiting the ALD through the adit plug, the mine drainage is directed to an in-line oxidation system, which utilizes hydraulic pressure within the mine to passively drive two jet (venturi) pumps. Passage of the mine drainage through the jet pumps aerates the water to promote oxidation of ferrous iron. The aerated drainage is introduced to a lined oxidation/precipitation pond (approximate volume 2,300 m^3) where iron flocs are allowed to settle. The decanted drainage is directed to a sand filter bed to further remove suspended iron flocs, and the drainage is then directed down a 10 cm diameter, 1.6 km pipeline to the constructed wetland treatment system.

The lined constructed wetland cells cover an area of approximately 0.8 ha and are designed to polish remaining trace metals (e.g., zinc) in the Mike Horse Mine drainage. The initial cell (Cell A) of the wetland consists of (1) a small, lined precipitation pond, (2) a downward-flowing sand filtration bed, and (3) a subsurface flow gravel wetland. Upon issuing from an open limestone channel (OLC), drainage from the Anaconda Mine (approximately 10 L/min) is commingled with the Mike Horse Mine drainage in this initial cell. A primary purpose of the initial cell is further iron removal, given that iron flocs may tend to plug the subsurface flow gravel wetlands.

The commingled drainage within the initial cell is collected in perforated piping, passed through a hydraulic control structure with adjustable weir, and then introduced into the first of two larger subsurface flow, anaerobic gravel wetlands (Cells 4 and 5). An adjustable hydraulic control structure separates these cells as well. The purpose of the structures is to allow for hydraulic gradient manipulations within the wetlands, as may be needed to account for long-term decreases in porosity and hydraulic conductivity potentially associated with metals retention and plant root-mass proliferation. Each of the subsurface flow wetlands is planted with Carex (spp.).

Drainage from the lower subsurface flow wetland cell is directed to a free-water, aerobic wetland cell, also planted with Carex (spp.). The primary function of the aerobic cell is to re-oxygenate the treated waters and to evolve hydrogen sulfide gas prior to discharge to the Blackfoot River.

4.0 TREATMENT SYSTEM PERFORMANCE

As noted above, discharge water from the Mike Horse Mine contains elevated concentrations of several metals including iron, zinc, copper, cadmium and lead. Iron and zinc occur at the highest concentrations with iron ranging from 60 mg/L to 180 mg/L seasonally and zinc concentrations ranging from 20 mg/L to over 150 mg/L. The discharge water pH ranges from 5.5 to 6.0 with total $CaCO_3$ alkalinity less than 100 mg/L and total acidity ranging from 200 mg/L to 350 mg/L. Sulfate concentrations in the Mike Horse Mine discharge are typically 2,000 mg/L to 3,000 mg/L. Prior to plugging of the Mike Horse 300-level adit, the mine discharge flow ranged seasonally from approximately 1.6 L/sec to 9.5 L/sec. Since the adit was plugged in late 1995, the adit flow has been regulated between 1.6 L/sec and 4.4 L/sec. Table 1 includes a summary of Mike Horse Mine discharge water characteristics.

4.1 *Water Chemistry Trends Through Treatment System*

Operational water quality monitoring has been performed approximately monthly since the treatment system was brought on-line in October 1996. A network of monitoring sites was established to reveal treatment efficiencies for individual components of the treatment system, as well as overall treatment system performance. Monitoring parameters were selected to assess removal of major contaminants of concern in the adit discharge water, and to provide information on metals removal processes within the treatment system.

TABLE 1. Historic Mike Horse Adit Discharge Water Chemistry

Parameter	Units	May 22, 1996	October 23, 1996
Flow	liter/sec	4.40	1.95
Temperature	Centigrade	8.1	7.5
pH	S.U.	5.76	5.72
Specific Conductance	μmhos/cm	4,000	3,670
Dissolved Oxygen	mg/L	1.14	1.37
Eh	millivolts	+343	+373
Total Alkalinity as $CaCO_3$	mg/L	7.5	16
Total Acidity	mg/L	364	196
Cadmium	mg/L	0.25	0.25
Copper	mg/L	3.0	2.3
Iron	mg/L	68	96
Ferrous Iron	mg/L	66	82
Lead	mg/L	0.37	0.35
Manganese	mg/L	86	76
Zinc	mg/L	132	114

Metals removal efficiencies have varied seasonally since the constructed wetland treatment system was brought on-line in October 1996. Metals removal rates through July 1998 for iron, copper, and lead have averaged greater than 90%, while cadmium removal rates have averaged 86%. The removal rate for zinc has varied seasonally with summertime removal rates as high as 96% achieved in September 1997 and winter removal rates, when the water temperature in the wetlands drops below 1 degree centigrade, in the range of 20%.

Water chemistry trends through the treatment system on July 22, 1998 are shown in Figure 3 and monitoring sites are described in Table 2. As shown in Figure 3, dissolved oxygen and metals concentrations undergo significant changes through the individual treatment system components. Dissolved oxygen concentrations change from 2.51 mg/L at the Mike Horse Mine (site ILS-B), to 7.07 mg/L at the pretreatment pond discharge (site MHPD-01) due primarily to aeration through the jet pumps. The dissolved oxygen concentration then decreases through anaerobic wetland cells 4 and 5 to 0.69 mg/L at site HCS-4/5 and 0.30 mg/L at site HCS-5/6. Dissolved oxygen then increases to 5.61 mg/L through aerobic wetland cell 6 (site AW-003A) prior to discharge to the Blackfoot River. The pH increased from 5.45 at the Mike Horse Adit (ILS-B) to a high of 6.71 at site HCS-5/6.

TABLE 2. Operational Monitoring Sites for the Mike Horse/Anaconda Adits Water Treatment System- UBMC

Site	Description
ILS-B	Mike Horse Adit Discharge.
MHPD-01	Mike Horse Pretreatment Pond Discharge.
HCS-4/5	Hydraulic Control Structure Between Wetland Cells A and 4.
HCS-5/6	Hydraulic Control Structure Between Wetland Cells 4 and 5.
AW-003A	Discharge of Constructed Wetland Treatment System to Blackfoot River.

Site Code	pH (s.u.)	SC (µmhos/cm)	DO (mg/L)	Cadmium (mg/L)	Copper (mg/L)	Iron (mg/L)	Manganese (mg/L)	Lead (mg/L)	Zinc (mg/L)
ILS-B	5.45	3110	2.51	0.24	4.23	177	67.2	0.759	82.1
MHPD01	4.99	3060	7.07	0.24	1.41	20	65.2	0.01	82
HCS 4/5	6.54	3080	0.69	0.005	0.007	0.31	42	< 0.003	37.2
HCS 5/6	6.71	3030	0.3	na	na	na	na	na	na
AW-003A	6.06	3010	5.61	< 0.001	0.008	0.24	26.5	< 0.002	14.0

na- not analyzed

All metals tested as total recoverable concnetrations

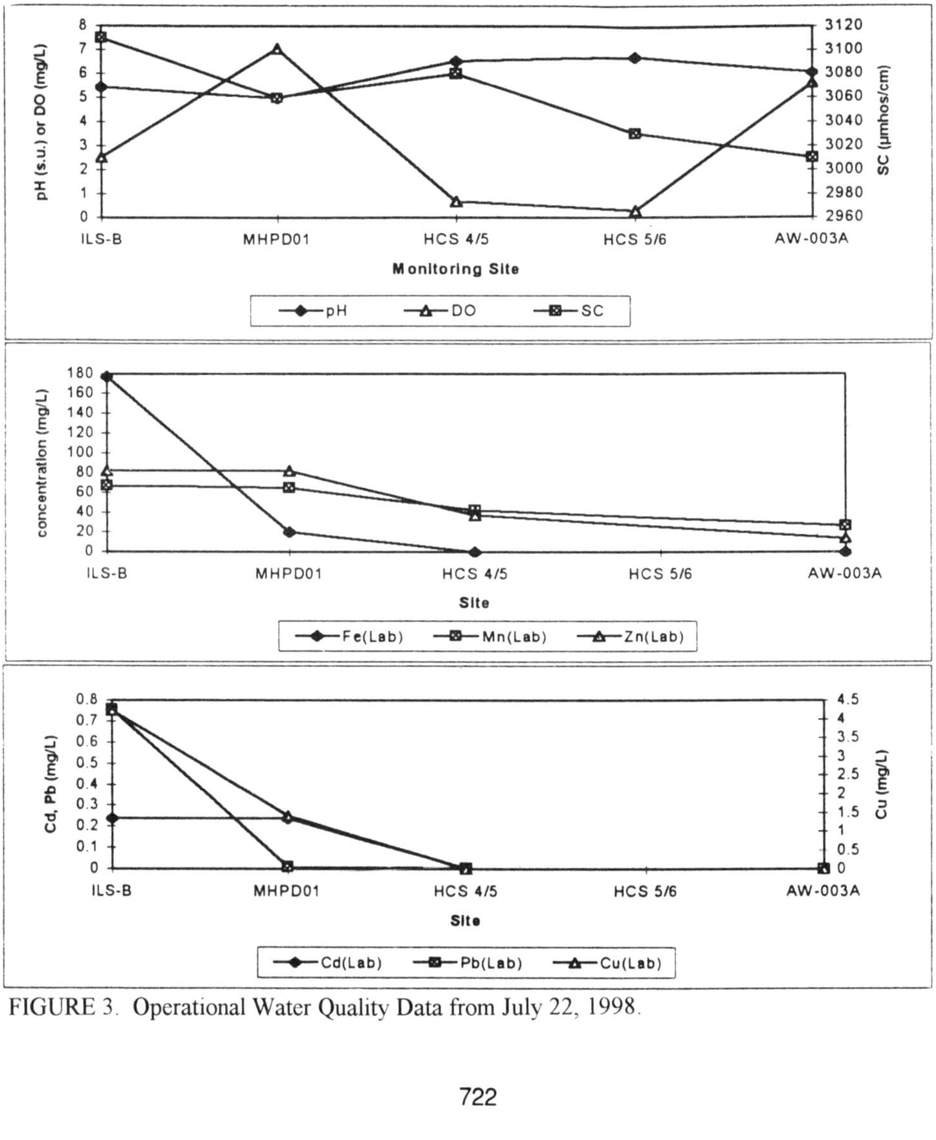

FIGURE 3. Operational Water Quality Data from July 22, 1998.

Total recoverable concentrations of iron decrease through the Mike Horse pretreatment pond from 177 mg/L at site ILS-B, to 20 mg/L at site MHPD-01. The iron concentration decreased to 0.31 mg/L at site HCS-4/5 due predominately to passage through the sand filter bed located between the pretreatment pond and the wetland cells (Figure 2). Zinc concentrations remain relatively stable through the pretreatment pond and sand filter bed. Zinc concentrations do decrease through the wetland system however, from 82 mg/L to 14.0 mg/L. This equates to a removal efficiency of 83% for zinc.

Both cadmium and lead concentrations decreased to below analytical detection limits through the treatment system on July 22, 1998. Lead concentrations decreased through the pretreatment pond from 0.76 mg/L at site ILS-B to less than 0.002 mg/L at site MHPD-01. Cadmium concentrations decrease from 0.24 mg/L at site ILS-B to less than 0.001 mg/L at site AW-003A, the constructed wetland treatment system discharge. Copper concentrations decrease from 4.23 mg/L at site ILS-B to 0.008 mg/L at site AW-003A. Sulfate concentrations on July 22, 1998 decreased from 2,570 mg/L at site ILS-B, to 2,280 mg/L at site HCS-4/5 between anaerobic wetland cells 4 and 5, to 2,220 mg/L at site AW-003a, the constructed wetlands discharge. Other data collected on July 22, 1998 includes dissolved organic carbon concentrations of 2 mg/L and 3 mg/L at sites HCS-4/5 and HCS-5/6, respectively, and 0.3 mg/L and 0.16 mg/L for sulfide at sites HCS-4/5 and HCS-5/6, respectively.

5.0 REFERENCES

Hydrometrics, Inc., 1992. Surface Water & Sediment Monitoring Program Upper Blackfoot River, Lewis & Clark County, Montana. Prepared for ASARCO Incorporated, Wallace, Idaho. June.

McCulley, Frick & Gilman, Inc., 1993. Master Plan for Remedial Actions at the Upper Blackfoot Mining Complex. Report to Asarco/ARCO. October.

Menges, J., 1997. Investigation of Temporal Changes of Heavy Metal Concentrations in Sediments and Water of the Blackfoot River, Montana. Master's thesis prepared for the Department of Geology, University of Montana, March.

Moore, J. N., 1990. Mine effluent effects on non-point source contaminants in the Blackfoot River, Montana. Prepared for Lewis and Clark City-County Health Department and the Montana Department of Health and Environmental Sciences, Water Quality Bureau, Helena, Montana. University of Montana Geology Department, Missoula, Montana.

PTI Environmental Services, 1993. Historical Data Report, Upper Blackfoot Mining Complex. Prepared for Asarco and ARCO, August.

PTI Environmental Services, 1994. Upper Blackfoot Mining Complex, Phase I Data Report. April.

Spence, L.E., 1975. Upper Blackfoot River Study, a pre-mining inventory of aquatic and wildlife resources. Montana Department of Fish and Game, Helena, Montana.

Tailings and Mine Waste '99 © 1999 Balkema, Rotterdam, ISBN 90 5809 025 6

Evaluation of wetland treatment for the removal of metals from mining impacted groundwater

D. McCarthy & R. Bullock
ARCO Environmental Remediation, L.L.C., Anaconda, Mont., USA

J. Pantano
ARCO Exploration and Production Technology, Anaconda, Mont., USA

ABSTRACT: Three pilot scale treatability tests are currently underway in Butte, Montana to test the applicability of constructed wetland technology to treat groundwater with elevated metal concentrations. The first treatability test consists of several configurations of subsurface flow constructed wetlands. The second treatability test consists of open water cells separated by highly conductive treatment walls. The third treatability test has the same configuration as the second, with the incorporation of a lime addition system. The initial results of these treatability studies are the focus of this paper.

1.0 INTRODUCTION

Mining within the Upper Clark Fork River Basin over the past 100 years has resulted in impacted surface and groundwater quality throughout portions of the basin. Underground and open pit mining operations along with urban development have resulted in elevated metal and nutrient concentrations throughout a portion of the alluvial and bedrock aquifers in the community of Butte, Montana. The Silver Bow Creek/Butte Area Superfund Site located in Butte (Silver Bow County), Montana was listed on the National Priorities List (NPL) in 1983.

The alluvial aquifer stretches along Blacktail Creek, which joins with Metro Storm Drain to form Silver Bow Creek (see Figure 1). From 1985 through 1998, many source control measures have been taken throughout the site such as removal, capping, and revegetation. In 1998, the removal of 1.8 million cubic yards of mine tailings along a one-mile length of Silver Bow Creek within the project site was completed. As part of this removal action, 1.2 million cubic yards of tailings below the water table were removed, and therefore is

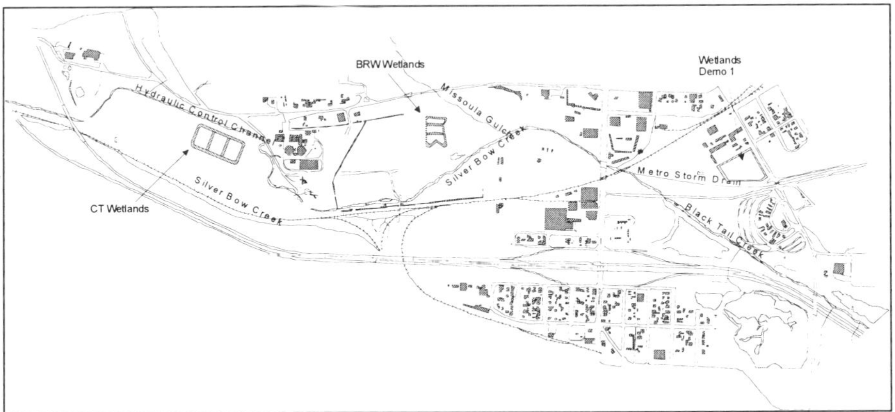

Figure 1. Treatability Study Location Map (Butte, Montana)

anticipated to result in improved alluvial groundwater quality. Two years of groundwater monitoring began in 1998 and will continue into 2000 in an effort to investigate the quantity and quality of groundwater that will be collected at the lower end of the site for eventual treatment.

Groundwater treatment is anticipated to be a large component of the final site remedy. Currently available conventional treatment methods such as lime precipitation would result in large amounts of sludge production and high operation and maintenance costs. With the limited amount of space available in and around Butte for a treatment and sludge disposal facility, additional treatment technologies were sought to treat groundwater with reduced sludge production resulting in lower operation and maintenance costs. Wetland treatment technology was selected for investigation because of its potential to minimize chemical usage and residual production through the use of natural processes.

Wetland treatment to remove metals has historically involved many different processes, including the following:

- Sulfate reduction to form metal sulfide precipitates;
- Biological modification of water pH, oxidation/reduction potential, and buffering; and
- Physical reduction of suspended metals and metal precipitates.

In addition to these processes, chemical modification of the incoming water may be utilized to initially reduce the concentrations of metals to allow the wetland biological and physical processes to complete the metal removals to the necessary levels. This type of "hybrid" wetland treatment technology is believed to result in significantly lower sludge production and O&M costs than a conventional lime precipitation treatment plant. For these reasons, it was decided to attempt a number of wetland technology treatability studies to investigate the effectiveness, implementability and relative costs of this type of treatment alternative. The initial results of these treatability studies are the focus of this paper.

2.0 OVERVIEW

Three treatability studies were developed to investigate the different wetland treatment types and configurations. The following sections give a brief background and description of each treatability study. The results are described in more detail for each treatability study in the subsequent sections.

2.1 Wetland Demonstration Project 1

Wetland Demonstration Project 1 (WDP1) was the first pilot scale treatability study. The WDP1 investigation was initiated in 1992, and WDP1 was constructed in 1996 to evaluate, compare, and document the effectiveness of differing configurations of subsurface flow wetland technology. Three different subsurface flow configurations were evaluated: horizontal flow gravel substrate, horizontal flow gravel/organic substrate, and vertical upward flow gravel/organic substrate

2.2 Wetland Demonstraticn Project 2 – Butte Reduction Works

Wetland Demonstration Project 2 – Butte Reduction Works (referred to as The Butte Reduction Works Wetlands or the BRW Wetlands) was developed in 1997 to evaluate the physical and biological processes within a surface flow wetlands without chemical addition as a comparative baseline to the Colorado Tailings Wetlands (see following section). The secondary objective was to document the effects of infiltration upon the underlying impacted alluvial aquifer. The configuration of this wetland treatment system includes three open water surfaces separated by highly conductive subsurface treatment walls.

2.3 Wetland Demonstration Project 2 – Colorado Tailings

Wetland Demonstration Project 2 – Colorado Tailings (referred to as The Colorado Tailings Wetlands or the CT Wetlands) was developed in 1997 at the same time as the BRW Wetlands to evaluate the effectiveness of surface flow wetlands with lime addition. An important objective of this treatability study is to develop/refine design components for a full-scale system, including the amount of lime addition necessary to optimize biological and physical reactions, the optimal hydraulic configuration, the physical and chemical characteristics of the sediment, and the management of accumulated sediment. The configuration of this wetland treatment system is similar to the BRW Wetlands in that it includes three open water surfaces separated by highly conductive subsurface treatment walls. The significant difference between the two systems, is that the CT Wetlands includes the addition of calcium hydroxide prior to the first cell to modify the incoming water chemistry and to precipitate a portion of the incoming metal load.

3.0 WETLAND DEMONSTRATION PROJECT 1

3.1 Background

The purpose of Wetland Demonstration Project 1 (WDP1) is to investigate the effectiveness, implementability, and cost of using subsurface wetland treatment cells for treatment of groundwater. Four types of subsurface wetlands are being investigated:

- Cell 1 – Horizontal flow, anaerobic, gravel substrate, no organics added;
- Cell 2 – Horizontal flow, anaerobic, gravel with 20% compost substrate;
- Cell 3 – Vertical flow, anaerobic, gravel with 50% compost substrate; and
- Cell 4 – Same as Cell 1, only at half-size to decrease residence time.

Influent water to the four anaerobic cells is drawn from Metro Storm Drain (MSD) near Kaw Avenue, which currently serves as an interception trench for alluvial groundwater in the area. The influent water is pumped from a sump within MSD to a constant head surge pond. The surge pond feeds each of the four subsurface flow cells through subsurface pipes.

The effluent from the two wetland cells with gravel only substrate drains to one aerobic polishing cell (Cell 6); while the effluent from the two wetland cells with gravel and compost substrate drains to another aerobic polishing cell (Cell 7). (Note that there is no Cell 5.) Both of these aerobic cells then drain to a sump, which is then pumped back into MSD. All of the cells are lined with high-density polyethylene for experimental control. (See Figure 2.)

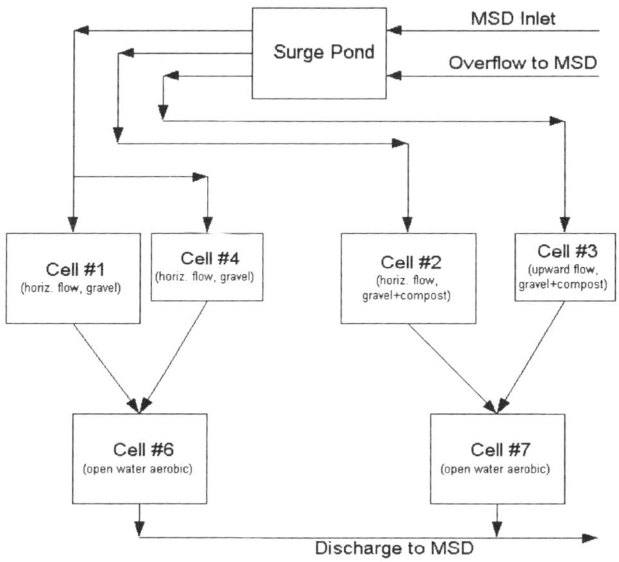

Figure 2. WDP1 Layout

3.2 *Methods*

Water chemistry, hydraulic characteristics, and vegetation are monitored periodically. The cells are designed to be sampled with instrumentation and conventional sampling techniques. Each of the anaerobic subsurface wetland cells has sampling ports which allow samples to be drawn along the length of the cells and at different depths within the cells. Flowmeters monitor the flow into and out of each cell, while a weather station records the precipitation, relative humidity, and wind conditions. Influent and effluent samples are collected on a monthly basis for analysis of pH, Eh, SC, DO, H_2S, metals, anions, and carbon. Samples from the interior of the cells are taken on a quarterly basis and analyzed for the same parameters.

Hydraulic parameters are monitored through periodic measurement of water level throughout the cells and through the use of tracer testing.

3.3 *Results and Discussion*

Water quality data from the cells indicate good removal of metals. Tables 1 and 2 show mean dissolved concentrations of arsenic, cadmium, copper, lead, and zinc for the influent and effluent of each cell.

Table 1. WDP1 Gravel Only Cells Water Quality Summary

Analyte	Mean Concentrations (dissolved)			
	Influent Conc. (ug/L)	Anaerobic Cell 1 Effluent (ug/L)	Anaerobic Cell 4 Effluent (ug/L)	Aerobic Cell 6 Effluent (ug/L)
Arsenic	6	27	13	25
Cadmium	42	1	2	1
Copper	188	13	14	15
Lead	3	3	3	3
Zinc	12176	572	3347	1516

Table 2. WDP1 Gravel+Compost Cells Water Quality Summary

Analyte	Mean Concentrations (dissolved)			
	Influent Conc. (ug/L)	Anaerobic Cell 2 Effluent (ug/L)	Anaerobic Cell 3 Effluent (ug/L)	Aerobic Cell 7 Effluent (ug/L)
Arsenic	6	23	32	32
Cadmium	42	<1	<1	<1
Copper	188	12	58	36
Lead	3	3	3	3
Zinc	12176	361	88	242

For copper, the influent water ranges from 100 to 500 μg/L. For the gravel only substrate Cells 1 and 4 effluent, copper concentrations are consistently below 10 μg/L and typically below 5 μg/L. The gravel and compost cells showed similar reductions. In both cases, the aerobic cells following the anaerobic cells resulted in no significant change to the copper concentrations.

Zinc also showed good removal from influent concentrations of approximately 12,000 μg/L down to less than 100 μg/L at times for both the gravel only and gravel+compost cells. Once again, the aerobic polishing cells did not seem to modify the concentration of zinc significantly. At times, zinc removal efficiency decreased, possibly due to freezing of the top surface of the cells which resulted in a smaller cross sectional flow area and thus a lower retention time. The data indicate that zinc appears to be the metal most easily influenced by modifications to the system.

The data indicate that the reduced conditions developed within the anaerobic wetland cells resulted in initial dissolution of arsenic from the substrate. This phenomenon appeared to decrease over time (slightly more so for the gravel only cells) to where the influent and effluent arsenic concentrations are now approximately the same.

Hydraulic conductivities within the wetland cells have been calculated since the start of the project from the measured water levels and known cell geometry. All wetland cells have shown a reduction in calculated hydraulic conductivity over time. Both of the gravel only cells began at greater than 10,000 ft/day and prior to any cell repairs were measured around 1000 ft/day. Cell 2 (horizontal flow gravel+compost) began at greater than 1000 ft/day and was recently measured around 200 ft/day. Measurements of water levels have also shown that conductivity appears to be lowest at the influent end of the cells. This may be due to deposition of suspended sediments or precipitates within the substrate; however, it is more likely due to plugging of geotextile wrapped around the influent distribution pipes. To remedy the problem, the inlet and effluent ends of each horizontal flow cell were excavated and the geotextile wrap was removed. Subsequent water level data indicate that restrictions to flow were greatly reduced upon the removal of the geotextile wrap.

Soon after the project began, flow through Cell 3 (vertical flow gravel+compost) began to slow from 5 gpm down to approximately 1 to 2 gpm. This may partially be attributed to the increased amount of organics in this cell (50% by volume); however, the water levels within the cell seem to indicate that the inlet distribution pipe, or the geotextile fabric surrounding it, may be plugged.

Flow rates through each of the four anaerobic cells were initially set at 5 gpm. After an initial period, the maximum amount of flow allowed by the system hydraulics was allowed to flow through the system. This was generally around 10 gpm for the horizontal flow cells. As described previously, Cell 3 declined continuously from 5 gpm down to around 1 to 2 gpm. Cell 2 has shown a continuously decreasing trend in flow and is currently around 5 gpm. Cells 1 and 4 at times decrease in flow, but following periodic cleaning of flowmeters and pipes, seems to remain around 10 gpm.

3.4 Conclusions/Future Investigations

In summary, data gathered over the past two years indicate significant reductions in metal concentrations but a limited ability to handle higher flow rates. Though hydraulic conductivity has been difficult to evaluate, there seems to be a decline over time in hydraulic conductivity. Data collected following the removal of the geotextile wrap indicate that most of the decline in hydraulic conductivity can be attributed to plugging of the geotextile wrap. Therefore, for the horizontal flow cells, any reduction in substrate hydraulic conductivity appears to be minimal enough to allow the cells to remain operational. Though the data is limited because the geotextile wrap could not be removed from Cell 3, the vertical upflow configuration with 50% compost substrate may be limited to low treatment flow rates due to the reduced hydraulic conductivity and difficulties with inlet distribution pipes. Though additional data is being collected and must be analyzed, this initial overview indicates that the horizontal subsurface flow systems may be appropriate to certain types of low flow water sources. Design factors and cost details are still being developed based upon the ongoing data collection and interpretation.

Future attempts at subsurface flow wetland treatment systems may benefit from some of the lessons learned during this pilot scale attempt.

- Small PVC pipes plug with biological matter and sediments over time even with periodic cleaning. Larger pipes to transmit water with frequent cleanouts, or surface channels to transfer water would result in lower maintenance costs.
- Filter fabric was placed around many of the influent and effluent distribution pipes and also between certain layers of substrate. Experience has shown that this filter fabric becomes plugged with biological matter and sediment, hindering flow or causing preferential flow paths through the substrate. Do not use filter fabric unless necessary for structural reasons.

- Nutating disc flowmeters were initially utilized to measure influent and effluent flowrates. These types of flowmeters wore quickly and required significant maintenance. Other types of non-mechanical flowmeters such as magnetic flowmeters will reduce long-term maintenance costs.
- The future engineering challenge will be to design a water distribution system with cell inlets and outlets that distribute and collect flow uniformly without plugging.

4.0 WETLAND DEMONSTRATION PROJECT 2 – BUTTE REDUCTION WORKS

4.1 Background

Wetland Demonstration Project 2 – Butte Reduction Works (the BRW Wetlands) was constructed to evaluate the effectiveness, implementability, and costs associated with an open water wetland type system. Innovations from previous open water wetlands include the use of treatment walls, which separate the open water cells. The layout of the BRW Wetlands consists of three open water cells separated by two treatment walls. The first treatment wall consists entirely of highly permeable river rock (d_{50}=6 inches). The second treatment wall consists of 90% river rock and 10% compost by volume.

The treatment walls serve as subsurface flow zones where physical, chemical, and biological processes may take place. Filtration through the permeable material should reduce the amount of suspended metals making it through each open water cell. Additionally, as vegetation establishes on the walls, they may serve as anaerobic zones for potential sulfate reduction and metal sulfide precipitation.

The source of water for the BRW Wetlands is Missoula Gulch baseflow. The majority of the Missoula Gulch drainage basin consists of urban storm sewers. The water in Missoula Gulch flows perennially, but is uncharacteristic of groundwater within the area. For this reason additional studies are underway to determine the source of the Missoula Gulch baseflow; however, it has been utilized to date as the source of water for the BRW wetlands due to its elevated metal concentrations and proximity to the site.

The Missoula Gulch baseflow enters two sedimentation/retention basins prior to entering the BRW Wetlands. The influent flow rate is typically between 50 and 150 gpm. Under dry conditions, the flow to the wetlands is reduced significantly and has resulted in complete dryout of the wetlands on one occasion for a period of weeks in early 1998 as the sedimentation/retention basins were put into service. Currently, during storm events, a control device is used to divert only a portion of the flow from Missoula Gulch into the BRW Wetlands.

Because the BRW Wetlands are elevated above the water table, they infiltrate most of the influent water. Monitoring of the open areas, groundwater wells, and piezometers are conducted periodically to investigate the interaction with groundwater and the metal removal effectiveness.

4.2 Methods

Monitoring wells and piezometers are located around the BRW Wetlands and within the treatment walls to allow measurement of water levels. Readings of Eh, pH, SC, temperature, and dissolved oxygen are taken on a daily basis for the influent, free water surfaces, and treatment walls. Weekly water quality samples are taken at each of these locations and from some of the monitoring wells. Additional water quality samples are taken on a monthly basis from additional monitoring wells. Depending upon the location of the sample, it may be analyzed for alkalinity, chlorophyll A, dissolved inorganic carbon, heterotrophic plate counts, anions (ion chromatograph), metals (ICP), non-purgeable organic carbon, and total suspended solids. Measurement of flow into the wetlands is via a 60° trapezoidal flume with continuous stage recorder.

4.3 Results and Discussion

As expected, monitoring of the open water cells indicates no discernable change in elevation between the three open areas. This indicates that the hydraulic conductivity of the treatment walls is more than adequate to convey the anticipated treatment flow rates. There has been no measurable change to the hydraulic conductivity of the treatment walls.

Measurement of water levels in nearby groundwater wells and piezometers indicate that the wetlands are serving as a groundwater mound. Such would be expected since the entire influent water source is currently infiltrated within the wetlands. Ongoing monitoring of surrounding wells will help determine the interaction of the wetlands with the alluvial groundwater system. Under a full-scale wetland scenario in Butte, infiltration of water from the wetland system would not cause significant concern since the alluvial groundwater that surrounds the wetlands would be captured and routed back into the wetlands for treatment. Under a full-scale scenario, such large infiltration losses are not expected since the wetlands will likely be located near or within the water table.

Water quality associated with the BRW Wetlands may best be evaluated by looking at influent concentrations compared to the concentration in the last free water cell. If the water were to be discharged rather than infiltrated, the water in this free water cell would most closely represent the effluent water quality.

Initial data from surrounding groundwater wells indicate a positive effect upon the underlying impacted alluvial groundwater. Monitoring results from downgradient wells show one to two orders of magnitude lower metal concentrations than upgradient wells.

Table 3. BRW Wetlands Water Quality Summary for Dec. 18, 1996 to Sept. 1, 1998

Analyte	Mean Influent Conc. (ug/L)		Mean Effluent Conc. (ug/L)		Percent Removal
	Dissolved	Acid Soluble	Dissolved	Acid Soluble	Acid Soluble
Arsenic	4.2	23.6	5.1	6.8	71%
Cadmium	39.4	45.2	25.0	25.2	44%
Copper	328.8	1,535.5	84.7	140.8	91%
Lead	3.2	235.8	2.7	10.3	96%
Zinc	11,272.9	13,645.2	6,995.2	7,858.3	42%
pH		7.6		7.9	-----

Arsenic typically enters the BRW Wetlands between 10 and 100 µg/L (acid soluble) and on the average is reduced to less than 10 µg/L. Cadmium is mainly in the dissolved phase as it enters the wetlands and a significant percentage appears to be removed through the wetlands system.

Influent copper concentrations are typically around 1000 µg/L (acid soluble) and are reduced by nearly an order of magnitude to around 100 µg/L. Dissolved copper concentrations are also typically reduced through the wetlands system. Influent dissolved copper concentrations of between 200 and 900 µg/L are reduced to around 100 µg/L.

Most of the zinc appears to be in the dissolved phase as it enters the BRW Wetlands. Influent dissolved or acid soluble concentrations of zinc are typically around 10,000 to 15,000 µg/L. Effluent dissolved or acid soluble zinc concentrations vary dramatically between 2000 µg/L and the influent concentration. Zinc appears to be the most difficult metal to remove because the solubility of zinc remains in the parts per million range until pH is above 9 to 9.5 for a period of time.

The principle mechanisms for the decreases in metal concentrations are believed to be 1) pH modification by changes in CO_2 via biological processes (photosynthesis) and abiotic processes (groundwater supersaturated with CO_2 with respect to atmosphere), 2) flocculation and settling, 3) sorption, 4) co-precipitation, and possibly 5) metal sulfide precipitation in micro-environments within the treatment walls.

4.4 Conclusions/Future Investigations

Because monitoring is still ongoing, it is premature to begin forming conclusions regarding this data. However, additional data is currently being collected and analyzed to further understand this type of surface flow wetland system and to develop accurate design criteria and cost estimations.

5.0 WETLAND DEMONSTRATION PROJECT 2 – COLORADO TAILINGS

5.1 Background

Wetland Demonstration Project 2 – Colorado Tailings (the CT Wetlands) was constructed to evaluate the effectiveness, implementability, and costs associated with an open water wetland system with lime addition. Though the geometry of the system is slightly different, the layout is essentially the same as for the BRW Wetlands: three open water cells separated by two treatment walls. The major difference is in the water source to the wetlands and the incorporation of a lime addition system.

The source of water for the CT Wetlands is alluvial groundwater captured within the Hydraulic Control Channel (HCC) constructed downgradient of the impacted alluvial aquifer. The HCC was constructed to capture the majority of contaminated alluvial groundwater to minimize adverse impacts to Silver Bow Creek. Under a full-scale scenario, this is most representative of the actual water that would be treated with either wetlands or conventional treatment methods. Prior to July 1998, the HCC was still under construction and carried runoff water from upgradient Silver Bow Creek and Butte's sewage treatment plant effluent. In early July, construction of the HCC was completed, the sewage treatment plant effluent was removed from the HCC, and runoff conditions subsided allowing water to be drawn as the influent source to the CT Wetlands. Prior to this time, influent water for the CT Wetlands had been drawn from a number of sources believed to be representative of the water that would be found in the completed HCC. Because of ongoing construction around the CT Wetlands, influent water was drawn from open water ponds or construction dewatering trenches near the HCC. Additionally, the source of water would at times be interrupted causing operational problems for the CT Wetlands.

Up to July of 1998, the lime addition system consisted of a mechanical water wheel lime addition machine. The incoming water would cause a water wheel to rotate which augered a specified amount of pelletized CaO into the water. Problems with this system included incomplete dissolution of the CaO, difficulty in operation during winter conditions, and scaling of the auger. Attempts to increase the dissolution of the CaO were marginally successful, with a large amount of unreacted CaO being deposited within the first open water cell.

In July of 1998, a new lime addition system was put into operation, which included two calcium hydroxide slurry-mixing tanks and two metering pumps. This improvement has resulted in the lowering of the lime requirement and calcium precipitate deposited within the first open water cell. This will also allow for more consistent treatment of the water with less maintenance and repair downtime. Lime is currently added at a rate between 30 and 120 parts per million (ppm) depending upon the influent water chemistry.

The incoming flow rate to the CT wetlands has varied over time due to construction conditions and maintenance problems; however with the completion of most construction in the vicinity and recent modifications to the lime system, it is anticipated that the flow will remain relatively steady at 125 gpm. Effluent from the CT Wetlands currently discharges to an open water area immediately to the west of the system.

5.2 Methods

Measurements of Eh, pH, SC, temperature, and dissolved oxygen are taken on a daily basis for the influent, effluent, free water surfaces, and treatment walls. Water quality samples are analyzed weekly for alkalinity, chlorophyll A, dissolved inorganic carbon, heterotrophic plate counts, anions (ion chromatograph), metals (ICP), dissolved organic carbon, total organic carbon, and total suspended solids. Measurement of influent and effluent flow rates is currently via mechanical totalizing flow meters.

5.3 Results and Discussion

As for the BRW Wetlands, monitoring of the open water cells indicates no discernable change in elevation between the three open areas. This indicates that the hydraulic conductivity of the treatment walls is more than adequate to convey the anticipated treatment flow rates. There has been no measurable change to the hydraulic conductivity of the treatment walls.

Water quality associated with the CT Wetlands may be evaluated at a summary level by looking at influent concentrations compared to effluent concentrations. Table 4, below, shows the water quality currently being maintained at the CT Wetlands. Data after June 2, 1998 was specifically chosen for presentation since previous data is influenced significantly by frequent interruptions in treatment due to construction conditions and problems with the lime addition system.

Table 4. CT Wetlands Water Quality Summary for the period of June 2 to Sept. 14, 1998

Analyte	Mean Influent Conc. (ug/L)		Mean Effluent Conc. (ug/L)		Percent Removal
	Dissolved	Acid Soluble	Dissolved	Acid Soluble	Acid Soluble
Arsenic	8.4	43.2	10.3	11.2	74%
Cadmium	34.4	36.2	1.5	2.2	94%
Copper	286.1	839.0	22.0	35.6	96%
Lead	2.0	11.0	2.1	2.0	82%
Zinc	9,474.0	9,849.3	68.0	235.7	98%
pH		7.2		8.3	-----

Arsenic typically enters the CT Wetlands between 10 and 100 μg/L (acid soluble) and on the average is reduced consistently to around 10 μg/L.

Cadmium is mainly in the dissolved phase as it enters the CT Wetlands at concentrations of around 20 to 80 μg/L and is removed to below 5 μg/L.

Influent copper concentrations are typically around 800 to 1000 μg/L (acid soluble) and are reduced consistently to levels less than 50 μg/L.

Influent acid soluble zinc concentrations are typically around 10,000 μg/L. Effluent acid soluble zinc concentrations are reduced to less than 300 μg/L.

The principle mechanisms for the decreases in metal concentrations for this system are believed to be 1) pH modification by lime addition, 2) pH modification by changes in CO_2 via biological processes (photosynthesis) and abiotic processes (groundwater supersaturated with CO_2 with respect to atmosphere), 3) flocculation and settling, 4) sorption, 5) co-precipitation, and possibly 6) metal sulfide precipitation in micro-environments within the treatment walls.

5.4 Conclusions/Future Investigations

In summary, the initial data indicate that the open water wetland system with lime addition significantly reduces metal concentrations. It also appears that only minimal lime addition is required since the physical and biological activity within the free water surface modifies the water chemistry resulting in further metal reduction.

Additional investigations will continue at this facility to further understand seasonal effects upon the system and modifications to the lime requirement. Because much of the data collected to date is for water not entirely representative of the water that will be treated under a full-scale treatment scheme, it is difficult at this time to form solid conclusions regarding this particular treatment system. Continued monitoring throughout 1998, 1999, and a portion of 2000 using water from the Hydraulic Control Channel will allow for a better understanding of the capabilities of this type of wetland treatment system and will allow for the development of design criteria and detailed cost estimations.

Tailings and Mine Waste '99 © 1999 Balkema, Rotterdam, ISBN 90 5809 025 6

Selection of treatment process for UMTRA groundwater

Randolph B. Richardson & Randall M. Juhlin
MACTEC-ERS, Grand Junction, Colo., USA
Donald R. Metzler
US Department of Energy, Grand Junction, Colo., USA

ABSTRACT: The groundwater at the uranium millsite at Tuba City, Arizona Tuba City site is contaminated with nitrate, sulfate, and uranium. After a lengthy evaluation of processes for cleanup of the contaminated groundwater, an active pump-and-treat system was determined to be the most viable treatment alternative. Available treatment technologies were reviewed for applicability. Spray-enhanced solar evaporation was the most cost-effective, but it was rejected primarily because of concerns regarding the sustainability of groundwater extraction without injection of treated water. Processes that incorporated injection of treated water included distillation, chemical treatment with biological denitrification, and nanofiltration with biological denitrification. Distillation was found to offer more process advantages, and less uncertainties, than either of the other processes, as well as offering lower total operating costs over the expected project lifespan. Therefore, distillation was chosen as the preferred treatment alternative for Tuba City.

1 SITE DESCRIPTION

The Tuba City Uranium Mill Tailings Remedial Action (UMTRA) Project site is located within the boundaries of the Navajo Nation, approximately 5 miles east of Tuba City and 85 miles northeast of Flagstaff, at an elevation of approximately 5,100 feet. Land near the site is used only for occasional grazing; adjacent land is used for dry and irrigated farming and for residences. The terrain below the site slopes gently to the southwest towards Moenkopi Wash, an intermittent stream which is the principal water source for the Hopi villages of Upper and Lower Moenkopi.

Between 1956 and 1962, the uranium mill at the Tuba City site processed an average of 300 tons of ore per day using a sulfuric acid leach. The plant was reconfigured in 1962 to use sodium bicarbonate in an alkaline process, and from 1963 until the mill closed in 1966 an average of 200 tons of ore per day were processed. All mill tailings were placed as a slurry into unlined evaporation ponds at the site.

Surface remedial action at the site was initiated in 1988 and completed in 1990. A total of 1.4 million cubic yards of uranium mill tailings and associated materials, as well as surrounding windblown contaminated soils, were moved and stabilized in a 50-acre engineered disposal cell located at the site of the former mill. The site license, granted by the U.S. Nuclear Regulatory Commission in 1997, deferred groundwater cleanup to the UMTRA Ground Water project.

2 GROUNDWATER CHARACTERISTICS AND TREATMENT STANDARDS

The general requirements for contaminant levels in the ground water at UMTRA sites are specified in 40 CFR 192.04, Table 1. The Navajo Nation has proposed secondary cleanup standards for the Tuba City site, based on the National Secondary Drinking Water standards in 40 CFR 143, that included constituents not listed in 40 CFR 192. DOE has incorporated the suggested secondary cleanup standards as goals for aquifer restoration and treatment of extracted ground water.

The concentrations of all constituents in the Tuba City ground water have been monitored beginning in 1985 at 41 monitoring wells distributed across an area covering over one square mile between U.S. Highway 160 and Moenkopi Wash. Regular sampling of these monitoring wells enables the UMTRA groundwater project to measure the extent of the contamination as well as to monitor movement of contaminants in the aquifer. Samples from 21 of the 41 monitoring wells are presently showing concentrations of at least one constituent that are in excess of either the aquifer restoration standards of 40 CFR 192.04, Table 1, or the secondary cleanup goals based on the Navajo Nation recommendations. Median and maximum concentrations of all contaminants of concern in the contamination zone, as well as concentrations of other constituents that can affect the operation of treatment processes, are summarized in Table 1. Also shown in Table 1 are the treatment levels (maximum allowable concentrations) in the aquifer at the conclusion of treatment. The treatment levels shown for molybdenum, nitrate, selenium, and uranium are treatment standards specified in 40 CFR 192; other treatment levels are project goals taken from the Navajo Nation recommendations.

Table 1. Contaminant and Treatment Levels in mg/L

Contaminant	Median	Maximum	Treatment Level
Alkalinity (as $CaCO_3$)	510	1,780	Not applicable
Calcium	443	1,071	Not applicable
Chloride	91	440	250
Magnesium	462	2,330	Not applicable
Molybdenum	0.16	0.6	0.1
Nitrate	951	2,330	44
Selenium	0.096	0.469	0.01
Sodium	219	615	20
Sulfate	2,257	7,590	250
Total Dissolved Solids	4,548	15,600	500
Uranium	0.404	1.38	0.044
pH	6.67	6.32 - 8.28	6.5 - 8.5

Total suspended solids and concentrations of organic contaminants are insignificant.

The volume of contaminated water can be estimated using the extent of the contamination area. Most of the contaminant concentrations exceed applicable standards only in confined areas, generally in the approximate location of the tailings ponds. The only contaminants which are distributed in high concentrations (in excess of applicable standards) across a significant area are nitrate, sulfate, and uranium. The volume of the contamination "plume" for each of these constituents is given in Table 2. The volume calculations assume a porosity of 0.25 and an average thickness of 86 ft of contaminated Navajo Sandstone aquifer.

Table 2. Summary of Volumes of Contaminated Ground Water

Contaminant	Concentration, mg/L	Area, sq ft	Volume, gallons
Nitrate	44	10.5×10^6	1.69×10^9
Sulfate	250	8.86×10^6	1.42×10^9
Uranium	0.044	4.84×10^6	0.78×10^9

The volume of contaminated water is 1.69 billion gallons, based on the volume of nitrate-contaminated water.

3 SELECTION OF GROUNDWATER REMEDIATION STRATEGY

An UMTRA groundwater site may use one of four possible compliance strategies:

(1) No remediation is required, because no contamination is present in excess of Maximum Concentration Limits (MCLs) or background levels.
(2) No remediation is required, because either Supplemental Standards or Alternative Concentration Limits (ACLs) are applicable.
(3) Remediation is required, and cleanup requirements can be met using natural flushing.
(4) Remediation is required, and active treatment must be used to meet cleanup requirements.

The concentrations of several contaminants at Tuba City exceeded both the MCLs and the background levels, so no-remediation option (1) was inappropriate.

Supplemental Standards could not be applied because the Tuba City groundwater did not qualify as limited-use water under the requirements of 40 CFR 192, and because there was no indication that excessive environmental harm would result from remediation. ACLs did not apply at Tuba City because concentrations of contaminants in five wells located outside the fenced area of the site exceed MCLs or health-based levels. Therefore no-remediation option (2) was also inappropriate.

Implementation of natural flushing requires that it can be demonstrated that one pore volume of background water will flush through the contaminated area within 100 years. Geologic modeling of the site estimated that the actual time required for natural flushing would be approximately 220 years, so the natural flushing option (3) was inappropriate. Therefore, the only groundwater compliance strategy which could be implemented for the Tuba City site was active remediation.

The capacity of the treatment system depends on the volume of contaminated water that is present, and the duration of the treatment. A goal of the UMTRA Groundwater project is that treatment systems are to be designed to complete the treatment process over a nominal lifetime of 20 years. The volume of water which must be treated by the treatment process is 1.69 billion gallons. A treatment duration of 20 years gives an average treatment system capacity of 160 gpm. The project team decided to implement the treatment system by first installing a treatment system having a capacity of 50 to 100 gpm, and increasing the treatment capacity to 200 gpm over a period of two to four years. Under this strategy, the total time required for remediation of 1.69 billion gallons would be approximately 16 years.

4 TREATMENT PROCESSES - INITIAL SCREENING

The first phase of the treatment selection process was to review available water-treatment technologies for potential applicability to the Tuba City site. The categories of water-treatment processes which were considered included evaporation systems, both simple and enhanced; distillation (evaporation and recondensation) systems; through-medium processes such as ion exchange; land application (land/plant treatment); biological unit processes; chemical treatment processes; and membrane separation processes, including reverse osmosis and nanofiltration.

An initial screening was used to eliminate processes which were clearly not suitable for the specific requirements of the Tuba City site. This section summarizes the results of that initial screening.

A solar pond treating the maximum extraction rate would require a surface area of about 43 acres, which was judged impractical. Use of a spray system can substantially reduce the size of the pond required, though this increases the complexity of the system and requires more maintenance and operator attention than simple solar evaporation. The major disadvantage of an evaporation process is that the water is lost to the aquifer. However, due to its low cost and relative simplicity of operation, spray-enhanced evaporation was selected for detailed evaluation.

Distillation recovers almost all of the water for injection back to the aquifer, and the product water is of very high quality. The Tuba City groundwater does not contain any volatile contaminants, so use of a distillation process is feasible. Energy requirements for distillation units can be greatly reduced by the use of "vapor recompression," making the operating costs of a distillation process more nearly competitive with those of other treatment processes. Considering the many other attractive features of the process, distillation was selected for detailed evaluation.

An ion exchange process would replace the nitrate and sulfate anions with chloride, which would report to the treated water. The project treatment goal for chloride concentration is 250 mg/L. Based on the median concentrations given in Table 1, use of ion exchange would increase the chloride concentration in the treated water to almost 2,100 mg/L, or over eight times the project treatment goal. This is obviously unacceptable, and therefore ion exchange was rated as an inappropriate remediation technology for Tuba City.

In-situ bioremediation, which uses bacteria to eliminate the hazards posed by COCs in the ground water, was evaluated by the Center for Radioactive Waste Management (CeRaM). Substrates injected into the ground water disperse through the contaminant plume to promote the growth of bacteria, which immobilize the COCs by reducing them to an insoluble form. The technology was eliminated from further consideration primarily because of difficulties in delivering the substrate to all areas of the aquifer and in maintaining a chemically reducing environment in the aquifer.

Ex-situ biological denitrification, either in a reactor or in a pond, is an attractive process for removing nitrate. However, biological desulfurization is not nearly as attractive, because it does not appear to be capable of meeting the treatment goal of 250 mg/L sulfate, and because of the problems associated with handling and disposing of the by-product H_2S. Thus, removal of sulfate must precede the biological denitrification process. Biological denitrification was retained for detailed evaluation as part of a complete treatment process utilizing other processes for removal of sulfate and uranium.

Land treatment systems use nitrate-contaminated water to irrigate tolerant indigenous rangeland plants and grasses. Pretreatment of the ground water before land application is required for removal of uranium, other trace metals and metalloids, and sulfate (at the concentrations present in the Tuba City groundwater). Also, due to the seasonal nature of land application, treating the volume of contaminated water at the Tuba City site within the project target time of 20 years would require a maximum irrigation rate of over 1200 gpm, with a planted area in excess of 140 acres, and an enormous pretreatment system for sulfate and uranium removal at the peak extraction rate. Thus, use of a land application system at the Tuba City site was considered impractical.

Chemical treatment typically uses precipitation, coagulation and flocculation, gravity settling, and filtration processes. For the Tuba City ground water, such a process would incorporate addition of an iron source as a flocculent to precipitate uranium, and a barium source to precipitate sulfate as barium sulfate. Chemical treatment is ineffective for removal of nitrate, so a denitrification process would be required downstream of the chemical process. DOE owns a 200-gpm chemical treatment facility that is currently in storage in Colorado. Use of this facility would greatly reduce the capital costs of the process. Accordingly, a process combining chemical treatment with biological denitrification was selected for detailed evaluation.

The most effective membrane separation processes are reverse osmosis and nanofiltration. The nitrate treatment standard of the Tuba City remediation project is almost certainly too low to be met by any membrane separation process. However, both reverse osmosis and nanofiltration can easily meet the Tuba City treatment goal for sulfate. Compared to reverse osmosis, nanofiltration has somewhat lower capital costs, significantly lower maintenance costs, and produces only about half as much reject water. Therefore, it is better suited as a pretreatment process for biological denitrification than is reverse osmosis, and a process combining nanofiltration with biological denitrification was selected for detailed evaluation.

5 FINAL DETERMINATION OF TREATMENT PROCESSES

The treatment processes that remained for detailed evaluation following the initial screening were Solar Evaporation with Spray Enhancement, Distillation, Chemical Treatment with Biological Denitrification, and Nanofiltration with Biological Denitrification. These processes were compared according to three criteria: *Effectiveness, Implementability,* and *Cost*. These criteria were divided into "sub-criteria" and each treatment process was evaluated based on its predicted performance as judged by each sub-criterion.

5.1 Effectiveness

Effectiveness was subdivided into three "sub-criteria": Conformance with 40 CFR 192, effect on the aquifer, and residual disposal.

Distillation was judged to be the best with regard to conformance with 40 CFR 192, because the treated water is of higher quality than that produced by any other treatment process. Nanofiltration with Biological Denitrification produces a better quality of water than Chemical Treatment with Biological Denitrification and was second choice. Solar Evaporation with Spray Enhancement does not produce any treated water for injection. Use of a injection system will enhance the chances for success of the remedial action, so solar evaporation was considered to have the lowest chance of meeting the regulatory requirements.

Distillation produces the most treated water of any of the processes and so was also considered to have the least effect on the aquifer. Chemical Treatment with Biological Denitrification was considered to be a slightly better choice than Nanofiltration with Biological Denitrification, because of the larger amount of reject water produced by the nanofiltration process. Solar Evaporation with Spray Enhancement, which would deplete the aquifer, is the poorest choice

Since residuals other than those from chemical addition did not vary significantly between the processes, ranking the processes according to residual disposal effectively ranked them in increasing order of how much chemical addition was required. Solar Evaporation with Spray Enhancement does not require any chemical pretreatment. Both of the processes utilizing Biological Denitrification were ranked below Distillation, because of the considerable amount of organic sludge generated by those processes. Chemical Treatment with Biological Denitrification requires a great amount of makeup chemicals in addition to the biological sludge and so was the poorest choice from this standpoint.

5.2 Implementability

Implementability was subdivided into five "sub-criteria": Ease of construction, ease of operation and maintenance, expected reliability, ability to handle changes in influent composition, and ability to handle increases in extraction capacity.

Distillation was judged easiest to construct, because commercial distillation units are often available as "packaged" systems requiring a minimum of on-site construction work. Solar Evaporation with Spray Enhancement must be completely constructed and installed at the site. However, the skill level required is not high, so it was rated as second choice. Nanofiltration with Biological Denitrification was third because, although the nanofiltration system would be a packaged system, the denitrification system would be a custom fabrication. Chemical Treatment with Biological Denitrification was judged to have the most difficult construction, because both of the two units would require considerable skill for proper assembly.

The Distillation unit will have relatively few moving parts and will be highly automated, so it should be very easy to operate and maintain. Solar Evaporation with Spray Enhancement will have many parts and will not be automated to a significant extent, but the skill level required to operate and maintain the spray-evaporation system would be relatively low. Nanofiltration with Biological Denitrification and Chemical Treatment with Biological Denitrification are comparatively complicated processes, requiring a high degree of skill for proper operation. The operator-intensive nature of the chemical treatment process made it the last choice from this standpoint.

Solar Evaporation with Spray Enhancement was judged to be most reliable, because, although it has a large number of spray nozzles which are prone to plugging, failure of a single spray nozzle or pump will not incapacitate the entire system. Distillation was judged likely to be highly reliable also, although a single failure will likely put the entire system out of commission. The two processes incorporating biological denitrification will likely be much less reliable; Nanofiltration with Biological Denitrification was judged likely to be more reliable than Chemical Treatment with Biological Denitrification because the nanofiltration process should be more reliable than chemical treatment.

The composition of the feed can vary significantly depending on which extraction wells are in service, as well as over the course of remediation as contaminants are removed. Solar Evaporation with Spray Enhancement is essentially totally insensitive to influent composition. The pretreatment system for Distillation will be designed to handle high carbonate loadings, and the distillation process itself should be able to adjust to variations in feed composition reasonably well. The two processes incorporating biological denitrification were judged less likely to be able to seamlessly handle changes in feed composition. The nanofiltration process is expected to be less sensitive to changes in feed composition than the chemical treatment process, so Nanofiltration with Biological Denitrification is preferable to Chemical Treatment with Biological Denitrification.

The ability to handle increases in extraction capacity will become important if the volume of contaminated water is found to be significantly higher than the current estimate, or if the timetable for remediation must be accelerated. Nanofiltration with Biological Denitrification was judged best able to handle such increases, because the nanofiltration units are modular and extra units can be added for relatively modest cost. The biological denitrification units are relatively easy to construct, so addition of more denitrification capacity would not be overly difficult. Increasing the capacity of the Distillation process would require adding more distillation units, which would be relatively easy but very expensive. Increasing the capacity of the Solar Evaporation with Spray Enhancement system would require addition of a complete second pond and spray system. The existing chemical treatment system is sized for 200 gpm, cannot be easily increased, and is in no sense modular or prefabricated. So increasing the capacity of the Chemical Treatment with Biological Denitrification system would require that an additional chemical treatment system would have to be designed, procured, and installed.

5.3 *Cost*

Cost was considered from two points of view: *Capital Cost*, the amount of money required to design, procure, and install the treatment system; and *Total Project Cost*, which includes both the capital cost and the total operating and maintenance (O&M) cost required over the lifetime of the project.

The total cost of the remediation system includes the capital and O&M costs for the extraction and injection systems as well as the treatment process. The Distillation, Chemical Treatment with Biological Denitrification, and Nanofiltration with Biological Denitrification processes utilized identical extraction and injection systems, with a total capacity of just over 200 gpm. The Solar Evaporation with Spray Enhancement process required a totally different extraction system, which did not incorporate injection and had a maximum sustainable estimated extraction rate of only 115 gpm. Thus, the total time required for remediation was estimated at 26 years for the Solar Evaporation with Spray Enhancement process, vs 16 years for the other three processes.

The estimated capital, annual O&M, and total lifetime costs for each of the four treatment systems, including the extraction and injection systems (as applicable), are given in Table 3.

Table 3. Estimated Costs, in Millions of U.S. Dollars, for Tuba City Treatment Processes

Injection Option	Treatment	Capital	O&M	Total
No Injection	Solar Evaporation with Spray Enhancement	4.90	1.13	17.92
Injection	Distillation	7.78	2.05	26.59
Injection	Chemical Treatment with Biological Denitrification	3.59	5.83	58.41
Injection	Nanofiltration with Biological Denitrification	5.85	2.40	28.15

The additional extraction wells required for Solar Evaporation with Spray Enhancement account for much of its capital cost. However, even though its O&M costs must be figured over a longer time than any of the other treatment processes, these costs were so low that this process still has by far the lowest total project cost. Chemical Treatment with Biological Denitrification offers the lowest initial capital outlay, largely because it utilized an existing DOE-owned chemical treatment facility. However, from a total project cost standpoint, it was a distant fourth, due to high consumption of barium chemicals and to the labor-intensive nature of the process.

Distillation is very capital-intensive. Making matters worse, the remediation is to be "phased in," with an initial capacity of 100 gpm which would later be expanded to 200 gpm, so the cost estimates required purchase of two 100-gpm systems rather than a single 200-gpm system. However, the O&M costs for the distillation process were relatively low, because the process is highly automated and requires little in the way of expensive chemical pretreatment. The most expensive single line item in the Distillation O&M costs is electricity to operate the two distillation units, which averaged about \$550,000 per year based on the average Tuba City cost of 11.5 cents per kilowatt-hour.

6 SELECTION OF PREFERRED TREATMENT SYSTEM

The results of the evaluation process, as described above, are summarized in Table 4, which lists each of the ten evaluation "sub-criteria" and ranks the four treatment alternatives in descending order of preference for each of these "sub-criteria," with 1 meaning that the process was the first choice and 4 meaning that it was the last choice.

Table 4. Results of Treatment Selection Process

Sub-criterion	Solar Evap	Distillation	Chem/Bio	Nano/Bio
Conformance with 40 CFR 192	4	1	3	2
Effect on the Aquifer	4	1	2	3
Residual Disposal	1	2	4	3
Ease of Construction	2	1	4	3
Ease of Operation and Maintenance	2	1	4	3
Expected Reliability	1	2	4	3
Ability to Handle Changes in Influent Composition	1	2	4	3
Ability to Handle Increases in Extraction Capacity	3	2	4	1
Initial Capital Outlay	2	4	1	3
Total Project Cost	1	2	4	3

"Solar Evap" = Solar Evaporation with Spray Enhancement
"Chem/Bio" = Chemical Treatment with Biological Denitrification
"Nano/Bio" = Nanofiltration with Biological Denitrification

As mentioned above, Chemical Treatment with Biological Denitrification requires less capital than the other alternative processes because it utilizes the existing DOE-owned chemical treatment facility. However, it is the fourth choice for seven of the ten evaluation sub-criteria, including having by far the highest total project cost. Thus both from the technical and cost viewpoints, Chemical Treatment with Biological Denitrification is the least desirable treatment process for Tuba City.

Nanofiltration with Biological Denitrification is the only treatment process which does not rank last on any of the sub-criteria under evaluation. However, it is the alternative of first choice only in regards to the ability to handle increases in extraction capacity. This ability will matter only if the timetable for treatment is significantly accelerated or if the volume of contaminated water proves to be much greater than the present estimates. Both of these possibilities are considered remote; the calculation of contaminated volume is considered conservative, since it assumes that the entire thickness of the aquifer is contaminated if it is contaminated at any vertical point, and the remediation timetable already significanly exceeds regulatory requirements. Nanofiltration with Biological Denitrification is the third choice for eight of the ten evaluation sub-criteria, and therefore overall it is the third choice of the four alternatives.

Solar Evaporation with Spray Enhancement is the preferred technology for four of the eleven evaluation sub-criteria, including overall project cost; spray evaporation is expected to cost about two-thirds as much as the next-least-expensive process over the lifetime of the project. However, Solar Evaporation with Spray Enhancement also has significant drawbacks. Hydrologic modeling suggested that, without injection of treated water, "dry spots" may develop within the aquifer, where further extraction of contaminants will not be possible, making it difficult if not impossible to meet aquifer restoration standards. Also, the flushing effect of treated-water injection is expected to be of beneficial value in enhancing the effectiveness of the pump-and-treat process, and injection at the boundaries of the plume will significantly retard the spread of contaminants beyond their present extent. Without injection, the possibility that the treatment system will have to remain in operation beyond the current estimated lifetime (already 10 years longer for this process than for the others because of its lower maximum extraction rate) is considered high. Also, the site stakeholders have indicated a clear and strong preference for technologies which return as much water as possible to the aquifer, after as much of the total contaminants as possible have been removed. Solar Evaporation with Spray Enhancement does not satisfy this preference.

Distillation is first choice for four of the ten criteria, and is second choice for five of the other six. Only in terms of immediate capital outlay required does distillation rank last among the four processes. The Navajo and Hopi tribes have indicated their particular interest in implementing distillation technology for this project. Advances in distillation technology have virtually eliminated the high energy costs which were traditionally associated with distillation processes, making Distillation a very attractive and cost-competitive treatment option.

If cost was the overriding consideration of the Tuba City ground-water remediation program, solar evaporation with spray enhancement, with a total project cost of about two-thirds that of the next least expensive alternative, would be the best choice. However, the realities of the Tuba City situation argued for a different choice. The technical uncertainty as to whether a treatment process that did not incorporate injection of treated water could meet treatment standards, and the expressed wish of DOE to satisfy stakeholders' desires to preserve the integrity of the aquifer to the extent possible, obviate against selection of solar evaporation. Distillation more than satisfies the regulatory requirements of 40 CFR 192, and will produce a high-quality effluent that will recharge the aquifer with a minimum of loss while containing and preventing the spread of the contaminant plume. Distillation will produce the highest-quality effluent, and is also expected to return more of the water to the aquifer than any other treatment alternative. Distillation is the second choice economically, and is the preferred process from an overall technical standpoint. Therefore, Distillation was chosen as the preferred treatment technology for the Tuba City ground-water remediation program.

7 PROPOSED TUBA CITY GROUNDWATER TREATMENT PROCESS

The extraction and injection system consists of an approximately equal number of extraction and injection wells, and an infiltration gallery north of the disposal cell. Recovery well locations are proposed to be in the areas of maximum contaminant concentrations to achieve rapid reductions in contaminant mass. Injection wells are located along the downgradient margin of the plume to control its migration. Other injection wells and the infiltration gallery are located within the plume to enhance flushing and rinsing.

Extracted ground water will be pumped to the feed pretreatment system, where acid is added under automatic pH control to reduce alkalinity in the water. Effluent from the feed pretreatment system will be pumped directly to the distillation system. This will consist of one or more self-contained units with a feed capacity of between 200 and 220 gpm. The units will be instrumented to permit continuous operation with remote monitoring capability.

The distillation system will produce two product streams: a concentrate, or brine, that contains all the contaminants, and a clean treated water. The brine, which is expected to average 5 percent or less of the total feed, will be pumped to solar evaporation ponds for final concentration. The dry sludge from these ponds will be removed and disposed of at a suitable landfill as needed. The treated water, which will average at least 95 percent of the total feed, will be pumped to an effluent holding tank prior to injection by gravity flow.

8 IMPLEMENTATION OF THE PROPOSED TREATMENT PROCESS

During the late summer and early fall of 1998, pilot tests of the distillation process will be conducted at the site, using existing extraction wells and ponds. Construction of the first phase of the extraction and injection systems is expected to begin in the summer of 1999, with remediation beginning with the startup of the first distillation system in late fall. This first phase system is expected to have a capacity of 50 to 100 gpm. Over the next two to four years, additional extraction and injection wells will be constructed, and a second distillation system will be installed, to bring the total capacity of the remediation system to 200 gpm.

Tailings and Mine Waste '99 © 1999 Balkema, Rotterdam, ISBN 90 5809 025 6

Treating heavy metals in mine soil and waste using Envirobond™ binder

R. Jensen & Ali Sogue
Rocky Mountain Remediation Services, L.L.C., Golden, Colo., USA
Steven Schurman
Goldrush Casino and Mining Corporation, Arvada, Colo., USA

ABSTRACT: This paper presents the results of the remediation of a former mining site in Central City, Colorado, one of the Clear Creek Superfund projects. The Envirobond™ binder was successfully deployed to solidify and stabilize 3,000 tons of metals contaminated mine waste and soils, reducing the toxicity level substantially below TCLP RCRA limits. Mine waste containing high levels of arsenic, lead, zinc, and mercury were treated with Envirobond™. Typical heavy construction equipment was used to incorporate the binder with the soil in 500-ton lifts. A large lay down area of 5000 sq. ft. was used to layer and mix all of the soil. Final cost of the treatment was much less than for standard industry solidification processes and was significantly less than offsite disposal options.

1 INTRODUCTION

Rocky Mountain Remedial Services (RMRS) has developed and deployed an effective process that provides for safe, simple handling and effective permanent stabilization of contaminated soil and soil-like radioactive and hazardous wastes. A low cost, multi-purpose binder (Envirobond™) that uses well-established technology and does not produce secondary waste is the key to the success of the process. RMRS has built upon existing company and parent company (Morrison Knudsen Corp. and BNFL, Inc.) expertise to successfully deploy the technology.

The Envirobond™ binder was used in a field application to immobilize heavy metals in soil that was contaminated by mining operations that took place more than 50 years ago. The Envirobond™ In-Place Stabilization/Compaction process was used to stabilize arsenic, lead, and zinc. The primary objective of the project was to stabilize the waste to levels that would meet the EPA's criteria for releasing the site for development. A secondary objective was to reduce the TCLP leachable levels to regulatory, (including the Universal Treatment Standard), levels. The demonstration was conducted at the client's site in Central City, Colorado. Equipment included a front-end loader, a road grader, a tractor-tiller, spreading equipment, a water truck, and a sheep's foot field compactor. Fly ash was delivered and spread from a truck with a unique distribution device that alleviates a potential dusting problem.

2 EXECUTION OF THE GREGORY GULCH PROJECT

The Gregory Gulch site is a former mining site. It is a designated Clear Creek Superfund site in Central City, Colorado. The site has been inactive since the early 1900's and construction projects and others have distributed heavy metals from the mining operations throughout the area. Most of the soil in the surrounding area contains heavy metals, and some is at elevated levels. Piles of contaminated mine soil/waste with levels of arsenic, lead, zinc, and mercury that exceeded the EPA threshold levels were on the site. In order to proceed with development of the contaminated properties, it was necessary to treat or remove the contaminated piles of soil.

The various sources of potential contamination were studied to establish the nature and extent of their environmental impact. Remedial actions were required only if the excavated soil was been found to exceed the following established EPA threshold levels for arsenic, lead, zinc and mercury. Success criteria were based on the RCRA Standards for these metals. They are arsenic – 5mg/kg, lead – 5 mg/kg, zinc – No standard, and mercury – 0.2 mg/kg. The pH must be above 5.5 and below 8.

Two alternatives were evaluated for remediating the site. The first was to excavate the waste piles, and other contaminated areas, and transport them to a disposal site in Denver, Colorado. There are two serious disadvantages to this alternative. The cost of transportation and disposal is high, and after the waste leaves the site, the owner remains liable for any further contamination that may occur during transportation, or at the burial site for years to come.

The second alternative was on-site treatment. The major disadvantage to this alternative was that the use of traditional cement, lime, or other similar additives will add large amounts of material to the waste, and some of these treatments are ineffective on some types of metals. Some metals such as zinc cannot be treated without adding large volumes of additives. With cement or lime, the heavy metals may not leach out, but they are still available to future environmental upsets because they are not permanently bound to the chemical additives. For these reasons, even at a higher cost, the EPA was inclined to allow only the excavation and removal of the soil.

The second alternative is much more attractive with a binder that does not add weight or volume. Use of the Envirobond™ binding agent adds less than 2 wt. % to the volume, and when combined with compaction, significantly reduces the volume. Furthermore, Envirobond™ chemically binds the metals so that they are not only physically stabilized, but they are incorporated with a chelating bond that cannot be penetrated under severe even severe conditions. It is not soluble, and the treated soil hardens to form a cap over the treatment area.

Costs are reduced when compared to cementation or excavation and hauling. With cementation, the additional bulking of the waste can easily add a 100% volume increase. The cost of materials handling and mixing is also higher. With excavation and hauling, there are additional transportation and disposal costs, and the excavated material would have to be replaced with clean fill dirt. All of these factors add cost that is avoided with Envirobond™.

2.1 Treatment and Storage of the Tailings

Three sites on the property were of concern. These sites included the south bank of Gregory Gulch above the existing Dynasty rip rap, the tailings pile located on the Leavitt Lode – Jackson property line and soil requiring treatment when the foundations for two historic houses were excavated. Impacted soil from all sites was excavated with a track hoe and pushed to the treatment area with a D-8 tractor. After soil removal, the bank along Gregory Gulch was sloped and covered with large rocks to prevent erosion.

Approximately 3,000 tons of mine waste was treated with Envirobond™ and fly ash. The procedure consisted of putting down approximately 500 tons of mine waste in a 10 to 15 inch lift. Six lifts were treated. Envirobond™ and fly ash were applied to the surface of the soil on each lift. The Envirobond™ was spread using a garden type spreader pulled with a tractor. A final application of Envirobond™ in solution was sprayed on the surface.

Fly ash was used increase the pH. Class F fly ash had been used successfully in the laboratory for use in this application, and it was selected for use in the field. Class F fly ash costs less, and was readily available. A covered truck with a live bottom was used to deliver the fly ash. An attachment was designed and attached to the back end of the truck to spread the fly ash. In this way, the truck was driven over the soil releasing the ash in a continuous manner that ensured good distribution over the soil to be treated. The attachment was designed in such a way that the fly ash is delivered without excess dusting.

A grader-mounted ripper was used to mix the Envirobond™ and fly ash into the soil. The grader latter furrowed the soil back and forth with the grader blade to complete the mixing process. After treatment and mixing, the lift was sampled. The lift was then compacted with a vibrating roll compactor. After the target soil had been moved to the treatment area, the treated soils were capped with one to two feet of clean soil. The soil capping was gently sloped to drain all incidental rainfall away from the treated material and seeded to reduce erosion.

3 RESULTS

The results of the test were very successful. In addition to meeting the EPA's standards for Toxic Characteristic Leaching Procedure (TCLP), the results of the TCLP testing were also less than the more stringent Universal Treatment Standards (UTS). Table 1 shows the results. Arsenic, lead, and zinc exceeded the EPA's threshold level and were the contaminants of concern. Untreated soil was given the TCLP test, and lead and zinc exceeded the standards without treatment. Arsenic was present at 4 ppm, which is just below the RCRA standard. After treatment with Envirobond™, all three metals were below the UTS standards. This project demonstrates the versatility of Envirobond™. Less than 4 wt. %, Envirobond™ and fly ash, was added to the volume of soil, and it was mixed using field equipment. The TCLP results from the treated soil were well below the RCRA limits set for the metals of concern, as well as the more stringent UTS standards.

Table 1. On-site full scale in-situ treatment of mine tailings in Central City

	Arsenic	Lead	Zinc	Mercury
Total Metals (ppm)	136	1270	1270	5.9
TCLP Extract Before Treatment (ppm)	0.07	91.7	108	<.02
TCLP Extract After Treatment (ppm)				
Sample 1	<.05	<.05	1.5	
Sample 2	<.05	<.05	1.0	
Sample 3	<.05	<.05	3.9	
Sample 4	<.05	<.05	4.5	
Sample 5	<.05	<.05	4.2	
Sample 6	<.05	<.05	4.7	
Sample 7	<.05	<.05	3.3	
Average	<.05	<.05	3.3	
TCLP Regulation CFR 261.24 (a)	5	5	NA	0.2
Current Universal Treatment Standard (UTS)	5	0.37	5.3	0.2

Approximately 2.8 wt. % fly ash was blended into the 3,000 tons of mine waste for pH adjustment. The average final pH for the 3,000 tons was 6.9.

The project also demonstrated that it is easy to prepare Envirobond™ for use in field applications. Other projects had shown that Envirobond™ could be applied in a wet form, or dry as in the Gregory Gulch project. The Envirobond™ does not add bulk to the waste, which is especially important when the waste is transported to a disposal facility. Finally, the project also used the treated soil to form a base for future construction. The treated soil was successfully layered, mixed and compacted to meet the proctor specifications for construction.

As this was a CERCLA site, all of the activities were completed with EPA Region 8 approval. The objective of the soil removal plan was to isolate contaminated soil that exceeded threshold limits and manage them in accordance with the Record of Decision (ROD). The work plan submitted to the EPA anticipated the removal of all contaminated soil to an open foundation where it was to be treated with Envirobond™ and fly ash. The EPA approved this work plan. The successful results were reported to the EPA in a report prepared for the EPA titled "Goldrush Superfund Site Final Report", Gregory Gulch, Central City, Colorado, September 9, 1998. The client is now seeking to have the EPA remove the lands subject to EPA Superfund action order CERCLA-VII-97-73, from the Superfund National Priority List.

4 DISCUSSION

4.1 Process Chemistry

The process chemistry of the proprietary additives and their interactions (see Figure 1) with the waste forms and phosphate binding agent are outlined below.

Phosphate Binder Reactions

- The Envirobond™ has a polyphosphate chain that binds with the metal ion (including radionuclides) forming an insoluble phosphate metal complex. The Envirobond™ also adheres the waste particles together which contributes to the structural integrity of the product.
- The metal ion binds to the oxygen sites on the phosphate ligand. Phosphate in the chain rotates freely about the P–O–P bond. This mobility allows the metal ion to be linked to more than one site on the polyphosphate chain. (i.e., if the oxidation state is 2+, then the metal binds to two sites; if the oxidation state is 3+, then the metal binds to 3 sites and so on.)
- The reaction proceeds at ambient temperatures, and does not produce secondary waste streams.

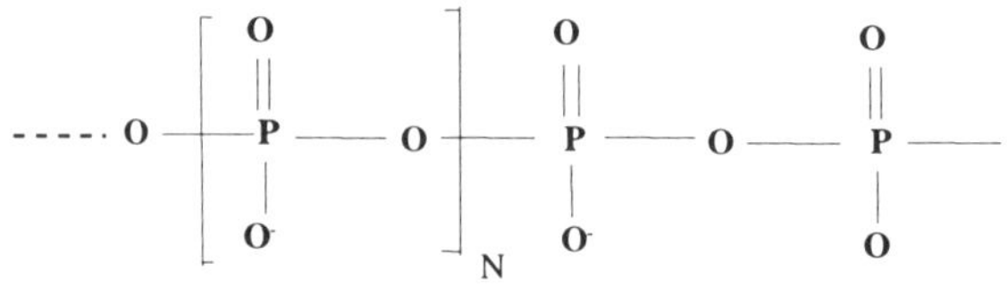

N = multiple units of phosphate structure

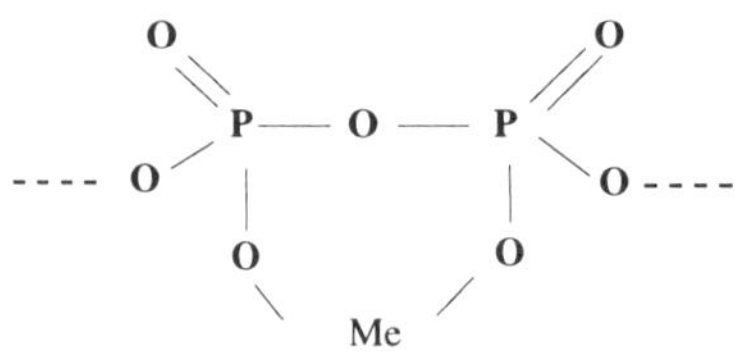

The metal ion binds to the O- sites, the example of a Metal 2+ to the phosphate is shown above

Figure 1. Chemical reactions in the proprietary Envirobric™ process

The heavy metals and resulting material is not soluble and typically cures to a hard, non-friable product. Its final physical properties can be enhanced with compaction. In this application, the treated waste was compacted in place using a vibrating roll compactor. In other applications, RMRS has formed the treated material into bricks using a rammed earth brick maker. Additional applications of the technology are discussed below.

4.2 Current and Future Applications of Envirobond™

4.2.1 Current Applications

Currently, Envirobond™ has been used in two basic ways. The first is described in this report. The second involves the stabilization of sludge, soil, or fine particulate materials such as mill tailings. In those applications, the treated material is mechanically compacted to form a rammed earth brick referred to as an Envirobric™. The machine was designed for mass production of sand/clay construction blocks. Rammed earth bricks differ from conventional construction bricks because they are produced under extremely high compaction forces, and they are not fired. The bricks possess very high compressive strength, (3000 to 6000 p.s.i.), which have very low porosity. The bricks are about 10 in. x 14 in. x 4 in., each weighing approximately 35 pounds.

This application is designed to produce a product that is easily handled, shipped, and stored. It also reduces the volume of the waste up to 40%, which cuts the cost of disposal dramatically.

4.2.2 Future Applications

In-situ Capping
In-situ capping is aimed at the thousands of abandoned mill sites throughout the U.S. A solidified layer of Envirobond™ with fly ash would be compacted on top of the pile to form a solidified layer. The surface could then be armored with riprap.

Heap Leach Process
In a second application, thousands of tons of mine rock waste, such as copper mine waste piles could be treated using a heap leach process. This process would take advantage of the solubility of Envirobond™ and use it as a treatment solution. Envirobond™ would be introduced to the pile by strategically placed sprinklers.

On-Site Stabilization of Mill Tailings
In a third application, mill tailings mixed with Envirobond™ as a preventive environmental measure would preclude the discharge of the heavy metals from the tailings into groundwater. By reacting the metals with Envirobond™, it may be possible to store the tailings without an expensive tailing liner, which generally leak and release metals to the groundwater over time.

Metals Stabilization
A fourth application involves a scenario where the process would be done in a similar fashion to the heap leach described above. Dissolved Envirobond™ is introduced into near surface soils containing metal contamination from industrial processes such as chrome plating facilities, lead recovery sites, aluminum processing sites, and numerous DOD sites. The Envirobond™ could be used in conjunction with a passive groundwater barrier (PGB) to flush metals located above the groundwater table to the PGB. The result could totally remediate the site from the contaminate source to the groundwater plume.

Mine Waste Water Treatment
Using the existing application of Envirobond™ described above; metal sludge from mine water flocculation treatment plants could be treated and formed into Envirobrics™, if desired. By treating the metal sludges with Envirobond™, they could be disposed of on-site, (perhaps even in the mine), rather than transporting to an off-site disposal facility.

Preventative Treatment
Finally, preventative treatment applications are also possible. Two applications are possible. In the first, soil below the bentonite clay impermeable barrier at mixed waste disposal sites would be treated as a precautionary method to stabilize existing heavy metals, and stand as a preventive barrier to any potential metals that leak through the barrier. An extra layer of protection would be formed with the Envirobond™ that would also react with metals that leak through the physical barrier. The second application envisions pretreating soil below the HDPE liner system used at mines and mills in the same way as the disposal sites. In this way, any metals contamination that leaks is treated in-situ before it enters the groundwater system.

Tailings and Mine Waste '99 © 1999 Balkema, Rotterdam, ISBN 90 5809 025 6

Utilization of dredge material-based grouts for mineland reclamation

B.E.Scheetz
Materials Research Laboratory, The Pennsylvania State University, University Park, Pa., USA
C.R.Schantz
Consolidated Technologies Incorporated, Blue Bell, Pa., USA

ABSTRACT: Large, unfilled mine sites remain in Pennsylvania as pre-act scares from coal mining for the past 150 years and more acutely from the era of World Wars I and II. Many of these mine sites are contained within Commonwealth lands that are part of the public trust. Abandoned minelands and acid mine drainage represents Pennsylvania's largest environmental problem. Closure of these 250,000 acres is currently impractical because of the enormous volumes of fill materials needed and, consequently, the cost for reclamation is estimated in 1976 dollars as $15 billion. A potential source of materials for applications in mineland restoration is derived from the annual, maintenance dredging of coastal and inland waterways. New York/New Jersey, for example, yearly removes 7 million tons of dredge material. These silt and clay-size particles wash into the harbor from surrounding uplands and sometimes contain organic and inorganic contaminants. A demonstration program has been initiated where fly ash and a chemical activator are mixed with de-watered dredge material to form a chemically stabilized and physically solidified "controlled low strength structural fill" with a compressive strength of 100 to 200 psi and a permeability of 10^{-5} to 10^{-6} cm/sec which is suitable for application as structural fill in mineland reclamation/remediation.

1 BACKGROUND

1.1 *Bark Camp abandoned mine site demonstration project*

The Bark Camp Mine Site Demonstration Project is an abandoned site in Clearfield, Co., PA where extensive coal mining occurred on leased property owned by the Commonwealth of Pennsylvania (Scheetz et al. 1995). With the termination of mining in 1984, the responsibility for the environmental consequences of the mining has reverted to the Commonwealth. The site is located near Penfield in northern Clearfield County and consists of 1400 acres of PA Game Commission, Bureau of Forestry and Department of Environmental Protection land. The challenges at the site are: a) acid mine drainage (AMD) from deep mining amounting for up to 0.18 million gallons per day from Bark Camp #1, and a much smaller flow (25,000 gal/day) from Bark Camp #2; b) two open deep mines; c) an accumulation of refuse from coal processing that occurred on the site; and d) an exposed high wall remaining as the result of surface mining. Bark Camp Run traversing the site supports a native fishery but receives an influx of AMD as it passes through the site rendering the stream severely degraded.

The restoration of the Bark Camp site will require: a) the reclamation of the high wall; b) encapsulation of the tipple refuse; c) injection into the deep mines; and d) remediation of the Bark Camp stream. The high wall at Bark Camp consists of approximately 11,000 lineal feet of exposure extending 40 feet high to a bench and then to an additional 60+ feet above this bench to the hill side. Tipple refuse estimated at approximately 40,000 tons of coal screening materials and other refuse is present on the site plus additional refuse material was originally used to fill in the Bark Camp Run valley for construction of the tipple site. These screenings act as an additional source of further AMD development.

Reclamation and remediation activities utilizing a fly ash-based grout have successfully demonstrated the efficacy of this approach. Tipple refuse was added in lifts to cementitious grouts as the grouts were placed against the high wall in order to stabilize and recontour the slope. The refuse was placed so that it is intimately encapsulated in the grout thus keeping it in direct contact with a material with neutralization potential and prohibiting water from contacting the refuse. As the remediation of the high wall was completed it was covered with soils and planted. To date, as part of this fly-ash grout demonstration, approximately 100,000 tons of fly-ash grout have been placed, soil applied and revegetated.

Finally, the open, void space in the deep mines will be closed by pressure – injecting a cementitious grout. The Bark Camp #1 mine is extensive, stretching for over six miles with a wide lateral expanse. In contrast, the Bark Camp #2 mine is much smaller, extending for nearly a mile and a half with much less lateral expansion.

Studies show that from 6-10 million tons of grout will be required to complete the restoration of the Bark Camp Site. This volume is so large that the use of ordinary construction materials is cost prohibitive. However, many alumino-silicate materials, which are produced in very large volumes as waste products, are available. Fly ash produced from the combustion of coal in utility power plants and fluidized bed combustion facilities served as pozzolanic materials for phase I of this demonstration project. Several approaches for using cementitious materials, based on coal combustion by-products in the remediation/reclamation of acid producing sites, have been demonstrated. Schueck, et al. (1993, 1994, 1996), Fontana (1993), Scheetz et al. (1993, 1994, 1997, 1998) and Zhao (1995) have reported on the use of ashes from fluidized bed combustion in the remediation of tipple refuse piles. In this approach, the naturally cementitious ashes were pressure grouted in the refuse piles. In this case, by-product alkaline earth additions were used to activate the latent pozzolanic potential of the coal ashes. The demonstration at Bark Camp has recently entered into phase II of the study in which stabilized dredge material is being used for structural fill to reclaim and remediate the exposed high wall.

The Bark Camp Demonstration Project will continue at the site for 10-15 years before restoration is complete.

1.2 *Dredge material*

Dredge material is a "giga-scale" waste product in the same class of material as fly ash and slag that is generated by industrial activity and, because of its composition and extremely large volume, generates significant environmental and economic challenges related to its disposal. Dredge material is an unconsolidated, earthen material taken from harbors and rivers that has particle sizes that range from clay to silt and occasionally to sand (see table 1 for classification). These materials form from erosion of surrounding upland areas and are transported by rivers and streams. Nationally, approximately 500 million tons of sediments are dredged annually. As an example, the New York/New Jersey harbor, a naturally shallow water harbor, accumulates sediments at the rate of one foot per year, which amounts to the need for annual maintenance dredging of approximately 7 million tons. On the other hand, areas along the Savannah River in Georgia accumulate sediment at the rate of 12 to 15 feet per year.

Table 1. Size Classification of Sediments

Size	**Fragment**
2 mm	
	Sand
1/16 mm	
	Silt
1/256 mm	
	Clay

Sediments accumulate in aquatic lowlands and, in similar fashion, chemical constituents which may pose a threat to the environment and the health and the welfare of the public, may accumulate along with the sediments. Until 1992, most of the dredged material from the New York/New Jersey region was acceptable for ocean disposal. Changes in federal and state regulations now restrict ocean disposal of certain dredged sediments containing low to moderate levels of chemical constituents that may pose a threat to the environment and human health in an aquatic environment. These changes have caused the volume of sediments removed from the region to temporarily decrease dramatically in recent years. Additionally, these restrictions on ocean disposal have created a critical need for *long-term* upland management options for large quantities of dredged sediments.

Prior to these present day environmental regulations, PCB's and other chemicals were routinely discharged into this nations waterway and found their way into the sediments of many harbors; one of the most notably was the Hudson River estuary. With the secession of uncontrolled environmental discharge, very little, if any toxic materials are now recorded during these maintenance-dredging operations involving ocean disposal. With regard to the dredge material utilized for mine reclamation, the contamination levels in the dredge material are significantly below those levels classified as hazardous or toxic under state or federal laws. However, because of the possibility of contaminants, extreme care is being taken to ensure that no contamination is incorporated into this demonstration project.

2 DREDGE MATERIAL TESTING PROTOCOL

Before any materials are removed from the harbor, the targeted dredge material area is sampled to below the depth of sediment removal. The sampling protocol has been established by a bi-state agreement between New York and New Jersey and signed-off by the Commonwealth of Pennsylvania. If any areas within the dredge material site are found to contain elevated levels of contamination, that area is re-surveyed on a finer grid scale to identify and isolate the contamination. This material will be specially handled and transported to a hazardous materials disposal site (Tavolaro, et al. 1987).

Once approved, the dredge material is recovered and transferred to a process treatment facility where it is stabilized in order to condition it for transfer to the Bark Camp Demonstration Site in central Pennsylvania. After the stabilization, it is again sampled for confirmatory purposes and placed on rail cars. At the Demonstration Site, additional fly ash and a chemical activator are added to the stabilized-dredge material and it is sampled and analyzed a third time before placement as a structural fill.

The Bark Camp Demonstration Project is being conducted pursuant to specific Quality Assurance & Quality Control (QA/QC) measures as required by the various state regulatory agencies involved. These measures are implemented throughout the project to provide assurances to protect the public health, welfare, and the environment. In addition to the many day-to-day operational QA/QC measures implemented, an extensive testing protocol is undertaken to physically and chemically characterize and monitor the dredge materials and industrial by-products used to create the manufactured fill product.

Quality Control measures for this project included characterizing the chemical and/or physical properties of the raw dredge material and any additives utilized in the treatment process prior to commencing dredging. The physical and chemical properties of the additives are determined from testing, and analytical results provided by the generators of these materials. Vendors supplying coal ash products are required to demonstrate that the ash materials meet the Pennsylvania Department of Environmental Protection (PADEP) Module 25 chemical criteria for ash placement in mine reclamation. Additionally, Material Safety Data Sheets (MSDS) that are available for these materials are provided and kept on file in the administrative office at the Bark Camp facility.

The chemical analysis protocol for the dredge material intended for Bark Camp consists of three stages. These include the core sampling and analyses of in-situ dredged material samples (Stage I); the sampling and analysis of the dredge material at the portside, offloading facility, which is intended to confirm that the material being shipped is similar to the in-situ materials

(Stage II); and the sampling and analysis of the treated materials at the Bark Camp facility to assure that the manufactured materials comply with the criteria established in by the PADEP (Stage III).

The sampling and analysis of the dredge material utilized at Bark Camp is conducted in accordance with the requirements of the PADEP Beneficial Use Order (BUO) issued for the Bark Camp Demonstration Project. The BUO includes the requirements of the New Jersey Department of Environmental Protection (NJDEP) guidance manual entitled, "The Management and Regulation of Dredging Activities and Dredged Material in New Jersey's Tidal Waters," October, 1997. The New Jersey manual (referred to as the "Guidance Document") specifies sampling and analytical requirements for upland disposal and beneficial re-use of dredged materials in the State of New Jersey. The manual specifies sampling procedures and frequency requirements, target analyte lists, analytical test methods to be used, and acceptable method detection limits for in-situ sediment samples.

A dredge site specific Sampling and Analysis Plan (SAP) for the in-situ sediment is prepared for and submitted to the NJDEP and PADEP's review and approval. Individual core samples of the in-situ sediment are taken to the proposed project depth plus allowable over-dredge. Composite samples are prepared from the individual core samples. The individual core and composite samples are subjected to the analysis specified in the Guidance Document and the approved SAP.

Bench tests utilizing the dredge material from the in-situ testing and various percentages of proprietary additives are performed to simulate the creation of the manufactured fill. The bench test product samples are analyzed in order to chemically and physically characterize the manufactured fill and to determine the ability of the fill from each mix design tested to stabilize chemical constituents found in the in-situ dredge material.

The analytical and test results for the Stage I in-situ sediment samples are submitted to the PADEP and state of origin of the dredge material for their respective review and approval.

Quality Assurance measures (Stages II and III) for the Bark Camp project are implemented to confirm that the chemical and/or physical properties of the pre-amended dredge material transported to Bark Camp and the manufactured fill were similar to that of the in-situ sediment sample properties.

Stage II testing occurs at the portside facility, and is performed to confirm that the pre-amended dredged material is physically and chemically similar to the dredge material sampled in Stage I. This confirmatory sampling is performed pursuant to the BUO, at a frequency of one composite per 25,000 cubic yards of dredged material. The analytical and test results are reported to the PADEP for its review and information.

The final stage (Stage III) of the QA process is performed after the final amendment of the dredge material at the Bark Camp facility. One (1) composite sample of the manufactured fill is obtained pursuant to the BUO requirement of one composite sample per every 25,000 cubic yards of material. The composite sample is chemically analyzed and geotechnically tested pursuant to the specific requirements specified in the BUO for the manufactured fill. The analytical and test results are reported to the PADEP for its review and information.

The dredge material testing protocol includes, but is not limited to, total analysis as well as TCLP and/or SPLP analysis for RCRA constituents (organic and inorganic), PCB's and dioxins and furans. Dredge material accepted and utilized at Bark Camp may not contain levels of chemical constituents rendering it classified as 'Hazardous' or 'Toxic' material. Additionally, the manufactured fill must possess certain physical properties including a permeability of not more than 10^{-5} cm/sec, and a compressive strength of 35 psi or greater.

3 DREDGING AND PROCESSING OPERATIONS

The states of New Jersey and New York have offered their support of the Bark Camp Demonstration project. The Port Authority of New York and New Jersey (PANYNJ) is responsible for assuring that the channels and harbors serving these states are maintained and navigable. The PANYNJ, in cooperation with state government, has committed to designating specific dredging locations within the harbor for the Bark Camp demonstration project. The initial dredging location within the harbor designated for the Bark Camp project was the City of Perth Amboy Municipal Marina, located on the Arthur Kill waterway, in Middlesex County, New Jersey. Figure 1 is a map of the New York/New Jersey harbor with the Perth Amboy Marina located. The Perth Amboy Marina dredging project is representative of how future projects associated with the Bark Camp Demonstration project will be conducted, and is described below.

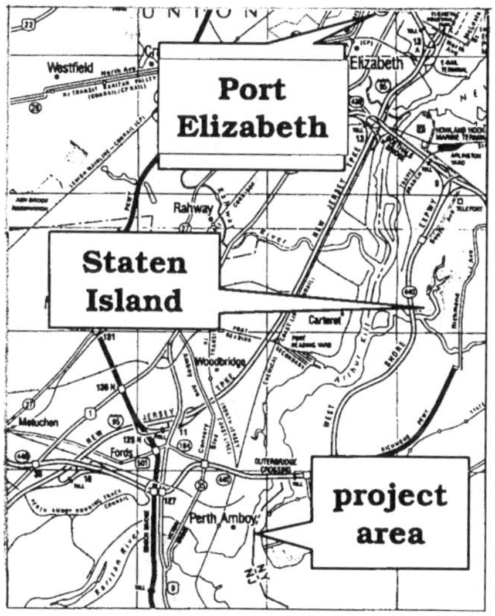

Figure 1. Map sketch of the New York/New Jersey harbor noting dredge material location.

For this project, dredging was conducted with a computerized, environmental, closed clamshell bucket. Experience has shown that this system result in higher solids to water ratio dredge material than is achievable by other techniques. The placement of the bucket can be controlled to +/- 6 inches, and each bucket's location can be tracked.

Off-loading of the dredged material was performed at Consolidated Technologies, Inc.'s transfer and processing facility located on the Arthur Kill waterway in Elizabeth, New Jersey. Off-loading was accomplished utilizing a 50-ton crane equipped with a clamshell bucket. As necessary, the loaded barges were moored at the Elizabeth facility to allow the sediment to settle for dewatering of the barges prior to off-loading. Water decanted from the loaded barges was pumped through a particulate filter to portable frac-tanks. After an adequate period of settling, the water in the tanks was tested for compliance with discharge criteria contained in the Water Quality Certificate. Upon confirmation, the decanted water was discharged from the tanks to the Arthur Kill waterway.

The dredge material was off-loaded into a large receiving hopper and through a series of screens to separate debris from the sediment. The debris was staged for transport and disposal at an alternate, approved disposal facility. The dredge material was placed into a pugmill where it was mixed with coal fly ash to pre-amend (physically stabilize) the material for transport to Bark Camp. The raw material was solidified to a solids ratio of 40/60 to ensure that no free liquids were present in the material. From the pugmill, the pre-amended material was loaded into gondola rail cars for transport to the Bark Camp Facility.

At the mine site, the dredge material was off loaded from the rail cars by a backhoe and transported to the processing site by truck. Here, it was again processed through a pugmill with additional fly ash and a lime activator to attain a water-to-solids ratio of approximately 30/70. On a demonstration basis, approximately 1,000 to a maximum of 4,000 tons per day can be processed in this manner. The activated, processed dredge material is allowed to stand for 24 to 48 hours for a pre-cure before being transported to the highwall where it is placed and roller compacted into place.

Manufactured Fill Being Placed at Bark Camp

Manufactured Fill Being Placed at Bark Camp

Figure 2. Photographs of the placement of the dredge material grout against a highwall.

4 DREDGE MATERIAL TO CONTROLLED LOW STRENGTH CEMENT FOR STRUCTURAL FILL

The cementitious solids (i.e. manufactured fill) that results from the process just described relies upon the pozzolanic reactions between the fly ash, the alumino-silicates in the dredge material and the lime activator. The properties of the resulting solids can best be described as a "controlled low strength fill." The ultimate properties of the cementitious solids, however, are dependent upon the varying nature of the dredge material and the fly ash used in the process.

Two physical/chemical characteristics are monitored for this project, unconfined compressive strength and permeability. Strengths range from 115 to as much as 300 psi in 28 days and permeabilities range from 10^{-5} to 10^{-7} cm/sec.

5 CONCLUSIONS

Innovative approaches, such as the one described here, when used for the remediation of abandoned mineland, represents a win-win-situation for all parties concerned. Moreover, as long as the proper precautions are taken to ensure that no labile contaminants are present in the cementitious grouts, it will represent a very important approach to the solution of several, major, environmental problems. The need for extremely large volumes of low-cost structural fill materials can be met by many of the "giga-scale" materials produced by industry, such as fly ash and the dredge material that is the subject of this publication. The availability of this material on a continual basis has the potential to significantly impact the needs within Pennsylvania for low-cost fill. At the same time, it would eliminate health and safety issues in the 250,000 acres of abandoned mines and restore vast areas of land for recreational use. Furthermore it would remove extremely large volumes of materials from landfill disposal.

REFERENCES

Fontana, Christopher F. Large 1993. Volume use of fly ash for mine reclamation. *M.S. thesis, The Pennsylvania State University, University Park.*

Scheetz, Barry E., Michael J. Menghini, Rodger J. Hornberger, Thomas D. Owens and Joseph Schueck 1997. Beneficial use of coal ash in anthracite and bituminous mine reclamation and mine drainage pollution abatement in Pennsylvania. *Proc. Air & Waste Management Association, Toronto June 1997.*

Scheetz, Barry E., Michael J. Menghini, Rodger J. Hornberger, Thomas D. Owen, Joseph Schueck and Ernest Giovannitti 1998. Beneficial use of coal ash in mine reclamation and mine drainage pollution abatement in Pennsylvania. *Tailings and Mine Waste '98:* 795-805. Rotterdam: Balkema.

Scheetz, B.E., Richard Reifsnyder, and M.R. Silsbee 1994. Design of fly ash based grouts for acid mine drainage abatement. *16th Annual Meeting of the Association of Abandoned Mine Land Programs, Park City, 18–21 September 1994.*

Scheetz, Barry E., Michael R. Silsbee, Christopher Fontana, Xiaogong Zhao, and Joseph Schueck 1993. Properties and potential applications of large volume use of fly ash-based grouts for acid mine drainage abatement. Timothy C. Richmond (ed), *Proc. 15th Annual Abandoned Mine Land Conference:* 64-76.

Scheetz, B.E., M.R. Silsbee and J. Schueck 1995. Field applications of cementitious grouts to address the formation of acid mine drainage. *Sudbury '95, Conference on Mining and the Environment, Sudbury.*

Scheetz, Barry E., Michael R. Silsbee and Joseph Schueck 1998. Acid mine drainage abatement resulting from pressure grouting of buried bituminous mine spoils. *Tailings and Mine Waste '98*: 859-870. Rotterdam: Balkema.

Schueck, Joseph, Terry Ackman, and Barry E. Scheetz 1994. Acid mine drainage abatement using fluidized bed combustion ash grout after geophysical site characterization. *International Land Reclamation and Mine Drainage Conference and the Third International Conference on the Abatement of Acidic Drainage, Pittsburgh, April 24-29, 1994.*

Schueck, Joseph, Mike DiMatteo, Richard Jones, Terry Ackman, Mike Silsbee, and Barry E. Scheetz 1996. Water quality improvements resulting from fbc ash grouting of buried piles of pyritic materials on surface coal mines. *1996 Annual Meeting of the American Society for Surface Mining and Reclamation (ASSMR), Knoxville, April 1996.*

Schueck, Joseph, and Barry E. Scheetz 1993. Acid mine drainage abatement from small, buried piles of tipple refuse using fluidized bed fly ash grout. Timothy C. Richmond (ed), *Proc. 15th Annual Abandoned Mine Land Conference:* 77-103.

Tavolaro, J.F., M.A. Paula, and R.M. Carrato. 1987. Investigations of subaqueous borrow pits and disposal sites for contaminated *dredge material* from New York Harbor. T.R. Patin (ed), U.S. Army Corps of Engineers, Water Resources Center, *Management of Bottom Sediments Containing Toxic Substances: Proc. 13th US/Japan Experts Meeting, Baltimore, 3-5 November 1987:* 308-325.

Zhao, Xiaogong 1995. Cementitious grouts based on fluidized bed combustion ash for environmental restoration. *M.S. thesis, The Pennsylvania State University, University Park.*

Reprocessing, utilization, and treatment

Tailings and Mine Waste '99 © *1999 Balkema, Rotterdam, ISBN 90 5809 025 6*

Survey of three hard-rock acid drainage treatment facilities in Colorado

M.Willow & C.tenBraak
SRK Consulting, Lakewood, Colo., USA

ABSTRACT: Three water treatment facilities receiving acid-rock drainage from historic hard-rock mining operations in the Rocky Mountains of Colorado were surveyed for both cost and process. The purpose of the survey is to familiarize other potential ARD impacted mining operations with treatment processes, methods, and relative costs for systems currently employed for treating ARD. The facilities examined included: the Argo Tunnel water treatment facility in Idaho Springs, CO; the Eagle Mine water treatment facility near Minturn, CO; and the Leadville Mine Drainage Tunnel treatment plant, located in Leadville, CO. Each of the facilities was evaluated for the following criteria: influent and permitted effluent quantity and quality; process design; capital construction costs; operation and maintenance costs; and regulatory rationale and overview. The three facilities are currently achieving their respective regulated discharge water quality criteria but the processes and associated costs are quite varied. Two of the three facilities are operated under CERCLA.

1 INTRODUCTION

1.1 *What is Acid-Rock Drainage?*

Acid-rock drainage (ARD) occurs as a result of natural oxidation of sulfide minerals upon exposure to air and water. Subsequent biological and chemical reactions generate sulfuric acid and mobilize heavy metals associated with the particular hard-rock ore, waste rock, and/or tailings from mining operations. The oxidation reactions may be accelerated by biological activity. Both operational and abandoned mine works, as well as some natural systems, can contribute ARD to the surrounding environment.

Infiltration from snowmelt, rainfall, and groundwater into mine workings can perpetuate the contamination of downstream aquatic resources if this drainage is allowed to reenter the surface environment. High concentrations of sulfate, iron and additional base metals such as lead, zinc, copper, cadmium and manganese may result. Release of ARD to the environment can not only pose an immediate threat to the fauna and flora of a region, but, left unchecked, often impacts human activities such as recreation, irrigation, and livestock watering for the duration of ARD generation.

1.2 *Regional History and Impacts*

Acid-rock drainage is estimated to directly impact nearly 7,000 kilometers (~4,350 miles) of streams and rivers and over 23,000 hectares of lakes and reservoirs throughout this country, while indirectly contributing to contamination of 20,000 kilometers (~12,400 miles) and 70,000 hectares, respectively (Kleinmann, 1989).

While the predominant source of ARD in the eastern United States is coal mining, the principle generating source in the western United States is hard-rock metal mine workings whose drainage often contains metals such as cadmium, lead, nickel, copper and zinc as well as iron and manganese (Wildeman, 1991).

Emerick et al. (1988) suggests that over 2,000 kilometers of streams and rivers in Colorado may be exceeding water quality standards for heavy metal concentrations due to abandoned mining operations. As ore veins are mined, the mineral surfaces become exposed to oxygen and water, especially in the waste rock and tailings piles which are commonly discarded outside mine portals. If, during operation, a mine becomes flooded with infiltrating surface and ground water, additional shafts and adits may be constructed to facilitate drainage or to lower the water table in order to expose deeper deposits. More potentially acid-generating minerals are then exposed to oxidation. The resulting drainage is typically acidic with higher than normal concentrations of metals, and ultimately discharges into the closest surface and groundwater drainage systems. Typical hard-rock acid drainage sources have been investigated in the Argo Tunnel in Idaho Springs, the Eagle Mine in Minturn, Colorado, and the Leadville Mine. Historically, drainage from these sources degraded the quality of the downgradient streams and rivers. Through the efforts of local, state, and federal agencies, significant steps have been taken to improve the quality of these drainage sources and reduce their impacts on the surrounding environment.

1.3 *ARD Treatment Techniques*

It has been estimated that nearly $1 million is spent each day on ARD prevention and abatement by mining companies throughout the U.S. (Kleinmann and Hedin, 1993). The typical ARD wastewater treatment processes employed by active mines are typically composed of three phases: 1) chemical oxidation, 2) lime/caustic soda softening, and/or 3) sulfide addition, though the last method is infrequently used. The objectives of these treatment processes is to eliminate acidity, precipitate heavy metals, and remove potentially harmful substances such as arsenate or antimonate.

The chemical oxidation phase primarily targets iron, manganese, and aluminum in an attempt to precipitate these metals as hydroxides through the addition of oxidizing reagents such as chlorine and potassium permanganate. The pH is raised from around 2 to 7.5. Sedimentation and filtration are required in order to remove the solid phase precipitates prior to water discharge into receiving waters. The addition of lime, caustic soda, and/or soda ash raises the pH to nearly 10.5, thus facilitating the nearly 100-percent removal of dissolved cadmium, copper, lead, nickel, and zinc as hydroxides and carbonates. Again, a secondary sedimentation or filtration step is usually necessary. The introduction of sulfides, typically in the form of hydrogen sulfide (H_2S), at a pH of 8.5, results in heavy metals combining with the aqueous bisulfide ion and precipitating a solid sulfide phase. All three treatment processes are somewhat inefficient and/or potentially expensive. The addition of sulfides also requires constant monitoring to avert a low pH that could cause the release of H_2S gas, and to reduce effluent concentrations of H_2S.

Each process and/or chemical has advantages and disadvantages under each acid mine drainage treatment scenario. According to the British Columbia Acid Mine Drainage Task Force (1989), the main disadvantages of chemical treatment are: it requires continuous operation and maintenance; the relatively high risk of equipment or power failure makes back-up or contingency measures necessary; and the sludge generated during treatment must also be removed and handled, with each chemical generating a different type of sludge residue.

The following sections describe the use of some of these chemical treatment processes at three ARD sources in the Rocky Mountains of Colorado: the Argo Tunnel; the Eagle Mine, and the Leadville Mine Drainage Tunnel. For each case, a brief history of the site is provided, followed by a discussion of the drainage treatment systems employed and the government agencies and rationale for the site.

2 THE ARGO TUNNEL

2.1 *Background*

The Argo Tunnel is located along Clear Creek and Interstate-70, approximately 30 miles west of Denver, Colorado. It is part of a larger Superfund study area which encompasses nearly 400-square miles of the Clear Creek drainage basin. The site was placed on the National Priorities List (NPL) of Superfund sites in September 1983, enabling both the Colorado Department of Public Health and Environment (CDPHE), the lead agency, and the U.S. Environmental Protection Agency (EPA) to prioritize and initiate cleanup of the historic mining properties that were contributing ARD to Clear Creek. The Argo Tunnel was identified as one of a number of sources of ARD in the area which were impacting the Clear Creek system, resulting in aquatic habitat impairment as well as disruption of human recreational and commercial activities. In addition, risks to human health were increased from exposure to heavy metals, primarily lead, arsenic, and cadmium, in the water and soil.

One part of the overall clean-up plan called for the construction of the Argo Tunnel water treatment plant to treat one of the identifiable point sources of pollution. Construction on the treatment plant began in 1996 and continued through 1997. Continuous operation of the plant began in April, 1998.

The State of Colorado and the EPA jointly own and operate the Argo Tunnel water treatment plant. It is administered by the State, as lead agency, under the Federal Comprehensive Environmental Response, Compensation, and Liability Act (CERCLA) or (Superfund) Program.

2.2 *Drainage Quality and Treatment Process*

The quality of the drainage from the Argo Tunnel is presented in Table 1. Estimated flows from the tunnel range from a low of 150 gallons per minute (gpm) to a seasonal high of around 550 gpm, with a yearly average flow rate of approximately 250 gpm. The design capacity for the treatment system is approximately 700 gpm.

Table 1: Argo Tunnel Drainage Water Quality and Regulated 30-Day Average Post-Treatment Discharge Limits

Constituent	Argo Tunnel Drainage (mg/L)	Discharge Limit (mg/L)
pH	2.0 3.0	6.5 – 9.0
Aluminum	15 - 35	1.0
Arsenic	0.02 0.3	0.050
Cadmium	0.05 – 0.3	0.003
Copper	4.0 – 7.0	0.017
Iron (Total)	75 - 200	1.0
Lead	0.005 – 0.15	0.0016
Manganese (Total)	65 – 150	1.0
Nickel	0.15 – 0.35	0.059
Zinc	25 – 50	0.125
Sulfate	1,000 – 3,000	none
Total Dissolved Solids	2,500 – 3,750	none

Zinc and manganese is the primary constituents of concern in the Argo Tunnel drainage. Manganese is perhaps one of the most difficult elements to remove from acid-rock drainage. The primary mechanism for treatment of the Argo Tunnel drainage is sodium hydroxide (NaOH) precipitation (see Figure 1 for process diagram). Flow is currently split between the two process trains, with each side receiving approximately 180 gpm.

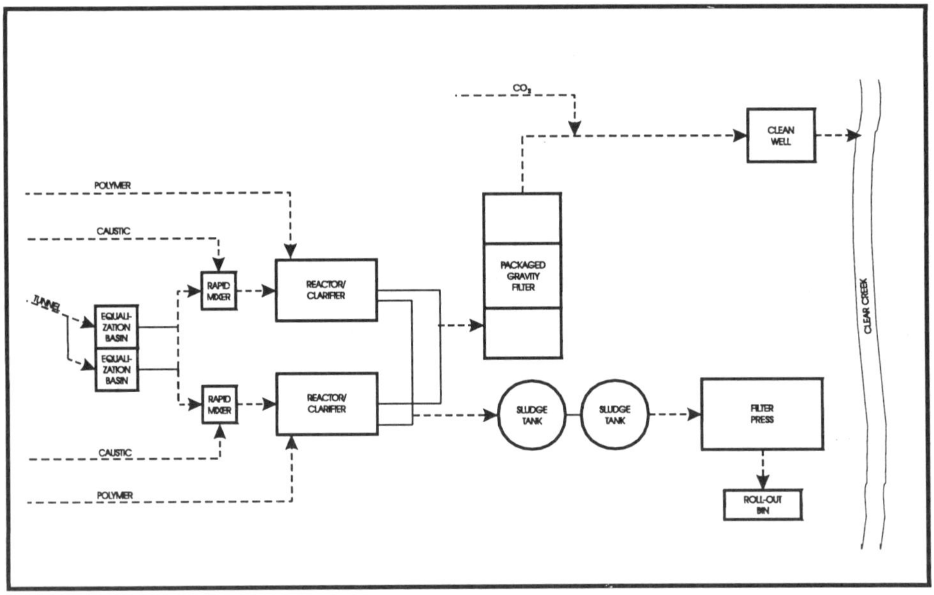

Figure 1: Process Diagram for Argo Tunnel Water Treatment Plant.

The capital costs associated with the feeding and handling of NaOH are generally lower than for other neutralization reagents, though the chemical unit costs may be higher in most cases. Because of the solubility of sodium salts, the total dissolved solids of the effluent tend to increase, though this has not occurred at the Argo Tunnel treatment facility. As a result, less sludge is typically generated. However, the sludge produced during NaOH precipitation is generally more difficult to dewater than lime precipitation sludge. The introduction of calcium in the lime increases the Ca:Mg ratio. It has been observed that sludge with a Ca:Mg ratio less than 2 is more difficult to dewater while a sludge with a Ca:Mg ratio greater than 5 is more easily dewatered (AWWA, 1990).

A polymer is added to enhance flocculation of the hydroxide precipitates in the clarifier. Overflow from the clarifier is routed to a gravity filter to remove any unsettled solids. Carbon dioxide (CO_2) is then added to adjust the pH of the effluent to acceptable levels before it is discharged to the cleanwell and ultimately Clear Creek. Underflow from the clarifier is routed to holding tanks. The solids content of the clarifier sludge underflow is approximately 3 percent. Filter presses are used to dewater the sludge and increase the solids content to approximately 18 percent. It was originally anticipated during the initial design period that a higher solids content could be achieved. However, this was not realized once operations began. The result is the generation of as much as three to four times as much sludge as expected. This required the introduction of a second filter press to accommodate the increased volumes of sludge. The dewatered sludge is disposed of at a regional landfill.

2.3 *Costs*

The Argo Tunnel water treatment plant cost approximately $5 million to construct. Because the plant has only been in operation since April of this year, exact costs associated with operation and maintenance (O&M) are not available. However, O&M costs for the facility are estimated to be on the order of $980,000. The majority of these costs, $410,000, are associated with labor costs for the approximately 10 full-time employees who operate and maintian the plant 24 hours a day. The cost of NaOH is approximately $190,000 per year, with the remaining reagents averaging about $79,000 per year. The remaining costs, $310,000, are for general operation of the facility, including sludge disposal, which is approximately $45/ton of dewatered sludge.

3 THE EAGLE MINE

3.1 *Background*

The Eagle Mine is located between Gilman and Minturn, in Eagle County, approximately eight miles southwest of Vail and one hundred miles west of Denver. The site includes the Eagle Mine workings, the town of Gilman, eight former mine tailings piles, Rock Creek Canyon below Highway 24, and approximately 14 waste rock piles.

The history of mining at the site began in the late 1870's. Gold and silver were mined there until the early 1900's, when economic lead and zinc deposits were discovered. From 1919 to 1979, an underground mill was operated to extract lead and zinc. Subsequent copper and silver extraction continued until 1984 when mining was halted, and the underground workings were allowed to flood with groundwater and surface infiltration. The State of Colorado subsequently filed notice and claim against the mine for "natural resource damages'" resulting in the mine being placed on Superfund's NPL in 1986. The State believed that significant threats to local aquatic resources and human health was posed by lead, zinc, cadmium, arsenic, and manganese associated with the mining wastes and discharge from the underground workings. Drainage from the mine was entering the Eagle River, which flows northwesterly through the site and past the town of Minturn.

An attempt was made as early as 1989 to reduce the amount of ARD entering the Eagle River by sealing off the underground mine. As water levels rose within the mine workings, the hydraulic head also increased and resulted in the increased seepage of underground ARD into the Eagle River. A water treatment plant was constructed in 1990 and brought on-line to collect mine seepage, groundwater at the main tailings pile, and run-off from various tailings areas which were contributing heavy metals into the river.

While the treatment facility is owned and operated by Viacom and McLaughlin Water Engineers, Ltd., respectively. The State of Colorado has regulatory authority over the operation under the Superfund program.

3.2 *Drainage Quality and Treatment Process*

As part of the treatment program, water levels in the underground mine must be kept at an elevation that minimizes the direct discharge of acid-rock drainage to the Eagle River. Pumping is used to suppress the water table in the mine. A surge pond holds surface runoff and drainage pumped from the mine prior to treatment. Use of a surge pond allows for continuous pumping of mine drainage, even during periods of treatment plant maintenance.

Annual average inflow to the Eagle Mine water treatment plant has increased from 250 gpm in 1996 to 400 gpm in 1998, ranging from 0 to 450 gpm. The design capacity of the facility is 200 gpm. The quality of the drainage from the Eagle Mine is presented in Table 2. As noted, the treatment plant is currently operating at twice the design capacity. This is largely due to the fact that the influent water quality has been improving with time, thus reducing treatment requirements.

Table 2: Eagle Mine Water Quality and Regulated 30-Day Average Post-Treatment Permit Limits

Constituent	Eagle Mine Drainage (mg/L)	Permit Limit (mg/L)
pH	5.1	6.5 – 9.0
Cadmium	0.25	0.008
Copper	1.08	0.15
Iron (Total)	320	9.8
Lead	0.07	0.12
Manganese (Total)	300	31
Zinc	130	0.4
Sulfate	4,000	none
Total Suspended Solids	70	20

*Note: Influent and permit limit information for aluminum, arsenic and nickel were not reported.

Similar to the Argo Tunnel, manganese is the primary constituent of concern in the Eagle Mine drainage. Unlike at the Argo Tunnel, however, calcium hydroxide [$Ca(OH)_2$ or lime] is used as the primary neutralization reagent. Figure 2 presents the treatment process employed at the Eagle Mine.

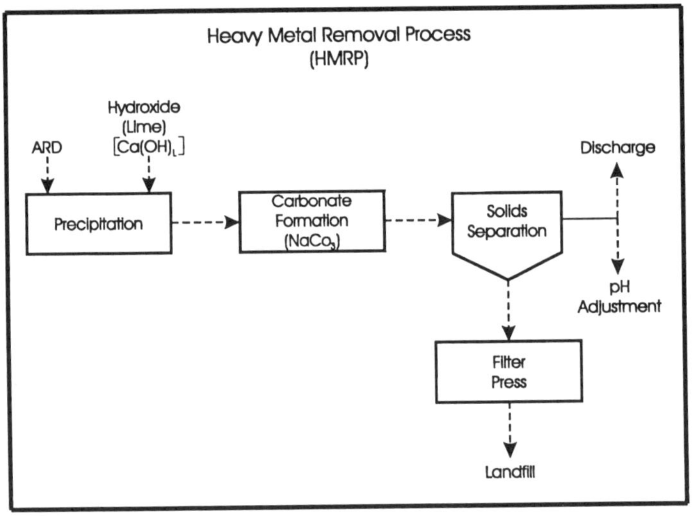

Figure 2: Process Diagram for the Eagle Mine Water Treatment Plant

Lime is generally less costly than caustic. However, use of lime at the Eagle Mine has resulted in gypsum scaling associated with the excess calcium introduced to the water. This has led to subsequent maintenance problems and an increase in total suspended solids in the treated water (MWE, 1998). To help reduce some of the scaling, sodium carbonate ($NaCO_3$ or soda ash) is added directly to the waste water stream in a second stage of the precipitation process. Soda ash generally has the same neutralizing capacity as lime or caustic, but adds carbonate to the system. In some cases, it is used to remove metals (e.g., cadmium and nickel) as carbonate species instead of hydroxides. An anionic polymer is also added to assist flocculation.

The pH of the clarifier overflow is adjusted prior to discharge to the Eagle River. Plate and frame presses are used to dewater the sludge underflow from 4 – 8 percent solids to a solids content of 50 percent. The dewatered sludge is disposed of in an on-site landfill at a cost of approximately $15 per cubic yard (cy).

3.3 *Costs*

At a capital cost of $600,000, the Eagle Mine treatment plant is the least costly of the three facilities reviewed. Annual O&M is approximately $756,000, which include $276,000 for all reagents and $400,000 for labor and expenses for five full-time employees. The remaining costs, including replacement parts, laboratory analyses, supplies, utilities and sludge disposal, averages approximately $80,000 per year.

4 LEADVILLE MINE DRAINAGE TUNNEL

4.1 *Background*

Leadville is located approximately 120 miles west of Denver. At 10,152 feet, it is the highest incorporated town in the United States. Mining for gold, silver, copper, zinc, manganese and lead in the area began in 1859. While the Leadville Mine Drainage Tunnel is not regulated under Superfund, the EPA did place several portions of Colorado's Leadville Mining District on its NPL in 1983. Together with the State of Colorado and mining companies, prioritized cleanup of the historic mining properties was initiated in the hope of reducing contamination in the Arkansas River watershed.

Starting in 1989, EPA and the U.S. Bureau of Reclamation initiated a plan to treat drainage from the Leadville Mine Drainage Tunnel, one of two principle point-source discharges which drain many of the underground workings in the district. The other point-source discharge, the Yak Tunnel, is more than four miles long, and discharges almost 210 tons of metals per year. These metals are currently being treated/recovered at the American Smelter and Refining Company (ASARCO) treatment plant. Both treatment plants operate on a continuous basis in an attempt to improved the water quality in the Arkansas River. Only the Leadville Mine Drainage Tunnel treatment facility is presented in this paper. The Leadville Mine Drainage Tunnel treatment plant was brought on line in 1992.

4.2 *Drainage Quality and Treatment Process*

Discharge from the Leadville Mine Drainage Tunnel occurs at an annual average rate of approximately 1,600 gpm. A bulkhead at the tunnel entrance allows for equalization and regulation of the discharge to the plant, though the flow can reach a maximum of 2,000 gpm.

The quality of the tunnel drainage, along with the plant's regulatory discharge permit limits are presented in Table 3.

Table 3: Leadville Mine Drainage Tunnel Water Quality and Daily Maximum Post-Treatment Permit Limits

Constituent	Tunnel Drainage (mg/L)	Permit Limit (mg/L)
pH	5.9	6.5 – 9.0
Cadmium	0.02	0.0044
Iron (Total) (30-day Average)	1.4	1.0
Manganese (Total) (30-day Average)	1.8	1.0
Zinc	3.6	0.127
Sulfate	370	none

The primary treatment process utilizes CO_2 stripping followed by precipitation and solid contact clarification using a 50-percent NaOH feed solution. A schematic of the treatment process is provided in Figure 3.

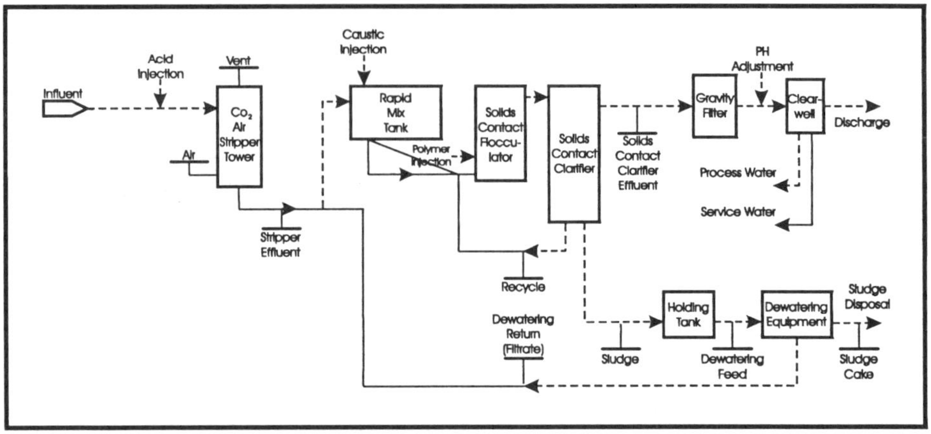

Figure 3: Process Diagram for Leadville Mine Drainage Tunnel Water Treatment Plant.

Gravity filtration is used on the clarifier overflow to remove residual solids. The pH is then adjusted prior to discharge. The sludge underflow, at a solids content of approximately 3 – 5 percent, is dewatered to 35 – 45 percent solids using plate and filter presses. Non-hazardous sludge is transported to an appropriate landfill, approximately 150 miles away in Colorado Springs. A closer disposal facility would greatly reduce sludge disposal transportation fees.

As a note, the equipment used in the Leadville Mine Drainage Tunnel treatment facility is the same equipment used at the Argo Tunnel facility, though on a slightly larger scale.

4.3 *Costs*

At $7.5 million to construct, the Leadville Mine Drainage Tunnel water treatment plant was to most expensive of the three facilities reviewed. It should be noted, however, that the average influent flow rate is at least four times greater than the Argo Tunnel and Eagle Mine systems. In addition to the actual capital costs associated with the plant, another $7.5 million was expended on litigation leading to construction.

Annual O&M costs for the facility average $800,000. $220,000 is expended on treatment reagents, and $182,000 is expended on the four full-time employees at the site. The cost of sludge disposal is approximately $50/cy for non-hazardous material.

5 CONCLUSIONS

All three treatment facilities currently meet their respective effluent discharge requirements. Table 4 presents a line-item comparison of the three facilities with respect to flow rate, capital cost, O&M, and unit treatment costs.

Both the Eagle Mine treatment facility and the Leadville Mine Drainage Tunnel treatment facility have been in operation for more than six years. The Argo Tunnel treatment facility is the newest of the three plants, yet treats the least amount of water for the highest cost.

While the water from the Eagle Mine contains the highest concentrations of metals, the costs associated with metal removal are greatest at the Leadville Mine Drainage Tunnel treatment plant where the concentration of metals is lowest. This is primarily due to the lower permit limits coupled with the high flow from the tunnel.

Table 4: Comparison of Water Treatment Facilities

Facility	Avg. Flow Rate (gpm)	Capital Cost	Annual O&M Costs	Cost / 1000 gal Treated	Cost / lb. of Metal Removed *	Sludge Disposal
Argo Tunnel	250	$5.0 M	$0.98 M	$7.45	$2.88	$45/ton
Eagle Mine	400	$0.6 M	$0.76 M	$3.60	$0.61	$23/ton
Leadville	1,600	$7.5 M	$0.80 M	$0.95	$23.0	$75/ton

* 'Metal Removed' based on difference between average influent chemistry and regulated permit limit for Al, As, Cd, Cu, Fe, Pb, Mn, and Zn.

6 ACKNOWLEDGMENTS

The technical and financial information contained in this paper was provided by representatives of the participating facilities. To this end, the authors wish to thank Mr. Ron Abel and Ms. Mary Scott of the Colorado Department of Public Health and Environment; Mr. Jim McLaughlin of McLaughlin Water Engineers, Ltd. and the Eagle Mine; and Mr. Brad Littlepage of the U.S. Bureau of Reclamation for their assistance and technical information on their respective water treatment facilities.

REFERENCES

American Water Works Association (AWWA). 1990. Water Quality and Treatment: A Handbook of Community Water Supplies. Fourth Edition. New York: McGraw-Hill, Inc.1195 p.

British Columbia Acid Mine Drainage Task Force. 1989. Draft acid rock drainage technical guide: Volume 1. Prepared by Steffen Robertson and Kirsten (B.C.), Inc., Vancouver, B.C., Canada. August.

Emerick, J.C., W.W. Huskie, and D.H. Cooper. 1988. Treatment of discharge from a high elevation mine in the Colorado Rockies using an existing wetland. Pp. 345-351. In: Proceedings of a Conference on Mine Drainage and Surface Mine Reclamation, 1: Mine Water and Mine Waste. United States Department of the Interior, Bureau of Mines Information Circular IC-9183.

Kleinmann, R.L.P. 1989. Acid mine drainage: U.S. Bureau of Mines researches and develops control methods for both coal and metal mines. Engineering and Mining Journal. 190: Pp. 16I-16N.

Kleinmann, R.L.P., and R.S. Hedin. 1993. Treat mine water using passive methods. Pollution Engineering. 25(13): Pp. 20-22.

Wildeman, T.R. 1991. Drainage from coal mines: chemistry and environmental problems. Pp. 499-512. In: D.C. Peters, (ed.), Geology in Coal Resource Utilization. American Association of Petroleum Geologists, Tulsa, Oklahoma.

Tailings and Mine Waste '99 © 1999 Balkema, Rotterdam, ISBN 90 5809 025 6

Phosphogypsum uses in agriculture and it's impact on the environment

J.E.Rechcigl & I.S.Alcordo
University of Florida, Range Cattle Research and Education Center, Ona, Fla., USA
R.C.Littell
Statistics, University of Florida, Gainesville, Fla., USA
C.E.Roessler
University of Florida, Ona, Fla., USA

ABSTRACT: Phosphogypsum is a by-product of the manufacture of phosphoric acid from phosphate rock and is a potential source of sulfur and calcium for crops. There are currently more than 700 million Mg of phosphogypsum in Florida alone stacked in waste piles and an additional 30 million Mg produced annually. A field study was conducted to determine whether addition of phosphogypsum to bahiagrass would increase production and quality. Results indicate that addition of up to 4.0 Mg/ha phosphogypsum increased bahiagrass yields, protein content, and *in vitro* digestibility of forage. This study has demonstrated that phosphogypsum can be used as an alternative source of sulfur and calcium for forage crops.

1 INTRODUCTION

Sulfur deficiencies in plants have been reported in more than 45 states, including Florida. Although sulfur is usually considered a secondary plant nutrient, it still needs to be viewed as one of the major nutrients essential for crop growth along with nitrogen, phosphorus, and potassium. Sulfur is required by plants for the synthesis of certain amino acids which are required for protein production. If sulfur is limiting, forage quality, as well as quantity, will be reduced. In fact, sulfur deficiencies are often confused with nitrogen deficiency. While in less severe cases of sulfur deficiency visual symptoms may not always show up, crop yield and quality can be adversely affected.

Until recently, little attention has been given to the need for sulfur fertilization in Florida and other parts of the country. This is understandable since in the past low analysis fertilizers contained sulfur impurities sufficient to meet the nutrient requirements for crop production. However, fertilizer manufacturing technology has now become highly advanced and consequently high analysis fertilizers, such as triple superphosphate and diammonium phosphate, are free of sulfur impurities. As a result, sulfur deficiencies are becoming more pronounced and widespread throughout the world. Coarse textured soils, such as those commonly found in Florida, may also exhibit sulfur deficiencies because of their very low nutrient holding capacity.

It is important to note that sulfur fertilization will increase yields and improve the quality of crops only if the plants are deficient in sulfur. The sulfur status of a crop is best determined by having plant tissue samples analyzed for sulfur. Tissue analysis is better correlated to crop yield than soil tests for sulfur. For grasses, the level of sulfur in plant tissue should range from 0.2 to 0.5 percent. If the level of sulfur falls below 0.2

percent, sulfur deficiency is indicated and the grass should respond to sulfur fertilization.

Over the years, we have demonstrated that addition of sulfur can increase production of harvested forages, such as bahiagrass (*Paspalum notatum* Flugge), by as much as 25 percent and protein by 1.2 percent. In these studies, the sources of sulfur were ammonium sulfate and potassium sulfate which are relatively expensive. Bahiagrass, which is an important forage crop in Florida, is grown on nearly five million ha, exceeding production of all other improved grasses combined. To provide sulfur for this land area would be a considerable expense.

There is a need to find an alternative economic source of sulfur which would be more affordable to growers than traditional sulfur fertilizers. Phosphogypsum, which is primarily gypsum ($CaSO_4.2H_2O$), is a by-product of the wet-acid production of phosphoric acid from rock phosphate, is a potential low cost source of sulfur and calcium for forages and other crops. In Florida alone, there are more than 700 million Mg of phosphogypsum stored in waste stacks, with 30 million tons being added to the stacks annually. In the entire country the total amount of phosphogypsum in stacks is estimated at seven billion tons. Until now, phosphogypsum has had little commercial use because it contains low levels of radium (8-30 pCi Ra-226/g), raising concern over its potential harmful effects.

This paper presents the results of the agronomic effects of phosphogypsum, applied at agronomic rates, on bahiagrass.

2 MATERIALS AND METHODS

Over the past ten years, we have evaluated the agronomic and the environmental impact of phosphogypsum use on pastures in Florida. The studies were conducted at the University of Florida Range Cattle Research and Education Center at Ona, Florida. The analysis for the phosphogypsum used in the study are shown in Table 1.

Yearly phosphogypsum rate of 0.4 Mg/ha and one-time rates of 2.0 and 4.0 Mg/ha were applied to long-established bahiagrass plots located on a Myakka fine sand soil. No-phosphogypsum plots serve as controls. Bahiagrass forage was harvested monthly from May until December in order to assess the influence of phosphogypsum on forage production and quality. Soil samples were collected annually to a depth of 1 meter. Water samples were collected after each heavy rain to a depth of 1.5 meter. Forage, soil samples , and groundwater samples were analyzed for various plant nutrients. Fluorides were also determined in forage and groundwater samples.

3 RESULTS AND DISCUSSION

3.1 *Forage yield*

Regardless of the rate or time of application, phosphogypsum tended to increase regrowth and mature (hay) bahiagrass yields by approximately 20 percent (Figure 1) over the 3-year period. Significant increases in regrowth and hay yields were noted for the 0.4 Mg/ha as well as at higher rates, for at least two years, and over the 3-year period for all rates. Other studies have also shown that addition of phosphogypsum, mined gypsum, or other sources of sulfur can increase forage production when sulfur is deficient (Alcordo and Rechcigl, 1993).

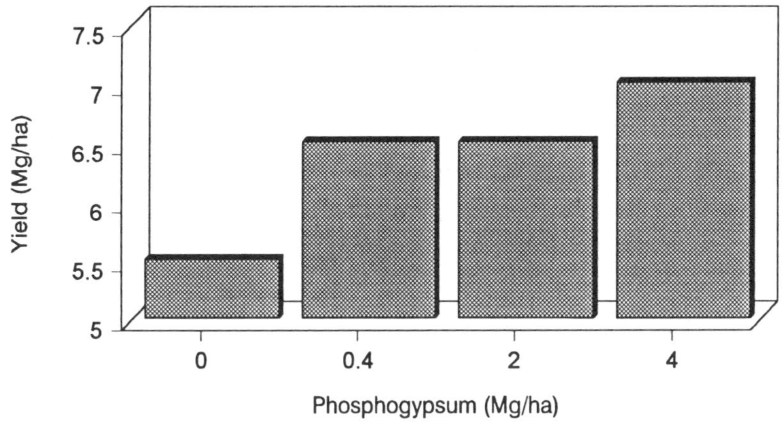

Fig. 1. Influence of phosphogypsum on bahiagrass yields.

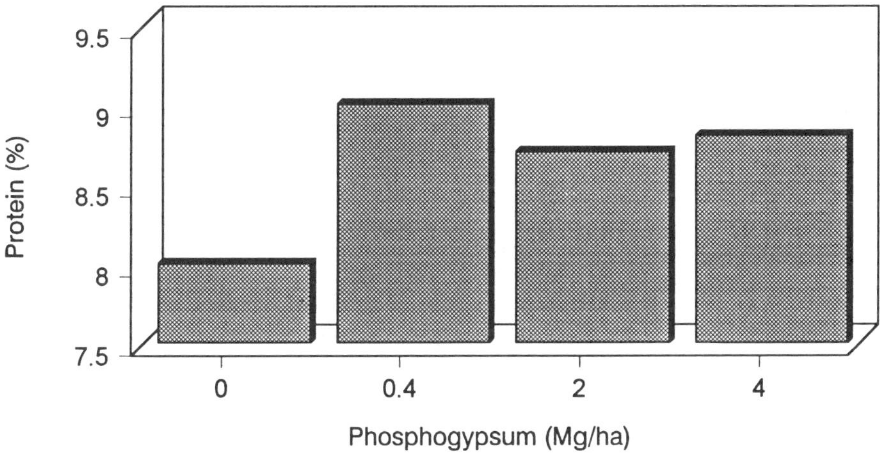

Fig. 2. Influence of phosphogypsum on bahiagrass protein content.

3.2 *Forage quality*

Phosphogypsum tended to increase crude protein content of the mature bahiagrass forage, by as much as 1% in all years and over the 3-year period (Figure 2), and the digestibility, by as much as eight percentage units (Figure 3) in some individual harvests during the first year (1990). This is in agreement with other studies showing that addition of sulfur will increase the nutritive value of forages on sulfur deficient soils. Increases in both digestibility and protein content of forage are known to increase the weight gains in livestock. Phosphogypsum increased the sulfur (Figure 4) and calcium (Figure 5) content of the bahiagrass tissue. The calcium content ranged from 0.42 to 0.60%. Sulfur content ranged from 0.18 to 0.40% for the 0 and 4.0 Mg phosphogypsum/ha treatments, respectively.

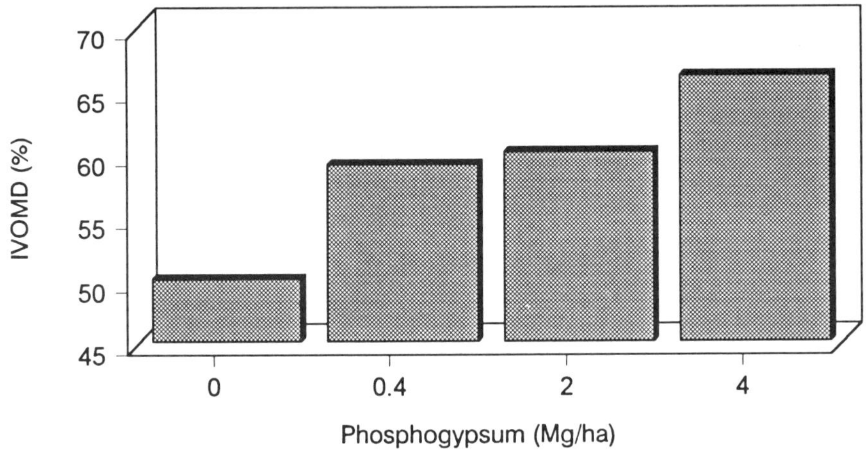

Fig. 3. Influence of phosphogypsum on bahiagrass digestibility.

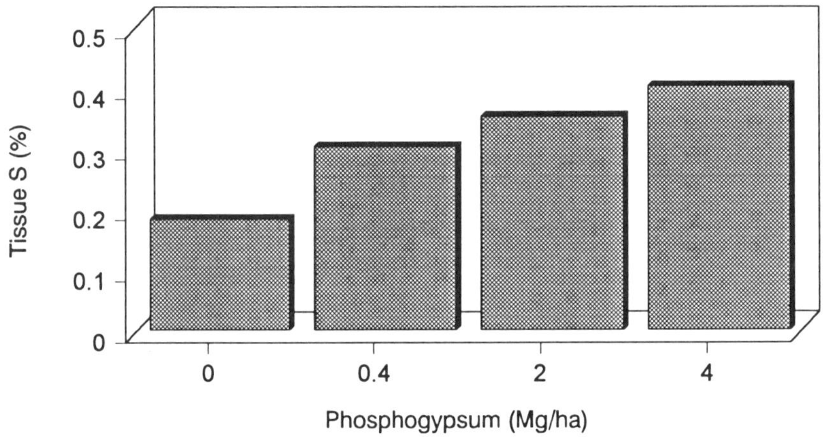

Fig. 4. Influence of phosphogypsum on bahiagrass tissue sulfur content

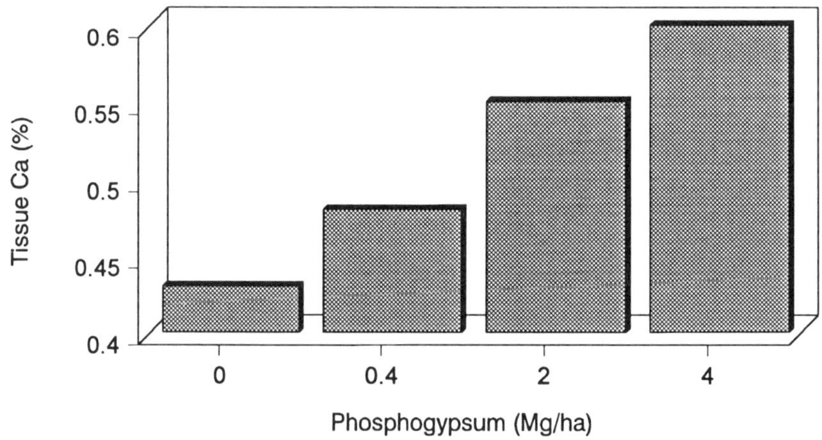

Fig. 5. Influence of phosphogypsum on bahiagrass tissue calcium content.

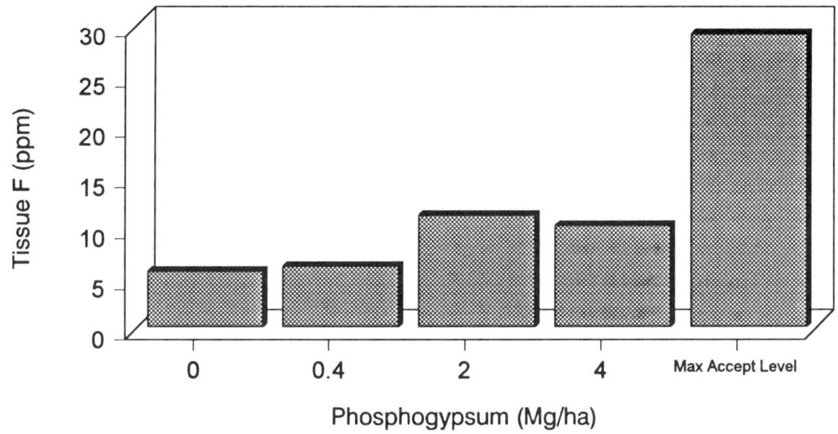

Fig. 6. Influence of phosphogypsum on bahiagrass tissue fluoride content.

Phosphogypsum also slightly increased the fluoride content of the bahiagrass tissue from 7.2 mg/kg for the control to 8.6 mg/kg for the 4.0 Mg phosphogypsum/ha (Figure 6). However the content was well below the 30 ppm maximum acceptable level for livestock intake. The low levels of tissue fluoride found in this study are of some importance since high levels of fluoride may bring about the loss of teeth in cattle.

This study demonstrates that phosphogypsum can increase both the yield and quality of bahiagrass, and possibly other forage grasses. This can, in turn, lead to greater livestock weight gains and increased stocking rates, resulting in increased profits of ranchers. Phosphogypsum may, thus, be a viable and an economical source of sulfur and calcium for forage production.

REFERENCES

Alcordo, I. S. and J. E. Rechcigl. 1993. Phosphogypsum in agriculture: a review. Advances in Agronomy. 49:55-118.

Alcordo, I. S. and J. E. Rechcigl. 1995. Phosphogypsum and other by-products gypsums. In: Soil Amendments and Environmental Quality. J. E. Rechcigl (ed.). pp. 365-425. Boca Raton, FL: CRC/Lewis Publishers.

Tailings and Mine Waste '99 © 1999 Balkema, Rotterdam, ISBN 90 5809 025 6

Characteristics and utilization potential of the mining refuse from the Hamtae Coal Mine

Daesuk Han
Korea Institute of Geology, Mining and Materials, Taejon, Korea

ABSTRACT: Coal mine refuse is one of the three solid by-products of the coal industry in the Republic of Korea. Little attention has been paid to the refuse as a potential replacement of natural material.

A laboratory investigation on the refuse material collected from the disposal site of the Hamtae Coal Mine was undertaken including gradation, Atterberg limits, specific gravity, compaction characteristics, CBR, permeability, freeze/thaw durability, wet/dry durability, petrography, particle shape, leachate quality, sulfate content, pH, etc. The test procedure and material characterization are discussed in the paper.

The refuse was classified as GW-GM according to the Unified Soil Classification System, containing 5.1% fines by weight. Based on all the test results, the refuse may be utilized as a sub-base material for road and parking lot.

1 INTRODUCTION

Coal, anthracite, is the only resource that can be used as an energy-producing fuel in the Republic of Korea. Coal mine refuse is one of the three by-products of the coal industry in the country, the other two resulting from the combustion of the pulverized coal mainly used for electric power generation and the briquette chiefly utilized for cooking and heating. Little attention has been paid to the refuse as a replacement of conventional construction material that becomes scarce.

The Hamtae Coal Mine is situated in the Samcheok Field, one of the major coal fields in the northeastern part of the country. It was opened in 1954 and closed in 1993 due to not only the increase of mining cost but also the decrease of coal consumption; about 60,000 metric tons per year were produced in the mid 1980's.

A main laboratory investigation on the coal refuse from the Hamtae Coal Mine was undertaken before 1993 and a further research on it was carried out in 1997 for the following purposes: (1) to find out its physical, engineering, and chemical characteristics and (2) to evaluate its utilization potential as a construction material.

2 LBORATORY INVESTIAGATION

A laboratory testing program included tests for grain-size distribution, Atterberg limits, specific gravity, compaction characteristics, California bearing ratio, freeze/thaw durability, wet/dry durability, permeability, point load strength, petrography, particle shape analysis, leachate quality, sulfate content, pH, and loss on ignition.

2.1 *Sampling*

About 200 kg of sample was collected from the main disposal site (Figure 1) having about 300,000 m^3 of refuse. Particles larger than 3-inch sieve opening were not collected as they were not included in the testing program, the amount being determined in field at about 9% by weight.

Figure 1. Refuse disposal site of the Hamtae Coal Mine. Note the refuse below the dotted line.

2.2 *Sample preparation*

The coal refuse collected was first oven-dried at 105°C and then divided into 20 groups of particles by using a riffle box in order to have all kinds of laboratory tests be performed on the samples with a very similar gradation, each weighing about 10 kg. A riffle box was used whenever necessary during the course of the subsequent laboratory testing.

Throughout this and subsequent chapters, the term, "waste material No. 1" refers to the particles passing a 3-inch sieve, while that of "waste material No. 2" does to the refuse consisting of the grains smaller than 3/4-inch sieve opening; the latter one was used for most of the laboratory tests.

2.3 *Physical characteristics*

The grain-size distribution test for the waste material No. 1 was carried out in accordance with ASTM D 422. The material was wet-sieved using 1.5-inch (38.1 mm), 3/4-inch (19.0 mm), 3/8-inch (9.5 mm), No. 4 (4.75 mm), No. 20 (2.35 mm), No. 40 (0.42 mm), No. 100 (0.15 mm), and No. 200 (0.075 mm) sieve. The waste material No. 2 also was wet-sieved using all the above-mentioned sieves but 1.5-inch and 3/4-inch one. Grain-size distribution curves were constructed using the test results (Figure 2). The composition of coarse and fine particles are presented in Table 1.

The Atterberg limits for the waste material No. 1 were conducted in accordance with ASTM D 4318, being determined at the values presented in Table 1.

The specific gravity of the waste material No. 2 was undertaken following the Gas Jar Method of BS 1377: 1975, Test 6(A). The test result also is presented in Table 1. The apparatus for the Gas Jar Method is shown in Figure 3.

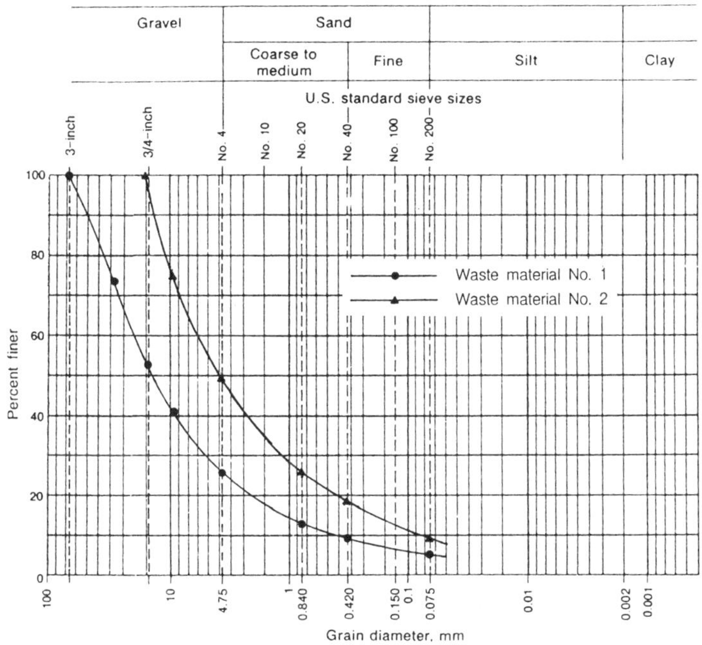

Figure 2. Grain-size distribution curves for the waste material Nos. 1 and 2.

Table 1. Physical properties for the waste material Nos. 1 and 2.

Sample No.	Gradation			Atterberg limits, %			Specific gravity
	Gravel	Sand	Fines	LL	PL	PI	
Waste material No. 1	74.0	20.9	5.1	20.8	19.0	1.8	-
Waste material No. 2	51.0	40.0	9.0	-	-	-	2.64

Figure 3. Gas Jar and mechanical shaker.

2.4 *Engineering characteristics*

The moisture-density relationship for the waste material No. 2 was determined in accordance with the procedure specified in ASTM D 698 Method D, "Standard Test Method Using 2.5-kg Rammer and 304.8-mm Drop"; a 152-mm diameter mold was used. A diagram was constructed illustrating the relationship between dry density and moisture content (Figure 4). Based on the figure, the maximum dry density and optimum moisture content are 2.168 g/cm^3 and 7.3%, respectively.

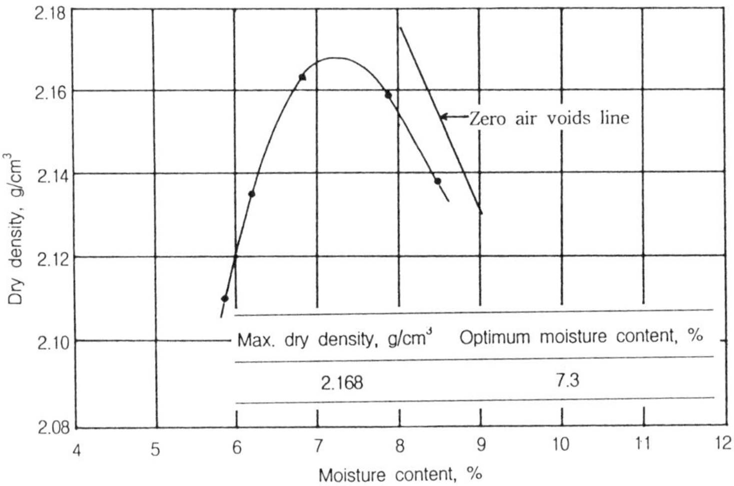

Figure 4. Moisture-dry density relationship for the waste material No. 2.

The waste material No. 2 was compacted to its respective optimum moisture-density condition to prepare two specimens for the determination of California bearing ratio. One specimen was tested immediately after preparation in accordance with ASTM D 1883; the other was soaked until swelling ceased prior to determining its CBR. The platen speed and annular surcharge weight employed were 1.27 mm/min and 4 kg, respectively. For the soaked specimen, swell was monitored; the swelling ceased within 24-hour. Table 2 presents the test results..

Table 2. CBR for the unsoaked and soaked specimens of the waste material No.2 and the specimens subjected to freeze/thaw and wet/dry cycles.

Unsoaked		Soaked			Freeze/thaw cycles		Wet/dry cycles	
Penetration, mm	CBR, %	Penetration, mm	CBR, %	Swell, %	Penetration, mm	CBR, %	Penetration, mm	CBR, %
2.5	43.1	2.5	16.6	0.94	2.5	26.3	2.5	11.5
5.0	42.1	5.0	21.5		5.0	31.4	5.0	15.6

Two specimens also were compacted to its optimum moisture-density condition for the determination of their durability. One was used to evaluate its freeze/thaw durability. The specimen was subjected to two cycles of freezing and thawing; each cycle consisted of placing it in a temperature-controlled chamber, lowering the temperature of the chamber from 21 °C to –15 °C during a 5-hour period and maintaining at that temperature for 9 hours. After that stage the temperature was raised to 21°C during another 5-hour period and maintained at that temperature for 9 hours. After the second freeze/thaw cycle, the specimen was tested under 1.27 mm/min platen speed and 4 kg annular surcharge for its CBR. The other was tested for its wet/dry durability. It was subjected to two cycles of wetting and drying, each cycle comprising submerging it in water at room temperature for 24 hours followed by oven drying at 105°C for 24 hours. After the second cycle, it was soaked for 48 hours prior to determining its CBR. The CBR test was undertaken under the platen speed and annular surcharge mentioned above. The test results appear in Table 2.

Using the waste material No. 2, a specimen was prepared at its optimum moisture-density condition following ASTM D 698 Method C and tested for its permeability in accordance with a generally accepted procedure of falling head permeability test. The coefficient of permeability at 20°C was determined at 6.0×10^{-6} m/s. For the waste material No. 1, its coefficient of permeability was estimated using the Hazen's formula, $k = 0.01\ (D_{10})^2$ m/s; the value was 5.0×10^{-3} m/s.

Point load strength test was conducted in accordance with the procedure recommended by ISRM for the fragments selected among the particles of the waste material No. 1 which are coarser than 3/4-inch sieve opening. The test results are presented in Table 3. Unconfined comperessive strength is approximately equal to 22 times $I_{S(50)}$. The unconfined compressive strength computed by the use of the relationship also is listed in Table 3.

Table 3. Point load strength and unconfined compressive strength for the particles coarser than 3/4-inch sieve opening.

Rock type	Point load strength, $I_{S(50)}$, MPa		Unconfined compressive strength, $22 \times I_{S(50)}$, Mpa	
	Range	Average	Range	Average
Sandstone	2.76 – 19.90	11.07	60.72 – 437.80	243.44
Shale	3.60 – 6.71	5.25	79.20 – 147.62	115.50
Miscellaneous	1.90 – 14.71	8.31	27.57 – 319.22	182.82

2.5 *Petrography and particle shape*

The particles between 3-inch and 3/4-inch sieve opening were collected from the waste material No. 1. They were grouped into five rock types, sandstone, shale, coaly shale, coal, and miscellaneous rock; their weight percentage was determined (Table 4). All the groups were then subjected to shape analysis. The analysis was conducted on the shape classification of BS 812: Part 1: 1975, rounded, irregular, angular, flaky, elongated, and flaky and elongated. The results of the particle shape analysis also are presented in Table 4.

Table 4. Rock type and shape analysis for the particles between 3-inch and 3/4-inch sieves.

Rock type	Composition, %	Particle shape, % by weight			
		Angular	Flaky	Elongated	Flaky & elongated
Sandstone	57.52	31.15	13.64	9.65	3.08
Shale	25.25	7.86	9.90	6.47	0.92
Coaly shale	1.69	0.44	1.16	0.09	-
Coal	1.16	0.49	0.11	0.56	-
Miscellaneous	14.48	7.14	3.99	2.16	1.20

2.6 *Chemistry*

The leachate sample of the waste material No. 2 was obtained by means of the Toxicity Test Extraction Procedure of the U.S. Environmental Protection Agency. After extraction, the leachate was analyzed with the aid of an inductively coupled plasma atomic emission spectro-

Table 5. Results of the leachate quality test on the waste material No. 2.

Element	Concentration, ppm	*Maximum permissible concentration, ppm
Al	1.61	
As	<0.1	5
Ba	0.54	100
Cd	<0.01	1
Cr	0.0015	5
Cu	0.11	
Fe	4.23	
Mn	6.22	
Mo	<0.01	
Ni	0.11	
Pb	0.071	5
Zn	0.26	

*EPA Hazardous Waste Toxicity Criteria (100 times drinking water standards).

meter. The concentration of 12 elements is listed in Table 5.

The sulfate content of the waste material No. 1 was determined for the particles passing No. 4 sieve in accordance with the ion-exchange method of BS 1377: 1967, Test 10 (water-soluble sulfates in soils), the value being 0.28 g/liter. The ion–exchange column and constant head apparatus for this test method is shown in Figure 5.

The determination of pH for the waste material No. 2 was conducted using an electric pH meter and following the Electrometric Method, BS 1377: 1975, Test 11(A). The material was crushed to pass through a 3.35-mm sieve and a sample of 30 g was taken for the determination. The pH value was determined at 6.6.

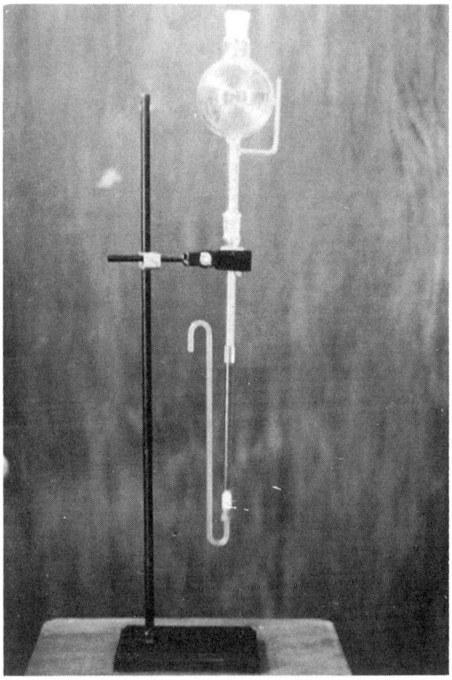

Figure 5. Ion-exchange column and constant head apparatus.

The particles passing No. 4 sieve was crushed to pass No. 20 (0.85 mm) sieve and oven-dried at 80°C. A sample of 20 g was taken from the oven-dried material and heated to 800°C for about 5 hours. The loss on ignition was 11.56%.

3 DISCUSSION

Based on the results of gradation and Atterberg limits, both of the waste material Nos. 1 and 2 are classified as GW-GM according to the Unified Soil Classification System.

Grain-size analysis was conducted on the waste material No. 2 before and after it was compacted for its optimum moisture-density relationship to see the degree of particle breakdown; the material did not suffer extensive degradation.

According to the permeability classification of Terzaghi and Peck (1967), the waste material No. 1 is characterized by high degree of permeability, while the waste material No. 2 is of low degree in permeability.

Both of the material Nos. 1 and 2 are not frost-susceptible when their grain-size distribution curves (Figure 2) are compared with the limits of frost susceptibility shown in Figure 6.

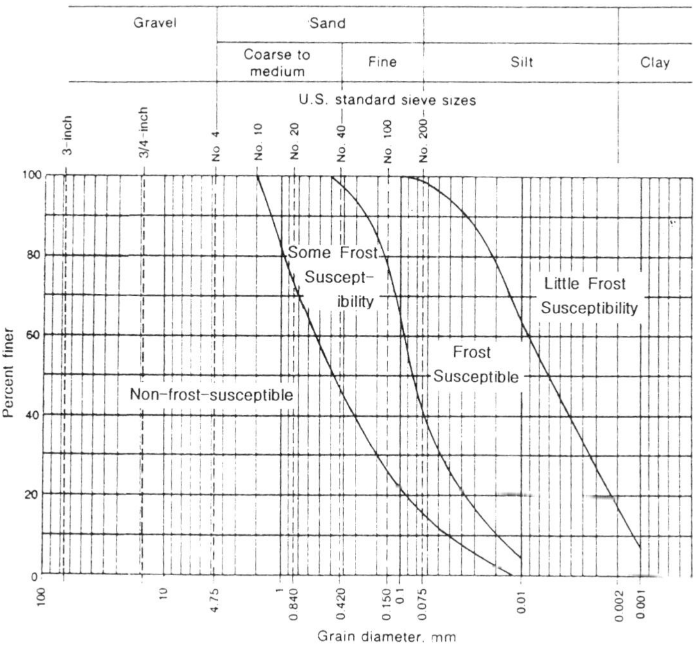

Figure 6. Limits of frost susceptibility according to Armstrong and Csathy (1963).

The CBR value of the soaked specimen is about half of that of the unsoaked one. The CBR value of the specimen subjected wet/dry cycles also is about half of that of the specimen subjected to freeze/thaw cycles; this result indicates that the waste material No. 2 is more durable under freeze/thaw condition than under wet/dry one.

The main rock type of the particles passing 3-inch sieve and retained on 3/4-inch sieve is sandstone; it comprises about 58% of the composition. Of the percentage, angular shape occupies about 31%.

The water-soluble sulfate usually found in soils are sodium sulfate (Na_2SO_4) and magnesium sulfate ($MgSO_4$). The sulfate content of the waste material No. 1 is not as high as to require sulfate-resisting Portland cement for buried concrete.

According to the pH value measured, the waste material No. 2 is slightly acidic; this degree of acidity does not appear to cause corrosion of metals.

4 CONCLUSION

On the basis of the rating of the performance of soils in accordance with CBR number (The Asphalt Institute, 1962), the CBR number for the soaked specimen is rated as "good", while that for the specimen subjected to wet/dry cycles as "fair".

The results of the leachate quality test indicate that the coal refuse is environmentally acceptable. It should be emphasized that the concentration of elements in the leachate from compacted refuse would be reduced over those obtained from the leachate of the loose sample.

Considering all the available information, it is concluded that the coal refuse may be utilized as a sub-base material for road and parking lot, especially in the coal mining regions where in general conventional construction materials are scarce and economically disadvantageous over coal refuse materials.

REFERENCES

American Society for Testing and Materials 1980. *Annual Book of ASTM Standards.* Section 4, Vol. 04.08.

Armstrong, M.D. & T.I. Csathy 1963. Frost design practice in Canada. *Highway Research Record, No. 33*: 170-201.

Bowles, J.E. 1978. *Engineering properties of soil and their measurement.* McGraw-Hill: New York.

British Standards Institution 1975. *Methods of test for soils for civil engineering purposes, BS 1377.*

British Standards Institution 1975. *Methods of test for soils for civil engineering purposes, BS 812: Part 1.*

Han, D. 1989. *Geotechnical characteristics of the coal-related wastes in the Republic of Korea.* Korea Institute of Energy and Resources Research Report KR-89-(B)-6: 27-49.

Head, K.H. 1980. *Manual of soil laboratory testing vol. 1.* Pentech Press: London.

Head, K.H. 1982. Manual of soil laboratory testing *vol. 2.* Pentech Press: London.

ISRM Commission on Testing Methods 1985. Suggested method for determining point load strength. *Int. J. Rock Mech. Min. Sci. & Geomech. Abstr.* Vol. 22, No. 2: 51-65.

Lambe, T.W. & R.V. Whitman 1979. *Soil mechanics, S.I. version.* Wiley: New York.

McQuade, P.V., W.J. Head & R.B. Anderson 1981. *Investigation of the use of coal refuse-fly ash compositions as highway base course materials.* Federal Highway Admission Offices of Research & Development Report No. FHWA/RD-80/129.

Terzaghi, K. & R.W. Peck 1967. *Soil mechanics in engineering practice.* Wiley: New York.

The Asphalt Institute 1962. *The asphalt handbook.*

Case histories

Tailings and Mine Waste '99 © 1999 Balkema, Rotterdam, ISBN 90 5809 025 6

Closure of a cyanide tailings pond

S.B. Patton
Department of Mining Engineering, Montana Tech of the University of Montana, Butte, Mont., USA
H. Gerbrandt
Department of Environmental Engineering, Montana Tech of the University of Montana, Butte, Mont., USA
C. Wassmann
Montana Tech, Butte, Mont., USA

ABSTRACT: In late 1994, low concentrations of cyanide were detected in a domestic well in the town of Pony, Montana, and at a nearby spring. The cyanide is believed to have leaked through the liner of a tailings impoundment at the site of the Pony Custom Gold Mill. The Montana Department of Environmental Quality has responsibility for the site and has begun reclamation efforts by removing hazardous chemicals from the site and dewatering the tailings pond. Further reclamation of the site required engineering design involving the dam, pond, and the surrounding slopes. Of particular interest is the manner in which the cyanide-processed tailings will be handled. The tailings will be encapsulated inside the current pond liner and will be buried under several feet of soil. Successful completion of the encapsulation includes: minimization of leachate escaping from the liner, minimization of differential settling, and minimization of ponding water on the liner.

1 INTRODUCTION

Pony, Montana is a small town of 110 people nestled in the Tobacco Root Mountains west of Bozeman Montana. (Refer to Figure 1) Many of the landowners in the area are weekend residents. The Chicago Mining Company (CMC) built a custom flotation and cyanide vat leach mill to recover gold from ore in 1989. The mill site lies between Pony Creek and Willow Creek approximately ¼ mile above the town of Pony. As shown in the topographic map in Figure 2, the site consists of a tailings impoundment dam approximately 70 feet high that was designed to hold 190,000 tons of tailings. The tailings impoundment is lined with two layers of 40 mil VLDPE with a geonet between the two liners. A leak detection sump, monitoring wells, mill buildings, ore and topsoil stockpiles are also on the site.

No permit was issued from the Department of Environmental Quality Hard Rock Bureau because the proposed mill fell under the small miner's exclusion. The Montana Small Miner Exclusion (SMES) is not a permit but an exclusion from obtaining a full operating permit if the following conditions are met:

- disturbance is less than 5 acres or processing is less than 36,500 tons per year
- water is not contaminated
- human and animal life are protected
- an annual summary report and map showing disturbance is submitted
- compliance with the Noxious Weed Act is maintained

No closure bond was required. New Hard Rock Mining regulations were passed in 1989 that essentially required a limited operating permit with bonded closure if a SMES operator was using cyanide in the recovery process. Only that portion of the mine in which cyanide was used required the operating permit. The permit application required in moderate detail:

- environmental baseline information
- operating and reclamation plans

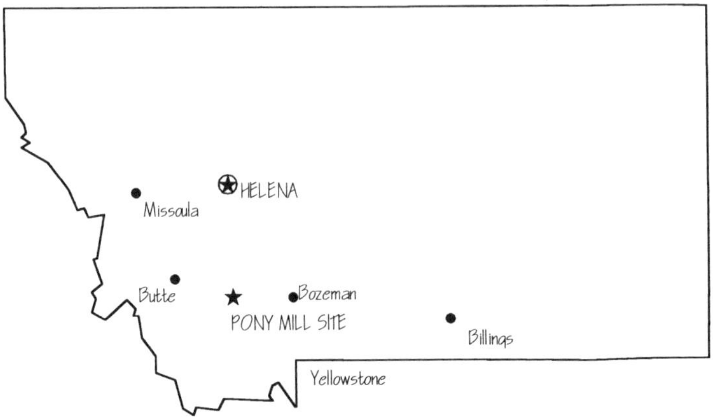

Figure 1. Pony site location map.

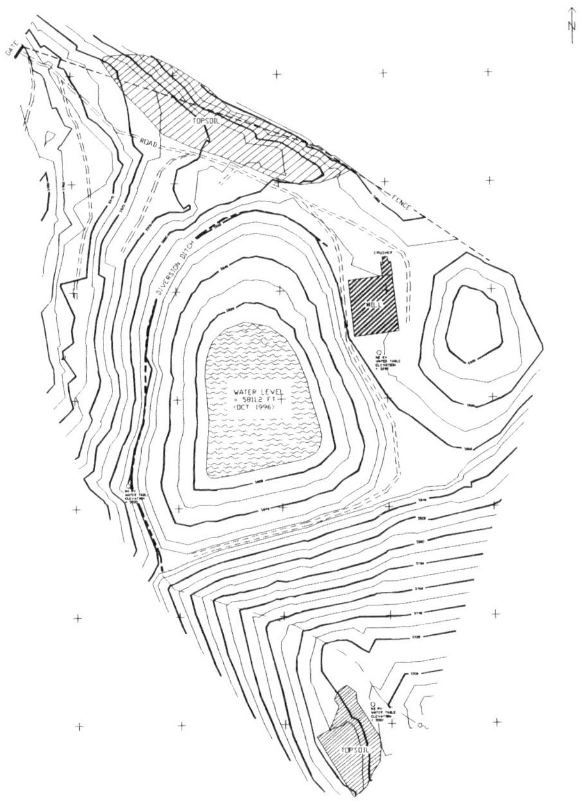

Figure 2. Pony Mill Site topographic map.

A bond of the full closure cost is required, should the state have to close the site. This applied to sites not in full operation by June 1, 1990. Since that time, the laws have changed to require reclamation bonding for all mine and mill sites. The Pony Gold Mill poured a gold and silver bar the last week of May 1990, one week prior to the June 1st deadline. Although exempt from mine operations permitting, the Pony Mill was subject to permitting requirements under the Ground Water Protection Act. Where infiltration or discharge is likely to occur, the mine is required to obtain a Ground Water Discharge Permit. There was potential discharge to the ground water from the lined tailings impoundment. The site was classed as zero discharge for surface water.

In January of 1990, after several battles with environmental protection groups, the Montana Department of Health and Environmental Sciences (DHES) issued Montana Groundwater Pollution Control System (MGWPCS) Permit No. 0059 to the Chicago Mining Company for the Pony site. So onerous was the battle that when in September of 1990, the tailings impoundment was determined to be leaking; newspaper accounts by Kirkpatrick (1990) stated that Bob Lee, President of the CMC, felt that the tailings impoundment liner was purposely sabotaged by a gunshot from planes taking aerial photographs. The Montana Department of State Lands issued a cease operation order but failed to succeed in stopping the operation. Since the CMC did not have a permit the regulatory authorities could not tell them to cease operation. The mill operated intermittently during 1990 and 1991. In January 1994 the DHES revoked the groundwater discharge permit for failure to monitor for 8 consecutive quarters. CMC abandoned the site and filed for bankruptcy. Great American Gold (GAG) assumed ownership of the site. In November and December of 1994, cyanide in concentrations of 0.007 ppm was found in a local resident's well and at the spring at the toe of the tailings impoundment. In July of 1995 responsibility for the Pony Mill Site was transferred to the Montana Department of Environmental Quality (DEQ). The DEQ took CMC to court to force payment of closure costs, stating that CMC wasn't really in operation when the first bar was poured. The trial was held in a county that favors the mining industry, and the State lost.

Although the site is very small and not even listed as a major problem with the DEQ (Winegar, 1997), the site has created a very visible black eye on the record of the Montana mining industry. Mining opposition groups are quick to point out the Pony site as an example of why mining operations should not be permitted, even though the laws have been changed to prevent the reoccurrence. In an unprecedented effort by the reputable mining companies in the state, funds were raised and in-kind services offered by the mining operators and affiliated groups to assist the state with the estimated $250,000 cleanup. Legislative action was required to limit the liability of those companies who volunteered services.

2 ENVIRONMENTAL AUDIT

In the first step toward closure of the site, the Water Quality Division of the DEQ performed an environmental audit of the proposed closure plans. All potential impacts to the physical or human environment were determined to be either no impact or minor. Minor impacts were to the water quality, vegetation cover, aesthetics, human health, and industrial activity. Six alternative actions were analyzed (Webster, 1996):

1) Maintain the Leak Detection System
2) Repair and Maintain the Liner
3) Construct an Internally Drained Grass Lined Basin
4) Construct an Internally Drained Wetland Basin
5) Reestablish Drainage Through Partial Removal of the Dam
6) Reestablish Drainage Through Total Removal of the Dam

Alternative 3 was selected as the preferred reclamation alternative. This alternative amounts to stabilization of the tailings mass by addition of on-site material from the dam. Folding the liners from the side slopes over onto the stabilized mass like a burrito would follow, encapsulating the tailings. This lowers the dam to a safer height and minimizes the amount of water that can enter the tailings. The DEQ, with the help of local mining companies, removed the hazardous chemicals from the mill and free water in the pond to a land application area nearby. Montana Tech was responsible for the engineering design of this alternative. This included the full spectrum of engineering services from survey of the site to revegetation planning. This paper specifically addresses the design of the encapsulation of the tailings. The tailings consolidation, water balance and drainage layer design are described here. The complete design is available in Wassmann (1998).

3 SITE DESCRIPTION

The site contains a mill building, tailings impoundment, office trailers, heavy equipment, crushed ore and topsoil piles. The dam is a 70 foot high centerline earthern dam with a 360 foot downstream face at a 2.5:1 slope, a 22 foot wide crest that is 600 feet long, and a 200 foot long upstream slope of 1.85:1. The pond contains a double liner system of two layers of 40 mil HDPE liners with a geonet between. A leak detection sump, made up of a 2-foot layer of gravel between two 8-inch layers of sand with two sampling ports, is located at the bottom of the pond. There are 23,000 cy of tailings in the pond at a maximum depth of 21 feet. The cyanide concentrations run from 6.56 ppm (February 1996) to 48.9 ppm (February 1998). The impoundment was constructed in a small basin with steep natural slopes on either side. The area receives 15 inches of precipitation per year. Native grasses and sagebrush grow on the undisturbed areas.

4 ENCAPSULATION DESIGN

Reclamation of the Pony Mill Site involves encapsulating the tailings in the existing pond liner, forming an internally draining basin. Before the tailings can be encapsulated they must first be sufficiently consolidated. This allows for both the operation of heavy equipment on the tails and the removal of water. Two ways of consolidating the tailings were considered, (1) completely dewater the tails and (2) remove the freestanding water and add soil to the tailings. Crushed ore from on-site stockpiles is not considered to be free from acid generation potential therefore it to will be placed inside the encapsulation.

A grain size distribution analysis was run on several samples of tailings. As expected the tailings are more than 50% clay material. The d_{50} is 0.08 mm. The porosity was determined to be 40% based on the mass before and after oven drying.

The addition of dam material to the tailings serves two purposes, to support vehicular loads and aid in consolidation. The quantity of material that will be mixed in with the tailings is based on the minimum volume capable of both adsorbing the excess water and supporting equipment load. 200 cubic yards of excess water includes water left standing or in the sump and water released during consolidation. Since the tailings will be wrapped in the liner, differential settling could pose problems for the drain design. The dam material is poorly sorted gravel with a porosity of 25% and a specific yield of between 15 and 20%. The field capacity is approximately 4.5%. With a porosity of 25%, the material would be 18% saturated at field capacity. The moisture content is assumed to be 2% for this design. The initial saturation is taken to be 8%. Additionally, a conservative assumption is made that the dam material can support machinery at 70% saturation. The resulting available capacity for water in the material would then be 60% of the pore space.

Dam material will be applied to the tailings at a minimum thickness of 2 feet in order to provide a surface for the operation of equipment. For two feet of depth, a volume of 4000 cubic yards is required to cover the entire surface of the pond. Since the material can adsorb a maximum of 60% of its porosity, a volume of 4000 cubic yards of material can hold 400 cy of water.

The tailings are submerged and assumed to be 100% saturated. The porosity of the tails is 40%. The coefficient of consolidation is unknown. The effective stress on the tailings was determined by distributing the weight of overlying materials over a square foot area. Void ratios were determined using information from tests run on Pegasus Gold's Montana Tunnels Mine tails (Jefferson City, MT). The Montana Tunnels tails are finer than the Pony tails. Estimates using the Montana Tunnels data should be conservative. This method for determining the settlement is considered to be the best alternative in lieu of consolidation tests.

The percent settling can be determined by the following equation:

$$\%\text{settling} = (VR_i - VR_f)/(1+VR_i)$$

where: VR_i is the void ratio before and VR_f is the void ratio after consolidation. Multiplying this ratio by the depth gives the actual settling in feet for each interval.

Up to 3 feet of settlement is expected in the deepest part of the pond. A total of 155 cy of excess water will be expelled from the tailings. This volume is well within the holding capacity of the 2 feet of material to be added. An additional 3,500 cy of dam material will be used to construct a crown at the point of greatest settling. The crown will be graded 2% towards the outer edges of the pond, and the overall slope will be 2% towards the drain in the core of the dam.

5 WATER BALANCE

The Hydrologic Evaluation of Landfill Performance (HELP) Model was the method initially chosen to balance water in the system. The model allows the input of numerous layers of varying thickness and water conductivity and storage parameters with accompanying precipitation data. Ten years of precipitation storm events were applied. Since the liner sump had been pumped three times, the HELP model was calibrated by equating the pumped volumes to liner leakage rates. Liner defects, installation quality, and number of holes per area were adjusted to match the sump pump volumes. However, calibration resulted in unrealistic near perfect liner quality, and HELP model results were abandoned for the water balance effort. Instead, a simple mass balance was used to analyze the potential for leachate generation from the encapsulation.

After the liners have been folded over the tailings and dam material, a gravel drainage layer will be placed over the encapsulation. A drainage layer is recommended to ensure that any infiltrating water will migrate away from the encapsulated tails. The HELP model was utilized to determine the thickness of drain layer required. Six inches proved adequate. No cushion was specified between the drain layer and the top of the "burrito" because the folded liners include 4 or 6 layers (two from the west side, two from the east side, and possibly two from the north side). A filter fabric layer is specified to overly the drain layer in order to prevent the migration of fines into the drain. The encapsulation design is illustrated in Figure 3.

Any water entering the encapsulated tails must first pass through 4 feet of subsoil before reaching the triple liner. Water infiltrating the full four feet will be intersected by the gravel drainage layer and directed to the drain in the core of the dam. It is unlikely, even during a major storm event, that water will infiltrate the multiple layers of liners.

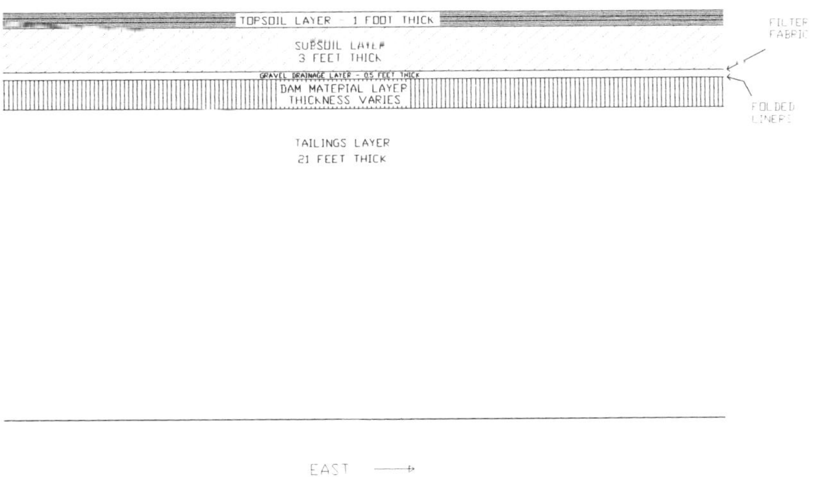

Figure 3. Pony mill site encapsulation design

6 SUMMARY

Although a small problem from an environmental standpoint, the Pony Mill site represents a black eye to the mining industry. The closure and monitoring of the site should help to heal the rather large public relations wounds caused by such a small operation.

This collaboration between the Montana DEQ and Montana Tech of the University of Montana will result in permanent closure to an ill-run operation and restore the integrity of the area groundwater.

REFERENCES

Kirkpatrick, D., 1990, "Miner defies state," Montana Standard, Butte, MT, Sept. 22.

Wassmann, C, 1998, *Pony Mill Reclamation Project Engineering Design Report*, submitted to MT DEQ March 1, 1998.

Webster, T., 1996, Pony Mill Maintenance and Reclamation, MT DEQ Environmental Assessment, May 21, 1996.

Winegar, B., 1997, Personal Communications, February.

Tailings and Mine Waste '99 © 1999 Balkema, Rotterdam, ISBN 90 5809 025 6

Reclamation of a coal mine waste disposal facility to a recreation lake

Donald E. Stump
Office of Surface Mining, US Department of the Interior, Pittsburg, Pa., USA

ABSTRACT: Federal regulations require the coal mining industry to cover coal waste impoundments with a minimum of 1.2 meters of nontoxic, noncombustible material as part of the reclamation plan. However, Federal regulations also allow the coal industry to request alternate reclamation plans as an experimental practice. In Pike County, Kentucky an experimental practice is currently being evaluated by State and Federal regulators that would create a recreation lake at the site of a coal waste slurry impoundment in lieu of traditional reclamation using four feet of cover.

Construction of the coal waste slurry impoundment began in 1986. The preparation plant supplying slurry to the impoundment ceased operations in 1993. In 1995 an experimental practice application was submitted to State and Federal regulators that revised the postmining land use to a recreation lake. The embankment is 43 meters high from the upstream toe to the spillway crest. The lake will have a surface area of 8.4 hectares, will contain 1,000,000 cubic meters of water, and have a shoreline of 1,800 meters.

The experimental practice was approved and the monitoring period began in April 1998. Monitoring of lake water levels, water quality, seepage through the embankment, and embankment stability will continue for a minimum of two years before responsibility for the lake can be transferred to the County and State Authorities.

1 INTRODUCTION

Regulations require coal waste impoundments to be reclaimed and vegetated prior to abandonment. A request was submitted to the Office of Surface Mining (OSM) to reclaim a coal waste slurry impoundment as a recreation lake instead of using typical reclamation plans in Pike County, Kentucky (Figure 1). This request to allow an alternate reclamation plan was submitted to the regulatory authorities as an experimental practice. The experimental practice has been approved and the site is currently being inspected and monitored. A final decision on the success of the experimental practice will be made at the end of the monitoring period.

2 ORIGINAL PLAN

The original coal waste slurry impoundment began construction and operations in 1986 and had an anticipated completion date of 2005. The original plan included the construction and operation of a coal refuse disposal facility for the disposal of both slurry and coarse refuse. The facility upon completion would require that the entire valley above the impoundment embankment would be filled with refuse and the drainage permanently diverted.

Figure 1: Location Map

The site is located in Pike County, Kentucky (Figure 1). Hillside slopes in the vicinity vary from 40 to 70 percent. The coal waste slurry impoundment has a drainage area of 130 hectares with a maximum topographic relief of 340 meters. Second and third growth timber comprise the major portion of vegetative cover throughout the watershed.

Geologic mapping shows the area to be underlain by sedimentary rocks of the Breathitt Formation which represents the Pennsylvania geologic period. The Breathitt Formation consists of cyclic sequences of sandstones and shales with lesser amounts of coal and under clays. The average rock strata dip is roughly 9 meters per kilometer to the northwest.

The Pond Creek coal seam is located beneath the impoundment area. Commercial mining of the Pond Creek coal seam reportedly began about 1912. The Pond Creek coal seam was mined extensively throughout the area in the 1940's and 50's using room and pillar method of mining. Rock borings indicate the Pond Creek coal seam thickness of 1.5 to 2.0 meters.

An underground coal mine in the Pond Creek seam is about 150 below ground level at the impoundment embankment. The coal pillars beneath the embankment area were left in place, resulting in a more or less uniform foundation condition. The average percent coal extraction beneath the embankment area is approximately 60%.

The disposal facility was designed to be constructed in four phases. Phase I was an earth starter embankment with Phases II, III and IV utilizing coarse coal refuse to raise the top of the embankment to a final elevation of 432 meters (M.S.L.).

The initial work included construction of a sediment structure downstream of the disposal area. Followed by clearing and grubbing of the area for embankment construction. All topsoil was removed and stored from areas disturbed.

Phase I of the embankment was approximately 138 meters long with a top width of 10 meters and a maximum height of 42 meters. The side slopes were 3H to 1V upstream and 2.5H to 1V downstream. The structure had a top elevation of 390 meters. The Phase I embankment was composed of silty clay and clay borrow material. Phase I required 264,000 cubic meters of borrow material, 40,000 of which was coarse refuse for the downstream berm. A 32 meter wide berm was provided on the downstream slope at elevation 372 meters in order to provide stability for phase 1 construction. A cutoff trench extends into competent bedrock along the centerline of the embankment and had side slopes of 1H to 1V. Phase I included a stream diversion conduit, spillways, diversions.

Phase II of construction raised the embankment to elevation 409 meters using coarse refuse. Phase II required 628,000 cubic meters of coarse refuse and 50,000 cubic meters of soil. The Phase II embankment was approximately 225 meters long with a top width of 15 meters and a maximum height of 65 meters. Embankment slopes were 2.5H to 1V on both upstream and downstream slopes. A 5 meter thick clay layer was continued from the Phase I elevation to the top of the embankment. A layer of filter fabric separates the soil and coarse refuse and aid in preventing piping.

Phase III and IV were in the original plan but were never constructed. Phase III would have raised the embankment to elevation 432 meters using coarse refuse with a lesser amount of durable rock. Phase IV would have also utilized coarse refuse. Phase IV was to be placed on the downstream face of the embankment. The final configuration was to be approximately 210 meters wide at the top and have downstream slopes of 2.5H to 1V.

The abandonment plan was to be initiated at the time the slurry impoundment is filled to capacity. This would be done by placing coarse refuse in the reservoir area until a grade is achieved which will allow the Phase III spillway to act as a permanent diversion channel. At this point, the entire reservoir area was to be sloped at approximately 2.5% and would allow the area to be revegetated.

The subsidence potential was evaluated for the surcharge load being added to the area. Little relevant literature was found. Drill holes into the mine show it is open and not collapsed. The company safety officer inspected the underground workings below the area in May 1985 and found isolated areas of small roof falls, the mine was essentially dry with isolated shallow water areas, and no significant pillar deterioration or floor heaving.

3 REGULATIONS

Two of the Federal agencies involved in regulating coal waste slurry impoundments are the Mine Safety and Health Administration(MSHA) and the OSM. The primary concerns of MSHA is to ensure the safety of mine company personnel and public safety, while OSM is concerned with reclamation, environmental impacts, and public safety.

OSM was established by Public Law 95-87, the Surface Mining Control and Reclamation Act of 1977. In Section 711 (Experimental Practices) of the Act it states "In order to encourage advances in mining and reclamation practices or to allow post-mining land use for industrial, commercial, residential, or public use (including recreational facilities), the regulatory authority with approval by the Secretary may authorize departures in individual cases on an experimental basis from the environmental protection performance standards promulgated under sections 515 and 516 of this Act. Such departures may be authorized if (i) the experimental practices are potentially more or at least as environmentally protective, during and after mining operations, as those required by promulgated standards: (ii) the mining operations approved for particular land-use or other purposes are not larger or more numerous than necessary to determine the effectiveness and economic feasibility of the experimental practices; and (iii) the experimental practices do not reduce the protection afforded public health and safety below that provided by promulgated standards." It was under this authority that the request was made to revise the original coal waste slurry impoundment reclamation plan to create a recreation lake.

The experimental practice request for the coal waste slurry impoundment focused on allowing departure from two of OSM's regulations (30 CFR 817.83(c)(4) and 30 CFR 817.84). In 30 CFR 817.83 (c)(4) it states that " Following final grading of the refuse pile, the coal waste shall be covered with a minimum of 4 feet (1.2 meters) of best available, nontoxic and noncombustible material, in a manner that does not impede drainage from the under drains." In 30 CFR 817.84 : Coal Mine Waste : Impounding Structures it states that "Such structures may not be retained permanently as part of the approved post mining land use."

4 EXPERIMENTAL PRACTICE

In January 1995 an experimental practice request was submitted for a coal waste slurry impoundment in Pike County, Kentucky (Figure 1). At the time of the request the water elevation in the impoundment was being held at 396 meters. The proposed water level would be 404 meters. Since the upstream toe of the embankment is at 361 meters, this would make the impoundment 43 meters high (Figure 2).

The finished impoundment will have a surface area of approximately 8.4 hectares and will contain 1,000,000 cubic meters of water. The lake will have 1,800 meters of shoreline making a Shore Development Index of 1.82. The Shore Development Index is an index of the

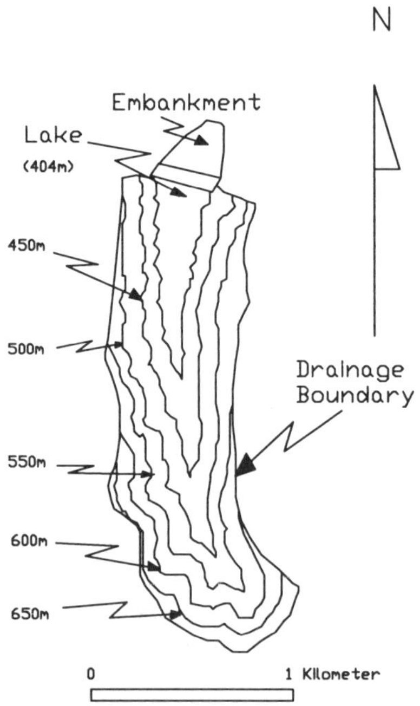

Figure 2: Plan View

irregularity of the shoreline with higher numbers being more favorable. A circle would have the lowest index possible, 1.0.

The experimental practice construction will include earthwork, revegetation, initial site development, and future development. The earthwork, revegetation and initial site development will be done as part of the experimental practice. The future development will by done by the Pike County Court.

Earthwork and Revegetation will include widening road to the site, grading top and face of refuse fill, cover refuse with alternate topsoil material, revegetation of project area, and modify spillway. Road drainage control is designed to handle a 100-yr, 24-hr storm.

Initial site development will include cleaning and painting existing entrance gate and install sign, construct caretaker's residence, rehabilitate maintenance building, connect to existing water system, stock impoundment with fish, pave roads, install guardrails, construct picnic shelter, install security lighting system, construct sidewalks, and construct chain link fence.

The future development by the Pike County Court will include construction of basketball/tennis courts, develop hiking trails, construct lavatory structure, install water lines, construct playground, construct volleyball court, and construct fishing pier.

The mine refuse to be revegetated has been tested and is non-toxic and non-acid producing. An alternate topsoil material capable of sustaining an acceptable vegetated cover has been identified within the permit and will be placed over the refuse to a depth of 0.6 meters.

Water quality of the lake inflow and outflow is of good quality (exceeds drinking water standards) and is predicted to remain more than adequate to support a variety of aquatic life and to provide an aesthetically pleasing environment.

As a public facility there must be enough population available to justify construction and maintenance. Populations within ten and thirty miles radius of the facility were evaluated. There are an estimated 29,994 and 208,069 people within a 16 and 48 kilometer radius respectively. There are an estimated 2,868 and 21,086 licensed fisherman within a 16 and 48 kilometer radius respectively. These numbers support the prediction that the facility would get considerable use.

Local support for the construction of the recreation lake has been documented in letters from the Pike County Fiscal Court, the KY Department of Fish and Wildlife, and the Pike County Judge.

The KY Department of Fish and Wildlife Resources has agreed to create and manage a fishery resource in the impoundment including initial stocking, potential habitat improvements, monitoring water quality in regard to the health of aquatic organisms, and ongoing management of fish species.

5 REVIEW

The review of the experimental practice application covered many aspects, including water level requirements, water quality, embankment stability, hillside stability, breach analysis, flood routing through the facility, overburden loading of the abandoned mine under the area.

Based on the review a siphon was required to be installed for emergency drawdown and an emergency action plan was required to be in place because during the monitoring period the water level was planned to be maintained at the permanent pool level.

The recreation lake is a Class "C" structure and the review found the impoundment has been designed and constructed to be in compliance with applicable MSHA regulations. The proposed facility is also in compliance with applicable OSM regulations.

6 MONITORING

The monitoring required for the experimental practice includes groundwater levels in the embankment, discharge quality, discharges from the mine below the area, underdrain flows. The monitoring period was set at a minimum of two years. The monitoring period of the experimental practice was not allowed to start until water was flowing over the spillway. In April 1998 water was flowing over the spillway and the monitoring period began.

. Water samples are being collected quarterly at the inflow into the slurry pond from natural stream, standing water in the pond along the north shoreline, the discharge from underdrawn, the discharge from the sediment pond, and the spillway outflow. The parameters being analyzed for are pH, acidity, alkalinity, conductivity, total iron, total manganese, sulfate, total aluminum, dissolved oxygen.

Quarterly reports are being submitted to OSM and the State with lab analyses, field observations, and field work conducted. Two major reports will also be prepared and submitted. The first report will be at the time of Phase I bond release request and will include the results of the water and soil analyses to be used to determine additional work which may be necessary to create a suitable public recreation area. The second and final report will be submitted prior to a Phase III bond release request and will include information pertaining to the success or failure of the experimental practice. To be included in the report are sediment loads into the lake, general water quality, uses of the lake, stocking activities, existing fish and wildlife species, results of any creel surveys, results of any KY Dept. Of Fish and Wildlife Studies, success of revegetation efforts, and recommendations.

7 FUTURE DECISIONS

The monitoring data is critical to provide support for the future decisions that will be made concerning the success of the experimental practice and the planned transfer of the recreation lake to the authorities of Pike County, Kentucky. The final decision will involve several regulatory agencies and will be based on the success of the vegetation, the suitability of the lake for public use, the release of reclamation bonds, compliance with effluent requirements, and embankment stability.

Tailings and Mine Waste '99 © 1999 Balkema, Rotterdam, ISBN 90 5809 025 6

Summitville Mine site – Wide reclamation design

M.J.Gobla
US Bureau of Reclamation, Denver, Colo., USA

ABSTRACT: The U.S. Bureau of Reclamation has prepared a reclamation plan for the Summitville Mine Superfund Site that will minimize the continued generation of poor quality stormwater runoff by revegetation of 300 acres of disturbed surfaces. Challenging aspects of the site include surface elevations above 11,000 feet, steep slopes, and wide-spread occurrence of acid-generating soils and rocks. The reclamation design includes the reduction of steep waste banks, construction of permanent drainage channels, and limestone neutralization of disturbed surfaces. Limestone riprap will be used in steep drainage channels as a form of passive treatment. Revegetation using a native seed mix will allow natural succession to proceed.

A value engineering study was performed by an independent team of mine reclamation experts from government and industry to review the design and recommend alternative methods. Integration of the value engineering proposals into the reclamation design has lowered project costs and enhanced project quality.

1 INTRODUCTION

The Summitville Mine Superfund Site is an open-pit gold mine that was abandoned in December of 1992 when the operating company declared bankruptcy. The Environmental Protection Agency (EPA) initiated emergency response actions to prevent the release of cyanide solutions from the heap-leach pad. The U.S. Bureau of Reclamation (BOR) was engaged by the EPA to provide engineering design, contract administration, and construction management services for the project. The Site was added to the National Priorities List in June of 1994. Transition from EPA management to State-lead management was recently initiated by the Colorado Department of Public Health and Environment (CDPHE).

1.1 Operable Unit 4, Site-Wide Reclamation

Engineering for State-lead Operable Unit 4 (OU-4) "Site-Wide Reclamation" was initiated in 1997 when the BOR was engaged by the CDPHE to prepare reclamation plans for the unit. Although the EPA-led cleanup efforts have made significant progress in site remediation, non-point sources of acid rock drainage (ARD) generation remain. Major goals are to minimize non-point sources of ARD, minimize water treatment for non-point source contaminated waters, and enhance and improve site biota habitat. The major design components are described along with the results of an independent Value Engineering Study that was conducted prior to completion of the reclamation design.

Water treatment requirements are expected to decline as the reclamation work is completed and vegetation is reestablished. Annual monitoring and maintenance will be conducted at the site for a period of up to ten years to evaluate the reclamation, reseed as needed, repair erosion features, and perform other maintenance.

2 SITE CONDITIONS

Challenging aspects of the Summitville site include a remote high-altitude site with a harsh climate, steep north facing slopes, wide-spread occurrence of acid generating soils and rocks, and presence of underground mine workings and cultural resources. The site presents significant reclamation challenges that had a major influence upon the reclamation design.

2.1 *Location and topography*

The Summitville Mine is located about 2 miles south of Del Norte, in southwestern Rio Grande County, Colorado. Located within the San Juan Mountains of southern Colorado, the site is approximately two miles east of the Continental Divide.

The disturbed areas of the site are situated between 11,200 feet and 12,200 feet in elevation. The site lies on the north facing slope of South Mountain. The crest of South Mountain reaches an elevation of 12,555 feet. Flat ground is mainly confined to the lower portions of the site which are occupied by wetlands, the historic Summitville townsite, and some of the mine buildings. Except for the backfilled mine pits, most of the disturbed areas are in steep terrain characterized by 33% or greater slopes.

2.2 *Climate*

The site is characterized by harsh winter weather with cool and wet summers. Summer rain and hail storms are frequent during July and August. Freezing temperatures are normally experienced from October through May and also occur infrequently during the summer season. The site receives an average of 55 inches of annual precipitation, most of this is in the form of snow (389 inches/year). Located on a north facing slope surrounded by mountain peaks, direct sunlight on a majority of the site is limited from mid-morning to mid-afternoon during the short summer growing season.

Frequent storms cause "white out" conditions in the spring and fall, while summer brings intense lightening, heavy downpours, and mud. Poor weather often shuts down reclamation activities at the site as workers seek shelter and heavy equipment stops because of poor visibility and lost traction. For example, heavy rains in July and August of 1997 delayed the placement of the heap-leach pad clay cap and required the addition of another season to the reclamation contract.

2.3 *Geology*

The site lies on the northern margin of the Platoro caldera complex within the San Juan volcanic field. The mine is situated in the Tertiary age, Summitville caldera on the north slope of the South Mountain vent-dome complex (Lipman, 1974). The Tertiary rock units are quartz latite, quartz latite porphyry, and andesite (Steven and Ratte, 1960).

The principal rock type at the site is quartz latite which exhibits varying stages of hydrothermal alteration from propylitic alteration at the margin of the deposit to an intensely altered, vuggy silica zone at the innermost core of the mineralized structures. Most of the latite contains visible iron-pyrite and it rapidly breaks down into a soft clay-rich soil upon exposure to weathering.

Following emplacement of the South Mountain quartz latite volcanic dome, the mineral deposit developed over the course of at least six different events (Gray, Coolbaugh, and Plumlee, 1993). The events are 1) early acid-sulfate alteration, 2) copper sulfide and gold mineralization, 3) widespread hydrothermal brecciation, 4) deposition of base-metal-sulfide bearing barite veins, 5) emplacement of small-scale kaolinite matrix breccias, and 6) supergene oxidation and secondary sulfide enrichment.

The andesite is most prevalent in the vicinity of the historic townsite area. A hard black rock, the andesite is not considered a source of acid rock drainage unless it has been intruded by the younger latite.

The wide-spread occurrence of pyrite-bearing latite is the reason for the pervasive ARD problem at Summitville. Practically every disturbed surface generates ARD laden with copper and other heavy metals.

2.4 *Cultural resources*

Under CERCLA, protection of cultural resources, as mandated by the National Historic Preservation Act of 1966 and amendments, is required. A Cultural Resource Survey (Clark, 1997) has been performed for the Summitville Mine Superfund Site. The cultural resources include the historic Summitville townsite lying below the work area, and five sites within or immediately adjacent to the proposed reclamation work. The reclamation plans have been designed to avoid all but one of the cultural resource sites. Sites in or near to the work areas will be protected by orange construction barrier fencing.

2.5 *Underground mine workings*

Historic underground mine workings are present throughout much of the site. The reclamation plans avoid work around the underground mine openings. Underground mine maps have been included in the construction documents to inform the contractor of the potential hazards.

An open stope filled with snow is suspected to be the source for drainage from the Iowa adit. The stope was evaluated by the BOR for closure. Difficult access, loose sloughing walls, a failing crown pillar, and location beneath an unstable overhanging highwall, resulted in an estimated $200,000 worth of site safety work being required prior to the closure operation. The excessive cost has led the CDPHE to delay the stope closure. It will not be pursued during the Site-Wide Reclamation project.

3 PREVIOUS RECLAMATION WORK

Previous cleanup activities conducted under EPA leadership include water collection and treatment, chemical drum removal, waste stabilization, grading, topsoil replacement, seeding of approximately 100 acres of land, and salvage operations. Over $100 million has been expended at the site; more than 55 % of the costs are for water treatment.

3.1 *Heap-leach pad*

The pad has been pumped down and rinsed with water to reduce the cyanide levels. The gold recovery process plant was modified and used to destroy the cyanide compounds in the rinsate. The water was then further treated to reduce the heavy metals content.

Dike 1, an embankment across Cropsy Creek that supports the heap-leach pad structure has been rehabilitated. A toe drain has been installed and additional fill has reduced the embankment slope thereby enhancing the geotechnical stability of the structure.

The heap-leach pad has been capped with a geocomposite-clay liner to prevent saturation of the pile and to minimize ARD generation from the sulfide minerals contained in the pile.

3.2 *Cropsy Waste Pile*

In order to reduce the cost of water treatment at the site, response actions were taken to reduce or eliminate ARD generated from several sources (Ketellapper, Cressman, and Carmody, 1995). The Cropsy Waste Pile was removed from Cropsy Creek where it had covered a number of active seeps and springs that were forming ARD. The waste was placed in the mine pits.

3.3 *Mine pits*

The bottom of the North Pit and South Pit were covered with a compacted-clay drainage barrier and then covered with a neutralizing layer of lime-kiln dust. Mine wastes from the Cropsy Waste Pile and the historic mill tailings from other areas of the site were backfilled into the pits. A geocomposite-clay liner was placed on top of the South Pit backfill.

3.4 *Summitville Dam Impoundment*

In 1994 and 1995 the existing tailings pond was modified and enlarged into a water storage dam. Renamed the Summitville Dam Impoundment, the facility is utilized to collect and store ARD from the site. The water is pumped back to water treatment plant where metals are removed using conventional lime precipitation treatment methods. A high-density sludge circuit (sludge recirculation) has reduced water treatment sludge volumes and disposal costs.

3.5 *Adit Plugs*

Excavation of the mine pits encountered old underground mine workings and lowered the ground surface to within a few hundred feet of the level of the historic Reynolds adit. The ARD issuing from the Reynolds adit increased in quantity while the water quality declined. In 1994 hydraulic plugs were placed in the Reynolds and Chandler adits. This helped re-establish the original groundwater levels in South Mountain and reduced the quantity of ARD requiring collection and treatment. Negative aspects of the plugs are the reactivation of natural acid seeps and springs issuing from the lower portions of the site, leakage around the plugs, and the maintenance requirements. A hydraulic adit plug is in effect a dam; failure must be prevented by routine monitoring and maintenance. Annual inspections are performed to observe seepage and record water pressure acting on the plug.

3.6 *Salvage operations*

Salvage activities have been conducted by a salvage contractor, and by Rio Grande County. The crushers and conveyors have been removed from the site and some of the crusher structures have been partially dismantled. The generators, generator building, and pumphouse have also been removed for reuse elsewhere. Because of the remote nature of the site, most of the heavy steel structures have no commercial value as scrap. The transportation costs exceed the scrap steel value.

3.7 *Revegetation*

Approximately 100 acres of disturbed surfaces were amended, topsoiled, seeded, and mulched during 1998. The revegetation work was mainly associated with the completion of the remediation of the heap-leach pad and adjacent disturbed areas in the Cropsy Creek valley.

4 SITE-WIDE RECLAMATION DESIGN

Specifications and over seventy design drawings were prepared over the course of one year. Construction is scheduled to commence in 1999 with completion expected in 2002.

4.1 *Reclamation goals*

The goals for OU-4 are directed at water quality. They seek to remove, reduce, stabilize and/or contain non-point sources of acid rock drainage to prevent further releases to the site; eliminate or minimize non-point source impacts to aquatic receptors in Wightman Fork, the Alamosa River, and the Terrace Reservoir; eliminate or minimize potential water treatment for non-point source contaminated waters; and enhance and/or improve on-site biota habitat.

4.2 *Reclamation resources inventory*

The site and surrounding area were evaluated for the availability of reclamation materials. Adequate amounts of topsoil (470,000 yd^3) have been stockpiled at the site in 13 piles. The topsoil will require limestone amendment to raise the pH from its current value of around 3.5.

Rock for riprap and bedding has been located in an andesite outcrop located adjacent to the site. The andesite will be quarried to supply 130,000 yd^3 of riprap and gravel bedding for construction of permanent ditches.

The nearest source for limestone for soil neutralization is at Bonanza, Colorado located about 80 miles from the site. Organic matter is also needed as a soil amendment. It is available in the San Louis Valley about 50 miles from the site in the form of mushroom compost. Sawdust from local lumber mills will be blended with the mushroom compost to increase the bulk and reduce the overall cost of the organic amendment.

4.3 *Reclamation design principals*

The design team formulated a number of design principals that were used to guide the engineering activities to meet the reclamation objectives. The principals employed include stabilization of landforms, control of surface stormwater runoff, improvement of runoff water quality, reduction of water treatment requirements, and minimization of project impacts. Grading, water diversion, construction of permanent drainage ditches, soil amendment, and revegetation were identified as major activities needed to reclaim the site. To minimize project impacts the plans also must minimize additional disturbance to vegetated areas, protect cultural resources, control safety hazards, and minimize dust and sediment generation.

4.4 *Demolition*

Over fifty structures were identified and evaluated as to function and value. Fourteen structures will be demolished to make way for the OU-4 Reclamation activities. Other facilities are needed for the ongoing monitoring, maintenance, and water treatment activities at the site. The structures requiring demolition are the dormitory, administration building, a concrete tank containment structure, lime silo, and the crushing and conveying facilities. The crushing and conveying facilities are massive steel structures and include a 750-ton bin, secondary crusher building, four conveyor galleries, two ore surge bins, an agglomeration silo, and the number 5 transfer station. The demolition debris will be consolidated in the Upper Mine Pit and buried under a minimum earth cover of six feet. Concrete foundations will be buried in place.

4.5 *Grading*

Over 450,000 yd^3 of mine wastes will be removed from steep road embankments to reduce them to a 3H:1V or flatter slope. This will enable topsoil replacement and revegetation of the road fills. Backfill derived from the embankment excavations will be used to backfill road cuts and some mine pit highwalls to a 3:1V slope also thereby expanding the proposed areas of revegetation. Other grading will be conducted to eliminate water holding surface depressions, remove acidic materials from natural watercourses, and establish permanent drainage pathways. The 700-foot-high mine highwall cut into South Mountain will remain.

4.6 Drainage control

Surface water flows will be carried from the site in a system of ditches sized to pass the 100-year storm event under revegetated conditions. Pipes were avoided because of the harsh site conditions and anticipated maintenance problems. One exception is the use of HDPE pipe for road culverts. Plastic pipes must be used because of the aggressive acid conditions. Most of the galvanized pipe culverts installed by the mining company have corroded and collapsed.

The ditches were originally designed as riprap and geomembrane lined structures. As a result of the Value Engineering study, the ditch designs have been modified. All ditches on the North Waste Dump will be geomembrane lined to reduce water infiltration into the dump. Ditch construction is now dependent upon water flow velocity. Flat gradient ditches (velocities under 4.5 ft/sec) are grass-lined, medium gradient ditches (velocities from 4.5 to 10 ft/sec) are lined with grass and flexible channel liner, and steep ditches (velocities above 10 ft/sec) are riprap lined. Over five miles of drainage ditches will be constructed.

Limestone riprap will be placed in the steep ditches as a form of passive treatment. ARD from most metal mines will coat limestone drains with iron and aluminum hydroxides rendering them ineffective. Recent studies in Eastern coalfields show that if high water velocities are present, the oxide scales are scoured off and passive treatment can be achieved

(Ziemkiewicz, and others, 1997). Steep gradients and long ditch runs make Summitville a favorable site for this type of passive treatment.

4.7 Revegetation

The BOR has been assisted by the Colorado State University and the USDA Forest Service's Rocky Mountain Research Center in Logan, Utah in formulating the revegetation techniques for the site. The CDPHE engaged revegetation experts from the Colorado State University to conduct revegetation experiments. Following the initial greenhouse studies, 32 test plots were placed upon the top of the North Waste Dump (Redente and Richard, 1998). Based upon this work, peer reviews and suggestions by the Forest Service, and cost investigations by the BOR, a revegetation procedure has been developed for the site.

Subsoil (mine wastes) will be amended to a 1 foot depth with 40 tons/acre of organic material, and neutralized with a variable rate (50 to 150 tons/acre) of limestone. The limestone rates are based upon a 1997 site survey that included over 150 acid-base accounting tests. Six inches of topsoil will be replaced and amended with 20, 30, or 80 tons/acre of limestone depending upon the particular topsoil stockpile source.

Nitrogen and phosphate fertilizers will be added to the hydroseeding/hydromulching operation. A long slope (450 feet) on the North Waste Dump will receive a combination of erosion protection blankets and natural-fiber check dams to prevent erosion prior to vegetation establishment.

Approximately 300 acres will be revegetated using a six-species native grass seed mix. The selected species are tufted hairgrass - *deschampsia caespitosa*, alpine timothy - *phleum alpinum*, alpine bluegrass - *poa alpina*, spike trisctum - *trisetum spicatum*, slender wheatgrass - *agropyron trachycaulum*, and western yarrow (a forb) - *achillea millifolium*. All of the species are native to the site, acid tolerant, and will allow natural succession to take place. It is expected that woody species will begin to invade the vegetation stands after two years of growth. This time period is needed for soil conditions to sufficiently developed to allow the woody species to establish.

Two species, *phleum alpinum* and *trisetum spicatum* are not available for purchase, therefore a seed collection and seed farming program is being implemented under the direction of Dr. Edward Redente of the Colorado State University. Seed will be collected from the site and nearby mountain areas. Seed will also be grown at the site with the first planting in reclaimed areas in the fall of 1998. Although the two named species are the main target of the seed harvesting efforts, other native species will be collected as available and be included in the seed furnished to the reclamation contractor.

5 VALUE ENGINEERING STUDY

In the United States, Federal agencies are required to conduct value engineering (VE) studies on construction projects costing over $1 million. The Department of the Interior has a stricter standard by requiring VE studies on projects costing over $100,000. Considering the magnitude of the Summitville mine site-wide reclamation project, a VE study was considered a worthwhile investment. The study cost $40,000 and resulted in the identification of nine proposals offering a total savings of $3.3 million over the design team's plans.

5.1 *The Federal value engineering program*

The principal objective of value engineering within the Federal Government is to realize better value from the expenditure of funds. Better value is defined in several ways such as achieving project objectives at a lower cost, providing a higher quality product for similar costs, increasing project durability, or lowering maintenance and operational costs. In the VE process it is project life cycle costs that are considered. Life cycle costs are construction, operation, maintenance, and replacement costs.

There is no guarantee that a VE study will reduce the cost or increase the quality of a particular project. Federal experience with a number of projects has demonstrated that the VE process does produce cost savings and quality enhancements far exceeding the expenses incurred from conducting the VE studies.

5.2 *The Summitville VE team*

Selection of the VE team is an important aspect of the process. Ideally, the team members should not be associated with the project so they can provide an unbiased review of the project designs. In the case of Summitville, three of the eight team members had been involved with ongoing reclamation work at the site, however, none of them had participated in the preparation of the Site-Wide Reclamation design.

Team members are sought that have expertise and practical experience in the principal disciplines encompassed by the project under consideration. In this case, mining engineering, mine reclamation, geology, geotechnical engineering, hydrology, civil engineering, and construction management were identified as the main fields of expertise involved. Specialists were sought with extensive experience and broad backgrounds, all but two of the team members selected had 20 years of experience in their respective fields. The team composition included two mining engineers, two geologists, a geotechnical engineer, a civil engineer, a construction manager, and a value engineer. Affiliations were one from the EPA, one from the CDPHE, two mining industry consultants, and four team members from the Bureau of Reclamation.

5.3 *The VE process*

The VE process practiced at the Bureau of Reclamation uses a combination of techniques such as functional analysis, cost analysis, brainstorming, and benchmarking to identify and analyze alternative methods of achieving the project requirements. Team members are not required to be VE experts, the value engineer directs the team and leads them through the study process. Although the VE team has access to the design staff, most of the study is conducted in private to ensure that an uninhibited critical review of the project is conducted.

The study follows a job plan that provides a reliable, structured approach to reaching conclusions and solutions. Using the design documents, the VE team identifies the component features of the project, their critical functions, governing criteria, and associated costs. Next, a functional analysis identifies the primary and secondary functions of the design components. The functions are then related in graphic form. A cost breakdown identifies the percentage of the total project cost associated with each project component.

With the initial tasks accomplished the VE team begins formulation of preliminary responses. Using brainstorming techniques the team suggests alternative ideas and solutions to perform project functions at a lower cost or with an increase in long term value, consistent with project requirements. The ideas are evaluated, analyzed, and prioritized. The best ideas are then developed to a level suitable for comparison with the current design. A draft VE report is prepared and the ideas, criticisms, and formal proposals are presented to the design team for discussion. The report is further refined and finalized.

5.4 *Value engineering proposals*

Nine project proposals resulted from the Summitville VE study as follows:

1. Develop an on-site quarry to produce riprap and other project rock products.
2. Replace gentle gradient rock-lined drainage channels with grass-lined channels.
3. Modify the channel armor; delete geomembrane lining in selected channel reaches.
4. Integrate access road revegetation with the revegetation of the adjacent disturbed areas.
5. Replace the CDP culvert fill with an inexpensive fill material.
6. Expand the revegetation planting to include shrubs & trees to increase diversity.
7. Pump low quality water back into the underground mine workings.
8. Install horizontal drains in seep areas to concentrate flows and dry out the surface.
9. Construct passive wetlands type systems to treat poor quality water.

5.5 *Implementation*

Federal regulations require that all VE proposals be evaluated by the design team. Each agency must submit an annual report summarizing the results of VE studies and amounts of cost savings. If specific proposals are not implemented, the reasons for rejection must be explained in the annual report.

The design team is currently involved with implementation of the proposals. Proposals 1 through 4, and modified versions of proposals 5 and 6, are being implemented through revisions to the reclamation plans.

Proposal 8 has been declined by the design team as being difficult to implement and having potential negative impacts to the reclamation process. Drilling into the seeps could increase the amounts of ARD flowing from these sources to the point where additional water treatment would be required. Over a longer period, the drains would be likely to clog up with ferrocrete precipitates thus eventually returning to the current situation of numerous small acid seeps.

Proposals 7 and 9 are beyond the scope of OU-4. Proposal 7 is being retained for consideration under the Groundwater Operable Unit for Summitville.

Studies are already underway for proposal 9 in the form of tests to evaluate an organic-substrate bioreactor type of wetland system that would employ sulfate reducing bacteria to treat acid seeps. One challenge to such a system is the inefficient performance in cold climates such as that found at Summitville. A possible solution is to cover the system with a thick insulating blanket of earth, however, this would significantly increase construction and maintenance costs.

5.6 *Conclusions*

Final quantities and revised cost estimates are not complete; however, an estimated savings of at least $2 million in construction costs will be achieved as a result of the VE study. The savings are net and include deductions for the study costs of $40,000 and the re-engineering costs of $80,000. Overall, $1.1 million has been expended on engineering and plan preparation for OU-4. While the VE study increased overall engineering costs by 11 percent, a return of $16 dollars in construction costs saved for every dollar invested in the VE process has been achieved.

The VE process provided a more intense and different type of project scrutiny than normally derived from peer reviews and public comment. The depth of experience and expertise of the VE team served to identify weak points in the design, validate other aspects of the project plans, and provide creative alternatives that fulfill the client's needs at the greatest value. The study also served to gain public acceptance that realistic, innovative, and cost-effective reclamation designs have been prepared for the site. Other mine reclamation projects could benefit from a VE study similar to the process described in this paper.

REFERENCES

Bureau of Reclamation (1998) *Value engineering final report, OU-4 Site-Wide Reclamation Summitville Mine Superfund Site Colorado.* BOR Technical Service Center, Denver, Colorado, 41 p.

Clark, B. J. (1997) *A survey of cultural resources in the vicinity of the Summitville Superfund Site, Rio Grande County, Colorado - draft report.* SWCA, Inc. Environmental Consultants, Denver, Colorado, 58 p.

Gray, J. E., Coolbaugh, M. F., and Plumlee, G. S. (1993) *Geologic framework and environmental geology of the Summitville acid-sulfate mineral deposit.* U.S. Geological Survey, Open File Report 93-677, 1993.

Ketellapper, V. L., Cressman, J. E., and Carmody, C. (1995) Cropsy Waste Pile, Beaver Mud Dump, Cleveland Cliffs and Mine Pits response action. In Posey, H. H., Pendleton, J. A., and Van Zyl, D. (eds), *Proceedings Summitville Forum '95.* Colorado Geological Survey Special Publication No. 38, pp. 121-126.

Lipman, P. W. (1974) *Geologic map of the Platoro Caldera area, southeastern San Juan Mountains, southwestern Colorado.* U.S. Geologic Survey, Map I-828, 1 sheet.

Redente, E. F., and Richard, C (1998) *Reclamation alternatives at Summitville Mine Superfund Site: greenhouse and field studies.* Colorado State University, Fort Collins, Colorado, 56 p.

Steven, T. A., and Ratte, J. C. (1960) *Geolgy and ore deposits of the Summitville District, San Juan Mountains, Colorado.* U.S. Geological Survey Professional Paper 343, Washington, DC.

Ziemkiewicz, P. F., Skousen, J. G., Brant, D. L., Sterner, P. L., and Lovett, R. J. (1997) Acid mine drainage treatment with armored limestone in open limestone channels. *Journal of Environmental Quality*, Vol. 26, p. 560-569.

Tailings and Mine Waste '99 © 1999 Balkema, Rotterdam, ISBN 90 5809 025 6

Author index